INTERNATIONAL MILITARY AND DEFENSE ENCYCLOPEDIA

INTERNATIONAL MILITARY AND DEFENSE ENCYCLOPEDIA

Volume 3

G–L

Editor-in-Chief
Col. Trevor N. Dupuy, USA (Ret.)

Executive Editor
Col. Franklin D. Margiotta, USAF (Ret.), Ph.D.

Managing Editors
Mr. Curt Johnson
Col. James B. Motley, USA (Ret.), Ph.D.

Associate Managing Editor
Mr. David L. Bongard

Brassey's (US), Inc.
A Division of Maxwell Macmillan, Inc.

Washington · New York

Copyright © 1993 by Brassey's (US), Inc.
All rights reserved. No part of this book may be reproduced, stored in a retrieval system, or transmitted in any form or by any means—electronic, electrostatic, magnetic tape, mechanical, photocopying, recording, or otherwise—without permission in writing from the publisher.

Brassey's (US), Inc.

Editorial Offices	*Order Department*
Brassey's (US), Inc.	Brassey's Book Orders
8000 Westpark Drive	% Macmillan Publishing Co.
First Floor	100 Front Street, Box 500
McLean, Virginia 22102	Riverside, New Jersey 08075

LIBRARY OF CONGRESS CATALOGING-IN-PUBLICATION DATA

International military and defense encyclopedia / editor-in-chief,
  Trevor N. Dupuy . . . [et al.].
    p.   cm.
  Includes index.
  ISBN 0-02-881011-2 (set)
  1. Military art and science—Encyclopedias. I. Dupuy, Trevor
Nevitt, 1916–
U24.I58 1993
355'.003—dc20
                                                          92-33750
                                                             CIP

ISBN 0-02-881063-5 (vol. 3)

10 9 8 7 6 5 4 3 2 1

PRINTED IN THE UNITED STATES OF AMERICA

INTERNATIONAL HONORARY ADVISORY BOARD

Field Marshal the Lord Bramall, GCB, OBE, MC, JP
Former Chief of the Defence Staff
United Kingdom

Professor Samuel P. Huntington, Ph.D.
Former President, American Political Science Association
Director, Institute for Strategic Studies, Harvard University
United States of America

Mr. Masimichi Inoki
Chairman, Research Institute for Peace and Security
Japan

General David C. Jones, USAF (Ret.)
Former Chairman, Joint Chiefs of Staff
United States of America

General P. X. Kelley, USMC (Ret.)
Former Commandant, United States Marine Corps
United States of America

Baron Ewald von Kleist
Organizer, Wehrkunde Conferences
Federal Republic of Germany

Field Marshal S. H. F. J. Manckshaw, MC
Former Chief of the Army Staff
Republic of India

Professor Charles C. Moskos, Ph.D.
Northwestern University
Chairman, Inter-University Seminar on Armed Forces and Society
United States of America

General Paik Sun Yup, ROK Army (Ret.)
Former Chairman, Joint Chiefs of Staff
Republic of Korea

Admiral of The Fleet Sir William Stavely, GCB, DL
Former First Sea Lord
United Kingdom

General Wu Xiuquan, CPLA
Member, Standing Committee of the Central Advisory
Commission of the Communist Part of China
People's Republic of China

EDITORIAL BOARD AND SUBJECT EDITORS

Maj. Gen. Hassan El Badri (Ret.)
Former Director, Nasser Military Academy
Egypt

Capt. Roger W. Barnett, USN (Ret.), Ph.D.
The National Institute for Public Policy
United States

Maj. Gen. Jibrail Bitar, Syrian Army (Ret.)
Military Historian
Syria

Col. Trevor N. Dupuy, USA (Ret.)
Author and Media Military Analyst
United States

Col. Kenneth Evans, USA (Ret.), Ph.D.
Vice President, System Planning Corporation
United States

Maj. Gen. Johannes Gerber (Ret.), Ph.D.
Former Deputy Commanding General of a German Army Corps
Germany

John G. Honig, Ph.D., USA
Former Deputy Director, Weapon Systems Analysis, Office of the Army Chief of Staff
United States

Mr. Curt Johnson
Military Historian
United States

Rear Adm. William C. Mott, USN (Ret.)
General Counsel and Director, National Strategy Information Center
United States

Brig. Gen. Adrian von Oër (Ret.), Ph.D.
German Army
Germany

Brig. Gen. Howard T. Prince II, USA (Ret.), Ph.D.
Dean, Jepson School of Leadership Studies, University of Richmond
United States

Brig. J. H. Skinner (Ret.), M. Litt. (Oxon)
British Army
United Kingdom

Lt. Col. John F. Sloan, USA (Ret.)
Defense Intelligence College, Washington, D.C.
United States

Maj. Gen. Perry M. Smith, USAF (Ret.), Ph.D.
Author and Cable Network News Military Analyst
United States

Professor Claude E. Welch, Jr., Ph.D.
SUNY, Buffalo; Former Editor of Armed Forces and Society
United States

Lt. Gen. Franz Uhle-Wettler (Ret.), Ph.D.
Former Commandant, NATO Defense College
Germany

ASSOCIATE EDITORS

Col. John R. Brinkerhoff, USA (Ret.)
Former Deputy Assistant Secretary of Defense for Reserve Affairs
United States

Lt. Col. Lewis Sorley, USA (Ret.), Ph.D.
Author
United States

Lt. Cdr. James T. Westwood, USN (Ret.)
Defense Analyst
United States

Maj. Charles F. Hawkins, USAR
President, Data Memory Systems, Inc./Historical Evaluation and Research Organization
United States

CONTENTS

Abbreviations and Acronyms	xi
International Military and Defense Encyclopedia	
Volume 1: A–B	*1*
Volume 2: C–F	*436*
Volume 3: G–L	*1041*
Volume 4: M–O	*1587*
Volume 5: P–S	*2083*
Volume 6: T–Z	*2657*
Index *(Volume 6)*	2985

ABBREVIATIONS AND ACRONYMS

AA	antiaircraft	APO	army post office
AAA	antiaircraft artillery	AR	Army Reserve (US)
AASR	advanced airborne surveillance radar	AR	repair ship(s)
AAM	air-to-air missile	Arg	Argentina
AAW	antiair warfare	ARM	antiradiation (antiradar) missile
AB	airborne	armd	armored
ABD	airborne division (Sov)	ARNG	Army National Guard (US)
ABM	antiballistic missile(s)	arty	artillery
about	the total could be higher	AS	submarine depot-ship(s)
ac	aircraft	ASAT	antisatellite weapon
ACDA	Arms Control and Disarmament Agency	aslt	assault
ACE	Allied Command, Europe	ASM	air-to-surface missile(s)
ACM	advanced cruise missile	ASRAAM	advanced short-range air-to-air missile
ACV	air cushion vehicle/vessel	ASROC	antisubmarine rocket
AD	air defense	ASTT	antisubmarine TT
adj	adjusted	ASUW	antisurface-unit warfare
AE	auxiliary(ies), ammunition carrier	ASW	antisubmarine warfare
AEF	auxiliary(ies), explosives and stores	AT	tug(s)
AEW	airborne early warning	ATBM	antitactical ballistic missile
AF	stores ship(s) with RAS capability	ATF	advanced tactical fighter
AFB	air force base	ATGW	antitank guided weapon(s)
AFV	armored fighting vehicle	ATK	antitank
AFR	Air Force Reserve (US)	ATTU	Atlantic to the Urals
AGHS	hydrographic survey vessel(s)	Aust	Australia
AGI	intelligence collection vessel(s)	avn	aviation
AGM	air-to-ground missile	AVT	aviation training ship
AGOR	oceanographic research vessel(s)	AWACS	airborne warning and control system
AGOS	ocean surveillance vessel(s)	BA	Budget Authority
AH	hospital ship(s)	BB	battleship
AI	artificial intelligence	bbr	bomber(s)
AIFV	armored infantry fighting vehicle	bde	brigade(s)
AIP	air-independent propulsion	bdgt	budget(s)
AK	cargo ship(s)	Be	Belgium
ALCM	air-launched cruise missile(s)	BMD	ballistic missile defense
amph	amphibious/amphibian(s)	BMEWS	ballistic missile early warning system
AMRAAM	advanced medium-range air-to-air missile	bn	battalion(s)/billion(s)
ANG	Air National Guard (US)	BSAG	battleship surface attack group
AO	tanker(s) with RAS capability; area of operations	bty	battery(ies)
		Bu	Bulgaria
AOE	auxiliary(ies), fuel and ammunition, RAS capability	BVR	beyond visual range
		CAD	computer-aided design
AOT	tanker(s) without RAS capability	CAP	combat air patrol; civil air patrol
AP	passenger ship(s); armor piercing	CAS	close air support
APC	armored personnel carrier(s)	Cat	Category
APDS	armor piercing, discarding sabot missile	cav	cavalry
API	armor piercing, incendiary	cbt	combat

Abbreviations and Acronyms

CBW	chemical and biological warfare	excl	excludes/excluding
CC	cruiser(s)	exp	expenditure
CCM	counter-countermeasures	FAC	forward air control
Cdn	Canada	FAE	fuel-air explosive
CCP	Chinese Communist Party	fd	field
CD	civil defense	FEBA	forward edge of the battle area
cdo	commando	FF	frigate(s)
CEP	circular error probable	FFG	frigate(s) with area SAM
CG	SAM cruiser(s)	FGA	fighter(s), ground-attack
CGF	Central Group of Forces (Sov)	FH	frigate(s) with helicopter
CGH	CG with helicopters	FLIR	forward-looking infrared radar
CGN	nuclear-fueled CG	flt	flight(s)
cgo	freight aircraft	FMA	foreign military assistance
Ch	China (PRC)	Fr	France
C^3I	command, control, communications, and intelligence	ftr	fighter(s) (aircraft)
CINC	commander in chief	FW	fixed-wing
CLOS	command line-of-sight	FY	fiscal year
COIN	counterinsurgency	GA	Chinese Integrated Group Army
comb	combined/combination	GB	Sarin (chemical agent)
comd	command	GBU	guided bomb unit
COMINT	communications intelligence	GCI	ground control intercept
comms	communications	GD	Soman (chemical agent)
CONUS	Continental United States	GDP	gross domestic product
coy	company(ies)	Ge	Germany
CP	command post	GLCM	ground-launched cruise missile
CPSU	Communist Party of the Soviet Union	GNP	gross national product
CV	aircraft carrier(s)	gp	group(s)
CVBG	carrier battle group	GP	general purpose
CVN	nuclear-fueled CV	GPS	global positioning system
CVV	V/STOL and hel CV	Gr	Greece
CW	chemical warfare	GW	guided weapon(s)
CY	current year	hel	helicopter(s)
Cz	Czechoslovakia	HARM	high-speed antiradiation missile
DD	destroyer(s)	HQ	headquarters
D Day	day operation begins	Hu	Hungary
DDG	destroyer(s) with area SAM; destroyer(s) with hel	HWT	heavy-weight torpedo(es)
		hy	heavy
def	defense	ICBM	intercontinental ballistic missile(s)
defn	definition	IFF	identification friend or foe
det	detachment(s)	IFS	International Financial Statistics
DEW	directed-energy weapon; distant early warning	IG	inspector general
		IMF	International Monetary Fund
div	division(s)	imp	improved
Dk	Denmark	incl	includes/including
ECM	electronic countermeasures	indep	independent
ECR	Electronic combat and reconnaissance	Indon	Indonesia
		inf	infantry
ELINT	electronic intelligence	INF	intermediate-range nuclear forces
elm	element(s)	IR	infrared
EMP	electromagnetic pulse	IRBM	intermediate-range ballistic missile(s)
engr	engineer(s)	Is	Israel
EOD	explosive ordnance disposal	It	Italy
eqpt	equipment	k	kilobyte
ESM	electronic support measures	KE	kinetic energy
est	estimate(d)	kg	kilogram(s)
EW	electronic warfare	kHz	kilohertz

km	kilometer(s)	mob	mobilization
KT	kiloton(s)	mod	modified/modification
LAMPS	Light airborne multipurpose system	mor	mortar(s)
LAW	light antitank weapon	mot	motorized
LCA	landing craft, assault	MPS	marine pre-positioning squadron(s)
LCAC	landing craft, air cushion	MR	maritime reconnaissance/motor rifle
LCM	landing craft, mechanized	MRASM	medium-range air-to-surface missile(s)
LCT	landing craft, tank	MRBM	medium-range ballistic missile(s)
LCU	landing craft, utility	PSC	principal surface combatants
LCVP	landing craft, vehicles and personnel	psi	pounds per square inch
LHA	landing ship(s), assault	PSYOPS	psychological operations
LHX	light helicopter, experimental	R&D	research and development
LKA	assault cargo ship(s)	RAS	replenishment at sea
log	logistic	RCL	recoilless launcher(s)
LORAN	long-range air navigation system	RCS	radar cross-section
LPD	landing platform(s), dock	RDF	rapid deployment force
LPH	landing platform(s), helicopter	recce	reconnaissance
LSD	landing ship(s), dock	regt	regiment(s)
LSM	landing ship(s), medium	RF	radio frequency
LST	landing ship(s), tank	RL	rocket launchers
lt	light	Ro	Romania
LWT	light-weight torpedo(es)	ROM	read-only memory
m	million(s)	ro-ro	roll-on, roll-off
MAC	Military Airlift Command (US)	RPG	rocket-propelled grenade
MAAG	military assistance and advisory group	RPV	remotely piloted vehicle(s)
MAD	mutual assured destruction; magnetic anomaly detection	RV	re-entry vehicle(s)
		SAC	Strategic Air Command (US)
maint	maintenance	SACEUR	Supreme Allied Commander, Europe
MBT	main battle tank(s)	SACLANT	Supreme Allied Commander, Atlantic
MC&G	mapping, charting, and geodesy	SAH	semi-active homing
MCC/I/O	mine countermeasures vessel(s), coastal/inshore, offshore	SALT	Strategic Arms Limitation Treaty
		SAM	surface-to-air missile(s)
MCMV	mine countermeasures vessel(s)	SAR	search and rescue; synthetic aperture radar
MCR	Marine Corps Reserve (US)		
MD	Military District(s)	SDI	Strategic Defense Initiative
M Day	mobilization day	SES	surface-effect ship(s)
mech	mechanized	SF	Special Forces
med	medium	SGF	Southern Group of Forces (Sov)
medevac	casualty transport/air ambulance	SHAPE	Supreme Headquarters Allied Powers, Europe
MEF/B/U	Marine Expeditionary Force(s)/Brigade(s)/Unit(s) (US)		
		SIGINT	signals intelligence
MFO	Multinational Force and Observers	sigs	signals
MG	machine gun	SLAR	side-looking airborne radar
MHC/I/O	minehunter(s), coastal/inshore/offshore	SLBM	sea- or submarine-launched ballistic missile(s)
MHz	megahertz		
MICV	mechanized infantry combat vehicle(s)	SLCM	sea- or submarine-launched cruise missile(s)
mil	military		
MINURSO	UN Mission for the Referendum in Western Sahara	SLEP	service life extension program
		SLOC	sea line of communication
MIRV	multiple independently targetable re-entry vehicle(s)	some	up to
		SOP	standard operating procedure
misc	miscellaneous	Sov	Soviet
Mk	mark (model number)	Sp	Spain
ML	minelayer	SP	self-propelled
MLRS	multiple-launch rocket system	spt	support
MMW	millimeter-wave radar	SQ	superquick fuze

Abbreviations and Acronyms

sqn	squadron
SRAM	short-range attack missile(s)
SRBM	short-range ballistic missile(s)
SS(C/I)	submarine(s) (coastal/inshore)
SSB	ballistic-missile submarine(s)
SSBN	nuclear-fueled SSB
SSGN	SSN with dedicated nonballistic missile launchers
SSM	surface-to-surface missile(s)
SSN	nuclear-fueled submarine(s)
START	Strategic Arms Reduction Talks
STOL	short take-off and landing
STOVL	short takeoff, vertical landing
SUGW	surface-to-underwater GW
Sw	Sweden
Switz	Switzerland
sy	security
t	tons
TA	Territorial Army (UK)
tac	tactical
TAC	Tactical Air Command (US)
TADS	target acquisition and designation system
TASM	tactical air-to-surface missile
TD	tank division
T&E	test and evaluation
TERCOM	terrain contour-matching guidance
tk	tank(s)
tkr	tanker(s)
TLE	treaty-limited equipment
TOE	table of organization and equipment
TOW	tube-launched, optically tracked, wire-guided antitank missile
tps	troop(s)
tpt	transport(s)
trg	training
TT	torpedo tube(s)
Tu	Turkey
UHF	ultra-high frequency
UN	United Nations
UNAVEM	UN Angolan Verification Mission
UNDOF	UN Disengagement Observer Force
UNFICYP	UN Force in Cyprus
UNIFIL	UN Interim Force in Lebanon
UNIKOM	UN Iraq/Kuwait Observer Mission
UNMOGIP	UN Military Observer Group in India and Pakistan
UNTSO	UN Truce Supervisory Organization
URG	underway replenishment group(s)
USGW	underwater-to-surface GW
USMC	US Marine Corps
UUGW	underwater-to-underwater GW
UV	ultraviolet
veh	vehicle(s)
VERTREP	vertical replenishment
VHF	very high frequency
VIP	very important person(s)
VLS	vertical launch system(s)
V(/S)TOL	vertical(/short) takeoff and landing
WGF	Western Group of Force (Sov)
WP	Warsaw Pact
wpn	weapon
Yug	Yugoslavia

Designations of Aircraft and Helicopters listed in Country Articles

Type	Name/designation	Origin	Maker
AIRCRAFT			
A-3	Skywarrior	U.S.	Douglas
A-4	Skyhawk	U.S.	MD
A-5	Fantan	China	Nanchang
A-6	Intruder	U.S.	Grumman
A-7	Corsair II	U.S.	LTV
A-10	Thunderbolt	U.S.	Fairchild
A-36	Halcón (C-101)		
A-37	Dragonfly	U.S.	Cessna
AC-130	(C-130)		
AC-47	(C-47)		
Airtourer		NZ	Victa
AJ-37	(J-37)		
Ajeet	(Folland Gnat)	India/U.K.	HAL
Alizé		France	Breguet
AlphaJet		France/Ge	Dassault/Breguet/Dornier
AM-3	Bosbok (C-4M)	Italy	Aermacchi
An-2	"Colt"	USSR	Antonov
An-12	"Cub"	USSR	Antonov
An-14	"Clod"	USSR	Antonov
An-22	"Cock"	USSR	Antonov
An-24	"Coke"	USSR	Antonov
An-26	"Curl"	USSR	Antonov
An-30	"Clank"	USSR	Antonov
An-32	"Cline"	USSR	Antonov
An-124	"Condor" (Ruslan)	USSR	Antonov
Andover	[HS-748]		
Atlantic	(Atlantique)	France	Dassault/Breguet
AS-202	Bravo	Switz	FFA
AT-3		Taiwan	AIDC
AT-6	(T-6)		
AT-11		U.S.	Beech
AT-26	EMB-326		
AT-33	(T-33)		
AU-23	Peacemaker [PC-6B]	U.S.	Fairchild
AV-8	Harrier II	U.S./U.K.	MD/BAe
Aztec	PA-23	U.S.	Piper
B-1		U.S.	Rockwell
B-52	Stratofortress	U.S.	Boeing
BAC-111		U.K.	BAe
BAC-167	Strikemaster	U.K.	BAe

Abbreviations and Acronyms

Type	Name/Designation	Origin	Maker
BAe-146		U.K.	BAe
BAe-748	(HS-748)		
Baron	(T-42)		
Be-6	"Madge"	USSR	Beriev
Be-12	"Mail' (Tchaika)	USSR	Beriev
Beech 50	Twin Bonanza	U.S.	Beech
Beech 95	Travel Air	U.S.	Beech
BN-2	Islander, Defender, Trislander	U.K.	Britten-Norman
Boeing 707		U.S.	Boeing
Boeing 727		U.S.	Boeing
Boeing 737		U.S.	Boeing
Boeing 747		U.S.	Boeing
Bonanza		U.S.	Beech
Bronco	(OV-10)		
Buccaneer		U.K.	BAe
Bulldog		U.K.	BAe
C-1		Japan	Kawasaki
C-2	Greyhound	U.S.	Grumman
C-4M	Kudu (AM-3)	S. Africa	Atlas
C-5	Galaxy	U.S.	Lockheed
C-7	DHC-7		
C-9	Nightingale (DC-9)		
C-12	Super King Air (Huron)	U.S.	Beech
C-18	[Boeing 707]		
C-20	(Gulfstream III)		
C-21	(Learjet)		
C-22	(Boeing 727)		
C-23	(Sherpa)	U.K.	Short
C-42	(Neiva Regente)	Brazil	Embraer
C-45	Expeditor	U.S.	Beech
C-46	Commando	U.S.	Curtis
C-47	DC-3 (Dakota) (C-117 Skytrain)	U.S.	Douglas
C-54	Skymaster (DC-4)	U.S.	Douglas
C-91	HS-748		
C-93	HS-125		
C-95	EMB-110		
C-97	EMB-121		
C-101	Aviojet	Spain	CASA
C-115	DHC-5	Canada	De Havilland
C-117	(C-47)		
C-118	Liftmaster (DC-6)		
C-119	Packet	U.S.	Fairchild
C-123	Provider	U.S.	Fairchild
C-127	(Do-27)	Spain	CASA
C-130	Hercules (L-100)	U.S.	Lockheed
C-131	Convair 440	U.S.	Convair
C-135	[Boeing 707]		
C-137	[Boeing 707]		
C-140	(Jetstar)	U.S.	Lockheed
C-141	Starlifter	U.S.	Douglas
C-160		Fr/Ge	Transall
C-212	Aviocar	Spain	CASA
C-235		Spain	CASA
CA-25	Winjeel	Aust	Commonwealth
Canberra	(B-57)	U.K.	BAe
CAP-10		France	Mudry
CAP-20		France	Mudry
CAP-230		France	Mudry
Caravelle	SE-210	France	Aérospatiale
CC-109	(Convair 440)	U.S.	Convair
CC-115	DHC-15		
CC-117	(Falcon 20)		
CC-132	(DHC-7)		
CC-137	(Boeing 707)		
CC-138	(DHC-6)		
CC-144	CL-600/-601	Canada	Canadair
CC-18	F/A-18		
CF-116	F-5		
Cheetah	[Mirage III]	S. Africa	Atlas
Cherokee	PA-28	U.S.	Piper
Cheyenne	PA-31T [Navajo]	U.S.	Piper
Chieftain	PA-31-350 [Navajo]	U.S.	Piper
Chipmunk	DHC-1		
Citabria		U.S.	Champion
Citation	(T-47)	U.S.	Cessna
CJ-5	[Yak-18]	China	
CL-215		Canada	Canadair
CL-44		Canada	Canadair
CL-601	Challenger	Canada	Canadair
CM-170	Magister [Tzukit]	France	Aérospatiale
CM-175	Zéphyr	France	Aérospatiale
CN-235		Sp/Indon	CASA/IPTN
Cochise	T-42		
Comanche	PA-24	U.S.	Piper
Commander	Aero-/Turbo-Commander	U.S.	Rockwell
Commodore	MS-893	France	Aérospatiale
Corvette	SN-601	France	Aérospatiale
CP-3	P-3 Orion		
CP-121	S-2		
CP-140	Aurora (P-3 Orion)	U.S.	Lockheed
CT-4	Airtrainer	NZ	Victa
CT-39	Sabreliner	U.S.	Rockwell
CT-114	CL-41 Tutor	Canada	Canadair
CT-133	Silver Star [T-33]	Canada	Canadair
CT-134	Musketeer		
Dagger	(Nesher)		
Dakota		U.S.	Piper
Dakota	(C-47)		
DC-3	(C-47)	U.S.	Douglas
DC-4	(C-54)	U.S.	Douglas
DC-6	(C-118)	U.S.	Douglas
DC-7		U.S.	Douglas
DC-8		U.S.	Douglas
DC-9		U.S.	MD
Deepak	(HT-32)		
Defender	BN-2		
DH-100	Vampire	U.K.	De Havilland
DHC-1	Chipmunk	Canada	DHC
DHC-2	Beaver	Canada	DHC
DHC-3	Otter	Canada	DHC
DHC-4	Caribou	Canada	DHC
DHC-5	Buffalo	Canada	DHC
DHC-6	Twin Otter	Canada	DHC
DHC-7	Dash-7 (Ranger, CC-132)	Canada	DHC
DHC-8		Canada	DHC

Abbreviations and Acronyms   xvi

Type	Name/Designation	Origin	Maker
Dimona	H-36	Ge	Hoffman
Do-27	(C-127)	Ge	Dornier
Do-28	Skyservant	Ge	Dornier
Do-128		Ge	Dornier
Do-228		Ge	Dornier
E-2	Hawkeye	U.S.	Grumman
E-3	Sentry	U.S.	Boeing
E-4	[Boeing 747]	U.S.	Boeing
E-6	[Boeing 707]		
EA-3	[A-3]		
EA-6	Prowler [A-6]		
Electra	(L-188)		
EC-130	[C-130]		
EC-135	[Boeing 707]		
EMB-110	Bandeirante	Brazil	Embraer
EMB-111	Maritime Bandeirante	Brazil	Embraer
EMB-120	Brasilia	Brazil	Embraer
EMB-121	Xingu	Brazil	Embraer
EMB-312	Tucano	Brazil	Embraer
EMB-326	Xavante (MB-326)	Brazil	Embraer
EMB-810	[Seneca]	Brazil	Embraer
EP-3	(P-3 Orion)		
Etendard		France	Dassault
EV-1	(OV-1)		
F-1	[T-2]	Japan	Mitsubishi
F-4	Phantom	U.S.	MD
F-5	-A/-B: Freedom Fighter; -E/-F: Tiger II	U.S.	Northrop
F-6	J-6		
F-7	J-7		
F-8	J-8		
F-8	Crusader	U.S.	Republic
F-14	Tomcat	U.S.	Grumman
F-15	Eagle	U.S.	MD
F-16	Fighting Falcon	U.S.	GD
F-18	[F/A-18]		
F-21	Kfir	Israel	IAI
F-27	Friendship	Nl	Fokker
F-28	Fellowship	Nl	Fokker
F-35	Draken	Sweden	SAAB
F-84	Thunderstreak	U.S.	Lockheed
F-86	Sabre	U.S.	N. American
F-100	Super Sabre	U.S.	N. American
F-104	Starfighter	U.S.	Lockheed
F-106	Delta Dart	U.S.	Convair
F-111		U.S.	GD
F-172	(Cessna 172)	France/U.S.	Reims-Cessna
F/A-18	Hornet	U.S.	MD
Falcon	Mystère-Falcon		
FB-111	(F-111)		
FH-227	(F-27)	U.S.	Fairchild/Hiller
Flamingo	MBB-233	Ge	MBB
FT-5	JJ-5	China	CAC
FT-6	JJ-6		
FTB-337	[Cessna 337]		
G-91		Italy	Aeritalia
G-222		Italy	Aeritalia
Galaxy	C-5		

Type	Name/Designation	Origin	Maker
Galeb		Yug	SOKO
Gardian	(Falcon 20)		
Genet	SF-260W		
GU-25	(Falcon 20)		
Guerrier	R-235		
Gulstream		U.S.	Gulfstream Aviation
Gumhuria	(Bücker 181)	Egypt	Heliopolis Ac
H-5	[Il-28]	China	Harbin
H-6	[Tu-16]	China	Xian
H-36	Dimona		
Halcón	[C-101]		
Harrier	(AV-8)	U.K.	BAe
Harvard	(T-6)		
Hawk		U.K.	BAe
HC-130	(C-130)		
HF-24	Marut	India	HAL
HFB-320	Hansajet	Ge	Hamburger FB
HJ-5	(H-5)		
HJT-16	Kiran	India	HAL
HPT-32	Deepak	India	HAL
HS-125	(Dominie)	U.K.	BAe
HS-748	[Andover]	U.K.	BAe
HT-2		India	HAL
HU-16	Albatross	U.S.	Grumman
HU-25	(Falcon 20)		
Hunter		U.K.	BAe
HZ-5	(H-5)		
IA-35	Huanquero	Arg	FMA
IA-50	Guaraní	Arg	FMA
IA-58	Pucará	Arg	FMA
IA-63	Pampa	Arg	FMA
IAI-201/-202	Arava	Israel	IAI
IAI-1124	Westwind, Seascan	Israel	IAI
IAR-28		Ro	IAR
IAR-93	Orao	Yug/Ro	SOKO/IAR
Il-14	"Crate"	USSR	Ilyushin
Il-18	"Coot"	USSR	Ilyushin
Il-20	(Il-18)		
Il-28	"Beagle"	USSR	Ilyushin
Il-38	"May"	USSR	Ilyushin
Il-62	"Classic"	USSR	Ilyushin
Il-76	"Candid" (tpt) "Mainstay" (AEW) "Midas" (tkr)	USSR	Ilyushin
Impala	[MB-326]	S. Africa	Atlas
Islander	BN-2		
J-2	[MiG-15]	China	
J-5	[MiG-17F]	China	Shenyang
J-6	[MiG-19]	China	Shenyang
J-7	[MiG-21]	China	Xian
J-8	[Sov Ye-142]	China	Shenyang
J-32	Lansen	Sweden	SAAB
J-35	Draken	Sweden	SAAB
J-37	Viggen	Sweden	SAAB
JA-37	(J-37)		
Jaguar		Fr/U.K.	SEPECAT
JAS-39	Gripen	Sweden	SAAB
Jastreb		Yug	SOKO
Jet Provost		U.K.	BAe

Abbreviations and Acronyms

Type	Name/Designation	Origin	Maker
Jetstream		U.K.	BAe
JJ-6	(J-6)		
JZ-6	(J-6)		
KA-3	[A-3]		
KA-6	[A-6]		
KC-10	Extender [DC-10]	U.S.	MD
KC-130	[C-130]		
KC-135	[Boeing 707]		
KE-3	[E-3]		
Kfir		Israel	IAI
King Air		U.S.	Beech
Kiran	HJT-16		
Kraguj		Yug	SOKO
Kudu	C-4M		
LIM-6	[MiG-17]	Poland	
L-4	Cub		
L-18	Super Cub	U.S.	Piper
L-19	O-1		
L-21	Super Cub	U.S.	Piper
L-29	Delfin	Cz	Aero
L-39	Albatros	Cz	Aero
L-70	Vinka	Finland	Valmet
L-100	C-130 (civil version)		
L-188	Electra (P-3 Orion)	U.S.	Lockheed
L-410	Turbolet	Cz	LET
L-1011	Tristar	U.S.	Lockheed
Learjet	(C-21)	U.S.	Gates
Li-2	[DC-3]	USSR	Lisunov
LR-1	(MU-2)		
Magister	CM-170		
Marut	HF-24		
Mashshaq	MFI-17	Pakistan/Sweden	PAC/SAAB
Matador	(AV-8)		
MB-326		Italy	Aermacchi
MB-339	(Veltro)	Italy	Aermacchi
MBB-233	Flamingo		
MC-130	(C-130)		
Mercurius	(HS-125)		
Merlin		U.S.	Fairchild
Mescalero	T-41		
Metro		U.S.	Fairchild
MFI-15	Safari	Sweden	SAAB
MFI-17	Supporter (T-17)	Sweden	SAAB
MH-1521	Broussard	France	Max Holste
MiG-15	"Midget" trg	USSR	MiG
MiG-17	"Fresco"	USSR	MiG
MiG-19	"Farmer"	USSR	MiG
MiG-21	"Fishbed"	USSR	MiG
MiG-23	"Flogger"	USSR	MiG
MiG-25	"Foxbat"	USSR	MiG
MiG-27	"Flogger D"	USSR	MiG
MiG-29	"Fulcrum"	USSR	MiG
MiG-31	"Foxhound"	USSR	MiG
Mirage		France	Dassault
Missionmaster	N-22		
Mohawk	OV-1		
MS-760	Paris	France	Aérospatiale
MS-893	Commodore		
MU-2		Japan	Mitsubishi
Musketeer	Beech 24	U.S.	Beech
Mya-4	"Bison"	USSR	Myasishchev

Type	Name/Designation	Origin	Maker
Mystère-Falcon		France	Dassault
N-22	Floatmaster, Missionmaster	Aust	GAF
N-24	Searchmaster B/L	Aust	GAF
N-262	Frégate	France	Aérospatiale
N-2501	Noratlas	France	Aérospatiale
Navajo	PA-31	U.S.	Piper
NC-212	C-212	Sp/Indon	CASA/Nurtanio
NC-235	C-235	Sp/Indon	CASA/Nurtanio
Nesher	[Mirage III]	Israel	IAI
NF-5	(F-5)		
Nightingale	(DC-9)		
Nimrod		U.K.	BAe
O-1	Bird Dog	U.S.	Cessna
O-2	(Cessna 337, Skymaster)	U.S.	Cessna
OA-4	(A-4)		
OA-37	Dragonfly		
Orao	IAR-93		
Ouragan		France	Dassault
OV-1	Mohawk	U.S.	Rockwell
OV-10	Bronco	U.S.	Rockwell
P-2J	[SP-2]	Japan	Kawasaki
P-3		Switz	Pilatus
P-3	Orion	U.S.	Lockheed
P-95	EMB-110		
P-149		Italy	Piaggio
P-166		Italy	Piaggio
PA-18	Super Cub	U.S.	Piper
PA-23	Aztec		
PA-24	Comanche	U.S.	Piper
PA-28	Cherokee	U.S.	Piper
PA-31	Navajo	U.S.	Piper
PA-34	Seneca	U.S.	Piper
PA-44	Seminole	U.S.	Piper
PBY-5	Catalina	U.S.	Consolidated
PC-6	Porter	Switz	Pilatus
PC-6A/B	Turbo Porter	Switz	Pilatus
PC-7	Turbo Trainer	Switz	Pilatus
PC-9		Switz	Pilatus
PD-808		Italy	Piaggio
Pembroke		U.K.	BAe
Pillán	T-35		
PL-1	Chien Shou	Taiwan	AIDC
Porter	PC-6		
PZL-104	Wilga	Poland	PZL
PZL-130	Orlik	Poland	PZL
Q-5	"Fantan" [MiG-19]	China	Nanchang
Queen Air	(U-8)		
R-160		France	Socata
R-235	Guerrier	France	Socata
RC-21	(C-21)		
RC-47	(C-47)		
RC-95	(EMB-110)		
RC-135	[Boeing 707]		
RF-4	(F-4)		
RF-5	(F-5)		
RF-35	(F-35)		
RF-84	(F-84)		
RF-104	(F-104)		
RF-172	(Cessna 172)	France	Reims-Cessna

Abbreviations and Acronyms

Type	Name/Designation	Origin	Maker
RT-26	(EMB-326)		
RT-33	(T-33)		
RU-21	(King Air)		
RV-1	(OV-1)		
S-2	Tracker	U.S.	Grumman
S-3	Viking	U.S.	Lockheed
S-208		Italy	SIAI
S-211		Italy	SIAI
Sabreliner	(CT-39)	U.S.	Rockwell
Safari	MFI-15		
Safir	SAAB-91 (SK-50)	Sweden	SAAB
SC-7	Skyvan	U.K.	Short
SE-210	Caravelle		
Sea Harrier	(Harrier)		
Seascan	IAI-1124		
Searchmaster B/L	N-24		
Seneca	PA-34 (EMB-810)	U.S.	Piper
Sentry	(O-2)	U.S.	Summit
SF-37	(J-37)		
SF-260	(SF-260W Warrior)	Italy	SIAI
SH-37	(J-37)		
Shackleton		U.K.	BAe
Sherpa	Short 330, C-23		
Short 330		U.K.	Short
Sierra 200	(Musketeer)		
SK-35	(J-35)	Sweden	SAAB
SK-37	(J-37)		
SK-50	(Safir)		
SK-60	(SAAB-105)	Sweden	SAAB
SK-61	(Bulldog)		
Skyvan		U.K.	Short
SM-1019		Italy	SIAI
SM-601	Corvette		
SNJ	T-6 (Navy)		
SP-2H	Neptune	U.S.	Lockheed
SR-71	Blackbird	U.S.	Lockheed
Su-7	"Fitter A"	USSR	Sukhoi
Su-15	"Flagon"	USSR	Sukhoi
Su-17/-20/-22	"Fitter"	USSR	Sukhoi
Su-24	"Fencer"	USSR	Sukhoi
Su-25	"Frogfoot"	USSR	Sukhoi
Su-27	"Flanker"	USSR	Sukhoi
Super Etendard		France	Dassault
Super Galeb		Yug	SOKO
Super Mystère		France	Dassault
T-1		Japan	Fuji
T-2	Buckeye	U.S.	Rockwell
T-2		Japan	Mitsubishi
T-3		Japan	Fuji
T-6	Texan	U.S.	N. American
T-17	(Supporter, MFI-17)	Sweden	SAAB
T-23	Uirapurú	Brazil	Aerotec
T-25	Neiva Universal	Brazil	Embraer
T-26	EMB-326		
T-27	Tucano	Brazil	Embraer
T-28	Trojan	U.S.	N. American
T-33	Shooting Star	U.S.	Lockheed
T-34	Mentor	U.S.	Beech
T-35	Pillán [PA-28]	Chile	Enaer
T-36	(C-101)		
T-37	(A-37)		
T-38	Talon	U.S.	Northrop
T-39	(Sabreliner)	U.S.	Rockwell
T-41	Mescalero (Cessna 172)	U.S.	Cessna
T-42	Cochise (Baron)	U.S.	Beech
T-43	(Boeing 737)		
T-44	(King Air)		
T-47	(Citation)		
TB-20	Trinidad	France	Aérospatiale
TB-30	Epsilon	France	Aérospatiale
TC-45	(C-45, trg)		
T-CH-1		Taiwan	AIDC
Texan	T-6		
TL-1	(KM-2)	Japan	Fuji
Tornado		U.K./Ge/Italy	Panavia
TR-1	[U-2]	U.S.	Lockheed
Travel Air	Beech 95		
Trident		U.K.	BAe
Trislander	BN-2		
Tristar	L-1011		
TS-8	Bies	Poland	PZL
TS-11	Iskra	Poland	PZL
Tu-16	"Badger"	USSR	Tupolev
Tu-22	"Blinder"	USSR	Tupolev
Tu-26 (Tu-22M)	"Backfire"	USSR	Tupolev
Tu-28	"Fiddler"	USSR	Tupolev
Tu-95	"Bear"	USSR	Tupolev
Tu-126	"Moss"	USSR	Tupolev
Tu-134	"Crusty"	USSR	Tupolev
Tu-142	"Bear F"	USSR	Tupolev
Tu-154	"Careless"	USSR	Tupolev
Tu-160	"Blackjack"	USSR	Tupolev
Turbo Porter	PC-6A/B		
Twin Bonanza	Beech 50		
Twin Otter	DHC-6		
Tzukit	[CM-170]	Israel	IAI
U-2		U.S.	Lockheed
U-3	(Cessna 310)	U.S.	Cessna
U-7	(L-18)		
U-8	(Twin Bonanza/Queen Air)	U.S.	Beech
U-9	(EMB-121)		
U-10	Super Courier	U.S.	Helio
U-17	(Cessna 180, 185)	U.S.	Cessna
U-21	(King Air)		
U-36	(Learjet)		
U-42	(C-42)		
U-93	(HS-125)		
UC-12	(King Air)		
UP-2J	(P-2J)		
US-1		Japan	Shin Meiwa
US-2A	(S-2A, tpt)		
US-3	(S-3, tpt)		

Type	Name/Designation	Origin	Maker
UTVA-66		Yug	UTVA
UTVA-75		Yug	UTVA
UV-18	(DHC-6)		
V-400	Fantrainer 400	Ge	VFW
V-600	Fantrainer 600	Ge	VFW
Vampire	DH-100		
VC-4	Gulfstream I		
VC-10		U.K.	BAe
VC-11	Gulfstream II		
VC-91	(HS-748)		
VC-93	(HS-125)		
VC-97	(EMB-120)		
VC-130	(C-130)		
VFW-614		Ge	VFW
Victor		U.K.	BAe
Vinka	L-70		
Viscount		U.K.	BAe
VU-9	(EMB-121)		
VU-93	(HS-125)		
WC-130	[C-130]		
WC-135	[Boeing 707]	U.S.	Boeing
Westwind	IAI-1124		
Winjeel	CA-25		
Xavante	EMB-326		
Xingu	EMB-121		
Y-5	[An-2]	China	Hua Bei
Y-7	[An-24]	China	Xian
Y-8	[An-12]	China	Shaanxi
Y-12		China	Harbin
Yak-11	"Moose"	USSR	Yakovlev
Yak-18	"Max"	USSR	Yakovlev
Yak-28	"Firebar" ("Brewer")	USSR	Yakovlev
Yak-38	"Forger"	USSR	Yakovlev
Yak-40	"Codling"	USSR	Yakovlev
YS-11		Japan	Nihon
Z-43		Cz	Zlin
Z-226		Cz	Zlin
Z-326		Cz	Zlin
Z-526		Cz	Zlin
Zéphyr	CM-175		

HELICOPTERS

Type	Name/Designation	Origin	Maker
A-109	Hirundo	Italy	Agusta
A-129	Mangusta	Italy	Agusta
AB-...	(Bell 204/205/206/212/214/etc.)	Italy/U.S.	Agusta/Bell
AH-1	Cobra/Sea Cobra	U.S.	Bell
AH-6	(Hughes 500/530)	U.S.	MD
AH-64	Apache	U.S.	Hughes
Alouette II	SE-3130, SA-318	France	Aérospatiale
Alouette III	SA-316, SA-319	France	Aérospatiale
AS-61	(SH-3)	U.S./Italy	Sikorsky/Agusta
AS-332	SuperPuma	France	Aérospatiale
AS-350	Ecureuil	France	Aérospatiale
AS-355	Ecureuil II		
ASH-3	(Sea King)	Italy/U.S.	Agusta/Sikorsky
AUH-76	(S-76)		
Bell 47		U.S.	Bell
Bell 204		U.S.	Bell
Bell 205		U.S.	Bell
Bell 206		U.S.	Bell
Bell 212		U.S.	Bell
Bell 214		U.S.	Bell
Bell 406		U.S.	Bell
Bell 412		U.S.	Bell
Bo-105	(NBo-105)	Ge	MBB
CH-3	(SH-3)		
CH-34	Choctaw	U.S.	Sikorsky
CH-46	Sea Knight	U.S.	Boeing-Vertol
CH-47	Chinook	U.S.	Boeing-Vertol
CH-53	Stallion (Sea Stallion)	U.S.	Sikorsky
CH-54	Tarhe	U.S.	Sikorsky
CH-113	(CH-46)		
CH-118	Bell 205		
CH-124	SH-3		
CH-135	Bell 212		
CH-136	OH-58		
CH-139	Bell 206		
CH-147	CH-47		
Cheetah	[SA-315]	India	HAL
Chetak	[SA-319]	India	HAL
Commando	(SH-3)	U.K./U.S.	Westland/Sikorsky
EH-60	(UH-60)		
EH-101		U.K./Italy	Westland/Agusta
FH-1100	(OH-5)	U.S.	Fairchild-Hiller
Gazela	(SA-342)	France/Yug	Aérospatiale/SOKO
Gazelle	SA-341/-342		
H-34	(S-58)		
H-76	S-76		
HA-15	Bo-105		
HB-315	Gavião (SA-315)	Brazil/France	Helibras/Aérospatiale
HB-350	Esquilo (AS-350)	Brazil/France	Helibras/Aérospatiale
HD-16	SA-319		
HH-3	(SH-3)		
HH-34	(CH-34)		
HH-53	(CH-53)		
Hkp-2	Alouette II/SE-3130		
Hkp-3	AB-204		
Hkp-4	KV-107		
Hkp-5	Hughes 300		
Hkp-6	AB-206		
Hkp-9	Bo-105		
Hkp-10	AS-332		
HR-12	OH-58		
HSS-1	(S-58)		
HSS-2	(SH-3)		
HT-17	CH-47		
HT-21	AS-332		
HU-1	(UH-1)	Japan/U.S.	Fuji/Bell
HU-8	UH-1B		
HU-10	UH-1H		
HU-18	AB-212		
Hughes 269		U.S.	MD
Hughes 300		U.S.	MD
Hughes 369		U.S.	MD

Type	Name/Designation	Origin	Maker	Type	Name/Designation	Origin	Maker
IAR-316/-330	(SA-316/-330)	Ro/France	IAR/Aérospatiale	SA-316	Alouette III (SA-319)	France	Aérospatiale
Ka-25	"Hormone"	USSR	Kamov	SA-318	Alouette II (SE-3130)	France	Aérospatiale
Ka-27	"Helix"	USSR	Kamov	SA-319	Alouette III (SA-316)	France	Aérospatiale
KH-4	(Bell 47)	Japan/U.S.	Kawasaki/Bell	SA-321	Super Frelon	France	Aérospatiale
				SA-330	Puma	France	Aérospatiale
KH-300	(Hughes 269)	Japan/U.S.	Kawasaki/MD	SA-341/-342	Gazelle	France	Aérospatiale
				SA-360	Dauphin	France	Aérospatiale
KH-500	(Hughes 369)	Japan/U.S.	Kawasaki/MD	SA-365	Dauphin II (SA-360)		
				Scout	(Wasp)	U.K.	Westland
Kiowa	OH-58			SE-3130	(SA-318)		
KV-107	[CH-46]	Japan/U.S.	Kawasaki/Vertol	SE-316	(SA-316)		
				Sea King	[SH-3]	U.K.	Westland
				SH-2	Sea Sprite	U.S.	Kaman
Lynx		U.K.	Westland	SH-3	(Sea King)	U.S.	Sikorsky
MH-6	(AH-6)			SH-34	(S-58)		
MH-53	(CH-53)			SH-57	Bell 206		
Mi-1	"Hare"	USSR	Mil	SH-60	Sea Hawk (UH-60)		
Mi-2	"Hoplite"	USSR	Mil	Sioux	(Bell 47)	U.K.	Westland
Mi-4	"Hound"	USSR	Mil	TH-55	Hughes 269		
Mi-6	"Hook"	USSR	Mil	TH-57	SeaRanger (Bell 206)		
Mi-8	"Hip"	USSR	Mil	UH-1	Iroquois (Bell 204/205)		
Mi-14	"Haze"	USSR	Mil				
Mi-17	"Hip"	USSR	Mil	UH-12	(OH-23)	U.S.	Hiller
Mi-24	"Hind"	USSR	Mil	UH-13	(Bell 47J)		
Mi-25	"Hind"	USSR	Mil	UH-19	(S-55)		
Mi-26	"Halo"	USSR	Mil	UH-34T	(S-58T)		
Mi-28	"Havoc"	USSR	Mil	UH-46	(CH-46)		
Mi-35	(Mi-25)			UH-60	Black Hawk (SH-60)	U.S.	Sikorsky
NAS-332	AS-332	Indon/France	Nurtanio/Aérospatiale	VH-4	(Bell 206)		
				Wasp	(Scout)	U.K.	Westland
NB-412	Bell 412	Indon/U.S.	Nurtanio/Bell	Wessex	(S-58)	U.S./U.K.	Sikorsky/Westland
NBo-105	Bo-105	Indon/Ge	Nurtanio/MBB	Whirlwind	(S-55)	U.S./U.K.	Sikorsky/Westland
NH-300	(Hughes-300)	Italy/U.S.	Nardi/MD	Z-5	[Mi-4]	China	Harbin
NSA-330	(SA-330)	Indon/France	Nurtanio/Aérospatiale	Z-6	[Z-5]	China	Harbin
				Z-8	[SA-321]	China	Changhe
OH-6	Cayuse (Hughes 369)	U.S.	MD	Z-9	[SA-365]	China	Harbin
OH-13	(Bell 47G)						
OH-23	Raven	U.S.	Hiller				
OH-58	Kiowa (Bell 206)						
OH-58D	(Bell 406)						
PAH-1	(Bo-105)						
Partizan	(Gazela, armed)						
RH-53	(CH-53)						
S-55	(Whirlwind)	U.S.	Sikorsky				
S-58	(Wessex)	U.S.	Sikorsky				
S-61	SH-3						
S-65	CH-53						
S-70	UH-60						
S-76		U.S.	Sikorsky				
S-80	CH-53						
SA-315	Lama [Alouette II]	France	Aérospatiale				

Source: International Institute for Strategic Studies. 1991. *The Military Balance, 1991–1992*. London: Brassey's.

Note: The use of [square brackets] shows the type from which a variant was derived. "Q-5 . . . [MiG-19]" indicates that the design of the Q-5 was based on that of the MiG-19.

(Parentheses) indicate an alternative name by which an aircraft is known—sometimes in another version. "L-188 . . . Electra (P-3 Orion)" shows that in another version the Lockheed Type 188 Electra is known as the P-3 Orion.

Names given in "quotation marks" are NATO reporting names (e.g., "Su-27 . . . "Flanker").

When no information is listed under "Origin" or "Maker," take the primary reference given under "Name/designation" and look it up under "Type."

G

GENDARMERIE

The term *gendarmerie* is the collective form of *gendarme*, derived from the French *gens d'armes*, meaning "men at arms" or "armed men." A gendarmerie is a national military force that has both civil and military functions. There are gendarmeries in several European nations; all trace their origin to the French Gendarmerie Nationale, a type of military police that was established when the Maréchausée, the military police of the Ancien Régime, was reorganized in 1791.

By 1798, this gendarmerie was placed under civil authority, the Ministry of the Interior. It was later reorganized by Napoleon I, who increased its strength, gave it greater civil responsibilities, and stationed it in each *département* in France. Despite these changes, by the end of the Napoleonic era the gendarmerie's duties had again become largely military, as its primary responsibilities were to combat enemy forces, suppress internal rebellion, and maintain discipline in the army.

Gendarmerie in Germany

Under Napoleon's rule, a gendarmerie was established to ensure law and order in the French-occupied German territories. The Prussian Gendarmerie was established in 1812. Although it was subordinated to a civilian authority—the Ministry of Interior—its structure was military in that its officers and men were drawn from the ranks of the army, its arms (rifles, bayonets, and pistols) were similar to those of the army, and it was under the command of a general.

The Prussian Gendarmerie was very successful in protecting the population against the retreating French in the wars of 1813 and 1815. Other German states, including Bavaria, Saxony, and Baden, emulated this Prussian force, and the gendarmerie concept was reflected in the constabularies that were established in the German colonies. As of 1871, there was no single national gendarmerie; rather, the several gendarmeries were subordinate to their respective districts. Nonetheless, they were occasionally used for military purposes during maneuvers, because there were no military police. Thus, a close connection between the civil gendarmeries and the military was continued.

This connection was severed effectively after Germany's defeat in World War I. The victorious Allies insisted that the gendarmerie be disassociated from the Ministry of War and that police jurisdiction be transferred to the states. The local and state police no longer underwent military training and they had neither military equipment nor a military structure.

In 1934, the National Socialists re-established the central authority for police functions in Berlin. The gendarmerie was re-established, to include a military organization, and remained subordinate to the Ministry of the Interior. After World War II, the gendarmerie continued to exist in Baden-Wurttemburg, Hesse, and the Rhein Palatinate, where it was placed under the authority of the local governments.

Other Contemporary Gendarmeries

The gendarmerie system of training, organizing, and equipping men as soldiers to perform police duties for civilian authorities can be seen in several nations. Spain has its Guardia Civil; Italy, its Carabinieri; and the Netherlands, its Maréchausée and Rijkswacht. In Austria, the federal gendarmerie is a uniformed, armed constabulary that is stationed outside larger towns and in the nation's rural areas. In the French-speaking cantons of Switzerland, the gendarmerie is a part of the canton police. Belgium's gendarmerie, established after the kingdom gained its independence in 1830, is a modern force that performs specialized police tasks. The Dutch Maréchausée dates back to 1814, when it was established by King William I. Following World War I it was almost exclusively responsible for the civilian sector, but when the police system was reorganized in 1945, the government assigned all military police duties to the Maréchausée Corps and created a civilian police force. This has been changed slightly, and today the Maréchausée has partial responsibility for the civilian sector.

Nature of Contemporary Gendarmerie

While the nature, organization, duties, and equipment of gendarmeries vary considerably from nation to nation, certain general observations are nonetheless valid.

Gendarmerie and the Military

Until the Napoleonic era there was no clear distinction between the gendarmerie and the military, since soldiers were used as a police force to maintain order within a given state. The close affinity between the two was due to the gendarmerie's practice of accepting into its ranks reliable veteran soldiers who were no longer suitable for military service. These men brought with them established military concepts of discipline and control, which contributed to the hybrid character of the gendarmerie—it performed civilian police duties as a powerful military organization. The taking in of ex-soldiers coupled with the military structure of the gendarmerie, made the gendarmerie a quasi-military organization that could be employed for security functions, guarding prisoners, traffic control, medical assistance, mail delivery, arbitration in military-civilian disputes, and other military duties.

Organizational Control and Structure

Generally, in peacetime, a nation's gendarmerie is subordinate to the Ministry of the Interior and civilian authorities, while in wartime and periods of national emergency it becomes subordinate to the nation's army.

Because of the military influence, a gendarmerie is organized hierarchically, with a clearly defined command structure. A military influence also is evident in the organization's recruitment, appointment, training, and operations, which are subject to strict regulations. Normally, a gendarme's duties are performed within divisions in response to orders from higher command. Gendarmes may be assigned duties in detachments or may be assigned to independent, individual billets. The major source of recruitment is the army's noncommissioned officer (NCO) corps, and applicants who have passed the army's NCO examination are also eligible for service in the gendarmerie. Recruits are trained in schools and units and are housed in barracks.

In addition to its regular police units, the gendarmerie also has specialized units, which may include air police, antiterrorist commandos, mountain police, river police, security police, border guards, and honor guards.

Functions

Contemporary gendarmeries perform military and civil duties. They are invested with comprehensive police jurisdiction and are subordinate to civilian authority. The tasks of the gendarmeries are maintaining law and order and providing adequate security for the nation. Thus, the jurisdiction of France's Gendarmerie Nationale extends over both military personnel and civilians. It performs general police duties, criminal investigations, and administrative police duties and has court responsibilities. The Dutch Maréchausée provides protection for the royal family; performs military police duties involving the army, air force, and navy; and guards the nation's borders. Additionally, it would provide support to the nation's civilian departments to maintain law and order during a period of unrest. It is also responsible for special duties that may be assigned to it by law or cabinet resolution.

Given the capabilities of the gendarmeries and their jurisdiction that extends into the military realm, the jurisdictions of the gendarmerie and the military police may overlap. Generally, in wartime a nation's military police have considerable jurisdiction extending into the civilian sector. Among the duties common to military police are providing traffic control; rounding up military stragglers; caring for refugees and evacuees; providing medical and nuclear, biological, and chemical (NBC) warfare assistance; providing security; and combating saboteurs. Some of these duties are also performed by gendarmeries; jurisdictional conflicts are settled by the state.

Equipment

Because of its paramilitary nature, a gendarmerie often has equipment similar to that of a military organization, with firepower that surpasses the average police force. Most have modern, sophisticated telecommunications systems that provide excellent command and control. Some detachments are provided with small arms and mortars. The gendarmerie also has a wide assortment of vehicles, including automobiles and motorcycles, armored vehicles, helicopters, motor boats, and torpedo boats, which provide excellent mobility along the nation's land frontiers and sea coasts and in its airspace.

Summary

Since its modern origins in 1791, the gendarmerie—a unique force that has both military and civilian aspects and functions—has been popular in Europe. The consolidation that will occur among many European nations in 1992 may affect the national composition of the several European gendarmeries. The relaxation of many European borders may well create situations that are ideal for drug smugglers and terrorists—threats to which the gendarmerie normally responds. One thing appears certain: the national gendarmeries or similar organizations will continue to provide the vital services afforded by today's gendarmeries.

Stephan Link
Karlheinz Böckle

SEE ALSO: Internal Security and Armed Forces; Internal Security Forces; Paramilitary Forces.

Bibliography

Blankenstein, W. 1931. *Die Preussische Landjaegerei im Wandel der Zeiten.* Erfurt: W. Blankenstein.
Böckle, K. 1987. *Feldgendarmen, Feldjaeger, Militaepolizisten: Ihre Geschichte bis Heute.* Stuttgart: Motorbuch Verlag.
Harnischmacher, R., and A. F. Semerak. 1986. *Deutsche Polizeigeschichte.* Stuttgart: Kohlhammer Verlag.
Steinhauser, A. 1900. *Geschichte des Grossherzoglich Badischen Gendarmeriekorps.* Karlsruhe: Braunsche Hofbuchdruckerei.
Wilhelm, T. M. 1980. *200 Jahre Deutsche Polizei.* Lübeck: Verlag Polizei.

GENERAL WAR

The term *general war* implies direct military conflict between the superpowers. Such conflict encompasses not only the forward-based armed forces and resources of the powers, but their territories as well. The cause and objective of general war lie in the offensive and active enforcement of vital interests of the powers concerned.

General war involves the direct employment of strategic offensive nuclear weapons (i.e., intercontinental ballistic missiles [ICBMs], submarine-launched ballistic missiles [SLBMs], and nuclear weapons carried by heavy bombers). Such direct employment is considered limited as long as intrawar deterrence prevails; otherwise it is unlimited.

General war may reach its highest intensity of all-out nuclear war either immediately at the onset of conflict in the case of a surprise attack, or at the final stage of gradually or progressively increasing escalation. Such all-out nuclear war means the use of strategic offensive nuclear weapons, unlimited in terms of target selection and effects on the targets.

General war is distinct from total war, which prevailed at the end of World War II after a long period of development, starting with the wars of revolutionary and Napoleonic France, and resulting in a new quality at the end of the war. The means of that new quality of war was the introduction of nuclear armament; its cause, the developing antagonism between the United States and the Soviet Union and their respective political clients.

Whereas total war was characterized by the total mobilization of personnel and materiel resources for war, general war is characterized by the unlimited use of all weapons available, in particular strategic offensive nuclear weapons.

Definition

The term *general war* is defined as "armed conflict

- between major powers,
- in which the total resources of the belligerents are employed,
- and the national survival of a major belligerent is in jeopardy." (DoD/JCS 1979)

The above definition of general war is, in the strategic debate in the United States, criticized as being too narrow and unprecise on the following grounds:

1. Only regional powers or pacts can be understood as "major powers." A war between such states or powers, in which the superpowers, the United States and the Soviet Union, are not directly involved, cannot be defined as general war, but must instead be defined as *limited war* in terms of geography and weapons used.

   Consequently, the *Glossary of Strategic Terms* of the U.S. Army Command and General Staff College widens the above definition of general war by adding, "Commonly reserved for a showdown between the United States and the USSR and featuring nuclear weapons."

2. The term *general war* should never be mistaken for the historic concept of *total war*, a major characteristic of which was the total mobilization of all personnel and materiel resources. General war needs to be carried out with all the forces *available* (Kissinger 1957), and such a conflict between the United States and the Soviet Union would presumably be over before the total resources could be employed.

   The employment of all resources available, as postulated in the definition in the *Dictionary of Military Terms* (U.S. Dept. of Defense 1984), appears only conceivable in the case of a protracted war.

The term *general war* must consequently only be identified: with a direct "genocidal showdown" (Collins 1973) between the United States and the Soviet Union in which the unlimited use of strategic offensive nuclear weapons against strategic targets (i.e., against targets on the territories of both powers) will have to be understood as the most relevant defining criterion of the phenomenon of general war.

As far as application of the definition is concerned, it is not of primary importance whether the unlimited use, in terms of target selection and effects on the targets, of strategic offensive weapons happens at the beginning of an escalatory phase or at its end. Potentially different courses of general war necessitate a differentiation between

- the self-reliant phenomenon of a direct military conflict between the United States and the Soviet Union, and
- the phenomenon of war caused by escalation:

  1. of a limited war in which the United States and the Soviet Union may be involved indirectly or directly, or
  2. of a regional conflict, which may in its course be widened to include either superpower.

Strategic offensive nuclear weapons are mostly reserved as a means of deterrence and thus for the covering of strategic targets in case of general war, whereas substrategic nuclear weapons are essentially implements of limited war. Strategically important targets are those on the territories of either superpower, the carriers of strategic offensive nuclear weapons, and the assets of strategic defense. That differentiation, acknowledged in principle, between strategic and nonstrategic nuclear weapons should serve to reduce the accelerating factor of the momentum of escalation.

General war can thus be understood as *central war*, in which *central systems*, "which the US considers central to the strategic nuclear relationship between the US and the USSR" (U.S. Arms Control and Disarmament Agency 1974), are in the end employed without limit. Such unlimited use of strategic nuclear weapons in general war may occur in the form of a surprise attack at the onset of the conflict or following an escalatory process.

In its abstract form, the process of general war may be subdivided into two phases of differing duration:
1. The initial phase would be the time period when intrawar deterrence was still credible and effective, a state that implies the possibility of a limited use of strategic offensive nuclear weapons against strategic targets.
2. The end of the escalatory phase of general war would be marked by actual general nuclear response, in which case the Single Integrated Operations Plan (SIOP) would be executed by order of the U.S. president. That final stage of escalation of general war would imply "all-out nuclear war."

Historical Dimension of General War

The current perception of the term and contents of *general war* in the United States started with the strategic air war and its doctrine in World War II. Yet the new dimension of warfare that resulted from the availability and possible use of nuclear weapons was not recognized in the beginning, because such weapons were not available until the final weeks of the war. The atom bomb was soon recognized, however, as a means to achieve efficient economy of forces in strategic warfare. Rather than using fleets of "flying fortresses," a single aircraft carrying a nuclear warhead was sufficient to achieve destruction equivalent to that achieved by strategic forces as exercised in World War II.

It was presumed in the political and strategic thinking in the United States after 1945 that a future war could only take place in the form of *general war* caused by a large-scale Soviet offensive in Europe. Because of the limited range of the atom bomb carriers of the time, such general war could not threaten targets on the territory of the United States.

Strategic Air War

In the United States, efforts were made to evaluate a general strategic doctrine from the success of the strategic air wars against Germany and Japan in World War II. This doctrine was exclusively oriented toward general war, then presumed to be unavoidable in the event of superpower conflict. The intrinsic exclusion of intermediate stadia between total peace and general war resulted in a drive toward exclusively worst-case thinking, and consequently to the neglect of American preparations for challenges on lower conflict levels, although realistically these could be expected. The conviction prevailed that a future war would be a global strategic conflict between the United States and the Soviet Union in which it would be mandatory to destroy the vital installations of the attacker by nuclear means. With the increasing destructiveness of the nuclear arsenal under development, the challenge justifying the use of such weapons, according to American views, would have to be the greater. In doing so, the United States blocked its own political actions because Soviet actions were always interpreted to be of such dimensions that general war was the only form of reaction considered, in particular by the military leadership in the United States. Lesser reactions were not considered "worthwhile" (Kissinger 1957).

When the limited war in Korea broke out in 1950, the United States was militarily and theoretically unprepared. The Korean War did not correspond to the current strategic thinking in the United States and thus became a doctrinal challenge for the inclusion of possible stages of minor conflict intensity into a revised theory and practice of the planning for war.

The introduction of nuclear weapons at the end of World War II made previous theories of war and strategy obsolete. The continued, and already traditional, doctrine of strategic bombing (attacking the social and industrial centers of the opponent) was considered the most effective use of the atom bomb, oriented toward achieving the breakdown of the opponent from within, which should obviate the necessity of achieving victory on the battlefield (Freedman 1989).

In the United States, strategic nuclear bombing in general war was conceived as the perfect technological development in support of what was becoming the traditional American method of war (Baylis et al. 1987).

Douhet

The theoretical basis for strategic nuclear bombing as the intrinsic criterion of general war can be traced to the form of warfare during World War I that steadily encompassed totality. One of the central topics of the strategic debate after 1918 was that of "independent" air forces and, consequently, an "independent" air war to be directed primarily against the civilian population.

These ideas for the offensive use of air forces were further developed to a consistent doctrine by the Italian general Giulio Douhet (1869–1930) in his book *Il dominio dell' aria* (1921). His ideas were shared in the United States by William Mitchell and in Great Britain by Hugh Trenchard. The doctrine had a strong influence on the developing theory of air warfare in the 1920s and 1930s and, consequently, led to strategic bombing (Millis 1956).

According to Douhet, it should be the primary objective of the air offensive to achieve air supremacy (if necessary, to first reduce the existing air supremacy of the opponent), the maintenance of which would further his victory. Following the achievement of air supremacy, civilian and economic targets on the opponent's territory should be completely destroyed during a second phase, facilitating the breakdown of social order. In the interest of survival, the population would then force termination of the war before completion of the mobilization of land and sea forces (Fuller 1961).

The doctrine of Douhetism (primarily based on intuition rather than on analysis) still influenced the conceptions of strategic nuclear bombing after World War II, although it had already led to fatal consequences (at least in the eyes of some Allies, as well as the victims) in the decisions of the Allies during World War II.

General War and Strategic Doctrine in the United States

American strategic doctrine after World War II was not a result of consideration of the potential use of the atom bomb, rather it represented an updated version of warfare that had evolved during World War II. The doctrine was exclusively oriented toward the waging of general war, in which victory should be achieved by reliance on the industrial potential of the United States and the means of strategic air forces.

The aims of a major war between the superpowers (at the core of the doctrine and its practical application) should be—as in World War II—the achievement of unconditional surrender. The achievement of that goal appeared feasible only by total defeat of the opponent in general war rather than as a consequence of limited war.

Unconditional surrender implies that the victor takes upon himself the responsibility for the defeated party, including a general build-up after war. Since the consequences of general war would likely be disastrous to the victorious party as well, victory in such a case would have little historical relevance (Kissinger 1957).

MASSIVE RETALIATION

With the Soviet Union's development of a nuclear armament able to threaten targets on the territory of the United States, there came a need to rethink the doctrine of general war. That doctrine had been founded on the theory of total war oriented toward a regional conflict in Europe or Asia.

The basic maxims of the new thinking of the nuclear age, which also modified the concept of general war, comprised:

- the impossibility of defense,
- the vulnerability of large cities,
- the attractiveness of surprise attack, and
- the necessity of a credibly assured capacity for counterstrike

The cold war made it necessary for U.S. strategic thinkers to abandon the one-dimensional perception of purpose and employment of nuclear weapons. That perceptual revision led also to a different concept of general war, to the acceptance of the possibility of a limited war between the superpowers, and thus to a complex interaction between the conduct of the cold war and the strategic importance of enhanced strategic and substrategic nuclear armaments (Freedman 1989).

The conception of a rather traditional strategy of general war in the years after World War II had failed to prevent the Korean War, and had not contributed to optimum efficiency of conventional warfare therein. Whereas deterrence against general war continued to be guaranteed to a large extent by the available, and increasing, nuclear potential of both sides, the problem of limited war remained unsolved.

These developments resulted in a demand for the concentration of American efforts on the defense of the continental United States, and the shaping of a capacity for massive retaliation.

That premise required a division of strategic functions between the United States and its nonnuclear allies. The United States became responsible for the organization of the nuclear components of general war; its allies became responsible for the handling of potential limited and local attempts at aggression. Agreement in principle along these lines eliminated the possibility of armed forces that could be used in all potential conflicts and conflict areas. Deterrence consequently rested upon the capacity of the United States for massive retaliation.

Warning to the potential enemy of the concept of massive retaliation did not imply either abandoning defense or instant resort to general war in any case. Conflict control, and dominance in the escalation process, would be guaranteed by keeping the initiative in the conduct of conflict and by controlled escalation in response to direct threats. The deterrence resulting thereby should prevent war.

At the end of the 1950s, general war was in principle perceived as strategic aerial warfare distinct from the historic phenomenon of war. The dominance of strategic airpower predetermined and consequently simplified the target planning and target selection within the system.

The question of the use of land and naval forces in general war, however, remained unsolved; that is whether they could effectively operate during the intensive strategic aerial war, and whether they could achieve a victorious termination of the war following a successful strategic nuclear exchange (Brodie 1959). In the structuring of the armed forces, as in strategy and doctrine, the utmost attention was devoted to the question of whether a general war should be perceived as a short or long war.

Nuclear exchange could be executed within a short period. The duration of war at that time was assumed to depend on the existence of retaliation forces, which at that time were still vulnerable. As soon as they could become invulnerable, a general war of longer duration would become conceivable because the weapon systems concerned would no longer need to be used immediately in order to be safeguarded from counterforce attack.

Congruent with the problem of the duration of general

war was the question of whether a strategy of destruction or of attrition should be applied. The former corresponded to a short, devastating war, whereas the latter would have been preferred in cases where a gradual weakening of the power of the opponent was intended (Morgenstern 1961).

FLEXIBLE RESPONSE

The doctrine of massive retaliation was gradually replaced by that of flexible response. Initial massive retaliation was supplemented by a credible adequate option for the conduct of limited war and thus for expanded deterrence. In a similar way, the differentiation between general war and limited war after the Korean War, which demonstrated the reality of limited war, was introduced into the strategic doctrine of the United States in the early 1960s.

The strategic doctrine of flexible response, which evolved in the 1970s, requires: (a) a highly survivable second-strike force for deterrence against general war; (b) flexibly employable armed forces to cover the remaining spectrum of potential forms of attack; and (c) the securing of a permanent ability to control the conduct of conflict at all levels, which is regarded as imperative for the prevention of automatic escalation.

The concept of flexible response requires an ability to conduct war successfully on any level, by conventional or nuclear means, without triggering unlimited general war. Intrawar deterrence should enforce conflict limitation; only its failure would make escalation to all-out nuclear war conceivable.

Preconditions for the efficacy of flexible response are:
1. the effective securing of strategic stability in the upper range of the conflict spectrum by secured second-strike capacity, requiring above all a credibly high survivability of the strategic offensive potential, as well as protected (and thus survivable) command, control, and communications systems. That should obviate the necessity for early employment of the directly threatened strategic offensive weapons in order to prevent them from early destruction by the opponent. As a consequence, the opponent should be forced to respect vital American interests, thereby forcing conflicts to remain at lower levels.
2. The existence of "sufficient" armed forces, able to deny any success to the opponent at any of the conflict levels chosen by him.

The political aim lies in achieving and maintaining crisis stability to prevent escalation to general war or an immediate and direct American-Soviet military exchange in the form of all-out nuclear war if deterrence fails.

The doctrine of flexible response, based originally on the nuclear supremacy of the United States and counterforce options, should prevent general war during the first phase of a superpower conflict because the development of equivalent Soviet nuclear armament removed the secured strategic supremacy of the United States and its dominance of escalation control in the case of conflict. Thus, strategic nuclear balance became the basis for flexible response. That balance, however, can only be achieved by arms control in concert with the Soviet Union (i.e., by SALT and START).

Central to the importance of general war is the imperative to safeguard an assured destruction capability by each of the components of strategic offensive armament (ICBMs, SLBMs, heavy bombers) and their total capacity to achieve flexible response as well as to perform unlimited execution of the strategic target plan. But even in the case of a regional conflict involving U.S. allies, American forces, and nonstrategic nuclear weapons, the strategic offensive nuclear weapons would primarily be reserved to cover those strategic targets (i.e., targets on the territories of either power) that have been selected to be destroyed in a general war between the United States and the Soviet Union.

Such general war is conceivable as (a) a controlled development starting as limited general war, which by failure of intrawar deterrence, escalates to unlimited general war, or (b) as a surprise attack in the form of all-out nuclear war from the beginning (Kahn 1965).

It is the primary objective of the American strategic nuclear potential to produce deterrence. If that deterrence should fail, not only in a regional conflict but also in a conflict between the United States and the Soviet Union, strategic offensive nuclear weapons are regarded as a militarily usable means for the conduct of general war.

As already mentioned above, in general war a distinction would have to be made between a limited and an unlimited phase, in correspondence with the escalation model.

Summary

General war reflects a strategic problem for which no historic counterpart can be found. The nuclear potential of intercontinental range and large-scale destructive capacity causes risks far exceeding any potential success, thereby allowing little space for politico-strategic nonrationality. As a consequence, for the first time in history, the political-military emphasis is placed on efforts toward the prevention of war.

Central aspects found in the theory of general war concern:

- the role and applicability of nuclear weapons,
- the future importance and form of nuclear warfare, and
- the complex of primary subject or object, respectively, of threat (Head and Rokke 1973).

The definition of deterrence and defense, the two central objectives of the strategy of general war, is of crucial importance to the United States as one of the powers directly involved. It is the objective of deterrence to deter the opponent from military aggression by making costs and risks exceed the potential gains that could be expected. It is the objective of defense to minimize American costs and risks in case of a failure of deterrence. Whereas deterrence is aimed at the potential intentions of the opponent, defense is conceived as reducing the ability of the opponent to cause damage.

General war, in which the United States and the Soviet Union would be involved with their total territories, requires consequently above all the counterforce use of armed forces as a means of damage reduction in order to reduce one's own costs and to cause the opponent to restrict escalation via such damage reduction (Snyder 1975).

RUDOLF HECHT

SEE ALSO: Airpower, Strategic; Arms Control and Disarmament; Deterrence; Douhet, Giulio; Flexible Response; Limited War; Nuclear Employment Policy and Planning; Nuclear Powers; Nuclear Theory and Policy; Spectrum of Conflict; Total War.

Bibliography

Baylis, J., K. Boost, et al. 1987. *Contemporary strategy*. 2 vols. New York: Holmes and Meier.
Brodie, B. 1959. *Strategy in the missile age*. Princeton, N.J.: Princeton Univ. Press.
Collins, J. M. 1973. *Grand strategy: Its principles and practices*. Annapolis, Md.: U.S. Naval Institute Press.
Douhet, G. 1942. *The command of the air*. Trans. D. Ferrari. New York: Coward-McCann.
Freedman, L. 1989. *The evolution of nuclear strategy*. New York: St. Martin's Press.
Fuller, J. F. C. 1961. *The conduct of war, 1789–1961: A study of the impact of the French, industrial, and Russian revolutions on war and its conduct*. New Brunswick, N.J.: Rutgers Univ. Press.
Head, R. G., and E. J. Rokke, eds. 1973. *American defense policy*. Baltimore, Md.: Johns Hopkins Univ. Press.
Kahn, H. 1965. *On escalation: Metaphors and scenarios*. New York: Praeger.
Kissinger, H. 1957. *Nuclear weapons and foreign policy*. New York: Harper.
Lider, J. 1977. *On the nature of war*. Farnborough, U.K.: Saxon House.
Millis, W. 1956. *Arms and men: A study of American military history*. New Brunswick, N.J.: Rutgers Univ. Press.
Morgenstern, O. 1961. *The question of national defense*. New York: Vintage Books.
Snyder, G. H. 1975. *Deterrence and defense: Toward a theory of national security*. Westport, Conn.: Greenwood Press.
U.S. Arms Control and Disarmament Agency (ACDA). 1974. *SALT lexicon*. ACDA Publication No. 71. Washington, D.C.: ACDA.
U.S. Army. 1982–83. *Glossary of strategic terms*. Ft. Leavenworth, Kans.: U.S. Army Command and General Staff College.
U.S. Department of Defense. Joint Chiefs of Staff (JCS). 1984. *Department of Defense dictionary of military terms*. JCS Publication no. 1. Washington, D.C.: U.S. Government Printing Office.

GENERALSHIP

As used originally, the term *generalship* involved those duties, responsibilities, and attributes required of a general officer during the conduct of active campaigns. In more recent times, the term has come to encompass all the functions carried out by general officers in modern armies including, but not limited to, directing wartime activities. Most theorists include the functions of leadership, management, and command of military organizations larger than brigades. Additionally, the modern analyst usually includes such activities as high-level staff responsibilities: analysis, planning, and organization and training of military organizations in peacetime. Less often included, but frequently assumed as components of generalship, are such functions as liaison with and advice to the civilian leadership of governments and even functioning as a civil government official in some cases. According to W. Duane's 1810 *A Military Dictionary:*

> *The natural qualities of a general*, are a martial genius, a solid judgment, a healthy robust constitution, intrepidity and presence of mind on critical occasions, indefatigability in business, goodness of heart, liberality, a reasonable age; if too young, he may want experience and prudence; if too old, he may not have vivacity enough. His conduct must be uniform, his temper affable but inflexible in maintaining the police and discipline of an army.
>
> *Acquired qualities of a general* should be secrecy, justice, sobriety, temperance, knowledge of the art of war from theory and practice, the art of commanding, and speaking with precision and exactness, great attention to preserve the lives and supply the wants of the soldiers, and a constant study of the characters of the officers of his army, that he may employ them according to their talents.... His experience inspires his army with confidence, and an assurance of victory; and his good qualities, by creating respect, augment his authority.... He ought to be fond of glory, to have an aversion to flattery, to render himself beloved, and to keep a strict discipline and regular submission.

History of Generalship

Much of the early study of generalship relied on historical analysis of famous generals of antiquity, often drawing up lists of "Great Captains" who were assumed to display all the attributes necessary for success as general officers. Such lists usually began with Alexander and included such luminaries as Hannibal, Caesar, Gustavus Adolphus, Marlborough, Eugene, Frederick the Great, Napoleon, and Wellington. Often the compilers of such lists drew conclusions as to characteristics or traits necessary for generalship.

In addition to the Great Captains school of generalship, virtually every military theorist to the present day has tried to define the qualities generals should have. Sun Tzu, Machiavelli, Clausewitz, and Jomini are among the better-known theorists whose aphorisms and advice on generalship have been most widely cited.

The study of leadership from the historical perspective reached its zenith in the period following World War I, when dozens of veterans of that war wrote their reminiscences and advice on leadership and generalship. The scientific study of human behavior accelerated in the period before World War II, and the behavioralist approach to the study of leadership surged during and after that war. With the ascendancy of behavioral science approaches in the period following World War II, the concept of attributes or traits of leadership became more and more discredited and the historical study of leadership and generalship declined as a discipline.

As of 1990, behavioral scientists and historians have been moving toward more of a synthesis on generalship, and, in fact, often incorporate one another's findings into their literature. Few of these studies would be likely to use one approach to the exclusion of the other.

Modern Theories of Generalship

Both scientific and historical studies of generalship tend to identify a mixture of psychological, intellectual, and physical components necessary for success as a general officer. These components are drawn from the functions of leadership and management, at which a general must be adept; often the separate function of command is included.

Until recent times, it would have been impossible for a general to be successful without being an effective leader; even in 1990 it is extremely rare to find a successful general officer who would not be seen first as an accomplished leader. Leadership in this context is the process of working with people, developing their attitudes and values so that they will attain an appropriate goal. Given the size of the organizations generals lead, they can personally lead only a few soldiers at a time—their staffs and soldiers in their proximity, for example. As a consequence, generals must often exert leadership through subordinate layers of command. The ways in which generals have successfully or unsuccessfully exerted such leadership are often featured in historical studies of generals. Anecdotes such as Napoleon's calling experienced veterans by name to encourage units before battle and Eisenhower talking to camouflage-painted paratroopers before the Normandy invasion are contrasted with Haig's refusal to visit the slaughter in the trenches of World War I. Scientific studies of senior leadership likewise stress the importance of a general officer's imprinting his personality on his unit and describe ways in which it can be accomplished.

The study of generalship in earlier times seldom would have stressed management skill as being a talent equal to leadership skill as a requisite for success. Such is not the case at the beginning of the 1990s—management is usually considered of equal and in some respects of paramount importance for effectiveness as a general officer. In retrospect, it would seem that this was always the case. Studies have shown that Alexander's operational concepts were predicated in large part on his logistic requirements. Napoleon's acumen in the same field has been analyzed as an essential component in many of his successes, just as his logistic problems contributed immeasurably to his failure in Russia.

Management is less difficult to define and analyze than are the more intangible ramifications of leadership. The management skills of the military are more closely related to those of the world of business and bureaucracy than is military leadership to its other forms. The latter fact may partially account for the stress on management as a component of generalship in recent times. At any rate, few theorists argue that it is less than a *sine qua non* for effective generalship. Management deals with the scientific planning, apportionment, and use of resources of all kinds. In modern warfare, for which World War II has been the predominant case study since 1945, the successful apportionment and use of resources by the Allies brought their success on the battlefields of Europe and the Pacific, just as resource problems and logistic shortcomings contributed to the ultimate failures of both Germany and Japan. In that war, the management accomplishments of dozens of generals who never commanded forces in the field contributed as much to ultimate victory as did the better-known field commands of generals such as Omar Bradley, Bernard Law Montgomery, and Carl Spaatz. As a result, management threatened at times to eclipse leadership in the study of generalship in the decade following World War II.

Command is the third component that most authorities would recognize as a function of generalship, although some theorists subsume it in the other two. Some successful generals might never exercise command in the traditional sense—such generals as Brehon Somervell, who apportioned resources for the American army in World War II, never commanded units on campaign, for example. Command encompasses the overlapping functions of leadership and management and focuses on the effectiveness of the unit or organization.

When one moves from the functions of generalship to the qualities, attributes, or traits needed by general officers, there is much less agreement among theorists. From the chroniclers of the Great Captains to the 1990s, each analyst has his or her own list of requirements without which a general is bound to fail—such items as intellect, imagination, knowledge, creativity, judgment, experience, objectivity, *coup d'oeil*, communication skills, character, temperament, willpower, ambition, physical and moral courage, integrity, perseverance, boldness, resilience, tenacity, adaptability, human relations skills, technical skills, physical fitness, good health, youth, and luck. Some of these are clearly contradictory—a youthful general is unlikely to have a great deal of experience or refined judgment, for example. Some, such as character, are virtually undefinable or unidentifiable, however real they may be. Still others are self-explanatory or not particularly helpful. From a historical perspective, one could quickly name generals lacking in several of these traits, although it would be hard to find many successful generals who displayed none of them. Nonetheless, keeping the lack of agreement among the experts in mind, a few of these attributes are worth discussing.

Almost all students of generalship would agree on intelligence as a critical attribute of a successful general. Intellect gives the capacity to grasp concepts and to improvise solutions. When combined with imagination, it allows for vision and creative problem solving. As a Russian general stated, "A battle takes place twice: first in thought, then in reality." Napoleon expressed similar thoughts, and the elder Moltke said that no plan survives the first engagement, implying that a talent for improvisation is invaluable to any general.

Intellect enhances the ability to make the most of experience. Frederick the Great cited his mule who had experienced many campaigns but knew no more of war than after his first. Through study of military theory, history, and contemporary doctrine, the general with intelligence and imagination can learn not only from his own experiences but from the experiences of others. In generalship, intellect implies common sense—the native good judgment that allows for discerning evaluation of alternatives.

Most writers on generalship mention strength of character as an essential attribute. Character, one of the less easily definable traits, refers to the aggregate of moral qualities distinguishing an individual's development, usually becoming more fixed with maturity. Thus a general's character would tend to manifest itself in all aspects of his personality and in all that he does. It is most often remarked on in such towering individuals as George Washington, Robert E. Lee, and George Marshall, but it can be identified, if not precisely measured, in lesser mortals.

Courage, both physical and moral, is most evident in campaigning, but is equally essential for a general in less hazardous situations. Examples such as Ulysses Grant riding the front lines unperturbed by the bullets whizzing around him or of "Vinegar Joe" Stilwell marching out of Burma with his defeated troops, heartening the soldiers they commanded, come most easily to mind. In many respects, however, they are overshadowed by such displays of moral courage as Eisenhower ordering the Normandy landings against heavy odds and with the possibility of catastrophic losses, yet assuming full personal responsibility. Napoleon referred to this sort of moral courage as "two o'clock in the morning courage."

In virtually all modern armies, the soldiers who make up the army are the general's most valuable and irreplaceable resource, and his greatest challenge both to lead and to manage. For this reason, all studies place a high premium on a general's skills in human relations and communication. Anecdotes abound of how a general could assess his soldiers—their strengths, weaknesses, morale, and capabilities—and then impose his will on them through his ability to communicate with them. Whether in the stirring rhetoric of Douglas MacArthur, the earthy and calculated profanity of George Patton, or the back-to-the-wall bluntness of Walton Walker in the Pusan Perimeter, the effective general evokes reserves of tenacity and spirit that his soldiers often do not know they possess.

The technical skills required of a general in modern war are more difficult to assess. Alexander's prowess with the horse and weapons of his day were a prerequisite for success in the warfare he practiced. A modern general's skill with the weapons of war is more problematical. Although no one insists that a general must—or could—be able to operate all the weapons of a modern battlefield, a general must thoroughly understand their capabilities and limitations. It is likely that he will have been expert with some of the weapons to have risen to his position. Yet some commanders in modern armies have been criticized as being preoccupied with their own technical expertise at the expense of the vision required to employ the weapons and army as a team. As modern information-processing capabilities become more sophisticated and refined, the danger of fixating on technology rather than capabilities will pose an increasing risk.

The health and physical fitness of generals have always been important, but they have received additional attention in studies of generalship. George Marshall's biographer reported that, as Chief of Staff of the American army in World War II, Marshall had to relieve more generals for health or fitness reasons than for any other. Some authors add mental fitness as a component of physical fitness; certainly history reveals many generals who lost the edge on their opponent for what can only be explained as lack of mental toughness.

Training of General Officers

Until the latter part of the twentieth century, there has been no systematic method to train general officers. Since the American difficulties in the Vietnam War, more attention has been given to how general officers should be selected and trained for their responsibilities. Leadership institutes have been established, senior leadership manuals written, and both behavioral and historical studies of generalship have proliferated. The U.S. Army has developed methods for evaluating and training operational general staffs separately from their units. These efforts promise more precise scientific analysis of generalship in the future.

K. E. HAMBURGER

SEE ALSO: Command; Leadership; Management; Span of Control.

Bibliography

Barnett, C., ed. 1989. *Hitler's generals*. New York: Grove Weidenfeld.
Bausum, H. S., ed. 1986, 1987, 1988. *The John Biggs Cincinnati lectures in military leadership and command*. Lexington, Va.: VMI Foundation.
Creveld, M. van. 1985. *Command in war*. Cambridge: Harvard University Press.
Fuller, J. F. C. 1936. *Generalship: Its diseases and their cure*. Harrisburg, Pa.: Military Service.
———. 1957. *Grant and Lee: A study in personality and command*. Bloomington: Indiana Univ. Press.
Rommel, E. 1953. *The Rommel papers*. New York: Harcourt, Brace.
Slim, W. J. 1961. *Defeat into victory*. New York: David McKay.
U.S. Department of the Army. 1987. *Field manual 22-103: Leadership and command at senior levels*. Washington, D.C.: Department of the Army.
Wavell, A. 1941. *Generals and generalship*. New York: Macmillan.
Zais, M. M. 1985. *Generalship and the art of senior command: Historical and scientific perspectives*. Thesis. U.S. Army Command and General Staff College. Ft. Leavenworth, Kans.

GENEVA CONVENTIONS

For more than 100 years the multilateral treaties called the Geneva Conventions have created and defined broadly accepted rules of international law designed to protect—insofar as possible—sick and wounded combatants, military medical units and personnel, prisoners of war (PWs), and civilian war victims. The first Geneva Convention was concluded in 1864. It was inspired by a Swiss businessman, Henri Dunant, who proposed it in reaction to the suffering of the wounded he had witnessed in 1859 on the battlefield at Solferino. Through the stimulus of the all-Swiss International Committee of the Red Cross (ICRC), the law of Geneva was revised and expanded, usually as a result of the experience gained in war, in 1899, 1906, 1929, 1949, and 1977.

The first several conventions dealt solely with the protection of sick and wounded combatants on land. Coverage was extended to prisoners of war in 1929, and to sick, wounded, and shipwrecked at sea and to civilians in 1949, when four conventions were concluded to replace all earlier ones. These four conventions were supplemented in 1977 by the conclusion of two protocols, the first dealing with international armed conflicts and the second with civil wars. The four Geneva Conventions of 1949 are accepted by virtually all nations; as of mid-1991 there were 164 states party to them, and the general principles of those conventions are clearly part of customary international law binding on all states. Protocol I at the same time had 102 states party to it, and Protocol II had 92.

The conventions envision that each party to a conflict will appoint a "protecting power"—a neutral state that agrees to look after the interests of that party. The representatives of the relevant protecting power are to safeguard the interests of that party by observing the application of the conventions, reporting deficiencies, and helping to resolve disagreements concerning interpretation or application of the conventions. Unfortunately, since 1949 protecting powers have almost never been accepted by all parties to an armed conflict.

Protection of Sick, Wounded, and Shipwrecked Military Personnel

The first Geneva Convention of 1949 deals with the sick and wounded on land, and the second with the sick, wounded, and shipwrecked at sea. The basic principles of these conventions are that sick or wounded military personnel must be respected and protected in all circumstances and given the medical treatment they require, without any adverse distinction based on nationality, religion, race, sex, political opinions, or any similar criteria. The parties to the conflict are obliged to search for and collect the wounded and sick and bury the dead and, in all cases, to record the names and other relevant information about such persons. With respect to enemy personnel, this information must be sent promptly to a central information agency (usually the central tracing agency maintained in Geneva by the ICRC) and to the relevant protecting power that looks after the interests of the enemy. Protocol I strengthened these provisions by creating a right to recover the remains of one's dead. These improvements resulted largely from the experience of the Vietnam War, in particular from American concern at the almost total absence of information concerning the fate of its military personnel who were missing in action and Vietnam's subsequent efforts to bargain for the release of remains of the dead.

The first and second Geneva Conventions of 1949 provide that military medical units and personnel are not to be attacked and are to be identified by the red cross sign, flag, and armband. If captured by an enemy, military medical personnel are to receive all the protections of PWs, but they are not PWs and are to be released except when needed to care for sick and wounded PWs. Protocol I extends similar protection to civilian medical personnel and units and prohibits reprisals against both military and civilian medical units and personnel.

Hospital ships must meet the requirements set forth in the second convention, including prior notification of their designation and characteristics to all parties to the conflict. They cannot be used for any other purpose during the armed conflict, and they enjoy immunity from attack or capture. Ambulances and other medical transports on land are not to be attacked when recognized. Medical aircraft, however, were protected by the first and second conventions only "while flying at heights, times and on routes specifically agreed upon between the belligerents concerned." This unrealistic rule has been changed by Protocol I in an effort to provide meaningful protection to medical aircraft—a matter of considerable humanitarian importance. The survival rates of wounded military personnel vary inversely with the time required to move them to medical units where they can receive medical attention. If medical units are moved close to the battle site to expedite such treatment, the dangers to which the units are exposed obviously are heightened. Increasingly, medical aircraft—primarily helicopters—retrieve wounded from the battle site for prompt removal to medical units located far enough away to be reasonably safe. Consequently, the protection of medical aircraft from attack has become a matter of great concern.

Protocol I has faced up to this task by establishing a series of rules. The most important rule is that medical aircraft, when they are recognized as such, are not to be attacked; no agreement between the parties to the conflict is required. The protocol also creates new means of identification—flashing blue lights, a special radio signal on designated frequencies, and a radar transponder—that should significantly increase the probability that such aircraft will, in fact, be recognized.

Protection of Prisoners of War

The third convention provides detailed and extensive rules for the treatment of PWs. At the outset, it makes clear that the capturing state is responsible for the treatment of PWs and may not transfer their custody to another state unless it first satisfies itself of that state's willingness and ability to comply with the convention. During the Vietnam War, the United States transferred all PWs captured by U.S. forces to the custody of South Vietnam and maintained advisers at all South Vietnamese PW facilities to ensure that the standards set by the convention were adhered to. The convention requires humane treatment at all times and prohibits reprisals against PWs. It requires the prompt evacuation of PWs from danger areas and the provision of free medical care, and it prohibits any form of coercion to obtain information from PWs beyond names, ranks, birth dates, and serial numbers.

PWs must be interned in camps on land (a reaction to the infamous prison ships of the Napoleonic Wars). These camps must be away from the combat zone or unhealthy areas and marked as PW camps. The enemy is to be notified of the locations of such camps, and the presence of PWs may not be used to render certain points immune from military operations.

The third convention sets forth requirements as to quarters, food, clothing, hygiene, and medical attention; religious, intellectual, and physical activities; and disciplinary and penal sanctions. The convention permits the detaining power to utilize the labor of PWs below officer rank, but requires that such labor be paid and limits the type of work that can be ordered; work with a military purpose or dangerous or humiliating work may not be assigned. PWs are entitled to send and receive letters and cards and receive relief parcels.

A vital device to ensure oversight of the well-being of PWs and of compliance with the convention—as with the other three conventions—is the protecting power. However, parties to a conflict may not always grant the rights of a protecting power; frequently even the ICRC has been denied access to PWs and other war victims. Protocol I attempts to remedy this situation; it establishes a procedure involving the ICRC designed to bring about the appointment of protecting powers or of a substitute international body like the ICRC, and makes it more embarrassing for a party to refuse.

Another subject dealt with in the third convention that in practice has given rise to difficulties is the repatriation of PWs. The convention requires the repatriation during hostilities of seriously sick or wounded PWs; this has frequently been ignored, for example, in the eight-year war between Iran and Iraq. The convention also requires that all PWs be "released and repatriated without delay after the cessation of active hostilities." At the close of hostilities in Korea in 1953, the United Nations Command was not prepared to repatriate by force Korean and Chinese PWs who objected to such repatriation on ideological or political grounds. Similar situations have since arisen at the ends of other conflicts. It is now generally accepted that forcible repatriation is not required by the convention, provided that an appropriate protecting power or the ICRC is enabled, through a private interview with each PW who rejects repatriation, to verify that the PW's decision has been freely made and is not the result of coercion.

Protection of Civilians

The fourth convention establishes many important protections that must be accorded to "protected persons," persons who find themselves in the hands of a party to the conflict or an occupying power of which they are not nationals. Virtually all inhabitants of occupied territory are likely to be "protected persons," as are enemy aliens in the territory of a party to the conflict. The convention reflects the experiences of the Second World War in which the inhabitants of many nations suffered greatly as a result of German and Japanese occupation policies and practices.

The convention deals with many subjects. It states that protected persons are entitled to respect for their persons, their honor, their family rights, their religious practice, and their customs. They must be treated humanely without any adverse distinctions based, in particular, on race, religion, or political opinion. They may not be detained at points in order to immunize such points from attack. No physical or moral coercion may be used against them in an effort to obtain information. Collective punishments, reprisals, and the taking of hostages are prohibited.

In occupied territories, the convention imposes many restrictions upon the occupying power. For example, the rights of protected persons in occupied territory cannot be affected by any annexation or other change in the status or government of that territory. Deportation of protected persons from occupied territory is prohibited, as is the transfer by the occupying power of parts of its own civilian population into the occupied territory. The occupying power must ensure adequate food and medical supplies for the population of the occupied territory. Extensive protections are established for protected persons who are subject to prosecution by an occupying power, including notification to the protecting power and the right (in most cases) of representatives of the protecting power to attend the trial. Finally, the convention includes detailed provisions regulating the treatment of protected persons who are interned by a party to the conflict, whether in its own territory or in occupied territory. Protocol I has added a few additional protections for persons in the power of a party to the conflict—most important is a listing of fundamental and minimum guarantees for persons who benefit from no better treatment, such as a party's own nationals.

With a few exceptions relating largely to medical and relief activities and family reunification, the fourth convention affords protection only to civilians who are in the power of an enemy. Thus, civilians in their own territory are not protected unless such territory is occupied. Protocol I has extended the protection of civilians. It provides that the civilian population may not be made the object of attack, nor may individual civilians so long as they take no direct part in hostilities. The protocol prohibits indiscriminate attacks, including target-area bombardment within cities or other similar concentrations of civilians and any attacks that may be expected to cause civilian losses disproportionate to the anticipated military advantage. The protocol also provides civilian objects with some protection from attack. It prohibits starvation of civilians as a method of warfare and protects objects indispensable to the survival of the civilian population. It prohibits the use of methods or means of warfare intended or expected to cause widespread, long-term, and severe damage to the natural environment. Significant restraints are imposed on attacks against dams, dikes, and nuclear electrical generating stations, to prevent the release of dangerous forces (water or radioactive particles) and consequent severe losses among the civilian population.

Civil War

Article 3, common to all four of the 1949 conventions, is the only provision of those conventions dealing with civil wars. It applies to armed conflicts "not of an international character" and lists a few "minimum" requirements that must be met. Article 3 states that persons taking no active part in the hostilities, including prisoners and the sick and wounded, must be treated humanely without any adverse distinction. The following acts are specifically prohibited with respect to such persons: violence to life and person; the taking of hostages; outrages upon personal dignity; and the passing of sentences and carrying out of executions without previous judgment by a regularly constituted court that affords all the generally recognized judicial guarantees. The article adds that the wounded, sick, and shipwrecked must be collected and cared for and that the ICRC may offer its services to the parties to the conflict.

Protocol II adds considerably to the substantive and procedural protections contained in Article 3. Perhaps most significantly, it prohibits collective punishments, acts of terrorism, slavery, and pillage; and it expands the protections due to persons accused of criminal offenses related to an armed conflict, in particular by prohibiting the ex post facto imposition of criminality, establishing a presumption of innocence and the right not to testify against oneself. The protocol also sets minimum standards for the treatment of persons detained for reasons related to the civil war, and it adopts a few of the provisions from Protocol I concerning the protection of civilians. Unfortunately, Protocol II establishes a relatively high threshold of applicability, requiring sustained and concerted military operations and control of territory by hostile forces. This threshold seems likely to allow almost any government, faced with a war of rebellion or succession, to find excuses to declare the protocol nonapplicable.

Future Prospects

The four 1949 conventions are accepted nearly universally, although they are not always fully complied with in practice. In particular, their principal oversight mechanism—the protecting power—has been virtually unused; at least one party to each armed conflict since 1949 has persistently refused to permit such scrutiny of its compliance with the conventions. For similar reasons, the ICRC has frequently found its access to PWs and protected persons denied or at least restricted. While Protocol I attempts to make it more difficult for a party to an armed conflict to refuse to accept a protecting power or a substitute, it is unclear how effective the protocol will be in that regard. Moreover, the protocol is not yet accepted universally; it remains to be seen whether the ideological objections raised by certain Western governments can be overcome.

In 1986 the Reagan administration decided not to send Protocol I to the Senate for advice and consent to ratification, largely because of objections to two provisions. One provision recognizes wars of national liberation as conflicts to which the protocol would apply in situations where the liberation movement is prepared to accept all the obligations of the 1949 conventions and the protocol. The other treats as legitimate combatants irregular soldiers in occupied territory who carry arms openly in certain defined circumstances. Spokesmen for the Reagan administration complained that these provisions would give aid and comfort to the PLO and other terrorist groups. Most NATO members reject the Reagan administration rationale and have ratified the protocol. France has refused to accept the protocol on different grounds, apparently its many prohibitions of reprisal and the French concern about any possible effect it might have on the use of nuclear weapons.

In the end, whether Protocol I will become as universally accepted as the four 1949 conventions may well turn on whether it is seen to improve compliance with the law. In that regard, the protocol establishes a new, fifteen-member International Fact-Finding Commission, headquartered in Geneva. It would be empowered to investigate alleged serious violations of the conventions and the protocol, when requested to do so by a state that has accepted in advance the competence of the commission, and when the alleged violations have been committed by another state that has accepted its competence. When, in a surprise move, the Soviet Union ratified Protocol I in 1989 and accepted the competence of the commission, new possibilities opened for improving compliance with the law. The commission was established and its members elected in 1991.

In civil wars, the sensitivities of governments to the granting of any status to rebel or secessionist movements seem likely to continue to deter them from acknowledging the applicability of either common Article 3 or of Protocol II. The main hope for restraining the inhumanity of such wars is that the rules of Article 3 and at least some of those in Protocol II may be accepted in practice as norms of customary international law.

GEORGE H. ALDRICH

SEE ALSO: Civilian Populations in War; Civil War; Hostage; Morality and War; Prisoner of War; Refugees.

Bibliography

Aldrich, G. H. 1991. Prospects for United States ratification of additional protocol I to the 1949 Geneva Conventions. *American Journal of International Law* 85:1–20.
Bothe, M., K. J. Partsch, and W. Solf. 1988. *New rules for victims of armed conflicts*. The Hague: Martinus Nijhoff.
Levie, H. S., ed. 1987. *The law of non-international armed conflict*. Dordrecht, The Netherlands: Martinus Nijhoff.
———. 1979. *Protection of war victims*. Dobbs Ferry, N.Y.: Oceana.
Roberts, A., and R. Guelff, eds. 1989. *Documents on the laws of war*. Oxford: Clarendon.
Schindler, D., and J. Toman, eds. 1981. *The laws of armed conflicts*. Alphen aan den Rijn, The Netherlands: Sijthoff and Noordhoff.
Sofaer, A. D. 1988. The rationale for the United States decision. *American Journal of International Law* 82:784–87.

GENGHIS KHAN [ca. 1167–1227]

Genghis Khan, the "Master of Thrones and Crowns," created the largest land empire in world history and passed it on, intact, to his sons and grandsons. Genghis Khan created not only the Mongol Empire but also the Mongol nation itself. Before Genghis Khan's rise to supremacy, the Mongol people were divided into numerous tribes and clans. After Genghis Khan, there was but one Mongol people, one Mongol nation.

Birth and Ancestry

The future Genghis Khan was born circa 1167. His father, Yesugei, was a minor chieftain of the Kiyat clan. Yesugei had gained a small following as a result of his success in battle. When his first son was born, Yesugei, following custom, named the boy Temujin after a captured Tatar chieftain.

Temujin's mother was Houlun, whom Yesugei had abducted from her husband, a member of the Merkit tribe, shortly after her wedding. Yesugei and Houlun had three other children besides Temujin.

Although the son of a chieftain, Temujin's lifestyle differed little from that of the common folk. He had, however, something the common folk lacked, an illustrious ancestry: the royal line of the Mongols, the Borjigin (Blue-Eyed Men).

Temujin's great-great-grandfather, Kaidu Khan, had been the first Mongol king. His grandfather, Kabul Khan, and his great-uncles Ambakhai Khan and Khutula Khan, had also borne the royal title. However, the Mongol kingdom was short-lived. Antagonism between the Mongols and the Tatars and the foreign policy of the Kin dynasty of North China, undermined the Mongol kingdom. Both Kabul Khan and Ambakhai Khan died at the hands of the Kin, after their capture by the Tatars. With the death of Khutula Khan, the Mongol kingdom disintegrated.

Early Life

Such was the fragility of tribal unity among the Mongols that, when Yesugei died, the clans that had gathered around him quickly deserted his widows and children and left them alone and destitute. Temujin was only 9 years old when his father died, leaving him and his brothers and half-brothers to glean a living from the steppe.

Nature was not their only enemy. Steppe bandits were a constant threat. The greatest danger to Temujin, however, came from his cousins the Taychiuts, once followers of Yesugei. The Taychiuts captured Temujin, kept him prisoner in their camp, and forced him to wear a cangue, a heavy wooden collar. Faced with slavery or, more likely, death at the hands of his cousins, Temujin escaped with the help of Sorkhan-shira and his family.

As they survived the perils of the steppe and grew,

Temujin and his brothers were able to stabilize the precarious situation of their family. Temujin also was able to gather a small following of other destitute families. More important, with the improvement in his family's fortunes, Temujin was able to renew the friendship forged between Yesugei and Toghrul Khan, king of the Keraits, a Turkic tribe. With the support of Toghrul, Temujin pressed his claim to the hand of Borte, to whom he had become betrothed shortly before his father's death. In due course, Temujin and Borte were married.

Temujin's fortunes took a turn for the worse when a large Merkit raiding party, greatly outnumbering Temujin's followers, attacked the Mongol camp. Temujin was forced to flee, leaving Borte at the mercy of the Merkits. The Merkit raiders gave Borte to a relative of the man whose brief marriage to Houlun had ended with her abduction by Yesugei.

Temujin sought the aid of Toghrul Khan in rescuing Borte. Toghrul gladly agreed to help in the rescue, as did Jamukha, khan of the Jalairs. Jamukha had been Temujin's *anda* (blood brother) when the two were children. The combined Mongol-Kerait-Jalair forces easily scattered the Merkits and freed Borte. Nine months later she gave birth to Juchi, whom Temujin acknowledged as his first-born son despite questions concerning the boy's paternity.

Supremacy over the Mongols

The successful rescue of Borte and the renewal of the Kerait alliance increased Temujin's stature among the Mongols. In 1190, the princes of the Kiyat clan—including Prince Altan, son of the last Mongol king, Khutula—asked Temujin to accept the title of king. As a member of a junior branch of the clan, Temujin at first refused, suggesting instead either Prince Altan or one of the princes of the Jürkin, the senior line. When the other princes refused, Temujin agreed to become king.

The Mongol princes knew the Kerait alliance was a personal one between Toghrul and Temujin, not between Keraits and Mongols. Without the alliance the newfound importance of the Mongols would evaporate. Therefore, the princes elected Temujin to be their leader in war and in the hunt. Temujin, however, conceived his role differently.

Temujin realized that without a strong leader the Mongol peoples were fated to remain dispersed and disunited, prey not only to each other but to outside forces as well. To create a single, united people—a nation—Temujin had to neutralize the dissident factions within his own kingdom and then defeat and incorporate into the Mongol kingdom the other ethnic Mongol tribes.

WAR AGAINST THE TATARS

The Kin dynasty of North China followed the traditional Chinese policy of setting one group of barbarians against another. When the Mongols became too strong, the Kin allied with the Tatars to neutralize the Mongol menace. With the Tatars now a threat, the Kin reversed themselves and offered an alliance to Temujin.

Temujin realized that the alliance with the Kin would be temporary, and, given the nature of their foreign policy, the Kin might quickly switch their support back to the Tatars. Therefore, the Mongols, in alliance with the Keraits, had to bring the maximum force possible against the Tatars in the shortest possible time. In a series of lightning moves, Temujin defeated the Tatars and crippled their war-making potential by slaying all Tatar males taller than the linchpin of a wagon wheel.

PURGE OF THE JÜRKIN PRINCES

Temujin declared the Tatar War a national war of vengeance. The Tatars had helped bring about the deaths of both Kabul and Ambakhai. Temujin's own father, Yesugei, had died poisoned by a Tatar. Therefore, he considered the war to be just punishment for the crimes of the Tatars.

When Temujin marshaled his forces for the war, the Jürkin princes were absent. After the war, Temujin called his uncles to account for their treason. The Jürkin princes paid with their lives, smothered to death lest their royal blood be spilled.

END OF THE KERAIT ALLIANCE

The defeat of the Tatars and the purge of the Jürkin princes greatly strengthened Temujin. However, his very success caused concern among his allies, the Keraits and Jalairs. Jamukha Khan convinced Toghrul that Temujin had designs on the Kerait kingdom. There followed numerous combinations—of Jalairs, Merkits, Naimans, Tatar remnants, and Keraits under the vacillating Toghrul—against Temujin.

FOUNDING OF THE MONGOL EMPIRE

After years of warfare, including a near-disastrous war with the Keraits, Temujin emerged victorious. In the spring of 1206, a general council (*khuraltai*) was held. The Mongol princes not only reaffirmed Temujin's kingship, but also conferred upon him a new title: Genghis Khan (*Chinggis Khahan*), "King of the Sea-Surrounded Land." Genghis Khan decreed that, henceforth, all his subjects—including Merkits, Tatars, Keraits, Naimans, and Oiyats—would be known as Mongols, members of one Mongol nation (*ulus*).

Structure of the Mongol Empire

The Mongol Empire was an absolute monarchy. Genghis Khan held all executive, legislative, and judicial powers. Although ruthless toward his enemies, Genghis Khan valued loyalty and friendship above all else. Men who betrayed their leaders to Genghis Khan were summarily executed. A victim of the all-too-common Mongol tendency to betray and prey upon one another, Genghis Khan gave short shrift to traitors but magnificent rewards to friends and loyal companions.

COMPANIONS OF GENGHIS KHAN

Long before his rise to unchallenged power, Genghis Khan had attracted warriors to his cause. His first companion was Boguechi, who as a youth accompanied an equally youthful Temujin in tracking down stolen horses. Boguechi remained with Temujin to become one of his "Four Heroes." Tribal origin was no obstacle to entry into the emperor's retinue. Besides Boguechi, the other Four Heroes were Mukhali of the Jalair; Chilaan of the Taychiuds, a son of Sorkhan-shira; and Borokul of the Jürkin, one of the emperor's adopted brothers. Khubalai of the Barulas, Jelme of the Uriyangkhadai and his younger brother Subotai, and Jebe the Merkit were feared as the "Four Hounds" of Genghis Khan. To these men, to his brother Kasar, and to his four sons by Borte—Juchi, Chagatai, Ogadai, and Tuli—Genghis Khan delegated responsibility for the administration of the empire and command of the Mongol army.

YASSA

The governance of the empire was regulated by a series of laws decreed by Genghis Khan and enumerated in the *Yassa*. The *Yassa* established the Mongol Army on a permanent footing. In its regulations concerning marriage, theft, and quarrels between his subjects, Genghis Khan sought to end the internal conflicts that had plagued his people for centuries. With the expansion of the Mongol Empire, the *Yassa* became the supreme law for all subject peoples.

Expansion of the Mongol Empire

Political and economic factors explain the expansion of the Mongol Empire in the 21 years between its creation and the death of Genghis Khan. Military campaigns and the resulting booty were necessary to channel the Mongol tendency to raid and plunder—each other if no one else was available—a tendency that, if not controlled, would have destroyed the empire from within.

The campaigns against the Kin of northern China solved another crucial security problem. The Kin posed a direct threat to the very existence of the Mongol Empire through their foreign policy toward the barbarians. The only way to neutralize the threat posed by the Kin was the total destruction of their empire.

Mongolia lay astride the traditional caravan route between China and the Middle East, the famous Silk Road. Control of the Silk Road gave Genghis Khan the power to collect transit taxes from Chinese, Persian, and Arab merchants. The Mongol conquest of the kingdoms of Hsi-Hsia and Kara-Khitai strengthened Mongol control of the Silk Road.

Death of Genghis Khan and His Legacy

Genghis Khan died in August 1227. In obedience to the wishes of Genghis Khan, his third son, Ogadai, was elected emperor by the princes and high-ranking officers of the empire. The Mongol Empire, instead of disintegrating, as had the Mongol kingdom of earlier times, continued to flourish under the sons and grandsons of Genghis Khan. Charged by Genghis Khan with the task of perpetuating and expanding the empire, his successors eventually conquered China, Korea, Iran, present-day Iraq, most of Russia, parts of Poland and Hungary, Indochina, and Burma. The Mongol army, led by veteran generals such as Subotai, annihilated the armies of countries with manpower resources far greater than that of the Mongols.

The most enduring legacy of Genghis Khan was the Mongol nation itself. Although the Mongols were later subjugated by Manchu and Russian, the concept of Mongol nationhood, as conceived by Genghis Khan, survived. In Mongol folk religion, Genghis Khan still holds an exalted position as an incarnation of the Everlasting Blue Sky and the special protector of the nation. So strong is Mongol identification with Genghis Khan that the Mongol people celebrated his 800th birthday despite the official disapproval of the Mongolian Communist Party and the displeasure of the Soviet Union.

LAWRENCE D. HIGGINS

SEE ALSO: Mongol Conquests.

Bibliography

Boyle, J. A. 1958. *The history of the world conqueror by 'Ala-ad-Din 'Ata-Malik Juvaini*. 2 vols. Manchester: Manchester Univ. Press.
Grousset, R. 1966. *Conqueror of the world*. Trans. M. McKellarand and D. Sinor. New York: Orion Press.
Haenisch, E. 1941. *Die geheime geschichte der Mongolen* [The secret history of the Mongols]. Leipzig: O. Harrassowitz.
Heissig, W. 1980. *The religions of Mongolia*. Berkeley: Univ. of California Press.
Kuo-yi Pao. 1965. *Studies on the secret history of the Mongols*. Bloomington, Ind.: Indiana Univ. Press.
Lamb, H. 1940. *The march of the barbarians*. New York: Literary Guild of America.
Legg, S. 1970. *The heartland*. New York: Farrar, Straus and Giroux.
Martin, H. D. 1977. *The rise of Chingis Khan and his conquest of North China*. New York: Octagon Books.

GEOGRAPHY, MILITARY

Military geography is that part of military science that deals with the characteristics of the area of operations as they relate to military missions and forces. Military geography is the application of the geographic method of analysis to military problems.

Geography

Geography is the science that deals with the spatial distribution of phenomena at or near the surface of the earth. Geography is used to explain the patterns and relationships of natural and human phenomena: people and the artifacts of their cultures, animals, vegetation, climate, oceans, and landforms. These patterns and relationships are significant only in terms of a problem to be solved. Geography may be broken down into branches according to the type of problem being addressed. Economic geography covers the spatial distribution of factories and ports and the movement of goods and dollars among various locations. Human geography covers the way in which people live and relate to their environment. Urban geography covers the relationships among the different parts of cities. Political geography covers patterns among nation-states or other political entities. Physical geography covers the natural features of the earth's surface, such as rivers, mountains, plains, storms, and soil. Finally, geography is concerned with regions defined by the nature and scope of the problem being addressed. The geographic method of analysis seeks to define appropriate regions and explain the spatial distribution of phenomena in the region in the context of a problem.

Military Geography

Military geography is applied in regions defined by the missions of the military forces. Military geography is subdivided into four major branches: terrain analysis; theater analysis; geopolitics; and topical military geography. The first three of these may be related to the tripartite division of military art as follows:

Level of warfare	Scope	Branch of military geography
Strategy	Global	Geopolitics
Operational art	Theater of operations	Theater Analysis
Tactics	Battlefield	Terrain Analysis

Terrain Analysis

Terrain analysis is used to determine the effect of the natural and manmade features of an area of operations on tactical military operations. It includes consideration of natural phenomena such as landforms, relief, drainage patterns, vegetation, animal and insect life, and surface materials. It includes consideration of works of man such as buildings, roads, railroads, airfields, dams, pipelines, and cultivation, but it does not usually consider humans in the area of interest. Terrain analysis may also include consideration of weather and climate.

Terms other than *terrain analysis* are also used to describe the application of military geography at the tactical level. *Terrain appreciation* is often used interchangeably with terrain analysis, although it implies a narrower and deeper study of the landforms of the area. *Terrain intelligence* implies more emphasis on basic data compilation than on mission-oriented analysis. *Military topography* was used formerly to mean the study of landforms from a military viewpoint, but now it refers primarily to map making and map reading. *Topography* still means the landforms of an area.

Terrain analysis is mission oriented. The area of operations is defined by the mission, and the significance of a terrain feature will vary depending on the nature of the mission. For example, a hill or a river has a different significance if the mission is to defend rather than to attack.

Terrain analysis is dynamic. The evolving military situation constantly changes missions and viewpoints, and thus constantly changes the significance of the terrain for the military commander. Changes in military weapons and technology can also alter the relative significance of terrain features. A river, which was a formidable barrier, can become a trivial problem with the introduction of improved combat bridging equipment. A distant target, which was not worth considering, can become important with the introduction of long-range weapons. The terrain itself also changes. It is modified by natural forces such as erosion and earthquakes, and it is also modified by man's construction of roads, airfields, and bridges. It is modified by military operations from the effects of troop movements, artillery fire, air strikes, and demolition of structures.

The military components of terrain analysis are:
1. Obstacles: terrain features that slow down or stop momentarily the movement of either enemy or friendly forces.
2. Fields of fire: the tendency of an area to facilitate or hinder direct fire by flat-trajectory weapons and missiles.
3. Observation: the tendency of an area to permit or deny visual or sensor detection of the enemy.
4. Concealment: the tendency of an area to facilitate avoiding observation by the enemy.
5. Cover: the tendency of an area to afford protection against being hit by enemy direct-fire weapons and missiles.
6. Routes of communication: roads or paths for movement of troops and vehicles.

An essential element of terrain analysis is the definition and interpretation of the spatial relationships among terrain features. For example, if the military mission is to seize a crossroads, several factors must be noted: the precise location of the crossroads, the distance and direction of the crossroads from the military unit, the characteristics of the intervening ground, and the relative position of terrain features (i.e., whether they will help or hinder the unit in accomplishing its mission). All of these factors must be taken into account along with the capabilities of the unit's weapons, equipment, and troops, to determine not only the time it will take to accomplish the mission, but also whether the mission can be accomplished at all.

Terrain analysis also varies according to the level at which it is carried out. Smaller units use a greater level of detail than larger units. For an individual rifleman or gunner, for example, a single tree or a small hill or hollow offering cover and concealment are of paramount interest. At the rifle company level, however, a clearing or the next ridgeline are the significant terrain features and fields of fire the primary concern. At the infantry or tank battalion level, the commander's interest is in obstacles such as villages, forests, or streams, while at the division level routes of communication may be the most important features.

Terrain is evaluated differently by different commanders, depending on the type and extent of influence the terrain will have on their units. A tank company commander, for example, with great organic tactical mobility, will draw conclusions about an area different from the conclusions drawn by an airborne rifle company commander with few or no vehicles. An air assault division commander with several hundred helicopters will draw different conclusions about the nature of his area of operations than would an armored division commander with several hundred tanks. Each commander must perform terrain analysis that is appropriate for the mission, role, and circumstances of the unit.

THEATER ANALYSIS

Theater analysis, or strategic area analysis, is the application of military geography at the level of operational art. Theater analysis is used to describe the influence on military operations of the characteristics of an actual or potential theater of war.

Theater analysis, unlike terrain analysis, does include humans in its consideration of natural and manmade features in the theater of operations. The occupancy patterns of human activity, consisting of towns, agricultural areas, roads, railroads, and airfields are of interest, as well as landforms, drainage, vegetation, and climate.

There is also a difference in scale between theater analysis and terrain analysis. Theater analysis takes into account the entire area of operations; terrain analysis has a more restricted or localized viewpoint. To the division commander, a river is an obstacle either to be crossed or defended. To the theater commander, the same river is only part of a total pattern of drainage indicating likely defensive positions or avenues of approach for an offensive. The principles are the same, and the influence of the river is likely to be similar, but the scale is different.

Theater area analysis also tends to be less mission-oriented than terrain analysis. At the theater and army group levels, missions are generally stated in broad terms, significant mission changes occur infrequently, and the planning cycle may be several weeks or months. For an army corps, the planning cycle may be several days or a few weeks, for a rifle company it may be measured in minutes.

Theater analysis tends to be more predictive than terrain analysis. Estimates used in the planning process at the theater headquarters must predict the impact of the area of operations on operations several weeks or months in the future. Terrain analysis deals with the immediate impact; theater analysis, with future impact.

The intelligence sections of theater headquarters or major land, air, or naval headquarters in the theater are responsible for theater analysis. In peacetime, the primary activity of theater analysis is compilation of data on the physical and human characteristics of the theater of operations. This includes descriptions of landforms and underlying geology, climatic data, distribution of vegetation and fauna, and demographic data. Special studies are made of trafficability, highway networks, railroads, ports, navigation channels and straits, airfields, airways, pipelines, power and communications networks, urban and built-up areas, and other characteristics of interest.

Theater analyses provide the basis for planning military operations in the theater. If the winter will be severe enough to require special clothing, that will have to be taken into account in planning the campaign. If the cloud cover will restrict air operations, that has to be considered in planning the kind and amount of air force units to be employed. If the terrain in a particular location is unsuitable for tanks, that should be taken into consideration when organizing the forces for combat. The presence of civilians has implications for nuclear and conventional fire support planning. The impact of refugees on the movement of troops and supplies may cause a diversion of resources. All of these factors must be taken into account in theater analysis.

Finally, the spatial perspective is an essential element of theater analysis. Because the distances are greater and the times longer, the interaction of space-time factors with military forces becomes more important at the level of operational art (theater analysis) than at the tactical level (terrain analysis). This is particularly true for selection of targets for battlefield or long-range interdiction, major defensive positions, or offensive axes of advance. Properly conducted, the theater analysis constitutes a complete application of geographic method to the military problem.

GEOPOLITICS

The application of military geography at the strategic or global level is called geopolitics. Geopolitics integrates political, diplomatic, sociological, economic, and military considerations into an overall strategic approach. Geopolitics is concerned with relative power among nations and coalitions. It includes consideration of the foundations of national power: population, industry, commerce, financial status, internal stability, resources, and national will, as well as military forces.

The essence of geopolitics is consideration of the size, shape, location, and characteristics of nations with respect to one another. History offers numerous examples of the importance of location and terrain. Poland, a nation between two great powers, but without natural lines of defense, has suffered repeated invasions. Switzerland has remained neutral and untouched through several major wars in its alpine bastion. The United States, secure from invasion and remote from Europe, needed only a small navy and an even smaller army from 1865 to 1917. Japan, lacking a large land area and raw materials, sought security by expanding into China and Southeast Asia.

Geopolitics recognizes tension between nations that are maritime powers and those that are land-based powers. Alfred Thayer Mahan advanced the concept of maritime power based largely on the experience of the British, who, invulnerable to invasion from the European continent, ruled a global empire for 140 years by virtue of a superior navy and a substantial merchant marine. In 1904 Sir Harold Mackinder identified the plains of Russia as the Heartland of Europe and predicted eventual global supremacy for the ruler of the Heartland. In Germany before World War II, Karl Haushofer bolstered German war aims by asserting that a combination of Germany, Russia, and Japan was unbeatable. Hitler's invasion of the Soviet Union, against Haushofer's advice, forced a coalition between the maritime power of the British Empire and the United States and the land power of the Soviet Union, which led ultimately to the defeat of Germany in 1945. In 1943 Nicholas Spykman advanced the concept of the Rimlands in opposition to the Heartland concept. According to Spykman, a combination of the economic and industrial superiority of the Rimlands—the United States, Western Europe, and the nations of the Pacific basin—would be more powerful than the Soviet Heartland.

Geopolitics is a major element of military strategic thinking in the nuclear age. Geopolitical concepts of relative location and power are important in maintaining a global balance of power by coalitions between the superpower and the medium powers. Geopolitical ideas underlie the current debate in the United States between advocates of a maritime strategy and adherents of a coalition (land-based) strategy. Geopolitics helps to understand how future changes in the relative power of nations will affect potential military operations.

TOPICAL MILITARY GEOGRAPHY

Topical military geography covers a particular, well-defined type of phenomena (a topic) on a worldwide basis. The major military applications of topical geography are:

Environmental studies. Environmental studies of climate, vegetation, and fauna are important in providing the correct equipment and training for military forces to be employed in various parts of the world. Troops employed in arctic regions need to be trained and equipped differently from troops employed in low-latitude deserts. This kind of topical military geography is often employed in the research and development process as new equipment, clothing, and supplies are developed.

Military geology. Geology is a scientific discipline dealing with the nature of the rock formations underlying the surface of the earth. Military geology provides a sound basis for protective construction for cover against conventional or nuclear explosions. It is also used to locate sources of water. Military geology is sometimes considered to be separate from military geography.

Geodesy. Geodesy is the science of global earth measurement that allows the precise location of points on the surface of the earth. Surveying, geodesy on a smaller scale, has been important in military operations since the introduction of field telegraphy allowed the use of indirect fire control with artillery. The advent of nuclear weapons and very long-range missiles has increased the importance of knowing the exact location of potential targets.

Military topography. Military topography originally meant the study of the impact of landforms on military operations, but the term now applies to the making and, particularly, the reading, of maps. Topographic maps are a representation of a portion of the earth's surface, usually of a land area, and include a means to represent altitude or relief. Relief is the difference between the high points and low points of landforms in the area. Relief is shown on a topographic map by contour lines that connect points of equal elevation, by color tinting, by shading, or by hachure marks to depict mountains and other elevated landforms.

Cartography. Cartography is the science of making maps, including topographic maps, aerial charts, and naval charts. Aerial charts provide a representation of the land or sea surface, emphasizing recognizable landmarks and information on navigational aids and airfields. Navigational charts provide a representation of sea or ocean areas, coastal areas, water depth in coastal areas, and hazards and aids to navigation.

Influence of the Area of Operations on Warfare

The characteristics of the area of operations have had enormous influence on the nature of combat and warfare throughout the ages. Generally, the confluence of terrain and technology on the battlefield has determined tactics. The nature of the theater of operations has been the primary basis for campaign planning and execution. Time and distance relationships among various regions of the world along with considerations of resources and statecraft have determined strategy. The nature of the terrain and weather affects all forms of warfare. The most obvious influence is on land warfare, but air and naval warfare are also influenced by the nature of the surface of the earth.

Armies fight on and must conform to the nature of the earth's surface. Therefore, military commanders and planners must appreciate the interaction of men and equipment with the terrain and weather. Streams and forests are both barriers and avenues of advance and supply;

mountains are both barriers and bastions; gaps and passes historically have had great military significance; hills are easy to cross and easy to defend; dry plateaus resemble the sea and favor rapid, mobile warfare; wet plateaus are rugged and make it difficult for military forces to move rapidly; and the boundary between land and sea—the coasts—are important for amphibious warfare and for access to the interior. Man-made structures have become increasingly important both as objectives and as defensive positions. Intimate knowledge of landforms and how to take advantage of them is a valuable tool for accomplishing a military mission.

Air and naval forces do not fight on land, but they are both dependent ultimately on land bases, although the nuclear-powered aircraft carrier and its nuclear-powered escorts may operate for extended periods of time without returning to base. So it is important to understand what influence the land will have on military operations by air and naval forces. Air forces fight from and over land, and their concerns in the area of operations are the suitability of the land for airfields and the adequacy of supporting roads and railroads. The nature of the terrain also influences the tactics and munitions that are appropriate to attack ground targets, particularly for low-flying aircraft such as helicopters and close-support attack aircraft. Despite great advances in high-technology navigation systems, aviators may have to rely sometimes on recognition of terrain features to locate themselves and their targets.

Navies also must learn the lay of the land. The nature of coastlines determines the availability of safe harbors and anchorages, for storms remain a dangerous foe for ships at sea. The distance of naval combat from supporting bases is still a major factor in planning and implementing naval warfare, although the time that a fleet can remain at sea without resupply has increased. The configuration of the ocean bottom is a major factor in undersea warfare. Even space warfare would be influenced by the wind patterns of the atmosphere and the shape and nature of the earth below.

WORLD WAR II

World War II was a truly global war, which necessitated an appreciation for military geography by both sides. Although the major land battles early in the war were fought in Europe, ultimately the Allies also conducted major ground campaigns in the Middle East, the Mediterranean area, China, Burma, the southwest Pacific, the central Pacific, and Manchuria. The nature of the war in each of these major theaters was dictated by the relative priority of the theater for resources and by the terrain and climate. Combat occurred in frozen mountains in Italy, on hot deserts in North Africa, across the stormy North Atlantic, on the vast ocean expanses of the Pacific, and in the moist tropical rain forests of Burma. The Allies had to produce uniforms and equipment, tactics, and techniques suitable for operations under these varied conditions. That they did this successfully reflects the best use of military geography until the Persian Gulf War of 1990–91.

On the tactical level, there were pluses and minuses. The Allies planned carefully for the breakout of armed forces from the landing areas secured by the Normandy invasion in June 1944. However, the planners had paid insufficient attention to the implications of the local hedgerows. These were formidable stone walls overgrown with thick vegetation and were characteristic of that part of France. It took a field expedient, attaching blades to tanks to clear out the hedgerows, to free the British and American armies to move to the Rhine.

Earlier, in May 1940 the Germans and the French did take note of a terrain feature—the Ardennes Forest—but drew different conclusions. To the French the Ardennes was an obstacle impassable to vehicles and worthy only of a light defense; to the Germans the Ardennes was an avenue of attack capable of handling the main effort of the German blitzkrieg.

In North Africa the Germans under Rommel and the British under Montgomery each adapted their operations to the realities of the desert, but in different ways. Rommel operated on a shoestring with rapid mobility and improvisation to take advantage of the ease of movement over most of the area. Montgomery adopted a mobile defense backed with air superiority, which could strike the Germans and Italians almost at will. Ultimately, the British prevailed when the Germans could not resupply their fighting forces.

In Russia, the Soviets learned from their earlier war with the Finns and adapted their clothing and tactics to snow and ice while the Germans froze and bogged down in the mire.

In the Pacific, the Japanese underestimated the ability of the U.S. construction troops to carve airfields and ports out of what was thought to be impassable and unusable terrain. This lack of appreciation of the capability of 1940s technology to alter the terrain cost the Japanese heavily as they were repeatedly outflanked during General MacArthur's island-hopping campaigns.

At the Ardennes Forest late in 1945 weather played an important role in the ability of the Allies to hold back the last desperate offensive by the Germans. The U.S. troops holding out in Bastogne were cut off from supplies and air support for several days because of bad weather, but when the clouds cleared, the U.S. and British planes were able to do their job and help defeat the Germans decisively. Terrain and weather were important elements during World War II.

KOREAN WAR

The invasion of South Korea by North Korea brought on some of the fiercest land fighting of the modern era. Korea was an infantryman's war. The navy supported and shelled, and the air forces attacked the North Koreans almost unchallenged. They both played important roles in the war, but the nature of the terrain in Korea was such that the outcome had to be decided on the ground.

Except for the western plain and small areas on the coast, Korea is a mountainous country with high relief. Relief is a measure of the difference in elevation between high points and low points and indicates the ruggedness of the land. Korea is a land of steep slopes, long ridges, and narrow valleys. Initially, the North Koreans took advantage of the roads and railroads to move swiftly southward with their tanks and trucks.

The U.S. forces, thrown unprepared into the breach, at first failed to appreciate the significance of the terrain; they moved in the valleys, which invited ambush and defeat. Adding to their misery was the cold and snowy weather for which the U.S. troops were also unprepared, with respect to both clothing and tactics. The North Koreans and the Chinese, however, took advantage of the terrain and weather. They fought at night in the worst weather, keeping to the ground and moving along the ridgelines to bring plunging fire to bear on the road-bound Americans in the valleys below. Fighting desperately, the U.S. and Republic of Korea (ROK) forces gradually adapted to the terrain and weather. They learned to fight on the ridges and make sure they had the high ground. They used artillery to attack North Korean troops dug into defensive positions high in the mountains. They learned how to live and fight in cold weather. Eventually, they fought their opponents to a military and political draw. The early stages of the Korean War illustrate the consequences of a lack of attention to military geography. Things that should have been known were not, and lives and battles were lost as a consequence.

VIETNAM WAR

The terrain and weather of Southeast Asia had a major impact on the Vietnam War, and the time and distance of the area of operations from the United States influenced the strategy and the outcome of the war. The long distances involved increased the difficulties of establishing and maintaining the logistical pipelines of supplies and replacements. While the materiel problem was solved by the application of massive resources, the distance of Vietnam from the United States made the problem appear remote and may have contributed to the loss of public support that eventually ended the war without the United States having achieved its strategic objectives.

The nature of the theater had a definite influence on campaign planning. The three major regions of Vietnam where U.S. and allied troops fought were the northern coastal plain, the Central Highlands, and the Mekong Delta. At the outset the U.S. campaign plan was to find and defeat decisively the North Vietnamese forces while simultaneously conducting counterinsurgency operations against the Viet Cong. This led the United States to distribute its forces across the nation more or less in proportion to the population rather than to the threat. Key terrain was defined tactically and not for the entire theater, and the option of closing off the border between South and North Vietnam extending into Laos was not pursued aggressively or with overwhelming force. Although the United States consistently won on the battlefield, the campaign turned into a war of attrition, which was won by North Vietnam as the United States lost its will to fight.

At the tactical level, the rugged and mountainous terrain in the Central Highlands and the heavy rainfall in the Mekong Delta slowed the tempo of operations and made it difficult for the United States to bring to bear fully its advantage in modern weapons. The thick vegetation of the triple canopy rain forest diminished the effects of air attacks and made bomb damage assessment difficult. The vegetation also offered the attacking North Vietnamese and Viet Cong the advantage of concealment, which they used to great effect in ambushes. The United States responded by removing the protective vegetation, cutting down the trees and killing them with chemical defoliants.

Fighting a new kind of warfare without fronts, the U.S. and South Vietnamese forces tried to overcome the terrain with new technology—primarily helicopters. This worked to a certain extent and made it possible for the U.S. and South Vietnamese forces to win almost all tactical engagements. After losing a few early battles conclusively, however, the North Vietnamese refused to stand and fight in decisive battles. They appreciated the nature of the terrain and took advantage of it to build combat power slowly and steadily under the concealment offered by the terrain until they had sufficient force to defeat the South Vietnamese forces in conventional combat. The U.S. and South Vietnamese forces fought well and won most of the time but in the end were defeated by the terrain and the will of their opponent.

PERSIAN GULF WAR

The Persian Gulf War took place in a region entirely different from Southeast Asia. The terrain in Kuwait, Saudi Arabia, and Iraq is desert with low elevations, little vegetation, and little rainfall.

Kuwait is 7,000 miles from the United States, but the United States had for several years been building a capability to project its armed forces rapidly to just such a remote location. Airlift, sealift, and pre-positioned equipment and supplies were on hand when President Bush decided to commit U.S. forces to defend Saudi Arabia and then to free Kuwait from Iraq. The strategic time and distance factors, however, did cause great anxiety for the U.S. commanders who were forced to wait for several weeks until the U.S. and coalition forces were sufficient to defend Saudi Arabia and another several weeks until enough forces had been assembled in the theater to take the offensive.

From a theater viewpoint, the man-made features of Saudi Arabia were critical to the success of the U.S. and coalition forces. Saudi Arabia had constructed modern airfields, ports, and roads in the northern area near Kuwait and Iraq in anticipation of this contingency. The availability of these facilities was crucial to the success of the U.S. buildup and resupply operations. If these facilities had not been available, it would have been much more difficult—and perhaps impossible—for the United States to have done what it did. The shallow seas of the Persian Gulf and the coastal islands off the coast of Kuwait made naval operations difficult and contributed significantly to the decision to make an amphibious operation a feint rather than a real attack.

Tactically, the lack of cover and concealment for the Iraqi forces was very important. The U.S. and coalition aircraft, having beaten the Iraqi air forces, could attack ground targets that could not hide. Although there were some problems due to unfavorable weather, in general the area was ideal for air operations. The mobility afforded ground vehicles in the desert areas west of Kuwait made possible the gigantic single envelopment that struck deep into Iraq and then turned to cut off the Iraqi Republican Guard from behind. Although the Iraqis sought to create artificial obstacles, they found it a difficult task lacking terrain favorable for the defense. Finally, modern high-technology tank guns and missiles were at their best in the flat terrain and able to fire accurately at long ranges over excellent fields of fire. The success of the United States and the coalition against Iraq is evidence of sound appreciation of the terrain and weather in the area of operations and illustrates the application of military geography at its best.

Overall Characteristics of Military Geography

Military geography is mission oriented. The mere compilation of data on a military theater or area of operations does not constitute an application of military geography. The essential nature of the geographic process comes into play only when the spatial relationships and impacts of the area are interpreted in light of a mission. If the mission changes, the effect of the features of the area of operations changes also.

Military geography is part of the commander's planning process. The planning process begins when a new mission is received from higher headquarters. After the commander has analyzed and elaborated on the mission, the next step is to make an estimate of the situation. The estimate of the situation includes consideration of the mission, friendly forces, enemy forces, and a military geographic analysis of the characteristics of the area of operations. Alternative courses of action to accomplish the mission are drawn up and evaluated. The commander decides upon a course of action and issues orders to that effect. The products of the planning process are missions for subordinate units. The receipt of these new missions at lower levels in turn initiates a new cycle of military planning, including additional estimates of the situation and geographic analyses of new areas of operations.

The area of operations is the geographic region defined by the military mission. It comprises the area directly influenced by the forces and weapons under the control of the commander. It includes the area occupied by the opposing enemy force, terrain features designated as objectives to be seized or held, area held by adjacent friendly units, and the support area of the military organization itself. The commander needs to know everything about the area of operations, and is also interested in major events in a larger area of interest, which includes the area of operations. The commander may assign reconnaissance resources to report on the area of interest or request intelligence on the area of interest from higher headquarters.

Military geography is an element of military intelligence and the responsibility of the staff intelligence officer. The compilation of data on natural and man-made terrain features and on human factors in a theater or area of operations usually is performed by intelligence sections and organizations. In intelligence terms, combat intelligence corresponds to terrain analysis, and strategic intelligence corresponds to theater analysis.

Military geography is three-dimensional. Warfare takes place in the oceans below the earth's surface and in the atmosphere above the earth's surface, as well as on the surface. Time and distance factors and the importance of relative positions take on different meanings when it is possible to deliver munitions by aircraft or missile or observe from a satellite. Submarines, aircraft, and satellites in earth orbit, with the possibility of manned space stations, must be considered in making the estimate of the situation. The area of operations is in reality a three-dimensional volume ranging from below the surface of the earth to the outer boundary of inner space.

Military geography uses the latest in modern technology and methodology. Earth satellites are used in geodesy and cartography. Aerial and space photography is used in cartography, terrain analysis, and theater analysis. Computers are used to calculate time and space factors and target locations, and to compile, manage, and analyze geographic data. Modern methods of mathematical, statistical, and spatial analysis are applied to military geographic problems.

Problems of Military Geography

There is a general lack of knowledge about military geography. All military forces employ the various elements of military geography, but most of them do not realize that they are using military geography. Terrain analysis is always part of tactical doctrine. Theater analysis is an accepted part of the intelligence process. Geopolitics is employed in politico-military strategic studies. These applications of military geography often are used without an appreciation for the spatial viewpoint of geographic analysis.

Military geography is seldom taught as a unified discipline. Terrain analysis and theater analysis are taught in military schools as part of courses on tactics or intelligence, and the elements of geopolitics are taught at war colleges. The essential appreciation of geography as a discipline unified by the process of spatial analysis, however, has been lost.

The inclusion of area analysis in the intelligence staff function has the advantage of providing a sponsor for this activity. It has the disadvantage of reinforcing the tendency of commanders and operations officers to concentrate on enemy and friendly forces and ignore or relegate to secondary importance the characteristics of the area of operations.

Value of Military Geography

Military geography has had substantial impact on military combat in the past. The Russian winter played a part in defeating both Napoleon and Hitler. The defensible landforms (*cuestas*) north of Paris helped France stop the invading Germans in 1914. The hedgerows of Normandy stalled the Allied advance in 1944. The vastness of China thwarted Japanese attempts at military domination in the 1930s. The nature of the Persian Gulf area made it possible for U.S. technology to crush the Iraqi army overwhelmingly in short order. It is logical to believe that military geography will also have a substantial impact on combat in the future.

The value of military geography is that it integrates the effects of the area of operations by a process of spatial analysis. Military geography does not live up to its potential because it is seldom applied by trained geographers or by military personnel who understand geographic method. Even so, the ideas and concepts of geography are now, more than ever, essential to the planning and conduct of military operations.

JOHN R. BRINKERHOFF

SEE ALSO: Deception; Lines of Communication; Mahan, Alfred Thayer; Mapping, Charting, and Geodesy; Maps, Charts, and Symbols, Military; Theater of War.

Bibliography

Able, R. F., M. G. Marcus, and J. M. Olson, eds. 1992. *Geography's inner worlds: Pervasive themes in contemporary American geography*. New Brunswick, N.J.: Rutgers Univ. Press.
Faringdon, H. 1986. *Confrontation: The strategic geography of NATO and the Warsaw Pact*. London and New York: Routledge and Kegan Paul.
Gray, C. S. 1988. *The geopolitics of super power*. Lexington, Ky.: Univ. of Kentucky Press.
Hartshorne, R. 1962. *Perspective on the nature of geography*. Association of American Geographers. Chicago: Rand McNally.
James, P. E., and C. F. Jones, eds. 1954. *American geography: Inventory and prospect*. Association of American Geographers. Syracuse, N.Y.: Syracuse Univ. Press.
O'Sullivan P., and J. W. Miller. 1983. *The geography of warfare*. New York: St. Martin's Press.
Rosen, S. J. 1977. *Military geography and the military balance in the Arab-Israeli conflict*. Jerusalem: Hebrew Univ.
Zoppo, C. E., and C. Zorgbibe. 1985. *On geopolitics: Classical and nuclear*. NATO Advanced Science Institutes Studies. Dordrecht: Martinus Nijhoff.
U.S. Department of the Army. 1972. *Field manual 30–10: Military geographic intelligence (terrain)*. Washington, D.C.: Government Printing Office.

GEORGIA

Georgia, one of the fifteen former republics in the Union of Soviet Socialist Republics (USSR), declared its independence from the Union on 9 April 1991. On 26 May 1991, Zviad Gamaskhurdia was elected president of the republic with 87 percent of the popular vote. By December, however, opposition to Gamsakhurdia's policies led to fighting between his supporters and those who would end what they called his dictatorial rule. The president fled the capital on 6 January 1992. It is not known whether Georgia will eventually join the eleven Soviet republics who formed the new Commonwealth of Independent States on 21 December 1991. For the next several years, relations between Georgia and the rest of the world are likely to remain unsettled. Over time, new structures and patterns will emerge in economics, trade and commerce, politics and government, finance, manufacturing, religion, and virtually all aspects of human life. New arrangements must be devised for dealing as a sovereign state with the new Commonwealth of Independent States (which is itself undergoing frequent change) and with the world outside the boundaries of the former Soviet state. If the history of the Soviet Union since 1985 is any guide, we can expect dramatic surprises and dynamic change.

An important question for the world is how Georgia and the Commonwealth will organize and provide for security. Most of the Soviet Union's armed forces stationed in Georgia are likely to be withdrawn. Also of great concern is the disposition of nuclear weapons, the security of these weapons, the command and control of their potential use, and compliance with arms control agreements entered into by the former Soviet government. The world can only hope these issues are settled amicably.

It will be years before the above issues are resolved for Georgia and some time before events settle down into more routine and measurable patterns. No accurate description of this new country's policies, defense structure, and military forces was available to be included in this encyclopedia. Only time will reveal the future of Georgia as a separate sovereign state. The reader is thus referred to the historic information contained in the article "Soviet Union," and to the latest annual editions of the *Military*

Balance, published by Brassey's (UK) for the International Institute of Strategic Studies; the *Statesman's Year-Book* published by the Macmillan Press Ltd and St. Martin's Press; and the *World Factbook*, developed by the U.S. Central Intelligence Agency, and published commercially by Brassey's (US).

F. D. MARGIOTTA
Executive Editor

GERMANY, FEDERAL REPUBLIC OF

Having been divided for more than 40 years following the end of World War II, East and West Germany were reunited on 3 October 1990. The Federal Republic of Germany is now facing the costs of that transition, including transfer of billions of dollars to its new eastern states and commitments of large sums to the Commonwealth of Independent States (formerly the Soviet Union) and Eastern Europe.

Power Potential Statistics

Area: 356,910 square kilometers (137,803 sq. mi.)
Population: 76,877,400
Total Active Armed Forces: 476,300 (0.620% of pop.)
Gross Domestic Product: US$1,157.2 billion (1990 est.)
Annual Defense Expenditure: US$47.1 billion (4.7% of GDP, 1990 est.)
Iron and Steel Production:
  Crude steel: 41.073 million metric tons (1989)
  Pig iron: 32.777 million metric tons (1989)
Fuel Production:
  Coal: 71.428 million metric tons (1989)
  Coke: 5.7 million metric tons (1987)
  Crude oil: 13.94 million metric tons (1988)
  Natural gas: 14,783 million cubic meters (1988)
Electrical Power Output: 580,000 million kwh (1990)
Merchant Marine: 598 vessels; 5,029,615 gross registered tons
Civil Air Fleet: 239 major transport aircraft; 647 usable airfields (312 with permanent-surface runways); 4 with runways over 3,659 meters (12,000 ft.); 86 with runways 2,440–3,659 meters (8,000–12,000 ft.); 95 with runways 1,220–2,440 meters (4,000–8,000 ft.).

For the most recent information, the reader may refer to the following annual publications:
The Military Balance. International Institute for Strategic Studies. London: Brassey's (UK).
The Statesman's Year-Book. New York: St. Martin's Press.
The World Factbook. Central Intelligence Agency. Washington, D.C.: Brassey's (US).

History (1945–55)

Three historical developments predating the establishment of the Federal Republic of Germany in May 1949 have colored its defense policies from the beginning: (1) the experience of the first German republic (Weimar Republic, 1918–33) with the military tradition it inherited from the Prussian/German national state (the Second Reich), which existed from 1871 to 1918; (2) the foreign and security policies of National Socialist Germany (the Third Reich, 1933–45); and (3) the cold war, which developed between the Western powers (Britain, France, and the United States) and their World War II ally, the Soviet Union, from 1945 to 1949. The cold war provided the impetus for the existence and the rearming of West Germany; the experiences of Weimar and the Nazi years determined, in large measure, the shape of both the new republic and its armed forces.

When the armed forces of the United States and the Soviet Union met on the Elbe River in April 1945, they symbolized the determination of their political leaders to preclude the possibility of a renewed German drive for hegemony in Europe. Berlin's victory in the Franco-Prussian War (1870–71) laid the foundation of the German national state proclaimed at Versailles in 1871; its defeat in the First World War (1914–18) led to the fall of the Hohenzollern dynasty and establishment of the Weimar Republic, but not to the end of the national state. In 1945 the victorious Allies viewed the unity of that state, its physical size, and its industrial potential, as the causes of the war thrust on Europe by Adolf Hitler in 1939. The plans the Big Three adopted for the occupation of Germany at the 1945 Yalta Conference reflected the resolve of Britain, the Soviet Union, and the United States—a resolve shared by their French ally—to deprive Germans of their war-making potential.

There were generally three components to the occupation policies carried out by the four allies from 1945 to 1947: the constriction of Germany into borders substantially less than those of 1937; the transfer of ethnic Germans out of the countries of Eastern Europe; and the decentralization of political and economic authority in the four occupation zones of post-1945 Germany. These measures, agreed on in London (1943–45) and approved at Yalta, were collectively a means to curb the resurrection of a dominant political and industrial German power in the heart of Europe.

The territorial constriction of the Reich began with undoing its aggrandizements of 1938–40: the four Allies re-established the independent Austria (under four-power occupation until 1955) created in 1919 and annexed by the Third Reich in 1938; they returned to Czechoslovakia the Sudetenland, whose annexation by Berlin had been approved by Britain and France at the 1938 Munich Conference; and they restored to France the two provinces of Alsace and Lorraine, annexed by Germany in 1871 and again in 1940 (after having been French from 1918 to 1940). They did not stop, however, at returning Germany to its 1937 borders. The Soviet Union annexed East Prussian territory, including the historic German city of Koenigsberg; and, as the Soviet frontier moved westward at the expense of Poland, the western Polish border was also pushed westward, into the heart of historic Germany. The old Hansa city of Danzig, under League of Nations administration in the interwar years, became the Polish city of Gdansk.

The Germany occupied by the four victorious powers in 1945 was, therefore, substantially reduced in size from the state of 1937, when its borders were essentially those of 1918 and 1871. These territorial changes were purportedly provisional, pending the conclusion of a peace treaty ending World War II. The second component of allied policy, however, the expulsion of Germans from their historic homelands in Bohemia and east of the Oder-Neisse rivers, was a good indication of the Allies' determination not to allow the restoration of a German state with its 1937 borders. The forced transfer of ethnic Germans out of Czechoslovakia and post-1945 Poland reflected the unwillingness of Prague and Warsaw to tolerate irredentist German minorities. Czechoslovakia had ample experience with the nearly five million Sudeten Germans, whose rights had been guaranteed by minority treaties in 1919, and their demands in 1938 to go "home to the Reich" (where, as subjects of the Hapsburgs until 1918, they had never been). The policies of the Third Reich had definitively ended the possibility of tolerance for large German minorities after 1945. Germans in Silesia and East Prussia were forced westward into the new Germany of the four occupation zones.

What they found there has been described as a "zero hour" in German history. By the 1980s, the phrase had come under much criticism: Germans, like every other nation, were precluded from shaking off their past. However, the victorious Allies were determined to reshape the course of German history and to undo much of the legacy of the national state founded in 1871. In that sense, 1945 was a zero hour, in which the third component of allied policy, the decentralization of German political and economic power, began to take shape. Had East-West allied unity prevailed, it is difficult to predict what would have happened to Germany; arguably, instead of the two German states founded in 1949, four or more smaller German states in the heart of Europe might have emerged from four-power occupation.

Allied unity did not prevail, however. Soviet policies in the countries liberated from Nazi occupation by the Red Army in 1945 led, from 1945 to 1948, to increasing Western concern that all of Europe would fall under Moscow's control. The West's fears of Stalin's intentions were confirmed by the February 1948 communist coup against the Benes-Masaryk government of Czechoslovakia, despite Prague's obvious determination to be responsive to Soviet foreign policy concerns (including declining to participate in the 1947 Marshall Plan conference). Moscow's refusal to allow free elections in Poland, the establishment of communist governments under Soviet control throughout Eastern Europe, the threat of a communist victory in the Greek civil war, and the possibility of communist electoral victories in France and Italy had already led, by 1947, to the United States' enunciation of the Truman Doctrine and the Marshall Plan. American foreign policy adopted as its principal goal the "containment" of Soviet power and communist ideology.

In Germany, the East-West struggle for the future of Europe culminated in 1948–49 with the Soviet blockade of the Western Allies' land routes to Berlin. Divided at Yalta into four sectors of occupation, paralleling the zones of Germany as a whole, the city of Berlin was entirely surrounded by the Soviet zone. Air and land access to their occupation sectors had been guaranteed to the Western powers at Yalta and Potsdam in 1945. Convinced that it would not be possible to reach a constructive compromise with the Soviet Union in the four-power Allied Control Council (ACC), the three Western powers in the spring of 1948 determined to push ahead with economic reforms, specifically the issuing of a new currency, in their zones and sectors. The Soviet walk-out from the ACC and the imposition of the Berlin blockade set the stage for a test of will between the two sides. Three years after the end of World War II in Europe, the economic potential of Germany had suddenly become of crucial importance in the East-West balance of power. The threat in Europe was no longer from Berlin, and the role of Germans in postwar Europe appeared in a new light.

Throughout the winter of 1948–49, in the shadow of the Western Allies' airlift of food and fuel to the western sectors of Berlin, representatives of the states (*Laender*) of the three western zones of occupied Germany met in Bonn to draft a new constitution for a western German state. The Basic Law (*Grundgesetz*), seen as provisional until the German people could freely draft a constitution for Germany as a whole, established a state based on four principles: democracy, federalism, the rule of law, and the social contract. Adopted in May 1949, it did not provide for armed forces or a defense establishment. The three Western Allies retained responsibility for the defense of the Federal Republic. Moreover, they did not relinquish the quadripartite rights and obligations established toward Berlin and Germany as a whole at the 1945 Yalta Conference, as a result of their common defeat of the Third Reich.

The political climate of 1949–50, however, was changing rapidly. Three events over the course of fourteen months determined the future of West Germany's defense policies: the signature in Washington, D.C., of the North Atlantic Treaty by the United States, Canada, and ten Western European countries six weeks before the adoption of the Basic Law; the establishment of the German Democratic Republic (GDR) in October 1949, as a response to the West German state; and the invasion of South Korea in June 1950 by North Korean forces. Virtually overnight, with the Korean invasion, the United States' assessment of the possibility of a Soviet invasion of Western Europe changed. Containment, which had had an economic aspect (the Marshall Plan) designed to promote the rapid economic recovery of postwar Western

Europe and a political aspect (the North Atlantic Treaty) meant to assure Western Europeans that the United States would not remain aloof to threats to their security, now acquired a military dimension. In the fall of 1950, the United States and its allies began to establish the institutions of the North Atlantic Treaty Organization (NATO); American troops would be stationed in Europe in peacetime.

However, the U.S. government had also come to the conclusion that the defense of Western Europe required the rearming of the Federal Republic. Much to France's dismay, the United States began to press for a decision to reestablish a (West) German army; Paris responded with the proposal for a European Defense Community (EDC), in which West German armed forces would be fully integrated, modeled on the European Coal and Steel Community (ECSC) then being set up. Within the Federal Republic, the U.S. demands met with resistance; only five years after the defeat of the Third Reich, the remilitarization of Germany was an issue that split the government of Chancellor Konrad Adenauer and produced the first popular German peace movement. The Soviet threat and the prospect of equality in Western foreign policy councils impelled Bonn toward rearmament, but the recent German past made the prospect unattractive.

The debate over West German rearmament stretched on through the early 1950s, as Western Europeans drafted and then debated the constituent treaty of the EDC. The Korean War ended in June 1953, three months after Stalin's death and five months after the inauguration of a new American president: Dwight D. Eisenhower, who had been the first NATO Supreme Allied Commander in Europe (SACEUR). Until the summer of 1954, when Pierre Mendès France became prime minister with a mandate to end the Indochina War, France was involved in a disastrous attempt to maintain its colonial empire in Asia. His government ended the war in Vietnam with the 1954 Geneva Accords, but Mendès was less successful in convincing the French National Assembly to ratify the EDC treaty.

The failure of EDC led immediately to the search for another framework within which to accomplish West German rearmament. At Paris in the fall of 1954, the Western Allies and the Federal Republic agreed that Bonn would establish an army, navy, and air force (collectively, the *Bundeswehr*) under West German command in peacetime. In time of war, command of the *Bundeswehr* would pass to NATO's SACEUR (the air force's [*Luftwaffe*] mission is fully integrated, even in peacetime). Bonn renounced the acquisition of chemical and nuclear weapons, and London agreed to maintain the British Army of the Rhine (BAOR) in West Germany. The Federal Republic joined the 1947 Brussels Treaty powers (France, Britain, and the Benelux countries) in a rechristened Western European Union (WEU) and in 1955 became the fifteenth member of NATO. (Greece and Turkey had joined in 1952.)

In the Federal Republic in 1954–55, ten years after the defeat of the Third Reich and its *Wehrmacht*, the Adenauer government and the West German parliament (*Bundestag*) began the difficult task of creating the Federal Republic's armed forces. With the defeat in 1945 had come the discrediting of the Prussian-German military tradition; ten years later, the political leadership of the Federal Republic had to redefine the role of the armed forces in West German society. The principles on which the *Bundeswehr* was established reflected the experience of the Weimar Republic with its army (*Reichswehr*) and the legacy of the Third Reich.

Politico-Military Policy and Strategic Problems (1955–90)

The first generation of political and military leadership in the Federal Republic had personal experience with the Weimar and Nazi years. Consequently, that leadership set out to create armed forces appropriate to the second German republic. Members of the *Bundeswehr*, officers and men, were to be "citizens in uniform," who did not relinquish their right to participate in the West German political process because of their military service or career. Burned by the concept of an "apolitical" army that had never accepted the Weimar Republic but had sworn personal allegiance to Adolf Hitler, the Federal Republic sought instead to create an army subordinate to elected civilian authority and committed to the democratic state established by the Basic Law.

The amended articles of the Basic Law that established the *Bundeswehr* reflected the German past. Preparations for a war of aggression were expressly forbidden (Art. 26), but the Federal Republic could join a collective security organization to maintain the peace (Art. 24). The federal government alone was granted authority to raise armed forces (Art. 73) for defensive purposes (Art. 87a). General male conscription was established (Art. 12a) with provision for conscientious objection (Art. 4). The *Bundeswehr* was made subject to the laws and legal decisions of the democratic state. Oversight into its activities was placed in civilian hands: the federal government, the *Bundestag*, and the defense minister, who, in peacetime, has overall command of the armed forces. In a situation in which the Federal Republic is under attack (*Verteidigungsfall*), strictly defined by the Basic Law (Art. 115a), command passes from the defense minister to the chancellor (Art. 115b) and, through him, to NATO's SACEUR. Provisions for maintaining democratic civilian control during a military crisis were also fully laid out in the Basic Law (Art. 115c–e).

Within this constitutional framework developed armed forces with a unique mission. The role of the *Bundeswehr* was to deter, through its conventional capabilities, an attack on the Federal Republic and its allies and, should deterrence fail, to defend and reestablish the integrity of West German and NATO territory. It had no other security or foreign policy role and could not be deployed outside the NATO area. By the 1980s, West German security policy, as one aspect of Bonn's foreign policy, was based on three principles: renunciation of force (except in self-defense), international cooperation, and the pursuit of peaceful change through negotiation. In the words of the 1983 West German Defense White Paper, "Safeguarding security through balance is accompanied by the readiness for dialogue and cooperation with the East."

Bonn's defense policy reflected the goals of West German foreign policy in several ways. The Federal Republic after 1969 was one of the most outspoken advocates of the approach to East-West relations outlined in NATO's 1967 Harmel Report on the Future Tasks of the Alliance. The Harmel Report added a third "d" to NATO's role: *detente* joined *deterrence* and *defense* as a task of the Alliance in a nuclear age. The NATO report spelled out the common East-West interest in survival, which was the conceptual foundation of strategic arms control negotiations in the early 1970s, as well as the *Ostpolitik* of Willy Brandt's social-liberal coalition government (1969–74).

West German defense policy was grounded in the principles of that *Ostpolitik*, all of which were contained in the 1975 (Helsinki) Final Act of the Conference on Security and Cooperation in Europe (CSCE). They included renunciation of force, respect for existing borders, free movement of people and ideas, and good faith fulfillment of international obligations. At the CSCE follow-up conferences held from 1977 to 1989, the Federal Republic was one of the most energetic supporters of implementing the CSCE principles by broadening human contacts, establishing a regime for the reporting and observing of military maneuvers, and promoting economic and political cooperation throughout Europe.

The third component of Bonn's foreign and defense policy was the development of increasing cooperation on all levels—political, economic, and military—with West Germany's Western European neighbors, especially France. In the European Community (EC) in the late 1980s, Bonn, Paris, and the other ten member countries were focused on implementation of the "EC 92" program established by the 1985 Single European Act. Bilaterally, the two countries finally began to implement the defense provisions of the 1963 Franco-German Treaty of Friendship and Cooperation. At the end of 1988, they established a joint Franco–West German brigade in the Federal Republic, under the rotating command of officers from each army. The chancellor and the French president also held talks on the role of French nuclear forces in assuring the security of the Federal Republic.

Finally, West German security policy was NATO policy. Since 1955, Bonn had defined the *Bundeswehr*'s role, deployment, weapons systems, and structure in conjunction with its NATO allies. West Germany was well aware of the critical NATO part played by its armed forces. The 1983 Defense White Paper noted that, "The weight carried by the Federal Republic of Germany in the Alliance is decisively determined by the German contribution to the common defense. The *Bundeswehr* gives the Federal Republic of Germany a say and influence in the Atlantic Alliance." Bonn developed that influence by maintaining, through conscription and real increases in defense expenditures, the major conventional force on NATO's central front. In the 1980s, it used that influence to encourage its allies to pursue an East-West dialogue, and to deflect criticism of its own energetic pursuit of cooperation with the states of the Warsaw Pact.

East-West cooperation seemed essential to West Germans for a number of reasons, one of which was the precarious strategic situation of the Federal Republic. Warsaw Pact tanks were stationed a little more than an hour from Hamburg; its fighter-bombers were five minutes flying time away. Across the narrowest central part of the Federal Republic, the inner-German border was less than 160 kilometers (100 mi.) east of the Rhine. As the 1983 Defense White Paper stated, "There is no sector in Western Europe where its depth meets the military strategic requirements."

West Germany's strategic situation was complicated by its numerous borders. In addition to the Warsaw Pact states of the GDR and Czechoslovakia, the Federal Republic bordered neutral Austria and Switzerland, the Baltic and North seas, and five NATO nations (Denmark, the Netherlands, Belgium, Luxembourg, and France). Given the short distances involved and the nature of modern weapons, the defense policies and military preparedness of all of these states directly affected the Federal Republic.

Finally, Bonn faced the same "particularly heavy handicap," in the words of the 1983 White Paper, as all European members of NATO: 3,000 miles of Atlantic Ocean separating them from the major power of their Alliance. Unlike the Soviet Union, in time of crisis the United States expected to resupply and reinforce its troops in Western Europe, and to ship strategic raw materials, notably oil, through hostile waters and airspace. As a result, keeping open Atlantic sea lanes was as critical to West German security as it had once been to beleaguered Britain. Modern warfare made it equally essential to maintain NATO's European airfields in order to receive reinforcements and resupply from the United States.

Political-Military Revolution (1989–91)

Like much else based on the threat that had confronted NATO for 40 years, West German foreign and defense policy began a year of revolutionary change on 9 November 1989 with the opening of the Berlin Wall and inner-

German border. Less than twelve months later, the GDR ceased to exist and German unification became a reality (3 October 1990). In Moscow on 12 September 1990 the four World War II allies, the Federal Republic, and the GDR, concluding the "Two Plus Four" negotiations begun in May, signed the Treaty on the Final Settlement with Respect to Germany that ended the Big Four's rights and responsibilities in Berlin and in Germany as a whole. Under the terms of that agreement, the Soviet Union agreed to withdraw all of its armed forces stationed in Berlin and the GDR by the end of 1994 (Art. 4). The Federal Republic undertook to reduce the personnel strength of the *Bundeswehr* to 370,000, with a ceiling of 345,000 on land and air forces (Art. 3), the drawdown to begin with the entry into force of the treaty limiting Conventional Forces in Europe (CFE). On 19 November 1990 the 22 member states of NATO and the Warsaw Pact signed the CFE treaty at the Paris CSCE summit.

The new Federal Republic of Germany was larger by 108,330 square kilometers (41,020 sq. mi.) and 17 million people. In a restructuring that took place before unification, the GDR in early 1990 created the five *Laender* that acceded to the Federal Republic on 3 October: from north to south, Mecklenburg-Western Pommerania, Saxony-Anhalt, Brandenburg, Saxony, and Thuringia. Reunified Berlin retained West Berlin's place in the *Bundestag* and *Bundesrat*, but with the direct elections and full voting rights that the western half of the city had never had. For the first time, the young men of West Berlin, like their countrymen in the new *Laender*, became eligible for conscription into the *Bundeswehr*.

The armed forces faced two immediate tasks: integration of the GDR's National People's Army (*Nationale Volksarmee*, NVA) into the *Bundeswehr*, and coordination with the Red Army and local civilian authorities of the withdrawal of all Soviet forces stationed in the five new states at the time of unification. On the day of unification, the NVA was not the force it had been a year before, when the GDR celebrated its fortieth, and last, anniversary. NVA service had been required of all males between the ages of 18 and 25, with no provision for conscientious objection. The GDR's army, navy, and air force had had a total personnel strength of 172,000, with an extensive reserve system.

In the eleven months from November 1989 to October 1990, however, much of that force had ceased to function. The *Bundeswehr*, which with unification incorporated 18,000 of 70,000 NVA officers and men into its structure provisionally (subject to individual review of their skills and security record by the end of 1992), thus faced the task of reducing its strength in the 1990s from 486,000 in late 1990 to the agreed maximum of 370,000 by 1995. In January 1992, the Defense Ministry announced plans to destroy the 7,000–8,000 heavy weapon systems, mainly tanks and armored vehicles, inherited from the NVA, while retaining other NVA equipment (tents, blankets, medicines, etc.) for use in natural disaster and humanitarian relief.

In 1991, the *Bundeswehr* began to implement the withdrawal of Soviet forces from Germany, scheduled to be accomplished in phases by the end of 1994: 30 percent each year in 1991, 1992, and 1993, and 10 percent the final year. When done, the formidable task will have relocated 337,800 military personnel, 163,700 dependents, and 44,700 civilian employees, as well as millions of tons of materiel, including more than 4,000 main battle tanks and 8,000 armored vehicles. German financial assistance to the withdrawal effort amounts to over DM 12 billion in direct outlays (including DM 7.8 billion for housing in the former Soviet Union) and DM 3 billion in interest-free loans. Over 154,000 personnel and 680,000 tons of materiel had been withdrawn by the end of November 1991.

While the *Bundeswehr* could thus look back on its first year of service to a united Germany as a period in which it had made a good start in carrying out the commitments of the Two Plus Four and CFE treaties, the future of German armed forces in the 1990s was far from certain. On the heels of unification, the Persian Gulf War brought to the forefront of the German political debate the question of *Bundeswehr* deployment "out of area" (i.e., outside the North Atlantic area as defined by the Alliance's constituent treaty), whether for United Nations enforcement and peacekeeping actions or in fulfillment of other international obligations. The federal government maintained that such deployment required a constitutional amendment—an interpretation of the Basic Law that was far from unanimous; and it was not at all clear, at the end of 1991, that such an amendment would pass the *Bundestag* with the necessary two-thirds majority. Germany also experienced an even more painful debate on the nature of its NATO obligations, with several cases of serving *Bundeswehr* personnel refusing deployment to Turkey during the war. In early 1992, deployment through the United Nations of *Bundeswehr* medical corps (*Sanitaetsdienst*) personnel to Cambodia raised new questions of morale and logistics for the German armed forces.

In the course of the year, the number of conscientious objectors in the Federal Republic doubled, from 74,000 young men who had refused conscription in 1990, to 150,000 in 1991. For the first time in the country's history, its political and military leaders began to consider seriously the option of a professional army. If opposition to both military service and "out of area" deployment remained strong—as appeared likely—in the 1990s, there was a strong prospect that the *Bundeswehr* would be totally reorganized as a professional, volunteer force before the end of the decade.

Military Assistance

The Federal Republic does not receive military assistance. It is one of only two NATO countries (the United States is the other) providing defense aid to its allies: to Portugal since 1978, to Greece and Turkey since 1964. Eighty percent of this aid is in the form of new materiel, 20 percent comes from *Bundeswehr* overstock. A one-time aid package to Turkey to modernize its armored equipment included 77 Leopard I tanks. By 1985, the total value of West German defense deliveries to Portugal was DM 292 million; to Greece, DM 802 million; and to Turkey, DM 2.92 billion.

The Federal Republic also provides foreign military assistance to 40 non-NATO countries, with a total value by 1985 of DM 750 million. From 1961 to 1985, over 1,400 soldiers from 55 developing countries had received training with the *Bundeswehr* in West Germany.

Defense Industry

As one of the leading industrial nations of the world, the Federal Republic has a modern defense industry. Approximately 40 percent of the *Bundeswehr*'s budget in the 1980s went to weapons research, development, procurement, and maintenance. In 1985 that amounted to DM 18.9 billion, 85 percent of which went to West German industries. Research and development received about 4 percent (DM 700 million) of the total.

Nevertheless, defense comprised on average in the 1980s only about 3.4 percent of West German industrial production. The defense share of the total in individual industries ranged from 1 percent to 2 percent of the automotive industry, to 10 percent of shipyard production, and over 50 percent of the aerospace industry's annual output. Approximately 250,000 jobs were involved in West German defense production in the mid-1980s, about 1 percent of the total in the Federal Republic.

In addition to purely West German research, development, and weapons production, Bonn initiated joint projects with several of its NATO allies in the 1980s. They included the second-generation antitank helicopter (PAH-2) with France and the European Fighter Aircraft (EFA) with Britain, France, Italy, and Spain, but the future of both these projects was called into question in 1992, as Germany began to restructure its weapon procurement policies along with the *Bundeswehr*'s size and missions. Other joint projects in which West Germany industry participated included the short-range (ASRAAM) missile with Britain and the Netherlands, the Patriot air defense system with the United States, and the Roland missile with France. The *Bundeswehr*'s Bremen-class frigate was a product of West German–Dutch cooperation, and in the late 1980s Bonn was developing plans with Britain, Canada, France, Italy, the Netherlands, Spain, and the United States for a new NATO frigate in the 1990s.

Defense Structure

In keeping with the civilian control of the armed forces established by the Basic Law, the *Bundeswehr* has no general staff. Its senior officer is the inspector general. The armed forces are divided into the army, navy, and air force. Their administration (*Bundeswehrverwaltung*) is a civil service; in 1990, it had 138,000 employees, with a reduction of at least 50 percent planned in the 1990s. In the *Bundeswehr* itself, women serve only in the medical corps; the Basic Law (Art. 12a) expressly forbids their training in the use of weapons.

Officers and men fall into three categories: conscripts, long-term volunteers (*Zeitsoldaten*) serving up to twelve years, and career soldiers. The ratio of each category varies from service to service. In 1985, 53 percent of the army was composed of conscripts, 37 percent were long-term volunteers, and 10 percent were career soldiers. In the navy, conscripts were only 27 percent of the total, while long-term volunteers (51 percent) and career soldiers (22 percent) were an overwhelming majority of its total strength. Finally, the air force also had a majority of long-term volunteers (47 percent) and career soldiers (17 percent), with conscripts making up 36 percent of the *Luftwaffe*'s manpower. A reorganization of the Bundeswehr is currently underway to achieve the required manpower level of 370,000 by 1995. Current planning calls for the *Bundeswehr* in that year to have an army of 255,400 men, an air force of 82,400, and a navy of 32,200. Of these, 125,943 (army), 25,074 (air force), and 7,983 (navy) will be conscripts, with the rest career or professional officers and soldiers. These will include 22,245 officers, 82,882 NCOs, and 24,330 soldiers (army); 12,026 officers, 37,260 NCOs, and 8,040 airmen (air force); and 5,429 officers, 13,158 NCOs, and 5,630 sailors (navy).

(For an explanation of the abbreviations and symbols used in the following section of military statistics, see the list of Abbreviations and Acronyms in each volume.)

Total Armed Forces

Active: 476,300 (203,000 conscripts; 5,000 active reserve trg posts, all services, and 7,500 interservice staff, not listed below). Terms of service: 12 months.
Reserves: 1,009,400 (men to age 45, officers/NCO to 60): Army 857,000, Navy 28,000, Air 124,400.

ARMY: 335,000 (163,300 conscripts).
Field Army: (219,000) 3 Corps, 12 div.
I Corps (NORTHAG): 3 armd, 1 armd inf div.
II Corps (CENTAG): 1 armd, 1 armd inf, 1 AB, 1 mtn div.
III Corps (CENTAG):
  2 armd, 1 armd inf div.
  1 armd inf div (LANDJUT).
  (Armd div with 2 armd and 1 armd inf bde; armd inf div with 2 armd and 1 armd bde; mtn div with 1 armd, 1 armd inf and 1 mtn bde; all with 1 armd recce bn, 1 arty regt (1 bn each: 18 FH-70, 18 203mm, 16 110mm MRL), 1 AD regt (with 35mm Gepard), 1 avn sqn; AB div with 3 AB bde.)
Corps Tps: 4 SSM bn each with 6 Lance; 3 AD comd (each 1 regt with 36 Roland).
1 Roland SAM bn.
1 AD arty bn with Gepard 35mm.
3 avn comd each 1 lt (48 UH-1D), 1 med tpt (32 CH-539), 1 ATGW hel (56 Bo-105 HOT) regt.

Territorial Army (cadre: 64,600 in peacetime).
Command Structure: 3 Territorial Comd (linked with NATO cmd) 5 Military Districts, 28 Military Regions, 76 Sub-regions: Units (eqpt holding only unless stated).
10 Home Defense bde, 5 with 2 armd, 2 armd inf, 1 arty bn plus full log spt (at 50–60% in peacetime) (two assigned to field army div). 5 with 1 armd, 2 armd inf, 1 arty bn.
1 German/French bde (Ge units incl 1 mech inf, 1 arty bn; 1 SP ATK coy).
15 Home Defense regt with 3 mot inf bn, 18 120mm mor.
150 Home Defense coy, 300 Security pl.
Eastern Command (51,400):
2 Military Districts, 14 Military Regions, 45 Sub-Regions.
6 Home Defense bde; 3 inf, 4 arty bn.
Equipment:
MBT: 7,000: 648 M-48A2G (Territorial bn), 2,054 Leopard 1A1 (1,258 to be upgraded to A5), 2,024 Leopard 2, 1,725 T-54/-55, 549 T-72M.
Light tanks: 143 PT-76 (CFE: HACV).
Recce: 410 SPz-2 Luchs, 56 TPz-1 Fuchs (NBC), 84 Wiesel, 1,262 BRDM-1/-2.
AIFV: 3,254: 2,104 Marder A1/A2 (to upgrade to A3), 1,150 BMP-1/-2.
APC: 10,327 (CFE 4,208): 889 TPz-1 Fuchs (CFE 344), 2,276 M-113 (CFE-537), 2,115 BTR-40 (CFE 1,073), 293 BTR-50 (CFE 123), 2,165 (CFE 1,402) BTR-60, 1,175 BTR-70, 685 BTR-152, 729 MT-LB.
Total arty: 4,579.
  Towed arty: 1,486: 105mm: 39 M-56, 192 M-101; 122mm; 335 D-30; 397 M-1938 (M-30); 130mm: 175 M-46; 152mm: 137 D-20; 155mm: 211 FH-70.
  SP arty: 1,263: 122mm: 374 2S1; 152mm: 95 2S3; 155mm: 573 M-109A3G; 203mm: 221 M-110A2.
  MRL: 556: 110mm: 204 LARS; 122mm: 260 Cz RM-70, 59 BM-21, 227mm: 33 MLRS.
  Mortars: 1,274: 120mm: 490 Brandt, 499 Tampella on M-113, 210 M-120, 75 2B11.
SSM: 26 Lance launchers (incl 2 in store).
ATGW: 1,975 Milan, 205 TOW, 316 RJPz-(HOT) Jaguar 1, 162 RJPz-(TOW) SP, AT-3 Sagger (incl BRDM-2SP), AT-4 Spigot, AT-5 Spandrel.
RCL: 106mm: 99 (in store).
ATK guns: 85mm: 64 D-48; 90mm: 121 JPz-4-5 SP (CFE HACV); 100mm: 267 T-12.
AD guns: 3,072: 20mm: 1,766 Rh 202 towed; 23mm: 295 ZU-23, 131 ZSU-23-4 SP; 35mm: 432 Gepard SP; 40mm: 204 L/70; 57mm: 244 S-60.
SAM: 658 Fliegerfaust 1 (Redeye), SA-7, 226 SA-4/-6/-8/-9. 163 Roland SP.
Helicopters: 207 PAH-1 (Bo-105 with HOT), 186 UH-1D, 109 CH-53G, 97 Bo-105M, 138 SA-313, 10 SA-318, 8 Mi-2, 28 Mi-8 (T/TB), 49 Mi-24.
Marine: (River Engineers): 36 LCM, 12 PCI (river) (.

NAVY: 37,600 (5,200 Eastern Command) incl naval air, 8,900 conscripts. Bases: Glücksburg (Maritime HQ) and three main bases: Wilhelmshaven, Kiel and Warnemünde. Other bases with limited support facilities: Baltic: Eckernförde, Flensburg, Olpenitz, Neustadt, Rostock (Eastern Command HQ), Peenemünde. North Sea: Borkum, Bremerhaven, Emden.
  Deployment: 1 DDG, 1 FF, 1 AO in Mediterranean on 3-month roulement with some gaps (about 50% cover).
Submarines: 24:
  18 Type 206/206A SSC with Seeaal DM2 533 mm HWT (8 conversions to T-206A complete).
  6 type 205 SSC with DM3 HWT.
Principal Surface Combatants: 14:
Destroyers: 6:
  DDG: 3 Lütjens (mod US Adams) with 1 × 1 SM-1 MR SAM/Harpoon SSM launcher, 2 × 127mm guns; plus 1 × 8 ASROC (Mk 46 LWT, 2 × 3 ASTT.
  DD: 3 Hamburg (ASUW) with 2 × 2 MM-38 Exocet, 4 × 533mm TT (SUT), 3 × 100mm guns.
Frigates: 8 Bremen with 2 Lynx hel (ASW/OTHT), 2 × 2 ASTT; plus 2 × 4 Harpoon; plus 4 Bützow (Parchim1) with 2 × 12 ASW RL, 4 × ASTT ex-GDR but not to be incorporated into Ge Navy.
Patrol and Coastal Combatants: 45:
Corvettes: 5 Thetis (ASW) with 1 × 4 ASW RL, 4 × 533 mm TT.
Missile Craft: 40:
  10 Albatros (Type 143) PFM with 2 × 2 Exocet, and 2 × 533mm TT.
  10 Gepard (T-143A) with 2 × 2 Exocet.
  20 Tiger (Type 148) PFM with 2 × 2 Exocet.
Mine Warfare: 54:
Minelayers: 2 Sachsenwald (600+ mines).
Mines Countermeasures: 52:
  10 Hameln (T-343) comb ML/MCC.
  6 Lindau Troika MSC control and guidance, each with 3 unmanned sweep craft.
  11 converted Lindau (T-331) MHC.
  5 Schütze (T-340/-341) comb ML/MSC.
  18 Ariadne/Frauenlob MSI.
  2 MCM diver spt ships.
  Plus 5 Kondor II MSC ex-GDR but not to be incorporated into Ge Navy.
Amphibious: Craft only: some 20 LCU.
Support and Miscellaneous: 48:
Underway support: 4: 2 Spessart, 2 Eifel AO.
Maintenance/Logistic: 32: 2 AR, 4 Rhein SS/MCMV spt, 6 small (2,000t) AOT (incl 1 ex-GDR), 8 Lüneburg log spt, 2 AE, 8 tugs, 2 icebreakers (civil); plus 5 log spt ex-GDR to be deactivated by Dec 1991.
Special purpose: 9: 3 AGI, 2 trials, 3 multi-purpose (T-748), 1 trg.
Research and survey: 3: 1 AGOR, 2 AGHS (civil-manned for Ministry of Transport).

NAVAL AIR ARM:
4 wings, 9 sqn: 2 wings with Tornado; 1 MR/ASW wing with Atlantic, Lynx; 1 SAR/liaison wing with Do-28, Sea King.
FGA: 3 sqn with Tornado.
FGA/Recce: 1 sqn with Tornado.
MR/ELINT: 2 sqn with Atlantic.
Liaison: 1 sqn with Do-28/D228.
ASW: 1 sqn with Sea Lynx Mk 88 hel.
SAR: 1 sqn with Sea King Mk 41 hel.
Equipment: 118 cbt ac, 19 armed hel.
Aircraft:
  Tornado: 104 (72 FGA, 24 FGA/recce, 8* trg) plus 5 in store.
  Atlantic: 19 (14 MR, 5 ELINT).
  Do-28: 18 (16 SAR, liaison; 2 environmental protection).
  Do-228: LM: 1 (environmental monitoring).
Helicopters:
  Sea Lynx Mk 88: 19 (ASW).
  Sea King Mk 41: 22 (SAR).

Germany, Federal Republic of

Missiles:
  ASM: AS-12/-20/-30/, Kormoran, Sea Eagle.
  AAM: AIM-9 Sidewinder.

AIR FORCE: 103,700 (12,200 Eastern Command) (30,800 conscripts).
Tactical Command (GAFTAC).
5 air div: 2 tac, 2 AD, 1 mixed Eastern Division (note, no op use of ex-GDR ac, except perhaps MiG-29).
FGA: 11 wings, 21 sqn: 5 wings with Tornado; 2 with F-4F; 4 with Alpha Jet.
Fighter: 2 wings with F-4F.
Recce: 2 wings with RF-4E.
EW: 1 trg sqn with HFB-320 Hansa Jet.
SAM: 4 wings (each 6 sqn) Patriot, being deployed; 9 wings (each 4 sqn) HAWK; 12 sqn Roland being deployed, 2 wings (each 2 sqn) SA5.
Radar: 2 tac Air Control Commands: 10 sites, 3 remote radars; 4 sites; 10 remote radars in Eastern division.
AAM: Sidewinder.
ASM: AS-20.
Transport Command (GAFTC).
Transport: 3 wings: 4 sqn with Transall C-160, incl 1 (OCU) with C-160, Do-28.
  1 special air mission wing with Boeing 707-320C, VFW-614, CL-601, Do-28 ac; UH-1D hel (VIP).
Helicopters: 1 wing: 3 sqn; plus 1 det with UH-1D (liaison/SAR).
Eastern division: 1 special mission sqn with Il-62, Tu-134, Tu-154, L-410S, Mi-83, 1 tpt sqn with An-26, 1 tpt/SAR sqn with Mi-8T/mil.
Training Command:
FGA: 1 det (Cottesmore, UK) with Tornado; 1 OCU (Beja, Portugal) with Alpha Jet.
Fighter: OCU (George AFB, Alabama) with F-4E.
Training: NATO joint pilot trg (Sheppard AFB, Texas) with T-37B, T-38A; primary trg sqn with P-149D.
Liaison: base flt with Do-28D.
Equipment: 638 cbt ac (44 trg (overseas)); plus 336 for disposal, no attack hel.
Aircraft:
  F-4: 231. -F: 152 (FGA, ftr); -E: 7 (OCU, in US); RF-4E: 72* (recce).
  Tornado: 223 (161 FGA, 19* ECR, 24* OCU, 19* in trinational trg sqn, (in UK)).
  MiG-29: 20 (ftr), -UB: 4 (trg).
  Alpha Jet: 164 (146 FGA, 18* wpn trg (in Portugal)).
  Transall C-160: 84 (tpt, trg).
  Boeing 707: 4 (VIP). CL-601: 7 (VIP). Do-28-D2: 60 (6 VIP, 54 tpt/liaison). Do-228: 1 (tpt). HFB-320: 7 (tpt). Il-62: 2 (tpt). L-410-S: 4 (VIP). T-37B: 35. T-38A: 41. Tu-134: 3 (tpt). Tu-154: 2 (tpt). VFW-614: 3 (VIP).
Helicopters:
  UH-1D: 109 (105 SAR, tpt, liaison; 4 VIP).
  Mi-8: 20 (SAR tpt).
  Mi-8S: 6 (VIP).
  Mi-2: 25 (civil rescue).
Missiles:
  ASM: AS-20, AGM-65 Maverick.
  AAM: AIM-9 Sidewinder.
  SAM: 216 HAWK launchers; 68 Roland launchers. 48 Patriot launchers, 24 SA-5 launchers.
Aircraft of former GDR air force:
Fighter: MiG-21: 251; MiG-23: 58; Su-22: 27; (all TLE will be disposed/sold in accordance with CFE).
Transport: Tu-134: 3; Tu-154: 24; Il-62: 3; An-26: 12; L-410S: 4 (further op use is under consideration), L-39: 52; An-2: 1, (no op use intended).
Helicopters: Mi-2: 24 (in use for civilian rescue); Mi-8: 52 (further op use is still under consideration).

PARAMILITARY
Federal Border Guard (Ministry of Interior): 25,600; 5 cmd (constitutionally has no combat status). Eqpt: MOWAG SW-1/-2 APC; 2 P-149D, 1 Do-27 ac; Bo-105M, 32 Alouette II, 13 UH-1D, 10 Bell 212, 22 Puma hel. Major reductions in force are expected in the 1990s, as a result of the scheduled abolition of border controls within the EC, German unification, and the demise of the Warsaw Pact and the Soviet Union.
Coast Guard: 550; 1 inshore tug, 8 PCI.

FOREIGN FORCES
NATO:
  HQ Northern Army Gp (NORTHAG).
  HQ Central Army Gp (CENTAG).
  HQ Allied Air Forces Central Europe.
  HQ Allied Land Forces Jutland and Schleswig-Holstein (LANDJUT).
  HQ Allied Command Europe Mobile Force (AMF).
  HQ 2 Allied Tactical Air Force (2 ATAF).
  HQ 4 Allied Tactical Air Force (4 ATAF).
Belgium: 22,800; 1 corps HQ, 1 div HQ; 1 armd, 1 mech inf bde (NORTHAG).
Canada: 7,000; 1 mech bde gp. 1 tac hel sqn.
  1 air div with 2 FGA sqn (CENTAG/4 ATAF).
France: 43,700; 1 corps HQ, 2 armd div.
  Berlin: (2,700), 1 armd, 1 inf regt.
Netherlands: 5,700; 1 armd bde (NORTHAG)
United Kingdom: 63,400; 1 corps HQ, 3 armd div, 13 ac sqn (NORTHAG/2 ATAF).
  Berlin: (2,800), 1 inf bde.
U.S.: 222,500. 1 army HQ, 2 corps HQ; 2 armd, 2 mech div; 1 armd. 1 air force HQ; 2 air div (CENTAG/4 ATAF).
  Berlin: (4,300), 1 inf bde.
USSR: 338,000. Army: 1 Gp, 5 Army HQ, 7 TD, 6 MRD. Air: 1 Air Army HQ; 8 FGA, 9 ftr regt.

Future

Although formally united, western and eastern Germany are still dramatically different. Western Germany is a leading exporter with a well-developed market economy dominated by manufacturing and service industries. Its highly skilled and urbanized population has high living standards and benefits from a comprehensive social welfare system. In contrast, eastern Germany is struggling to make the transition from an inefficient command economy to one responsive to market demands; industrial production in the new federal states was down 50 percent in early 1991 compared with the same period in the German Democratic Republic the year before. Restructuring the eastern economy will take a major effort of political will in the 1990s and significant amounts of funding that will increase pressures to reduce defense spending and restructure the *Bundeswehr*.

EDWINA S. CAMPBELL

SEE ALSO: European Communities; NATO; Prussia-Germany, Rise of; Western Europe; World War I; World War II.

Bibliography

Abenheim, D. 1989. *Reforging the iron cross.* Princeton: Princeton Univ. Press.
Campbell, E. S. 1989. *Germany's past and Europe's future: The challenges of West German foreign policy.* McLean, Va.: Pergamon-Brassey's.
Die Bundeswehr in der Demokratie, 1955–1985. 1986. Bonn: Presse- und Informationsamt der Bundesregierung.
International Institute for Strategic Studies. 1991. *The military balance, 1991–1992.* London: Brassey's.
Ireland, T. 1981. *Creating the entangling alliance.* Westport, Conn.: Greenwood Press.
Kelleher, C. M., and Gale A. Mattox, eds. 1987. *Evolving European defense policies.* Lexington, Mass.: Lexington Books.
McGeehan, R. 1971. *The German rearmament question.* Urbana: Univ. of Illinois Press.
Sommer, T. 1974. "Wiederbewaffnung und Verteidigungspolitik." In *Die Zweite Republik,* ed. R. Loewenthal and H.-P. Schwarz, pp. 580–603. Stuttgart: Seewald Verlag.
Weissbuch 1985: Zur Lage und Entwicklung der Bundeswehr. 1985. Bonn: Bundesminister der Verteidigung.
White book 1983: The security of the Federal Republic of Germany. 1983. Bonn: Federal Minister of Defense.

GHANA, REPUBLIC OF

Ghana, a small nation on the west coast (Gulf of Guinea) of Africa, is a major supplier of the world's cocoa. Ghana remains a member of the Commonwealth, although it emerged as the leader of the nationalist Pan-African movement during the early days of its independence. Ghana has had several democratic constitutions, but it is currently ruled by a military government.

Power Potential Statistics

Area: 238,540 square kilometers (11,020 sq. mi.)
Population: 15,427,200
Total Active Armed Forces: 11,900 (0.077% of pop.)
Gross National Product: US$5.8 billion (1990 est.)
Annual Defense Expenditure: US$23 million (0.5% of GNP, 1988 est.)
Iron and Steel Production: none
Fuel Production: none
Electrical Power Output: 4,110 million kwh (1989)
Merchant Marine: 4 vessels; 52,016 gross registered tons
Civil Air Fleet: 6 major transport aircraft; 9 usable airfields (5 with permanent-surface runways); none with runways over 3,659 meters (12,000 ft.); 1 with runways 2,440–3,659 meters (8,000–12,000 ft.); 7 with runways 1,220–2,440 meters (4,000–8,000 ft.).

For the most recent information, the reader may refer to the following annual publications:
The Military Balance. International Institute for Strategic Studies. London: Brassey's (UK).
The Statesman's Year-Book. New York: St. Martin's Press.
The World Factbook. Central Intelligence Agency. Washington, D.C.: Brassey's (US).

History

In 1871 the Tanti Confederation was formed on the Gold Coast by Akan kings seeking incorporation into the British Empire. The Tanti Confederation was destroyed within the year by the formation of the Ashanti Confederacy, a violent nationalist movement that opposed incorporation into the empire. In 1896 British troops challenged the Confederacy and entered what is now Ghana to maintain the cocoa trade. Ashanti warriors resisted the British until 1911 when Ghana was annexed by Britain. Nationalist pressure continued, however, and resulted in the implementation of the 1925 constitution and a long string of concessions by Governor Guggeisberg.

Participation in government by a newly emerging class of educated black Africans resulted in a massive boycott of the cocoa industry in 1937. The boycott, and worries of nationalist activity by returning veterans of the Commonwealth's war effort, prompted Governor Sir Alan Burns to draft a new constitution in 1943. The Burns Constitution went into effect on 1 March 1946 and was declared by the governor to be a move toward self-government.

Dissatisfaction with the Burns Constitution resulted in the formation of the United Gold Coast Convention (UGCC) in 1947. On 28 February the nationalist movement took a new turn. Until that day, protests had been essentially nonviolent and had consisted primarily of boycotts and peaceful marches. On one such march on 28 February, protesters marched through the streets of Accra to the seat of government at Christianburg Castle. Nervous policemen opened fire killing six and sparking two days of rioting that killed 29, injured 237, and destroyed £2 million worth of property.

After another constitution and elections scheduled for 1951, Kwame Nkrumah split with the UGCC and formed the Convention Peoples Party (CPP). A series of nationwide strikes enforced with violence by the CPP, and an election victory in February 1951, gave the CPP the leverage needed to negotiate for independence. In the 1954 election, the CPP continued to gain strength, defeating the more radical National Liberation Movement and capturing 71 of the 104 legislative seats.

On 6 March 1957 Ghana gained its independence from Britain. The road to independence, a revolution in every sense of the word, was comparatively nonviolent and was influenced only slightly by the use or threat of use of military force.

Politico-Military Background and Policy

The participation of over 70,000 men in the Commonwealth war effort in World War II, at a time when the entire population of the Gold Coast Colony was only 4 million, not only produced the deleterious economic effects that accelerated political change, but also broadened the political views of the returning veterans and established a strong military tradition. The newly elected president of an independent Ghana soon exploited the prevailing militarism in his country. Nkrumah mobilized the CPP youth groups federated as the Committee of Youth Organizations (CYO). The CYO was officially replaced by the Young Pioneers in 1957, but continued to act as Nkrumah's strong arm against political opposition.

Parliament soon transferred power to Nkrumah with a series of laws that included the State Secrets Act and the Security Service Act of 1963. The establishment of the Security Service completed the institutionalization of totalitarian control.

Nkrumah's Security Service was based on the military's Criminal Investigation Department (CID). The British-trained CID were soon retrained by Soviet and Eastern European intelligence specialists.

After an assassination attempt on 2 January 1964, Nkrumah became concerned over the reliability of the military and acted to create Ghana's "Special Force." The creation of Nkrumah's private unit soon drained the most capable soldiers from Ghana's army, and infuriated the British-trained senior officer corps. On 24 February 1966, 600 men of the Ghanian army launched a coup d'état. Immediately after the coup, a national liberation council was formed for the purpose of re-establishing democracy in Ghana. On 24 September 1979, a new democratic constitution took effect.

Corruption and a failing economy again prompted military intervention into the affairs of state and on 31 December 1981, Flight Lt. Jerry John Rawlings seized power and formed the seven-member Provisional National Defense Council. Opposition political parties are currently (1990) banned, although Rawlings promises an eventual return to democracy.

The role of the military is primarily the maintenance of Rawling's control over Ghana. Ghana's military has also been called upon to secure the national borders to prevent the entry of illegal aliens. Foreign workers and refugees from neighboring countries have in the past been a severe burden on the Ghanian economy.

Several international treaties currently govern Ghanian politico-military policy. Ghana became a signatory to the partial test ban treaty in 1963, the Geneva Protocol in 1967, the biological warfare convention in 1975, and the Enmod Convention in 1978.

Strategic Problems

Ghana's neighbors, Burkina Faso, Côte d'Ivoire, and Togo pose no significant threat to Ghana. Ghana's most significant defense problem appears to be the control of its 200-nautical-mile territorial sea.

Defense Industry

Ghana has no domestic defense industry.

Alliances

Ghana is a member of the Commonwealth, the UN, and the OAU. Foreign trading partners include the United Kingdom, the United States, and Nigeria. Foreign military aid amounts to only US$0.3 million.

Defense Structure

The ruling Provisional National Defense Council maintains complete control over the armed forces of Ghana.

Officers are educated in the United Kingdom and the United States. Military equipment is purchased from a variety of sources, including Britain, France, Germany, and India.

Ghana's armed forces are supported without conscription, and are composed of the army, the air force, and the navy.

(For an explanation of the abbreviations and symbols used in the following section of military statistics, see the list of Abbreviations and Acronyms in each volume.)

Total Armed Forces

Active: 11,900

ARMY: 10,000
2 Command HQ:
  2 bde (comprising 6 inf bn (incl 1 trg, 1 UNIFIL), spt units).
  1 recce bn (2 sqn).
  1 AB force (incl 1 para coy).
  1 indep inf bn.
  1 arty 'regt' (mor bn).
  1 fd engr regt (bn).
Equipment:
Recce: 3 Saladin, 3 EE-9 Cascavel.
AIFV: 50 MOWAG Piranha.
Mortars: 81mm: 50; 120mm: 28 Tampella.
RCL: 84 mm: 50 Carl Gustav.

NAVY: est 1,100 Commands: Western and Eastern.
Bases: Sekondi (HQ, West), Tema (HQ, East).
Patrol and Coastal Combatants: 4:
Coastal: 2 Achimota (Ge Lürssen 57-m) PFC.
Inshore: 2 Dzata (Ge Lürssen 45-m) PCI.

AIR FORCE: 800. 18 cbt ac, no armed hel.
COIN: 1 sqn with 4 MB-326K†, 2 MB-339.
Transport: 3 sqn: 1 VIP with 3 Fokker (2 F-27, 1 F-28); 1 with 3 F-27, 1 C-212; 1 with 6 Skyvan.
Helicopters: 2 Bell 212 (VIP), 2 Mi-2, 4 SA-318.
Training: 1 sqn with 10 Bulldog 122†, 12* L-29.

FORCES ABROAD
Liberia: about 1,500 forming part of ECOWAS force.
UN and Peacekeeping:
Afghanistan/Pakistan: (OSGAP): 1 adviser.
Iraq/Kuwait (UNIKOM): 8 observers.
Lebanon (UNIFIL): 1 inf bn (980).

PARAMILITARY
People's Militia: 5,000: part-time force with police duties.

Future

With an increasing literacy rate, which is currently at 30 percent, and growing public dissatisfaction with the Rawlings regime, democratic reforms seem likely. It is almost a certainty, however, that the Ghanian military will play a significant role in politics for years to come with the precedent it has established in this young nation.

LEE A. SWEETAPPLE

SEE ALSO: Colonial Empires, European; Commonwealth of Nations; Organization of African Unity; Sub-Saharan Africa.

Bibliography

Apter, D. E. 1963. *Ghana in transition.* New York: Atheneum.
Bourret, F. M. 1960. *Ghana: The road to independence, 1919–1957.* Stanford, Calif.: Stanford Univ. Press.
Bretton, H. L. 1968. *The rise and fall of Kwame Nkrumah: A study of personal rule in Africa.* New York: Praeger.
Hunter, B., ed. 1991. *The statesman's year-book, 1991–92.* New York: St. Martin's Press.
International Institute for Strategic Studies. 1991. *The military year-book, 1991–1992.* London: Brassey's.
James, C. L. R. 1977. *Nkrumah and the Ghana revolution.* Westport, Conn.: Lawrence Hill.
Omari, T. P. 1970. *Kwame Nkrumah: The anatomy of an African dictatorship.* New York: Africana.
Stockholm International Peace Research Institute. 1987. *Yearbook: World armaments and disarmament.* London: Oxford Univ. Press.
U.S. Central Intelligence Agency. 1990. *The world factbook.* Washington, D.C.: Government Printing Office.

GIAP, VO NGUYEN [1912–]

Senior General Giap (Fig. 1), five-star former commander in chief of North Vietnam's army, as well as its defense minister, is the only North Vietnamese general widely known in connection with the Indochina-Vietnam wars. Giap (also known as Anh Van, or Tran Van Lam) won his place in world history by leading the Viet Minh peasant forces that decisively defeated U.S.-supported French Mainland, Foreign Legion, Colonial, and collaborating Vietnamese regular and auxiliary forces in Dien Bien Phu in 1954, thus winning the First Indochina-Vietnam War (the Viet Minh War).

Subsequently, Giap brought sufficient continual military pressure on U.S., Free World (including Asian-Pacific allies), and South Vietnamese armed forces. This pressure combined with the results of North Vietnamese political-diplomatic maneuvering forced the governments of the United States and its allies to retire their troops from the field in 1973. Demoralized, unsupported South Vietnamese forces were then overrun by North Vietnamese troops in 1975. Giap's self-taught strategic, logistical, and tactical abilities, combined with North Vietnamese political-diplomatic maneuvering and Sino-Soviet sup-

Figure 1. Vo Nguyen Giap. (SOURCE: Presumed to be a North Vietnamese official photo)

port, enabled Eastern peasants to defeat Western professional military commanders, their citizen-soldiers, and their doctrine, to win the Second Indochina-Vietnam War (the Vietnam War).

Giap's determination and willingness to spill his troops' blood enabled Third World socialist forces to overcome the will of democratic powers' East-West alliances. The Western nations' complex political maneuvering, economic predominance, technological superiority, massive military strength, and immense firepower fell to this one-time history teacher's able organizational talents, his cunning, and his single-minded will power. While Giap is now a larger-than-life legend, American critics are downplaying his strategic abilities and his ideas on "People's War" by focusing on the huge losses suffered by his troops and their tactical defeats. But Senior Gen. Vo Nguyen Giap's military-political achievements—necessarily coupled with those of a wily and tenacious Ho Chi Minh—are and will remain monumental.

Giap the Individual

Vo Nguyen Giap was born in 1912 (or 1911 or 1909—many published "facts" about Giap are questionable) in Quang Binh / Vinh (or Thanh Hoa) in the then-French protectorate of Annam. Officially from a peasant family, he was reportedly the son of a hard-working, low-ranking mandarin scholar who sacrificed much to have his son educated.

Giap began working at age 14, immediately becoming involved with the Tan Viet Nationalist revolutionary party. In 1930 he took part in demonstrations against the French, earning two or three years in jail on Puolo Condore Island—where he read extensively. Giap then attended the French-operated Lycée National (with classmate and future adversary Ngo Dinh Diem), assisted by a future father-in-law. Later, while studying law at Hanoi University, he was first a journalist and then a high school history teacher, becoming known for his interest in Napoleon and for his inflammatory speeches. Awarded a less-than-M.A. degree in three years, Giap failed his fourth-year entrance exam, although some believe he received doctorates in political science and in law.

Giap was persuaded to embrace communism by Truong Chinh, later chief ideologue of the Vietnamese communists. Giap married his first wife, Minh Thai, in 1934, and they joined the Indochina Communist Party. In 1936 he was a founding member of the Democratic Front. Giap and Phan Van Dang then published anti-imperialist newspapers. Giap published his first book in 1938, coauthoring with Truong Chinh *The Peasant Question*. Regarded by the French as a subversive document and a guide to revolution, it was seized and destroyed.

When communism was outlawed in France in 1939, the French repressed the Indochinese Communists; and Giap went to China with Pham Van Dang, later premier of North Vietnam. There he became allied with Ho Chi Minh, leader of exiled Vietnamese Communists. He helped start the Viet Minh "League for the Independence of Vietnam" at the Tsin Tsi Conference in 1941. During World War II Giap worked in the mountains of Vietnam, gathering information to sell to the Republic of China and eliminating local opposition to communism. (Giap did not hesitate to eliminate challengers, including fellow Viet Minh.) In December 1944 he commanded the 34-person armed propaganda "Brigade" for National Liberation, which evolved into the People's Army of Vietnam (PAVN). This group reportedly attacked the French on 19 December 1946 in retaliation for earlier French shelling of Hanoi, thus initiating the Viet Minh War.

For a short time Giap was interior minister of the newly declared independent Democratic Republic of Vietnam but was soon promoted by Ho Chi Minh to general and commander in chief. He became defense minister in 1946, and in that year married his second wife, Dan Thai Ha, daughter of the minister of education, Dang Thai Mai. Giap often appeared in public with her and they had two children.

In Giap's early days of command he was, to Western eyes, a slight and rumpled figure—seeming strangely out of place near the tall French officers of the army he would eventually defeat or striding in review of his peasant soldiers. Later, the stern face that glared from the cover of *Time* magazine better fit Western ideas of a powerful military figure. Throughout Giap's career he has been *Nui Lua*, the "snow-capped volcano," as Ho Chi Minh referred to the calm exterior that barely concealed his seething interior. Occasionally he has exploded violently—once resigning as defense minister.

Forceful and arrogant, imaginative and impatient, energetic and ambitious, Senior General Giap has an analytic mind. Reported to have an "encyclopedic knowledge of military history" (Bowman 1985), he first received military training in 1940 from the Chinese Communists, later supplemented by reading Western and Eastern military works and studying Chinese Communist warfare against Japanese invaders. Giap alleges that he was influenced by the ancient Chinese strategist Sun Tzu, and by the early Vietnamese guerrilla warrior Trang Hung Dao, who defeated Kublai Khan's Chinese in 1287. The deaths in French jails of his first wife and child, his father, two sisters, and others of his family, plus the guillotining of his sister-in-law, strengthened Giap's hatred of the French and his determination to prevail. Although less well known and perhaps less sophisticated than Ho Chi Minh—even needing someone like Ho to fully succeed—Giap nevertheless approaches him in historical significance.

Giap's Accomplishments

Employing the philosophy of many successful military commanders, Giap urged his forces not to lose sight of the main objective of fighting—destruction of the enemy. He believed more in armed struggle than political struggle (in contrast to his former coauthor and subsequent doctrinal archrival, Truong Chinh), although conceding the need for both. Focusing on frontal attacks rather than the guerrilla tactics he had employed so well, Giap seemed first to follow the direct, "real warfare" philosophy of Germany's Clausewitz rather than the indirect approach offered by Sun Tzu. But his complex thoughts possibly parallel the in-between combat philosophy of Japan's Musashi.

An anonymous U.S. analyst (Giap's Use . . . 1969) noted that the classic principle of war, "economy of force," was used by great offensively minded commanders of history: Napoleon, Jackson, Lee, Rommel, Bradley, and Giap—and that Giap's application fitted his time. He organized only a few main force units and used them to attack isolated posts—first French, then American. But his irregular organizers were protected by guerrillas; they shunned contact, and concentrated on the enemy's rear. The reaction was what Giap expected: Counter–main force units were organized and fielded. The attention of enemy commanders was held rigidly to the relative handful of Giap's main forces—while the enemy's rear crumbled. From late 1946 until 1949, the French thought the lull in military activities meant they had won the war; they were not finding anyone to fight. Lulls while fighting the Americans in 1968 and 1969 were similar, the analyst stated. (At one time during the American phase, 95 percent of the combat elements of the U.S. expeditionary force was dispersed in territory where only 5 percent of the population was located.) Giap's most important force was conquering the real battlefield (the people) while his "economy of force" units (regular battalions) kept U.S. forces occupied elsewhere. Giap did not, however, neglect his assumed need for the final "big push."

Giap's determination to use periodic, unrelenting uniformed force—at first defeated again and again by the French—took him to the heights of conventional military success at Dien Bien Phu. There he brought artillery into terrain considered impossible by the French, establishing so vigorous a siege (with a four-to-one artillery dominance) that faraway U.S. supporters following the situation decided not to interfere by providing the troops, air strikes, or nuclear weapons requested by the French. The French surrendered.

Giap's willingness to learn about his foe, without regard to the number of men he lost, was demonstrated at the 1965 battle of Ia Drang. There he continued to send troops into battle against U.S. forces to see how the Americans fought, in order to develop a strategy and tactics to counter them. If his critics are correct, Giap did not learn. He continued to use frontal attack combat tactics (which had been successful at Dien Bien Phu) at the siege of Khe Sanh, in the Tet Offensive, and in the Eastertide Offensive. This reportedly caused his removal from office (Zabecki and Montpelier 1988; Summers 1985). He was replaced as vice chairman of the National Defense Council in 1971. In 1975, Chief of Staff Van Tien Dung (a proponent of political pressure) succeeded Giap as commander in chief. In 1977–78, when Vietnam initiated the Third Indochina War, Giap attempted to dissuade his colleagues from the proposed Soviet-style over-the-border overt invasion and occupation of Kampuchea, but his advice was ignored. He retired as minister of defense of the Socialist Republic of Vietnam in 1980, and from the Politburo in 1982. His work shifted from military commander and strategist to leader of an effort to improve the Vietnamese economy and future prospects through the use of science and technology. Giap's change of emphasis and offices, following the death of Ho Chi Minh (whom he once was expected to succeed), may be related as much to the loss of his patron as to arguments about strategy (Gurtov 1970).

Senior Gen. Vo Nguyen Giap was highly respected by his opponents as well as his fellow Communists for his policy, strategic (especially logistical), and tactical accomplishments. The West's most knowledgeable scholar of North Vietnam and Indochina reported (Pike 1966, p. 36) that Giap had redefined Mao Tse-tung's "three stages" into the National Liberation Front's (NLF) "third generation," and that "Mao-Giap became to revolutionary warfare what Marxism-Leninism is to Communist theory." Pike has provided two careful evaluations (1966, pp. 49–51; 1987, pp. 339–43), concluding that Giap cannot be regarded as a military genius (except, perhaps, a logistics genius). Despite his achievement of tactical skills and his energy, audacity, and meticulous planning, Giap was a competent but not brilliant commander, a first-rate military organizer (once past the stage of innovative conceptual work), and a meticulous planner. As a strategist, Giap was at best a "gifted amateur." Nevertheless, organization of the "Legion of Porters" and architecture of the logistical miracle of the Ho Chi Minh Trail, with its ubiquitous bicycles that moved tons of supplies, shows that this user of "floating bridges" and the willing backs of his people surely was a consummate logistician.

But more significant than specific incidents of successful strategic or tactical surprise, superb logistic movements, or other strictly military accomplishments in battle (despite costly failures through frontal combat in major engagements) are the global implications of Giap's ultimate victories over Western powers and their Eastern allies. To the nations and peoples of today's developing countries (who first were shown the possibility of victory by the East when turn-of-the-century Japan defeated the West—the Russian fleet at Tsushima), Giap demonstrated clearly the feasibility and the specific means for seemingly weak "resistance movements" to overcome apparently overwhelmingly strong "great powers." Richard Nixon notes (1978, p. 269) that Giap stated in 1965: "The war against South Vietnam was a model for the communist movement around the world; if such a style of aggression could succeed there, it could work elsewhere."

Giap's Future

Vo Nguyen Giap will always provide an example of how to use military thinking to achieve political ends, not just a general who was important in battle for a brief period in history. Although the target of political enemies and of whispering campaigns, Giap remains a "national treasure" of Vietnam. He seems to have overcome a reported deadly illness, and continued to write and be written about (Pike 1987, p. 343; Davidson 1988). With a continuing stream of his own publications, and others writing about him and his philosophy, both Giap's accomplishments and his purported failings will provide clear examples of military thought put into practice.

DONALD S. MARSHALL

SEE ALSO: Ho Chi Minh; Mao Tse-tung; Musashi, Miyamoto; Vietnam and Indochina Wars.

Bibliography

Bowman, J. S., ed. 1985. Vo Nguyen Giap. In *The Vietnam war: An almanac*. New York: World Almanac.
Davidson, P. B. 1988. *Vietnam at war—The history 1946–1975*. Novato, Calif.: Presidio Press.
Dobbs, C. 1988. Vo Nguyen Giap. In *Dictionary of the Vietnam War*, ed. J. S. Olson. New York: Greenwood Press.
Fall, B. B. 1966. *Hell in a very small place: The siege of Dien Bien Phu*. Philadelphia: J. B. Lippincott.

Giap, Vo Nguyen

Giap's use of the "economy of force" principle. 1969. LORAPL Papers, Indochina Archive, Univ. of California, Berkeley.

Giap, V. N. 1962. *Dien Bien Phu.* Hanoi: Foreign Language Publishing House.

———. 1967. *Big victory, great task.* New York: Praeger.

———. 1975. *Unforgettable days.* Hanoi: Foreign Language Publishing House.

———. 1976. *How we won the war.* Philadelphia: Recon.

———. 1979. *War for national liberation.* Hanoi: Su That Publishing House.

Gurtov, M. 1970. *Some recent statements by North Vietnamese leaders: General Giap.* Rand Document D-20026-ARPA/AGILE. Santa Monica, Calif.: Rand Corp.

Jenkins, B. 1972. *Giap and the seventh son.* (P4851) Santa Monica, Calif.: Rand Corp.

Nixon, R. 1978. *The memoirs of Richard Nixon.* New York: Grosset and Dunlap.

O'Neill, R. J. 1969. *General Giap: Politician and strategist.* New York: Praeger.

Pike, D. B. 1966. *Viet Cong.* Cambridge, Mass.: M.I.T. Press.

———. 1987. *PAVN: People's Army of Vietnam.* Novato, Calif.: Presidio Press.

Summers, H. 1985. Giap, Vo Nguyen. In *Vietnam war almanac.* New York: Facts on File.

Zabecki, D. T., and R. P. Montpelier. 1988. Unlimited expense account. *Vietnam* 1(2):43–49.

GONZALO DE CÓRDOBA
[1453–1515]

Gonzalo (Gonsalvo) Fernández de Córdoba y Aguilar played a key role in the transition from medieval to modern warfare. He developed the Spanish infantry to the point where it replaced heavy cavalry as the principal combat element. His carefully planned campaigns, which coordinated infantry, cavalry, and artillery, replaced the medieval cavalry duel. He fitted his infantrymen with light armor and steel helmets and armed them with short swords instead of cumbersome Swiss halberds, or with harquebuses instead of crossbows. He divided his men into captaincies, which he combined into larger units, prototypes of regiments or Spanish *tercios.* In Italy his inspiring leadership in wars against France forged his poorly paid soldiers into a disciplined army that outmoded Renaissance condottiere warfare.

Early Life

Born in the family castle of Montilla, 40 miles south of Córdoba, the son of Pedro Fernández de Córdoba, Gonzalo and his brother Alfonso de Aguilar were carried into battle virtually from infancy, against the Muslims of Granada or against their own cousins, the counts of Cabra. As boys they had suits of chain mail. Gonzalo represented his family at the court of Princess Isabella of Castile and her husband Prince Ferdinand of Aragón, but he withdrew in 1473 in accordance with the politics of the marquis of Villena, his brother's father-in-law. Gonzalo returned to Isabella's service in 1479 after she had gained his release from captivity by his Cabra cousin. Commanding a hundred knights of the Order of Santiago and funded by his brother against Alfonso V of Portugal, he helped Isabella gain the Castilian throne.

Gonzalo emerged as an outstanding commander in the campaigns against Granada (1481–92) in which feuding Castilian vassals united and fought under royal leadership to finish the Reconquest, ending the 700-year Muslim presence in Spain. The queen rewarded him by making him a knight commander of Santiago and giving him twelve towns.

Gonzalo's Italian Campaigns (1495–97 and 1500–1504)

Commanding an expedition to protect Naples against the invasion of Charles VIII of France, Gonzalo lost his first battle against the French at Seminara because of poor coordination with his Italian allies. Afterwards he concentrated on training and organizing his men, refusing challenges for a field battle with D'Aubigny, the French commander in Calabria. By July 1496, however, when the allied army assembled at Atella, Gonzalo had gained control in Calabria with the guerrilla-like tactics he had used in the Granada wars. His men hailed him as "The Great Captain." Speed and skill in taking fortresses at Atella and Ostia further enhanced his reputation.

In 1500 Gonzalo returned to Italy and nearby waters as captain-general of a large armada, ostensibly to support a Venetian war against the Turks, but Ferdinand had already agreed with Louis XII of France to conquer and divide the Kingdom of Naples. Again Gonzalo, with Pedro Navarro and Antonello da Trani, demonstrated brilliant siegecraft in taking the Turkish fortress of Saint George on Kephallenia (Cephalonia).

Although the Franco-Spanish conquest of Naples in 1501–1502 presented few difficulties, major warfare soon broke out between the two victors. For nine months Gonzalo, headquartered in Barletta, restricted himself to small operations and raids launched from a line of advanced fortified garrisons. Finally, reinforced and supplied, he assembled his forces for a grueling march to his chosen battlefield near Cerignola on a rising slope among vineyards. Arriving late in the afternoon of 28 April 1503, the Great Captain established a defensive base of operations, ordering his exhausted men to dig a long trench and build a parapet behind it with the dirt they extracted. The French, under the viceroy, the duke of Nemours, arrived at dusk, and attacked at nightfall. The Franco-Swiss infantry stumbled into Gonzalo's trench, and were then hit by harquebus fire from behind the parapet. A French cavalry charge fared no better and Nemours was killed. Gonzalo then led his Italian-Spanish cavalry in a counterattack, turning the battle into a rout.

Fabrizio Colonna, an Italian commander in Gonzalo's army, said disdainfully that a mere ditch had won the battle. In fact, that battle changed warfare forever. Gonzalo's soldiers halted their pursuit of the retreating Frenchmen to mutiny for pay. He placated them with appeals to national pride and allowed them to sack the city of Bari, the Castel Nuovo in Naples and even his own house. But later he hanged the leaders of the mutiny.

Louis XII sent another army to Italy, and Gonzalo, outnumbered three-to-one, fought another campaign that year, this time in the valley of the Garigliano River. Again, with headquarters in San Germano, he established garrisons to hold a defensive perimeter while avoiding a major confrontation. The Roccasecca garrison diverted the French approach to San Germano, and the two armies faced each other across the river in a war of attrition. The marquis of Mantua, who had replaced the duke of La Trémouille as French commander, retired and the marquis of Saluzzo replaced him. After the French army withdrew from the flooded river banks to winter quarters, Gonzalo assembled his army and attacked during a terrible rainstorm on 27–28 December. He surprised and completely defeated the enemy in this Battle of the Garigliano.

Gonzalo's Loyalty to Ferdinand "the Catholic"

King Ferdinand made Gonzalo viceroy of Naples, duke of Terranova, and rewarded him generously, but he also surrounded him with spies and bureaucrats. In consequence the beleaguered viceroy presented royal officials with the famous "accounts of the Great Captain," including fictitious sums paid to the clergy for prayers. After Isabella's death in 1504, her daughter Juana inherited Castile. The future of the union of Castile and Aragón was in grave jeopardy. Many of the great vassals of Castile, including Gonzalo's kinsmen, turned to support Juana and her husband, Philip of Hapsburg. As a consequence, threatened King Ferdinand was still less trustful of his viceroy in Naples. The Venetians and the pope had already offered him command over their armies, and rumors circulated that he intended to make himself ruler of Naples.

However, Castilian Gonzalo remained loyal to the Aragonese king, illustrating in his own life the emergence of transcending national loyalty. Relieved of his Italian command in 1507, Gonzalo returned peacefully to Spain. Denied his promised reward, the mastership of Santiago, he withdrew to his Castilian estates. When Gonzalo's nephew, the marquis of Priego, revolted and Ferdinand pulled down Montilla Castle as punishment, the uncle counseled submission. In 1512, after Spain's defeat at Ravenna, the king called on the Great Captain to lead a new Italian expedition, but then reneged. Gonzalo accepted this disappointment, too, and remained a loyal Spaniard. He died three years later on 1 December 1515 from malaria, which he had contracted during his long stay in southern Italy.

PAUL STEWART

SEE ALSO: Arab Conquests; Cavalry; Italian Wars.

Bibliography

d'Auton, J. 1889–95. *Chroniques de Louis XII.* Editions de la Société de l'histoire de la France, vols. 245, 250, 264, 278. Paris: Renouard.
de Gaury, G. 1955. *The grand captain.* London: Longmans.
Lojendio, L. M. 1952. *Gonzalo de Córdoba, el gran capitán.* Madrid: Espasa-Calpe.
Pieri, P. 1952. *Il Rinascimento e la crisi militare italiana.* Torino: Einaudi.
Quatrefages, R. A. 1977. A la naissance de l'armée moderne. *Mélanges de la casa de Vélasquez* (Paris) 13:119–59.
Rodríguez Villa, A., ed. 1908. *Crónicas del Gran Capitán.* Nueva biblioteca de autores españoles. vol. 10. Madrid: Balliére.

GOVERNMENT, MILITARY

There are two major connotations of the term *military government.* The first, which defines the term as the periodic interference of a nation's military leadership in its civil affairs, is more accurately defined as a usurpation of a nation's power by members of the armed forces. The second definition applies to the establishment of a military jurisdiction over a defined region by a controlling power. Military government may be semipermanent or it might be short term. In the former, it is in place over the long term; in the latter, its purpose is to permit the transition from wartime chaos and devastation to orderly civilian government.

Historically, colonial powers—Great Britain, France, Italy, and others—have had considerable experience in military government as a component of their colonial policies. In recent times, however, the most significant military governments have been established by the United States. Since World War II, four major military governments—Germany, Austria, Japan, and Korea—have been administered by the United States. Because the United States has had the greatest amount of experience in military government in modern times, and because through its influence in the United Nations and other international forums will probably greatly influence any military governments established in the future, the following focuses on the U.S. experience in military government.

The United States did have experience in military government prior to World War II. The U.S. Army had established such governments in various territories (e.g., Florida, Louisiana, New Mexico, and California) as the United States established its control across the North American continent. In addition, the army, Treasury Department, and the navy had all administered to Alaska, and the U.S. Navy maintained a military government on Guam from 1898 until 1950. Nonetheless, there were several factors that made the post–World War II U.S. military governments unique.

The first was that Austria, Germany, and Japan had

been enemies and had engaged in war under regimes that were considered hostile to democratic interests. Thus, the military governments that administered these countries had to be involved in psychological reorientation on a significant scale. In its most famous sense, in Germany, this type of program was called "denazification," but a similar program was pursued in Austria; in Japan, it was called "demilitarization."

The second factor was that in each case—Germany, Austria, Japan, and Korea—the nations had been destroyed by war. Thus, unlike previous U.S. military governmental experiences, these postwar governments had to administer significant economic recovery programs to bring the nations back from wartime devastation.

Third, the scale of the problem was unique. Never before had the United States established military governments over so many people with such severe political, military, social, and economic problems. Thus, the size of the U.S. military governmental effort was unprecedented.

Military Government in the Post–World War II Era

Traditionally, U.S. military governments served two functions. On the frontier, they were transitory and provided for government and law in sparsely developed regions until adequate numbers of people had settled and civil institutions were established. Such governments often existed in response to the Indian threat in the new territories, using defense as the justification for their existence.

Military governments of a more permanent nature were established in U.S.-controlled territories such as Guam, the Pacific islands, the Panama Canal Zone, and Alaska. These were condoned because they provided for the most efficient administration, and, as in the first case, were often linked to U.S. defense interests.

GUAM

The second type has greater relevance to the contemporary situation because it provided the U.S. experience in military government on which postwar U.S. military governments were developed. Guam is of particular relevance because it spanned the period from 1898 to 1950, thereby encompassing both the prelude and the implementation of the postwar governments.

The U.S. administration on Guam was intimately linked to U.S. policy concerning Micronesia. The strategic importance of both was stressed by Alfred Thayer Mahan in his discussions of seapower. Mahan attached great importance to the Panama Canal as a waterway vital to U.S. national interests. Concerning defense of the canal, Mahan recommended looking to its approaches. In respect to the canal's eastern defense, he said that the United States must control a Caribbean island on the canal's eastern approaches and argued for possession of Cuba. While the United States subsequently opted for Puerto Rico, the security of the canal on its eastern side was established nonetheless. Concerning the western approaches, Mahan's argument was considerably more aggressive. He had discussed the importance of seapower projection elsewhere in his works, and applied these postulations to this portion of his canal discussion. He said that the place to defend the canal was not off the coast of the Americas, but rather, in the Far East, on a latitude roughly equivalent to that of the canal. He called for a U.S. base on the Philippines to defend U.S. interests. However, while very much under Mahan's influence, subsequent U.S. policy expanded on his argument by attaching great importance not only to the Philippines but also to Guam, Hawaii, and Micronesia.

The administration of Guam amounted to a colonial government, and the pattern was set during the period from 1898 to 1914. The acquisition of Guam was a part of the greater theme of Spanish defeat and the elimination of Spanish influence from the region. The strategic value of Guam as a coaling station was immediately apparent to both the United States and Japan.

On 12 January 1899 the Navy Department appointed Capt. Richard P. Leary the military governor of Guam, although he had no previous experience in civil affairs. At least three aspects of the situation in Guam would influence subsequent U.S. military governments. The first was that Guam had no mineral resources or industries. In short, the populace was not self-sustaining and relied immediately on the United States for relief and assistance. (This same situation—large numbers of destitute people looking to the U.S. military government for support—would reappear in postwar Germany, Austria, Japan, and Korea.)

The second aspect was the arrogant, somewhat racist approach the United States took toward Guam's affairs. These negative aspects of U.S. administration would haunt future U.S. military governments, particularly in Panama, Japan, and Korea, because while the subjugated peoples initially had to rely on U.S. largess, as recovery occurred, U.S. arrogance would become more and more difficult to accept.

The third problem concerned balancing colonial (primarily military) priorities with the establishment of "democratic values" among the people of Guam. On the one hand, the United States possessed Guam and had a military government on the island for perceived strategic interests. On the other hand, there was a moral imperative in the United States against developing a colonial empire, and it demanded that the government accomplish "good works"—specifically, the inculcation of democratic values in the indigenous population and the improvement of living standards. Such achievements would reflect the beneficence of U.S. rule and help to justify what otherwise would be unjustifiable to the American public.

The Japanese threat was perceived throughout the 1920s and 1930s, and was justified, given Japanese conquests that included occupying the island in World War

II. In the postwar years, Guam was an element in the U.S. containment policy, and it later provided a base for U.S. ballistic missile submarine (SLBM) patrols in the Pacific that were targeted against the Soviet Union. It was not until the 1980s that the range of U.S. SLBMs was of such magnitude that ballistic missile submarine deployments from Guam were no longer necessary.

Such perceived U.S. defense interests were a constant concerning the three problems discussed above. The first problem, the standard of living, improved under U.S. administration, and probably benefited from U.S. aid to the island. Yet the same defense interests exacerbated the second and third problems. The desire of the people of Guam for a greater say in the government was considered secondary to U.S. defense interests and affected the third problem—U.S. attitude toward the islanders. In this light, such incidents as the 1925 attempt by the people of Guam to gain greater self-government were denied as the U.S. Navy was firmly committed to defending the status quo.

Adm. Chester Nimitz was appointed governor of Guam when it was retaken from the Japanese, and U.S. policy was to reinstall the status quo military government that had existed before the war. The navy's refusal to reinstall the Guamanian Congress with the rights that it enjoyed before the war and the U.S. appropriation of significant territory on Guam for new military bases incensed the people. Both President Harry Truman and Secretary of State George Marshall favored a transfer of power from the navy to the Guamanian Congress. In spite of this, the navy refused to concede its position, resulting in a crisis during the administration of Adm. Charles A. Pownall as governor. The situation resulted in a Guamanian victory when, on 21 July 1950, the Organic Act of Guam became law. The result was the creation of Guam, U.S.A., and the granting to the people of Guam all the privileges of U.S. citizenship in addition to territorial government.

PANAMA CANAL ZONE

The Panama Canal Zone presented a special case because it was not a true military government. Administration of the zone was entrusted to two closely related U.S. agencies: the Panama Canal Company and the Canal Zone government. The governor of the zone was appointed by the president and was supervised by the secretary of the army. U.S. control of the zone existed from 4 May 1904 to 1 October 1979. During this time the zone was a "little America"; the United States was concerned with the canal's administration and limited its involvement in Panamanian affairs to the concern that whoever was in power would support continued U.S. presence in the zone. The result was that two different Panamas developed. One was plagued with corruption and abuses, while the other, the zone, prospered. The disparity between the two, and the fact that the canal cut a swath through the nation of Panama, were chronic problems that may be resolved if Panama assumes full control of the canal on 31 December 1999, as scheduled.

Relevance of Pre–World War II Military Governments

The military governments in Guam and Panama did provide the U.S. military with experience in military government, but the importance of this experience should not be overemphasized. Guam was a navy, not an army, experience, and Panama was a special case, not typical of traditional army-administered military government. Furthermore, they were long-term military governments for strategic purposes, so U.S. defense interests prevailed over the interests of the local inhabitants. Third, the relationship of the native populations with the United States at the start was benign. Finally, the military governments in Panama and on Guam represented only a very small, low-priority part of the total U.S. military effort. The post–World War II military governments differed in three out of four respects. While they were motivated in part by U.S. defense interests, they were intended for the short term, they often concerned populations that had been hostile to the United States, and they represented a considerable part of the total postwar U.S. military effort.

PREPARING FOR POSTWAR DEMANDS

The goal of the United States in World War II was the unconditional surrender of both Germany and Japan. This meant that there would most likely be considerable destruction in both nations, that there would be military occupations to administer the nations in the immediate postwar period, and that a new goal—the reorientation of the nations' political and value systems to ones more acceptable to the United States—would be pursued. Prior to World War II, the United States had pursued an isolationist path, preferring not to be involved in the affairs of others. In addition, U.S. military governments up to that point had not had to deal with the kind of problems that would be present in postwar Germany, Austria, and Japan. In short, prior U.S. efforts in running military governments did not provide the experience needed to govern postwar Germany, Austria, and Japan.

But the U.S. approach was farsighted, and preparations were made during the war for the postwar governments. During the war, military government was simply a mission of the Civil Affairs Division of the Department of the Army—one of many. However, it was realized that in the post–World War II period, it would be *the* major activity of the Civil Affairs Division. As a result, a School for Military Government was begun at the University of Virginia on 11 May 1942, and an under secretary of the army was designated as the primary administrative and staff person. His job was to be solely concerned with military government in all places. Developing adequate military governments required detailed economic, social, and political planning, with due consideration of all legal requirements—a monumental planning effort. Accomplishment of this task involved years of effort by thousands of staff personnel. In short, it amounted to a very significant, very demanding new mission for a U.S. military that had only nominal prior experience in such matters.

Legal Issues

One of the first issues to be decided by U.S. wartime planners was the legal grounds under which the military governments would be established. The U.S. military's previous experiences in governing either had pre-dated the existing laws pertaining to military governments, which were established in 1907 (Guam), or occurred (as in the cases of Panama, Alaska, and elsewhere) under conditions other than the termination of war.

The codes pertaining to war, the U.S. Army's definitions, and relevant international law define military government as a "form of administration by which the occupying power exercises governmental authority over occupied territory." The concept was established as a principle of international law in Article 42 of the Fourth Hague Convention of 1907. The U.S. Army expanded on this definition by noting that an occupying power exercises executive, legislative, and judicial power over an occupied territory. Defined as the law of belligerent occupation, it is derived from that branch of international law known as the law of war. Technically, the law of military occupation takes effect when the actual battle line has been passed and the enemy population is clearly within the territory belonging to the invading army. Usually, the transition is marked by a military proclamation that suspends the local laws and substitutes occupation statutes.

Martial law is military authority exercised in accordance with the laws and usages of war. It contrasts with military oppression, which is the abuse of the power that martial law confers. Similarly, the law of belligerent occupation imposes definite obligations on the victor as well as the vanquished. Such occupation, if it is harsh, is to be terminated within a year. Finally, in establishing a military government, the invader can choose one of two administrative patterns: he can deal directly with the civilians of a defeated state, or he can work through the civilian structures of the local government, thereby establishing indirect control.

Clearly, the existing law concerning military government was inadequate to cover the unprecedented postwar occupation experiences, and was modified by the postwar experiences. Policy definitions were expanded accordingly, and this was most evident in the evolution of the term *civil affairs*. Initially, its definition described a condescending and dictatorial relationship between the occupation armies and the local population, and the legal rights and obligations of both. Since a reorientation of civilian values and views and a defense of these same civilians against the new threat of communism were among the goals of the U.S. military governments, the changes pertaining to civil affairs amounted to toning down the hostility in the definition.

Occupation of Germany

The U.S. military government in Germany existed from 9 May 1945 until the establishment of the Federal Republic of Germany in September 1949. The initial U.S. military governmental policy was hard-line and stressed denazification, demilitarization, decartelization, decentralization, and democratization—"the five d's." The goal was to divest Germany of those aspects that were perceived as inimical to world interests. As the occupation progressed, however, the United States was confronted with several unforeseen problems, including famine, economic stagnation, a gross disruption of communications, a severe housing shortage, a very strong black market, and a very unstable currency. As these problems were perceived and addressed, U.S. military governmental policy was modified accordingly.

By late 1945, the perception of postwar realities was a sufficient reason to reconsider the U.S. position. First, the talents of many ex-Nazis were needed in the new Germany. Second, the number of ex-Nazis far exceeded the number that had been anticipated earlier. Third, there was the question of the limits of denazification—that is, how far did the United States intend to push the program. As a result, the earlier policy goals were gradually modified by policy revisions that reduced the severity of the initial program. After 1947, the U.S. policy was to punish those Nazis who had been involved in the atrocities of the Third Reich.

Concerning the issue of whether to administer to the public directly or to work through German organizations, the U.S. Army's policy in Europe strongly favored the latter. U.S. commanders were urged to find Germans to run the local governments. The regulations were developed in such a way that the alternative—field commanders personally fulfilling the governmental functions—was considered an additional burden that was not to be assumed unless it was absolutely necessary.

Concerning decartelization, the initial U.S. program was modified significantly. From 1945 to 1947, a stern decartelization program was in place, but this was reversed and, with the Marshall Plan, the military government put mechanisms into operation that would allow the German economic miracle of the 1950s.

Decentralization was accomplished to a certain extent through bolstering local and regional power and uniting the territories into a federation. Likewise, democratization was accomplished through education.

Equally important, however, was that the United States learned through experience. The five d's that had been defined during the war were recognized as unrealistic because they had not foreseen or taken into account some critical and very real problems that existed in postwar Germany. From 1947 onward, the United States shifted the emphasis of its military governmental policy from accomplishing the five d's to developing political, economic, and social institutions that would address the problems of the postwar scene.

U.S. Military Government in Japan

The U.S. military government that existed in Japan from 1945 until 1951 contrasted vividly with the military government in Germany. Whereas the German model emphasized maximum use of German civilians and provided on-scene commanders with considerable decision-making power, the Japanese model was so highly centralized that no subordinates in the field were allowed to make decisions that varied from the overall pattern established by Gen. Douglas MacArthur (Supreme Commander, Allied Powers). MacArthur's policy was forceful, imaginative, and dynamic. His brand of democracy—which included extension of the franchise, women's rights, land reform, a constitution, and ideas about popular sovereignty and civil rights—was farsighted and popular with the Japanese. In essence, he sought to retain what he considered positive in Japanese culture while transforming those militaristic, authoritarian aspects that he considered to be pernicious.

The initial goals of the U.S. military government in Japan were demilitarization, democratization, and development of a peacetime economy. The early months of the occupation saw the United States move strongly against Japanese militarist institutions. The armed forces were demobilized, the State Shinto was disestablished, and nationalist organizations were abolished. A new constitution, strongly influenced by MacArthur, was approved on 3 November 1946 and went into effect on 3 May 1947.

While MacArthur's approach might be defined as forceful, even dictatorial, and possibly egotistical, it was also effective. When Japan regained its sovereignty in 1952, the democratic changes that MacArthur had encouraged were already taking root and continued to develop in an independent Japan. Likewise, in respect to the economy, U.S. policy came to support the reestablishment of the major Japanese companies that had previously existed while encouraging the modification of their policies and views.

As in the German experience, the U.S. military government in Japan learned through practical experience. The result was impressive: the reconstruction of Japan as a world economic power, but one divested of the militaristic traits of the past.

Korea

U.S. occupation forces maintained a military government in Korea from war's end until 15 August 1948, when the Republic of Korea was inaugurated. Henceforth, it existed as a legitimate entity, and for many years was ruled by the authoritative Syngman Rhee, who, on many occasions, was certainly a match for the authoritarian MacArthur. Throughout the war, U.S. civil assistance was provided through a complex United Nations system, and a highly defined military government such as those in postwar Germany and Japan, did not exist.

Panama (December 1989)

The U.S. invasion of Panama, Operation Just Cause (December 1989), showed that U.S. policy concerning military government had matured considerably. At no time was a military government established. Rather, Manuel Noriega's successor, President Guillermo Endara, was sworn in shortly before the invasion commenced. And throughout the invasion and in its aftermath, Endara was considered by the United States to be the legitimate head of Panama.

Accompanying the U.S. combat forces invading Panama were contingents of civil affairs personnel, and immediately upon arrival, they began organizing assistance for those Panamanians who had been dislocated by the war. A significant portion of these personnel were reservists—sheriffs, educators, engineers, medical professionals, and a host of other professionals. In addition to providing for the immediate needs of the people, they assisted U.S. embassy personnel in their civic programs and developed aid and assistance programs that could be accomplished in future years by reserves serving their annual active-duty requirements. They also fulfilled an important law enforcement function by developing a course of instruction for prospective Panamanian police officers.

U.S. Army combat personnel also fulfilled police and law enforcement duties in 1990 when they accompanied Panamanian police officers on patrols and when they participated in dragnets that were conducted to incarcerate criminals during the looting and riots that followed U.S. intervention. In January and February 1990, these reserve efforts were completed and the results were given to embassy personnel. The U.S. civil affairs effort, for the most part, was concluded pending the approval of the additional aid and assistance programs, and the reserves were withdrawn.

The significance of the Panamanian experience was that the United States achieved the goals, needs, and purposes of a military government while working through a foreign government. Thus the goals were achieved without the problems that a military government would have created.

Forecast

The traditional tenets of military government and belligerent occupation that existed before World War II have generally remained valid. However, the experiences in Germany and Japan have imposed a contemporary reality that has considerably modified traditional assumptions. First, the traditional goals, most accurately placed in the realm of high strategy or even philosophy—denazification, Japanese demilitarization, and so forth—have given way to defining more realistic, more tangible aims. The postwar experience demonstrated that the goals that are perceived during a war may or may not have validity in the postwar peacetime period. In addition, if they do have validity, their precedence and importance may have to be adjusted to meet postwar realities.

Second, it has been realized that military government in occupied territories involves a much broader range of occupations and professions than it did in the past. The U.S. Army has devoted much attention to civil affairs in an attempt to be as prepared as possible for the future demands for military government.

Third, it has been realized that there are limits to what a military government can expect to accomplish. Civil affairs personnel must be knowledgeable in a nation's culture, economy, and political tradition in order to be effective. Here, again, there has been considerable progress.

Finally, as the experience in Panama demonstrated, the issue of military government is not as clearly defined as it was in the past. Today's U.S. Army Civil Affairs forces are prepared to provide a variety of economic, social, and political assistance programs. The emphasis today, however, is to maintain a low-visibility presence and to let civil affairs personnel work with existing indigenous political, economic, and social organizations to achieve their goals.

BRUCE W. WATSON

SEE ALSO: Civil Affairs; Civil-Military Cooperation; Civil-Military Relations; Hague Conventions.

Bibliography

Coles, H. L. 1964. *Civil affairs: Soldiers become governors.* Washington, D.C.: Office of the Chief of Military History, Department of the Army.
Maga, T. P. 1988. *Defending paradise: The United States and Guam 1898–1950.* New York: Garland Press.
U.S. Department of Defense. 1985. *Field Manual 41-10: Civil affairs operations.* Washington, D.C.: Department of the Army.
Watson, B. W., and P. G. Tsouras, eds. 1990. *Just Cause: The U.S. military intervention in Panama.* Boulder, Colo.: Westview Press.

GRAECO-PERSIAN WARS

The military encounters between the Persian Empire and the Greek city-states in the early part of the fifth century B.C. were among the most pivotal events of human history. As a result of the eventual Greek victory, the culture of Classical Greece would flourish and have a profound effect on subsequent human history. The major source of information on this time is the Greek historian Herodotus, who, about the middle of the fifth century B.C.—a generation after the wars ended—compiled a lengthy account of the wars. There are other sources that can be used to add to and correct the account of Herodotus, but details remain uncertain.

Causes

The Persian Empire was at its most powerful and most expansionist during the reign of Darius I (521–486 B.C.), who may have had designs on Greek territory. The genesis of the Graeco-Persian wars can be found in the revolt of the Ionian Greeks in 499. Ionian Greek city-states on the coast of Asia Minor had been absorbed into the Persian Empire a generation earlier, but in 499 they broke into open revolt and sought aid from the older city-states of the Greek mainland. Assistance came from only two cities: Athens and Eretria. The revolt was eventually crushed, and Darius seems to have resolved to punish Athens and Eretria for the aid they offered to the rebels. But even without vengeance as a motive, Darius had incentive to conquer Greece (Fig. 1). His empire contained a large number of different ethnic groups, including the Greeks of Asia Minor, and to rule over some of the Greeks but not all would be a constant source of danger in the future. Thus, the attempted conquest of Greece by the Persian Empire may be viewed as simple expansionism, as a desire to more efficiently incorporate ethnic peoples into the empire, as simple vengeance, or, and more likely, as a combination of these.

It should be noted that throughout this period contact between Greeks and Persians was regular and frequent. Greece was rife with political factionalism, and throughout the campaigns the Persian emperor had the advice of Greek expatriates who resided at his court. It is likely that

Figure 1. Greece in the fifth century B.C.

the preparations by each side were well known to the other.

Campaign of 490 B.C.

In the summer of 490 B.C., Darius sent an invading force under the command of his generals Datis and Artaphernes by ship from Asia Minor across the Aegean Sea. Herodotus suggests that there were approximately 600 ships and 25,000 to 30,000 men, including a sizable cavalry contingent. Some modern historians, however, doubt the Persian land troops much exceeded 10,000 in number. After receiving the submission of the small islands along the way, who were in no position to resist, the Persian force landed on the island of Euboea off the coast of Attica and laid siege to the city of Eretria. Although assistance was sought from Athens, and promised, Eretria was eventually betrayed by some of its inhabitants, and the Persians entered the city and destroyed it. The Eretrians were killed or sold into slavery. The Persian fleet then made a landing at Marathon in northeast Attica, about 42 kilometers (26 mi.) from Athens. This particular spot was chosen on the advice of an exiled Athenian tyrant named Hippias, who believed it would be an ideal place for the deployment of cavalry. In the account of the battle, however, little use seems to have been made of the cavalry.

Battle of Marathon (12 September 490 B.C.)

A force of about 10,000 heavily armed hoplites (armored spear or pikemen), mostly Athenian, was at Marathon to meet the Persians, but the Persians had no difficulty disembarking their troops and making camp.

The battle plans of either side are difficult to determine. The Persians were apparently in no hurry to engage the Athenian force and may have expected to simply overawe the opposition with their numbers. The Athenian force under Miltiades seemed intent on just blocking the land route to Athens, although it was easily possible for the Persians to re-embark and sail around Attica to Athens (which they eventually did). The Athenians also appeared to be in no hurry to engage, and several days were spent with each side observing the other and repositioning their forces. Finally, apparently as the Persians began to re-embark some of their troops, Miltiades ordered an attack. The Persians were surprised, not only by the unexpected charge, but also by the fact that 10,000 Greeks in heavy armor were bearing down on them in dense array. The shock was effective, as was the Athenian tactic of strengthening the wings at the expense of the center. The Persians seem to have driven back the Athenian center but were in turn enveloped by the wings. A rout ensued, and the casualty figures are striking: 6,400 Persian dead, 192 Athenian dead. The remnant of the Persian force re-embarked and sailed around Cape Sunium to Athens, only to find that Miltiades had marched his victorious army overland and was waiting for them on the beach. The Persians did not attempt a landing; instead, they abandoned the effort and sailed back to Asia.

Marathon marked the end of the first campaign in Greece proper, but there was little doubt that another attempt would be made. The next invasion attempt was delayed ten years, however, by internal problems within the Persian Empire.

Campaign of 480 B.C.

The Persian invasion of 480 B.C. was several years in preparation. Herodotus gives specific information, which modern scholars find difficult to accept. We are told that the Persian force numbered 1.7 million infantry, 80,000 cavalry, and 1,200 warships with crews of approximately 200 each, for a total of about two million fighting men. There were in addition more than 3,000 smaller vessels and transports, and numerous support personnel and camp followers, a total of nearly 4.2 million. This force was prepared under the direction of the Persian king Xerxes (486–465 B.C.), who had succeeded his father, Darius. Herodotus describes in detail the method of mustering and counting the troops as well as the considerable logistics involved in transporting and feeding a force this size. Nevertheless, the great German military historian Hans Delbrueck considered Herodotus' figures to be a gross exaggeration, and estimated that Xerxes's army "cannot have numbered more than some 75,000 warriors, including the allied Greeks" (Delbrueck 1975, pp. 35–36). The route of march was through Asia Minor, across the Hellespont (a bridge was built of ships lashed together and covered), and then through what is now northern Greece, reaching the objective, Athens, from the north. The land force was to stay near the coast; the fleet would sail alongside.

The Greeks had undertaken some advance planning as well. For the first time in Greek history, a large number of Greek city-states put aside their quarrels with one another and their insistence on complete independence and formed an alliance, usually referred to as the Hellenic League. Thirty-one states were involved, under the nominal leadership of Sparta, which was generally agreed to be the strongest military power among them. The land forces available to the Greek alliance probably numbered about 70,000, although they were never all mustered in one place, and were commanded by King Leonidas of Sparta. In Athens, the statesman and general Themistocles had convinced the Athenians to build up their navy, which now numbered nearly 300 ships and was by far the strongest naval force among the Greeks. Although the nominal command of the naval forces was conferred on a Spartan, the practical command lay with Themistocles. Like the Persians, the Greeks saw the necessity of coordinated land and sea activity.

After considering all the options, the Greeks decided to post a blocking force at the pass of Thermopylae in central

Greece. The easiest passage from northern Greece to southern Greece and the only practical way to move a large army was along the coastal road, which at Thermopylae traversed a defile that, at its narrowest, was only about 13 meters (40–50 ft.) wide between the steep, rugged mountains and the sea. (Changes in the seacoast have substantially widened the pass in subsequent centuries.) It was a place where the superior numbers of the Persians would provide little advantage. Not far offshore, the island of Euboea provided a convenient base for blocking the accompanying Persian fleet, which would either have to go around the island or sail through the Euripos channel; in either case, it could be outflanked and attacked in detail by taking the other route. Therefore, the Greek fleet, which numbered about 400 ships, positioned itself at the northern end of the island of Euboea near a place called Artemisium, a location close enough to Thermopylae to maintain contact with the land army.

Battle of Thermopylae

The Greek forces at Thermopylae numbered about 10,000 heavily armed troops; only 300 were Spartans. Sparta had sent only the king and his bodyguard because of religious festivals at home (a similar situation had prevented them from aiding the Athenians at Marathon in 490). It is doubtful that the Greeks ever expected to do more than delay the Persian advance by their stand at Thermopylae. In that, they succeeded. The Persian movement was halted for perhaps as long as two weeks, during which time many of the cities to the south were evacuated in an orderly manner. The Greek stand at Thermopylae not only cost Xerxes valuable time—it was already September and winter would soon be upon the Persians—but it seems he also took heavy casualties both on land and sea.

For several days the Persians assaulted the Greek position, only to be driven back by the Greek soldiers; the 300 Spartans under King Leonidas particularly distinguished themselves. At the same time, the Persian fleet was unable to dislodge the Greek fleet from the narrow waters off Artemisium. There were no victories at sea for the Greeks but no defeats either. Although the naval encounters were indecisive, they were costly for the Persians, who had no intimate knowledge of the waters and weather patterns, and suffered severe losses from shipwrecks during storms. The land battle at Thermopylae was finally won by the Persians when the Greek position was turned by a small Persian force that had been guided along an undefended mountain path by a Greek traitor. This came as no surprise to King Leonidas; it was just a question of when, not if, the Persians would find the path. When he learned from his scouts that the position was no longer tenable, Leonidas ordered most of the Greek force to withdraw and remained with his 300 Spartans and a few other troops to buy a little more time. Leonidas and his Spartans fell to the last man. Once the land battle had been lost, the Greek fleet retreated from its exposed position and regrouped off the island of Salamis near Athens.

Naval Battle of Salamis

The land army of Xerxes marched south into central Greece while his fleet sailed around the promontory and arrived in the Saronic Gulf, near Athens. Xerxes received the submission of some of the smaller towns that were still inhabited, but most of the cities had been evacuated and were plundered and burned. Athens, whose inhabitants had fled, was sacked and burned.

The Greek land forces took up positions on the narrow Isthmus of Corinth to prevent a Persian advance into the Peloponnesus. The land forces saw no further action, however; the decisive battle of this second and final Persian campaign was a naval battle in the Straits of Salamis, between the island and the mainland. The subsequent Greek success at Salamis was due entirely to the efforts of Themistocles. The Greeks were disunited and fearful, and there was great sentiment for fleeing, either to avoid the Persians or to protect one's own family, city, and territory. Themistocles had detailed knowledge of the local topography, sea conditions, and winds. He managed to convince the Persians by secret messages that their best hope was to trap the Greeks in the narrow straits of Salamis and annihilate them there; while at the same time he persuaded the Greeks that their best chance for a victorious battle was at Salamis. The Greek ships, obscured from Persian view by the irregularities of the coast on the island of Salamis, were lying in ambush for the Persian fleet as it entered the narrow channel on 23 September. Persian superiority in numbers was irrelevant. Just after the battle got under way, the winds and the current changed, causing the Persians to run afoul of one another and making the Greek task of ramming and boarding easier. As at Thermopylae, the Greeks gave battle in a location where superior numbers could not be used to advantage. Half of the Persian ships were sunk or captured; the Greeks lost only 40 ships.

The Sequel

With the defeat at Salamis, Xerxes had little choice but to withdraw most of his army and navy from Greece. A force so large could not be maintained in a hostile land for any length of time, particularly without command of the sea. He left an army in Greece under his general Mardonius and withdrew the remainder into Asia. Mardonius's army suffered many deprivations during the ensuing winter, and what was left of it was annihilated in a battle at Plataea in the spring of the following year. Only a few stragglers made their way back to Persia. That spring (479 B.C.) the Athenian fleet sailed through the islands of the Aegean and to the coast of Asia Minor to liberate the Greek cities remaining under Persian rule.

There was never again a Persian invasion of Greece,

and the entire episode may be viewed as the genesis of the Athenian Empire of the fifth century B.C.

JANICE J. GABBERT

SEE ALSO: History, Ancient Military; Persian Empire.

Bibliography

The primary sources are Herodotus, as well as Aeschylus's play *The Persians*, and the biographies by Plutarch of Themistocles and Aristides. A full treatment of modern scholarship on the subject can be found in *Cambridge Ancient History*, Vol. 4 (Cambridge, 1926, rev. 1964).

Adcock, F. E. 1957. *The Greek and Macedonian art of war*. Berkeley: Univ. of California Press.
Delbrueck, H. 1975. *History of the art of war within the framework of political history*. Vol. 1, Antiquity. Trans. W. J. Renfroe, Jr. Westport, Conn.: Greenwood Press.
Ferrill, A. 1985. *The origins of war: From the Stone Age to Alexander the Great*. London: Thames and Hudson.
Frost, F. J. 1980. *Plutarch's Themistocles: A historical commentary*. Princeton, N.J.: Princeton Univ. Press.
Hammond, N. G. L. 1967. *A history of Greece to 322 B.C.* Oxford: Clarendon Press.
Pritchett, W. K. 1971–85. *The Greek state at war*. 4 vols. Berkeley: Univ. of California Press.
Rodgers, W. L. [1937] 1964. *Greek and Roman naval warfare*. Reprint. Annapolis, Md.: U.S. Naval Institute Press.
Warry, J. 1980. *Warfare in the classical world*. New York: St. Martin's Press.

GRANT, ULYSSES SIMPSON (1822–85)

Figure 1. *Ulysses S. Grant*. (SOURCE: U.S. Library of Congress)

Ulysses Simpson Grant (Fig. 1), the chief architect of the Federal forces' (or Union) victory over the Confederacy in the American Civil War, rose from an inauspicious prewar business career to become one of the most prominent leaders in American military history.

Early Life and Pre–Civil War Activities

Grant was born on 27 April 1822 at Point Pleasant, Ohio, the first of six children of Jesse Root and Hannah Simpson Grant. Christened Hiram Ulysses Grant, when he entered the United States Military Academy at West Point, New York, he altered his name to Ulysses Simpson Grant, the name provided by the congressman who applied for his appointment. Growing up in Ohio, Grant led a comfortable frontier life, attending school and working on his father's farm. In 1839, his father, unbeknownst to Grant, secured for him an appointment to West Point. Grant accepted it without enthusiasm and went on to become a mediocre cadet with a distaste for military life and no intention of remaining in the army. He achieved distinction at West Point primarily for his accomplished horsemanship, but he also collected a substantial number of demerits and at graduation in 1843, ranked only 21st academically in a class of 39.

In September 1843, Grant joined the 4th Infantry Regiment in St. Louis. One year later, the unit assembled with Gen. Zachary Taylor's Army of Occupation in Corpus Christi, Texas. Taylor moved against the Mexican army in May 1846, and Grant served under him in battles at Palo Alto and Resaca de la Palma (8–9 May 1846) and Monterrey (21–23 September 1846). Grant participated in Gen. Winfield Scott's 1847 Vera Cruz–Mexico City campaign, along with numerous other future Civil War leaders, both Confederate and Federal. He saw action at Vera Cruz (9–29 March 1847), Cerro Gordo (17–18 April 1847), Churubusco (20 August 1847), Molino del Rey (8 September 1847), and the storming of the hilltop fortress of Chapultepec (13 September 1847), where he was breveted captain for bravery. On 16 September 1847, Grant was commissioned a first lieutenant, the rank he held at the end of the Mexican War (1848).

Grant returned to the United States in 1848 and married Julia Dent, whose acquaintance he had made while stationed in St. Louis. In 1854, after duty in New York, Michigan, the Pacific Northwest, and California—enduring separation from his family and suffering from excessive drinking while on the Pacific Coast—Grant resigned with the rank of captain. Until 1860, he lived with his wife and children in Missouri, where he tried his hand at farming, real estate, and other occupations. In 1860, he started work with a family-owned business in Galena, Illinois, and remained there until the outbreak of the Civil War.

Civil War Generalship

Grant trained a company of Galena militia and then worked in the Illinois Adjutant General's Office in Springfield, the state capital. He tendered his services to the army in May 1861 and was ignored. He was, however, soon appointed colonel and made commander of the 21st Illinois Volunteer Infantry Regiment. Promoted to brigadier general in August 1861, Grant took command of the District of Southeast Missouri, with headquarters in Cairo, Illinois. On 7 November 1861, he commanded Federal forces in an inconclusive battle at Belmont, Missouri. In February 1862, he scored major victories in capturing Forts Henry (6 February) and Donelson (16 February), actions that earned him promotion to major general of Volunteers and the nickname "Unconditional Surrender" from a message he sent to the Confederate commander at Fort Donelson. Although temporarily removed from command due to failed communications with his superior, Gen. Henry W. Halleck, Grant regained his position as head of the Army of the Tennessee in March. On 6–7 April 1862, a Confederate army under Gen. Albert Sydney Johnston surprised Grant's troops encamped at Shiloh (Pittsburg Landing), Tennessee, and pushed them back until Grant rallied his troops and then, reinforced, drove the Southerners from the field in severe fighting. After Shiloh, Grant again temporarily lost command of the Army of the Tennessee to Halleck, but resumed it in July 1862 and, following victories by Maj. Gen. William S. Rosecrans at Iuka (19 September) and Corinth (3–4 October), initiated his campaign to open the Mississippi River and divide the Confederacy.

Cooperating with the river gunboats of Rear Adm. David D. Porter, Grant conducted a long campaign, beset by natural obstacles and Confederate resistance. Finally, in April 1863, he moved his forces south of Vicksburg, crossed the river, and marched overland to Jackson, Mississippi. He defeated Confederate general John C. Pemberton's army at Champion's Hill (16 May 1863). Grant then invested Pemberton's army at Vicksburg and forced the surrender of the Confederate garrison on 4 July. With the Federal victory at Port Hudson on 9 July, the Union effectively controlled the entire Mississippi River.

Grant earned his promotion to major general in the regular army with his victory at Vicksburg and received command of the newly created Military Division of the Mississippi on October 1863. His next actions came at Chattanooga, Tennessee, where he took command of the besieged Federal army (October) and won a decisive victory at Lookout Mountain–Missionary Ridge (24–25 November 1863).

President Abraham Lincoln was greatly impressed by Grant's achievements; he brought Grant to Washington, promoted him to lieutenant general, and appointed him general in chief in March 1864. Although Grant exercised command over all Federal armies, he remained in the east with Gen. George G. Meade's Army of the Potomac for the rest of the Civil War. Using the telegraph, he directed the overall Union army war effort, sending Gen. William T. Sherman through Georgia and the Carolinas, while directly overseeing the crucial campaign in Virginia. His drive on Richmond was carried out in accordance with a simple but effective strategy and was characterized by brutal fighting with Gen. Robert E. Lee's Confederate army at the Wilderness, Spotsylvania, Cold Harbor, and the attack on Petersburg (May–June 1864). Grant continued undeterred from his goal to wear down Lee and laid siege to Richmond and Petersburg. The Virginia campaign thus entered a new stage of fighting, foreshadowing the static trench warfare of World War I. After a nine-month siege, the overwhelming Federal strength and dwindling Southern resources forced Lee to evacuate his defensive line around the Confederate capital and march westward (April 1865). Grant pursued, sending part of his army to cut off the Confederate retreat. Lee surrendered at Appomattox Courthouse, in Virginia, on 9 April. Lee's capitulation doomed the Confederacy.

Postwar Activities

In 1866, Grant was promoted to general of the army and, in 1868, following an increasingly turbulent political relationship with Pres. Andrew Johnson, accepted the nomination by the Republican party as candidate for president of the United States. Grant was elected in 1868 and served two terms. He appointed a number of unscrupulous or incompetent officials to his administration posts, and his presidency suffered for it. Financial scandals such as the Black Friday episode (1869), the Whiskey Ring (1872), and the Credit Mobilier crisis (1873) rocked his terms in office. Grant was politically naive and not personally involved in the corruption. Negative sentiment prevented his renomination for a third term in 1880. Moving to New York City in 1881, he invested his life's savings with a banking firm and lost it when the business went bankrupt due to mismanagement by fraudulent partners. This financial catastrophe coincided with the onset of throat cancer, the disease that eventually killed him.

Grant spent the remainder of his life writing his autobiography, proceeds from which restored his family's finances. He completed the highly acclaimed literary work only four days before his death on 23 July 1885.

Grant achieved his place in history as a "Great Captain" from his accomplishments during the Civil War, both in the western and eastern theaters. In the Vicksburg campaign he proved to be an able tactician and strategist. In Virginia, his determination and persistence were most evident. After the intense battle at Spotsylvania, he demonstrated a strong will in continuing to press Lee by stating that he would "fight it out on this line if it takes all summer" (Dupuy 1969, p. 109). Lincoln held Grant in particularly high esteem. When urged by influential politicians to replace him, Lincoln refused with the response, "I can't spare this man. He fights!" (Heinl 1966, p. 135). Grant's great determination and personal courage remained through his final years.

BRIAN R. BADER

SEE ALSO: Civil War, American; Lee, Robert Edward; Scott, Winfield.

Bibliography

Dupuy, T. N. 1969. *The military life of Abraham Lincoln: Commander-in-chief.* New York: Franklin Watts.
Fuller, J. F. C. 1929. *The generalship of Ulysses S. Grant.* New York: Dodd, Mead.
——— 1957. *Grant & Lee: A study in personality and generalship.* Bloomington: Indiana Univ. Press.
Heinl, R. D., Jr. 1966. *Dictionary of military and naval quotations.* Annapolis, Md.: U.S. Naval Institute Press.
Lewis, L. 1950. *Captain Sam Grant.* Boston: Little, Brown.
Grant, U. S. 1952. *Personal memoirs of U. S. Grant.* Ed. with notes and an introduction by E. B. Long. Cleveland: World.
Porter, H. 1897. *Campaigning with Grant.* New York: Century.

GREECE (Hellenic Republic)

Greece is a mountainous country in southeastern Europe, with a long recorded history. In ancient times, Greece was the home of the first constitutional democracies, the most famous of which was Athens.

Power Potential Statistics

Area: 131,940 square kilometers (50,547 sq. mi.)
Population: 10,174,400
Total Active Armed Forces: 158,500 (1.558% of pop.)
Gross Domestic Product: US$76.7 billion (1990 est.)
Annual Defense Expenditure: US$3.7 billion (5.5% of GDP, 1990 est.)
Iron and Steel Production:
  Crude steel: 0.890 million metric tons (1986)
  Pig iron: 0.160 million metric tons (1986)
Fuel Production:
  Coal: 37.11 million metric tons (1986)
  Crude oil: 1.451 million metric tons (1986)
  Natural gas: 65.72 million cubic meters (1986)
Electrical Power Output: 36,420 million kwh (1989)
Merchant Marine: 958 vessels; 21,585,048 gross registered tons
Civil Air Fleet: 35 major transport aircraft; 79 usable airfields (60 with permanent-surface runways); none with runways over 3,659 meters (12,000 ft.); 20 with runways 2,440–3,659 meters (8,000–12,000 ft.); 22 with runways 1,220–2,440 meters (4,000–8,000 ft.).

For the most recent information, the reader may refer to the following annual publications:

The Military Balance. International Institute for Strategic Studies. London: Brassey's (UK).
The Statesman's Year-Book. New York: St. Martin's Press.
The World Factbook. Central Intelligence Agency. Washington, D.C.: Brassey's (US).

History

The earliest city-building civilization in Greece was that of the Minoans, centered on Crete and the islands, which flourished from 1800 to 1400 B.C. The Mycenean culture followed, forcibly replacing the Minoans, and lasted until about 1150 B.C. The archaic stage of classical Greek civilization developed by 900 B.C. For most of the ancient period, up through the period of Roman conquest (about 150 B.C.), Greece was a land of small independent city-states, frequently at odds and often at war. Despite these conflicts, this was the most brilliant cultural and artistic period of classical Greek civilization, reaching its height between 550 and 300 B.C.

Subsequently, Greece passed through periods of Macedonian, Roman, and Byzantine domination or outright rule; the last 250 years of Byzantine rule (A.D. 1204–1460) also saw Western European influences introduced by members of the Fourth Crusade. When the Ottoman Turks conquered the Peloponnese in 1460, their domination of mainland Greece was complete, and they governed the country for the next 369 years. By the late eighteenth century, a modern Greek nationalist movement had arisen, influenced by European romantic nationalism. The Greek War of Independence began in 1821, and was waged with great bitterness and cruelty on both sides until Greek independence was recognized by the Sultan at the Treaty of Adrianople (14 September 1829).

The concept of the *Megali Idea* (Great Idea), bringing together all Greeks under one Greek state, brought Greece into conflict with the Ottoman Empire but also resulted in gradual territorial expansion. At independence, Greece's northern frontier extended from the Gulf of Volos to the Gulf of Arta. The Ionian Islands were added in 1864, Thessaly in 1881, and Macedonia, Crete, and the Aegean Islands at the end of the Balkan Wars in July 1913. Greece, which sided with the Allies in World War I after much internal political wrangling, gained Western Thrace from Bulgaria in 1918. Greece also launched a major effort to bring western Asia Minor under its control just after World War I, but the war with Turkey which this effort provoked resulted in a costly Greek defeat in 1921–22, and the deportation of 1.3 million Greeks from the Turkish mainland. Absorbing these refugees posed enormous problems.

The aftermath of this defeat resulted in a republican government in 1924, but King George II returned to the throne in 1935. Greek interwar politics were dominated by a power struggle between monarchists and republicans, only resolved by the dictatorship of Gen. Ioannis Metaxas (1936–41). Greece entered World War II when

Italian troops invaded the country from Albania in October 1940. The Greeks drove them back, but other Balkan developments brought German intervention. Despite some British aid, the Greeks were outmaneuvered and outfought by the Germans, who overran the country in three weeks (6–27 April 1941), and Crete fell to a German airborne assault (20–31 May). German and Italian occupation troops never established complete control over the countryside because of Greek partisans, and the Germans withdrew in October 1944 as Soviet forces swept into Yugoslavia.

After World War II the communists, who had played a major role in partisan activities, twice tried to gain control of Greece, in winter 1944–45 and during the civil war of 1946–49 (which left 80,000 dead). With British and later U.S. aid, the government survived, and Greece entered the North Atlantic Treaty Organization (NATO) in 1952 as a democracy. After over a decade of stable conservative government, Col. George Papadopoulos mounted a successful coup (21 April 1967) to forestall election of a socialist government. During its seven years in power, the military suppressed civil rights and abolished the monarchy (confirmed in a 1974 plebiscite), but interference in Cyprus provoked the Turkish invasion (20 July–August 1974) and produced the restoration of civilian government (24 July).

Greece became the tenth member of the European Community on 1 January 1981. Later that year (18 October), Greek voters elected a socialist government with the victory of the Panhellenic Socialist Party (PASOK), led by Andreas Papandreou. PASOK won the June 1985 elections, but continuing scandals involving Papandreou and his government led to a prolonged political crisis in 1989. Elections in summer 1989 failed to produce a clear majority for either PASOK or their conservative New Democracy Party rivals; New Democracy entered into a peculiar alliance with the communists to form a caretaker government which cleaned up the worst of PASOK's scandals. A new election a few months later produced a similar result, and New Democracy has formed a minority government.

Politico-Military Background and Policy

Despite Greece's membership in NATO, much of Greek military planning is directed toward possible conflict with Turkey. The animosity between the two countries goes back to the Greek War of Independence, and so is both deep-seated and bitter. The most likely spark for conflict would be the situation on Cyprus. The ouster of President Makarios in early 1974, paving the way for a Cyprian government favorable to union with Greece, is the event that prompted the Turkish invasion. Relations between Turkey and Greece have improved since the mid-1980s, and there are some hopeful signs of a *rapprochement*.

Coupled with the disputes with Turkey is Greek unhappiness with the U.S. bases on Crete and the Greek mainland. Particularly under the Papandreou government, the Greeks made no secret of their resentment of U.S. aid to Turkey, and renegotiation of U.S. basing rights could easily become derailed. For its part, the United States (and some other Western European nations) were uneasy over Greece's rather close relations with some terrorist groups and radical Muslim states like Iran, Libya, and Syria. These issues all served to complicate Greek relations with its neighbors, with fellow NATO members, and with the United States.

Greece spends, by NATO standards, a high proportion of its limited budget on defense. Every Greek male in good health is liable for about two years of military service on reaching age 18 (21 months in the army, 25 months in the navy, 23 months in the air force). After their active service, former enlisted personnel have a reserve obligation through age 40; officers have reserve obligations to age 50.

Strategic Problems

The geographical position of Greece, near the Black Sea and the Middle East, and in a commanding position over the Eastern Mediterranean, make the country important to NATO's plans for the defense of southern Europe. Although the 1,000-kilometer (620-mi.) northern frontier is mountainous, the routes across it run perpendicular to the border, so that the mountains are only rarely an obstacle to attack. The narrowness of northeastern Greece is a second weakness, as geography precludes strategic depth. In an effort to remedy this weakness, much of Greece's army is deployed in the north in peacetime.

The most serious problem, however, is the long-standing dispute with Turkey. Greece is still adamantly opposed to the Turkish presence on Cyprus, and although the two countries have not fought since the early 1920s, war over Cyprus nearly broke out in 1963, 1964, and 1974. Until some permanent settlement can be reached, NATO's southern flank will remain relatively weak because of poor Greco-Turkish relations.

Military Assistance

Greece receives considerable military aid from the United States. Despite frequent Greek complaints about the amount of aid given to Turkey by the United States, the quantities of aid are, despite occasional fluctuations, generally similar. U.S. military aid to Greece during 1988 came to $313 million in loans and $81.1 million in grants, a total of $394.1 million. This was a decrease from the mid-1980s (in 1984, loans of $500 million, $1.4 million in grants), but represents a significant increase from the $140-million loan and $31-million grant (total $171 million) received from the United States in 1979. Greece receives smaller quantities of aid from other NATO countries, including US$41.8 million from West Germany in 1988.

Defense Industry

Greece has a limited but significant and growing domestic armaments industry. The most significant recent product is the Leonidas tracked APC, which has so far only been employed in Greece. Greek shipyards produce many of the Hellenic navy's ships, especially support vessels and small combatants. Greek factories also produce artillery and small arms ammunition, small arms, electronic equipment, and support vehicles. Aircraft and helicopter components are built in Greek factories as well, as are RPVs and other equipment. Most of the factories engaged in defense work are partially, if not wholly, state-owned.

Alliances

Greece has been a member of NATO since 20 September 1951, but Greek forces have not participated in joint NATO exercises since 1981. Greece signed a mutual security treaty with Yugoslavia and Turkey on 9 August 1954; while this was a remarkable step of reconciliation among old adversaries, the practical utility of that treaty is probably limited. Greece is also a member of the United Nations (UN), and since 1 January 1981 has been a member of the Common Market and the European Community. Greece's principal bilateral Mutual Defense Cooperation Agreement provides for U.S. use of bases on Greek soil. The most recent agreement was concluded in November 1990.

Defense Structure

The Hellenic army, navy, and air force have been part of an integrated Defense Ministry since 1950. An armed forces commander in chief is responsible to the Minister of Defense, and the commander in chief also heads a staff composed of the chiefs of staff of the three services.

(For an explanation of the abbreviations and symbols used in the following section of military statistics, see the list of Abbreviations and Acronyms in each volume.)

Total Armed Forces

Active: 158,500 (125,800 conscripts, 4,200 women). Terms of service: Army up to 19, Navy up to 23, Air Force up to 21 months.
Reserves: some 406,000 (to age 50). Army some 350,000 (Field Army 230,000, Territorial Army/National Guard 120,000); Navy about 24,000; Air about 32,000.

ARMY: 113,000 (100,000 conscripts, 2,200 women).
Field Army (82,000): 3 Military Regions.
1 Army, 4 corps HQ.
2 div HQ (1 armd, 1 mech).
9 inf div (3 inf, 1 arty regt, 1 armd bn) 2 Cat A, 3 Cat B, 4 Cat C.
5 indep armd bde (each 2 armd, 1 mech inf, 1 SP arty bn) Cat A.
2 indep mech bde (2 mech, 1 armd, 1 SP arty bn), Cat A.
1 marine bde (3 inf, 1 lt arty bn, 1 armd sqn) Cat A.
1 cdo, 1 raider regt.
4 recce bn.
10 fd arty bn.
6 AD arty bn.
2 SAM bn with Improved HAWK.
2 army avn bn.
1 indep avn coy.
Units are manned at 3 different levels: Cat A 85% fully ready; Cat B 60% in 24 hours; Cat C 20% ready in 48 hours.
Territorial Defense: (31,000):
Higher Mil Comd of Interior and Islands HQ.
4 Mil Comd HQ (incl Athens).
1 inf div.
1 para regt.
8 fd arty bn.
4 AD arty bn.
1 army avn bn.
Reserves (National Guard): 34,000. Role: internal security.
Equipment:
MBT: 1,879 (154 in store) (CFE: 1,725): 396 M-47, 1,220 M-48 (299, 110 A2, 212 A3, 599 A5), 154 AMX-30, 109 Leopard 1A3.
Light tanks: 198 M-24 (CFE: HACV).
Recce: 48 M-8.
AIFV: 96 AMX-10P (CFE).
APC: 1,995 (517 in store) (CFE: 1,478): 110 Leonidas, 114 M-2, 403 M-3 half-track, 372 M-59, 996 M-113.
Total arty: 1,908.
  Towed arty: 875: 105mm: 18 M-56, 469 M-101; 140mm: 32 5.5-in; 155mm: 271 M-114; 203mm: 85 M-115.
  SP arty: 299: 105mm: 76 M-52; 155mm: 48 M-44A1, 51 M-109A1, 84 M-109A2; 175mm: 12 M-107; 203mm: 28 M-110A2.
  Mortars: 107mm: 602 M-30, 132 M-106 SP; plus 81mm: 690.
ATGW: 394: Milan, TOW (incl 36 SP).
RCL: 90mm: 1,057 EM-67; 106mm: 763 M-40A1.
AD guns: 20mm: 101 Rh 202 twin; 30mm: 24 Artemis 30 twin; 40mm: 227 M-1, 95 M-42A twin SP.
SAM: 42 Improved HAWK, Redeye.
Aircraft: 3 Aero Commander, 2 Super King Air, 20 U-17A.
Helicopters: 8 CH-47C, 77 UH-H/AB-205, 1 UH-1N/AB-21Q, 11 OH-58/AB-206, 14 OH-13HS.

NAVY: 19,500 (11,400 conscripts, 900 women); Bases: Salamis, Patras, Soudha Bay.
Submarines: 10: 8 Glavkos (Ge T-209/1100) with 533mm TT; 2 Katsonis (US Guppy) with 533mm TT.
Principal Surface Combatants: 18:
Destroyers: 11:
  7 Themistocles (US Gearing) (ASW) with 1×8 ASROC, 2×3 ASTT, 1 with AB-212 hel; plus 3×2 127mm guns, 4 with 2×4 Harpoon SSM.
  1 Miaoulis (US Sumner) with AB-212 hel, 2×3 ASTT; plus 3×2 127mm guns.
  3 Aspis (US Fletcher) with 2×3 ASTT; plus 5×533mm TT, 4×127mm guns.
Frigates: 7:
  2 Elli (NI Kortenaer) with 2 AB-212 hel, 2×3 ASTT; plus 2×4 Harpoon
  4 Aetos (US Cannon) with 2×3 ASTT.
  1 Aegeon (Ge Rhein AD) with 2×3 ASTT, 2×100mm gun.
Patrol and Coastal Combatants: 36:
Missile Craft: 16: 14 Laskos (Fr Combattante) PFM, 8 with

4× MM-38 Exocet, 6 with 6 Penguin 2 SSM, all with 2×533mm TT; 2 Stamou, with 4×SS-12 SSM.
Torpedo Craft: 10: 6 Hesperos (Ge Jaguar) PFT with 4×533mm TT; 4 No 'Nasty' PFT (with 4×533mm TT.
Patrol: 10:
  Coastal: 2 Armatolos (DK Osprey) PCC.
  Inshore: 8: 2 Tolmi, 6 PCI (.
Mine Warfare: 16:
Minelayers: 2 Aktion (US LSM-1) (100–130 mines).
Mine Countermeasures: 14: 9 Alkyon (US MSC-294) MSC; 5 Atalanti (US Adjutant) MSC.
Amphibious: 12:
  1 Nafkratoussa (US Cabildo) LSD: capacity 200 tps, 18 tk, 1 hel.
  2 Inouse (US County) LST: capacity 400 tps, 18 tk.
  5 Ikaria (US LST-510): capacity 200 tps, 16 tk.
  4 Ipopliarhos Grigoropoulos (US LSM-1) LSM, capacity 50 tps, 4 tk.
  Plus about 55 craft: 2 LCT, 8 LCU, 11 LCM, some 34 LCVP.
Support and Miscellaneous: 13: 2 AOT, 4 AOT (small), 1 trg, 1 AE, 5 AGHS.

Naval Air: 15 armed hel.
ASW: 1 hel div: 3 sqn: 2 with 14 AB-212 (11 ASW, 3 ECM); 1 with 4 SA-319 (with ASM).

AIR FORCE: 26,000 (14,400 conscripts, 1,100 women).
Tactical Air Force: 7 cbt wings, 1 tpt wing.
FGA: 10 sqn:
  3 with A-7H.
  4 with F-104G.
  2 with F-16.
  1 with F-4E.
Fighter: 9 sqn:
  2 with F-4E.
  3 with F-5A/B.
  1 with RF-5A.
  2 with Mirage F-1CG.
  1 with Mirage 2000E/D.
  Deliveries of Mirage 2000 continue.
Recce: 2 sqn: 1 with RF-4E; 1 with RF-104.
MR: 1 sqn with HU-16B.
Transport: 3 sqn with C-130H, YS-11, C-47, Do-28, Gulfstream.
Liaison: T-33A.
Helicopters: 3 sqn with AB-205A, AB-206A, Bell 47G/OH-13H, AB-212.
AD: 1 bn with Nike Hercules SAM (36 launchers). 12 bty with Skyguard/Sparrow SAM, twin 35mm guns.
Air Training Command:
Training: 4 sqn: 1 with T-41A; 1 with T-37B/C; 2 with T-2E.
Equipment: 375 cbt ac (plus 73 in store), no armed hel.
Aircraft:
  A-7H: 52. 47 (FGA); plus 4 in store; TA-7H: 5 (FGA).
  F-104: 96. F-104G: 70 (FGA); plus 24 in store; TF-104G: 8* (FGA); RF-104G: 18 (recce).
  F-5: 90. -A: 60 (36 FGA, 24 ftr); -B: 10 (6 FGA, 4 ftr); plus 13 (10 -A, 3 -B) in store. RF-5A: 20, plus 3 in store.
  F-4: 37. -E: 32 (17 FGA, 15 ftr; plus 17 in store); RF-4E: 5 (recce).
  F-16: 36 (FGA/ftr); plus 4 in store.
  Mirage F-1: 36. CG: 32 (ftr); BG: 4* (trg), plus 4 in store.
  Mirage 2000: 28. -E: 24 (ftr); -D: 4* (trg), plus 4 in store.
  HU-16B: 12 (8 MR; 4 being updated). C-47: 8 (tpt).
  C-130H: 11 (tpt). CL-215: 14 (tpt, fire-fighting). Do-28: 15 (lt tpt). Gulfstream I: 1 (VIP tpt). T-2: 36* (trg). T-33A: 56 (liaison). T-37: 25 (trg). T-41: 20 (trg). YS-11-200: 6 (tpt).
Helicopters: AB-205A: 15 (tpt); AB-206A: 1 (tpt); AB-212: 4 (tpt); Bell 47G: 11 (liaison).
Missiles:
  ASM: AGM-12 Bullpup, AGM-65 Maverick.
  AAM: AIM-7 Sparrow, AIM-9 Sidewinder, R-550 Magic.
  SAM: 36 Nike Hercules; 40 Sparrow.

FORCES ABROAD
Cyprus: 2,250. 2 inf bn and officers/NCO seconded to Greek-Cypriot forces.
UN and Peacekeeping:
Iraq/Kuwait (UNIKOM): 7 observers.

PARAMILITARY
Gendarmerie: 26,500; MOWAG Roland, 15 UR-416 APC, 6 NH-300 hel.
Coast Guard and Customs: 4,000; some 100 patrol craft, 2 Cessna Cutlass, 2 TB-20 Trinidad ac.

FOREIGN FORCES
U.S.: 1,950. Army (250); Navy (200) facilities at Soudha Bay. Air (1,500) 2 air base gp.

Future

Like its Balkan neighbors and Turkey, Greece is not yet a fully industrialized country, but it will continue to move rapidly in that direction over the next twenty years. In political terms, Greece would gain considerable benefit from strengthening its ties to Turkey, despite long-standing Greek dissatisfaction with Turkish policies. A solution to the Cyprian situation would help make this possible, because it would ease the Greek memories of the 1974 Turkish invasion. Greece's geographic position relative to both the Balkans and the eastern Mediterranean, and its growing economic power, will almost certainly produce pressure from other NATO countries to fully participate in the alliance, especially as PASOK has failed to maintain power.

DAVID L. BONGARD
TREVOR N. DUPUY

SEE ALSO: Alexander the Great; Byzantine Empire; European Communities; History, Ancient Military; Peloponnesian War; Turkey; Western Europe.

Bibliography

Clogg, R. 1979. *A short history of modern Greece.* New York: Cambridge Univ. Press.
Hunter, B., ed. 1991. *The statesman's year-book, 1991–92.* New York: St. Martin's Press.
International Institute for Strategic Studies. 1991. *The military balance, 1991–1992.* London: Brassey's.
Kousoulas, D. G. 1974. *Modern Greece: Profile of a nation.* New York: Scribner's.
Woodhouse, C. M. 1976. *The struggle for Greece, 1941–1949.* London: Faber.

GRENADA, U.S. INTERVENTION IN

The events that triggered the U.S. intervention in Grenada, code-name Urgent Fury, began long before the 24 October 1983 presidential directive. Beginning in 1974, when Grenada gained its independence from Britain, corruption and political repression were trademarks of successive Grenadian prime ministers. The execution of Prime Minister Maurice Bishop on 19 October 1983, however, set into motion a series of events that would culminate in the largest U.S. military operation since the Vietnam War.

The intervention was overwhelmingly supported by the American public but condemned internationally and likened by some to the Soviet's 1979 invasion of Afghanistan. The decision to launch Urgent Fury, however, proved to be a sound one, both militarily and politically. Documents captured by U.S. forces revealed that almost every communist satellite regime under Soviet influence had representatives and observers, at some stage, in Grenada. These individuals were helping to transform the island into a major military camp, as revealed by U.S. forces' discovery of artillery, antiaircraft weapons, ammunition, armored personnel carriers, and rocket launchers. Moreover, a democratically elected government was subsequently restored to Grenada, an island that for years had suffered under autocratic, corrupt, and ruthless regimes. For Cuba and the Soviet Union, the intervention was a serious blow.

By 15 December 1983, less than two months after they were ordered into action, U.S. combat forces were withdrawn from Grenada. Some maintained that the victory re-established the professional competence of the U.S. military, which had been lost in Vietnam. However, others pointed out that Grenada was not a flawless military operation but had come close to being a military disaster. William S. Lind, a staff aide to former Senator Gary W. Hart, was highly critical of the operation. The press coverage of his allegations was partly responsible for demands for a congressional investigation, but an investigation was forestalled by the Chairman, Joint Chiefs of Staff's written responses to the allegations.

Overall, U.S. military operations in Grenada succeeded beyond expectations against a small and marginally equipped group of defenders. According to former Secretary of Defense Caspar W. Weinberger, it was a "complex operation in which we saved a lot of lives and a lot of misery." The operation showed the ability of the U.S. military to mount a swift, well-executed attack on short notice. More important, it demonstrated U.S. public support for the use of selective military force to accomplish limited political-military goals.

History

The island of Grenada was discovered by Columbus in 1498. Thereafter, it passed through Spanish and French hands and finally came under British ownership during the eighteenth century; England governed the island until the late 1950s. During the early 1960s, Britain sponsored two attempts to form her Caribbean colonies into a single federation. Failing in this, the individual islands were given independence within the British Commonwealth system, starting with Grenada in 1974. Grenada's first prime minister was Sir Eric Gairy, whose administration was known for its corruption and political repression. In March 1979, Gairy was overthrown in a bloodless coup led by Maurice Bishop. Bishop took power as head of Grenada's Provisional Revolutionary Government (PRG) in the name of his New Jewel Movement Party (NJMP), an acronym that stood for the Joint Endeavor for Welfare, Education, and Liberation.

The New Jewel Movement

Bishop and his NJMP faced both political and economic problems but saw a solution in the example of Cuba. Over the next few years, Bishop invited increasing assistance from both Cuba and other communist states. This, plus Bishop's disinterest in holding elections, brought PRG into conflict with U.S. foreign policy in the region, a situation aggravated by the PRG's announcement of its most ambitious project—construction of a new international airport at Point Salines. The airport was to be built by Cuban workers and feature a 2,750-meter (9,000-ft.) runway. Its purpose was to improve Grenada's sagging tourist trade, because the existing airport could accept only twin-engine air traffic. U.S. military forces, however, captured documents revealing that the Cubans had planned to use the airport as a staging area to airlift supplies to their troops in Africa and as a refueling stop for Soviet planes en route to Nicaragua.

By late summer 1983, the NJMP had split into two factions. One faction, led by Bishop, wanted closer ties with the West. The other faction, led by Deputy Prime Minister Bernard Coard, wanted to retain Grenada's communist connections and to speed up the country's conversion to a Marxist state. (By the time of the U.S. intervention, Grenada had agreements with the Soviet Union, Vietnam, Czechoslovakia, North Korea, Cuba, and East Germany for supply of modern sophisticated military equipment and technical logistical assistance.) The crisis between the two factions came to a head on 13 October, when Coard, with the PRG military's backing under Gen. Hudson Austin, ordered Bishop to step down from office for failure to carry out the orders of the NJMP's Central Committee. Bishop and a number of his ministers were placed under house arrest and subsequently executed on 19 October. A 24-hour curfew was then imposed and notice given that violators would be shot on sight.

Events In Grenada: 19–28 October 1983

On the night of 19 October, General Austin announced the formation of a 16-man Revolutionary Military Council

(RMC) with himself as head. No mention was made of Coard; however, he would serve as an adviser to Austin. Over the next four days, a number of prominent citizens were arrested.

On 21 October, the leaders of six small nations (Dominica, St. Lucia, St. Vincent, Montserrat, St. Kitts-Nevis, and Antigua) composing the Organization of Eastern Caribbean States (OECS) met to consider collective action against Grenada. They voted to intervene militarily to restore order to the region. However, because none of the six possessed the necessary forces, they recognized that an appeal for assistance would have to be made to nonmember states Jamaica and Barbados, regional neighbors, and to the United States.

The following day, the Caribbean Community (CARICOM) met. A majority of the delegates supported the idea of intervention, if the RMC would not peacefully accept a CARICOM fact-finding commission and a Caribbean Peacekeeping Force.

On 23 October, the British deputy commissioner in Barbados and two U.S. diplomats met with Grenadian officials. The diplomats found the officials "obstructionist and uncooperative." (Early that morning, 241 U.S. Marines stationed in Beirut had been killed by the explosion of a bomb-laden truck driven by a suicide driver. After an all-day meeting between President Reagan and his advisors on the situation in Beirut, attention shifted to Grenada. The president expressed concern over the possibility of U.S. hostages because of the 1,000 U.S. citizens on Grenada, most of whom were students and faculty of St. George's University Medical School, an American-run institution.) The next day, two additional U.S. diplomats arrived on Grenada. Grenadian officials demanded six hours' advance notice for any evacuation flights and warned the diplomats they could not guarantee the safety of foreigners. U.S. officials became concerned that the RMC was planning to use the safety issue of American citizens as a bargaining chip.

A formal request for assistance under Article 8 of the OECS charter was presented to Jamaica, Barbados, and the United States on 23 October. President Reagan met with senior U.S. officials the next day and that evening signed the directive committing U.S. military forces into action. Early on the morning of 25 October, a combined force of 7,000 military troops from the United States, Jamaica, Barbados, and the six OECS states landed on Grenada by sea and air. By Friday, 28 October, the force had secured all significant military objectives and had defeated the People's Revolutionary Army and Territorial Militia and captured about 600 Cuban construction workers. Forty-nine Russians, seventeen Libyans, fifteen North Koreans, ten East Germans, and three Bulgarians were found on the island and quickly repatriated.

Justification For U.S. Action

The Grenada operation was the first time since the Dominican intervention of 1965 that U.S. forces had been ordered into combat in the Caribbean. Elaborate justification for the action was given by senior U.S. officials. President Reagan's 25 October announcement of the "rescue mission" outlined the main reasons for the operation: to ensure the "personal safety" of between 800 and 1,000 U.S. citizens on Grenada, to "forestall further chaos," and "to assist in a joint effort to restore order and democracy" there. He also strongly emphasized that the operation had been mounted in response to "an urgent, formal request" from several eastern Caribbean states.

Secretary of State George Shultz, the U.S. representative to the Organization of American States, Ambassador J. William Middendorf, and United Nations Ambassador Jeane Kirkpatrick also elaborated upon the president's statement and provided further justification for the intervention: the lack of governmental authority on the island, international law requirements, regional security concerns, and the reality of the use of military force in world politics.

U.S. Domestic Reactions to the Intervention

Despite the Reagan administration's efforts to portray the Grenada operation as a humanitarian rescue mission, a response to an urgent request for help by friendly democratic neighbors, and the successful foiling of a Soviet-Cuban colony, the U.S. public supported the intervention because it was swift, conclusive, and relatively free of cost. Several polls affirmed the American public's overwhelming support of the operation. Many Americans believed that the intervention demonstrated that the United States had overcome the "Vietnam syndrome," that is, the reluctance to use military power to defend U.S. interests and values.

Initial congressional reaction, however, was largely negative, due in part to the administration's failure to consult Congress prior to the intervention. But as the medical students began returning from Grenada and as U.S. forces uncovered evidence of Cuban and Soviet weapons, congressional criticism diminished markedly and shifted toward other issues associated with the military operation. Two such issues were the restrictions that had been placed on journalists prohibiting them from going on the island until several days after the operation had begun, and the accidental U.S. bombing of a mental hospital that claimed several patients' lives.

International Reactions

Although U.S. domestic opinion supported the intervention, international opinion did not. From 25 to 27 October, the United Nations Security Council debated U.S.

and OECS actions. Of 63 countries that spoke regarding the intervention, only the United States and the OECS states defended the action. On 28 October, by a vote of eleven "yes," one "no," and three "abstain," a UN resolution was passed that "deeply deplored the armed intervention in Grenada" as "a flagrant violation of international law and of the independence, sovereignty and territorial integrity of that state." An identical resolution was introduced in the General Assembly on 2 November and was passed without debate with over 100 states voting "yes."

The European allies of the United States also expressed initial disapproval of the intervention. But as the Reagan administration continued to produce evidence showing that the U.S. students had been in danger and that Grenada had become a Soviet outpost, European opposition began to soften.

U.S. Military Performance

The United States deployed elite military units to Grenada that included airborne, ranger, and special operations forces, and marines. By the time hostilities were officially declared over, these forces equated to nine infantry battalions. (Grenada was divided roughly in half, with the Marines responsible for the northern part of the island and the Army for the southern part.) Missions that were accomplished included the seizure of key objectives and the securing of the island with minimum casualties and destruction. American ground units were supported by a naval carrier battle group complete with ground attack aircraft, naval gunfire, air force and helicopter gunships, army artillery, and massive logistical support. The forces opposing U.S. military units consisted of about 750 Grenadian troops and 600 Cuban "construction workers," all of whom were of questionable morale, with no air or naval forces, no tanks, no artillery, and only a few outdated antiaircraft guns.

Although Urgent Fury was a success, U.S. military operations were not without their problems, most of which remain classified. However, based on open source material, the most highly publicized were the bombing of U.S. troops by their own aircraft, intelligence failures, the lack of maps, and problems with special operations.

A report of lessons learned from the U.S. intervention was submitted to the Chairman, Joint Chiefs of Staff on 6 February 1984 by Adm. Wesley L. McDonald, the commander of Task Force 120, the combined headquarters that controlled military operations. The report identified a number of problems: the lack of understanding of interservice close air support procedures, the confusion caused by the lack of maps, the need for senior liaison officers from all commands at the controlling headquarters, the lack of equipment and preplanning to deal with prisoners of war, and the lack of experience with communications systems.

In addition to these problems, Admiral McDonald said that there was confusion with the insertion of the Caribbean Peacekeeping Force; air assets were "not always properly controlled"; "helicopters are highly vulnerable to well-aimed ground fire"; and "medevac operations at night became a great concern because Black Hawk pilots had not been trained to land on seaborne helicopter platforms and were denied permission to land." Although some have maintained that there were "unforgivable blunders" in intelligence, Admiral McDonald's report stated: "Available basic intelligence was generally adequate for overall planning purposes" and that the estimate of enemy strengths "was within an acceptable range of uncertainty."

Costs and Implications

Almost a decade after the end of the Vietnam war, the U.S. intervention in Grenada reintroduced the use of military power into U.S. foreign policy. However, Urgent Fury cost the lives of eighteen U.S. servicemen killed in action. A total of 116 were wounded in action. Nine U.S. helicopters were either destroyed or damaged. American citizens evacuated from the island totaled 599.

Of the approximately 600 Cubans on Grenada, 24 were killed in action and 59 wounded. No official attempt was made to separate Grenadian casualties into military and civilian totals; the combined total was 45 killed and 337 wounded.

The loss of Grenada was a severe blow to Cuban prestige worldwide and, more directly, to its plans in the region. The exposure of Grenada's secret agreements with East Bloc countries revealed the subversion that was underway, which, if left uncontested, would have threatened both regional and U.S. interests.

Militarily, Urgent Fury demonstrated the readiness, capability, and professionalism of the U.S. armed forces. The entire operation was planned and executed in less than a week. While operating under stringent rules of engagement, U.S. forces accomplished their mission. When the operation was over, Americans felt a renewed sense of pride in their military forces.

In the final analysis, the U.S. intervention in Grenada revealed that the use of military power, tempered by effective and determined leadership, remains a valid tenet of international politics.

JAMES B. MOTLEY

SEE ALSO: Air Assault; Airborne Land Forces; Amphibious Forces; Dominican Republic: 1965 Crisis; Land Warfare; Low-intensity Conflict: The Military Dimension; Panama, U.S. Invasion of; Special Operations Forces.

Bibliography

Adkins, M. 1989. *Urgent Fury: The battle for Grenada.* Lexington, Mass.: D. C. Heath.

Dunn, P. M., and B. W. Watson, ed. 1985. *American intervention in Grenada.* Boulder, Colo.: Westview Press.
Motley, J. B. 1984. Grenada: Low-intensity conflict and the use of U.S. military power. *World Affairs* 146 Winter 1983–84, (3):221–38.
Russell, L. E., and M. A. Mendez. 1985. *Grenada 1983.* London: Osprey.
Seabury, P., and W. A. McDougall, eds. 1984. *The Grenada papers.* San Francisco, Calif.: Institute for Contemporary Studies.
Schoenhals, K. P., and R. A. Melanson. 1985. *Revolution and intervention in Grenada.* Boulder, Colo.: Westview Press.
Weinberger, C. 1990. *Fighting for peace: Seven critical years in the Pentagon.* New York: Warner Books.
JCS replies to criticism of Grenada operation. *Army* (August):28–37.

GRENADE

Grenades provide soldiers on the battlefield with a cheap but lethal weapon suitable for use in a variety of situations, both on the attack and in defense. There are two main types of grenades: hand grenades, which the soldier throws by hand; and others that are fired from attachments to the soldier's rifle.

Hand grenades are used primarily by infantry soldiers and special troops (rangers, commandos, and parachutists). They are small in size, light in weight, and easily thrown by hand. The many types of hand grenades include defensive, offensive, smoke, antitank, illumination, and tear gas.

Rifle grenades are designed to be fired without the need for additional equipment. Again there are many types; they include antipersonnel, antitank, smoke, illumination, and dual-purpose antipersonnel.

Hand Grenades

Table 1 sets forth the main characteristics of the two primary types of hand grenades, those designed for defensive and those for offensive use.

Defensive hand grenades are used by infantry soldiers who throw them and then remain under cover, thereby causing heavy casualties among attacking forces.

TABLE 1. *Characteristics of Defensive and Offensive Grenades*

CHARACTERISTIC	DEFENSIVE	OFFENSIVE
Effectiveness	Danger area 5–15 m	Morale effect
Range of throwing	25–30 m	up to 40 m
Mechanism	Time-fuzed (3.6–4.5 sec.)	Impact-fuzed
Fragmentation	About 3,500 steel balls, each 2.5–3 mm in diameter	Very light fragments causing considerable blast effect
Weight	Heavy (309–486 gm)	Light (220–309 gm)

Offensive hand grenades are used by assaulting infantry. They throw these grenades while continuing to move forward, seeking to daze and demoralize the defenders while creating suitable conditions for gaining temporary advantages and pressing home the assault. Such grenades are used in assaulting buildings, strongpoints, and other defensive positions.

Multipurpose (defensive and offensive) hand grenades have characteristics such that they can be used in both defensive and offensive action. For use in the defensive mode, a steel fragmentation casing may be fitted over the grenade's body. For offense only the basic grenade is used, primarily for its blast and shock effect.

Smoke-generating grenades are made of tinned steel and are capable of producing smoke in different colors (typically white, green, red, yellow, or gray), which lasts for one to two minutes. Some smoke grenades contain white phosphorus and produce not only smoke but also incendiary effects.

Antitank hand grenades are used at close ranges (15–20 m). These are antiarmor grenades that can pierce steel up to 75–100 mm thick. The effective fragments have a range of 20 meters (approx. 65 ft.). Such grenades are used by infantry from protected defensive positions against armored vehicles, and may also be effectively employed by guerrillas and by militia units engaged in street fighting.

Illuminating grenades are designed to illuminate the surrounding area up to a radius of 100 meters (328 ft.). Burning time is about 45 seconds. They are used for observation of a particular area.

Tear gas (riot control) hand grenades are cylindrical grenades with a body made of tinned rolled plate. They can be thrown about 40 meters (approx. 130 ft.). The fuze is initiated to puncture a gas cartridge, which produces a stream of tear gas for about one minute. These grenades are used mainly for control of civil disturbances.

Rifle Grenades

Antipersonnel rifle grenades come in two types, each described below.

Short-range antipersonnel rifle grenades have a maximum range of 300 meters (approx. 985 ft.) and produce a large number of fragments. The affected area is 20 meters (approx. 65 ft.) in diameter. Such a grenade is light, weighing not more than 341 grams, which enables the soldier to carry a number of them without limiting his mobility. It can be fired from a standing or prone position because of its light recoil. The weapon may be used effectively against small groups of enemy in the open, against buildings and bunkers, and in defilade positions.

Long-range antipersonnel grenades are fitted with a rocket propellant that ignites automatically when the grenade is fired, thus extending the range up to 650 meters (approx. 2,150 ft.) and the dangerous area up to a 40-meter (130-ft.) diameter. They can be fired from a standing or prone position, holding the rifle at a 45-degree angle, enabling the soldier to engage hidden targets behind barriers and in effect allowing the rifle to perform the function of a light mortar.

Antitank rifle grenades can pierce steel to a depth of 25 centimeters (approx. 10 in.). Their weight does not exceed 350 grams, so a soldier can carry several without impairing his mobility. They are fitted with a marked sight graduated from 50 meters to 200 meters (165 to 650 ft.), which is the maximum range of this grenade. They are noted for accuracy, have very low recoil, and can be used successfully to cover the distance beyond antitank hand grenade range (more than 50 m) and the dead area that antitank missiles cannot reach (out to 200 m).

Smokegenerating rifle grenades produce a persistent and intense opaque smoke. They are effective for some 80 seconds at ranges of 50 to 300 meters (165 to 985 ft.) from the firer. The grenade weighs about 470 grams. A soldier can launch up to five grenades per minute. Versions are available that produce red, green, yellow, or white smoke, so that the resultant smoke can be used for either spotting or screening.

Illuminating rifle grenades produce an intense yellow illumination for about 30 seconds as a flare slowly descends by parachute after having been deployed at an angle of 80 degrees and a height of 100 meters (328 ft.). The grenade weighs 420 grams.

Dual-purpose antipersonnel grenades are designed so that they can either be thrown by hand or launched from a rifle with the use of a regular ball cartridge. For the latter purpose the tail of the grenade is fitted with a bullet trap. The grenade is also time-fuzed, making air bursts possible. Maximum range is 225 meters (approx. 740 ft.). Each round weighs 615 grams and produces about 500 fragments. The weapon may be employed either offensively or defensively by changing the ball of the grenade in accordance with the desired effect.

SAMIR HASSAN SHALABY

SEE ALSO: Ammunition; Firepower; Mortar; Munitions and Explosives Technology Applications; Small Arms.

Bibliography

Chamberlain, P., and T. Gander. 1976. *Allied pistols, rifles and grenades.* New York: Arco Books.
———. 1977. *Axis pistols, rifles and grenades.* New York: Arco Books.
Hogg, I. V. 1977. *The encyclopedia of infantry weapons of World War II.* New York: Thomas Y. Crowell.
Jane's Infantry Weapons 1983–84. 1983. London: Jane's.
Rosser-Owen, D. 1986. *Vietnam weapons handbook.* Wellingborough, U.K.: Patrick Stephens.

GROUND DEFENSE OF AIR BASES

An air base exists only to launch combat sorties. An enemy will try to stop sortie generation by initiating air and ground attacks. The objective of ground defense of air bases is to keep enemy ground forces from slowing the generation of sorties by preventing enemy attacks on the air base or minimizing their impact. Air base defenses support and shield the conduct of the air operations; their primary mission is not to defeat enemy ground elements. Air base defenses succeed when enemy impact on air base operability is minimal. The only viable measure of success is the number of effective sorties flown from the base, not the number of enemy casualties on and around the perimeter. The defenders win when they deny enemy approaches to the perimeter and/or swiftly repel attacks.

The Need for Ground Defense

To be able to function at war, the entire air base must also be able to fight in order to protect and sustain itself. An air base retains its operability through a unified process. The primary objective of air base operability should be to facilitate and ensure generation of a maximum number of effective sorties during the war. The role of the air base ground defense is to prevent enemy forces from having a degrading effect on base operations. The protection of the base is accomplished by a combination of damage prevention, base resurrection, and force-containment measures.

Ground attack on air bases was a useful tool against the airpower supported by the base. For example, ground attack on NATO air bases was presumed to constitute a critical element of the Soviet strategy for a future war in Europe. The Soviets expected their on-ground operations against NATO air bases to compensate for possible setbacks of aerial operations and to contribute to success at the combined-arms operational level. The neutralization of NATO air bases could be achieved through the cumulative effect of several dedicated, local combat actions, mostly by small subunits. Most NATO air bases were extremely vulnerable to the forms of attack that the Soviets could have employed.

Airpower has a major influence on modern warfare despite the relatively small numbers of aircraft and crews used by the belligerents. Determined offensive use of airpower has proven virtually unstoppable, provided the attacking forces are willing and able to withstand the required levels of attrition.

The major potential constraint of airpower is the requirement that aircraft start from, and return to, fixed air bases where they are serviced and maintained. As aircraft become more powerful and complex, their dependence on larger, more complex air bases increases tremendously. Attacking enemy air bases is a most effective and efficient method to preempt and prevent the use of airpower without having to challenge the flexibility and performance of the aircraft themselves. Attacking air bases is not only an integral component of the quest for air superiority, but also an alternative means by which a force with inferior airpower can balance that inferiority.

Using one's own aircraft to hit the enemy's air bases is

the traditional method. As air base fortifications improve, aerial strikes are no longer sufficient. Long-term, persistent pressure on the enemy air base is crucial. Airpower, which is characterized by brief and swift, albeit powerful, strikes, is incapable of maintaining uninterrupted pressure on point-targets during a war. Since most air bases are beyond the range of artillery, special forces and raiding elements (including irregular forces and terrorists) are the primary method for attacking air bases on the ground. A few detachments of raiding elite forces can maintain enough protracted pressure on an air base to have a major effect on its sortie generation capability.

Precedents

WORLD WAR II

British and Soviet ground attacks on German air bases during World War II still constitute a source of valuable and viable lessons on the objectives and effectiveness of such operations.

The German bombing of Britain created a need for innovative action. Late in 1940 (especially after the German raid on Coventry on 14 November), it became clear that the Royal Air Force (RAF) night fighters and the electronic jamming effort were not sufficient to prevent the German KG100 from relying "on the beams" to accurately mark targets for night bombings. The British sought drastic solutions. The KG100 operated from Meucon air base in south Brittany. The British learned that the aircrews, who were housed in nearby Vannes, traveled in two buses from the village to the air base prior to the strikes. The Air Ministry instructed the Special Operations Executive (SOE) to ambush the buses and kill the specially trained aircrews. Five French operatives of SOE were dropped on the evening of 15 March 1941 under the cover of an air raid. They discovered that the Germans were taking extra precautions, that the KG100 aircrews traveled in pairs in small cars, and that the Germans patrolled the roads. Consequently, they aborted the raid.

The British in North Africa. The effectiveness of special forces as an integral component of the air war was demonstrated in the Western Desert by the British Long Range Desert Group (LRDG) and the Special Air Service (SAS). Since September 1940, the British had relied on special operations to compensate for lack of offensive capabilities, including airpower, in North Africa. Regardless of the actual damage achieved by the first raids, the Luftwaffe allocated large numbers of sorties to locate and attack LRDG patrols, sorties that were sorely missed by Rommel.

British special operations against air bases became highly successful during the raiding partnership of the LRDG and SAS. The first week of 1942 saw the impact of raiding forces in a surprise attack. The LRDG-SAS conducted a series of relentless raids on rear air bases in the Sirte area in Libya. In one week they destroyed 90 aircraft on the ground with small bombs, blew up installations and fuel and ammunition depots, and attacked the enemy air and ground crews, causing heavy casualties. During this week, Paddy Mayne, later the commanding officer of the 1st SAS Rgt, personally destroyed 47 aircraft in a single raid, more than the wartime record of any Western fighter ace. By the end of 1942, his record was over 100 aircraft destroyed. By July 1942, the LRDG-SAS had destroyed 143 aircraft and by the late summer nearly 250, a score surpassing that of the most aggressive RAF squadron in the Middle East.

Special raids could solve specific problems for the RAF. As an example, the Germans' Me-109F outperformed anything the RAF had in the Western Desert. To solve this problem, in a single night raid the LRDG-SAS destroyed with machine-gun fire fifteen Me-109Fs—most of the operational German fleet available in the theater—without suffering any casualties. It also destroyed large quantities of aviation fuel and ammunition and killed air and ground crews in raids during 1942.

The primary outcome of the persistent anti–air base campaign by the LRDG-SAS and the RAF was residual impact on the German and Italian ability to use airpower. RAF Marshal Lord Tedder explained that by mid-1942, "the enemy has been forced to adopt a degree of dispersal so extreme that this of itself severely handicapped his operations." The LRDG-SAS raids not only severely hurt the Luftwaffe's already tenuous lines of communications but compelled the Luftwaffe to waste a large number of sorties and already scarce fuel in reaction to these raids. The continued raids on the air bases and lines of communication (LOC) had a cumulative effect on the aircrews, significantly reducing the effectiveness of their performance due to physical and mental exhaustion. Consequently, at the most crucial stages of the Desert Campaign, the German and Italian air forces could realize but a small fraction of their airpower potential.

The Soviet experience. During the Great Patriotic War (GPW), the Soviets conducted systematic, large-scale irregular warfare behind the German lines. Partisan operations reached maturity in support of the Soviet offensives in 1943. Luftwaffe air bases, fuel and ammunition supply lines, and dumps were priority objectives. The most effective raids were directed at the ground lines of communication. The partisans also marked and attacked air bases whenever possible. According to Soviet data, partisans destroyed 1,100 aircraft, compared with the 57,000 aircraft claimed by the Soviet Air Force (VVS), and the total of 77,000 aircraft the Germans lost on the Soviet front to all causes.

The cumulative impact of the partisan attacks on the Luftwaffe's fuel supply system and air bases was telling. Not only did the Luftwaffe suffer from an acute shortage of fuel, but existing stockpiles were often misused. The continued redeployment and wide dispersal of aircraft between air bases, because of partisan and VVS air base attacks, required that every airfield be able to receive, maintain, and supply larger numbers of aircraft than its regular deployment. Emergency fuel stockpiles were built in every air base, denying the Luftwaffe the use of large quantities of fuel that reached the front line at the most crucial stages of its operations. Aircraft that were used extensively to patrol roads and railways and to hunt partisans were not available for battle use at the front.

In order to defend against the partisan and VVS attacks, air defense formations and regional defense units, comprised of older soldiers, were assigned to the air bases. The Luftwaffe field divisions, which had been organized to defend the Luftwaffe major deployment perimeters, were taken over by the army to be used as regular divisions against the Soviet offensives. As the partisan attacks intensified, the antiaircraft troops were also used for guard duties. Consequently, as the Soviet air and ground pressure grew, the German air and ground defense system failed. There were too few regional defense soldiers to establish static perimeters around air bases, and the older soldiers could not cope with the highly mobile partisans. Air defense deteriorated because the antiaircraft troops could not cope with both ground and air defense duties due to exhaustion and attrition. The Luftwaffe attempt to reduce the damage by a further dispersal of air bases and installations strained the system to collapse.

The combination of the disruption of fuel supply by the partisans and the increasing attacks on air bases affected the Luftwaffe well beyond the actual physical damage. The Soviet attacks forced the Luftwaffe to disperse, consume, and stock fuel at a time when fuel shortages were caused by interdiction of supplies. The failure of air base defense, exacerbated by the growing dispersal, forced the allocation of sorties to air base and supply line defense, further straining the Luftwaffe and its fuel crisis. The net outcome was a drastic reduction in Luftwaffe participation in the major land operations. This concept of building pressure on air bases and their supply lines served as the basis of the Soviet approach to the neutralization of airpower, with the emphasis shifting toward special force operations.

POST–WORLD WAR II

U.S. air bases in Vietnam were repeatedly attacked by enemy ground units beginning on 1 November 1964, when 81mm mortars were used to shell Bien Hoa. After the 1968 Tet Offensive, highly trained sappers (engineers) of the North Vietnamese Army (NVA) conducted the raids, relying heavily on intelligence and observation. Four types of attacks were employed against U.S. air bases. Stand-off attacks by mortars and rockets were the most frequent. Second came sapper attacks and penetration efforts. Less frequent were mass attacks and sabotage by agents inside. In establishing air base defense, the U.S. Air Force (USAF) had to overcome challenges that are still valid. Air base defense was coordinated with the local authorities, which severely limited the freedom of fire and timeliness of reaction. The sole exception was Da Nang where U.S. Marines, later augmented by ARVN (army of the Republic of Vietnam) troops, installed a 45-kilometer (28-mi.)-deep defensive perimeter with free-fire zones and anti-infiltration lethal barriers.

In most air bases, defense was based on a three-zone system that included static and mobile elements deployed in sectorized defense in depth. It was designed to contain and destroy enemy penetrations on base but away from the runway area. This defense system relied on lethal barriers, routine patrols, fire sources in bunkers, and quick-reaction mobile forces. In 1968, this system proved insufficient to cope with a determined sapper thrust. Consequently, U.S. Army gunships were integrated into the defense plan to provide quick-reaction firepower. Subsequently, ARVN artillery, airborne companies, and Vietnamese Air Force (VNAF) controlled light tanks were added to the air base defense in order to provide the required mobility and firepower beyond the capabilities of the USAF Security Police squadrons. The radius of the perimeter patrolled and covered by sensors eventually reached an average of 29 kilometers (18 mi.). Despite all efforts, including airborne patrols and fields of sensors, the USAF never had reliable tactical warning on impending NVA attacks. Consequently, all the NVA penetration attempts were blocked only by fierce fighting inside the air bases in which massive firepower and out-of-base U.S. reinforcements were brought into use. The USAF security was incapable of preventing and/or preempting the stand-off fire attacks.

Even small-scale ground attacks on air bases can be decisive in local wars. In July 1976, during the Entebbe rescue mission, a small Israeli detachment destroyed the Ugandan Air Force on the far side of the airfield. Consequently, the Israeli C-130s could operate safely without fighter cover even after the surprise factor was lost.

By inserting a single battery of G-5 155mm howitzers into the UNITA (*Uniao Nacional para a Independencia Total de Angola*) rear, the South Africans changed the aerial balance of forces in southeast Angola without using their own aircraft (by choice). For a few months in late 1987 and early 1988, the South Africans neutralized the only major air base in the area—Cuito-Cuanavale—with sporadic shelling, preventing the Cuban Air Force from deploying tactical aircraft to the area.

Current Approaches to Air Base Defense

At present, there is a sharp difference between the approaches of the West and the East to attacking and defending air bases. The West is preoccupied with sudden attacks on enemy airpower such as characterized operation Barbarosa, Pearl Harbor, and the Six Day War. Attention is aimed at conducting and countering air attacks on air bases.

In the East, by contrast, there is an increased interest in attacking and defending air bases on the ground. It is a major component of the revival of the nonnuclear strategic deep offensive. Soviet military writings and exercises suggested that the Soviets would have preferred not to use surface-to-surface missiles (SSMs) for surprise attacks, wishing to avoid confusion with nuclear SSMs and unnecessary risks of escalation and retaliation. There was a growing emphasis in the former USSR on the use of special forces for attacks on key objectives in the rear of the enemy, including air bases. These trends were reflected in the assignments of specialist and/or Afghan veteran senior officers, helicopter pilots, and special forces commanders and senior NCOs to the Group of Soviet Forces-Germany (GSFG, currently known as the Western Group of Forces or WGF) and the Group of Soviet Forces-Central (GSFC, formerly based in Czechoslovakia), as well as the kind of air base attack exercises conducted there.

THE NATO NATIONS

It has been suggested that the West's preoccupation with surprise attacks and short (nuclear) wars has had a direct effect on NATO's air base operability. There has been an overall neglect in establishing defensive measures against forms of attack associated with protracted conventional wars. Air base defense and operability have a low priority in NATO—an alliance constrained by budget factors and virtually dependent on airpower to balance the inferiority in land forces. The overall defense of air bases and related installations (radars, communication facilities, etc.) is a component of the general defense of the rear, which is entrusted to local territorial and auxiliary forces. There are only small security forces on the air bases themselves.

For example, United States Air Force, Europe (USAFE) air bases are to be defended by a small, dedicated Security Police (SP) force supported by Law Enforcement (LE) policemen and untrained reinforcements whose primary role is as lookouts on the perimeters. Most people on USAFE air bases are unarmed. The SPs are supposed to protect key objectives on the base, conduct preventive patrols, and engage and destroy intruding enemy forces. At their present composition, the SPs are overstretched, undertrained and under-armed for such demanding tasks. In recent years there has been a growing awareness in the U.S. Air Force of the vulnerability of its air bases, and concrete steps are being taken to improve the combat performance of the SPs. Since October 1987, SPs have received ground combat skills training by the army at Fort Dix, New Jersey. However, the defense of USAFE air bases in a protracted war is still largely based on the timely arrival of army reinforcements from the United States and on a protective shield to be provided by local territorial forces. The lessons of World War II clearly demonstrated that territorials and other older-age auxiliary and guard forces are no match for special forces. The deep perimeters, proven so critical in Vietnam, are nowhere to be found in Europe.

Within NATO, the United Kingdom is an exception to the rule; it pays attention to its own extensive offensive experience and to the other side's capabilities. The defense of RAF air bases is provided by the dedicated RAF regiment with heavy combat equipment and its own territorial forces. The British Territorial Forces include several elite units with strong cadres of regulars such as the 21st and 23d SAS Regiments. The RAF regiment is divided into five squadrons, each committed to the defense of a cluster of installations. In 1987–88, the regiment's combat vehicles included 38 Scorpions, 19 Scimitars, 113 Spartans, and 7 Sultans. To ensure unity and centrality of command and control, the RAF regiment is also in charge of the point air defense, which is provided by nine squadrons each based on Rapier surface-to-air missiles (SAMs) and supporting weapons. The tactical approach of the RAF regiment is to counter the excellence, surprise, and flexibility of raiding forces with superiority of firepower, speed, and protection provided by armored vehicles. Similar lessons were learned and implemented by the USAF in Vietnam, where ARVN armored units were used for air base defense.

ISRAEL

Israel pays close attention to the defense of its air bases even though they are located in the midst of a committed rear with a military-trained population. Air base defense is organized as if all the other defenses had collapsed or did not exist at all. Each base has an integral infantry battalion with all its heavy weapons and mechanized equipment in addition to well-armed lookouts in towers on the perimeters and other key points. At war, these defenses are further augmented by highly trained reservists, many of them veterans of combat units. The point air defense battalion, which includes dual-role automatic guns, is also an integral part of the air base forces. Moreover, all the manpower on base, from technicians to cooks, are armed and are trained in basic infantry tactics. They are expected to protect their installations in case of an enemy penetration. To reduce the vulnerability to air base attacks, major components of the air force can deploy to auxiliary and forward strips. The ground crews are armed and trained to defend these locations against ground threats even in a case where the dedicated defense force fails to link up or establish a proper perimeter. Even with the proper defenses deployed, the ground crews are entrusted with the defense of their own facilities and installations.

THE SOVIET UNION

There was a growing awareness in the former USSR of the significance of air base defenses. The Soviets emphasized that all ground crews and aircrews must be armed and trained to defend their air bases and/or forward sites. Providing air bases with credible defenses against raiding forces was considered a critical question of combat training. In Afghanistan the entire base was actively involved in the protection of the forward installations with the airmen standing in bullet-proof vests holding submachine guns, almost indistinguishable from motorized riflemen. Experience in Afghanistan also pointed to the strategic potential and lethal effectiveness of properly conducted special operations. Each base had a dedicated defense force of at least a motorized rifle company. The ground crews were armed and detailed for the defense of their installations and aircraft. The air base defense was under the command of a pilot, because only pilots could properly assess the significance of various installations and the priorities in organizing defense.

The Soviets conducted repeated two-side live exercises, pitting the elite forces and the air base defenders against each other in realistic conditions. In such exercises, pilots were pulled away from flying to protect their assets. The Soviets also conducted extensive and diversified exercises for attacks on, and seizure of, air bases by elite formations, ground and heliborne raiding forces, chemical weapons, and roving combined-arms reinforced battalions and tank battalions. Their objectives were destruction of the soft areas of the base and the conduct of fire and denial attacks on the runway areas if possible, with emphasis on harassment and aggregate damage rather than decisive results. Operating under a contamination scenario was practiced under realistic conditions, using diluted and harassment agents, in training air bases maintained by the Chemical Troops.

Principles of Air Base Defenses

It is impossible for the overextended modern armed forces to provide all-around dedicated forces for defense of key installations such as air bases. The quality of the former Soviet special forces and their realistic approach to attacking air bases made air base defense a challenge well beyond the capabilities of security police supported by paramilitary and rudimentarily trained territorial forces. The prime objective of air base attacks is to render the enemy militarily useless (neutralization), not to destroy the base. This objective is attained by concentrating on the base's soft areas and the building of pressure and terror among its personnel. To survive, the entire air base must protect itself. One need not send everyone to the trenches; pilots are still more effective in the cockpit. However, the wealth of quality manpower in air bases should be used for the defense of their own installations.

The growing significance of ground attack on air bases is reflected in the former USSR's preoccupation with the security of its own bases. Reconnaissance-intelligence and assault-*desant* (landing) raiding detachments challenged the defenses of former Soviet air bases—a combination of motorized rifle subunits and armed ground crews—in realistic exercises that clearly demonstrated the sophistication and effectiveness of the attackers.

At present, most Western dedicated air base defense forces—such as the USAF SP—are expected to carry out reactive tactics, to engage and fight the best soldiers the other side has rather than to strike and repel them. Because of the location of most NATO air bases, their defenders are deprived of the depth of perimeter and suppressive firepower that proved so crucial in Vietnam for the containment of enemy penetrations. Nor are the defenders—with the exception of the RAF regiment—equipped with firepower and mobility to compensate for their lack of surprise, initiative, and depth. Although most Western air bases now are extremely vulnerable to ground attack, existing procedures provide a sound basis for the establishment of effective defenses.

Accumulating experience and expertise point to some key elements of air base defense. There is no substitute for the establishment of a centralized oversight headquarters that can ensure the most effective use of the existing assets. Reliable protection is achieved primarily through the establishment of realistic expectations from the defenses—namely, the speedy resurrection of air base operability and the containment and repelling of enemy penetrations—and the organization of the entire air base population to meet the challenge.

Only the combat training and arming of all the air base personnel will ensure the availability of defensive firepower throughout the base. In clashes with raiding forces, the dedicated defense forces, not the entire defending force, constitute the determining factor. Therefore, those dedicated forces are provided with the firepower and mobility required to overwhelm enemy intruders. Sensors, obstacles, and barriers can support and assist the defending forces, but they cannot replace or substitute for a maneuvering force.

Yossef Bodansky

SEE ALSO: Artillery, Tube; Missile, Surface-to-air; Mortar; Sensor Technology; Special Operations; Special Operations Forces; Vietnam and Indochina Wars; World War II.

GROUND RECONNAISSANCE

Ground reconnaissance, a military function and craft, is the examination or observation of an area or specific location to gain militarily significant information. Information gained can be of a tactical or a strategic nature and may include enemy strengths, activities, and dispositions; the nature of terrain; obstacles to movement; results of friendly actions; and likely routes of approach. Strategic ground reconnaissance may be aimed at learning a target population's attitudes, social conditions, economic and political activities, and receptivity to friendly forces. Ground reconnaissance may be accomplished by combat action, overt observation, or stealth; the latter method is often referred to as "scouting." Ground reconnaissance is of ancient origin and has always been important to military commanders. The absence of an adequate ground reconnaissance can be fatal to a plan's execution, a commander's performance, or the conduct of war itself. For much of recorded history, there was little distinction made between reconnaissance and what would later become known as intelligence. Although air and satellite reconnaissance have greatly extended the ability of military staffs to determine an area's status and the condition of an enemy force, ground reconnaissance is expected to retain its prominence and importance in military affairs.

Strategic and tactical ground reconnaissance differ in several ways. Tactical ground reconnaissance is almost always conducted just prior to or in the midst of combat operations. Strategic ground reconnaissance may be conducted during peacetime as well as during war. The two forms of ground reconnaissance also differ in scope. Strategic ground reconnaissance may reveal general demographic patterns, such as how a population is settled within a large terrain corridor. The tactical reconnaissance product may deal only with the disposition of buildings in a particular village or the relation of a road junction to a hill. Information produced by these two forms of reconnaissance is used in different ways. The results of strategic ground reconnaissance are likely to be used in assessments that can influence both policy choices and national strategic options. On the other hand, tactical information may be of use only to an infantry company commander. Normally, there is also a distinction in permanence; strategic products may remain current for decades while information derived from a tactical ground reconnaissance may only be useful for a matter of hours.

Ideally, information gained from ground reconnaissance is integrated with other information to produce a refined, final intelligence product. But on occasion, ground reconnaissance is the sole basis for action. Both strategic and tactical ground reconnaissance are often accompanied by efforts to acquire information through technical means: satellite surveillance systems with photographic, radar, infrared, and electronic intercept sensors; air photography; and supplementary sources such as human intelligence and research of pertinent literature. When all of this information is integrated, analyzed, and combined in a single document it is usually called "finished" intelligence. Therefore, the product of ground reconnaissance can be an integral part of finished intelligence, information ready for either the policymaker or the military field commander. However, the results of ground reconnaissance may be so timely, persuasive, and important that a commander may take dramatic action on the basis of one reconnaissance report alone. Such a report was made in 1945 providing American commanders with the surprising knowledge of light defenses around the Remagen Bridge over the Rhine River. Discovery of the German failure to destroy this bridge and the bold Allied action in exploiting this weakness probably shortened the Second World War in Europe by several weeks.

The very nature of ground reconnaissance—a trusted subordinate reporting from the actual scene of interest—can be far more convincing to a commander than a photograph that might not discriminate between actual weapons systems and decoys, an agent who could be operating for the other side, or an electronic surveillance report that may only reflect sound enemy communications discipline. For these reasons, ground reconnaissance will likely be valued over many other forms of intelligence gathering for years to come.

Although ground reconnaissance has its advantages, it also has some distinct drawbacks. In most cases, ground reconnaissance is a much slower process than air or satellite reconnaissance. In addition, the latter two forms can cover much wider expanses of territory and can place terrain and troop dispositions deep in the enemy rear under friendly surveillance. These areas are likely to be beyond the practical reach of friendly ground reconnaissance units. However, a sound ground reconnaissance performance can result in taking prisoners, thus gaining information about the enemy's will to fight. Furthermore, a skilled member of a reconnaissance unit may be able to determine enemy strengths, habits, and past activities on a battlefield simply by reading tracks, noting terrain marks, and observing refuse left by opposition forces. Ground reconnaissance forces may be used, on the one hand, to provoke enemy reactions, reactions that might reveal strengths and weaknesses as well as unit dispositions. On the other hand, a poorly executed ground reconnaissance can reveal one's own intentions. If an attacker concentrates his reconnaissance forces at a likely point of attack, the defender may determine his opponent's design in time to reinforce a critical weak point.

Through the years a number of patterns have developed around ground reconnaissance activities, and future shaping trends can be discerned. Several distinct types of ground reconnaissance have evolved, each with its own characteristics. The historic use of light cavalry to conduct reconnaissance was almost universal, but industrialized nations began replacing these units with armor organizations in the twentieth century. Standard, well-known defensive techniques have been established to deny an opponent the opportunity to gain potentially damaging information through ground reconnaissance. Thus this field of military lore has developed like others, a cat-and-mouse game of moves and countermoves. Eastern nations (China, Vietnam, etc.) have often concentrated reconnaissance missions, training, and capabilities in foot-mobile units. On occasion, they have accomplished combat reconnaissance by use of their clandestine intelligence organizations.

Several evolutionary trends are emerging. Until recently, ground reconnaissance was shifting toward units using helicopters. Helicopters, of course, speed the process of moving to and returning from a target area. Often these units land and recover reconnaissance troops who have performed their classic role on foot. However, helicopters are becoming increasingly vulnerable to shoulder-fired infantry antiaircraft missiles. Highly trained special operations forces are gradually being given more and more reconnaissance tasks, often of a strategic nature. In a different kind of change, the information explosion, growth in the knowledge industry, and ever-expanding computer-assisted means to manage information are making quiet research and mundane staff work an increasingly valuable tool in strategic ground reconnaissance.

Historical Background

The function of ground reconnaissance has been associated with military affairs and warfare since the earliest recorded times. The Biblical phrase for ground reconnaissance was "spying out the land." Moses used these words to describe the task for some of his followers in the land of Canaan. The word *reconnaissance* has evolved from the Old French word for "recognize"; this implies that the hearer of a reconnaissance report would know an area although he had never laid eyes on it. Early descriptions of warfare made no distinction between intelligence and reconnaissance. Indeed, the distinction was only firmly established in military literature during the latter part of the nineteenth century and the early part of the twentieth.

The simple requirements of receiving a general's instructions, proceeding to the area in question, and returning to render the anticipated report put a premium on speed. It is therefore not surprising that the reconnaissance craft has been associated with cavalry organizations. The Mongols were particularly adept at reconnaissance, sending specially picked riders hundreds of miles into an adversary's territory well in advance of an approaching army. The reports from these cavalrymen formed the primary basis for a Mongol commander's decision. This type of ground reconnaissance lasted well into the twentieth century. It was exhibited by camel-borne raiders coming from Saudi Arabia in formations of as many as 2,000 riders to pillage and loot Iraq as late as 1929. The techniques used by these warlike bedouin tribes included a general mounted reconnaissance to establish the locations of probable grazing areas, a more specific search maybe a month later, and a detailed reconnaissance the night before a raid on hapless shepherds and their flocks. These traditional reconnaissance techniques, witnessed and recorded by twentieth-century British military authorities, are believed to be little different from the methods used by reconnaissance elements of Genghis Khan's great army hundreds of years before.

Most modern-day distinctions among various types of tactical reconnaissance and counterreconnaissance techniques and the current fundamentals of strategic ground reconnaissance can be dated from the nineteenth century. Napoleon was noted for the organized collection of written material and maps. He regularly gave his ambassadors detailed tasks for acquiring specific information and he made systematic observations during his own travels. All of this data and information were developed on regions where Bonaparte believed future military campaigns might occur. In effect, he was conducting strategic ground reconnaissance and was rarely without a considerable wealth of vital information for contingencies. Later, military attaches, collection agencies, and large, complex intelligence staffs would duplicate the basic procedures Napoleon developed.

The Gettysburg campaign during the American Civil War provides examples of reconnaissance and counterreconnaissance techniques that today's professional soldier can easily recognize. In the early summer of 1863 the Confederate commander, Gen. Robert E. Lee, directed his cavalry commander, Gen. J. E. B. Stuart, to screen the initial Confederate move northward. Stuart executed his screening mission well, placing his horsemen between the Union army and the advancing Southerners. Stuart's cavalrymen largely prevented their own line of widespread outposts and patrols from being penetrated by Union reconnaissance forces. Stuart's screen thereby secured Lee's flank and provided him with information on Union movements. Despite the fact that the Union cavalry under Gen. Alfred Pleasonton had just been reorganized into balanced teams of cavalry, horse artillery, and rapidly moving infantry, the Union commander, Gen. George Meade, learned little of Lee's move from Pleasonton's force. However, hilltop Union observation and signal posts along Lee's route of advance provided Meade with another form of reconnaissance—battlefield observation. As a result of this and other sources, the Union commander was well advised of Lee's progress. When Stuart went on his futile and ill-advised raid toward Washington, Lee was deprived of a substantial part of his reconnaissance force. Moreover, Stuart's late arrival at the Gettysburg battlefield may have contributed to Lee's imperfect understanding of Union dispositions. The lack of an adequate reconnaissance is considered by some to be a major factor in Lee's defeat, and a better use of reconnaissance figured heavily in the survival of Meade's Army of the Potomac.

Modern Ground Reconnaissance

Twentieth-century reconnaissance forces perform many of the same tasks their forebears did, but their accouterments and weapons are quite different. Horse cavalry proved to be extremely vulnerable to modern firepower during World War I, a conflict that saw the rapid development of air reconnaissance. The airplane seemed to make ground reconnaissance a thing of the past. However, the radio-equipped tank and lightly armored wheeled vehicle were also created. With the growing reliability and improved cross-country mobility of these vehicles, ground reconnaissance reclaimed its importance on World War II battlefields.

It retained its prominence in the various wars of the Middle East, the Vietnam War, and beyond. Although some armies favor thin-skinned, speedy armored vehicles for reconnaissance tasks, others, like the U.S. Army, are partial to the use of combined arms teams with main battle tanks, self-propelled artillery, and mechanized infantry formations. This heavy type of organization is capable of conducting stand-up battles with substantial enemy formations, allowing American commanders to readily determine an opponent's actual strength and weapons positions. In addition, a hefty reconnaissance formation can easily brush aside or roll through light resistance, contributing to the rapid advance of the main body. Also, heavy armored cavalry forces can confuse an attacking opponent by portraying an apparent main line of a defense when, in fact, the attacker is actually probing a relatively thin cavalry screen, well in advance of the primary defense position.

The penalty for this type of weighty reconnaissance structure, of course, is that it is every bit as expensive, every bit as heavy and difficult to move strategically, and just as difficult to maintain as a standard armor or mechanized infantry formation. In the latter stages of the Vietnam War, however, when Hanoi increasingly sent regularly organized infantry units to South Vietnam, the heavily equipped American armored cavalry units proved particularly useful in a reconnaissance-in-force role. Ranging about the flat, thinly forested area north of Saigon known as War Zones C and D, and in the Central Highlands region of South Vietnam, these units covered vast areas of territory in a short amount of time, often discovering North Vietnamese forces. In most instances, the speed of reinforcement that these armor forces exhibited ensured that communist units were quickly outnumbered, outgunned, and overcome. In the rare cases where the North Vietnamese held the upper hand, the cavalrymen completed their reconnaissance-in-force role by maintaining contact with their enemies until artillery and air-supported infantry formations could be transported to the contact site.

Without a substantial industrial base to equip large armor formations, some Eastern Hemisphere nations became adept at foot-mobile reconnaissance. During the Korean War, China's army used "line crossers," reconnaissance troops who in some instances were dressed as South Korean peasants or soldiers. Line crossers infiltrated United Nations positions during darkness and returned to their own headquarters with detailed, tactically significant information. During the Vietnam War, North Vietnamese and Vietcong troops mimicked their Chinese counterparts but improved on the technique by recruiting and training local civilians, some of whom worked at U.S. or South Vietnamese installations. North Vietnamese Army sappers, especially, won the respect and admiration of their adversaries by their daring reconnaissance feats. Stripped almost naked, these brave soldiers would crawl through barbed wire and mine fields nightly, weeks before an attack, in order to accurately locate communications bunkers, ammunition stocks, and weapons positions. When the communist assault was finally launched, Americans were invariably surprised at the precision of North Vietnamese mortarmen and the unerring way the Northerners found the critical installations. This proficiency was, of course, attributable to well-executed, painstaking ground reconnaissance.

The Vietnam War also brought about wide-ranging developments in the use of helicopters for ground reconnaissance. Early in the war, Americans experimented with a technique that was initially known as the "eagle flight." In time, the eagle flight was transformed into a standard U.S. Army organization with well-tested doctrine. These organizations, "aero scout" platoons, usually had three helicopter-borne rifle squads with accompanying helicopters armed with machine guns and rockets. Rifle squad members were often landed or let down by ropes into difficult, heavily forested terrain to conduct detailed searches while the armed helicopters hovered overhead, ready to assist the foot soldiers should an enemy unit be discovered. Upon finding an enemy unit, the infantrymen would try to maintain contact with their adversaries. If outnumbered, the scouts had to break off the engagement and be recovered by transport helicopters. The aero scout platoon leader then called in larger infantry units, air strikes, and artillery fire. These units were perhaps misnamed as scout units because they rarely gained information through stealth.

Even less stealthy was the use of the armed helicopters in a reconnaissance-by-fire role. In this type of action, air cavalry units would fly over a suspected enemy area and fire their rockets or machine guns at likely hiding places, hoping to draw return fire from communist units. Occasionally, Vietcong and North Vietnamese units would rise to the bait, firing at the probing armed helicopters. Normally this type of reaction came from an erroneous belief on the part of the communists that their hiding spots had been discovered by the Americans. On being fired upon, the air cavalrymen would bring in enormous amounts of firepower.

Americans have considerable twentieth-century experience in classic ground reconnaissance using stealth, particularly in Asia. The World War II Burma campaign was characterized by a brilliant use of reconnaissance on the part of the Office of Strategic Services Detachment 101. Operating deep in the Burmese jungle, the detachment provided allied air forces with detailed, accurate information on heavily camouflaged and concealed Japanese troop dispositions and facilities. During the Korean War, U.S. Army units employed and directed upwards of 20,000 Koreans mostly in intelligence-gathering roles deep in the rear of Chinese and North Korean lines. Based on islands off the coasts of North Korea, these organizations supplied UN forces with a steady stream of target information for naval bombardment and air strikes and produced hundreds of prisoners of war for interrogation by American and South Korean intelligence officers.

This traditional type of reconnaissance role was repeated during the Vietnam War. From 1962 until 1972, American Special Forces units recruited, trained, and led thousands of Southeast Asians on reconnaissance missions. Normally, the tasks for these units of mixed nationality involved only the acquisition of information, not combat. In these cases, weapons were used only in self-defense, a means to break contact and escape capture. The steady expansion of these Special Forces reconnaissance units was accompanied by the battlefield activation of American Long-Range Reconnaissance Patrol units. Operating under the control of corps headquarters, these all-U.S. forces provided corps commanders with timely information on communist dispositions in the latter phases of the war.

Some of the world's special operations forces are trained and occasionally used in strategic ground reconnaissance. While most reconnaissance patrols led by the American Special Forces during the Vietnam War were gathering information in South Vietnam for tactical purposes, others were employed in more of a strategic role. For instance, U.S. Special Forces teams assigned to the Studies and Observation Group of the U.S. Military Assistance Command Vietnam (MACV-SOG) conducted extended-duration missions in Laos, defining the extent, organization, and functioning of the Ho Chi Minh Trail, Hanoi's primary line of communication to South Vietnam. These reconnaissance teams were brought to Laos either by parachute or by helicopter and their explorations along the trail complex sometimes lasted for as much as two weeks. During the Falklands/Malvinas War, the British Special Air Service (SAS) was employed in a somewhat different fashion, keeping Argentine forces under constant surveillance prior to the start of British landings and the subsequent recapture of the islands. Although details are sketchy, the former Soviet Spetsnaz forces are believed to have had a strategic ground reconnaissance role among their other duties. Members of these units are reported to have visited Allied installations in Western Europe, gathering information for their wartime tasks. Targeted facilities were believed to include command and control bunkers, ammunition storage sites, North Atlantic Treaty Organization missile units, and critical airfields.

While strategic ground reconnaissance may on occasion be accomplished by dramatic and daring operations in war, most of the world's activity in this esoteric field is carried on in peace by large bureaucracies. Both military attaches and the employees of civilian intelligence services are daily engaged in acquiring volumes of information of military utility in potentially hostile countries or in regions where combat operations might be anticipated. The type of information collected includes details about airfields—details that would be of use whether the facility were to be either used or destroyed. For example, it is important to know the thickness of the concrete forming the runways and taxiways. Knowledge of that thickness is essential so that appropriate ordnance can be chosen for the airfield's destruction; it is equally important when a plan involves the facility's capture and use, because insubstantial construction might rule out the basing of heavy cargo aircraft. Ground commanders need to know the availability and volume of potable water supplies in a region. The construction and security arrangements of headquarters, storage sites, port facilities, and other such facilities have to be examined and calculated or estimated. The demands of amphibious operations require the examination of beaches for a determination of slopes, soil and sand characteristics, navigation hazards, and typical wave heights. Bridges must be examined for their ability to support heavy armored vehicles and logistic traffic. All of this information and much more are being collected daily by many nations conducting strategic ground reconnaissance in preparation for military operations.

Principles of Tactical Ground Combat Reconnaissance

While the armed forces of some nations may proffer little attention to formal principles of ground tactical reconnaissance, there is general agreement among professionals on what leads to the successful performance of such missions. These principles include: orientation on the objective, rapid and accurate reporting, retention of the freedom to maneuver, gaining and maintaining contact, and rapid development of a contact. Of course, the principles do not apply to reconnaissance missions conducted with stealth, the true scouting tasks.

Orient on the objective. This principle merely states the seemingly obvious necessity of remaining fixed on the physical object of the mission. If the reconnaissance mission is to determine the suitability of a road as an avenue of approach, the roadway, its bridges, and the surrounding commanding terrain must be thoroughly examined for obstacles or enemy forces. However, if the mission deals with the location and description of an enemy force, a reconnaissance unit must orient on the opponent's formation even if it is moving. The reconnaissance leader must position his unit where it can best observe the enemy and survive.

Develop rapid and accurate reporting. Although reconnaissance is a broad function, its prime purpose is usually the development of information, and information is useless unless it is presented to the decision maker in an accurate and timely fashion. Complete reconnaissance reports normally answer the questions: who, what, when, where, and how.

Retain the freedom to maneuver. This principle is a reminder for reconnaissance leaders to position their units so that escape can be effected if a superior enemy force is encountered. Normally, a reconnaissance mission does not include decisive engagements. The purpose is most often to gather information so other friendly elements can win in battle.

Gain and maintain contact. This rule of thumb applies only to reconnaissance units whose task is to locate and temporarily fix enemy forces so that other, more powerful friendly forces can be brought to bear.

Develop the situation rapidly. The emphasis here is to make the maximum use of time and resources. Combat reconnaissance leaders should not be deterred by distractions, minor obstacles, or scanty resistance.

Types of Tactical Ground Reconnaissance Operations

There are a number of different types and forms of ground reconnaissance operations. It is not surprising that some types include ancillary tasks of preventing an enemy's reconnaissance of friendly forces.

Scouting. Usually accomplished on foot, scouting involves obtaining information by stealth. Ideally, the scout sees without being seen. And, since the word *scout* stems from an Old French word meaning "to listen," a scout presumably listens without being heard.

Route reconnaissance. Armored cavalry units are well suited for this task. The product of the route reconnaissance is detailed information on every possible obstacle to the use of a route by a friendly force. The route may be a road or a general direction of advance. Obviously, enemy strengths, unit compositions, and locations are of vital importance, but so too are bridge capacities and unoccupied but commanding terrain features.

Screening. This reconnaissance task is usually accomplished by placing outposts or widely separated units in front or on the flank of a larger force. The purpose is to prevent the larger force from being surprised. It might be accomplished either during movement or in static situations. One task of the screening force is to prevent enemy reconnaissance forces from penetrating the screen, thereby depriving an opponent of useful information. A large screening force, perhaps two or more armored cavalry regiments, might be called a covering force.

Zone reconnaissance. Considered by many professionals to be the most time-consuming and painstaking type of reconnaissance, zone reconnaissance requires the reconnaissance unit to examine all ground intervening two boundaries, usually terrain features such as rivers or ridgelines. Reports concern roads, enemy activity, obstacles, and other pertinent information for the entire designated area.

Area reconnaissance. Area reconnaissance examines a particular region, often a suspicious town site or forest. This type of reconnaissance is often conducted by dismounted armored cavalry units or the infantry squads of an aero scout platoon.

Commander's reconnaissance. This is the reconnaissance of an area, route, objective, or prospective defensive position by a unit commander.

Engineer reconnaissance. This type of reconnaissance is usually accomplished in friendly held territory. Military engineers conduct reconnaissance operations to assess future construction work or road or bridge repair, or to gather information to advise a commander on the tactical or logistical feasibility of a course of action.

Reconnaissance in force. A reconnaissance operation conducted in force seeks to obtain information and to challenge and test enemy forces. Units conducting this type of mission are normally prepared to seize the initiative and capture critical terrain, overpower light opposition, and exploit success. Along a broad front, a reconnaissance in force may consist of simultaneous, strong probes at several points. The great danger in this kind of reconnaissance is that the probing force can be trapped, pinned down in decisive combat, and destroyed before help arrives.

River reconnaissance. A mission of this type usually involves determination of likely crossing sites. The emphasis here is on steepness of banks, structure of the river bottom to identify shallow, fordable bodies of water, and typical current speeds. In the latter case, it should be noted that knowledge of the average current speeds of European rivers has been instrumental in the design of amphibious combat vehicles.

Beach reconnaissance. A sound beach reconnaissance is a time-consuming task, but skilled reconnaissance teams have conducted this type of mission during the hours of darkness and under the noses of enemy forces. The purpose is to gain critical information for possible amphibious operations. The slope and composition of the beach, wave heights, underwater obstacles, nearby commanding terrain, enemy reinforcement times, and other types of information are vital to the work of amphibious warfare planners.

The Future of Ground Reconnaissance

Although it is impossible to predict the future, some distinct trends can be discerned in the realm of strategic reconnaissance; yet at the same time, the development of tactical reconnaissance seems to be reaching a crossroads. In the main, the future of strategic ground reconnaissance appears to lie in the continued evolution of special operations forces and increasingly effective staff research and information management. The growth of both civilian and military intelligence bureaucracies will probably continue, and that growth will undoubtedly be accompanied by mounting data banks of militarily significant information. It is unlikely that acquisition efforts will diminish; thus each region of the world that can conceivably be exposed to military operations will be subjected to constant scrutiny by overt and covert collection agents from many countries. Concurrently, the globe is undergoing a rapidly accelerating explosion of information: books, magazines, videotape, travel tips, travelers' tales, maps, interviews, and news reports. These are only some of the apparently countless sources in the information flood. Some knowledge industry products are useful to military staffs interested in terrain, targets, and facilities—all part and parcel of strategic ground reconnaissance. Thus, the collection, analysis, and cataloging of open source information is likely to continue expanding. This information revolution has developed side by side with the steadily improving computer-assisted information management field. The latter greatly facilitates efficient storage and responsive retrieval of militarily useful information. The desktop publishing phenomenon, together with quick computer-drafted graphics, makes rapid publication of easily understood intelligence products feasible. Accordingly, this rather pedestrian, staff-labor-produced part of strategic ground reconnaissance can be expected to grow in importance, use, and value.

Another aspect of strategic ground reconnaissance, the use of special operations forces to acquire important concealed or protected information, may also be expected to grow. In part, the reason for this growth is that the distinction between war and peace will probably remain blurred. Declarations of war seem to be a subject for historical study, so some military tasks, including strategic ground reconnaissance, are constantly underway. For example, Soviet Spetsnaz troops were probably not the only special operations forces who visit their planned wartime targets wearing civilian attire. This condition, a legacy of the Cold War, has resulted in the virtual elimination of another distinction, the once-sharp division between soldier and clandestine agent. The use of special operations forces to conduct strategic ground reconnaissance missions either in combat or before fighting has begun can be expected to continue apace.

The expansion of strategic ground reconnaissance may be accompanied by a reduction in mounted tactical ground reconnaissance. The primary reason for this possibility is that the tools for such reconnaissance—tanks and helicopters—are becoming increasingly vulnerable to hand-held antitank and antiaircraft missiles. From their initial battlefield appearance in the 1970s, shoulder-fired or man-portable, passive seeker missiles have rapidly gained effectiveness. In the late 1990s and the early years of the twenty-first century, these weapons can feature spoof-proof multiple seekers, tandem warheads, and variable flight profiles. The coming infantry antitank missile, for example, may seek its prey by sensing heat, millimeter wave emissions, acoustic signals, or magnetic distortions. The path of this missile will probably be programmed so that it weaves its way to the target, spoiling the aim of counterweapons. It will likely maneuver for a top or rear attack approach in the last few meters of flight, depending on the thickness of the tank's armor. This approach selection will probably be a gunner's option.

The foot soldier may also be equipped with shoulder-fired antiaircraft weapons of vastly improved effectiveness. Some idea of the impact of these weapons on a battlefield was glimpsed during the war in Afghanistan. The weapon used there, the American Stinger, was a relatively early and crude model. By the mid-1990s, the U.S. Stinger and its Russian equivalent may have more than just simple infrared sensors, and they will probably be able to discriminate between decoy flares and their intended targets.

These approaching innovations may shift the emphasis of tactical ground reconnaissance back to foot-mobile scouts and away from mounted reconnaissance elements. Whatever the future brings, the importance of ground reconnaissance in all its forms can be expected to continue well into the twenty-first century and beyond.

ROD PASCHALL

SEE ALSO: Air Reconnaissance; Cavalry; Fog of War; Intelligence, Tactical; Principles of War; Tactics.

Bibliography

Burgess, W. H. 1990. *Inside Spetsnaz: Soviet special operations*. Novato, Calif.: Presidio Press.
Chambers, J. 1976. *The devil's horsemen: The Mongol invasion of Europe*. New York: Atheneum.
Chandler, D. 1966. *The campaigns of Napoleon: The mind and method of history's greatest soldier*. New York: Macmillan.
Cleaver, F., et. al. 1955. *UN partisan warfare in Korea, 1951–1954*. Baltimore, Md.: Johns Hopkins Univ. Press.
Davidson, P. B. 1988. *Vietnam at war: The history*. Novato, Calif.: Presidio Press.
England, J. W. 1987. *Long range patrol operations: Reconnaissance, combat and special operations*. Boulder, Colo.: Paladin Press.
Gavin, J. 1955. Cavalry and I Don't Mean Horses. *Armor* 63 (3).
Glubb, J. B. 1960. *War in the desert: An RAF frontier campaign*. New York: W.W. Norton.
Gugeler, R. A. 1954. *Combat actions in Korea*. Washington, D.C.: Association of the U.S. Army.
Hastings, M., and S. Jenkins. 1983. *The battle for the Falklands*. New York: W.W. Norton.
Jones, A. 1987. *The art of war in the Western world*. Urbana, Ill.: Univ. of Illinois Press.
Peers, W. R., and D. Brelis. 1963. *Behind the Burma road: The story of America's most successful guerrilla force*. Boston: Little, Brown.
Pike, D. 1986. *PAVN: People's Army of Vietnam*. Novato, Calif.: Presidio Press.
Stanton, S. L. 1985. *Green Berets at War: U.S. Army Special Forces in Southeast Asia, 1956–1975*. Novato, Calif.: Presidio Press.

Stevens, P. H. 1969. *Search out the land: A history of American military scouts.* Chicago: Rand McNally.
Sumner, E. M. 1944. *Modern reconnaissance: A collection of articles from the Cavalry Journal.* Harrisburg, Pa.: Military Service.
U.S. Army. 1981. *Field manual 17–95: Cavalry.* Washington, D.C.: Government Printing Office.
———. 1982. *Field manual 17–35: Aero scout procedures.* Washington, D.C.: Government Printing Office.
———. 1986. *Field manual 100–5: Operations.* Washington, D.C.: Government Printing Office.
Wagner, A. L. 1893. *The service of security and information.* Kansas City, Kans.: Hudson-Kimberly.

GUATEMALA, REPUBLIC OF

Guatemala is the northernmost of the six Central American republics. Before the arrival of Europeans, this mountainous area was the heartland of the brilliant civilization of the classic Mayan between 250 B.C. and A.D. 800.

Power Potential Statistics

Area: 108,890 square kilometers (42,042 sq. mi.)
Population: 9,329,200
Total Active Armed Forces: 39,600 (0.425% of pop.)
Gross Domestic Product: US$11.1 billion (1990 est.)
Annual Defense Expenditure: US$113 million (1% of GDP, 1990 est.)
Iron and Steel Production:
 Crude steel: 0.009 million metric tons (1986)
Fuel Production:
 Crude oil: 0.255 million metric tons (1986)
 Natural gas: 1.9 million cubic meters (1986)
Electrical Power Output: 2,594 million kwh (1990)
Merchant Marine: 1 vessel; 4,129 gross registered tons
Civil Air Fleet: 10 major transport aircraft; 381 usable airfields (11 with permanent-surface runways); none with runways over 3,659 meters (12,000 ft.); 3 with runways 2,440–3,659 meters (8,000–12,000 ft.); 19 with runways 1,220–2,440 meters (4,000–8,000 ft.).

For the most recent information, the reader may refer to the following annual publications:
The Military Balance. International Institute for Strategic Studies. London: Brassey's (UK).
The Statesman's Year-Book. New York: St. Martin's Press.
The World Factbook. Central Intelligence Agency. Washington, D.C.: Brassey's (US).

History

To a great extent, Guatemala's history has been the product of the division of Guatemalan society into two nearly equal parts: the Spanish-speaking ladinos (Indians, often of mixed Spanish and Indian blood, who have adopted Spanish customs and language) and the rest of the Indian population, mostly Mayans, who are generally farmers and agricultural laborers, illiterate, and often desperately poor. This bifurcated society is a direct legacy of Guatemala's conquest by Spanish conquistadors in the early sixteenth century, a society in which the dominant ladino element has often exploited the Indians.

Guatemala proclaimed its independence from Spain on 15 September 1821 in the aftermath of the Spanish revolution of 1820. Guatemala was at first one of the member provinces of the Federation of Central America, but in 1838 it broke away and became fully independent. During much of Guatemala's history, the country has been dominated by a series of strongmen and dictators; the periods of peaceful rule by law have been few and all too brief. The policies of these autocrats have varied; some, like Justo Rufino Barrios and Jorge Ubico, were honest and dedicated to Guatemalan national welfare. Others, like Manuel Estrada Cabrera, were unscrupulous and greedy.

The fall of the Ubico regime in 1944 led, after the presidency of Juan José Arévalo, to the election of former army officer Jacobo Arbenz Guzmán in 1950. Arbenz, a progressive who favored land reform and the nationalization of foreign-owned industries, alarmed the U.S. government. The American press unjustly painted him a Communist, in part for threatening the holdings of the United Fruit Company and the Industrial Railways of Central America (IRCA). Isolated from the United States, he turned to other sources of support, including the Soviet Union, but was overthrown in June 1954 by an invading exile army based in Honduras and supported by the U.S. government.

Alarmed at guerrilla successes and disgusted by government policy, progressive army officers overthrew the last in a series of venal general-dictators, Fernando Romeo Lucas García, in March 1982, and installed as president Gen. José Efraín Ríos Montt (who had lost a fraudulent election in 1974). Ríos Montt peacefully yielded power to Marco Vinicio Cerezo Arévalo in January 1986, following free elections (which Ríos Montt and the army had organized) in November and December 1985. At the end of Cerezo's term, Jorge Serrano was elected president in 1990.

Politico-Military Background and Policy

Despite the aura of democracy afforded by elections, the army has dominated Guatemalan politics for a decade. An ongoing Marxist guerrilla movement in the countryside has resulted in an often brutal counterinsurgency campaign, mostly directed at the Indian population, in which the army has taken politically unpopular steps. The army's relationship to the rest of the Guatemalan body politic is further complicated by the activities of right-wing "death squads," often either sponsored by the army or composed of military personnel. These groups have killed not only perceived guerrilla sympathizers but also labor leaders, teachers, students, lawyers, priests, peasant organizers, and others presumed to harbor sympathies for the left. Although the government maintains that this activity has ceased, such groups remain active, and many people still "disappear."

In theory, all Guatemalan males reaching age 18 are liable for 30 months of military service. Since over 80,000 men reach military age each year, and since the armed forces need only about one-eighth that number, military service is in fact selective. The early 1980s saw reports of army "press gangs" in rural areas forcibly recruiting Indian teenagers, in part a reflection of the relative unpopularity of military service among ladinos. This practice has fallen off since the 1985 election.

Strategic Problems

Aside from the ongoing guerrilla war inside Guatemala, the country faces relatively few security problems. There is, however, an ongoing territorial dispute with Belize, formerly British Honduras. Guatemala claims sovereignty over Belize, and moved troops to the border in 1972 and 1977. However, the presence of British troops in Belize has tempered Guatemalan moves in that direction.

Ex-President Cerezo worked diligently but with mixed success for a general peace settlement in Central America. In 1987, he was the architect of a proposed compromise accord for Nicaragua, and this concern for regional security is a noteworthy development.

Military Assistance and Armaments Supply

Until the late 1970s, the United States provided Guatemala with extensive military aid. Between 1950 and 1980, Guatemala received US$27.7 million in foreign military sales, coupled with US$16.2 million through the Military Assistance Program. These levels dropped in response to repressive measures undertaken by the Lucas García government but have resumed, especially after President Cerezo took office in early 1986, reaching a total of US$9.4 million in 1988. The bulk of Guatemala's materiel comes from the United States, although some items are bought from Israeli and Swiss sources. All commissioned officers are graduates of the Escuela Polytécnica (Polytechnical School) and many have studied at military schools in France, West Germany, Italy, Spain, and especially the United States.

Alliances

Guatemala is a member of the United Nations, the Organization of American States (OAS), and the Organization of Central American States (ODECA). It has no formal military alliance with the United States, but like most of its neighbors has extensive informal defense arrangements.

Defense Structure

The armed forces are officially titled Ejército de Guatemala, or Army of Guatemala. The army administers and provides logistical support for the air force and the navy, although the air force has some autonomy. The president is the commander in chief of the army, and exercises his control through the minister of defense.

(For an explanation of the abbreviations and symbols used in the following section of military statistics, see the list of Abbreviations and Acronyms in Volume I.)

Total Armed Forces

(National armed forces are combined; the army provides log spt for navy and air force.)
Active: 39,600. Terms of service: Conscription; selective, 30 months.
Reserves: Army 4,500 (trained), Navy (some), Air 200.

ARMY: 37,000
HQ: 19 Military Zones.
3 Strategic Reserve bde (each 2 bn).
39 inf bn.
1 engr bn.
Equipment:
Light tanks: 10 M-41A3.
Recce:† 10 M-8, 10 RBY-1;
APC: 9 M-113, 7 V-150 Commando, 18 Armadillo.
Towed arty: 75mm: 10 M-116; 105mm: 4 M-101, 8 M-102, 56 M-56.
Mortars: 81mm: 55 M-1; 107mm: 12 M-30; 120mm: 18 ECIA.
RL: 89mm: 3.5-in. M-20.
RCL: 106mm: M-40A1.
AD guns: 20mm: 20 incl 12 Oerlikon;

NAVY: 1,200 incl 600 marines. Bases: Puerto Barrios, Santo Tomás de Castillas (Atlantic), Puerto Quetzal, Puerto San José (Pacific)
Patrol Craft, Inshore 8: 1 Kukulkan† (US 'Broadsword' 32-m) PFI, 7 (plus some 17 boats.
Amphibious: craft only.

Marines: (600); 1 bn.

AIR FORCE:
1,400; 16 cbt ac, 6 armed hel. Serviceability of ac is perhaps less than 50%.
COIN: 1 sqn with 8 Cessna A-37B, 8 PC-7.
Attack helicopters: 3 Bell 212, 3 Bell 412.
Transport: 1 sqn with 6 C-47, 3 F-27, 2 G-222, 6 IAI-201, 1 Super King Air (VIP).
Liaison: 1 sqn with 3 Cessna 170, 1 -172.
Helicopters: 1 sqn with 9 Bell 206, 5 UH-1D/-H, 3 S-76.
Training: 6 T-41
Tactical Security Group: 3 coy; 4 M-3A1 White scout cars.

PARAMILITARY
National Police: 8,000.
Treasury Police: 2,100.
Territorial Militia (R) (CVDC): est. 500,000.

OPPOSITION
Unidad Revolucionaria Nacional Guatemalteca (URNG): some 900–1,000; coalition of 4 groups: Ejército Guerrillero de los Pobres (EGP): 300. Partido Guatemalteco del Trabajo (PGT). Fuerzas Armadas Rebeldes (FAR): 300. Organización del Pueblo en Armas (ORPA): 400.

Future

Guatemala will not achieve real stability and internal order until the socioeconomic problems of the nation are solved (or at least treated) and the violent excesses of both

Guatemala, Republic of

left-wing and right-wing radicals are curbed. That achievement will require a strong, self-confident, and popular civilian government and a depoliticized military. The election of President Cerezo has been a worthy beginning, but there is still much to be done.

DAVID L. BONGARD
TREVOR N. DUPUY

SEE ALSO: Central America and the Caribbean; South America.

Bibliography

American University. 1983. *Guatemala: A country study.* Washington, D.C.: Government Printing Office.
Hunter, B., ed. 1991. *The statesman's year-book, 1991–92.* New York: St. Martin's Press.
International Institute for Strategic Studies. 1991. *The military balance, 1991–1992.* London: Brassey's.
Schlesinger, S., and S. Kinzer. 1982. *Bitter fruit: The untold story of the American coup in Guatemala.* Garden City, N.Y.: Doubleday.

GUDERIAN, HEINZ [1888–1951]

Heinz Guderian was born on 17 June 1888 in Kulm, East Prussia. His father was an army officer and an aristocrat. In 1900 Heinz entered the cadet school at Karlsruhe in Baden, transferring in 1903 to the Main Cadet School at Gross Lichterfeld outside Berlin.

An eager and highly intelligent student, Heinz easily passed his cadet training and was assigned to the 100th Hannover Jaeger Battalion, then his father's command. On 27 January 1908, he was commissioned a lieutenant in the battalion.

In 1912 Guderian was selected to join a newly formed radio company. During this assignment Guderian learned to appreciate the importance of effective radio communications, and the flexibility they provided in the exercise of command.

In 1913 Guderian was selected to attend the War College. In October of that same year he married Margarete Goerne. The War College was closed upon the outbreak of war in 1914, and Guderian was given command of a wireless station in the headquarters of the 5th Cavalry Division in the Third Army.

World War I

In April 1915 Guderian was transferred to the headquarters of the Fourth Army to serve as assistant signals officer. He was promoted to captain and, in April 1917, was transferred to the General Staff of the 4th Infantry Division in Champagne, France.

Except for one month when he commanded the 2d Battalion, 14th Infantry Regiment, Guderian continued to hold staff positions from division to army level throughout the war. He was assigned in February 1918 to the Army General Staff and in October to the prestigious Operations Branch as general staff officer.

Interwar Years

In May 1919 Guderian was assigned to the staff of the "Iron Division" in Latvia. The "Iron Division" was a Freikorps unit involved in halting Bolshevik expansion from Russia into Latvia and East Prussia. In June, after the withdrawal of German troops from the Baltic states, Guderian returned to Hannover and a staff appointment with his old battalion. At the end of 1919 he was selected as an officer in the newly formed Reichswehr, the small 100,000-man defensive force allowed Germany under the terms of the Versailles Treaty.

In January 1922 Guderian was assigned to Reichswehr headquarters and posted to the new Office of the Inspectorate of Transport Troops. He threw himself into the job with his customary eagerness and intensity. Along with the study of motorized troops, Guderian developed an interest in the tank and read all the available works on tank development and tactics, most notably those by Fuller, Liddell-Hart, and Martel.

Guderian quickly made himself a leading tank expert. He began to supplement his income as well as to disseminate his theories by writing articles for various military periodicals on the uses of tanks and troop transports on the battlefield.

In October 1924 Guderian was assigned to the 2d Division as instructor of tactics and military history. He used his new position to develop further his concepts of mobile warfare. In 1927 he was promoted to major and in October was reassigned to Reichswehr headquarters. Posted back to the Transport Section, he was given the specific tasks of studying motorized troop transport and of instructing members of his section in the operational uses of tanks.

Guderian worked zealously at developing an armored (panzer) force for the army. In February 1930 he was given command of a motor transport battalion and immediately began transforming it into an armored reconnaissance battalion. Despite resistance to his ideas, Guderian persevered in his belief that fast, mobile armored units were the wave of the future. In October 1931 he was promoted to lieutenant colonel and offered the post of chief of staff by his old friend, General Lutz, now inspector of Transport Troops. Lutz had always been a supporter of Guderian's ideas, and the two of them began to work toward not only the development of a panzer force but also its acceptance by the High Command.

Under Lutz, Guderian was given free reign to experiment and to develop his theories of mobile warfare. By the time of Hitler's accession to power in 1933, Guderian had already laid the groundwork for the building of Germany's panzer forces. By 1934 the first panzer division was created and Guderian was promoted to colonel and made chief of staff to the Commander in Chief, Panzer Troops. By 1935 three panzer divisions had been created and Guderian, although still only a colonel, was given command of the 2d Panzer Division at Wuerzburg.

In 1936 Guderian's book, *Achtung! Panzer*, was published. In it he detailed his theories on using massed armored formations, supported by artillery and air attacks, to breach the enemy's lines and then to fan out into their rear areas. Once through, the panzers would use their mobility to exploit the breach further, while accompanying motorized infantry would clear the breach and hold it open for the follow-on forces. Guderian emphasized the importance of panzer mobility. As long as the panzers kept moving, they could keep the enemy off balance and deny it a chance to concentrate.

Later in 1936 Guderian was promoted to *generalmajor* (brigadier general). In 1938, as a result of the Blomberg-Fritsch crisis, Lutz was forced into retirement by Hitler. Guderian was then promoted to *generalleutnant* (major general) and made acting commander of the XIII Corps. In November he was promoted to *general* (lieutenant general) of Panzer Troops and succeeded Lutz as commander in chief of Mobile Troops.

By January 1939 Guderian had five panzer divisions operational and a sixth on the way. He was also preparing the four light divisions for quick conversion into panzer divisions if the need should arise. Shortly afterward, Guderian was finally given the command he desired most: commander in chief, Panzer Troops.

World War II

On 22 August 1939, Guderian was given command of the XIX Corps for the planned invasion of Poland. The XIX Corps was composed of the 3d Panzer and the 2d and 20th Motorized divisions and had the important role of spearheading the German drive across the neck of the Polish Corridor. The German offensive opened on 1 September, and Guderian, leading his corps from the front, proceeded to put his theories to the test. By the evening of 4 September, Guderian's troops had crossed the corridor and by the 16th had taken Brest-Litovsk, inside the allocated Russian zone of influence, effectively ending the campaign.

On touring the battlefield, Hitler was amazed and impressed by the destruction wrought by the panzers. Guderian had proven his case for the effectiveness of massed, highly mobile panzer forces. He was awarded the Knight's Cross in October during the victory celebrations in Berlin.

During the preparation and planning stages of Case Yellow (code name for the planned invasion of France) Guderian helped Gen. Eric von Manstein, Chief of Staff of Army Group A, persuade Field Marshal Gerd von Runstedt, Army Group A commander, of the soundness of a surprise armored thrust through the Ardennes Forest. Guderian also encouraged Manstein to approach Hitler with this strategic concept.

Upon acceptance of Manstein's plan, Guderian's XIX Corps was assigned the difficult operation of spearheading the drive across the Meuse River. The XIX Corps was now composed of the 1st, 2d, and 10th Panzer divisions and the Grossdeutschland Regiment, whose battalions alternately served as Hitler's personal escort. Guderian's corps was one of three that formed Panzer Group Kleist. Kleist's group contained five panzer divisions—half the entire German panzer force.

On 10 May, Guderian's corps drove through the Ardennes in eastern Belgium and smashed into the French lines. By the 13th he had taken Sedan and forced a crossing of the Meuse. On the 20th his panzers reached Abbeville on the French coast and by the 23d had taken Calais and Boulogne before receiving Hitler's order to halt. Guderian was furious at not being able to finish the destruction of the French and British troops, now trapped with their backs to the sea, but Hitler was adamant. When the Germans renewed their offensive southward into France on 5 June, Guderian drove to the Swiss border before halting again. Guderian's success was a direct result of his tactics of mobility, command from the front, and an excellent working relationship with the Luftwaffe unit that supported the panzer group.

The French surrendered unconditionally on 22 June; for his part in the victory Guderian was promoted to *generaloberst* (general) on 19 July. Guderian then returned to the task of raising and training new panzer divisions. Within the year he had increased the number of panzer divisions to twenty but had not increased the strength of the panzer force. Much to Guderian's displeasure Hitler had ordered the creation of the new divisions but had equipped them by taking panzers from the existing divisions. This was to become a standard practice for Hitler. Combined with the inability of the German automotive industry to keep pace with the necessary production requirements, it severely weakened the further development of the panzer forces.

At the beginning of the Russian Campaign (22 June 1941) the Wehrmacht boasted four panzer groups. Guderian was given command of the 2d Panzergroup (later called Panzerarmee Guderian and then 2d Panzer Armee). This, with General Hoth's 4th Panzergroup, formed the spearhead of General von Bock's Army Group Center.

Guderian launched his panzers into Russia with the same skill and tenacity as in Poland and France. Forming a pincer with Hoth's corps, Guderian raced across Russia, slicing deep into enemy territory. Guderian and Hoth met on 27 June outside Minsk, having covered more than 320 kilometers (200 mi.) in five days and encircling nearly 300,000 Russian troops. They repeated this tactic at Smolensk on 27 July, trapping another 100,000 Russians.

In August, Hitler shifted the axis of advance from Army

Group Center to Rundstedt's Army Group South. Guderian was incensed that Hitler could not see the ramifications of not pushing on to take Moscow and only against his better judgment moved his army south to support the drive on Kiev. On 15 September, Guderian linked up with Kleist near Lokhvista, completing the encirclement of over 600,000 Russians.

Guderian was then ordered to turn north and resume the advance toward Moscow. Although he succeeded in penetrating south beyond Moscow, the onset of winter and the exhaustion of both men and equipment convinced Guderian that the city could not be taken and that the army's only recourse was to retreat to prepared winter positions. Hitler, however, ordered the army to stand fast regardless of the losses from exposure. Guderian, Rundstedt, and Bock opposed this idea and withdrew their forces in spite of Hitler's orders. On 26 December, Guderian and the others were relieved of their commands and assigned to the reserve officers' pool.

Guderian remained literally unemployed until February 1943 when Hitler, realizing that the military situation demanded all of his experienced officers to stem the tide of German reversal, recalled him to act as inspector general of Panzer Troops. Guderian immediately buried himself in the duties of his new position which involved the training, organization, and development of the panzer forces.

After the abortive assassination attempt on Hitler's life on 20 July 1944, Guderian was appointed to replace General Zietzler as chief of the General Staff of the Army High Command (OKH) on 21 July. One of his first duties was to act as chairman of the army tribunal that investigated those implicated in the assassination conspiracy. The tribunal had the responsibility of turning over hundreds of the accused to a People's Court for trial and summary execution. Guderian found the entire affair disgusting but was virtually trapped by his position as chief of staff.

He did use his new position to fight for an increase of troop strength on the eastern front. Guderian predicted that the Russians would launch a major offensive that would result in a collapse of the German lines if they were caught unprepared. He spared no pains to challenge Hitler on this issue whenever necessary. When the Russian offensive began in December, Guderian's predictions came true. With the Allies closing in on Germany from east and west, Guderian urged Hitler to make peace with the western powers. Hitler had had many violent disagreements with Guderian in the past but this was too much. He dismissed Guderian for reasons of failing health on 21 March 1945. Guderian retired to the Tyrol to await the end and was taken prisoner by American forces on 10 May. He was held in captivity as a war criminal until June 1947 when the charges against him were dropped and he was released.

In 1951 his memoirs, *Errinerungen eines Soldaten*, were published. In them he brutally lambasted the dilettante mentality of the High Command that resulted in Germany's defeats on the eastern front. Guderian died on 15 May 1951 in Schwangau bei Fussen.

A brilliant tactician, organizer, and commander, Guderian was the mastermind behind the creation of the panzer forces, the cutting edge of the German blitzkrieg.

VINCENT B. HAWKINS

SEE ALSO: Blitzkrieg; Manstein, Erich von; Mechanized Warfare; World War II.

Bibliography

Guderian, H. 1952. *Panzer leader*. London: Michael Joseph.
Keegan, J. 1973. *Guderian*. New York: Ballantine Books.
Pfannes, C. E. and V. A. Salamone. 1980. *The great commanders of World War II*. Vol. 1, *The Germans*. New York: Kensington.
Wistrich, R. 1982. *Who's who in Nazi Germany*. New York: Bonanza Books.

GULF WAR, 1991

Iraq's invasion of Kuwait on 1 August 1990 triggered a series of events that led to one of the largest-scale conflicts of the modern era. Although the actual fighting lasted only 43 days, it involved massive air and armored operations and the widespread use of new military technologies ranging from stealth attack aircraft to modern tank fire-control systems with thermal imaging sights.

The war also marked a major shift in East-West relations and within the developing world itself. It became a contest between a regional superpower, under the leadership of an ambitious dictator, and a broad coalition of United Nations forces, led by the United States and Saudi Arabia and operating in a political context where the United States had the political support of the Soviet Union. As such, it may well have been the first conflict of the post–cold war era.

Iraq's Invasion of Kuwait

Unlike most conflicts, Iraq's invasion of Kuwait was an act of naked aggression with little political justification or sophistication. In the period before the invasion, Iraq claimed that Kuwait was violating its oil quotas and improperly draining oil from the Rumalia oil field—a large reservoir largely in Iraq but whose southern tip is in Kuwait. In fact, Iraq had never agreed to a quota of its own, and most of Kuwait's modest production from the Rumalia field had gone to sales that aided Iraq during its war with Iran (1980–88). Further, Kuwait had provided Iraq with billions of dollars in aid during that war, and had offered both to cut its exports and halt production from the Rumalia field before the Iraqi invasion.

Iraq also undermined any justifications for its actions during the first days of the invasion when it first claimed to be supporting a nonexistent uprising by prodemocratic Kuwaiti forces, stated that it was withdrawing from the country but then moved toward outright annexation, and sent its forces to Kuwait's southern border with Saudi Arabia.

While Iraq then claimed it was simply liberating territory stolen from it by Britain, these claims had equally little historical justification: Iraq had no claim to Kuwait as a successor state because modern Iraq had been created by Britain after the collapse of the Turkish Empire at the end of World War I. Even the Turkish Empire had had an uncertain claim to Kuwait, since it exercised only limited or dual jurisdiction over the area, and Kuwait had normally existed as a small independent Bedouin settlement on the Persian Gulf coast. Kuwait's boundaries as a city-state were set by the British in the 1920s, and only in reaction to the threat of a Saudi invasion.

The true causes of Iraq's invasion were a mixture of economic problems and the ambitions of Saddam Hussein. Under Saddam's leadership, Iraq had continued to expand its military machine after the cease-fire in August 1988. It did so even though it had obtained more than US$60 billion worth of arms during 1980–88, and the war had cost Iraq as much as one-third of its gross domestic product. Further, Iraq was spending additional billions on missiles and biological, chemical, and nuclear weapons. At the same time, Iraq spent billions on ambitious civil development projects like the reconstruction of Basra and Al Fao.

This saddled Iraq with a foreign debt of some US$80–100 billion at a time when oil prices were depressed and there was a significant world surplus of oil exports. As a result, Iraq could not continue to pay for its military machine, could not meet its debt payments, and experienced steadily greater problems in giving its people the kind of economic development and reconstruction they expected at the end of the war with Iran.

The invasion of Kuwait thus offered Saddam Hussein a means of distracting Iraq's population, a potential source of vast wealth, and the strategic asset of a deep-water port on the Persian Gulf. Kuwait's Fund for the Future had investments worth more than US$100 billion. Kuwait was capable of adding at least 2 million barrels a day of oil to Iraq's exports of roughly 3.5 million, and it offered the opportunity to increase Iraq's total oil reserves from 100 billion to 198 billion barrels (a total of nearly 25% of the world's total reserves). At the same time, it placed Iraqi forces on Saudi Arabia's border and within easy striking range. Even if Iraq did not attack Saudi Arabia's nearby oil fields and oil facilities, this strategic position gave it political and military leverage over nations that possessed an additional 28 percent of the world's total reserves.

International Reactions and the Forging of a U.S.-Led Military Coalition

The success of Iraq's invasion depended, however, on the reaction of its neighbors, the United States, and other regional and world powers. Saddam Hussein seems to have calculated that neighboring states like Saudi Arabia would be too frightened to act and that the United States would either not send forces or not be willing to go to war. As it turned out, he fatally miscalculated the reaction of his neighbors, the United States, and the other nations of the world.

Instead of paralysis, Saudi Arabia immediately gave the Kuwaiti government-in-exile its full support and consulted with the United States. Rather than be intimidated when Iraq moved its divisions to the Saudi border, in position to invade Saudi Arabia, Saudi Arabia sought outside military aid. Further, Saudi Arabia immediately obtained the support of other Gulf Cooperation Council states—Bahrain, Oman, Qatar, and the United Arab Emirates.

Pres. George Bush of the United States also acted immediately to check Iraqi aggression. Consulting with France, Britain, and many of the same allies that had supported joint naval action in the Persian Gulf in 1987 and 1988, President Bush sent a delegation to Saudi Arabia that pledged the commitment of massive military forces to defend Saudi Arabia. At the same time, President Bush took immediate action to freeze Iraqi assets and to obtain UN support for a naval blockade of Iraq and an embargo on all Iraqi imports and exports other than medicine and food for humanitarian purposes. If Saddam Hussein counted on what he perceived to be weakness demonstrated by the U.S. withdrawal from Vietnam and Lebanon, he was proved totally mistaken. On 7 August, less than a week after the first Iraqi troops entered Kuwait, the United States announced it would send land, air, and naval forces to Saudi Arabia.

Most of the rest of the world proved equally decisive. Britain, France, the other members of the North Atlantic Treaty Organization (NATO), Japan, most Eastern European nations, and the Soviet Union immediately joined in condemning Iraq's actions. While the Soviet Union jockeyed for political position and made its own efforts to seek Iraqi withdrawal from Kuwait, it consistently supported the United States in the United Nations and never gave Iraq any support for its actions. Most of the remaining Arab world proved equally firm. On 3 August 1990, the Arab League Council voted to condemn Iraq and demand its withdrawal from Kuwait. Egypt and Syria strongly opposed Iraq and sent military forces to defend Saudi Arabia and liberate Kuwait. So did other Arab states including Algeria. Only Jordan, Libya, Mauritania, the PLO, the Sudan, and Yemen gave Iraq significant political support during any point of the crisis.

The shift toward cooperation between East and West had an equally important impact in allowing the United Nations to take unprecedented action against Iraq. On 2 August 1990, the Security Council voted 14 to 0 (Resolution 660) to demand Iraq's immediate and unconditional withdrawal from Kuwait. The Security Council then passed resolutions that ordered a financial and trade embargo against Iraq (6 August), declared Iraq's annexation of Kuwait null and void (9 August), demanded that Iraq free all the foreign hostages it had taken (18 August), established an international naval blockade (25 August), halted all air cargo shipments (25 September), declared Iraq liable for all war damages and economic costs (29 October), and authorized the nations allied with Kuwait "to use all necessary means" if Iraq did not withdraw from Kuwait by 15 January 1991 (29 November).

For the first time since the Korean War, the United Nations was allowed freedom of action in checking an aggressor. As a result, Iraq suffered a complete naval and economic blockade, could not export oil, and lost any access to arms imports. Its economic and military strength was severely undermined, and it was forced to deploy a steadily increasing portion of its best forces to defend the Saudi-Kuwaiti border and its border with Saudi Arabia. By the UN deadline, Iraq had sent 545,000 men and 12 armored (heavy) and 31 mechanized infantry (light) divisions to the Kuwaiti theater of operations.

Force Ratios at the Beginning of the Conflict

Time did not favor Iraq or Saddam Hussein. His military inaction during the months of diplomatic maneuvering—1 August to 15 January—gave the powers who opposed him time to build up a massive international force. The United States built its forces from a few tactical air squadrons, which it deployed to Saudi Arabia shortly after the Iraqi invasion, to a massive land-sea-air force of 527,000 men and women, including over 110 naval vessels, 2,000 tanks, 2,200 armored personnel carriers, 1,800 fixed-wing aircraft, and 1,700 helicopters.

The U.S. forces were the largest element of what became a 38-nation coalition that included major contributions by other nations. Saudi Arabia contributed 118,000 troops, 550 tanks, 179 aircraft, and over 400 sites for artillery bases. Equally important, it made its modern air bases and military infrastructure available to the other nations of the coalition. Britain contributed naval forces, 43,000 troops with 170 tanks, an armored division, and 72 combat aircraft. Egypt contributed 40,000 troops with 2 armored divisions and 250 tanks. France contributed 16,000 troops with tanks, helicopters, a light armored division, and combat aircraft. Syria contributed 20,000 troops and 2 divisions. Other allied nations, such as Canada and Italy, contributed air and naval forces, and Oman, Qatar, and the United Arab Emirates deployed a significant portion of their small forces.

The end result was the largest set of opposing forces since the Korean War, and the largest mix of modern armor and air units since World War II. While experts still argue over the exact numbers involved, U.S. reports after the war indicate that, in toto, the coalition had well in excess of 600,000 land troops to Iraq's 545,000; 3,360 tanks to 4,230; 3,633 artillery weapons to 3,110; 4,050 other armored vehicles to 2,870; 1,959 helicopters to 160; and some 2,700 aircraft to 770. Moreover, the coalition had a massive technological advantage in virtually every category of weaponry, munitions, communications, and command and control. It also had an effective monopoly over photo, signal, and electronic intelligence.

Air Phase of the War

The 1991 Gulf War began on 17 January 1991 when U.S.-led air units launched a devastating series of attacks on targets in Iraq. These targets included command and control facilities, communications systems, air bases, and land-based air defenses. The war began when AH-64 Apache attack helicopters knocked out Iraq's forward radar system. The United States then used F-117 stealth attack fighters, which flew 31 percent of the attacks during the first day and attacked even heavily defended targets such as downtown Baghdad with complete impunity. They also involved the first significant use of sea-launched cruise missiles and a wide range of precision-guided weapons.

As early as the third day of the war, the coalition air forces were able to shift their attacks from Iraq's main air defenses to such strategic targets as key headquarters, civil and army communications, electric power plants, and Iraq's plants and facilities for the production of biological, chemical, and nuclear warfare.

The coalition also took full advantage of its monopoly on long-range reconnaissance, photo and signal intelligence from satellites, electronic intelligence aircraft, refueling capability, air control and warning aircraft (AWACS), and sophisticated targeting aircraft like the JSTARS. This gave it further advantages in both air-to-air and air-to-ground combat.

The advantage in air-to-air combat became clear during the first days of the war. Iraq had 770 combat aircraft, 24 main operating bases, 30 dispersal bases, and a massive network of some 3,000 surface-to-air missiles when the coalition attacked. Coalition air forces, however, were so superior that Iraq was unable to win a single air-to-air engagement and lost a total of 35 aircraft in air-to-air combat. By the end of the first week of the air war, Iraq ceased to attempt active resistance in the air, and Iraqi aircraft began to flee to Iran in hopes that Iran would return the aircraft and pilots after the war. Iraq halted even token efforts to use its aircraft in combat after the fourteenth day of the air war, and Iraq's land-based air defenses then proved vulnerable to electronic warfare, infrared and other countermeasures, and antiradiation missiles throughout the rest of the war.

At the start of the second week of the air war, coalition air forces shifted their focus and began attacking the Iraqi field army in the Kuwaiti theater of operations. This phase of the air conflict lasted for the next 26 days. The coalition was able to use its AWACS and refueling capabilities, maintain extremely high sortie rates, and concentrate up to 600 aircraft in the air over a country as small as Kuwait.

It then used a mix of highly sophisticated attack fighters like the F-15E and F-16 with the LANTIRN night targeting system to launch precision-guided weapons against Iraqi armor and artillery. Other lighter aircraft—the AV-8B, F-18A/B, A-10, and AH-64—also played a major role in striking Iraqi army targets in Kuwait.

These strikes were backed by bombers like the B-52 and F-111, using a mix of guided and unguided ordnance, and heavy attack fighters like the Tornado. These aircraft were able to keep Iraq's air bases suppressed and to conduct massive strikes on land targets. Typical bomber targets included Republican Guard concentrations, key supply and communications facilities in the rear and border area, dug-in artillery positions, and Iraq's defensive barriers and positions near the border area.

Almost from the outset of the war, Iraq realized it had no way to retaliate against the coalition's attacks except to launch its long-range modified Scud missile. Iraq began these missile strikes by attacking Israel and Saudi Arabia on the second day of the war and persisted in them until the cease-fire. Iraq launched a total of 40 Scud variants against Israel and 46 against Saudi Arabia, but these missiles never succeeded in doing major military damage. They also failed to provoke Israel into retaliating against Iraq, largely because the United States rushed Patriot defense missiles to both Israel and Saudi Arabia; the Patriot's ability to hit most incoming Scuds provided a vital boost in public confidence. Israeli restraint may have played a key role in ensuring that the Arab members of the coalition continued to support it throughout the war.

In retrospect, the only major impact of the Scuds was to force the coalition's air forces into a massive game of hide and seek in trying to kill Iraqi Scud units. Even here, however, the Scuds also imposed new costs on Iraq. While the attacks on the Iraqi Scuds diverted sorties from other targets, they increased the damage to Iraqi targets outside the Kuwaiti theater of operations. The only strategic damage Iraq was able to inflict on Kuwait and the coalition was to set fire to some 600 Kuwaiti oil wells. These fires, however, did nothing to affect allied air operations or slow the pace of the war.

By the time the ground war began at 0400 hours on 24 February 1991, Iraqi ground forces had been hit by more than 40,000 attack sorties. While the resulting damage estimates are controversial, coalition airpower claimed to have destroyed or severely damaged: all of Iraq's nuclear reactor facilities, eleven chemical and biological weapons storage facilities and three production facilities, 60 percent of Iraq's major command centers, 70 percent of its military communications, 125 ammunition storage revetments, 48 Iraqi naval vessels, and 75 percent of Iraq's electric power–generating capability. Logistic supply to the theater had been cut by up to 90 percent, and the U.S. command estimated that at least 1,300 Iraqi tanks, 800 other armored vehicles, and 1,100 artillery pieces had been destroyed from the air.

These air attacks continued throughout the air-land phase of the war that followed. By the cease-fire of 28 February, coalition air forces had dropped a total of 88,500 tons of ordnance, of which 6,520 tons were precision-guided weapons. A total of 216 Iraqi aircraft had been destroyed, along with nearly 600 aircraft shelters. Coalition air forces had also destroyed 54 bridges or made them inoperable—playing a major role in cutting off Iraqi land forces from their final route of escape along the Tigris north of Basra during the last days of the war.

Air-Land Phase of the Battle

Once the ground phase of the war began, it proved to be extraordinarily quick and decisive. The coalition not only attacked a gravely weakened Iraqi army and had a massive advantage in intelligence and virtually every area of tactical technology, it had vastly superior tactics. The coalition forces used the "air-land battle" concept, which the United States had developed to meet the Warsaw Pact's most modern forces in Europe, against an Iraqi force that was equipped with modern weapons but had trained and organized to fight a relatively static trench war against an Iran that lacked significant airpower.

While coalition land forces did not enjoy a significant superiority in weapons strength and manpower over Iraq, they did consist largely of highly motivated professionals. In contrast, the majority of Iraqi forces were poorly trained conscripts who seemed to have poor morale and little motivation. It is impossible to determine how much of this weakness stemmed from having to invade an Arab "brother," poor leadership and organization, or the coalition's air attacks. All three combined to undermine Iraqi military capabilities.

Iraq's one "success" in the land war occurred long before coalition land forces began to liberate Kuwait. Several Iraqi brigades made a brief incursion into the Saudi border town of Khafji on 29 January. But the town had been evacuated, and the Iraqi forces were driven back the next day by Saudi, Qatari, and U.S. Marine forces. The net result was that Iraq did more to reveal its weaknesses than advance its own cause.

The attack on Khafji also did nothing to keep coalition land forces from making a massive shift from positions along the coast and to the south of Kuwait, to areas near the Iraqi-Saudi border to the west of Kuwait. These shifts later allowed coalition land forces to drive deep into Iraq and Kuwait.

The shifts of land forces began on 17 January, the same day the air phase of the war began. They involved massive logistic and movement difficulties, but they eventually positioned the U.S. Marine Expeditionary Force and Saudi, Syrian, and Egyptian forces where they could drive north from the center of Kuwait's southern border toward Kuwait City. At the same time, French and U.S. forces in VII Corps moved far to the west, where they could launch an attack to cut off southern Iraq from Baghdad and then drive around Kuwait to move against Basra from the west. British and U.S. Army forces in XVIII Corps moved to areas on the Saudi-Iraqi border just west of Kuwait.

In a move he later called his "Hail Mary play," the allied commander, Gen. Norman Schwartzkopf, was able to position two full armored corps along the Iraqi border to the west of Kuwait without the Iraqis detecting these movements. He was able to keep them undetected through the use of special forces and extensive frontline patrols, and because coalition airpower denied Iraq any air reconnaissance capability, even near its own border.

This element of surprise played a key role when the air-land battle began on 24 February 1991 (Fig. 1). Two major simultaneous attacks quickly crossed Iraqi defensive positions. The first consisted of Pan-Arab (Saudi, Kuwaiti, Qatari, and Omani) forces and U.S. Marine forces attacking on a broad front from the northern "notch" in the Saudi-Kuwaiti border to the coast and penetrating the Iraqi defenses along the southern Kuwait border. These forces advanced as far as half the distance to Kuwait City within twelve hours. This attack was aided by the fact that many of Iraq's forces were kept pinned down by a U.S. deception operation that convinced Iraqi commanders that U.S. Marine amphibious forces might strike at any point along the coast.

The second attack occurred at the far western edge of coalition positions along the Iraqi-Saudi border. The French 6th Light Armored Division and one brigade of the 82d Airborne Division drove 90 miles north to seize an airfield at Al-Salman. The U.S. 101st Airborne Division then launched the largest air assault operation in military history, and U.S. heliborne forces moved first to a forward logistic base in Iraq and then to positions near Samawah on the Euphrates. This attack cut Iraqi forces off from the main routes from Basra to Baghdad that run south of the Euphrates.

The coalition attacked along a third major line that same afternoon. The 3d Armored and 24th divisions of XVIII Corps drove across the Iraqi border and toward the Wadi al-Batin and the western approaches to Kuwait City. Immediately to the east, U.S. and British land forces also advanced into Iraq. The 1st U.S. Division forced a breach in the Iraqi defenses that was rapidly exploited by the 1st British Armored Division and the 2d Cavalry Regiment and 1st and 3d Armored divisions of the U.S. Army. This advance rapidly turned into a deep thrust against the Republican Guard forces west of Kuwait City, north of Kuwait, and west of Basra. Finally, Egyptian and Saudi forces, backed by Syrian fire support, launched a fourth attack on Iraqi positions to the east of the gorge of Al-Batin on the 25th.

These attacks, and the relentless air attacks that had preceded them, quickly shattered the remaining organization, morale, and war-fighting capability of most of the Iraqi army, while the Republican Guards remained pinned down outside Kuwait. As a result, coalition forces were able to drive up through Kuwait to strike positions south and west of Kuwait City, while VII and XVIII corps forces moved deep into Iraq, to positions south of a line drawn from the border to Nasiryah.

On 26 February, forces of VII and XVIII Corps closed on the Iraqi Republican Guard forces and reserves defending Basra in the longest sustained armored advance in history (Fig. 2). They destroyed the key Republican Guard divisions holding the area just north of the Kuwaiti border. Other coalition forces, including the 1st and 2d Marine Divisions, reached positions on the edge of Kuwait City and began fighting for control of the international airport. These advances took place despite extraordinarily bad weather, which created substantial amounts of mud and interfered with air cover.

The war ended with a devastating series of engagements where Iraqi forces were able to put up only limited resistance. The thermal sights and superior fire-control systems of coalition tanks allowed them to achieve massive kills against Iraqi armor, backed by lethal systems like the AH-64 attack helicopter and the Multiple Launch Rocket

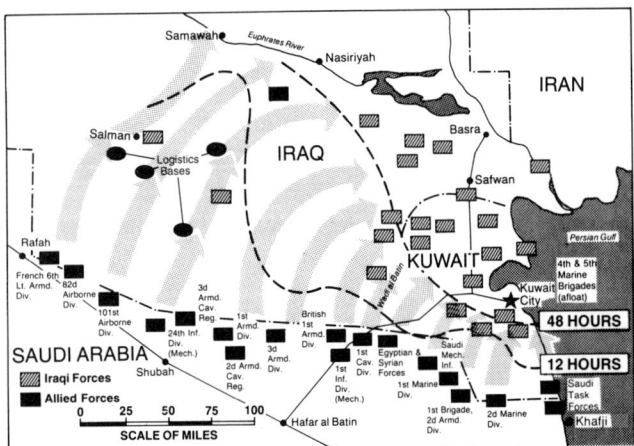

Figure 1. The first two days of the air-land battle, 24–25 February, 1991 Gulf War. (SOURCE: AUSA 1991)

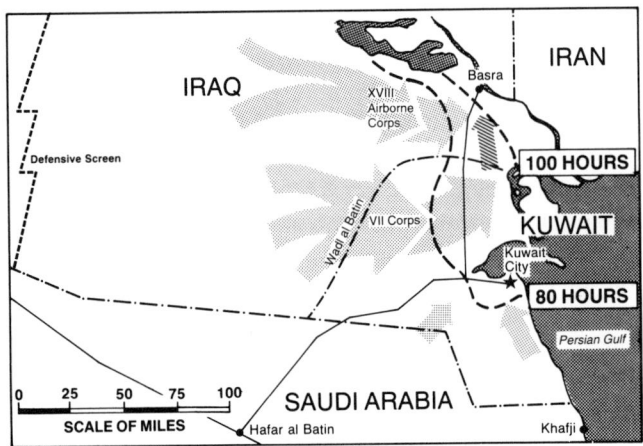

Figure 2. The final push of the 1991 Gulf War, 26–27 February. (SOURCE: AUSA 1991)

System. The coalition's vastly superior intelligence and night vision devices, combined with the use of new navigation aids that provided precise location data from global positioning satellites, gave its land forces control of both the desert and the night.

From 27 February to President Bush's order to halt operations on 28 February, the remainder of Iraq's 43 divisions in the theater were destroyed or rendered ineffective. The VII Corps shattered the remainder of the armored (heavy) Republican Guard divisions; the XVIII corps moved close to Basra. British and U.S. Army forces liberated northern Kuwait. Saudi and other Arab forces liberated Kuwait City, and U.S. Marine forces secured the southern and western outskirts of Kuwait City.

The scale of the coalition's success in fighting a 1,000-hour air battle and the 100-hour air-land battle that followed is indicated by the fact that coalition land forces succeeded in reaching every major objective ahead of schedule and with far fewer casualties than their commanders dared to hope for. They achieved a rate of advance so fast that many units did not bother to halt at their intermediate objectives.

The scale of the coalition's success is also indicated by U.S. estimates that coalition forces had destroyed nearly 4,000 Iraqi tanks, more than 1,000 other armored vehicles, and nearly 3,000 artillery weapons. In contrast, the coalition suffered combat losses of four tanks, nine other armored vehicles, and one artillery weapon. Although coalition aircraft flew a total of 109,876 sorties by the end of the war, the coalition lost only 38 aircraft—the lowest loss rate of any air combat in history and less than the normal accident rate per sortie in combat training. The difference in manpower losses is even more astounding, although no precise estimates are possible. U.S. intelligence issued rough estimates after the war that 100,000 Iraqi soldiers died in combat. Allied killed—less casualties to friendly fire—totaled less than 200.

Political and Military Consequences of the Conflict

It will be years before all the lessons of the 1991 Gulf War are fully analyzed and its strategic political consequences are clear. It is, however, already apparent that the coalition victory had a number of complex impacts.

The war substantially shattered Iraq's military organization. While substantial Iraqi forces still remain, largely composed of units never deployed to the Kuwaiti theater of operations, these are incapable of offensive action against neighbors like Iran, Turkey, and Syria, or against Kuwait and Saudi Arabia as long as they have American backing. Iraq has little chance of intimidating the conservative southern Gulf states, and a new and far more stable balance of power has been established within the region.

Elsewhere in the Middle East, the war had more ambiguous effects. It critically weakened a military threat to Israel, but it has also demonstrated Israel's vulnerability to missile attacks and its potential vulnerability to weapons of mass destruction. It has further shattered the myth of Arab unity without bringing any new concept of regional order in its place. It created new opportunities for an Arab-Israel peace initiative, but Palestinian and Jordanian alignment with Iraq also made some aspects of such negotiations more difficult.

The war ended with the United States emerging as the leader of a 38-nation coalition and as a pre-eminent military power that took only 43 days to inflict one of the most decisive military defeats in military history. Nearly two decades after the U.S. withdrawal from Vietnam, and almost a decade after U.S. withdrawal from Lebanon, the reputation of U.S. military forces has been decisively restored. The United States demonstrated a combination of strategy, tactics, readiness, training, weaponry, and manpower quality that clearly had no equal. At the same time, it showed that the United States was heavily dependent on time to deploy its powers and access to the ports, air bases, and facilities of nations in the forward area. Further, it showed that American military freedom of action was heavily dependent on an international consensus and allied military support.

The implications for the future of East-West relations were equally complex. The coalition victory came at a time when the Soviet Union was already in a deep political and economic crisis and drifting toward dissolution. The Warsaw Pact ceased to exist a few months after the cease-fire, and the Soviet Union transformed itself into the Union of Sovereign States only a year after Iraq first invaded Kuwait. The bipolar world that existed from 1945 to 1990 had begun to vanish without creating any clear movement toward the "new world order" that President Bush had mentioned in some of his speeches before the war.

Like most victories, the war also produced new issues and uncertainties. There is no doubt that the coalition scored a major victory in grand strategic terms. It liberated Kuwait, it destroyed Iraq's ability to invade or use military pressure against its neighbors, it destroyed most of Iraq's capability to build and use weapons of mass destruction, and it forced Iraq to agree to cease-fire terms that promised to steadily weaken its military capabilities for years to come.

The importance of superior technology was evident. Advanced weapon systems provided the coalition forces with a clear-cut advantage over Iraq, a nation which itself was equipped with some very modern Western systems. The war, which received unprecedented television coverage, marked the dawn of a new technological era. Precision-guided munitions (PGM) proved immensely effective. Cruise missiles, antiballistic missile defenses, advanced reconnaissance systems, F-117 stealth aircraft, and Apache helicopters were all used successfully for the first time in major combat.

Land and naval forces played key roles in the U.S.-led coalition's strategic plan. Airpower, however, provided the decisive element. For 38 days, allied aircraft methodically attacked Iraq's offensive machine, leaving a shattered, demoralized, and disorganized army to be mopped up in 100 hours by coalition ground forces. Factors contributing to the successful application of airpower against Iraq included:

- Highly accurate navigation and weapon delivery systems that could deliver PGMs to within one meter (3 ft.) of their target.
- Stealth technology, which returned the element of surprise to air warfare.
- Night attack systems to maintain pressure around the clock.
- Surveillance and intelligence-gathering systems, space systems, and tactical reconnaissance aircraft that provided coalition commanders with theater-wide situational awareness.

The importance of technology and airpower should not be overstated, however. The extent to which the early collapse of Iraq's air defense system was due to allied technology as opposed to human factors remains unanswered. And it should be noted that the war was conducted in terrain that has historically favored air operations. But the bottom line is that high-technology weapon system dramatically increased the effectiveness of coalition forces.

The war did not, however, destroy Saddam Hussein's control over Iraq. It left the rivalry between Iraq and Iran intact and caused new tensions between the southern Gulf states and Yemen and the Sudan. It did not bring more liberal or democratic regimes to any state in the region. The Arab portions of the coalition that defeated Iraq did not hold together, and no new security structure arose in its place. Enforcing the terms of the cease-fire rapidly became a major challenge, and Iraq constantly attempted to cheat on its terms in every area—from recognition of Kuwait's sovereignty to preventing the destruction of its weapons of mass destruction. Rather than marking the end of history, the legacy of the 1991 Gulf War was one of creating further sources of instability in a radical period of change.

© ANTHONY H. CORDESMAN

SEE ALSO: Air-Land Battle; Airpower, Strategic; Airpower, Tactical; Arab League; Coalition Warfare; Desert Warfare; Precision-guided Munitions; United Nations.

Bibliography

Association of the U.S. Army (AUSA). 1991. *Special report: The U.S. Army in Operation Desert Storm, An overview.* Arlington, Va.: AUSA Institute of Land Warfare.
Blackwell, J. 1991. *Thunder in the desert.* New York: Bantam Books.
Friedman, N. 1991. *Desert victory: The war for Kuwait.* Annapolis, Md.: U.S. Naval Institute Press.
U.S. Department of Defense. 1992. *Conduct of the war report.* 15 February.
Woodward, B. 1991. *The commanders.* New York: Simon and Schuster.

GUN, AERIAL

By the end of World War II, the aerial fighter forces of all major combatant nations were using cannon and heavy machine guns up to 0.5 inches. Such a weapon was then sufficiently effective against fighter aircraft. The Germans were employing 30mm and 50mm cannon, mainly to destroy heavy bombers, but two or three hits by 20mm shells were generally thought sufficient to down a fighter aircraft.

During the Korean War of 1950–53, where large-scale air battles took place between U.S. F-86 Sabres and Soviet-built MiG-15s, the Sabre was armed with six half-inch heavy machine guns delivering 110 rounds per second. The half-inch projectile had excellent ballistic qualities, coupled with the high muzzle velocity of the gun; hits were often scored on MiG-15s at ranges of 375–470 meters (400–500 yd.). Unfortunately for United Nations forces, the MiG-15 was a tough airplane that often survived such hits. The armament of the MiG was one 37mm and one or two 23mm cannon. The Soviet guns had a relatively low muzzle velocity and a slow rate of fire. The time of flight of their shells was also comparatively long, which made it difficult for them to score hits on an evading target.

By the late 1950s many experts were convinced that fighter guns were obsolete due to the advent of air-to-air missiles (AAMs). The new weapons provided the capability of following an evading target and hitting it at long range with a warhead large enough to inflict lethal damage. Many studies of fighter armament dating from that period ignored the gun completely. The spin-stabilized rocket was considered to be the only viable alternative to the AAM. One officer who disagreed with this approach was the commander of the U.S. Air Force's 8th Tactical Fighter Wing, who said in 1965 that "a fighter without a gun is like a bird without a wing."

Importance of the Aerial Gun

The aerial gun is a very reliable close-range weapon for fighter aircraft. It is often considered solely a defensive weapon that would be used only in a dogfight with hostile aircraft. For this purpose a well-designed fighter aircraft should carry one or more guns that have the highest possible instantaneous rate of fire coupled with the highest possible muzzle velocity. Today some types of combat aircraft carry only two missiles; the problem with such an approach is that in a major engagement these missiles would soon be expended. Also, fighters armed only with missiles and no guns sometimes find themselves in dogfights at ranges too short to permit launching their missiles. In such situations, being armed with guns would both permit self-defense and provide an additional weapon for use if further attack opportunities arise.

The general belief of military experts during the period

1955–70, that aircraft guns were obsolete, has now come to be viewed as quite mistaken. A modern fighter mounting an internal gun is a close-support asset, since such guns have a secondary role for use in ground strafing.

Mounting the Aerial Gun in Combat Aircraft

Aircraft guns, like electronic countermeasure (ECM) jammers, can be mounted either internally or conformally or in a pod. Internal guns, designed along with the aircraft in which they are to be mounted, can be very efficient, causing low drag and leaving all external hard points free for other uses. The use of external gun pods is fundamentally undesirable unless it is known that the gun will seldom be needed. Such mountings use up a pylon, upset aircraft trim, and make accuracy difficult to attain. Most U.S. fighters mounting the M61 gun have it placed well off the centerline. In the F-15, firing the gun automatically applies just enough left rudder to hold the aircraft on target.

Gun pods have recently been seen on the various MiG-23 and MiG-27 dedicated attack versions, despite the fact that these aircraft already have a multibarrel internal gun with great hitting power mounted on the ventral centerline. The addition of the external pods seems to allow the guns to be tilted down to about -20 degrees for firing at ground targets without the necessity of diving. Such fire cannot be sustained on one target and would normally sweep along at the ground speed of the aircraft, so accuracy could never be high no matter what sighting system is used.

Almost all modern Soviet aircraft have at least one internal gun. The SU-24 long-range interdictor has two of different calibers, probably one 23mm and the other 30mm. For stealth aircraft, internal mounting of the guns is infinitely preferable to any kind of external mounting; the supercruise advanced tactical fighter (ATF) is almost certain to have its guns internally mounted.

Types of Aerial Gun

The standard gun of nearly all U.S. fighters is the General Electric M61, originally designed as Project Vulcan in 1949–52 and still in service with no planned replacement more than 30 years later. This gun is usually used in 20mm configuration with six barrels, producing a rate of fire of 6,000 rounds per minute. Some versions have hydraulic or electric drive, while others are self-powered. Most current versions use a linkless feed system.

For aerial combat, smaller caliber guns now appear most desirable, since larger calibers lack the high muzzle velocity that is needed to perform effectively in aerial combat. The next generation guns in all countries are likely to have much shorter (so-called "telescoped") square-section propellant charges. Another probable development is elimination of the cartridge case, or at least its replacement by a case that will be consumed when the charge is fired. This would offer numerous advantages, including cost reduction, weight reduction, reduced ammunition bulk, elimination of case-induced stoppages, and elimination of problems involved in getting the spent cartridge case to separate cleanly from the aircraft or the alternative necessity of storing it on board.

Air-to-air Operations

The approach stage of aerial combat begins when the target is detected, either visually or by a fighter's radar. It concludes when the fighter has occupied the key position for attack or entered into the zone of probable attack. Attack is the decisive stage of aerial combat. It is carried out by a single aircraft or a group of fighters and consists of sighting, maneuvering, and firing. The attack phase commences when the fighter is in the zone of probable attack, which is that part of the airspace relative to the target within the limits of which the fighter is able to sight and take the target under fire.

The traditional gun attack is from astern with an overtaking speed of between 50 and 150 knots. But the historical record shows that the aces of past generations took far more head-on shots than is generally realized. It may be assumed that the high speeds of modern fighters invalidate the head-on gun attack. Perhaps they can still be made, even at supersonic speeds, but with present-day closing speeds on the order of 450–900 meters per second, that appears to be a doubtful proposition. What the head-on gun attack does at such speeds is to hang a curtain of shells in front of the approaching enemy aircraft. This form of attack appears unlikely to achieve consistently good results.

The procedure for air-to-air weapons employment under modern circumstances should therefore be long-range missiles first, then closure to visual range to employ heat-seeking homing missiles and guns. But when it comes to aerial combat there are no hard and fast rules. Guns and missiles do not exist in isolation, but are merely the cutting edge of an integrated weapons system. Heat seekers close the gap between the medium-range weapon and the gun. They are essentially visual-distance weapons that have consistently produced better results in combat than radar homers. But when the enemy is very close, only the gun remains.

Gun Weaponry of Fighter Aircraft

The gun as an air-to-air weapon is essentially a compromise. Most cannon are now based on the revolver principle developed by Germany in the closing years of World War II. In those days a fighter carried four cannon. The additional weaponry developed since then has so reduced the space available for cannon that modern aircraft carry only two or even one. It is interesting to note that the Tornado F2 has only one gun, whereas the interdiction strike version of the same aircraft has two. This configuration was decided upon by the British air staff primarily because the F2, built to kill at long stand-off ranges, was thought unlikely to be in close-in dog-fighting.

An aerial gun's recoil is enormous, so much so that it quickly slows the aircraft. Such guns are typically mounted two degrees nose down so that the barrels come onto the aircraft's centerline as they fire. Table 1 shows the variety of aerial guns mounted in different aircraft.

In 1917 many victories in aerial combat were scored at ranges of less than 30 meters (98 ft.). By 1945, 140 meters (460 ft.) was considered close range. Now anything less than 275 meters (900 ft.) is considered almost suicidal.

Modern fighters are unlikely to be knocked down by a single hit. The aerial gun therefore needs the highest possible rate of fire, to pump out the greatest number of shells, thereby increasing the chance of scoring multiple hits. Since no more than two or three hits are likely to be scored on an evading or rapidly crossing target, the shell needs to be as lethal as possible, which means large. Finally, in order to be effective the shell needs to have the highest possible muzzle velocity and the best possible ballistic characteristics.

The problem of airborne fire control involves one very important element, the high velocity of the weapon carrier itself. The muzzle velocity of a gun and the velocity of the aircraft may both be great. Therefore the vector of motion of the projectile, which is the resultant of the vector of motion of the projectile and of the aircraft, is greatly affected by the aircraft's speed.

Aerial Guns Mounted in Helicopters

The first guns mounted on helicopters were rifle-caliber machine guns on gimbal or pintle mounts. Such weapons were aimed by hand. In the Korean War, machine guns were mounted in the doorways of helicopters. This is still a common practice despite obvious shortcomings in terms of vulnerability, limited arc of fire (to one side only, for example), and blockage of the doorway. In general all guns aimed by hand from helicopter doorways have extremely poor accuracy, even using tracer ammunition.

Helicopters may be expected to play an increasingly important role in air-to-air combat against both airplanes and other helicopters. But until recently helicopter guns were almost ignored as aerial combat weapons, although

TABLE 1. *Aerial Guns Mounted in Various Aircraft*

DESIGNATION	TYPE OF AIRCRAFT	AERIAL GUNS
UNITED KINGDOM		
Aerospace Harrier	Tactical attack and reconnaissance	Under-fuselage strakes, each replaceable by a 150-round pod containing one 30mm Aden gun
Sea Harrier FRS.1	Multirole fighter	Normally fitted with two 30mm Aden MK4, each with 150 rounds
FRANCE		
Dassault-Breguet Alpha Jet	Light strike/reconnaissance	Detachable belly fairing housing one 30mm DEFA or 27mm Mauser cannon with 125 rounds
Mirage 2000	Interceptor and air superiority fighter	Two 30mm DEFA 5-54 cannon, each with 125 rounds
Super Etendard	Strike fighter	Two 30mm DEFA 5-53 cannon, each with 125 rounds
UNITED STATES		
Fairchild Republic A-10 Thunderbolt II	Close support attack	One 30mm GEGAU-8/A high velocity high-energy gun with 1174 rounds
General Dynamics F-111	All-weather attack, FB, strategic attack, tactical ECM jammer	One 20mm M61A-1 gun with 2084 rounds
Grumman F-14 Tomcat	Multirole fighter	One 20mm M61A-1 cannon
F-4 Phantom II	Multirole fighter and all-weather interceptor	One 20mm M61 multibarrel gun; virtually all versions can carry the same gun in external centerline pod
McDonnell Douglas F-15 Eagle	Air superiority fighter with secondary attack role	One 20mm M61A-1 gun with 940 rounds
FORMER USSR		
Mikoyan/Guervich MIG-21	Fighter	One GP-9 comprising one 23mm GSH-23 gun 200 rounds
MIG-27	Ground attack	One 23mm six-barrel gun on centerline with approximately 500 rounds
Sukhoi Su-7	Ground-attack fighter	Two NR-30 guns in the wing roots, each with 70 rounds
Su-24	All-weather attack and reconnaissance	23mm or 30mm multibarrel gun with over 1000 rounds
Kfir-C2	Multirole fighter and attack	Two 30mm DEFA 5-53 cannon, each with 150 rounds

they were regarded as useful for employment against personnel and soft-skinned vehicles.

There is no inherent problem in mounting small guns at the sides of a helicopter, either bare (using ammunition provided from within the fuselage) or in a streamlined pod. High-powered guns pose problems of installed weight, recoil force, and severe muzzle blast effects, as well as the problem of muzzle flame, which at night can destroy the crew's night vision adaptation. In the past, shock-absorbing mountings have been used to reduce recoil forces transmitted to the fuselage; but such mountings introduce inaccuracies in aiming that are multiplied by the whip of long cannon barrels. This was not a significant problem when the weapons involved were little better than scatter guns, but with today's all-weather precision aiming systems much more has to be done to point the barrels accurately.

Almost all guns mounted in U.S. and Soviet attack helicopters are fitted in powered turrets. Such mountings, or at least precision aiming under remote power control, appear to be the preferred form of installation under current circumstances.

Electric power is often used to drive rotary cannon; many types of ammunition, including the standard U.S. M50 series of 20mm, have electrical priming instead of percussion. A few helicopters have used guns with limited pivoting—in one plane only—to overcome the difficulty of tilting the whole helicopter in pitch. An example is the M621, carried on the right side of the Gazelle, with elevation limits of $+6°/-4°$. Sometimes in a small helicopter there are problems in reconciling the heavy recoil forces with aiming accuracy. In the AH-64 Apache, the gun is mounted with powered elevation $+11°/-60°$ and traverse 110 degrees to left and right.

Some chin turrets have unrestricted all-around traverse, although usually this is limited by the twist of the ammunition feed. One possible advantage of mounting the turret right in the nose is that it eliminates the danger of a crew's being injured by the turret in a crash landing. Otherwise, at high rates of descent, the complete turret may be pushed up into the fuselage.

KHEIDR K. EL DAHRAWY

SEE ALSO: Ammunition; Ballistics; Close Air Support; Firepower; Gun Technology Applications; Helicopter; Munitions, Aerial.

Bibliography

Friedman, R., et al. 1985. *Advanced technology warfare.* London: Salamander Books.
Gunston, B. 1983. *Modern air combat.* London: Salamander Books.
———. 1985. *Warfare of the future.* London: Salamander Books.
Gunston, B., and M. Spick. 1986. *Modern fighting helicopters.* London: Salamander Books.

GUN, ANTITANK

After the Allies introduced the tank in battle in 1916, German soldiers had to find a means to negate it. Some types of German machine-gun ammunition were found to be effective against the relatively thin armor of the early British Mark I tank, and artillery and mortars of the day also proved effective when fire was accurately placed. The first gun specifically designed to defeat tanks, however, was a modified German Mauser—a high-powered, bolt-action rifle. Increased in size to fire a 13mm bullet at 900 meters (3,000 ft.) per second, it could be used by infantrymen and proved sufficient to defeat the armored vehicles used in World War I.

In the years between World War I and the onset of World War II, many gun designers and manufacturers attempted to produce weapons capable of destroying tanks by using the armor-penetrating power of a kinetic-energy round. The British "Boys Rifle," Mark 1, named for one of its principal designers, Captain Boys, is one such example. Likely the best weapon of its kind, the Boys Rifle was completed in 1936 and fielded in 1937. It fired a 0.55-inch steel-cored bullet at 990 meters (3,250 ft.) per second and could penetrate armor to a thickness of 20 millimeters (0.8 in.) at a distance of 250 meters (825 ft.). Since the average tank armor at the time was about 15 millimeters (0.6 in.) thick, the Boys Rifle stood an excellent chance to defeat any tank of the day.

Tank armor, however, was greatly improved in the few years just prior to the German blitzkrieg attack against the Allies in May 1940, and small-caliber antitank guns were no longer very effective. The effectiveness of large-caliber guns was also problematic until the introduction of the hollow-charge explosive warhead. This new warhead, made with a cone of explosive lined with metal such that on detonation the blast is focused and a jet of the molten metal is propelled at speeds of about 6,000 meters (20,000 ft.) per second, provided a means to effectively engage more thickly armored tanks. Hollow-, or shaped-, charge explosive technology was adapted to both antitank guns and rockets.

Of the numerous guns used in an antitank role in World War II, the most famous was Germany's "88." Originally developed for use against aircraft, the 88mm artillery piece proved deadly effective against all but the heaviest, most thickly armored tanks. The United States had no equivalent, and the British 6-pounder was underpowered. Only the Soviet 76mm gun came close.

During World War II, Swedish weapons designers began work on a recoilless antitank gun. Unlike the breech-loading, recoil-operated antitank guns that fire an explosive shell from a cartridge, recoilless guns have the propellant charge housed in the weapon itself. On firing, a disk of plastic (or a similar substance) at the base of the propellant case ruptures, permitting the rearward exhaust of propellant gas and other counterweight material (e.g., water, iron filings). The mass and velocity of rearward exhausting material imparts a balancing forward velocity on the mass of the shell, allowing recoilless operation.

Recoilless guns are substantially lighter than their recoil-operated cousins. A typical recoilless system can weigh about 18 kilograms (40 lb.) compared to a typical recoil system that can weigh thirteen times as much. Both types of systems permit effective engagement of targets to similar ranges—out to about 1,000 meters (3,300 ft.).

One of the more successful recoilless antitank guns, developed in the mid-1970s, is the Swedish weapon, the RCL Carl-Gustaf M2 (discussed below). It fires an 84mm projectile to an effective range of 450 meters (1,485 ft.) against tanks, and 1,000 meters (3,300 ft.) against troops in the open.

Because of their light weight, relative simplicity, and ease of manufacture, recoilless weapons have provided interesting possibilities in the conduct of warfare. One example is a disposable gun, manufactured by the firm of Raikka in Finland, and discussed later in this section.

Antitank guns of both types are in use in most militaries around the globe. The following discussion provides a variety of examples of contemporary systems, the predecessors of which first saw action in World War I.

Miniman Light Antiarmor Weapon (Sweden)

The Swedish Miniman is a one-shot throwaway, recoilless gun issued to infantrymen to provide them with an effective defense against close-in tanks. It arrives in the forward area with its projectile already in place. The user has only to estimate range and speed, cock the firing mechanism, lay on the target, and fire.

The weapon's barrel is made of filament-wound fiber. An attached label provides illustrated instructions for applying the correct lead to a moving target. The gunner need only judge the range (up to 150 meters or 495 ft.) to the nearest 25 meters (82½ ft.), and the target's speed as very slow, slow, or fast, according to parameters provided.

The HEAT (high-explosive antitank) shell includes a distance tube at the front, made of alloy, which establishes the stand-off distance; the shell body with copper liner and a shaped charge of octol; and a stabilizing tube of light alloy, in the rear of which are four slots forming flaps that, forced out by gas pressure, form four fins to stabilize the shell.

When this shell strikes a target, the body is compressed, the piezo pushes the firing button forward, the firing rod goes rearward, and the pin ignites the primer. The resulting flame travels down an ignition transmission line to the igniting and propelling charges. The shaped-charge warhead is initiated from the rear. The resulting jet can penetrate up to 340 millimeters (13.6 in.) of armor plate.

The 74mm weapon weighs 2.9 kilograms (6.4 lb.) loaded and has a range of 150 meters (495 ft.) against moving targets and 250 meters (825 ft.) against stationary ones.

M72 LAW (Lightweight Multipurpose Assault Weapon) (United States)

The U.S. M72 is light, short when configured in the carrying mode, and expendable after firing. Its small size makes it easy to carry, while its low weight—3.2 kilograms (7 lb.)—does not add appreciably to the existing considerable load of infantry soldiers. The M72's multipurpose capability is also unique in this class of antiarmor weapon systems. Its accuracy, safety, and reliability give the combat soldier a highly effective short-range assault weapon. On both operational grounds and considerations of cost effectiveness, this weapon is impressive, including its performance, acquisition cost, and training and logistics factors.

SPG-9 73mm Recoilless Gun (USSR)

The SPG-9 is a lightweight antitank gun normally carried by two men, crewed by four men, and mounted on a tripod for firing. It can be towed using a small two-wheeled carriage. It is used by motorized rifle battalions of the former Soviet army and is also in service in Bulgaria, East Germany, Hungary, and Poland. The weapon fires a fin-stabilized round with a HEAT warhead. The projectile is also given a slow spin inside the barrel by means of offset holes in the launching charge. The propellant charge is carried in a case attached behind the fins, thus making for a very long round. Not only does this system produce a high muzzle velocity, but the projectile is subsequently rocket-assisted, further increasing velocity to some 700 meters (2,310 ft.) per second.

The launcher weighs 47.5 kilograms. The weapon is normally quad-mounted in sets of four tubes. The system has a maximum range of 1,300 meters (4,290 ft.) and can penetrate more than 390 millimeters (15.6 in.) of armor.

Raikka Recoilless Gun (Finland)

The firm of Raikka has developed a novel and interesting series of antitank recoilless guns. In their arrangement, the barrel is a plain tube into the center of which is inserted the cartridge. These barrels may be either smoothbore or rifled. On firing, the shell goes forward and an equivalent weight of another substance is blown backward, thus balancing the recoil. Raikka offers a range of such guns in calibers of 41, 55, 81, 120, and 150mm. The 41mm and 55mm weapons are man portable; the 81mm size comes in both man-portable and mounted versions; and the larger calibers are mounted. The 120mm version has a fin-stabilized high-velocity APDS round (HVAPDS [FS]). It is claimed that this projectile can achieve velocities of 1,500 meters (4,950 ft.) a second and is effective beyond 1,000 meters (3,300 ft.) against main battle tanks.

Multiple versions are produced, and their characteristics differ very greatly from one to the other. The smallest man-portable weapon, the 41mm, weighs 3 kilograms (6.6 lb.) and has an effective range of 200 meters (660 ft.). At the upper end of the spectrum, the 150mm mounted weapon weighs 1,200 kilograms (2,640 lb.), uses a round weighing 42 kilograms (92.4 lb.), and has an effective range of over a kilometer. Two different versions of the 120mm weapon weigh 1,500 kilograms (3,300 lb.) and have ranges of 1.5 kilometers (4.95 ft.) or greater.

It might appear that the firm is a little late in producing these guns since most armies are phasing out the recoilless principle as a type of main launcher of antitank projectiles, but it may yet turn out that there is more to these guns than at first appears.

85mm Antitank Gun D-48 (USSR)

The 85mm antitank gun D-48 was originally given the Western designation 100mm field gun following its first appearance in the 1955 May Day parade in Moscow. Subsequent investigation determined that the weapon was in fact a special high-performance 85mm gun designed by the FF Petrov design bureau as a replacement for the 100mm field gun M1944 (BS-3) for use in antitank warfare.

The D-48 is a towed weapon served by a six-man crew. It has a range of 18,970 meters (20,867 yd.) and can fire at a sustained rate of eight to nine rounds per minute or a maximum rate of fifteen rounds per minute. The gun has a 54-degree traverse.

The ammunition for the D-48 was developed for high performance by necking down the cases for 100mm to accommodate a new 85mm projectile. Two basic forms of projectile were developed, a full-caliber hard-core high-velocity armor-piercing (HVAP) projectile and at least one type of high-explosive round. Soviet references mention an armor-piercing (AP) projectile weighing 9.3 kilograms (20¼ lb.), an HE projectile weighing 9.7 kilograms (21.3 lb.), and a muzzle velocity of over 1,000 meters (3,300 ft.) per second. If the AP projectile is fired at that velocity, the corresponding HVAP would be fired at nearly 1,200 meters (3,960 ft.) per second. Estimated penetration could then be about 190 millimeters (7.6 in.) for AP and 240 millimeters (9.6 in.) for HVAP at a range of 1,000 meters (3,300 ft.) and zero degrees incidence.

The D-48 was replaced in the Soviet antitank elements of artillery formations during the mid-1960s by the 100mm antitank gun T-12, a very long barreled gun with the entire recoil system located over the breech area. A single castor wheel assists in bringing the weapon into and out of action; during travel the breech is clamped between twin box-section trails. the D-48 can be fitted with an infrared night vision device. It is most likely that the D-48 served as the basis for the 86mm gun D-70 mounted on the ASU-85 airborne assault gun.

85mm Antitank Gun Type 56 (People's Republic of China)

The Type 56 is towed by a four-by-four truck and served by a crew of six to eight men. Its maximum range is 15,650 meters (10.7 mi.) or, with HEAT ammunition, 970 meters (0.6 mi.). The gun can traverse 54 degrees, elevate to +35 degrees, and depress to −7 degrees. It features a double-baffle muzzle brake and a recoil system employing a hydraulic recoil buffer and hydropneumatic recuperator. The breech mechanism is a semiautomatic vertical sliding block and the carriage a split tubular trail. The weapon can sustain a rate of fire of fifteen to twenty rounds per minute. Its traveling weight is 1,750 kilograms (3,850 lb.).

ASU-85 85mm Air Transportable Self-Propelled Antitank Gun (USSR)

ASU was the Soviet designation for airborne assault gun. This weapon is deployed only with air assault divisions and is air transportable in the AN-12 Cub aircraft. Many of the automotive components for ASU-85 are identical to those used in the PT-76 light amphibious tank, although the ASU-85 has no amphibious capability. It can, however, ford to a depth of 1.1 meters (3.6 ft.).

The ASU-85 has infrared night vision equipment and is probably fitted with an NBC (nuclear, biological, and chemical) defense system. Its main armament consists of an 85mm gun mounted in the glacis plate, slightly offset to the left of the centerline. The weapon has a double-baffle muzzle brake and a fume extractor. Mounted coaxially to the right of the main armament is a 7.62mm SCMT machine gun; some later versions have been observed with a 12.7mm DSHKM machine gun mounted on the roof for antiaircraft defense. Mounted over the main armament is an infrared searchlight that moves in elevation and traverse with the gun. The main weapon can elevate to +15 degrees, depress to −4 degrees, and traverse to 12 degrees. Both elevation and traverse are manual.

The main armament, the D-70 85mm gun, appears to be a variant of the D-48 85mm towed antitank gun. Served by a crew of four, it fires AP, HE, or HVAP projectiles at a muzzle velocity of over 1,000 meters (3,300 ft.) a second. Some systems have been fitted with a bank of smoke dischargers over the top of the hull at the rear; these can launch smoke grenades over the frontal arc of the ASU-85. The vehicle has a manual transmission and is powered by a 240-horsepower diesel engine (some later production models may have a 280-hp engine) and can attain maximum road speeds of 45 kilometers (27 mi.) per hour.

90mm Antitank Gun Models 50 and 57 (Switzerland)

The Swiss army currently employs two towed antitank guns, the Model 50 Panzerabwehrkanone (or Pak 50, for short) and the Model 57 (or Pak 57). Both weapons have a two-wheel split trail carriage with a small fixed spade mounted at the end of each trail. The Model 50 is fitted with a flat shield with its lower portion sloping forward at an angle of about 45 degrees; the Model 57 has a smaller curved shield. The Model 50 is towed muzzle-first with the split trails folded into the vertical position to reduce overall length of the weapon. In the firing position the towing attachment is removed and mounted on the rear of the shield. On the Model 57 the complete barrel, mount, and shield are swung through 180 degrees so that they rest over the trails. Both weapons are fitted with a day and night sight. They are served by a crew of five and can be towed by a jeep or Land Rover–type vehicle.

The Model 50 fires a HEAT projectile weighing 1.95 kilograms (4.3 lb.), the Model 57 a 2.7-kilogram (5.9-lb.) HEAT projectile. These will penetrate 250 millimeters (10 in.) of armor. Both have a muzzle velocity of 600 meters (1,980 ft.) per second and a maximum range of 4,000 meters (13,200 ft.), a normal rate of fire of six rounds per minute and a maximum rate of fire of twenty rounds per minute. In 1983, the Swiss Ministry of Defense announced that new HEAT ammunition for these weapons had been developed by Bofors of Sweden and that production was being undertaken under license in Switzerland by the federal ammunition factory at Altdorf. The Model 57 is also fitted with an American 12.7mm ranging machine gun, added after the weapon was introduced into service with the Swiss army.

90MM MECAR KENERGA 90/46 ANTITANK GUN (BELGIUM)

The Kenerga 90/46 antitank gun features low recoil, an approximately 6-ton trunnion load, and light construction, making it particularly well suited for use with light armored vehicles and a small gun carriages. Its semiautomatic breech is of the classical horizontal sliding block type. The recoil system consists of a two-stage muzzle brake that reduces recoil energy by some 70 percent, an optimized hydraulic recoil cylinder, and a hydropneumatic recuperator. A shield is available to provide a degree of crew protection against small arms and shrapnel. The weapon can fire a range of projectiles, including APFSDS-T, HEAT-T, HESH-T, WP smoke, and canister. It can achieve a maximum rate of fire of ten rounds per minute and a sustained rate of seven rounds per minute. Its range of traverse is 54 degrees.

100MM ANTITANK GUN T-12 (USSR)

The smooth-bore T-12 is the replacement for the 85mm antitank gun D-48. The fin-stabilized nonrotating APFSDS and HEAT projectiles fired by the T-12 resemble those fired by the 115mm gun of the T-62 main battle tank. The weapon fires: an APFSDS projectile weighing 5.5 kilograms (12.1 lb.) having a muzzle velocity of 1,500 meters (4,540 ft.) per second that is capable of penetrating 406 millimeters (16.2 in.) of armor at zero degrees incidence and 500 meters (1,650 ft.) range; and a HEAT projectile weighing 9.5 kilograms (20.9 lb.) that develops a muzzle velocity of 990 meters (3,267 ft.) per second and can penetrate 400 millimeters (16 in.) of armor at an incidence of zero degrees.

The T-12, served by a crew of six, has a split trail carriage and a gun shield. It utilizes a pepperpot muzzle brake. The weapon can traverse through 27 degrees, elevate to +20 degrees, and depress to −10 degrees. Its rate of fire is ten rounds per minute. Maximum effective range is 900 to 1,200 meters (2,970–3,960 ft.).

The T-12 reached full operational capability with the Soviet army in 1965. A slightly modified version, known in the West as the T-12A, was called the MT-12 by the East German army. It differs from the basic T-12 in that it weighs slightly more and uses a different tire specification. The T-12 was issued in the Soviet army on the basis of 12 guns (two batteries, each with six guns) in the antitank battalion of the motorized rifle division and 36 guns (two battalions, each having 18 guns) in independent antitank regiments.

100MM ANTITANK GUN TYPE 86 (PEOPLE'S REPUBLIC OF CHINA)

The Type 86 antitank gun appears to be a modification of the 85mm field gun Type 56, incorporating a new ordnance and recoil system. The Type 86 gun is smooth-bored and features a distinctive muzzle brake. The breech mechanism has a vertical sliding block. The sighting mechanism is located to the left of the gun: a night vision device is also available. The carriage uses split trails, with a castor wheel on the left trail leg to assist in bringing the gun into action. A shield with sloping sides is also provided.

The overall weapon weighs 3,600 kilograms (8,052 lb.) in traveling configuration. It has a maximum range of 13,650 meters (8.5 mi.) and a traverse of 50 degrees. Although the Type 86 is primarily an antitank weapon, it can also be used in an indirect fire role, although potential range is limited by the maximum gun elevation of +38 degrees. The maximum rate of fire is quoted as eight to ten rounds per minute.

105MM NORICUM ANTITANK GUN ATGN 105 (GERMANY)

The Noricum ATGN 105 was originally developed as a mobile trails mount for the long recoil Noricum 105mm system and its associated ammunition, but the manufacturer realized that there might be a market for a 105mm antitank gun using existing 105mm L7/M68 tank gun ammunition. The result was the development of this gun, mounted on a split-trail, single-axle, two-wheeled carriage. The split trail has cranked, box-type trail legs of welded steel that are fitted with fixed trail spades. For towing, the barrel can be rotated through 180 degrees and folded over the trail legs. Recoil length is 590 millimeters (23.6 in.) and the weight of the complete gun 3,900 kilograms (8,580 lb.).

The ATGN 105 can fire all standard NATO 105mm L7/M68 tank gun ammunition, including HEAT and APDS, as well as the Noricum NP 105 A2 APFSDS round. Developed jointly with Hirtenberger GmbH, that projectile is made of Tungalloy 176 FA, uses a long rod penetrator, and develops a muzzle velocity of 1,485 meters (4,900.5 ft.) per second.

106mm Self-Propelled Recoilless Gun Type 60 (Japan)

Following successful employment of jeep-mounted recoilless rifles by the U.S. Army in the Korean War, the Japanese Ground Self-Defense Force issued a requirement for a self-propelled tracked recoilless gun, specifying that it was to be equipped with two recoilless rifles that could be raised hydraulically (with additional manual controls for emergency use) and have a crew of three, a loaded weight of 5,000 kilograms (11,000 lb.) an air-cooled diesel engine, and a maximum road speed of over 55 kilometers (33 mi.) per hour.

Prototype development and testing of the system in the mid- to late 1950s were fraught with difficulties, including engines overheating and steering mechanisms behaving poorly. After modifications, the weapon was accepted into service in 1960. Subsequent improvements have been made to increase engine power from 110 horsepower at 2,300 rpm to 150 horsepower at 2,800 rpm. Something on the order of 250 Type 60s have been produced with production ending in fiscal year 1979.

The type 60 is armed with two 106mm recoilless rifles that have an effective range of 1,100 meters (3,360 ft.). A 12.7mm spotting rifle assists with aiming the parallel-mounted recoilless rifles, further assisted by a 75-centimeter (30-in.) stereo rangefinder and an infrared night sight. The Type 60 has a three-man crew consisting of driver, loader, and commander and can carry a combination of eight HEAT or HE rounds. Special tracks give the vehicle a cross snow capability, but there is no amphibious capability or NBC protection.

Jagdpanzer Kanone JPZ 4–5 Self-Propelled Antitank Gun (West Germany)

When the West German army was formed, it issued a requirement for a 90mm self-propelled antitank gun—or tank destroyer, as the Germans normally call this kind of weapon. The first vehicle intended to meet this requirement was basically a Swiss-designed HS-30 armored personnel carrier with a 90mm gun ball-mounted in the glacis plate. In trials, this approach did not meet with success, so development of a new system began. This time a chassis was developed that could be used for a number of basic vehicles, including the Jagdpanzer Rakete, a reconnaissance tank (which was subsequently developed to the prototype stage but not put into production); an MICV, which eventually became the Marder; and various other supporting vehicles. Three series of prototypes of the self-propelled antitank gun were built between 1960 and 1965. The prototypes were followed by 750 production vehicles completed between 1965 and 1967, half being built by Hanomag and half by Henschel.

The weapon is served by a crew of four. It has a combat weight of 27,500 kilograms (60,500 lb.). Powered by a 500-horsepower diesel engine, it has a road range of 400 kilometers (240 mi.) and a maximum road speed of 70 kilometers (42 mi.) per hour.

The vehicle's 90mm gun, which fires the same ammunition as the M47 and M48 tanks, is mounted in the glacis plate, slightly offset to the left of center. The barrel is fitted with a double-baffle muzzle brake and a fume extractor. It can elevate and depress to +15/−8 degrees and traverse through 30 degrees. Both elevation and traverse are manual. The gun fires HEAT-T and HESH-T projectiles of a maximum effective range of 2,000 meters (6,600 ft.) at a maximum rate of fire of twelve rounds per minute.

Mounted coaxially with the main armament is a 7.62mm MG3 machine gun. A similar weapon mounted at the commander's station for antiaircraft use can also be removed and fitted at the loader's station. Mounted on the rear of the fighting compartment are eight electrically operated smoke dischargers that fire forward over the front of the vehicle. Mounted over the main armament is an infrared and white searchlight that can be removed and stowed at the rear of the hull when desired. The Jagdpanzer Kanone is also equipped with infrared night-vision equipment and a nuclear-biological-chemical defense system. The basic vehicle can ford to a depth of 1.4 meters (4.6 ft.) without modification, and to 2.1 meters (6.9 ft.) with use of a fording kit. Some 160 of these systems have been rebuilt to a Jaguar 2 configuration and fitted with the Hughes TOW ATGW system with AN/TAS-4 night sight. Some have also been used in the role of an observation post with the 90mm gun removed.

Conclusion

Although their importance in antiarmor fighting—once dominant—has diminished, antitank guns will remain in service around the world for the foreseeable future. Many military commanders recognize the tactical capability of antitank guns to complement other antiarmor systems, and that their effect in urban fighting can be greater than that of weapons that must be guided to a target and have relatively slower projectile speeds. They also tend to be less expensive to produce and field, and so will remain popular with countries of limited resources or warfighting requirements.

Alaa El Din Abdel Meguid Darwish

See Also: Armor; Armor Technology; Mechanized Warfare; Missile, Antitank Guided; Rocket, Antiarmor; Tank.

Bibliography

Hogg, I. V. 1977. *The encyclopedia of infantry weapons of World War II*. New York: Thomas Y. Crowell.
Jane's armour and artillery 1988–89. 1988. London: Jane's.
Reid, W. 1976. *Arms through the ages*. New York: Harper and Row.

ADDITIONAL SOURCES: *Armada International; Military Technology; NATO's Sixteen Nations.*

GUN TECHNOLOGY APPLICATIONS

Gunpowder for use as a propellant was known in China as early as the eleventh century, and according to Franciscan friar Roger Bacon, has been known in western Europe since 1249. About 60 years later, the Europeans learned how to harness that explosive energy to accelerate a projectile. References to the "handpipe" in the city of Aachen chronicles of 1338 describe a device still recognizable as a gun (Fig. 1).

Since the first appearance of that primitive fire-pot, a wide variety of firearms has been developed, from slow match to flintlock and from small revolver to the heaviest siege guns. All are based on the same principle: the tube is loaded with powder and projectile, and the ignition charge is actuated by means of sparks, causing deflagration that propels the projectile.

During the nineteenth century, firearms development progressed enormously. The impact-sensitive percussion cap was introduced in 1807, to be succeeded by the paper cartridge, breech loading, and the metal case with center firing. The final step on the road to automatic weapons was taken in the United States by Richard Gatling, who, in 1862, developed the first operable machine gun. This was followed in 1884 by the recoil-operated Maxim and in 1892 by the Hotchkiss gas-pressure-driven system. Cast bronze muzzle-loading guns were then replaced by steel breech-loaders with rifled barrels; the recoil gun subsequently led to a higher rate of fire. Black powder was replaced by the smokeless and residue-free nitrocellulose powder in 1884. Table 1 lists a variety of modern guns, from handguns to heavy artillery, categorized according to size, form, and function. Some of the features and processes associated with operating a firearm are shown in Table 2.

TABLE 1. *Types of Modern Guns*

CATEGORY	EXAMPLES
Handguns	Revolver, pistol, rifle, machine pistol, machine gun
Machine guns	Revolver, Gatling gun
Automatic guns	Recoil-operated, gas-pressure operated, mass locked weapons (blow back), externally powered
Artillery guns	Field howitzers, field guns, mountain guns, fortification guns, naval guns
Tank guns	Tank guns
Antiaircraft guns	Antiaircraft guns
Special guns	Mortars, recoilless guns

Basic Operation and Assembly of a Gun

The gun barrel (Fig. 2) is the tube in which gas pressure accelerates the projectile. The projectile is propelled through the muzzle with a certain velocity and in a certain direction. The rear end of the tube is closed gas-tight, and ammunition is loaded through the breech there. Between the projectile base and the rear of the tube is the "chamber" that holds the propellant and ignition device. The ignition device is actuated through the breech.

A "round" of ammunition is inserted into the chamber through the breech. The breech is then closed, and the igniter element is actuated by a mechanical blow or electrically. This actuates the primer charge, evenly igniting the propellant. The time between the firing pin impact and the start of propellant deflagration is less than one millisecond in small-caliber weapons and up to tens of milliseconds in large-caliber systems. Ignition transforms the propellant charge into gas.

The pressure created by the hot propellant gas expanding between the tube rear end and the projectile base drives the projectile with increasing velocity down the tube until it exits through the muzzle. The characteristic in-bore pressure distribution (Fig. 3) from ignition until total propellant transformation into gas is controlled by contrary facts:

1. Transformation velocity increases at rising pressure, resulting in further pressure increase.
2. The projectile converts gas pressure into kinetic energy, which results in pressure decrease. Simulta-

TABLE 2. *Functional Sequence of Guns*

Gun	Ammunition transport, chambering, closing, firing, opening, case removal
Ammunition	Propellant ignition, in-bore acceleration, ballistic flight, effect on target
Aiming means	Target assignment, target acquisition, aiming, observation, correction
Mount	Gun ammunition transport, weapon mounting, absorption/deflection of firing forces

Figure 1. Firing of a handpipe. (SOURCE: City Chronicles of Aachen, City Museum, Aachen, Germany)

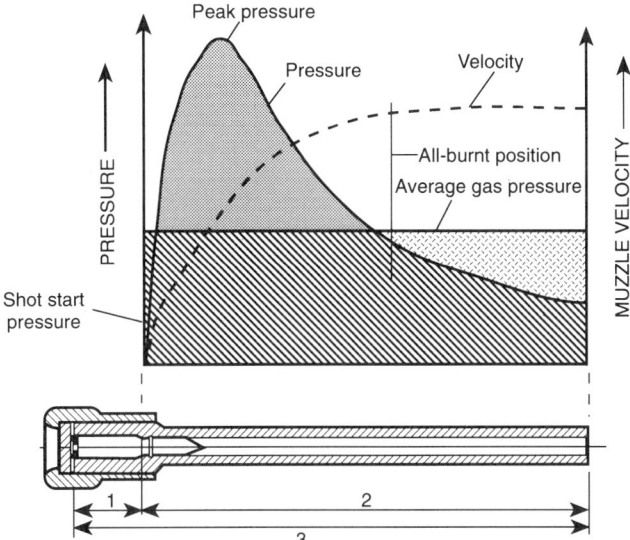

Figure 2. Schematic view of a loaded tube—typical pressure/time, velocity/time curves. (SOURCE: Rheinmetall)

neously, as the projectile moves down the tube, the space available for the gas to expand increases, resulting in further pressure decrease.

Normally, the in-bore gas pressure rises to a peak over the projectile travel and then drops until final combustion is achieved. Thereafter, pressure keeps dropping according to a polytropic expansion.

Ideally, propellant deflagration should be completed when the projectile is as far toward the muzzle as possible. The acceleration of the projectile takes place over a very short distance, but creates up to 10^5 g-loads. Projectile velocity at the muzzle can reach more than 1,700 m/sec. (about 1 mi./sec.) at maximum gas pressures in excess of 6,000 bar and temperatures of up to 4,000 degrees K.

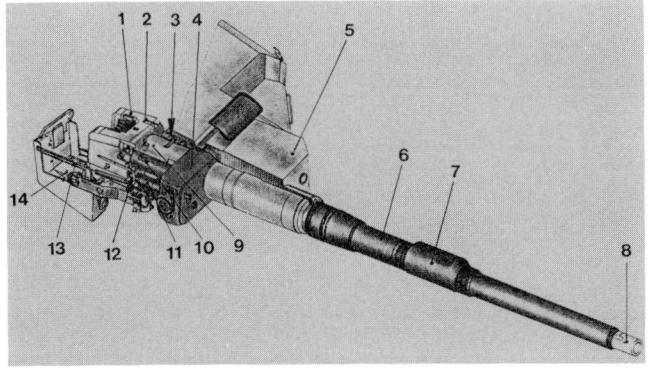

1. Breech-block with wedge
2. Buffer
3. Barrel brake, LH
4. Cradle
5. Shield with sealing
6. Thermal sleeve
7. Bore evacuator
8. Tube
9. Cradle tube
10. Recuperator
11. Emergency trigger
12. Barrel brake, RH
13. Recoil indicator
14. Spent case box

Figure 3. Smooth-bore tank gun (120 mm). (SOURCE: Rheinmetall)

The chemical energy released when the propellant is ignited is consumed as follows:

Projectile motion	33%
Frictional losses	3%
Propellant gas motion	4%
Heat loss to gun and projectile	20%
Heat retained by propellant gases	40%
Total propellant energy released:	100%

Accordingly, only one third of the chemically bound propellant energy is used for actual projectile propulsion; the rest is lost primarily to gas dynamics and heat build up.

BARREL

The gun barrel consists of the tube, the breech system, the firing mechanism (and, if necessary, the muzzle brake), the bore evacuator, and the thermal shroud.

Tube. The tube is divided into two areas: at the rear is the chamber, which tapers at the front to enclose the cartridge case neck; and the narrower tunnel of the bore, through which the projectile is propelled. The inner diameter of the bore, measured in millimeters, is the caliber of the gun. The bore is rifled along its length—that is, it has grooves that extend helically toward the muzzle to impart spin to the projectile as it travels down the bore. This spin stabilizes the projectile in its trajectory.

Gun bores may also be smooth; these are simpler and easier to produce, and have been used in mortars to fire relatively slow-moving, fin-stabilized rounds. Since the 1960s, smooth-bore technology has also gained the edge in tank guns, because modern high-speed, armor-penetrating projectiles can develop their full potential only when fin stabilized (see Fig. 3). In addition, the smooth bore surface allows higher thermal loads (i.e., peak gas pressure and temperature) and less erosion (i.e., wear life) when fired.

The tube wall thickness is determined by the gas pressure load and should decrease toward the muzzle. Tube rigidity must be high enough to avoid any uncontrollable oscillations (longitudinal, lateral, twisting) that might have negative influence on firing accuracy.

With the rapid advances in weapons technology during the nineteenth century, cast bronze, cast iron, and cast steel tubes became obsolete. Today, forged or centrifugally cast monobloc tubes are used for smaller-caliber weapons. For high-performance guns, blanks are specially produced using extremely hard heat-treated vacuum-melted steel alloys with optimized properties. Special tube designs have been developed, such as built-up (multilayered) tubes, monobloc autofrettage tubes with interchangeable inner liners (where only the worn-out liner is replaced), and easily and quickly replaceable tubes.

A tube is subjected to many stresses during system operation, including:

- Radial and tangential stresses due to gas pressure
- Longitudinal stresses due to rotating band pressure
- Torsional stresses due to projectile rotation
- Bending stresses due to the tube's own weight

Radial and tangential stresses exerted during firing are unevenly distributed across the tube diameter from the inside to the outside. This is compensated by autofrettage, which loads and evenly distributes residual stresses through the tube wall when firing. This technology facilitates (1) higher peak gas pressures, (2) better barrel dimensioning, and (3) very high fatigue life.

Tube heating can be a problem in rapid-firing guns, and special measures must be taken to carry off heat since barrel wear grows as the temperature increases, as does the risk of propellant "cook-off." Preventive measures are water-cooled tubes, radiating ribs, fast and easy tube interchange, and heavier tubes with higher heat absorption capability.

Thermal shrouds (sleeves) are used on especially long tubes because the tubes tend to bend at higher temperature differences due to one-sided (solar) heating or cooling; this reduces accuracy. Thermal shrouds prevent such temperature influences.

A bore evacuator is a device that removes residual in-bore propellant smoke through the muzzle, thus avoiding blow-back into the crew space when the breech is opened for reloading. In closed weapon stations, the bore evacuator or smoke absorber is mounted on the tube. The bore evacuator uses the ejector principle. As soon as the projectile has passed the absorber holes, which are canted towards the muzzle, part of the propellant gas fills the absorber pot and builds up pressure. As pressure in the tube decreases, these gases expand back into the bore and also wash out residues from the tube rear through the muzzle.

Tube service life is determined by wear life and fatigue life. Wear life is reached when mechanical drive and propellant gas erosion deteriorate a gun so that it can no longer meet performance requirements. Wear life depends on the type of projectile, muzzle velocity, propellant charge, tube temperature, rate of fire, and so on. Wear can be reduced by hard chromeplating of the bore surface, use of propellant additives, and (in small-caliber weapons) cold forming of rifling. Drawn or rifled tubes have, at identical stress loads, a shorter wear life than smooth-bore tubes.

Fatigue life is reached when materials fatigue and can result in spontaneous tube destruction. This is caused by microscopic cracks at the materials' grain boundaries that grow and eventually attain "critical crack depth." Autofrettage increases the weapon's life up to three times the original number of rounds. Highly stressed guns (tank guns) usually reach *wear* life sooner, whereas less stressed howitzer tubes reach *fatigue* life earlier. Highly stressed tubes usually are designed so that they can rapidly be exchanged; examples are machine gun tubes with rapid locks and tank gun tubes with bayonet thread between breech block and tube.

Breech (bolt). A certain sequence of operations is necessary to fire multiple rounds:
1. Chamber the cartridge
2. Lock the breech or bolt
3. Actuate the firing mechanism
4. Fire
5. Unlock the breech or bolt
6. Extract the empty cartridge case from chamber
7. Store the actuating energy for breech or bolt
8. Eject the spent cartridge case
9. Chamber the new cartridge

Automatic weapons perform these steps automatically; other weapons may be partly automated so that certain operations have to be performed manually.

Actuation of the breech or bolt system facilitates the repetition of rounds being fired. Its tasks are:

- Gas-proof sealing of the tube rear
- Initiation of propellant ignition
- Absorption of gas pressure
- Extraction and ejection of cartridge case
- Chambering of new cartridge for next round (in automatic weapons only)

Breech or bolt systems can be quite different and are determined by rate of fire, caliber, and ammunition type. Breech systems include the closed, screw, and wedge types.

The simplest form of breech is the tube closed at one end. This closed breech form is as old as the gun itself and today is used only in mortars. At the closed end the mortar has a fixed firing pin onto which the round and its percussion cap drop.

In the mid-nineteenth century, the screw-type breech was developed, which made breech loading possible. A bayonet-threaded plug is manually or mechanically swung into the bore axis and screwed into the tube. Due to the use of an elastic obturator, this breech system is well suited to firing ammunition with bulk charges. This design depends on an obturating mushroom that protrudes into the chamber and, using gas pressure, compresses the elastic obturator.

The wedge-type breech developed later in the nineteenth century is a rectangular block that, by simply moving in a breechblock recess vertically to the bore center axis, closes or opens the rear tube end for loading. The wedge-type breech is simple in design and operating sequence and, in connection with a ring-obturator, can also be used to fire ammunition without a sealing metal case. The ring obturator consists of a base ring plus an actual obturator that seals the breech by being forced against the barrel wall and wedge when under pressure.

Generally, ammunition is manually chambered in both

breech types. Chambering may be partly automated using loading aids (flick-ramming). In the future, fully automated chambering due to automatic loaders will gain ground even with large-caliber guns.

The simple operating sequence of the wedge-type breech is very fast and suitable for automation: driven by the tube recoil after firing, the breech opens, ejects the spent case, retorques the closing mechanism, and automatically closes after the new round is chambered.

Automatic Weapons

Very early revolvers were first tested in the Middle Ages. By about 1715, Puckel had developed the basic principle of the revolver weapon; following invention of the percussion cap in 1807, Samuel Colt used Puckel's principle in designing his own famous revolver in 1835. Introduction of the metal-cased cartridge in the mid-nineteenth century led to the development of the automatic weapon with constant ammunition feeding. Over time, numerous variations emerged.

In 1862, the rapid-fire gun developed by Richard Gatling was presented. It was a multibarrel system with a rotating barrel arrangement driven manually by a crank. This was the first automatic rotary action cannon in which the cylinder was reloaded during firing. Hiram Maxim followed in 1884 with the first self-actuating machine gun with a linear oscillating bolt system operating on the recoil principle, and in 1892, J. M. Hotchkiss developed the first gas-operated automatic weapon. Only at the end of World War II did the German company Mauser develop the Colt-equivalent revolver gun with a single fixed tube.

Modern automatic weapon developments are based principally on the three main types of bolt action: rotating, linear, and lateral action. The functional round-to-round sequence of an automatic weapon (including the drive for simple ammunition feeding) employs energy that is provided either from outside (external power), manually, or mechanically; or from the propellant charge (self-driven), by tube recoil, case base pressure, or gas pressure.

The simplest form of external drive is found in the repeating rifle. Here, the bolt is cycled manually. The Gatling gun derived its energy from a manually powered crank. Modern machine guns use an external motor, either electric, pneumatic, or hydraulic.

The self-actuating method uses energy derived from the propellant charge; this is obtained via tube recoil (see Fig. 4), directly from gas pressure, tapped from the barrel, or direct via the case acting on the bolt.

The main advantages of externally powered systems are smooth and uniform bolt motion and a controllable rate of fire due to drive speed. A disadvantage is the need for external energy, plus the fact that Gatling guns do not immediately reach their full rate of fire.

The main advantages of self-powered systems are independence from external energy, high and immediately

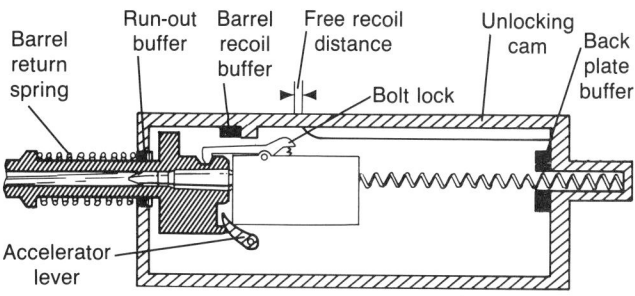

A. Start of cycle

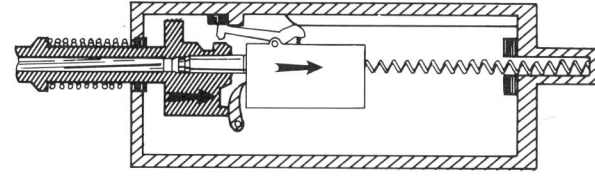

B. Bolt unlocked; acceleration starts

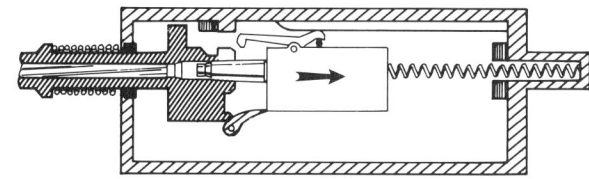

C. Acceleration completed

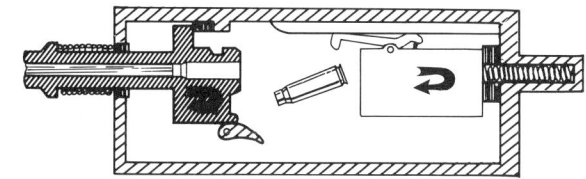

D. Barrel rebounding from recoil buffer; bolt rebounding from back plate buffer

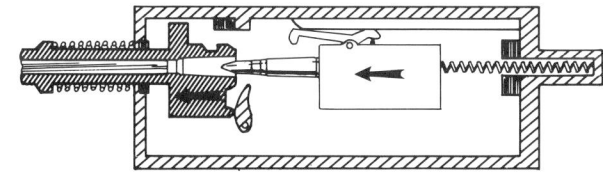

E. Bolt loading fresh cartridge

Figure 4. Recoil loader with short recoil. (SOURCE: C. J. Marchant Smith and P. R. Haslam, *Small Arms and Cannons*, London: Brassey's, 1982, p. 145)

available rate of fire, and sufficient energy in gas pressure operation for belt pull without affecting the rate of fire. Disadvantages are rough bolt motion and the difficulty controlling a low rate of fire.

ROTATING BOLTS

Revolver/Gatling principle. Weapons with a rotating bolt system have a chambered cylinder that holds the cartridges. In revolvers and revolver guns, this loaded cylinder rotates at the end of a single tube (gas sealing), whereas in the Gatling gun, each chamber is affixed to its own tube.

The rotating cylinders are loaded by transport stars and cartridge pushers. The pushers also extract the spent cases after firing. The rotating bolt action and the arrangements of multiple chambers facilitate a high rate of fire. Rotary guns differ in both their cylinder cycling and their drive mode. While the Gatling, due to its rigid cylinder/barrel assembly connection can fire while rotating, the revolver gun must be at rest when fired. The revolver gun taps the gas used to cycle the cylinder and feeds ammunition from the rigid tube. This driving mode creates problems in a Gatling gun because of the rotating tube assembly. Therefore, Gatling guns are mostly externally powered, either electrically, hydraulically, or pneumatically.

Roller breech. The roller breech or tilt cylinder is a special form of the rotary bolt principle especially suited for caseless ammunition. The roller breech pivots with the chamber back and forth from firing position to feeding position. It is gas-pressure-operated and seals the chamber via telescoping sleeves. This new functional principle was designed by Heckler and Koch especially for the caseless ammunition of their G11 rifle.

Linear bolt. The essential features of this bolt are that it moves horizontally to chamber a cartridge, locks with the tube, ignites the cartridge, and ejects the spent case as it recoils. The energy required for this bolt movement can be provided either externally or through self-actuation. In the latter case, the bolt is forced backwards by gas pressure of the round being fired; this simultaneously torques a spring that provides the return action. As the bolt motion is buffered and diverted, a new cartridge is inserted, and the returning bolt pushes it into the chamber and locks. In externally powered systems, the closing spring is not necessary, since the bolt motion is forced.

Most automatic weapons designed for multipurpose applications and mounts, from machine pistol to heavy guns, operate with the linear bolt action. Only when higher rates of fire (more than 1,000 rounds/min.) are required is the linear bolt replaced by the faster revolver or Gatling principle. The main difference between these two systems is their motion and functional sequence: the steps in revolver and Gatling guns are staggered but parallel while they are in sequence in a linear moving action.

Lateral bolt. The lateral bolt is a special version of the linear bolt in which a feeder transports the cartridge laterally and a laterally moving wedge locks the tube. This version is especially suitable for self-operating guns because of its smooth performance.

BOLT ACTUATION

There are four basic means of bolt driving: the recoil loader with long and short tube recoil, the case-base loader using the cartridge case with simple or delayed action bolt, the gas pressure loader that taps gas directly from the tube, and the externally powered gun.

The tube recoil loader has a tube that is loosely seated in the weapon housing. When the gun is fired, gas pressure forces the locked tube and bolt backwards. With long tube recoil, the tube pushes the bolt backwards to the return point, whereas with the short tube recoil, an additional accelerator is used to drive the bolt backwards. Short recoil systems often use a recoil booster at the muzzle to increase the recoil force and improve the rate of fire; this also is called reverse muzzle brake action. The recoil loader is a very robust design; however, the rate of fire can be varied only by design measures.

The simplest automatic weapons are driven by the cartridge case base and are "semilocked" by the dynamic mass of the bolt. There are two types of case-base loaders: the simple dynamic mass bolt with high mass for locking, and the delayed action bolt with a lightweight mass. The latter type compensates for the missing weight necessary for secure locking by energy transfer, assisted by an accelerator for part of the two-part bolt. Because of the interference during the acceleration phase, this locking mode is also called "semirigid." The simplicity of this drive and locking mode makes case-base loaders attractive for use in machine pistols and machine guns.

The gas-pressure loader taps a portion of the propellant gas directly from the tube and uses it to activate a piston, which unlocks the bolt from the receiver housing and drives the bolt backwards. Gas-pressure loaders have a number of advantages: the timing of actions and the amount of gas (nozzle cross-section) can be adapted to system functions and the rate of fire can—within a certain range—easily be varied. Most modern automatic weapons are driven by this method.

Externally powered guns with linear bolts use a chain or other mechanical means to actuate the to-and-fro motion of the bolt. The motor drives the bolt throughout the whole cycle. Advantages of this system are its variable rate of fire based on the drive speed and the uniform and function-adequate controllable bolt movement (feeding, simple and safe locking, case removal, etc.); one disadvantage is its dependency on external energy.

FUNCTIONAL SAFETY

To avoid hazards in weapon system and handling, safety features are provided that prevent unintentional firing (trigger) and ensure that the tube and bolt are securely locked during firing.

Locking and ignition. A number of methods are available to lock the bolt either directly to the tube (in recoil-operated systems) or to the housing (in fixed tube systems). The semilocking system—the simplest system—uses the kinetic mass effect of battery firing. Semirigid locking also uses the kinetic mass effect, but the bolt is lighter and more complex. The missing mass is replaced by transformed recoil momentum. Both semilocking and semirigid locking facilitate the very simple bolt actuation via the cartridge case base. Rigid or form-fit locking can be achieved by screw-type lock, supporting lugs, support rollers, balls, or similar supporting elements, or by a lateral bolt. Rigid locking provides a high degree of safety. The supporting-lug bolt locks quickly and is favored in high-rate-of-fire systems. Under certain conditions, the bolt tends to jump, due to the front-end impact on the tube rear, causing premature unlocking. By employing special delay elements, this danger can be avoided.

Triggering system. The trigger is manually or electromagnetically (solenoid) actuated to initiate firing. Ignition of the cartridge is usually performed by a spring-loaded firing pin or electric current. Prevention of unintentional firing is of the utmost importance and ignition must start only after the bolt is securely locked. In systems with lateral bolt function, the bolt is held by the trigger in the ready-to-release rear position during firing interruption.

Closed bolt weapon systems "park" the bolt in the rear position, whereas in the open bolt mode, the bolt is locked in the forward position and cartridge ignition is separately controlled. Open bolt systems are used only when utmost precision in rate of fire and accuracy is required.

AMMUNITION FEEDING

Automated ammunition feeding systems called "loaders" (in large-caliber weapons) are required in single-shot weapons to increase the rate of fire, minimize crew size, and compensate for munition weight, or allow for remote weapon system control. The degree of automation in single-shot weapons feeding covers the range from loading aids to remote-controlled fully automatic loaders.

Feeding systems for automatic weapons are considered an integral, functional part of the weapon. Cartridges can be fed in limited number by magazine; in the case of a larger supply, they can be contained in a "linkless feed system" or be fed linked in belts. If transport distance is not excessive, the feed mechanism (which transports a cartridge into chambering position) is driven mostly through tube recoil, weapon recoil, or gas pressure. Modern multipurpose weapons are equipped with alternating feeders, allowing different types of cartridges to be selected for different missions.

Carriages or Mounts

The "gun" comprises the actual ordnance together with the carriage or mount. The carriage or mount consists of an array of assemblies to support the ordnance when firing. The carriage absorbs or deflects occurring forces as well as mounting the ordnance and aiming and training the tube. It may also be used to transport ordnance and ammunition.

Carriages/mounts and their assembly groups can be configured differently, depending on their use: for ground- or surface-based guns, there are static or semistatic, mobile or self-propelled mounts; for guns based on aircraft, the mounting platform aims the gun system.

The main assembly of a carriage or mounting, the superstructure (Fig. 5), consists of the top carriage, the elevating and traversing gears, the balancing gears, the cradle, the recoil system, and the sighting device. These elements are used to mount and train the gun, and to absorb and deflect the firing stresses. The main parts of the basic structure are the bottom carriage, the trials and articulation system, the platforms, the spades, and the wheels, axles, and so forth. Of course, handguns and rifles have the simplest mounts: the user.

The carriages of more complex gun systems (e.g., artillery pieces or the main guns of tanks) incorporate subsystems to aid in tube elevation, gun traverse and lay, and gun stabilization. These subsystems are respectively known as cradles, laying gear, and stabilization units.

Cradles support the mass of the gun tube at rest and during firing and facilitate tube elevation (or depression) about its trunnions. As the trunnions provide the pivot point for the tube, their location and the use of counterweights are important considerations in gun design, since artillery weapons are generally used for indirect high- or low-angle fire, and tank guns are used more in direct fire roles.

Laying gear may comprise various combinations of gears, threaded spindles, or hydraulic cylinders, and is used to adjust the elevation and traverse of the gun so that the trajectory of the projectile will intercept the sighted target. Gun stabilization systems, on the other hand, permit accurate laying of a gun while mounted on a moving platform, usually ships or armored vehicles. More complex, modern stabilization subsystems use ballistic computers and compensate for elevation, roll, and transverse motion.

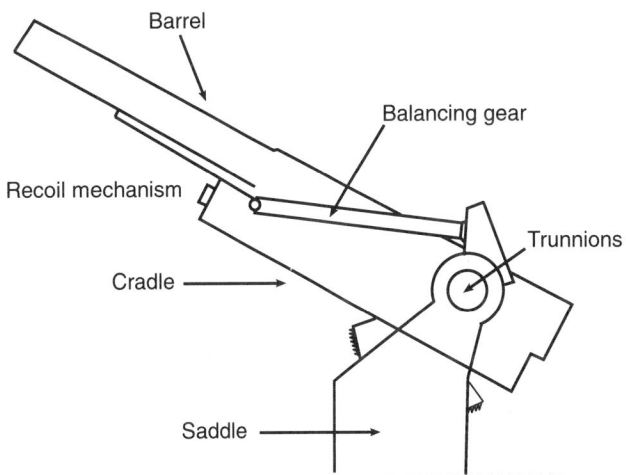

Figure 5. Superstructure of the conventional gun. (SOURCE: J. W. Ryan, *Guns, Mortars and Rockets*, London: Brassey's, 1981)

Absorption/Deflection of Firing Forces

Recoil force. The gas pressure accelerating the projectile must, as a counterforce acting on the breech, be absorbed by the mount. The dimensions of the mount must be adequate to absorb and deflect the recoil forces. The forces at the elevating mass are transmitted to the saddle via trunnions, elevating gear, and equilibrators. In systems with a recoiling tube, the recoil forces acting on the tube are transformed into reverse acceleration of the tube—recoiling mass. The kinetic energy of the recoiling mass is absorbed during recoil travel by a hydraulic tube recoil brake.

Recoil energy acting on the mount can be reduced further by increasing the weight of the recoiling mass, using muzzle brakes, and employing out-of-battery firing. Special-purpose guns may use low-recoil or recoilless systems.

The recoil system consists of a recoil brake, the recuperator, and a slider rail in the cradle. A recoil system must provide for uniform and vibration-free action and, when firing, a secure resting of the mount to avoid delays in firing sequence due to extensive readjustments. In general, only hydraulic brakes are used. They are connected with the recoiling mass and the cradle. Inside the brake unit, hydraulic fluid is pressed through an opening in the piston and converted into heat. By design, hydraulic brakes can be tuned to the exact loads of a mount, depending on the length of recoil and the elevation, so that an almost constant mount load is maintained throughout the recoil travel. After firing, the recuperator brings the tube back into rest position. The tube recoil energy is pneumatically or hydropneumatically accumulated.

The braking force to be absorbed by the mount is inversely proportional to the recoiling mass, that is, at identical recoil force the mount is less stressed by a heavier recoiling mass. To minimize mount stress during firing the principle of out-of-battery firing also can be applied: gas pressure is countered by the forward impulse of the recoiling mass in single-shot systems, the interrupted counterrecoil in automatic cannons (the next round is fired before the weapon is back in battery position), or the forward impulse (kinetic energy) of the bolt mass. Accordingly, the mount loads or the mass of the bolt can be smaller.

Muzzle brakes significantly reduce mount stresses by deflecting the gases expanding through the muzzle over impact surfaces, thus pulling the tube forward. Muzzle brakes with an efficiency of 35 percent are most practical, since greater efficiencies would create considerable crew hazards from sound pressure and dust.

Recoilless guns use propellant gas effects to compensate for recoil force. The tube is open at the rear end, where a nozzle is attached, and part of the gases are exhausted through this nozzle. When the impulse of these gases is equal to the projectile impulse, there is no tube recoil and no recoil system is required (which makes the tube much lighter). One disadvantage is that rear safety is restricted because of the exhaust gases especially because overall pressure must be higher to propel the projectile. Physically, the recoilless gun is a transitional step to a rocket, which carries propulsion and housing and is self-propelling.

Projectile Propulsion

The metal case contains and protects the cartridge (which consists of projectile, propellant charge, and igniter) and securely seals the breech (or bolt) when the gun is fired. The case is made of brass, steel, or aluminum; it must be gas-pressure resistant and have the elasticity required for obturation and sufficient resilience for extraction.

The main disadvantage is that it is inconvenient: although it was a prerequisite for developing rapid-fire weapons, the case is used only once, and it is expensive, heavy, and voluminous, and it must be removed after firing. Various configurations of cartridges are known: the conventional cartridge with metal case, the cartridge with combustible case (and stub case, as in the Leopard 2 main battle tank), caseless ammunition, the mortar grenade, and the split-loading case (howitzer ammunition).

Although the first patents for systems without metal cases date back to 1891, fully or semicombustible cases or even caseless solutions have only recently been achieved. The combustible case—a mixture of paper, felt, and nitrocellulose—is (for the 120mm smooth-bore gun—the main armament of modern Western tanks) still equipped with a metal stub case to provide absolutely safe obturation at high operating gas pressures.

Caseless ammunition is still being developed. Such munitions are cheaper, up to 50 percent lighter, up to 30 percent smaller, and allow a higher charge density. Furthermore, eliminating empty case removal facilitates the round-to-round cycle of operations considerably.

Until recently, efforts to develop caseless ammunition have failed because of problems such as obturation, cook-off (self-ignition in hot chambers), rigidity, and the difficulty in removing an unspent, chambered cartridge. But some developments show promise: for example, the cartridge for the G11 Heckler and Koch automatic rifle manufactured by the German company Dynamit Nobel.

The simple mortar grenade functions fairly well as caseless ammunition; because it is muzzle loaded, no rear obturation problems exist. In-bore obturation between the projectile and the tube wall is solved by means of a specially designed obturating ring that is forced against the tube walls by gas pressure. Range, aside from tube elevation, is adjusted by incrementally adding propellant charge units.

Large-caliber artillery guns normally use separately loaded (split) ammunition. Projectile, propellant charge, and primer are separately loaded because the amount of propellant needed depends on the range and elevation (in modern howitzers, up to ten bags or modules may be needed). Moreover, the individual projectile alone can be exceedingly heavy.

Future Technology

Modern weapons that use standard propellant powder are technologically mature; developmental quantum jumps can hardly be expected. However, with an efficiency of over 30 percent, their ease of handling, and their ability to immediately fire without prewarming, they are good and reliable machines.

However, one disadvantage has not yet been eliminated: the mean gas pressure is only 40 to 50 percent of the true peak pressure. Thus, the launcher and the projectile must be designed to withstand very high peak pressures. A lower peak pressure with a tendency toward a constant pressure/time curve would be very desirable.

The search for alternate solutions to the conventional powder gun began during World War II. Today, intensive efforts are under way to achieve chemical propulsion with liquid propellants and electric propulsion.

Liquid Propellants

Compared with standard propellant powder, liquid propellants (LP) for projectile acceleration have certain advantages in system technology, such as a continuously overlapping range in howitzers and mortars because of infinitely variable propellant injection, and the elimination of the cartridge case in tank guns, which permits the protected storage of LP, use of the same propellant for all types of rounds, more favorable silhouettes for main battle tanks, and so forth.

Liquid propulsion systems under development include propulsion with liquid monopropellants containing fuel and oxidizer in one liquid, and propulsion with liquid bipropellants, which store oxidizer and fuel separately.

Two versions of propellant injection are known: bulk loading, in which the propellant is pumped into the chamber prior to firing, and regenerative loading, in which the propellant is injected into the chamber during combustion. The regenerative mode may require greater technical efforts, but the defined injection during combustion gives it the advantage of a more favorable pressure curve development tending toward even pressure distribution. At constant projectile acceleration at lower peak pressure, the stress loads on mount and projectile would be significantly reduced.

Electric Gun

Higher projectile velocities are required in air defense because of ever-shorter engagement times, for increased penetration to defeat the modern tank targets with kinetic energy projectiles, and so on. But the projectile velocity achievable with a current powder gun physically cannot be increased significantly. New methods for propulsion must be found in order to keep up with the development of potential targets. One option is the electric gun, which achieves considerably higher velocities than conventional guns.

The first patents were granted early in this century to the Norwegian physicist Kristian Birkelund. However, so far all efforts to develop a weapon system have failed because of a lack of suitable energy storage. Still, test systems have accelerated smaller projectiles to a velocity of over 4,000 m/sec. A usable system will require additional research on the storage of sufficiently high energy, pulse-forming, launchers, or projectile configurations for those velocity regimes.

Acceleration Methods

For the most part three alternative acceleration methods are being investigated: the coil gun, the rail gun, and the electrothermal gun. The coil and rail guns accelerate the projectile by electromagnetic force. They differ in the way the magnetic field is generated. The electrothermal gun accelerates the projectile in a similar way to a conventional gun—that is, by high gas pressure: a plasma burner creates an electric arc that vaporizes a suitable material, such as polyethylene, thus creating a high-pressure plasma. These principles or a hybrid electro-chemical solution will probably be the future technology applied in hypervelocity guns.

Ingo Wolfgang Weise

SEE ALSO: Ammunition; Artillery, Tube; Automatic Weapon; Explosives Technology, Conventional; Machine Gun; Materials Technology; Munitions and Explosives Technology Applications; Space Weapons; Technology and Warfare.

Bibliography

Canby, C. 1964. *Geschichte der Waffe*. Switzerland: Editions Rencontre and Erik Nitsche International.
Handbook on weaponry. 1982. Rheinmetall. Frankfurt am Main: Brönners Druckerei Breidenstein GmbH.
Lee, R. G. 1981. *Introduction to battlefield weapons systems and technology*. London: Brassey's.
Marchant Smith, C. J., and P. R. Haslam. 1982. *Small arms and cannons*. London: Brassey's.
Ryan, J. W. 1981. *Guns, mortars and rockets*. London: Brassey's.

GUNBOAT DIPLOMACY

Gunboat diplomacy is the use or threat of limited naval force, other than as an act of war, in order to secure advantage or avert loss, either in the furtherance of an international dispute or against foreign nationals within the territory or the jurisdiction of their own state. Only governments with warships can resort to gunboat diplomacy as an instrument of peacetime coercion. However, the warships need not be actual gunboats (which are less often employed in recent years than they were in the twenties or thirties), and the big navies do not always get their way. If the conditions are right—which they seldom are for anyone—gunboat diplomacy is an instrument many different countries can employ.

Some people wrongly suppose that gunboat diplomacy is a practice that flourished only in earlier eras. During the 1970s, 23 governments used or threatened limited naval force. They can be described, without implying any moral or political judgment, as assailants. Forty-eight governments were victims, some of them—as was also true of the assailants—more than once. This pattern has continued.

Political Categories of Gunboat Diplomacy

Limited naval force is *definitive* if it creates a fait accompli that the victim cannot resist and that forces the victim to choose between acquiescence and escalation. *Purposeful* force tries to threaten or inflict such damage that the victim agrees to do what the assailant wants rather than undergo further suffering. *Catalytic* force, more often threatened than employed, tries to resolve or even to forestall disputes indirectly by deploying visible power that excites alarm and encourages prudence or even concessions. The true function of *expressive* force—although this is not always admitted—is to provide an outlet for emotion in the government employing it and to appease public opinion at home.

Except in the case of definitive force, measuring the degree of success achieved by any particular threat or use of limited naval force can be difficult. However, gunboat diplomacy is always deemed a failure if it leads to war, to which it should be an alternative and never a prelude. Nothing done in war counts as gunboat diplomacy, unless naval action is taken against allies or neutrals.

DEFINITIVE FORCE

Gunboat diplomacy with definitive force tries to resolve a dispute by removing the cause: taking or sinking an offending ship; clearing a minefield or destroying a fort; rescuing prizes, captives, or those in danger; maintaining or breaking a blockade; or seizing territory. In 1968 North Korea put a stop to American seaborne electronic espionage off its coast by overpowering the USS *Pueblo* and making hostages of the crew. In 1974 a Turkish amphibious expedition established Turkish ascendancy—which has proved durable—over a substantial part of Cyprus that had been dominated by the Greek-speaking majority of the population. In 1982 the initial success of the Argentine amphibious Operation Rosario—which captured the Falkland Islands/Islas Malvinas—ended disastrously because the victim, Britain, reacted by escalating into war. Very few international disputes lend themselves to resolution by the definitive use of limited naval force. When the conditions are right, however, definitive force is often the most effective way of employing gunboat diplomacy.

PURPOSEFUL FORCE

Gunboat diplomacy with purposeful force tries to induce the victim to do something (or to stop doing it) or to refrain from even starting to do it. It is a less direct and less reliable expedient than definitive force because its success depends on how the victim decides to respond to pressure. In such a conflict the size of the naval forces available to the two sides may be less important than their relative political motivation. The fishery dispute between Britain and Iceland began in 1958 and, with intervals of quiescence, lasted until 1976. Because they were allies, they employed purposeful, but severely limited, force. Icelandic gunboats harassed British trawlers and tried to arrest them. British frigates obstructed the gunboats, sometimes even bumping them aside, but did not shell them. Nobody won a naval victory, but Iceland prevailed by making the waters it claimed hopelessly unprofitable for British fishing. Its success was the result of greater political fortitude—fishing mattered more to Iceland. The country persevered until the British caved in completely. In the eighties, other countries—notably Argentina, France, Greece, Iran, and the United States—made use of limited naval force, sometimes more violently and on a larger scale, but none for so long or with such success as Iceland.

CATALYTIC FORCE

Gunboat diplomacy with catalytic force is not so much a response as a readiness to respond to an essentially formless menace. There may actually be a crisis or the government of a naval power may believe that a crisis could be prevented by the presence of a fleet in the appropriate sea. Since 1979, for instance, the United States has maintained one, sometimes two, and occasionally even three carrier battle groups in the Indian Ocean in case the local conflict in the Persian Gulf should widen or escalate. Few countries are capable of using or threatening limited naval force on anything approaching this scale, although Britain, France, and the former Soviet Union also manifested this kind of naval "presence" in the second half of the twentieth century. Experience so far has not clearly established it as a cost-effective expedient. Nevertheless, naval powers will always be tempted to deploy their ships where they might be needed.

EXPRESSIVE FORCE

In the expressive use of limited naval force, governments try to lend extra emphasis to some disputed claim or to manifest their own indignation by ostentatiously deploying or maneuvering warships. These ships are expected to make a striking gesture, not to take effective action. In March 1986, for instance, Spain dramatically reasserted its sovereignty demands by sending the carrier *Dedalo* into the territorial waters of Gibraltar and deploying some helicopters.

Naval Categories of Gunboat Diplomacy

These political purposes may be pursued by a variety of naval methods, the choice depending on the resources available and the circumstances of the dispute. Where serious opposition is not expected, a single warship may be enough for a *simple ship* operation. In other cases a *superior ship* or ships may be needed to awe or overpower. Sometimes a *simple fleet* may be required, not to fight, but for tasks needing numbers: to patrol a coast, for instance, and to intercept gunrunners or other undesirable vessels without significant combat potential. Where resistance is likely, a *superior fleet* must be used, and the same factor marks the difference between *simple* and *opposed amphibious operations*.

Simple and Superior Ship and Fleet Operations

Once a regular routine in Chinese rivers or on Caribbean coasts, simple ship operations today are much less frequent because most victims are better able to resist coercion. The French sloop *Albatros* did sink the trawler *Southern Raider* in 1986, and a Philippine gunboat forced a Kenyan tug into harbor in 1987, but neutral navies trying to protect their merchant vessels during the long Gulf War of the eighties nearly always needed warships that were superior to those of the belligerents. Spain sent a simple fleet to Ceuta and Melilla in 1975 in response to Moroccan pressure on those enclaves, and the Soviet Union sometimes deployed a simple fleet merely to demonstrate its interest in what the U.S. Navy was up to: off the coast of Korea, in the Indian Ocean, or in the Mediterranean. American practice is to deploy a superior fleet even in those disputes where naval opposition seems improbable.

Simple Amphibious Operations

In the first half of the twentieth century, few years passed without a simple amphibious operation, but the practice became less frequent as victims improved their political organization and their coastal defenses. Nevertheless, in 1953 Argentina effected a landing of brief duration on the disputed British territory of Deception Island. In 1954 South Korea more lastingly occupied the Takeshima Islands claimed by Japan. In 1963 Cuban forces seized some of their dissident citizens from a British island in the Bahamas. In 1976 Vietnam occupied the Spratley Islands, which China, the Philippines, and Taiwan all thought belonged to them. A traditional kind of rescue was carried out by an unusual ship in 1986: the Royal Yacht *Britannia*, which sent in its boats to evacuate British and other expatriates endangered by civil war from a beach in South Yemen. Nor were these the only instances.

Opposed Amphibious Operations

The modern trend, however, is toward the opposed amphibious operation, the riskiest and most difficult application of gunboat diplomacy. In 1983 even little Grenada had some defensive potential, prompting the United States to deploy a carrier battle group as well as an amphibious group in support of the 7,000 soldiers and marines landed to change the government of Grenada. Forcing changes in the policy or composition of the victim's government has been a frequent motive for amphibious operations, the largest of these having occurred in 1927, when various European powers joined Japan and the United States in sending 35 warships and 40,000 troops (Britain provided the largest contingent) to Shanghai.

The Nature of Limited Naval Force

Limited force is a political concept. The threat or use of force does not cease to be limited merely because many warships are involved or because these include such capital ships as strike carriers. The landing of troops, the use of aircraft, the bombardment of targets ashore, or the sinking of ships at sea does not necessarily deprive acts of force of their limited character. Moreover, gunboat diplomacy has no "safe" level or category of actions that will never provoke any victim into escalation to war. The acceptance—by the assailant, by the victim, by third parties—of an act of force as limited rather than an act of war does not depend on objective or generally valid criteria or on motives, morals, or international law. Only the result matters. An assailant contemplating the use of gunboat diplomacy must give fresh and specific consideration to the political circumstances of every contingency as and when it arises.

Even Lenin's famous question—Who? Whom?—is not a sufficient guide. Both superpowers have been the victims of relatively insignificant assailants. In 1971 Ecuadorean warships seized eight U.S. fishing vessels on the high seas and did not release them before a fine of half a million dollars had been paid for fishing in a 200-mile zone not then recognized by the United States. In 1977 the Argentine navy fired on seven Soviet trawlers (inflicting casualties) before arresting them in what Argentina claimed as a 200-mile zone.

Britain went to war in 1982 to recover the Falklands/Malvinas, which Argentina had seized in what was meant to be a definitive act of limited naval force. Planners in Buenos Aires might have noted that five islands, groups of islands, or parts of islands had changed hands in similar fashion and without adverse consequences during the seventies. Iran took Greater and Lesser Tumb at the mouth of the Persian Gulf; China seized the Paracels; first South Vietnam and then the Vietnamese Liberation Navy captured the Spratleys; Indonesia occupied East Timor; and Turkey seized a substantial portion of Cyprus. This last case—an opposed amphibious operation of fair size—may have been regarded in Buenos Aires as an encouraging precedent, for Britain had military bases on Cyprus and treaty obligations there, but took no forcible action.

Any assailant contemplating the threat or use of limited naval force has three questions to answer: What is the desired terminal situation, the happy ending to the dispute? Can it be achieved, regardless of the likely reactions of the victim or its allies (actual or potential), by limited naval force? If a risk of war exists, is it small enough—and are the stakes so high—that the risk is worth running? Often these questions are considered only after the fleet has put to sea. The substantial naval forces Britain deployed during and after the evacuation of Aden in 1967 did not actually do anything, although Britain's Arab clients lost the struggle for power ashore. The task force (including 3 carriers) that the United States assembled in the Sea of Japan after the capture of the USS *Pueblo* also did not do anything. The precise purpose of the prolonged French naval presence (2 carriers among the 18 warships) off Djibouti during and after the 1977 referendum in that colony remains obscure. One of the great advantages of navies is that warships can hover on the high seas while governments ponder how to employ them and, if the problem remains insoluble, the ships can quietly change their course.

The Evolution of Gunboat Diplomacy

The many changes in how limited naval force is usually threatened or employed for political purposes have misled many politicians, journalists, and academics into believing that the days of gunboat diplomacy are over. It has not been eliminated, however; it has merely evolved. The original class distinction between assailants, who were drawn from a small number of naval powers, and victims, who tended to be largely defenseless and often rather chaotically administered coastal states, has been eroded. Today gunboat diplomacy is a game that may involve almost any country with a seacoast in either role. This major change is the result of two kinds of continuing growth. The first is the number, the nationalism, and the administrative efficiency of independent states; the second is the diffusion of cheap and effective weapon systems for use at sea or in coastal defense.

One comparison is enough to demonstrate how, internationally speaking, gunboat diplomacy has ceased to be the accepted privilege of the elite and has become a practice open to wider use and exposed to general criticism. In 1900 the United States, Russia, Japan, Britain, France, Italy, Germany, and Austria—some of them usually rather prickly rivals—joined in an ambitious but successful opposed amphibious operation to lift the siege of the foreign legations in Peking and to punish the Chinese government for permitting it. At the time victory over superior forces (the Chinese had modern, German-built destroyers and artillery to defend the Taku forts at the mouth of the river) was accounted less remarkable than the political success of international cooperation in upholding the rule of law.

In 1983, by contrast, other than local countries in the Caribbean area, only Chile, El Salvador, and Israel expressed support for American intervention to remove the government of Grenada whose "atrocious acts" even President Fidel Castro of Cuba (hitherto a staunch supporter) had condemned. To escape the censure of the United Nations, which was less concerned with the domestic conduct of member states than with their sovereign equality and independence, the United States needed its veto.

In 1900 the world included 37 states that were nominally independent, even if some of them—for instance, China and Morocco—had to submit to a degree of outside interference that their governments would today consider intolerable. In 1987 the United Nations had 159 members, most of them enjoying substantial independence, exercising a measure of administrative control, and able to appeal to a sense of conscious nationalism among their citizens. Being politically incapable of effective resistance—as most potential victims were at the beginning of the century—was an even greater weakness than their lack of effective coastal defense. Modern technology—mines, radar, and missiles, for instance—has now made this defense cheaper than seaborne attack. Where once a gunboat and its landing party would have sufficed, a task force is now sometimes necessary.

In contrast, when more favorable opportunities arise for simpler kinds of limited naval force, nearly 50 navies have some potential for exploiting them. More states can now resist gunboat diplomacy, but more can also employ it. What has changed most during this century is not the relevance of limited naval force but the degree of political judgment and discrimination needed for its successful application. This trend is reflected in the tendency of presidents and prime ministers to make the decisions once left to consuls and gunboat captains.

The Future of Gunboat Diplomacy

As a specialized form of coercive diplomacy, the political application of limited naval force is a peacetime expedient. It can be used in someone else's war—as it was during the thirties in the Spanish civil war and during the eighties in the Gulf War—but seldom in one's own. The only year without an instance was 1944, when all the world was at war.

The violent peace of recent decades has made even those governments most anxious to avoid war impatient with the frustrations of diplomatic negotiation and the indignity of bargaining under duress. From one end of the political spectrum to the other, the nations of the world have increasingly resorted to coercion—particularly at home but sometimes abroad—in the second half of the twentieth century. Unfortunately, the people of the world have no reason to expect any reversal of this tendency or any lessening in the violence of the peace.

Naturally the level of violence will fluctuate in the future, as it has in the past. Conflict will probably continue to be more frequent and more extensive on land than at sea. The occasions when limited naval force would be appropriate are likely to be as rare in the future as they have been in the past. Nevertheless, gunboat diplomacy will continue. It may even become the main employment of navies. Since 1945 there has been one real naval war, half a dozen rather one-sided naval contributions to operations on land, and more than 200 political applications of limited naval force.

JAMES CABLE

SEE ALSO: Amphibious Warfare; Coast Defense; Limited War; Naval Warfare; Warship.

Bibliography

Booth, K. 1977. *Navies and foreign policy.* London: Croom Helm.
Cable, J. 1981. *Gunboat diplomacy 1919–1979.* London: Macmillan; New York: St. Martin's Press.
———. 1985. *Diplomacy at sea.* London: Macmillan; Annapolis, Md.: U.S. Naval Institute Press.
———. 1989. *Navies in violent peace.* London: Macmillan.
Dismukes, B., and J. McConnell. 1979. *Soviet naval diplomacy.* New York: Pergamon Press.
Dunn, P., and B. Watson. 1985. *American intervention in Grenada.* Boulder, Colo.: Westview Press.
Jonnson, H. 1982. *Friends in conflict: The Anglo-Icelandic cod wars.* London: Hurst.
Luttwak, E. 1974. *The political uses of sea power.* Baltimore, Md.: Johns Hopkins Univ. Press.
O'Connell, D. 1975. *The influence of law on sea power.* Manchester, U.K.: University Press.
Panikkar, K. 1953. *Asia and western dominance.* London: Allen and Unwin.

GUSTAVUS ADOLPHUS
(Gustavus II) [1594–1632]

Gustavus Adolphus (Fig. 1), called the "Father of Modern Warfare" by military historians, was known as the "Lion of the North" during his time. A skilled statesman and conscientious ruler, he is most famous as an innovative early modern military leader.

Early Years: Scandinavian and Baltic Campaigns

Gustavus was born in Stockholm on 9 December 1594. His grandfather, Gustavus I (1496–1560), was the first of the Swedish royal house of Vasa. While still a teenager Gustavus Adolphus successfully led Swedish troops against Dan-

Figure 1. Gustavus Adolphus. (SOURCE: U.S. Library of Congress)

ish invaders. In 1611, at the age of 17, he succeeded his father, Charles IX, as Gustavus II Adolphus.

With the crown he inherited three persistent conflicts with Denmark, Poland, and Russia. He quickly took the field, and with skillful tactical command and careful strategic judgments, he was able to obtain reasonable treaties with Denmark (Treaty of Knäred in 1613) and Russia (Treaty of Stolbovo in 1617). He then focused on a more serious challenge—King Sigismund III of Poland. The Polish king, who claimed the Swedish throne, withstood several military defeats and agreed only to a series of truces. Initially, Gustavus needed the truces to begin the reforms in the Swedish army for which he became famous. However, his continued war with Poland was handicapped by his increasing interest in the Thirty Years' War in Germany, and by the very effective command of Polish forces under Stanislaw Koniecpolski.

Satisfied that he had effectively established Sweden as a power among the Scandinavian and Baltic states, and had secured sufficient security for his homeland, Gustavus agreed to a six-year truce with Poland, signed at Altmark on 26 September 1627. He then prepared to go to war against the Emperor Ferdinand of Hapsburg, whose forces had largely defeated the armies of the Protestant states of northern Germany. Gustavus' brilliant achievements in the Swedish phase of the ongoing Thirty Years'

War are best appreciated by understanding his reforms of the Swedish army.

Formation of the Modern Army in the Age of Gunpowder

Gustavus Adolphus' legacy to military art was far more significant than his conquests, which were too extensive for Sweden to hold. As a monarch, he inherited his command, but he had to develop his military force. The warrior-king had a unique advantage; he played the part of both combat commander and logistic administrator. Gustavus' interest and authority spanned the spectrum of military organization and employment. He could introduce innovations with the full knowledge and control of their implementation—rarely the case in military history.

Seventeenth-century military institutions were struggling to adapt to the use of gunpowder. In addition, it was essential for a small nation with limited resources, like Sweden, to be efficient in warfare. Military reform was given high priority by Gustavus, who saw both his country and religion threatened by larger powers. Gustavus was an avid student of military history. He understood the essence of mobility plus force, and he sought to establish these elements in each of the arms on the battlefield.

Gustavus built upon the concepts of the great Dutch general, Maurice of Nassau. What made Gustavus' contribution unique was his vision. He did not just improve muskets and cannon, he also addressed the structure of field formations and logistics. One of his most effective reforms was to emphasize the use of cavalry as a shock weapon.

Gustavus restructured the infantry formations to enhance the firepower of the muskets. He reduced the battle formations to six lines in various postures (standing, stooping, and kneeling) while delivering their fire. He redesigned artillery pieces both to limit the logistics of supporting varied calibers and to enhance the mobility of some of the field pieces. He developed battlefield artillery as far as the manufacturing techniques of his time allowed, and directed the manufacture of light guns (some made of metal-reinforced leather and wood) to allow rapid horse-drawn or manhandled movement. He had shot and powder packaged in cartridges for both artillery and infantry.

He did not focus on weapons alone, but emphasized training, discipline, and professionalism—such as making his artillerymen military personnel, rather than civilian specialists. He trained his army to emphasize coordinated employment of infantry, cavalry, and artillery on the field. His reforms stretched into logistics and troop discipline.

Although Gustavus' reforms sustained the Swedish army for some time following his death, they fell short of a self-perpetuating institution. In the long run, the inheritors of Gustavus' military ideas were other great leaders, in other nations. Bernhard of Saxe-Weimar, who took command of the Swedish army at the battle of Lützen after the king fell, introduced Gustavus' ideas into the French army and, in particular, to the great French general, Turenne. Gustavus' cavalry shock tactics were adopted in France by Turenne, implemented in England by Cromwell, and later, fully exploited by the Duke of Marlborough. Gustavus' use of artillery in an effective mobile balance with other arms was never fully achieved until Napoleon, who benefitted from newer casting technology and the reforms of Gribeauval.

Campaigns in the Thirty Years' War

In the early phases of the war the military successes of the Imperial (i.e., Holy Roman Empire) forces in Germany posed a threat to Swedish power in the Baltic Sea. Gustavus also saw a threat to the Protestant cause, but religion was not the primary motive behind most of the actions at this time. In fact, the French, under royal minister Cardinal Richelieu, subsidized a considerable part of the Protestant king's campaigns against the Catholic Hapsburgs.

Delegating much of the duties as regent of the Swedish Kingdom to his chancellor, Count Axel Oxenstierna, Gustavus led his army of 16,000 to northern Germany in June 1630. Gustavus was fortunate in that soon after his arrival, one of the ablest commanders of the Imperial forces, Albrecht von Wallenstein, was dismissed. Thus, Gustavus did not have to confront him during the critical early phase of his campaign. In a short time, Gustavus drove the Imperial forces out of Pomerania and marched south.

On 3 April 1631 his army stormed Frankfort-on-Oder, but was unable to prevent the sack of Magdeburg by the Imperialists on 20 May. Though his army was weakened by a lack of supplies, Gustavus defeated the Imperialist army under Count Tilly near Werben in late July. When the Imperialist army started to lay waste to Saxony, Gustavus secured support from Brandenburg and Saxony, and on 17 September 1631, again confronted the Imperial forces under Count Tilly, at Breitenfeld, near Leipzig.

The Battle of Breitenfeld was the first test of Gustavus' new army against the deeper, less maneuverable formations of the Imperialists. Gustavus defeated the initial attack of the Imperial army on his right and then directed his forces to deal with the collapse of his Saxon allies on his left wing. He led a cavalry attack that captured a large portion of the Imperial artillery. The Imperialist army's inflexible formations collapsed under the intense Swedish artillery bombardment and massed cavalry charges. Tilly, who had been wounded, withdrew his army, which had suffered heavy losses (approximately 11,000). The Saxons and Swedes lost only about 4,500 men.

Gustavus then undertook a campaign to consolidate his position in northern Germany. He seized the strategically important town of Mainz on 22 December 1631, and in the following year marched south toward Vienna. Tilly had formed another army and attempted to prevent the

Swedish army from crossing the Lech River. In the ensuing Battle of the Lech (15–16 April 1632), Gustavus won a decisive victory and Tilly was killed.

The Emperor, recognizing the severity of the threat, recalled Wallenstein to succeed Tilly in command of the Imperial forces. While Wallenstein lacked the finely honed army that Gustavus possessed, he was a clever strategist and highly capable tactician. Wallenstein quickly formed an army and moved to prevent Gustavus from joining up with the Saxons. For weeks the two armies maneuvered around Nüremburg.

Finally the two armies met in the Battle of the Alte Veste. Neither commander allowed himself to be lured into an engagement that the other attempted to develop. Gustavus realized that his supply line was too vulnerable to maintain such a southern position and made a series of futile attacks against Wallenstein's well-defended position on 31 August–4 September 1632. Failing in this, Gustavus disengaged and moved north.

Wallenstein followed Gustavus to Saxony. There, on 16 November 1632, the two commanders engaged in the bitterly fought Battle of Lützen. It was here that Gustavus' recklessness led him into the enemy's formation during a cavalry melee and he was killed. The Swedish army rallied to gain this last victory for their fallen leader, and rushed forward in mass to overpower the Imperial army.

Gustavus Adolphus was a popular leader, and imparted a structure and reputation to the Swedish army for some time after his death. His excellent leadership qualities included the selection and training of highly capable subordinate commanders. While these commanders ensured excellent terms for Sweden in the Treaties of Westphalia, which ended the Thirty Years' War, they were not monarchs nor did they have the genius of Gustavus.

Significance

The integrated, combined arms employment of forces was the most advanced of Gustavus Adolphus' concepts. He showed his grasp of the integrated effect of combined arms as it is understood today. Further, he demonstrated fully the traditional concepts of streamlined organizational structure for effective command and control, and appreciated the value of rapid, aggressive action. The significant thrust of many of his innovations was to give firearms more potential in the offense.

His personal leadership qualities were exemplary, both as a monarch and as a field commander. He elicited the highest morale from his troops and love from his subjects.

ALBERT D. MCJOYNT

SEE ALSO: History, Early Modern Military; Sweden; Thirty Years' War.

Bibliography

Ahnlund, N. 1983. *Gustavus Adolphus*. Trans. M. Roberts. Westport, Conn.: Greenwood Press.

Dodge, T. A. 1895. *Gustavus Adolphus*. 2 vols. Boston: Houghton Mifflin.

Dupuy, T. N. 1969. *The military life of Gustavus Adolphus, father of modern war*. New York: Franklin Watts.

H

HAGUE CONVENTIONS

The Hague Conventions were a series of international agreements that resulted from the two Hague Conferences held in the Hague, the Netherlands, in 1899 and 1907. These conventions played a significant role in the development of the international rules of warfare, and created the Permanent Court of Arbitration (popularly known as the Hague Tribunal). This court is viewed by many as the primary achievement of the Hague Conferences, although its importance was eclipsed by the World Court (under the League of Nations) after World War I, which in turn was superseded by the International Court of Justice in 1945. As of 1973, 71 countries had adhered to the convention regarding the Permanent Court of Arbitration, and more than twenty international disputes had been administered.

First Hague Conference (1899)

The first Hague Conference was called by Czar Nicholas II of Russia in his rescript (or edict) of 14 August 1898—an invitation to the "civilized" nations of the world to a world peace conference to discuss ways to maintain the peace generally, and specifically to attempt to limit the rapid increase of armaments. Nicholas' main reason for calling the conference was to ease the financial burden required to maintain the arms race; at the time, armaments were being developed and produced "at such a pace that armed peace was more burdensome than actual warfare had been a generation before" (Trueblood 1914).

In January 1899, Czar Nicholas sent out a proposed program of topics for discussion at the upcoming Hague Conference that included: (1) a limitation on the expansion of armaments, (2) application of the Geneva Convention of 1864 to naval warfare, and (3) revision of the laws and customs of land warfare. Many believed that his decision to call the conference was influenced by the strength of the international peace movements of the time and by the precedents that had been established both at the international conference held in Berlin to discuss the Congo (15 November 1884) and at the Pan-American Conference held in Washington, D.C. (2 October 1889).

The first Hague Conference (18 May–29 July 1899) was attended by representatives from 26 nations: the European nations, the United States and Mexico from North America, and China, Japan, Persia, and Siam. The primary issue of the talks—arms limitation— met with strong opposition (especially by Germany, which was rapidly increasing its armaments), and was not resolved. Primary achievements were three conventions, and several declarations and resolutions:

Conventions:

- improved rules for the conduct of "civilized" warfare on land
- application of the 1864 Geneva Convention to maritime warfare
- adoption of the Convention for the Pacific Settlement of International Disputes, which created the Permanent Court of Arbitration

Declarations:

- five-year prohibition on throwing projectiles or explosives from balloons
- prohibition on the use of projectiles designed to spread asphyxiating gases
- prohibition on the use of expanding bullets (dum-dums)

Resolutions were agreements among the nations that certain matters (such as the rights and duties of neutral powers, limitations on the caliber of weapons, the bombardment of forts and cities) should be brought up for discussion at a subsequent conference.

Many peace activists declared the conference a failure because it did not achieve the primary goal; others felt that it was a success, claiming that the most important achievement of the first conference was the proposal to hold a second conference. The final act of the first conference was the declaration that the nations of the world were "animated by a strong desire to concert for the maintenance of the general peace," and "resolved to promote by their best efforts the friendly settlement of international disputes" (Trueblood 1914).

Period Between the Conferences

Not only had the first Hague Conference failed to achieve its primary goal, it was followed quickly by two bloody wars—the Boer War (1899–1902) and the Russo-Japanese War (1904–1905). The latter occurred despite efforts by U.S. Pres. Teddy Roosevelt to convince Japan and Russia to submit their dispute to the newly formed Court of

Arbitration (Russia refused arbitration). In fact, the Court of Arbitration was not used until 1902, when it settled a dispute between the United States and Mexico; the court was used only three more times prior to 1907.

Second Hague Conference (1907)

The second Hague Conference (15 June–18 October 1907), proposed by President Roosevelt but officially convened by Czar Nicholas, was attended by representatives from 44 nations: all the members who had attended the first conference plus representatives from all the nations of Central and South America. This conference was recognized as "the first general representative assembly of the world" (Trueblood 1914).

During the second conference, the three conventions adopted in the first conference were strengthened, and eleven more conventions (mainly regarding naval warfare and the actions of neutral powers in both land and naval warfare) were adopted. Again, however, no agreement was reached on arms limitations. The additional conventions limited or defined such actions as the:

- use of force to recover contractual debts,
- responsibilities and actions of neutral powers,
- use of submarine contact mines (the use of submarine mines in the Russo-Japanese War made restrictions on their use a matter for discussion in the 1907 conference),
- bombardment by naval forces,
- rules concerning the opening of hostilities,
- conversion of commerical vessels into warships,
- limitations on the right of seizure, and
- establishment of an international prize court.

In addition to the conventions, several nonbinding *voeux* (lit., vows) were adopted, including one to hold a conference in 1915—a conference that never took place because of the outbreak of World War I.

Assessment

Critics said that both Hague conferences had failed, claiming that restrictions had not been placed on armaments, that the arbitration clauses of 1899 were only paper agreements, and that the Permanent Court of Arbitration was meaningless. This view seems justified when one considers that the conclusion of the second conference was followed by the outbreak of World War I and that only fourteen cases had been submitted to the Permanent Court of Arbitration prior to that.

Others, however, saw the two conferences as the start of an era of progress toward international cooperation and understanding, claiming that the conferences were a necessary precursor of the formation of the League of Nations, which was created after the war. They viewed accomplishments with regard to specifics (e.g., dum-dum bullets and asphyxiating gases) as less significant than the fact that the nations of the world thought it important to attempt to legislate international disagreements and to codify the laws of war.

JOHN W. HOPPER

SEE ALSO: Attache, Military; Bombardment, Strategic; Geneva Conventions; Hostage; Law, Military; Prisoner of War.

Bibliography

Choate, J. H. 1913. *The two Hague Conferences.* Princeton, N.J.: Princeton Univ. Press.
New Encyclopedia Britannica. 1990. *New encyclopedia Britannica.* 15th ed. London: Britannica.
Scott, J. A. 1918. *The Christian nations and the Hague.* Lawrence, Kans.: Univ. of Kansas Press.
Scott, J. B. 1972. *The Hague Peace Conferences of 1899 and 1907: A series of lectures delivered before the Johns Hopkins University in the year of 1908.* New York: Garland.
Trueblood, B. F. 1914. *The two Hague Conferences and their results.* Washington, D.C.: American Peace Society.

HAITI, REPUBLIC OF

Haiti occupies the western third of the island of Santo Domingo (Hispaniola), which it shares with the Dominican Republic. A former French colony, Haiti has been independent since 1804 and so is the second-oldest state in the Western Hemisphere. Unfortunately, Haiti is also one of the poorest countries in the Americas, and most of its people survive as subsistence farmers.

Power Potential Statistics

Area: 27,750 square kilometers (10,700 sq. mi.)
Population: 6,636,800
Total Active Armed Forces: 7,400 (0.112% of pop.)
Gross Domestic Product: US$2.7 billion (1990 est.)
Annual Defense Expenditure: US$34 million (1.5% of GDP, 1988 est.)
Iron and Steel Production: none
Fuel Production: none
Electrical Power Output: 264 million kwh (1990)
Merchant Marine: none
Civil Air Fleet: 4 major transport aircraft; 10 usable airfields (3 with permanent-surface runways); none with runways over 3,659 meters (12,000 ft.); 1 with runways 2,440–3,659 meters (8,000–12,000 ft.); 4 with runways 1,220–2,440 meters (4,000–8,000 ft.).

For the most recent information, the reader may refer to the following annual publications:
The Military Balance. International Institute for Strategic Studies. London: Brassey's (UK).
The Statesman's Year-Book. New York: St. Martin's Press.
The World Factbook. Central Intelligence Agency. Washington, D.C.: Brassey's (US).

History

Inhabited by Arawak Indians before the arrival of Europeans, the island of Hispaniola was visited by Columbus during his first voyage (December 1492). Although the Indian population disappeared by the late 1500s, Spanish settlement in the area was sparse. French buccaneers settled on western Hispaniola in the mid-1600s, and French claims were formally recognized by the Treaty of Ryswick in 1697. By the late eighteenth century, large plantations worked by black slaves produced valuable exports of sugar, coffee, cocoa, indigo, and dyewoods. The 32,000 whites and 24,000 free mulattos lived well, but conditions for the nearly 500,000 slaves were harsh and demanding.

News of the French Revolution sparked a major slave uprising in August 1791. The ensuing chaos in Haiti triggered a three-way struggle for the country between France, Britain, and Spain in 1793. France decreed emancipation for the slaves in Haiti on 4 February 1794, and thereby helped secure its hold on Haiti. In May 1801 Pierre-Dominique Toussaint l'Ouverture, a former slave, proclaimed a constitution. In reaction, Napoleon sent an army to Haiti under Gen. Charles Leclerc to crush Toussaint later that year. Toussaint was captured, but Leclerc's army was decimated by fever and disease, and fears among black and mulatto troops that Napoleon planned to reinstate slavery led to renewed revolt. Leclerc died of fever in summer 1803, and his successor surrendered the remnants of the army in November.

Haiti's independence was proclaimed on 1 January 1804, but this brought neither political stability nor prosperity. Most of Haiti's history consists of a succession of vicious autocrats and violent revolutions, with occasional periods of relative stability. Some of the rulers were colorful, like former slave Henri Chistophe who, as Henri I, left behind him the magnificent palace of Sans Souci and an impressive, if militarily useless, citadel near Cap-Haitien. Haiti ruled all of Hispaniola from 1821 to 1844, when the Dominican Republic successfully revolted.

The assassination or overthrow of six Haitian presidents in four years (August 1911–July 1915) led to U.S. occupation partly triggered by fears of German influence. The U.S. presence, which ended only in 1934, gave Haiti stable government and sound fiscal administration, but little else. Although government in the twentieth century has been more stable than before, it is still autocratic and sometimes violent. The most lasting political influence since 1934 was the rule of the Duvaliers, first Francois "Papa Doc," and then his nephew Jean-Claude, derisively referred to as "Baby Doc." Jean-Claude was chased into exile in the face of a popular uprising in 1986. In late 1990, Father Jean-Bertrand Aristide was elected president, but a military coup in 1991 forced him to leave the country.

Politico-Military Background and Policy

Throughout Haitian history, the military has served as one of the pillars of the state. Since Haiti has faced few external threats (the last real foreign war dates to 1855), the real purpose of the armed forces is to support the government and maintain internal stability. This point is underscored by the army's duties as police force, customs agents, and firemen.

Under the Duvaliers, an important mechanism of internal control and repression was the dreaded *Tonton Macoutes*, a paramilitary secret police force that employed strong-arm tactics to suppress dissent. Their effectiveness was enhanced by deliberate efforts to draw on popular beliefs concerning *voudou*, the local African-derived folk religion, to enhance the nefarious and ruthless character of the *Macoutes*. Since 1986, most members of the *Macoutes* have kept a low profile, but there are reports that the military leadership may be employing some of them as soldiers or police.

Strategic Problems

Haiti's capacity to defend itself against any significant foreign aggression is doubtful. Fortunately, it is not likely to face any such external threats. The domestic political situation, however, is potentially much more dangerous. The appalling poverty in which most Haitians live has never been addressed by any Haitian regime (including the U.S. occupation), and long-term political stability will depend on economic progress for the mass of the people.

Military Assistance

Haitian governments have often received significant military aid from the United States, a tradition dating to the independence struggle with France in the early 1800s when American merchants smuggled arms and supplies to the rebels. U.S. military assistance totaled US$1.6 million in 1987 and US$1.8 million in 1988, but U.S. dissatisfaction with the progress of democratization has affected aid levels, and will probably continue to do so. The United States must tread a fine line when pressuring an uncooperative government, so as not to undermine its capabilities lest the government collapse and be replaced by a worse one.

Defense Industry

Haiti's overall industrial base is extremely small, and the country produces no weapons. It relies on imports from Europe and the United States for all its military needs.

Alliances

Haiti has no formal security ties with any other countries, although there is an informal security relationship with the United States. Haiti is a member of the Organization of American States (OAS) and the United Nations.

Haiti, Republic of

Defense Structure

The unsettled nature of current Haitian politics and the frequent changes of regime that have taken place in the years since Jean-Claude Duvalier's ouster make it impossible to describe the formal constitutional lines of authority in the national defense structure. It is safe to say, however, that the recent heads of government (most of them military officers) have exercised close and direct control over the armed forces.

(For an explanation of the abbreviations and symbols used in the following section of military statistics, see the list of Abbreviations and Acronyms in each volume.)

Total Armed Forces

Active: 7,400.

ARMY: 7,000 (has police/gendarmerie, fire-fighting, immigration, etc., roles); 1 defense unit (3 inf, 1 hy wpn coy); 9 military departments (27 coy).
Equipment:†
APC: 5 M-2, 6 V-150 Commando.
Towed arty: 75mm: 5 M-116; 105mm: 4 M-101;
Mortars: 60mm: 36 M-2; 81mm: M-1.
ATK guns: 37mm: 10 M-3A1; 57mm: 10 M-1.
RCL: 57mm: M-18; 106mm: M-40A1.
AD guns: 20mm: 6TCM-20, 4 other; 40mm: 6M-1.

NAVY: est. 250 (Coast guard). Base: Port au Prince.

Patrol Craft: 1 PCO (ex-US tug), boats.

AIR FORCE: 150; no cbt ac, no armed hel.
Transport: 1 Baron, 1 DHC-6.
Training: 3 Cessna 150, 1 -172, 5 SF-260TP, 1 Twin Bonanza.

Future

Haiti faces severe problems. As a nation, it has not yet been able to create a satisfactory government. Further, the pressures of overpopulation and diminishing resources (the rate of deforestation is extremely high) ensure that most Haitians live in conditions of appalling poverty and squalor. This situation is not impossible, however, because Haiti possesses considerable unexploited natural resources, including coal, gold, antimony, sulphur, nickel, and gypsum, all of which could contribute a great deal to a nearly moribund subsistence agriculture economy. In the absence of a stable and responsible government, however, those resources will remain unexploited.

DAVID L. BONGARD
TREVOR N. DUPUY

SEE ALSO: Central America and the Caribbean; Dominican Republic; Organization of American States.

Bibliography

Dupuy, T. N., et al. 1980. *Almanac of world military power.* 4th ed. San Rafael, Calif.: Presidio Press.
Hunter, B., ed. 1991. *The statesman's year-book, 1991–92.* New York: St. Martin's Press.
International Institute for Strategic Studies. 1991. *The military balance, 1991–1992.* London: Brassey's.

HALSEY, WILLIAM F. ("Bull") [1882–1959]

Adm. William Frederick Halsey, Jr., U.S. Navy, nicknamed "Bull" Halsey by the press, was a major U.S. naval leader in the Pacific theater during World War II (Fig. 1). He directed the Guadalcanal, Solomon Islands, and Philippine Islands campaigns, and the controversial battle of Leyte Gulf. A great proponent of sea-based airpower, he had a positive effect on the scope and operational development of carrier airpower in World War II.

Early Life and Career

Halsey was born on 30 October 1882 in Elizabeth, New Jersey. He came from a naval tradition, as his father had graduated from the U.S. Naval Academy in 1873 and retired from the navy as a captain. Graduating from the Naval Academy in 1904, Halsey served on the battleship *Kansas* as that ship steamed around the world as part of President Theodore Roosevelt's Great White Fleet.

As a navy commander during World War I, Halsey commanded first the destroyer *Benham* and then the *Shaw*. He was awarded the Navy Cross for his wartime service. Following the war his various assignments included duty at the Naval War College in Newport, Rhode

Figure 1. Admiral William F. Halsey. (SOURCE: U.S. Library of Congress)

Island, and the Army War College in Washington, D.C. His attention then turned to naval aviation; he qualified as a naval aviator in 1935 at the age of 53. He subsequently commanded the carrier *Saratoga* and then the Pensacola Naval Air Station. In the spring of 1940, Admiral Halsey was designated Commander Aircraft Battle Force, commanding all the carriers of the Pacific Fleet.

World War II

On 7 December 1941, Halsey's force, consisting of the carrier *Enterprise*, three cruisers, and nine destroyers, was 240 kilometers (150 mi.) west of Oahu when the Japanese attacked Pearl Harbor. Refueling at Pearl Harbor, Halsey's force sortied on 9 December in search of enemy submarines. In January 1942, Admiral Nimitz ordered a carrier force under Halsey's command to attack Japanese bases in the Gilbert and Marshall islands, and on Wake Island and Marcus Island. In April Halsey's title was changed to Commander Carriers Pacific, and he delivered Lt. Col. James Doolittle's B-25 bombers to their launch point in the Pacific Ocean, from which they sortied to bomb Tokyo on 18 April.

Halsey fell ill and was unable to participate in the Battle of Midway in June 1942. He returned to duty in October and became commander of the South Pacific Force and Area. He directed his forces in victories off the Santa Cruz Islands and Guadalcanal in October and November.

After the Japanese evacuated Guadalcanal on 9 February 1943, Halsey planned and directed an offensive up the Solomon Islands that lasted throughout 1943 and into 1944. His strategy was to bypass the Japanese strongpoints, including Rabaul, sealing them off with airpower and leaving their garrisons stranded. U.S. forces then constructed new air and naval bases in less strongly defended locations several hundred miles closer to Japan. In November 1943, carrier-based aircraft pounded Rabaul while aircraft from Bougainville continued to destroy enemy air forces. By 25 March 1944, Rabaul was neutralized and the war in the South Pacific had ended.

BATTLE OF LEYTE GULF

Halsey was made Commander Third Fleet in June 1944; in August he directed carrier support for the Philippines campaign, the timetable of which he had moved up after his probing perceived weak enemy defenses. In reaction to the American landing on Leyte Island on 20 October, the Japanese committed their entire fleet against American forces. This fleet was divided into three forces: the southern, central, and northern. The southern force, consisting of battleships and cruisers, was to pass through Surigao Strait and rendezvous with the powerful central force that was to move through San Bernadino Strait and come around Samar from the north. The northern force, built around four carriers, was to lure Halsey's carrier force northward, away from Leyte Gulf. As combat began, forces under Adm. Thomas Kincaid engaged the southern force, sinking several ships and sending the remainder fleeing. Halsey's aircraft had engaged the central force and he overestimated the damage his planes had incurred. When the northern force was discovered moving southward, Halsey moved his forces northward to engage them, thereby leaving San Bernadino Strait unguarded. While Halsey's forces engaged and sank all four Japanese carriers in the battle off Cape Engano, the powerful Japanese central force left San Bernadino Strait unopposed and approached the northern entrance to Leyte Gulf undetected. It then attacked Kincaid's escort carrier forces off Samar Island and the ensuing action was among the bloodiest and most gallant in U.S. naval history. Since the Japanese lacked airpower, the Americans were able to defeat their fleet, which had more than ten times the U.S. naval firepower. The Japanese fleet was damaged so severely in the battle that it could no longer mount an offensive, but in the battle's aftermath Halsey was strongly criticized for leaving the beachhead undefended and open to enemy attack.

In the final months of the war, Halsey's forces launched strikes on Japanese installations on other Philippine islands, Formosa, Okinawa, and the Chinese coast. Following the bombing of Hiroshima and Nagasaki, Japan surrendered unconditionally on 14 August 1945. The official surrender took place on 2 September on board Halsey's flagship, *Missouri*, anchored in Tokyo Bay.

Halsey was promoted to fleet admiral in December 1945, retired from the Navy in 1947, and died 16 August 1959 on Fishers Island, New York.

Halsey's Significance

Halsey was an aggressive risk-taker who welcomed hazardous situations. He was an early supporter of airpower, and stressed its flexibility, importance, and effectiveness. He had great confidence in his carrier-based forces and the effect that they could have in war. He believed that such airpower could establish command of the air at a required place and time and thereby establish the conditions necessary for successful amphibious operations.

Halsey spent his final years defending his decision at Leyte Gulf, and that battle has been replayed continually in U.S. war games and strategy sessions. The correctness of his decision is still disputed, reflected in the fact that the navy has not named a class of ships after him, while a class of destroyers was named after Admiral Spruance, who participated in the Battle of Midway. His supporters point out that he had a talent for selecting the brightest and most capable officers to be on his staff, for rarely going against his staff's recommendations, and for defending his staff when it was criticized. He was extremely popular with the press and the navy's rank and file, and has gone down in history as one of America's most colorful admirals.

BRUCE W. WATSON

SEE ALSO: Airpower, Tactical; Fleet; Maritime Strategy; Naval Warfare; South and Southeast Asia; World War II.

Bibliography

Halsey, W. F., and J. Bryan, III. 1976. *Admiral Halsey's story.* New York: Da Capo Press.
Keating, L. A. 1965. *Fleet admiral: The story of William F. Halsey.* Philadelphia: Westminster Press.
Merrill, J. M. 1976. *A sailor's admiral: A biography of William F. Halsey.* New York: Crowell.
Potter, E. B. 1985. *Bull Halsey.* Annapolis, Md.: U.S. Naval Institute Press.
Reynolds, C. G. 1978. *Famous American admirals.* New York: Van Nostrand Reinhold.
Spiller, R. J. 1989. *American military leaders.* New York: Praeger.

HANNIBAL BARCA [247–183 B.C.]

Hannibal Barca, commonly referred to as Hannibal, was a Carthaginian general renowned for his daring attack on Rome using an indirect approach across the Alps during the Second Punic War (218–202 B.C.). Also known as the "Father of Strategy," Hannibal successfully commanded Carthaginian armies in the Siege of Saguntum (219), at the River Ticinus (Ticino) (218), and at the Trebbia in 218. He won another victory at Lake Trasimene in 217 and destroyed eight Roman legions at Cannae in 216. Hannibal campaigned vigorously against the Romans in southern Italy, but although he won many battles, he was unable to force Rome to sue for peace. Scipio Africanus' invasion of Africa in 204 forced Hannibal's recall to defend Carthage, but he was defeated by Scipio at Zama in 202 and was forced into exile. Greatly feared by his Roman enemies, all that is known about him is derived from Roman accounts.

Early Career

Hannibal was born in 247 B.C., the son of Hamilcar Barca, a member of a distinguished Carthaginian noble family. Hamilcar had, following Carthage's defeat in the First Punic War (264–241 B.C.), directed Carthaginian expansion in Spain. Hamilcar took his son and his son-in-law, Hasdrubal, to Spain with him in 237. Popular legend has it that Hannibal swore an oath before he left for Spain never to be a friend of Rome. Over the next nine years he was almost constantly at his father's side, gaining a thorough education in war. Hamilcar was drowned while on a campaign in 228, and Hannibal returned to Carthage to complete his education while command of the army in Spain fell to Hasdrubal.

Hannibal the General

Hannibal returned to Spain in 224 and took command of the cavalry in Hasdrubal's army. Hasdrubal was assassinated in 221, just after he had concluded a treaty with Rome that set the Ebro River as the boundary between Roman and Carthaginian Spain. Hannibal was chosen to succeed his brother-in-law as commander in Spain, and over the next two years he led two effective campaigns to subdue the northwest part of the country between early 221 and autumn 220. He won the respect and loyalty of his soldiers; Roman chroniclers noted that "he never required others to do what he could not and would not do himself."

In spring 219, Hannibal moved his army against the port of Saguntum. Although this city lay south of the Ebro, Roman envoys warned him that any attack upon it would lead to war. Hannibal prosecuted the siege with his accustomed vigor, and Saguntum fell in late autumn that year after an eight-month siege. The following spring, Rome declared war on Carthage, and Hannibal marched northward. He led about 60,000 troops, including some 80 elephants, across the Ebro and into Gaul in July 218. Continuing toward Italy, Hannibal eluded the army of P. Cornelius Scipio near Massilia and headed up the Rhone Valley, making an epic crossing of the Alps in September and October. Hannibal soon acquired allies among the Gallic tribes of northern Italy and defeated a force of Roman cavalry and light troops at the River Ticinus (Ticino) in November. The following month, he defeated the main Roman army of T. Sempronius Longus at the River Trebbia.

Early the following spring, Hannibal moved his army (now numbering about 40,000) south and evaded the army of G. Flaminius near Pisea in March 217 before turning and destroying Flaminius' army in an ambush at Lake Trasimene in April. Hannibal led his army south, passing west of Rome, hoping to find allies among the Greek subject cities of southern Italy. The Romans, frustrated at their inability to defeat the nimble army of Hannibal, raised a force of sixteen legions (four consular armies totaling about 80,000 men) and placed it under the command of both consuls, G. Terentius Varro and L. Aemilius Paulus. Hannibal met this formidable army at Cannae on 2 August 216 and, employing his superiority in cavalry, encircled the Roman center and nearly annihilated it, inflicting some 55,000 casualties.

Despite this great victory, strategic success eluded Hannibal, as Rome refused to make peace and continued to field new armies. Even the success at Cannae failed to induce Rome's allies to desert it in any numbers. Hannibal continued to campaign vigorously in southern Italy. Although his first attack on Rome was repulsed by M. Claudius Marcellus at Nola (215), he had more success at Capua, the Silarus River, and Herdonia in 212. A second attempt on Rome in 211 also failed, but Hannibal blunted a Roman offensive against his bases, winning battles at Second Herdonia and Numistro in 210. Hannibal's position in Italy was severely damaged by the loss of his valuable base at Tarentum to Q. Fabius Maximus in 208, but he still retained the edge in the field.

Rome Ascendant

The following year (207), Hannibal's younger brother, Hasdrubal, led a reinforcing army from Spain to Italy, following in Hannibal's footsteps. Unfortunately for Hasdrubal, his messengers to Hannibal were captured by the Romans. Thus forewarned, Consul G. Claudius Nero left a force to cover Hannibal's army and hastened north. At the River Metaurus in summer 207, Hasdrubal's army was destroyed and he was killed by the forces of Nero and M. Livius Salinator. Hannibal heard of the disaster only when his brother's head was hurled into his camp.

Although he realized that his cause was nearly hopeless, he withdrew to Bruttium and continued the struggle. In 204, P. Cornelius Scipio (the son of the general mentioned above), capitalizing on a series of successes in Spain, invaded Africa. Although he was forced to lift the siege of Utica soon after, he won notable success in a series of battles with Hasdrubal Gisco and Syphax in early 203. Scipio won a great victory at Bagbrades in early autumn 203 and captured Syphax. The desperate Carthaginian Senate recalled Hannibal and his brother Mago (leading a small army in Liguria) to Africa to defend Carthage. Hannibal led a new army, some 48,000 strong, built around his veterans from Italy, into the field in early 202. He met Scipio's army on the plain of Zama in early March. Despite some initial success, and the stout resistance of his Italian veterans, Hannibal's army was overwhelmed and defeated when the Numidian cavalry of Massanissa, having defeated Hannibal's own cavalry, fell on the Carthaginian rear. For this success, Scipio gained the honorific "Africanus."

After Zama

As the leading Carthaginian general, Hannibal played a major role in convincing his more bellicose countrymen that peace was preferable to the destruction of Carthage. Elected Suffete (a post similar to consul) in 196, his moderate politics had aroused distrust and suspicion in both Carthage and Rome, and he was exiled at Roman insistence. He took shelter at the court of King Antiochus III the Great of Syria and raised a naval squadron for him from the cities of Phoenicia. He attacked Rhodes but was defeated at Eurymedon, off Side, by Eudamus of Rhodes and L. Aemilius Regilus in 190.

Following Antiochus' defeat by Rome in 188, Hannibal fled first to Crete and then took shelter in Bithynia. There, to avoid capture by the Romans, he took poison in 183. As he lay dying, he is alleged to have said, "Let us release the Romans from their long anxiety, since they think it too long to wait for the death of an old man."

Evaluation

Hannibal was certainly the ablest general Carthage ever produced. He was resolute in adversity, eager to seize opportunities when they arose, but cautious when he needed to be. He was a supremely gifted strategist, and as a tactician he was able to coax remarkable performances from his disparate forces. The information about Hannibal is drawn entirely from the Romans, who in general both feared and hated him.

LEE A. SWEETAPPLE

SEE ALSO: History, Ancient Military; Punic Wars; Roman Empire; Scipio Africanus.

Bibliography

De Beer, G. R. 1969. *Hannibal: Challenging Rome's supremacy.* New York: Viking.
Dodge, T. A. 1891. *Hannibal.* Boston: Houghton Mifflin.
Dupuy, T. N. 1969. *The military life of Hannibal.* New York: Franklin Watts.
Livy (Titus Livius). 1965. *The war with Hannibal.* Trans. A. de Selincourt, ed. B. Radine. Baltimore, Md.: Johns Hopkins Univ. Press.
Polybius. 1922–27. *Histories.* 6 vols. Trans. W. R. Paton. Cambridge, Mass.: Harvard Univ. Press.
Proctor, D. 1971. *Hannibal's march in history.* Oxford: Basil Blackwell.

HELICOPTER

The military helicopter has become a valuable weapon in the arsenals of the world's military forces. No other aircraft has the unique qualities of hovering and vertical takeoff and landing.

For centuries men have dreamed of rotating-wing flight, but it was not until the early days of World War II, when the German navy first listed the helicopter as a military machine, that the dream became a reality. Early helicopters were so fragile, however, that they never played an important military role. Postwar military rivalries stimulated further helicopter development and early experiments in using them for observation, reconnaissance, search and rescue, and medical evacuation, especially during the Korean War.

The French, during their war in Algeria, were the first to use the helicopter as an armed aircraft, and by the mid-1950s British carriers had launched the first helicopter assault during the Suez War.

During the Vietnam War, helicopters really came of age. U.S. Army helicopters performed the first airmobile combat actions, and other missions such as heavy lift, heliborne assault, minesweeping, antisubmarine patrol, and vertical replenishment were developed and perfected. The lessons of that war, and experience during conflicts in the Middle East, Afghanistan, and the Falkland Islands/Islas Malvinas, led to further helicopter development and the formidable combat machines that are in wide use today.

Military helicopters today may be classified into two groups, based on their characteristics, capabilities, and roles—battlefield and naval helicopters. These powerful ships can save lives under adverse conditions, reconnoiter battlefields in total darkness, destroy the heaviest main battle tanks, hunt down and kill submarines, and engage in aerial combat with their counterparts in the enemy's arsenal.

Origin

Although the helicopter is considered by many to be a modern invention, the rotating-wing concept of flight has been pursued for hundreds of years. (The name *helicopter* is derived from two Greek words, *heliko*, meaning helical, and *petron*, wing.) Indeed, it was long considered a more likely avenue to successful flight than the fixed-wing concept, and the first studies of helicopters were conducted well before the first airplanes. Leonardo da Vinci made the earliest description of a helicopter in his 1483 sketch of an aerial-screw machine.

Over the next four centuries many models were designed, but it was not until the Breguet and Cornu trials in France that primitive helicopters able to lift a man were constructed. On 29 September 1907 at Douai, France, Volumand was lifted clear of the ground (although to a height of only 2 ft.) by an elaborate Gyroplane built by Louis Breguet. This aircraft was totally uncontrollable, requiring four assistants to steady it, so what resulted could not really be called free flight. Paul Cornu, a bicycle maker from Lisieux, is officially recognized as having made the first vertical free flight when he succeeded in remaining airborne for about twenty seconds at a height of 30 centimeters (1 ft.), using his powered machine, the "flying bicycle," on 13 November 1907.

Between that time and the mass production of Igor Sikorsky's R-4B helicopter in 1942, many experimental helicopters were flown with various degrees of success in Europe, the United States, and the Soviet Union. Some of these efforts constituted significant milestones on the long road to the modern helicopter.

The first major achievement was the invention of the "autogyro" by the Spaniard Juan de la Cieva in the early 1920s. The autogyro was basically a wingless airplane in which an autorotating or windmilling rotor performed the function of the wings. Although the autogyro could not take off and climb vertically, hover in the air, or perform other such useful maneuvers, it was a great contribution to the development of the rotating wing. The principles of autorotation and automatic blade accommodation to varying loads, first demonstrated in the autogyro, provided the technological breakthrough that led to the modern helicopter.

Helicopter development was given new impetus when Louis Breguet solved the problem of stability by using coaxial contrarotating rotors fitted in his open-fuselage helicopter. This machine made its first flight on 26 June 1936, setting a series of world records for helicopters: an altitude of 157 meters (518 ft.), a flying time of 1 hour, 2 minutes, 5 seconds over a distance of 44 kilometers (27.32 mi.) and speed in level flight of 44.8 kilometers per hour (27.77 mph).

The German Anton Flettner was also a leader in the transformation of autogyros into the modern helicopter. In 1939, he developed a machine that the German navy planned to use for antisubmarine patrolling. The first two-seat Fl282 Kolibri flew near the end of 1941. By the following year, it was operational on those warships that could provide suitable platforms, escorting convoys in Baltic and Mediterranean waters. It was thus the first military helicopter in the world, although fielded in limited numbers.

A jet helicopter also appeared in Germany during World War II. Built by Friedrich von Doblhoff in 1942, it was never mass-produced. In fact, the last of four prototypes was captured by American forces.

Meanwhile, the Russian emigré Igor Sikorsky resumed research in vertical flight, producing the VS-300 in the summer of 1939. After a successful test flight of that machine, Sikorsky built the R-4 two-seater in 1942. A total of 130 of these were produced during the war, laying the foundation for the helicopter industry in the United States. Following the success of the R-4, Sikorsky set to work on the R-6, which subsequently established world distance, duration, and altitude records by flying 625 kilometers (387 mi.) in 4 hours, 53 minutes, and to 1,524 meters (5,000 ft.). The R-6 was also put into mass production. Sikorsky also built the R-5, a two-seat observation helicopter, which made its first flight on 18 August 1943. The U.S. Army Air Force ordered 24 preproduction YR-5As and 100 R-5As. The achievements of Sikorsky made him truly the pioneer of the helicopter industry.

Flight and Control

The helicopter is an aircraft in which lift, propulsive thrust, and control are produced by one or several sets of blades that rotate about a vertical axis (the main rotor). Lift is produced by the continuous rotation of these variable-pitch blades, which operate under power to thrust the air downward, thus increasing the pressure below the blades and decreasing the pressure over them, to generate lift.

As with fixed-wing aircraft, the amount of lift varies with the angle of the blades relative to the airflow. Thus, the pitch of the main (horizontal) rotating blades, which can be adjusted, determines the amount of lift generated. Increasing the pitch angle and the rotating speed of the main blades increases the lift, because the increased angle of attack and air velocity both generate an intense reduction in pressure above the front of the blades while also increasing the pressure along the underside of the leading edge. The result is sufficient lift to overcome the helicopter's weight. Since the rotation of the main rotor creates a torque reaction that tries to drive the helicopter fuselage in the opposite direction, a small tail rotor (which rotates in the vertical plane), which pushes the tail sideways, is needed to counteract the torque.

In hovering flight, all blades are set to the same angle of incidence at all times. Their angle of attack is equal to the angle of incidence, reduced by an angle proportional to the vertical (downward) velocity of the air moving through the rotor disk. Hovering is achieved when the produced lift equals the helicopter's weight.

Horizontal flying is achieved by tilting the rotor disk to propel the helicopter forward, backwards, or sideways. In vertical flight, the blades must rotate in the horizontal plane so that their lift is always perpendicular to the rotor disk, which has only a vertical component. If the rotor disk is tilted, then a portion of the lift generated will act to propel the helicopter in that direction.

The pilot controls the helicopter with the collective pitch lever, the cyclic pitch lever, and foot pedals. The collective pitch lever controls changes in the direction of flight by changing the angle of attack of the blades at different points in their cycle of rotation. Thus a change in the angle of roll and pitch is achieved. Foot pedals are used to control the direction of the helicopter by varying the thrust of the tail rotor, thus causing the craft to yaw clockwise or counterclockwise.

History of Development

Generally speaking, the development of helicopters has proceeded from a first generation of multipurpose machines to a current third generation of mission-specific craft.

Western Nations

Compared with fixed-wing aircraft, the helicopter lacks performance. It has less speed, a lower ceiling, and a shorter range. Its load-carrying capacity is relatively small. It is also noisy, uncomfortable, and fatiguing to fly. It is, moreover, mechanically complex, has little "g" tolerance, is inherently unstable, and is expensive for what it does.

Offsetting these shortcomings is the fact that the helicopter can carry out a wide range of missions. Its hovering capability allows it to do things that no fixed-wing aircraft can attempt. No other system can match its ability to conform to terrain contours during flight. Thus it has the ability to skim the surface, to climb vertical cliffs, and to leapfrog obstacles. It is more versatile than the fixed-wing aircraft and, as a result, over the past 40 years has brought about a revolution in almost every branch of warfare.

From its humble origins in the 1940s and early 1950s, the helicopter has developed to the extent that it can now fulfill nearly all aerial functions that do not require long range, high speed, and high altitude. Apart from strategic bombing, the helicopter can thus participate in the majority of aerial warfare roles.

The development of the military helicopter was stimulated by a series of post–World War II conflicts, especially those in Algeria, Indochina, Korea, and Vietnam. During World War II, the Soviet Union, Germany, Japan, and the United States all used rotary-wing aircraft for different missions. Soviet autogyros were employed for reconnaissance and to drop propaganda leaflets. The Germans used the Fl282 Kolibri for antisubmarine patrolling. The Japanese also used autogyros flown from an aircraft carrier to hunt submarines. The U.S. Army Air Force used its R-4s almost exclusively for search and rescue. The helicopter at this point was still fragile, and the early types constituted essentially raw technology waiting to be improved and exploited.

That development came quickly after the war, and, as improvements made the helicopter more reliable, the military embraced the changes. It was not long before the helicopter was being widely used for such varied roles as search and rescue, transport, medical evacuation, wire laying, reconnaissance, surveillance, and support of amphibious assault.

This postwar development of the helicopter as a military system began with the Indochina War at the end of the 1940s. The French Army developed medical evacuation and observation techniques using the Hiller 360, and troop and cargo carrying with the Westland-Sikorsky WS-51 Dragonfly and WS-55 Whirlwind. The British also employed helicopters during operations in Malaya, where the Royal Air Force and the Royal Navy used the WS-51 and the WS-55 for similar missions while learning to cope with the difficulties of operating helicopters in tropical climates.

During the Korean War (1950–53), U.S. armed forces gained extensive experience in helicopter operations. That war saw the further refinement of the medical evacuation and observation roles using the Bell H-13 Sioux. The Sikorsky H-19, in addition to medical evacuation and liaison missions, was used for troop transport. By November 1951, the utility of the military helicopter had become so apparent that Gen. Matthew Ridgway, then commanding the United Nations forces in Korea, asked for four helicopter transport battalions. Hundreds of helicopters were eventually used during the conflict by all branches of the American armed forces.

By the mid-1950s, military planners had become convinced that the helicopter could be successfully employed in assault operations. On 6 November 1956, during the Suez War, 22 craft launched the first helicopter commando assault at Port Said. Operating from two British carriers stationed offshore, sixteen Whirlwinds and six Sycamores managed to move 415 men and 25 tons of stores and to put them ashore in 90 minutes. Not only were commandos landed by helicopter, but airfield defense elements of the Royal Air Force were also ferried ashore to secure Gamil airfield.

Meanwhile, the French army had also rapidly grasped the military significance of the helicopter. During the fighting in Algeria, the French further developed the use of helicopters in mountainous terrain, arming their new turbine-engine Alouettes and larger helicopters with machine guns and rockets for use in ground attack. By the time a cease-fire was reached in 1962, the French had concentrated no fewer than 600 helicopters in Algeria, including 380 troop-carrying craft of the H-34 and Vertol H-21 types, 25 medium craft of the S-55 and H-19 types, and about 200 light helicopters, mainly of the Alouette type. While these numbers indicate extensive French use of the helicopter in the Algerian campaign, these heliborne operations were not overly successful. Algerian forces soon learned about the noisy craft and quickly developed effective techniques for helicopter-baiting and trapping.

The true military transition to helicopters came about in the U.S. armed forces as a result of the Vietnam War. The arrival of 32 U.S. Army H-21 helicopters in Saigon at the end of 1961 signaled the beginning of a new era of airmobility. Within days of their arrival, these helicopters were committed to the first helicopter assault conducted in Vietnam. With the subsequent buildup of U.S. forces there, the number of helicopters increased tremendously, reaching several thousand by the late 1960s.

The tactical concept of "air cavalry" units—forces taken into battle by helicopter and provided with organic helicopter-mounted fire support—came as a result of the Howze Board study of 1962. The board's final report shaped the future of U.S. Army helicopter use, setting forth new concepts of battlefield mobility centered around the developing technology of the modern military helicopter. The farsighted report compared helicopter development with that of fixed-wing aircraft and predicted similar development in the capabilities and combat power of rotary-wing aircraft. It further predicted that transport helicopters would need the protection of armed helicopters, which would thus provide airmobile accompanying fire support. These insights paved the way for establishment of the U.S. Army's 1st Cavalry Division (Airmobile). That unit, which was deployed to Vietnam in the summer of 1965 with its 428 helicopters, became the most tactically mobile division in military history.

In addition to troop carrying, the Bell UH-1, a second-generation utility transport helicopter, incorporated the second change in the helicopter's role to emerge from Vietnam. The reliability and capability of the "Huey," as the UH-1 became known, made possible the changes in tactical use of the helicopter that took place during the Vietnam War. Soon after conversion of the Huey to a tactical or battlefield troop transport, it was armed with twin fixed forward machine guns, and then with traversing M60 machine guns mounted on fuselage outriggers and manned by door gunners. By the end of 1962, the first twenty armed UH-1s, mounting factory-fitted Emerson Electric turret-mounted machine guns and rocket racks, had arrived in Vietnam. This marked the change of helicopters from an exclusively support role to an offensive system, one capable of operating as a weapons platform.

With the advent of air cavalry operations in Vietnam, a U.S. Army UH-1 company was formed around nine armed UH-1s (called gunships) and sixteen unarmed troop carriers (nicknamed "slicks"). As successive models of these aircraft were developed, it was noted that the UH-1B and UH-1C, which were being used increasingly in the gunship role, were slower than the newly arriving UH-1D troop carriers. Thus, the need for a more advanced helicopter became urgent.

The U.S. Army requirement was defined as an advanced aerial fire support system (AAFS). It called for a top speed of no less than 420 kilometers (260 mi.) per hour and the ability to conduct a ten-minute hover in its main mission profile—all while fitted with heavy armament. Under the pressure of urgent war needs, the Bell 209/AH-1G Huey Cobra was accepted as an interim AAFS. However, the Cobra was essentially only a very extensive modification of the basic Huey. Its design incorporated earlier dynamic components in support of the armed helicopter mission requirement.

The first Huey Cobras arrived in Vietnam in August 1967. As originally armed, the Cobra carried a 7.62mm six-barrel rotary Minigun mounted in an Emerson Electric TAT-102A (the designator referred to a tactical armament turret). This was soon replaced by the XM-28 armament system, which featured twin 7.62mm Miniguns, a pair of XM-129 40mm grenade launchers, or one of each. Under its stub wings, the Cobra could carry rocket or Minigun pods.

The Cobra was supposed to be an interim replacement for the UH-1B until an optimum helicopter fire support system could be developed. The AH-56 Cheyenne was to be that optimum system. The Cheyenne attempted to combine the most advanced techniques in helicopter dynamics and construction with equally advanced avionics and armament systems, thus equipping it to operate in all weather at high speed and with formidable weaponry. The result was a machine that was big, heavy, and expensive.

As the Huey Cobras proved highly successful, and U.S. Army pilots were becoming increasingly skilled in operating them, the urgent need for an immediate and much more expensive follow-on system diminished. The success of the interim system led to two other developed versions, a two-engined over-water model for use by the U.S. Marine Corps, called the AH-1J Sea Cobra, and the AH-1S, equipped with eight Hughes TOW antitank missiles. In 1966–67, the U.S. Navy also formed a heavily armed helicopter squadron using the UH-1B, basing it aboard amphibious ships and floating craft to provide close support for naval operations conducted in the Mekong Delta.

Offshore, the Navy helicopters were used for heavy lift work and other maritime missions, such as oceanographic survey and at-sea replenishment of warships, in addition to their early roles of medical evacuation, reconnaissance, liaison, and search and rescue. Apart from the Vietnam requirement, the U.S. Navy also developed helicopter units for use in antisubmarine and mine countermeasures operations.

Based on the lessons learned in Vietnam, the U.S. Army stated a new requirement for an advanced attack helicopter, one capable of all-weather, day and night operations from front-line bases, able to carry a heavy weapon load, and so constituted that it could survive in the nap-of-the-earth environment. The Hughes YAH-64 realized these requirements, but it took ten years of development before the first AH-64 Apache became operational. This system reflected the technological improvement of the third-generation helicopter. Those advances had an immediate effect on the battlefield environment in which the new ship was going to operate, and also on the equipment with which it was provided.

The Apache is equipped with very sophisticated target acquisition and advanced weapons systems, as well as night and adverse weather operations equipment. In addition to the TADS/PNV (target acquisition and designation/pilot's night vision) systems, it is armed with a 30mm chain gun and can also carry sixteen laser-homing Hellfire antitank missiles, which have a range of up to 10 kilometers (6.2 mi.). These missiles can be launched using the Apache's own laser target designator or any other ground- or air-based designator.

Some Apaches are equipped with an airborne target handover system, which communicates the precise location and characteristics of a target from one helicopter to another, or from a helicopter to a ground station, using a coded data pulse. This system could, for example, allow a ground laser designator or a scout helicopter to mark a target for a Hellfire missile launched from an Apache whose crew has not even acquired the target.

In addition to the attack helicopter, U.S. helicopter development in the years following Vietnam featured three other thrusts: upgrading the existing fleet, building more-advanced light multirole helicopters, and attempting to build more capable heavy-lift helicopters.

The conversion of the Bell OH-58A to the D (scout) version and the model 406 CS (combat scout) are typical examples of the upgrading effort. The modified systems have been provided with a more powerful engine, a mast-mounted sight-producing TV/FLIR (forward-looking infrared) pictures, a laser rangefinder/designator, a new integrated multiplex cockpit control and display system, an airborne target handoff subsystem, and other advanced avionic systems. The upgraded system is intended to provide scouting/reconnaissance and surveillance support for both attack helicopter and ground forces. The combat scout version can carry four TOW-2 antitank missiles or a mix of 70mm rockets, 7.62mm or 12.7mm machine guns, air-to-air missiles, and other pylon-mounted weaponry.

The S-70 (UH-60 Blackhawk; see Fig. 1) illustrates the second direction of development: the light multirole helicopter. The UH-60, as a third-generation utility transport, provides the U.S. Army with improved hot-weather performance, reduced daily and routine maintenance requirements, and upgraded survivability, crashworthiness, and troop and cargo capacity. It is also designed to operate in low-level nap-of-the-earth environments. The success of this version has led to derivatives (SH-60 B and F models) for use in naval roles.

The third direction of development, heavy lift, has had two aspects: the heavy transport, such as the S-65 (CH-53E Super Stallion), and the flying crane (Boeing Vertol XCH-62).

The need for improved assault transport and heavy-lift capabilities in the latter stages of the Vietnam War drove development in those areas, although it took about ten years to deliver the first CH-53E to the U.S. Marine Corps. This helicopter can carry 55 troops or seven standard cargo pallets. It has a maximum internal payload of 13,635 kilograms (30,000 lb.) with a range of 185 kilometers (115 mi.), and a maximum sling-load capacity of 16,365 kilograms (36,000 lb.). It has also been developed in the MH-53 MCM version for mine countermeasures operations.

The development of the flying crane was driven by a stated U.S. Army requirement for a heavy-lift helicopter capable of moving a 40,000-pound sling load. Such a capability would enable the system to move 70 percent of the army's equipment, including its 16-ton M551 Sheridan light reconnaissance tank.

Western European countries have also continued their contribution to development of helicopter design and em-

Figure 1. A Special Forces Group exits from a UH-60 Blackhawk helicopter. (SOURCE: U.S. Department of Defense)

ployment. They have drawn on their unique experience in an attempt to close the gap between their own helicopter development and those of the United States and the former Soviet Union, although that gap still remains.

Since the early 1950s, many helicopters have been built in Western European countries, either under license or based on their own designs. There have also been joint programs of helicopter development and production.

Western European helicopter development has concentrated on light and medium multirole systems. Some versions of each have also been developed for use in the naval role. The only exceptions to this focus on multirole systems have been the Agusta Mangusta, a promising attack helicopter in service with the Italian army; its derivative, the Tonal, which is still under study; and a fruitless joint program (PAH-2/HAC-3G/HAP) conducted by France and the German Federal Republic.

The light multirole development is represented by 2- to 5-ton helicopters such as the French SA 342 Gazelle and SA 365 Panther, the German BO 105 and the German-Japanese BK 117, the Italian A 109, and the British Lynx. Some versions of these are armed with antitank missiles, rockets, and machine guns, while others have been converted to the naval role and could be provided with antiship missiles, antisubmarine torpedoes, or antisubmarine warfare (ASW) equipment.

With their improved avionics and armament and a troop-carrying capacity of four to ten men, these helicopters could be employed in several roles as antitank, scout, general reconnaissance, observation, liaison, medical evacuation, ASW, or antisurface vessel systems.

The French AS 332 Super Puma and the British-developed versions of the Sea King/Commando are typical examples of the medium multirole helicopter. With their payload capacity of 12 to 28 troops, avionic systems, and armament, these helicopters have been used in both battlefield and naval roles. Their payload capability could be employed in helicopter assault and logistical missions, or, in the naval version, for antisubmarine or antisurface vessel operations.

As a result of the Falklands/Malvinas operation, the HAS-2 Westland Sea King was converted to meet a Royal Navy requirement for aerial early warning. This version has been equipped with a Thorn-EMI Search water radar, with the antenna housed in a semiretractable radome mounted on the side of the cabin.

The A 129 Mangusta is the first European helicopter specifically designed as an attack antiarmor and scout helicopter. Like the Apache, the Mangusta is a twin-engined helicopter with a crew of two seated in tandem. It is designed to be difficult to detect and is provided with such protective systems as active jamming countermeasures and various passive sensors (radar/laser threat warners and the like). The Mangusta is designed to be invulnerable to 12.7mm high-explosive incendiary projectiles, and nearly so to hits by 23mm shells. It has high crashworthiness, with crew survival assured at crash impacts up to about 11 meters (36 ft.) per second.

The Mangusta is armed with the TOW antitank missile and 7.62mm rockets, which is the Italian choice. But it can also carry other antitank missiles (Hellfire, TOW, HOT, Trigat), air-to-air missiles (Mistral, Sidewinder, Stinger), pod-mounted 20mm cannon, and 81mm rocket launchers. Mangusta is also equipped with advanced avionic systems, which are managed, as are the other onboard systems, by the integrated multiplex system (IMS). IMS-collected data are recorded and displayed on a multifunction display, and both the pilot and gunner can work interactively with the system.

SOVIET UNION

The former Soviet Union, once the world's second largest helicopter operator, also worked to exploit experience and war lessons in developing the design of its helicopters and their military use. The development of Soviet helicopters before Vietnam reflects their interpretation of post–World War II conflicts in the context of Soviet doctrine, strategy, and tactics.

For two full decades following World War II, Soviet helicopters were regarded as useful vehicles for troop lift and for administrative and logistical tasks, but hardly as weapons platforms. In the early 1950s, the Soviet view was that helicopters could under certain circumstances be employed in tactical heliborne assaults, a view probably derived from analysis of two small-scale American assaults conducted in the Korean War. In 1951, fifteen months after the outbreak of that conflict, the Soviet Union had only the Mi-1 Hare in service. At that point, Stalin gave his famous directive to Mil and Yakovlev. Mil was to build a single-engine, single-rotor 12-passenger helicopter, while Yakovlev was to develop a twin-engine, twin-rotor 24-passenger helicopter. The new types became the Mi-4 Hound and the Yak-24 Horse, respectively.

One role specified for the new helicopters was lifting airborne troops ahead of the main body of the army to capture key objectives such as bridges, road junctions, and airfields. This employment had the further advantage of carrying these troops over natural obstacles and minefields. The new concept was tested in exercises and judged to be valid.

As the doctrine of Soviet armed forces began to embrace nuclear war, it became clear that troop movement would have to be conducted much more quickly. The payload and range capabilities of neither the Hare nor the Hound met the new requirements of being able to lift heavy army weapons. Mil offered the heavy-lift solution in its huge Mi-6 Hook and its derivative, the Mi-10 Harke flying crane. The Hook entered service in the early 1960s, providing a maximum payload capability of 12,000 kg. This huge helicopter, the first in the world in its heavy class, made it possible to carry a large number of troops and a wide range of equipment, including armored vehicles and heavy guns, and to land them simultaneously in a selected area behind enemy lines. Thus, before the Vietnam War had even begun, the Soviets had accepted the helicopter as an unarmed battlefield vehicle.

Meanwhile, the Mi-4 was developed to operate with the Soviet navy in a shore-based ASW role, using a search radar mounted under its nose. Later machines also carried magnetic anomaly detection equipment, sonobuoys, markers, flares, and antisubmarine torpedoes. In this role, the Mi-4 was known in the West as the Hound-B.

Viewing the war experience and lessons derived from use of the armed Mi-4 by the Egyptians in Yemen and UH-1 employment by the Americans in Vietnam, the Soviets further developed their basic version of the Mi-4. A close-air-support version equipped with a gun turret for a 12.7mm machine gun under the nose and air-to-surface rockets was reported in 1968. But the main role of the Hound continued to be that of a troop carrier; eventually, it was replaced in that role by the Mi-8 Hip, which had a payload of 24 troops.

The extensive use of helicopters in Vietnam convinced Soviet leaders of the potential of both armed and unarmed helicopters to increase troop mobility. As the Soviets believed that victory goes to the side that has the greater firepower, they became convinced that all battlefield vehicles, including helicopters, had to be able to fire on the enemy.

To put this concept into action, the Soviets armed the Mi-8C, their standard troop carrier (which went into production in 1964), with four pods that each could accommodate 16 or 32 57mm rockets. Four 250-kg bombs could be substituted for the four pods. This armament proved suitable for neutralizing enemy air defenses when supporting heliborne troops in assault operations. Subsequently, several other versions of the Mi-8 were produced for different roles.

While the Hip J and Hip K are dedicated for electronic countermeasures operations, the Hip E is a heavily armed attack variant that carries a triple stores rack on each side of its fuselage. Up to 192 rockets can be mounted in six pods, plus four Swatter antitank missiles. In addition, a 12.7mm machine gun is fitted on a flexible mount in the nose. The Mi-17 Hip H, a new development of the Hip family, was introduced in 1981 to meet Soviet and export requirements for an advanced assault transport helicopter.

By 1973, production of a specialized antisubmarine version, the Mi-14 Haze, had also been initiated to replace the aging Mi-4. Operational equipment of the Mi-14 includes a chin-mounted search radome, a retractable sonar unit, and a towed magnetic anomaly detector. Torpedoes and depth charges can also be carried.

Meanwhile, a third generation of Soviet helicopters was under development. In 1969, about two years after arrival of the first Cobra in Vietnam, armed clashes took place on the border between Soviet and Chinese forces. Since the frontier with China is some 7,000 kilometers (4,350 mi.) long, Soviet commanders were faced with the problem of concentrating their tanks at the critical points within a few hours. A flying tank, heavily armed and with high speed, was needed. This implied development of a specially designed attack helicopter with heavy weapons and some armor protection, rather than a utility helicopter with some strap-on armament. That conclusion led to a remarkable change in the Soviet view of both design and use of the military helicopter. Successful American employment of the Huey Cobra in Vietnam also influenced the Soviet outlook, as did Soviet maneuver experience in integrating armed helicopters into combined-arms formations to enhance those elements' flexibility and ability to achieve surprise. The success of that concept further confirmed the requirement for an armed assault and antitank helicopter.

The design of a two-seat dedicated attack helicopter in the mold of a Huey Cobra was advocated, but priority was given to an assault troop capability with integrated weapons. Mil designed the basic layout of the new type around the dynamics of the Mi-8, providing space for eight combat-equipped soldiers and a crew of four. By early 1973, the Mi-24 Hind-A, as the new type was designated, had entered active service. The armament of Hind-A included four AT-2 Swatter antitank missiles; four pylons for mounting podded rockets, bombs, or other munitions; and a single 12.7mm machine gun in the nose, slaved to a chin-mounted sight.

The availability of the more powerful 2200 SHP TV3-117 engines, which replaced the original 1700 SHP Istov TV2-117 power plants, enabled the Hind to exceed Soviet expectations as a heavily armed battlefield helicopter, so the decision was taken to develop it into a dedicated attack helicopter. Different versions have since been produced, the most notable being the last three, designated Hind D, E, and F.

The Hind D is similar to Hind A, but with TV3-117 engines and heavily armored accommodation for a two-man flight crew. It is optimized for the gunship role, while still providing a transport capability. Under the nose, a four-barrel 12.7mm machine gun provides air-to-air and air-to-surface capability. Unlike the Hind A, which does not have much armor protection, the D version is armored against small-arms fire from the ground. Critical components, including the crew, have been protected by replacing aluminum with steel and titanium plating.

Hind E was developed in the same manner as Hind D, but with modified wing-tip launchers for four AT-6 Spiral laser-homing antitank missiles and an enlarged under-nose sensor pod. The four-barrel 12.7mm machine gun found in Hind D has been replaced in the E version by a twin-barrel 23mm cannon.

By 1982, the Hind F had entered service with Soviet forces. It is generally similar to the E version, but with a twin-barrel 30mm cannon replacing the 23mm cannon of the E Model.

Apart from their armament, all Hind versions have all-weather navigational aids, including a projected map display. Most versions have an electro-optical (low-light-level TV) sensor. The Hind A has an optical gunsight, while the D and E versions have an impressive group of sensors, including radar and low-light-level TV and, since 1982, forward-looking infrared. ECM/IRCM (electronic countermeasure/infrared countermeasure) warning and jamming systems have also been installed.

Paralleling the development of these troop-carrying and attack helicopters, the Soviets undertook development in the heavy-lift class of an updated successor to the Mi-6, one capable of day and night operations with increased capability. These efforts produced the Mi-26 Halo, which was specifically designed to meet the heavy-lift requirement in both battlefield and underdeveloped regions.

The Halo can carry a normal payload of more than 90 combat-equipped troops, or 20,000 kilograms (440,000 lb.) of cargo internally or sling loaded. This payload capability, as well as the Halo's wide cabin, enable it to carry a variety of important weapon systems. Heavy loads can either be winched into the cabin or slung, using built-in winches and television monitoring.

OTHER COUNTRIES

From the design point of view, other countries have made less of a contribution to helicopter development, as they have mainly bought the helicopters they needed or built them under license. Nevertheless, certain countries, including those of East and Southeast Asia and the Middle East, have made some contribution to helicopter development. Based on their experience and war lessons, they have moved in two directions: improvement of methods of helicopter employment, especially in antitank, heliborne commando assault, and electronic warfare roles; and design development to meet their own particular requirements.

Future Prospects

The helicopter in its present form probably will always be limited in speed and agility, compared with fixed-wing aircraft. In the long run, development of the rigid coaxial advanced blade rotor concept or the stopped-rotor concept may offer a suitable solution to this persistent problem. The best that has been done to this point is to minimize the existing defects by improvements in aerodynamic and structural efficiency.

The American LHX (light helicopter, experimental) program reflects the present design capability and affordability of the U.S. effort. This program derived from the need to have a more advanced and competent helicopter, one that could compete with the new Soviet attack helicopter, the Mi-28 Havoc, and the emerging threat of the Hokum, a dedicated air-to-air combat helicopter. A new helicopter was also needed that could survive in the modern battlefield environment, which is lavishly populated with advanced air defense weapons.

As defined by the U.S. Army, the LHX would come in two light versions (scout/attack, or SCAT, and utility); accommodate a single pilot; be extremely versatile and maneuverable; have no less than 70 percent common components between the two versions; have a range of 2,030 kilometers (1,260 mi.) without refueling; carry several weapons systems and integrated targeting means; have all-weather and night vision flight capability; include nuclear, biological, and chemical protection; provide a 40 percent reduction in operating and support costs; and have a unit production cost not greater than US$6 million.

As it is envisaged, the exterior appearance of the LHX/SCAT version would incorporate stealth technology. It would have an all-composite airframe and a fully integrated cockpit incorporating complete offensive and defensive electronics systems, integrated through high-speed data buses. An electro-optical target acquisition and designation system would provide a thermal image output directly to the pilot's helmet-mounted display for navigation and target search. That would be combined with a digital map system. The basic armament would consist of four Hellfire and two Stinger missiles, plus a 20mm cannon with 750 rounds of ammunition.

The LHX/Utility version is required to replace the Huey in all its transport versions, carrying a squad of six to eight men or internal/external cargo. The first LHX prototypes were expected to fly in 1991.

Another type based on a different concept, the Boeing V-22 Osprey was to enter service with the U.S. armed forces in the early 1990s. This system integrates the tilt-rotor concept. The V-22 is in fact really more of an airplane than a helicopter. It is fitted with very large propellers that can be tilted up to lift the machine without forward speed. Compared with the conventional helicopter, it is less efficient in the hovering mode, but faster in cruising flight, where it is lifted by its wing and pulled along by its propellers.

The capabilities of the V-22, including a cruising speed of over 480 kilometers per hour (300 mph), quieter operation with less vibration than the conventional helicopter, and a payload of 24 combat-equipped troops or external cargo of 4,545 kilograms (10,000 lb.), make it an attractive replacement for the Boeing Vertol CH-47 and CH-53 in their transport role. The V-22 is expected to enter service a few years earlier than LHX, as its production deliveries to the U.S. Marine Corps were expected to begin in 1991.

Partly for economic reasons, the Western European allies have more modest plans than the United States for future helicopter employment. There are in prospect three joint programs: the PAH-2/HAC-3G/HAP, involving France and the German Federal Republic; Tonal, sponsored by Italy, the United Kingdom, the Netherlands, and Spain; and EH 101, a joint Italy-U.K. project.

Negotiation of the French-German program began in

1975 and was aimed at defining an antitank helicopter for army use. The helicopter was to be built in three versions: PAH-2 (Panzerabwehr Hubschrauber, second generation) and HAC-3G (Helicoptere Anti-Char, third generation) as antitank helicopters for German and French use, respectively; and HAP (Helicoptere d'Appui et Protection) as an escort and support helicopter for the French.

In 1987, after more than eleven fruitless years, it was decided to develop a single antitank version instead of the three versions initially contemplated. The new choice was a twin-engine helicopter crewed by two men seated in tandem. It was to be capable of night combat and have two-hour endurance. In addition to a mast-mounted sight and European visionics, it would be provided with eight antitank and four air-to-air missiles. If the program proceeds, the first helicopters will be delivered in 1997.

Until now, the Tonal has been no more than a study launched in 1986 by four countries: Italy, the United Kingdom, the Netherlands, and Spain, and intended to produce a multirole combat helicopter (antitank, scout, and antihelicopter) to be operational by the mid-1990s. It would be a derivative of the A 129 Mangusta, with the primary modification being incorporation of a new weapon system. This would probably be the long-range "fire-and-forget" version of the third-generation Trigat antitank missile.

The third Western European program is the EH 101, which reflects a joint effort of the British Westland and Italian Agusta companies to build a new multirole helicopter able to satisfy the needs of the sponsoring countries for a replacement of the SH-3 Sea King. The EH 101 would be produced in two military versions, naval AWS and utility, and civil versions. The naval version will have the ability to operate from frigate-sized ships in a high-sea state. It will also be equipped with search radar and fully integrated avionics management and mission systems. The utility EH 101 will be capable of carrying up to 28 combat-equipped troops or nearly 6 tons of cargo. Loads of up to 6,818 kilograms (15,000 lb.) could be carried on the external hook. As the EH 101 has already made its initial flight, it is expected to enter active service before the other two versions have completed development.

The former Soviet Union, in addition to its superiority in heavy-lift helicopters, has once again gone further than Western nations in the development of combat helicopters. The new Mi-28 Havoc, the first Russian helicopter specifically designed for the attack role, will provide forces with not only an improved antitank capability, but an antihelicopter capability as well. The Havoc, designed in the late 1970s on the basis of experience gained in employment of the Mi-24, has—as a result of dropping the troop lift role—improved maneuverability and agility and a greatly reduced cross-section.

It is believed that the Havoc has day/night vision and radar sensor and sighting capabilities. Its armament includes a ventrally mounted 23mm gun turret and mix of AT-6 Spiral antitank and SA-14 air-to-air missiles mounted on four stub-wing pylons. Infrared suppressors and infrared decoy dispensers are also fitted. The Havoc has apparently been in service since early 1988.

While the Havoc was designed as an attack helicopter with a potential antihelicopter capability, the Hokum, a forthcoming Kamov helicopter, is dedicated to the antihelicopter (helicopter hunt) role. Some commentators believe that the Hokum will give the Russians a significant rotary-wing air-superiority capability.

Like its Kamov predecessors, the Hokum will have twin engines driving coaxial rotors, but with a slimmer fuselage. Its speed in level flight is believed to reach 350 kilometers per hour (217 mph). Many reports indicate that the Hokum is armed with a multibarrel cannon (installed in the lower nose) and air-to-air missiles (carried on the stub-wing hard points). It is also provided with protective systems, such as infrared suppression equipment, armor, and other survivability features. According to a statement about the Hokum capability attributed to the U.S. Department of Defense, "the system has no current Western counterpart."

The widespread use of attack helicopters, with their advanced antitank missiles and capability of nap-of-the-earth flight, has increased the threat they represent and made it difficult and expensive to counter them with fixed-wing aircraft. Fighters will have difficulty in discovering, to say nothing of engaging, a helicopter operating near the ground and exploiting natural and artificial cover.

Thus, a specially designed helicopter hunter may be the best answer to the threat of the attack helicopter. The Russian Hokum is, thus far, the only system specifically designed for this role. The United States hopes that the LHX program will produce a system that can counter the Havoc and Hokum. The Western nations, lacking any system capable of performing this new role, and not having any under development, have sought to improve their antihelicopter capabilities by arming their current antitank helicopters with air-to-air missiles for self-protection. Based on the continuing development of both Russian and Western helicopters, there is reason to believe that an all-weather helicopter hunter will soon become a reality, pending only solution of the problem of integrating such a weapon system into the reconnaissance air defense network within the framework of airspace control.

Conclusion

As with the earlier development of fixed-wing aircraft, the helicopter is evolving from general-purpose to mission-specific systems. In addition to the multipurpose helicopter, there are already helicopters specifically designed for transport, heavy lift, scouting, attack, and—forthcoming—the helicopter hunter.

Under pressure from the world's military organizations, with their continuing thirst for more advanced capabilities of vertical flight, it is expected that the helicopter and its competitor, the tilt-rotor aircraft, will continue to emulate the dramatic technological advances achieved by fixed-wing aircraft. Then, like the aerial battles between fixed-wing aircraft of previous wars, future wars will see helicopter battles in abundance.

GABR ALY GABR

SEE ALSO: Air Assault; Airborne Land Forces; Casualties: Evacuation and Treatment; Helicopter, Battlefield; Helicopter: Military Missions; Helicopter, Naval; Machine Gun; Missile, Air-to-air; Missile, Antitank Guided; Mobility.

Bibliography

Apostolo, G. 1984. *World encyclopedia of civil and military helicopters.* London: Willow.
Armitage, M., and R. Mason. 1985. *Air power in the nuclear age 1945–84, theory and practice.* London: Macmillan.
Campbell, C. 1984. *Air warfare, the fourth generation.* London: Hamlyn.
Chant, C., et al. 1977. Helicopter warfare. In *The encyclopedia of air warfare,* pp. 240–247. London: Salamander.
Gunston, B., and M. Spick, 1986. *Modern fighting helicopters.* London: Salamander.
Heath, E. 1983. *Soviet helicopters: Design, development and tactics.* London: Jane's.
The illustrated encyclopedia of aviation. 1979. London: Cavendish.
Polmar, N., and F. Kennedy. 1981. *Military helicopters of the world.* London: Arms and Armour.

HELICOPTER, BATTLEFIELD

The battlefield helicopter is an important component of the weapons of modern warfare. It now performs a variety of jobs formerly performed only by transport and fighter-bomber aviation, as well as jobs that are unique to the helicopter. Battlefield helicopters can carry troops and their equipment—even heavy vehicles and light tanks—beyond the front and into the enemy's rear areas. They can render fire and general support for ground troops as well as hunt enemy helicopters. The diverse use of battlefield helicopters has greatly increased the mobility of ground troops.

Contemporary battlefield helicopters are of numerous types, from light observation and liaison ships to heavy lift and flying cranes, from unarmed utility carriers to lavishly armed attack systems. This article deals with the types of battlefield helicopters; their characteristics, roles, and missions; and the tactics they use to carry out such missions in combat.

Types

Battlefield helicopters are those helicopters that carry out their missions in the battlefield area, both over friendly terrain and behind the enemy's front lines. In most countries that possess a helicopter force, they are typically formed into units of the army and air force. According to their missions, capabilities, equipment, and armament, battlefield helicopters are classified as transport/general purpose and heavy lift, multirole, or attack.

TRANSPORT/GENERAL PURPOSE AND HEAVY LIFT

The helicopters in this group have spacious, well-equipped cabins for accomplishing the lift mission plus external load capabilities. The payload of ships of this type varies greatly. Medium transports such as the SA 330 Puma and the Boeing Vertol 114 (CH-47D) can lift 16 to 44 fully equipped soldiers or about 2,976 to 12,600 kilograms (6,614–28,000 lb.) of externally slung cargo. Heavy lift helicopters, typified by the Soviet Mi-6 "Hook" and Mi-26 "Halo," have a lift capability of 70 to 100 soldiers or 12,000 kilograms (26,400 lb.) of internal cargo or 20,000 kilograms (44,000 lb.) of externally slung cargo. Helicopters in the transport/general purpose and heavy lift category usually accomplish one or more of the following missions: helicopter assault, troop and cargo lift, aerial resupply, heavy lift, search and rescue, and medical evacuation.

MULTIROLE HELICOPTERS

These are ships that can perform more than one mission, either with the provision of additional equipment or by being manufactured in different versions. Examples are the Mil Mi-8 Hip, the Sikorsky S-70, the Westland Lynx, and the Aerospatiale 365. Some armed versions of multipurpose helicopters are considered attack helicopters and will be discussed in that category.

According to their capabilities and equipment, multirole helicopters can carry out more than one of the following missions: assault, scouting and target acquisition, search and rescue, electronic warfare, chemical reconnaissance, mine laying and clearing, airborne command and control, and others.

ATTACK HELICOPTERS

These are armed ships capable of attacking and destroying ground targets such as tanks and artillery. The category includes both armed versions of multirole helicopters and helicopters specifically designed for the attack role.

Modified light multirole helicopters include such systems as the SA 342 Gazelle and the Hughes 500 MD armed with antitank missiles, rocket pods, and heavy machine guns. Heavier modified types include the Bell 209/AH-1S Cobra and the Mi-24 Hind. These helicopters, designed from the outset as attack ships, are intended primarily for the antitank role.

In addition to the attack role, the armed multirole helicopter can carry out a variety of missions, such as general reconnaissance, scouting, observation, control of artillery fire, and liaison. The specifically designed attack helicopters, represented by such systems as the AH-64 Apache and the Mi-28 Havoc, are intended mainly for the attack role, although the latter is able to engage enemy helicopters as well.

The armament of helicopters in this group varies according to type but generally includes antitank missiles (wire- or laser-guided), unguided rockets mounted in pods, and either cannon or machine guns. Equipment can include advanced navigation systems, target acquisition/designation sights, laser spot trackers, laser range finders/designators, FLIR (forward-looking infrared), and computerized fire control systems as well as night vision sensors, direct view optics, and television cameras. Infrared jammers, radar jammers, chaff dispensers, and passive radar warning receivers are also usually provided.

Roles, Missions and Tactics

The battlefield helicopter is now regarded as an important weapon in its own right. It has an offensive capability even though at times it is used in a defensive situation. Its use, roles, and missions depend on the military doctrine of the user nation, the helicopter's capabilities, and the battlefield situation—especially the air defense environment. The key roles are fire support, general support, helicopter assault, and antihelicopter defense. (The primary characteristics of modern and advanced battlefield helicopters expected to be in service in the 1990s are shown in Tables 1 through 3.)

FIRE SUPPORT

As with fixed-wing aircraft, helicopter attack gunships with advanced weapons systems can provide for effective fire support for friendly ground troops. The Western allies use attack helicopters mainly in the antitank role, while the Soviet Union and its allies made wider use of such systems to provide fire support for tank and motorized infantry units as well as for heliborne forces. These helicopters, integrated into combined arms groups, would have attacked enemy defensive areas, supported the advance of friendly troops, provided flank security, disrupted the movement of enemy reserves, and destroyed enemy tanks.

Attack helicopters are expected to play a major role in future conflicts, whoever the contending parties may be, but they are very vulnerable to air defense weapons, which requires them to take measures for secure completion of their missions.

During offensive operations, helicopter fire support is essential to create favorable conditions for breaking through an enemy's system of defenses, which have evolved to the point where they are highly resistant to conventional attack.

The helicopter, along with fixed-wing aircraft, works to accomplish air attack against enemy reserves, overrun tactical nuclear weapon delivery systems, and pursue withdrawing enemy forces. While fixed-wing aircraft are dealing with assigned targets in the operational depth of the enemy defenses, attack helicopters concentrate their efforts on the enemy's tactical zone of defense, mainly against enemy tanks, antitank weapons, artillery, and rocket launchers.

As the offensive develops, attack helicopters must secure the commitment of second-echelon formations and protect their flanks against enemy tanks and antitank weapons.

In defensive operations, attack helicopters usually participate in preventive strikes or counterpreparations against enemy first-echelon tanks, launching their attack

TABLE 1. *Characteristics of Assault/Transport and Heavy Lift Helicopters*

PARAMETER	AEROSPATIALE AS332 SUPER PUMA M	BOEING VERTOL 114 CH-47D CHINOOK	MIL MI-17 HIP H	MIL MI-26 HALO	BOEING VERTOL XCH-62/CH62
Origin	France 1978	U.S. 1979	USSR 1981	USSR 1977	U.S. 1975
Type	Medium transport	Transport	Assault transport	Heavy airlift transport	Heavy airlift
Engine	2 1,780 shp Turbomeca Makila 1A	2 3,750 shp Lycoming T55-L-712	2 1,900 shp Istov TV3-117 MT	2 11,400 shp Lotarev D-136 turbines	2 8,000 shp Allison T701
Rotor diameter	15.3+ m (51+ ft.)	18 m (60 ft.)	20.9– m (70– ft.)	31.5 m (105 ft.)	27.5 m (92 ft.)
Payload	22 troops or 4,500-kg (9,900 lb.) cargo	44 troops or 12,712-kg (27,966 lb.) cargo sling-loaded	32 troops or 4,000-kg (8,800 lb.) internal cargo	100 troops or 19,842-kg (44,092 lb.) internal or external cargo	55 troops or 18,000-kg (40,000 lb.) cargo sling-loaded
Maximum speed	294 kph (184 mph)	293 kph (183 mph)	248 kph (155 mph)	293 kph (183 mph)	Not available
Ceiling	5,905 m (19,684 ft.)	4,500 m (15,000 ft.)	4,920 m (16,400 ft.)	4,530 m (15,100 ft.)	Not available
Range	630 km (394 mi.)	387 km (242 mi.)	944 km (590 mi.) with reserve tanks	800 km (500 mi.)	Not available

TABLE 2. *Characteristics of Multirole Helicopters*

Parameter	Mil Mi-24 Hind D	Agusta 109 A (MK 11)	Westland Lynx (AH MK1)	Sikorsky S-70 (UH-60A Blackhawk)	Aerospatiale SA 365 Panther	MBB-Kawasaki BK 117 (A-3 M)
Origin	USSR 1971	Italy 1975	U.K. 1976	U.S. 1978	France 1984	Germany/Japan 1985
Type	Armed assault	General purpose multirole	General purpose multirole	Combat assault transport	Day/night attack/transport	Multirole armed
Engines	2 2,200 shp Istov TV3-117	2 420 shp Allison 250-C20B	2 900 shp Rolls Royce BS360 Gem	2 1,543 shp GE T700	2 838 shp Turbomeca TM333	2 550 shp Avco Lycoming LTS 101-650 B-1
Rotor diameter	16.7 m (56 ft.)	10.8 m (36 ft.)	12.6 m (42 ft.)	16.2 m (54 ft.)	11.7 m (39 ft.)	10.8 m (36 ft.)
Maximum weight	10,913 kg (24,250 lb.)	2,579 kg (5,732 lb.)	4,715 kg (10,478 lb.)	7,410 kg (16,468 lb.)	4,068 kg (9,040 lb.)	3,175 kg (7,055 lb.)
Maximum speed	318 kph (199 mph)	309 kph (193 mph)	280 kph (175 mph)	358 kph (224 mph)	278 kph (174 mph)	246 kph (154 mph)
Ceiling	4,425 m (14,750 ft.)	4,500 m (15,000 ft.)	Not available	5,700 m (19,000 ft.)	5,904 m (19,680 ft.)	1,350 m (4,500 ft.)
Range	746 km (466 mi.)	662 km (414 mi.)	677 km (420 mi.)	597 km (373 mi.)	746 km (466 mi.)	490 km (306 mi.)
Capacity	4 crew plus 8 troops	2 crew plus 6 troops	2 crew plus 10 troops	2 crew plus 11 troops	Pilot plus 8 troops	2 crew plus 10 troops
Armament	4-barrel 12.7mm MG(E), 2-barrel 23mm cannon (F), 2-barrel 30mm cannon, 4 rocket pods and 4 ATM	8 TOW or AAMs; scout 2 podded 12.7mm and 2 pintle-mounted 7.62mm machine guns	8 TOW missiles, or podded maching guns, rockets, etc.	2 7.62mm M60 machine guns + 12 Hellfire missiles or gun (rocket pods or M56 mine dispenser)	8 HOT missiles or podded rockets, machine guns, other	1 12.7mm gun + HOT or TOW missiles or podded rockets, machine guns, other

from behind the friendly front line. With the onset of the enemy offensive, attack helicopters direct extensive fires against advancing enemy tanks and self-propelled guns, concentrating their efforts on filling gaps between friendly formations and protecting their flanks until reserves and second echelons can get into position.

Attack helicopters usually work in cooperation with the army antitank reserve so as to increase antitank capabilities in the threatened direction, seeking to stop advancing enemy tanks at a certain line of defense. Should the enemy seek to bypass the antitank position by a turning maneuver, the attack helicopters move to hinder enemy movement so that friendly antitank reserves can move to new blocking positions in the path of the enemy advance. Attack helicopters may also strike the enemy's flanks to force him in a certain direction where he will become more vulnerable to ground antitank weapons.

While counterattacking in this manner, attack helicopters participate in fire preparation, just as they do in the offense, and provide security for the commitment of the counterattack forces against enemy tanks and antitank weapons.

In both offensive and defensive operations, coordination and cooperation with friendly fixed-wing aircraft, antitank weapons, missile troops, artillery, and air defense weapons are essential. Strikes by attack helicopters and friendly ground weapons are coordinated through distribution of targets and the timing of strikes. Friendly maneuver elements must mark their positions with visual aids to assist their own attack helicopters. Artillery observation posts may also use lasers to illuminate targets for attack helicopters.

When supporting ground troops, attack helicopters usually operate from forward airfields located some 48 kilometers (30 mi.) behind the front lines. Resupply landing sites are also prepared and stocked at about half that distance from the front but out of enemy artillery range. In planning attack helicopter combat actions in offensive operations, special attention must be given to the nature of the terrain so as to select covered waiting and combat areas. In defensive operations, such areas, as well as ambush/attack positions, are selected to the front and on the flanks. The combat areas selected must provide for covered maneuvers between the ambush/attack positions. Attack helicopters typically deploy in small groups, although never less than two together, in combat/waiting areas that provide an unobserved route to preselected ambush/attack positions.

Table 3. *Characteristics of Attack Helicopters*

Parameter	McDonnell Douglas AH-64A Apache	Agusta A129 Mangusta	Mil Mi-28 Havoc (est.)
Origin	U.S. 1975	Italy 1983	USSR ca. 1982–83
Engines	2 1,690 shp GE T700-GE-701	2 952 shp Rolls Royce Gem 2MK 1004D	2 2,200 shp Istov TV3-117
Rotor diameter	14.4 m (48 ft.)	11.7 m (39 ft.)	16.7 m (56 ft.)
Weight	7,949 kg (17,665 lb.)	3,671 kg (8,157 lb.)	9,027 kg (20,060 lb.)
Maximum speed	376 kph (235 mph)	269 kph (168 mph)	368 kph (230 mph)
Ceiling	6,150 m (20,500 ft.)	2,355 m (7,850 ft.)	Not available
Range	606 km (379 mi.)	744 km (465 mi.)	238 km (149 mi.)
Armament	1 30mm cannon plus 16 Hellfire missiles or 76 2.75-in. rockets in pods	8 HOT or TOW missiles plus podded 7.62mm or 12.7mm machine gun or rockets, other	Antitank and air-to-air missiles, 23mm gun turret

Helicopter fire support may be provided in several different forms, including preplanned strikes, on-call close support, free hunt, and ambush. Preplanned strikes are used mainly in the phases of fire/air preparation and commitment of the counterattack force in defensive operations. On-call fire support and the free hunt can be used in all phases of operations to hinder and repel the enemy while on defense, or to overcome his most highly resistant positions in the offense. The ambush is used mainly while in defensive positions, independently, or in cooperation with ground antitank weapons along the direction of the enemy's approach and on his flanks. The ambush may also be used to repel an enemy counterattack during offensive operations.

Helicopters launch their weapons in two ways, from the hovering position and from dive attack. Missiles can be launched in either way, but the dive attack is usually used for rocket launching and gun firing. The hovering method is used where there is cover in the attack position. In open areas, dive attack is preferable, especially in desert areas where hovering causes sand clouds that interfere with visibility and reveal the helicopter's position.

Typical combat formations used by attack helicopters are organized around scout groups and strike groups. The scout group is used for reconnaissance of the combat area, target acquisition, and assignment and coordination of combat actions of strike groups. The scout group may select the attack positions or use previously selected positions. The scout group may also call for artillery and fixed-wing aircraft support.

In addition to their armament, scout group helicopters are usually equipped with advanced avionics systems such as night vision, target acquisition and designation, and navigation and artillery fire control systems. The scout group comprises approximately one third of the combat formation force, the strike group constitutes the main body of the combat formation.

When delivering preplanned or on-call strikes, attack helicopters fly directly from advanced airfields or resupply sites to the assigned or resupply sites to the assigned combat area, timing it so that they arrive just after suppression of enemy air defense weapons in the target area by friendly artillery or fixed-wing aircraft. The scout element arrives a few minutes before the strike force, which waits under cover in the combat area.

After the scout group reconnoiters the target area, the strike group moves to attack positions as directed by the scouts. Upon acquiring visual contact with its targets, the strike group launches its attack as directed; usually this begins with firing missiles from the hovering position (unless in desert terrain), followed by rockets and guns fired using dive attacks. To launch a dive attack, the helicopter performs a rapid climb followed by a brief attack run, then aims and fires its rockets while at the same time spraying the target area with gunfire so as to suppress return fire; this is followed by a hard down break and disengagement.

When engaged in the free hunt, attack helicopters traverse an assigned area, searching for enemy advanced/reconnaissance tanks that may be seeking a way around friendly antitank positions or trying to approach from the flank. Attack helicopters typically use combat flight formations when in the free hunt area. While a search-and-strike pair flies in a zigzag manner at minimum altitude and optimum speed, the leader brings up the rear at a distance that ensures good visual contact and mutual fire support. The zigzagging and frequent changes of position in the formation maximize the chance of spotting concealed targets. The flight leader suppresses enemy antiaircraft weapons on the flanks of the combat flight formation. Then, having contact with the enemy targets, the strike pair launches the attack according to the flight leader's instructions.

In the ambush, attack helicopters deploy to an assigned waiting/combat area that has been selected because it provides an unobserved run to the planned ambush positions into which they will move just before launching their attacks. Once in the ambush position, the helicopters observe advancing enemy tanks and select their targets. If the helicopters have mast-mounted sights, they will be able to operate with only the sights visible to the enemy.

To launch missiles (such as the Hughes BGM-71 TOW),

the helicopters may need to break cover, with either a pop-up or a pop-sideways maneuver, to allow the missiles to clear the cover. This prelaunch maneuver may not be necessary if the missile is designed to be launched at an angle so that it will clear the cover. During the missile's flight, the helicopters resume their former positions with just the sight exposed, waiting until the missiles strike their targets. Then they repeat the attacks until such time as enemy fire approaches their positions.

With the mast-mounted sight, helicopters armed with the laser-homing Hellfire missile can even keep their rotor behind cover while launching their weapons in a "fire-and-forget" mode.

GENERAL SUPPORT

The helicopter general support role comprises a group of missions that include general reconnaissance, scouting and target acquisition, chemical reconnaissance, artillery spotting, search and rescue, aerial command post, laying of smoke screens, mine sowing, and logistic and electronic warfare support.

The rapid changes and confusion typical of the battlefield environment may make it necessary to use helicopters for reconnoitering the situation along the front, to carry out scouting and target acquisition, and to insert ground reconnaissance teams into critical areas.

The helicopter's involvement in these missions will depend on the specific battlefield situation. For example, it would not be wise to use a helicopter for reconnaissance missions over enemy troops in a well-defended front with numerous antiaircraft weapons, especially when there is a suitable alternative such as the remote piloted vehicle (RPV), which is cheaper and less vulnerable under those circumstances. But in such missions as scouting and target acquisition, as well as chemical reconnaissance, there are few alternatives to the helicopter. On a front that is heavily infested with air defenses, scouting might be carried out from behind the friendly front line beyond the range of enemy air defense weapons. Insertion of chemical reconnaissance teams into contaminated areas, however, would require helicopters to operate behind enemy lines.

Scouting is essential for security, for directing attack helicopters to their targets, and for suppressing enemy antiaircraft weapons with artillery or fixed-wing aircraft strikes. The importance of scouting increases when the mission is to protect the flanks and to provide security for the commitment of second-echelon formations.

When the friendly force is engaged in exploiting chemical or nuclear strikes, helicopters are used to insert reconnaissance teams in the contaminated areas. Samples of the air may be collected during the helicopter's flight over the contaminated area, and survey teams can be landed within 15 minutes after the chemical or nuclear strike. Forward observers can perform their mission from helicopters behind friendly front lines.

Search and rescue, on the battlefield usually the location and extraction of bailed-out and crashed air crews, has long been one of the helicopter's missions. Helicopters assigned to this mission are maintained at an advanced state of readiness, equipped with search and rescue equipment, and on call around the clock.

The helicopter may also be used as an airborne command and control post. These helicopters must have wide cabins to provide a suitable place for the commander and his staff and special equipment for command and control. Such helicopters can also be used by commanders to control live-fire maneuvers and for radio relay.

Helicopters may also be used efficiently to lay smoke screens by using on-board pyrotechnic generators or by dropping smoke pots at low altitude. A single helicopter, such as the Mi-8 Hip, can, using some 24 BDSL-15 smoke pots, produce a 5-kilometer (3-mi.) smoke screen that will last for 15 minutes. These screens can be used to cover such friendly combat activities as hovering missile attacks, landing of assault helicopters behind enemy lines, and the maneuvering of ground elements.

Battlefield helicopters are also used for mine-sowing and mine-clearing operations. Two Hip helicopters, each carrying about 400 antitank mines, can lay their load in less than fifteen minutes while cruising at 25 kilometers (15 mi.) per hour. The importance of the helicopter's mine-sowing capability increases in critical situations, such as when it is necessary to hamper the rapid advance of enemy troops and tanks, or to rapidly emplace coastal defenses against impending amphibious operations. Helicopters can also clear mine fields by towing a device to detonate the mines.

Helicopter logistical support comprises mainly transport, supply, and medical evacuation missions. Such missions can be carried out behind friendly lines or across enemy front lines. Helicopters are used for logistical support behind friendly lines only when time is critical or when the helicopter is the only means that can complete the mission. Behind enemy lines, the helicopter's logistical support is provided mainly to heliborne assault troops or to troops encircled by enemy forces. In such cases, strong air protection must be provided for the logistical helicopters.

Electronic warfare support is usually carried out by helicopters from behind the friendly front line. Some helicopter versions, such as the Hip-K and EH-60A, are specially equipped for electronic warfare and intelligence.

HELICOPTER ASSAULT: THE SOVIET VIEW

Many nations expect to use heliborne operations extensively in future conflicts. Their aim is to increase the mobility and rate of advance of friendly troops, doing this in particular by attacking the enemy rear and thereby loosening and disrupting his system of defense. Many nations have thus organized heliborne forces, usually consisting of units of airborne troops and air-lifted motorized infantry units.

With their unique capabilities, helicopters play different roles in such heliborne operations. They can carry troops to an assigned area behind enemy lines without the complications of a parachute drop and the special training it requires. In addition, helicopters can provide the heliborne assault with most of the air support it requires, whether fire support or general support.

Employment of the helicopter assault has special importance in jungle terrain and on mountainous or desert battlefields where troop mobility is limited by the road network and combat involves fierce fighting to seize communications, road hubs, and key mountain passes.

Depending on the purpose of the operation, the strength of the heliborne force, and the depth of its penetration, helicopter assaults may (according to Soviet doctrine) be classified as operational, tactical, or special, as follows:

1. An operational assault is usually carried out by a heliborne brigade force to assist attacking troops in accomplishing operational missions according to the general concept of the operation.
2. A tactical assault is carried out by a heliborne force, varying in size from company to battalion, to accomplish tactical missions in combat of a field army's first-echelon divisions.
3. A special assault is usually carried out by a ranger force, varying in size from a squad to a company, to accomplish reconnaissance and sabotage missions. Special assaults may be carried out to both tactical and operational depth of enemy defenses.

In general, the helicopter assault aims to complete one or more of the following combat tasks: destruction of enemy tactical nuclear delivery means; seizure of river bridgeheads; seizure of vital territory; disruption of enemy reserves, command and control networks, and logistics systems; and exploitation of nuclear or chemical strikes.

A heliborne assault is initiated from a departure area where the heliborne force is loaded. The departure area is a suitable landing area for helicopters, usually 30 to 50 kilometers (18–30 mi.) behind the friendly front line and within the zones protected by antiaircraft weapons. It should include shelters, camouflage, and so on.

To minimize their vulnerability in the open departure area, especially when operating in desert terrain, helicopters arrive at the departure area to board the attacking force only 10 or 15 minutes before loading. The loading is then usually carried out under the protection of fighter aircraft.

Helicopters take off from the departure area in flights or squadrons, although night take-offs are usually carried out in flights to minimize the requirement for lights on the ground and aboard the helicopters.

Under the protection of fighters, the helicopters combat formation flies to the landing zone at very low altitude and in two groups, the landing group and the support group. The former makes the heliborne force landing, while the second includes scout and attack helicopters.

The scout helicopters fly far enough ahead of the landing group to permit reconnoitering the situation along the flight route and at the landing zone. Attack helicopters maintain contact with the landing group, flying ahead and along the flanks of its formation.

As the landing group approaches the enemy front line, friendly artillery and fighter-bombers, with coordinated fires, suppress enemy antiaircraft weapons along the flight route. When they reach the enemy lines, the attack helicopters, flying ahead and on the flanks of the landing group, suppress the revealed enemy antiaircraft weapons.

When they approach the landing zone (5 to 10 minutes before the helicopters land), the fighter-bombers suppress antiaircraft weapons in the landing zone and destroy any enemy tanks and artillery near it (out to a distance of approximately 15 kilometers [9 mi.] from the landing sites). Meanwhile, the density of friendly fighters in the air alert zones is increased so as to protect the combat operations of both the helicopters and fighter-bombers.

The landing zone should have been cleared of resistance by the fighter-bombers before arrival of the helicopters. If not, the attack helicopters must accomplish this, clearing the landing zone before arrival of the transport helicopters.

The landing group approaches the landing zone with the assistance of its navigational aids and the guidance provided by the scouting helicopters. As the transport helicopters approach the landing zone, the intervals between flights and squadrons are reduced so as to minimize the time needed for landing and disembarking and also the space required for landing sites.

In general, an area of 500 square meters (600 sq. yds.) is sufficient to land a medium transport helicopter brigade of 36 helicopters simultaneously, although the area must be doubled for a night landing. A helicopter brigade needs three landing sites for each squadron. Upon arriving at the landing sites, the transport helicopters land in squadrons using a direct approach. Disembarking is carried out as quickly as possible so the helicopters stay on the ground for the shortest time.

While the troops are disembarking and the heliborne force is assembling, the scout helicopters fly over the landing zone, in a radius of not more than 5 to 10 kilometers (3–6 mi.) from the landing sites, searching for any enemy weapons that might threaten the landing and assembly; if any are located, the scouts direct attack helicopters to destroy them.

After landing and disembarkation have been completed, the helicopters return to their home airfields by means of a less vulnerable flight route. In the case of an airborne raid, however, the helicopters must return to pick up the raiding party as soon as the mission has been accomplished.

Antihelicopter Defense

It is widely believed that the best counter to the helicopter is an air combat helicopter, or helicopter hunter, armed for the task and faster and more agile than the enemy ship. A helicopter such as this, with a well-trained crew, will be able to tackle enemy battlefield helicopters in their own ultra-low-level environment, accomplishing its mission by either offensive or defensive combat actions.

Offensive combat action is aimed primarily at detecting and destroying the multithreat attack helicopter, which implies the need for wide and continuous reconnaissance activity. The defensive concept, on the other hand, means self-protection and the protection of other friendly helicopters and tank units against both enemy helicopter hunters and attack helicopters, respectively. This is the approach now receiving the most emphasis, especially by Western allies.

Self-protection against enemy helicopters seeking duels can be achieved by arming friendly helicopters with effective long-range weapons and air-to-air missiles, as well as by providing such helicopters with protective systems. But unarmed friendly helicopters, and those that do not possess suitable air-to-air weapons, will need aerial protection against the enemy helicopter hunter. This can be achieved by escorting friendly helicopters during their flight and mission accomplishment, using for this purpose suitable other helicopters having air-to-air combat capabilities, supported by electronic countermeasures, and competitive with the enemy helicopters.

Protection of friendly tank units against enemy attack helicopters can be accomplished by helicopter patrols carried out in the expected direction of the enemy's approach and along the flanks of the tank units to be protected. Such patrols must be carried out at a distance of not less than one-and-a-half times the missile range of the enemy helicopters. In the offense, this protection will be even more necessary during the phases of repelling counterattacks and commitment of the second echelon. In the defense, the need for this air protection will increase when moving tank reserves and during commitment of the second echelon to conduct a counterattack.

Compared with fixed-wing aircraft, a helicopter's aerial combat is more difficult. In order to engage and destroy an enemy helicopter, three important and difficult steps must be accomplished successfully even before firing or launching missiles. These are detection, identification, and getting into firing position.

In the modern battlefield environment, helicopters can survive only by sticking to the terrain and using natural and artificial cover. During low-level flight, both visibility and observation are considerably reduced. This means that, without airborne radar, detection and identification will take place mainly by surprise and at short range. Thus, the pilot will not have time to assume a firing position before he is sighted by the enemy.

At short distances—from 1,000 to 1,500 meters (3,300–4,950 ft.)—reaction speed is the decisive factor. If the friendly pilot succeeds in gaining a good firing position before being detected by the enemy, he will then have a good chance of destroying his opponent. Otherwise, he must resort to speed and to vertical and horizontal maneuvers to overcome the enemy, a more difficult task under those circumstances.

Several maneuvers have been developed for use in aerial combat by helicopters. They include acceleration and deceleration in horizontal flight, dive attack and pulling out, turns in dive attack followed by wing-overs with high-pulled turns, horizontal backward flight, upward flight and spirals under acceleration, and yoyos at low altitude with side slip and rapid transition to stationary hover. Such maneuvers require a helicopter specifically designed and armed for the mission.

Future

With the continual development of the helicopter's capabilities, its importance as a potent force on the battlefield will increase. Advanced helicopters such as Havoc, Apache, and the forthcoming Hokum and LHX/SCAT (light helicopter experimental/scout attack), with their impressive speed, lethal weapons, and great agility and flexibility, will play a decisive role in future conflicts.

As reliance on attack helicopters increases for antiarmor battles, the indicators point toward the helicopter hunter as a good response. As long ago as September 1979, Major General Belove foresaw the future of the helicopter when he wrote in *Soviet Military Review* that, "like tank battles of the past wars, a future war between well-equipped armies is bound to involve helicopter battles."

GABR ALY GABR

SEE ALSO: Air Assault; Airborne Land Forces; Airlift; Air Reconnaissance; Army Aviation; Helicopter; Helicopter: Military Missions; Rotary Wing Aircraft Technology.

Bibliography

Apostolo, G. 1984. *World encyclopedia of civil and military helicopters*. London: Willow.
Chant, C., et al. 1977. Helicopter warfare. In *The encyclopedia of air warfare*, pp. 240–47. London: Salamander.
Gunston, B. 1984. *Modern fighting aircraft*. London: Salamander.
———, and M. Spick. 1986. *Modern fighting helicopters*. London: Salamander.
Illustrated encyclopedia of aviation. 1979. London: Cavendish.
Jane's all the world's aircraft 1986–1987. 1986. London: Jane's.
Jane's all the world's aircraft 1987–1988. 1987. London: Jane's.
Polmar, N., and F. Kennedy. 1981. *Military helicopters of the world*. London: Arms and Armour.
Soviet helicopters: Design, development and tactics. 1983. London: Jane's.

ADDITIONAL SOURCES: *International Defense Review; Military Technology*.

HELICOPTER: MILITARY MISSIONS

This article provides a discussion of combat helicopter operations from the early days of the pioneers in the U.S. Air Force (USAF) to the present, and a presentation of how the Royal Air Force (RAF) and the air force of the former Soviet Union utilize helicopters to accomplish military missions.

Inasmuch as the primary missions for USAF helicopters fall into two categories—special operations and combat search and rescue—it is fitting that the first combat use of the helicopter was the rescue of a downed aircrew that had been engaged in a special operation in Burma. In April 1944 an L-1 liaison plane crashed behind enemy lines, carrying a pilot and three passengers belonging to the "Chindit" raiders of British Maj. Gen. Orde Wingate. On 21 April, 1st Lt. Carter Harman flew a tiny YR-4 helicopter to the crash site and recovered one of the wounded men. The underpowered helicopter could lift only one passenger at a time from the crash site. Accordingly, it took two days to complete the four rescue sorties.

The Helicopters of the U.S. Air Force

Since 1944 the air force has operated nine different types of helicopters representing a total of 29 separate and distinct models. The following is a brief description of each of those helicopters.

THE R-4

The R-4 helicopter was built by Sikorsky and first flew in January 1942. It had a fabric-covered box fuselage resembling that of a conventional light aircraft. Even the rotor blades were fabric covered. A later version of the R-4 incorporated a 200-horsepower engine, but even that improvement left the helicopter severely power limited.

THE R-5 (H-5) SERIES

The XR-5 first flew in August 1943. This aircraft was twice the size of the R-4 and carried twice the useful load. It had a maximum speed of 120 kilometers (75 mi.) per hour. Unfortunately, it was designed with a narrow tandem cockpit that could accommodate only two persons.

The USAF operated these helicopters mainly in two search and rescue (SAR) variants, the H-5D and the H-5G. (Prior to the establishment of the air force as a separate service, the type symbol assigned to helicopters had been *R* for rotary wing. In the new air force this was changed to make *H* the exclusive designation for helicopters.)

THE H-19 SERIES

The H-19 series shared an all-metal construction and a three-bladed main rotor system. The H-19 was designed, built, and delivered in less than a year from contract agreement. The first aircraft flew in November 1949, and the air force began taking deliveries in 1950.

The most popular variant was the H-19B. A total of 266 H-19Bs were built, some for foreign transfer.

THE H-21 SERIES

The H-21 "Workhorse" was manufactured by Piasecki. It was a twelve- to twenty-place helicopter with two main rotors located fore and aft of the fuselage. The USAF purchased a total of 214 of these helicopters for both transportation and rescue roles.

THE H-43 SERIES

The basic H-43 had a box-shaped fuselage upon which were mounted two intermeshing rotors. The term for this configuration is *synchropter*. The rotors were of spruce; control was accomplished by moving servo flaps mounted on the trailing edge of the wooden blades, thereby actually warping the blades and changing their aerodynamics. The H-43 had small wheel-skis fitted to the landing gear, enabling it to operate from hard or soft surfaces.

The original model (H-43A) had a reciprocating engine but the two most popular variants (H-43B and H-43H) were turbine-powered. The H-43B was the world's first turbine-powered military helicopter.

THE H-3 SERIES

All of the USAF H-3 series aircraft were derived from the Sikorsky S-61R series and have a longer but narrower fuselage, a hydraulically operated rear ramp, and a fully retractable tricycle undercarriage.

The helicopters have been in the USAF inventory since 1963 and continue in active service today. The variants of the series were identified as CH-3B, CH-3C, CH-3E, and HH-3E. The HH-3E is the SAR variant and incorporates a rescue hoist and an aerial refueling system. This is the version first called the "Jolly Green Giant."

THE H-1 SERIES

The UH-1F, a considerably modified version of the Bell 204-B "Huey" helicopter, first flew in 1964. Although it had the small cabin of the 204-B, the tail boom was extended to accommodate the larger (48-ft. diameter) main rotor. A total of 146 of these helicopters was delivered by the end of the contract in 1967. They were used almost exclusively in the security support role for the ICBM sites of the Strategic Air Command.

The UH-1P was a modification of the UH-1F helicopter. It had hard points added so that it could accommodate weapon mounts for miniguns and rocket pods. It was used both as a troop carrier and as a gunship in special operations squadrons in Southeast Asia and Central America. In its gunship configuration, it carried two 7.62mm miniguns, 9,000 rounds of 7.62mm ammunition, and two rocket pods, each of which held seven 2.75-inch rockets.

The UH-1N is a military version of the Bell 212 helicopter. It was slightly larger than the UH-1F but the really significant change was in its powerplant. The UH-1N used the Pratt & Whitney "Twin-Pac" consisting of two PT6 turboshaft engines, each of which was capable of producing 900 shaft horsepower. The USAF bought 79 of these "Twin Hueys" with delivery beginning in 1970. The first of these helicopters went directly to Vietnam to replace the rapidly dwindling fleet of UH-1Ps.

The HH-1H was a variant of the army UH-1H, modified for local base rescue. It was bought to replace the aging HH-43 fleet in that role.

THE H-53 SERIES

The H-53 was a natural follow-on to the H-3 series. In the rescue role, the H-53 allowed a crew to fly farther, fly faster, and loiter longer to complete a rescue. The C or cargo variant gave the air force its first heavy-lift helicopter capability.

All of the USAF H-53s were originally built as either HH-53B/C SAR variants or as CH-53C heavy-lift cargo models. The rescue versions had defensive armament and aerial refueling systems. The cargo models were not air-refuelable and were virtually empty inside to maximize their internal and external cargo-carrying capabilities.

Under the USAF Pave Low III program, nine HH-53s were modified for night and all-weather search and rescue operations. The prototype of this highly modified helicopter first flew in 1975. Specialized equipment on this model included a stabilized forward-looking infrared (FLIR), an inertial navigation system, a projected map display, and terrain-avoidance and terrain-following radar. In 1980, after the failed attempt to rescue the American hostages in Iran, management of Pave Low helicopters was transferred within the USAF from rescue to special operations to enhance the USAF hostage-rescue capabilities. In 1986, the HH-53 designation of the Pave Low III was changed to MH-53 to more accurately reflect the special operations mission. Also in 1986, the decision was made to upgrade all 41 of the Air Force H-53s to a new Pave Low III "enhanced" standard that carries the designation MH-53J. The new modifications include even more sophisticated avionics plus titanium armor plating and .50-caliber machine-gun mounts. Also included in the enhancement package is the Navstar global positioning system (GPS) terminal. The GPS has wide-ranging applications to all kinds of aircraft, but in the helicopter it may be able to provide an aid to hovering that will allow the helicopter to establish a hover with 1-meter precision and hold it for three minutes. All USAF H-53s were to be modified to the MH-53J standard by early 1990.

THE UH-60 SERIES

The UH-60 was purchased by the air force as an interim remedy for its shortfall in rescue helicopters during the early 1980s.

The UH-60A was bought in the basic army UH-60 configuration with the addition of a de-icing system, a rescue hoist, and air-transportability kits. Ten of them were purchased by the air force and delivered in 1982 and 1983.

Since 1987, all H-60s acquired by the USAF have been of the MH-60G or Pave Hawk standard. That aircraft has an air refueling system, an auxiliary fuel tank, precision tactical low-level navigation equipment, and improved communications and weapon systems. Current plans call for the air force to buy approximately ten MH-60Gs each year for the next several years to add to the current rotary wing capabilities of both the special operations and combat search and rescue forces. (In 1986 the air force instituted a modification program called Credible Hawk, which modified all of the UH-60A helicopters to the MH-60G Pave Hawk standard.)

Korea

By the end of World War II Sikorsky had replaced the R-4 series of helicopter with the larger and more capable R-5. The newly established U.S. Air Force ordered 45 of these helicopters in the H-5G and H-5H configurations. Unfortunately, these were the only air force helicopters available when the five years of peace ended. On 25 June 1950, the day that units of the North Korean People's Army crossed the 38th parallel into South Korea, the only U.S. Air Rescue Squadron (ARSq) in the Pacific with assigned helicopters was the 3d ARSq in Japan.

In addition to its fixed-wing assets, the 3d ARSq had nine H-5 helicopters. Fewer than four weeks later, the first of those H-5s were on the ground in Korea and were conducting medical evacuation (medevac) missions at the battlefront. By the middle of August, six of these nine H-5s were operating in Korea.

Two significant dates in the annals of the H-5 in Korea are 15 February and 23 March 1951. On the first date, United Nations troops were pinned down at Chipyong-ni, 32 kilometers (20 mi.) east of Seoul. Six H-5s joined the fray, carrying badly needed blood plasma and medical supplies to the battle zone and wounded infantrymen away from it. In two days of nonstop action the H-5 crews evacuated 52 badly wounded soldiers.

On 23 March, evacuation began of more than 300 wounded and injured paratroopers from the Munsan-ni drop zone just south of the 38th parallel. Seven H-5s were sent into the battle, and they performed great feats during the six-day siege. They flew a total of 147 sorties into the surrounded drop zone and picked up the last of the wounded on 29 March.

Even in this impressive display of helicopter capabilities, the shortcomings of the H-5 became obvious. Just four days earlier, a test model Sikorsky YH-19 had been brought to Korea for a priority mission to recover a Russian-built MiG-15 jet fighter that had crash landed in North Korean territory. That mission completed, the YH-19 joined the H-5s in their evacuation effort at the Munsan-ni drop zone. The YH-19 could evacuate wounded four times as fast as the H-5, carrying eight litter patients on each trip out. On the return flight to the drop zone, the YH-19 could carry in 1,500 pounds of medical supplies.

It was not until February 1952 that the 3d ARSq began receiving H-19s to replace the H-5s as they were lost or worn out. These original H-19s were equipped with floats for water landings, but most rescue recoveries were made by hovering over the survivor and lowering a rescue device by hydraulic hoist. The H-19 had a combat radius of 192 kilometers (120 mi.) compared with the 136-kilometer (85-mi.) range of the H-5. This made a significant difference in the H-19's suitability for overwater rescue work.

In this environment of nearly unchallenged air superiority, the H-19 became one of the real success stories of the Korean conflict. In July H-19s were used to rescue 710 UN soldiers who were stranded in exposed forward positions by flood waters. On 12 April 1953 an H-19 crew rescued Capt. Joseph C. McConnell, Jr., after he parachuted into the Yellow Sea. This rescue allowed Captain McConnell to continue his flying exploits and to become the leading jet ace of the Korean War.

The H-5 and H-19 helicopter crews played a major role in the impressive statistics of the air rescue service (ARS) in Korea. In the final accounting, the ARS helicopter crews had rescued 846 U.S. and UN aircrew members from behind enemy lines, recovered 86 airmen from friendly territory, and performed front-line medical evacuation for 8,598 wounded soldiers.

Ten Years of Peace

One of the most impressive and daring feats of this period came while the Korean War was still raging in the other hemisphere. In July 1952, the crews of two USAF H-19s demonstrated the aircraft's ability to self-deploy to Europe by flying from Westover Air Force Base, Massachusetts, to Wiesbaden, Germany. The flight, appropriately named Operation Hopalong, covered 6,374 kilometers (3,984 mi.) in seven legs and required 52 hours of flying time. The longest leg, Greenland to Iceland, covered 1,344 kilometers (840 mi.) and established a nonstop world's record for helicopters.

In 1952 the Air Force took delivery on the first model of another new rescue helicopter, the Piasecki H-21. The H-21 went on to have a distinguished career in the air force that spanned nearly two decades. Besides being used in many base support roles, the H-21 found a niche for itself in arctic search and rescue. It was indeed the workhorse of the Alaskan Air Command from 1956 until its retirement in the early 1970s. In September 1967 the worst flood in Alaska's history hit Fairbanks. The H-21s from both Eilson AFB (Fairbanks) and Elmendorf AFB (Anchorage) were called into action. The helicopters flew day and night searches of a 240-kilometer (150-mi.) radius around Fairbanks. They carried hundreds of persons to the safety of higher ground. When they were not carrying evacuees, the H-21s airlifted 30,000 doses of typhoid serum and almost 100,000 pounds of supplies and equipment to Fairbanks.

The H-43B was intended to be the primary aircraft in the new scheme consolidating all LBR detachments under the single manager of ARS. By late 1961 every major USAF base in the world had an ARS local base rescue unit. The LBR concept had the H-43 providing rescue coverage within a 120-kilometer (75-mi.) radius of its home air force base and fire rescue coverage within 24 kilometers (15 mi.). For fire rescue, the helicopters used a fire suppression kit (FSK). This kit was carried under the H-43 while it orbited in anticipation of a precautionary or emergency landing. In the event of an actual fire, the H-43 would deploy the FSK and two firefighters upwind of the fire. The H-43 would then use its directional downwash to blow a path through the flames for the firefighters to penetrate to the crashed aircraft. The objective was not to extinguish the fire but rather to provide emergency access to possible survivors.

Vietnam

The total rotary wing fleet available to ARS in 1961 comprised 17 H-43As, 69 H-43Bs, 58 obsolete H-19Bs, and 4 H-21Bs. The force was configured for peacetime search and rescue in the immediate vicinity of USAF bases. The most capable helicopter, the H-43B, had no weapons, no armor, and a pitifully short 120-kilometer (75-mi.) operation radius of action. That was the force available when the conflict in Southeast Asia began to produce a pressing need for a combat search and rescue force.

SEARCH AND RESCUE

The first air force helicopters to arrive in Southeast Asia were the two HH-43Bs sent to Nakhon Phanom, Thailand, in June 1964. In October 1964, two permanent rescue detachments were established at Da Nang and Bien Hoa in South Vietnam. Each detachment was equipped with three of the more powerful HH-43F models. The combat radius of the HH-43F was 192–224 kilometers (120–40 mi.) as compared with the 120-kilometer radius of the HH-43B. This increase allowed the Da Nang–based helicopters to reach into North Vietnam for the first time. By January 1965, HH-43s were operating out of three more detachments in Thailand, bringing their operating locations in Southeast Asia to a total of six.

In July 1965 the first two CH-3C helicopters, on loan to ARS from the Tactical Air Command, arrived in Thailand. This aircraft's range of approximately 800 kilometers (500 mi.) gave the ARS their first capability to penetrate into North Vietnam from Thailand without having to land enroute to refuel. Four months later the first of the even longer-range and more powerful HH-3Es began arriving for duty in Thailand.

The next major breakthrough in USAF helicopter capability came with the advent of aerial refueling from fixed-wing tanker aircraft. In December 1965, the feasibility of that maneuver was proven when a modified Air Force CH-3C made ten successful probe/drogue hookups with a Marine Corps KC-130F tanker. The USAF ordered twenty of its HC-130 Hercules aircraft and seventeen HH-3E helicopters modified for air refueling operations. On 31 May and 1 June 1967, two USAF HH-3E helicopters were flown nonstop from Brooklyn Naval Air Station, New York, to Paris where they landed during the Paris Air Show. During the 31-hour, 6,720-kilometer (4,200-mi.) trip, the helicopters were refueled nine times by five HC-130P tankers from Eglin Air Force Base in Florida. Just three weeks later, the procedure was operationally tested in Southeast Asia; by September 1967, aerial refueling was being used routinely on combat rescue missions.

In September 1967 the first two of seven HH-53Bs arrived in Thailand to join the rapidly growing rescue force. Two years later the first of many HH-53Cs joined the fight in Southeast Asia. These HH-53B/Cs were probably the finest rescue helicopters in the world. Five of these HH-53s and one HH-3E were used in the raid on the Son Tay prisoner-of-war camp near Hanoi in November 1970. As good as the HH-53 was, it was no match for the concentration of North Vietnamese antiaircraft weapons in the vicinity of Hanoi in the later stages of the war. During the B-52 raids on North Vietnam in December 1972, none of the many American airmen shot down were recovered from the heartland due to the intensity of the opposition.

Between 1964 and August 1973, the combined forces of the Aerospace Rescue and Recover Service (ARRS) were credited with saving 3,883 lives. All but a handful of those were saved by air force helicopters

SPECIAL OPERATIONS

From 1967 to 1973, USAF special operations squadrons (SOSs) provided helicopter support to U.S. Army special forces conducting cross-border operations into denied areas of Southeast Asia. The organizations were the 20th and 21st Special Operations Squadrons. The 20th SOS was located in Nha Trang, Vietnam, and had a forward operating base at Ban Me Thuot. It also had a detachment in Udorn, Thailand, which was known as the Pony Express. The Vietnam-based part of the squadron operated the Bell UH-1P helicopter while the Pony Express operated a mixture of UH-1Ps and Sikorsky CH-3Cs. The 21st SOS was located at Nakhon Phanom, Thailand, and operated the CH-3E.

A reorganization of forces in 1969, due mainly to the attrition of UH-1Ps in the 20th SOS, closed the Pony Express detachment. The detachment's CH-3s were transferred to the 21st SOS and the UH-1Ps were sent to the 20th SOS in Vietnam. In August 1970, the 21st SOS received its first CH-53C and began a transition from the H-3 to the H-53 that took sixteen months to complete. In September 1970, the 20th SOS received the first of its UH-1N "Twin Hueys" to replace the rapidly diminishing supply of UH-1Ps.

SOUTHEAST ASIA FINALE

In the space of 32 days in the spring of 1975, USAF helicopters were involved in three of the largest operations of the nearly twelve years they had been in Southeast Asia.

Operation Eagle Pull—The evacuation of Phnom Penh. On 12 April 1975, USAF and U.S. Marine Corps helicopters evacuated 291 persons from the Cambodian capitol just as the city fell to the Khmer Rouge forces. The Marine CH-53s transported all of the actual evacuees while ARRS HH-53s removed the combat control team and the Marine Security Force at the conclusion of the evacuation. While this was going on, nine CH-53s from the 21st SOS orbited just north of the city in case they were needed.

Operation Frequent Wind—The evacuation of Saigon. On 29 April 1975, seven CH-53s of the 21st SOS and two ARRS HH-53s joined an armada of helicopters assigned to evacuate the personnel from the American Embassy in Saigon to ships waiting offshore. The nine USAF helicopters made four round trips each and carried a total of 1,479 evacuees to safety. One of the ARRS helicopters was engaged by a heat-seeking missile and escaped the threat by firing a flare to decoy the missile.

Rescue of the crew of the Mayaguez. The last major USAF helicopter operation in Southeast Asia began on 12 May 1975, when Cambodian naval forces captured the American-registered cargo ship *Mayaguez* and held its crew captive on Koh Tang Island.

Before sunrise on 15 May the assault on the island began, using eight of the fifteen H-53s available at U-Tapao to ferry the Marine assault force. The five CH-53s and three HH-53s in the first assault wave met unexpectedly intense opposition. By a little after eight o'clock that night the fight was over, and the crew and vessel were safe. The intervening hours, however, had seen some of the most furious fighting ever to involve USAF helicopters. During the day some 230 men had been landed on the island and withdrawn. Total U.S. casualties were fifteen killed, three missing and 49 wounded. The helicopter toll was enormous. Of the fifteen H-53s involved in the fight, four were destroyed and nine were damaged.

Mayaguez to 1989

In the years since the *Mayaguez* incident, the USAF helicopter force has seen considerable restructuring due to both mission changes and fiscal constraints. In September 1981, the Aerospace Rescue and Recovery Service was credited with having saved its 20,000th life since the inception of the service in 1946.

The number of helicopters in the active air force inventory dropped from 237 in 1984 to 171 in 1989. The Air Force bought the HH-1H to replace the HH-43 in the LBR role. All HH-43s have been retired from active service. The air force reallocated the UH-1N and HH-1H to take over the SAC missile site support role. All UH-1Fs, UH-1Ps, and TH-1Fs were retired by mid-1987.

The H-3 fleet is contracting as the helicopters become more difficult and expensive to maintain. Approximately 50 of the CH/HH-3s remain in the active inventory; their average age now is 22 years.

The drone recovery mission, for which the CH-3 was originally purchased, was inactivated in 1987. In seventeen years of using the Mid-Air Recovery System (MARS) to snag and safely return missiles from above the Gulf of Mexico, that single H-3 detachment had retrieved more than 3,000 drones at an estimated savings to the U.S. government of over a billion dollars.

The Future of U.S. Air Force Helicopters

Light-Lift Helicopters

The future of the USAF light-lift helicopter fleet remains in doubt. The UH-1Ns and HH-1Hs continue their role in supporting SAC missile sites and the two air force survival schools, but whether these will remain USAF missions is still in question.

Medium/Heavy-Lift Helicopters

The shift in emphasis from combat search and rescue to special operations has had a significant impact upon the medium- and heavy-lift helicopter units. Nearly all of the H-60 and H-53 assets in the current inventory are designated as special operations aircraft, and all of the H-53s are being modified to the MH-53J Pave Low III standard. That leaves the classic air force helicopter mission of combat search and rescue without assets. The one bright star is the MH-60G Pave Hawk helicopter. The procurement process continues with most new helicopters earmarked to form the nucleus of a new and vital rescue force.

The CV-22 Osprey

This is a tilt-rotor, multimission aircraft based on the Bell XV-15 design. It is designed to have the vertical-lift characteristics and performance of a helicopter and the speed of a fixed-wing aircraft. The USAF has specified a requirement for 55 of the long-range version (designated CV-22A) of this aircraft for special operations. The USAF currently classifies the CV-22 as a fixed-wing airplane, but its acquisition will definitely affect the helicopter fleet. If it replaces the MH-53J in the role of long-range support for special operations, a significant number of the very capable Pave Low helicopters might be returned to the combat search and rescue mission.

Royal Air Force Helicopters

Special Operations

The RAF first used the helicopter to combat an insurgency in 1950 when they deployed a flight of four Westland-Sikorsky Dragonfly HC2s to Malaya to counter the communist bandits. The Dragonfly was built in England by Westland under license from Sikorsky. It was a variant of the Sikorsky S-51 and was similar to the USAF H-5. In 1954, these forces were joined in Malaya by Sycamore helicopters. The Sycamore was the production version of the first British-designed helicopter.

The first major test of RAF helicopters in a counterinsurgency role was in Cyprus in the late 1950s. Here the Sycamore carried the brunt of the missions as the colonial war often manifested itself in terrorist acts against the British forces. Night operations and rope insertions were first tried during the two years of counterterrorism operations in Cyprus.

In 1962, yet another counterguerrilla operation in the Far East saw yet another new British helicopter tested in combat. When the Sultan of Brunei appealed to Britain for help in maintaining order, the RAF responded by sending its giant Bristol Belvedere helicopters from their base in Singapore to Borneo. This initial positioning flight required an impressive 640-kilometer (400-mi.) nonstop over-water sortie by a large number of helicopters. Although it took only one month to disperse these insurgents, the RAF involvement in the area had four more years to go.

When the new Federation of Malaysia was formed in 1963, Indonesia reacted violently and infiltrated guerrilla forces into Malaysia all along their 1,600-kilometer (1,000-mi.) common border. The RAF deployed a large helicopter force to support the Malaysian troops. Included in this force were the new Wessex helicopters. The Wessex was another Sikorsky product (S-58), built in Britain under license by Westland. In the three years of operations in the Asian jungles, the Wessex proved its worth in both the violent counterinsurgency operations and the newly evolved techniques of nation building in the campaign for the "hearts and minds" of the local population. So successful was the helicopter in this new and more gentle side of COIN operations that the Indonesian insurgents never found widespread support from the Malay villagers.

In 1969, the counterinsurgency effort in Northern Ireland began, and it continues to this day. The RAF helicopter effort there was originally carried entirely by the Wessex but, in the early 1970s, the new Westland-Aerospatiale Puma was added to the force. The Puma was larger and faster than the Wessex. It could move up to twenty troops at speeds of up to 240 kilometers per hour (150 mph). This allowed for a more rapid insertion of larger groups of security forces.

SEARCH AND RESCUE

The RAF helicopter force has been a cornerstone of the British search and rescue system since the early days of the tiny Sycamore. The helicopter that carried the RAF SAR banner for almost three decades, however, was the Westland Whirlwind. The Whirlwind was still another example of a Sikorsky product (S-55) built in the UK under license by Westland. The Whirlwind was in active service from the early 1950s until the early 1980s. The RAF bought 42 of the HAR 10 version of the Whirlwind, which specialized in mountain and offshore rescue. That variant significantly upgraded the basic S-55 by replacing the piston engine with the Rolls Royce Gnome turboshaft engine.

Another SAR stalwart in the RAF inventory is the Wessex. It entered military service in the early 1960s and is still in active service. The newest SAR platform in RAF service is the Westland Sea King HAR 3 helicopter. This is another Sikorsky-Westland license package based upon the Sikorsky S-61 helicopter. The Sea King is similar to the USAF CH-3 helicopter, but it is a much more capable rescue machine.

Soviet Air Force Helicopters

SPECIAL OPERATIONS

Contrary to what might be the popular conception, there is little in Soviet history or military doctrine to support the concept of helicopters used in the special operations role. Many would contend that the Soviet use of helicopters in Afghanistan constituted a special operation, since the opposition was an insurgent guerrilla force. On the contrary, although the helicopter was the single most effective weapon the Soviets used against the Afghan rebels, it was used almost exclusively in a conventional air assault role.

The use of helicopters to insert a reinforced rifle battalion deep behind enemy lines is a classic air assault mission. When the force to be inserted, however, is a so-called special purpose brigade whose mission is reconnaissance, sabotage, infiltration, or any other form of subversion, then the mission meets the Western definition of a special operations infiltration. That mission was definitely a part of Soviet military doctrine.

That doctrine also supports heliborne air assaults that fit the Western definition of direct action missions or raids. Here, helicopters are used to insert specialized strike forces behind enemy lines for a single, specified task or target, and then rapidly return those forces to their own territory.

The helicopters used to accomplish these special operations would almost certainly be the Mi-8 Hip to carry the troops and equipment and the Mi-24 Hind to provide cover and support for the insertion. In the case of small teams, the Mi-24 Hind might be used as the troop-carrying vehicle itself.

SEARCH AND RESCUE

Soviet secrecy and the selective release of Soviet publications made research into this aspect of military doctrine nearly impossible. Much of what follows is inferred rather than documented.

The Soviets did establish an active air rescue service. The *Soviet Military Encyclopedia* defines that service and states that "search and rescue operations are performed by airplanes, helicopters, ships, vessels, and ground facilities equipped with radar search apparatus and rescue equipment."

All evidence points to the Mi-8 Hip as the primary rescue helicopter. It is comparable in speed and size to the USAF's HH-3E "Jolly Green Giant" and is equipped with an electrically operated hoist installed in the door.

Whether the Soviet air forces had dedicated SAR units assigned to the Afghanistan campaign is beyond the scope of unclassified research. The Soviets were even less inclined to publicize their crashes and combat losses in that conflict than they had been in previous cases. It is known, however, that the first Soviet military pilot to be awarded the "Hero of the Soviet Union" for military action in Afghanistan was a helicopter pilot; he received the award for rescuing another helicopter crew that was downed in a combat situation.

HELICOPTER AIR-TO-AIR

This topic goes beyond the scope of comparing and contrasting the three air forces and how they utilize helicopters in special operations and search and rescue missions. However, any current article on Soviet helicopters would be incomplete without the mention of this new air-to-air mission, which has the potential of determining the victor on future battlefields.

The logic the Soviets used to validate their requirement for an air-superiority helicopter is elegant in its simplicity. The cornerstone of Soviet strategy for a high-intensity conventional war in Europe was the rapid movement of troops supported by fast-moving tanks. The greatest threat to that strategy came from the antitank attack helicopters of NATO such as the AH-64 Apache, the Mangusta and the MBB Bo-105. If the Soviets' strategy was to remain viable, they had to be able to counter the threat of those attack helicopters. The most promising way to defeat NATO's helicopters was with counterair helicopters.

To meet the need for counterair helicopters, the Soviets designed and built at least one, and probably two, new helicopters dedicated to the antihelicopter role.

The Mi-28 Havoc attack helicopter is similar in appearance to the AH-64 Apache. Like the Apache, it can carry antitank guided missiles and was assigned to perform that air-to-ground role. These missiles, coupled with the anticipated Havoc speed of 214 knots, make the helicopter a formidable air-to-air combatant as well.

The Kamov Hokum may not share the multiple-mission personality of the Havoc. It has not yet been observed carrying antitank missiles. Belief is that this helicopter is intended for the single purpose of air-to-air combat. It will carry air-to-air missiles and will fly at speeds up to 250 knots. Coupling such speed with a day/night all-weather combat capability, the Hokum can provide a significant rotary-wing air-superiority capability.

As of 1990, no helicopter had ever been shot down by another helicopter. In fact, there is no record of any helicopter-to-helicopter fire fight. The Western air forces, having given the former Soviets a commanding lead in both the technology and the techniques of antihelicopter warfare, seem to think that that situation will continue into future conflicts.

PHILIP STINSON

SEE ALSO: Helicopter; Korean War; Rotary-Wing Aircraft Technology; Special Operations; Vietnam and Indochina Wars.

Bibliography

Boyne, W. J., and D. S. Lopez. 1984. *Vertical flight: The age of the helicopter.* Washington, D.C.: Smithsonian Institution Press.
Cardoza, A. A. 1987. Soviet aviation in Afghanistan. *Proceedings*, February pp. 85–88.
Cockerham, S. G. 1987. A half-century of helicopters. *Military Review*, September, pp. 82–83.
Gunston, B. 1981. *Military helicopters.* New York: Arco.
Polmar, N., and F. D. Kennedy, Jr. 1981. *Military helicopters of the world.* Annapolis, Md.: U.S. Naval Institute Press.
Wragg, D. W. 1983. *Helicopters at war: A pictorial history.* New York: St. Martin's Press.

ADDITIONAL SOURCES: *Air Force Magazine; Airman; Air University Review.*

HELICOPTER, NAVAL

Since their first military use with the German navy during World War II, naval helicopters have developed in three aspects: their characteristics and capabilities, the equipment that has been installed on them, and the missions they have been enabled to perform.

It is no longer necessary to have aircraft carriers in the fleet to support helicopter operations: modern helicopters can operate from the small decks of destroyers and even frigates. As a result, many nations have been able to afford naval helicopters, since they have not had to undergo the expense of procuring helicopter carriers. Only in the case of amphibious operations are specialized helicopter carriers needed to support helicopter assault. For other roles, such as fire support and general support, naval helicopters need only small platforms.

Today's naval helicopters comprise a wide variety of multirole and transport types, in addition to attack helicopters. This article deals with those types, their characteristics, the roles and missions they carry out, and how they fulfill those tasks.

Types

Naval helicopters are characterized by their ability to accomplish over-water missions as well as different roles in amphibious operations. The naval helicopter may be specifically designed as a naval type, like the Soviet Kamov Ka-25 Hormone and Ka-27 Helix and the American Sea Sprite, or it may be a naval version of a multirole helicopter specially modified and equipped for naval missions, such as the American Sikorsky S-70L Seahawk.

According to their capabilities, missions, equipment, and armament, naval helicopters may be classified as transport and assault, multirole, or attack. In the major world powers, naval helicopter units are usually organic to the navy, but in other countries, naval helicopters sometimes are only operationally subordinated to the naval commander.

Transport and assault helicopters should have the capability for considerable lift, as well as wide cabins suitable for carrying either a number of troops or cargo loads, or for being equipped with heavy naval equipment. They usually carry out one or more of the following missions: helicopter assault, troop and cargo transport, aerial resupply and replenishment at sea (Fig. 1), medical evacuation, or heavy lift. The payload capabilities of transport and assault helicopters vary from 25 fully armed troops or 2,268 kilograms (5,000 lb.) of cargo, as in the Westland Commando, to 37 fully armed troops or 3,629 kilograms (8,000 lb.) of cargo, the capacity of the S-65 (CH-53 ND) Sea Stallion, to the heavy lift capability of the CH-53E Super Stallion, which can lift 55 fully armed troops or 16,330 kilograms (36,000 lb.) of cargo.

Multirole helicopters comprise those types which can carry out different missions, either by virtue of additional equipment or because they are manufactured in different versions. The Westland Lynx, Sikorsky S-61 and S-70L, Aerospatiale Dauphine, and Kamov Ka-27 are good examples of the multiversion, multirole helicopter. Depending on their equipment and armament, multirole helicopters can carry out more than one of the following missions: antisubmarine warfare (ASW); antisurface vessel (ASV) operations; over-the-horizon (OTH) missile targeting; maritime patrol, search, and rescue (SAR); minesweeping; airborne early warning (AEW); electronic warfare support; and coastal reconnaissance. In addition to their advanced systems, mission equipment may include surveillance radar, magnetic anomaly detection (MAD), dipping sonar, sonobuoys, and guided missile sights. For mine sweeping, the helicopter is usually equipped with towing equipment, towed mine hunting sonar, and acoustic minesweeping gear. Special airborne early-warning radar is provided for use in that role: for example, that used by the Sea King MK2 helicopter in service with the U.K.

Figure 1. A CH-46 Sea Knight helicopter drops supplies on the deck of a hospital ship during vertical replenishment. (SOURCE: U.S. Department of Defense)

TABLE 1. *Characteristics of Naval Assault, Transport, and Heavy Lift Helicopters*

PARAMETER	BOEING VERTOL V-107 (CH-46E)	SIKORSKY S-80 (CH-53E)	BELL/BOEING VERTOL (MV-22A)
Origin	U.S. 1975	U.S. 1974	U.S. 1988
Type	Tactical transport	Heavy transport	Assault transport
Engines	2 1,870 shp GE	3 4,380 shp GE	2 6,000 shp Allison
Rotor diameter	15.2 m (50 ft.)	24.1 m (79 ft.)	11.6 m (38 ft.)
Payload	26 troops or 3,175-kg (7,000-lb.) cargo	55 troops or 16,330 kg. (36,000 lb.) cargo sling-loaded	24 troops
Maximum speed	270 kph (168 mph)	315 kph (194 mph)	630 kph (391 mph)
Ceiling	3,960 m (13,000 ft.)	3,780 m (12,400 ft.)	Not available
Range	371 km (230 mi.)	500 km (310 mi.)	3,900 km (2,418 mi.) ferry

Royal Navy. The multirole helicopter, depending on its mission, may be armed with antiship missiles or with antisubmarine torpedoes and depth bombs (charges).

As with battlefield helicopters, the naval attack helicopter is a modified or specifically designed machine for providing fire support. In the naval role, such fire support is accomplished in both over-water and amphibious operations. For over-water operations, the multirole helicopter is a type specially equipped for antisurface vessel operations. But in amphibious operations, a ship-borne attack helicopter is usually committed to the mission, as in the case of the Bell 209 (AH-1) Sea Cobra and Super Cobra in service with the U.S. Marine Corps.

Roles, Missions, and Tactics

Naval helicopters are usually involved in three main roles: fire support, general support, and helicopter assault. The first two roles can be carried out in both over-water and amphibious operations, while the assault role is usually part of an amphibious operation, or sometimes of commando raids. The ways in which these roles are performed depend on the nation's military doctrine, the nature of the threat, the size of the available force, and the tactical situation.

FIRE SUPPORT MISSIONS AND TACTICS

Fire support from naval helicopters includes that provided during over-water missions, such as antisubmarine and antisurface vessel operations, and over-land missions, as when helicopter fire supports combat actions in amphibious or helicopter assault operations.

Antisubmarine operations are performed by naval helicopters and fixed-wing aircraft, as well as by surface and subsurface vessels. While fixed-wing aircraft, utilizing their superior speed and range, sweep the sea far from the surface force, helicopters maintain a barrier closer in as part of the ASW mission of protecting and ensuring the safe and timely arrival of the surface fleet.

Exploiting their hovering and vertical take-off and landing capabilities, naval helicopters also play an offensive role as the most important element in antisubmarine operations, where they can operate from land or from carrier or other ship bases. In the conduct of antisubmarine operations, naval helicopters usually accomplish two complementary missions: search for enemy submarines, then destroy them. Even if the enemy submarine is only prevented from launching its weapons, the mission may be considered successful.

In the search for enemy submarines, naval helicopters use three practical means of detection: radar, magnetic anomaly detection (MAD), and sonar. Visual detection of a submarine is entirely a matter of luck. To have some

TABLE 2. *Characteristics of Naval Multirole Helicopters*

PARAMETER	AEROSPATIALE DAUPHINE 2	MIL MI-14 HAZE	SIKORSKY S-70 (SH-60B)	KAMOV KA-27 HELIX	WESTLAND LYNX 3	EUROPEAN INDUSTRIES EH-101
Origin	France 1975	USSR 1976	U.S. 1979	USSR 1980	U.K. 1984	Italy/U.K. 1987
Type	Surface attack/patrol	Amphibious MCM/ASW	Multipurpose/ASW	(A) ASW; (B) ASST; (C) SAR/utility	ASW/ASV	ASW/utility
Engines	2 700 shp Turbomeca Ariel	2 2,000 shp Istov TV2-117	2 1,690 shp GE T 700	2 2,225 shp Istov TV3-117V	2 1,115 shp Rolls Royce Gem 60	3 1,725 GE T700-GE-401
Maximum weight	4,100 kg (9,039 lb.)	14,000 kg (30,864 lb.)	9,183 kg (20,244 lb.)	11,000 kg (24,250 lb.)	5,900 kg (13,000 lb.)	13,000 kg (28,660 lb.)
Maximum speed	297 kph (184 mph)	231 kph (143 mph)	234 kph (145 mph)	250 kph (155 mph)	306 kph (190 mph)	279 kph (173 mph)
Ceiling	Not available	4,500 m (14,765 ft.)	5,791 m (19,000 ft.)	6,000 m (19,685 ft.)	Not available	Not available
Range	882 km (547 mi.)	800 km (497 mi.)	1,100 km (683 mi.) ferry	800 km (497 mi.)	620 km (385 mi.)	556 km (345 mi.)
Armament	(M) 4 AS.15TT ASMs	Torpedoes, depth charges	2 MK-46 (or EX-50 ALWT) torpedoes	2 ASW torpedoes or ASV missiles	2 MK-46, MK-96, or Sting Ray torpedoes, depth charges, or 4 Sea Skua antiship missiles	4 torpedoes or antiship missiles

chance of detecting a surfaced submarine, it is essential to be in the right place at the right time, since the surfaced submarine constitutes such a small target in the vastness of the sea. Radar detection, too, can be accomplished only in the case of a partially or fully surfaced submarine.

Since a submarine is a large metal object, one which distorts the earth's magnetic field, its presence can also be detected by a very sensitive device having the ability to detect the magnetic field distortion created by the submarine. This device, called a magnetic anomaly detector, is an extremely short-range detection device, which means that it is normally used in the attack rather than as a passive detection device.

The most widely used means of submarine detection is acoustic sonar. This device, which may be either active or passive, relies on dunking sonar or disposable sonobuoys. Active buoys operate somewhat like underwater radar, sending out sound waves. If these expanding waves strike a submerged object, a portion of the signal energy is reflected. This reflected signal can then be heard by the sensitive buoy's hydrophone. Passive buoys listen for underwater sounds, then measure the direction of their sources. In the case of the modern submarine, which is specially designed to be very quiet, passive buoys are less effective. This gives even more importance to active buoys, which can measure both range and direction.

With fixed-wing aircraft providing the outer layer of antisubmarine defense, naval helicopters carry out their protective missions at the near approaches to the defended surface force. Using active buoys, the helicopters usually lay a screen across the path of the surface force, renewing that screen successively as the force progresses.

The rear of the surface force is also covered by a sonobuoy screen. Since the enemy submarine, approaching from the rear, must make suitable speed to enable it to close on the surface force, it cannot run as silently, which means that passive sonobuoys can be used in this application. The rear screen thus is laid just astern of the surface force, where it remains active for several hours while a protective net is formed of overlapped circles of detection.

Helicopters using dunking sonars provide effective flank coverage. According to the specific situation, either both flanks can be covered or only the one deemed to be threatened. Antisubmarine helicopters using this technique typically work in pairs, thereby maintaining continuous underwater contact while moving from one position to

TABLE 3. *Naval Attack Helicopter*

PARAMETER	BELL 209 (AH-1) SEA COBRA/SUPER COBRA
Origin	U.S. 1980/1983
Type	Two-seat ground support
Engine	(T) 1970 shp Pratt & Whitney T400-WV-402
Rotor diameter	14.6 m (48 ft.)
Maximum weight	6,350 kg (14,000 lb.)
Maximum speed	277 kph (172 mph)
Ceiling	3,795 m (12,450 ft.)
Range	577 km (358 mi.)
Armament	1 20mm M197 3-barrel gun, TOW missiles (W can carry 8 TOW or Hellfire), or gun/rocket pods

another. The two helicopters can thus set up an efficient overlapping pattern of search.

Passive dunking sonar is the preferred means of detecting a running submarine; since it does not alert the enemy vessel, the helicopter using such sonar need not cope with the target submarine's evasive action. But against a silent runner, or in critical situations—as when the enemy submarine is nearing torpedo range of the surface force—it is essential to use active dunking sonar to gain contact with the silent runner, or at least to warn the attacking submarine that it is being tracked and may come under attack. This usually forces the enemy submarine to withdraw for its own protection instead of completing its attack.

As soon as one of a pair of patrolling helicopters has gained contact with an enemy submarine, the other—using its MAD, and guided by the helicopter in contact—marks the exact course of the submarine and determines its approximate speed by laying a line of smoke floats by day or flares by night. The attack can then be launched using either depth charges or homing acoustic torpedos.

ANTISURFACE VESSEL OPERATIONS

ASV operations consist of a group of combined and coordinated combat actions carried out by friendly naval aviation (both helicopters and fixed-wing aircraft), surface vessels, and submarines. While the fixed-wing aircraft and submarines accomplish their missions at the far approaches of the enemy surface vessels, helicopters and their base ships carry out their missions at the near approaches of these vessels.

As with ASW, ASV operations comprise two complementary missions: searching for, and then destroying, the enemy vessel. Search is conducted mainly by radar, and both active and passive detection are used. Although visual detection capability is poor even in good weather, it cannot be entirely discounted.

Since the radar has a very short range at low altitudes, the helicopter must gain height to provide a suitable range of detection. If the target's location is accurately known and there are no large ships in the assigned area, a straight pass at medium altitude, using radar, will be sufficient to detect the target. To search a wide area, and when the target ship is underway, a different technique is required.

Wide-area search is conducted by one or two helicopters which comb the area in several parallel and opposing passes using overlapped radar screening. The length of the leg flown is based on a calculation of slightly less than half the time the expected target would take to cross the scan width at full speed.

Once a target is detected, the methods of attack used depend on the type of antiship missile carried and its homing system. With missiles that use internal midcourse guidance, once the helicopter has acquired a target, it can set its radar on standby mode and drop below the target's radar detection zone, where it feeds attack data to the missile's autopilot. If two helicopters are working as a team at wide separation, the detecting helicopter transmits the attack data to its partner, which usually lurks unseen. In this case, the partner will have a good chance to launch its missile using a dog-leg approach that ensures a surprise attack from an unexpected direction. The detecting helicopter can also transmit attack data to friendly surface vessels. A system such as this enables surface vessels to launch their missiles at targets located over the horizon.

OVER-LAND FIRE SUPPORT

Over-land fire support is usually provided by naval helicopters in the course of amphibious operations. These missions are intended to weaken the resistance of enemy defenses and secure the success of amphibious operations, and also to assault enemy tanks that may threaten the landing.

In coordination with naval gunfire and strikes by fixed-wing aircraft, naval attack helicopters thus usually launch their attacks against two main targets: enemy weapons emplaced in and around the landing area and enemy armor reserves that may threaten the landing/assault force. Naval helicopter fire support is usually conducted in one of two forms: planned combat actions, such as aerial preparation for a landing, and on-call combat actions, such as close fire support. For planned strikes, attack helicopters launch their attacks on the basis of reconnaissance information provided by other sources, whereas in on-call operations they locate their own targets by means of direct reconnaissance or communication with their own contact team accompanying the landing or assault force.

The tactics and weapons used in such operations depend on the tactical situation and the nature of the targets encountered. Naval attack helicopters launch their antitank missiles from a hovering position—except in desert areas, where too much dust would be generated by hovering. In the desert, it is preferable to use a dive attack to launch the helicopter's weapons.

GENERAL SUPPORT MISSIONS

These missions contribute to the overall success of naval operations. They include mine countermeasures, missile decoying, and airborne early warning (AEW), in addition to the range of other missions similar to those carried out by battlefield helicopters.

Some naval helicopters, such as the RH-53D MCM (mine countermeasures) and the Mi-14 B Haze, are specially equipped for mine sweeping, but missile decoying is carried out by multirole helicopters. When they spot an enemy antisurface vessel missile, these helicopters can deploy chaff and flares to decoy the missile away from friendly ships.

The experience of the Falkland Islands (Islas Malvinas) conflict in 1982 confirmed the need for airborne early warning to accompany the fleet and to operate far beyond the range of friendly fixed-wing AEW aircraft. When there is no available aircraft carrier capable of accommodating fixed-wing AEW aircraft (such as the Grumman E-2C Hawkeye), it is essential to look for an alternative, as the U.K. Royal Navy has done with the modified AEW Sea King.

The airborne radar with which the Sea King has been fitted enables it to patrol the fleet at a moderate altitude and a range of 23 miles for about 45 minutes. Published reports indicate that the helicopter's radar has a full 360-degree scan and that it can detect bomber-sized targets at medium altitude out to a distance of about 200 miles and low-level cruise-missile-type targets out to about 46 miles.

HELICOPTER ASSAULT

Naval helicopter assault is carried out in the same manner as battlefield helicopters, except for a few naval modifications, summarized as follows: (1) the heliborne force is formed mainly of the marine force rather than of airborne or infantry troops; (2) the assault usually is performed within the framework of an amphibious operation and according to its concept; (3) the assault is initiated from carrier bases directly rather than from departure areas; (4) enemy coastal defenses represent the enemy front line; and (5) landing sites are chosen in the coastal area or behind the coast line, depending on the concept of the amphibious operation and the mission of the heliborne force.

Future Prospects

With the continuing development of naval helicopters and their equipment, as well as of naval armament, the future will see more adaptation of helicopters for meeting naval requirements. The forthcoming tilt rotor technology, which has been demonstrated in the Bell-Boeing V-22 Osprey, forms an important competitor to the traditional helicopter in the naval field. Its cruising speed of more than 300 miles an hour and its dramatically lower operating costs in comparison with the traditional helicopter provide a serious challenge, particularly in terms of the transport/assault and search and rescue missions, in which the time factor is critical. Naval helicopters in the decade of the 1990s may be expected to follow the development cycle of fixed-wing aircraft in response to the dramatic growth of new technology.

GABR ALY GABR

SEE ALSO: Aircraft, Military; Airpower; Antisubmarine Warfare; Close Air Support; Helicopter; Helicopter: Military Missions; Helicopter, Battlefield; Naval Auxiliary and Support Ships; Naval Warfare; Radar; Seapower; Seapower, British.

Bibliography

Apostolo, G. 1984. *World encyclopedia of civil and military helicopters.* London: Willow.
Gunston, B., and M. Spick. 1986. *Modern fighting helicopters.* London: Salamander.
Jane's all the world's aircraft 1986–1987. 1986. London: Jane's.
Jane's all the world's aircraft 1987–1988. 1987. London: Jane's.
Jordan, J. 1983. *Modern naval aviation and aircraft carriers.* London: Salamander.
Polmar, N., and F. Kennedy. 1981. *Military helicopters of the world.* London: Arms and Armour.
Soviet helicopters: Design, development and tactics. 1983. London: Jane's.
Sweetman, B. 1981. *Soviet military aircraft.* Novato, Calif.: Presidio Press.
Warner, O., et al. 1975. *The encyclopedia of sea warfare.* pp. 215–23. London: Hamlyn.

HIERARCHY OF COMBAT

An understanding of warfare is facilitated by using a hierarchy of combat to relate all aspects of waging war from the highest activities of governments to the lowest details of tactical combat. The hierarchy of combat in common use today is as follows: war, campaign, battle, engagement, action, and duel. Also hierarchical in nature is a threefold conceptual division of the process of waging war, namely tactical, operational, and strategic.

Hierarchy of Combat Definitions

War is armed conflict, or a state of belligerence, involving military conflict between two factions, states, nations, or coalitions. Hostilities between the opponents may be initiated with or without a formal declaration that a state of war exists. A war is fought for a stated particular political or economic purpose or reason, or to resist an enemy's efforts to impose domination. A war can be short, sometimes lasting a few days, but is usually lengthy, lasting for months, years, or even generations.

A *campaign* is a phase of a war involving a series of military activities or operations related in time and space and aimed toward achieving a single, specific, strategic objective or result in the war. A campaign may include a single battle, but more often it includes a number of battles, connected over a protracted period of time or a considerable distance, but within a single theater of operations or delimited area. A campaign may last only a few days or weeks but usually lasts several months or even a year.

A *battle* is combat between major forces, each having opposing assigned or perceived operational missions; each side seeks to impose its will on the opponent by accomplishing its mission, while preventing the opponent from achieving his. A battle starts when one side initiates combat and ends when one side accomplishes its mission or when one or both sides fail to accomplish the mission(s). Land battles are often parts of campaigns, and modern battles last from a few days to several weeks. Naval battles tend to be short and, in modern times, decisive.

An *engagement* is combat between two forces, neither

larger than a division nor smaller than a company, in which each has an assigned or perceived mission, which begins when the attacking force initiates combat and ends when the attacker has accomplished its mission, or ceases to try to accomplish that mission, or one or both sides receive significant reinforcements, thus initiating a new engagement. An engagement is often part of a battle. An engagement usually lasts only one or two days; although it may be as brief as a few hours, it is rarely longer than five days.

An *action* is combat between two forces, neither larger than a battalion nor smaller than a squad, in which each side has a tactical objective, and which begins when the attacking force initiates combat and ends when the attacker seizes the objective, or one or both forces withdraw, or both forces terminate combat. An action is often part of an engagement and is sometimes part of a battle. An action lasts for a few minutes or a few hours, and never for more than a day. An action is often called a *skirmish*.

A *duel* is combat between two individuals or between two mobile fighting machines, such as combat vehicles, combat helicopters, and combat aircraft, or between a mobile fighting machine and a counterweapon. A duel starts when one participant attacks the other and ends when one or both participants are unable to continue the action, usually by ceasing fire voluntarily. A duel is almost always part of an action and lasts only a few minutes at most.

Hierarchy of Combat Examples

To cite some historical examples, the Normandy campaign of World War II is generally reckoned to have lasted from the Allied landing on the beaches in the early hours of 6 June 1944 until the end of German resistance in the Falaise Pocket about 21 August, a total of 76 days. Numerous battles occurred during that time, including the battle of the beaches from 6 to 10 June, the American capture of Cherbourg between 21 and 27 June, and the several British offensives in the Caen sector, particularly Operation "Goodwood," of 18–20 July. Engagements during the Normandy campaign included the counterattack of the 21st Panzer Division against Juno and Sword beaches on the night of 6–7 June and the valiant effort of the 15th (Scottish) Infantry Division to capture the Odon bridges during Operation "Epsom" on 26 and 27 June. Actions, often difficult to identify in the midst of larger operations, include the 2d American Ranger Battalion's capture of the empty German gun battery at Pointe du Hoe on the morning of 6 June. Another example is the capture and defense of the Bénouville and Ranville bridges by five platoons of the 2d Oxfordshire and Buckinghamshire Light Infantry which landed in gliders near the village of Hérouvillette near Caen on 6 June; the action continued until they were relieved by troops from the 5th Parachute Brigade of the 6th Airborne Division.

Tactical, Operational, and Strategic Levels of Warfare

The conceptual relationship between tactics, operations, and strategy is closely related to the hierarchy of combat. Tactics, as distinct from operations or strategy, may be defined as the technique of deploying and directing military forces—troops, ships, or aircraft, or combinations of these, and their immediate supporting elements—in coordinated combat activities against the enemy in order to attain the objectives designated by strategy or operations. The combat operations of units smaller than divisions or brigades are usually considered tactical in nature, as are duels, actions, engagements, and some battles.

The operational level is an intermediate category of military activity between strategy and tactics. It was originally developed by the German Army during World War I to provide an intermediate planning concept between the performance of battlefield activities (tactics) and the multifront or multitheater aspects of waging war (strategy). The Germans saw operations as a lower level of strategy, pertaining to the activities of large ground forces, usually armies or army groups, in a discrete theater of combat. During World War II, Soviet armed forces adopted this concept, modifying it and applying it in somewhat more rigid fashion to the activities of armies and army groups. The Soviets call their modified operational concept Operational Art, and the U.S. Army adopted a concept similar to that of the Soviets in 1982. The combat activities of divisions, corps, armies, and army groups are operational, and the operational level of war therefore covers large-scale battles and campaigns.

Strategy, in its broadest sense (national strategy), is the planning and management of a nation's total available resources—economic, social, political, and military—in order to achieve the goals of national policy, and in wartime, to maximize the chances of victory. In a narrower military sense, strategy is the art and science of planning and directing military movements and activities so as to achieve victory or, in the words of Clausewitz, "the use of engagements to attain the object of war." Properly used, then, the term "strategic" should apply only to the broader goals of a war and describes the activities of entire national armed forces.

The concept of the operational level has not won universal acceptance among military analysts. An older division has persisted, dividing the tactical level, which takes place on the battlefield or in the presence of the enemy, from the strategic level, which takes place outside of battle itself. This view, which still holds some influence, maintains that the identification of an "operational level" of warfare adds an unnecessary degree of complexity to the analysis and conduct of warfare. Other analysts, however, contend that German advocacy and study of the operational level of war, and the particular style of leadership and command needed to work within it, contributed to the qualitative advantage the German army enjoyed during both world wars. The general trend, which has become particularly noticeable in NATO during the last few years, favors the inclusion of the operational level, thanks largely to the recognition of the crucial role that operational art plays in the Soviet concept of warfare. Recent commentary in the United States has linked the operational level with the doctrine of "maneuver warfare" and with the U.S. Army's Air-land Battle concept for waging conventional warfare.

Examples of Levels of War

To return to the Normandy campaign, the decisions taken by the Allies to invade France and expand their lodgement there to wage their major land campaign against Germany in Western Europe were both strategic decisions, concerning the establishment of goals. The allocation of resources and the choices of time and place were operational decisions, concerning the methods of employing the available resources to achieve those goals. The operational decisions therefore include, on the Allied side, such choices as landing six divisions by sea and three other divisions by air. By comparison, the Germans never really resolved the operational-level dispute between Rommel, who favored the forward deployment of mobile reserves, and Rundstedt, who wanted the reserves kept back from the coast. Finally, the conduct of battle was a matter of tactics. On the day of invasion (6 June 1944), the German placement of defenses and beach obstacles and the Allied use of naval gunfire support and engineer teams in the initial waves to clear beach obstacles, as well as the British use of specially equipped armored vehicles, were all matters of tactics.

DAVID L. BONGARD

SEE ALSO: Operational Art; Strategy; Tactics.

Bibliography

Carlson, K. G. 1987. Operational level or operational art? *Military Review* 67 (10):50–54.
DeVries, P. T. 1983. Soviet operational formation for battle: A perspective. *Military Review* 63 (2):2–12.
Dick, C. J. 1984. Soviet operational concepts: Part I. *Military Review* 65 (9):30–45.
———. 1984. Soviet operational concepts: Part II. *Military Review* 65 (10):4–19.
———. 1988. Soviet operational art: Part I: The fruits of experience. *International Defense Review*, July, pp. 755–761.
Dupuy, T. N., J. R. Brinkerhoff, C. C. Johnson, and P. J. Clark. 1986. *Handbook on ground forces attrition in modern combat*. Fairfax, Va.: Hero Books.
Dupuy, T. N. 1986. *Understanding war: The historical basis for a theory of combat*. Fairfax, Va.: Hero Books.
Glantz, D. M. 1983. Maneuver and the operational level of war. *Military Review* 63 (2):13–34.
Hines, J. G., and P. A. Petersen. 1984. The Soviet conventional offensive in Europe. *Military Review* 64 (4):2–29.
Runals, S. E. 1987. A different approach. *Military Review* 67 (10):44–49.
Sullivan, G. R. 1987. Learning to decide at the operational level of war. *Military Review* 67 (10):24–33.
U.S. Department of the Army. 1986. *Field manual 100-5: Operations*. Washington, D.C.: Government Printing Office.

HINDENBURG, PAUL VON [1847–1934]

Paul von Hindenburg was a German field marshal during World War I and president of Germany during the Weimar Republic (Fig. 1). Exhibiting strength of character and a strong sense of duty, he was the most popular figure in Germany during the First World War and in the 1920s. Controversy surrounds his dependence upon more brilliant subordinates during the war and his selection of Adolf Hitler as German chancellor in 1933.

Life Prior to the First World War

Born in Posen, Prussia, on 2 October 1847 with the full name of Paul Ludwig Hans Anton von Beneckendorff und von Hindenburg, he was from an old Prussian aristocratic ("Junker") family, whose line went back to the Teutonic Knights of the Middle Ages. Upon achieving fame in the

Figure 1. Paul von Hindenburg. (SOURCE: U.S. Library of Congress)

Hindenburg, Paul von

First World War, he used the surname of von Hindenburg, although prior to that he was listed in the army records under the surname of von Beneckendorff und von Hindenburg.

He became a cadet at age 11, and served with distinction as a junior officer in the Austro-Prussian War (1866) and the Franco-Prussian War (1870–71), and was awarded the Iron Cross. Hindenburg was then selected to attend the school of the German General Staff, the Kriegsakademie. As a member of this elite and powerful general staff, his assignments alternated between staff duties and troop duties. He retired as a general in 1911. There is some suspicion that his retirement was hastened by his performance as a commander during a maneuver exercise in 1908, during which he "defeated" the opposing forces commanded by the kaiser. In later years, however, Hindenburg denied that this incident had affected his decision to retire.

The First World War

The outbreak of war in 1914 provided Hindenburg with the opportunity to return to active service. The German plan to win the war quickly, the Schlieffen Plan, called for German forces to hold the slower-to-mobilize Russians in the east, while the bulk of German forces swept through Belgium to conquer France. As the war began, Gen. Max von Prittwitz, the commander of German forces in the east (the Eighth Army), panicked when the Russians massed two field armies on the border and advanced into East Prussia. The German high command relieved Prittwitz and his chief of staff, and named Hindenburg the new commander. On the train specially assigned to carry him east to his new command, Hindenburg met his newly appointed chief of staff, Maj. Gen. Erich Ludendorff, for the first time. Under the Imperial German system, the chief of staff was more a partner of the commander than a subordinate; the two shared responsibility for the performance of the unit. The meeting of Ludendorff and Hindenburg was the beginning of a command team that would eventually direct the entire war effort of Germany.

Hindenburg was deliberate, calm, and self-effacing. Ludendorff was brilliant, intense, and subject to extremes of temperament. Working together, the two men complemented each other superbly. They demonstrated their collective abilities immediately by reversing the situation on the eastern front. With two Russian armies entering East Prussia (but not cooperating with each other), the H-L team (to use an abbreviation that Winston Churchill devised) concentrated first against the southern army and annihilated it at Tannenberg, then massed against the other army and drove it out of East Prussia. The victories were the result of cooperative efforts, not only of the H-L team, but also with significant contributions from Lt. Col. Max Hoffmann of the operations staff and from the impetuous I Corps commander, Gen. Hermann K. von François. But the object of public adoration was Hindenburg. He received fan mail and presents from an adoring public and the nation's highest decoration from the kaiser. To his credit, Hindenburg never allowed this adulation to go to his head.

After Tannenberg, Hindenburg assumed command of all German forces on the eastern front and was promoted to field marshal. With Ludendorff at his side, he directed more spectacular victories over the Russians, driving them deeper into Russia. Germany, having failed to conquer France in 1914, now fought a two-front war. The H-L team had bitter disputes with the chief of the General Staff, Gen. Erich von Falkenhayn, who directed the national war effort and believed that the war would be won by concentrating efforts in the west. The H-L team felt that Falkenhayn was both denying adequate resources for their efforts in the east, and interfering with their conduct of operations. When Falkenhayn's plans to win the war proved unsuccessful by August 1916, Hindenburg was named to replace him as chief of the General Staff. Ludendorff, as first quartermaster general, accompanied Hindenburg to the new post.

The H-L team thus assumed control of the entire German war effort. The kaiser was an indecisive figurehead, and no German politician appeared on the scene to lead the nation. By default, the military leaders assumed responsibilities that were well beyond their scope and experience. Hindenburg was so popular that, by threatening to resign, he and Ludendorff could impose their will on national policy, even forcing the resignation of the chancellor. The policies advocated by the H-L team included resumption of unrestricted submarine warfare (which brought the United States into the war) and aggressive territorial demands (which eliminated any hope of a negotiated peace).

The End of the War

In 1917 the German armies fought a masterful defensive campaign on the western front and drove Russia to internal collapse on the eastern front. The year 1918 was the year of decision. Germany had to destroy the Western Allies before the full military power of the United States could be brought to bear. The great German offensives of the spring of 1918 were largely the result of Ludendorff's efforts. They achieved initial success for which Hindenburg was awarded a national decoration that had not been bestowed since the Napoleonic Wars. But the flaws in Ludendorff's conduct of the operations, the exhaustion of Germany, and the resilience of the Western Allies caused Germany's last gamble to fail. By late 1918 it was clear that Germany could not win the war. In the bitterness of that realization, Ludendorff became the object of widespread German resentment, and resigned. Ludendorff and many others felt that Hindenburg had callously allowed Ludendorff to absorb the blame, while the field marshal remained serenely at his post. Even when the kaiser abdicated, Hindenberg remained as chief of the General Staff, supervising the withdrawal of German forces into Germany, where they crushed uprisings of radicals and restored order. Hindenburg's support for the new and fragile republic was crucial, for he was the symbol of continuity and order for the psychologically distraught German people. The field marshal's influence was also critical in the painful decision to accept the terms of the Treaty of Versailles (although Hindenburg's name was never publicly associated with that acceptance). With his popularity still high, Hindenburg retired in 1919.

Political Career

Despite his lifelong commitment to the German monarchy, Hindenburg agreed to run for president of Germany in 1925, at the age of 77. He entered politics from a genuine sense of patriotic duty, but his lack of political experience and his old age did not serve him well in the treacherous political climate of Germany during the interwar years. By the early 1930s, Hindenburg was the only national figure whose popularity exceeded that of the rising Adolf Hitler. Hindenburg defeated Hitler in the presidential election of 1932. Like many other respectable Germans, Hindenburg mistakenly believed that extremists like Hitler could be tamed or manipulated by the more reasonable elements of German society, never realizing that Hitler was always one step ahead of those elements in his plans to undermine and revolutionize German society. Political pressure forced Hindenburg to name Hitler chancellor in 1933. Hindenburg died on 2 August 1934, not living to see the results of his misjudgment. He was buried as a national hero.

Assessment

Hindenburg was an able military commander. He was not merely a figurehead in the First World War; his was the final decision in several critical situations. He bore himself extremely well under great pressure, persevering under the uncertainties of battle better than Ludendorff. His military career demonstrates the importance of combinations of personalities in modern war, a factor that is often overlooked in societies that emphasize and analyze individual contributions in isolation. Hindenburg also never allowed his enormous popularity to cloud his thinking. However, as he grew older in the murky political climate of the 1920s and early 1930s, he deferred too often to the last opinion he had heard. Like so many of his countrymen, Hindenburg never grasped that a determined megalomaniac could bend Germany to his own will. The tragedy of Field Marshal Paul von Hindenburg is that his outstanding service in the First World War gave him the popularity and respect to stand as the last obstacle to Hitler, a final role in which the aged field marshal failed.

TIMOTHY T. LUPFER

SEE ALSO: Germany; Generalship; Ludendorff, Erich; Strategy.

Bibliography

Craig, G. A. 1964. *The politics of the Prussian army.* Oxford: Oxford Univ. Press.
Dupuy, T. N. 1970. *The military lives of Hindenburg and Ludendorff of imperial Germany.* New York: Franklin Watts.
———. 1977. *A genius for war.* Fairfax, Va.: Hero Books.
Goerlitz W. 1953. *The German general staff.* New York: Praeger.
Hindenburg, P. von. 1920. *Out of my life.* London: Cassell.
Lupfer, T. T. 1981. *The dynamics of doctrine: The changes in German tactical doctrine during the First World War.* Ft. Leavenworth Kans.: Combat Studies Institute.
Wheeler-Bennett, J. W. 1936. *Wooden titan: Hindenburg in twenty years of German history.* New York: William Morrow.

HISTORY, ANCIENT MILITARY [Prehistory–A.D. 476]

The military arts underwent an impressive development from prehistoric times to the fall of the Roman Empire. Large, fully integrated armies moved out of Egypt and Mesopotamia in the ancient Near East, and in the Roman Empire armies of Napoleonic proportions defended the frontiers. During the reign of Theodosius the Great (A.D. 379–395) the Roman army numbered about 500,000; this compared with the army of 600,000 that Napoleon I led into Russia. Many of the armies of antiquity were as large as or larger than the armies of the Middle Ages and Early Modern times.

Prehistoric Warfare

The origins of war go back to prehistoric times. Some anthropologists have argued that war is a creation of civilized man and that in prehistoric times there was no armed aggression of man against man. Archaeological discoveries of the twentieth century, however, have produced ample evidence that prehistoric man did wage wars. No one knows when men were first organized into armies, but they appeared at least by Neolithic times and possibly earlier. When Narmer moved down the Nile and united the two kingdoms of Egypt at the beginning of Egyptian history, he did it with an organized army, at least according to the stone relief sculpture that has survived from those early times. When man first learned how to write, he had wars to write about.

The earliest man-made weapons were chipped stone tools and fire-hardened spears. There is very little evidence for the bow and the sling in Paleolithic times. The famous cave paintings of France and Spain (20,000–30,000 years old) contain few scenes of human figures and none that unequivocally depict man killing man. Most of the pictures are peaceful and show beautiful animals, many of which are now extinct.

Although some scholars believe that the bow may be 50,000 years old, there is no definite evidence for it until the Late Paleolithic, around 12,000 B.C. A cemetery site discovered in excavations during the building of the Aswan dam on the Upper Nile has skeletons with what appear to be arrowheads imbedded in the bones. The first clear evidence of the use of the bow is from the Neolithic period in the Eastern Mediterranean (8,000–7,000 B.C.). Neolithic cave paintings from Spain depict man fighting man in an organized fashion, deployed in column and line. The sling is illustrated in drawings found at Çatal Hüyük in modern Turkey.

The most impressive evidence for warfare in the Neolithic period are the ruins of the great walls and fortifications of prehistoric settlements, the most famous of which are Jericho and Çatal Hüyük. The massive stone walls of Jericho even had towers. At Çatal Hüyük there were no surrounding walls, but the dwellings were all built with contiguous walls. The only entrance into them was up ladders through holes in the roof, and the ladders could be pulled up if there was an attack. Since the walls of the settlement were contiguous, there was in effect an outside wall, and if an attacker managed to get through it, he would find himself inside a single room. It was an ingenious form of military architecture.

In the Neolithic some villages contained a few thousand inhabitants. Jericho's population is estimated at 2,000; and the number of fighting men at 500 to 600. The wall enclosed an area of about ten acres and was 700 meters (765 yd.) long. It has been suggested that the new developments in warfare are what caused man to settle down in the Neolithic villages and discover agriculture; man had to live in permanent settlements behind strong walls for his own protection against the new missile weapons of the bow and the sling.

Ancient Egypt and Babylonia

Civilization first appeared in the river valleys of Egypt and Mesopotamia in the fourth millennium B.C. In the course of Egyptian history there were many wars, and much is known about the army of the pharaohs from literary and pictorial evidence. Although Egypt was well protected by deserts and relatively isolated from the other centers of civilization in Syria-Palestine, Asia Minor, and Mesopotamia, the pharaohs always had an army. Under warlike pharaohs such as Tuthmosis III and Ramses II, the mobile field army numbered around 20,000. It consisted of infantry armed with spears, swords, and bows, and of fast, light horse-drawn chariots. At the Battle of Kadesh in 1285 B.C. Ramses II marched several hundred miles out of Egypt into Syria where he fought the King of the Hittites in a major battle. Although the Egyptian army was caught in an ambush as it marched up to Kadesh and was hit hard in the flank, Ramses personally rallied his panic-stricken troops and led them to a tactical victory over the Hittites. They were not able to take the fortified city, however, and the campaign ended in a strategic stalemate.

In Babylonia the war chariot appeared much earlier than in Egypt. There is pictorial evidence that it appeared at least as early as the third millennium although the earliest ones were four-wheeled carts drawn by asses. The so-called Standard of Ur also depicts infantry pikemen in formation. From early times in Lower Mesopotamia, organized armies fought one another as the Sumerian cities competed for precious territory in the rich, fertile valley of the Tigris and Euphrates rivers. In the last half of the third millennium, Sargon the Great sent armies on raiding expeditions as far away as Asia Minor, and by the time of Hammurabi in Babylon (ca. 1750 B.C.) the military empire of the Amorite kings was extensive. Because Mesopotamia was relatively open to attack in comparison with isolated Egypt, military developments were faster and more effective. The last two great Mesopotamian empires, the Assyrian and Persian, both conquered Egypt.

Assyria and Persia

The army of the Assyrian Empire (900–612 B.C.) was one of the finest in the history of the ancient Near East. Numbering at least 150,000 men, it contained infantry, cavalry, chariots, skirmishers, and special forces (such as sappers, siege troops, mountain troops, infantry, cavalry, and chariot shock forces). Assyrians excelled in siege warfare. Walled and fortified cities were unable to stand against the Assyrian army. Although the Assyrians did develop cavalry, chariotry retained its pride of place in the armed forces. The logistical system of the Assyrian army was advanced enough to permit the army to move vast distances of up to 1,600 kilometers (1,000 mi.) and to fight effectively at the end of the march.

The rise of Assyria corresponds with the beginning of the Iron Age. Although iron had been used earlier as an ornamental metal, it was not until the early first millennium B.C. that it became available on a wide enough scale to have an impact on warfare. Iron is one of the earth's plentiful metals, much more so than the tin that was necessary for making bronze weapons in the earlier period. As a result the Assyrians were able to build a mighty arsenal.

Eventually, however, the Assyrian Empire fell, as much the result of internal rebellion as of external threat. Nineveh was destroyed in 612 B.C. After an interlude of about one-half century, during which the Neo-Chaldaeans of Babylonia and the Medes dominated the eastern Mediterranean, Cyrus the Great (559–530 B.C.) wrested control for his Achaemenid dynasty and launched the Persian Empire with extensive conquests.

The Persians introduced two new ingredients into the warfare of the civilized states around the east coast of the

Mediterranean. Although the Assyrians had used some cavalry, they relied mainly on chariots. Persia abandoned the chariot for the horse cavalry. Also, although the Persians were not sailors, they organized the first great fleet of triremes to dominate the sea. It was composed of ships provided by their allies, mainly the Phoenicians and also some of the Greek cities along the coast of Asia Minor. It is uncertain whether the trireme was invented by Phoenicians or Greeks, but it was the Persians who actually organized the first large fleet. With it they were able to dominate the Aegean and the Syria-Palestine coast. Under Cyrus's successors, the Persian Empire expanded to include all the ancient Near East from Egypt to India. It extended across the Bosporus into European Thrace, and by 500 B.C. the expansion of Persia brought the empire into contact with the Greek cities of Athens and Sparta.

Early Greek Warfare

Despite Homer's epic poems, the *Iliad* and the *Odyssey*, little is known about Greek warfare in the Bronze Age. The Trojan War remains semilegendary after more than a century of excavation in modern Turkey and Greece. The Minoans and Mycenaeans were familiar with the war chariot and also with Bronze Age warships. Homer implies that they fought duels as champions in the fashion of Achilles and Hector, but it is possible, and even likely, that they fought in formation.

In the Greek Dark Ages, when Homer lived, the duel of champions may have been common, but by 700 B.C. the Greek cities had adopted a new style of fighting with heavy infantry called the hoplite phalanx. Greek warriors wearing helmet, shield, breastplate and greaves, carrying a 6-foot thrusting spear, fought in a close-order formation eight ranks deep, the phalanx. The warrior was called a hoplite, meaning a shield bearer. This new heavy infantry dominated the warfare of classical Greece. Individual champions were powerless against it. The Spartans became the foremost practitioners of this style of fighting, which required considerable discipline because it was essential that the individual warrior hold his place in the line. The most important thing was not to panic and run. As one Spartan mother is supposed to have told her son, "Come home with your shield or on it." In other words, it was better to die in action and to be carried home on the shield than to throw it away and desert in the face of the enemy. Because the Greek hoplites fought with thrusting spears, they engaged in shock, or hand-to-hand combat, which requires courage and discipline.

In the period from 700 to 500 B.C., the Greeks were developing their distinctive type of warfare while the Persians (in the last half of that period) were conquering their empire, including the Greek states of Ionia on the coast of Asia Minor. In 499 those Greek states revolted against Persia, and Athens agreed to come to their aid. The Persians were too strong, however, and by 494 they had reconquered the Ionians and decided to invade the Greek mainland to protect the northwest frontier of their empire.

The Persian Wars

In 490 Darius, King of Persia, sent an expeditionary force across the Aegean against Athens. It landed at Marathon, about 40 kilometers (25 mi.) across the Attic Peninsula from Athens. The Athenians probably surprised the Persians by sending their army out to meet them in the field rather than defending the walls of their city. Considering the Persian superiority in siege warfare this proved a wise decision on the part of the Athenians. The Athenian general Miltiades, after occupying a position in the foothills surrounding the plain of Marathon so that the Greek army could not be attacked by the Persian cavalry, moved out against the enemy after thinning his line in the center so that he would not be outflanked by the larger Persian force. The Athenian army numbered about 10,000 and the Persian about 20,000. When the two armies were about a kilometer apart, Miltiades ordered the Athenians to attack on the run in order to reduce the amount of time they would be under the fire of the Persian archers. The Persians advanced against the weak center of the Greek line, but as they did so, the heavier Greek wings wheeled around and caught the invaders in a double envelopment. The Persians were driven back to the sea, and they withdrew from mainland Greece in their fleet.

In 480 Xerxes, successor to Darius, invaded Greece again. This time the Persian king personally accompanied the army, and it was the full army of Persia, not just an expeditionary force. It was so large, perhaps 150,000 to 250,000 strong, that no Greek army could hope to stand against it. The Athenian leader, Themistocles, realized that the Persian army was too large to live off the land in Greece and that it would have to be supported by the fleet. Therefore, if the Greeks could defeat the Persian fleet, they could force the army to withdraw. And Themistocles believed that it would be much easier to defeat the Persian fleet than to defeat the army. On the sea the Greeks would be outnumbered but not so much as on the land.

The Spartans, who had been given overall command of the Greek forces in the war, wanted to defend a wall they were building across the Isthmus of Corinth. Since Athens and a few other Greek cities were north of that line, however, the Greeks decided to stop the Persians farther north, in narrow constricted areas where the Persians would not be able to take advantage of their superiority in numbers. The main stand came at Thermopylae where the Spartans under their King Leonidas fought to the death in a three-day battle before the Persians finally breached the Greek position by learning of an alternate route that took them behind the Spartans. On the same three days, the Greek fleet fought a losing battle in the nearby waters of Artemisium. After this defeat on land and sea it was necessary to evacuate Athens, and the Greek fleet fell back to Salamis to help with the evacuation. The Persians moved up to offer battle, but in the narrow and constricted waterway that was favorable to the Greeks, they were decisively defeated.

King Xerxes then moved the bulk of his army back into Asia because the Greeks now controlled the western Aegean and were in a position to cut off the Persian line of retreat across the Hellespont. He left an army of about 50,000 in Greece under General Mardonius, but it was defeated in spring of 479 in a battle at Plataea. According to tradition, on the same day the Greek fleet defeated the Persian fleet again, this time in the eastern Aegean at Mycale. This marked the end of the Persian War; the Greeks had repelled a full-scale invasion by the Persian Empire.

The Peloponnesian War

Until the last half of the fifth century B.C., the Greeks relied almost exclusively on heavy infantry, the hoplite phalanx. There were few cavalry, skirmishers, or light infantry. During the Peloponnesian War between Athens and Sparta, skirmishers and light infantry made an appearance, but cavalry was very limited until the fourth century.

When Sparta invaded Athens in 431, the Athenian leader Pericles responded by having the Athenians abandon their fields and fall back behind the safety of their walls. Periclean strategy dictated that under no circumstances should the Athenians meet the well-disciplined and trained Spartan army in a pitched battle on land. Instead, the Athenians used their fleet to feed their city since they controlled the sea-lanes for the grain trade from the Black Sea, and they also circumnavigated and attacked the coast of the Peloponnese. Because the Athenians had greater financial resources, they could outlast Sparta in a war of attrition.

Except for a plague that swept through the city in 429 (among its victims was Pericles), the first few years of the war went well for Athens, culminating in a victory at Pylos (425) where the Athenians forced some Spartan hoplites to surrender. Under the leadership of the demagogue Cleon, however, the Athenians abandoned Periclean strategy and became more aggressive on land. This led ultimately to a major defeat at Amphipolis (422). In the following year, both sides, wearied by the conflict, agreed to end the war by the Peace of Nicias, named after the Athenian who negotiated it. It was supposed to last for 50 years, but only lasted about seven.

During the interval between the first and second phases of the war, the Athenians mounted a Sicilian expedition. A Greek city on Sicily asked Athens for help against a neighbor, closely allied with the large city of Syracuse, which in turn had good relations with Corinth and Sparta.

In Athens an ambitious young leader, Alcibiades, nephew and ward of Pericles, saw the invitation to intervene in Sicily as an opportunity to weaken the Peloponnesian League that Sparta dominated. In 415 B.C. Athens sent a fleet of 134 warships, about 5,000 hoplites, 480 archers, 700 Rhodian slingers, 120 light infantry, and 30 cavalry to Sicily. Unfortunately, the Athenians broke the principle of unity of command by naming three equal leaders—Alcibiades, Nicias, and Lamachus. Alcibiades, the one most likely to succeed, was removed from the expedition at the start because of a religious scandal.

In Sicily the Athenians moved against Syracuse and became bogged down in a siege that lasted until the summer of 413. Before it was over, Athens sent an additional 73 triremes and 5,000 troops, but the large armada failed to breach Syracusan defenses and was eventually totally destroyed. The losses numbered nearly 40,000 men, including the crews of the 200 ships that were also lost. The hoplite phalanx was outstanding on a level field of combat, but not particularly good for siege warfare. Syracuse was simply too far away from the center of Athenian vested interests, and Athens overcommitted its forces.

In the following year Sparta decided to renew the Peloponnesian War. The King of Persia, who had lost control of the Aegean to Athens after the Persian wars, saw an opportunity to regain much of Persia's former position, and he promised the Spartans strong financial support for the duration of hostilities. From 412 to 404 the war was fought on the sea, and Athens was often victorious in battle, but the Persians always helped the Spartans to rebuild their fleet, and eventually, at the Battle of Aegospotami in 405 the Spartan commander Lysander destroyed the Athenian fleet and blockaded the Athenian harbor at Piraeus. In March 404 Athens surrendered and the war was over. Sparta became the dominant city of the Greek mainland, and the King of Persia regained control of the Greek cities of Asia Minor.

Heavy infantry, the traditional force of the Greek states, had not been the decisive factor in this war. The need for lighter troops, for effective siege units, and for a navy was becoming widely recognized. What the Greeks lacked, and what the ancient Near Eastern states had long produced, was an integrated army, one composed of heavy and light infantry, skirmishers, heavy and light cavalry, and special forces. Exclusive reliance on the hoplite phalanx was beginning to decline.

Sparta and Thebes

The most important battles of the first half of the fourth century, the Battles of Leuctra and Mantinea between Sparta and Thebes, were fought by phalanxes. In them the Theban commander, Epaminondas, defeated the vaunted Spartan phalanx by stacking his own left wing in a formation 50 shields deep against the best troops on the Spartan right. In previous phalanx battles, the full force of the fighting had been felt all along the line and there had been no attempt at turning the flanks of an army (although it sometimes happened by accident). But Epaminondas deliberately massed his strength on one wing and used it against the best troops of the enemy. In that way Thebes won at Leuctra in 371 and again at Mantinea in 362, using the same tactics. At Mantinea, however, Epaminondas was killed in the fighting, and his death signaled the end of the Theban hegemony in Greece.

Significant changes were beginning to occur in Greek warfare. The simple catapult was invented in Syracuse in 399 B.C., and by the 350s the sophisticated and powerful torsion catapult was in use. At the beginning of the century the famous march of Xenophon and the Ten Thousand into the heart of Persia nearly to Babylon taught the Greek mercenaries a great deal about the art of logistics and the importance of cavalry and skirmishers. The Greeks were beginning to realize that light troops had their uses and that the best army was an integrated one. The first person to create such a force was Philip II of the northern kingdom of Macedon (359–336 B.C.).

Philip and the Macedonian Army

At the very start of his reign Philip reorganized the Macedonian army and fashioned one of the greatest fighting forces in the history of war. The premier arm of the new army was its cavalry. Macedonian aristocrats were horsemen, and the best of them were organized into the King's Companions, heavy cavalry that served as shock troops. They fought in a wedge formation and carried lances made of cornel wood. In weight and size the Macedonian lance was similar to the Napoleonic lance of modern times, but the Macedonian one had an iron point on both ends. For reconnoitering, skirmishing, and light cavalry, Philip used Thracians and Paeonians. His Thessalian allies provided another large contingent of heavy cavalry, and altogether the Macedonian cavalry numbered about 5,000.

In addition, Philip created the Macedonian phalanx which was heavier than the hoplite phalanx. This infantry fought sixteen deep, and the phalangites carried two-handed pikes (*sarissas*) that were probably about 4 meters (13 ft.) long. Front rankers wore very heavy armor. The formation was not very mobile, and it had a specialized tactical function—to meet and pin down the enemy line while the Macedonian cavalry attacked the flanks of the enemy or penetrated gaps in the opposing line. This represented the introduction of so-called hammer-and-anvil tactics in the Greek world. The phalanx was the anvil against which the surrounding cavalry, the hammer, smashed the enemy forces. The king trained these troops intensively to face in any direction by wheeling in an arc, and he could double the length of his line by stepping the last eight men in the files to the right or the left and moving them up. Although the usual battle order was in phalanx, the men were trained to fight in other formations, too, and Alexander (Philip's son) used them in sieges and mountain warfare.

Philip also used light infantry and skirmishers on a wide scale. *Peltasts* could perform as either: they were armed with javelins and could fight in a line or individually. There were units of archers and slingers in the Macedonian army. Philip also revolutionized logistics in the Greek world. He abandoned the use of carts and required his men to carry their own supplies, and he reduced the number of servants the troops were allowed. He was able to move his army an average of 25 kilometers (15 mi.) per day, and for short periods much more than that. Until the advent of the railroad in the nineteenth century, no armies exceeded these speeds. The new mobility increased the strategic range of Macedonian power. Alexander marched all the way to India, an impossibility for earlier Greek armies. Finally, Philip trained his new Macedonian army with the rigor that is today devoted to commandos, sending them often on training marches of nearly 65 kilometers (40 mi.) under full pack. His mobile, integrated army became, under his son Alexander, one of the world's finest armies.

Philip was able to dominate Greece with his army, and at the Battle of Chaeronea in 338 he defeated Athens and Thebes, formed the League of Corinth, and planned to lead the Greeks on a great crusade against Persia to free the Greek cities of Asia Minor. However, he was assassinated in 336 on the eve of his departure, and the young Alexander became king.

Alexander the Great

Alexander became one of the best generals in the history of warfare. He was an inspirational leader of men, a legend in his own lifetime, whose feats even today are remarkable. After devoting two years to securing his position on the throne of Macedon, he was ready in 334 to launch the crusade against Persia. He never turned back until he reached India about ten years later.

Almost immediately after crossing the Hellespont with his army he met the Persians at the Granicus River where he led a charge across a stream into the heart of the Persian defenders and breached their line. The Persians tried, but failed, to kill Alexander, and after his victory he marched triumphantly down the coast of Ionia in Asia Minor freeing the Greek cities as he went. Most opened their gates to him, although he was forced to overcome resistance at Miletus and Halicarnassus.

Then, in 333 B.C., the young king went through central Asia Minor to the city of Gordium where he untied the famous Gordian Knot and announced that he would become, as the legend promised, the Lord of Asia. As he moved down into Tarsus, the Persians tried unsuccessfully to stop him in the narrow mountain pass at the Cilician Gates. Alexander turned the corner from Asia Minor down the Syria-Palestine coast. He had decided to conquer all Persia, and first he needed to neutralize the Persian fleet by capturing its land bases around the eastern Mediterranean because his fleet was no match for the Persians'.

Darius III, King of Persia, marched the full Persian field army (perhaps 200,000 to 500,000 strong) in behind Alexander as the Macedonians advanced down the coast. Alexander immediately turned around and charged back against the Persians at Issus, where they fought a great battle. The Persians lost much of their numerical advantage by using the coastal road since the terrain forced them into a narrow area between mountains and sea. They took a strong defensive position along the Pinarus River, facing south, but since the ground was hilly on their left, Darius massed all his cavalry on his right.

Alexander, with a smaller army of about 45,000, approached in traditional formation with cavalry on both wings. He normally fought with the right cavalry and put the left wing of his army under the command of his senior general, Parmenio. In any event, Alexander advanced against the Persians so rapidly that he had to move from line of column into line of battle in the face of the enemy. As soon as the battle began, he charged with his rightwing cavalry into the left wing of the Persians and breached a gap in their line. Then he wheeled against the Persian center, where Darius personally fought, and threatened the rear of their line. The Persians panicked, including Darius, who fled from the field, abandoning his chariot, armor, and members of his family who had accompanied him on the expedition. In defeating the main army of Persia, under Darius, Alexander had won a major victory (November 333 B.C.).

The Macedonians continued their march down the coast, and the cities of Lebanon surrendered to them—Sidon, Byblos, Beirut—but the city of Tyre offered only neutrality. It was an island city about a kilometer off the coast with splendid defenses including strong walls, in some places about 45 meters (150 ft.) high. Alexander decided to build a mole, or walkway, across the sea to the walls of the city. The siege proved extremely difficult and lasted for almost seven months, from January to July of 332. Alexander was helped when the Phoenician fleets, which had formerly served Persia, defected to him. Eventually he used those warships against the walls of Tyre, and after several disappointments the Macedonians stormed that city, opening the road to Egypt.

After a brief siege at Gaza, Alexander entered Egypt without resistance and spent the winter of 332–331 there. Having gained control of the entire eastern coast of the Mediterranean, the King of Macedon was ready to strike into the heart of Persia against Darius. The Persian king had mobilized a new army and was waiting for Alexander on the plains of Babylon. The Macedonian had followed the Fertile Crescent north, and when he crossed the Euphrates River he decided to continue eastward toward the Tigris River rather than meet Darius on terrain that permitted the Persians to take advantage of their numerical superiority. Darius had no choice but to march up the Tigris, and the two armies met (September–October 331 B.C.) on the field near the village of Gaugamela.

The Battle of Gaugamela (often called the Battle of Arbela, which was the largest city in the vicinity; about 110 kilometers [70 mi.] away) proved decisive in the war for Persia. The Macedonians were outnumbered at least three or four to one. Alexander extended his line in echelon formation on the left wing, where he ordered Parmenio to fight a holding action. On the right he looked for an opportunity to lead a charge. When the Macedonians were on the verge of defeat, after their left wing had been enveloped, Alexander attacked through a gap that developed on the Persian left, wheeled around (as he had done at Issus), and moved against the rear of the Persian center. Again, Darius panicked and fled. Although victorious, Alexander was forced to help the beleaguered Parmenio on the Macedonian left, and Darius escaped. The full Persian army, however, was never reassembled.

Alexander then marched unopposed into Babylon and Susa, the Persian capital. After some opposition he also took the ancestral shrines at Persepolis, which he put to the torch in revenge for what the Persians had done to Athens during the Persian Wars of the fifth century B.C. Darius was killed in 330 by his own nobles, and Alexander spent the years 329–327 fighting a guerrilla war for control of northeastern Iran. Finally he was able to move into India, which had also once been part of the Persian Empire (the area known as the Punjab in modern Pakistan).

The Punjab was so far from the Greek world that Macedonians knew about it only vaguely in myth. When they arrived at the tributaries of the Indus River, Alexander thought he had discovered the source of the Nile because there were crocodiles in the rivers, and the Nile was the only river he knew where they existed. While in India, a shipment of equipment arrived from Macedonia, a tribute to Alexander's logistical system.

The main opponent in the valley of the Indus was a king named Porus, who took a strong position on the east bank of the Hydaspes River (the modern Jhelum River). He defended it with an Indian army of 30,000 infantry, 4,000 cavalry, 300 chariots, and, perhaps most important, some 200 war elephants. The elephants were effective—Alexander's men and horses were not trained to fight them, and horses unaccustomed to elephants are frightened of them (see Fig. 1). Since he could not move directly across the river against Porus, Alexander marched under cover of darkness 27 kilometers (17 mi.) upstream, crossed the river, and marched down against the Indians. Porus wheeled his army around and rushed to meet the invaders. Alexander won the great cavalry battle that developed on the Macedonian right and Indian left, gained control of the Punjab, and became friends with his former foe, Porus.

By this time Alexander had decided that he wanted to march to the end of the earth. Greeks believed that an

Figure 1. Elephants and camels were used in the time of Alexander, who originally saw them in use in the Indian Army. (SOURCE: Iconographic Encyclopaedia)

ocean stream flowed around the world, and Alexander heard that there was only one more river valley to conquer before reaching it, but after he reached the Hyphasis (modern Beas), his army refused to go farther. After weeping in his tent for three days, the king finally agreed to take the men home. By one modern estimate they had marched over 27,000 kilometers (17,000 mi.) since leaving Macedonia. They returned down the Indus, and at the mouth of the river, Alexander took part of the force overland across southern Persia; the rest went under the command of Nearchus by sea in a fleet the army built. Both forces suffered severely, but Alexander and Nearchus made it back to Babylon.

Alexander died there in 323 B.C. at the age of 32. His conquests are among the most extensive in history, and his generalship was superb. Characterized by decisiveness and speed, the quality of his command has rarely, if ever, been exceeded. In strategic thought he made few questionable decisions, and he was tactically successful as well. The most common criticism of his generalship is that he exposed himself to personal injury too often and recklessly, which, although true, is also part of what made him such an inspirational leader. His integrated army was as well trained as any in premodern times. Discipline was firm, and the quality of his senior officers was brilliant.

Alexander had a tremendous impact on warfare. From his time forward, armies were able to strike hundreds or thousands of miles from their home base. Ancient Near Eastern armies had also performed long-distance operations, and they too were integrated and had good logistical support. What the Macedonians added was outstanding heavy infantry and rigorous discipline. As a result of his invasion of India, one of Alexander's legacies was the introduction of elephant warfare into the Mediterranean; a new way of fighting for the Greeks, and through them the Romans.

Early Rome

When Alexander died in Babylon, the Romans in Italy were just starting to become a world power. According to legend, their city had been founded on 21 April 753 B.C., by its first king, Romulus. Late in the period of the monarchy, Rome adopted a phalanx army similar to those used at the same time in Greece. The Roman army successfully defended the fledgling republic during the fifth century B.C. against attacks by the neighboring hill people, and around 400 B.C. they won their first war against a major civilized state, the Etruscan city of Veii. Shortly afterward, however, Gauls from the Po valley in northern Italy sacked Rome (390 B.C.) and humiliated the army. It was probably at that time that the Roman military hero Camillus, who had defeated Veii, reorganized the phalanx army into legions.

The Roman legion was arranged in maniples, each consisting of 120 men (except for the maniples in the rear ranks which were only 60 strong). Altogether the legion numbered about 4,200. In addition to the legions, Rome's allies provided auxiliary units equal in manpower to the legions. Generally, the Roman army was strong in infantry and weak in cavalry. It normally depended on mercenaries for archers, mounted and unmounted. Discipline in the Roman army was notoriously severe—decimation is the most famous example.

The Roman army usually fought in three lines. Those in the first line were called *hastati* and *velites*. The *velites* were the youngest (1,200 in all), and they served as skirmishers with swords, javelins, and circular shields. The *hastati* and the troops in the second line, the *principes*, carried the short sword (*gladius*), two javelins, and the oval shield (*scutum*). There were 1,200 of each. In the third line stood the veterans, the *triarii*, 600 strong, who carried a pike instead of javelins. Each line wore a distinctive uniform. The cavalry squadrons attached to the legion numbered 300.

Normally, when a consul went into the field he took two legions plus an equal number of allied troops, nearly 20,000 men altogether. Sometimes in a grave crisis consular armies were doubled in size. All Roman legionaries were conscripts, and they were also property owners. The Romans did not want to arm the landless poor.

After conquering most of the Italian peninsula south of the Po valley, the Romans faced the Greeks of southern Italy. Greeks had been in Sicily and southern Italy for centuries, and they had always fought with the neighboring Italian hill people. In the 280s B.C., one of the Greek towns, Thurii, asked the increasingly powerful Romans to help them against attacks. Rome responded by sending troops and angered the Greek city of Tarentum, which believed that southern Italy was Tarentum's sphere of influence. Tarentum was bold enough to oppose Rome because a king from the Greek world, Pyrrhus of Epirus, offered to help the Tarentines.

Pyrrhus was an ambitious Hellenistic king and general with an army of about 25,000 professional troops trained to fight in the Macedonian style. He crossed to Italy in the spring of 280 B.C. and defeated the Romans at the Battle of Heraclea. The following year he defeated them again, at

Asculum, making good use of his elephants in both battles. His losses were heavy in both battles, and after the second he is supposed to have exclaimed, "Another such victory and I am lost." His problem was that he could not besiege Rome as long as Rome's Italian allies remained loyal because they would mobilize a relief army and pin him down around the walls of the city. Thus, although he had won great tactical victories, he was far from winning the war strategically. Despondent, he accepted an invitation to help the Greek cities of Sicily in a war against Carthage and left Italy. Although he was successful against Carthage, his Greek allies eventually turned against him, and he returned to Italy in 275 to resume the war with Rome. He was defeated at the Battle of Beneventum and left Italy to return to Epirus. He became King of Macedon and died a few years later in a street fight in the city of Argos. During this time the Romans organized the Greek cities of Italy into the Roman alliance system.

The Wars with Carthage

The conquest of Greek Italy brought Rome into direct conflict with the Carthaginian Empire of North Africa. Until the early third century B.C. Rome and Carthage had been friendly because they had no conflicts of interest. The traditional enemies of both states were the Etruscans and Greeks. Rome finally defeated both, and brought them into the Roman confederation. Because the Greeks of Italy had close ties with the Greeks of Sicily, Rome soon became involved in Sicilian affairs. However, Carthage occupied the western end of Sicily, and within ten years of the war with Pyrrhus, Rome was at war with Carthage.

The first Carthaginian War (or Punic War; *Punic* was the Latin word for Phoenician, and Carthage had originally been a Phoenician colony) became a war of attrition, lasting an entire generation from 265–241 B.C. Carthage was a major naval power, and Roman strength was in infantry, but because the Romans had to fight overseas, they built a fleet. Knowing they were inferior to the Carthaginians in standard naval battle maneuvers, the Romans added a large gangplank onto a special mast near the prows of their vessels. Attached vertically, it could be released to fall onto the deck of the enemy ship as it approached. The gangplank had a spike in its end to pierce the Carthaginian deck; it was called a crow (in Latin, a *corvus*, because the spike resembled a crow's beak). The Romans then sent their legions across the crow to board the Punic warship. With this new invention the Romans inflicted heavy losses on the Carthaginians.

The Roman forces surprisingly fared better on sea than on land. The Carthaginian army dug in on Sicily, and the Roman army could not dislodge them. In 256 B.C. the Romans used their fleet to carry the war to North Africa against Carthage, but the army, under the Consul Regulus, was defeated outside the city, and Regulus was captured. Although the Romans resumed the war at sea, they suffered heavy losses in storms, either because their ships were made unseaworthy by the heavy extra mast for the crow, or simply because they lacked naval experience. In 242 both states built new fleets, and the Romans won. In the following year Carthage agreed to give up Sicily, and a few years later Rome seized Sardinia and Corsica.

Hannibal and the Second Punic War (218–201 B.C.)

In the First Punic War Carthage had been defeated but not destroyed. Its North African empire was still intact, and Hasdrubal, the best Carthaginian general, decided to pursue Punic designs on Spain. Spain was far from Rome, and the Romans paid little attention to Carthaginian activities there. Rome fought the Gauls of the Po valley in the 220s and sent troops across the Adriatic to protect the shipping of the Greek cities of southern Italy against piracy. The Carthaginians under Hamilcar, and eventually under his son Hannibal, strengthened their position in Spain and grew strong on the mineral resources of that country. In 219 Hannibal put Saguntum (a Spanish city allied to Rome) under siege, although he knew it would lead to renewed war between the two major powers.

In 218 the Romans prepared for war. Determined to avoid the pitfalls of the earlier conflict with Carthage, they wanted a quick victory; they sent out two armies, one to Sicily and the other, under P. Cornelius Scipio, to Spain. Hannibal, however, achieved strategic surprise with his march across southern France and over the Alps into Italy. Although he suffered heavy losses crossing the Alps, he caught the Romans off guard and defeated them in three major battles in the first three years of the war. The Carthaginians overwhelmed the Roman army at the Trebia in December of 218 and at Lake Trasimene in 217. Following these defeats, the Romans, under the dictator Fabius Maximus, followed a strategy of exhaustion against Hannibal (popularly misnamed "Fabian tactics") and refused to offer conventional battle. Hannibal faced the same dilemma as had Pyrrhus earlier in the century. Carthage could not put Rome under siege as long as the allies in Italy remained loyal. Hannibal needed decisive victories in the field in order to incite Rome's allies to defect. Fabius hoped to deny Hannibal that opportunity, but Fabian tactics were not popular in Rome, and the Senate decided on a bold attempt to defeat Hannibal in Italy.

In 216 a large army (possibly 80,000 strong) met the Carthaginians in southern Italy at Cannae. There Hannibal put out strong cavalry on both wings, placed his weakest infantry in the center in a crescent formation, and stationed his best veterans on both sides of the crescent. The Romans drove back the weak troops in the center, but the Carthaginian wings wheeled around while their cavalry easily defeated Roman cavalry and then hammered the rear of the Roman line, catching the entire Roman army in a classic double envelopment. The defeat at Cannae was the darkest moment of the war for the Romans.

Philip V of Macedonia declared war the next year, and the Italian city of Capua defected to Carthage. Even Syracuse on the island of Sicily abandoned a long and loyal Roman alliance.

Hannibal, however, could not offer Capua or Syracuse much protection since he did not want to pin his own army down in a besieged site. The Romans retook these cities in the next few years, and in Spain, where they had sent an army at the outset of the war, a young general named Scipio won a total victory. In 205 Scipio won election to the consulship on a promise to carry the war to North Africa against Carthage. He left in 204 and a year later the Carthaginians recalled Hannibal from Italy—he had won many great battles, but he had not achieved strategic victory.

In North Africa at the Battle of Zama in 202 Scipio (later known as Africanus) defeated Hannibal, and in 201 Carthage accepted the terms Scipio imposed. Carthage lost all of her empire and was reduced to the status of a city; Rome took Spain and the Kingdom of Numidia (roughly modern Algeria), and Carthage became a Roman ally. Carthage was also forbidden to wage war without Roman permission. In geopolitical terms, Rome became dominant in the central and western Mediterranean and emerged as the leading power in the world.

Rome quickly demonstrated that power. First the legions humbled the Macedonian phalanx at the Battle of Cynoscephalae in 197, and then defeated the army of Antiochus the Great, ruler of the Seleucid Empire, at the Battle of Magnesia in 189. The Ptolemies of Egypt quickly adopted a peaceful policy toward Rome, and the legions of the west became the master of the world. Around 100 B.C. the Roman general Marius reorganized the Roman army and left an indelible stamp on Roman history.

Marius, Sulla, Pompey the Great, and Caesar

Marius abandoned the draft and property requirement for service in the army and called for landless volunteers. He promised them booty and land grants as a reward for service, rewards that were guaranteed by the commanding general rather than the state. The result was that the army became more attached to its general than to the government, and it was not long before Roman armies were willing to march on Rome itself if the general ordered it. Marius also streamlined the logistical supply system, following Alexander's father's example, by making Roman troops carry most of their own supplies. The army became known as Marius' Mules. He also made some tactical changes, organizing the legions around cohorts made up of six centuries each, and each century consisted of 80 men. There were ten cohorts in a Roman legion, and on paper each legion numbered slightly more than 5,000 men, although legions, like modern military divisions, were rarely up to paper strength. Some of Caesar's legions later were only about 2,000 strong. But the new legions of Marius, now highly politicized, were also extremely effective on the battlefield. It is at this time that they began to acquire the famous legionary standards and names. Marius is sometimes credited with creating a professional army; although this is not so, he did lay the foundations for it.

The armies of the Late Roman Republic made tremendous conquests and played a major role in Roman politics, often placing their generals in control of the government. Pompey the Great and Julius Caesar are the most famous examples of military dynasts. Pompey defeated Mithridates in Asia Minor and added Syria to the Roman Empire. Caesar conquered all Gaul. Then in a great civil war Caesar defeated Pompey and became dictator. Caesar's generalship was characterized by decisiveness (or Caesarspeed), personal bravery, and firm discipline. He also doubled legionary pay. In war there are many paths to popularity, but generosity and victory are perhaps the best.

Augustus and the Grand Strategy of Preclusive Security

The loyal support of the legions, however, was not enough to prevent Caesar's opponents in the Senate from assassinating him on the Ides of March, 44 B.C. Eventually, after a major military confrontation between his successors, Octavian and Mark Antony, Octavian emerged as the sole ruler of Rome and became known as the Emperor Augustus. It was Augustus who completed the transformation of Roman troops, begun by Marius, into a standing professional army.

After the victory over Antony and Cleopatra at Actium (31 B.C.), Augustus demobilized the great army that had developed during the period of the Second Triumvirate, reducing his forces to 28 legions (and an equal number of auxiliary troops). He pursued extensive conquests, especially along the northern frontier of the Rhine and Danube rivers and stationed most of the legions along that line. Only a few, particularly in Egypt and Spain, were kept for internal police purposes.

The emperor also regularized the length of service. Legionaries served for 20 years, members of the Praetorian Guard (an elite force of about 7,000 men stationed in Italy) for 16, and auxiliaries for 25. There were also two fleets, one stationed at Ravenna and the other in the Bay of Naples. Pay, bonuses, and retirement benefits also were controlled by the government, and a new military treasury was created for the pension system.

The imperial defense system was characterized by economy of force. The army, after the loss of three legions across the Rhine at the Teutoburg Forest in A.D. 9, was only 25 legions strong, and with auxiliaries numbered around 250,000 to 300,000 men. For an imperial frontier of several thousand miles this force was small and efficient. Training and discipline made the Roman imperial army one of the finest in the history of the world. Under Augustus and the Julio-Claudians, Romans used "client kings" or satellite states to help in the defense of the frontiers, but by the end of the first century A.D. there were about 30 legions, and the client kingdoms had been incorporated as provinces. In the second century the Romans had a fully developed system of preclusive security protecting the imperial frontiers which were rigidly defined. Hadrian's Wall in England is the most famous example. Some analysts have argued that the flaw in Roman grand strategy was that the legions were all deployed on the frontier and there was no central reserve.

The Collapse of the Third Century and the Recovery under Diocletian and Constantine

In the third century A.D. military defense of the Roman Empire broke down in a half century of civil war (235–84 A.D.), sometimes referred to as the period of the Barracks Emperors. Legion fought legion to place its own commander on the throne, and as the frontiers were abandoned by their defenders, barbarians poured into the empire while rebel legionary generals seized large territories for themselves. There were times when the emperor in Rome ruled only over Italy and parts of North Africa. One emperor, Valerian, was captured by the Persians (260) and never heard from again. Another was killed fighting the Goths. Barbarians raided Athens. The economy of the empire collapsed as the coinage was devalued to almost nothing but worthless metal.

Surprisingly, Rome recovered under the forceful Emperor Diocletian (284–305). He managed to reconquer almost all the former territory of the empire and to restore the system of preclusive security.

Constantine the Great (d. 337) made a major modification to imperial grand strategy. Abandoning preclusive security, he turned instead to a system of defense in depth and created a large mobile army, strong in cavalry, stationed near the imperial residence, wherever it happened to be. To do this, he had to weaken the frontier forces, but his system worked well for a time, and the empire remained powerful. In the end, however, the results were catastrophic, as the frontier garrisons deteriorated in their now secondary strategic role. Infantry was generally neglected while cavalry became more important. Associated with this development was the increased use of barbarian mercenaries which led to a deterioration in training and discipline. While Constantine the Great lived, Rome remained dominant in the entire Mediterranean area, but under his successors in the fourth century, Roman armies began to suffer great defeats.

The Fall of the Roman Empire

The first major loss came in 363 when the Emperor Julian the Apostate led an invasion of Persia. The Sassanid Persians had replaced the Parthians in Mesopotamia in the third century, and the new Persian Empire became a formidable opponent. During Julian's invasion the Persians adopted a scorched earth strategy, and Julian, although he advanced to the Persian capital, Ctesiphon, in Lower Mesopotamia, withdrew. On the return he was killed in a skirmish with the Persians, and the new Roman Emperor, Jovian, was forced to negotiate a humiliating settlement.

Fifteen years later the Romans suffered an even greater defeat against the barbarian Visigoths at Adrianople (378 A.D.). The emperor of the Eastern Roman Empire, Valens, had agreed two years before to let the Visigoths cross the Danube into the empire. They had been driven in panic against the imperial frontier by the appearance of the Huns from the steppes of Asia into the region around the Black Sea. Because Roman authorities mistreated the Visigoths when they made the crossing, the barbarians went to battle. Valens met them at Adrianople in 378, rushing there to get the glory before the Western Emperor Gratian, who was nearby, could arrive with reinforcements. On that Balkan battlefield the army of the Roman Empire, arriving little by little and engaging before it was fully deployed into line of battle, was badly beaten on a hot August afternoon. Valens was killed.

Theodosius the Great (379–95) took over in the East after Adrianople and by the time of his death had control of all the empire, East and West. Theodosius adopted a policy of appeasement and negotiation with the Persians and the barbarians, believing that he needed time to rebuild Roman military strength. He agreed to let the Visigoths stay in the empire as federated allies under their own kings and bearing their own arms. Never before had Rome made such a settlement with a Germanic tribe.

When Theodosius died, the Visigoths, under their new King Alaric, again attacked. The empire had been permanently divided between East and West on Theodosius' death; his son Honorius became emperor in the West and Arcadius in the East. Since Honorius was not yet 10 years old, Theodosius appointed General Stilicho as regent in the West. At this period the West was weaker than the East. The East was richer, and the walls of Constantinople on the European side of the Bosporus protected the capital and the rest of the Eastern Empire from invasion since the barbarians had no naval power for operations in the Eastern Mediterranean. So Alaric led the Visigoths first into the Balkans and then against Italy in the West. As long as Stilicho lived, he was able to keep Italy safe, but after his death in 408 Alaric put the city of Rome under siege. It fell to the Visigoths and was sacked for three days late in August, 410 A.D.

Earlier, in 407, Vandals, Suebi, and Alamanni had crossed the Rhine into Gaul and Spain. The entire western half of the empire was hard pressed, but the government of Honorius, now centered in the fortress of Ravenna on the northern Adriatic, held firm. Under a new military hero, Constantius, much of the West was retaken for a brief time, but Britian had been lost permanently in the crisis of 408–410, and the Visigoths had been allowed to settle in Aquitaine in southwestern Gaul.

Finally, in 429 Vandals moved from Spain into North Africa, and took Carthage in 439. The loss of North Africa, a major source of grain and money, was a strategic blow to the emperor in the West. In the meantime Burgundians and Alans had also settled in Gaul. The government of the Emperor Valentinian III (425–55) tried to keep things together, and with the help of the able General Aëtius did as much as possible under the circumstances. Militarily, the barbarian tribes relied heavily on infantry rather than cavalry, as is commonly believed, since only aristocratic tribesmen had horses. Generally, the Roman cavalry was better than the barbarian, but Roman infantry had declined drastically, and by the mid-fifth century it was not as good as the Germanic forces. In siege warfare and in naval warfare the barbarians were inferior to the Romans, but without good infantry Rome could not hold the western empire together. In the East this deterioration did not occur, mainly because the eastern army had not been so thoroughly barbarianized, and because discipline and training were much more rigorous.

The last great victory for Rome came in 451 at the Battle of Châlons when Aëtius organized an alliance of Romans, Visigoths, and Alans against Attila the Hun who had invaded Gaul. After this triumph, however, the alliance broke up, and on the deaths of Aëtius (454) and Valentinian III (455) the western empire nearly disappeared. For a generation it labored on under barbarian generals who ruled through puppets, but in 476 the last Roman emperor in the West, Romulus Augustulus, was deposed, and the western Roman Empire ceased to exist.

ARTHER FERRILL

SEE ALSO: Alexander the Great; Assyria, Military History of; Attila the Hun; Caesar, Julius; Graeco-Persian Wars; Hannibal Barca; Peloponnesian Wars; Persian Empire; Punic Wars; Roman Empire; Scipio Africanus; Tuthmosis III.

Bibliography

Campbell, J. B. 1984. *The emperor and the Roman army, 31 B.C.–A.D. 235*. Oxford: Clarendon Press.
Cheesman, G. L. 1914. *The auxilia of the Roman Imperial army*. Oxford: Clarendon Press.
Connolly, P. 1981. *Greece and Rome at war*. New York: Prentice-Hall.
Engels, D. W. 1978. *Alexander the Great and the logistics of the Macedonian army*. Berkeley, Calif.: Univ. of California Press.
Ferrill, A. 1985. *The origins of war: From the Stone Age to Alexander the Great*. New York: Thames and Hudson.
———. 1986. *The fall of the Roman Empire: The military explanation*. New York: Thames and Hudson.
Grant, M. 1974. *The army of the Caesars*. New York: Scribner.
Greenhalgh, P. A. L. 1973. *Early Greek warfare: Horsemen and chariots in the Homeric and Archaic Ages*. Cambridge, England: Cambridge Univ. Press.
Humble, R. 1980. *Warfare in the ancient world*. London: Cassell.
Keppie, L. J. F. 1984. *The making of the Roman army: From Republic to Empire*. Totowa, N.J.: Barnes & Noble Books.
Lazenby, J. F. 1985. *The Spartan army*. Warminster, England: Aris and Phillips.
Luttwak, E. N. 1976. *The grand strategy of the Roman Empire from the first century A.D. to the third*. Baltimore, Md.: Johns Hopkins Univ. Press.
Rodgers, W. L. 1964. *Greek and Roman naval warfare: A study of strategy, tactics, and ship design from Salamis (480 B.C.) to Actium (31 B.C.)*. Annapolis, Md.: U.S. Naval Institute Press.
Schulman, A. R. 1964. *Military rank, title, and organization in the Egyptian New Kingdom*. Berlin: B. Hessling.
Speidel, M. 1984. *Roman army studies*. Amsterdam: J. C. Gieben.
Starr, C. G. 1960. *The Roman Imperial navy, 31 B.C.–A.D. 324*. New York: Barnes & Noble.
Turney-High, H. H. 1971. *Primitive war: Its practice and concepts*. Columbia, S.C.: Univ. of South Carolina Press.
Warry, J. G. 1980. *Warfare in the classical world*. London: Salamander Books.
Watson, G. R. 1969. *The Roman soldier*. Ithaca, N.Y.: Cornell Univ. Press.
Webster, G. 1969. *The Roman Imperial army of the first and second centuries A.D.* London: Black.
Yadin, Y. 1963. *The art of warfare in biblical lands in the light of archaeological study*. New York: McGraw-Hill.

HISTORY, EARLY MODERN MILITARY [1453–1789]

In the mid-fifteenth century, while European princelings battled over a fragmented continent that was still trying to pull itself out of feudalism, an imperial era was reaching its crest throughout the rest of the world. Great empires were raised up and torn down by cavalry armies that swept through Asia. Magnificent pyramids and golden cities were built to mark the growing domains of the Aztecs and the Incas. Tribal chiefs were becoming kings in Africa. In this otherwise vital age, the fall of Constantinople, the last symbol of the great Roman Empire, to the Turkish sultan in 1453 appeared to mark the impotence and vulnerability of Europe.

From this nadir, however, the kings of Europe rose up, and in less than a century it was their empires that were on the advance. Their civilization, their religions, and their rule spread across the oceans, and the first true world empires were born. By 1789, Europeans dominated, ruled, or enslaved many of the other peoples of the globe. This imperial explosion was fueled by a complex military and technological revolution that began with these simple ingredients: 10 parts sulphur, 15 parts charcoal, and 75 parts saltpeter—the most popular formula for gunpowder.

Advent of Gunpowder Weapons

Gunpowder was not a European invention and was used against Europeans, rather than by them, for centuries. The stone walls of Constantinople, for example, which had shielded Europe from Islam for eight centuries, crumbled in 1453 when the sultan, Mohammed II, trained fourteen batteries of cannon on them. The largest of those five dozen guns was built, ironically, by a European, who cast the weapon in the European city of Adrianople. The gun, named Urban's Bombard after its creator, could hurl a 270-kilogram (600 lb.) stone ball over 1,600 meters (1 mi.). The huge bronze monster required about 60 kilograms (134 lbs.) of gunpowder for each such shot, and took so long to clean, load, and lay that it could only be fired seven times a day. The weapon, like many gunpowder engines of its day, frequently misfired. One such misfire burst the barrel and took the life of the gun's designer during the siege.

While Mohammed II's batteries pounded the imperial city, the Byzantine emperor, the eleventh Constantine to rule, had only a handful of small cannon mounted on his walls. The imperial artillery relied on catapults, ballistae, and dart throwers, which the first Constantine would have found familiar. The greatest city in Europe, moreover, could only muster 8,000 defenders for its population of 100,000. The sultan had more soldiers with him than the emperor had people in his capital.

The defense was desperate, brave, ingenious, and hopeless, as Mohammed's guns tore great gaps in the walls. Still, the defenders forced back the flood of feudal levies the Turks sent against them, and did not finally succumb until Mohammed sent in his elite corps of infantry, 12,000 Janissaries. After a day of looting by the Turkish horde, the Janissaries restored order, and Mohammed entered his city. With that procession, the last ancient empire fell.

Gunpowder was not a surprise to the Byzantine defenders. The explosive mixture had been in use in China since the start of the fourth century, about the same time that Constantinople was laid out by the engineers of Constantine the Great, the Roman emperor. Its reputation as a weapon in sieges in Asia was known in Europe by the ninth century, according to church manuscripts. By the twelfth century, Chinese inventors were building crude cannons, and Moorish invaders brought gunpowder weapons into Spain in 1118.

A century and a half later, small guns were being used in sieges throughout Europe, and by the time Urban cast his monster bombard for Mohammed, even larger bombards were in use by the rulers of Ghent, Scotland, and Muscovy. Cast-iron cannon balls and crude carriages were designed to replace the stone shot and immobile platforms of earlier guns, and within a century of the fall of Constantinople, a French inventor had built limbers that finally made guns mobile instruments of war that could be brought to the battlefield, rather than relegated to static siege work.

The proliferation of artillery swept aside the last remnants of feudalism in Europe. No castle, no matter how strongly built or stoutly defended, could withstand the thunder and lightning of artillery. The Europeans, once instructed in this new fact of life by the Turks, were quick to learn that lesson. Less than 40 years after Islamic power wiped out Europe's oldest empire, Spanish guns forced the surrender of Islam's oldest regime, Granada.

In 1481, Mulay Abdul Hassan, sultan of Moorish Spain, refused to pay tribute to Ferdinand and Isabella, the Christian rulers of Castile and Aragón. When asked for silver tribute, he replied, "Our mint at present coins nothing but blades of scimitars and heads of lances." Unfortunately for him and his subjects, the Spanish mints hired the best gunsmiths from France, England, Germany, and Italy, and one by one the Moorish castles and cities fell. On 2 January 1492, Mulay's successor, Boabdil, knelt before Ferdinand and Isabella, the walls of Granada breached by their guns. "These are the keys to the last relics of Arab empire in Spain," Boabdil told the Spanish rulers. "Thine, O sire, are our trophies, our kingdom and our person," Boabdil said as he gave up his crown. "Such is the will of Allah."

The "will of Allah" was heard in the blast of cannon both at Constantinople and at Granada, and at a hundred castles and cities from Scotland to Japan over the next 50 years. From one end of the Old World to the other, gunpowder crumbled principalities and built empires.

Into the Field

Cannon were expensive to cast and required miniature armies of specialists and laborers of their own just to move, use, support, and protect them. Feudal lords and minor nobles could neither remain safe behind ancient walls nor afford the expense of the new weapons. War became too costly for anyone but the wealthiest king, and his need for money to pay for war led the royal houses of Europe to demand taxes rather than military service from their vassals and subjects.

Some feudal armies, or at least feudal components of armies, would live on until the seventeenth century, especially in eastern Europe and the Ottoman Empire. In the 200 years after the fall of Constantinople, however, war would be decided as it had been during the age of Rome, in contests between increasingly larger and more professional armies that faced each other not from behind walls but across open fields.

While artillery forced armies out from behind castles and into the field, it was not mobile or powerful enough on its own to win, or even to be a decisive factor in, those field battles. The gunpowder revolution, however, did not stop. Although many artillery guildsmen worked to make bigger or more mobile guns, others put their talents to work making smaller, lighter weapons that individual soldiers could carry into battle. By the end of the fifteenth century, corps of handgunners could be found in most major armies. Although training men to use these short-range, slow-firing, cumbersome, inaccurate, and unreliable weapons was not difficult, their ineffectiveness led the more advanced armies of the age to retain their reliance on the proven weapons of antiquity: the bow, the pike, and the lance. That reliance, however, was about to change.

From the decline of the Roman Legion in the fourth century to the early fourteenth century, regular, reliable infantry had all but disappeared from the battlefield. Cavalry won the field battles for nearly a thousand years, and even at the time of the Siege of Constantinople, the horseman was still supreme from the shores of the China Sea to the Danube. The heirs of Tamerlane rode over the armies of the Delhi sultanate in India. Ming horsemen and the riders of the steppeland khanates ruled the Asian mainland. Turkish cavalry armies, for example, swept through the Balkans and North Africa in the years following the fall of Constantinople, and by 1477 the Ottoman horsemen were within sight of the defenses of Venice. Horsemen ruled the world, except in Europe.

The chivalry of Europe, the power of the feudal age, had fallen to the Welsh longbow and the Swiss and Flemish pikes, weapons wielded by common soldiers. The death knell of the knight was sounded by the Lombard citizen pikes at Legnano in 1176, and was heard again at Courtrai, Bannockburn, Morgarten, and a score of other battles throughout western Europe over the next 300 years.

The longbow, however, remained the weapon of a small handful of professionals, and except for limited use in Burgundy and Italy, was found exclusively in the small armies of the English kings. The pike, however, was the universal weapon that let the Swiss cantons and the Italian city-states and many cities and towns in between win their freedom from feudal lords. However, the masses of pikemen, the phalanxes of the fifteenth century, soon found themselves vulnerable to the long reach of gunpowder.

By the start of the Italian Wars (1494–1559), European generals had begun to experiment with lighter, long-range hand weapons called harquebuses. Select bodies of infantry armed with these weapons, the precursor of the musket, accompanied the pikemen into battle. Protected from cavalry by their own pikemen, they could shoot and kill the opposing pikes and hold off enemy crossbowmen or cavalry. Supported by artillery and their own cavalry, the combination of pike and shot dominated the battlefield in Europe and, shortly thereafter, the world. The use of gunpowder in weapons that a single man could wield gave rise to two very important and revolutionary factors that would forever change the face of war: the supremacy of infantry and the need for professional soldiers.

Infantry: The Queen of Battle

Despite the drubbing received from pikes, longbows, and harquebuses in the late sixteenth and early seventeenth centuries, the European nobility still preferred the saddle to shoe leather as a means of going into battle. The foot soldier, who had been looked upon as little more than an encumbrance or target for the last thousand years, now became the weapon of victory, and most nobles understood that. Even the aristocratic gendarmes of Catholic France, for example, were wise enough to bring Swiss pikemen with them into the fray against their Huguenot enemies during France's eight Wars of Religion (1562–98).

The foot soldier became the centerpiece around which a battle was planned. As feudal levies had proven worthless against the disciplined pikemen of Switzerland and the Italian cities, a new type of infantry was needed. In the 100 years following the fall of Constantinople, most major powers in Europe would form or hire units of full-time soldiers, the first regular infantry to march in western Europe since the fall of Rome.

In 1496, Ferdinand and Isabella raised the Spanish Infantry of the Ordinance, which by 1505 had been organized into *colunelas*, or regiments, that mixed pikemen and harquebusiers into battlefield units. Under able captains like Gonsalo de Córdoba, this infantry won battle after battle in Italy, defeating the pike masses of the Swiss at Cerignola (21 April 1503), annihilating the French on the Garigliano (30–31 December 1503), and winning other victories at La Motta (October 1513), La Bicocca (27 April 1522), and Pavia (24 February 1525).

Twenty of these regiments were formed initially, but by 1534 these had evolved into even larger units—brigades known as *tercios*. At first, the proportion of pikemen to harquebusiers was large, but the officers discovered that increasing the number of harquebusiers increased the effectiveness of the unit on the battlefield. Consequently, they gave more and more men harquebuses, until by 1550 just over one-half of the men carried firearms while the rest carried the long pike. By 1566, when the duke of Alva took his Spanish army into the Netherlands, the *tercios* included musketeers as well. The Spanish infantry (most of whom were from the Italian, German, and Walloon areas ruled by Spain) replaced the Swiss as the dominant force on the European battlefield, and they would retain that reputation for 150 years, until their defeat by the French at Rocroi on 19 May 1643.

France followed the Spanish model in 1531 when King Francis I created four "legions" of infantry, units that consisted of six "ensigns," each of 600 pikemen, 300 harquebusiers, and 100 halberdiers. The legion was bested by the *tercio* in several battles, notably at Ceresoles in 1544. In that fight, the French were saved at the last minute when their commander, Francis, prince of Enghien, turned the Spanish flank with his cavalry, showing that cavalry could still play an important battlefield role when properly combined with infantry and cannon.

Although not as successful as the Spanish *tercios*, the French infantry continued to grow, and two of the legions, named for the provinces of Picardy and Champagne, continued to exist in the French army without interruption until the revolution of 1789.

Other regular infantry forces also appeared in Europe. From the German states came mercenary pike units, with companies of harquebusiers and pikemen. Known as *landsknechts*, these units fought neither for king nor cause, but for cash. Similar mercenary infantry and cavalry units were raised by the condottiere captains of Italy. The Italian political philosopher Niccolo Machiavelli proposed a new type of Roman Legion, which he called the *battaglione*, in 1525, which besides 9,000 men carrying pike, sword, and harquebus, would also have included detachments of cavalry and artillery, as well as its own supply corps. Machiavelli's ideas were never implemented by the Italian states, but some of his theories were adapted by the generals of a later age.

The Ottomans, of course, had their regular infantry, the Janissaries. By the time of the unsuccessful Ottoman siege of Malta in 1565, there were 40,000 Janissaries, armed with sword, bow, and harquebus. They carried the banners of the Turks from the shores of the Persian Gulf to the waters of the Red Sea, and from the deserts of Arabia to the Pillars of Hercules. Their march into Europe proper reached its crest at the walls of Vienna (1529 and 1683). For nearly 250 years after they had captured Constantinople, the Turks repeatedly tried to take Vienna, the capital of the Holy Roman Empire. The Janissaries' greatest defeat, however, was self-inflicted as the corps became embroiled in dynastic politics and was eventually purged. As a unit, however, it would remain part of the Turkish army until destroyed by other Turkish forces in 1826.

A similar fate befell the first professional infantry of eastern Europe, the Russian Streltsi. These regular infantry units of the czar were formed in the mid-1500s, about the time the Janissaries were trying to storm Malta and the Spanish had won the last battles of the Italian Wars. Armed with large axes and harquebuses, these professional soldiers became a class, with service in the unit hereditary. At first paid in saltpeter and lead, the Streltsi grew rich and influential and, like the Janissaries, had to be wiped out by the "new" regular infantry that came into being in the later years of the seventeenth century.

The rise of infantry was not confined to the world west of the Volga and the Caspian, however, but at least for a brief time the foot soldier built empires for eastern warlords. Babur, king of Kabul, conquered most of India with an army of harquebusiers supported by mobile artillery. Chinese, Korean, and other Asian armies also raised infantry units that relied on firepower.

In the mid- to late 1500s, Oda Nobunaga and his generals, Toyotomi Hideyoshi and Tokugawa Ieyasu, unified Japan with saltpeter as well as samurai. Nobunaga developed the concept of continuous infantry volley fire, in which an entire rank fired in a single blast, then stepped back while another rank came up and fired. Although the Spanish experimented with a march-countermarch fire system at about the same time, the volley tactic would not become common in Europe until John of Nassau used it in 1594, nearly 20 years after Nobunaga's use at Nagashino in 1575. Japanese firearms were lighter than those used in Europe and were just as important to victory. One of the most decisive battles in Japanese history, the battle of Sekigahara (the Barrier Field) on 21 October 1600, for example, was won by infantry firepower. The victor, Tokugawa Ieyasu, had learned his trade as a general under Nobunaga. The shogunate he established would last nearly a quarter of a millennia.

The military might of the shoguns, however, could not be effectively extended off their island. Despite invasions in 1592 and 1598, even nearby Korea proved too difficult for Hideyoshi's forces to conquer. Warlords, like Hideyoshi, learned that an invincible army needs an invincible navy if it is to conquer over water. That lesson was also being learned in Europe, as gunpowder, combined with sail power, brought a revolution at sea.

Wooden Ships and Iron Guns

Gunpowder went to sea in the early fourteenth century. The largest ships in the Genoese navy reportedly were armed with a few guns in the early 1300s, but the first time cannon were fired at sea was in 1340 at the naval battle of Sluys in the Hundred Years' War (1337–1453). The fight at Sluys, like most sea battles up to that date, resembled a land battle fought on water. The French fleet was chained together in three immobile lines in the estuary. Ahead of the line were floating traps made from small boats filled with timber, sharp sticks, and dirt. Five of the ships, however, had light cannon. Positioned ahead of and on the left of the first line so as to sweep the approaches, these five ships had an open field of fire on the advancing English. Their guns, however, were not very powerful, horribly inaccurate, and slow to load. Their effect on the battle could not have been very great, as four of the five cannon-armed vessels were taken and the French lost the battle. The battle at Sluys was won by hand-to-hand fighting as armored knights and foot soldiers boarded enemy vessels. There were no maneuvers, and the battle looked a lot like the storming of a castle. Still, the victory gave England command of the Channel for 30 years.

Since antiquity, naval battles had been fought like land battles: masses of ships powered by oar closed with and tried to outflank one another while bowmen fired to kill enemy crewmen. For more than 2,000 years, from Salamis (480 B.C.) to Lepanto (1571), sea battles were decided by ramming and boarding.

By the mid-fifteenth century, however, ship design began to change. While the oar-powered galley and longboat still formed the mainstay of naval power, larger, taller sailing vessels were being built in the Atlantic and Baltic port cities of northern Europe and in the shipyards of China.

The Chinese were pioneers in seapower and in using navies to protect their might over long distances. A Muslim eunuch named Cheng Ho, for example, took his fleet of seagoing junks from Ming China to subjugate Sumatra, Ceylon, and much of the Indonesian archipelago in a series of great expeditions between 1405 and 1415. Over the next twenty years, Cheng Ho raided the coasts of the Indian Ocean, exacting tribute from Hormuz, on the Persian Gulf, and Mecca, on the Red Sea, in the early 1430s.

At about the same time, northern European shipyards were turning out relatively large merchantmen of up to 1,000 tons, while the merchant cities of Italy were building even larger vessels, called *uscieri*. Like Cheng Ho's junks, these ships depended solely on sail power and were designed for long-distance travel. The "great cogs" of the Baltic could carry up to 500 fighting men and, like the *uscieri*, were fitted with light cannon. By 1450, Venice and Genoa mounted scores of light cannon, called *petararae*, on their ships. The gun was not cast; instead, it was "built up," made of a tube that was reinforced with rings or strips of rope or, later, iron. A *petarara* fired a ball about the size of a man's fist and was designed to tear sails and kill men, not sink ships. One English vessel of the late fifteenth century, the *Regent*, mounted over 200 guns.

The effectiveness of these ships against oared galleys was demonstrated at the Siege of Constantinople in 1453. In one battle, over 100 Turkish galleys tried to force their way into the harbor, but were blocked by ten Byzantine sailing ships. Although the galleys had a few small guns, they could not be raised to fire up at the larger ships, and the Turks were raked by plunging fire.

Later in the siege, three big Genoese ships, escorting a grain carrier, tried to reach the city. They were surrounded by 75 Turkish galleys, led by the sultan's most experienced admiral, Baltoghlu. In the short time between the defeat at the harbor and the battle with the convoy, Baltoghlu had made many changes in the Turkish fleet. Ships' guns had been elevated and remounted, walls of shields were erected along the sides of galleys, and a picked force of Janissaries was chosen to lead the boarding parties.

The Turks repeatedly swarmed around the Christian ships, which were becalmed when the light breeze changed direction. From morning until sunset, Baltoghlu's sailors tried to take the big ships, but each time they were repulsed by fire from above. The admiral even ordered his flagship to ram the grain carrier, but he, too, was repulsed. The four large Christian warships were lashed together and held the Turks at bay until a fresh night breeze carried them safely to Constantinople. Like the English admiral John Byng three centuries later, the luckless Baltoghlu was disgraced and stripped of command. The Turkish admiral, unlike Byng, who was executed for losing the Battle of Minorca in 1756, was allowed to live, but only after he was given 100 lashes with a heavy golden rod.

Despite the proven power of the great sailing ships, the Turks, and most Mediterranean powers, continued to rely on galleys. Although some ships were powered by convicts or slaves, most rowers were paid professionals and were treated the same as any other member of the crew. These fast ships and their skilled soldiers did not disappear from naval history quietly, but went out with one last flourish of glory under two fifteenth-century admirals, Don John of Austria and the Muslim corsair Khair ed-Din.

Born a Greek Muslim, Khair ed-Din (also known as Barbarossa because of his red hair), was perhaps the greatest of the galley admirals. As dey of Algiers and vassal of the Ottoman sultan, Khair ed-Din scoured the Mediterranean, Aegean, and Adriatic seas throughout the early 1500s. He used his galleys for many different missions, including the conquest of North African states like Tunis and European seaports like Nice, which fell to him in 1543. His navy also harried Christian shipping and performed other complicated strategic missions, such as the evacuation of Spanish Moors (Moriscos) in 1533. Khair ed-Din died in 1546, 25 years before Don John led the last great galley fleet to victory at Lepanto.

Over 500 galleys of various sizes met off Lepanto in the Gulf of Corinth on 7 October 1571. Most of the ships, on both sides, had three to five light cannon mounted in their bows. The six largest ships, giant galleasses sent by Venice, also mounted guns along their sides. Many of the larger ships carried harquebusiers. Despite the guns, the battle was still one of ramming and melee. By the end of the day, nearly 200 Turkish ships had been sunk or captured, and nearly 20,000 rowers, soldiers, and sailors had been lost by the Ottomans. Although 15,000 Christians were killed or wounded, only thirteen of their ships were lost. The battle ended Turkish seapower forever, but despite the triumph of the Christian galley fleet, Lepanto was the last major naval action fought under oar.

The Armada

By the end of the fourteenth century, sails had long replaced oars in westernmost Europe, as they had in China. Christopher Columbus and Vasco da Gama had led sailing fleets across the Atlantic and down along the coast of Africa. Portugal's admirals reached China and Japan by the early 1500s, and Spanish explorers had penetrated the Pacific. None of these expeditions could have been made by galley fleets. In one of the rare galley versus sail battles of the era, a Turkish galley fleet sailing out of the Red Sea was forced back to port by a Portuguese sailing squadron. It was a defeat as important as Lepanto, as it forced the

Ottomans away from Africa and opened up that continent, and the Asian sea routes, to the Europeans. Portuguese admirals like Francisco de Almeida and Afonso de Albuquerque fought off Muslim squadrons for 50 years throughout the first half of the sixteenth century, securing Indian Ocean sea-lanes for Portugal.

Portugal needed a navy to maintain an empire that reached from its Brazilian colonies to its Japanese trading posts. So did Spain, whose growing empire stretched to the Philippines. The Iberian ships were large two-, three-, and later four-masted sailing vessels capable of carrying cargo, crew, and the provisions for the crew great distances. The largest of these were the royal galleons of Portugal and the East India galleons of Spain. They also mounted a wide variety of cannon, ranging from small swivel guns to hull-rending culverins.

Other European powers also built navies, both for exploration and trade (Spain and Portugal) and for privateering (England). England had had a standing navy since Henry VII (1485–1509), who built five warships, among them the *Regent*, and who gave the navy a home base at Portsmouth.

Henry VIII established construction and repair yards along the Thames and enlarged both the navy and its ships. He also replaced the light weapons of his predecessor's reign with much larger, ship-killing guns. The heavier weight of these guns meant that they could no longer be mounted on the fore and stern castles or rails of the ships; for balance they had to be set on the upper or main deck of the vessel. The guns were mounted on carriages, or trucks, and portholes were cut in the sides so that the guns could be slid out and fired, then retracted. One of the first of this new type of broadside-gunned ship was Henry VIII's *Mary Rose*, built in 1509. It foundered and sank in 1545, but Henry and his shipwrights did not give up.

Although the fleet had decayed under Queen Mary (1553–58), Queen Elizabeth's admiral, John Hawkins, rebuilt and revitalized the fleet and its bases. Between 1577 and the year the Armada sailed (1588), Hawkins increased Britain's naval power to such an extent that he was able to put 197 armed ships in English waters that year. Only one of those, moreover, was oar powered.

The English also built flush-deck and race-built ships, which were much lower in the water than the great galleons. The lower English ships appeared small when seen next to the Spanish galleons, but many, like the *White Bear* and the *Triumph*, each of which was over 1,000 tons, were larger than many of the galleons, which averaged about 500 to 700 tons.

England waged a commercial war at sea against Spain and the Spanish treasure fleets. In April 1587, another of Elizabeth's admirals, Sir Francis Drake, sailed into Cádiz harbor. His squadron of seventeen ships fought off ten galleons, burned or captured 37 merchantmen, and then sailed off to seize the Azores. On the way home, he captured the Portuguese treasure ship *San Felipe*. Unable to protect the sea-lanes or his coasts from English raiders, King Philip II of Spain decided to strike the English in their den; he ordered his commanders to prepare an invasion of the British Isles.

Although Julius Caesar, Canute of Denmark, and William of Normandy had all crossed from the continent to England under oars, the Spanish fleet that was collected and built for that same goal in the years after Lepanto contained only eight ships propelled by oars. Four of those, moreover, were galleasses that depended on sails as their principal means of propulsion. Although Philip's chief planner, Alvaro de Bazan, Marquis de Santa Cruz, had commanded a squadron of galleys at Lepanto, he successfully argued for the creation of a sailing fleet for the invasion. He brought shipwrights and gunsmiths from all over Europe to Spain's Atlantic ports for the construction of a new fleet, the Armada. Santa Cruz recognized the value of the cannon used by Drake. Over 100 of the guns mounted on Armada ships were, in fact, made in England. A cannon foundry in the Forest of Dean made 100 guns for the warships of Spain, and merchants in Bristol shipped nine vessels loaded with long-range culverins to Spain for the Armada.

By the time it was ready to sail in May 1588, the Armada had over 130 warships, half of them classified as galleons or great ships. They mounted over 2,400 guns among them and were manned by 26,000 sailors and soldiers. The fleet even included 200 large boats that were carried aboard the big ships for the purpose of disembarking troops and guns. These were the sixteenth-century equivalent of modern landing craft.

Santa Cruz had died in February 1588, and Philip chose the richest, most powerful, and perhaps least-qualified man in Spain to succeed him, the duke of Medina Sidonia. Other experienced commanders were either passed over, relegated to subordinate positions, or detached to other theaters. When the fleet sailed in May 1588, moreover, its destination was not England, but the Netherlands, where it was to escort the duke of Parma's army across the Channel.

Through a combination of Spanish luck and bickering among English commanders, Medina Sidonia sailed into the Channel unopposed. On 30 July the entire Armada stood outside Plymouth, where two-thirds of the main English fleet lay at anchor, unprepared for battle. Rather than emulate Drake's Cádiz raid, Medina Sidonia remained on course. Over the next two weeks, the 130 ships of the Armada and the 118 warships sent after them by the English navy fought a series of running battles. Drake and Hawkins tried to stand off at long range with their guns, but during four battles in the Channel sank only two galleons, one of which was lost because of fouled rigging. The Spanish, largely outranged, tried to close and board (at one point Medina Sidonia's flagship closed to within 100 meters [110 yds.] of the English line), but they too did little damage and reached Calais as planned on 6 August.

Although the Armada fight was the first time ships mounting broadside guns engaged in a large-scale fleet battle, the decisive moment came not at sea but in port, when English fireships were sent among the Armada. Although actual damage was minor, many Spanish captains cut their anchor cables to make way hurriedly, and Medina Sidonia's carefully ordered fleet degenerated into ad hoc squadrons and groups of stragglers.

Medina Sidonia managed to collect two dozen of the large galleons into a battle squadron, but after ten hours of combat on 8 August, both his fleet and the English had used up almost all of their powder and shot. The Armada sailed north, chased at a distance by the English, who after four days broke off the action and headed home. The Armada, in squadrons and by individual ship, tried to sail around England and Ireland. Half of the fleet was lost at sea. Of the 67 ships that made it back to Spain, nearly half were so badly damaged that they had to be broken up to repair the rest.

Spain built three more armadas, two of which were wrecked by storms. The fourth, sent to sea three years after Philip's death, landed a small army in Ireland in 1601, but to no avail. The two weary successors to Philip and Elizabeth made peace in 1604.

World Empires Made at Sea

Spanish seapower, although still great in terms of numbers of ships, never fully recovered. The English, who took credit for its demise, did not supplant the Spanish at sea, and were themselves nearly swept from the waves by the Dutch, whose navy was essentially a huge fleet of armed merchantmen. The Dutch, who gained their freedom from Spain in the 1630s, quickly built up a trading empire that brought spice from Asia and furs from the Americas. Colonies and trading posts from New Amsterdam (New York) to Nagasaki, Japan, served as bases for the trading navy, which in turn made Holland the warehouse of the world.

Civil wars in England (1642–51), unrest in France (1648–53), and a general malaise in the Iberian kingdoms gave the Dutch a head start toward building an empire and a fleet. When the rival navies finally began a twenty-year war for mastery of European waters and Atlantic trade routes, it was the Dutch, initially, who had the advantage, in terms of experience, numbers of ships, and money. By the middle of the century, Holland had over 10,000 sail and 168,000 seamen upon the water.

As the more maneuverable English had an advantage against the galleons, so Holland's lighter, flatbottomed ships often gave the Dutch an advantage over the English, French, and Spanish. Most of the battles between and among these forces in the seventeenth century were fought in coastal, tidal, or shoal-filled waters, where the Dutch were more maneuverable and less likely to run aground.

Naval battles of the period were contests of broadsides, but at first still resembled the melees of galley warfare. Fire ships were used to break up enemy formations in the same way that later navies would use torpedo boats. The Dutch, for example, took fire ships into the sea fight of Southwold Bay (Sole Bay) in 1672. By the end of the era, however, the tactical order of line abreast, used by galleys, was supplanted by the line ahead formation. This allowed the increasingly longer ranged guns to be best brought to bear in massive fleet broadsides.

England, France, and the Netherlands fought four naval wars in the second half of the seventeenth century, and although the Dutch fleet under De Ruyter once sailed up the Thames, it was the English who eventually triumphed.

Rather than armed merchantmen, the fleets of the late seventeenth century relied on ships designed solely for war. These vessels, known as ships of the line, mounted from 50 to as many as 130 guns of standard calibers. The bigger ships could even outgun coastal forts. The old iron trucks were replaced by wooden gun carriages, usually made of cedar or elm, and systems of ropes and pulleys allowed the guns to be fired, hauled in, reloaded, run out, and fired again in minutes. Unlike the Armada's guns, some of which had to be loaded by men leaning over the rails, these guns could be serviced from behind the protective sides of the warships.

The Dutch won most of the European naval actions, but the English kept coming back out to sea. They took the war to the colonies, moreover, and gained an empire. The commerce of England eventually surpassed that of Holland, and the Dutch, worried over the continental ambitions of France, made peace.

The French had an overseas empire, but they were not convinced of the advantages of maintaining a large bluewater navy. By the 1670s, the French decided to lay up their large warships and fight a war of privateers, preying on commercial vessels for profit and economic harm rather than contest the growing English mastery of the sea.

France produced its own class of courageous and innovative seamen, such as the privateer admiral Jean Bart and the fleet commander Tourville. Under Louis XIV, France made another, unsustained, bid for seapower, and built a new battle navy that by 1689 was nearly as large as the fleets of England and Holland combined. Unfortunately for France, England and Holland united. Despite inconclusive victories like Beachy Head (1690), the capture of an English treasure fleet from Jamaica, and the skillful retreat of La Hogue in 1692, the French decided again to break up their navy into commerce raiding. By the turn of the century, the French had all but disappeared from the sea, and the navies of England and its allies were able to sail anywhere at will.

This command of the sea allowed England to seize islands and raid ports, ravage coasts and capture strategic outposts such as Gibraltar, which fell to English forces in

1704. Spain, under the aggressive and talented royal minister Alberoni, tried to reestablish itself as a seapower, but the fleet was crushed at Cape Passaro in 1718, and the terms of peace imposed on it in 1719 forced Madrid to destroy the dockyards where the new navy was being built.

By 1740, the English fleet of nearly 100 ships of the line and an equal number of frigates (smaller and faster than ships of the line, mounting 20 to 40 guns), outnumbered that of its next largest rival, France, by two to one. That ratio grew as France and England went to war in the 1740s and 1750s. Although the English suffered a few defeats, such as Byng's failure at Minorca, the French were again forced to break up the navy and rely on a raiding strategy. At the end of the Seven Years' War (1756–63), the English navy outnumbered the French by three to one in ships of the line alone. Although French privateers captured or sank 2,500 English merchantmen during the war, over 8,000 remained to bring troops, supplies, and goods from one end of the growing British Empire to the other. Nearly 250 of the privateers had been captured by war's end, and by 1761 not a single French ship of the line was under sail. French commerce, moreover, was driven from the seas.

Command of the oceans allowed England to move armies at will, and not just in Europe. British forces captured the cities of rival European overseas empires. Quebec was taken from France in 1759, and Havana was wrested from Spain in 1762. Seapower also gave the British the opportunity to challenge the power of more distant kingdoms and to win an even larger empire in India. While mastery of the sea gave the English, as it had the Spanish before them, the ability to reach across the ocean to build empires, those empires were won, and held, on the ground, and that was the job of the professional soldier.

Cannon, Cavalry, and Muskets

As the ship of the line became the main unit of mastery at sea, so the regiment became the instrument of victory on land. The regularity of warfare, both in Europe and overseas, and the increasing size of the armies needed to win those wars, led to the formation of standing armies of professional soldiers. Although all the powers of Europe relied on mercenaries or men who were forced or induced to join the colors to flesh out their armies, the central core of the Spanish, French, Swedish, and other important forces of the period were professional combat units of horse and foot.

The combination of pike and harquebus pioneered by the Spanish *tercios* became the model for all European armies by the turn of the seventeenth century. By 1618, at the start of Europe's last—and worst—great religious war, muskets had begun to replace the harquebus, giving infantry units a greater and more sustained reach across the battlefield. The 4–6-meter-long (4½–6½-yd.-long) spears carried by the pikemen, less an offensive weapon in an age when a musketeer could kill a man at 60 to 100 paces, were, however, still needed to protect the musketeer while he reloaded. Muskets easily became fouled by black powder, and a light drizzle could render the powder useless or put out the match that was used to ignite the powder in the musket pan. While the musket was put to use as a club in close fighting, particularly by the Irish regiments that fought for France and Spain, the pike was still needed to ward off cavalry and the opposition's own pike masses.

Throughout the Thirty Years' War (1618–48), European arms, tactics, and organizations went through a series of changes, most of them brought about by firepower. Artillery became more mobile, and light guns were used with great effect by Sweden's warrior king Gustavus Adolphus and his imitators. These light cannon, manned by a small crew, could keep pace with the advancing infantry, while the heavy guns remained comparatively immobile once in place. Gustavus at first used small guns made of copper tubes bound with iron hoops, cords, plaster, and leather, but he discarded these in favor of a cast-iron piece that fired a four-pound shot. It weighed only 200 kilograms (500 lbs.) and could be moved about by a two-horse team and a three-man crew.

The Swedish king attached a pair of these guns to each of his regiments and used them to plow great gaps in the massive squares of Imperial infantry that he faced at Breitenfeld, Lützen, and a dozen other battles of the Thirty Years' War. Accompanying light guns soon became standard equipment in the infantry regiments, a practice that continued throughout the musket era. As guns and muskets improved, the numbers of pikes dwindled in proportion, and generals and soldiers traded cold steel for hot powder.

For a time, even the cavalry of Europe came to rely on firepower. Pistol-armed horsemen had made their appearance in the French Wars of Religion half a century before, when Protestant cavalry combined sword and pistol in their duels with Catholic lancers. Large units of armored cavalry armed with pistols rode onto the battlefields of Germany in the early 1600s and performed a maneuver called the caracole, in which rank after rank of horsemen came forward, fired at their opponents, and wheeled to the side to let the rank behind come forward and fire.

This complete turnabout in tactics from shock to missile weapons by horsemen was short-lived, however, as Gustavus, England's Oliver Cromwell, and other generals raised new regiments of cavalry that were trained to charge at the gallop, wielding heavy steel swords. Lines of cavalry riding to the charge with swords remained the trademark of the heavy cavalry for 300 years, until the middle of the nineteenth century, when rifled muskets and quick-firing artillery were able to break up the horse regiments at ranges of several hundred meters.

The clumsy lance of the chivalric era was still carried by gendarmes and other heavy horse at the start of the seventeenth century, but by the end of the Thirty Years' War even they had discarded it in favor of the sword. Light horse units and eastern cavalry in general still carried a light lance, but even the Polish winged hussars stuck an axe, mace, or sword, and sometimes a brace of pistols, in their gear. The bow all but disappeared in the west, but remained an important tool of war from the Volga to the Yalu. The Manchu Banner cavalry, which overran Ming China in the 1600s, for example, fought with lance and bow, as did the Moghul cavalry in India. Japanese samurai also continued to use their bows.

Light cavalry also made a comeback on the European battlefield. The unarmored cavalry on their small ponies, such as the Hungarian hussars, performed the duties of scouting, patroling, and foraging for the army. On the battlefield, they usually kept to the far flanks, but were often used in harassing skirmishers, annoying flanks, and riding down already routed infantry units. European generals also experimented with units called dragoons, which moved and performed some of the duties of the light cavalry, but which usually fought on foot with muskets, firelocks, or carbines.

The use of such a variety of forces required the skill, timing, and discipline found only among well-led regular forces. The Stuarts were taught that lesson by Cromwell's New Model army, which, even when outnumbered, won impressive victories, like Naseby (14 June 1645). By the end of the Thirty Years' War, even the *tercio* had fallen to the more advanced combined arms tactics of men like the French general Louis duke of Enghien (later prince of Condé). The final day for the *tercio* came at the battle of Rocroi on 19 May 1643, when 18,000 Spanish infantry were broken by Enghien's army. The French cavalry defeated the Spanish horse, leaving the mighty *tercios* stranded. Still, the Spanish fought off a succession of French horse and foot attacks, until late in the day, when the duke brought his mobile guns up close and broke open the pike squares. Like the loss of the Armada, the destruction of the *tercios* was a defeat from which Spain could not recover, and Spanish dominance of land warfare was ended.

Spain was not the only empire whose sun had begun to set in the mid-1600s. By the end of the century, a new age had come about in Europe. With the last great gasp of Turkish power hurled back from the gates of Vienna by John Sobieski and his Polish cavalry in 1683, the Ottoman Empire began a rapid decline. By the end of that decade, the Turks had been forced out of Hungary, the Ukraine, and much of Greece and were in internal turmoil following a Janissary revolt and dynastic upheaval. Relieved of Ottoman pressure, another power in the east was also on the advance. Russia, whose cossacks had reached across Asia to Siberia and the borders of Manchu China by 1689, also turned westward, where it became embroiled in a series of wars with Poland and Sweden. By 1700, the last of the old empires had been beaten down, and those that rose to take their places engaged in a century-long world war.

The Rise of France

If any nation emerged victorious from the 30 years of destruction that left one-third of the population of central Europe dead or displaced, it was France. For the rest of the century following Enghien's victory at Rocroi, the French army was considered the best in Europe. Under Louis XIV it was certainly the largest and the best organized. By the 1660s, Louis had over 160 regiments of horse and foot in his army. Nearly a quarter of a million men marched under his royal banner.

Louis's generals also standardized and professionalized the French artillery. In the mid-seventeenth century, there was no standard artillery piece. The Spanish, for example, had over twenty different calibers of artillery, although they at least had them classified into four basic groups, which was far more ordered than most of their rivals. Louis's artillery regiment and his inspector of artillery, Claude du Metz, standardized the guns into six land and seven naval classes, categorized by weight of shot. Older guns were refounded, and new brass pieces were cast. Artillery would not become truly standardized until the French introduced their Système Vallière in 1732, the year that French gunners finally received military rank.

At the start of the 1700s, France was the leading power in Europe. Louis XIV had the largest army on the continent, and France itself was protected by a series of fortifications designed by the king's chief engineer, Vauban. The new style of fortress built by Vauban was designed to offset the power of artillery that had rendered stone castles obsolete. Combinations of stonework and dirt made the relatively low walls more resistant to the pounding of iron shot, while star-shaped gun positions, called bastions, were laid with an eye to establishing overlapping fields of long-range cannon fire.

As Vauban himself demonstrated, almost any fortress, even one modeled on his own designs, could be taken if the besieger had the time, manpower, will, and skill to do so. The Vauban fortress was not impregnable, nor was it intended to be. It was designed to slow an invasion and to allow small forces to buy time. While an invader was busy digging trenches and mines, foraging for food and otherwise tied down in a siege, the opposing state could marshal an army of its own and march to the rescue. Such a rescue might involve a field battle, but more likely it could be accomplished by cutting off supply lines or threatening to lay siege to the invader's own bases. Tied to their depots, armies could not leave fortresses in their rear, unless they detailed large bodies of men to besiege or at least mask them. Such detachments would greatly weaken an invader, again helping the state under attack to even the odds.

Eighteenth-century warfare, however, was hardly the staid, bloodless chess match that the war of fortresses would suggest. In 1704, the English commander John Churchill, Duke of Marlborough, marched an Anglo-Dutch army from the Netherlands to Bavaria, knocked France's Bavarian ally out of the war and joined up with an Austrian Imperial army under Prince Eugene of Savoy, to win a bloody battle at Blenheim, near the banks of the Danube. At Blenheim, in August 1704, nearly 35,000 men, on both sides, were killed or wounded. Losses were even heavier at Malplaquet, another of Marlborough's victories, five years later.

The principal instrument of death in those battles, and in half a hundred other major engagements fought across Europe in that decade-long war, was the infantry musket. Unlike the clumsy matchlocks of the Thirty Years' War, the musket of 1700 was a flintlock weapon, a much more reliable and faster-to-load firearm than the matchlocks of the previous century. Louis XIV raised the first full regiment to be equipped with flintlocks in 1670, and they were the standard weapon throughout Europe by 1699. The last pikes disappeared in the West just about that time, although a few pike companies were still found in the Austrian, Russian, and Swedish armies in the first years of the century.

Infantry tactical formations went from men massed ten deep to more compact, linear formations three to six ranks deep. The troops fought in closely packed lines, with the purpose of bringing massive firepower to bear on other closely packed lines. Most firefights were fought at under 100 meters (110 yds.), and the slaughter was dreadful. When the mobile cannons were brought to the front by their regiments, the bloodletting was even worse.

Infantrymen were equipped with plug bayonets, which could be forced into the muzzle of a musket, and later with ring and socket bayonets (a Vauban invention), which could be fitted around the muzzle, thus permitting them to fire their muskets with bayonets fixed. However, cold steel was used more frequently by the cavalry, which continued to slash about at one another with heavy swords. The firepower of good, disciplined infantry gave them little to fear from cavalry approaching from the front, but the linear tactics dictated by the weapons made them more vulnerable on the flanks. While the infantry could still form a hollow square and hold off the horse with musket fire and a hedgehog of bayonets, an army whose cavalry was driven off the field could soon find itself in the position of the Spanish at Rocroi or the French at Blenheim.

The Great Wars for Empire

The eighteenth century saw the first true world wars in that great powers fought in several theaters of conflict. French, British, and Spanish fleets and armies, and the forces of their allies, fought each other in Europe, the Americas, India, and off the coasts of Africa and Asia. While most of the wars were initiated by dynastic squabbles in Europe, they mushroomed into world conflicts that decided or influenced the destiny of millions.

Some of these wars, however, were purely European in nature, such as the Great Northern War between Peter the Great of Russia and Charles XII of Sweden. Heir to Gustavus's tough little professional army, Charles triumphed over every army Peter sent against him for nearly a decade. The czar, however, learned from the Swedes, and when he met Charles on the battlefield of Poltava, in southern Russia in 1709, the weight of Russian numbers and artillery won the day. Russia replaced Sweden as the Baltic power, and for the next century would expand to the south and west at the expense of the Ottomans, the Poles, and the Swedes.

No longer threatened from the West, Russia also grew eastward, and became a great land empire and a power whose influence was felt from the privy council chamber in London to the Manchu court in Tientsin. Russia even launched its own bid for an overseas empire, with expeditions that sailed to Alaska and the Pacific coast of North America.

Unlike the Great Northern War, the wars initiated by Prussia's Frederick the Great involved coalitions that included nations with overseas empires. While his victories over Austrian, French, and Russian forces led to a battlefield legend for the Prussian disciplinarian and his army, battles fought outside of Europe by the allies and enemies of Frederick had at least as great an effect on world history.

In the two Frederickian wars of mid–eighteenth-century Europe (the War of Austrian Succession [1740–48] and the Seven Years' War [1756–63]), France and England sent fleets and armies to battle around the world. From the forests of New York to the plains of Delhi, small European regular forces, backed by colonial militia and local allies, fought and won huge empires that were out of all proportion in numbers of people and areas of land to the size of the armies engaged. The actions of a few thousand English troops on the Plains of Abraham (Quebec, 13 September 1759) and the mango grove at Plassey (23 June 1757) in India, ensured the ascendancy of Britain's world empire. Like Cortez's conquistadors, who conquered the Aztec Empire of Mexico 240 years before, those small numbers of professional, disciplined European firepower infantry proved more than a match for hordes of poorly trained feudal troops, even when, like the Nawab's army at Plassey, they also had gunpowder and courage to spare.

The World Turned Upside Down

The revolution in land and sea warfare that by 1763 had helped create the British Empire, the Prussian dominance of Central Europe, and the Russian colossus in the east did not end in the mid-eighteenth century (Fig. 1). The changes in warfare that had crumbled the last ancient empire at Constantinople in 1453 also contributed to the political and military revolution that came about in the closing decades of the eighteenth century.

MARK G. MCLAUGHLIN

SEE ALSO: American Revolutionary War; Seapower, British; Charles XII, King of Sweden; Condé, Louis II de Bourbon, Prince de; Cromwell, Oliver; Eugene, Prince of Savoy-Carignan; France, Military Hegemony of; Frederick the Great; Gonzalo de Córdoba; Gustavus Adolphus; Italian Wars; Manchu Empire; Marlborough, John Churchill, Duke of; Moghul Empire; Musashi, Miyamoto; Ottoman Empire; Peter the Great; Prussia and Germany, Rise of; Saxe, Hermann Maurice, Comte de; Seven Years' War; Spanish Empire; Suffren de St. Tropez, Pierre André de; Suvorov, Aleksandr Vasil'evich; Turenne, Henri de La Tour d'Auvergne, Vicomte de; Vauban, Sébastien Le Prestre de; Washington, George.

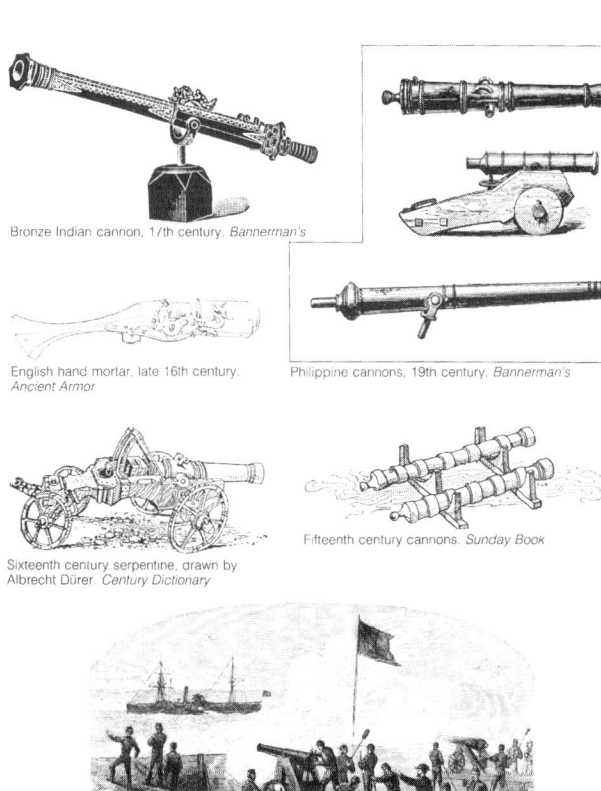

Figure 1. Examples of fifteenth- through nineteenth-century cannon and a sixteenth-century mortar. (SOURCE: Hart Picture Archives)

Bibliography

Beeching, J. 1983. *The galleys at Lepanto*. New York: Scribner's.
Braudel, F. 1973. *The Mediterranean and the Mediterranean world in the age of Philip II*. 2 vols. Trans. S. Reynolds. New York: Harper and Row.
Chandler, D. G. 1974. *The art of warfare on land*. London: Hamlyn.
———. 1976. *The art of warfare in the age of Marlborough*. New York: Hippocrene.
Cipola, C. M. 1963. *Guns, sails, and empires*. New York: Minerva Press.
Corvisier, A. 1979. *Armies and societies in Europe 1494–1789*. Trans. A. T. Sidall. Bloomington, Ind: Indiana Univ. Press.
Duffy, C. J. 1975. *Fire and stone: The science of fortress warfare 1660–1860*. New York: Hippocrene.
———. 1988. *The military experience in the age of reason*. New York: Atheneum.
———. 1985. *Frederick the great: A military life*. New York: Atheneum.
Dupuy, R. E., and T. N. Dupuy. 1977. *The encyclopedia of military history*. New York: Harper and Row.
Fuller, J. F. C. 1967. *A military history of the western world*. New York: Minerva Press.
Guilmartin, J. F. 1974. *Gunpowder and galleys: Changing technology and Mediterranean warfare at sea in the sixteenth century*. Cambridge: Cambridge Univ. Press.
Hale, J. R. 1983. *Renaissance war studies*. London: Hambledon Press.
Hogg, I. V. 1975. *Fortress: A history of military defense*. New York: St. Martin's Press.
Koch, H. W. 1981. *The rise of modern warfare*. Greenwich, Conn.: Bison Books.
Lawford, J. 1976. *Cavalry*. New York: Bobbs-Merrill.
Leach, D. E. 1973. *Arms for Empire*. New York: Macmillan.
Machiavelli, N. 1965. *Chief works and others*. Trans. A. Gilbert. Durham, N.C.: Duke Univ. Press.
Mahan, A. T. 1918. *The influence of sea power upon history, 1660–1763*. Boston: Little, Brown.
Mallet, M. E. and J. R. Hale. 1984. *The military organization of a renaissance state: Venice, c. 1400–1617*. Cambridge: Cambridge Univ. Press.
Mattingly, G. 1959. *The armada*. Boston: Houghton Mifflin.
Oman, C. W. C. 1937. *A history of the art of war in the sixteenth century*. New York: E. P. Dutton.
Parker, G. 1972. *The army of Flanders and the Spanish road*. Cambridge: Cambridge Univ. Press.
———. 1988. *The military revolution: Military innovation and the rise of the West, 1500–1800*. Cambridge: Cambridge Univ. Press.
Roy, I. 1972. *The Hapsburg-Valois wars and the French wars of Religion: Selections from the commentaries of Blaise de Monluc*. Hamden, Conn.: Archon Books.
Parkman, F. 1968. *The Seven Years' War*. New York: Harper-Torch.
Rogers, H. C. B. 1968. *Battles and generals of the civil wars*. London: Seeley.
Taylor, F. L. 1929. *The art of war in Italy, 1494–1529*. Cambridge: Cambridge Univ. Press.
Vale, M. G. A. 1982. *War and chivalry: Warfare and aristocratic culture in England, France, and Burgundy at the end of the Middle Ages*. Athens, Ga.: Univ. of Georgia Press.
Wedgewood, C. V. 1961. *The Thirty Years' War*. Garden City, N.Y.: Anchor Books.

HISTORY, MEDIEVAL MILITARY
[ca. A.D. 450–1450]

The medieval era (ca. A.D. 450–1450) is often regarded by military historians as devoid of significant developments, apart from the military system of the Mongols. This is unfortunate because numerous military developments occurred, many of them closely related to social changes that influenced the evolution of modern European military institutions. The decline and collapse of large empires in India, China, and the Mediterranean basin during the fourth century A.D. was followed by a prolonged period of political chaos. This situation endured longest and was most severe in Europe, but India remained politically fragmented for most of the ensuing ten centuries, and even China saw long periods of disorder and conflict between imperial regimes. By the mid-fifteenth century, though, Europe had achieved sufficient political organization and technological superiority to begin spreading its influence, marking the advent of the early modern era.

Dark Ages, A.D. 350–800

The collapse of the western Roman Empire in Europe had more profound results than the end of the Gupta Empire in India or the chaotic and sometimes anarchic period in China between the fall of the Later Han and the rise of the T'ang. Principal among these results was the change in the social fabric that followed the advent of the Germanic kingdoms in western Europe. The literate, urban society of the Roman Empire was replaced in less than a century by a collection of haphazardly organized successor-kingdoms ruled by kings who were little more than chiefs, and whose authority over their subjects was tenuous. Most cities in these kingdoms decayed, trade waned, and public order broke down as brigandage became common and quarrels among the new aristocracy became small wars. Perhaps most important, military service became linked to the ownership of land (the roots of manorialism); loyalty to the state or the emperor was replaced by personal, practical ties to a person's immediate overlord (the origins of feudalism). Some of these trends, like the decline of urban life and trade, and the beginnings of feudalism, were apparent as early as the mid-fourth century, but the collapse of the Roman Empire in the west in the fifth century hastened their development.

In terms of military organization and the conduct of warfare, these changes meant an end to the sophisticated warfare of Rome, and a return to days when campaigns were often expeditions for plunder, battles happened accidentally, and armies were usually poorly organized masses of armed men. Much the same could be said of the armies of the very late Roman Empire, however. The exception to this decline was in the Eastern Roman Empire where, with a more urban population, society did not regress even under the pressures of barbarian raids and crises of imperial succession nearly as badly as those in the west. The Eastern Empire's army, though at first staffed largely by *foederati* (allied barbarian soldiers serving for pay), retained most of the professionalism and technical expertise of the old Roman army.

A clear example of this is the Eastern Roman conquest of Italy, the so-called Gothic War of 534–54, undertaken by Justinian I's able generals Belisarius and Narses, less than 60 years after the fall of the Western Empire. Though handicapped by Justinian's notorious (and ultimately shortsighted) stinginess, Belisarius waged his campaigns with energy and speed, using his army's superior tactical skill to defeat much larger armies. At the Battle of Casilinum in 554, Narses' 18,000 men decisively defeated 30,000 Franks and Goths under the Frankish chieftains Buccelin and Lothair. Narses deployed his cavalry and archers on both flanks of his army, and when the Franks had committed themselves to the attack, he brought the wings of his army together, surrounding the Franks. After softening them up with arrows, Narses' cavalry charged and destroyed the Frankish army.

While the armies of the Germanic kingdom in western Europe were less organized than those of the Romans, they were not simply mobs. In theory all the adult male members of a tribe (around which each of the successor-kingdoms was originally created) were liable for military service, but in practice monarchs rarely summoned more than a small portion of the available manpower for service in a particular campaign. This self-imposed limitation occurred because a long campaign with an army based on universal compulsory service of adult males would wreak havoc on the subsistence agricultural economies of the barbarian kingdoms.

As military service was a significant economic burden (soldiers in the Germanic kingdoms were expected to show up with their own weapons, horses, and supplies when they were summoned), most of the kingdoms soon came to rely on their aristocracies, and those men who could be supported by them, to provide their armies. This practice kept armies relatively small, but it meant that the economic dislocation caused by raising troops was minimized, an important consideration in an economy just above the subsistence level. The tactics of these armies were simple: attack was an all-out charge, especially for the cavalry forces of the Goths and Lombards, and a defense was made by a dense mass of footsoldiers, all generally facing the enemy. Battles were usually confused and bloody, and commanders served less as tactical directors (battles in this period were nearly impossible to control once they began) and more as rallying points and champions. The strength of a soldier lay not in his tactical skill but in his personal prowess as a fighter.

This era lasted for approximately three centuries after the fall of the Roman Empire in the west and, although more primitive than the preceding or succeeding eras, it was not always lacking in sophistication, supplied by such monarchs as Theodoric the Great.

The Arab Conquests and the Rise of the Byzantine Empire, 630–800

The rise of Islam and the astoundingly rapid Arab conquest of North Africa, the Levant, and nearly all of Spain within a century of Muhammad's death in 632 had the most profound effect on military developments in Europe since the fall of the Roman Empire two and one-half centuries before. The Arab armies of the early Caliphate were primarily light troops, often poorly armed, but what they lacked in materiel they made up for in morale and esprit de corps. The Arab armies were also gifted with several brilliant generals, such as Khalid ibn al-Walid and Saad ibn ali-Waqqas, commanders who understood not only the strengths and weaknesses of their own troops but also those of their enemies. The best of the Arab generals also showed a clear grasp of strategy. Their campaigns employed the desert as a screen for their movements, a level of strategic sophistication almost entirely lacking in western Europe.

The most immediate result of the Arab conquests in Europe, other than the loss of Spain and destruction of the Visigothic kingdom, was the metamorphosis of the Latinized Eastern Roman Empire into the Greek-oriented Byzantine Empire, whose attentions were drawn to the east. The emperor who presided over this metamorphosis was Heraclius, a ruler and general of rare talent and ability. He became emperor (or "took the purple," as Byzantine chroniclers put it) as a result of a successful coup d'état against the notorious tyrant Phocas in 610. Heraclius came to the throne of an empire beset by Persian and Avar enemies, but his energy and ability enabled him to regain the initiative. Despite the loss of Egypt, the Levant, and most of Asia Minor he carried the war into the Persian homeland in a series of campaigns begun in 622.

The failure of the Persian siege of Constantinople, coupled with the effects of Heraclius' continued successes in Armenia, Anatolia, and Assyria, compelled the Persians to make peace in 628. Over the next few years, Heraclius busied himself restoring the Empire's prosperity after twenty years of war. Heraclius' restorations were interrupted by Arab attacks on the Levant, which began in the mid-630s. Despite valiant efforts, he was unable to halt the Arab advance, and when he died in 641, weary and despondent, the Empire had lost Egypt and all of the Levant as far north as the Taurus mountains. The people in these areas were too tired of war to resist for long, and were further separated from the Imperial government by long-standing religious differences.

Heraclius' most important work, though, was his expansion of the ideas of Emperor Maurice (whom Phocas had murdered to seize the throne) on military administration, and his creation of the first *themes*, or military provinces, in Asia Minor in the 620s. The troops stationed in and recruited from each *theme* were to provide its defense, and were under the command of a *strategos* (general), who also acted as governor of the *theme*. The thematic system was patterned after the militarized administration of the exarchates of Ravenna and Africa, and represented a final break with the remnants of Diocletian's administrative system. The thematic system also provided the basis for Byzantine territorial defense and the army until the eleventh century, when the combination of the loss of Anatolia, the decreasing size of individual themes (and a corresponding increase in number), and the disappearance of the soldier-freeholder at last destroyed it.

Western Europe felt the Arab onslaught later than the East; Spain was not invaded until 711. Perhaps the most famous military incident of this period of Arab conquests was the Battle of Tours in October 732. Charles Martel led an army of heavily armed and armored Franks against a large force of Arab raiders under their commander Abd er-Rahman, the governor of Muslim Spain. Despite repeated fanatical attacks, the light Arab cavalry was unable to make an impression on the well-armored Frankish troops, who had dismounted to fight. When the Arabs found that their commander had been killed in the fighting, they abandoned the field. The Battle of Tours showed that while the Arab armies were not invincible, they could be defeated only through careful and skillful generalship. Encounters with Arab and other Muslim armies in subsequent centuries showed that European commanders often lacked that quality.

Second Barbarian Assault: Vikings and Magyars, 800–1000

Western Europe, soon after stopping the Arab advance, was faced with two new threats: the Magyars from the east, and the Vikings from Scandinavia to the north. The Viking attacks began first, during the first decade of the ninth century. Although at first they only conducted raids for plunder, the warriors from the north began to settle along the coasts they raided, establishing networks of strongholds in the mid to late ninth century. Most of the Vikings who attacked England were Danes (the area they dominated was known later as the Danelaw), while the Norwegians (or Norse) operated primarily in Ireland and Scotland, and both Danes and Norse harassed the French coast. The Swedes took a different course and pressed deep into Russia, eventually arriving on the Black and Caspian seas, and even attacking Constantinople, albeit unsuccessfully.

Since the first Viking attacks were simply pirate raids, true armies of Vikings did not appear until the latter half of the ninth century. The physical prowess of the Vikings and their evident delight in violent activities did much to enhance their reputation as warriors, and the news that the Norsemen were coming was sometimes enough to empty a district of its population. Viking tactics, like those of their European opponents, were simple, but generally effective. These included the "shield wall," where defenders overlapped their shields for additional protection, and the use of *berserkers* (literally, "those without armor"), who worked themselves into a killing frenzy before a battle.

Viking generals also used ruses and stratagems, and their cunning and resourcefulness earned them many triumphs over more sophisticated opponents. Viking campaigns were at first constrained by the availability of water transport routes that their longships used to gain access to territories for pillage. When the Vikings later established local bases, they were able to mount some of their troops on horseback. While the Norsemen always fought dismounted, they used horses to increase their strategic mobility.

The strain of resisting the Viking depredations wrecked, or at least badly damaged, what little central government had developed in western Europe since the late fifth century. Authority naturally devolved onto those local leaders who could organize resistance or at least provide refuges for people fleeing Viking raids. In this fashion, especially in England and France, the Viking raids hastened the growth of feudalism. By the beginning of the tenth century, many of the Danes and Norse had settled in the lands they raided and often accepted the authority of the native rulers, if only in the same unruly fashion as other feudal magnates. The situation was different in England. By the end of the reign of Alfred the Great, the Danes and Anglo-Saxons had reached a happy accommodation. Alfred had also created a navy, which he and his successors used to guard their coast, to transport their troops, and to fend off enemy raiders.

While northern and western Europe endured the depredations of the Vikings, eastern Europe faced the onslaught of the Magyars, a nomadic people of Finno-Ugric ancestry. These people, defeated by an alliance of Patzinaks (Pechenegs) and Bulgars in southern Russia about 895, settled soon after in the Hungarian plain under their chieftain Arpad. There they allied with the remnants of Charlemagne's old enemies, the Avars. Even before settling in Hungary, the Magyars had raided eastern Germany, and they soon began again, raiding Germany and northern Italy, ranging as far afield as Rheims and Lyons. The highly mobile, bow-armed forces of the Magyars at first easily eluded the heavily-armed and rather clumsy German cavalry.

The Germans responded, however, and worked better to protect their lands and to improve their armies. Henry the Fowler, then King of Germany, defeated a Magyar army at the Battle of Allstedt (near Erfurt) on 15 March 933. The Magyars, chastened by this reverse and frustrated by increasingly effective German frontier defenses, reduced their raiding activity during the next two decades. In 954, though, they assembled an army of at least 50,000 and swept through Bavaria into Franconia. German Emperor Otto I made a treaty with them and helped them move into Lorraine. The Magyars ravaged much of northeastern France, then crossed the Great St. Bernard Pass to conduct further pillage in Italy before crossing the Carnic Alps back into Hungary. In 955 the Magyars tried again, and advanced into Bavaria with another army of 50,000. They were besieging Augsburg when Otto I's army approached and compelled them to raise the siege on 9 August. The next day, the Magyars gained the advantage in the first clash, but Otto arrayed his heavy cavalry in line and delivered a charge that broke the more lightly armored Magyars and drove them from the field. Otto ordered a pursuit, and inflicted such losses on the Magyars that 75 years passed before another Magyar army invaded Germany.

The decentralization of power necessary to withstand these barbarian raids had effectively ended the last remnants of Carolingian state structure on the continent. Even under the well-organized Carolingians some decentralization was unavoidable given the state of communications at the time, and the strain of resisting raids nearly eliminated central government. Widespread raids could be resisted successfully only at a local level, and any local leader who defended his region against Viking or Magyar raiders was unlikely to surrender his authority to a distant monarch he had never seen and who had done little to help him. This strengthening of local authority, and the sporadic efforts of the kings to regain their paramount position, was the most important characteristic of medieval politics.

By 1000, feudal society was well-established in most of western and central Europe. The situations in England and Italy were a little different. In England, the limited size of the realm and its well-defined boundaries made the task of the monarchy easier, and England was better-administered and more tightly-organized than other European kingdoms. In Italy, the high level of urbanization, and the economic power of the cities (few had more than 10,000 people even in the late 1200s) limited the power of rural feudal magnates. Most feudal lords in northern and central Italy had become clients of neighboring towns by the mid-twelfth century, and many actually moved to the towns.

The Normans, the *Reconquista*, and the Crusades, 1000–1200

The Normans

When the Norse chieftain Rollo, or Rolf the Granger (so-called because he was reputedly too tall to ride a horse), in 911 accepted the offer of King Charles the Fat of France to settle in the lands at the mouth of the Seine River in exchange for forgoing future raids on French lands, he founded a people who became famous for producing military adventurers. The region was soon called Normandy after its new inhabitants, who adopted many French customs, including their methods of making war. By 1050, following the same impetus that had driven their Norse forefathers to seek their fortunes outside Scandinavia, the Normans had gained considerable land and power in southern Italy.

Over the course of the next fifty years, the Normans achieved their three greatest military successes: the conquests of southern Italy, Sicily, and England. The conquest of England by William the Conqueror in 1066 is perhaps the best known of these, but is not necessarily the most spectacular. It is, however, illustrative of Norman capabilities. Harold Godwinson's Saxon army, which fought William's Normans at Hastings in October 1066, was a capable force built around a core of professional heavily armed mounted infantry called *huskarls*. The bulk of the army was militia, though, and like many militia forces, the Anglo-Saxon army was not well-disciplined. The decisive point at Hastings came when the Normans feigned a retreat (a difficult maneuver, and indicative of notable tactical skill; it was also a favorite Byzantine tactic, and may have come to William's attention through Norman involvement in southern Italy) and enticed Harold's militia to make a hasty attack. As the Anglo-Saxons charged down off Senlac hill, the Normans rallied and launched a carefully timed counterattack, routing much of the militia and forcing the remnants of Harold's army back to the hilltop, where they were surrounded and destroyed by the Normans.

While William's Normans conquered England, other Normans under the adventurer turned state-builder Robert Guiscard were gaining control of southern Italy. The conquest of Sicily by Robert's brother Roger, begun in 1060, was not completed until 1091, delayed by the demands of Norman interests on mainland Italy, where Robert was involved in conflicts with the pope, the Byzantine Empire, and the quasi-independent feudal magnates of southern Italy. In 1071 Robert captured Bari, eliminating the last Byzantine foothold in Italy. He then annexed the Duchy of Amalfi in 1073, and took Apulia in 1076. In 1081, he captured Corfu and prepared to invade the Byzantine Empire itself. He led his army into what is now Albania and besieged Durazzo, but he was opposed by the able and resourceful Byzantine Emperor Alexius II Comnenus. Robert eventually captured Durazzo (February 1082) despite the earlier loss of his fleet to a combined Venetian-Byzantine fleet. He left his son Bohemund to direct the advance overland against Thessalonica in the spring, while he responded to an urgent plea from Pope Gregory VII and returned to Italy. Bohemund had the worst of the Battle of Larissa against Alexius (1083), and a second naval encounter off Corfu was indecisive (1084). However, the death of Robert on Cephalonia (17 July 1085) compelled Bohemund to return to Italy to see to his father's dynastic interest.

Robert Guiscard's conquest of southern Italy and Sicily laid the foundations for a state that survived more than a century. During its existence it guarded the development of a remarkable culture, combining Norman, Byzantine, and Arab elements. This Sicilian Kingdom, as it is often called, also dominated the central Mediterranean, its fleets controlling the Ionian and Tyrrhenian Seas. It provided a powerful counterweight to the ambitions of the Holy Roman Empire in northern Italy, as it was also a strong ally of the papacy. The Sicilian Kingdom weakened in the latter half of the twelfth century, and in 1194 the throne was usurped by the Hohenstaufen Emperor Henry VI, who had a claim to the Sicilian throne through marriage.

THE *RECONQUISTA*

The Arab invasion of Spain, which destroyed the Visigothic Kingdom in the early eighth century, did not completely conquer Iberia. While most of Spain succumbed to the Arabs, the foothills of the Pyrenees, as well as Galicia and the northern coast along the Bay of Biscay, maintained a precarious independence. Charlemagne's conquest of the "Spanish March" along the modern French border in the northeast gave sorely needed help to the Christians in Spain. By 1000, five Christian states had arisen in Spain: (from west to east) the kingdoms of León, Castille, Navarre, Aragón, and the County of Barcelona. Barcelona was absorbed by Aragón soon after, and in 1034 the Christian cause gained an advantage by the dissolution of the Caliphate of Córdoba. Muslim-Arab Spain broke up into more than a dozen small states, which were (like their Christian rivals) as ready to fight each other as to fight their common enemy.

This situation was a great benefit to the Christian cause, which was further aided by the periodic arrival of knights from southern France, eager to fight the Muslims. The Christian reconquest of Spain, known in that country as the *Reconquista*, continued for over two centuries. The Christian progress southward was interrupted by the Almoravids, a tribal Islamic sect, who had conquered Morocco and responded to an appeal for help from the Muslim princes of Spain in 1085. Their general Ibn Tashfin defeated Alfonso VI of Castile at Zallaka, north of Badajoz, in 1086. Ibn Tashfin, though, was less successful in his campaigns against Rodrigo Díaz y Bivar, better known as El Cid, and his forces suffered heavily even though they had some successes against the Cid's armies. After Ibn Tashfin's death, Almoravid power waned and the Christians resumed their advance.

The Almoravids were overthrown by the Almohades, another tribal Islamic reform movement, in the mid-twelfth century. The Almohades also succeeded in blocking Christian conquests. By the time the Almohades conquered Morocco in 1147, the Christian states had taken both Toledo and Lisbon, and controlled the northern half of the Iberian peninsula. Over the next half-century, the Christian kingdoms made little progress in their struggle to drive out the Moors. However, on 6 July 1212, the Almohades suffered a crushing defeat at the Battle of Las Navas de Tolosa, and within forty years the kingdom of Castile (which had merged with León in 1230) ruled half of Spain. Castile blocked Navarre from contact with Muslim Spain, left Aragón in control of the northeast and east, and pushed southward to overrun the rest of Spain except for the Kingdom of Granada in the southeast. Farther west, the county of Portugal, which had become independent from León in 1140, forced the Muslims south of the Tagus River.

The warfare of the *Reconquista* was distinct from medieval warfare in northern Europe in several ways. First, the ruggedness of Iberian terrain limited the deployment of cavalry, and Spanish armies made wider and more effective use of footsoldiers than their French or German contemporaries. Second, the Spanish cavalry was notably more lightly equipped than its northern counterparts. In part, this was a result of climatic conditions, but it was also a response to the Muslim armies, whose horsemen were skirmishers and horse-archers. Most Spanish cavalrymen were *genitors*, men mounted on small, agile horses and armed with javelins, a sword, and often a shield. Many of the military lessons of the *Reconquista*, however, were ignored outside Spain.

The Christian states of Spain consolidated the gains won at Las Navas de Tolosa by the mid-thirteenth century, but because of civil strife and war among the Christian states, the kingdom of Granada remained independent for more than two centuries. King Ferdinand of Aragón and Queen Isabella of Castile brought the resources of their kingdom together, and between 1481 and 1492 conquered Granada. During this time, they undertook a major reform of their army (1483–87) and created a permanent national army. One of the leading commanders of the new army was Hernandez Gonzalo de Córdoba, later famous for his victories over the French in southern Italy. The conquest of Granada marked the entry of Spain into the ranks of major European powers.

The Crusades

The Crusades, as a great manifestation of popular religious feeling, were more than a series of military expeditions. Yet even considered as merely military operations, the Crusades were interesting for several reasons. The idea of the "crusade" probably arose in Spain, as part of the warfare between Christian and Moor for control of the Iberian peninsula. The First Crusade was launched in 1096 and led by Norman and Flemish barons who had been inspired by the preaching of Pope Urban II. Over the ensuing 150 years, there were eight crusades, but only two, the First (1096–99) and the Third (1189–92), were successful. The First Crusade captured Jerusalem and brought most of what is now Lebanon, Israel, and coastal Syria under Christian rule. Some of the Crusader states thus created did not last long: the county of Edessa was destroyed in 1144, and frontier warfare against Muslim raiders was constant.

After the fall of Jerusalem to the Turkish-Egyptian monarch Saladin in 1187, the Crusader states (which the French called *Outremer*, "[the land] across the sea") waned, dwindling to a narrow coastal strip from Tripoli south to Jaffa and eventually to a few coastal enclaves. Despite military reversals, Europeans learned much of military value from the Crusades. The Crusaders gained from their Muslim adversaries an appreciation for light cavalry and its use in irregular warfare, the importance of scouting, ambushes, and surprise, and the importance of combining shock and missile action to gain success in battle. Most importantly, the Crusades introduced eastern, and especially Byzantine, fortification techniques to western Europe. These included concentric rings of walls and round towers, both previously unknown in the west.

Feudal Warfare and the Laws of War

The Middle Ages were an unusually violent period, marked not only by warfare between kings, but by warfare among a king's subjects, and between kings and their subjects. The frequency with which feudal lords resorted to war made warfare almost constant. The concept of war as a manifestation of private feud between two rival lords produced the concept of *dampnum*. Since it was usually impractical for one lord to strike at another directly, a lord instead injured his rival by destroying his crops, ruining his fields and villages, and killing or carrying off his serfs and livestock. This concept of striking at a lord through his subjects was called *dampnum*, and its widespread use was the main reason for the relative destructiveness of medieval warfare. It also tended to transform campaigns into semiorganized brigandage.

Some authorities, notably the church and royal governments, desired to limit the mayhem, and this desire, coupled with the medieval "laws of war" produced the idea of "just war." The medieval laws of war were not written statutes, but a compilation and distillation of traditions governing the exchange of envoys, the status of hostages, the treatment of prisoners, the granting of safe passage, the implementation of truces, and the like. A just war was characterized by three conditions: it was either a war against barbarians or infidels (in which case the laws of war among Christians did not generally apply), a war against rebels or outlaws undertaken by a legally constituted authority, or a defense against aggression. A war begun under any other conditions was unjust, and a war undertaken as part of a private quarrel was particularly unjust.

These conditions were applied most rigorously to lesser magnates, or to those who owed allegiance to a monarch. Matters were less clear-cut when dealing with disputes between monarchs, and the concept of just war at that level was honored more in the breach than the observance.

The Mongol Conquests and the Mongol System of Warfare, 1190–1280

The Mongols, who conquered most of Asia and half of Europe in less than 100 years, were the most efficient and ruthless war-makers of the Middle Ages. They were brought from obscurity to greatness by their chieftain Temujin, later called Genghis Khan. Genghis transformed the army of the nomadic Mongols from a simple force of horse-archers into a highly organized military force, adding armored lancers, an important shock element, to increase combat power. The Mongols were, by virtue of their harsh environment, hardy warriors, and under the command of a skilled general like Genghis, or his lieutenants Chepe Noyon and Subotai, an army of such men could achieve wonders.

Temujin unified the tribes of Mongolia over a sixteen-year period (1190–1206) and accepted the accolade-name of Genghis Khan ("Perfect War Emperor" or "Supreme Emperor") from the tribes. Over the next ten years, the Mongols overran most of northern central Asia, absorbing the empires of the Western Hsia, Chin, and Kara-Khitai. Genghis conquered the Khwarezmian Empire between 1218 and 1224, and by the time of his death in 1227 ruled an empire that stretched from the Urals to the Pacific and reached south into India and touched the Persian Gulf. Genghis' successors completed the conquest of the Chin (1231–34) and Sung Empires in China (1234–79), and between 1237 and 1241 invaded eastern Europe, easily conquering Russia, Poland, and Hungary before the death of Khan Ogotai compelled them to withdraw. This was their high-water mark in Europe, but while they never again passed west of the Ukraine, they remained in Russia for three centuries.

Even the decline of the Mongol state, and its fragmentation into four major parts (the Khanate of the Golden Horde in Russia, the Il-Khan state in Persia, the Chagadia or Ili state in central Asia, and the Khanate of the Great Khan in Mongolia and China) did not end Mongol military power. The career of Tamerlane (Timur-e-lenk, or Timur the Lame), a Tartar who became chief of the Jagatai Mongols, is the greatest example. For over twenty years, from 1381 until his death in 1405, Tamerlane and his armies, operating from his capital at Samarkand, were the terror of the Middle East. Less an empire-builder than a plunderer, he conquered Persia and parts of central Asia, and invaded India (1398–99), Syria (1400), Iraq (1401), and Anatolia (1402). When in Iraq he sacked Baghdad, and in his invasion of Anatolia so mauled the Ottoman Turks at Angora (30 June 1402) that it took them years to recover.

The High Middle Ages and the Origins of Modern Warfare, 1150–1450

THE ENGLISH

By the middle of the twelfth century, feudalism and the manorial system had reached their greatest level of sophistication. In feudalism, an individual placed his services at the disposal of a more powerful person in exchange for that person's protection and access to his economic resources (the English word *lord* is derived from the Anglo-Saxon *hlaford*, or loaf-giver, meaning someone who provided food to his followers and supporters). The manorial system was a parallel to feudalism, and served to tie the holding of land (a fief) to obligations for military service (varying widely, but often fixed at a biblical 40 days). The gradual recentralization of political authority, which began to reach noticeable proportions in the twelfth century, made many rulers realize how little control they had over the military establishment. Their military vassals, in turn, became less eager to leave their homes and families for a campaign, even if for only six weeks.

The result of these two trends was the development of a money payment in lieu of service. Called shield money, or *scutage* in England, it allowed the knights and men-at-arms to remain happily at home, and it allowed the rulers to hire armies for as long as they wanted and as long as they could afford the expense. This practice began in England about 1176, where it was especially popular because of the difficulties of providing garrisons for the extensive Plantagenet holdings on the continent. The development of *scutage* indicates a revival of a money economy from one based on barter, which in turn shows renewed trade and a growth of urban centers. The institution of *scutage* proved so popular that by the time of King John's war against Philip II Augustus of France early in the first decade of the thirteenth century, more than half the troops in his army were mercenaries. The practice spread to France and the Low Countries, and eventually to Germany.

The widespread use of *scutage* in England led to another innovation in the late thirteenth century. King Edward I, exasperated by the continual political disarray in Wales, invaded and subjugated that country in a series of wars in the 1270s, 1280s, and 1290s. These campaigns were long-term operations, often lasting for several months, and they took place far enough from Edward's English bases that the traditional military institutions, even fortified by *scutage*, were inadequate to provide him with the necessary forces to conquer, and then garrison, his new domains.

Edward began two practices to give him the forces he needed. First, he instituted muster and review to raise footsoldiers: each county was to muster all its militia, essentially every free male in reasonable health from his late teens to his early sixties. The county officials (usually the sheriff) were then to review these men and select a number of them (specified by the king) to serve on a campaign. They were supposed to select the most eager and able men, arm them from the county armory, and send them to the army's concentration point. These men were paid regularly, usually three pence a day for footsoldiers (a rate comparable to what a skilled or semi-skilled worker would earn), and they were often provided with rations. Nonmilitary specialists like coopers, miners, carpenters, and blacksmiths were also raised in this fashion. Although the cost was high, England was able to manage the administrative burden imposed by this system, and the payoff in military efficiency was considerable.

Second, Edward I also began the practice of raising troops through indentures. An indenture was a written agreement between the king and one of his subjects (usually a man-at-arms or a knight) to provide the king with a specific number and type of soldiers, at a particular date and place, and for a specific period of time (at first three months, later raised to six, and then to a year). The indenture also specified how the men were to be equipped, how much they were to be paid, and so forth. The king agreed to pay a set sum of money to the subject, and the subject promised to get the specified troops to the right place at the right time.

These two methods of raising troops enabled Edward I and his successors to raise and maintain what amounted to standing armies, removing military service from the realm of traditional duties and obligations and placing it on an economic footing. Since all soldiers, including those holding indentures, were awarded a share of any plunder gained, or ransom received for captives, military service was popular, and raising troops was not a problem.

Edward and his grandson Edward III refined this system, so that when Edward III began the war with France in 1337 (the Hundred Years' War), the English possessed a military system superior to any in Western Europe. Edward III even began to issue indentures for supplies (mainly food), and began centralized, government-supervised manufacture of weapons, mainly at the Tower of London. These weapons were used to equip the troops raised by muster and review (a system that survived until 250 years later). This military system provided the English monarchy with armies that—while small (rarely more than 12,000 men, and often half that)—were cohesive, determined, and manned by a large proportion of experienced professionals, skilled not only in fighting, but also in the art of making war.

THE *CONDOTTIERI*

While the English were refining their systems of indenture, and muster and review, the Italians in the city-states of northern and central Italy were finding that warfare was costly and inefficient when waged by citizen levies. By the middle decades of the fourteenth century, most Italian city-states had begun making contracts (*condotta*) with independent soldier-entrepreneurs called *condottieri*. These contracts specified the number and types of troops, the duration of the contract, and the rates of pay. Many of the early *condottieri* were foreigners, including Germans, Frenchmen, and the notable English soldier Sir John Hawkwood, who went to Italy when he was left unemployed by the Peace of Brétigny in 1360.

The *condottieri*, like most mercenaries, were more interested in getting paid than in killing their fellow *condottieri*, who happened for the moment to be enemies. Consequently, Italian warfare during the last century of the Middle Ages was characterized more often by *ruses de guerre*, clever stratagems, and a great deal of maneuver than by actual fighting. However, Machiavelli's claim that the *condottieri* were militarily inept and commonly fought battles where armies numbering thousands suffered only a few fatal casualties is untrue.

The *condottieri* did professionalize the conduct of war. Mercenary bands, once assembled, tried to remain together, and soldiers who fought together for years developed considerable skill, unit cohesion, and efficiency. The officers in particular began to think and write seriously about the methods of waging war and fighting battles. These trends produced well-disciplined, well-organized forces, and increasingly sophisticated battle tactics. Further, the city-states that hired *condottieri* found it worthwhile to retain the same troops even in times of peace. This provided greater security for the states and, more importantly, created stronger ties between the mercenary and his employer. By the middle of the fifteenth century, the largest Italian states (Venice, Milan, Florence, and the Papal States) had all developed permanent standing armies of *condottieri*. These soldiers were not like their predecessors, who had been free agents; the *condottieri* of the mid-1400s were usually tied to a particular employer, and were *condottieri* only in the sense that they were hired through contractors, rather than directly by agents of the state. Those few *condottieri* who were still independent, like Federigo da Montefeltro or Sigismondo Malatesta, had their own principalities as power bases.

This institutional change in the status of the *condottieri* paralleled changes in warfare in Italy, especially in tactics and force structure. Early *condottieri* armies were dominated by the mounted man-at-arms, although some of the foreign *condottieri* like Sir John Hawkwood continued to rely heavily on infantry. The dominance of mounted troops became more widespread between 1390 and 1430 under Alberico da Barbiano and his protégés, Muzio Sforza and Braccio da Montone. Sforza, like his patron, favored the employment of troops in large bodies, unleashed at just the right moment in a mass attack. Braccio da Montone, in contrast, believed in operating in smaller units, which could maintain steady pressure on the enemy and be relieved by fresh troops when they became exhausted. Montone's system also allowed junior officers to exercise greater initiative. These rival "systems" and their followers, named "Sforzeschi" and "Bracceschi" after their patrons, dominated Italian military practice in the middle decades of the 1400s. Both systems had merit under different conditions.

By the later years of the 1400s, especially after the Peace of Lodi in 1454, the advent of infantry firearms began to change the face of battle. The campaigns of the late fifteenth century showed increased use of field entrenchments, and armies employed higher proportions of infantry, light cavalry, artillery, and engineers (Fig. 1). These changes, contemporary with the last stages of evolution from the old, entrepreneurial *condottieri* to the new, state-controlled mercenaries, foreshadowed the organization of French and Spanish armies in the early sixteenth century, an Italianization of European warfare that is often overlooked.

THE SWISS

The political history of the Swiss people began in the late thirteenth century, but their importance as a military force in central Europe was not immediately apparent. The Hapsburgs, feudal overlords of the Swiss cantons, were their first military opponents. In a series of conflicts lasting nearly 200 years, the Swiss Confederation (at first only the three cantons of Uri, Schwyz, and Unterwalden) freed themselves from Austrian domination.

The Swiss won their first victory at Morgarten on 15 November 1315 when their army of 1,500 spearmen and archers ambushed and routed an Austrian army of about 8,000 on a narrow battlefield between mountains and a lake. Seventy years later, an Austrian army 6,000 strong under Leopold III of Swabia was defeated at Sempach on 9 July 1386 by 1,600 Swiss. Although the Austrians fought dismounted and enjoyed initial success, they were soon exhausted by their exertions and fell victim to the more agile (and less heavily armored) Swiss halberdiers and pikemen. The Swiss warred again with Austria in 1415, 1460, and 1499. It was not until the Treaty of Basel on 22 September 1499 that the Swiss at last won independence.

During this period, the Swiss had also fought other neighbors; the French, Burgundians, and Milanese went to war with the Swiss between 1339 and 1450. Most of these clashes served to enhance the Swiss military reputation. A case in point was the Battle of St. Jakob, near Basel, fought on 24 August 1444. A French army of 30,000 men under Dauphin Louis (later King Louis XI) invaded Swiss territory as part of an alliance with Frederick III of Austria. This army was opposed by a Swiss force of 1,500 that fought to the last man, inflicted 3,000 casualties, and so disordered the French army that the French withdrew. The French were so impressed with the Swiss determination and ferocity that they changed sides and raided Frederick's domains in Alsace.

The Swiss armies, which performed so well in battle, differed in several ways from typical European armies of the day. First, since the troops were raised from a population of freeholders and town-dwellers of modest means, most of them were lightly armored infantrymen; Swiss armies contained few if any cavalrymen. The weapons of the infantry soldier changed over time. Initially, about one-third of the men used bows or crossbows, and the rest spears or halberds (a spear with an axe-blade at the end). As encounters with cavalry became more common, and as Swiss discipline and tactics improved, they adopted the pike in increasing numbers, eventually employing an 18- to 21-foot version. Swiss pikemen fought in large, dense columns and were trained to move at relatively high speed. Records of the period often refer to the speed with which Swiss troops moved to the attack, and the fearsome impact of their charge.

Fearsome as these pike phalanxes were, they were not invulnerable. They could be overwhelmed by superior numbers, as happened at St. Jakob, and they could be beaten by clever tactics and astute combat leadership. A Swiss army invaded the northern territories of Milan in 1422, as part of a sporadic border conflict that continued for years. About one-third of the Swiss army was pike-armed, while most of the rest had halberds, and only a small portion were armed with crossbows. At Arbedo, near Bellinzona, the invading Swiss met a Milanese army under the *condottiere* Francesco Bussone di Carmagnola on 30 June. Carmagnola's mounted attack was repulsed with heavy losses and, reverting to Hawkwood's tactics of 40 years before, he ordered his men-at-arms to dismount and fight on foot, covered by the fire of his crossbowmen. The dense blocks of Swiss troops suffered heavily from the Milanese crossbowmen, and hastily withdrew. So sobering was this defeat for the Swiss that they did not fight Milan again until 1478.

The strength of the Swiss lay in their effective combination of offensive power, from their halberds and pikes, with the ferocity and determination for which they were renowned. The latter characteristic was a product of strong traditions of community and local loyalty: no Swiss soldier would dare let his comrades down, and the *esprit de corps* that resulted gave Swiss armies remarkable cohesion. Swiss tactics, while aggressive and remarkably effective in many circumstances, did have limitations. Their pike columns could move rapidly, but were at a disadvantage on uneven ground, and were vulnerable to skirmishers. A lack of appreciation for troops equipped with crossbows or firearms also showed tactical narrow-mindedness. Still, the Swiss were formidable opponents; even after they abandoned an aggressive foreign policy in the 1520s, their

Figure 1. The infantry carried long spears to kill the horses, forcing the riders out of action. (SOURCE: Iconographic Encyclopaedia)

mercenaries were in high demand, noted for their zeal and steadfastness.

The End of Medieval Warfare: The French Army of the *Compagnies d'Ordonnance*, 1445–90

The Hundred Years' War was dominated for the first 90 years (1337–1427) by the tactical superiority of the English. The French, despite numerous efforts to reorganize their armies and regularize their armament and quality, were unable to match the English in the field unless they were led by an unusually talented commander like Bertrand du Guesclin. After the defeat at Agincourt, where an English army of 6,000 under King Henry V defeated a French army of over 30,000 on 30 October 1415, French military fortunes declined steadily. By the time of Henry V's death in 1422, the English and their Burgundian allies held Paris and most of France north of the Loire River.

The advent of Jeanne d'Arc (Joan of Arc) and her leadership of the relief of Orléans from the Duke of Bedford's siege in 1429, provided the spark the French needed for revitalization. Within six years, the Burgundians had switched sides, the French had retaken Paris, and the English were on the defensive. This success, and the victories of the ensuing ten years, were accomplished by mercenary bands led by experienced professional soldiers. Many of the soldiers and their commanders were foreigners, and many of them were not far removed from brigands. In 1445, taking advantage of one of the periodic truces during the war, King Charles VII decided to place the army on a permanent footing.

On 26 May 1445, the King declared the *Grande Ordonnance*, which established a standing army. The new law granted official sanction to certain mercenary captains and made them royal officers. The soldiers of the *compagnies* were organized into lances, each of which contained a man-at-arms, a squire, two archers, and two pages (one for the archers and one for the man-at-arms and the squire). Each member of the lance had one horse, and the French authorities counted each lance as four combatants and six horses (although there were often extra horses, and some lances were understrength in manpower). The lances, which at first totalled about 1,800, were grouped into companies of 30 to 100 lances. The company commanders were paid on the basis of the number of lances fit and present, so royal inspection was frequent and precise.

Initially, the soldiers and officers of the *compagnies* were a surprisingly cosmopolitan group, only two-fifths being native Frenchmen. Further, many of the first group of leaders of the *compagnies* had unsavory careers, but transforming them into royal officials placed them under royal control, where their unruliness could be contained. To a large degree, the creation of the *compagnies d'ordonnance* paralleled the transformation of the *condottieri* from military entrepreneurs to state employees and, as a manifestation of the growing authority governments exercised over their territories and populace, serves to mark the end of the Middle Ages and the beginning of Early Modern Europe.

Meaning of Medieval Warfare

The military history of the medieval period is marked by two major characteristics. First, the control of military institutions was decentralized; it rested in the hands of local leaders rather than central or national governments. Second, these institutions were based on tradition and informal agreement, not on written law or formal decree. Similarly, command and organization were generally informal and irregular, based on groups of personal followers, or on fighters drawn from a particular region. Rank in the modern sense was virtually unknown, and a man who was commander of an army in one campaign could find himself, without having suffered disgrace, commanding a mere company in the next.

The regularization of these haphazard methods of making war, which began to spread through Europe in the last half of the fifteenth century, effectively ended medieval warfare. Regularly raised standing armies, firmly under the control of the central government, were a far cry from the feudal hosts of the eleventh or twelfth centuries. If nothing else, they could stay in the field for more than the traditional 40-day span of feudal obligations. Nevertheless, several crucial facets of medieval warfare had longlasting effects. The tradition of command by titled aristocrats lasted well into the modern era, and the traditions of chivalry and culture developed by the knightly class during the period had a long effect on European culture and military traditions.

DAVID L. BONGARD

SEE ALSO: Arab Conquests; Attila the Hun; Byzantine Empire; Charlemagne; Crusades; Edward I; Edward III; Feudalism; Genghis Khan; Gonzalo de Córdoba; Hundred Years' War; Khalid ibn al-Walid; Mongol Conquests; Normans; Persian Empire; Ruses and Stratagems; Tamerlane; Turkic Empire; Vikings.

Bibliography

Allmand, C. T., ed. 1973. *Society at war: The experience of England and France during the Hundred Years' War.* Edinburgh: Oliver and Boyd.
Beeler, J. H. 1971. *Warfare in feudal Europe.* Ithaca, N.Y.: Cornell Univ. Press.
Bloch, M. 1961. *Feudal society.* Trans. L. A. Manyon. Chicago: Univ. of Chicago Press.
Burne, A. H. 1956. *The Agincourt War: A military history of the latter part of the Hundred Years' War from 1369 to 1453.* London: Eyre and Spottiswoode.
———. 1955. *The Crécy war: A military history of the Hundred Years' War from 1337 to the Peace of Bretigny.* London: Eyre and Spottiswoode.
Contamine, P. 1984. *War in the Middle Ages.* Trans. M. Jones. London: Basil Blackwell.
———. 1972. *Guerre, état, et société à la fin du Moyen Age.* Paris: La Haye, Mouton.

Glubb, J. B. 1963. *The great Arab conquests*. London: Hodder and Staughton.

Hewitt, H. J. 1966. *The organization of war under Edward III, 1338–62*. Manchester: Univ. of Manchester Press.

Keegan, J. 1976. *The face of battle*. New York: Viking Press.

Keen, M. H. 1965. *The laws of war in the Late Middle Ages*. London: Routledge and Kegan Paul.

Kwantern, L. 1979. *Imperial nomads: A history of Central Asia, 500 to 1500*. Philadelphia: Univ. of Pennsylvania Press.

Mallet, M. E. 1974. *Mercenaries and their masters: Warfare in Renaissance Italy*. Totowa, N.J.: Rowman and Littlefield.

Norwich, J. J. C. 1967. *The Normans in the South, 1016–1130*. London: Longmans.

Oman, C. W. C. 1924. *The art of war in the Middle Ages*. 2 vols. London: Methuen.

Ostrogorsky, G. 1969. *History of the Byzantine State*. Trans. J. Hussey. New Brunswick, N.J.: Rutgers Univ. Press.

Runciman, S. 1964–67. *A History of the Crusades*. 3 vols. Cambridge and New York: Cambridge Univ. Press.

Smail, R. C. 1956. *Crusading warfare (1097–1193)*. Cambridge: Cambridge Univ. Press.

Trease, G. 1971. *The condottieri*. New York: Holt, Rinehart and Winston.

Turnbull, S. 1985. *The book of the medieval knight*. London: Arms and Armour Press.

Vale, M. G. A. 1981. *War and chivalry: Warfare and aristocratic culture in England, France, and Burgundy at the end of the Middle Ages*. Athens, Ga.: Univ. of Georgia Press.

HISTORY, MILITARY

At its core, military history is the analytic review of wars, campaigns, battles, and military institutions, including their economic, political, and social foundations and effects, and relationships between military and civil authorities.

A distinction may be made between professional, academic, and popular military histories.

At its most fundamental level, professional military history (or "operational history" as it is sometimes called) examines past conflicts in order to derive insights that might guide decision makers in preparing for contemporary or future warfare. In that capacity, it moves in the direction of military theory. To the extent that theoretical touchstones like the principles of war are subordinated to historical example, they belong in the realm of didactic military history rather than undiluted doctrine.

It is possible, while often not easy, to distinguish military history from general history. When the latter deals with armed conflict and military forces, these subjects have traditionally been treated as extraneous to the normal flow of events. As such, they are not discussed extensively.

Military history, as understood here, highlights such factors as the strategy, tactics, and logistics of a campaign; skill and knowledge of the commanders; characteristics or influence of weapons systems; and the composition and combat capabilities of the units engaged. Auxiliary, noncombatant services, such as advances in military medicine, the administration of field hospitals, and the prevention of tropical diseases, are also topics covered in military history studies, as are topics as diverse as arrangements for provisioning an army in the field, effects of various transport systems upon rates of advance, and techniques and devices for signals communications.

The military historian also analyzes the instruments that nations devise to ensure success in defending or advancing vital national interests. Such matters have long been considered by academic historians to be unworthy of attention, but this view has changed in the last 35 years.

In addition to critical accounts of wars, campaigns, and battles, military history includes institutional and administrative studies of armed forces that analyze their bureaucratic structure, managerial efficiency, and relationship to government and society. The social and economic background of enlisted personnel and officers and the cohesion of units also may be discussed in monographs, which do not necessarily include accounts of battlefield activities.

The academic variety of military history in particular includes assessments of the effects of war upon peoples and civilian institutions, participants and neutrals alike. The nature of modern "total war," harnessing all of the social and economic resources of the nation and endangering its very survival, has brought military history out from the constricted descriptive mode and into a broader context—what is now known as "war and society" studies. Academia has accepted the latter variety of military history. The concluding section of this article shows that some war and society scholars have strained so hard to avoid the "stench of cordite" that subscribers to military journals publishing their work may well wonder whether the editors have forgotten that combat is the *ultima ratio* of armies, navies, and air forces. On the other hand, the better operational historians are now borrowing from related disciplines to present a more rounded view of the relative military effectiveness of competing societies and economic systems.

On another level are unit histories. These are compiled from orders, dispatches, phone or radio message transcripts, logbooks, unit war diaries, citations, awards, and personal diaries of the fighting men. The least useful to military history are those composed simply to serve as souvenirs or memorials. As such, they may take the form of scrapbooks or captioned photo albums, patching together the bits and pieces of rosters, documents, and memorabilia with a cursory chronology of events, including sports and social festivities. The best of this type may be valuable as source material on daily routine and morale-building mechanisms. When the designated unit historian is also a practiced military historian, and affiliated with the unit, he may relate the unit's activities to events. British specialists like C. T. Atkinson, Sir Frederick Maurice, H. C. Wylly, and Dudley Ward have produced the best examples of this genre. Even then, the writer commissioned to memorialize the unit may hesitate to include incidents reflecting unfavorably upon the unit's performance. The British, French, and German models usually comprise valuable minihistories of the various wars, whereas American divisional, regimental, and battalion annals, with a few notable exceptions, tend toward the cut-and-paste commemorative approach.

Another variant, official history, has its own special credibility problems. The origin and development of this subgenre is discussed in the next section. Since the second quarter of the nineteenth century, defense establishments of major powers have commissioned official histories to distill lessons from past conflicts; to objectively record the role of the nation's armed forces, their triumphs, and errors; and to instill a sense of continuity and tradition among the troops. The last is an extension of the unit history's role to embrace the entire fighting force. Up until World War II, there was too often a built-in bias: to prove the correctness of pre-war (or current) doctrine and to gloss over the mistakes of certain commanders or statesmen. The 1939–45 conflict produced official histories that employed the talents of trained historians who were asked to do their best to ascertain exactly what had transpired on the battlefield regardless of who won or lost.

Military memoirs and biographies highlight the military career of the subject. The subject's early life, training, emotional experiences, and so on, are presented only as background to his martial life. A few memoirs, such as those of Confederate artillerist Edward Porter Alexander and Wehrmacht Field Marshal Manstein, are depersonalized to the extent that they read like operational studies of the memoirist's campaigns rather than vivid personal recollections. A fascinating hybrid between the memoir and the troop orientation lecture can be found in Erwin Rommel's collection of World War I battle vignettes describing the exploits of his mountain infantry company on the Italian front. Finally, works such as G. F. R. Henderson's *Stonewall Jackson and the American Civil War* are really strategical studies of an epoch's prevailing doctrine as seen through the eyes of a preeminent practitioner and theorist.

A good deal of the "popular" military historical writing done for the enthusiast or "buff" is devised more to entertain than to inform. This ephemeral literature includes patriotic or macho tales of heroic deeds, a phenomenon of the nationalist ferment in the latter half of the nineteenth century that persists today in magazines appealing to would-be mercenaries, both armchair and active, and gun collectors. There is also the heavily illustrated coffee-table book or magazine article that is little more than extended captions for color pictures of uniforms, decorations, accoutrements, weapons, and fanciful battle scenes.

However, one should not dismiss the output of hobbyists as irrelevant to the advancement of the form. The best of the semiamateur magazines written for (and often by) subscribers who are recreational war-gamers, modelers, collectors of militaria, and fanciers of military dress are thoroughly researched and documented. Dedicated and diligent buffs often add insights to their limited field that trained historians might utilize.

Origins of the Genre and Early Practitioners

From classical times through the seventeenth century, history was viewed as the doings of kings, nobles, ministers, and cabinet members. Because in ancient times warfare was endemic and preparation for it all-consuming, the written record is dominated by the martial exploits of rulers and lords. Scribes were not self-consciously writing military histories, but simply reflecting the dominant theme of societies organized for and preoccupied with war. Many of these works are didactic in that they set out to demonstrate the military genius of the subject ruler or the weakness resulting from the treacherous counsel of his or her internal enemies. Although one may learn of the nature of ancient armies and navies as well as their techniques and stratagems from these court histories, it was not until the advent of the "scientific" military historians at the end of the nineteenth century that students of military affairs had reliable guides to guide them through the "minefields" of the half-mythological tales of ancient and medieval chroniclers.

The histories of the wars of antiquity have come down to us through the writings of the Old Testament scribes and aristocratic intellectuals such as Herodotus, Thucydides, Xenophon, Polybius, and Arrian. Although their subject was warfare, they were writing for posterity rather than for the edification of future warriors.

Military biography and memoirs are among the earliest military histories. Ancient and early medieval historians wrote the "life and times" of great rulers who were, perforce, commanders as well. Although statesmanship was stressed, this often was seen as the diplomatic face of war, anticipating late nineteenth-century writers who looked beyond the battlefield to the cabinet chambers and ministries.

The Old Testament oral tradition, as written centuries after events, has left a sufficiently detailed record of bloody encounters among the great and small powers of the Fertile Crescent to enable modern military critics to reconstruct many of the pivotal battles among the Hebrews and their neighbors.

Herodotus, considered the father of history, wrote of the Persian wars in an epic-heroic vein. The history of Thucydides presents an analytic and critical treatment of the Peloponnesian Wars. Thucydides demonstrates a professional familiarity with strategy, tactics, and military organization and a scholar's concern with weighing conflicting sources in fixing relative numbers and positions as well as with explaining the reactions of the troops in human terms.

The campaign histories of Julius Caesar are replete with the dynamics of the various engagements but are obviously biased to present the most favorable picture of the author/protagonist/historian, and the student of the conquest of Gaul or the Roman civil wars must consult independent sources from archaeology and bardic sagas to obtain objective perspectives of Caesar's version of the events.

The medieval and Renaissance periods produced manuals on warfare, such as Maurice's *Strategicon*, Leo VI's *Tactica*, and Machiavelli's *The Art of War*, as well as renewed interest in Vegetius' *De Re Militari*. One must not overlook the ancient Chinese general Sun Tzu's treatise, *The Art of War*, which has served as a guide to Asian warlords for two millennia. But these were surveys of contemporary practice with an occasional snippet of relevant history rather than histories proper. To the extent that historical examples were included, they were often anachronistic, like those medieval miniatures depicting Joshua's Israelite warriors clad in suits of gothic armor. The authors viewed the ancient commanders as their contemporary colleagues in arms. Nonetheless, their historical examples help to illuminate both earlier and contemporary methods.

Although far from military history, the chronicles of the medieval and early Renaissance campaigns, such as the reminiscences of Joinville about St. Louis during the Seventh Crusade (1248–54) and Froissart's account of the first phase of the Hundred Years' War, are generally the only sources available. While conveying the authentic atmosphere of the times and presenting vivid portraits of the leading personalities, these chatty narratives are poor guides to details of battles or the size and composition of armies.

This is true not only of European writers but also of those in the Middle East, where the rise of Islam was writing a new and important chapter in military history that would be ignored in Europe until modern times due to a lack of militarily astute witnesses and the paucity of nineteenth-century scientific historians interested in sifting the dross from the early Arab commentators' accounts. The early Muslim historians, eager to record the great events of the Arab conquests of the seventh through ninth centuries, set out the various conflicting versions of the campaigns without evaluating their relative merits. They left this to their reader, with the admonishment, "And Allah knows best."

The Renaissance in the Mediterranean littoral of Europe witnessed a rediscovery of the texts of Vegetius and Maurice as well as the wars and campaigns that inspired them. Machiavelli's *Art of War* (1521) was more concerned with contemporary practice than the study of the military past. However, it did examine and embellish the arguments of the Roman historian Livy in support of a citizen army of native troops rather than mercenaries like the *condotierri*. Machiavelli's treatise also mines the works of Vegetius, Frontinus, and Polybius to lay out the precise organization, battle formations, logistical arrangements, campsites, chain of command, and so on of the Roman armies. His purpose was to hold up that ancient fighting machine as an ideal model for the forces of the Florentine city-state. He dismissed the impact of modern artillery upon his Roman paradigm, since he estimated, erroneously, that the guns of his day were not much more mobile or effective than were the Roman siege engines.

Transition from War Histories to War Studies

It was not until the Enlightenment of the late seventeenth and eighteenth centuries that the "Great Captains" began to lay out their military maxims, based on concrete experience, so as to instruct the novice officer in the ways of war. The works of Marshal de Saxe, Chevalier de Folard, Guibert, Count Raimondo Montecuccoli, Henry Lloyd, and Frederick the Great—part military memoir and part distilled wisdom—are not so much military histories as rules and guidelines (practical hints for day-to-day troop management rather than principles of war) that assume that the reader is familiar with the political and technological context. Thus it was the immediate experience of contemporary wars rather than an idealized past that provided the ore for the metal of military wisdom.

Lloyd's *A History of the Late War in Germany* (1766) is widely regarded as the most influential elucidation of embryonic principles of war in a historical context. He admonished the reader that the best strategy was the slow and cautious advance along a short and well-protected line of communication of an undivided force that would, if possible, avoid battle. If a fight became necessary, Lloyd's intention was to lure a fragmented and ill-prepared enemy to attack well-prepared defensive positions. Lloyd was concerned more with preconceived notions of the character and qualities of physical force and the best methods of commanding troops than he was with recording accurately the progress of the Seven Years' War.

Perhaps the first truly modern military historian, as Peter Paret has pointed out, was Gerhard von Scharnhorst, who in 1797 published a history of the War of the First Coalition. Scharnhorst's avowed purpose was to identify and analyze crucial differences between the old monarchies and revolutionary France as demonstrated in their ways of war. Paret notes the distinction between Scharnhorst's treatment and that of other writers who found the key to French victory in such special techniques as the *levée en masse* or skirmishing. The pioneering German military educator noted instead that France enjoyed a more favorable strategic position, greater numbers, a unified political and military command, and stronger motivation. He also traced the origins and development of the French army's organization and methods and analyzed the strategy of the French revolutionary commanders as derived from the politico-military context. As an object lesson for his own Prussia of the early 1800s, Scharnhorst identified the psychological and social factors that generated superior moral force in a free political system that encouraged the participation of the common folk. Scharnhorst, who was the real founder of the Prussian German general staff system that dominated warfare for a century, was the first of a distinguished line of Prussian German military historian-theorist-generals. His major successors were Moltke and Schlieffen.

Emergence of Professional Military History in the Nineteenth Century

Scharnhorst's pupil, Karl von Clausewitz, who wrote from 1802 to 1832, is best known for his magisterial treatise *On War*. Besides distilling profound and widely applicable military truths (*not* readily memorized maxims) from the campaigns of Napoleon, he cautioned the would-be didactic military historian to examine his selected illustrative historical examples in depth to make certain that the circumstances do not invalidate the lesson derived. Book 7 of this work briefly surveyed the methods of making war employed by different European societies since ancient times. It concluded that one should not seek a single theory of war valid for all times, but that several different theories were necessary to account for the distinctive features of each epoch. The only immutable generalizations that could be sustained were those governing the psychological bases of human behavior under stress. His own study of the Thirty Years' War demonstrated how the men of the early seventeenth century behaved on the battlefield according to the dictates of economic and technological circumstances, as well as their political and religious motivations and mind-sets.

Next, Clausewitz's contemporary and fellow veteran of the Napoleonic Wars, Antoine Jomini, drew upon a thorough survey of 30 campaigns of Frederick and Napoleon in order to deduce maxims and principles to guide apprentice military leaders. In that sense, Jomini's work was a logical extension of Lloyd's précis mentioned above. Much has been made of Jomini's deleterious effect upon British, Continental, and American commanders of the nineteenth century who slavishly followed his precepts without taking into account conditions of terrain, social background of the troops, and the character of the political and military leadership. However, Jomini qualified his published lectures (*Précis de l'art de la guerre*) by recognizing the importance of the personality of the general and the esprit of the men, as well as differences in the political impetus among wars. Since his works were clearly written and well organized, the newly established military educational institutions formed under the aegis of German-model general staff systems adopted the books of Jomini and his imitators as texts.

In Britain, a new precision was attained in the writing of military history with the publication of Sir William Napier's *History of the War in the Peninsula* (1828–1840). Based upon an exhaustive perusal of orders, dispatches, war diaries, and his own combat experiences in the campaign, Napier leavened his battle narrative with observations that fell between the precepts of Jomini and the intangibles of Clausewitz. So far as British military education was concerned, several prominent Staff College professors adapted Jomini's approach to their own uses, notably Maj. Gen. Patrick MacDougall, whose *Theory of War Illustrated by Numerous Examples from Military History* (1856) was a digest of Jomini, liberally sprinkled with examples from Napier, and represented the first in a line of British casebooks on military history. The most well known and durable of the didactic surveys was that done by Gen. Sir Edward Bruce Hamley, who provided the British Staff College (founded in 1858 after the exposure of British army shortcomings in the Crimean War) with its first "scientific" text consisting of heavily illustrated principles covering a wide range of problems confronting the apprentice field commander. His *The Operations of War Explained and Illustrated*, first published in 1866, went through seven editions (the last in 1922), taking Hamley's thematic campaign analyses through World War I. The book not only amplified Jominian precepts but also concurred with the Swiss theoretician's view of history as a treasurehouse to be plundered for proof of predetermined rules, rather than a body of knowledge to be studied for whatever insights it may happen to reveal. Hamley differed with MacDougall in questioning the relevance of combat predating the Seven Years' War to contemporary practice.

Late nineteenth-century British military historian George F. R. Henderson, also a lead instructor at the Staff College, granted that *Operations* was an excellent teaching device but believed Hamley omitted crucial factors: the commander's leadership qualities and troop morale. While Henderson continued the emphasis upon tactical battle analysis, his work in this area is exemplary, bringing the student into the center of the action and to the seat of decision. This utilizing the vantage point of a single commander, with only that knowledge he possessed, was called the "applicatory" method of studying military history. Henderson emphasized the value of analyzing the neglected American Civil War at a time when most of European and America's service schools were digesting and regurgitating lessons from the great Prussian victories over Denmark (1864), Austria (1866), and France (1870). Henderson exploited fully his insights into human nature and the power of personality in war in his 1898 biography of Stonewall Jackson, which is in reality a personalized operational study of the first half of the Civil War in the eastern theater.

Part of the reason that Europeans were transfixed by the German Wars of Unification was the mystique of the German general staff under the direction of the great strategist, soldier's schoolmaster, and military historian Helmuth von Moltke. The latter's perfection of the general staff system included the creation of a historical section.

This team of official researchers was entrusted with compiling the authorized, definitive histories of all of Germany's wars. These projects were the forerunners of all official history and, much as the German war directorate was copied by the leading nations of Europe, so was their history-writing function. Until the end of World War I, official history as written by the various armed forces was technically precise, highly detailed, and careful neither to stain national heroes nor to question the correctness of current military doctrine.

Since the study of warfare had been preempted by the military establishment, German academic scholars generally shied away from it. There were a few notable exceptions. German professional historians, led by Otto von Ranke, originated the modern methodical and scientific study of the past. In the mid-1800s, one of them, Friedrich Wilhelm Rüstow, dared to investigate the history of infantry (*Geshichte der Infanterie*) using the new critical techniques.

Inspired by Rüstow's independent analytical war studies and prompted by Clausewitz's dictum that war was but an extension of politics into a violent medium, Hans Delbrück's contribution to military history is arguably the most significant development in the field. Delbrück is generally considered the archetypical military historian. A highly trained and skilled professor of history, Delbrück was no mere armchair pedant. He had seen action in the Franco-Prussian War and was appointed as tutor to Waldemar, son of Crown Prince Frederick, where he was given wide access to army leaders and combat archives.

Appointed as Privatdozent to the University of Berlin in 1881, Delbrück began to concentrate upon the evolution of warfare as an extension of diplomatic intercourse. His authorized life of the great Prussian commander in the Napoleonic Wars Field Marshal Graf August Neidhardt von Gneisenau required an extensive grounding in Napoleonic strategy, for which he explored the works of Clausewitz and fused that great thinker's insights with the methods of Ranke and Rüstow. By the 1890s, Delbrück was making his mark as a master of military historical research. His contributions have been summed up by Richard Bauer as threefold:

1. Exposure and correction of many legends concerning the size of armies participating in ancient and medieval battles.
2. Division of all military strategy into the strategy of exhaustion (*Ermattungsstrategie*) and the strategy of annihilation (*Niederwerfungsstrategie*). The former was the preferred method in the eighteenth century, especially by Frederick the Great, and consisted of weakening the enemy gradually by a series of maneuvers without necessarily engaging his forces.
3. Explanation of the events of a war by going beyond mere descriptions of battlefield operations and considering the constraints imposed by technological and geographical limitations as well as the political intentions of the national leaders directing the war.

We can mark the beginning of the war and society trend with the advent of Delbrück, although Delbrück himself focused more on military organization, strategy, and tactics than upon social context as it has become known in the 1990s.

A contemporary of Delbrück, Otto Hintze, also placed warfare in its general context but found a new analytical touchstone in the relationship between military organization and political structure. In the process, he advanced the case for military history's relevance to the patterns of cultural and institutional development.

In England, Sir Charles Oman and Spenser Wilkinson, civilian scholars, advanced the study of military history by applying Delbrückian methods to, respectively, the wars of the Middle Ages and the military education of and influences upon the young Napoleon Bonaparte. The multivolume tour de force of John Fortescue, the pre-eminent historian of British land forces, is a masterpiece of traditional military history, but demonstrates the hold that battle narrative still exerted upon institutional studies.

Nineteenth-century French military history was principally the province of the officer-pedants attached to the general staff or instructors in the *Ecole de Guerre Supérieure*. Following the disaster at Sedan in 1870, the spirit and genius of Napoleon I was evoked to guide soldiers. A search for ways to revive the Napoleonic spirit permeated the prescriptive historical reviews churned out by the high command. "Drum and trumpet" histories, emphasizing élan, predominated. Several notable Napoleonic scholars rose above the hero-worship of their fellow officers to produce fine critiques of the master's mind and method. The works of Palat, Bonnal, Camon, Vachée, Colin, and Pierron have enduring value.

American military historians covering land warfare worked under the shadow of Jomini. John Bigelow's 1895 primer on strategy was a Jominian survey of the Civil War, though his 1910 study of Chancelorsville represents, along with G. F. R. Henderson's tactical studies, the perfection of the campaign history. Matthew Forney Steele's 1909 West Point history text, *American Campaigns*, was the standard text for the institution's course in military art and engineering for almost 50 years. It also kept a "principles" scorecard for each major engagement from colonial times through the Spanish-American War.

It was only in the realm of naval history that American strategic thinkers broke new ground. Alfred Thayer Mahan's 1890 survey of seapower's influence upon world affairs in the age of sail also sought out eternal verities of naval warfare. But Mahan looked deeper for his principles of seapower than the landpower theoreticians' themes and found links between geographical configuration, demographic distribution and evolution, industrial infrastructure, political makeup, and the "control of the sea-lanes." His Darwinian racist credo and emphasis upon seeking decisive action between battle fleets on the high seas limited his influence to the first quarter of the twentieth century, yet his accent upon broader social and economic factors endures.

History Lessons for the Era of Total War

In the aftershock of World War I's slaughter, two British veterans of the trenches re-examined warfare to find formulas that might overcome the offense-defense deadlock.

J. F. C. Fuller searched the past to exhume its concealed "science of war," based upon immutable principles. His work in the 1920s is linked to the enumeration of discrete principles of war that found their way into the British field service regulations and the United States's field manual *Operations*, now designated *FM 100-5*. Fuller went beyond Jomini's list to evoke Darwin's evolutionary theories and devise a "law of military development," which held that armies must adapt to changes in their environment. Fuller turned to history to verify the natural laws of military science he had already divined. Much as Hamley did before him, Fuller plundered the historical data for insights without going into the primary sources. The result was lively and illuminating military history (especially his *Military History of the Western World*), albeit often wrong in the particulars.

Sir Basil Liddell Hart found his key to the puzzle of the World War I killing grounds in the "strategy of the indirect approach." In order to amass an inventory of examples proving that the indirect approach had been the determinant of success, Liddell Hart examined numerous battles throughout history, using the evidence selectively and, like Fuller, shunning primary sources or incidents that might contradict his hypothesis.

In short, both Fuller and Liddell Hart enriched the literature with provocative ideas and inspired much-needed investigation into contemporary doctrines. In their use of history they represent perhaps the last of the intellectual heirs of Jomini and Hamley.

While the British Committee of Imperial Defence supervised a meticulously detailed official operational history of the Great War, its editors were constrained to protect the reputations of much-decorated blunderers. The Americans did not hazard to analyze their brief participation. Rather than waste precious defense dollars on staff pedagogues, the U.S. War Department was content to assemble the various orders, corps, and division after-action reports and let the documents speak for themselves. The story of individual battles and campaigns was consigned to monographs that served as in-house teaching aids at the army's War College and at the Infantry School. In the mid-1930s, a new sophistication in prescriptive small-unit action studies was attained with the U.S. Army Command and General Staff College's *Infantry in Battle*.

The New Military History: 1945–2000

The Second World War inspired novel approaches in military history. At one level, official history had to shed its mask of self-exoneration and achieve greater depth and breadth than the up-the-hill, down-the-hill chronicle. The British version certainly bettered its World War I counterpart, especially in the "Grand Strategy" series, but did not match the ingenuity of the American method. The latter was developed by S. L. A. Marshall, a journalist-historian, who headed a talented academic team commissioned by the U.S. Army to write its combat history while it occurred. Marshall combined a survey of war diaries and orders and dispatches with group after-action debriefings to commence an innovative procedure in oral history. He extended this method to interrogations of captured German officers, who were assisted by access to captured documents.

A second level of analysis proceeds directly from Marshall's spotlight on small-unit interaction and cohesion under stress. As University of Chicago military historian Walter Kaegi has observed, it was sociologists and political scientists rather than historians who carried the study of men under fire to its next logical step: research on armies as social institutions, as bureaucracies, as political entities with internal rivalries and self-interests, and on their interactions with civilian agencies. Trailblazers of this aspect of war and society analysis include Samuel Huntington, Morris Janowitz, and S. E. Finer.

The sociological study of unit cohesion during World War II was embraced by erudite conflict historians in their investigations of the citizen at war in earlier periods. In France, Phillipe Contamine improved upon Oman's survey of medieval warfare by tracking religious, class, cultural, and ritual influences from village and manor to the field of battle. R. C. Smail did the same with crusading warfare for English-speaking readers.

In Britain and the United States, Michael Howard, John Shy, and Peter Paret brought new insights from the social sciences to bear upon multidimensional conflict studies, particularly the impact of the anticolonialist struggles in the Third World. Paul Kennedy has provided a holistic framework upon which to hang his explanation of the rise and fall of British naval mastery, which he later expanded to encompass the symbiosis between capital and military power since the Renaissance, updating Sombart and Nef to this effect.

Some practitioners of combat-oriented military history have been calling for a return to utilitarian considerations of military institutions as fighting instruments. Utilitarian military pedants believe that the scholarly approach toward military organizations and their impact on society has lost sight of the fact that their primary mission is combat or the threat of combat. Accordingly, since the mid-1980s there has been a fusion of the pristinely civilian outlook with the old emphasis upon battle. Exemplary

American works are the U.S. Command and General Staff College's review of *America's First Battles* (meaning the American baptism of fire in each of its wars) and Ohio State University's three-volume project on military effectiveness, a concept that offers an excellent vehicle for combining the two approaches in a historical context.

Operations research (OR), or systems analysis, is another spin-off from World War II–stimulated progress in the social sciences that affected the writing of military history. This was an essentially algebraic procedure to solve special combat problems requiring a scientific approach. It was first devised to hunt submerged U-boats that evaded air and sonar search, and expanded to cover such questions as artillery fire plans, proper force mixture for different contingencies, bombing target priorities, and so forth. Its origins stretch back to William Lanchester's 1916 speculations upon the "force multiplier" effects of aircraft in combat. The British and American defense establishments adapted the method to solve postwar dilemmas. The harmful effects of OR upon American strategy in Vietnam somewhat discredited the technique as a planning tool and led to the revival of historical analogy in military education and planning.

Successful applications of OR concepts to professional war-gaming also have enriched the historian's craft. One example of this is American Col. Trevor Dupuy's quantified judgmental method. Applying statistically driven formulas covering such aspects as relative weapons lethality, effects of surprise, field fortifications, and troop density, Dupuy has isolated the factors explaining combat results over a wide range of historical battles. While critics have quibbled over the degree of precision one may expect, the method has enriched historical debate over what makes for success in battle.

Soviet historians have essentially utilized a variation of this method in their calculation of "norms" for artillery concentrations, armor dispersal, and the like for their World War II engagements.

Since the late 1960s, the vast expansion and refinement of the popular military hobby trade, including accurate scale models and diorama construction by gifted amateurs and the rise of recreational war-gaming, has created a demand for guides and authentic background information. Sometimes denigrated by academic military historians as mere buffs or enthusiasts, scale modelers and war-gamers often have made real contributions to military history.

A quick survey of the magazines catering to the hobbyists reveals a spectrum of professionalism ranging from fascination with uniform and equipment minutiae to erudite order-of-battle analysis. Necessarily, the modeling magazines are preoccupied with details of appearance, but this does not necessarily mean that the broader implications are ignored. Those directed to the more advanced creators of museum-quality reproductions often feature articles offering the historian hard-to-find information on matters pertaining to topics like unit insignia, weapons development and variants, equipment, camouflage techniques, and military costume.

There is considerable interaction between the hobby war-gaming industry and defense establishments, at least in the United States and Great Britain. The names of prominent Sandhurst and U.S. Army or Naval War College analysts appear regularly in the better historical simulations journals.

JAMES BLOOM

SEE ALSO: Ardant du Picq, Charles J. J. J.; Art of War; Caesar, Julius; Clausewitz, Karl von; Fuller, J. F. C.; Jomini, Antoine Henri; Mahan, Alfred Thayer; Moltke the Elder; Operations Research, Military; Principles of War; Saxe, Hermann Maurice, Comte de; Scharnhorst, Gerhard Johann David von; Science of War; Search Theory and Applications; Social Science Research and Development, Military; Sociology, Military; Sun Tzu; Systems Analysis.

Bibliography

Bauer, R. H. 1942. Hans Delbrück. In *Some historians of modern Europe*, ed. B. Schmitt. Chicago: Univ. of Chicago Press.
Best, G., B. Bond, D. Chandler, J. Childs, J. Gooch, M. Howard, J. C. A. Stagg, and J. Terraine. 1984. What is military history? *History Today* 34:5–15.
Dupuy, T. N. 1984. *The evolution of weapons and warfare*. Fairfax, Va.: Hero Books.
Earle, E. M., ed. 1943. *Makers of modern strategy from Machiavelli to Hitler*. Princeton, N.J.: Princeton Univ. Press.
Gooch, J. 1980. Clio and Mars: The use and abuse of military history. *Journal of Strategic Studies* 3:21–36.
Howard, M. 1962. The use and abuse of military history. *R.U.S.I. Journal* 117:4–10.
James, R. R. 1966. Thoughts on writing military history. *R.U.S.I. Journal*, May.
Jessup, J., and R. W. Coakley, eds. 1979. *A guide to the study and use of military history*. Washington, D.C.: U.S. Army Center of Military History.
Jones, C. 1982. New military history for old? War and society in early modern Europe. *European Studies Review* 12:97–108.
Luvaas, J. 1982. Military history: Is it still practicable? *Parameters* 12(1):2–14.
Millis, W. 1961. *Military history*. Washington, D.C.: Service Center for Teachers of History.
Millett, A. R. 1977. The study of American military history in the United States. *Military Affairs* 41:58.
Paret, P. 1971. The history of war. *Daedalus* 100:376–96.
———, ed. 1986. *Makers of modern strategy from Machiavelli to the nuclear age*. Princeton, N.J.: Princeton Univ. Press.

HISTORY, MODERN MILITARY
[ca. 1792–present]

Although the point at which modern warfare began is debatable, a good case can be made for the period of the French revolutionary wars. Dynastic conflicts for limited objectives faded, and the rise of nationalism dramatically changed the conduct of war. Organization at the national level and an approach toward total war marked the conflicts of the late eighteenth and early nineteenth centuries.

Naval warfare became highly organized and disciplined as wooden sailing ships reached their zenith. Battles were fought with heavier guns when large ships of the line appeared as precursors to the battleships of a later age.

Wars of the French Revolution

The revolution had a tremendous impact on the French army. Seventy percent of its officer corps and most of its trained men were lost in the chaos, leaving the French armed forces incapable of operating on the old model. In 1791, a new set of tactical concepts and reforms provided speed and flexibility in deployment for action.

European monarchs, alarmed at the spread of revolutionary ideals, reacted vigorously. By 1792, French volunteers and remnants of the Royal Army were in combat with the disciplined, professional armies of Prussia and Austria. The disorganized French forces were repeatedly defeated. Fortunately for the new government, the committed Prussian and Austrian forces were insufficient to subjugate France.

By 1793, France faced a coalition of Prussia, Austria, the Netherlands, Spain, and Great Britain and began drafting hundreds of thousands of men, many of whom were poorly trained and meagerly equipped. Nevertheless, the masses of men, revolutionary zeal, new tactics, and poor coordination of the coalition forces, along with the advantage of fighting from interior lines, enabled the French to hold their own.

In March 1796, 26-year-old Napoleon Bonaparte took command of a newly structured French army in Italy. His impact on military thought during this war-prone era was enormous. The old French officer corps was uprooted, the army rebuilt, and the obscure Corsican rose to leadership and later became emperor of France. Warfare was not greatly affected by new technology, though some innovations did appear, including Cugnot's invention of a steam-powered vehicle to haul cannon and the development of canned food to improve military diets.

Napoleon's initial campaigns in northern Italy were considered by some analysts to be his most brilliant. Even as he "learned his trade," he drove the competent but aging Austrian generals out of Italy in a series of short, sharp battles and maneuvers (April 1796–March 1797). Following an interlude in the Mideast, where he conquered Egypt but was checked in Acre, Palestine, he returned to France to face a revolt in the Vendée and the continuing war with Great Britain and Austria, the latter having reconquered almost all of Italy by 1799.

Napoleon took over the government as first consul and prepared for more vigorous warfare. Since Great Britain could not be attacked directly, he targeted Austrian forces in Italy via Switzerland. Although nearly defeated at Marengo (14 June 1800), the French forced a temporary peace on Vienna.

Great Britain, along with Turkish troops, forced the French out of Egypt. Finally, in 1802, the Treaty of Amiens was signed and lasted fourteen months. Friction between France and England continued and, in May 1803, France declared war, seized Naples, occupied Hanover, and prepared for a channel crossing to invade Great Britain. British seapower proved too formidable and the invasion was never attempted.

Meanwhile Prime Minister William Pitt crafted another coalition, and by 1805, Russia, Austria, Naples, Great Britain, and Sweden had mobilized half a million men to fight the French. That autumn, Napoleon destroyed a major Austrian army at Ulm (17 October) before Russian forces arrived. Even in victory, however, a naval battle at Cape Trafalgar where a French-Spanish fleet was destroyed (21 October) ended any threat to Britain's sea supremacy.

Subsequently, at Austerlitz (2 December 1805), a Russian-Austrian army of 84,000 was shattered by Napoleon's army of 71,000. Vienna again sued for peace and the czar's army was sent reeling back into Poland.

On 7 August 1806, Napoleon secretly goaded the Prussians into a war declaration, and on 8 October, his columns crossed into Saxony. At the twin battles of Jena-Auerstädt (4 October 1806), the Prussian army was beaten and further shattered during its disordered retreat. Napoleon's month-long "lightning war" dragged on for six months, sustained by Prussia's iron-willed Queen Louise. Napoleon became master of most of Europe but there were still Russia and Great Britain to deal with. He tried an embargo to damage the British economy, but it failed.

Russian forces opened a winter offensive in northern Poland but were forced back after an indecisive battle at Eylau (8 February 1807). Napoleon won a notable victory at Friedland (14 June) and peace was finally achieved that month at a summit meeting between Czar Alexander and Emperor Napoleon. The French then faced a festering war on their southern flank as Portugal resisted invasion; however, shortly thereafter, Napoleon took over Spain, his nominal ally, and placed his brother on the throne.

The Spanish, with British backing, resisted, but were unable to repulse the French in the field. Widespread irregular warfare broke out; eventually, the Spanish, with the help of British forces under the Duke of Wellington, pushed the French out of Spain. In 1809, Austria reopened hostilities and Napoleon went on the offensive to win a costly victory at Wagram (5–6 July). A quarter of his men (32,000) were casualties, but the Austrians fared even worse. Napoleon's peace terms were accepted; however, this campaign was the last of his victories.

After two and a half years of relative peace, Napoleon again collided with the Russian Empire. In mid-1812, he assembled half a million men, many of whom were unwilling allies, and marched eastward. The Russians retreated until they reached Borodino, where they fought Napoleon almost to a draw in a bloody (about 80,000 casualties) twelve-hour battle (7 September). The Russians abandoned the field and the crippled French army marched on to Moscow.

Napoleon dallied in Moscow for several weeks and then began a disastrous retreat during which most of his invasion force was destroyed (October–December 1812). Following this fiasco, he created a new army to combat his persistent foes: Prussia, Austria, and Sweden. Campaigning began in Germany in April 1813, and his new French army fought well (although not up to 1809 standards), particularly when led by "the master." Napoleon's subordinates were less successful and more often targets of enemy attacks.

A major defeat at the Battle of Nations, at Leipzig on 16–19 October 1813, ended French hopes for a trans-Rhine empire. Napoleon refused a peace proposal that would have made the Rhine and the Alps France's frontier. The allies invaded France, captured Paris in March 1814, and forced the emperor's abdication.

Exiled to the island of Elba, he returned a year later, rallied his followers, drove out the French king, Louis XVIII, and began the "Hundred Days," in which he tried to reconstitute his empire. The allies relentlessly opposed these attempts. He was defeated at Waterloo by an allied Anglo-Prussian army (18 June 1815), which put an end to the great Napoleonic wars.

The Post-Napoleonic Era

Peace was not universal. In South America, revolutions led by Simón Bolívar, José de San Martín, Antonio José de Sucre, and others demolished the decrepit Spanish Empire between the years 1808 and 1828. Elsewhere, anti-Manchu uprisings in China, along with Great Britain's colonial wars in India, the French conquests in North Africa, local wars in the western Sudan, and a Greek revolt against the Turks, kept warfare a means of solving ethnic, religious, and political conflicts. Strife in Italy led to the ousting of Austrians, and in 1848 there were widespread revolts of one kind or another in Europe.

During this period, weaponry was marginally improved while warfare characteristics remained generally unchanged. A Greek revolt, beginning in 1821 and threatened by an Egyptian army victory, brought European intervention (1827) as an allied fleet of British, French, and Russian ships, in the last major battle between wooden sailing vessels, annihilated Turkish and Egyptian fleets at Navarino (20 October 1827).

The Crimean War (1854–56) reflected the most changes in warfare since, perhaps, the advent of gunpowder. During this conflict, France, Great Britain, and later Sardinia came to the aid of the Ottoman Empire, which was on the verge of destruction by Russian attacks. On the naval side, steam overtook the fighting sail vessel. Armored floating batteries were used by the allies against the fortress of Kinburn (16 October 1855), presaging the widespread use of armor plate on naval vessels. Other innovations included use of a railroad to carry supplies from a landing area to the allied camps, the presence of war correspondents at the battlefields, and a reduction of mortality rates among the sick and wounded through the efforts of British nurse Florence Nightingale.

The siege of Sevastopol (8 October 1854–8 September 1855) was marked by hard-fought and confused battles with sustained bombardment of the city. The allies forced the Russians to withdraw and a peace settlement was concluded on 30 March 1856, thus preserving the Ottoman Empire.

The year 1859 was notable for several reasons. First, the French produced a steam-powered armored frigate, *Gloire*. It had a wooden hull, but its vitals were protected by armored plating. Second, during the Wars of Italian Independence, the battle of Solferino, fought by 120,000 Austrians and 118,000 French and Piedmontese (24 June 1859), forced the defeated Austrians to cede all of Lombardy to an emerging united Italy. Another major development that grew out of the carnage of Solferino was the formation of the International Red Cross. Swiss philanthropist Jean Henri Dunant witnessed the battle and was horrified at the poor care the wounded received. His post-battle activism led to the founding of the organization, which has had a strong and continuing impact on the conduct of war, stimulating treaties governing the treatment of civilians in war zones, the treatment of prisoners, and the conduct of hostilities.

The War Between the States (1861–65)

Considerable new battlefield technology appeared during the American Civil War, including electrically fired land and naval mines, new forms of heavy artillery, a balloon observation corps, Monitor-type naval vessels (turreted armored warships), breech-loading and magazine rifles, and the Gatling gun (an early form of machine gun). The most important technology associated with this war was not really new at all: the railroad and the minié ball (previously used in combat by the Austrians).

The minié ball was a conoidal bullet with a hollowed-out base that was used in muzzle-loading rifles in standard infantry equipment. When the powder charge ignited, the hollow base expanded, making a gas-tight fit and forcing the soft lead to engage the rifling of the barrel. This spin-stabilized bullet maintained a relatively accurate trajectory over a long distance. In this way, every infantry musket (technically rifles, although still called muskets) became an accurate, long-range weapon infinitely superior to the inaccurate, smooth-bore muskets of the past. (Patch and ball rifles had long been used in warfare, but their slow rate of fire limited them on battlefields.)

Rifled muskets drastically changed battlefield tactics

and campaign strategies. Field fortifications could no longer be charged with impunity unless attacks were made with overwhelming forces. Also, artillery could no longer unlimber at 270-meter (300-yd.) ranges and open fire on opposing infantry. The cannoneers would be wiped out by musketry.

Railroads had been around longer than the minié ball but, prior to 1861, had made only small contributions to warfare. Now their logistical importance became so great that battles and campaigns were planned around rail support. In addition to transport for men and supplies, the trains themselves could serve as semimobile artillery: gun tubes were sometimes mounted on rail trucks (e.g., the train-mounted mortars and guns used at the siege of Petersburg 1864–65).

Ultimately, the greater economic, demographic, and manufacturing base of the North provided the margin of victory in what was largely a war of attrition. Here, the United States Navy also played an important role, blockading the South and intercepting war materiel. Naval action included the first battle between ironclads—the *Monitor* and the *Virginia* (9 March 1862)—and later the widespread employment of this type of ship. The war ended with the surrender of Confederate general Robert E. Lee at the Appomattox Courthouse on 9 April 1865.

The Franco-Prussian War

After the Napoleonic wars, Prussia continued to improve its military staff system and reserve structure and began to move toward unifying Germany. Schleswig-Holstein was quickly wrested from Denmark in 1864. Austria was defeated in the Seven Weeks' War in which the two armies—some 170,000 on both sides—collided at Königgrätz (Sadowa) on 3 July 1866.

The next step toward unification was to defeat France. Both parties to the Franco-Prussian conflict wanted war: the Prussians to eliminate French influence and unify Germany, and Napoleon III, who hoped to regain the grandeur of the empire of Napoleon I.

On 19 July 1870, war was declared. The French had superior rifles and a "secret weapon," the *mitrailleuse*. Their mobilization, however, was chaotic and tactics obsolete. The mitrailleuse, a manually cranked precursor of the machine gun, had little impact because, to maintain secrecy, training was limited, and the guns were inappropriately used as artillery rather than as infantry weapons. The better organized and trained Prussian forces had superior artillery, better logistics support, and a better command structure. The French were beaten in every major engagement as the Prussians drove one French army into Metz to be captured later (27 October with 173,000 men), and then captured Napoleon with his entire army of 83,000 at Sedan on 2 September. France thereupon declared itself a republic and resolved to carry on the struggle and ordered "war to the bitter end."

The Prussians surrounded Paris, initiating a famous siege that brought several innovations to warfare, including the use of balloons for communications and the use of microphotography to send messages attached to trained carrier pigeons. The Prussians also added a footnote to military intelligence history. They tapped an underwater telegraph line leading out of Paris and carried out what may have been the first signal intelligence intercept operation ever conducted in Europe (telegraph messages had been intercepted and decoded in the American Civil War). With besieged Parisians nearing starvation, the French government capitulated on 28 January 1871.

Colonial Wars, Rebellions, Spanish-American War, China, Boer War

Following the Franco-Prussian War, hostilities among major powers quieted as if bracing for the upcoming storm of World War I. There were, however, a host of revolutions, colonial wars, and minor conflicts throughout the world.

The Spanish-American War of 1898 was largely determined by two naval battles in which one Spanish fleet was destroyed at Manila in the Philippines (1 May) and another at Santiago, Cuba (3 July). Spanish soldiers ashore had superior personal weapons, including Mauser rifles and smokeless powder (developed by the French), but were quickly beaten. Spain capitulated and signed the Treaty of Paris (10 December 1898) and the United States emerged as a world power.

China was in almost continual warfare during the last half of the nineteenth century. As the Manchu dynasty weakened, the Sino-Japanese War, which ended in April 1895, provided victorious Japan with the island of Formosa, the Pescadores, a huge indemnity, and other considerations.

More widely publicized was the Boxer Rebellion of 1899, which was initiated by a xenophobic Chinese secret society and culminated in a siege of the foreign legation quarter in Peking (20 June–14 August 1900). Japanese, French, British, Italian, U.S., and Russian military forces (later joined by a German unit) relieved the legations, then fought with the Chinese for a year before peace was imposed (12 September 1901).

The Boer War (or South African War, 1899–1902) was a revolt of descendants of Dutch settlers in South Africa against British incursions. Early Boer successes were largely due to their accurate fire from magazine-fed rifles with smokeless powder. The British rapidly reinforced their troops, deploying 450,000 men against 90,000 Boers. Outnumbered, the Boers turned to guerrilla warfare, again with initial success; the British countered with effective, although harsh, antiguerrilla tactics. The war ended in May 1902 with the Boers accepting British sovereignty.

New, or relatively new, to warfare was the use of steam tractors to haul cannon and supplies, the use of armored and armed trains, and the large-scale employment of guerrilla and antiguerrilla warfare tactics such as concentration camps to segregate the civilian population from combatants. This tactic, pioneered by the Spanish in Cuba, presaged such later twentieth-century conflicts as Vietnam and Afghanistan.

Russo-Japanese War (1904–1905)

The Russo-Japanese War was, in many ways, a dress rehearsal for World War I. The naval battles included the use of breech-loading guns, mines, armor-piercing shells, wireless communications, and torpedoes at sea. On land were barbed wire, machine guns, and field artillery with recoil mechanisms. Naval actions and Japanese squadrons defeated Russian naval detachments; the Port Arthur squadron surrendered to besieging Japanese ground forces (2 January 1905) and a Russian squadron from the Baltic was destroyed at sea at Tsushima (31 May 1905). Although the Japanese were successful, they were hard-pressed to maintain the level of conflict as Russian ground reinforcements came east on the Trans-Siberian railroad.

A U.S. offer to act as mediator was accepted and the war ended with the Treaty of Portsmouth, N.H. (5 September 1905). Japan's sphere of influence in Korea was recognized and both parties agreed to evacuate Manchuria.

World War I

The causes of World War I have been studied and argued for years. It is clear, however, that the great powers of Europe were, in the early 1900s, psychologically and militarily ready for combat, although few had any inkling of the tremendous bloodletting that lay ahead. Nationalism, Pan-Slavism, and desires for glory or revenge had turned Europe into a tinderbox that needed only a spark to set it ablaze. That spark came at Sarajevo, Serbia, on 28 June 1914 when Archduke Francis Ferdinand, heir to the Austrian throne, was assassinated by Serbian radicals. In the aftermath, Europe began massive and complex mobilizations; war was inevitable.

Desperate, last-minute attempts to preserve peace failed and war began with the Central Powers—Germany, Austria-Hungary, Turkey, and Bulgaria—opposed by the Entente headed by Russia, France, and the United Kingdom.

CHARACTERISTICS OF THE GROUND, NAVAL, AND AIR WAR

Ground warfare was markedly different from that of previous wars. Masses of manpower were employed on a scale never seen before. Europe's mass armies resulted from several factors, including long-term population growth; industrialization, which freed more manpower for combat; and nationalism, which sanctioned popular support for conscription. Most changes, however, involved technology. The large-scale, defensive use of machine guns, for example, led to development of vast field fortifications, miles in depth, that spread from the English Channel in the north to the Swiss-French frontier in the south. Combatants were supplied with magazine rifles, and artillery was equipped with recoil mechanisms and no longer bucked out of position with each discharge. Field telephones and to a lesser extent radios improved communications. Later in the conflict new ground force weapons, the tank and poison gas, appeared. One material-of-war on the western front was barbed wire, a relatively recent agricultural development now adapted to warfare. Thousands and thousands of miles of wire obstacles were strung before entrenchments, thus contributing to the static nature of combat on the western front.

Naval warfare was characterized by all-big-gun battleships (dreadnoughts), with ever larger guns, improved torpedoes and mines, better fire control systems, electrical shipboard communications, and wireless ship-to-shore and ship-to-ship communications. Oil-fired, turbine-driven ships became more common and, for the first time, submarines dominated the force at sea.

At the war's beginning, fixed-wing aircraft were few in number, balky, and limited to short-range reconnaissance missions. By war's end, thousands of aircraft from armies on both sides had taken to the skies. Range and speed improved many times over and new classes of aircraft were phased into combat so quickly that an aircraft design often had a useful life of only a few months. Machine guns appeared after a period in which planes were unarmed. Bombers emerged along with other specialized aircraft. Most aerial operations were tactical and aimed at supporting ground combat operations. Some longer range strikes on cities, however, were initiated by the Germans who used dirigibles and later long-range bombers. By the end of the war, the Allies were attacking distant communications centers and planning major strikes on German cities.

At the war's beginning, Germany had dirigible (zeppelin) capability and used airships extensively until incendiary machine-gun bullets exposed their vulnerability to fire. France and the United Kingdom also developed dirigibles, using them in various roles. Observation balloons were especially useful for artillery spotting on the western front.

Aerial warfare never impacted greatly on naval operations, but dirigibles and fixed-wing aircraft (both seaplanes and land-based) were important for reconnaissance.

WESTERN FRONT

Imperial Germany, knowing that the forces of the Central Powers would be greatly outnumbered by those of the Allies after all combatants were fully mobilized, determined to maximize its initial advantages: interior lines, better mobilization planning and execution, and better transportation. The strategy was to hold the slowly mobilizing Russians at bay while knocking France out of the war before the Russians could become a significant factor or before the British made large ground forces available in France. The concept was to sweep through Belgium, concentrate fast-moving German columns on the right wing of the advance, and envelop French forces with a vast turning movement. It was a strategy that demanded the German mobilization advantage not be squandered by drawn-out peace negotiations. On 3 August, Germany declared war and invaded Belgium.

The initial success failed partly because German generals did not rigorously adhere to the plan, partly because the small British Expeditionary Force (5 divisions, 70,000 men) became an unexpected obstacle. There was also the monumental calmness of the French commander, Gen. Joseph Joffre, who refused to panic or allow his forces to panic. Most important was a Russian offensive that forced the Germans to transfer badly needed troops to the east.

The Germans were halted at the First Battle of the Marne on 6–10 September, and shortly afterward both sides began to entrench along a 800-kilometer (500 mi.) front. At first the fortifications were simple breastworks, but these were steadily elaborated with extensive dugouts, barbed-wire obstacles, and defenses that defied attempts at breakthrough. The western front remained static until 1918.

In an attempt to break the stalemate, the Germans unleashed a chemical weapons attack at Ypres on 22 April 1915, killing 5,000 men, which opened a 6.4-kilometer (4-mi.) gap in Allied lines. The gap was soon sealed off by Allied reinforcements, but the precedent had been set for the use of gas for artillery preparation.

The British secretly developed armored tanks that protected crews from machine-gun fire when crushing barbed wire. Poor tactics and muddy battlefields limited their effectiveness at Flers-Courcellette (15 September 1916), but they met with greater success in later battles, most notably the use of 350 of the tanks at the Second Battle of the Marne (15–17 July 1918).

Eastern Front

In August, czarist armies, unready but responding to Allied pleas to relieve German pressure in the west, advanced into East Prussia. Germany was forced to shift troops to the east and recalled aging Gen. Paul von Hindenburg to duty to save the situation. Ably seconded by Erich Ludendorff, he soundly defeated the Russians at Tannenberg (26–31 August 1914).

Russian attacks against the heterogeneous armies of the Austro-Hungarian Empire were more successful, yet despite German reinforcements, by 1915 all fronts were in retreat.

Other Fronts

The Japanese entered the war in 1914, attacking German colonies in the Far East; Turkey joined the Central Powers operating initially against Russia. Balkan, German, Austrian, and Bulgarian forces battled Serbia as the forces of the Central Powers overran much of that country. In October–November 1915, Romania joined the Allies but was defeated in December 1916. In Greece, King Constantine was deposed by pro-Allied forces, and war was declared on the Central Powers. The Allies sent an army to Salonika, where they defeated Bulgarian forces.

Britain occupied Egypt, and in Africa, British and French forces captured the German colonies of Togoland (1914), Kamerun (1916), and German Southwest Africa (1915). In German East Africa, Colonel (later general) von Lettow-Vorbeck fought vastly superior Allied forces, successfully operating in the field until after Germany itself surrendered.

In the Mideast, British troops carried the war to the Ottoman Empire, crossed into Palestine from Egypt, and attacked Mesopotamia to secure the Basra oil fields. Arabian allies, revolting against the Turks, were supplied with money, arms, and advisers. Among the latter was then-captain T. E. Lawrence, brought to fame by American journalist Lowell Thomas as "Lawrence of Arabia."

Allied naval and ground forces tried to open the Dardanelles in 1915 as a supply route to Russia, but they were stopped at Gallipoli by stubborn Turkish resistance on land and sea. A British army was forced to surrender at Kut in Mesopotamia (29 April 1916) after a five-month siege and the failure of relief efforts.

The Naval War

Germany was unable to directly challenge the strong British navy but did send several surface raiders to attack Allied shipping. After initial successes, the raiders were hunted down and destroyed. A German squadron destroyed a British one off Coronel, Chile, but was itself later destroyed off the Falkland Islands (Islas Malvinas). The most important naval battle was fought off Jutland on 31 May 1916 when the main British and German battle fleets engaged. The British incurred the most damage, including three battle cruisers. The German fleet, however, was driven back and never again challenged the British.

The Allies faced their greatest threat from German submarines. German U-boats did enormous damage to British merchant shipping until the institution of the convoy system and the development of such antisubmarine warfare (ASW) weapons as sonar, along with evasive tactics to reduce the damage from underwater warfare.

Final German Offensive

On 15 July 1918, Ludendorff launched an all-out offensive with three armies. By the evening of 17 July, massive Allied counterattacks had forced the Germans to retreat. On other fronts, the war was going badly for the Central Powers. Bulgarians asked for peace in September and the Austro-Hungarian Empire was fragmented beyond repair. Turkey, having lost Palestine, Mesopotamia, and Syria, asked for terms in November. Germany fought on until 11 November, when an armistice was signed.

World War I, the "War to End All Wars," was over. Eight million men had died, and the old order in Europe was destroyed forever.

Between the Wars

During the 1920s, there was prosperity in the United States until the 1929 stock market crash ushered in the Great Depression. With the election of Franklin Delano Roosevelt as president in 1933 came the devaluation of the dollar and a greater federal role in industry and farm production. New Deal programs were aimed at reducing unemployment.

America became isolationist as it saw that World War I had not brought peace, democracy, or disarmament, and European nations bickered as before. President Woodrow Wilson's Fourteen Points had been discarded and his concept of the League of Nations, adopted by many nations but refused in the United States, lacked authority to do more than issue verbal reprimands. Also, the United States believed that the two oceans would protect Americans from foreign belligerents.

In Russia, the November 1917 revolution, led by Vladimir Lenin, brought communism to power. After Lenin's death in 1924, Joseph Stalin became the dictator, later crushing all opposition by eliminating hundreds of thousands of intellectuals, army officers, and civil servants in the Great Purge of 1936–39. The fate of those purged was execution or labor camps in Siberia. Stalin's goal was to industrialize the USSR, nationalize steel and electrical plants, and seize private farms for use as collectives.

In the Far East, Japan and Korea (taken over by Japan in 1910) needed more territory to sustain the 65 million people, who increased by a million each year. The Japanese economy was hurt by the 1929 crash and it needed iron, coal, oil, and cereals in order to survive. Japanese forces overran nearby Manchuria in China in late 1931 and early 1932. When condemned by the League of Nations, Japan withdrew membership in 1933, signed an alliance with Germany and Italy in November 1936, and on 7 July 1937 declared war on China. Japanese armed forces attacked Peking, Nanking, Canton, and Shanghai. Entire village populations were slaughtered, and in the infamous Rape of Nanking, 20,000 were massacred in six weeks (13 December 1937–25 January 1938). Movietone News and graphic photography on the front pages of newspapers portrayed the horrors of war with the classic photograph of a Chinese child crying on the railroad tracks in Nanking.

In Europe, the civil war in Spain began as a rebellion by generals hostile to the Frente Popular and the Republican government (18 July 1936). The war ended in March 1939 with the victory of the rebels under dictator Gen. Francisco Franco. Adolf Hitler and Benito Mussolini backed the rebels, while the Soviet Union under Joseph Stalin and 45,000 volunteers from 53 countries under the banner of the International Brigade joined the Republican cause. Tens of thousands of civilians died, hundreds of thousands of soldiers were killed, and 350,000 people fled the country. Stalin bolstered the republic with arms and advisers and Hitler tested his weaponry.

In January 1933, Adolf Hitler became chancellor of the German Reich with the promise of righting the spiteful and selfish wrongs of the Versailles Treaty. At first he used ruses to avoid the clause that prohibited the manufacture of modern military equipment, such as tanks, planes, and submarines. Marshal Hermann Göring established a League of Air Sporte, which actually trained future Luftwaffe pilots. Companies prepared prototypes of fighter planes and tanks for mass production.

In 1935, Hitler restored compulsory military service; France and Great Britain did nothing except express indignation. Meanwhile, Germany's arms industry flourished with steelworks, metallurgical and chemical factories, and the production of airplanes, tanks, and armored vehicles. Strategic roads and airfields were constructed and unemployment dropped. In March 1936, Hitler sent troops into the demilitarized zone of the Rhineland. Inaction by Britain and France emboldened Hitler, who supported Benito Mussolini's invasion and conquest of Ethiopia (October 1935–May 1936), and in March 1938, the Nazi dictator annexed Austria. In Czechoslovakia, which was carved out of the Austria-Hungary Empire in 1919, the Sudeten region, with its industries and large mineral resources and inhabited largely by Germans, was Hitler's next demand. In September 1938 at Nuremberg, annexation on threat of invasion was demanded. On 28 September 1938, in a conference with Hitler and French prime minister Edouard Daladier at Munich, British prime minister Neville Chamberlain refused to commit British troops "for this small nation," and promised "peace in our time." In March 1939, Nazis entered Prague, and Czechoslovakia was erased from the map.

Poland was next. Hitler demanded Danzig and free passage of the corridor that linked the port of Danzig with the rest of Poland. On 24 August 1939, Germany signed a secret nonaggression pact with Stalin's Russia and on 1 September, 2,000 Luftwaffe planes bombed Polish cities while troops and hundreds of tanks overwhelmed the country in a blitzkrieg ("lightning war"). France and Britain declared war but Poland fell in early October as Soviet forces occupied its eastern half.

World War II

WEHRMACHT OFFENSIVE IN WESTERN EUROPE

France had built the Maginot Line in 1932, a gigantic maze of underground works and barracks, all linked by tunnels and supplied with electric power, food, and ammunition. This great fortification was intended, in future war, to spare France the high casualties of World War I. Their guns, however, were short ranged and only limited air defense was provided. Most important was the fact that it stopped at the Belgian border from which the Germans would invade France. Conversely, Germans had the Stuka fighter-bomber that could make deadly attack dives, with the pilot-operated siren heightening the fear below.

In May 1940, using blitzkrieg tactics, German forces invaded the Netherlands, Belgium, and Luxembourg, smashing through defenses at Sedan in three days. British and French troops retreated to the Channel port of Dunkirk, and every available small boat from England sped to the rescue of 250,000 troops. The massive rescue saved 200,000, but the 50,000 left behind were either killed or became prisoners of war. The Germans crossed the Somme and Seine rivers, and on 10 June, Italy declared war on France. Italian advances were successfully repelled.

Meanwhile, the French army fell apart; hundreds of thousands of civilians filled roads with cars, carts, bicycles, and walking casualties, even as the government left Paris for Bordeaux. Paris was declared an open city and Marshal Henri Philippe Pétain told his people by radio to stop fighting. Although Pétain signed an armistice with Germany, Gen. Charles de Gaulle left for London on 17 June to organize a resistance movement.

Britain Alone

In July 1940, as the Nazi flag flew in France, Belgium, Denmark, and Norway, Prime Minister Winston Churchill in London gave his famous "Blood, toil, tears, and sweat" speech. Britain's ports were fortified and miles of trenches and shelters built to guard against invasion. But first, Hitler had to control the skies.

Battle of Britain. For 57 consecutive nights, 200 Luftwaffe bombers dropped explosives and incendiary and time bombs, leaving London in flames and many towns in the south almost completely destroyed. Still, the Germans failed to wreck British fighter bases, and the Royal Air Force and antiaircraft guns inflicted heavy damage on the German air force. (By 31 October 1940, the Luftwaffe had lost nearly 1,750 of the 2,200 planes initially employed.) Hitler saw that invading Britain was out of the question and canceled invasion plans on 12 October. Nevertheless, Air Marshal Göring persisted, sending 500 bombers carrying 600 tons of bombs to destroy the city of Coventry on 14 November.

The courageous Allied pilots were British, Norwegian, Dutch, Belgian, Polish, and French. Radar (radio detection and ranging) enabled the defenders to know when and where the Germans would strike. Britain girded for a long fight with the manpower and resources of its dominions (Australia, New Zealand, South Africa, and Canada), its colonies (in India, Kenya, and Nigeria), and its protectorates (such as Egypt) activated. Raw materials and arms were shipped from the United States.

Vichy France. Germany annexed Alsace and part of Lorraine and occupied the northeastern two-thirds of France. Southern France was declared a "free zone," headed by Marshal Pétain. He and Pierre Laval passed the Law on Jews and closely collaborated with Hitler. In November 1942, following Anglo-American landings in French-controlled North Africa, the Germans crossed into the "free zone," thus controlling the entire country.

Barbarossa

The Barbarossa plan (named for the 12th-century German emperor) ignored the nonaggression pact of 1939 in which Hitler and Stalin divided up Europe. Hitler had no fear of the Red Army after the fiasco in Finland, when the Soviets' mass attack on so small a nation during the Winter War (30 November 1939–13 March 1940) revealed poor leadership, coordination, and judgment. The 105-day war offered the go-ahead to invade Russia and reach Moscow, the Volga River, and the Caucasus Mountains before winter. Beginning on 22 June 1941, 3 million men, 3,000 tanks, and 3,000 planes crushed the surprised Soviet first line of defense. By 10 July, the Germans had advanced 400–800 kilometers (250–500 mi.) into Soviet territory. The Soviets lost hundreds of tanks; tens of thousands of men died and 300,000 were taken prisoner.

Stalin's defense during retreat was to destroy transportation, cattle, cereals, and oil and encourage guerrilla (partisan) attacks on the invaders. During these battles, the effectiveness of the Russian T-34 medium tank was an unwelcome surprise to the Germans. Hitler ordered some troops south to conquer Caucasia's oil fields and coal mines in the Donetz Basin. Others headed for Leningrad. On 30 August 1941, the Germans isolated Leningrad for a 900-day siege. The Soviets dug 480 kilometers (300 mi.) of trenches, put up 32 kilometers (20 mi.) of barricades, and built 15,000 blockhouses. Conditions during the siege were appalling, starvation was rampant, and many died of the cold. People ate crows, cats, and dogs and burned books and furniture for heat. Even so, half a million men, women, and children died before the siege was lifted in January 1944.

In October 1941, German forces totaling 1.5 million men, 1,800 tanks, and 1,500 planes pushed toward Moscow—320 kilometers (200 mi.) distant. The goal was to reach there before winter, but the "fifth element," weather, sided with the Soviets. Autumn rains and snowstorms stalled or halted the German advance, first with seas of mud, then with numbing cold. Reinforcements from the Far East, along with trenches, shelters, and antitank ditches hastily built by Muscovites, plus Marshal Georgi K. Zhukov's orders that "anybody leaving his post will be shot," helped even the odds. On 2 December, the Germans were within 24 kilometers (15 mi.) of Moscow, but they were exhausted, cold, and without sufficient food and fuel. On 6 December, the well-equipped and well-led Red Army launched a counteroffensive and pushed the Germans back. In retreat, they abandoned trucks, tanks, and other vehicles. The German losses in killed, wounded, or taken prisoner totaled 800,000.

PEARL HARBOR

At 7:50 A.M., Sunday, 7 December 1941, Japanese torpedo bomber and dive-bomber planes, backed by fighters, attacked Pearl Harbor, Hawaii, where most of the U.S. Pacific fleet was concentrated. The first bombs caught the Americans by surprise, and they were unable to respond. A second wave of 170 planes finished the job, and by 9:45 A.M. the attack was over. The damage: 2 battleships were sunk outright, 6 were damaged; 159 planes were destroyed; 2,334 Americans were killed and 1,341 wounded. The three U.S. aircraft carriers in the Pacific were at sea at the time and so were unhurt. The submarine base and shipyard, along with the vital underground oil storage tanks, also survived unscathed.

President Roosevelt addressed the nation by radio and referred to the attack as "a day which will live in infamy." Congress immediately voted funds, and four days later, Germany and Italy declared war on the United States. War was now on a truly global scale.

The Japanese captured the American possession of the Philippines in May after an unexpectedly difficult six-month campaign, and Gen. Douglas MacArthur, who had left for Australia in a submarine in March, vowed, "I shall return." In rapid succession, the Japanese captured the British colonies of Hong Kong in December 1941 and overwhelmed Dutch Indonesia in March 1942, Malaya and Singapore in April 1942, and Burma in May 1942.

The Americans soon struck back. On 18 April 1942, Gen. Jimmy Doolittle's "Tokyo Raiders" took off from the flight deck of the carrier *Hornet* and bombed Tokyo. A Japanese effort to seize strategic Port Moresby in southeastern New Guinea was foiled in the Battle of the Coral Sea (7–8 May 1942). The U.S. carrier *Lexington* was sunk, but the Japanese lost the light carrier *Shoho*. The fleet carrier *Shokoku* was damaged and the carrier *Zuikaku* lost so many aircraft that it, too, was out of action.

The decisive Japanese naval defeat came at Midway. On 5 May 1942, 6,000 soldiers and sailors, 40 large warships, 8 aircraft carriers with 500 planes, and 2 additional ships—all commanded by Adm. Isoroku Yamamoto—headed for Midway atoll, 1,440 kilometers (900 mi.) from Japan. The capture of Midway would give the Japanese protection against any surprise attack from the east while simultaneously threatening the U.S. west coast. American intelligence sources had broken the communications code used by the Japanese navy. This enabled Adm. Chester Nimitz to read Japan's plans.

On 3 June, Yamamoto's forces launched an air attack preparatory to troop landings. There was extensive damage, but the following day, waves of American dive bombers along with torpedo planes appeared, swooping down on the Japanese air and sea fleet. They destroyed rows of enemy planes readying for a strike on the flight decks, and set three of the four carriers aflame. Later, the fourth carrier was sunk by another U.S. air attack. Without air protection, Yamamoto ordered a retreat. His navy had lost a dozen ships, including a heavy cruiser and four of Japan's six fleet carriers; 250 planes were shot down. More than 3,500 men were killed, including 100 of Yamamoto's best airmen. The Americans lost one carrier, the U.S.S. *Yorktown*, which had been damaged at the Coral Sea battle. The Battle of Midway was mourned as a major defeat in Japan and showed that the aircraft carrier was the key to success in the Pacific.

BATTLE OF THE ATLANTIC (3 SEPTEMBER 1938–
8 MAY 1945)

On 3 September, the day war was declared, German submarines sank the British liner *Athenian*. After the fall of France, the German U-boat blockade of Britain, under the command of Adm. Karl Dönitz and operating out of Brest, Lorient, Saint-Nazaire, and La Pallice, sank more ships than were built during that time. The British protected their shipping with convoys shepherded by escorts equipped with sonar; losses remained high in the Atlantic and elsewhere as 400 German submarines operated freely from Greenland to the Gulf of Mexico and roamed the east coastal waters of the United States and from the coast of Africa to South America. In five months, the United States lost 505 ships, of which 112 were oil tankers. The Germans increased their U-boat production from 60 in 1939 to 250 in 1942. They often operated at night in groups known as "wolf packs." They surfaced, fired torpedoes on convoys, then submerged. They hid in Norway's fjords to attack convoys heading for the USSR; in the Mediterranean, they attacked convoys sailing to the Middle East and Egypt. Allied convoys began sailing with warships and small aircraft carriers, employing improved detection procedures and depth charges, which finally reversed Allied losses. During 1943–44, U-boats suffered heavy losses; by the war's end, the Allies had sunk hundreds of German U-boats and 85 Italian submarines.

THE MEDITERRANEAN

Mussolini invaded Albania in 1939 and in September 1940 launched an attack on British-held Egypt from Italian-held Libya. Two months later, Italy attacked Greece. The Italians were poorly equipped and trained and their arms were often mediocre. The Greeks inflicted a humiliating rout on Italian forces there (December 1940) and the British mauled them in Libya, but Gen. (later field marshal) Erwin Rommel's Afrika Korps saved the Italians from total disaster. Provoked by an anti-Axis coup in Yugoslavia, the Germans invaded Greece and Yugoslavia in April 1942. Both countries surrendered within three weeks and British Commonwealth forces sent to aid Greece were evacuated, some to Crete.

On 20 May, German paratroopers carried out the first large-scale airborne operation of the war as more than 20,000 paratroopers landed on Crete. By month's end the British had withdrawn and the island was in German hands.

In Libya, Rommel—nicknamed the Desert Fox—tangled with British, Canadian, New Zealand, Australian, Indian, and Free French forces in the desert. In the summer of 1942, Rommel overcame the British and occupied Tobruk, crossing the Egyptian border and opening the road to Alexandria and Cairo. At El Alamein, however, the British put up a solid defense and strengthened their position against the Afrika Korps (July–September). Rommel's forces were exhausted and their equipment worn out; in November, the British took the initiative. After a grueling battle (23 October–4 November 1942), Rommel retreated to Tunisia. North Africa was lost. By then, British and American troops had landed in Morocco and Algeria in Operation Torch (8–11 November 1942).

THE HOLOCAUST

In 1942, the Germans rounded up Jews by the thousands and deported them or packed them into ghettos. The Germans pillaged occupied countries for food and money and rationed everything. Food was a major problem as was manpower for factories. "Volunteers" from France, Belgium, and Holland were sent to Germany to work. In France, work service was mandatory. Special units (SS Einsatzkommando) were sent to Russia in 1941 to kill Jews. The "final solution," begun on 20 January 1942, involved shipping Jews to concentration camps where old people and children went directly to the gas chambers and able-bodied men were put on work details and given starvation rations until they died. Six million Jews perished in the genocide.

Although the Jews suffered the most, Gypsies and other ethnic minorities, homosexuals, and political and religious dissidents perished by the hundreds of thousands in death camps like Sobibor, Auschwitz, Treblinka, Bergen-Belsen, and Buchenwald.

STALINGRAD

The conquest of the USSR was critical, although the Wehrmacht was halted outside Moscow and Leningrad. Elsewhere, the industrial center and major port of Stalingrad spread over 40 kilometers (25 mi.) along the right bank of the Volga River. By the end of August, the German army of Gen. Friedrich Paulus was on the verge of success. The northern part of the city of 800,000 was aflame from bombings, and buildings had been reduced to rubble. Defenders formed groups and fought the enemy in bitter, costly street battles, often hand-to-hand, but by November, Paulus's forces controlled the city's center. Hitler ordered the offensive to continue even though Paulus reported his forces were weakened by the cold and snow.

Marshal Zhukov planned a large offensive, bringing together a million men, 6,000 tanks, and 3,000 planes. He organized them in pincers, and struck on 14 November 1942. Within five days he had surrounded 300,000 Germans; neither the Luftwaffe nor German armies from elsewhere could help Paulus's beleaguered Sixth Army. On 24 January 1943, Paulus informed Hitler of the desperate situation; 150,000 had already died. Hitler refused to allow surrender. On 31 January, Paulus and his staff were captured, German resistance stopped, and 99,000 were taken prisoner. Thousands of dead Germans were piled between railroad tracks and set afire. In Germany, there were two days of mourning.

During the Battle of Kursk (5–17 July) one of the largest tank battles in history was fought. The Wehrmacht deployed 3,200 tanks and 5,000 guns on a 256-kilometer (160-mi.) front. The Red Army's superior firepower resulted in a massacre, and after two days of fighting, hundreds of armored vehicles were destroyed and tens of thousands of men killed or wounded. The Germans were driven back hundreds of kilometers by partisans and guerrillas, obliterating any future offensives in Russia.

LOGISTICS

World War II was fought as much in factories as on the battlefield. Soldiers used highly sophisticated, costly equipment that was replaced when the old wore out or became outdated. Britain began the war with 185,000 soldiers and by mid-1940 had a million and a half. Those between the ages of 18 and 50 were mobilized and trained with broomsticks instead of guns. Women were recruited for passive defense or auxiliary forces.

In Germany, the Wehrmacht increased from 5.6 million in 1940 to 9.5 million in 1943. Albert Speer, Germany's minister of equipment and ammunition, was also responsible for armament and war production, increasing fighter plane production from 771 at the war's beginning to 12,740 during the last six months of conflict. In 1939–44, Britain had 70,000 planes and tripled its tonnage of ships. In the USSR, 1,300 factories were dismantled and transferred to the Ural region. Germans considered the Soviet T-34 tank the best of World War II. Until 1942, the Allies were the only ones who had radar, and by the end of 1944, the Luftwaffe became the first to use jet fighter planes. The Messerschmitt Me 262 was an effective fighter aircraft and the V-1 and V-2 rockets caused great damage in Britain during the last eleven months of war.

America's goal on 6 January 1942 was to produce 60,000 planes and 45,000 tanks that year, and 125,000 planes and 75,000 tanks the following year. Within four years, the United States built 65,000 landing craft, 320,000 pieces of artillery, 15 million small arms, and more than a million trucks, and produced 4 million tons of ammunition. The United States shipped aid to Stalin via Iran, building two ports linked to Teheran by a railway 1,100 kilometers (700 mi.) long. The United States supplied 35 percent of the arms used to fight Germany and 85 percent of the arms used against Japan.

At the war's beginning, there were 7 to 8 million Americans unemployed. All this changed as the young, elderly, and women moved from country to cities where defense industries provided jobs. Blacks left the south for California and the northeastern cities. Sweden designed a gun that required 450 hours of work in Sweden but only 10 hours on American assembly lines. Shipbuilders put Liberty ships to sea in as little as twelve days and, at full capacity, Boeing built six B-29 bombers every day.

THE PACIFIC

The war in the Pacific was one against distances. The Americans crossed the longest distances ever imposed on vessels and merchant ships in a major war. For the Japanese forces, sea routes had to be protected from U.S. submarines, planes, and ships so that raw materials (metals, rubber, and copper) could reach factories in Japan. In the Pacific, economic strength counted as much or more than individual bravery, and Japan had only one-tenth of U.S. industrial and technical strength and power. Americans had sophisticated radar and antiaircraft guns on vessels and by mid-1942 were producing more and better aircraft than the Japanese. In the Solomon Islands during 1942–44, dependable supplies of modern planes and ships ensured U.S. victory.

Admiral Yamamoto was shot down during a flight over the Pacific on 7 August 1943, and a year later, the Americans launched an assault on the Marianas Islands. Japanese forces in their island garrisons often fought to the death, preferring suicide to surrender. During the reconquest of the Philippines from October 1944 to August 1945, the Japanese used bomb-loaded suicide planes—"kamikazes"—for the first time, and they were at first horribly effective. In the February 1945 Iwo Jima campaign, soldiers carrying explosives around their belts threw themselves against U.S. command posts or under planes on the ground. Improved tactics and warning gradually reduced the effect of aerial kamikazes, but the fanaticism disturbed U.S. strategists. The battle of Okinawa in the spring of 1945 was fanatical in the extreme. Only a few Japanese surrendered, preferring suicide to the dishonor of defeat. After three months of continual fighting, 100,000 Japanese were dead, including 7,000 kamikazes. U.S. losses were 12,000. The final step was to invade Japan itself, with 60 million people defending the homeland.

NORMANDY

On 5 June 1944, Allied convoys sailed across the English Channel toward the French coast. D-Day, code-named Operation Overlord, was planned by Gen. Dwight D. Eisenhower, with British Field Marshal Bernard L. Montgomery coordinating the movement of all Allied forces. Three million Allied soldiers and thousands of vessels were involved. British admiral Louis Mountbatten built artificial ports that were towed across the Channel and assembled off the landing beaches, enabling 40,000 tons of equipment and 6,500 vehicles to be unloaded each week. Gas came by pipeline across the Channel.

From midnight to dawn of that day, 10,500 British and American planes bombed the Normandy coast, and frogmen cut the barbed wire. At 2:00 A.M., 27,000 parachutists landed along the coast. On the so-called Longest Day, ten divisions with arms and supplies landed in Normandy. The British captured Bayeux on 8 June and linked with the Americans. The liberation of France had begun.

The Germans suffered a humiliating defeat. In the ensuing campaign, more than 500,000 of them were killed, wounded, or taken prisoner. Battlefields were littered with equipment and the rotting corpses of men and horses. On 15 August 1944, the Allies landed in the south of France and, with the help of Free French Forces, joined up in Burgundy a month later with the Normandy forces.

On 19 August, Paris revolted and was liberated with the help of the 2d Armored Division of Free French Forces under General Leclerc and American forces. An attempt by German officers to assassinate Hitler on 20 July failed, and 5,000 conspirators were executed. A counteroffensive in the Ardennes, beginning on 16 December 1944, resulted in the bitter and costly month-long Battle of the Bulge—a German defeat.

YALTA

Churchill, Roosevelt, and Stalin met at Yalta 2–10 February 1945 to plan the future. The USSR agreed to enter a war against Japan and the future occupation of Germany was settled, along with defining the role of the United Nations. As Germans retreated everywhere, Allied bombings had left their cities in ruins and 600,000 civilians dead. At the same time, 10 million refugees fled from the Soviets. Desperate for food, civilians found road and rail traffic at a standstill; trains had been bombed and no canals or rivers could be used.

THE END

Although V-2 rockets, Messerschmitt 262 jets, and sophisticated submarine equipment, including the snorkel, gave hope to the Germans, atomic scientists building nuclear weapons were the worst threat. Technology could not be perfected in time and countries fell one by one: Romania on 23 August 1944, Bulgaria on 28 October; Poland was liberated in January 1945 and Yugloslavia in April of that year.

On 7 March 1945, U.S. forces crossed the Rhine River, and in the east, the Russians approached the outskirts of Berlin. On 25 April, Russian and U.S. troops met on the banks of the Elbe. By the end of April, the Soviets had a million men and 15,000 guns encircling Berlin. Hitler entered his bunker with its 15-foot-thick walls and, on 30 April, committed suicide. On 8 May, in the French city of Rheims, Germany signed the unconditional surrender; on the following day, the ceremony was repeated in Berlin.

HIROSHIMA

After the fall of Okinawa, the U.S. blockaded Japan and during July 1945 bombed Tokyo every day. On 17–18 July, six waves of 1,500 bombers destroyed Osaka and Yokohama as the Japanese swore they would fight to the death to defend national soil. President Roosevelt died on 12 April 1945, and Harry S Truman became the U.S. president. Estimates were that it would cost a million U.S. casualties to conquer Japan proper, although these figures may have been too high. A secret weapon was needed to force Japan's surrender.

On 6 August 1945 at 8:15:17 A.M., the B-29 bomber *Enola Gay* opened her bomb-bay doors. Suspended from a parachute, a device descended on Hiroshima. Fifty-one seconds later, at almost 600 meters (2,000 ft.), it exploded. A gigantic mushroom cloud rose more than 14,900 meters (50,000 ft.) into the sky. Tens of thousands of people were buried in the rubble; many vanished and more were burned or mutilated. Fires were everywhere and the destruction was great. The atomic bomb killed 130,000, with survivors suffering irreversible changes in their bodies.

The USSR attacked Manchuria, and the United States dropped a second atomic bomb on Nagasaki, killing 35,000 and wounding 60,000. In a radio broadcast on 16 August, Emperor Hirohito announced to an astonished nation that Japan had surrendered. The six-year war was over.

Korea

Japan's defeat in World War II led to the partition of Korea along the 38th parallel. The cold war between the United States and the Soviet Union was the implied cause of the conflict in Korea in 1950–53.

It began on 25 June 1950 when the armies of North Korea, a protégé of the Soviet Union and Red China, invaded South Korea across the 38th parallel of latitude. The United Nations Security Council condemned the aggression, and on 27 June, President Truman ordered U.S. land, air, and sea forces to the defense of South Korea.

Eventually, fifteen UN members joined the U.S. "police action": Australia, Belgium, Canada, Colombia, Ethiopia, France, Great Britain, Greece, Holland, Luxembourg, New Zealand, the Philippines, South Africa, Thailand, and Turkey, with medical aid coming from Denmark, India, Italy, Norway, and Sweden.

At first, the Communists pushed back the UN forces nearly into the sea at Pusan. The United States was surprised and ill-prepared, while the People's Korean Army (PKA) had been well supplied by the Soviets with modern military equipment, including 150 T-34 tanks, 100 Yak fighters, and other heavy military equipment and artillery. There had been recruitment and training by Soviet advisers, and the PKA's 135,000 men were divided into seven infantry divisions with 122mm howitzers and self-propelled 76mm guns.

The Republic of Korea (ROK) forces were only an upgraded paramilitary constabulary of 95,000 men, with mortars and light artillery but no tanks. On 10 July, Gen. Douglas MacArthur was appointed commander in chief of the United Nations command and on 13 July Lt. Gen. Walton H. Walker (one of Patton's corps commanders in Europe during World War II) established his Eighth U.S. Army in Korea (EUSAK). On 23 December, General Walker was killed in a jeep accident and replaced by Lt. Gen. Matthew B. Ridgway. In the face of near defeat, Walker had ordered a defensive perimeter to protect the vital port of Pusan. There were 47,000 U.S. combat troops and 45,000 ROK troops holding the lines until reinforcements could be brought in. Names like "The Bowling Alley" appeared in news headlines to describe battles for the Naktong Bulge and the defense of Taegu. The ultimate success of the Pusan perimeter was due to the inability of the PKA to resupply its rapidly advancing troops. Conversely, U.S. forces were comparatively well supplied with men and materiel.

The Inchon landing had been carefully planned by MacArthur months in advance. A seaborne attack from the enemy's rear gained control of enemy communications in Seoul. Strategically, all north-south roads and rail links passed through the city, and to capture it proved politically significant to ROK forces. On 15 September 1950, the following forces participated in the massive Inchon operation: the 1st U.S. Marine Division from the U.S. mainland, the 7th U.S. Infantry Division stationed in Japan, 30 landing craft (LSTs) from Japan, the 5th Regiment of the 1st Marine Division from the Pusan perimeter, South Korean troops, the 7th Regiment U.S. Marines of the Sixth Fleet from the Mediterranean, and 230 ships from navies of the United States, Great Britain, Australia, Canada, New Zealand, South Korea, and France. On 25 September, aerial reconnaissance reported that PKA troopers were leaving Seoul, but some units remained behind to fight a desperate rearguard action. Three days later, UN forces had complete control of the city. Throughout the entire operation, fighting was furious and bloody, but the victory was decisive and took just thirteen days.

Of the 165,000 North Koreans who invaded the South, not more than 25,000 to 30,000 made it back to the North. Far East Air Forces (FEAF) under the command of Lt. Gen. George E. Stratemeyer provided essential air cover and close support.

Following the success of the Inchon operation, MacArthur ordered an invasion of North Korea and by late November UN troops occupied most of the country. But there were ominous signs of large-scale intervention by

Chinese communist forces. Despite the lateness of the season and the Chinese communist threat, MacArthur launched an operation on 24 November to clear all of North Korea. Less than a day later, a massive Chinese offensive, employing 180,000 troops in eighteen divisions, outmaneuvered and nearly overran the UN troops. The 1st Marine Division was almost surrounded in the mountains around the Chosin Reservoir and had to fight its way out in the bitter cold through eight Chinese divisions over nearly impassable mountain terrain to reach the port of Hungnam for evacuation (27 November–9 December). The division's commander, Gen. Oliver Smith, was quoted as saying, "Retreat, hell! We're just attacking in a different direction!" General Ridgway eventually established a solid front line, running roughly along the 38th parallel.

UN forces had been driven from North Korea; over the next seven months, the front seesawed back and forth with vicious attacks and counterattacks (January–July 1951). The intent was to influence the intermittent peace talks at Panmunjom (November 1951–July 1953). On 27 July 1953, after a costly repulse of yet another Chinese communist offensive (10–30 June 1953), an armistice was signed. Neither the North nor South Korean regimes is reconciled to the permanent division of the nation, and the demilitarized zone (DMZ) remains heavily armed and patrolled. North and South Korea were admitted to the United Nations in 1991, along with the three Baltic nations and two South Pacific nations.

Vietnam (1961–75)

In July 1950, one month after the outbreak of the Korean War, the United States sent a Military Assistance and Advisory Group (MAAG), along with three DC-3 Dakotas, to bolster the French-supported puppet government of Emperor Bao Dai in Saigon. It was the beginning of the longest war in American history, but its roots went back to World War II, when President Roosevelt declared that Vietnam should not be returned to France as a colony. During the cold war, President Truman favored neutrality and withdrew American advisers of the OSS (later the CIA), who had been working with the guerrilla forces of Ho Chi Minh fighting the Japanese.

The French Indochina War of 1946–54 ended with the fall of the French fortress of Dien Bien Phu (7 May 1954). Indochina was divided into four parts: Laos, Cambodia, and two Vietnams divided along the 17th parallel. Ho Chi Minh's forces were in the north, and the Saigon government controlled the south. In 1956, a civil war broke out between the Communists in the north (Vietminh) and the "democratic" U.S.-backed nationalist forces in the south. Fighting was mainly guerrilla, the Viet Cong fighting the Army of the Republic of Vietnam (ARVN). In 1961, U.S. forces were authorized to fight alongside ARVN units, but they had virtually no authority over their ARVN counterparts, made no battlefield decisions, had no disciplinary role, and were unable to remove incompetent commanders. Americans advised ARVN forces on everything from battlefield tactics and logistics to communications and intelligence. They also worked with the CIA, AID, and USIA and helped with health matters, medicine, finance, and agriculture. The Green Berets (Special Forces) trained the South Vietnamese in commando tactics and worked with the ethnically different Montagnards, who disavowed loyalty to South Vietnam but nevertheless remained faithful to the Americans.

On 2 August 1964, North Vietnamese patrol boats reportedly attacked two U.S. destroyers in the Gulf of Tonkin. In hindsight, they probably did not actually attack the ships in international waters. At any rate, President Lyndon Johnson received congressional authority to repel armed attacks, and U.S. bombing raids began in earnest. American ground forces were sent to fight as allies alongside ARVN forces. North Vietnamese army units marched continually down the "spiderweb upon spiderweb," as the Ho Chi Minh Trail in Cambodia was called. Teaming up with the Viet Cong, their goal was to liquidate the enemy. By late 1967, there were half a million U.S. troops in Vietnam, and together with ARVN forces, they began "search and destroy" missions employing "free fire zones" coupled with a pacification policy to win the "hearts and minds" of the Vietnamese people. Meanwhile, there was continual bombing of Viet Cong targets and supply dumps in Cambodia.

In January 1968, the North Vietnamese Army (NVA) and the Communist-formed National Liberation Front (NLF) based on the Viet Cong launched the Tet Offensive against 36 cities. Intelligence had underestimated the strength of the Tet offensive—although the NLF and NVA forces were defeated and suffered heavy casualties, the U.S. and ARVN forces were also badly mauled. In the United States, antiwar demonstrations captured the headlines and by 1968–69 a policy of Vietnamization, in which the Vietnamese would do more of the fighting, gradually replaced the active participation of U.S. troops. Most of the 500,000 Americans were to be returned home, leaving behind the air force to continue air support.

On 30 March 1972, the NLF and NVA launched another major offensive, crossing the demilitarized zone (DMZ) at the 17th parallel and capturing a South Vietnamese province. After heavy combat, ARVN forces halted the offensive with the help of U.S. air support (May). At that point, the United States mined the Haiphong harbor and other North Vietnamese ports. Peace talks, which had been going on from time to time, broke down, and in December 1972, President Richard Nixon ordered eleven days of intensive bombing of North Vietnamese cities.

Eventually, peace talks were resumed and a cease-fire agreement was reached on 27 January 1973. Fighting continued, however, with each side accusing the other of violations. In 1974, the ARVN began withdrawing troops from distant outposts, and the NLF quickly took them over. In January 1975 the expected offensive began. The NLF gained the central highlands; when the South Vietnamese government evacuated the northern cities of Quang Tri and Hue, it was clear the struggle was nearing the end. Southern coastal towns and villages were abandoned; soldiers and civilians fled, and the remaining U.S. forces escaped by sea and air in ignominious defeat. On 30 April 1975, U.S. Marines serving as a rear guard were flown out of Saigon from the rooftop of the American embassy. Two hours later, Gen. Duong Van Minh, who had replaced President Thieu, announced Saigon's unconditional surrender in a radio and television broadcast. General Minh, who had led the coup that deposed Ngo Dinh Diem twelve years earlier (1–2 November 1963), appealed to all Saigon troops to lay down their arms and avoid further bloodshed. Fearing reprisals, ARVN troops hastily discarded uniforms and changed into civilian clothes.

Two hours later, North Vietnamese and Viet Cong troops rolled into the city on tanks, armored vehicles, and camouflaged Chinese-built trucks. The red, blue, and yellow-starred flag of the NLF was raised over the presidential palace; North and South Vietnam were formally reunited as the Socialist Republic of Vietnam on 2 July 1976.

Israel

The state of Israel was created in the aftermath of World War II to provide a homeland for survivors of the Holocaust. In 1947, the UN voted to replace the British mandate with a sovereign Jewish state in a part of Palestine. Arabs were opposed to this, and in May 1948 invaded Israel. In the end, after seven months of intermittent but bitter fighting, Israel occupied disputed areas in Palestine, and 400,000 Palestinian Arabs settled in refugee camps in neighboring Arab countries. In 1956, the nationalization of the Suez Canal by Egypt's president Gamal Abdel Nasser precipitated another crisis, leading to Israel's conquest of Sinai and the Gaza (29 October–6 November). In the end, Israel turned over the Gaza Strip, all of Sinai, and Sharm-el-Sheik to the United Nations forces.

The Six-Day War (5–10 June 1967), triggered by Nasser's militant nationalism, left Israel in possession of Sinai, the West Bank, Gaza, and the Golan Heights, but the basic causes of conflict were still unresolved. Egypt and Syria attacked Israel (6 October 1973), but at war's end (25 October), Israelis had gained the upper hand. Although Israel and Egypt signed a peace treaty after U.S.-moderated negotiations (26 March 1978), peace remained elusive. An Israeli invasion of southern Lebanon (June–August 1982) displaced the Palestine Liberation Organization (PLO) presence temporarily, but military action again failed to produce a lasting peaceful settlement.

Unrest continues in the Mideast as Israel swells the population of the West Bank and Gaza Strip with Jewish immigrants from the former USSR and elsewhere. The Palestinian Intifada—an independence movement—continues to be at odds with Israeli policies.

Gulf War, 1991

When his troops invaded Kuwait on 1 August 1991, Iraqi leader Saddam Hussein apparently expected neighboring Arab states to acquiesce to his aggression. Instead, the local wrath and concern that he provoked rapidly grew into worldwide condemnation. Only the PLO and a handful of countries—Jordan, Libya, Mauritania, the Sudan, and Yemen—gave Iraq any significant political support during the crisis. The opposing international coalition of 38 nations gathered under the UN aegis provided a textbook demonstration of the application of collective security.

After several months of threats and standoffs during Operation Desert Shield, the war began in earnest after Iraq ignored UN demands that it withdrew unconditionally by 15 January 1991. On 17 January the first phase of the air war began when U.S.-led air units launched a devastating series of attacks on targets in Iraq. The coalition quickly achieved air superiority and then proceeded to focus on targets of military significance within Iraq. These included command, control, and communications facilities; radar and surface-to-air missile sites; power-generating plants; transportation facilities (e.g., airfields, runways, bridges, and major roadways); and nuclear, biological, and chemical weapons development and production facilities. Toward the end of the air war (after about four weeks), coalition airpower began to concentrate on Iraqi forces in the field in preparation for the ground war. An intensive series of air strikes was launched, using precision-guided weapons against Iraqi armor and artillery, backed by B-52 and F-111 bombers and heavy attack fighters like the Tornado. Coalition airpower destroyed or incapacitated all of Iraq's nuclear reactor facilities, eleven chemical and biological weapons storage facilities and three production facilities, 50 percent of Iraq's major command centers, 70 percent of its military communications, 48 Iraqi naval vessels, and 75 percent of Iraq's electric power-generating capability. Iraq's only means of retaliation was to launch its long-range modified Scud missiles against Israel and Saudi Arabia and to set fire to about 600 Kuwaiti oil wells. The Scud attacks were largely ineffective, while the oil fires did nothing to affect allied air operations or slow the pace of the war.

With Iraq distracted by air bombardment (from both land and sea forces) and by the threat of an attack by amphibious forces stationed in the Gulf, coalition land forces moved secretly from the southeast Kuwait-Saudi border 320–480 kilometers (200–300 mi.) westward to the Iraq-Saudi border. The element of surprise played a key role when the ground war began on 24 February 1991.

Within 48 hours, coalition troops from the south swept through Kuwait and established positions south and west of Kuwait City, while those forces that had secretly moved far to the west launched a deep strike northward into Iraq.

On 26 February, in the longest sustained armored advance in history, U.S. forces destroyed key Republican Guard divisions holding the area north of the Kuwaiti border, while other coalition forces began fighting for control of Kuwait City. A devastating series of engagements over the next two days virtually destroyed Iraq's remaining forces. British and U.S. Army forces liberated northern Kuwait, Saudi and other Arab forces liberated Kuwait City, and U.S. Marine forces secured the southern and western outskirts of Kuwait City.

The war was quick and decisive, due to the weakened and demoralized state of the Iraqi army and the superior technology and tactics wielded by the coalition's motivated, professional forces.

Summary

The 1991 Gulf War marked a major change in modern history and war. By the late twentieth century, many nations, both industrialized and developing, had acquired the modern equipment of war, including weapons of mass destruction. At the same time, with the dissolution of the Soviet Union, the long confrontation between the superpowers began to wind down. As of mid-1992, the threat of massive nuclear warfare has been significantly reduced, yet the threat of nuclear, chemical, and biological warfare at a lower level of intensity may be increasing dramatically. New states have been created and new power blocks and political/military alliances will continue to form; universal peace still lies in the future.

Relatively new features in conflict, primarily since World War II, are the increased use of peacekeeping forces to separate combatants and enforce truces and the use of multinational forces under UN or other aegis to restrain aggression (e.g., the Korean War, the 1991 Gulf War). The era of East-West polarity is apparently over, an era in which virtually every conflict of any magnitude involved and mirrored, in some way, the interests of the superpowers and the cold war. Although there may be more local conflict and tensions in the future, global conflict seems to be less of a danger.

During the past 200 years, military affairs have changed from relatively primitive strategy and weaponry to sophisticated computer-assisted long-range planning, "invisible" bombers, the Strategic Defense Initiative, and nuclear weaponry. Communications and airlift capabilities have brought conflicts into instant focus, but peace settlements and long-term resolution of ethnic, religious, economic, and political conflicts remain elusive. Environmental and social issues, migrations, and immigration are probable trouble areas of the future.

ELOISE H. PAANANEN
DWAYNE ANDERSON

SEE ALSO: Arab-Israeli Wars; Boer Wars; Civil War, American; Civil War, Russian; Civil War, Spanish; French Revolutionary–Napoleonic Wars; Gulf War, 1991; Korean War; Russo-Japanese War; Vietnam and Indochina Wars; World War I; World War II.

Bibliography

Allinson, A. R. 1926. *The war diary of Frederick III 1870–1891.* Westport, Conn.: Greenwood.
Bailer, S., ed. 1969. *Stalin and his generals.* New York: Pegasus.
Calvocoressi, P., and G. Wint. 1972. *Total war: The story of World War II.* New York: Pantheon.
Catton, P. 1962. *The Army of the Potomac.* 3 vols. New York: Fairfax.
Chandler, D. G. 1966. *The campaigns of Napoleon.* New York: Macmillan.
Dupuy, R. E., and W. H. Baumer. 1968. *Little wars of the U.S.* New York: Hawthorne.
Engle, E., and A. Lott. *America's maritime heritage.* Annapolis: U.S. Naval Institute Press.
Engle, E., and L. Paananen. 1973 and 1992. *The winter war.* New York: Scribner's, and Harrisburg, Pa.: Stackpole.
Esper, G., and Assoc. Press. 1983. *The eyewitness history of the Vietnam War.* New York: Associated Press.
Fest, D. C. 1974. *Hitler* (Eng. transl.). New York: Harcourt Brace Jovanovich.
Harbottle, T. 1971. *Dictionary of battles.* New York: Stein and Day.
Humble, R. 1974. *Napoleon's peninsular marshals; a reassessment.* New York: Taplinger.
Moore, R. 1965. *The Green Berets.* New York: Crown.
Nevins, A. 1950. *The emergence of Lincoln,* vols. 1 and 2. New York: Scribner's.
Noyes, A. H. 1934. *Europe—Its history and its world relationships.* Boston: Heath.
Pierre, M., and A. Wieviorka. 1987. *The Second World War* (Eng. trans.). Morristown, N.J.: Silver Burdett Press.
Rees, D., ed. 1984. *The Korean War: History and tactics.* New York: Crescent.
Smith, S. E., ed. 1966. *The United States Navy in World War II.* New York: Morrow.
Warner, O. 1973. *Great battle fleets.* London: Hamlyn.
Welsh, D. 1982. *The USA in World War 2: The Pacific theater.* New York: Galahad.

HO CHI MINH [1890–1969]

In many senses Ho Chi Minh is the "father of his country"; he seems to have meant even more to Vietnam than Mao Tse-tung did to China or Joseph Stalin did to the Soviet Union. Ho Chi Minh ("He who Enlightens"), also known as Nguyen Ai Quoc ("Nguyen the Patriot") as well as by a host of other aliases, led Vietnam for some 30 years. He was president from 1945 to 1969 and chairman of his party's Central Committee. More than just a Vietnamese patriot, Ho was one of Asia's foremost anticolonialists and one of the twentieth century's most influential leaders of communist movements.

The Early Years—The Wanderer

Ho was born Nguyen That Thanh (or Nguyen Sinh Cung) on 14 (or 19) May 1890 in Huong Tru hamlet, Nghe An Province, Central Vietnam—the French Protectorate of Annam, later to be North Vietnam. He was, like his colleague Vo Nguyen Giap, the son of a poor mandarin scholar. (His father, Nguyen Sinh Sae, left the family to become an itinerant teacher.) After a dreadful childhood, Ho managed to attend grammar school in Hue from age 14 to 18; later, as Van Bu, he taught school in various villages and also worked in Saigon. In 1911 (or 1912), Ho (now called Bu) became a cook on a French freighter and started his 30-year odyssey. Before he returned to Vietnam, he had visited such U.S. ports as Boston and New York. After sailing around the world for three years, he lived in Brooklyn for a year. Moving from the United States to London, he worked from 1915 to 1917 as a gardener, sweeper, waiter, photo retoucher, and stoker.

In the midst of World War I, Ho moved to France, where he remained until 1923. Then known as Nguyen Ai Quoc (Nguyen the Patriot), he organized a group of Vietnamese living in Paris. At Versailles, in 1919, Ho presented an eight-point petition to representatives of the Great Powers (Fig. 1). His demands centered on equal rights for Vietnam from the colonial French. There was little response from the great powers, but Ho became a hero to the Vietnamese.

Using his Nguyen Ai Quoc pseudonym, he was a founding member of the French Communist party on 30 December 1920. At that time, he was both speaking and writing on his primary ideological notion that anticolonial nationalism and socioeconomic revolution are inseparable. In 1922 (or 1924), impressed by Soviet victories—and tracked by French security—he went to Moscow to study Marxist doctrine at the School of Oriental Workers. He also met with the leaders of the Soviet Union. In 1925, he was assigned to Canton, China, as a Comintern agent. While in southern China, he formed the Vietnamese Revolutionary Youth League to campaign for Vietnamese independence. This led to the forming of the Indochinese Communist party (Lao Dong) in Hong Kong in 1930, with an amalgam of Vietnamese, Lao, and Khmer communist groups. The group received instructions from Moscow based on the Communist Internationale (Comintern). This activity led to Ho's imprisonment by the British in 1930. After his release in 1933, he traveled widely in China, Thailand, and elsewhere in Asia. In 1936, the French released his comrades, who fled to China. Ho returned to Moscow in 1938, remaining as one of the notably few Comintern agents to survive the Stalinist purges.

The Middle Years—Back in Vietnam

In 1940, the Japanese invaded Indochina. Ho aligned himself with the Allied powers, believing that Japanese domination was no better than French imperialism. Using the name Ho Chi Minh, he returned to Vietnam in May 1941 and organized the Vietminh (Vietnamese Doc Lap, Long Minh, or Dong Minh Hui—League for Vietnamese Independence). Initially, this was to be a coalition of all anti-French Vietnamese groups. Ho also chaired the eighth plenary session of the Indo-Chinese Communist party.

In 1942, Ho Chi Minh went to China again to ask the Chinese Nationalist government for help in operating against the Japanese, but he was arrested as a Communist agent and jailed until 1943, when the Chinese released him so he could organize an anti-Japanese resistance movement in Vietnam. Ostensibly copying Stalin, he proclaimed his movement "nationalist"—not Communist—anti-Japanese *and* anti-French.

During World War II, Ho cooperated with the Allies against the Japanese—and with the French, supposedly disbanding the Indo-Chinese Communist party in 1940. As Stalin was looked upon and called "Uncle Joe," so Ho Chi Minh became Vietnam's much loved "Uncle Ho." He cultivated a successful relationship with the United States through OSS agents who were acting against the Japanese in Indochina. In 1945, Bao Dai, the emperor of Vietnam who had cooperated with the Japanese, abdicated in Ho's favor.

By the end of the war, Ho was the most well known indigenous leader in Vietnam. On 2 September 1945, he declared Vietnamese independence, borrowing language from the U.S. Constitution as part of his declaration, and attained political power. In 1946, the French returned to Vietnam. Although at the May 1946 conference at Fontainebleau, Ho was prepared to compromise and work toward independence within a French union (to be something like the British Commonwealth), by September the plan broke down. By late 1946, the Vietminh were fighting the French forces using guerrilla tactics.

Figure 1. Ho Chi Minh as a member of the French Socialist party at the Versailles Peace Conference, 1919. (SOURCE: U.S. Library of Congress)

The Later Years—Independence

In 1950, Ho formed and proclaimed the Democratic Republic of Vietnam, which was rapidly recognized by China, the USSR, and many Soviet-bloc countries. True independence was won in battle when Vo Nguyen Giap and his Vietminh forces beat the French decisively at Dien Bien Phu. There had been peace talks in Geneva in April, and after Giap's 7 May victory, a truce was signed on 20 July 1954 that separated Vietnam into North and South; Laos and Cambodia were independent. The subsequent South Vietnamese Diem regime did not collapse as anticipated, and the United States supported it with both military and economic assistance. Ho, who had consolidated his power in the north between 1954 and 1960, organized, in 1960, the National Liberation Front and began supporting the (Laotian) Pathet Lao and (Cambodian) Khmer Rouge. He gave up his secretary-general post to lead more symbolically as the head of state. But Ho drove on to achieve his goal of Vietnamese unification and national independence together with domination of the French-created Indochina. Even though he used brutality to implement his plans, Ho Chi Minh was and is still recognized as the "father of his country."

Ho died 2 September 1969 and was accorded the dignity of a Leninist, Maoist-style mausoleum. In May 1975, when the North Vietnamese Army and Viet Cong forces captured Saigon and South Vietnam, Ho was further memorialized when the city of Saigon was renamed Ho Chi Minh City—an empty honor since Saigon remains the city's name to most of the world, including many of the city's own residents.

Donald S. Marshall

See Also: Giap, Vo Nguyen; Vietnam and Indochina Wars.

Bibliography

Fall, B. 1968. *Ho Chi Minh on revolution.* New York: Signet.
Halberstam, D. 1971. *Ho.* New York: Random House.
Ho Chi Minh. 1966. *Prison diary.* Hanoi: Foreign Language Publishing House.
———. 1966–67. *Selected works.* 4 vols. Hanoi: Foreign Language Publishing House.
Lacoutre, J. 1968. *Ho Chi Minh: A political biography.* New York: Random House.

HONDURAS, REPUBLIC OF

Honduras, the poorest of the Central American nations, borders Nicaragua, Guatemala, and El Salvador. Honduras has been the subject of great controversy throughout the 1980s because of the presence of anti-Sandinista Nicaraguan guerrilla camps inside its borders.

Power Potential Statistics

Area: 112,090 square kilometers (67,254 sq. mi.)
Population: 5,319,800
Total Active Armed Forces: 17,500 (0.329% of pop.)
Gross Domestic Product: US$4.9 billion (1990 est.)
Annual Defense Expenditure: US$82.5 million (1.9% of GDP, 1990 est.)
Iron and Steel Production: none
Fuel Production: none
Electrical Power Output: 2,023 million kwh (1990)
Merchant Marine: 173 vessels; 527,481 gross registered tons
Civil Air Fleet: 9 major transport aircraft; 134 usable airfields (8 with permanent-surface runways); none with runways over 3,659 meters (12,000 ft.); 4 with runways 2,440–3,659 meters (8,000–12,000 ft.); 13 with runways 1,220–2,440 meters (4,000–8,000 ft.).

For the most recent information, the reader may refer to the following annual publications:
The Military Balance. International Institute for Strategic Studies. London: Brassey's (UK).
The Statesman's Year-Book. New York: St. Martin's Press.
The World Factbook. Central Intelligence Agency. Washington, D.C.: Brassey's (US).

History

Honduras was granted independence from Spain on 15 September 1821. Since independence, it has tottered between democracy and military rule as witnessed by its 136 coups and revolutions.

In 1821 Honduras became a member of the Central American Federation. By 1829 Honduran Francisco Morazan became the ruler of the Central American Federation when he overthrew the regime of Manuel Arce. Tensions between Arce's conservatives and Morazan's liberals grew during the Morazan regime, and by 1838 the Federation recognized its inability to govern such a diverse federation. On 31 May 1838 the Federation declared that individual provinces could establish independent governments. On 28 October 1838 Honduras officially seceded from the Federation, and by January 1839 had adopted its first constitution.

By 1840 General Francisco Ferrera had firmly entrenched his conservative government. In 1847 General Ferrera passed the reigns of power to Juan Lindo Zelaya. Ferrera's popularity declined after a costly and unsuccessful attempt to invade El Salvador and destroy what little remained of the Central American Federation. The Lindo years were marked by internal violence, and disputes with the British about debts and the ownership of the Bay Islands. Honduran sovereignty over the Bay Islands was finally recognized by Britain in 1859, during the regime of the liberal Trinidad Cabanas. From 1852 until 1891, the Liberal Party retained power in Honduras.

During the mid- to late 1800s, Honduran politics were greatly influenced by its more powerful neighbors, Guatemala and Nicaragua. From 1873 until 1883, Honduran presidents held power at the pleasure of the Guatemalan dictator General Justo Rufino Barrios. By 1891 Nicaraguan influence began to dominate the Honduran political scene when the Liberal party of Honduras was returned to power with the help of Nicaragua's dictator, José Santos Zelaya.

In 1906 Central America erupted into war when Honduras invaded Guatemala. Although repulsed, the Guatemalan invasion was followed in 1907 by an invasion conducted by Nicaraguan-supported Honduran exiles. (During this same period, El Salvador and Nicaragua also waged war.) The conflict in Central America was ended only after the United States intervened with marines and naval forces to protect the primarily U.S.-owned banana plantations in Honduras. U.S. pressure resulted in the Central American Peace Conference of 1907, and the General Treaty of Peace and Amity of 1907. The treaty of 1907 committed the signatories to prohibiting invasion attempts of other states from their territory by exiles. The treaty further established Honduras as a neutral state.

In 1908 U.S. president William Howard Taft identified the huge Honduran debt as the greatest factor of Honduran instability, and by 1909 the US$120 million Honduran debt was refinanced by U.S. entrepreneurs in exchange for control of the Honduran national railroad. The debt reduction appeared to have come too late, however, and Honduras continued in a state of political turmoil until its most recent return to civilian rule in 1978. The years prior to 1978 were highlighted by the "Soccer War" of July 1969 in which El Salvador briefly invaded Honduras, and by intermittent border clashes with Nicaragua spanning 70 years.

In 1978 continued tensions with El Salvador prompted the cancellation of travel permits to that country. Honduran internal problems were also compounded by a growing number of refugees from the civil war in Nicaragua, and the scene was set for yet another coup d'état. On 7 August 1978, General Policarpe Paz Garcia led a bloodless coup and formed a three-member military junta with the stated intention of an eventual return to democracy. During elections in 1980 and 1981 a legislature and president were chosen. On 20 January 1982 a democratic constitution went into effect. In 1985 a democratic transfer of power occurred once more, with the election of José Azcona Hoyo as president. During the Azcona administration, Honduras was the victim of limited invasions by Nicaragua in March 1986 and in April 1988. The Nicaraguan invasions were apparently aimed at the destruction of Nicaraguan Contra guerrilla camps. During the 1988 invasion by Nicaragua, a U.S. brigade task force was deployed to Honduras as a guarantee of Honduran security. Rafael Leonardo Callejas Romero was elected president in 1989 and sworn in on 26 January 1990.

Politico-Military Background and Policy

The precedent for political meddling by the Honduran military has been firmly established throughout Honduras's short history. As of 1989 civilian rule exists in Honduras, but only at the pleasure of a well-entrenched military power base. Under the constitution, the military retains the right to veto both legislation and executive acts; an ever-present threat of a military takeover tempers the actions of the Honduran government.

The Honduran government's tolerance of the Contra guerrilla camps inside its borders was mainly attributable to agreements by the United States to provide modern military equipment, training, and other security assistance. The Contra camps were also viewed as a means of moving thousands of Nicaraguan refugees out of Honduran cities and an already strained economy. The resultant growth of a U.S. military presence in Honduras is greatly resented by a deeply nationalistic and growing number of Hondurans.

Strategic Problems

The greatest strategic threat to Honduras is the size of the Nicaraguan military, which currently exceeds the size of the combined armies of all five other Central American nations. The 1990 election of Violeta Barrios de Chamorro as president of a noncommunist government in Nicaragua has ameliorated that threat, but it will remain until true stability emerges in Nicaragua.

Honduras is also fighting terrorism and a low-level insurgency conducted by several small guerrilla groups, the most active of which is the Popular Revolutionary Forces–Lorenzo Zelaya.

Military Assistance

The United States maintains 1,000 military personnel in Honduras on a permanent basis. In recent years U.S. troops have constructed airfields and roads in defense zones adjacent to the Nicaraguan border. U.S. commercial charter aircraft including C-130s and C-7s have been used to carry materiel inside Honduras. The exact amount of military aid delivered to Honduras is unknown, but reasonable estimates indicate over US$30 million dollars in aid during the 1980s.

Defense Industry

Honduras has no domestic defense industry.

Alliances

Honduras is a member of the Organization of American States and the United Nations. The security of Honduras was guaranteed in statements by U.S. president Ronald Reagan.

Defense Structure

The Honduran military includes the army, air force, navy, and a 5,000-man Public Security Force. Military reserves number approximately 60,000 men. Although the constitution calls for mandatory conscription with a term of service of 24 months, conscription actually takes place by means of random roundups of youth in an action referred to as "the harvest."

(For an explanation of the abbreviations and symbols used in the following section of military statistics, see the list of Abbreviations and Acronyms in each volume.)

Total Armed Forces

Active: 17,500 (reducing); (12,000 conscripts). Terms of service: conscription, 24 months.
Reserves: 60,000 ex-servicemen registered.

ARMY: 14,400 (10,500 conscripts).
10 Military Zones:
 4 inf bde (12 inf, 3 arty bn; 1 armd cav regt).
 1 arty bde (regt).
 1 territorial force, 1 SF, 1 engr bn.
Reserves: 3 inf bde.
Equipment:
Light tanks: 12 Scorpion.
Recce: 3 Scimitar, 1 Sultan, 72 Saladin, 12 RBY Mk 1.
Towed arty: 105mm: 24 M-101/-102; 155mm: 4 M-198.
Mortars: 400 60mm; 81mm; 120mm: 60 Brandt; 160mm: 30 Soltam.
RL: 84mm: 120 Carl Gustav.
RCL: 106mm: 80 M-40A1.
AD guns: 80: 20mm: incl M-55.

NAVY: 1,000 incl 500 marines (700 conscripts). Bases: Puerto Cortés, Puerto Castilla (Atlantic), Amapala (Pacific).
Patrol Craft, Inshore: 11: 3 Guaymuras (US Swiftships 31-m) PFI; 2 Copan (US Lantana 32-m) PFI (; 6 other PC (plus boats.
Amphibious: craft only; 1 Punta Caxinas LCT; plus some 3 ex-US LCM.

MARINES: (500); 1 bn.

AIR FORCE: some 2,100 (800 conscripts); 48 cbt ac, no armed hel.
FGA: 2 sqn: 1 with 13 A-37B; 1 with 10 F-5E, 2 -F.
Fighter: 1 sqn with 8 Super Mystère B2 (to be replaced).
Transport: 9 C-47, 1 C-123, 2 C-130A, 2 DHC-5, 1 L-188, 2 IAI-201, 1 IAI-1123, 1 -1124.
Liaison: 1 sqn with 4 Baron, 3 Cessna 172, 2 -180, 2 -185, 4 Commander, 1 PA-31, 1 PA-34.
Helicopters: 9 Bell 412, 4 Hughes 500, 7 TH-55, 8 UH-1B, 11 UH-1H.
Training: 4* C-101BB, 2 Cessna 180, 2 -185, 11* EMB-312, 2* T-33, 5 T-41A.

PARAMILITARY
Public Security Forces (Ministry of Public Security and Defense) 5,000: 2 bde.

FOREIGN FORCES
U.S.: Army: 800.
United Nations (ONUCA): elm.

Future

Economic development has been hampered by the necessity of a military buildup to counter Nicaragua, and widespread corruption and the investment of profits in foreign countries. Continued adverse economic conditions are likely to fuel a growing insurgency inside Honduras unless the current cycle of regional armament can be halted.

LEE A. SWEETAPPLE

SEE ALSO: Central America and the Caribbean; El Salvador; Nicaragua.

Bibliography

American University. 1983. *Honduras: A country study.* Washington, D.C.: Government Printing Office.
Hunter, B., ed. 1991. *The statesman's year-book, 1991–92.* New York: St. Martin's Press.
International Institute for Strategic Studies. 1991. *The military balance, 1991–1992.* London: Brassey's.
Stockholm International Peace Research Institute. 1987. *Yearbook: World armaments and disarmament.* New York: Oxford Univ. Press.

HOST NATION SUPPORT

Host nation support is the civil and military assistance rendered, in peace and war, by a host nation to forces of an allied nation that are located in, or in transit through, the host nation's territory.

In general terms, whenever an army, air force, or navy receives food, fuel, or any other logistic assistance from a friendly nation, it is receiving host nation support. The practice has been common for hundreds of years, but it is only since World War II (1939–45) that it has been recognized as a specific area of logistic planning.

The growth of the concept is best seen in the host nation support agreements implemented by nations of the North Atlantic Treaty Organization (NATO). From initial negotiations on storage, these have expanded to include the use of facilities and services and, eventually, the provision of military units to provide direct logistic support to incoming formations.

In the early days of warfare, host nation support was involuntary because armies tended to live off the land and simply take what they wanted from their enemies, allies, and even their own people. In the twentieth century, the demands of modern warfare, with its vehicles, artillery, and aircraft, made living off the land impractical. Large and elaborate schemes for provision of logistic support were incorporated into the armies themselves. Even so, there have always been demands for support made on the nations in which the fighting takes place. Usually, this involves the quid pro quo of support for help in fighting the war.

Background of Host Nation Support in NATO

It has always been the responsibility of a commander to provide for the logistic support of his force. This responsibility can never be discarded, but in NATO certain factors have forced nations to adopt a more flexible approach. As part of its commitment to the defense of Western Europe, the United States has undertaken to provide ten divisions and 30 fighter aircraft squadrons in Europe within ten days of the declaration of a crisis. Since not even a wealthy nation like the United States is able to afford the cost of maintaining such a large force permanently in Western Europe, only four divisions are in place. The remainder will travel from the U.S. mainland.

Although the United States can move the men and aircraft in the time scale required, it would take far longer to transport the heavy equipment and logistic support they need. Some way had to be found to alleviate this movement problem, and the first solution was to arrange for the heavy equipment and materiel required for the incoming six divisions to be pre-positioned in Europe.

Development of the Concept

The success of the storage program led military planners to examine other possibilities. Western Europe has many facilities and services of use to allied forces. Canada, the United Kingdom, and the United States reached agreement with their European allies on the provision of a variety of functions, including joint use of ports, airfields, and communication facilities, plus civil transport resources within Europe.

The next stage was an agreement between the United States and the Federal Republic of Germany. The Germans undertook to furnish 93,000 military reservists to provide combat service support for the incoming U.S. forces, including transport, maintenance, communications, and labor units. This agreement also included the provision of services by the German civil authorities.

Requirements for Successful Host Nation Support

The concept of host nation support is easy to grasp, but the planning required can be a long and laborious process. It offers obvious advantages in the efficient use of resources and facilities, full utilization of transport, and savings in manpower and equipment. Its success, however, depends on:
1. the willingness of commanders to entrust the logistic support of their forces to the authorities of another nation,
2. a firm commitment by both nations to the concept, and
3. detailed negotiation and planning at every level.

Before the events of 1989 in Eastern Europe, Warsaw Pact arrangements for host nation support were similar to, although probably more demanding upon the individual countries than, those of NATO. The Soviet concept of total war included the entire economy, industry, and population of the Soviet Union. This concept undoubtedly also applied to the other Warsaw Pact nations, which were expected to provide support for Soviet and other Pact forces operating in their territory as well as for their own forces.

N. T. P. MURPHY

SEE ALSO: Assistance, Military; Assistance, Mutual; Logistics, NATO.

Bibliography

Lawson, R. 1984. Wartime host nation support. *NATO's Sixteen Nations*, February–March.

HOSTAGE

A hostage is a person kept or given as a pledge for the fulfillment of certain agreements. Hostage taking dates back to the Roman tradition of educating the sons of allied monarchs in Rome in order to guarantee the ally's loyalty. In medieval England the families of knights lived in castles their king controlled as a means of guaranteeing the knights' loyalty in battle. In 1911 the Geneva Protocols outlawed reprisals against prisoners of war, thus differentiating the modern prisoner of war from the hostage.

In recent years the popularity of hostage taking has boomed as terrorist groups and outlaw nations have found that by taking hostages they are able to capture the attention of the entire world, if only for a few brief moments. The phenomenon of political terror and the helplessness felt by would-be adversaries of the superpowers has popularized hostage taking as a political weapon, and separates this crime from the category of common kidnapping for pecuniary reward.

The act of hostage taking on a large scale has been made logistically feasible with the growing popularity of long-range international air travel among government officials, businessmen, and tourists. This increase in foreign travel by citizens of targeted nations enables terrorists and hostile governments to seize hostages in areas beyond the reach of their nation's law enforcement agencies and military forces. International air travel has created a magnet for hostage taking in the form of air piracy because terrorists who have seized an aircraft can easily relocate to countries sympathetic to their cause. The capture and holding of hostages aboard an aircraft also complicates rescue operations and has, therefore, become the most popular form of modern hostage taking.

Although much recent attention has been focused on hostage taking, the phenomenon is not new. In 1192, Duke Leopold of Austria captured and held hostage King Richard I of England as he traveled through Vienna on his way home from the Crusades. The conditions for Richard's release consisted of limited control over his kingdom and a huge ransom that affected Richard's entire kingdom. Today's media ensure that the modern hostage taker will have a great effect on the hostage's home country, even though the hostage himself has far less political significance than a head of state.

The phenomenon of hostage taking was burned into the public memory with the massacre of Israeli athletes in Munich in September 1972. Then, in 1979, the world held its breath as Iran seized and held hostage the staff of the U.S. embassy in Tehran. Hostage takers with political goals had now discovered the power of the world media, and the foreign policy of the industrialized nations soon reflected a respect for the disproportionate power wielded by potential hostage takers.

The specter of hostage taking looms daily as nations and businesses spend billions on counterterrorism and hostage survival training. Physical barriers, evasive driving techniques training, personnel protection experts, and risk assessment consultants are all part of a thriving industry based on fear. The protection of American diplomats was estimated to have cost US$200 million in 1982. The fear of national embarrassment at the hands of terrorists has had other far-reaching effects. In June 1985 the U.S. Army issued special guidance procedures for all of its members entitled, "Code of Conduct Guidance for Personnel Subject to Terrorist Captivity." These guidelines came on the heels of the hijacking of TWA flight 727 and the murder of a U.S. Navy diver who happened to be a passenger. The U.S. Army's reaction included the implementation of annual terrorism awareness training for all personnel. In February 1989 the United States evacuated its embassy personnel from Afghanistan when a deteriorating military and political situation threatened a repeat of Tehran 1979.

The proliferation of hostage taking and the heralded example of the attempted, although unsuccessful, rescue of the Israeli athletes in Munich has prompted most of the world's armies to create special purpose forces trained primarily for the rescue of hostages. The success of such hostage rescue units is attested to by the rescue of Brig. Gen. James Dozier by an Italian counterterrorist squad on 28 January 1982. The rescue of General Dozier was a great success in the war against terrorism; however, the Italians had the rare luxury of conducting the operation on their home soil.

The emergence of state-supported terrorism has made it necessary for nations to rescue their citizens far from their national borders. The international cooperation required to accomplish rescue missions presents the greatest stumbling block, as the fear of terrorist retaliation drives the foreign policy of those nations that are needed for transit or staging areas. The military capability to conduct a long-distance rescue was dramatically illustrated by the Israeli raid on Entebbe in 1976.

Modern hostage taking has been shown to affect the world economy and the foreign policy of nations, but the victim taken hostage is also affected, even years after his or her release. In almost all cases of prolonged captivity, an unconscious emotional response to the stress of being held hostage has been shown to produce some degree of bonding between the captor and the captive. The term for captive–captor bonding is *Stockholm syndrome*.

The Stockholm syndrome was named after an incident that occurred in a Stockholm bank in which a failed robbery attempt became a 131-hour hostage drama. In the closing hours of the hostage negotiations, one of the victims began to side with his captors, and upon surrender the hostages voluntarily shielded the bodies of their captors from police gunfire. The captors were even visited in prison by their former captives, one of whom actually married one of the criminals.

Another graphic example of the Stockholm syndrome was Patricia Hearst's trial in the United States in which the Stockholm syndrome became the cornerstone of her defense against robbery charges. After being taken hostage by the Symbionese Liberation Army (SLA) terrorist group, Hearst is alleged to have been so overcome by the Stockholm syndrome that she assisted in a bank robbery while armed with an automatic weapon.

The bonding produced in hostage situations also affects the captors and is therefore used against them during hostage negotiations. Hostage survival training also depends heavily upon the Stockholm syndrome as a vehicle for obtaining better treatment during captivity and, ultimately, release.

Hostage survival training and extraordinary security measures promise to be with us for some time. Only after the world community unites against the scourge of terrorism will the specter of hostage taking abate. As long as the politics of fear prevails, even the strongest allies will think twice before cooperating with the other nations to prevent hostage taking.

LEE A. SWEETAPPLE

SEE ALSO: Crisis Management; Prisoner of War; Terrorism; Terrorism, Legal Aspects of.

Bibliography

Aston, C. C. 1982. *A contemporary crisis: Political hostage-taking and the experience of Western Europe.* Westport, Conn.: Greenwood Press.
Cassidy, W. L. 1980. *Political kidnapping: An introductory overview.* Boulder, Colo.: Paladin Press.
Mideast terror strikes Americans. 1985. *U.S. News* 24 (24 June):9–11.
Stanley, R. L. 1984. Hostage negotiations and the Stockholm syndrome. *Military Intelligence*, Oct.–Dec.
U.S. Army Command and General Staff College. 1984. *Terrorism counteraction.* FC 100-37, 15 July.

HUNDRED YEARS' WAR [1337–1453]

This conflict, certainly the longest and possibly the most famous of medieval wars, dominated the history of both England and France from 1337 to 1453, a period of 116 years. Before it was over, it had involved Scotland, Castile, Aragon, Burgundy, and the Holy Roman Empire as well as France and England. Fighting was not continual over the entire period, as the frequent truces lasted as long as twelve years.

The Causes of War and the Combatants

The causes of the war lay in questions of suzerainty and dynastic precedence, exacerbated by economics. King Edward III, eager for military glory, cited all of these in his declaration of war. Although on paper a war between France and England looked very one-sided because England's population of perhaps 3.25 million was outnumbered by France's 16 million, the worth of armies is not determined by numbers alone. Although the French could field armies of 30,000 men, these forces were badly organized and poorly led. The individual quality of the mounted men-at-arms and sergeants was high, but these troops did not fight well in units and held their humble infantrymen-cum-followers, who were generally poorly trained and indifferently armed, in disdain. The French also used crossbowmen, many hired from Italy.

The English army, although rarely able to field forces larger than 6,000 men, was much better organized than the French. The need for standing forces during the Welsh and Scottish wars of Edward I had led the English to raise, equip, and maintain soldiers on a long-term basis. These troops had high morale, and their armament was superior as well. The infantry Edward I raised was at first equipped with pikes, but by the 1290s many of the troops were armed with the Welsh longbow. Made of yew and about 1.8 meters (6 ft.) long, the longbow fired 0.9-meter (3 ft.) steel-tipped arrows (the famous clothyard shaft) a distance of up to 228.6 meters (250 yd.) and at a rate (in trained hands) of six to twelve per minute. Longbowmen in units of twenty *(vintennis)* and 100 *(centennis)* were organized to deliver controlled, accurate, and rapid missile fire at the direction of the commander. The English armies could do things in battle utterly beyond French capabilities.

Edward III's War

The first three years of war, waged in Flanders, consisted largely of futile maneuver. No battles were fought, and the principal effect of hostilities was to drain Edward III's treasury. However, Edward's great victory at the naval Battle of Sluys on 24 June 1340 ensured his control of the English Channel. The onset of a succession crisis in Brittany in 1341 opened a new theater of war, and Edward dispatched troops there to support the ducal claim of John de Montfort. An English army of 3,000 men under the earls of Northampton, Derby, and Oxford won a notable victory over the French army of about 12,000 under Charles of Blois at Morlaix on 30 September 1342.

French efforts against English Gascony had been frustrated by Edward's able lieutenant there, Henry de Grosmont, earl of Derby, who defeated the French at Auberoche in October 1345 and beat them again at Aiguillon the next year. Edward invaded France directly by landing in Normandy near Cherbourg with an army of 12,000 men in July 1346. He captured Caen and, after feinting toward Paris and maneuvering in Picardy, gave battle near Crécy. Edward's leadership, superior English weapons and organization, and French tactical incompetence led to the virtual destruction of the French army and gave Edward a great victory, despite French numerical superiority of more than five to two.

Following the repulse of a Scottish invasion of England at the Battle of Neville's Cross on 17 October 1346 and the fall of Calais to Edward in September 1347, the outbreak of the Black Death produced a truce that was renewed for seven years. War again flared in 1354, partly because of the ruthless and devious Charles the Bad of Navarre. Edward III's eldest son Edward, often called "the Black Prince" from the color of the armor he supposedly wore at Crécy, led the English armies. He undertook great raids, called *chevauchées*, through central France, in order to wreak such damage that the French king would be forced to submit. Regardless of the oft-demonstrated English battlefield superiority, the French remained unsubdued, even after a disaster like the Black Prince's victory at Poitiers on 19 September 1356, where he captured King Jean II le Bel (the Good).

Despite this success and the associated negotiating advantage, the French continued to resist. Edward led another invasion at the head of a large army in October 1359. Marching through northern France against minimal opposition, he compelled the French to accept the Treaty of Brétigny on 8 May 1360 and so gained all of Guyenne, as well as Calais, Guines, and a ransom of three million gold ecus for Jean II. The terms were never fulfilled, and after the end of the Breton War the French began an earnest campaign. Poor English leadership, the discontent of local nobles, and the inspired leadership of the French constable, Bertrand du Guesclin, led to considerable French success. English *chevauchées* in 1369, 1370, and 1373 were fruitless, and an English fleet was smashed by France's Castilian allies off La Rochelle in 1372. The death of the Black Prince in 1376, of Edward III the following year, and of du Geusclin in 1380 caused hostilities to wind down.

Henry V's War

The next three decades saw only nominal combat but included a disastrous English expedition to Ypres and Dunkirk in 1383 and a planned but unexecuted French invasion of England in 1385–87. Truces and peace negotiations in the 1390s almost ended the war, but Henry IV's displacement of Richard II in 1399, coupled with the outbreak of the Burgundian-Armagnac civil war in France in

1407, led to renewed war after Henry V succeeded his father.

Henry, the ablest leader on either side since du Guesclin, landed with an army of perhaps 7,000 men in Normandy in August 1415. He besieged and captured Harfleur on 14 September and marched for Calais. He hoped to avoid the formidably large French army on the way but was compelled to give battle at Agincourt on 22 October, and his army, reduced to perhaps 5,000 archers and 800 men-at-arms, defeated the ill-led and ill-organized French army of at least 25,000, mostly dismounted men-at-arms.

Campaigning vigorously, Henry conquered Normandy between 1416 and 1420. He also secured the Treaty of Troyes, which made him heir to the French throne in preference to Charles VI's son the dauphin, later Charles VII, on 21 May 1420. Henry's death on campaign on 31 August 1422, however, left the English government in the hands of his able but unruly and less brilliant brothers. Without Henry's enterprise, the English conquest of France slowed, although an Anglo-Burgundian army under the earl of Salisbury won a victory over the French at Cravant (31 July 1423) and the English defeated a Franco-Scottish army at Verneuil (27 August 1424).

The French Recovery

The turning point in the war came when the duke of Bedford, one of Henry V's brothers, led a small army to besiege the city of Orléans in 1428 and begin the conquest of the Loire Valley. Invigorated by the example of Jeanne d'Arc (Joan of Arc), the French were able to break Bedford's siege. Although the comte de Clermont failed to capture Sir John Fastolf's supply convoy at the famous Battle of the Herrings (12 February 1429), Fastolf and John Talbot, earl of Shrewsbury, were defeated by Jeanne d'Arc's army at Patay on 18 June. Although Jeanne d'Arc was captured and executed soon after, the defection of Burgundy in 1435 doomed the English cause, and they were slowly driven from their holdings in France, outnumbered and, for the first time, outfought and "outgeneraled."

In 1445, during a truce, French King Charles VII promulgated the *Grande Ordonnance* of 1445, which established a royal army composed of mounted men-at-arms, their pages and squires, and mounted archers, the first standing army in western Europe since the fall of Rome. This force, known as the *compagnies d'ordonnance* and backed by a large and efficient artillery train (largely the creation of the Bureau brothers), received its first test when the French invaded Normandy in 1449. An English army of 4,500 men under Sir Thomas Kyriell, sent to restore the situation, was destroyed at the Battle of Formigny on 15 April 1450 by two small French armies (a total of 5,500 men), under the count of Clermont and Artur de Bretagne, constable de Richemont. Turning their attention to Guyenne, the French conquered all of that area between 1451 and late 1452, except for the city of Bordeaux. The English dispatched an army under the earl of Shrewsbury to its relief, but Shrewsbury was killed and his army defeated when he launched a headlong and ill-considered attack on the fortified French camp at the Battle of Castillon on 17 July 1453. The fall of Bordeaux that October left the English holding only Calais and effectively ended the Hundred Years' War.

DAVID L. BONGARD

SEE ALSO: Edward I; Edward III; Feudalism; History, Medieval Military.

Bibliography

Allmand, C. I. ed. 1973. *Society at war: The experience of England and France during the Hundred Years' War*. Edinburgh: Oliver and Boyd.
———. 1976. *War, literature, and society in the late Middle Ages*. Liverpool: Univ. of Liverpool Press.
Burne, A. H. 1955. *The Crécy war*. London: Eyre and Spottiswoode.
———. 1956. *The Agincourt war*. London: Eyre and Spottiswoode.
Contamine, P. 1970. Les armées française et anglais à l'époque de Jeanne d'Arc. *Revue des Sociétés Savantes de la Haute Normandie: Lettres et Sciences Humaines*. 57:5–33.
———. 1972. *Guerre, état et société à la fin du Moyen Age*. Paris: La Haye, Mouton.
Cosneau, E. 1886. *Le Connétable de Richemont (Artur de Bretagne)*. Paris: A. Picard.
Curry, A. E. 1979. The first English standing army? Military organization in Lancastrian Normandy, 1420–1450. In *Patronage, pedigree and power in later medieval Europe*, ed. C. Ross. Gloucester and Totowa, N. J.: Rowman and Littlefield.
Fowler, K. A. 1969. *The king's lieutenant, Henry of Grosmont, First Duke of Lancaster, 1310–1367*. New York: Barnes and Noble.
———, ed. 1971. *The Hundred Years War*. New York: St. Martin's Press.
Hewitt, H. J. 1966. *The organization of war under Edward III, 1337–1362*. Manchester: Univ. of Manchester Press.
Oman, C. W. C. 1924. *The art of war in the Middle Ages*. 2 vols. London: Methuen.
Seward, D. 1977. *The Hundred Years War*. New York and London: Atheneum.
Solon, P. D. 1976. Valois military administration of the Norman frontier, 1445–1461: A study in medieval reform. *Speculum* 51:91–111.
Vale, M. G. A. 1982. *War and chivalry: Warfare and aristocratic culture in England, France and Burgundy at the end of the Middle Ages*. Athens, Ga.: Univ. of Georgia Press.

HUNGARY, REPUBLIC OF

Hungary, the ancient land of the Magyars, with its capital of Budapest situated where the Danube River turns south, sits at a strategic and cultural crossroads in Eastern Europe. It is strategic in the sense that transportation and communications corridors from the north and south, and the east and west, intersect or pass through the nation. Culturally and ethnically homogeneous, Hungary is surrounded by neighbors of dissimilar diversity—the rich ethnic soup that is the Balkans lies to the south; Czechs and Slovaks live across the northern border; to the east are Ukraine and Romania; Austria lies to the west.

The challenges that face Hungary are those that face all former Warsaw Pact nations: economic recovery, political stability, and ethnic peace.

Perhaps because of its ethnic homogeneity, combined with a sense of its vibrant history, Hungary has somewhat better adapted to the post-Communist challenges. Statistically, Hungary's economy is the strongest in the region. It not only has the lowest rate of inflation, but also the highest rate of export growth. Foreign investment in Hungary exceeds the combined total for all of Eastern Europe, except for the former East Germany. The one post–cold war election has seen a stable, three-party coalition government emerge under the guiding hand of Prime Minister Jozsef Antall. Despite such successes, and the respect with which the government is viewed internationally, difficulties remain. Strong ties to Germany evoke concern in some conservative-minded neighboring governments. In addition, there are more than 3.5 million ethnic Hungarians who are living in bordering countries where prejudice and discrimination are often the norm.

Power Potential Statistics

Area: 93,030 square kilometers (35,919 sq. mi.)
Population: 10,555,400
Total Active Armed Forces: 86,500 (0.820% of pop.)
Gross National Product: US$60.9 billion (1990 est.)
Annual Defense Expenditure: not available
Iron and Steel Production:
  Crude steel: 6.373 million metric tons (1988)
  Pig iron: 2.093 million metric tons (1988)
Fuel Production:
  Coal: 2.255 million metric tons (1988)
  Crude oil: 1.9 million metric tons (1989)
  Natural gas: 6,327 million cubic meters (1988)
Electrical Power Output: 30,400 million kwh (1990)
Merchant Marine: 17 vessels; 94,393 gross registered tons
Civil Air Fleet: 28 major transport aircraft; 90 usable airfields (20 with permanent-surface runways); 2 with runways over 3,659 meters (12,000 ft.); 10 with runways 2,440–3,659 meters (8,000–12,000 ft.); 15 with runways 1,220–2,440 meters (4,000–8,000 ft.).

For the most recent information, the reader may refer to the following annual publications:
The Military Balance. International Institute for Strategic Studies. London: Brassey's (UK).
The Statesman's Year-Book. New York: St. Martin's Press.
The World Factbook. Central Intelligence Agency. Washington, D.C.: Brassey's (US).

History

Hungary (*Hungavaria*, from the Latin) the "Land of the Huns and Avars" is located in Central Europe's Carpathian Basin. The Hungarians call themselves Magyars, after the Megyer tribe from which their ancestors selected their ruling prince, Almos. The ancestors of today's Hungarians lived (A.D. 520–880) between the Dnieper and Don rivers in what is today the Ukrainian SSR. They lived in a loose federation of some ten to twelve tribes, each tribe numbering about 50,000 to 60,000. They were well known to and chronicled by the Persians, the Arab Caliphate of Baghdad, and the Byzantine Emperor with whom they were frequently in a military alliance. Their closest neighbor, whose northwestern frontier they protected, was the Khazar Empire. Around A.D. 850, "the seven tribes"—Megyer, Nyek, Kurt-Gyarmat, Tarjan, Jeno, Ker, and Keszi—formed a nation. Their chieftains (Kende) held council in which they swore a "blood oath" of fidelity and elected Almos, Kende of the Megyers, as their Kagan or ruling prince. About 862, Arpad, Almos' son, married a Khazar princess. This princess' dowry included three Khazar tribes which brought the new nation the additional military force of 30,000 mounted archers.

Around 860 the Bulgarians invaded Byzantine territory, penetrating as far south as Thrace. The Magyars at this time were allied with the Byzantine Empire against the Bulgars, and with Arnulf, the Holy Roman Emperor (r. 887–899), against the Moravians under their King, Svatopluk (r. 892–894). They rendered military assistance to both, and in the process conquered Panonia Transdanubia, Transylvania, and the entire Carpathian Basin. Panonia had remained uninhabited since the Avar tribes were forced out through raids by the Franks under Charlemagne in 795. The occupation of the Carpathian Basin, which required moving some 600,000 people, cattle, horses, and related livestock about 800 kilometers (500 mi.), was finally accomplished by Arpad in 896. The Avar tribes in the northern Carpathians, and the scattered Szekely (Hunnish) tribes in Transylvania accepted Arpad's rule without resistance.

During the early part of the thirteenth century, the Kuhn tribes were forced out of their homeland around Lake Aral by the Mongols, and were allowed to settle in south central Hungary. The Kuhns were similar in tribal organization and ethnic lineage to the ninth century Magyars and were readily accepted. Two tribes of Pecsenegs settled into the northeastern Carpathians during the same period, completing the ethnic makeup of what today are known as the Hungarian people.

The Hungarians are not related in any way (linguistically, ethnically, or culturally) to any of their neighbors. In their lore, from the earliest times, they have considered themselves descended from the Huns of Attila. To a marked degree this was reinforced by their enemies in the

West who referred to them as Huns, as well as their on-again/off-again allies, the Byzantine imperial family, with whom the Hungarian royal family closely intermarried from earliest times.

Byzantium, in 895, was the cultural and commercial center of the known world, and its universities and libraries archived such books as the *Takitika* of Leo III, the Isaurian (r. 717–741), and the *De Administratio Imperio* of Constantine VII (r. 913–959). These works describe Hunnish military organization and tactics, language, and social structure as identical to that of the Magyars—an erroneous assumption. Since Magyar princes and military leaders (e.g., Bela IV was raised and educated in the East Roman capital) frequented the Byzantine court, they were naturally exposed to this interpretation of history.

The Magyars caused consternation throughout Europe. From 896 until the crowning of Stephen I as King of Hungary (Christmas Day, 1000) the Hungarian tribes conducted raids throughout Europe ranging from as far south as Naples to the Pyrenees in the West.

From Stephen I (r. 997–1038) until the death of the last Arpad king, Andrew III in 1301, the Hungarians maintained and solidified their national cohesion. Using their military strength and diplomatic skills, they managed to maintain their independence; becoming vassals to neither the Papacy nor the Holy Roman or Byzantine Empires.

The Mongol Invasion

The first great disaster to befall the Hungarian nation was the invasion of the Mongols under Batu Khan. On 10 April 1241, the Hungarian army, in an action at early dawn under the leadership of Ugrin Csaky, Archbishop of Kalocsa, successfully repelled the crossing of the Sajo River by the Mongol army, inflicting heavy losses on them. This resulted in a dangerous complacency in the Hungarian camp as the fearless Hungarians then perceived the Mongols as "beaten." The Mongols, in the manner of the Huns some 900 years earlier, moved at night and surrounded the camp. They attacked on the morning of 11 April 1241, causing such confusion that the entire Hungarian force of 20,000, never being allowed to group into an effective fighting unit, had to fight individually.

Realizing that all was lost, Archbishop Ugrin, Prince Kalman, and their retainers mounted their horses and threw themselves against the center of the Mongol force. They killed so many, including Batu Khan's Vizir, that the Mongols were put to flight. Only Batu Khan himself was able to halt the flight, crying out, "If we run, we'll surely die; if we must die, let us die here and now!"

It was at this precise moment that the tumens (units of 10,000 mounted archers) of Subotai Khan arrived and the Hungarian defeat was complete. After this defeat, although the king and some of his retainers managed to cut their way out of the encirclement and survive, the greater part of the country was devastated and depopulated. The Mongols remained only briefly; withdrawing abruptly with the death of the great Khan.

It is forever to the credit of Bela IV (r. 1235–70) that he managed to rebuild Hungary after the Mongols withdrew; he defeated invading Germans and Bohemians and repopulated the country with Germans and Slavs.

The Arpad Dynasty was followed in 1301 by the Anjou Kings, Robert I (r. 1307–1342) and Louis I. Louis I (r. 1342–82), the Great, in addition to being King of Hungary was also elected King of Poland.

The Regency and Beyond

With the regency of Janos Hunyadi (r. 1445–57), the Hungarians began to deal with the expansion of Islam in the Balkans, while still fending off repeated attempts by the Holy Roman Emperor and the Kings of Bavaria, Bohemia, and Poland to acquire Hungary as a fiefdom. The threat of Islam ended temporarily with the crushing defeat of the armies of Sultan Mohammed II by the Hungarian army under the leadership of Janos Hunyadi and his friend the Franciscan priest St. John Capistrano at Belgrade on 22 July 1456.

The election of Hunyadi's son, Mathias I (r. 1458–90), as King of Hungary gave the nation, now sorely pressed in on all sides, a leader of great military and political ability. Through the 32 turbulent years of his reign he managed to defeat repeated incursions by the Holy Roman Emperor, Frederick III.

Soon after Mathias' death, Hungary's decline began; ending with the disastrous defeat by the 300,000-man army of Sultan Suleiman the Magnificent at Mohacs. Following this defeat and the death of King Louis II (r. 1516–26), Hungary was divided among the Turks, the Hungarian Barons under the leadership of John I, Duke of Transylvania (King of Hungary, 1526–40), and Emperor Ferdinand I of Hapsburg (1526–64). From then on, the Hapsburgs, with the support of the Hungarian high nobility, ruled over Hungary—but as separately crowned kings of Hungary and not as Austrian emperors. Thus, Hungary managed to retain a degree of its historical independence.

Treaty of Trianon through World War II

The greatest of all disasters to befall Hungary, and that which is remembered by all Hungarians, is the Treaty of Trianon, 4 June 1920. This treaty divested Hungary of its 1,000-year-old frontiers. Hungary lost 71.4 percent of its territory, 63.5 percent of its population and 95 percent of its natural resources. It is ironic that, as a nation with a foreign monarch, Hungary had no vested interest in World War I from which this treaty resulted.

At the end of the war, Hungary's armies returned home

and demobilized without having suffered any great defeats. Following demobilization, Hungary elected a republican form of government. This government was quickly subverted by the more radical elements of the Hungarian Communist Party, who elected Bela Kuhn premier in March 1919.

While remaining in power, Bela Kuhn did try to reactivate the Hungarian army and defend Hungary from the allies—Serbia, France, and Romania—who occupied Hungary under the Treaty of Trianon. In the end, Bela Kuhn fled abroad and Hungary was partitioned.

The post–World War I period included the ouster of the Communist government and the establishment of a constitutional monarchy with Nicholas Horthy as regent (March 1920). This era also saw Hungary struggle to absorb the huge number of refugees from the territories lost under the Trianon Treaty and fight for the reestablishment of its national independence, free from Hapsburg rule.

The rise of Hitler's Germany, coupled with bitter feelings over the Treaty of Trianon and a national fear of Stalinist communism, brought Hungary reluctantly into the German-Italian Axis Pact. This culminated in Hungary's entry into World War II as an ally of Nazi Germany.

Post–World War II

The end of World War II brought with it the return to a republican form of government within the block of Soviet-occupied nations. By 1948, this government transitioned into a Soviet-style people's republic. Severe Stalinist and post-Stalinist rule by terror caused a bloody national uprising in 1956. Although the armed forces did not in any way support this attempt by the Hungarians to free themselves of Soviet rule, the Hungarian people managed to throw off Soviet occupation for five days, inflicting heavy casualties on the Soviets.

The uprising was quickly and brutally repressed, but the sacrifice of those who died was not in vain. A slow thaw in the controls exercised by the Soviet Union over Hungary came about—partly because of the fear of another unpleasant uprising, partly as a result of changes in the attitude of the Soviet Politburo after Stalin, and partly because of the affinity that has always existed between the Ukrainians (now occupying influential positions in Soviet government) and the Hungarians dating back to earliest times. After that time, Hungary became a model for economic development within the Soviet bloc.

As a result of Hungarian industry and the work ethic of its people, Hungary emerged from the crumbling Soviet Union in somewhat better shape than other nations. On 1 April 1991, the Warsaw Pact was dissolved, ceasing to exist as a military alliance; by the end of June all Soviet forces had been withdrawn from Hungary. Hungary has now embarked on a new phase of its long history.

Politico-Military Background and Policy, and Strategic Problems

With its central location relative to NATO and Eastern Europe, bulwarked by 7,767 kilometers (4,816 mi.) of rail and 29,796 kilometers (18,474 mi.) of excellent roads, Hungary remains militarily strategic. During the cold war, the country provided excellent staging and assembly areas for large ground and air forces that might threaten attack through northern Yugoslavia into northern Italy and southern France. In view of the ethnic warfare that has characterized much of Yugoslavia, the strategic position of Hungary continues to be important, although with a shift in potential significance.

Politically, however, Hungary's government is much more focused on economic recovery, and continued economic progress. The nation's military forces have been redistributed to provide a balanced presence in all parts of the country, instead of the previous westward orientation during the cold war. Military forces have also been cut, and conscription reduced by six months to a mandatory one year term of service.

Government policy, which must consider the feature of strong nationalist sentiment, relies on rhetoric about Hungary's glorious past, yet avoids confrontation with its neighbors along ethnic lines. Still, the possibility of ethnic-based conflict along Hungary's borders cannot be ignored.

Only two acknowledged disputes mar formal relations with Hungary's neighbors: a territorial dispute with Romania regarding Transylvania and a disagreement with Czechoslovakia regarding the Nagymaros Dam.

Military Assistance

Long a constituent of the former USSR, Hungary no longer receives military aid or assistance from any source.

Defense Industry

Hungary's defense industry is extremely modest, limited to production of look-alike versions of Soviet equipment (e.g., versions of armored personnel carriers). Some defense production will likely remain in order to maintain and service existing weapons and equipment. Electronic-based early warning upgrades to air defense deployments may be a feature.

A key aspect of Hungarian industry, however, is that it is one of the world's largest producers of gallium arsenide, the essential component of nuclear-resistant computer chip circuitry. Until recently, the Soviet Union was Hungary's major, and almost singular, customer for this important defense-related mineral.

Alliances

With the dissolution of the Warsaw Pact, many Eastern European nations sought to form agreements and alliances with their neighbors. One of the earliest, and most important, of such alliances is the Pentagonal, which spans membership of NATO, former Warsaw Pact countries, and neutral and nonaligned nations. Its original five members were Italy, Czechoslovakia, Hungary, Austria, and Yugoslavia. It became the Hexagonal with the admission of Poland in July 1991; Bulgaria and Romania have applied to become members.

Hungary and Poland have signed a bilateral treaty: a Military Co-operation Agreement, which allows for coproduction, procurement, and officer training and exchange; a symbolic statement was signed by Czechoslovakia, Hungary, and Poland at Visegrad (where the kings of these countries had signed a cooperation agreement in 1335). The statement concerned cooperation rather than competition in economic affairs, and addressed the need to consult on matters of security.

Defense Structure

Upgraded with more modern Soviet equipment prior to the dissolution of the Warsaw Pact, Hungary has substantial military means at its disposal. Recently, however, there has been some downsizing in terms of force strength, and one (of four) Scud brigades has been disbanded.

The Hungarian army of 66,000 members is organized into three corps headquarters, each with one tank division and three motorized rifle divisions, supported by one artillery brigade, one antitank brigade, and one air defense regiment. These forces are more evenly distributed over Hungary's geography than previously, and with increased emphasis on a balanced air defense.

One of the smaller air forces in Eastern Europe, the Hungarian air arm comprises about 20,000 members with over 100 combat aircraft and some in storage. Naval forces in landlocked Hungary are nonexistent, although there is a Danube flotilla of six patrol craft.

(For an explanation of the abbreviations and symbols used in the following section of military statistics, see the list of Abbreviations and Acronyms in each volume.)

Total Armed Forces

Active: 86,500 (45,900 conscripts). Terms of service: 12 months.
Reserves: 210,000: Army: 192,000; Air: 18,000 (to age 50).

ARMY: 66,400 (36,400 conscripts) incl Danube Flotilla.
1 Army, 3 Corps HQ: each Corps with 1 tk, 3 MR, 1 arty bde, 1 ATK, 1 SA-6 regt.
Army tps: 1 arty, 1 ATK, 1 SA-4 bde; 1 MRL, 1 AD arty regt, 1 AB bn.
(2 tk, 2 MR bde at Cat A, remainder at Cat B.)
Equipment:
MBT: 1,482: 62 T-34, (in store), 143 T-54, 1,139 T-55 (152 in store), 138 T-72.
Light tanks: 7 PT-76 (CFE: HACV).
Recce: 331: some 160 BRDM-2, 171 FUG D-442.
AIFV: 502 BMP-1 (8 in store).
APC: 1,261: 148 BTR-80, 40 FUG D-442, 1,042 PSZH D-944, 31 MT-LB. (Plus some 700 'look alike' types).
Total arty: 1,087:
  Towed: 591: 122mm: 234 M-1938 (M-30) (4 in store); 152mm: 49 M-1943 (D-1), 308 D-20 (6 in store).
  SP: 171: 122mm: 154 2S1 (3 in store); 152mm: 17 2S3 (12 in store).
  MRL: 122mm: 62 BM-21 (6 in store).
  Mortars: 120mm: 263 (9 in store).
SSM: launchers: 18 FROG-7, 9 Scud (all to be destroyed).
ATGW: 312: 117 AT-3 Sagger, 21 AT-4 Spigot (incl BRDM-2 SP), 174 AT-5 Spandrel.
ATK guns: 85mm: 162 D-44; 100mm: 106 MT-12.
AD guns: 23mm: 14 ZSU-23-4 SP; 57mm: 144 S-60.
SAM: 110: 18 SA-4, 44 SA-6, 44 SA-9, 4 SA-13; plus 240 SA-7, 54 SA-14.
Danube Flotilla:
MCMV: 6 Nestin MSI (riverine); boats.

AIR FORCE: 20,100 (9,500 conscripts); 111 cbt ac plus 34 in store, 39 attack hel.
1 air corps:
Fighter:
  2 regt with 60 MiG-21 bis/MF.
  1 sqn with 9 MiG-23MF.
  (Plus 34 MiG-21 bis/MF in store.)
Recce: 1 sqn with 11 Su-22.
Transport: 2 An-24, 12 An-26, 1 L-410, 1 Il-18.
Helicopters: 3 regt.
  1 ATK/tpt.
    2 sqn with 39 Mi-24.
    1 sqn with 23 Mi-8/-17;
  1 tpt with 25 Mi-8
  1 liaison with 35 Mi-2
Training: 25 *MiG 21, 3 *MiG 23, 3 *Su-22.
AAM: AA-2 Atoll.
AD: 1 bde: SAM regt, some 16 sites: 120 SA-2/-3/-5.

FORCES ABROAD
UN and Peacekeeping:
Angola: (UNAVEM II): observers.
Iraq/Kuwait: (UNIKOM): 7 observers.

PARAMILITARY
Border Guards: (Ministry of Interior) 18,000 (13,500 conscripts); 11 districts.
Construction Troops: (Ministry of Defense) 800 (675 conscripts).

Future

On balance, one can conjecture that Hungary has a reasonably bright future, although not without caveats. The government has done more than others to promote decentralized and market-oriented enterprises; however, opting for social stability, they have not taken aggressive measures to initiate sweeping market reforms, and insolvent state industries continue to be subsidized. Thus, while the economic picture is good overall, it is not as progressive as it might become. In general, the government and the people are probably content with this scenario, and gradual change and improvement are likely to characterize future growth. Also, there is the large former USSR market that must be reassessed. It seems that Russia, other former Soviet republics, and former Warsaw Pact countries will continue to be trading partners with Hungary, Ukraine and Poland in particular. Without communist domination, new markets can be addressed, especially in Western Europe.

Hungary seems well situated to address its economic and stability challenges. It is also well prepared to deal with ethnic divisiveness on its borders, but this is perhaps a greater challenge, and will require continued sure-handed diplomatic efforts at home and abroad.

SZABOLCS M. DE GYÜRKY
CHERYL DE GYÜRKY

SEE ALSO: Attila the Hun; Eastern Europe; Mongol Conquests; Soviet Union; Warsaw Pact.

Bibliography

Bodolai, Z. 1978. The timeless nation. Sydney, Australia: Hungarian Publishing Co.
Defense and foreign affairs handbook. 1988. Hungary, p. 427. Fort Bragg, N.C.
Dienes, I. 1972. *The Hungarians cross the Carpathians.* Budapest: Athenaeum Printing House (Hereditas, Corvina Press).
Dummerth, D. 1977. *Az Arpadok Nyomaban (The Arpad Dynasty).* Budapest: Panorama.
Kaleidoscope: Current world data. 1988. Hungary. Santa Barbara, Calif.: AB-CLIO. p. 4.
Military technology. January, 1988. *MILTECH.* p. 174.
Padanyi, V. 1956. *Dentumagyaria.* Buenos Aires: Transylvania Press.
Skinner, T., ed. 1988. Hungary, Magyar Nepkoztarsasag. In *International Yearbook and Statesmen's Who's Who,* p. 252. 36th ed. West Essex, England.
Zajti, F. 1939. *Magyar Evezredek (Hungarian Millennia).* Budapest: Fovarosi Nyomda.

I

IDENTIFICATION FRIEND OR FOE (IFF) TECHNOLOGY

War planners and field commanders have one basic objective in war: to kill, disable, or dissuade an enemy to the point of capitulation by efficient use of weapons and forces while minimizing one's own losses in each engagement. Hence, identification friend or foe (IFF) is a keystone technology for the planning and conduct of warfare, since it provides the capability to assure the most effective use of weapons at all ranges. Throughout history, combatants have taken actions to identify friends uniquely by the use of uniforms and insignia, even on some occasions disguising friends so they appear to the foe as "friends," in order to achieve tactical advantage. The cost and destructive power of late-twentieth-century warfare demand that IFF be given high priority. The performance and lead time of costly weapons introduced since World War II for land, air, and sea battle necessitate an understanding of the capabilities and limitations of IFF technology in planning a warfighting capability.

Whether or not overall superior technical performance can be a sufficiently effective force multiplier to overcome numerical superiority may well be determined by the effectiveness of IFF technology and its employment. For example, the U.S. development and production of the advanced medium-range air-to-air missile (AMRAAM) as a fire-and-forget capability for the late 1990s typify the search for beyond-visual-range (BVR) weapons intended to counter a numerically superior force. However, in the absence of effective BVR IFF technology, a rule of engagement specifying "shoot them down at extended range and sort them out on the ground" is not a winning strategy for selling state-of-the-art weapons programs. Nations throughout the world continue to develop, produce, or purchase weapons that significantly exceed the capability of the eventual user to perform a reliable identification function. The net result in a conflict of any scale may be constrained use of weapons, loss of friends due to fratricide, or increased chaos. Rules of engagement necessitate the positive identification of foes. For example, the United States will have approximately 2,400 F–16s in a worldwide inventory in the 1990s. It is therefore an enemy's objective to make U.S. identification of foes difficult, time-consuming, and uncertain.

There is no general-purpose, single technical solution for beyond-visual-range foe identification; ultimately the decision to engage a target is a function of the confidence level under a particular set of circumstances. "Positive" identification of a foe at a confidence level of 85 percent is considered good enough to engage a target beyond visual range. Since friendly pilots are subject to the possibility of hostile attack by friends, the identification process has as its first input the positive identification of friends.

Since there is no single technical solution for the complete IFF process, what is IFF technology? Over the years, the technology has been categorized as: (1) cooperative means that by design interact with each other, notably question-and-answer techniques (interrogators and transponders); (2) noncooperative means, which include all technical methods that provide information about a target that does not willingly emit useful signals and/or attempts to minimize emissions useful to the opponent's decision process; and (3) the correlation/fusion/decision process, which is the technical means to "weigh the evidence" and synergistically decide the target's identity under a particular set of circumstances.

IFF technology moves forward only incrementally for several reasons. It is driven by the need for interoperability with widely dispersed allies. It must adapt to the constant introduction of new weapons, some of which complicate or thwart the use of previously fielded noncooperative recognition techniques (NCRT) technology and the ability to perform fusion and high-confidence decision making at a level that provides real-time outputs to the pilot or weapon commander engaged in split-second deadly combat. By its nature, the quantity and quality of useful information available at or on an autonomous weapon platform is limited at best. Strategies to tie multiple-source information NCRT into a correlation/fusion/decision process linked to a high-performance electronic BVR "gun sight" are dependent on implementation of robust data links such as the Joint Tactical Information Distribution System (JTIDS). In the final analysis, the decision to fire a BVR missile at a target will be based on a real-time identification process that is far from perfect.

One of the most dangerous threats to friendly close-air-support aircraft is the cold and tired foot soldier behind the optical sight of a Stinger-type Missile. Relative to advances in computers, aircraft performance, radars, and so forth, IFF capabilities are only slightly improved from those of the early 1970s.

The identification process can be viewed as repeated subtraction from a large set of unknown targets at various ranges. The first priority is to subtract positively-identified friends. It would be convenient to classify all remaining unknowns at this point as enemy, but the history of fratricidal losses has proven this to be unacceptable. Consequently, the unknowns are treated as enemy, friends, or neutrals.

The identification of friends must be certain and quickly established. For this reason, cooperative technology is based on interactive questions and answers using equipment that has its roots in developments of World War II. The original friend IFF techniques were methods to alter radar returns. It was quickly determined, however, that an enemy could duplicate reflective devices and appear as a "friend" (spoofing). The change to independent radio question-and-answer systems operating at 1030/1090 MHz was thus adopted, and it sustains all evolutionary developments, including integration with air traffic control systems. Through the years other question-and-answer systems have been proposed; these have evolved to four standardized interrogations (not ciphered) consisting of two pulses of appropriately timed radio frequency energy and preformatted replies. These modes of operation provide military information, personal identification of each aircraft, and air traffic control information. The replies to these fixed modes of interrogation are assigned to an aircraft as an electronic insignia that must be accepted as genuine by a would-be gunner.

The system engineering of cooperative systems involves many considerations. The early radio systems could identify friends only as long as the user could keep the enemy from finding out how the system worked. (One generation of equipment incorporated explosive charges to dissuade the inquisitive.) By 1964 serious proposals for use of cryptographic techniques were being made so that installed question-and-answer systems could be used to identify friends long after the enemy had a copy of the equipment. The crypto security was intended to assure that an enemy would not appear as a friend unless he could obtain a key setting, and this would be changed each day. The conventional crypto systems of the day could be broken by an analyst who could collect large numbers of interrogation-reply pairs. The availability of transistor technology to designers in the mid-1950s made possible a system concept with many more interrogations and only sixteen replies. Thus, a "friend" decision could be made only if a reasonably high percentage of interrogations elicited correct replies—a decision that could be made in an automatic reply evaluator. Such a system is relatively easy to cipher but extremely difficult to decipher. By the time this U.S. system was fully developed and authorized for use in 1963, the crypto secure mode had become known as Mark XII, Mode 4. This system was offered to NATO allies, but it has not been widely implemented in NATO because some nations believe a more modern system is needed.

Vulnerability to foe exploitation, or to tactics that deny friends the use of the systems, is a serious concern that drives new developments. More robust cryptographic techniques were being developed for the Mark XV, which was intended to replace the Mark XII. The Mark XV program has been terminated, however, primarily for reasons of cost. A major factor in friends' failure to respond to interrogations is the placement and coverage of transponder antennas. Antenna performance is a compromise between antenna placement in relation to aircraft performance and radar-aspect angle of the potentially attacking weapon system. In an attempt to provide all-aspect coverage, the typical high-performance aircraft will have two omnidirectional antennas so as to provide an unconstrained "view of the world." Unfortunately such antennas have as much gain in the direction of a potential hostile SAM site as in the direction of a friendly air base.

Once friends have been identified by the interrogation and reply system, the IFF process focuses its resources on sorting out foes from the unknowns by the use of NCRT. NCRT technology is based on sensor signal analyses that range from the detailed examination of radar returns to the sorting and identification of target radio frequency emissions (altimeters, data links, etc.). This process is limited by the physical location and quality of the sensors, timeliness of the information, and the ability to correlate the information in a decision process. For example, if enemy aircraft are expected to emit a frequency-hopping spread spectrum signal in a voice communication mode, it would be unlikely that a small, single-seat friendly attack fighter would have an on-board capability to determine precisely the angle of arrival of such a signal. However, associating such a signal with any other unique signal or a number of specific previously unknown targets would dramatically increase confidence in foe identification by the autonomous defender.

In reality the cost and limited availability of weapons dictate that autonomous target engagement be minimized. In wartime scenarios the dynamics of weapons engagement dictate that the IFF process be integrated with the command, control, communications, and intelligence (C^3I) system to achieve synergism in detection, tracking, and attack tasking. The C^3I assets participating in the IFF system must correlate exploitable signals in a near-real-time process and pass high-confidence-level indicators to the weapon system taskers. The sensors, processors, and robust data links needed to support even a modest architecture are costly and make highly attractive targets for the enemy to attack. For example, ground or airborne adaptive main-beam nulling technology would be useful in the presence of interferers to determine precisely direction of arrival of enemy-emitted signals in the foe identification process. "Usefulness" is unfortunately not a persuasive rationale in the budgeting process and is a low-priority selling point.

The evolution of NCRT technology for positive identification follows two paths: (1) for weapons platforms, the focus is on making better use of the radar returns to extract more information, and (2) more robust fusion of multiple source information in C^3I centers, with the results forwarded in real time to the weapon tasker and/or platform gunsights. Advocates of the fusion-center approach point to success in the development of counter-C^3I systems, where the identification of enemy command posts is now accomplished through correlation of multiple sources of information. In the counter-C^3I scenario the time factor is measured in hours, whereas in the IFF case it is measured in minutes (at best) and ultimately in seconds, and identification must be over 85 percent certain with each target.

The use of NCRT technology in the time-constrained positive identification of foes will ultimately employ artificial intelligence algorithms to weigh the evidence and assign a confidence level to the decision. In a highly dynamic situation this probably will be far from an ideal process. Limited information (often deceptive) will burden the process, forcing automated decisions that in all likelihood will not have high confidence levels. Thus, high-confidence "positive" foe identification for the launch of a beyond-visual-range weapon in a "target-rich" conflict will be difficult at best. The deciding factor may be simply the possibility that firing first is the only assured tactic for survival. If this were to become a rule of engagement, NATO war games routinely demonstrate that as many as half of all NATO aircraft could be shot down as mistaken enemy. Thus, the 85 percent confidence level of positive foe identification may be overly optimistic.

War planners and IFF technologists face a serious challenge in their attempt to promote a major improvement in IFF capability, primarily because they offer an evolutionary product rather than the acquisition of a specific number of black boxes. There does not appear to be any breakthrough technology that by the year 2000 could raise the confidence level of positive foe identification above 85 percent. To achieve this level of performance would require commitment, planning, and financial resources devoted to an evolutionary development program in an era of reduced defense spending. It remains to be seen whether a reduction in long-term weapons acquisition will provide the opportunity to sell the concept of a more capable evolutionary IFF system as a high-payoff force multiplier.

Finally, although the identification issue has historically focused on aircraft, the advent of long-range antitank weapons has now extended the problem to the ground. Fratricide losses incurred in Operation Desert Storm have prompted the U.S. Army to begin the development process on techniques for battlefield IFF. The army is seeking to improve situation awareness as well as the identification of individual vehicles. The technical considerations described above for aircraft identification will generally apply to tanks as well, while affordability constraints will probably be more severe. Operationally, however, there is no avoiding the need for improved identification if the full effectiveness of modern antitank weapons is to be realized.

MICHAEL I. KELLER

SEE ALSO: Close Air Support; Intelligence, Tactical; Radar; Radar Technology Applications; Rules of Engagement; Surveillance and Target Acquisition Equipment.

Bibliography

Dickson, P. 1976. *The electronic battlefield.* Bloomington, Ind.: Indiana Univ. Press.
Price, A. 1977. *Instruments of darkness: The history of electronic warfare.* London: Macdonald and Jane's.

INDIA, REPUBLIC OF

India is the second most populous country on earth. Since gaining independence from Great Britain in 1947, India has become a major power in Asia.

Power Potential Statistics

Area: 3,287,590 square kilometers (1,269,339 sq. mi.)
Population: 857,960,200
Total Active Armed Forces: 1,265,000 (0.147% of pop.)
Gross National Product: US$254 billion (1990 est.)
Annual Defense Expenditure: US$9.2 billion (3.5% of GNP, 1991 est.)
Iron and Steel Production:
 Crude steel: 14.194 million metric tons (1988–89)
 Pig iron: 12.253 million metric tons (1988–89)
Fuel Production:
 Coal: 200 million metric tons (1989–90)
 Coke: 13.3 million metric tons (1986)
 Crude oil: 32.04 million metric tons (1988–89)
 Natural gas: 13,217 million cubic meters (1988–89)
Electrical Power Output: 245,000 million kwh (1990)
Merchant Marine: 308 vessels; 6,087,451 gross registered tons
Civil Air Fleet: 93 major transport aircraft; 288 usable airfields (198 with permanent-surface runways); 2 with runways over 3,659 meters (12,000 ft.); 57 with runways 2,440–3,659 meters (8,000–12,000 ft.); 88 with runways 1,220–2,440 meters (4,000–8,000 ft.).

For the most recent information, the reader may refer to the following annual publications:
The Military Balance. International Institute for Strategic Studies. London: Brassey's (UK).
The Statesman's Year-Book. New York: St. Martin's Press.
The World Factbook. Central Intelligence Agency. Washington, D.C.: Brassey's (US).

History

Although India's history goes back at least 1,500 years before the birth of Christ, its history as a modern sovereign nation begins only with its independence from Great

Britain on 15 August 1947. British influence in India began in the early 1700s as the British and French replaced the Dutch and Portuguese as the dominant European powers in the Indian Ocean basin. In an extended series of wars lasting until the 1840s (often waged through princely Indian proxies), the British gained effective political control of most of India, driving out their European rivals by 1805. For the next century, the British governed the subcontinent, both directly and through the local rulers they allowed to retain some power. In the process, the British gave India three great legacies: a magnificent railroad system, a foreign language which has proved a powerful unifying force, and effective legal and administrative machinery which, by the time of independence, was largely run by Indians.

Almost at the instant of independence, India was involved in a brief but bitter war with Pakistan, the neighboring new nation (formed from the Muslim-dominated areas of the Indian subcontinent) and former British colony. The war was fought over the province of Jammu-Kashmir, a predominantly Muslim area ruled by Hindus, and the struggle was exacerbated by the problems of masses of refugees (as many as 12 million) fleeing to shelter with their co-religionists. For fifteen years after that first war, India was at peace, pursuing a policy of domestic development and foreign nonalignment.

In 1962 a border dispute with the People's Republic of China led to attacks by Chinese troops on Indian border areas in Assam and Kashmir. The Indian forces, with inadequate training, leadership, and weapons, performed poorly, and the shock of that defeat led Prime Minister Jawaharlal Nehru to rearm India and expand its defense forces.

When India clashed with Pakistan for a second time over disputes in the Rann of Kutch and Kashmir in Augustand September 1965, Nehru's new policy proved its worth, and Indian forces performed adequately in both minor and major engagements. A third serious clash with Pakistan in December 1971, sparked by a separatist revolt in East Pakistan (which led to the establishment of Bangladesh), resulted in a decisive Indian victory, as Bangladesh became a close Indian ally. India took another step toward great-power status when it exploded an atomic bomb, euphemistically referred to in the Indian press as "a nuclear device," on 18 May 1974. Despite the alarm this event caused in Pakistan, Indian officials insisted that their nuclear project had peaceful, nonmilitary goals.

Politico-Military Background and Policy

For most of the quarter-century following independence, India followed a policy of nonalignment between the Soviet bloc and the West. This changed in 1971, when India concluded a mutual assistance treaty with the Soviet Union, and when relations with the United States became strained due to India's war with Pakistan that year. After that, India maintained close relations with the Soviets, and has had correspondingly poor relations with both China and Pakistan. Relations between the United States (Pakistan's other ally) and India have been variable. India has also developed close ties with the Himalayan border states of Nepal, Sikkim, and Bhutan; in the case of the latter two nations, India handles their foreign relations and defense responsibilities, tasks India undertook in the aftermath of the border war with China in 1962.

India's armed forces are manned entirely by volunteers. Recruitment is not a problem because many Indians view a career in the armed forces as honorable and comparatively well-paid. Volunteers may be as young as 17 years old, and serve for ten to fifteen years depending on the amount of specialized or technical training they receive. There are upper age limits for volunteers, based on the same criteria as length of service.

Strategic Problems

India's greatest strategic problem is the high cost of the defense effort as a proportion of the total government budget; India spends about 20 percent of the budget on defense. This is a major problem when the cost of development programs is considered, and, unless the economy grows rapidly, the defense spending increases must stop.

India has few geographic liabilities. The most serious is the narrow Assam neck, connecting oil-rich Assam and the old North East Frontier Area with the rest of India. Fewer than 100 kilometers (62 mi.) wide, the neck separates the Himalayas in the north from Bangladesh in the south, and is vulnerable to a Chinese attack. The ongoing Tamil insurgency on Sri Lanka is a potential problem as well, because of the support for the insurgency in the neighboring state of Tamil Nadu in extreme southeast India.

Finally, India faces some serious internal insurgent problems. Some of these, like Gurkha unrest in Assam and the long-festering tribal insurgency in the Naga hills, have recently ended. The most serious, though, is Sikh unrest in Punjab, an issue complicated by the many positions in army and government held by Sikhs, who constitute less than 2 percent of India's population. Radical Sikhs, unhappy with government policies toward them and desiring greater autonomy for their Punjab homeland, have gained considerable support from moderates since the Indian army stormed the Golden Temple complex in Amritsar on 6 June 1984. This act provoked the assassination of Prime Minister Indira Gandhi on 31 October that year. The accumulation of grudges and hatred between Sikh militants and Indian security forces has produced a bitter and bloody terrorist campaign in the Punjab. An end to this problem would require compromises on both sides, and that seems unlikely at this time.

Military Assistance

In the late 1980s, India received no formal military assistance from any other nation, although that had not always been the case. For the first few years after independence, for instance, Indian army forces were led with the assistance of British officers, because Indian officers lacked staff and command experience. For some time after, India received aid from the United States and Great Britain, although much of that was in the form of sales. In a sense, India has received aid from the former Soviet Union, in the form of sales of advanced equipment, like MiG-29 fighters, Mi-35 attack helicopters, and one Charlie-class guided missile submarine, which the Soviets did not ordinarily sell.

India does extend some military assistance to other nations. It provides some training and equipment to the Nepalese armed forces, and extensive support for Bhutan's small army. India has also become involved in the Sri Lankan civil war between minority Tamil and majority Sinhalese ethnic groups. In 1988 there were some 55,000 Indian troops on Sri Lanka, including four infantry divisions, operating against those Tamil rebels who had rejected the Indian-sponsored peace initiative of 1987. On a similar but smaller scale, India sent paratroops to the Republic of Maldives in early November 1988 to combat an attempted coup by Tamils from Sri Lanka.

Defense Industry

Although it imported many of its major weapons systems from the former Soviet Union, India has created a substantial armaments industry of its own in the 40 years since independence. The Indians build many of their own aircraft (often under license), naval vessels, and tanks (also under license) and other military vehicles. By early 1985, the Indian defense industrial establishment included 33 ordnance factories, nine public sector concerns, and nearly 40 research and development labs, employing a total of some 300,000 workers.

The ordnance factories, inherited from the British, produce items like ammunition, small arms, vehicles, and optical instruments for the armed forces and the police. The public sector concerns are larger industrial operations set up to research, develop, and produce major weapons systems. The oldest of these is Hindustani Aeronautics, Ltd. (HAL), which makes helicopters, jet fighters, interceptors, and trainers, most of these under license agreements with foreign companies. Three of the public sector operations are shipyards, and others produce electronics, armored vehicles, and other items. The research labs are centrally administered by the Department of Defence Research and Development, created in 1980.

As already noted, India has an extensive nuclear energy program, and has several nuclear power plants as well as facilities for producing more "devices." India also has a space program, which launched its first satellite on 18 June 1980, and has since launched several more. These launches have been made with Soviet rockets, but the Indians have some rocket capacity of their own.

Alliances

India's main alliance was the bilateral treaty of peace, cooperation, and friendship with the Soviet Union signed in August 1971. India is a member of the United Nations, and has in the past contributed troops for truce observation and peacekeeping forces. Finally, India is a member of the British Commonwealth, although that tie is more economic than political.

Defense Structure

Constitutionally, the president is supreme commander of the armed forces, but actual responsibility for defense rests with the cabinet, presided over by the prime minister, who is assisted by the standing Defence Committee and advised by the National Defence Council. The Minister of Defence heads the defense organization, and is responsible to both Parliament and the prime minister for the administrative and operational control of the armed forces. There are three services—army, navy, and air force—each with its own chief of staff.

(For an explanation of the abbreviations and symbols used in the following section of military statistics, see the list of Abbreviations and Acronyms in Volume I.)

Total Armed Forces

Active: 1,265,000.
Reserves. (obligation to age 60) Army 300,000. Territorial
  Army (volunteers) 160,000. Air Force strength n.k.

ARMY: 1,100,000.
HQ: 5 Regional Comd (= Fd Army), 10 Corps.
2 armd div (each 2/3 armd, 1 SP arty [2 SP fd, 1 med regt]
  bde).
1 mech div (each 3 mech [4/6 mech bn, 3 armd regt], 1 arty
  bde).
21 inf div (each 2–5 inf, 1 arty bde; some have armd regt).
11 mtn div (each 3–4 bde, 1 or more arty regt).
19 indep bde: 8 armd, 9 inf, 1 mtn, 1 AB/cdo.
5 indep arty bde.
5 AD bde.
4 engr bde.
These formations comprise: 56 tk regt (bn); 25 mech, 332 inf
  bn; 9 AB/cdo bn; 190 arty regt (bn): incl 1 hy, 5 MRL, 50
  med (11 SP), 69 fd (3 SP), 39 mtn; 29 AD arty regt; perhaps 10 SAM gp (3–5 bty each); 7 sqn, 25 flt, Air Observation; 6 ATK/tpt, 4 liaison hel sqn.
Equipment:
MBT: 3,100 (est. 500 in store): some 500 T-55, 900 T-72/-M1,
  1,700 Vijayanta.
Light tanks: 100 PT-76.
Recce: BRDM-2.
AIFV: 800 BMP-1/-2 (Sarath).
APC: 400 OT-62/-64, 50 BTR-60.

India, Republic of

Towed arty: over 4,000 incl: 75mm/76mm: 900 75/24 mtn, 215 Yug M-48; 88mm: 1,000 25-pdr (retiring); 100mm: 185 M-1944; 105mm: some 800 (incl M-56 pack), some 100 IFG Mk II; 130mm: 550 M-46; 140mm: 150 5.5-in (retiring); 155mm: 410 FH-77B.
SP arty: 105mm: 80 Abbot; 130mm: 100 mod M-46
MRL: 122mm: 80 BM-21.
Mortars: 81mm: L16A1; 82mm: M-43; 120mm: 1,000 M-43; 160mm: 200 M-43.
ATGW: SS-11-B1, Milan, AT-3 Sagger, AT-4 Spigot.
RCL: 57mm: M-18; 84mm: Carl Gustav; 106mm: 1,000+ M-40A1.
AD guns: 2,750: 23mm: 140 ZU 23-2, 75 ZSU-23-4 SP; 40mm: 1,245 L40/60, 790 L40/70; 94mm: 500 3.7-in.
SAM: 26 SA-6, 620 SA-7, 20 SA-8A/-B, SA-13, SA-16, 25 Tigercat launchers.
Helicopters: 9 sqn with 50 Chetak, 40 Cheetah, 30 Krishnar Mk2.
Reserves:
Territorial Army: 30 inf bn.
Deployment:
North—1 corps with 2 inf, 1 mtn div; 1 mtn, 1 indep inf, 1 indep arty bde. 1 corps with 4 inf div; 2 indep armd, 1 indep inf, 2 indep arty bde.
West—1 corps with 1 armd, 1 mech div; 1 corps with 2 inf div; 1 corps with 3 inf div.
Central—1 corps with 1 armd, 2 inf div, plus 3 indep div (2 inf, 1 mtn).
East—3 corps each with 3 mtn div.
South—1 corps with 4 inf div.

NAVY: 55,000, incl 5,000 Naval Air Force and 1,000 Marines.
Principal Commands: Western, Eastern, Southern. Subsidiary Commands: Submarine, Naval Air.
Bases: Bombay (HQ Western Cmd), Goa (HQ Naval Air), Lakshadweep (Laccadive Is), Karwar; Cochin (HQ Southern Cmd), Visakhapatnam (HQ Eastern and Submarines), Calcutta, Port Blair (Andaman Is).
Fleets: Western (based Bombay), Eastern (based Visakhapatnam).
Submarines: 17 SS: 8 Sindhughosh (Sov Kilo) with 533mm TT; 2 Shishumar (Ge T-209/1500) with 533mm TT; 7 Kursura (Sov Foxtrot) with 533mm TT; (incl some 3 in reserve).
Principal Surface Combatants: 28:
Carriers: 2:
 1 Viraat (UK Hermes) (29,000t) CVV.
 1 Vikrant (UK Glory) (19,800t) CVV.
 Air group typically:
  ac: 8 Sea Harrier fighter/attack. hel: 8 Sea King ASW/ASUW (Sea Eagle ASM).
Destroyers: 5 Rajput (Sov Kashin) DDG with 2×2 SA-N-1 Goa SAM; plus 4 SS-N-2C Styx SSM, 5×533mm TT, 2×ASW RL, 1 Ka-25 or 27 hel (ASW).
Frigates: 21:
 3 Godavari FFH with 2× Sea King hel, 2×3 ASTT; plus 4×SS-N-2C Styx SSM.
 2 Talwar (UK Whitby) FFH with 3×55-N-2A Styx, 1 Chetak hel.
 6 Nilgiri (UK Leander) with 2×3 ASTT, 4 with 1×3 Limbo ASW mor, 1 Chetak hel, 2 with 1 Sea King, 1×2 ASW RL; plus 2×114mm guns.
 6 Kamorta (Sov Petya) with 4 ASW RL, 3×533mm TT.
 3 Khukri (ASUW) with 4 Styx, hel deck.
 1 Betwa (UK Leopard) (trg).
Patrol and Coastal Combatants: 40:
Corvettes: 13:
 3 Vijay Durg (Sov Nanuchka II) with 4×SS-N-2B Styx SSM.
 5 Veer (Sov Tarantul) with 4× Styx.
 1 Vibhuti (similar to Tarantul) with 4× Styx.
 4 Abhay (Sov Pauk) (ASW) with 4×ASTT, 2×ASW mor.
Missile Craft: 9 Vidyut (Sov Osa) with 4× Styx.
Patrol, Offshore: 4 Sukanya PCO
Patrol, Inshore: 14: 12 SDB MK 2/3; 2 Osa PFI.
Mine Warfare: 22:
Minelayers: None, but Kamorta FF and Pondicherry MSO have minelaying capability.
Mine Countermeasures: 22:
 12 Pondicherry (Sov Natya) MSO;
 4 Bulsar (UK 'Ham') MSI;
 6 Mahé (Sov Yevgenya) MSI (.
Amphibious: 9: 1 Magar LST, capacity 200 tps, 12 tk, 1 hel; 8 Ghorpad (Sov Polnocny C) LSM, capacity 140 tps, 6 tk; plus craft; 7 Vasco da Gama LCU.
Support and Miscellaneous: 20: 2 Deepak AO, 1 AOE/R, 1 Amba (Sov Ugra) sub spt, 1 div spt, 2 ocean tugs, 4 AO sm, 4 Sandhayak and 4 Makar AGHS, 1 Tir trg.

Naval Air Force: (5,000); 46 cbt ac, 75 armed hel.
Attack: 1 sqn with 21 Sea Harrier FRS Mk-51, 3 T-60 trg.
ASW: 6 hel sqn with 26 Chetak, 7 Ka-25, 10 Ka-28, 32 Sea King Mk 42A/B.
MR: 2 sqn: 9 BN-2, 5 Il-38, 8 Tu-142M Bear F.
Communications: 1 sqn with 5 BN-2 Islander, Do-228 ac; 3 Chetak hel.
SAR: 1 hel sqn with 6 Sea King Mk 42C.
Training: 2 sqn: 6 HJT-16, 8 HPT-32 ac; 2 Chetak, 4 Hughes 300 hel.

Marines: (est. 1,000); 1 regt (2d forming).

AIR FORCE: 110,000; 630 cbt ac, 36 armed hel. 5 Air Comd.
Bombers: 1 lt bbr sqn with 9 Canberra.
FGA: 26 sqn:
 5 with 80 Jaguar IS.
 14 with 112 MiG-21 MF/PFMA.
 3 with 54 MiG-23 BN/UM.
 4 with 56 MiG-27.
Fighter: 17 sqn:
 4 with 74 MiG-21 FL/U.
 6 with 108 MiG-21 bis/U.
 2 with 26 MiG-23 MF/UM.
 3 with 54 MiG-29/UB.
 2 with 36 Mirage 2000H/TH.
Maritime attack: 8 Jaguar with Sea Eagle.
Attack helicopters: 2 sqn: 1 with 18 Mi-25; 1 with 18 Mi-35.
Recce: 3 sqn: 1 with 8 Canberra PR-57; 1 with 6 MiG-25R, 2 MiG-25U; 1 with 4 HS-748.
MR/Survey: 2 Gulfstream IV SRA, 2 Learjet 29.
Transport:
 Aircraft: 12 sqn:
  1 with 15 An-12B (being withdrawn);
  6 with 108 An-32 Sutlej;
  1 with 16 BAe-748;
  2 with 30 Do-228;
  2 with 24 Il-76 Gajraj;
 Helicopters: 11 sqn with 80 Mi-8, 50 Mi-17, 10 Mi-26 (hy tpt).
VIP: 1 HQ sqn with 2 Boeing 707-337C, 4 Boeing 737, 7 BAe-748.
Liaison: flt and det: 16 BAe-748, C-47.

Training: 24 BAe-748, 7 Canberra T-4/-13/-67, 120 HJT-16, 57 Kiran II, 20 HPT-32, 60 HT-2, 20 Hunter T-66, 5* Jaguar IB, 5 MiG-29UB, 44 TS-11 ac; 20 Chetak, 2 Mi-24, 2 Mi-35 hel.
Missiles:
  ASM: Akash, AM-39 Exocet, AS-7 Kerry, AS-11B (ATGW), AS-30, Sea Eagle.
  AAM: AA-2 Atoll, AA-7 Apex, AA-8 Aphid, AA-10 Alamo, AA-11 Archer, R-550 Magic, Super 530D.
  SAM: 30 bn: 280 Divina V75SM/VK (SA-2), SA-3.

FORCES ABROAD
UN and Peacekeeping:
Angola (UNAVEM II): 6 observers.
Central America (ONUCA): 22 observers.
Iraq/Kuwait (UNIKOM) 8 observers.

PARAMILITARY
National Security Guards: 5,000: anti-terrorism contingency deployment force. Comprises elm of the Armed Forces, CRPF, Border Guard.
Central Reserve Police Force (CRPF): (Ministry of Home Affairs) 100,000; Reserves: 250,000; 70 bn, internal security duties and army first-line reserves.
Border Security Force: (Ministry of Home Affairs) 140,000; some 120 bn, small arms, some lt arty, tpt/liaison air spt.
Assam Rifles: (Ministry of Home Affairs) 15,000; 36 bn, security within north-eastern states.
Ladakh Scouts: 5,000.
Indo-Tibetan Border Police: (Ministry of Home Affairs) 22,000; 22 bn.
Special Frontier Force: 10,000.
Central Industrial Security Force: (Ministry of Home Affairs) 55,000.
Defence Security Force: 30,000.
Railway Protection Forces: 60,000.
Provincial Armed Constabulary: 250,000.
National Rifles: (being formed, to be 10,000).

COAST GUARD: 2,800
Patrol Craft: 41: 9 Vikram PCO, 6 Tara Bai PCI, 5 Rajhans PFI; 13 Jija Bai PCI, 8 (.
Aviation: 3 air sqn with 9 Do-228, 2 Fokker F-27, 5 BN-2 Islander ac, 4 Chetak hel.

National Resistance Groups

The only major guerrilla/terrorist force operating in India as of autumn 1988 were the Sikh extremists in the Punjab. In the first six months of 1988, their operations had killed nearly 1,800 people. Many of the extremists were members of the traditional Sikh warrior brotherhood, the *Khalsa*, and were engaged in what they perceived as a holy struggle against Indian oppression and domination. They had assassinated then-Prime Minister Indira Gandhi in October 1984. Since then, they have been involved in frequent clashes with government security forces. Their numbers are difficult to estimate; many of its supporters are not full-time fighters, but there are at least several thousand active fighters. Government action against them is hampered by the considerable support they enjoy in the Punjab, where they have become popular with moderate Sikhs since the Indian army stormed the holy Golden Temple complex in Amritsar in June 1984, and thereby martyred the leading radical, Sant Jarnail Singh Bhindranwale. The crucial role Sikhs play in modern India is indicated by the fact that four of the six army generals who directed the June 1984 operations against Bhindranwale in the Golden Temple were Sikhs.

Future

India's size, population, relative wealth of natural resources, and military power give it the capability to become a major power. Balanced against this potential are considerations of national disunity (India is a patchwork of nationalities, ethnic groups, and religious sects) and the poverty in which so many Indians live. India's progress is further hampered by the lack of a nationwide sense of "Indian-ness," since many Indians still see themselves as Sikhs, Tamils, Mahrattas, Muslims, or Hindus first, and as Indians second. In addition, the territorial dispute over the Kashmir Province will continue to be a problem for India and Pakistan and will hinder improved relations between these two countries.

DAVID L. BONGARD
TREVOR N. DUPUY

SEE ALSO: Indo-Pacific Area; Moghul Empire; Oceania.

Bibliography

American University. 1985. *India. A country study*. Washington, D.C.: Government Printing Office.
Barnds, W. J. 1972. *India, Pakistan, and the great powers*. New York: Praeger.
Hardgrave, R. L. 1980. *India: Government and politics in a developing nation*. 3d ed. New York: Harcourt, Brace, Jovanovich.
Hunter, B., ed. 1991. *The statesman's year-book, 1991–92*. New York: St. Martin's Press.
International Institute for Strategic Studies. 1991. *The military balance, 1991–1992*. London: Brassey's.
———. 1988. *Strategic survey 1988–1989*. London: Brassey's.
Wolpert, S. 1977. *A new history of India*. London: Oxford Univ. Press.

INDIRECT FIRE

Indirect fire is one of three major procedures for placing fire effect from a weapon upon a target. The others are direct fire and unobserved fire.

In *indirect fire* the person (or crew, in the case of a crew-served weapon) aiming and firing the weapon does not see the target, but receives instructions for setting proper elevation (for range) and deflection (for direction) from an observer who can see the target. In *direct fire* the person or crew aiming and firing the weapon does see the

Indirect Fire

target at which the weapon is being fired. In *unobserved fire* the weapon's firer (person or crew) does not see the target, and there is no observer; instead, fire is placed on the presumed target area on the basis of information about its location that can range from a wild guess to extremely accurate information plotted precisely upon maps or firing charts.

The term *indirect fire* is sometimes confused with *elevation*, which is merely the angle of a tube or missile with respect to the horizontal plane; it has no relation to indirect fire.

Historical Background

For the first 500 years of gunpowder weapons, there was little difference between the process of aiming and firing an infantry musket and that used to fire artillery cannon, other than differences related to the size and mobility of the weapons. In the early days, cannon were usually emplaced in front of deployed troops and played a major role only in the first phase of the battle. The heavy, clumsy cannon could not keep up with the infantry as battle lines shifted. By early in the seventeenth century Gustavus Adolphus used horse-drawn field artillery, generally placed in line with the infantry and cavalry. Frederick the Great, in the eighteenth century, introduced horse artillery; the gunners and cannoneers rode horses into battle, enabling the artillery to keep up with the cavalry. Late eighteenth-century cannon had much longer ranges, and were much more accurate, than infantry muskets. Cannon placed on high ground could fire with relative accuracy at ranges up to 1,000 meters (0.6 mi.).

As early as the Civil War and Franco-Prussian War, ingenious artillerymen experimented with means of using observers to coordinate the fires of several guns, either to make the firing more efficient or to permit locating the guns out of sight of the hostile infantry. In his classic book, *Letters on Artillery,* Prince Kraft zu Hohenlohe Ingelfingen, pre-eminent Prussian artillery general of the Austro-Prussian and Franco-Prussian wars, briefly refers to examples of both. In fact, after describing a crude effort to "lay back" under cover, he commented, "I think myself that it is a complicated mode of working which will only in rare cases fulfill the expectations formed of it" (Hohenlohe 1898, p. 419).

Without benefit of the telephone, which had not yet been invented, Hohenlohe Ingelfingen was probably justified in that comment. Transmitting instructions from an observer by shouting from the top of a hill, or by means of signal flags, or by passing messages along a line of men, could never be really efficient. Yet in the Russo-Japanese War, less than 30 years after Hohenlohe Ingelfingen wrote those words, the use of indirect fire by means of field telephones was common practice, taken for granted. This we know from the commentary of a British observer of the Russo-Japanese War, General Sir Ian Hamilton, in *A Staff Officer's Scrap Book* (Hamilton, 1912).

After the invention of the telephone by Alexander Graham Bell in 1876, field telephones began to supplement, and then supplant, the telegraph as the principal means of military long-distance communication. Artillery no longer had to be placed out in the open, and exposed on top of ridge lines so that gunners could see their targets. Rather, the guns could be placed behind the shelter of the ridges and connected by telephone to an observer in an inconspicuous spot on the ridge or elsewhere. The observer, sending his commands by phone, would use indirect fire techniques to adjust the fire of the guns, which the enemy could not see.

Fundamentals of Indirect Fire

Figure 1 shows the basic elements of indirect fire. The guns are located behind a mask, usually a hill or ridge line. The target is identified by the observer in an observation post, located to facilitate surveillance of the target area where the enemy is located. He has communications by field telephone with the gun position. The battery commander or his executive officer have provided to the gun crews the necessary instructions so that all guns are aimed together at some imaginary point in the target area. The gunners have then been directed to train their sights on an aiming point visible to all of them, and then to record the deflection reading on their sights. This reading is the angular measurement (in degrees or mils) between the direction in which the gun tubes are pointed and the line of sight from each gun sight to the aiming point. (In the absence of a readily identifiable common aiming point, each gun crew may set out aiming stakes in some convenient direction from the gun, and the deflection they record is that when the sight is pointed at the aiming stake.)

As soon as the battery has gone into position, the observer begins to adjust fire on a target: he registers the

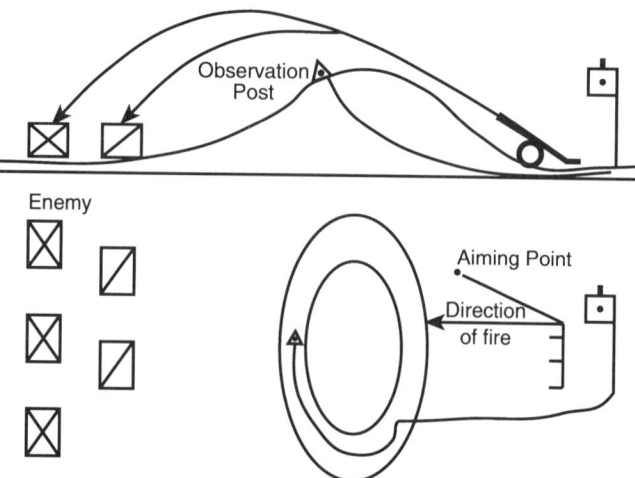

Figure 1. Electronic communications and indirect fire.

battery by firing on a *base point*—some prominent, fixed, readily identifiable, centrally located object in the target area, such as a house, a crossroad, a church steeple, or a lone tree. Usually only one gun of the battery will register on the base point. Once the observer is satisfied that the center of impact of the fire of that gun is on the target, he so informs the battery commander or executive, who will order the gunner to record *base deflection* and *base elevation*. The other guns will then be given the angular difference (for the registering gun) between the original recorded deflection and the base deflection; and they then each record their base deflections, which will automatically cause all of them to point in the same direction as the registering gun. Since the nonregistering guns are located nearby, all will use the same base elevation as that of the registering gun.

The artilleries of most nations have adopted the fire direction center concept introduced by the U.S. Army's artillery in the 1930s. This permits efficient coordination of fire of two or more batteries. While communication from the observer to the guns may be directly with the battery commander or executive at the battery fire direction center, it is usually with the battalion fire direction center.

When the observer sees a target, he reports its direction and range to the fire direction center. Usually by means of a firing chart, the angular measurement in deflection is calculated with respect to the *base line* (line between guns and base point), as is the angular difference in elevation, to permit the guns to fire at the range of the target. The guns (or one gun, the adjusting gun) will then fire at the elevation and deflection so measured and calculated. Even an expert observer can rarely estimate distances so accurately that the first round will hit the target. So the observer then adjusts the fire, giving corrections in range and direction after each round is fired until the center of impact is on the target. At that point he will call for *fire for effect*, and the battery commander (or the fire direction center) will instruct the battery (more than one battery, if appropriate) to fire a suitable number of rounds at the target. When the observer sees that the target has been destroyed or neutralized, he will inform the battery commander or fire direction center: *mission accomplished*.

As soon as fire for effect is begun, the fire direction center records the deflection and elevation for the battery's base piece for that target. Thus a future target can be engaged by shifting fire either from the base point or from the location of any target on which fire has been adjusted.

Conclusion

The indirect fire techniques discussed above have not changed fundamentally since the Russo-Japanese war nearly a century ago. The French, the British, the Germans, and the Americans added refinements in World War I. Innovations introduced by Brig. Gen. Charles P. Summerall, when he commanded the artillery brigade of the U.S. 1st Division in mid-1918, were brought to fruition in the 1930s, under his guidance as Chief of Staff of the U.S. Army, in the fire direction system mentioned earlier. The principal refinement of World War II was substitution of radio for telephone. Amazing new techniques have resulted from the introduction of computers into fire direction centers in the years since World War II, but these are merely refinements to a concept visualized by Hohenlohe Ingelfingen in the 1870s.

TREVOR N. DUPUY

SEE ALSO: Ammunition; Artillery; Artillery, Tube; Command, Control, Communications, and Intelligence; Franco-Prussian War; Frederick the Great; Gustavus Adolphus; Mortar; Russo-Japanese War.

Bibliography

Bidwell, S., and D. Graham. 1982. *Firepower: British army weapons and theories of war, 1904–1945*. London: George Allen and Unwin.
Hamilton, I. 1912. *A staff officer's scrap book*. London: Edward Arnold.
Hohenlohe Ingelfingen, Kraft zu. 1898. *Letters on artillery*. Trans. N. L. Walford. London: Edward Stanford.

INDONESIA, REPUBLIC OF

The Republic of Indonesia, with its capital in Jakarta, is the world's fifth most populous nation. It comprises about 13,500 islands that extend slightly less than 3,000 kilometers (1,860 mi.) along the equator between the Indian and Pacific oceans. Thus situated between continental Southeast Asia and Australia, it controls major Indo-Pacific sea lanes and constitutes a strategically critical geopolitico-military entity.

Less than half of Indonesia's islands are inhabited by its 190 million people, who are primarily Malays. The official language is Bahasa Indonesia, a deliberately developed, modified form of Malaysian; English is a second language. Many local dialects exist, the most widely spoken of which is Javanese. Nearly 90 percent of the people are Muslim. The national literacy rate is about 62 percent.

Indonesia has a tropical climate that features severe droughts, floods, and earthquake-generated tsunamis (tidal waves). Most of the country consists of coastal lowlands; the larger islands—Sumatra, Java, Borneo, Celebes, and New Guinea—have interior mountains with more moderate climate.

Rich in natural resources, particularly in petroleum, gas, and timber, Indonesia is a mixed economy. The most important sector is agriculture, including forestry and fishing, which accounts for more than one-fifth the GNP and employs more than 50 percent of the 67-million labor force. However, Indonesia continues to have difficulty feeding and employing its increasing population. Despite successful economic reforms, internal political disputes and official mismanagement have caused many of the people to lose confidence in the military-backed government.

By contrast, Indonesia's leaders have demonstrated skill in international diplomacy, having worked with some success to help resolve the Cambodian problem, and reaching agreement with Australia concerning use of the Timor Sea. Some international problems remain; these combined with internal unrest present challenges that will affect Indonesia and its Pacific and Indian ocean neighbors in the future.

Power Potential Statistics

Area: 1,919,440 square kilometers (741,096 sq. mi.)
Population: 182,269,800
Total Active Armed Forces: 278,000 (0.153% of pop.)
Gross Domestic Product: US$94 billion (1990 est.)
Annual Defense Expenditure: US$1.4 billion (1.8% of GNP, 1988 est.)
Iron and Steel Production:
  Crude steel: 1,337 million metric tons (1987–88)
Fuel Production:
  Coal: 7.793 million metric tons (1988–89)
  Crude oil: 66 million metric tons (1989)
  Natural gas: 1,787,000 million cubic meters (1988–89)
Electrical Power Output: 38,000 million kwh (1990)
Merchant Marine: 365 vessels; 1,647,632 gross registered tons
Civil Air Fleet: 216 major transport aircraft; 436 usable airfields (111 with permanent-surface runways); 1 with runways over 3,659 meters (12,000 ft.); 12 with runways 2,440–3,659 meters (8,000–12,000 ft.); 63 with runways 1,220–2,440 meters (4,000–8,000 ft.).

For the most recent information, the reader may refer to the following annual publications:
The Military Balance. International Institute for Strategic Studies. London: Brasseys's (UK).
The Statesman's Year-Book. New York: St. Martin's Press.
The World Factbook. Central Intelligence Agency. Washington, D.C.: Brassey's (US).

History

Civilizations had existed on Java and Sumatra for at least a thousand years when Columbus set sail in 1492 to seek the Spice Islands of Indonesia. From the seventh to the twelfth century a Buddhist kingdom centered on Sumatra; in the fourteenth century a Hindu kingdom had its capital in eastern Java. The temples of these early cultures and other structural remains are among the finest ancient stone art in the world.

Islamic peoples entered Indonesia in the twelfth century, and Islam gradually replaced Hinduism except on the island of Bali. As ancient kingdoms were broken into smaller states the region became unable to resist the spread of Western colonialism, first by the Portuguese in the sixteenth century. They were forceably supplanted by the Dutch in the seventeenth century; Portuguese presence remained only on the eastern half of the island of Timor and in a small enclave on its north coast. For 300 years the Dutch ruled the Netherlands East Indies, one of the world's richest colonial possessions, from which came most of present-day Indonesia.

The movement for Indonesian independence began early in the twentieth century and expanded through the two world wars. Its leaders were young professionals and students, some educated in Holland, and some—including Indonesia's first president, Sukarno—imprisoned there. Japanese occupiers in World War II encouraged nationalism and opened many administrative posts to Indonesians. Immediately after the war Sukarno and his followers declared independence, forming the Republic of Indonesia on 17 August 1945. Armed resistance to the Dutch followed, and on 27 December 1949 the Dutch formally transferred sovereignty to Indonesia except for its claim to western New Guinea.

Further conflict ensued over western New Guinea, termed West Irian, even as Indonesia became the United Nations' 60th member in 1950. The dispute, and occasional armed clashes, continued until a settlement was achieved in 1962; on 1 May 1963 West Irian (now Irian Jaya) was transferred to Indonesia. In 1969 an Act of Free Choice, supervised by the United Nations, confirmed the transfer by plebiscite.

Sukarno's troubles extended beyond wresting control of Indonesia from the Dutch. Despite the first national elections in 1955, parliamentary control proved difficult and the decade from the mid-1950s to the mid-1960s was characterized by violence and insurrection. Unsuccessful rebellions in Sumatra, Sulavesi, and other islands in 1957 were followed by a series of short-lived national governments.

In an attempt to establish control through an independent executive rather than by parliamentary rule, Sukarno imposed so-called Guided Democracy in 1959. This authoritarian regime aligned Indonesia with Asian communist states, boosting the importance of the Indonesian Communist Party (PKI), which gradually gained control and in 1965 sought to arm its followers as an official armed force. The Indonesian army leaders resisted this move, and on 1 October 1965 PKI-supported forces attempted to seize national power. PKI military units occupied key locations in Jakarta, and six senior Indonesian generals were kidnapped and murdered. The army defeated the coup attempt, and in Java and Bali thousands of communist followers were killed, leaving a lasting emotional turmoil that is still in evidence.

Sukarno tried to restore the position of the PKI, but his efforts, plus evidence of mismanagement and misconduct, rapidly eroded his popular support. By March 1966 he was forced to transfer key military and political power to Suharto, a leader who had rallied the country to defeat the coup attempt. A year later Suharto was named acting president; Sukarno died in June 1970.

Suharto shifted both foreign and domestic policies away

from Sukarno's traditional ideology, setting economic rehabilitation and development as primary goals. The army remained dominant, but was now advised by Western-educated economic experts. In 1968 Suharto was elected president in his own right; he has been reelected regularly every five years since then.

Since the attempted communist coup, Suharto has faced a major armed insurrection that stems from it and from the centuries-old Portuguese control of East Timor. In mid-1975, as Portugal struggled with its own internal political problems, the Portuguese withdrew from East Timor, ostensibly leaving it to its own government, but in reality opening the way for control by a well-armed Marxist faction, Fretilin.

Fretilin gained military superiority in East Timor's internal struggle, alarming the Indonesian government that a communist regime might be established on one of its borders. Responding to appeals from Timorese opponents of Fretilin, Indonesian troops intervened; they overcame Fretilin regulars and cleared the way for East Timor to become Indonesia's 27th province in mid-1976. Its first elected representatives were sent to Jakarta in 1982.

East Timor's status remains in dispute with Portugal, causing recent unrest. Armed force has been evident in clashes between the Indonesian military and elements of the population.

Politico-Military Background and Policy

The political value system of Indonesia tends toward authoritarianism and paternalistic rule, with emphasis on civil service employment, and more concern for prestige than for performance. This derives from historical Hindu-Buddhist and Islamic influences and from the efforts of Indonesian intellectuals who fought for freedom, finding aspects of Western tradition and political liberalism that could be incorporated into their own distinct concepts.

In practice the Indonesians have a military-backed government. Its major political party, known as Golkar (Golongan Karja, meaning Functional Groups), claims to speak for about 270 affiliated groups from all walks of life. Golkar is effectively a government party; its recent voting strength has been about 73 percent. Other political parties include the Indonesia Democratic Party (PDI), a federation of nationalist and Christian parties; and the Development Unity Party (UDP), composed of formerly Islamic parties. A residual Communist party of several thousand (less than 10% organized) is all that remains of the one-time party membership of 1.5 million; membership was banned after the bloodshed in 1966.

President Suharto is both leader and chief of state. He heads the executive branch and also the cabinet, which he selects. The unicameral legislature has a House of Representatives with 500 members, 100 of whom are appointed and 400 elected. A second body is the People's Consultative Assembly of 1,000 members, half of whom are selected through other than elective processes. It is this body that elects the president and the vice-president and in theory determines national policy. The highest judiciary is the Supreme Court. Suffrage is universal for those over 18 and for married persons regardless of age.

Indonesia's national military forces (TNI—Tentara Nasional Indonesia) provide for national defense and also fulfill a major sociopolitical role. Under Sukarno the armed forces gained political influence due to weak, ineffective, and poorly organized civil governments and adminstrations. So entrenched, the generals became difficult to dislodge. The military overall shows a distrust of civil authorities and of special interest or ideological groups. As a result, Indonesia's military has promoted the concept of a secular state in which all the social groups exist in peace.

It is understandable that the PKI found it difficult to infiltrate the armed forces, and the failed 1965 coup attempt might have been predicted. After this watershed politico-military event the TNI purged its ranks of officers involved in the coup, and under the leadership of Suharto combined army, navy, and air force into a unified command structure, with the army dominant and Suharto in control.

Although Suharto has acted to circumscribe the power of the generals, he has also taken steps to enhance their image among the population in an effort to reduce dissatisfaction with the military's multifaceted role in society. Despite Suharto's resolve to lower the military profile in Indonesia, it remains an entrenched political fixture. Suharto has urged that the military discipline themselves, behave as servants of the people, and refrain from inconveniencing villagers. Since political parties, villagers, and urbanites show no strong conviction to change conditions, a wary status quo will likely continue to characterize Indonesia's politico-military environment.

Internally, governmental policies continue to focus on issues of economics and stability, emphasizing the need to strike a balance between democracy and leadership but with recognition of Sukarno's mistakes. Internationally, Indonesia strives for a carefully maintained neutrality.

Strategic Problems

Perhaps the key strategic problem for Indonesia is feeding and employing its growing population. The average GNP growth each year from 1985 to 1989 was 4 percent, less than the 5 percent needed to absorb about 2.3 million new workers into the work force annually. This is relevant to Indonesia's place as East Asia's largest oil exporter and the only Asian member of OPEC, but with a population that still has a per capita income of less than US$500 per year.

Also important are internal stability issues relating to the Free Papua Movement (OPM) of Irian Jaya, of which there are about 600 members (about 100 of whom are armed); and the Revolutionary Front for an Independent

East Timor (Fretilin) with about 400 members. While at present these insurrections seem small and of little potential consequence, the East Timor problem remains an international dispute between Indonesia and Portugal; the UN has not recognized East Timor as a part of Indonesia.

In addition to being a major supplier of oil to Japan, Indonesia's strategic position astride Indo-Pacific sea lanes is of global significance since it has the potential to block passage of petroleum and gas products to Japan and other energy-dependent states. Should Indonesia return to Sukarno's concept of an "Archipelago Doctrine," it would create a giant block of sovereign territory of 4,800 by 1,600 kilometers (3,000 by 1,000 mi.).

Military Assistance

The amount of aid Indonesia has received, both military and economic, has varied with its evolving international relationships and political focus. During the Sukarno period U.S. military aid was about one-tenth that of the Soviet Union. During Suharto's rule, the United States replaced the Soviet Union as Indonesia's major military supplier, but on a smaller scale. For example, in 1988 the United States provided US$2.8 million in foreign military assistance; in the earlier period of 1958–65, Indonesia received US$1.2 billion in military aid from Soviet and Soviet-bloc countries.

For the years 1970–88 U.S. aid, including Export-Import Bank arrangements, totaled US$4.2 billion, and aid from other Western countries for about the same period came to US$19.8 billion. Aid from communist countries during this period totaled only US$175 million.

The diversity of military equipment in the armed forces provides an idea of the extent and flexibility of Indonesia's arrangements with military suppliers. Light tanks and reconnaissance vehicles from the United Kingdom, France, and the Soviet Union are in use; U.S. antitank recoilless rifles are inventoried, and U.S. aircraft fly army and air force missions. The Indonesian navy has submarines of German design and manufacture.

Defense Industry and Defense Structure

Indonesia has little in the way of defense industry. The manufacturing that is carried on is controlled by the military and tends to focus on ammunition, uniforms and field gear, ancillary equipment, unsophisticated repair parts, and small arms; a .30-caliber rifle is entirely made locally. Consequently, most major military end-items have been procured abroad.

Since the mid-1960s the United States has had perhaps the most influence on military thinking in Indonesia. In addition to being a supplier, the United States has provided military schooling for many of Indonesia's officers over the past three decades. This is in contrast with the Soviet influence during Sukarno's time, which saw few officers being trained in communist countries (Yugoslavia was a popular exception). Thus it is not surprising to find similarities in TNI organization, doctrine, and tactics to those of the armed forces of the United States.

The defense structure of Indonesia's armed forces relates to a doctrine that focuses on guerrilla warfare and considers the geographic layout of its vast island nation. Hence Indonesian armed forces are characterized as relatively small in size, lightly armed, and mobile units with a significant naval component.

As a result of restructuring in the late 1960s, the armed forces have a centralized command structure in which the army holds greater sway than the navy or air force. The structure is designed to functionally support the necessary battlefield functions of intelligence, operations, personnel, logistics, territorial affairs, and communications; the departmental functions of manpower, material, finance, education, legal affairs, and security; and nonmilitary affairs—sociopolitical development, civic mission, and functional groups—that distinguish it as a uniquely Indonesian institution.

Alliances

Consistent with its policy of nonalignment, Indonesia has shunned military alliances. It has, however, entered into military aid agreements with the United States and other nations and has made bilateral defense arrangements with neighboring states. Indonesia has also supplied a small contingent of troops to the UN peacekeeping forces in Vietnam (c. 1970) and more recently in the Persian Gulf states of Iraq and Kuwait (1990–91).

Indonesia became a member of the United Nations on 28 September 1950 and is also a member of the Association of Southeast Asian Nations (ASEAN), OPEC, the Association of Tin Producing Countries, and a number of other international, socioeconomic-related organizations.

(For an explanation of the abbreviations and symbols used in the following section of military statistics, see the list of Abbreviations and Acronyms in Volume I.)

Total Armed Forces

Active: 278,000. Terms of service: 2 years selective conscription authorized.
Reserves: 800,000: Army (planned): cadre units; numbers, strengths unknown, obligation to age 45 for officers.

ARMY: 212,000.
Strategic Reserve (KOSTRAD).
  2 inf div HQ.
  3 inf bde (9 bn).
  3 AB bde (9 bn).
  2 fd arty regt (6 bn).
  1 AD arty regt (2 bn).
  2 engr bn
10 Military Area Comd (KODAM)
(Provincial (KOREM) and District (KORIM) comd)
  65 inf bn (incl 4 AB).
  8 cav bn.
  8 fd arty, 8 AD bn.

8 engr bn.
1 composite avn sqn, 1 hel sqn.
SF (KOPASSUS): 2 SF gp.
Equipment:
Light tanks: some 125 AMX-13, 30 PT-76†.
Recce: 60 Saladin, 45 Ferret.
APC: 200 AMX-VCI, 45 Saracen, 200 V-150 Commando, 20 Commando Ranger, 140 BTR-40, 25 BTR-50.
Towed arty: 76mm: M48; 105mm: 170 M-101.
Mortars: 81mm: 800; 120mm: 75.
RCL: 90mm: 90; 106mm: 45.
AD guns: 20mm: 125; 40mm: 90 L/70; 57mm: 200 S-60.
SAM: 25 Rapier, 40 RBS-70.
Aviation:
  Aircraft: 1 BN-2 Islander, 2 C-47, 4 NC-212, 2 Cessna-310, 2 Commander 680, 18 Gelatik (trg).
  Helicopters: 10 Bell 205, 13 Bo-105, 10 NB-412, 10 Hughes 300C (trg).
Marine: LST: 1; LCU: 20 300t; 14 tpt.

NAVY: 42,000, incl est. 1,000 naval air and 12,000 marines.
Principal Commands:
Western Fleet (HQ) Jakarta/Tanjung Priok. Bases: Jakarta, Tanjung Pinang (Riau Is.), Sabang (Sumatra).
Eastern Fleet (HQ) Surabaya. Bases: Surabaya, Manado (Celebes), Ambon (Moluccas).
Military Sea Transport Command: (KOLINLAMIL): Controls some amph and tpt ships used for inter-island comms.
Submarines: 2 Cakra (Ge T-209/1300) with 533mm TT (Ge HWT).
Frigates: 17
6 Ahmad Yani (Nl Van Speijk) with 1 Wasp hel (ASW) (Mk 44 LWT), 2×3 ASTT; plus 2 with 2×4 Harpoon SSM.
3 Fatahillah with 2×3 ASTT (not Nala), 1×2 ASW mor, 1 Wasp hel (Nala only); plus 2×2 MM-38 Exocet, 1×120mm gun.
3 M.K. Tiyahahu (UK Ashanti) with 1 Wasp hel, 1×3 Limbo ASW mor; plus 2×114mm guns.
4 Samadikun (US Claud Jones) with 2×3 ASTT, (probably 3 in store).
1 Hajar Dewantara (trg) with 2×533mm TT, 1 ASW mor; plus 2×2 MM-38 Exocet.
Patrol and Coastal Combatants: 43:
Missile Craft: 4 Mandau PFM with 4×MM-38 Exocet SSM.
Torpedo Craft: 2 Singa (Ge Lürssen 57-m) with 2×533mm TT.
Patrol, Coastal: 3 Tongkak (Ge Lürssen 57-m) PFC.
Patrol, Inshore: 3 Yug Kraljevica, 8 Siliman (Aus Attack) PCI, 5 Bima Samudera PHM, 18 ⟨.
Mine Warfare: 2 Pulau Rengat (mod Nl Alkmaar) MCC (mainly used for coastal patrol).
Amphibious: 14:
6 Teluk Semangka LST, capacity about 200 tps, 17 tk, 4 with 3 hel (2 fitted as comd ships and 1 as hospital ship).
1 Teluk Amboina LST, capacity about 200 tps, 16 tk.
7 Teluk Langsa (US LST-512), capacity 200 tps, 16 tk.
Plus about 44 craft: 4 LCU, some 20 LCM, 20 LCVP.
(Note: 3 LST assigned to Mil Sea Tpt Comd.)
Support and Miscellaneous: 18: 1 Sorong AOR, 1 cmd/spt, 1 repair, 8 tpt (Mil Sea, Tpt Comd), 1 ocean tug, 6 survey/research.

Naval Air: (est. 1,000); 18 cbt ac, 15 armed hel.
ASW: 9 Wasp HAS-1 hel.
MR: 12 N-22 Searchmaster B, 6 Searchmaster L.
Aircraft: incl 4 Commander, 4 NC-212; 2 Bonanza F33 (trg), 6 PA-38 (trg).
Helicopters: 1 NAS-332F, *6 NBo-105, 2 SA-313.

Marines: (12,000): 2 inf bde (6 bn); 1 cbt spt regt (arty, AD)
Equipment:
Light tanks: 80 PT-76.
Recce: 20 BRDM.
AIFV: 10 AMX-10 PAC-90.
APC: 100: 25 AMX-10P, 75 BTR-50P.
Towed arty: 122mm; 40 M-38.
MRL: 140mm: BM-14.
AD guns: 40mm, 57mm.

AIR FORCE: 24,000: 81 cbt ac, no armed hel; 2 Air Operations Areas.
FGA: 2 sqn with 28 A-4 (26 -E, 2 TA-4H); 1 with 12 F-16 (8 -A, 4 -B).
Fighter: 1 sqn with 14 F-5 (10 -E, 4 -F).
COIN: 2 sqn: 1 with 15 Hawk Mk-53 (COIN/trg); 1 with 12 OV-10F.
MR: 1 sqn with 3 Boeing 737-200, 2 C-130H-MP, 4 HU-16.
Tanker: 2 KC-130B.
Transport: 4 sqn: 2 with 19 C-130 (9 -B, 3 -H, 7 -H-30); 2 with 1 Boeing 707, 7 C-47, 5 Cessna 401, 2 Cessna 402, 7 F-27-400M, 1 F-28-1000, 10 NC-212, 1 Skyvan (survey).
Helicopters: 3 sqn: 1 with 12 S-58T; 2 with 2 Bell 204B, 2 -206B, 10 Hughes 500, 12 NBo-105, 13 NSA-330, 3 SE-3160.
Training: 4 sqn with 40 AS-202, 2 C-47, 2 Cessna 172, 5 -207 (liaison), 23 T-34C, 10 T-41D.
Airfield Defense: 5 bn.

FORCES ABROAD
UN and Peacekeeping:
Iraq/Kuwait (UNIKOM): 7 observers.

PARAMILITARY
Police (POLRI): some 180,000: incl Police 'Mobile bde' (BRIMOB) org in coy: some 8,000 incl Police Coin unit (GEGANA); 3 Commander, 1 Beech 18, 7 lt ac; 10 Bo-105, 3 Bell 206 hel.
Marine: About 10 PCC and 15 PCI (armed).
Kamra (People's Security): 1.5m: some 300,000 a year get 3 weeks' basic trg. Part-time police auxiliary.
Wanra (People's Resistance): part-time local military auxiliary force under comd of Regional Military Commands (KOREM).
Customs: About 72 PFI ⟨, armed.
Maritime Security Agency: 4 PFC, 9 SAR PCI.
Military Sea Transport (Transport Ministry): 28 LSL (for Army).

OPPOSITION
Fretilin (Revolutionary Front for an Independent East Timor): some 150–200 incl spt; small arms.
Free Papua Movement (OPM): perhaps 500–600 (100 armed).

Future

Indonesia continues to adapt its economy to meet internal demands and international trade and commerce opportunities. GNP growth in 1989 and 1990 averaged over 7 percent, exceeding the assessments of observers in the West. While this is encouraging, inflation was also up for these years, from 6.5 percent to 10 percent.

Politically, the nation continues to evolve internally at a slow pace. The 1990 crisis of public confidence over the resurgence of corruption and favoritism will test the ability of an authoritarian government to deal with these problems and reassert its credibility with the public. Military forces and their leaders will have a major role to play in resolving the situation, either aided or detracted by the effects of rapidly shifting global economic, political, and military events.

This may not prove an easy chore. Economic change is intertwined with political stability, and the two can at times conflict. Managing change for economic improvement embodies a major opportunity for Indonesia; maintaining stability while doing so presents the most difficult challenge.

Charles F. Hawkins
Donald S. Marshall

See Also: Association of Southeast Asian Nations; Australia; Indo-Pacific Area; Malaysia; South and Southeast Asia.

Bibliography

American University. 1983. *Area handbook for Indonesia*. Washington, D.C.: Government Printing Office.
Crouch, H. 1978. *The army and politics in Indonesia*. Ithaca, N.Y.: Cornell Univ. Press.
Dahm, B. 1971. *History of Indonesia in the 20th century*. New York: Praeger.
Geertz, C. 1964. *The religions of Java*. Urbana, Ill.: Glencoe Press.
Glassburner, B., ed. 1971. *The economy of Indonesia*. Ithaca, N.Y.: Cornell Univ. Press.
Hunter, B., ed. 1991. *The statesman's year-book, 1991–92*. New York: St. Martin's Press.
International Institute for Strategic Studies. 1991. *The military balance 1991–1992*. London: Brassey's.
Murray-Brown, J. 1990. Indonesia. In *The Asia and Pacific review*. 11th ed. Saffron Walden, U.K.: World of Information.
Neill, W. T. 1973. *Twentieth century Indonesia*. New York: Columbia Univ. Press.
Population Reference Bureau. 1990. *Population data sheet, 1990*. Washington, D.C.: Population Reference Bureau.
U.S. Central Intelligence Agency. 1992. *The world factbook 1991–92*. Washington, D.C.: Brassey's.
U.S. Department of State. 1985. Indonesia. In *Background notes*. Washington, D.C.: Department of State.
Weinstein, F. B. 1976. *Indonesian foreign policy and the dilemma of dependence: From Sukarno to Soeharto*. Ithaca, N.Y.: Cornell Univ. Press.

INDO-PACIFIC AREA

The earth's single largest area with inherent geo-political significance is the Indo-Pacific area, often referred to politically and economically as the Asia and Pacific region.

The most significant geographic characteristic of the Indo-Pacific area is the connection of the Pacific and Indian oceans, which provides the means for travel and for transport of significant amounts of food, fuel, and other natural resources to all 60 nations of South, East, and Southeast Asia—and to the non-independent political entities throughout the Pacific and Indian oceans. Although these two bodies of water mingle within Indonesia without significant natural barriers, there are increasing possibilities of political barriers to free movement by sea and air throughout the region, with a potentially significant impact on the ability of nations to plan for contingency operations.

More critical for military purposes than the actual distances between locations in the Indo-Pacific area are the total number of "steaming days" required to move men and materiel over the constrained routings that may be available in time of conflict. Ship movement and resupply times over the Indo-Pacific area are often measured in terms of weeks—even months—rather than hours or days, and moving men and materiel by air may involve days rather than hours. The complex political relationships in the area make the actual time requirements far greater than what might be calculated by merely looking at maps and figuring the most direct routes.

Despite attempts by the U.S. military to reduce or divide the command and control responsibilities for the Indo-Pacific area, it continues to be the operational responsibility (approximately) of the United States Pacific Command (PACOM) and its Commander in Chief Pacific (CINCPAC).

Geo-Political Matters

The Indo-Pacific area incorporates six regions and subregions: Southeast, Central South, and Southwest Asia; the Indian Ocean; Northeast Asia; and Oceania. The ocean areas total some 239 million square kilometers (93 million sq. mi.), over 25 times the size of the United States. More specifically, the Indo-Pacific area is defined as that area from approximately 40 degrees east longitude (just off the coast of Africa, and including the Comoros and Madagascar) through 80 degrees west longitude (a line running through the Juan Fernandez Islands); and from about 40 degrees north latitude (to include Japan and Hawaii, and the western part of the Caspian Sea) through 60 degrees south latitude (to include the McDonald Islands in the Indian Ocean and the McQuarie Islands in the Pacific). The equator divides northern from southern subregions.

Two subregional parts of the Indo-Pacific area where the commingling natural and political borders of regions and subregions raise very serious politico-military–related issues are (1) Southwest Asia (covering most of what is often called the Middle East, where Mediterranean and Asian cultures fuse; the term stems from a time when East and Southeast Asia were called the Far East); and (2) Southeast Asia (where continental Asian and insular Austronesian societies and cultures mingle, and the Pacific waters become part of "Indonesia"—an area far larger than the present nation-state of Indonesia). In a larger sense, Southeast Asia reflects that area of historic contact where

Indian culture (particularly its religious facets) mixed with the Chinese, particularly in Indochina (Vietnam, Cambodia, and Laos).

Results of the clashes and subsequent intermingling of broadly related cultural groups in the area have been further complicated by more recently intrusive European colonial powers. The intrusions reflect expanded politico-economic interests, especially as they relate to fuels and their movement, and military force presence. This Indo-Pacific geo-political and cultural mix has been further affected by increasing Soviet military-economic expansion into Pacific and Indian ocean areas and the resultant strategic implications. Further complexity arises from newly changed politico-economic-military factors stemming from expansion of national boundaries which have been formalized by emerging Law of the Sea (LOS) agreements and varied national interpretations and usage of that "law." All of this is further influenced by the political, and especially the economic, effects of recent global moves away from communism toward market democracy.

Economic Significance

The primary economic factor underlining the politico-economic reality of the Indo-Pacific area is the location of the most significant known global reserves of petroleum and gas energy sources in the Middle East/Southwest Asia subregion. This factor will strengthen when results of ongoing oil field explorations in the South China Sea and archipelagos of the Pacific are known.

Another important economic factor for the area is the use of its waters for a large foreign fishing industry and for trade. Bulk cargo vessels, freighters, and commercial fishing and factory-processing boats use age-old trading routes to move goods from one national source to another, assisted by the entrepôt ports in Southeast Asia (Singapore) and Northeast Asia (Hong Kong and Macao). Use of these centers is now complicated by the pending turnover of Hong Kong and Macao to the People's Republic of China (PRC), and the continuing sovereignty issues among Taiwan/Republic of China, the PRC, and the "indigenous Formosans." Trading will also be influenced by recent changes from what were once "international waters" to more complicated politico-economic-defense zones provided for by LOS agreements and through the interpretations by signators and abstainers of these agreements.

Environmental Issues

Some serious ecological impacts from the increased trade in the region have been seen in pollution, the resultant endangerment of many marine species (detailed below under the "Ocean" sections), and the movements of biological pests and dangerous predators into new areas. Additional negative effects on the ecology can be expected from future commercial ocean-bottom mining and the testing and disposal of nuclear materials.

Political Significance

Economic implications and a demonstrated significant growth potential for many nations located in, on the shores of, or using transportation facilities and resources of the Indo-Pacific area are immensely complicated by political realities. Over the past 50 years, many of the region's political entities—often extremely small in land size and human population—have attained, or have been striving for, political independence. The newly formed nations have been ceded title to vast expanses of oceanic resources by LOS agreements, creating a nightmare for larger nation involvement. Technical provisions of treaties, and the influence of past relationships, are further complicated by ethnic, linguistic, and other divisive cultural issues. There are also increasingly strong political-military sentiments, both external and internal, for the establishment and maintenance of nuclear-free zones. (In particular, see entries for New Zealand and Oceania.

Influences resulting from the spread of various ideologies, the education in various "mother" countries of present leaders and future leaders of many new nation-states, the rise in the use of terrorism, and revolutionary conflict contribute to the likelihood of instability in the years ahead. This instability may lead to calls for politico-military intervention by external powers. Military forces would have to traverse what now has become a very complex set of subregions and individual political entities interconnected by sea lines of communication on (and airways above) the waters of the Indo-Pacific area.

Military Significance

A multitude of military forces are now in use in the waters of the Indo-Pacific area. Their presence makes for global as well as regional politico-military complexities which include: political calls for "nuclear-free" zones; moving military troops to control outbreaks of potential revolutionary and nationalistic activity in various Indo-Pacific islands and island groups, in addition to helping control insurgency elsewhere (e.g., Indian troops in Sri Lanka); increased Russian commercial and military activities in what was once an "American lake"; quarrels over mineral rights and fishing fleet intrusions; and the effects of student movements and politico-military change (e.g., China, Korea, and the Philippines), which may influence major powers' military deployments, weapons storage, and troop movement.

All of the above testify to an increasing potential for the need to move, and possibly deploy, military combat troops over the vast distances of the Indo-Pacific area—an area crisscrossed by interlinked lines of communication with barriers and zones of influence that will affect military planning. Such movement may require the use of both the Indian and the Pacific oceans. Accordingly, military planners must be aware of a host of geographic, economic, political, and other factors that may influence troop movement.

Indian Ocean

Almost eight times the size of the United States, the Indian Ocean is the third-largest ocean in the world, larger than the Arctic, but smaller than the Atlantic or Pacific. There are 66,526 kilometers (41,246 mi.) of coastline. Tributaries include, among others, the Arabian Sea, the Bass Straight, the Bay of Bengal, the Java Sea, the Persian Gulf, the Red Sea, the Strait of Malacca, and the Timor Sea.

The Indian Ocean floor is dominated by a mid-ocean ridge, subdivided by southeast and southwest ridges, and the "ninety east" ridge. The Java Trench is the deepest point, at 7,258 meters (23,951 ft.).

The Indian Ocean northeast and southwest monsoon seasons are from December to April and from June to October, respectively. There are tropical cyclones during May–June and October–November in the north Indian Ocean and during January–February in the south. Surface currents in the south Indian Ocean move counterclockwise, the reverse of those in the north; this is the result of low pressure over Southwest Asia from the hot, rising summer air, with southwest to northeast winds and currents. High pressure over northern Asia from cold, falling winter air produces northeast-to-southwest winds and currents.

Natural resources include oil and gas fields, fish, shrimp, sand and gravel aggregates, placer deposits, and polymetallic nodules. Industrial exploitation of these resources has raised environmental concerns about oil production in the Arabian Sea, Persian Gulf, and Red Sea.

The ocean's most significant economic function involves transportation of petroleum and trade goods. Ports include Bombay, Calcutta, and Madras in India; Columbo in Sri Lanka; Durban and Richard's Bay in South Africa; Jakarta in Indonesia; and Fremantle and Melbourne in Australia. Unlike the Pacific, no submarine communication cables have yet been installed in the Indian Ocean.

The most critical strategic areas include such chokepoints as: Bab el Mandeb, Strait of Hormuz, Strait of Malacca, the southern access to the Suez Canal, and Lombok Strait. The Indian Ocean also provides launch points and havens for nuclear-armed strategic missile submarines. Bases for ships of France, Russia, the United States, and other external navies are provided under various local country arrangements. The British Indian Ocean Territory provides a strategically significant logistical base and landing field for air- and seaborne U.S. armed forces.

Pacific Ocean

The Pacific Ocean is the largest ocean in the world, almost eighteen times the size of the United States, and larger than the Earth's total land surface. It covers one-third of the globe. The total area is some 165,384,000 square kilometers (63,574,000 sq. mi.), and there are 135,663 kilometers (84,111 mi.) of coastline. Tributaries include, among others, the Arafura Sea, the Banda Sea, the Bellingshausen Sea, the Bering Sea, the Coral Sea, the East China Sea, the Gulf of Alaska, the Makassar Strait, the Philippine Sea, the Ross Sea, the Sea of Japan, the Sea of Okhotsk, and the South China Sea.

The eastern Pacific sea floor is dominated by the East Pacific Rise; the western Pacific is dissected by deep trenches, which include the world's deepest trench, the Marianas Trench, 10,922 meters (36,043 ft.) deep. The coastal land masses that bound the Pacific Ocean make up the "Pacific Ring of Fire," which is so named because of the frequency of violent volcanic and earthquake activity.

There are thousands of islands in the Pacific Ocean, from low coral to high volcanic (and combinations of the two), particularly in the southwestern Pacific. These islands are subject to tropical cyclones (typhoons) in East and Southeast Asia from May to December, and especially from July through October. Tropical cyclones form south of Mexico and strike Central America and Mexico from June to October, especially in the months of August and September. The El Niño phenomenon occurs off the coast of Peru when tradewinds ease off and the warm equatorial countercurrent moves south—killing the plankton that is the primary source of food for anchovies. The anchovies move to better feeding grounds, but large numbers of marine birds starve.

Monsoons occur in the western Pacific during the summer, when moisture-laden winds from the ocean blow over the land; a dry season occurs in the winter when dry winds from Asian lands blow back over the ocean. Northern Pacific waters have a clockwise warm water circulation of currents; the South Pacific has a cool, counterclockwise system. In the north, ice forms in the Bering Sea and the Sea of Okhotsk in winter; ships are subject to superstructure icing from October to May in the extreme northern reaches. There is also a persistent fog covering the northern Pacific from June to December, which is hazardous to shipping. In the southern Pacific, ships contend with icing conditions from May to October; the maximum extension of ice up from Antarctica is in October, when icebergs threaten the southern shipping lanes.

Natural resources of the Pacific Ocean include: oil and gas fields, polymetallic nodules, sand and gravel aggregates, placer deposits, and fish. Both industrial fishing and oil and gas production take place in the area; the latter has led to oil pollution in the Philippine and South China seas. All of these activities have resulted in the endangerment of several marine species: dugongs, sea lions, sea otters, seals, turtles, and whales.

Major Pacific ports include Bangkok (Thailand); Hong Kong; Los Angeles, San Francisco, and Seattle (United States); Manila (Philippines); Pusan (South Korea); Shanghai (China); Singapore; Sydney (Australia); Vladivostock (Russia); Wellington (New Zealand); and Yokohama (Japan). Several submarine communication cable networks have their junctions in Guam and Hawaii. Major strategic chokepoints are: the Bering Strait; Panama Canal; Luzon Strait, and Singapore Strait.

Future

The twenty-first century is often referred to as the century of the Pacific. The economic significance of the Pacific Rim and the aggregate trade value of Pacific commerce to the United States, now exceeding that of the Atlantic, point to the increasing significance of the Pacific for the United States and the rest of the world.

However, the Pacific cannot be divorced from the Indian Ocean, especially regarding the intricate politico-economic issues of Southeast Asia. That region's political instability, combined with significantly increased Russian presence and off-shore naval and air movements in the Pacific, the competitive major power support for warring sides in the internal conflicts of island group political activities, and the politico-economic effects of LOS agreements, create the potential for greatly expanded military and defense activity at various levels of the spectrum of conflict in the Indo-Pacific area.

DONALD S. MARSHALL

SEE ALSO: South and Southeast Asia; Oceania.

Bibliography

Allen, P. M. 1987. *Security and nationalism in the Indian Ocean.* Boulder, Colo.: Westview Press.
Bellwood, P. 1979. *Man's conquest of the Pacific: The prehistory of Southeast Asia and Oceania.* New York: Oxford Univ. Press.
Bowman, L. W., and I. Clark, eds. 1981. *The Indian Ocean in global politics.* Boulder, Colo.: Westview Press.
Braun, D. 1983. *The Indian Ocean: Region of conflict or "peace zone"?* New York: St. Martin's Press.
Bunge, F. M., ed. 1983. *Indian Ocean: Five island countries.* Headquarters, Department of the Army DAPROM 550-154. Washington, D.C.: Government Printing Office.
Central Intelligence Agency. 1990. *The world factbook 1991–92.* Washington, D.C.: Brassey's.
Friis, H. R. 1967. *The Pacific basin—A history of its geographical exploration.* New York: American Geographical Society.
International Security Council. 1988. Security of the Asian-Pacific sea lanes. *Global Affairs* (Spring).
Kapur, A. 1982. *The Indian Ocean: Regional and international power politics.* New York: Praeger.
Kerr, A., ed. 1981. *The Indian Ocean region—Resources and development.* Boulder, Colo.: Westview Press.
Longhurst, A. R., and D. Pauly. 1987. *Ecology of tropical oceans.* Orlando, Fla.: Academic Press.
World of Information. 1987. *The Pacific guide.* 3d ed. Suffron Walden, U.K.: World of Information.
———. 1988. *The Asia and Pacific review: The economic and business report 1988.* Suffron Walden, U.K.: World of Information.

INFANTRY

Foot soldiers have been an important part of armies as long as wars have been conducted.

History

The armies of Sparta won undying fame fighting on foot with spears, and the legions of Rome conquered the peoples of the Mediterranean, the Near East, and Western Europe marching and fighting on foot. But in summer 378, the Roman emperor Valens attacked the Visigoth army of Fritigern outside Adrianople. Fritigern called on his nearby Alan and Ostrogoth allies, whose cavalry (armored horsemen) arrived while Valens' army was still deploying. The barbarian cavalry drove off the Roman cavalry, and then struck the Roman foot soldiers from flank and rear. The surrounded Romans were slaughtered, and Valens was killed. Adrianople marked the demise of the Roman foot soldier and demonstrated the tactical power of cavalry.

THE MIDDLE AGES

During the Middle Ages, armies underwent a fundamental change: the ancient military system composed of free men came under the feudal social system. Each feudal lord commanded his own army, which was built on a main striking force of armored horsemen (knights). As these knights were expensive to equip and train, they were supplied by wealthy landowners.

The decline of noble support for the knights coincided with the rise of *landsknecht* armies. (This initiative came from the Swiss confederates, who used well-trained foot soldiers to defeat armored knights.) The Germans gave the name *landsknecht* to this new version of the ancient Macedonian phalanx. The *landsknecht* forces were among the ablest foot soldiers of the Late Middle Ages.

DEVELOPMENT OF INFANTRY

The modern term *infantry* is a relic of the Middle Ages. A knight or man-at-arms was followed into battle by retainers, who usually fought on foot and were recruited from tenants of his estate. Because the knight was both commander and patron, these retainers were often called *enfants* or *infante*, "children." Later, *infantry* became a term applied to all foot soldiers.

During the seventeenth and eighteenth centuries, the Netherlands became the forerunner for establishing standing armies composed of mercenaries and recruited natives. The Dutch military structure developed by Maurice of Orange created a trained professional army with the infantry as the prime component.

The Dutch soldier, especially the infantryman, was subject to drill, coercion, and draconian penalties for the pettiest offenses. Individualism and initiative were not in demand in the common soldier; the combat drill aimed at exact execution of precise commands and perfection in the control of fire and maneuver. Nothing is more characteristic of the art of war of that time than the statement of a Prussian general who said: "It is true that 76 steps per minute have been prescribed for marching, but upon careful deliberation and frequent observation I have come to the conclusion that 75 steps per minute are even better."

Over time, the infantry was subdivided into groups of specialists. There were the grenadiers—hand-grenade throwers—who were considered "the core of the infantry." They were specifically selected and reliable assault troops, who in most cases were seasoned by long service. Musketeers originally were soldiers armed with matchlock muskets but the term eventually became the generic name for the common soldier.

In contrast to the musketeers, the pikemen, foot soldiers armed with a pike for close combat, were regarded as gentlemen. Fusiliers, originally artillery train guards, were equipped with flintlock muskets. Jagers, trained hunters and foresters employed as scouts and sharpshooters, were employed in irregular warfare, skirmishing, security tasks, and reconnaissance. The success of such volunteer riflemen in the American War of Independence led other nations—particularly France—to develop those troops as part of the "light infantry."

At the end of the eighteenth century, wars continued to be won by professional armies in battles with the proper employment of the infantry.

THE AGE OF REVOLUTION

When the masses of France were mobilized during the French Revolution, there was no time to train troops in the traditional way. However, like the American volunteers, the French troops had greater mobility and made better use of the terrain. Their use of columns echeloned in depth proved successful against the attacking armies who were arrayed in lines according to prescribed regulations.

The revolutionary demand for liberty, fraternity, and equality changed the principles of military command and control, and affected tactics as well as *esprit de corps*. The new *esprit* required superiors to treat their soldiers humanely and soldiers to be prepared to serve. The victories of revolutionary armed forces over the professional armies of the ruling order bear testimony to the success of this approach (again, the American War of Independence is an excellent example).

The rising nationalism connected with humane treatment by officers and noncommissioned officers contributed considerably to the success of the French troops and especially of the infantry. Napoleon's great victories strengthened that motivation.

The Prussians and the British who defeated the French at Waterloo had learned their lesson from Napoleon. The Prussian General Neithardt von Gneisenau described the light infantryman as follows:

> Ease in the individual, mobility, presence and agility of mind, skill in using all facilities as cover for himself or to the disadvantage of the enemy, jumping over ditches, fences, hedges and walls, running, correct target shooting, unnoticed creeping up and away, approaching and withdrawing with the speed of a horse, leaving no hiding place, no rock unnoticed to send the hitting bullet to the careless enemy.

Those demands are still valid today.

In his work *Vom Kriege*, Gen. Karl von Clausewitz emphasized the importance of the infantry when he discussed combat actions:

> An engagement consists of two essentially different components: the destructive power of firearms, and hand-to-hand, or individual, combat. The latter in turn can be used for either attack or defense (words employed here in an absolute sense, for we are speaking in the broadest terms). Artillery is effective only through the destructive power of fire; cavalry only by way of individual combat; infantry by both these means.

That is, the infantry is the only branch of service that combines both attack and defense, and it is thus (at least in the eyes of infantrymen) superior to both artillery and cavalry. As a logical consequence, Clausewitz stated that "Infantry is the main branch of the service; the other two are supplementary."

The reform of the armed forces in Europe at the beginning of the nineteenth century was largely a consequence of political development. With the end of absolutism and of mercenary armies, citizens were soldiers and soldiers were citizens. The professional situation changed for the military leader as well. Noble descent was no longer the key to advancement in the armed forces, but was superseded by knowledge and education in peacetime, and bravery in wartime. The increasing number of citizens available for military service and the continued improvement of weapons led to reorganization in all armies.

Moreover, the growth of technology in the armed forces was unstoppable. The introduction of the cone-shaped rifle bullet (the "Minié ball") in the midnineteenth century more than tripled the effective range of infantry fire, from under 250 yards to over 800, with a sharp increase in accuracy. By the 1860s (especially in the American Civil War), infantry firearms inflicted more than three-quarters of all battlefield casualties. The increased lethality of these firearms forced greater dispersion on the battlefield, from an average of one soldier per 200 square meters in Napoleon's time to one per 257 square meters in the American Civil War.

WORLD WAR I

The lethal potential of technology first became apparent during World War I. Infantrymen discovered, especially on the western front, that courage was no longer sufficient to decide the battle. The firepower of artillery and the deadly effect of machine guns forced the foot soldiers of the belligerents to take cover. World War I produced a new type of infantryman: the individual fighter specifically trained and equipped for his task. Soldiers began the war carrying rifle, bayonet, and pistol; by the end, they carried steel helmet, knapsack, spade, gas mask, and hand grenades. Flame throwers and satchel charges were supplemented by light and heavy machine guns and portable heavy weapons such as mortars and mine launchers.

The tank was first employed by the Allies in 1917 as an armored infantry escort vehicle. It indicated a solution to the stalemate of trench warfare. A successful attack against an enemy in trenches and behind deeply echeloned obstacles was possible only with specifically trained assault battalions after heavy preparatory fire, often lasting several days. This suggested that artillery, not infantry, played the decisive part in war. The attack of infantry formations in dense skirmish lines was abandoned. Neither the armies of the Central European Powers nor the Allied Forces could cope with the high losses of the offensives during the initial months of war.

The bitter experiences of the years 1914–18 introduced a new epoch of infantry tactics and training. British officers such as General Fuller and the former captain and military author Liddell Hart demanded the organizational combination of armored and infantry units. Liddell Hart asserted that the infantry must become mobile on light armored combat vehicles to ensure cooperation with the tanks. German general Heinz Guderian pursued that idea, and developed a balanced composition of personnel for the German armored (*panzer*) divisions.

WORLD WAR II

Until the outbreak of World War II, the basic idea of motorizing infantry divisions had been implemented only in individual cases. But the ever-increasing use of technology required more extensive logistical support. In the infantry divisions of World War I, more than 80 percent of the military personnel fought on the front lines; in World War II, there were as many as three supporters for every "fighter."

The use of mobile troops by the German Wehrmacht eventually influenced the course of the war. A combination of tanks and infantry formations in the initial phase ensured the success of the blitzkrieg campaigns. The motorization of combat troops allowed Germany to move large numbers of troops rapidly on the battlefield, thus achieving superiority. However, the blitzkrieg successes would have been impossible without the cooperation of the German air force.

On the other side, the Allies quickly learned their lesson. The achievements of the British infantry in the desert war in North Africa, the resilience and toughness of the Russian soldiers in the Battle of Stalingrad, the bloody fights of the American marines in recapturing the islands of the Pacific region from the determined Japanese unequivocally demonstrated General Montgomery's claim that "without infantry you can do nothing, absolutely nothing."

British Field Marshal Wavell wrote in *The Times* in April 1945

> We should be aware of three facts. Firstly, all battles and all wars are won by the infantryman in the long run. Secondly, the infantryman always carries the main burden of the fight, his losses are higher, he suffers to a greater extent from adverse conditions and from physical exhaustion than the other branches of service. Thirdly, the skills of the infantryman in modern warfare are less stereotyped and much more difficult to acquire than those of the other branches of service. With almost every step he takes, and with everything he does on the battlefield the infantryman must put his independence and intelligence to the test. Therefore, we should employ those among our personnel with the greatest intelligence and persistence in the infantry.

The most important changes that World War II wrought upon the infantry were caused by the mechanization of the armed forces and the ever-improving weapons technology. Compensation included changed tactics, improved protection, new organizational structures, and improved training. During World War I, chemical agents (gas) had become a new threat to the infantryman employed forward in the trench; the battle tank became the main enemy of the infantryman in World War II. Consequently, every effort was made to develop new armor-piercing defense weapons.

Special troops of different designations developed within the infantry branch. The *Grenadiere* and later *Panzergrenadiere* units of the German armed forces, equipped with armored personnel carriers, corresponded to the armored infantry of the U.S. Army. They had the task of protecting main battle tanks against enemy infantry and were to a certain extent able to fight in both mounted and dismounted modes. The Soviet Red Army had "desant" infantry units in their tank forces, which were mounted on battle tanks during attack operations and conquered terrain after a successful breach of breakthrough. Most armies employed specialized infantry including paratrooper, ranger (commando), and mountain troops. Another costly lesson was learned from the toll of lives paid primarily by front-line infantry. Human lives became valued, especially those of well-trained specialists; manpower increasingly would be replaced by technology and materiel. But the principal statement of the German regulation on *Truppenführung* ("military leadership command") remains valid, especially for the infantry: "In spite of all sophisticated technical means, man is the decisive element in battle."

Infantry Today

Today's infantry retains the virtues of the warrior, yet is more integrated with other branches. Modern war presents complex and diverse scenarios, such as fighting against regular military forces; combating guerrilla forces, partisans, and terrorists; civil war; and covert warfare. Moreover, it is subject to additional exogenous factors such as fighting in built-up areas, in the jungle, in the mountains, in the desert, or in large-scale amphibious operations. Although a "world war" between the great powers becomes more improbable, local and limited conflicts are increasingly frequent and virulent, requiring mobile, conventionally equipped task forces.

GERMAN INFANTRY

The German Federal Armed Forces are typical of a number of "special infantry forces" developed by several nations for almost every possible condition. The infantry comes under the collective term of armored and nonarmored combat troops. Rifle troops, mountain infantry troops, and parachute troops belong to the nonarmored combat troops, and the armored infantry troops belong to the armored combat troops. The term "infantry" itself is used in German service regulations almost exclusively in connection with the rifle troops. Each branch of service has specific tasks.

Armored infantry. The armored infantry troops are the most combat-effective and versatile infantry branch of service within the Field Army. They have the task of fighting against infantry and of combating light armored combat vehicles and main battle tanks. In addition, they can operate in close coordination with the armored troops.

The way armored infantry fight is influenced by armored personnel carriers, like the German Marder. Its cross-country and deep-fording capabilities permit employment of the armored infantry in many kinds of terrain. Its speed allows rapid concentration and dispersal of forces and thus provides additional protection against enemy weapons, especially against artillery fire. Its armor protects against hand weapons, to a limited extent against gunfire up to a caliber of 20mm, and against the effects of artillery shell fragments and blast. It also makes mounted combat possible even after nuclear, biological, or chemical weapons have been used. Armored infantry troops also carry 20mm secondary weapons which can use two types of ammunition and can be employed against armored and nonarmored targets. Long-range antitank missiles, machine guns, portable antitank weapons, and small arms complete the weaponry of the armored infantry.

Depending on mission, situation, and terrain, armored infantry troops may fight mounted or dismounted. In mobile operations, the infantry always fights in close proximity to its armored personnel carriers in order to make use of their mobility, firepower, and protection from enemy small arms fire. Armored infantry troops fight on foot to destroy the enemy in close combat during attack operations, or to defend decisive portions of terrain.

In conducting operations with various organizations, for example, brigades, combat groups, or regiments, the infantry forces must be organized so that they can be employed in mobile operations together with other mobile armored forces and antitank forces in attack, defense, and delay operations. The armored infantry forces are organized in battalions comprising a headquarters and supply company, two or three armored infantry companies, and an armored mortar company or a tank company. Battalions in turn make up armored brigades and armored infantry brigades.

Motorized infantry troops. Motorized infantry troops fight on foot, although they are equipped with armored or nonarmored vehicles to achieve greater mobility. Their primary task is to fight in built-up areas, in woods, and in swampy areas. Such terrain reduces the effectiveness of armored combat units and requires units with relatively large personnel strength.

The mission of motorized infantry troops is twofold. First, they must hold terrain from prepared positions and thus facilitate mobile operations of armored combat units during defensive operations. Second, they must dislodge the enemy from his positions, often engaging in close combat and often under the cover of darkness to create the preconditions for the employment of armored combat units in achieving a breakthrough and in pursuing the enemy.

To withstand attacks from armored combat vehicles, motorized infantry battalions are equipped with antitank rocket weapon systems. Motorized infantry units and armored infantry units employ the same small arms equipment. Because of improved air mobility, motorized infantry units, like parachute units, can be rapidly employed by transport helicopters as a mobile reserve. Motorized infantry forces are particularly suited to guarding and securing important facilities in the rear area and to defending them in case of enemy attack. They are also used in counter-raid operations against penetrated dispersed enemy forces or against SPETSNAZ (Soviet Special Forces). Motorized infantry normally operate on foot, often in difficult terrain, and rely on surprise attacks against the enemy.

Prospects

For centuries the infantry has adapted to new technical developments. In future conflicts, as in the past, the infantryman will play a decisive part. Infantry units must be optimally organized, equipped, and trained for numerous missions.

The specialization of the past generation will continue. Organization in smaller formations, units, and subunits will enable better command and control, and electronic command and control will play an important role.

Despite the most sophisticated technology, man will continue to be the main carrier of the fight. The individual soldier's power of resistance and ability will continue to be necessary. Both depend on hard, demanding training.

PETER BOLTE

SEE ALSO: Arctic Warfare; Artillery; Branches, Military; Cavalry; Combined Arms; Jungle Warfare; Land Warfare; Mechanized Warfare; Mountain Warfare; Rapid Reaction Force; Tooth-to-Tail: Land Forces; Trench Warfare.

Bibliography

Bohrmann, K. 1983. Schützenpanzer für die moderne Infanterie. In *Heere International*. Bonn: Verlag E. S. Mittler u. Sohn.
Clausewitz, C. von. [1834] 1934. *Vom Kriege*. Berlin: Vier Falken Verlag.
Liddell Hart, B. H. 1966. *Lebenserinnerungen (deutsche Ubersetzung)*. Düsseldorf: Econ-Verlag GmbH.
Heeresdienstvorschrift. 1988. *Das Panzergrenadierbataillon*. Bonn: BMVg.

INFORMATION FUSION

Information fusion has extensive implications for the military and impacts almost every facet of operations, from logistics and maintenance to the conduct of small unit tactical operations. It is not limited to sensor fusion, nor is it simply putting information from different sources on one display. It involves the acquisition, integration, filtering, correlation, and synthesis of useful information representations from diverse sources for the purposes of situation assessment, planning, detecting, verifying, diagnosing problems, making decisions, or improving system performance and utility. Information fusion is the conceptual process by which multisource and multimodal data are combined to produce useful and readily interpretable information representations. In particular, when many data sources are involved, it includes the idea of data reduction and synthesis of new information constructs. Another very significant perspective of information fusion is the concern for man-machine interaction, which can be described as the acquisition, modeling, processing, and presentation of information to the commander, operator, or analyst for the purpose of enhancing or aiding human decision-making performance.

Until recently, information fusion has been accomplished entirely by human operators, analysts, planners, and commanders. With the rapid evolution of electronics technology and system complexity, the data rates in computerized systems and sensors have been increasing by orders of magnitude. Hence, the quantity of battlefield information has increased to the point where effective utilization by military personnel may be impaired. It is essential that technology be applied to fuse the data acquired into more useful constructs and representations so as to reduce the quantity and improve the quality of information in the military decision-making process. Data must also be filtered to avoid human and/or system overload with irrelevant, outdated, redundant, or perceptually incomprehensible data.

This article begins with an examination of the military need and some military applications for information fusion. In addition, some of the diverse types of data, technologies, and methodologies used are mentioned, and some future concerns are briefly discussed. Only a limited number of examples can be given of the many known possibilities for the implementation and utilization of information fusion.

Military Need and Applications

MILITARY NEED

In the last hundred years, weapon, vehicle, and sensor systems have experienced an exponential increase in their power, range, speed, and accuracy. Operations in the future will develop faster and have much greater destructive capacity than in the past. Commanders will have considerably less time to make decisions and much more information on which to base decisions. The decisions they make will have greater impact and broader consequences than in the past. Without some form of automated information fusion, military personnel may be overwhelmed by information, which could in turn contribute to errors in judgment and delay of decisions, leading to catastrophic results.

MILITARY APPLICATIONS

Command, control, communications, and intelligence (C^3I) embodies the overall management of a battle. It encompasses functions for national and strategic operations, unit tactics, sensors, and weapons platforms such as tanks, ships, or tactical aircraft (e.g., the associate pilot). Some of the functions of information fusion in C^3I are to reduce the quantity; improve quality, reliability, and understanding; and increase the speed of distribution of information. In battle management, the volume of data is large and the time to process it may range from hours down to minutes. For high-performance weapons systems and platforms, the quantity of data is more limited, but the required speed of reaction can range from mere seconds to fractions of seconds. In addition to its role in transforming raw data into useful representations, information fusion should facilitate the prioritization and dissemination of information for C^3I. Many common military tasks, such as allocation of forces, intelligence production, situation and readiness assessment, planning supply and movement, and planning and rehearsing operations, have the potential to be improved by information fusion.

Communications are an essential component of C^3I and information fusion. Without them, data cannot be acquired and processed and commands cannot be used. The enemy will use highly sophisticated countermeasures (CM) and misinformation to disrupt communications and confuse the battle management process. Most information fusion research assumes unlimited capacity and bandwidth for communication, which is far from reality. Demand for communication resources increases rapidly as battles evolve. The technology of modern warfare and the proliferation of data sources such as sensors are creating rapid increases in the quantity and rate of data communication. Information fusion can ease this situation by reducing the quantity of data to be communicated and by optimizing sensor/source reporting schedules. It can also be used to improve the throughput and routing of communications networks and to increase survivability by providing critical and timely information on the electronic warfare environment and the condition of communication resources for network management and control.

Automatic Target Recognition (ATR) systems using single sensors have failed to perform as well as human-operated systems. The fusion of information from multiple sensors, collocated (local) or spatially distributed (global), should provide information that a single sensor does not provide; such fusion has already demonstrated improved ATR performance. Sensor information fusion is critical in the management of fire control for multiple weapon systems engaging multiple targets. Information must be distributed to each weapon system to assure that each engages the target it is most likely to destroy and that every possible target is engaged by at least one system. Embedded sensor information fusion for ATR potentially can be used to provide the performance necessary for autonomous precision-guided munitions. Sensor information fusion is also useful for navigation, nap-of-the-earth flight, and autonomous robotic vehicles.

Sources and Properties of Military Data

Information fusion systems must be able to correlate and integrate data obtained from all sources, and the sources and properties of military information are many and varied.

SOURCES OF DATA

The predominant sources of data processed on the battlefield are sensors, graphics, text, and databases.

There are large numbers of different types of sensors, such as infrared, visual, and radar-imaging sensors. Each frame of image data can contain a million bits of data and, at display rates of 30 frames per second, can require processing speeds of billions of operations per second (BOPS). In order to be useful, the image data must be registered (spatially and temporally), and individual sensor data must be correlated before the information can be fused. Features of interest in these data must be identified, located, and correlated among multiple sensors of varying types and displayed to commanders, operators, and analysts.

Another form of data that is similar to image data is graphics. Graphical data such as digital maps can contain on the order of a billion bits of data; for example, a 1:50,000 digital map contains 0.5 billion bits of information, excluding overlays. Information fusion must be capable of correlating features in graphics data with features in image data and their descriptions in text.

Other sources of sensor data include sonar; range and Doppler radar; threat warning receivers; and acoustic, seismic, magnetic, chemical, and biological sensors. Meteorological data may be considered as sensor data. In addition, humans can act as sensors, reporting data by communications traffic and verbal reports, which need to be fused into the information stream.

Information in the form of text is generated by messages, analyses, plans, and reports. The number of these sources increases dramatically before, during, and after battles. Automated means of extracting essential information from textual data and identifying the timeliness, reliability, criticality, sources, and destinations of the information is an important function of intelligent information fusion and processing.

There are numerous additional types of data typically stored in computer databases, such as the status of personnel, logistics, and maintenance. Position locating and reporting systems also generate data, which must be incorporated into the information fusion database. Accurate and timely automation of the acquisition, processing, and integration of all data sources will be an important aid to the commander's execution of battle management.

PROPERTIES OF DATA

Data may be in raw form as obtained directly from a source or may be preprocessed. Preprocessed data usually results in a compressed data set that expresses essential information more efficiently. For example, image data processing may identify objects and terrain features and represent them symbolically.

Time stamps and spatial references need to be acquired with data and the generation of information. Time is essential in the correlation and combination of information events that occur simultaneously or in a specific sequential time frame. Spatial references are needed to place information objects in accurate geographical locations. All information must be transformed to a common temporal (time) and spatial (location) framework.

Measures of accuracy, reliability, and confidence are required to resolve ambiguities and contradictions. Importance and criticality must be identified in order to establish priorities for processing and dissemination. Additionally, much research is concerned with the information-theory issues of data. For example, data needs to be classified as redundant (no new information) or unique (new information) for effective information fusion. Also, the information fusion operation will be dependent upon the information properties of the specific sensors and their data channels represented by the stored sensor models and their error characteristics.

Technologies of Information Fusion

Advanced information fusion systems, as they are currently conceived, are dependent upon a diverse set of technologies and disciplines. It is important to recognize that the design of useful, robust, high-performance systems requires an aggressive approach to interdisciplinary science and technology. Machine implementations of information fusion have only recently been realized, and these systems are less than adequate for future needs. In times past, hardware considerations and capabilities have dominated system design. This situation has been created by a perceived immediate need for an information fusion capability. However, robust quantitative design of systems has been hampered by the limited state of the art for encoding intelligence in machines. System models and implementations have been ad hoc, which has led to frustration when exact physical and mathematical analysis has been attempted. In the future, the system design process must include a focus on system functionality and performance that addresses a balance of the three major concerns—hardware, software, and human capability. This design process requires extensive continued research in the multiple technologies associated with the information fusion problem. Some of these technologies are discussed in the following paragraphs.

Sensor Fusion

The integration of data from multiple sensors, both locally and globally distributed, and from man-portable sensors to satellites is a complex and important application of information fusion. The goal of sensor fusion is to provide a unique information set that cannot be obtained from only one sensor. Two examples are the production of three-dimensional data, such as in the integration of image and range data, and the detection and identification of targets obscured or partially obscured at one sensor. Sensors employed in electronic support measures (ESM) and ATR systems contribute information critical to the survivability of forces and weapons platforms. For example, information obtained from infrared sensors and radar threat warning receivers must be fused to identify and locate threat signals and to determine their status (search, track, or missile launch for radar systems). This information will be utilized in intelligent systems to ascertain optimal evasive actions (maneuver, jamming, or deception) for survivability.

In the case of multiple sensors of the same kind, the data are similar. However, even this task is not necessarily a simple combination of data. Multiple sensors of the same type are normally geographically distributed, and they may be stationary or mobile. This means that the field of view and scale of the images they produce will vary. Also, moving sensors will have different translation and rotation functions. These factors must be evaluated and accounted for in the processing, or the processing must be robust—that is, independent of variations in these parameters.

In order to achieve useful and robust system operation, a number of factors must be carefully considered and incorporated into the sensor network design. Apart from the obvious concerns for sensor location, orientation, field of view, and motion, the sensors must be calibrated, and operational models of the sensors must be embedded in the system. The models and the processing must consider sensor sensitivity, resolution, noise, and error functions, such as geometric distortion. The data from each sensor must be spatially and temporally correlated with that from the other sensors. Before spatial correlation can be performed, images must be registered with high accuracy (on the order of 0.01%). This means that geometric displacement and distortion must be nearly eliminated.

Sensor fusion involving different kinds of sensors, such as acoustic, magnetic, radar, and imaging sensors, can be a more challenging problem than the integration of information from a set of identical sensors, since their properties and models can be vastly different. For example, acoustic information from a point sensor is a temporally modulated signal, whereas the data from an imaging sensor are spatially as well as temporally modulated. In this example, coordinate and bearing data are used to locate sound sources from acoustic sensors in the image, or composite image, fields of imaging sensors for the same time frame.

Implementations of heterogeneous sensor fusion require that the data from each sensor be treated differently according to each sensor's model or intrinsic operating characteristics (spectral sensitivity, transduction phenomena, etc.). Another task in this kind of implementation is to relate the sensor performance model to the existing battlefield environmental conditions. For example, infrared sensors are more sensitive to hot objects than visual sensors, and their performance is affected differently by such atmospheric conditions as wind, fog, and smoke.

Heterogeneous sensor fusion implementation usually has to be handled at a high level of abstraction and in an ad hoc manner so that the data can be processed as sensor-independent features, such as a particular kind of target/object in each data channel. Alternatively, the signal can be processed at the bit level using information theory. This approach is more complex but more exact, and more source information can be retained. Consequently, the value of the information fusion concept can be enhanced if sensor fusion is approached using communications, signal detection, or information theory. A major problem with this approach is how to encode the information, channel, and feature set to permit efficient, accurate, and reliable fusion.

Robust sensor information metrics and a systematic methodology (abstract or theoretical) are needed to make decisions about multisensor fusion. The answers to questions such as "What types of sensors should be used?" "Where should they be placed?" "How should their parameters be controlled?" and "How does one combine local and globally distributed sensor data?" must be available to make optimal use of sensor fusion.

INTELLIGENCE FUSION AS INFORMATION FUSION

The term *intelligence fusion* is used to capture an idea inherent in a military command and control environment. Sensor data are only part of the super-set of data commonly thought of as intelligence information (SIGINT, ELINT, COMINT, IMAGINT, HUMINT, etc.). The term *intelligence* implies that raw data have been processed to some extent, for example, intelligence reports. The ability to generate a comprehensive intelligence report is predicated on the effective use of information fusion. In fact, intelligence fusion is information fusion. Here, as with sensor fusion, the data from numerous sources must be integrated. Intelligence fusion typically operates on data on a higher level of the information hierarchy than sensor fusion. Intelligence fusion must deal with information in the form of speech, text, tables, database objects, maps, images, graphics, and so forth.

Progress is being made in text processing. Word processing software now typically includes a spelling checker and a thesaurus. There are also intelligent programs available to evaluate the quality, grammar, and education level of writing. However, progress in automated natural language understanding has been slow. The analysis of syntax and sentence understanding is limited to short sentences and does not readily include the context in which sentences occur or the history of the discourse. Current application of this technology is somewhat limited to text generated with a standard form and specified vocabularies.

Significant progress has also been made in speech recognition when system training is allowed for specific speakers. Some speaker-independent speaker recognition has been demonstrated for limited vocabularies. Some of the problems associated with military speech and speaker recognition are the high levels of noise and vibration and effects of stress on speech.

Considerable research also is being performed in the area of database management, which is highly relevant to intelligence fusion and information fusion in general. Relational database management is fairly mature. Current research topics include object-oriented database management and associative memories. Neural network research is also expected to contribute in this area.

ARTIFICIAL INTELLIGENCE

Artificial intelligence (AI) in its most general applied sense is concerned with automating functions normally accomplished by humans. In many instances, AI research and development are directed toward enhancing and extending human perceptual and commonsense reasoning abilities. AI is currently playing an important role in information fusion system development in the form of expert systems. Expert systems contain rules (inference, perception, and reasoning) with which to determine the importance, criticality, relevance, reliability, and association of data and to combine information based on those rules. Expert systems, however, have some serious limitations, and much current research is directed toward overcoming them. Current expert system implementations have been *ad hoc* and "hand-crafted" using an "expert" (pseudo-expert) and someone functioning as a "knowledge engineer" (sometimes the same person). Advanced AI systems involving new approaches, extended expert systems, and hierarchies of interacting expert systems require advances in fundamental understanding (information theoretics and the basis of intelligent behavior) and some degree of automation of the knowledge acquisition process.

Closely related to the problem of knowledge acquisition is the problem of knowledge representation. Knowledge is central to the idea of intelligence, and its usefulness is obviously related in some way to its representation. This perspective leads to the awareness that new data (knowledge) representations and associated databases are required for advanced intelligent systems.

DECISION TECHNOLOGY

As has been mentioned, the data providing the basis for information fusion may be incomplete, inexact, and ambiguous. Nevertheless, decisions must be made to facilitate useful information fusion. It is important to note that information fusion, AI, and decision technology itself are intimately related. All are based on the ability to reason over uncertainty and incompleteness. Furthermore, all must deal with the issue of conflict resolution. In automated systems, it is also necessary to consider the issue of commonsense reasoning, which is the human ability to make quick judgments of acceptable accuracy with limited resources. Attempts to automate human reasoning in a brute force (computational) way are doomed to failure because of the exponentially driven computational needs related to problem complexity. Therefore automated decision analysis requires techniques other than deterministic closed-form numerical computations. For nondeterministic systems, one must inevitably use some form of probabilistic analysis. Most approaches use Bayesian analysis, probably because it is well understood and mature. However, Bayesian analysis assumes that the data sources are independent, which is not usually the case, and requires precise probability, which implies large sample spaces, which are not always available. Also, the Bayesian approach does not readily permit the inclusion of ignorance, uncertainty, ambiguity, conflict, or partial belief, which are important factors in information fusion. Thus, modifications and extensions of Bayesian rule techniques have been investigated to cope with these inexact conditions.

The Dempster-Schafer theory is another approach. It uses probability intervals to compute two variables to represent a credibility interval, or interval measure of belief. The two variables are support (S), which represents the degree to which data support the hypothesis, and plausibility (P), the degree to which data do not deny the hypothesis. The difference, P − S, is a measure of the ignorance about the hypothesis. These measures enable us to identify degrees of belief, disbelief, and ignorance about the hypothesis. This rule is identical to the Bayesian when P = S, or exact point probabilities are known.

A third approach, which has gained some recognition, is called fuzzy logic, or fuzzy sets. It is an extension of classical set theory to sets that do not have well-defined edges, thus are fuzzy. This approach is useful in the analysis of data where accuracy exists in degrees. In classical set theory membership is binary, whereas in fuzzy set theory, members may have any value from zero to one. This enables the inclusion of data with an associated degree of belief or confidence. Additionally, fuzzy set theory has been shown to reduce the computational burden by an order of magnitude in some cases.

Advanced Information Processing

Advancing sensor technology has produced sensors of progressively smaller size with higher performance and lower cost. This advancing technology has contributed to an increasing number of sources, which in turn has led to a corresponding increase in data volume, data rates, and computation assets required for processing. This situation requires the development of parallel, multiprocessor architectures and fast, parallel algorithms to fulfill the performance requirements expected of information fusion systems. Many advanced information processing system implementations involve netted or distributed system architectures, and these are the subject of much study.

Space precludes an in-depth treatment of all possible variations on the concepts of parallel and distributed processing. For completeness, however, it is essential to mention that the ideas of hierarchical, asynchronous, adaptive, and highly interconnected processing nodes will play an important role in the future. Further, implementations called neural nets will provide new ways to acquire, process, store, and retrieve information; these highly interconnected and adaptive systems are expected to enable remarkable and robust implementations of real-time systems. The large volume of data that will be generated also requires development of efficient and distributed mass storage systems. Finally, some information fusion systems requiring extremely high performance will be designed as dedicated, embedded systems, while others will be designed for general-purpose processing.

Programming computers for AI, or manipulation of symbolic data and information, is not easily accomplished using common numerically oriented high-order languages such as Fortran, C, and Pascal. Languages such as Lisp and Prolog and their extensions have been developed for this purpose. These languages are designed to operate on narrative statements and to provide results as statements. Software development for parallel, distributed, and dedicated systems must be developed concurrently with the overall system design. Software development and maintenance make up one of the costliest components of total processing system development, and aggressive attempts are being made to automate standard and AI software development in the form of computer-aided software engineering (CASE).

Information Quality Measures

The data for information fusion come from multiple sources of widely varying inherent characteristics, such as bandwidth, sensitivity, and so forth, with outputs ranging from text to visual imaging. One of the problems in dealing with information from multiple sources is finding measures (e.g., accuracy, confidence) that are robust and consistent among all sources and across the boundaries of multiple levels of processing and integration. This is an extremely difficult task. For example, a fusion task may require combining data from an infrared image and a visual image. How does one measure, or compute, the degree of confidence in the data from each sensor and in the results of the combined information? These are questions being addressed by current research. Quantification of the complexity and information content (information theoretic issues) of the data from each sensor and the combined data is of particular interest.

Man-Machine Interface

Throughout this article the importance of the role of the human has been emphasized. Consequently, it follows that the man-machine interface (MMI) and human engineering must be important components of information fusion systems throughout the design process, from conceptualization to production and fielding. In addition to hardware and software, the MMI must be considered as a component of the whole system.

One of the aspects of MMI is the technology of interactive systems. This technology is concerned with the ability of the user to interact with the information fusion process and the derived information representations. The systems must permit the user to make queries, to retrieve specific information, and to determine the process by which the information was generated (the how and why). Also intrinsic in the idea of these information fusion systems is the need to synthesize information (data) in reduced sets of representation that are unambiguous and rapidly comprehensible to humans. The intelligent interactive information fusion system also must be capable of directing sensors or sources to acquire specific data in specific locations needed to fill information voids and resolve ambiguities and conflicts.

Another aspect of MMI is the visualization of information. This includes display technology and much more. Visualization, a technology in itself, involves the transformation of data into pictures or representations, both static and dynamic, in such a way as to convey a better understanding of the functional relationships of the information and to cue the observer to changes in the situation (variables) being monitored. Visualization technology includes the ability to generate graphics (a graphics generator) from extensive and multiple databases.

Human factors engineering is also an aspect of MMI. Here the objective is to include ergonomics in the design of machines to minimize human fatigue, facilitate MMI, and optimize system performance. An example is the use of speech understanding and synthesis to facilitate man's communication with machines by increasing the speed of interactions and freeing his hands for other tasks.

Future Trends

The need for intensive management aids for future rapid-paced battles and the performance goals of future notional information fusion systems require continued advances in sensor, information metrology, processing, and representation technologies. Current research will provide new technology, which must be rapidly and effectively utilized. One of the principal problems associated with information fusion will be the concurrent development of software to support new directions in system hardware and the MMI. Parallel processing, distributed nodes and databases, networks, and new data representations all require substantially new software implementations. This, coupled with the interactive nature of the systems and the real-time requirements, will lead to entirely new ways of thinking about system implementations. A systems level (hardware, software, and MMI) focus on research and development is required to realize affordable high-performance systems.

Future systems are likely to be strongly hierarchical. This multilevel structure will progress from the lowest level, a single source, to multiple globally distributed sources and from small units, or individual weapons systems, to theater, strategic, and national levels of operations. The databases and fusion processing will be distributed and accessible by multiple levels in the hierarchy. Accurate and flexible models of military systems will be essential. Future systems may even contain models of each individual commander's (friend and foe) beliefs and tactics to aid in predicting events.

Technology will provide an overwhelming quantity of information and the means to process it. Care must be taken to assure that the commander is provided with fused information optimally matched to his specific needs at specific moments in the battle. This will require particular attention to the application of knowledge engineering to military exercises and war-gaming simulations to extract the crucial elements and constructs of information needed to enhance battle management.

Finally, caution must be advised in the development of information fusion and decision support systems. In the hands of the opponent, such systems could be used to predict every action of the force using it. Also, the system could be unstable and promulgate self-destruction. For example, the stock market crash of 1987 has been attributed to widespread use of similar systems for recommending investment strategy. The commander must remain the ultimate decision maker and must understand the role information fusion and decision support systems will play in aiding battle management.

WILLIAM AUGUST SANDER III
DAVID WILLIAM HISLOP

SEE ALSO: Computers, Military Use of; Electronic Warfare Technology Applications; Mathematics and the Military; Research and Development Establishments and Policies; Robotics, Military.

Bibliography

Andriole, J. A., and S. M. Halpin. 1986. Information technology for command and control (introduction to special issue on Information Technology for Command and Control). *IEEE T-SMC* 16:762–65.

Flachs, G. M., J. B. Jordan, and J. J. Carlson. 1988. Information fusion methodology. *Proceedings of SPIE 1988 Symposium in Optics, Electro-Optics, and Sensors*, April.

Flachs, G. M., C. L. Beer, and D. R. Scott. 1989. A well-ordered feature space mapping for sensor fusion. *Proceedings of SPIE Conference on Sensors and Sensor Fusion*, March, Florida.

Garvey, T. D., and J. D. Lowrance. 1984. An AI approach to information fusion. *Journal of Electronic Defense* (7):31–41.

Goodman, I. R. 1987. A general theory for the fusion of data. *Proceedings of the First Tri-Service Data Fusion Symposium*, June 9–11, pp. 254–70. Washington, D.C.

Greer, T. H. 1985. Artificial intelligence: A new dimension in EW. *Defense Electronics* 17 (9):190–208, and 17 (10):108–28.

LaJeunesse, T. J. 1986. Sensor fusion. *Defense Science & Electronics* 5 (9):21–31.

MacRae, J. R., and C. D. Byrne. 1987. Connectionism applied to a real time expert system for tactical data fusion. *IEEE Expert Systems in Government 3d Conference*, October 19–23, pp. 66–71. Washington, D.C.

McKendall, R., and M. Mintz. 1988. Robust fusion of location information. *Proceedings of the 1988 IEEE International Conference on Robotics and Automation*, April 24–29, pp. 1239–44. Philadelphia.

McVicar, K. E. 1984. C^3I: The challenge of change. *IEEE T-AES* 20:401–13.

Mitiche, A., and J. K. Aggarwal. Multiple sensor integration/fusion through image processing: A review. *Optical Engineering* 25 (3):380–86.

Nahin, P. J., and J. L. Pokoski. 1980. NCTR plus sensor fusion equals IFFN or can two plus two equal five? *IEEE T-AES* 16:320–37.

Pecora, V. J. 1984. EXPRS: A prototype expert system using prolog for data fusion. *AI Magazine* 5 (2):37–41.

Rauch, H. E. 1984. Probability concepts for an expert system used for data fusion. *AI Magazine* 5 (3):55–60.

Sutherland, J. W. 1986. Assessing the artificial intelligence contribution to decision theory. *IEEE T-SMC* 16:3–20.

Taylor, E. C., H. M. Beebe, H. S. Goodman, and D. H. New-

ell. 1984. Man and machines: A synergy of tactical intelligence. *Defense Electronics* 16 (6):133–41.
Waller, L. 1989. Fuzzy logic: It's comprehensible, it's practical—and it's commercial. *Electronics* 62 (3):102–3.
Waltz, E. L. 1986. Data fusion for C³I systems. In *Defense Electronics*, eds. *C³I Handbook*, pp. 217–26. Palo Alto, Calif.: EW Communications.
———, and D. M. Buede. 1986. Data fusion and decision support for command and control. *IEEE T-SMC* 16:865–79.

INFRASTRUCTURE, MILITARY

In the military, the term *infrastructure* applies to all fixed and permanent installations, buildings, or facilities. Some examples are:

1. Barracks, camps, and bases built as separate installations or as part of a garrison complex to accommodate troops, their transport and equipment, and often their families; complete with amenities and utilities (applicable to all, except 7).
2. Headquarters, communication centers, and administrative facilities.
3. Hospitals, workshops, a wide range of storage accommodation, and other logistic installations, buildings, or facilities.
4. Port complexes and naval base installations.
5. Airfields and their support facilities.
6. Fuel pipelines and tanks.
7. Other installations providing utilities for military use (separate power stations, water, and sewage works).

National infrastructure is built, operated, and funded for the use of a nation's armed forces. Sometimes the forces of two or more nations share installations, buildings, or facilities by bilateral or similar agreements that specify access, utilization, maintenance, and funding. Within the North Atlantic Treaty Organization (NATO), certain specified categories of operational infrastructure are provided and financed on a special common, cost-sharing basis by member states for the use of their forces.

J. H. SKINNER

SEE ALSO: Installations, Military; Logistics, NATO.

Bibliography

Eccles, H. E. 1959. *Logistics in the national defense*. Harrisburg, Pa.: Stackpole.

INITIATIVE IN WAR

In military literature, the word *initiative* has two different meanings. As one of the principles of war, initiative means "freedom to act" toward an enemy. As a principle of military organization, command, and control, more specifically, it consists of decentralized decision making and command systems for the sake of assuring operational and tactical "flexibility," the key precondition for taking, maintaining, and regaining freedom to act.

Initiative as a Principle of War

As a principle of war, to take, maintain, or retake the initiative means to impose one's will on the enemy and achieve one's objective as if there were no powerful opponents.

Ruled by polarity, confrontation of opposite initiatives is a zero-sum game in which a loss on one side means an equal gain on the other.

Initiative does not depend, ultimately, on material factors, but on relative superiority in cleverness and will. One gains initiative by a better strategy or operational plan, and maintains it by an information-decision-action (IDA) process that is faster than the enemy's. A key precondition of initiative is the freedom to choose among several options and to impose the most favorable one upon the enemy, thus setting the rules of the game. Such is the case, for example, of escalation dominance (see below). But each deliberate choice of a particular option, as it happens in the escalation of conflict thresholds, will necessarily reduce flexibility and initiative: *electa una via non datur recursus ad alteram* (once one way is chosen, it's not possible to go back to the other.) Initiative does not necessarily imply hit first, although that is usually a result.

Military writers often confuse initiative with the offensive, considering the latter as the best, if not the most unique, way to assure initiative. Thus, offense at all costs and preemptive war or strike are judged intrinsically superior to defensive posture, even if national or common defense policy does not imply aggression as a possible means of obtaining objectives. This was the case for Soviet military strategy and doctrine; both based security on the offensive posture of the operational planning. Western military doctrine, on the other hand, bases security and dissuasion on operational defense, even if it includes forward defense, first use, and follow-on-forces attack (FOFA); but such a posture is often judged unnatural, a sort of tribute that military logic pays to the political principles of Western democracies.

But after a closer look at aggression, preemptive war or strike, offense, and attack, it often appears that they are all attempts to regain the initiative lost on a wider scale. It is frequently a lack of political or strategic initiative that drives one side to attack first. Also, the fact that one maintains initiative on the field does not necessarily imply that one has it at the other levels of the conflict, as was shown by Ludendorff's final offensives in 1918.

By being on the offensive, attackers show their intent and spend their reserves, thus paying cash while the defender pays on credit; as Clausewitz said, losing space and gaining time. Inconclusive offensives, such as those of the

Germans against the Allies in 1914 and Russia in 1941, or the Japanese attack on Pearl Harbor, changed into a loss of initiative for the attacker and into a corresponding gain for the defender.

Defense itself may assure initiative by disengagement, even before a counterattack, as the Chinese People's Liberation Army successfully achieved with the Long March, on the largest scale in modern military history.

In a more general way, however, threat, deterrence, indirect strategies, economic measures, coercive diplomacy, and limited use of force may assure political and strategic initiative in global conflicts more efficiently than wide-scale military operations.

Political self-restrictions in military operations are often held responsible for defeat, as happened in the Vietnam War. But war is an historical process that only politics can rule. In the ultimate analysis, military initiative at all levels, including tactical, will depend on political initiative.

Escalation Dominance

Escalation dominance means taking and maintaining initiative in nuclear war by both first use and intrawar deterrence capabilities. Prerequisites are the material and psychological capability of being the first to cross both theater and strategic nuclear thresholds as well as deterring retaliation by a second-strike capability or impeding it by nuclear defense. Material capabilities of escalation dominance consist of efficient surveillance, information, command, control and communication (SIC^3) systems, survivable nuclear sites, and long-range and stand-off delivery vehicles, as well as conventional defense capabilities, to gain the time necessary to decide to use nuclear weapons. Escalation dominance is the basis of extended deterrence and of such nuclear doctrines as "massive retaliation," "flexible response," "limited strategic options" (LSO), and "countervailing strategy."

In case of nuclear parity, the initiative of both first and second strike require that nuclear fire be controlled, graduated, and selective in order to assure firebreak and intrawar deterrence instead of victory. Said another way, escalation dominance relies not only on one's own SIC^3, invulnerability, or survivability to a surprise attack and second-strike capability, but also on a common rationality with the opponent. This implies that in case of nuclear parity, initiative and dominance may be exerted only through one's own escalation machine, which is not enough to master the opponent's will and strategy.

More recent technological developments in SIC^3 and battle management (BM) systems, as well as their lower vulnerability, have enhanced technological capabilities for escalation dominance. This, however, has become—or is perceived to be—less important because of recent progress in nuclear disarmament and conventional arms control.

Initiative in Combat Operations

At the operational and tactical level, initiative aims to impose on the enemy one's "action plan" in order to increase the efficiency of one's forces and to reduce that of the enemy. Any combat operation, at any level, consists of a clash of forces (material element) and a confrontation of will and cleverness (intellectual and psychological element). It is the last that gives organic unity to the first element and allows synergy between the elementary functions of the various components of its own force system.

The initiative is influenced not so much by the ratio of forces as by the ratio of the levels of "freedom to act" that the enemy can use. There is a kind of polarity here: the initiative consists of optimizing and increasing one's own "freedom to act" while contriving annulment of the enemy's. The force ratio affects the initiative only in the sense that the party with the material superiority has a greater number of choices than the enemy.

At the operational and tactical level, initiative gives an opportunity to utilize force components against enemy vulnerabilities. The initiative presupposes a constant aggressive attitude, but it can also allow delaying actions to gain time to withdraw forces into rear positions in order to avoid their destruction in the forward positions, and to provide defense in depth. Initiative includes stratagem, surprise, and deception or diversionary maneuvers.

Initiative on the battlefield depends on the level of command decentralization and on the level of authority given to subordinate commanders to freely adapt their actions according to the situation. The capability to react to unforeseen situations, due to enemy actions or changes in plan, is a basic aspect of the military profession. Initiative presupposes that orders will be related mainly to missions and not to detailed procedures on how to accomplish them.

Contrary to common belief, modern information technologies increase the importance of decentralization. Subordinate commanders, now that they have, in real time, the information relevant to their action, can avoid uncoordinated actions and can continue to monitor information during the operation. This allows more timely action and reaction and greater adaptability to unexpected changes. If there is no delegation of initiative to the subordinate commanders, the use of automated data processing (ADP) runs the risk of hindering the decision-making process because, after a certain point, the volume of information can slow down, and even paralyze, decision-making capability.

Initiative as Decentralized Command

Maintaining military initiative is much more difficult than seizing it. All military writers agree with Moltke's concept that a mistake in the original concentration of an army can hardly be rectified during the entire course of a campaign, and that no operational plan survives the first collision with the enemy's main body. Attrition and chance weaken the capability of maintaining initiative.

To deal with these constraints, typical military organizations have developed two different trends. One trend, that toward centralized command and detailed standard procedures, leaves nothing to chance and improves the capability of reacting against attrition, thus unifying the actions of the entire army as if the supreme or major commander were acting for subordinate field commanders. The other trend emphasizes the importance of giving freedom of action to field commanders, down to lower levels, to change orders and even vary standard procedures depending on their appreciation of the situation, to allow more flexibility to operations and tactics.

There is contradiction in the fact that concentration requires highly centralized command, whereas the movement of the different armies converging over separate routes in support of each other may be assured only by one that is decentralized.

As G. E. Rothenberg wrote, General von Moltke (the Elder) created "the Prussian General Staff as a unique instrument combining flexibility and initiative at local level with conformity to a common operational doctrine and to the intention of the higher command." The German army was the first to adopt a common fighting doctrine and to transform the higher command staff officer into a true "junior partner in command . . . who had the right, indeed the duty, to disregard what he considered unsound operational judgments" of his commander. Moltke also devised the concept of *Auftragstaktik*, "mission tactics, a command method stressing decentralized initiative within an overall strategic design."

British military reformers such as J. F. C. Fuller criticized Moltke's way of command on the assumption that it would require qualities too rare in subordinate commanders, and praised the example of Napoleon, who "led and controlled," instead of only managing the army.

A related problem in military literature was that of command at sea, a paradigm for the more general problem of the amount of autonomous initiative to be granted to theater or independent commanders, as in colonial wars or limited operations. In general, the amount of initiative tends to be in inverse proportion to the political or strategic importance of the mission, with no regard to the size of the forces committed. Cases of high political importance include the delegation of initiative to declare war and to resort to nuclear fire. In 1963, in order to be in a position to accomplish his mission, Gen. Lauris Norstad, then Supreme Allied Commander in Europe, requested unsuccessfully that SACEUR be given full authority by the president of the United States on matters relating to the employment of theater nuclear forces under his direct command.

Balancing centralized command and flexibility is also difficult because individual initiative conflicts with subordination and discipline. Classical tradition reports the case of a young Roman knight put to death by his father, the commanding general of the Roman army, for leaving the front line without permission even though he did so in order to kill an enemy single-handedly (see Livy, viii, 7). On the other hand, however, the Austrian Order of Maria Theresa was granted to officers who achieved victory by ignoring or deviating from their orders. The story is the subject of Kleist's Romantic tragedy *The Prince of Homburg*, presented for the first time in 1811. Just two years later, the Prussian general Hans D. H. Yorck von Wartenburg took the initiative as a field commander to conclude an armistice with the former enemy, the Russians, thus forcing his king to ratify a *fait accompli*. Yorck justified his rebellion by claiming that the king, residing in French-occupied Berlin, had no freedom of action. Much later, Nazi propaganda used Yorck's example as an instructive precedent for the *Reichswehr*'s attitude toward the Weimar Republic. Carl Schmitt made Yorck the prototype of military initiative in foreign policy, like De Gaulle or Salan. Some general staffs, such as that of the German, Austrian, French, and Japanese, have carried initiative too far when they have attempted to settle national interest independently of their respective civilian authorities.

VIRGILIO ILARI

SEE ALSO: Auftragstaktik; Principles of War; Tactics.

Bibliography

Aron, R. 1976. *Penser la guerre. Clausewitz.* Paris: Gallimard.
Ball, D. 1981. *Can nuclear war be controlled?* Adelphi Papers no. 169. London: International Institute for Strategic Studies.
Beaufre, A. 1966. *Stratégie de l'action.* Paris: Colin.
Gerasimov, I. A. 1985. Seizing and holding the initiative in combat. In *The Soviet Art of War*, H. Fast and W. F. Scott, eds., pp. 277–95. Boulder, Colo: Westview Press.
Rothenberg, G. E. 1986. Moltke, Schlieffen and the doctrine of strategic envelopment. In *Makers of Modern Strategy*, P. Paret, ed., pp. 296–325. Princeton, N.J.: Princeton Univ. Press.
Schmitt, C. 1963. *Theorie des Partisanen.* Berlin: Duncker and Humblot.
van Creveld, M. 1985. *Command in war.* Cambridge: Harvard Univ. Press.

INSPECTION AND SUPERVISION

Regardless of whether a military service is an army, navy, or air force, its ultimate purpose is to be ready to perform its mission. Likewise, whether a military entity is a combat or a support unit, its readiness and capability are based not only on the equipment it possesses but also on the ability of its personnel to maintain this gear and to use it expertly. Such use, in turn, is contingent on three major factors. First, the unit's personnel must have been properly educated and trained. Second, these personnel must have remembered and maintained the skills they have been taught. Third, they must have maintained their equipment in proper order. The combat readiness of an organization is a complex phenomenon. It can vary from day to day, it is in part contingent upon a unit's morale and attitude, and it can quickly deteriorate under poor leadership and supervision. The initial signs of such a deterioration are often seen in areas not vital to a unit's mission—uniforms look a bit less smart, living and working spaces are a bit less tidy. From this point, things can deteriorate quickly until finally matters and equipment vital to mission accomplishment are affected and the unit's capability to perform properly comes into question.

Supervision and inspection are two ways of ensuring that a unit achieves and maintains its capability. While the two are not mutually exclusive, in general terms, supervision has a greater influence on achieving a combat capability because it is directly linked with training less experienced personnel and judging and correcting their performance; inspections simply ensure that a capability or standard is maintained.

Supervision

Many military services define a supervisor as a trainee's reporting official and consider supervision to be of such importance that it is integral to a unit's rating and evaluation structure. Most military services conduct formal education and training programs to instruct newly inducted personnel in the performance of their general duties and responsibilities. After completing such schooling, the new trainee has acquired the general knowledge expected of a person in his grade, but he is still not intimately familiar with the systems of the unit to which he is then assigned. On-the-job training fills this void by supplementing general knowledge with detailed knowledge concerning the system or systems that a command has in its inventory. For example, in a navy, a gunner's mate will have been taught about the gunnery systems that the navy currently has on all of its ships. Then, when he reports to his ship, he will receive detailed knowledge about that ship's gunnery, including any idiosyncrasies inherent in those systems. This training amounts to the initial and most pervasive supervision that is provided at a command and most often is given by a person's immediate supervisor. It involves the direct transfer of knowledge from this supervisor to the trainee and it ensures that the proper, established techniques and operations are known and followed. Daily follow-up supervision ensures that the established procedures are followed correctly and regularly. This type of supervision is often defined as direct supervision, and a major standard in judging the performance of both military supervisors and trainees is how well this supervision transfers the necessary knowledge from one person to another.

A second category of supervision is indirect supervision. This is imparted to supervisors from their superiors, and from supervisors to trainees. Here the supervision may be less intense and constant and will include judging how well a trainee or a unit under a supervisor's aegis is performing the duties assigned. Such supervision allows for input from superiors and acts as a standard on how well the direct supervision is being administered.

Inspections

In essence, inspections are tests that judge the conditions existing in a given military unit against established military standards. They may be preflight tests of aircraft, prefiring tests involving artillery or missiles, or other maneuvers that test equipment operability. These tend to be mechanical and rote and will yield "go/no-go" decisions; these tests determine whether a piece of equipment is operating properly and can be used effectively. Other inspections test personnel readiness and capability. These involve all aspects of military life and occur at all levels of command.

LEVELS OF INSPECTIONS

Inspections may be formal, such as the United States Navy's commandwide operational readiness inspections (ORIs) and the United States Air Force's unit effectiveness inspections (UEIs), or they may be informal. Likewise, they occur at all levels in a military service's structure. In the case of direct supervision, an inspection might be informal, involving a check by the supervisor of the work that has been done by a subordinate to ensure that it has been done properly and that established procedures have been followed. On this level, inspections are an integral part of good supervision because they provide immediate evaluation of a person's daily efforts. From here, unit inspections (often conducted as exercises) determine that the people assigned to a unit can operate together and support each other effectively. Likewise, larger scale, commandwide inspections are held to make sure that all of the personnel in a command operate cohesively. The progression proceeds upward and can even involve interservice or international exercises and inspections that are held to ensure that a nation's or alliance's military forces are combat capable and can operate together effectively. In today's combined-arms combat environment, the importance of such interservice and international endeavors cannot be overstressed.

TYPES AND ASPECTS OF INSPECTIONS

It is not an exaggeration to say that there are inspections for every relevant aspect of military life. Frequent personnel inspections ensure that an individual's personal appearance—his grooming and cleanliness, and the smartness of his uniform—conforms to military standards. Inspections of an individual's property, called sea-bag inspections in the British and U.S. navies, ensure that all personnel have all the required uniforms, personal effects, and gear that they need to serve professionally. Berthing inspections ensure that a unit's living space conforms to military standards. Equipment and working space inspections make certain that established military standards have been met and that equipment has been maintained. In a broader sense, commandwide inspections may be held by either the commander or his superiors to assess the standards and capabilities of a command; on an even higher level, fleet- or armywide inspections can ideally assess how well entire forces are prepared to accomplish their missions.

Throughout an individual's professional military career, he or she will constantly encounter supervision and inspections. The former makes sure that all personnel assigned have been trained properly and that the performance from the lowest to the highest ranking individual is competent and conforms to military standards. The latter reinforces supervision, in that inspections will reveal areas of weakness in an organization's performance so that they may be corrected. Taken together, supervision and inspections are vital to ensuring that a unit or command is maintaining proper military standards and is capable of fulfilling its mission.

SUSAN M. WATSON

SEE ALSO: Combat Effectiveness; Education, Military.

Bibliography

Department of Defense, Department of the Air Force. 1989. *Air Force manual 11-1: Air Force glossary of standardized terms.* Washington, D.C.: U.S. Air Force.
Department of Defense, Department of the Army. 1986. *Field manual 22-5: Drills and ceremonies.* Washington, D.C.: Department of the Army.
Edwards, T. J. 1961. *Military customs.* 5th ed. Rev. by A. L. Kippling. Aldershot, Eng.: Gale and Polden.
Hering, P. G. 1961. *Customs and traditions of the Royal Air Force.* Aldershot, Eng.: Gale and Polden.
McCann, T. J. and G. LeBlanc, eds. 1981. *Basic military requirements.* NAVEDTRA 10054-E. Washington, D.C.: Government Printing Office.
Russell, E. C. 1980. *Customs and traditions of the Canadian Armed Forces.* Ottawa: Deneau and Greenberg for Department of National Defence.

INSTALLATIONS, MILITARY

Installations is a collective term for areas (temporary or permanent) adapted by construction for military use. They fall into the categories of combat support and administrative support—some serving both functions—and exist in various forms. They include the aerodrome or airbase, armory, arsenal, barracks, base, billet, camp, cantonment, center, commissary, college, depot, dump, exchange, facility, field, fort, garrison, headquarters, hospital, kaserne, laboratory, magazine, park, port, post, proving ground, range, shipyard, and station. Specific installations are described more fully, such as marine barracks, naval base, war or staff college, recruit depot, artillery park, rifle range, and air station.

Historically, military installations evolved in the context of the overall security or war system of which they were part. That is also true today. An understanding of this system as a whole, with its major components, provides insights as to when, where, and why military installations are required and when they are not.

Historic Development

The earliest installations with military significance were fortified villages and towns. Evidence of a walled settlement as early as 7000 B.C. has been found at the site of biblical Jericho. By 1000 B.C., the art of fortification was well-established in the Middle East and less sophisticated examples could be found in Europe and South and East Asia. Fortifications will be discussed here only in the broader context of military installations.

The security system of an early town organized solely for its own passive defense comprised the town walls with their gates, ditches, and towers; the interior stables and storehouses for livestock and provisions necessary to withstand a siege; a secure source of water; and facilities necessary for the fabrication and repair of weapons. As the citizenry was responsible for manning the defenses, there was no requirement for a separate military garrison. Such a system was simple and, other than the town walls, required little or no extraordinary construction. However, as social organization expanded beyond the city-state, so did the requirements for a corresponding security system and associated installational support.

Maintenance of a standing army required barracks and fields or exercise grounds for training. Fabrication and repair of arms and armor required shops and armories. Development of cavalry, chariots, and baggage trains required stables and parks. As the area to be defended expanded, so did the system for transporting forces from one area to another. Roads were built, and fortresses were developed to protect these lines of communication. These fortifications also served as garrisons and magazines for transiting forces. As armies began to travel and fight far from their home establishments, they required camps for their immediate defense and administrative support.

In the Roman military system, the camp (*castrorum metatio*) was developed into a transportable installation providing strategic reach as well as local tactical and administrative support. The method for erecting and organizing a camp on the march became an exact science (*castramentation*) and is described by Polybius and Vegetius, among others.

At the end of a march, the Roman army quickly erected its camp, complete with surrounding ramparts and ditch. Each soldier had specific responsibilities for camp construction and routinely carried excavating tools and two staves for the rampart palisade in addition to his arms, armor, and personal kit. The marching camp was sited on terrain favoring observation, defense, and drainage, and within reach of water, wood, and forage. If in contact with the enemy, the Roman army erected its marching camp under the protection of infantry and cavalry covering forces. According to Vegetius, a camp supporting six legions (55,000–60,000 men) would be generally square, 700 meters (765 yd.) on a side, and laid out in standard fashion. The surrounding ditch was about 4 meters (13 ft.) across and 3 meters (10 ft.) deep with a rampart raised an additional meter or more on the inside face of the ditch. The rampart was usually palisaded.

The marching camp was, in short, an advanced base. It served both as a fortified strongpoint for defense against attack and as a base of operations for security or offensive operations. It also served as the administrative garrison for the legions and as a repository for military stores that were required day to day and were accumulated for offensive or defensive operations. The most distant camps defined the frontiers of the empire and eventually were improved into permanent fortified garrisons with towers.

As a whole, the Roman military infrastructure comprised permanent fortified frontier camps; intermediate bases (according to Vegetius, fortified garrisons or strong, well-located cities provisioned in preparation for an impending campaign) as depots for livestock, grain, and engines of war; and a road network linking these nodes with Rome and the central Roman recruitment and provisioning regions. This system permitted the deployment and concentration of legions on interior lines at any point of threat or opportunity on the frontier of the Roman Empire. The marching camp provided en-route security and administrative support for the quick-marching legions and their baggage trains and also the means for extending the strategic reach—and frontiers—of the empire.

Like those of their Western precursors, the Eastern Roman military installations were systematic. However, they were organized on a different pattern. While the Roman system had, as its central virtue, mobility, the Byzantine system favored fortification. Fortifications were established in echelon to defend roads and waterways (such as exist today at the narrowest point of the Dardenelles) and to provide protection from invasion for the military defenders and for the local population.

As the effectiveness of centralized imperial government eroded—and with it the road network which facilitated the reach of law and legion—the Roman military establishment in Europe gradually disintegrated. As sovereignty decentralized, so did corresponding military needs, and fortification increased apace with feudalism. Either with armed forces or fortifications, perhaps both, the feudalistic lord was expected to protect the villages and serfs who provided his labor, sustenance, and military recruits. His seat of government was fortified (ranging from a castle to a fortified city) not only to protect him and his subjects (or the military forces established to defend those subjects) and their produce, but also to serve as a repository for the shops, armories, stables, barracks, and storehouses necessary to develop and maintain the weapons and military forces necessary for war. In addition to requirements for tactical defense, these fortresses were located in response to strategic and economic considerations. They were established close to economic freeways which fostered civilized intercourse—rivers, harbors, roads, and mountain passes through which passed the trade of the region and which also allowed military communications with allies and provided avenues of approach to and for enemies. The fortress, as in the earliest times, reappeared as the centralized military installation in a decentralized regional order.

However, as city-states and fiefdoms amalgamated into nation-states, the need for networks of military installations gradually increased. That has continued until the present time, and each twentieth-century nation and some alliances have evolved military support establishments similar in general outline to the Roman or Byzantine models. The fortress installation (e.g., Saipan, Singapore, Malta, and Gibraltar in World War II) occurs in this century only as a node within a much larger system.

Military and Naval Installations Today

Modern military installations, in the collective sense, are described by some as infrastructure, a term generally applicable to all fixed and permanent installations, fabrications, or facilities for the support and control of military forces (as defined by the Inter-American Defense Board and the north atlantic treaty organization). Infrastructure is developed in association with a transportation network and accompanying military support (termed variously as trains, mobile support echelon, etc.). As a whole, this aggregation comprises a military support system which has three functions: generating and maintaining military strength (force generation); projecting military strength (force projection); and supporting military forces in any contest at arms (combat operations support).

FORCE GENERATION

Force generation can be defined as the conversion of a nation's material and human resources into usable military power. Force generation includes the fabrication of military hardware, creation of war reserves, individual military training and education, formation and training of military units, maintenance of military hardware, and medical support. Each category has its unique requirements for military installations, most of which will be located in the national homeland. This part of an overall military support system may be called the homeland supporting establishment.

The construction of military weapons, equipment, vehicles, aircraft, and ships is undertaken by civil industry, by military arsenals, shipyards, and other facilities, or by some combination of the two. Nations without the means for fabricating their requirements must buy them on the international market or depend upon the resources of an ally. Military hardware fabrication requires more than factories and arsenals; it requires laboratories and proving grounds for basic and advanced research as well as for testing and evaluation.

While nations may depend upon civil facilities for creating weapon systems, munitions, and military stores, most nations establish military depots to store and preserve equipment and supplies held in readiness for war. Some nations develop underground depots as protection from air attack. At least one nation (the United States) has developed floating depots (mobile pre-positioning ships) not only for storage and preservation but also for the rapid movement of unit equipment and supplies worldwide.

Military individual training and education may be categorized as entry level and progressive. Recruit depots or training centers are established by each armed service for entry-level training of enlisted personnel. Some nations centralize their recruit depots for support of the service as a whole, while others decentralize their recruit training centers and affiliate them with the military organization (e.g., army division, regiment, navy ship) they are expected to serve. Entry-level officer training is often centralized at service military, naval, and air force academies. However, some nations augment this with decentralized officer training facilities affiliated with civil universities. Another approach is to organize a centralized, intense, short-term (months instead of the years associated with service academies) service school for screening and training officer candidates from the ranks or the civilian world. Progressive education for officers is usually centralized at service or joint schools (technical or tactical), staff colleges, and war colleges. Similar advanced technical and leadership education for enlisted personnel is usually provided at centralized service schools or training centers.

Installations for unit formation and training may take one of two general patterns. One pattern is a self-contained base which includes not only the barracks and administrative facilities but also the training areas (for maneuver, weapons firing, field exercises) for the units in garrison at the base. The other pattern is to provide barracks (or ship berthing), vehicle and equipment parks (or aviation hangars and runways), and administrative facilities at a military, naval, or air base and to conduct unit training in other areas. Most naval and air bases are developed along the latter pattern. Additional shore-based training facilities are sometimes created for air and shipcrews, and air and sea maneuver areas are established for centralized use out of the way of normal civil pursuits. Many European nations establish military bases, barracks, or kasernes for garrisoning army units with the expectation that field maneuvers will be conducted in nonmilitary areas once a year after harvest.

Most nations echelon the maintenance of military hardware so that minor and routine repairs are performed at maintenance shops near the supported forces. Major overhauls are performed at depots, facilities, and shipyards which provide centralized support. For naval overhauls, this requires dry dock facilities which can take the ship out of the water (Fig. 1). Repair ships and tenders are also used for naval repair and sometimes for military repair. These have the advantage of use to augment the homeland supporting establishment or deployment to augment combat operations maintenance capabilities.

Medical support, like hardware maintenance, is usually echeloned. Medical support facilities are decentralized for emergency and routine medical care to servicemen and their families at or near the installations, bases, and ships to which they are assigned. Definitive and specialized medical treatment is provided at centralized service or joint hospitals. Some nations depend entirely on civilian hospitals both for centralized specialized treatment and decentralized routine care. Others use civilian resources to augment military medical capabilities either routinely or to provide wartime expansion. Some nations maintain hospital ships which, like repair ships, may be used to expand homeland support capabilities or to augment combat operational support.

The homeland support establishment is the key not only for peacetime manufacture and support of national military capabilities, but also for expansion of these capabilities in time of war. Central to this is the existence of sufficient military infrastructure to support generation of additional military strength or the ability to adapt civil resources (e.g., factories, hospitals, repair shops, educational institutions) to the requirements for expansion. Most nations use a combination of both.

FORCE PROJECTION

Force projection may be considered in the context of moving military strength to the threatened area on the national borders or beyond and, in the case of maritime

Figure 1. A submarine base in Bangor, Washington. The pier is composed of support and administrative facilities and features a drydock that can accommodate Trident submarines. (SOURCE: U.S. Naval Photographic Center)

nations, transoceanic movement. In this context, infrastructure (ports, airfields, railyards, marshaling areas, intermodal transfer facilities, petroleum pipelines) facilitates movement of military forces, equipment, and supplies along land, river, sea, and air lines of communication. The objective of the force projection system is throughput, which may be defined as the timely flow of forces and material in the quantities and types necessary to conduct combat operations.

Military installations are located and designed for optimal throughput not only for routine peacetime needs but also wartime requirements. This may require the establishment by maritime nations of intermediate bases for bunkering ships, refueling airplanes, and en-route repair for oceanic transit. Although some of the installations comprising projection infrastructure may be purely military, much is civil in purpose and administration and is mobilized for military use only in an emergency or for training.

COMBAT OPERATIONS SUPPORT

The last category is combat operations support and includes both the installations which combat forces devise for temporary use and the more permanent facilities which are established for continuing support of military action in a theater of operations. These will differ in continental and maritime theaters.

In a continental theater, military installations in the forward part of the theater, sometimes termed the combat zone, are temporary in nature. Field hospitals, ammunition and fuel dumps, field maintenance facilities, and military field headquarters are typically found in a combat zone and are designed to be transported, erected, packed up, and moved in short order. Where possible, they are incorporated into existing civil structures using portable equipment. Where civil structures are unavailable, tents and field fortifications are used. Temporary airfields and equipment parks are cleared, used, and discarded as forces maneuver. Defensive installations may be temporary field fortifications developed by combat units and pioneer support, or they may be permanent structures or systems of structures integrating concrete, masonry, timber, and earthen construction. An example of the latter was the Maginot Line.

In the rearward sector of a continental theater—that area called the communication zone—infrastructure is more permanent in construction and is located to facilitate transportation throughput. The communication zone is structured for the reception and onward distribution of deploying forces, hospitalization and evacuation, major equipment repair, staging of combat equipment and supplies, and tactical and transport air operations. In this latter regard, air bases (e.g., main operating bases, forward operating bases) are echeloned between the communication zone and combat zone to provide the best air coverage while assuring physical security and access to the resupply of fuel, munitions, and parts. Infrastructure in the communication zone may be appropriated civil facilities (ports, hospitals, pipelines) or may be military construction (military airfields, arsenals, ammunition dumps). Without the terminal reception, storage, and distribution systems in place and functioning in the communications zone, rapidity of force projection is valueless; ships and aircraft will have to wait to unload, and disembarked supplies and equipment will quickly clog the ports and airfields.

In a maritime theater, the facilities and support located in a continental communications zone are contained in a network of advanced bases. These bases are tied to the homeland supporting establishment by secure lines of communication and extend strategic reach, much like the Roman camp. This is done by providing forward airfields to extend aviation range, by providing forward naval stores for reprovisioning and repairing forward-deployed ships, and by providing forward communication facilities for command and control.

Advanced bases may be permanent or temporary. If temporary, they may be constructed either by conventional construction or by assembling special support facilities such as hospital ships, repair tenders, floating dry docks, floating cranes, and floating piers or causeways for rapid development and operation. Advanced bases may be established to extend or protect a line of communications, to protect a strategic area or chokepoint, to conduct offensive operations in a forward area, or to build up logistic capabilities for future operations.

Future Developments

Looking to the future, military installations, both in form and function, will be shaped as much by civil development as by military need. For example, developments in the shipping industry—container ships and ultralarge petroleum tankers—influence what type of shipping and port configuration will be available in the years ahead. Probably the area which is most uniquely military is combat operations support infrastructure. In that category, the trends have been toward the expedited development of temporary facilities. Quickly erected airfields, hospitals, and port facilities are but a few areas where improvements have been made. Castramentation has indeed found its place in the twentieth century. Another trend has been the fortification or "hardening" of airfields, depots, shipyards, and headquarters against air and missile attack. Finally, the last trend has been the establishment of space installations for military communications and reconnaissance. It is in the latter regime, space, that the greatest strategic impact of military infrastructure development will be seen.

JAMES E. TOTH

SEE ALSO: Feudalism; Fortification; History, Medieval Military; Infrastructure, Military; Roman Empire.

Bibliography

Coakley, R. W., and R. M. Leighton. 1955 and 1968. *Global logistics and strategy.* 2 vols. Washington, D.C.: Government Printing Office.
Diehl, C. 1957. *Byzantium: Greatness and decline.* New Brunswick, N.J.: Rutgers Univ. Press.
Eccles, H. E. 1950. *Operational naval logistics.* Washington, D.C.: Bureau of Naval Personnel.
NATO. 1986. *NATO logistics handbook.* Brussels.
Polybius. 1980. *Polybius on Roman imperialism.* Trans. E. S. Shuckburgh. Lake Bluff, Ill.: Regnery Gateway.
Renatus, F. V. 1952. *The military institutions of the Romans.* Harrisburg, Pa.: Military Service.

INTELLIGENCE ANALYSIS, MILITARY

Military intelligence analysis is essentially the evaluation of foreign military activities and forces at all levels and in all potential situations, from tactical to strategic, performed as a basis for designing a national strategy. Induction and deduction, its reasoning techniques, are common to most analytic systems, but it is otherwise a highly specialized discipline.

This article focuses on present-day military intelligence analysis as generally practiced in major developed countries such as the United States, Great Britain, and France. Some consideration is also given to potential developments in intelligence analysis. (In the discussion of military capabilities, friendly countries are identified as "Blue" and enemy countries as "Red" throughout.)

Current Practice

A basic understanding of the contemporary practice of military intelligence analysis can be gleaned through an examination of its purposes, consumers, techniques, tools, and personnel.

PURPOSES

The ultimate purpose of military intelligence analysis is to assist a decision maker at any level to answer two fundamental questions: "How vulnerable is my country (or command)?" and "How vulnerable is my enemy?" These questions, of course, are complementary.

Although questions of vulnerability are phrased in rather absolute terms, they are often only one aspect of larger decisions concerning national strategy. Military intelligence about the potential strengths and vulnerabilities of nations supports a wide range of national options for action in conflict, among them deterrence, political or economic measures, moral suasion, or the direct application of military force. In addition to clarifying the existing dispositions and potentialities of military forces, military intelligence analysis also is responsible for identifying changing circumstances that require new decisions. Thus, military intelligence analysis is part of "feedback loops" both within the intelligence cycle and between the intelligence and policymaking communities.

CONSUMERS

Within the military sphere, intelligence analysis flows to several key functional groups, as described below.

Indications and warning centers. The activities of indications and warning centers, which enable Blue to detect and react to Red nuclear attack, require near-real-time intelligence reporting and quick-reaction analysis. Although such critical activities stress technical method, they also rely on a solid and intuitive understanding of Red's past and current behavior so that nuclear strike preparations are correctly identified.

Operating force commanders. The commanders of operating units require military intelligence analysis to support short-term decisions concerning the use of their forces in a battle area (ground, air, or sea). These commanders need to know the size, location, activities, and equipment of Red forces so that they can position Blue forces defensively or attack Red's vulnerable areas.

Contingency operations planners. Organizations for contingency operations planning develop concepts of operation and apportion forces in peacetime or wartime environments to anticipate circumstances in which a country might be required to use military force. To assist in this process, intelligence analysts make estimates based on Red plans, documents, and exercises, and on information provided by defectors. The intelligence analysts then provide descriptions of Red's most likely courses of action to the planners so that the planners can resolve issues concerning potential deployment and employment of Blue forces. Intelligence analysis is also used to assist in risk analysis—comparing Red's force capabilities with Blue's capabilities in potential areas of combat.

Research, development, and acquisition directors. Such users are concerned with the identification of basic technology, materials, and engineering required to make use of technology in a combat system, and with the processes required to produce combat systems. In the best of circumstances, intelligence analysis is based on examination of active Red systems that have been captured or obtained by other means. If Blue cannot obtain a Red system, then technical analysis is performed using a variety of methods. Radar systems, for example, can be analyzed from their emitted signals and other characteristics. Similarly, the vulnerability of tanks can be estimated crudely from photographs that show specific locations on the tank from selected angles. Such information is one element used in estimating the capability of friendly munitions to penetrate tank armor. The focus of Blue in analyzing Red technology is twofold: to develop or acquire Blue systems that can provide adequate protection against Red weapons and to develop or acquire Blue weapons that exploit the vulnerabilities of Red systems. Intelligence analysts thus provide the basic factors for Blue developers to use in computer models that simulate combat between Blue and Red at the system level.

Combat and training developers. These organizations formulate the doctrine, organizations, and tactics they deem most appropriate to conducting battles. Analysts provide these developers with intelligence estimates concerning Red systems, organizations, and tactics. They base these estimates on Red literature, training, exercises, and, if appropriate, actual combat operations. For example, the knowledge that many Arab officers were trained in Soviet institutions and that Arab forces were equipped with Soviet combat systems allowed U.S. military intelligence analysts to view operations conducted by Arab forces against Israel in 1973 for "developmental" purposes, as exemplary of Soviet tactics and equipment performance. U.S. analysts then could be more confident in making new estimates or reinforcing existing estimates of Soviet tactics and equipment.

National resource allocators. A country creates force structures that provide some assurance of either successfully defending against Red attacks or successfully attacking Red. The allocators range from military service staffs to heads of state and parliamentary organizations. Their confidence in, and risk assessment of, such forces are heavily dependent on the status of Red as depicted by military intelligence analysis. Intelligence analysts often face the pressures imposed by advocates and opponents of certain programs to "examine the data more closely or from a different perspective" in the hope of supporting particular viewpoints.

TECHNIQUES

Techniques used by the analyst are based on some combination of inductive and deductive reasoning, depending on the purpose of the analysis and what is already known. The deductive method starts with general information and uses logical reasoning to arrive at specifics. The inductive method starts with specific information and builds to a general case. Because military intelligence deals with many diverse areas, innumerable variables, large quantities of data, and a variety of confidence levels about reported data, both inductive and deductive methods are used simultaneously in the analysis process. These methods can best be highlighted by some simple examples.

Deductive methods. Suppose that Blue obtains a copy of a Red document that describes how a new weapon should be employed, and Blue had not previously seen evidence that Red had such a weapon in its inventory. Following *deductive* reasoning, Blue can conclude that (1) Red secretly built the weapon and equipped its forces; (2) Red is in the process of building facilities to produce such a weapon; or (3) Red has, or plans to enter, an agreement to obtain the weapon from another country. Alternatively, Blue can conclude that (1) Red sees some merit in having such a weapon in its inventory but has not yet decided to acquire it; or (2) Red is trying to mislead Blue through disinformation. Blue can make an estimate describing some or all of the options, depending on other evidence available. Alternatively, Blue can and usually does ask collectors to obtain more-specific information, such as trying to find factories or countries providing such weapons to Red forces. In the interim, Blue can publish an advisory notice of a potential Red capability.

Inductive methods. Suppose that Blue collectors report the presence of a large quantity of a specific type of chemical used as a primary ingredient in military rocket fuel as well as in other civilian industrial processes. Because Blue has not seen any evidence of a new civilian economic activity by Red, Blue inductively concludes that Red has rockets or plans to introduce them into its force. Blue then determines what other elements of rocket-force structure might be present and asks collectors to obtain further information to determine the extent and capability of Red's rocket forces.

"Fusion" of methods. Intelligence analysis normally is dependent on both deductive and inductive methods for success. The artful part of the process involves the recognition of "clues" from different collection means to put together an estimate of Red's capability and intentions. This process is sometimes called *fusion*. Successful fusion generally requires professional analysts combined with fortuitous circumstances and diligence on the part of both collectors and analysts. Although "scientific" methods—or more appropriately, technical processes and tools—are used (e.g., computer processing), fusion still is often based more on the "Eureka factor" than on scientific discipline. This statement has more pertinence as one moves from analysis of tangibles, such as the locations of combat units, to intangibles, such as Red's intention to attack Blue. Combat unit locations can be estimated fairly accurately using all-source intelligence processing (e.g., photography, communications signal intercepts, and radar signal direction finding). However, the unit's intentions are not as easily discernable solely from technical means, and less so if the enemy properly integrated deceptive measures with force operations.

TOOLS

Analytic tools are used with either inductive or deductive techniques. Tools can be as simple as paper templates, which can be used to match individual observations to an overall picture, or as complex as a large computer that sorts and analyzes millions of electronic signal emissions data in fractions of a second. Analysts use computer processing and analytic tools, such as mathematical equations, to correlate data rapidly or to narrow the choices to the most likely cases on which estimates can be based.

The capability of technical collection devices is increasing dramatically. High-speed digital information is rapidly replacing older formats, such as fixed-film photography. Electronic emissions data are collected on a wider front and in more depth. More data does not necessarily imply more intelligence, but improving the accuracy of collected data and the speed with which it can be communicated to an analyst can be beneficial if the analyst can use automated processing to assist in the sorting and matching functions. Otherwise, the analyst may spend an inordinate amount of time administering the data flow rather than analyzing the data. Cataloging, filing, and retrieving data are long, tedious tasks. Many times those functions are relegated to clerical or junior personnel, and the senior analyst may lose touch with the complete picture being presented by a variety of collectors.

In many cases, it is just as important to understand the timing of Red's activities as it is to know what those activities are. Some analysts currently have high-speed data exchange networks, but the networks do not include graphic displays of analyzed data except at the most sophisticated fixed-site facilities.

The British Army is supporting automated tactical intelligence with the Wavell system in its operational organizations. (Wavell is an automated data processing system for battlefield command and control, including the intelligence function. It is a mobile computer system mounted on a military truck, with visual display capability. It is designed to support corps, division, and brigade organizations.) However, most analyst support systems generally do not provide the ability to manipulate data in the sense of taking a "gaming" approach to analysis. The architectures of current graphics displays are such that data is extracted from structured computer files and is displayed in a fixed-format report or stylized graphic background. A gaming approach, using interactive automated graphics and true image processing, would allow the analyst to examine data from a variety of "what if . . ." perspectives. Such an approach gives the analyst a capability to display both Blue and Red forces on an interactive video display, to maneuver those hypothetical force representations with postulated operations plans, and to determine likely combat outcomes using a mathematical or judgmental combat adjudication methodology appropriate to the forces and circumstances being analyzed. Such a capability would also include the effects of weather and topography in either real or stylized fashion.

Analysts

Despite copious technical tools, the human intellect is still the most valuable tool for intelligence analysis. The value of analysis lies in being able to forecast events in sufficient time for a decision maker to act. In many cases, this requires an understanding of Red's behavior and the behavior of other countries. A professional analyst thus should know the history, language, culture, and geography of the area being studied and should have a solid grasp of world affairs.

The general orientations of a society and its educational system provide analysts in different countries with varying perspectives. The United States has spent billions of dollars in recent years on satellites to collect information on Soviet nuclear delivery systems and on large computers to analyze telemetry data containing millions of emissions. But it has not invested as much in the intellectual capital necessary to create a pool of analysts to understand the behavior of foreign countries, as evidenced by a recent Gallup survey that reported that American adults ranked seventh in geographic knowledge out of nine countries (Grosvenor 1988, p. 40).

The Soviet government, in contrast, placed great emphasis on historical research, as evidenced by the large quantity of general officers who wrote journal articles and books concerning historical events. This is partly because Soviet ideology emphasized historical patterns. Nonetheless, highly educated Soviet generals have also written doctrine and principles concerning current and future strategy.

Future Developments

Undoubtedly, computer processing capability will increase by orders of magnitude, in terms of speed and memory capacity, in the next ten years. Military intelligence analysis will benefit from advances in computer hardware, but it will also be greatly enhanced by several key software developments.

Among these developments are artificial intelligence techniques and relational databases. Both should help free analysts from more mundane, time-consuming tasks and provide more time to analyze behavior of threat forces. As the ability of technical collectors to gather ever-increasing amounts of data grows, the analyst can be overwhelmed with information. Therefore, more attention is being paid to providing user-friendly tools of analysis to enhance the effectiveness of military intelligence organizations.

Other techniques, such as image processing, can be used with artificial intelligence and relational databases to provide the capability of analyzing problems in multiple dimensions. Images in three dimensions (length, width, and altitude) could be correlated with time, movement, and other activities to provide a more realistic picture for analysts to work with. Such techniques should not be limited to analyzing traditional data collected by technical sensors about the movement of equipment. Multiple displays of information from several files could be displayed on one computer screen using simultaneous "windows." Image-processing techniques, used together with other analytic tools, could provide a robust environment in which to examine multiple aspects of foreign military capability simultaneously.

JOHN R. BONDANELLA

SEE ALSO: Intelligence Collection; Intelligence Cycle; Intelligence: Indications and Warning; Intelligence, Military; Models, Simulations, and War Games.

Bibliography

Allen, T. B. 1987. *War games.* New York: McGraw-Hill.
Bittman, L. 1985. *The KGB and Soviet disinformation: An insider view.* McLean, Va.: Pergamon-Brassey's.
Colby, W., and P. Forbath. 1978. *Honorable men: My life in the CIA.* New York: Simon and Schuster.
Druzhinin, V. V., and D. S. Kontorov. [1972] 1975. *Concept, algorithm, decision (a Soviet view).* Trans. U.S. Air Force. [Moscow: Voyenzidata.] Washington, D.C.: U.S. Air Force.
Grosvenor, M. 1988. World weary. *News Chronicle* (Thousand Oaks, Calif.), August 3, p. 40.
Kennedy, W. V., D. Baker, R. S. Friedman, and D. Miller. 1983. *Intelligence warfare: Penetrating the secret world of today's advanced technology conflict.* New York: Crescent Books.
Kurzweil, R. 1990. *The age of intelligent machines.* Cambridge, Mass.: MIT Press.
Lomov, N. A., ed. [1973] 1975. *Scientific-technical progress and the revolution in military affairs (a Soviet view).* Trans. U.S. Air Force. [Moscow: Voyenzidat] Washington, D.C.: U.S. Air Force.
Laquer, W. 1985. *A world of secrets: The uses and limits of intelligence.* New York: Basic Books.
Phillips, D. A. 1984. *Careers in secret operations: How to be a federal intelligence officer.* Bethesda, Md.: Stone Trail Press.
Prados, J. 1982. *The Soviet estimate: U.S. intelligence analysis and Russian military strength.* New York: Dial Press.
Rackham, P., ed. 1990. *Jane's C^3I systems.* 2d ed. Alexandria, Va.: Jane's Information Group.
Radziyevskiy, A. I., ed. [1965] 1976. *The dictionary of basic military terms.* [Moscow: Voyenzidat.] Washington, D.C.: U.S. Air Force.
Ranelash, J. 1985. *The agency: The rise and decline of the CIA.* New York: Simon and Schuster.
Richelson, J. 1987. *American espionage and the Soviet target.* New York: William Morrow.
Strong, K. 1971. *Men of intelligence: A study of the roles and decisions of chiefs of intelligence from World War I to the present day.* New York: St. Martin's Press.

INTELLIGENCE COLLECTION

Beneath the bland abstraction *intelligence collection* lie Byzantine networks of spies, technologies ranging from miniature electronic listening devices to satellite-borne thermal infrared image detectors, and painstaking research in everything from obscure technical journals to daily newspapers.

The basic task of all types of intelligence collection is to acquire information and provide it to intelligence establishments, which analyze and then distribute this information to appropriate civilian and military policymakers as "finished" intelligence. Intelligence collection is preceded by planning and direction by intelligence services, but essentially it is the primary and most fundamental step of the three-part intelligence cycle: collection, processing (analysis), and production (dissemination).

Effective intelligence collection requires highly skilled specialists and carefully developed intelligence infrastructures. Intelligence collection can best be understood by examining its basic sources and methods, the major intelligence services that perform these functions, and the implications of collection for present policy and future requirements.

Intelligence Sources and Methods

Within the collection process itself, information is derived chiefly from three major sources. First, human-source information comes from spies (i.e., agents), defectors, and other people. Second, technical intelligence comprises a diverse array of sources, reflecting the advances in various information-gathering technologies in recent years. And last, open-source information may come from technical publications, or it may be from widely available radio and television broadcasts, newspapers, or anything else in the public realm.

HUMAN INTELLIGENCE (HUMINT) COLLECTION

Human intelligence collectors may be spies or defectors; at a more mundane level, they may be travelers, refugees, or prisoners of war. In current intelligence practice, human sources are most effectively used where information is not communicated electronically or is protected by encryption or deception that cannot be unraveled before the information's intelligence value has perished. Human intelligence sources are especially important when intelligence information requires interpretation or is ambiguous in a way that requires on-the-scene observation and understanding. Assessments of plans, intentions, ideology, doctrine, training, and morale are not easily made by technical means.

Overt and clandestine dimensions of HUMINT. Human intelligence does not necessarily mean spying or clandestine collection. In fact, interviews of refugees or interrogation of prisoners of war have been major sources of intelligence information, as are the observations of front-line troops or reconnaissance patrols during wartime. Travelers frequently gain information of intelligence value during perfectly aboveboard activities; a businessman or scientist who tours a foreign manufacturing or research facility may pick up information by observation or conversation that, although not classified by the foreign power, is nonetheless highly sensitive or valuable for intelligence purposes.

Although a substantial amount of human intelligence-gathering may be overt, intelligence carried out by agents acting clandestinely remains the centerpiece of the human intelligence effort. Agents typically are used to gain access to a specific person or place that has information considered valuable enough to risk a death sentence or long imprisonment—the price of being caught conducting espionage.

Tradecraft. The sensitive methods of HUMINT are collectively termed *tradecraft*. They may involve the use of cover, secret writing, concealment devices, "dead drops," "cutouts," clandestine communications, and the "Moscow rules" publicized in spy novels.

Although seldom so well-equipped as the fictional James Bond, who sported an ejection seat in his Aston-Martin, espionage agents have and use a variety of technical tools. These include surveillance equipment, or "bugs"; cameras, radios, and radio direction-finding equipment; laser and other electro-optical detection instruments, and sensors that detect various emanations.

The most sophisticated and systematic use of electronics is in strictly technical applications, such as operating a "spy" satellite, but the line between the human intelligence and technical intelligence disciplines is sometimes difficult to draw. This is particularly true in covert military or paramilitary intelligence-gathering activities that rely principally on the activities of personnel but may employ unattended ground sensors, radio-jamming equipment, infrared photography, surveillance aircraft, and so on.

Advantages and disadvantages of HUMINT. Although human intelligence as a whole yields smaller quantities of data than do technical and open-source intelligence, it often provides the "golden nuggets" of information that enable an analyst or consumer to make sense out of information received from other techniques.

The principal drawback of HUMINT lies in human follies and frailties of the kind amply dramatized in espionage novels: agents can stop providing information; forget crucial details; be "turned" by a counterintelligence service to provide false information; or turn "rogue," becoming corrupt or uncontrollable, or both. Although such cases are exceptions, they can generate extremely unfavorable publicity, damage relations between countries, and have serious domestic political consequences. Of course, the dangers of failures in human intelligence are anticipated and guarded against, but the threat to nations and to the viability of their intelligence services remains real.

TECHNICAL INTELLIGENCE (TECHINT) COLLECTION

TECHINT is intelligence collected through electronic means; its two major divisions are signals intelligence (SIGINT) and imagery intelligence (IMINT). Each has an array of specialized subfields. And, of course, technology in the form of surveillance devices and systems pervades all sorts of intelligence activities.

SIGINT. Next to HUMINT, SIGINT is the oldest intelligence field, dating from telegraph wire–tapping in the American Civil War and coming of age with the radio and other electronic transmissions of the twentieth century. Signals intelligence was particularly important during World War II, when Britain and the United States pioneered decryption techniques that rendered vital German and Japanese coded radio traffic transparent to the Allies.

SIGINT today is a highly sophisticated endeavor, involving antennas and sensors—placed on the ground, on ships, in the air, and in space—gathering vast quantities of signals and other electronic emissions. Its subdisciplines include communications intelligence (COMINT), electronic intelligence (ELINT), and telemetry intelligence (TELINT). SIGINT and its associated subfields produce huge volumes of information—such copious amounts that, in fact, the information cannot all be deciphered immediately but must be taped for future interpretation.

Perhaps the greatest shortcoming of SIGINT is that it is often inconclusive by itself; it usually requires interpretation, decoding, and analysis. It also may be countered by effective signals security (SIGSEC) programs, a form of counterintelligence that concentrates on minimizing signals traffic and other clues to the nature of military and strategic activity.

An interesting example of both the strengths and weaknesses of information derived from communications signals was Project Gold, conducted by the U.S. Central Intelligence Agency and the British MI6 during the 1950s. This project began in the early 1950s, after Western intelligence agencies discovered that the teletype encryption system then used by both the West and the Soviets contained a flaw that permitted its decryption. Because the West also knew that most Soviet-bloc telephone and teletype circuits passed through Berlin, the Western agencies built a tunnel to tap Soviet messages and exploited this information freely from 1955 to 1956, when the Soviets uncovered the tunnel. It was not until 1961 that Western intelligence discovered that the Soviets had known about the project from the start through George Blake, a member of MI6 who was also a Soviet agent, and had used this knowledge to funnel useless or misleading information to the West.

IMINT. As practiced today, imagery intelligence is the newest and perhaps the most advanced of the major collection disciplines. Although reconnaissance photographs were taken from balloons in the nineteenth century and from aircraft during both world wars, contemporary reconnaissance has begun to exploit not only photographic intelligence (PHOTINT) but also a broader range of imagery intelligence collected and displayed digitally (i.e., by computer). In addition, imagery intelligence can be acquired by detectors sensitive to wavelengths not visible to the naked eye, from ultraviolet through the far infrared frequencies. These modalities are generally termed electro-optical intelligence (ELECTRO-OPTINT).

IMINT is significant because it often carries a convincing weight with its consumers. The influence of the U-2 photographs of Soviet missile installations in Cuba during the early 1960s was strong both in moving President Kennedy to action and in sustaining the U.S. case in the United Nations. IMINT may play an equally vital, if perhaps less volatile, role in assisting the monitoring of military forces and installations in pursuance of arms control treaties.

Like HUMINT and SIGINT, IMINT has its drawbacks, a principal one being the weather. The Eurasian landmass is covered by clouds up to 70 percent of the time, and other obscurants, such as smoke or air pollution, also may limit its effectiveness. In addition, IMINT can be countered by operations security (OPSEC) and cover, concealment and deception (CC&D) techniques, as was recently illustrated in the 1991 Persian Gulf War, when the Iraqi army placed dummy Scud missile launchers and inflatable "tanks" in the battle zone to draw fire away from actual installations. Sophisticated modern imaging techniques, such as thermal infrared detectors, can overcome obscurants and deception to a degree, but, at least as far as deception is concerned, advances there continue apace.

Other technical collection modalities. The technical intelligence field has still more specialized collection modalities. The acoustical properties of sound waves in water are of particular use in tracking ships and submarines; hence, collection has come to include acoustical intelligence (ACOUSTINT; or ACINT in U.S. Navy terminology), which measures and interprets acoustic waves radiated by a target.

Acoustic and seismic vibration-detecting devices, in the form of a system of unattended ground sensors, were developed for the McNamara Line, a proposal to monitor and interdict infiltration by communist forces across Vietnam's demilitarized zone. This line was never built, but the sensors were deployed along the Ho Chi Minh trail in a then-classified program known first as Muscle Shoals and later as Igloo White.

Other less-known collection methods include mapping, charting, and geodesy intelligence (MC&G), which is used extensively in the preparation of targeting information, tactical planning, and geographic intelligence. Esoteric collection methods include laser intelligence (LASINT); nuclear intelligence (NUCINT), which reads effects of radioactivity; radiation intelligence (RINT), which measures electromagnetic or radiated energy from sources other than nuclear detonations or radioactive sources. Finally, radar intelligence (RADINT) is derived from data collected by radar.

Open-Source Information Collection

In view of the vast resources devoted to clandestine information gathering, it is perhaps paradoxical that the bulk of intelligence collection is from open sources. (One estimate holds this figure to be 80 percent of all information collected.) This may be a reflection of the so-called information revolution, in which data of all kinds, including much sensitive data, increasingly appears in open publications.

One traditional open source of intelligence information in recent years had been the annual Moscow parade celebrating the anniversary of the revolution of October 1917. It was at this event that the Soviets frequently displayed new military equipment. Less formal, but often even more useful, are military maneuvers, which military attaches of all countries observe with interest. Other open sources are myriad: research of published works, translation of technical journals, newspaper and magazine articles, foreign radio and TV broadcasts; and increasingly, the monitoring of worldwide media (especially TV) coverage of ongoing events and crises.

Major Intelligence Services

Most countries have at least one or two intelligence services, and sometimes several, that conduct collection. These services work in the same fundamental way, as the basic techniques of intelligence collection do not vary greatly, but they all pursue different objectives in accordance with their own national security requirements and resources.

United States

Intelligence collection is parceled out among different organizations that comprise what is known as the "intelligence community." The Central Intelligence Agency (CIA) is the best known and was established in 1947 to coordinate the activities of the others. It is the paramount agency for HUMINT collection, a function carried out by its Directorate of Operations. The Defense Intelligence Agency (DIA), the pre-eminent military intelligence organization, established in 1961, is also involved in HUMINT collection abroad. The National Security Agency (NSA) is charged with SIGINT collection, and IMINT collection is carried out by specialized elements of the Defense Department. The American military services also conduct intelligence collection from human and technical sources to suit their own unique needs.

Commonwealth of Independent States/former USSR

The organization of intelligence in the former USSR remains in flux. The new Central Intelligence Service has replaced the former First Chief Directorate of the KGB in foreign collection, while the Chief Intelligence Directorate (GRU) of the General Staff is responsible for military intelligence. Both the KGB and GRU had their own HUMINT and SIGINT organizations, an arrangement quite unlike that of the United States. The GRU also has maintained the USSR's IMINT program. The former Soviet military services also collected intelligence for their own purposes. Further changes may occur with new developments in the CIS.

OTHER COUNTRIES' CIVILIAN INTELLIGENCE SERVICES

Intelligence services of other major powers plan and carry out collection. These include, among the best known: the United Kingdom's Secret Intelligence Service (SIS, otherwise known as MI6); France's General Directorate of External Security (DGSE); Germany's Federal Intelligence Service (BND); Israel's Central Institute for Intelligence and Special Duties (Mossad); and China's Ministry of State Security (MSS).

Nearly every other country has at least one intelligence service, and there is a tremendous variety among all of them in terms of professional competence, orientation, and resources. The only thing these services have in common is that they all collect HUMINT of one type or another; most of the world's intelligence services have limited resources and do not collect SIGINT or IMINT to the same degree as the major powers do.

The Present and Future of Intelligence Collection

An intelligence service's control of information gives it substantial leverage within its own government. Policymakers come to rely on the intelligence provided because it affects their ability to conduct an effective foreign policy or to craft an adequate defense program. Recognizing this importance, intelligence services can develop a proprietary view of the intelligence they collect and make available and are invariably vigilant to ensure that collection sources and methods remain secure.

CONTEMPORARY INTELLIGENCE PRIORITIES

Some of the major industrial countries spend heavily on technical collection programs. In the United States, it appears that several times as much money is spent on technical collection as on HUMINT (although specific figures remain classified), for high-technology SIGINT and IMINT sensors and platforms can be costly. Some have claimed that the U.S. competence in technical collection is more pronounced than its ability to conduct human-source collection—perhaps a reflection of a national affinity with technology.

In view of all the collection taking place, a major task for intelligence planners is how to best manage and sort the flow of data. More sophisticated collection systems can lead to more ambiguity for analysts and even a poorer quality of intelligence analysis if an analytical/production office is truly overwhelmed. A large influx of data from technical sources can, for example, monopolize an analyst's attention and perhaps result in oversight of significant facts contained in reporting from clandestine, attache, or diplomatic sources.

The former USSR had traditionally attracted about one-half of U.S. budget resources for intelligence collection. This priority should not be surprising, because the Soviet Union was the only country that could inflict catastrophic damage to U.S. territory within 30 minutes. The People's Republic of China and other communist nations have traditionally followed in terms of U.S. collection priorities.

The most difficult targets to collect against today are not necessarily communist countries or other police states, but terrorist organizations or other tightly knit conspiratorial groups such as narcotics traffickers. These groups are difficult to penetrate with human sources of information and often are not suitable for technical collection. Yet they can pose capital threats to any nation's security and require innovative methods of collection.

Every major power needs a wide-ranging and highly competent system of intelligence collection. In the case of the United States, Executive Order 12333, issued by President Reagan in 1981, states that the collection of accurate and timely information about the capabilities, intentions, and activities of foreign powers, organizations, or persons and their agents is essential to informed decision-making, and that this collection is a "priority objective" that will be pursued in a "vigorous, innovative, and responsible manner" consistent with the Constitution and applicable law and respectful of the principles upon which the United States was founded.

GUIDELINES FOR FUTURE COLLECTION

Regardless of the target, intelligence collection will always require continuity and consistency of effort. Whether it is through HUMINT, SIGINT, or IMINT, it should not be switched on and off at will. For the best results—compilation of meaningful intelligence products over time—the collection effort should be sustained and steady.

Another guide for future collection is that technical collection managers and collectors themselves should attempt to brief their customers on the capabilities and limitations of their collection systems. As these systems become increasingly complex, the need to understand them grows. This is a delicate matter, for the rules of compartmentation must also be respected; in this vein, the analyst or consumer of the intelligence should be given details only of what he or she needs to know to do the job.

As of the early 1990s, some influential U.S. policymakers have called for greater emphasis on HUMINT collection, with more specialists trained in foreign languages and cultures. Some see this as a continuation of a long rebuilding process after reductions in HUMINT capabilities in the late 1970s, and a renewed commitment to HUMINT may well prove necessary to best satisfy new requirements.

Intelligence collection agencies must be prepared to focus their efforts against an increasingly diverse array of targets. By the early 1990s, for example, the cold war had receded, but dangerous powers such as Iraq as well as new terrorist groups had emerged.

Likewise, intelligence collection agencies should stand

ready to focus on a host of topics, some of them nontraditional. These would include foreign economic systems, arms control monitoring, refugee flows, energy issues, multinational corporations, new science and technology issues, environmental topics, and extremist groups of every persuasion.

JAMES H. HANSEN
EDMUND R. THOMPSON

SEE ALSO: Attache, Military; Counterintelligence; Intelligence Cycle; Intelligence, Human; Intelligence, Imagery; Intelligence, Signal; Intelligence, Tactical; Reconnaissance and Surveillance.

Bibliography

Andrew, C. M. 1986. *Secret service: The making of the British intelligence community.* Sevenoaks, Kent: Sceptre.
Bamford, V. J. 1982. *The puzzle palace: A report on America's most secret agency.* Boston: Houghton Mifflin.
Burrows, W. E. 1986. *Deep black: Space espionage and national security.* New York: Berkeley Books.
Cline, R. S. 1976. *Secrets, spies, and scholars: Blueprint of the essential CIA.* Washington, D.C.: Acropolis Books.
Deriabin, P., and F. Gibney. 1959. *The secret world.* New York: Ballantine.
Dulles, A. W. 1965. *The craft of intelligence.* New York: Harper and Row.
Felix, C. [pseud.] 1963. *A short course in the secret war.* New York: Dutton.
Hood, W. 1983. *Mole.* New York: Norton.
Johnson, L. K. 1986–87. Making the intelligence "cycle" work. *International Journal of Intelligence and Counterintelligence,* Winter, pp. 1–23.
Kahn, D. 1967. *The codebreakers: The story of secret writing.* New York: Macmillan.
Kennedy, W. V., et al. 1983. *Intelligence warfare.* New York: Crescent Books.
Kent, S. 1949, 1965. *Strategic intelligence for American world policy.* Princeton, N.J.: Archon Books.
Lewin, R. 1978. *Ultra goes to war: The first account of World War II's greatest secret based on official documents.* New York: McGraw-Hill.
———. 1983. *The American magic: Codes, ciphers and the defeat of Japan.* New York: Farrar, Strauss and Giroux.
McCartney, J. 1988. Intelligence: A consumer's guide. *International Journal of Intelligence and Counterintelligence,* Winter, pp. 457–86.
Ranelagh, J. 1987. *The agency: The rise and decline of the CIA.* New York: Simon and Schuster.
Richelson, J. T. 1987. *American espionage and the Soviet target.* New York: William Morrow.
Taylor, J. W. R., and D. Monday. 1972. *Spies in the sky.* New York: Scribner's.

INTELLIGENCE CYCLE

The intelligence cycle is a five-step process by which raw intelligence data is converted into useful intelligence. The five steps are planning and direction, collection, processing, production, and dissemination.

The term *intelligence cycle* was created by the U.S. intelligence community. Because U.S. intelligence accounts for much of the intelligence collected worldwide, and because many other nations fashion their intelligence arrangement after that of the United States, the concept of the intelligence cycle is quite relevant, and the discussion will center around the intelligence cycle of the United States.

The intelligence cycle pertains to "the collection, collation, evaluation, analysis, integration, and interpretation of all collected information." It is the methodology necessary to produce the most accurate and complete intelligence.

The initial problem facing the intelligence gatherer concerns acquiring the type of information that is needed to fulfill a particular need. An important distinction is made between *information* and *intelligence information: information* is data that has been collected without any concern for a particular intelligence requirement and which therefore may or may not be of value; *intelligence information* is data that pertains to a particular problem but that has not been processed into intelligence. It is also referred to as *raw intelligence data* and is that part of the available information that is actually of intelligence value. The process by which such raw data becomes intelligence is the intelligence cycle.

Planning and Direction

In planning and direction, the first step, a determination is made that a need for intelligence exists, and a plan is devised to provide that intelligence. For example, an analyst monitoring an acknowledged problem situation might encounter a new phenomenon—an unexpected event, an unusual aircraft, troop, or ship movement, or a new weapon system. For decades, Soviet military organizational changes, operational movements, and new weapons systems created U.S. intelligence needs.

Alternatively, an entirely new threat might create intelligence needs that must be met in order to satisfy national security requirements. The situation in Kuwait in 1990 presented an entirely new national security problem for the United States. In the period leading up to the deployment of U.S. forces, hundreds of intelligence needs—information on enemy order of battle, force distribution, command, control, and communications—were identified. The existing database had to be expanded considerably as the United States moved from merely observing the actions of Iraq to contemplating deploying troops to confront them in crisis situations. Most complete and accurate intelligence was needed and did not automatically exist; it was created methodically by identifying and satisfying all the relevant intelligence needs.

Once intelligence needs are identified, a collection plan is prepared describing a scheme for collecting information from all available sources. The type of intelligence required will, in large part, determine the collection plan. Basic intelligence is factual general reference material readily available from a variety of sources. This includes information about the geography and about political, social, economic, and military structures; biographic and cultural data; and any information on the resources, capabilities, and vulnerabilities of a nation. This information provides a foundation for planning, policy making, and military operations. Cartographic intelligence is derived from maps and charts of the areas concerned. Foreign materiel (FORMAT) intelligence is information about both materiel (equipment) and its maintenance, operations, and support. Since most of the sophisticated military materiel in use today is produced by relatively few nations, it is also important to evaluate the equipment mix in order to assess the degree of compatibility of several systems from different sources.

Human intelligence (HUMINT) is provided by human beings, rather than by technical means. Imagery intelligence (IMINT) is derived from an analyst's interpretation of imagery. Photographic intelligence (PHOTINT) is a category of IMINT that consists of classified and interpreted photographic material.

Measurement and signature intelligence (MASINT) is scientific and technical information obtained by quantitative and qualitative analyses of metric, angle, spatial, wavelength, modulation, plasma, hydromagnetic, and other data, collected by technical sensors to identify any distinctive features associated with the source and to enable the subsequent identification and measurement of the same. MASINT includes but is not limited to acoustic intelligence (ACOUSTINT and ACINT), electro-optic intelligence (ELECTRO-OPINT), nuclear intelligence (NUCINT), radar intelligence (RADINT), radiation intelligence (RINT, derived from unintentional electromagnetic energy, but does not include energy emanating from nuclear detonations or radioactive sources), and debris collection.

Medical intelligence (MEDINT) is intelligence about natural and man-made environments that could affect the health of military forces. It includes general medical intelligence, which is concerned with biological and medical capabilities and health situations, and medical scientific and technical intelligence, which assesses and predicts technological advances that have medical significance, including defense against chemical, biological, and radiological warfare. This type of intelligence was crucial for U.S. forces in Operation Desert Shield, since the Iraqi chemical warfare capability was a significant component of the Iraqi threat.

Nuclear proliferation intelligence is intelligence relating to (1) the scientific, technical, and economic capabilities and programs and the political plans and intentions of a nation or organization that does not have nuclear weapons to acquire nuclear weapons or to research, develop, and manufacture nuclear weapons; or (2) the attitudes, policies, and actions that could incline a nation or organization with nuclear weapons to supply nuclear technology, facilities, or special materials to a nation or organization that does not have nuclear weapons. This type of intelligence was also crucial during Desert Shield, because if Iraq had a nuclear capability, the threshold of a possible war could be raised to the nuclear level.

Signals intelligence (SIGINT) includes any or all of the following: communications intelligence (COMINT), electronic intelligence (ELINT), and foreign instrumentation signals intelligence (FISINT). COMINT is derived from foreign communications obtained by other than the intended recipients. It does not include monitoring foreign public media or communications intercepts that are obtained in the course of counterintelligence investigations within the United States. ELINT is derived from foreign electromagnetic radiations that do not involve communications, including those from atomic detonations and radioactive sources. FISINT is derived from analyzing electromagnetic signals associated with testing and operationally deploying aerospace, surface, and subsurface systems that may have either military or civilian application. It includes but is not limited to telemetry, beaconry, electronic interrogation, tracking/fusing/arming command systems, and video data link signals. Telemetry intelligence (TELINT) is a subcategory of FISINT derived from intercepting, processing, and analyzing foreign telemetry.

Technical intelligence (TI) is intelligence on the characteristics and performance of foreign weapons and equipment. It is a part of scientific and technical intelligence and is distinct from order of battle.

IMINT may be of crucial importance in determining an enemy's order of battle, but will be of nominal benefit in determining its electronic warfare capability. COMINT, such as the intelligence derived from Enigma or the Japanese codes in World War II, can save thousands of lives because of the information that it provides concerning enemy force movements—but it may be of little value in determining the size of an enemy's ship or tank. Thus, in devising an intelligence collection plan, the correct source or sources must be tasked with collecting the information.

Once the correct intelligence collection sources are identified, the collection requirement is written as a task that asks them to collect and submit the necessary information. The associated collection plan defines the need, makes suggestions concerning how the information might be collected, and provides other advice and guidelines. This plan must be as complete as possible; vague, ill-defined, or incomplete plans might prompt the intelligence collector to amass information on the wrong requirement or collect information that only partially fulfills the intelligence need. Such errors, which require additional requests for information, are costly and inefficient. More important, they delay the production of intelligence or contribute to producing incomplete intelligence. Either situation can lead to disaster.

Funding to pay for the intelligence effort is then approved. Here, cost is the major factor, and the most efficient and economical collection source should be chosen. Orders in the form of collection requirements are issued, and these are sent to the appropriate collection units directing them to collect the information.

Collection

In this second step, the collection units gather the required information and deliver it to the production or processing agency. Here, too, there is an established procedure. Upon receiving the collection tasking, the collector first determines the means by which the information can be collected most efficiently and economically. Phone calls to subunits or allies will provide information for some requirements, while field collection operations may be required to fulfill others.

The necessary information is then collected. The collector screens the information and, based on his on-the-scene experience, discards anything that he knows to be false or irrelevant. He then formats the remaining material in the form that is most usable for the production or processing agency. In some cases, photographs are of great value, while other requirements might involve a large computerized database. Proper formatting can often save a great deal of time.

Processing

The third step, processing, is accomplished by the intelligence processing or production facility. Here the intelligence information is converted into the form that is most suitable for intelligence production. The processor often has access to more information than the collector, and his first duty is to screen the data for any information that he knows is false or irrelevant. The remaining information, which is material that appears to be relevant to the problem but which may or may not be accurate, is then prepared for analysis.

Production

Production converts the intelligence information into finished intelligence. It involves integrating the information with all other information that the analyst has concerning the subject. For example, the analyst may have information collected by HUMINT, COMINT, and MASINT. All of this is brought together, and anything that is obviously incorrect is discarded.

All the material is then analyzed. The collected information is reviewed to identify significant facts, which are compared and collated with other information, and logical conclusions are drawn. The information is then assessed. In assessment, the analyst determines the reliability and validity of each piece of information. Ideally, validity is determined through corroboration with other intelligence that has been collected from a different source. For example, information collected by HUMINT might be corroborated by information from a totally independent source, such as SIGINT. This permits an analyst to have much greater confidence in his data. Alternatively, the analyst may have conflicting information. In this case, the assessment becomes more sophisticated, as each piece of information is scrutinized for fallacies that will discredit it and permit the analyst to discard it. If conflicting data cannot be resolved, then the analyst will retain all the data and highlight the existence of conflicts, to diminish the user's confidence in the intelligence that is provided.

The material is then evaluated. Evaluation is the process of appraising what the information contributes to the achievement of the tasking—its pertinence or usefulness in terms of the intelligence need. The analyst judges the information that is available using two criteria: the validity of the source, and the accuracy of the information. A technical system that has a record of high reliability might be given a high grade, while information extracted from a government-owned newspaper that is known to practice disinformation might be considered as suspect. The analyst will also consider how and by whom the information was collected, determine whether he has other information that corroborates the data, examine the data in light of past patterns and occurrences, and use any other methods necessary to determine the degree of accuracy.

The resulting data is then integrated with all other intelligence concerning the problem. In this process, the results from different branches or offices might be brought together by a higher office to create a complete picture. For example, in naval intelligence, an analyst might be concerned only with operational intelligence. A higher office might take this operational intelligence and integrate it with scientific intelligence concerning naval construction. This additional intelligence might present a more complete picture, might resolve some questions that the operational intelligence analyst could not answer, or might even prove that some of the conclusions that the analyst made were incorrect. What may have seemed valid to an analyst concerned with only one facet of the problem may be revealed as inaccurate in a larger context. The remaining information is then interpreted with the aim of drawing inferences concerning enemy intentions and goals. Conflicting data are duly noted and any intelligence gaps or shortfalls are highlighted.

In the final phase of production, the analyst will prepare intelligence products, which present the intelligence in a form that the customer can best use. The mode of presentation (a message, a paper document, photography, schematics, viewgraphs, briefings) and the formatting of the information (statistics, narrative) are crucial: the information is useless unless the customer can understand and use it.

Dissemination

Finally, in the dissemination step, the finished intelligence products, in the appropriate oral, written, or graphic form, are distributed to departmental and agency intelligence consumers. Intelligence must be provided to all those organizations that need it. Most production centers have standard distribution lists, which include all organizations that usually have a need for the intelligence that the center produces. However, each item produced should be considered in light of potential consumers, because the effort that has gone into collecting, processing, and producing the intelligence is wasted if the result is not provided to those who need it.

SUSAN M. WATSON

SEE ALSO: Intelligence Analysis, Military; Intelligence Collection; Intelligence, General; Intelligence, Human; Intelligence, Imagery; Intelligence, Signal.

Bibliography

Clauser, J. K., and S. M. Weir. 1975. *Intelligence research methodology*. State College, Penn.: HRB-Singer.

Department of Defense, Defense Intelligence College. 1987. *Glossary of intelligence terms and definitions*. Washington, D.C.: Defense Intelligence College.

Hopple, G. W., and B. W. Watson. 1986. *The military intelligence community*. Boulder, Colo.: Westview Press.

U.S. Congress, Senate. 1976. *Final report of the Senate select committee to study government operations with respect to intelligence activities*. Washington, D.C.: Government Printing Office.

INTELLIGENCE ESTABLISHMENT

The term *intelligence establishment* usually connotes the broad array of organizations that carry out such intelligence functions as espionage, counterintelligence, technical collection, and analysis. In the United States, the term frequently used is *intelligence community*. In a slightly different sense, *intelligence establishment* is sometimes used to refer to a group of senior government officials who are intimately familiar with intelligence matters and who control the activities of those organizations with related functions. This article uses *intelligence establishment* in the former sense.

Possibly the first use of the term as a way to refer to the panoply of intelligence organizations serving a nation was in 1970 when Harry Howe Ransom published *The Intelligence Establishment*, which was primarily an examination of the U.S. intelligence establishment, although it also included some coverage of the British system. Ransom may have been influenced in his choice of title by *The Espionage Establishment* by David Wise and Thomas B. Ross, published only a few years earlier, which sketched the intelligence establishments of the United States, Great Britain, the Soviet Union, and the People's Republic of China (PRC). Those countries, and perhaps Australia, France, Israel, Italy, and Japan, continue to maintain the world's most robust intelligence establishments. The intelligence establishments of each of these countries are briefly outlined below.

United States

Until World War II, the U.S. intelligence establishment consisted mainly of the intelligence units of the armed forces, although until the early 1930s the State Department possessed a small office that sought to break the codes and ciphers foreign nations used in their diplomatic communications. The shock of the Japanese surprise attack on Pearl Harbor in 1941 is most often cited as the factor that convinced the U.S. government to create, after World War II, a peacetime intelligence establishment. This process began in the National Security Act of 1947, which created the Central Intelligence Agency (CIA). The process was furthered by a presidential executive order in October 1952, which consolidated the several armed forces communications intelligence entities, descended from the State Department unit of the 1930s and from military organizations, into a unified National Security Agency (NSA), now perhaps the most extensively staffed and budgeted of the agencies. Meanwhile, aerial photography and other means of overhead reconnaissance increased in importance, leading the United States to create in the summer of 1960 the still unacknowledged National Reconnaissance Office (NRO; discussed below). Then, an effort to consolidate military intelligence branches, reducing their dominance on national security matters, led to the formation in 1961 of the Defense Intelligence Agency (DIA). The army, navy, air force, and marine intelligence units continued their existence and ultimately regenerated, making moot the intentions of DIA's founders.

Specialized government agencies such as the Departments of Energy and Treasury also have certain intelligence functions. Finally, the Federal Bureau of Investigation (FBI), under the Department of Justice, had had the role of conducting counterintelligence investigations for the United States for several decades. All of these organizations form components of the American intelligence establishment.

CENTRAL INTELLIGENCE AGENCY

At the apex of the entire U.S. intelligence establishment sits the director of central intelligence (DCI), who is simultaneously the top official of the CIA. The DCI bears formal responsibility for coordinating the activities of the intelligence establishment as a whole and for managing and directing the CIA itself. The DCI's so-called community function has been hampered because the DCI lacks budgetary authority over such military intelligence components as NSA, NRO, and DIA, which continue to spend the preponderant part of the intelligence dollar (about 85% by the most recent available breakdown). Another obstacle to effective DCI coordination of the full intelligence establishment has been the lack of sufficient capability to monitor other elements of the bureaucracy. In recent years, the DCI has formed a special unit called the Intelligence Community Staff specifically to assist in managing the establishment.

The CIA remains the best-known component of the U.S. intelligence establishment. In its early years, the CIA assiduously carved out a role for itself in areas previously dominated by military intelligence and performed a variety of "services of common concern" for the intelligence establishment as a whole. More recently, the pendulum of power has seemed to swing the other way. The 1980s, for example, witnessed significant inroads by the military in peacetime covert operations and even espionage, traditionally CIA preserves. Similarly, the Iran-Contra affair involved a transfer to the National Security Council (NSC) staff of a CIA covert operation, and DCI William Casey was alleged to have attempted to create an "off the shelf" capability for covert operations outside the CIA.

In managing the Central Intelligence Agency, the DCI is assisted by a deputy director of central intelligence and an executive director. All three officials work from CIA headquarters at Langley, Virginia. Beneath these top managers are five directorates, each headed by a deputy director. The Directorate of Intelligence is responsible for producing intelligence reports, often called "finished intelligence," and for analytical inputs to national intelligence estimates. This unit theoretically uses the data of all segments of the intelligence establishment to produce authoritative products. The Directorate of Operations (DDO) conducts the CIA's espionage activities and its covert operations, utilizing a variety of offices or staffs that are either regional or functional in orientation. The Directorate of Science and Technology does analytical work on intelligence subjects of a highly technical nature and also conducts research and development work on new means of gathering technical intelligence, such as satellites or means of intercepting communications. The Directorate of Administration handles the CIA's housekeeping work. In 1990, the CIA added the Directorate for Planning, which is intended to help the organization anticipate the amount and kinds of intelligence gathering and analysis that will be required in the future. This unit should not be confused with the Directorate of Plans (DDP), the agency's covert action and espionage arm during its early years, which became the DDO in the 1970s.

NATIONAL SECURITY AGENCY

The National Security Agency (NSA), the largest component of the U.S. intelligence establishment, an element of the Department of Defense (DOD), is located at Fort Meade, Maryland. The NSA collects, digests, and summarizes everything that can be learned from the interception and decoding of foreign communications and electronic emissions. This agency is also responsible for the security of American codes and encipherment systems. Like the CIA, the NSA is subdivided into functional directorates with a combination of regional and functional subordinate units. Actual collection work on behalf of the NSA is carried out by the security groups or commands of each of the armed services and by certain instrumentalities managed by the air force for the National Reconnaissance Office. The manpower of these armed services collection units often matches that of the NSA central organization, which is itself roughly equivalent in size to the CIA. Thus, overall, the NSA far outstrips the CIA in manpower and also in budget.

NATIONAL RECONNAISSANCE OFFICE

Most amorphous as an intelligence establishment component is the National Reconnaissance Office (NRO), whose very existence the United States still refuses to acknowledge. The NRO contracts for, manages, launches, and then controls a variety of space satellites serving intelligence purposes ranging from secure communications to electronic intercept to overhead photography. In effect, the NRO conducts special space activities as a service of common concern for the rest of the intelligence establishment. The NRO is controlled on a daily basis by the undersecretary of the air force and is given policy direction by a committee consisting of that official, the DCI, and a representative of the secretary of defense. The NRO is believed to have the largest budget of any single component of the U.S. intelligence establishment.

PRODUCING INTELLIGENCE

All other U.S. intelligence agencies contribute to the establishment's major product and to its management by sitting upon the National Foreign Intelligence Board (NFIB). Finished reports are styled national intelligence estimates (NIEs) or special national intelligence estimates (SNIEs), depending upon whether they flow from a regular series or are the result of a particular request. In either case, the papers are written by members of the National Intelligence Council, a board of designated individuals each of whom represents the intelligence establishment's senior expert on some particular subject matter, such as the Soviet Union, China, or international terrorism. Two staff assistants help each of the national intelligence officers, who call upon all intelligence components to furnish the information used in their analyses. Draft NIEs and SNIEs are then discussed by the full National Intelligence Council, which is chaired by the deputy director of central intelligence. If approved, the drafts go forward to the NFIB, where agency representatives again review them. The State Department's Bureau of Intelligence and Research (INR), for example, might disagree with aspects of an NIE that have to do with foreign policy matters. It is the responsibility of the DCI to adjudicate such disagreements. The DCI may choose to accept amendments in the language of an NIE, to permit a dissenting intelligence agency to state its disagreement in the NIE, or to return the NIE for redrafting.

This process aims to produce intelligence that is responsive to the consumers of that commodity, specifically the National Security Council (NSC), which has the authority to direct the intelligence establishment under the 1947 law and the president. Other consumers are in Congress, especially in the Senate Select Committee on Intelligence and the House Permanent Select Committee on Intelligence, which have the task of legislative oversight over the intelligence establishment and also serve as clearinghouses for intelligence that may be provided to other interested members of Congress. In all, the U.S. intelligence establishment has grown into an exceedingly complex mechanism.

United Kingdom

During World War II, the Americans followed the British example in creating a number of U.S. organizations and in adopting several intelligence methods. In a sense, therefore, pride of place goes to the British in any set of profiles of international intelligence establishments. The United States benefits from an enormous array of additional resources, and the Soviet Union possesses a tremendous advantage in intelligence manpower, but the British have the distinction of accomplishing a very great deal despite relatively limited means.

Senior among the British organizations is the Secret Intelligence Service (SIS), sometimes also called MI-6, which is roughly the equivalent of the CIA although its origin formally dates back to 1907. The SIS conducts espionage, covert operations upon request, and maintains intelligence files. Technical intelligence in support of the SIS mission, such as from communications intercepts, is gathered by the Government Communications Headquarters (GCHQ), which, like the American NSA, is also responsible for the security of its own government's coded communications. Military intelligence files and reports are maintained and issued by the Defense Intelligence Staff under the director general of intelligence (DGI) at the Defense Ministry. Counterintelligence is the responsibility of the Security Service (also called MI-5) and the Special Branch of Scotland Yard, the British police force.

The figure known to history and literature as "C," the chief of the secret service, lacks the establishmentwide role of the director of central intelligence in the American system. Rather, C merely heads the SIS. Community management is accomplished by a Permanent Under Secretaries' Committee on Intelligence Services for foreign intelligence purposes and a Ministerial Group on Security for counterintelligence. At the working level, the Joint Intelligence Committee coordinates projects and approves appreciations, which are the British equivalent of national intelligence estimates. The London Signals Intelligence Board supervises the work of GCHQ, and an official Committee on Security does the same for counterintelligence components. The chief of secret services (C) chairs some of these units and is represented on all of them, but in the British system the senior intelligence adviser to the prime minister is a personal adviser, called the coordinator of intelligence and security.

Soviet Union

Although it may someday emerge that the very shadowy Chinese secret services actually lead the world in intelligence establishment manpower, this distinction was traditionally accorded to the Soviet Union. Intelligence was but one aspect of the work of the senior Soviet service, which was known as the Committee for State Security (Komitet Gosudarstvennoy Bezopasnosti, or KGB). Until very recently, the KGB handled not only foreign intelligence but also technical intelligence and both foreign and domestic counterintelligence, border and coast guard, personal security for Soviet leaders, anticorruption work, and nuclear weapons security. With such an array of activities, it is perhaps not surprising that KGB manpower was estimated at between 500,000 and 750,000 total.

The foreign intelligence function was handled by the KGB's First Chief Directorate, housed in a new headquarters building just outside the Moscow ring road, reportedly with space for some 15,000 employees. The directorate controlled a residency in each Soviet embassy abroad, and these units gathered material from open sources or conducted espionage. Until the recent disintegration of the Warsaw Pact, the First Directorate had access to, and sometimes even control over, the espionage resources of the services of Moscow's Eastern European allies. Technical intelligence in the Soviet system was gathered by the KGB's Eighth Chief Directorate from listening posts on Soviet bases; in Lourdes, Cuba; and aboard ships at sea. Counterintelligence and security functions fell within the bailiwick of the Second Chief Directorate.

Military intelligence functions in general were the task of the other Soviet service, the Chief Intelligence Directorate of the General Staff, which maintained a system of residencies in parallel to that of the KGB. The Chief Intelligence Directorate was familiarly known as the "cousin" of the KGB, more formally as the GRU (Glavnoye Razvedyvatel'noye Upravleniye).

As far as is known, the Soviets made little effort to produce finished intelligence such as the American national intelligence estimates or British appreciations. Rather, the Soviet intelligence establishment was called upon to furnish decision makers with raw intelligence reports. Although the KGB and GRU did produce intelligence studies or monographs, these are believed to have been relatively less important than a daily digest culled from the latest field reports and cables.

Coordination in the Soviet intelligence establishment was simplified due to the concentration of most functions within the KGB. Until the reforms that followed the abortive coup of 1991, the director of the KGB exercised full administrative and budgetary control over all aspects of Soviet intelligence save military intelligence. Even there, KGB officers with responsibility for operational security and ideological purity in overseas embassies had been known to wield substantial influence over GRU officials. The Soviet intelligence establishment worked theoretically for the Council of Ministers and in practice for the Politburo, of which the KGB director was a member during several periods of Russian history. The breakup of the Soviet Union in 1991 and the formation of the Commonwealth of Independent States will no doubt result in changes in the pattern of management over the former Soviet intelligence establishments.

People's Republic of China

In the wake of the Soviet upheavals of 1991, which forced the Soviet intelligence establishment to divorce ideology from espionage work, the most prominent intelligence establishment that remains strongly ideological is the Chinese. For more than six decades—since the 1930s—intelligence work has been the province of the Social Affairs Department of the Chinese Communist party. This organization worked against Chinese Nationalists during the Civil War, against the Japanese during World War II, and against the West during the cold war years. During the 1980s, the Chinese were cooperating with Western intelligence services, particularly the CIA, in monitoring Soviet strategic forces and in covert operations in Cambodia and Afghanistan.

Almost nothing is known of the internal organization of the Chinese intelligence establishment. Its locus in the hierarchy of the Party immediately suggests that intelligence policy coordination is carried out by the Party at the highest level, possibly directly between the Chinese leadership and the chief of intelligence. For many years, the man holding the latter post was K'ang Sheng, son of a landlord and one of the original Communist militants in Shandong province and Shanghai city. Until his death in December 1975, K'ang seems to have exercised complete control over Chinese intelligence, a span of half a century for which he truly merits the accolade "spymaster."

Change has crept into the Chinese system since the demise of K'ang. The Party's International Liaison Department and United Front Work Department appear to be major organs working with Communist and social democratic politicians and other sources. A Ministry of State Security, formed in 1983, handles internal security matters. Foreign intelligence appears to be divided among a network of research institutes, a unit of the Foreign Ministry, and agents under cover, including that of the New China News Agency. Analysis of foreign military capabilities is the province of the Military Intelligence Department of the Defense Ministry. There is no evidence in the open literature regarding overall manpower and budget levels for these assorted intelligence units.

France

The French foreign intelligence service has a long tradition among intelligence establishments, originating from military intelligence organizations under the Third Republic even before World War I. From that time, the French also operated a semiautonomous Service de Renseignements (SR), which responded to tasking from the Deuxième Bureau, the department of the French general staff concerned with intelligence matters. When Germany overran France during World War II, both the political and military units were subsumed in a single organization operating from England and later North Africa. After various name changes, internal reorganizations, and repatriation to France, this became the Service de Documentation Extérieure et de Contre-Espionage (SDECE), France's postwar intelligence service, with major divisions for administration and intelligence exploitation. The latter, SDECE's main operating unit, had sections for intelligence, counterintelligence, studies, and action. With manpower initially drawn primarily from military ranks, by the mid-1960s SDECE had shifted to a composition roughly two-thirds civilian and the rest military. The shift reversed itself in the 1980s, when the organization was renamed the Direction Générale du Sécurité Extérieure (DGSE), under which title it still operates today.

Internal security in the French system is handled by a unit known as Direction de la Surveillance du Territoire (DST), a national force akin to the counterintelligence unit of the American FBI. The DST and DGSE have had recurrent disputes over aspects of counterintelligence work. France's equivalent to the National Security Agency is the Groupement de Communications Radioélectriques (GCR). The DST and GCR have also fought over roles and missions in interception of radio messages from foreign agents. Estimates for the several French services in the 1980s suggest total intelligence manpower of about 7,000.

The president of France has ultimate authority over French intelligence services working through a special assistant on his personal staff. Day-to-day management falls to several cabinet-level committees including those focused specifically on intelligence, counterintelligence, and communications. The French services produce intelligence reports and monographs, but it is not known if they have an equivalent to American national intelligence estimates.

Italy

The Italian intelligence establishment can trace its origins to a service set up by the state of Piedmont within its Foreign Ministry as early as 1854. Military intelligence units were first set up the next year. An internal security unit existed by 1861. The Italian intelligence establishment grew slowly until World War I, received a great impetus at that time, and experienced further growth during the fascist government of Benito Mussolini. With a few exceptions, Italian intelligence proved largely ineffective during World War II and was then dismantled by Allied occupation authorities. Only in September 1949 did Italy regain a military intelligence unit called the Servizio Informazioni Forze Armati (SIFAR), renamed the Servizio Informazioni Difesa (SID) in 1965. In 1977, military and counterintelligence units were combined into a single Servizio per le Informazioni e la Sicurezza Militare (SISMI). A parallel domestic intelligence unit also exists, called the Servizio per le Informazioni e la Sicurezza Democratica (SISDE). In addition, military services operate their own intelligence sections and directly engage in communications intelligence activity. Assisted by a cabinet committee, the prime minister is responsible for supervision of the services and for intelligence policy.

Israel

Since 1960, when agents sent to South America successfully kidnapped Nazi war criminal Adolf Eichmann and brought him home for trial, Israel has had the reputation of having one of the best intelligence establishments. In fact, the Israeli services have their origins much earlier in the institutions created to assist the migration of Jews to Palestine before World War II. Today, the major foreign intelligence unit is called the Secret Intelligence Service (Mossad Letafkiddim Meyouchadim, or simply Mossad). The unit known as Counterespionage and Internal Security (Sherut Bitachon Klalt, or more familiarly Shin Beth), handles the functions it names. Military intelligence (Agaf Modiin) is the largest component of the Israeli intelligence establishment, which is managed by a Committee of Heads of Services (Va'adat Rasheri Hasherutim). The Israeli prime minister also has a special adviser on intelligence and terrorism and is capable of overriding lower level initiatives or setting his own intelligence policies. Available estimates of the size of the Israeli services suggest a total of about 10,000 on the overall size of the Israeli intelligence establishment.

Australia

Although it was originally established in 1952, the existence of the Australian Secret Intelligence Service (ASIS) was not officially acknowledged until 1977. Based upon the British model, the ASIS formed with British assistance, including a number of officers detached from the British SIS. An open source in the early 1980s estimated ASIS had more than 200 officers plus support staff. This small but high-quality unit evidently does both foreign intelligence and communications intelligence work. A central Joint Intelligence Organization (JIO) focuses on military matters, while counterespionage is the province of the Australian Security Intelligence Organization (ASIO). Formal responsibility for management of the Australian intelligence establishment was vested in a cabinet committee formed in 1958, which possibly due to political reluctance to become involved in intelligence issues, never met before 1976. There is no information on whether the cabinet group has convened since. Day-to-day activity is determined by a National Intelligence Committee (NIC) chaired by the director of the JIO and including the other agency directors. The Australian services provide reports and appreciations in much the same way the British services do.

Japan

This Far Eastern power has maintained military and naval intelligence organizations since the 1870s and 1890s, respectively. These appear to have performed indifferently during World War II. Allied occupiers insisted on the disbanding of Japanese intelligence organizations after 1945 and these only slowly regenerated. Under direct control of the prime minister, the Cabinet Research Office (Naicho), though small, is the major foreign intelligence unit. Each service component of the Japanese Self-Defense Force also maintains an intelligence unit. Within these the Annex Section of the Second Section, Investigation Division, Ground Self-Defense Force (Chosa Besshitsu) is Japan's electronic intelligence agency. Counterintelligence work is done by the Public Security Investigation Agency. A unit of the Ministry of Foreign Affairs and another within the Ministry of International Trade and Industry perform research and analysis within their fields of expertise. Little is currently known regarding Japanese procedures for managing the intelligence establishment.

JOHN PRADOS

SEE ALSO: Counterintelligence; Covert Action; Disinformation; Intelligence Collection; Intelligence Cycle; Intelligence, General.

Bibliography

Andrew, C. 1985. *Secret service: The making of the British intelligence community.* London: Heinemann.
Barnett, A. D. 1985. *The making of foreign policy in China.* Boulder, Colo.: Westview Press.
Barron, J. 1974. *The KGB.* New York: Bantam Books.
Faligot, R., and P. Krop. 1989. *La Piscine: The French secret service 1944–1984.* Oxford: Basil Blackwell.
Hohne, H., and H. Zolling. 1971. *The general was a spy.* New York: Coward, McCann.
Prados, J. 1988. *Presidents' secret wars.* New York: Quill Books.
Ransom, H. H. 1970. *The intelligence establishment.* Cambridge: Harvard Univ. Press.
Richelson, J. T. 1988. *Foreign intelligence organizations.* Cambridge: Ballinger Books.
———. *The U.S. intelligence community.* Rev. ed. Cambridge: Ballinger Books.

Ross, T. B., and D. Wise. 1967. *The espionage establishment.* New York: Random House.
Steven, S. 1980. *The spymasters of Israel.* New York: Macmillan.
Viviani, A. 1985. *Servizi Segreti Italiani.* 2 vols. Rome: Adnkronos.

INTELLIGENCE, GENERAL

Intelligence refers to information a government regards as important for advancing its military, foreign, and security interests. Often viewed solely as the cloak-and-dagger exploits of spies, intelligence encompasses a much broader range of activities. It applies to collecting, analyzing, and disseminating information (both so-called open-source information and secret material); to exploiting information—either by secretly influencing events in a manner beneficial to a government's interests (e.g., by disseminating *disinformation*) or by countering the intelligence activities of adversaries (*counterintelligence*); and to the governmental organizations that accomplish these tasks.

What constitutes "intelligence" may differ from government to government, according to the way a government—and hence its intelligence service—defines its interests and understands the threats to those interests. For example, a regime that regards international relations as an ongoing ideologically based struggle is likely to devote a higher priority to intelligence than one based on classic liberal principles that regards peaceful relations with other nations as the norm and war as a sporadic and theoretically preventable interlude. The former regime is likely to emphasize intelligence as a means of conducting the ongoing struggle during peacetime, whereas the latter may stress intelligence as a tool for anticipating adverse changes in the international environment.

More important in accounting for differences among intelligence agencies is the manner in which the government they serve defines its "security" interests. For example, totalitarian regimes are more likely to regard domestic opposition as a security threat, and their intelligence services concentrate on rooting out dissent. In contrast, liberal democracies usually are less preoccupied with internal threats to security, confine their domestic activities to detecting violent crimes that have been or are about to be committed, and concentrate on threats from abroad (including threats involving nonviolent crimes such as espionage).

Elements of Intelligence

Whatever their particular orientation, most intelligence services focus their work on the same four basic categories or elements: collection, analysis, covert action, and counterintelligence.

COLLECTION

Collection, as an intelligence term, refers to the gathering of data. In the intelligence craft, collection methods generally are geared toward amassing data about foreign governments, although some of the same methods may be used to collect information about dissident or terrorist groups that are seen as posing a threat to a regime's security interests. The principal methods of collection are (1) human intelligence (HUMINT); (2) technical intelligence, which comprises a plethora of specialized subdisciplines, the most prominent of which are signals intelligence (SIGINT) and imagery intelligence (IMINT); and (3) open-source information. Each of these methods has specific and valuable applications, but the proper choice or mixture of methods is equally essential to effective intelligence collection.

HUMINT. HUMINT (or espionage) is information obtained from individuals who have authorized access to government information and who are willing to pass on information they are supposed to keep confidential. Among the more common motives for this betrayal of trust are money, ideological attraction to a foreign government or rejection of one's own, resentment at superiors for perceived injustice or thwarted ambition, or blackmail on the basis of some discovered indiscretion. Alternatively, an agent may engage in espionage for his native country by gaining an adversary nation's confidence as an emigre, a supposed defector, or an "immigrant" (with forged papers that conceal a true nationality and former residence).

SIGINT. SIGINT is information derived from the interception of an adversary's electromagnetic transmissions. SIGINT itself can be divided into a number of subcategories, depending on the nature of the transmissions involved. For example, the term *COMINT* describes communications intelligence: the interception of the adversary's communications in either original or encrypted form. COMINT can be particularly valuable in providing timely and reliable access to an adversary's internal deliberations and operational traffic. During World War II, for example, Allied breaking of German and Japanese ciphers (Ultra and Magic, respectively) were COMINT coups of major strategic significance. Other forms of signals intelligence include ELINT (electronics intelligence, the interception of electromagnetic signals transmitted by military hardware such as radars) and TELINT (telemetry intelligence, the interception of telemetric information from the adversary's test missiles, airplanes, or other military vehicles). TELINT is a prime source of information about missile characteristics and capability; the ability to collect and use TELINT (which requires, among other things, that it not be encrypted) is of major importance for the verification of strategic nuclear arms control agreements.

IMINT. IMINT comprises intelligence collection from images, usually derived from air- or spaceborn platforms.

Airplanes may fly either directly over the territory to be photographed (e.g., U.S. U-2 reconnaissance aircraft were able to violate Soviet airspace during the late 1950s because they could function at altitudes unreachable by Soviet air defense) or just outside the target country's borders or territorial waters. Both the United States and the former Soviet Union have in practice asserted the right to collect intelligence from space and have developed and used systems to do so. Although international law remains ambiguous on the legitimacy of space reconnaissance, it has been implicitly recognized in bilateral arms control agreements such as SALT I and II and the INF Treaty.

IMINT allows one to "search" an adversary's territory for signs of new military construction and deployments by photographing large areas at relatively low levels of resolution. In principle, one would want to photograph the adversary's entire territory periodically; in practice, it is often necessary to concentrate on areas in which military forces are more likely to be located. Once an important new development has been sighted, higher resolution pictures of facilities, weapons, or equipment of interest may be obtained. High-resolution pictures are particularly important for gaining insights into the capabilities and intended operation of new items of military equipment.

Open-Source Intelligence. Perhaps surprisingly, a great deal of intelligence collected in peacetime comes from sources open to the public—radio, television, newspapers, technical journals, and so on. The extent to which crucial information can be publicly available is also surprising; in the 1970s, it was possible to piece together, entirely from public sources, a fairly complete design for an atomic bomb. Such episodes aside, open-source information has gained prominence in recent years because of attempts by intelligence agencies to devise more sophisticated methods for using social and economic data as a barometer of societies' stability and political and military capacity.

Collection Issues. The possible methods of technical intelligence collection are limited only by the laws of physics and the imaginations of the intelligence services' technical staffs. Collection can be directed against any part of the electromagnetic spectrum (e.g., infrared waves that enable one to track sources of heat, such as missile exhaust); against sound waves (e.g., sonar or seismic detectors); against sources of radiation; and so on.

Yet, partly because reconnaissance from space and other forms of high-tech intelligence wizardry are often so dazzling, they can mislead observers into overestimating their utility. Satellite imagery cannot detect activity that takes place indoors or below the satellite's horizon; moreover, an adversary may be able to schedule activities at times when they cannot be photographed or may ensure that sensitive equipment is covered when it is vulnerable to observation. Finally, adversarial activity that takes place in unexpected areas or that can be disguised as routine may never be noticed unless hints are obtained by other means.

In practice, overestimation of the usefulness of contemporary technical intelligence was a particular problem in the United States during the mid and late 1970s and has since resulted in a backlash that has re-emphasized the importance of HUMINT. This view emphasizes that human agents can provide information about the intentions and strategic concepts of an adversary's leadership that would be otherwise unavailable; clues for interpreting other types of material, such as photo imagery; and code books and other cryptographic material (including the original texts of messages transmitted in encrypted form) to facilitate cryptanalytic work. HUMINT also is essential in collecting information about nongovernmental groups (such as terrorist organizations) that lack the types of communications that are susceptible to interception by technical means.

Another factor re-emphasizing HUMINT in the 1980s was a series of espionage coups—themselves largely the product of HUMINT—that gave the former Soviet Union a great deal of detailed information about the way in which U.S. signals and imagery intelligence systems function. Such information potentially enabled the Soviets to hide important information from technical systems or to deceive them by providing false signals for them to collect.

ANALYSIS

Analysis refers to the process of transforming collected intelligence data into a form usable by policymakers and military commanders. The end result, or "intelligence product," can take the form of a short memorandum, an elaborate formal report, a briefing, or any other means of presenting information. Analysis comprises a large variety of information-processing activities from the most technical (the decrypting of intercepted communications or the interpretation of telemetric signals to determine a missile's characteristics) to the most theoretical and speculative (predicting an adversary's future actions).

The importance of analysis in the intelligence process has been enhanced by the fact that crucial information often can be derived from public sources. In fact, much of the particular contribution of intelligence services can reside not so much in the covert collection of "smoking gun" information as in thoroughgoing analysis of data that may require resources (such as large computerized databases) beyond the reach of individuals. To exploit such types of information, intelligence agencies such as the U.S. Central Intelligence Agency (CIA) have recently developed sophisticated models for collecting and analyzing social and economic information, most of it from open sources.

Covert Action

Intelligence services not only collect and analyze information but also provide the means by which a government can exploit that information in a secret manner to advance its foreign policy interests. In the United States, such activity is termed "covert action" and is defined as activity conducted in support of national foreign policy objectives abroad that are planned and executed so that the role of the United States government is not apparent or acknowledged publicly. (Other regimes, however, whose intelligence services concentrate on internal threats to the ruling party may use analogous methods to disrupt and destroy domestic political opponents.) Thus, covert action covers a broad spectrum of activities, from the most pedestrian, such as secretly providing technical aid (e.g., with respect to security or communications) to a friendly foreign government, to the most spectacular, such as assassinating foreign leaders or fomenting revolutions; no standard categorization exists among intelligence students or practitioners.

Intelligence services that carry out espionage activities typically have been responsible for covert activity as well, because they have the necessary resources—primarily the human agents—and are well equipped to act covertly. In the United States, the CIA was established under the auspices of the National Security Council (NSC) in 1947. The NSC established a separate subagency (under the auspices of the Secretaries of State and Defense), the Office of Policy Coordination (OPC), in 1948 to carry out covert action, largely in support of the policy of "containment" of Soviet influence in Europe. However, the division of responsibility between the OPC and the CIA's clandestine intelligence collection office, the Office of Special Operations (OSO) proved unworkable—in some cases, they bid directly against each other for the services of the same agents—and the two offices were merged in 1951–52.

The integration of the OPC and the OSO under the management of Allen Dulles as the CIA's Director of Central Intelligence ameliorated considerable bureaucratic confusion. Yet some have argued that the covert action function should be separate from the rest of intelligence. Conceptually, it differs in that it focuses on affecting events rather than on understanding and predicting them. Furthermore, it is claimed that the objectivity of the intelligence analysis function will suffer if the same organization that proposes and runs a covert action program is responsible for predicting the likelihood of its success and for monitoring its actual progress.

Just as *covert action* originated as an American term, the term *disinformation* has its origins in the former Soviet Union (as *dezinformatsia*). Disinformation is intentionally false information disseminated for the purpose of affecting another country's political or military decision making. It may take the form of forgeries, rumors, official denials, false media broadcasts, and activities by "front groups." Although the Soviet Union has propagated such blatant ideological disinformation as the notion planted in an Indian newspaper that the United States was developing an "ethnic weapon" capable of killing only nonwhites, many regimes have used some form or other of disinformation, especially in wartime. In a tactical disinformation move, U.S. battle maps of troop concentrations on the Kuwaiti border, displayed on television in the 1991 Persian Gulf War, misrepresented the situation in that actual troop concentrations were considerably further west. On a more ideological level, the Entente Allies in World War I magnified German atrocities in Belgium to stir war sentiment.

Counterintelligence

Counterintelligence refers to activities undertaken, and information collected and analyzed, to protect one's nation (including one's intelligence activities) against the actions of hostile intelligence services. Therefore, the scope of counterintelligence is in a sense as broad as the scope of intelligence itself, although it often refers specifically to attempts to defeat an adversary's intelligence collection efforts. Its proper scope and nature tend to be controversial issues even among intelligence professionals.

Counterintelligence activities may be placed along a spectrum from the most passive to the most active. Passive activities include such things as security measures, which seek to block an adversary's access to sensitive information. These comprise personnel security programs to ensure the loyalty and discretion of personnel granted authorized access to information; physical security programs to prevent theft or removal of documents; and communications security measures (including encryption) to prevent an adversary from gaining useful information from one's communications. More active measures include surveillance of known officers of adversaries' intelligence service, to determine who among one's own nationals might be in contact with them, and double-agent operations in which one's own agent pretends to commit espionage for an adversary's service). The most ambitious and effective counterintelligence measure may be recruitment of a high-level official in an adversary's intelligence service (a "mole") to learn directly about espionage activities directed against one's own service.

A *deception operation*, another method of counteracting an adversary's intelligence services, focuses not so much on an adversary's collection activities as on its intelligence analysis. By concealing the true facts and presenting the adversary with misleading information (via double agents, simulated communication networks, plywood dummies realistic enough to fool the adversary's photo reconnaissance capabilities, etc.), deception operations attempt to lead the adversary to misinterpret the actual situation and, on the basis of that misinterpretation, to act contrary to its own interests and in furtherance of one's own.

Counterintelligence also includes the analysis of available information to determine whether one's own intelligence service has been penetrated or is being deceived by an adversary. Such evidence might include information about the adversary service (e.g., that it has obtained a copy of a secret document from one's own government); inferences drawn from its observable behavior (which may indicate foreknowledge on its part of one's own actions); and inferences drawn from occasions on which it has been able to thwart one's own intelligence activities (again indicating possession of some source of knowledge about them). Information provided by one's own agents must be scrutinized to determine if it could be material provided by the adversary for the purpose of deception.

In a broader sense, counterintelligence involves protecting a regime against threats from nongovernmental groups (such as terrorists, drug "cartels," or domestic political opponents seen as subversive of the regime) that operate clandestinely. A totalitarian party that regards part of its society (e.g., non-Aryans, the bourgeoisie) as the "enemy" may concentrate its intelligence activity on this sort of internal counterintelligence.

Intelligence and Policy

The ultimate purpose of intelligence is to provide the information that policymakers or military commanders need to perform their duties as well as possible. Thus, the question of the proper relationship between intelligence and policy has always been an important, if subtle, one.

The standard view in the United States, developed during the immediate post–World War II period, has been that intelligence should not be contaminated by the biases of policy (see Kent 1949). Intelligence analysis therefore should not take cognizance of its own government's internal foreign policy disputes or actions. Under this view, intelligence does not attempt to predict the future activities of its own government, although they may be major determinants of the international situation whose future course it is trying to assess.

An alternative view holds that intelligence requires not "absolute prediction" ("Ruritania will attack Fredonia in three weeks") but rather "contingent prediction" ("Ruritania's decision about attacking Fredonia will be influenced by the following factors, which in turn are subject to the following external influences"; see Kendall 1949). Contingent prediction considers that the future may be influenced by steps one's own government may take and must make explicit judgments about whether a proposed course of policy will achieve its objectives; thus, it runs the risk of being "contaminated" by the policy positions of the various departments and agencies with which the intelligence analysts must maintain close contact. Indeed, because the contingent prediction approach implies that intelligence analysis and formulation of policy alternatives go hand in hand, it is not clear whether this type of analysis should remain within the sole purview of the intelligence agencies or whether it should be the product of joint intelligence–policy working groups.

A related criticism of the standard view holds that if intelligence is supposed to provide warnings to policymakers of potential dangers to the nation's interests, a concomitant part of its job should be to point out potential opportunities to advance the nation's interests. This perspective also sees the intelligence analyst as a more active participant in the policy-formulation process than the standard view would envisage.

These differences of opinion illustrate the twofold role of intelligence in the policy process: it must both aid in the formulation of policy and be prepared to serve as a critic of policy (i.e., to present to policymakers facts and assessments that make them uncomfortable, and call their policies into question). The standard view emphasizes the latter function at the expense of the former, on the grounds that any involvement in policy will lessen intelligence's freedom to criticize that policy.

Intelligence Failure

Many recent academic studies of intelligence have concentrated on the question of why intelligence often fails to provide adequate warning of impending dangers or is otherwise mistaken in its assessments of particular situations.

Causes

In most cases, intelligence failures seem to have been principally failures of organization and analysis: the necessary clues were available to the intelligence service, but various institutional or intellectual factors impeded their timely and accurate evaluation. Among these causes are the following:

Lack of "all-source" intelligence analysis. One common cause of intelligence failure has been the absence of a single location within a government for unified analysis of intelligence information from all sources. Lacking such a location or agency, governments fail to put together important clues, and no one has the whole picture or context in which to make the proper inferences.

Low "signal-to-noise" ratio. Borrowing the language of information theory, some analysts have described a major problem in intelligence analysis as the low "signal-to-noise" ratio characteristic of intelligence collection—that is, the tiny amount of useful data compared to the large amount of insignificant, misleading, or false information (data that are deceptive either accidentally or as a result of deliberate action by the adversary). This phenomenon is related to the "cry wolf" syndrome, in which an important warning sign is ignored because, in the past, identical or similar warnings turned out to be false alarms.

Bureaucratic or political bias. Intelligence failure is perhaps most commonly attributed to bureaucratic or political bias—that is, the often subtle tendency in intelligence for individuals and organizations to skew analysis results toward their bureaucratic or political interests. A self-serving bias can prevent an individual from judging situations correctly and can lead him ultimately to act against his own best interests. Similarly, within governments, intelligence judgments can have large effects on organizational policy decisions and budgets; an opportunistic bias that may enable an organization to win a short-term intragovernmental struggle may dangerously distort the policy process as a whole.

The tyranny of conventional wisdom. Often, important bits of evidence are overlooked or undervalued because they conflict with the preconceptions of a government or some bureaucracy within it. However difficult it is for individuals to change their basic view of a situation on the basis of new evidence, it is even harder for bureaucracies to do so, because advocates of change are likely to be in the minority. Thus, evidence that is inconsistent with an accepted view is more likely to be forced to fit an existing theory by means of ad hoc explanations than to indicate the need for a new perspective. It is even harder to re-evaluate the conventional view if it corresponds to the view commonly held in the academic world.

"Mirror imaging." An additional problem arises from what some consider an inherent human tendency to anticipate and interpret the actions of others in terms of one's own motivations. This may easily lead to faulty predictions about what members of other cultures, with other ways of thinking about political and military affairs, might do in a given situation. This difficulty is compounded to the extent that academic comparative government theory tends to assume that a general theory can be developed that applies to diverse societies and cultures.

In the area of scientific and technical intelligence a similar problem exists, often referred to as the "not-invented-here" syndrome: many intelligence experts have difficulty believing that another country has made technical or scientific discoveries that their own country has regarded as infeasible or unpromising. This syndrome is likely to be strongest where a more technically advanced society is in competition with a less advanced one, since the former will be psychologically unprepared to believe that the latter was able to surpass it in any given area.

PREVENTION AND CURE FOR INTELLIGENCE FAILURE

Although the possibility of intelligence failure can never be completely ruled out (if only because one side's advances in intelligence collection and analysis will motivate its adversaries to improve their concealment and deception techniques), various proposals have been made to avoid some of the major causes of failure discussed above.

The creation of a central agency with access to all sources of intelligence information helps avoid the lack of all-source intelligence analysis. Making this organization independent of the major government agencies in charge of national security policy (and hence of their political and budgetary interests) may also alleviate the dangers of bureaucratic bias.

At the same time, a centralized service may be more subject to the conventional wisdom, especially if it tends to dismiss outside challenges to its ideas as reflecting merely the self-interest of the other bureaucratic "players." In addition, its judgment may be distorted by its own bureaucratic interests—for example, its status as a centralized agency depends on its ability to argue that it is a necessary counterweight to the biased views of others.

Various solutions to these problems exist. "Competitive analysis" makes all-source information available not only to the centralized intelligence agency but to other organizations as well, typically parts of the major national security bureaucracies. This generates a variety of perspectives, and "competition" among intelligence agencies ensures a more careful review of the evidence and a greater likelihood that faulty reasoning will be uncovered and corrected. Problems with this approach arise when the results of the analyses are presented. A desire to reach a unified judgment may lead to a final product, made up of several converged opinions, that has been diluted or hedged to make it acceptable to the competing agencies and is less useful to the policymaker than the precise views of any particular agency would have been. Alternatively, the policymaker may be confronted with a set of discordant views, which may lead him merely to choose the one most congenial to his own policies, so that the impact of intelligence on policy is lessened.

The "devil's advocate" solution envisages an official charged with the responsibility of challenging the conventional wisdom. In principle, this might alert intelligence officials to weaknesses in the accepted views and might make the analysts more careful than they would otherwise be. In practice, however, such routine contrariness would soon be ignored and its arguments shrugged off; it could easily degenerate into a meaningless exercise.

The intellectual vice of "mirror imaging" can best be countered by training programs that ensure that analysts receive a thorough grounding (including linguistic ability) in the areas of the world on which they will be working. In addition, nations with heterogeneous immigrant populations (such as the United States and Israel) have an inherent advantage if they are able to make use of their nationals who have first-hand knowledge of other countries, cultures, and languages. On the other hand, overemphasizing area expertise may lead analysts to ignore world-wide or cross-cultural trends and may result in surprises when sudden changes in direction occur.

Finally, addressing the question of intelligence failure requires setting reasonable standards of performance. The typical implicit standard is the accuracy of intelligence predictions, and, in particular, whether intelligence was able to warn policymakers of the imminent occurrence of important events. This standard tends to ignore the other types of assistance intelligence should be able to provide (e.g., an assessment of the underlying factors of a situation that would allow government officials to better evaluate the likely consequences of alternative policies). Moreover, in many cases—especially when one's own actions exert a major influence on the foreign behavior one is trying to predict—this standard may not be reasonable. Simple predictions that do not explicitly take into account the effects of one's own actions may be useless or misleading. For instance, the optimism of U.S. intelligence with respect to the durability of the Shah of Iran was based partly on past crises in which the Shah defeated opposition to his rule; in those past instances, however, the United States had provided important assistance to the Shah, which it did not do in the Shah's final crisis in 1978–79.

Intelligence and Democratic Government

In any regime, the governance of intelligence activities presents special problems because of the need to restrict access to sensitive information to the smallest possible number of people. In democracies, where parliamentary and public debate about public affairs plays a major role in the political process, this difficulty is even greater. In addition, domestic counterintelligence requirements often conflict with the more ordinary law enforcement procedures of the society, leaving it unclear what actions intelligence services are permitted to take in this area.

In the United States, this question was addressed in the context of the congressional investigations of intelligence of the mid-1970s. A major result was the establishment of congressional oversight committees with the statutory right to be kept "fully and currently informed of all intelligence activities." Similar developments, albeit of a much less ambitious nature, have occurred in West Germany and Canada. In the latter case, the oversight body is not a parliamentary committee but is chosen by the prime minister in consultation with the leaders of the opposition from among the Privy Councillors, a corps of senior government advisers.

The theory of congressional oversight as developed in the United States is that the advantages of democratic debate and secrecy can be combined: the intelligence committees—whose members, under this theory, should represent a broad spectrum of views—are seen as surrogates for the public, and their secret deliberations substitute for public debate. In this way, the presumed "pathologies" of the intelligence services (e.g., a technocratic approach that loses touch with the public's sense of basic values and hence eventually forfeits the confidence of the public; a "can-do" attitude that pursues the goals of the executive branch leadership in a single-minded way that ignores longer-term considerations) are kept in check, while secrecy is preserved.

Critics of congressional oversight have pointed to a number of disadvantages that they feel outweigh the benefits described above. Secrecy is in fact endangered, either via "leaks" that become more likely as more people get access to sensitive information, or because cases of executive–legislative disagreement come to be resolved through the political process, which necessarily involves debate and legislation. For example, once the issue of aid to the Nicaraguan resistance became a major source of contention between the Reagan administration and the Democratic leadership in the Congress, to be resolved in a series of "Boland Amendments" (Edward Boland, former chairman of the House Permanent Select Committee on Intelligence) to appropriation bills it was impossible for the "covert action" program to remain secret.

At the same time, Congress' right to be kept "fully and currently informed" means that intelligence data or analysis that raises questions about the president's policies can be used by his political enemies to attack those policies. This increases the political sensitivity of the intelligence product and increases the policymaker's incentive to attempt to influence it. Some intelligence officials have welcomed the relative independence from executive branch pressures that congressional scrutiny provides, but it seems unlikely that the increased political role of the intelligence product will be beneficial to the intelligence agencies in the long run.

ABRAM N. SHULSKY

SEE ALSO: Counterintelligence; Covert Action; Deception; Disinformation; Espionage, Legal Aspects of; Intelligence Analysis, Military; Intelligence Collection; Intelligence Cycle; Intelligence Establishments; Intelligence, Human; Intelligence, Imagery; Intelligence, Naval; Intelligence, Signal; Intelligence, Strategic; Intelligence, Tactical; Intelligence: Indications and Warning.

Bibliography

Betts, R. 1978. Analysis, war, and decision: Why intelligence failures are inevitable. *World Politics* 31:61–89.
Codevilla, A. 1980. Comparative historical experience of doctrine and organization. In *Intelligence requirements for the 1980s: Analysis and estimates*, ed. R. Godson. Washington, D.C.: National Strategy Information Center.
Dziak, J. 1988. *Chekisty: A history of the KGB*. Lexington, Mass.: D. C. Heath.
Kendall, W. 1949. The function of intelligence. *World Politics* 1:542–52.
Kent, S. 1949. *Strategic intelligence for American world policy*. Princeton, N.J.: Princeton Univ. Press.
Maurer, A., M. Tunstall, and J. Keagle. 1985. *Intelligence: Policy and process*. Boulder, Colo.: Westview Press.
Orlov, A. 1963. *Handbook of intelligence and guerrilla warfare*. Ann Arbor: Univ. of Michigan Press.
Robertson, K., ed. 1987. *British and American approaches to intelligence*. London: Macmillan.
Treverton, G. 1988. *Covert action: The limits of intervention in the postwar world*. New York: Basic Books.
Turner, S. 1985. *Secrecy and democracy: The CIA in transition*. New York: Harper and Row.

INTELLIGENCE, HUMAN (HUMINT)

Human intelligence (HUMINT) is any intelligence collected from human beings. Its purpose is to assess the intentions, motivations, policy, and capability for warfare of possible adversaries. HUMINT is important to the decision-making processes of political leaders. For example, in the summer of 1990, HUMINT gave President George Bush insight into the intentions of the Iraqi leader, Saddam Hussein, and the capability of his armed forces. The advantages of HUMINT are not likely to be rendered marginal by technological collection. Overhead satellites may locate missiles, but HUMINT is critical to determining whether a potential foe intends to use those missiles. HUMINT also helps intelligence analysts make sense of a jumble of information and fill in critical gaps in their data. Moreover, while a hostile nation may be able to conceal its formal communications, HUMINT will remain available.

The disadvantages of HUMINT are twofold: first, an agent can be turned and used to supply disinformation to his case officer; second, a case officer must have cover, which often involves his employment by another government agency or a private business, either of which may intrude upon his time available for intelligence. In that situation, he must spend long hours of work each day, often not coming home till late at night. This is not only physically exhausting for the case officer, but it creates tensions within his family life, further exacerbating his problems.

Overt and Clandestine Sources

HUMINT includes both overt and clandestine intelligence. The person, organization, or government from which overt intelligence is gathered may be aware that the collector is an intelligence officer or agency. For example, the Central Intelligence Agency (CIA) openly subscribes to the British publication *The Economist*, which has careful analyses of problems throughout the world. Clandestine intelligence is obtained and transmitted to an intelligence organization without the knowledge of the person or group responsible for guarding it. When Jonathan Jay Pollard passed on documents from the U.S. intelligence community to Israeli intelligence for photocopying, he was doing so without the consent or knowledge of the Office of Naval Intelligence, for whom he worked.

The overt sources of information are numerous. Foreign publications—newspapers, journals, and technical and scientific periodicals—are an obvious source. Intelligence services monitor the broadcasts of foreign radio and TV stations; the CIA's Foreign Broadcast Information Service performs this task for the United States. Embassies abroad engage in overt collection using Foreign Service Officers, defense attachés, FBI personnel assigned as legal attachés to do liaison work with the counterintelligence and criminal divisions of other nations, and employees from the U.S. Treasury, the U.S. Department of Commerce (to gather economic intelligence), and the U.S. Department of Agriculture. U.S. defense attachés seek out information on foreign military services and weapons systems, and are controlled by the Directorate of Attachés of the Defense Intelligence Agency (DIA). Defense attachés generally make no secret of their status as intelligence officers. Travelers—tourists, business people, and academics—sometimes volunteer information to the CIA's Domestic Contacts Division upon their return to the United States.

Because, in most countries, several different organizations collect clandestine information, coordination is vital to avoiding duplication. In the United States, the CIA coordinates the activities of all U.S. intelligence services abroad. To focus clandestine collectors' attention upon crucial areas, the Director of Central Intelligence (DCI) issues a list of topics of vital importance and the priorities attached to each of these issues. The DCI uses the Human Resources Committee to identify these requirements and then develops plans that assign each segment of the U.S. intelligence system the task of acquiring certain information.

CIA case officers are part of the Directorate of Operations, seeking out people abroad (usually foreign citizens) who have access to protected information, documents, or technical equipment of interest to the United States. Once someone accepts a clandestine mission from a case officer, he has become an agent. Usually, an agent is motivated by a mixture of factors: the desire for more money, adventure, or revenge upon his employer.

Defectors, refugees, and prisoners of war present another clandestine source of information. In U.S. intelligence, the CIA and the U.S. Air Force Office of Special Investigations maintain facilities both in the United States and abroad to process these people. Interrogation is performed by skilled linguists who are sensitive to subtle nuances that might be explored. The 250,000 German POWs who came home from the Soviet Union after World War II provided a rich source of information on an inaccessible country.

Another source of information is clandestine collection by the military. The U.S. Army devotes more personnel to HUMINT through its Intelligence and Security Command (INSCOM) than other U.S. military services, focusing upon the development of weapons systems, early warning, and order-of-battle information. (Order of battle refers to the strength, leadership, location, mission, and equipment of enemy forces.) The DIA coordinates the requests of military intelligence services for information.

The FBI also provides clandestine HUMINT. Although its best-known function is law enforcement, the FBI's Intelligence Division conducts counterintelligence (CI) in the United States. In this capacity, it seeks to prevent and combat sabotage, espionage, and subversive activities. FBI special agents often acquire information useful to collectors of HUMINT. The CIA and military intelligence services generally coordinate their domestic activities with those of the FBI.

Methods

When a case officer identifies a possible agent overseas, he must first assess this person for stability and reliability. For example, the potential agent may have access to targeted documents, but if he is unstable, he may brag to his friends or otherwise endanger the entire operation. The case officer confirms the reliability of his agents by making sure that they really have the access to the information or equipment that they claim. A case officer who wastes large amounts of money on a fraud also feeds misinformation to his own organization. Once an agent is recruited, the case officer or handler must train him in the craft of clandestine collection and guide him in the "acquisition of the target."

The case officer arranges and conducts meetings with his agent(s), checking to see that neither he nor his agent has been followed to their rendezvous, a process that may require considerable time. The case officer must be able to follow someone without being detected and also must know how to elude or at least detect hostile surveillance. They minimize direct contact with agents by using "dead drops," places where documents and messages can be left without being discovered. They must arrange for clandestine communications with their agents, and sometimes meet them in a safehouse, a house or apartment maintained for this purpose.

SKILLS REQUIRED IN HUMINT

The case officer may be a specialist in some field, such as Soviet affairs, but is usually a generalist, often the product of a liberal arts education. A case officer who shows a flair for management may become a station chief in the field or a regional administrator at headquarters. He must be able to learn to speak different foreign languages, since exposure in one country may necessitate reassignment to another. Above all, the case officer must have an intuitive understanding of his agents so that he can sense and respond to their needs and problems before they interfere with the agents' effectiveness.

In the search for information, the case officer utilizes wiretapping and lock picking for the purpose of surreptitious entry, which is considered criminal activity under ordinary circumstances. On occasion, they must open the safes of foreign establishments or secretly enter offices and homes; open and reseal letters and diplomatic pouches without leaving traces; plant hidden microphones ("bugging"), which usually involves surreptitious entry; engage in wiretapping and intercept facsimile messages; photograph documents of potential adversaries and sometimes those of friendly nations; and communicate with agents through special codes, secret writing ("secret ink"), or the microdot process, which reduces a message to the size of a period or a comma.

History of Espionage

In the United States, clandestine collection began with Gen. George Washington, who used HUMINT to defeat his stronger foe, the British. After the American Revolution, the U.S. government resorted to clandestine collection on a sporadic and underfunded basis, usually developing a HUMINT capability during wartime and then allowing it to shrivel into insignificance or disappear entirely when the conflict ended.

However, in 1882 the U.S. Navy began its Office of Naval Intelligence, and in 1885, the U.S. Army established its Military Intelligence Division. In the late 1800s, in their effort to modernize the U.S. Navy, the Office of Naval Intelligence collected designs and specifications of the latest European warships and the equipment and doctrine of European navies. This facilitated the navy's transition from sails to engines and from wooden sides to steel hulls. As the Spanish-American War approached, the Military Intelligence Division estimated the strength of Spanish Army units with reasonable precision, but skeptical field commanders used very little of their information.

When the United States entered World War I, Maj. Ralph H. Van Deman, aided by British intelligence officers, reorganized and expanded U.S. Army intelligence, focusing upon domestic security and counterintelligence rather than clandestine collection. At the end of the war, Van Deman's organization was severely curtailed and was often used by commanders during the interwar period for such functions as public relations.

In World War II, the Office of Strategic Services (OSS), created in 1942, was under the leadership of Maj. Gen. William J. Donovan, who had won the Medal of Honor in World War I. Donovan, aided by British intelligence, developed the Secret Intelligence (SI) unit within his organization to conduct HUMINT. President Truman disbanded the OSS at the end of the War, but its SI and Counterintelligence sections were transferred to the War Department as the Strategic Services Unit (SSU). When the Central Intelligence Group (CIG) was created in 1946, the SSU and seven of its stations in North Africa and the Middle East were transferred to this new intelligence organization. When the National Defense Act of 1947 established the CIA, the facilities of the CIG and FBI activities in Latin America were merged into this new organization.

Having set up puppet governments in Eastern European nations after World War II, the Soviet Union tried to exploit postwar instabilities in Western Europe, with the objective of installing pro-Soviet governments in France and Italy. The CIA and U.S. military intelligence sought to counter this by supporting organizations opposing Communist front groups. This required further development of clandestine collection. After the threat in Western Europe receded, Soviet initiatives shifted to the infiltration or overthrow of the fragile governments in the developing nations. This required both overt and clandestine collection by Foreign Service and CIA intelligence officers, since relatively little was known at that time about these new nations. As the internal situations in some of these nations stabilized, the emphasis in U.S. clandestine collection shifted to "denied areas," chiefly those controlled by Communist governments. However, non-Communist governments that were hostile to U.S. interests—Libya, for example—were also targets of clandestine collection. During the Carter administration, the Shah of Iran fell from power, one result of which was that American personnel in the embassy in Tehran were taken hostage by Iranian fanatics. This development, along with the activities of a revolutionary Marxist government in Nicaragua, returned the emphasis of U.S. HUMINT to the developing nations.

When Jimmy Carter became president in 1977, he appointed Adm. Stansfield Turner as Director of Central Intelligence (1977–81). Turner had a strong background in engineering, and his preference for collection by overhead satellites and signals intelligence (SIGINT) by the National Security Agency (NSA) was detrimental to HUMINT. Turner dismissed 800 employees of the Directorate of Operations, the Agency's HUMINT division (this number is a matter of dispute within the intelligence community). Turner's successor, William J. Casey, worked during his tenure (1981–87) to rebuild the CIA's HUMINT capability.

In Great Britain, clandestine collection goes back to at least Elizabeth I (1558–1603) whose secretary of state, Sir Francis Walshingham, operated a small secret service. Among other things, this unit uncovered evidence that sent Mary Stuart ("Mary, Queen of Scots") to the gallows. Since 1797, Parliament has been required to vote annually on the disposition of a Secret Service Fund. This money was used to finance British propaganda on the Continent, informants, secret operations, and diplomatic bribes. The Secret Service Fund was greatly diminished after the British defeated Napoleon at Waterloo in 1815, but when the Franco-Prussian War (1870–71) revealed a startling lack of information on the part of British military intelligence, the army's Topographical and Statistical Department was reorganized to include an Intelligence Branch (IB). The main purpose of the IB was to monitor Russia's threat to India and to aid the army as the British competed with other European powers for colonies in Africa from approximately 1775 to 1914. In 1886, the Naval Intelligence Department was established to gather information on such topics as the movement of foreign fleets, though it was severely limited by a lack of funds and personnel.

In 1900, during the Boer War in South Africa, a novelist, William Le Queux, created hysteria in England, claiming that German spies were operating on a mass scale in Great Britain in preparation for a German invasion of this now vulnerable nation. Since this was a period of intense German naval construction, Le Queux and other merchants of fear gained the support of people like the elderly Field Marshal Earl Roberts, thereby gaining respectability for his wild claims of this nonexistent danger. In reality, the German General Staff was concerned with fighting a two-front ground war in Europe against France and Russia, and did not have the capability, time, or interest to invade Great Britain. However, this imagined spy menace highlighted the inadequacies of British intelligence and led in 1907 to increased clandestine collection activity by military intelligence under Maj. James Edmonds.

When Rear Adm. Esmond Slade became Director of Naval Intelligence in 1907, he began to search for agents in Germany to monitor the progress of the rapidly growing German Navy. Another consequence of the spy mania in Great Britain was the founding of the Secret Service Bureau in 1909, which was soon divided into a home department, predecessor of MI5 (counterintelligence), under the leadership of Capt. Vernon Kell, and a foreign department, headed by Commander Mansfield Cumming. This latter unit developed into MI6, whose purpose is to collect intelligence abroad (HUMINT).

Because the foreign department was responsible to the Admiralty from 1910–14, Cumming's job initially was to gather evidence of the nonexistent German plan to invade England. Consequently, his organization's attention was diverted from a real threat: the German Army's Schlieffen Plan to defeat both Russia and France on the ground. Cumming's agents did, however, provide considerable information on the technical aspects of the German Navy, such as the speed of their U-boats, which was of use to the British fleet in World War I.

When World War I broke out, Cumming concentrated his case officers in neutral Holland and Switzerland, as well as Belgium and Russia. In Belgium, about 3,000 agents worked for Cumming's organization, chiefly reporting on the movement of German troop trains.

After World War I, Cumming located his intelligence officers throughout Europe in the Passport Control Offices of British embassies, where they were not welcome or especially productive. His replacement in 1923, Capt. Hugh ("Quex") Sinclair, had to administer MI6 on a greatly reduced budget. Sinclair's efforts in clandestine collection were paralleled by the Z organization of Col. Claude Dansey, set up to obtain intelligence on Nazi Germany. Its information was generally not useful.

No British intelligence organization predicted with any accuracy the timing of Hitler's invasion of the Rhineland in 1936 and of Austria in 1938. Upon Sinclair's death in 1939, he was succeeded by his deputy, Col. Claude Menzies, who merged the Z organization into that of MI6. As the German Army swept over Europe, the Passport Control Offices were closed, forcing MI6 to rely upon its intelligence officers in the neutral capitals of Europe—Madrid, Lisbon, Stockholm, Ankara, and Berne.

Ivan the Terrible established the Oprichnina in 1565 chiefly to destroy the czar's enemies among his own people, rather than to collect intelligence abroad. Political police organizations were periodically revived, and in 1881, the Okhrana was established to combat the rise of terrorism by an alienated intelligentsia. In 1884, the Okhrana created its Foreign Agency in Paris to penetrate Russian revolutionary groups throughout Europe, essentially an external surveillance (or counterintelligence) function. By approximately 1900, the Okhrana had developed a SIGINT division to intercept and decrypt the communications of other governments. This codebreaking required HUMINT, so that stolen diplomatic telegrams might be compared with the coded originals.

Russian military intelligence before World War I derived considerable information about the Austrian armed forces from Colonel Alfred Redl, a high-level Austrian intelligence officer and homosexual. In the winter of 1901–02, a Colonel Batyushin of Russian military intelligence, using blackmail and bribery, obtained a massive amount of information from Redl—including the Austrian plans for mobilization against Russia and Serbia.

Lenin established the Cheka shortly after the Bolshevik Revolution as a means to remain in power despite the miniscule support for his party. Under its first director, Felix Dzerzhinsky, the Cheka slaughtered hundreds of thousands of Soviet citizens in a campaign of calculated terror, until major opposition to the Bolsheviks was either destroyed or silenced. In 1920, Dzerzhinsky established a Foreign Department, the INO, one of whose purposes was clandestine collection of intelligence (HUMINT).

In the 1920s and 1930s, the OGPU, successor to the Cheka, concentrated its HUMINT efforts in Great Britain, where the class struggle was particularly intense. This work resulted in the acquisition of Cambridge University traitors Harold Adrian Russell ("Kim") Philby, Guy Burgess, Donald Maclean, Anthony Blunt, John Cairncross, Norman John (James) Klugmann, and Leonard Henry ("Leo") Long. Kim Philby penetrated MI6; Burgess and Maclean infiltrated the Foreign Office; Blunt was in MI5 during World War II; Cairncross reached senior positions in several branches of the British government; Klugmann influenced British policy towards Yugoslavia from his position in the Special Operations Executive (sabotage, guerrilla warfare, reconnaissance, and assassination); and Long served in British military intelligence, both during and after World War II.

In the 1920s, the OGPU also focused on the penetration of foreign embassies, especially those outside Europe, where security tended to be lax. Documents stolen from these embassies were compared to the ciphered versions, thus rendering considerable support to Soviet codebreakers.

From 1933–38, Josef Stalin conducted purges of the Communist Party, the Red Army, the GRU (military intelligence), and the NKVD, successor to the OGPU. Despite these purges, Soviet clandestine collection continued to score impressive achievements. Soviet intelligence easily penetrated the U.S. Embassy in Moscow during the 1930s, since security was almost nonexistent in this outpost. A U.S. military attaché, Col. Phillip R. Faymonville, had probably been suborned by the NKVD. In 1944, a U.S. Navy electrician found 120 hidden microphones in his first examination of this embassy. Soviet intelligence officers also established influential spy rings in the United States government in the late 1930s and during World War II. Their probable agents included Alger Hiss, a high-ranking State Department official; Harry Dexter White, a senior official in the U.S. Treasury; and Lauchlin Currie, an administrative aide to Pres. Franklin D. Roosevelt. During these war years, Soviet intelligence also stole Allied nuclear secrets from the Manhattan Project through the Julius Rosenberg spy ring, Harry Gold, and nuclear scientists Klaus Fuchs and Alan Nunn May.

In postwar Great Britain, Soviet HUMINT acquired an important asset in MI6, George Blake, and in Geoffrey Prime of the Government Communications Headquarters (GCHQ). In the United States, Soviet HUMINT had by 1960 penetrated the National Security Agency (NSA), whose primary mission is to protect U.S. codes and break those of other nations, through three agents: Bernon F. Mitchell, William H. Martin, and Staff Sergeant Jack E. Dunlap. From 1968–86, Soviet intelligence also penetrated U.S. naval codes through the spy ring of Chief Warrant Officer John A. Walker. Soviet HUMINT obtained data on U.S. spy satellite systems through Christopher Boyce, an employee of TRW, and his friend, Andrew Daulton Lee, as well as from a former CIA employee, William Kampiles. Soviet intelligence used a Canadian academic, Hugh G. Hambleton, to obtain NATO documents and guided the Norwegian Arne Treholt to steal secret documents from his Foreign Ministry.

HUMINT in the 1980s and 1990s

Ronald William Pelton, a former NSA employee, volunteered his services to the Soviets in January 1980. Pelton had a photographic memory and was able to dredge up enough material to compromise five NSA (SIGINT) collection systems. On the other hand, U.S. intelligence officers found it increasingly difficult during most of the 1980s to obtain information from foreigners, because the latter feared their names could be deduced from documents obtained from the U.S. government under the Freedom of Information Act. U.S. intelligence officers in the Soviet Union met increasing difficulties, since the

KGB has arrested and executed their agents in Moscow through the information supplied by the CIA traitor, Edward Lee Howard.

However, two major developments occurred in the late 1980s and early 1990s which may have a mixed effect on Soviet HUMINT capabilities. First, when the Soviet armed forces began to leave East European nations in 1990, intelligence services became independent of Soviet control, thus draining HUMINT resources from the KGB and GRU. Second, as the Communist Party of the Soviet Union lost control of the government, new political forces called for the reform of the KGB, a factor which may enable the KGB to transfer some resources from domestic security to clandestine collection abroad, perhaps strengthening the Soviet HUMINT capability. On the American side, intelligence collectors have received the additional tasks of aiding in the interdiction of narcotics coming into the United States and the penetration of terrorist groups.

KENNETH J. CAMPBELL

SEE ALSO: Cold War; Covert Action; Deception; Disinformation; Intelligence Collection.

Bibliography

Andrew, C. 1985. *Her Majesty's secret service.* New York: Viking.
Andrew, C., and O. Gordievsky. 1990. *KGB: The inside story.* New York: Harper Collins.
Bidwell, B. W. 1986. *History of the Military Intelligence Division, Department of the Army General Staff: 1775–1941.* Frederick, Md.: Univ. Publications of America.
Breckinridge, S. 1986. *The CIA and the U.S. intelligence system.* Boulder, Colo.: Westview Press.
Dorwart, J. 1979. *The Office of Naval Intelligence.* Annapolis: U.S. Naval Institute Press.
Hopple, G. W., and B. W. Watson. 1986. *The military intelligence community.* Boulder, Colo.: Westview Press.
Ranelagh, J. 1986. *The agency: The rise and decline of the CIA.* New York: Simon and Schuster.

INTELLIGENCE, IMAGERY

Imagery intelligence (IMINT) is obtained by describing, locating, and identifying objects, activities, installations, and terrain as represented on film or electronic display devices. An outgrowth of aerial reconnaissance—particularly of photography and the interpretation techniques developed to support it in World Wars I and II—IMINT has been evolving rapidly with each technological development of the space age. Present-day platforms for remote sensing (near-infrared, infrared, radar, etc.) enable the acquisition of data about objects or phenomena without physical contact. The advent of this technology has afforded an opportunity for numerous and diverse research efforts conducted in and among a variety of disciplines.

Early Developments

As early as 1794, French observers in the Battle of Fleurus used reconnaissance balloons to track the movement of Austrian troops. But perhaps the first actual airborne imagemaker was the American artist George Catlin, who painted the 1827 *Topography of Niagara* from a balloon, providing a remarkable near-vertical view that closely resembled later aerial photographs.

Airborne photography seems to have originated with the Parisian photographer Gaspard Felix Torunachon (working under the pseudonym Nadar). In 1858, at an altitude of 80 meters (260 ft.) above the Val de Bievre on the outskirts of Paris, he took the first aerial photograph, but it did not survive. On 13 October 1860, American photographer James Wallace Black ascended twice above Boston in a balloon, *The Queen of the Air.* Drifting at 370 meters (1,200 ft.), Black exposed six photographic plates, one of which survived. Photo interpretation was provided in the July 1863 issue of *Atlantic Monthly* by Oliver Wendell Holmes, the future jurist, who himself was a photographer and student of Black.

The American Civil War provided substantial impetus to technical imagemaking from balloons, although not photographic imagery as such. Professor Joseph Henry, secretary of the Smithsonian Institution, sent Thaddeus Sobieski Coulaincourt Lowe, a 29-year-old Ohio aeronaut, aloft on 18 June 1861. Lowe soon ascended with telegraphic equipment lent by the American Telegraph Company and transmitted descriptions of the encampments outside Washington. In other ascents, Lowe brought along Maj. Leyard Colburn of the 2d Connecticut Infantry, who sketched a map of a portion of Fairfax County, Virginia. Lowe's telegraphic dispatches and Colburn's maps were enormously useful to Union generals; in particular, Lowe's observations were described as "revolutionary" to the art of gunnery and may have marked the first instance of indirect fire by artillery.

During the rest of the nineteenth century, balloons remained the most reliable platforms for aerial reconnaissance, although inventors tried to take cameras aloft via kites, rockets, and pigeons. In June 1898, American balloonists confirmed the presence of the Spanish fleet in the harbor of Santiago, Cuba, and contributed to the victory at the Battle of San Juan Hill by directing artillery fire onto Spanish forces and delineating potential routes for the attacking American forces.

The airplane provided the next great advance in photographing the world from aloft. The first photograph known to have been obtained from an airplane was a motion picture taken in 1909 by Wilbur Wright over Centocelli, Italy. Soon afterward pilots everywhere were taking pictures of the earth with hand-held cameras.

World War I

Aerial photography leapt forward in World War I, although the technique of taking useful photographs from airplanes—especially in combat situations—required much refinement. World War I airplanes were vibrating, bouncing platforms. Cameras were mounted on the sides of the open-cockpit craft, and the slipstream made plate-changing a difficult operation. Nonetheless, during the war the British, French, Germans, and Americans developed automatic cameras that could take photographs in rapid succession over a prescribed area, without the aid of an aerial photographer. Pasted together, these photographs formed a mosaic that, properly scaled, could be made into an accurate battlefield map. Later, cameras were designed to take photos of overlapping areas. Two photos of the same area, taken from different angles, were called stereoscopic pairs; when viewed in a stereoscope, they provided three-dimensional views. Heights, widths, and lengths of objects and territory could be measured precisely.

Trench warfare played a major role in World War I, and photographic reconnaissance proved effective in identifying jump-off trenches, gun positions, and logistical areas and thereby in better predicting the timing and scale of enemy offensives. Interpreters soon established image "signatures" for specific pieces of military equipment: a set of characteristics, including size, shape, and texture, that could be measured and used to identify an object reliably. By war's end, reconnaissance pilots and photo interpreters were recognized as the "eyes of the Army." Col. Edward Steichen, commander of the Photographic Section of the American Expeditionary Forces and later famous as a civilian photographer, asserted that "at least two-thirds of all military information is either obtained or verified by aerial photography."

Between the wars, photo interpreters and photogrammetrists helped to plan transcontinental highways, chart air lanes, search for minerals and for oil and gas fields, construct large dams for electrical power, implement flood and erosion control measures, map national timber reserves, monitor the nation's crop and grazing lands, and plan the expansion of cities and towns.

World War II

The Germans, who had advanced substantially in optical and film technology, had developed outstanding military reconnaissance and photo-interpretation technology after World War I. That Germany recognized the importance of imagery intelligence in war was clearly illustrated in 1938, when General-Oberst (Colonel-General) Baron Werner Von Fritsch, chief of the German General Staff, commented, "The nation with the best photo interpretation will win the next war." The Germans had outstanding photo interpreters who analyzed aerial photographs before, during, and after blitzkrieg campaigns. (Luftwaffe photography taken of those campaigns is now available to the public at the Cartographics and Architectural Branch of the National Archives in Washington, D.C.)

But the British, too, had developed a photo reconnaissance and interpretation capability, and they enjoyed certain strategic advantages, primarily due to the efforts of a number of intrepid individuals. The most notable of these was Frederick Sidney Cotton, a color photography businessman, who had hidden cameras in his aircraft and on his frequent prewar trips to Germany had clandestinely filmed a number of German factories and military installations.

Once the war began, its demands for accurate and sophisticated information rapidly advanced the technologies, techniques, and applications of imagery intelligence. In the European theater, American photo interpreters were quickly integrated with British and other Western Allied photo interpretation units. The British established a major photo interpretation school at Dansfield House, Medmenham, a palatial mansion on the north bank of the Thames between Marlow and Henley. In January 1942, American officers trained there by the British were involved in organizing the U.S. Navy Photo Interpretation Center and School near Washington, D.C., and the U.S. Army center at Harrisburg, Pennsylvania.

Aircraft with increased speed, maneuverability, payload, and capability were being added to the reconnaissance inventory throughout World War II. Concomitant improvements were being made in aerial cameras, including continuous strip cameras, night flash photography, and infrared and radar photography. In particular, the trimetrogon camera system, combining vertical and oblique-angle pictures, was a revolutionary achievement for photographing unmapped combat areas and the complex logistical networks required to prosecute a war.

An important part of the advance in imagery intelligence was the development of photo-interpretation skill and personnel. A World War II Navy publication suggested the expanding scope of interpretation, noting, "A photo interpreter . . . must know the enemy's country economically and physically, its . . . industries, airfields, railways and other inland transportation, warships, shipbuilding, general shipping, radio, camouflage, gun installations and armored vehicles, bomb damage assessment, decoys and dummies."

Western Allied photo interpreters devised a standard system of three separate reporting phases for the photo-interpretation effort. *Flash*, or first-phase reporting, involved a quick interpretation of imagery and the "flashing" of intelligence information deemed vital or immediate that might affect tactical operations or combat. Second-phase reporting involved a close study of the photos and a report prepared on the subject of concern. Third-phase reporting involved detailed research conducted on a number of reports and preparation of a comprehensive report.

There were many photo-interpretation successes during World War II: The location of V-1 and V-2 sites, the finding and bombing of German battleships and cruisers, analysis of German radars, analysis of German jet and rocket aircraft, and the continued monitoring of the German missile and rocket installation at Peenemunde. The greatest photo-interpretation effort in that war was applied to preparations for Operation Overlord, the Normandy landings of 6 June 1944. Photo interpreters pored over thousands of aerial photos for days, pinpointing enemy beach defenses, troop dispositions, radars, and lines of communications and transportation to the beaches. It was the photo interpreters who selected, analyzed, and constantly monitored the beaches and airborne landing areas to be assaulted by Western Allied forces. More than 1,700 officers and enlisted men working around the clock studied daily takes of more than 85,000 negatives and prints. This task alone took more than a half-million man-hours.

Prosecution of the war in the Pacific presented entirely different circumstances from those of Europe. The geography was different, the enemy's base of operation was often beyond the range of Western Allied aircraft, and little reliable information was available about the islands to be assaulted. Hence, up-to-date information and maps for planning offensive operations were paramount needs. The "uncontrolled mosaic" image proved invaluable in meeting that need. To produce this mosaic, a reconnaissance plane first flew a predetermined track at a given altitude, photographing an assigned area. The film was developed, prints made and assembled, and a mosaic compiled. Photo interpreters then painstakingly studied the photos and extracted information on coastal defenses, airfields, supply dumps, communication centers, and so on.

Photoreconnaissance in the Pacific theater was critical for the conduct of an island-hopping war and assisted in the preparation of detailed studies about approaches and egresses for amphibious landings. Three-dimensional terrain models derived from aerial photography were used to brief aviators and ground and naval commanders prior to a landing. Photo interpreters also developed methods to determine underwater depths for invasion planning and provided information to the newly organized U.S. Navy Underwater demolition teams for the removal of mines, tetrahedrons, posts, wire, and a variety of other enemy-placed obstacles designed to impede beachhead landings. Massive photo-interpretation efforts were launched to support the invasions of the Philippines and Okinawa and in preparation for the invasion of Japan.

In the Pacific air battle, photoreconnaissance prepared target charts for aviators. Photo interpreters became so adept at locating and identifying enemy antiaircraft weapons that they also were often called upon to help determine the safest target approaches and departing routes for strike aircraft. Another important function performed by the photo interpreter was the preparation of bomb damage assessment (BDA) reports. These determined how many bombs actually hit their target and assessed their effectiveness.

The Postwar Period

After the war, President Truman authorized a massive effort by the armed forces to conduct the U.S. Strategic Bombing Survey, an evaluation of the bombing effects of the war. Its conclusion: "Although viewed with indifference and skepticism at the beginning of World War II, aerial photography ultimately became one of the most important single sources of intelligence in the Pacific War. [It played] an important part in more phases of military and naval operations than any other sources."

World War II not only greatly advanced the art and science of aerial photography and photogrammetry but also spurred the development and use of remote sensing devices, including color infrared, thermal infrared, and radar systems. In the postwar period, these detectors were applied to the development of airborne optical-mechanical scanners, radiometers, and spectrometers.

The Cold War

Successive Soviet technological breakthroughs in the immediate postwar period—among them the detonation in 1949 of a nuclear device—provided ample evidence that the USSR would become the principal military, political, and economic adversary of the United States. The U.S. intelligence community was perturbed over its ignorance of the details of Soviet nuclear capabilities and over repeated failures to procure the needed strategic intelligence about the Soviet Union. In late 1953, a U.S. military attache observed a heavy jet bomber (later designated the Bison) at a test and experimental airfield near Moscow. The Soviets clearly possessed both atomic weapons and the capability of delivering them to targets in the United States, and concern escalated at the highest levels of U.S. government.

On 27 March 1954, at a meeting of the Science Advisory Committee of the Office of Defense Mobilization, President Eisenhower voiced the need for better and more current intelligence on Soviet strategic capabilities. The president asked Dr. James Killian, Jr., "to direct a study of the country's technological capabilities to meet some of its current problems," particularly intelligence problems. Dr. Edwin Land was appointed to chair the Technological Capabilities Panel, which concerned itself with the development of film, lenses, cameras, and the platforms on which to position them over prescribed targets. The platform for the ultimate in reconnaissance, the U-2 aircraft, was designed by Clarence L. "Kelly" Johnson, and production was undertaken by Lockheed.

At the same time, the U.S. Central Intelligence Agency (CIA) hired a World War II photo intelligence expert, Arthur C. Lundahl, away from the U.S. Navy to establish its advanced photo-interpretation center. Lundahl was given a free hand in selecting his cadre of personnel. His admonition to them was simple: "to observe the inaccessible, to make that that is not known perceptible, and to do it in a humble and teaching manner."

THE ADVENT OF AERIAL SURVEILLANCE

The United States' increased capability for IMINT enabled President Eisenhower to suggest at the Geneva Summit Conference of 21 July 1955 that the United States and the Soviet Union negotiate a program of mutual surveillance "to ease the fears of war." Eisenhower suggested that each country should allow the aerial photographing of the other's territory and that they also provide each other with blueprints of all military installations. The proposal, labeled "Open Skies" by the press, was rejected by the Soviets, and Eisenhower subsequently approved what he characterized as "unilateral inspections for peace": operational U-2 missions over the Soviet Union.

The first U-2 flight over the Soviet Union took place on 4 July 1956. Subsequent missions criss-crossed the Soviet Union. Within a few months, these intelligence-gathering missions dispelled the rumor of the "bomber gap," "missile gap," and "megatonnage gap," revealing that Soviet capabilities were less advanced than had been feared (Fig. 1). On 1 May 1960, Frances Gary Powers's U-2 was downed near Sverdlovsk in the Soviet Union. An incensed Khrushchev demanded, but did not receive, an apology from President Eisenhower, thus causing a collapse of the Paris Summit Meeting on 16 May.

U-2s again proved central to international military affairs when they provided high-level reconnaissance pictures of what appeared to be offensive missile sites under construction in Cuba, precipitating the Cuban Missile Crisis of 1962. Imagery intelligence—particularly photo interpretation performed by Arthur C. Lundahl and the National Photographic Interpretation Center—played a crucial role in the crisis. Lundahl recalls his briefing of President Kennedy on the offensive missile sites:

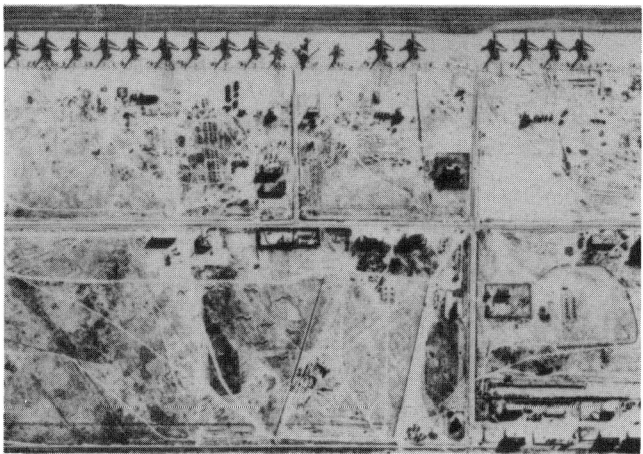

Figure 1. Photo of Soviet bombers, taken during U-2 flight in the 1950s. (SOURCE: U.S. Central Intelligence Agency)

I placed the enlarged photos in front of the President, along with a magnifying glass, then stood behind him and pointed out the highlights over the President's shoulder. I showed him the various pieces of equipment that supported the medium-range missiles; he looked up from the U-2 photos, turned in his chair, and said, "Are you sure of this?" I replied, "Mr. President, I am as sure of this as a photo interpreter can be sure of anything and I think you might agree that we have not misled you on the many other subjects we have reported to you."

The interpreted photos were shown on the floor of the United Nations General Assembly and did much to defuse the immediate threat of nuclear war and to ensure the eventual acceptance of photo intelligence as a technical means of monitoring international truce and peace agreements.

A decade later, reconnaissance and photo interpretation were called upon to play a significant role in monitoring a peace agreement. After the Arab-Israeli War of October 1973, the United States proposed that U-2s could be used to monitor the Egyptian-Israeli truce. U-2 missions proved so successful that an agreement between Egypt and Israel on 1 September 1975 provided for the continuation of aerial reconnaissance missions every seven to ten days over designated areas.

RECONNAISSANCE SATELLITES

During the 1960s, satellites again revolutionized the collection of intelligence information. Remote sensors placed in space by the National Air and Space Administration—especially the Earth Resources Technology Satellite (ERTS) and later the Landsat—were able to detect, record, and transmit details that cannot be seen by the human eye alone. Sophisticated imagery obtained from classified satellites has contributed to the confidence that a disarmament treaty could be monitored by "national technical means of verification."

The success of the 1991 Gulf War prompted Les Aspin, Chairman of the House Armed Forces Committee, in an address to the American Institute of Aeronautics and Astronautics on 1 May 1991, to state:

In the Gulf War our commanders knew where the friendly forces were and where the enemy was to an unprecedented degree. This was the result of a number of things: spy satellites, AWACS, J-STARS, airborne radars, global positioning satellites, aerial reconnaissance—all of these things put together allowed us to have an enormous knowledge of what was going on across the whole battlefield.

The satellite intelligence revolution meant that interpreters had to be sensitized to the new forms and patterns recorded on imagery acquired by these sensors. In fact, by the 1960s, the description "photo interpreter" had become anachronistic, and a new term, *imagery analyst*, was devised to convey the broad spectrum of image-producing technologies available to interpreters.

In addition to their uses for military and arms control intelligence, air- and space-based imagery collection techniques have been instrumental in discovering and measuring the earth's resources and in monitoring the dynamic and often harmful processes by which humans are damaging the biosphere. IMINT has turned outward as well: lunar and planetary probes armed with cameras and sensors have brought back volumes of important scientific data. The future holds even greater opportunities for increased knowledge, because sensor-collected imagery can now be digitized, supplementing expert human imagery interpretation with computer technology.

DINO BRUGIONI

SEE ALSO: Air Reconnaissance; Aviation, Military; Cuban Missile Crisis; Defense, Aerospace; Reconnaissance and Surveillance; Sensor Technology; Space, Military Aspects of; Surveillance and Target Acquisition Equipment.

Bibliography

American Society of Photogrammetry. 1983. *Manual of remote sensing*. Vol. 1. Falls Church, Va.: American Society of Photogrammetry.
Brugioni, D. A. 1984. Aerial photography: Reading the past, revealing the future. *Smithsonian Magazine*, March, pp. 150–61.
———. 1984. Photo interpretation and photogrammetry in World War II. *Photogrammetric Engineering and Remote Sensing*, September, pp. 1313–18.
———. 1985. Arlington and Fairfax Counties: Land of many reconnaissance firsts. *Northern Virginia Heritage*, February, pp. 3–8.
———. 1985. New roles for recce. *Air Force Magazine*, October, pp. 94–101.
———. 1986. Aerial photography and multisensor imagery: Opening new vistas on the world. *Renewable Resources Journal*, Autumn, pp. 6–10.
———. 1987. Naval photo intelligence in WW II. *U.S. Naval Institute Proceedings*, June, pp. 45–51.
Brugioni, D. A., and A. C. Lundahl. 1985. Aerial photography: World class disaster fighter. *The Information Society, An International Journal*, no. 4, pp. 327–45.
Brugioni, D. A., and R. F. McCort. 1988. The art of aerial photography. *Photogrammetric Engineering and Remote Sensing*, February, pp. 270–72.
Newhall, B. 1969. *Airborne camera*. New York: Hasting House.

INTELLIGENCE: INDICATIONS AND WARNING

All nations seek to determine the plans and intentions, especially those of a hostile nature, of their potential enemies. Warning that another nation is preparing to attack or to take some other adverse action is critical to policy makers and military commanders. History abounds with instances in which inadequate warning, or inadequate response to warning, led to disaster. Today, warning intelligence systematically seeks to construct continuous, comprehensive, and timely assessments of an adversary's plans and intentions, thus enabling responsible officials to take effective and appropriate countermeasures to forestall hostile action or lessen its impact.

Dimensions and Goals of Warning Intelligence

National leaders must anticipate a variety of dangers, not all of them military in the traditional sense. Coups or revolutions in other nations may have consequences as adverse as regular military operations; small groups of terrorists, guerrillas, or saboteurs can inflict serious damage on national interests; and even hostile trade measures may wreak havoc on a nation's economy, including its warmaking capacity. Intelligence today must deal with all of these potential threats.

As a specialized type of analysis, warning intelligence deals particularly with traditional military threats and especially with the threat of sudden or surprise attack. Yet in doing so, it must also consider other factors, especially political factors, for these may be equally important in assessing the military intentions of another country. In practice, then, warning intelligence becomes involved in forecasting conflicts almost anywhere on the globe that may directly or indirectly affect national interests.

TACTICAL AND STRATEGIC WARNING AND PREPAREDNESS

Both intelligence and military commands distinguish between tactical and strategic warning. *Tactical warning* is a short-term warning that attack is either under way or so imminent that it cannot be avoided. It is too late for intelligence analysis or new policy decisions; there is time only to alert military forces and perhaps to implement previously planned defensive measures or counteractions.

Strategic warning, the kind of warning with which national-level intelligence is primarily concerned, is any warning issued early enough that decision makers can consider the problem and respond to it. Response may be both political and military and may perhaps forestall the threat altogether. If the conflict cannot be avoided, adequate strategic warning may enable the nation to take measures that can greatly reduce the damage from the enemy's initial attack and offset the advantage of surprise. When strategic warning is totally lacking, tactical warning may also fail, either because military forces are not alert or because the tactical indications are ignored or misinterpreted.

Strategic warning and preparedness are closely linked, but they are not synonymous. Sometimes a firm warning from intelligence has not led to adequate preparedness, but military forces can be well prepared against a potential threat even when there has been no strategic warning.

Structure of Warning Intelligence

Warning intelligence as a separate specialty is a product of the cold war. In the United States, it began with the assignment of a few analysts, part-time, to the problem in the late 1940s. For 25 years after the outbreak of the Korean War, an interagency watch committee, supported by a full-time research staff and chaired by a senior CIA official, provided warning assessments at least each week for senior policy officials. Subsequent organizational changes after 1975 led finally to the appointment of a National Intelligence Officer for Warning. He and a small staff serve as the focal point for warning in the intelligence community and report directly to the Director of Central Intelligence.

Other major Western nations, particularly in NATO, have also set up procedures to facilitate systematic centralized warning, but probably none has devoted as much organizational effort or as many personnel to the problem as has the United States.

Despite the seeming centralization of the warning function in the United States, collection and analysis is diffused throughout the intelligence community and the military commands. All major agencies and military commands have alert, after alert, operations, or indications centers, and all are free to make their own threat assessments. In addition, the National Security Adviser and his staff, and other senior political and military officials, with their direct access to the president, may often be the most influential voices in warning.

For warning, the bureaucratic structure and the chain of command are less important than the assignment of experienced, dedicated analysts to the problem, competent and imaginative supervisors, and free communication between intelligence and policy makers.

Relation to Other Intelligence Specialties

Despite its importance, warning intelligence does not have high status compared with other intelligence specialties or subdisciplines. Because warning problems tend to be infrequent and extraordinarily diverse, analysts generally do not work on them exclusively in peacetime. There are thus relatively few warning specialists, and efforts to develop methodologies to assist analysts and decision makers in this difficult problem have been of only marginal value. Still, studies by historians and social scientists have shown that the same types of problems and errors recur in warning for virtually all nations. Thus, the potential for improving assessments would seem to exist in a rigorous training of analysts from many specialties in the lessons of history.

Of necessity, warning intelligence often overlaps or draws on other specialties. For example, the primary analysis of information on day-to-day developments is by so-called current intelligence. Similarly, the primary producers of long-term assessments of foreign military establishments are the basic intelligence and estimative processes.

Objectives

Although the lines between warning intelligence, estimates, current intelligence, and basic intelligence are somewhat ill-defined, warning intelligence still has crucial and specialized objectives.

Warning intelligence specifically, and largely exclusively, seeks to discover any information—however tenuous or fragmentary—that may suggest that an adversary is preparing for a hostile or surprise action. Fragments of warning information, usually called *indications*, may be military, political, economic, or technical. They may be high-grade data from reliable sources or low-grade information that has not been and may never be confirmed; reports whose significance, even if confirmed, is obscure and ambiguous; and data whose relevance to the enemy's intention is uncertain or in dispute.

Warning intelligence's primary objectives are to deal with indications in depth and to undertake a cumulative analysis of information over a period of days, weeks, and months to anticipate hostile action. Warning intelligence thus serves as a watchdog or devil's advocate, constantly prodding other intelligence elements to reexamine their data and be sure that nothing has been overlooked.

Processes and Problems of Warning Intelligence

Warning intelligence uses a variety of specialized procedures to analyze and disseminate intelligence, and it must confront a host of significant and sometimes unique problems.

Processes

Regardless of how an intelligence service is organized to cope with warning, the processes involve essentially the same interrelated steps or procedures.

Collection. The first and overriding responsibility of intelligence in any threatening situation is to ensure that collectors have been alerted and are functioning at maximum effectiveness. In fact, the collection of reliable and up-to-date information is obviously so vital to warning that many believe inadequate collection is the primary—or even sole—cause of warning failures. This is only partially true. Clearly, if collectors could have total access to an adversary's preparations, they could provide almost certain warning. Yet because this cannot be expected, and because time is always on the side of the aggressor, intelligence agencies cannot rely solely on collection.

One important task of collection is to provide as much information as possible concerning the sources of incoming data so that analysts are better able to weed out extraneous data (sometimes called *noise*) and disinformation and concentrate on the most reliable material. This is particularly true in a crisis, which usually leads to a vast increase in incoming data, much of which cannot be evaluated immediately or comes from obscure sources.

Selection of indications. From the vast quantities of incoming data, the warning process seeks to identify items that bear specifically on intentions and to discard the irrelevant. Indications are usually classed as *positive* (pointing toward the likelihood of hostile action) or *negative* (indicating a decreased likelihood of such action or revealing that an expected preparation has not been made). A lack of information should not be confused with a true negative indication.

Military data or indications are the primary element in warning. A certain minimum of preparation is essential for war, and in practice nations usually undertake a wide range of preparedness measures—for example, partial or full mobilization; redeployment of units; raising of combat readiness; extensive logistic preparations; readying of command, control, and communications; and military and civil defense precautions. Many military preparations are physically observable and to some extent measurable or quantifiable.

Warning intelligence seeks to identify in advance the specific key preparedness measures the adversary may be expected to take. Warning indicators may be derived from logic or history, from analysis of the adversary's military doctrine and exercises, or from a study of the enemy's behavior in past crises or conflicts. Warning concentrates not only on the number of positive indications but also on certain key actions that distinguish routine or relatively normal activity from actual combat preparations. Although many indications are ambiguous individually, some— indications of "high diagnostic value"—are critical to recognition of hostile action. The compiling of indications is thus both quantitative and qualitative.

Intelligence must also consider political indications, because the determination to wage war is fundamentally a political decision of the national leadership. Political indications are often difficult to pin down. Political preparations for war are generally not quantifiable and are much less predictable. Moreover, some of the most important political preparations and decisions can be totally concealed, and political deceptions are much easier to carry out than military deceptions.

Assessment of cumulative evidence. The diversity, subtlety, and complexity of warning indications means that warning intelligence cannot be a mere compilation of facts or a simple weighing of positive against negative indications. More facts or more positive indications do not necessarily provide more warning. In fact, too much positive evidence may suggest that the enemy is bluffing or seeking to gain an objective by threat or blackmail.

Evaluation of indications is a highly subjective, inferential, and inductive process. Individual indications may be contradictory and of varying reliability; it may not be possible to determine their validity, and the evidence will certainly be incomplete. Since many of the "facts" cannot be established with certainty, the inferences drawn from them are nearly always controversial. Often, individuals who have examined the same evidence as objectively as they can will come to opposite conclusions about what it signifies. And even if an intention has been correctly perceived, there is always the possibility that the adversary's leaders will change their minds.

In a crisis situation, an adversary almost certainly will be taking unusual steps to conceal its preparations; because time is critical, indications analysts must consider information that normally would not meet criteria for acceptance. For example, if intelligence discards unconfirmed but persistent reports of large-scale troop deployments (too low grade individually to be accepted by order-of-battle analysts), it will probably underestimate the threat.

The threat of hostile action calls for the imaginative integration of data from a wide variety of specialties, coupled with careful and reflective evaluation by experts. Otherwise, obscure but significant clues may be overlooked. For example, because military forces often are more directly responsive to the political leadership during a crisis, the import of military moves may be misinterpreted unless the political and military indications are considered as a whole and interwoven into a meaningful pattern.

Judgments for decision makers. Wars may result from long-planned deliberate aggression by a militaristic state; from long-festering border disputes; from nationalistic desires to regain territory or to unify a country (*irredentism*); from deep-seated antagonisms; and for ideological, religious, or economic reasons. They range from small-scale guerrilla operations to massive attacks. They may be planned in secret over months or years, or they may follow a period of rising tensions over some development or issue and involve a series of actions and reactions. Both sides may actually be seeking to avoid conflict, but miscalculations or misperceptions may culminate in a conflict that might have been avoided.

Decision makers obviously would like firm and unequivocal warning judgment of what an adversary will do, but the tenuous and often contradictory nature of the available evidence and the inherent uncertainties in assessing how foreign leaders may behave make such judgments difficult. Warnings, however phrased, generally must be qualified judgments or probabilities. Research and experience have shown that firm judgments are particularly difficult if there is a recently accepted estimate to the contrary; if the prevailing climate of opinion holds the action to be unlikely; if there are sharply differing opinions that intelligence minimizes to obtain a single, unanimous (but weak or ambiguous) judgment; and if the news is potentially so bad that there is reluctance to alarm the decision maker in the absence of firm proof.

Yet if it is rarely possible to arrive at a clear-cut, definitive assessment, intelligence still has a responsibility to provide the decision maker with the clearest possible assessment of the threat, particularly a realistic description of the adversary's military capabilities. Thus, when warning intelligence perceives a growing possibility of conflict, it will likely issue a series of assessments designed to keep policy makers and military commanders abreast of the situation and to apprise them of changes in the enemy's capabilities and apparent intentions.

Action by decision makers and commanders. Warning differs from most other aspects of intelligence in that it is useless unless it is heeded by responsible officials and results, when possible or desirable, in some action to forestall the threatened hostile action or reduce its impact.

Intelligence is only one of many factors that decision makers take into account in planning policy, and rarely is it the most important. Moreover, intelligence is not held in high esteem by many policy makers (and often by military commanders as well), most of whom consider themselves at least as competent, if not more so, to come to judgments about what their adversaries will do. They receive much (although far from all) of the same daily flow of information as does intelligence; also, they are often privy to other, tightly held information such as direct communications from the adversary's leadership and other heads of state, which affords them unique insights unavailable to their intelligence services. (These channels may also be used for high-level deception.) In a crisis, what decision makers usually want most from their intelligence is up-to-date and reliable basic data (especially accurate descriptions of the strength, dispositions, and equipment of opposing forces).

Occasionally, decision makers may perceive a threat more accurately than their intelligence services; they may disbelieve or disregard warnings from intelligence; or both may be essentially right, or wrong, or partially wrong. Even when the danger of hostile action has been correctly perceived, there may be valid reasons to limit or defer countermeasures, especially if they might appear provocative to the adversary and lead the latter to preempt; if they would be unduly alarming or expensive domestically; or if there is still a prospect that the issue can be resolved peacefully.

In the face of a clear military threat, however, political and military leaders can rarely afford to do nothing. Whatever the assessment of the adversary's intention may be, there is a time-honored military precept that the commander should prepare against the enemy's military capability.

Problems

A brief description of some of the inherent uncertainties and difficulties that plague the warning process from beginning to end will contribute to a better understanding of why threats are so often underestimated and why there are so many warning failures.

Gaps in collection. Even when collection is generally good, some critical military information will almost certainly be lacking. It is most unlikely that sufficient knowledge of the adversary's crucial decisions will be available to permit a definitive judgment as to whether attack is imminent, is still being debated, or has been deferred or rejected. Where collection is poor, basic intelligence on the adversary's order of battle may be limited and insight into decision making totally lacking. (This was the case in North Korea's attack on South Korea in June 1950.)

High volume of data. Even when critical data are lacking, a perceived crisis usually generates a flood of information. This vastly complicates the analytic problem and is a prime reason why potentially important information is "lost" or not considered in the assessment process.

Overreliance on highly classified sources. Even when these are unproductive or too slow to provide the needed data, many analysts are reluctant to credit less classified and readily available information until it can be confirmed by their time-tested reliable sources. Valuable and timely information thus may be ignored as being of insufficient quality.

Preconceptions. Time and again, nations have misjudged how their adversaries will act, believing that a resort to force would be "illogical," counterproductive, or unlikely to solve the problem. Many scholars of analysis believe that such preconceptions, which can blind analysts and policy makers alike to an objective examination of the evidence, are the single most important cause of warning failures. Sometimes, for example, the misjudgment results from "mirror-imaging," believing that another nation would act as we would.

Misperceptions of an adversary's commitment. Another common error in warning intelligence is the failure to appreciate the determination of another nation to obtain its objectives by any means, including force, if other means fail. An adversary's willingness to run high risks thus may be seriously underestimated. (In the Cuban missile crisis, for example, both the United States and the Soviet Union misjudged the other's commitment.)

Inadequate examination of the evidence. It is almost impossible to overemphasize the importance of meticulous and exhaustive research for warning. Retrospective analyses nearly always uncover numerous indications that were either overlooked or misevaluated. Inadequacies in research can result from many factors: failure to allocate sufficient or appropriate personnel to the problem; poor methodology, including inadequate attention to tenuous and seemingly low-grade data; failure to integrate military, political, and other data into a meaningful pattern; and poor communication among various research offices.

Even if the analysts have produced the requisite information and assessments, it will be useless unless the committee or working group that is to render judgment is also willing to undertake an exhaustive review of the evidence before reaching its conclusions.

Insufficient time. The greatest impediment to effective warning is often simply a lack of time—time to collect the needed data, to confirm it, to analyze it, to come to considered assessments, to warn the policy maker and military commander, and to mobilize and deploy defensive forces. The problems of warning, complex at best, are immeasurably compounded by the requirement, in many cases, to make judgments and to take appropriate countermeasures before crucial information is available and before the situation can be accurately assessed. Yet if appropriate countermeasures have not been taken and an enemy's forces are in a high state of readiness, the enemy may have the advantage of tactical surprise. (This was the Israelis' dilemma in October 1973, when they had only twelve hours' definitive warning that the Arabs would attack—too little warning to fully mobilize or to prevent the initial successes of the Egyptian forces.)

Compartmentalization and breakdown of communication. Important information may fail to reach those who need it. The reasons are manifold: overloading of the system; poor coordination among analysts or between intelligence and policy officials; or breakdown of communications facilities. Problems also arise from intentional compartmentalization of information—for example, information from particularly sensitive or diplomatic sources may be withheld from both intelligence and military personnel or military operational data may be denied to intelligence personnel. Any such communications gap impedes the warning process and impairs the preparedness of military forces. (It is generally conceded that such problems contributed to the U.S. lack of preparedness at Pearl Harbor.)

Underestimates of military strength and capabilities. Underestimation nearly always plays a part in rendering a nation open to successful surprise attack, in part because of the inherent difficulties of collecting intelligence under stringent time constraints. Because collection of accurate data is disrupted when forces are mobilizing or in motion, order-of-battle estimates are usually too low—sometimes much too low—or too long delayed. Even if a nation's overall estimate of enemy strength is fairly accurate, the adversary is often successful in deploying overwhelming strength at the point of attack.

Underestimates of the capability of supposedly inferior or poorly equipped forces are also not uncommon, nor are assumptions that superior technology will prevail (e.g., U.S. assumptions about the Chinese forces in Korea in 1950 and about the Viet Cong in the 1960s).

Timing of attack. Contrary to what many believe, predicting the timing of an attack is one of the most difficult and elusive aspects of warning. For a variety of political and military reasons, offensives may not occur on what is perceived as a "logical schedule"; the attack may be advanced or, more likely, delayed, sometimes for weeks or more. Particularly if the attack is repeatedly postponed or appears to be delayed after military forces are judged to be ready, there may be numerous false alarms that lead to disbelief in the final, authentic warning (the "cry wolf" syndrome), as in Hitler's repeatedly postponed offensive against Western Europe in 1939–40. Or there may be a general relaxation of vigilance or a rising expectation that attack will not occur at all, as in the seeming delay in the Chinese offensive in Korea in the fall of 1950 and in the Soviet invasion of Czechoslovakia in August 1968.

Weak or ambiguous judgments. Confronted with uncertainties and conflicting opinions, intelligence often seeks a compromise between those who consider the attack likely and those who do not. The result may be a weakly worded judgment that glosses over the differences, downplays the military threat, and fails to convey any clear warning to policy makers. Sometimes, also, if there is no demand for a new assessment, intelligence reporting may consist largely of a roundup of new information without any judgments about what it may signify.

If military analysts permit their preconception that an adversary will not attack to override their preparation of a clear and comprehensive assessment of capabilities, they may seriously undermine the value of collected intelligence. Even when intelligence analysts fail to reach agreement on the probability of the action or lack the resources to perform a full analysis, they may assist policy makers by providing a forthright and unvarnished description of the adversary's military strength and capabilities, which may carry an implicit warning.

Security, deception, and disinformation. Possibly the most unpredictable factor in warning is the potential effect of enemy deception and disinformation. All nations—even the successful practitioners of the arts of deception—appear vulnerable to ruses that have been used time and again. Even when strategic warning of the likelihood of attack has been ample, deception can be highly successful in misleading an adversary about the time, place, and strength of the attack. Probably the most notable example was the brilliantly successful Allied (primarily British) deception effort against the Nazis in the Normandy invasion in June 1944.

The Future of Warning Intelligence

Two major developments of recent years will profoundly affect the nature and quality of warning intelligence in the future.

First are the enormous improvements in collection and

international communications. The advent of high-quality satellite photography now permits intelligence to provide the commander a reasonably accurate and up-to-date assessment of another nation's normal order of battle, recent troop movements, and major logistic preparations. While such information is directly available to only a few major powers, many other nations are the indirect beneficiaries of this great technical achievement. All major nations also have greatly improved the speed of communications and processing and dissemination of other information. At the same time global communications, particularly international television, have opened up much of the world on a real-time basis, providing both decision makers and the public with immediate and constant coverage of many events abroad.

The second major development is the dramatic change in the once-communist world. The seeming threat of possible major conflict in Europe, which preoccupied warning intelligence for 40 years, has virtually vanished, but the disintegration of the Soviet Union has opened up dangerous prospects for instability and new types of conflict that could make the future more hazardous than the past.

Further, these changes do not mean that the warning problem has been solved or that the assessment of intentions is necessarily simplified. In the Kuwait crisis in 1990, accurate assessments of Iraq's high capability to invade did not result in a generally accepted judgment that it would do so. The relative openness of Soviet society in August 1991 and the presence in Moscow of many foreign diplomats, delegations, and news services did not provide a clue that a coup was imminent. Once it was under way, the superb television coverage was invaluable to Western decision makers, but the most expert and experienced political analysts could not, for at least 24 hours, even hazard a guess as to the likely outcome.

Thus, the main problem facing warning intelligence—which is to assess what foreign nations and leaders *are likely to do*, not just their capabilities—remains essentially the same. The world contains many unstable nations, messianic political leaders, and potential areas of conflict that will continue to challenge the best efforts of warning intelligence and decision makers in the years to come.

CYNTHIA M. GRABO

SEE ALSO: Deception; Ruses and Stratagems.

Bibliography

Betts, R. K. 1982. *Surprise attack: Lessons for defense planning.* Washington, D.C.: Brookings.
Daniel, D. C., and K. L. Herbig, eds. 1982. *Strategic military deception.* New York: Pergamon Press.
Herzog, C. 1975. *The Arab-Israeli wars.* London: Arms and Armour.
Hinsley, F. H., et al. 1979–88. *British intelligence in the Second World War: Its influence on strategy and operations.* 4 vols. New York: Cambridge Univ. Press.
Jervis, R. 1976. *Perception and misperception in international politics.* Princeton, N.J.: Princeton Univ. Press.
Knorr, K., and P. Morgan, eds. 1983. *Strategic military surprise: Incentives and opportunities.* New Brunswick, N.J.: Transaction Books.
May, E. R., ed. 1985. *Knowing one's enemies: Intelligence assessment before the two world wars.* Princeton, N.J.: Princeton Univ. Press.
Wohlstetter, R. 1962. *Pearl Harbor: Warning and decision.* Stanford, Calif.: Stanford Univ. Press.

INTELLIGENCE, MILITARY

Military intelligence (hereafter generally referred to as intelligence) is a primary function in war; it runs throughout all command echelons and the entire conflict spectrum. Intelligence is absolutely essential for the rational conduct of war, and as such is one of the oldest functions in the political/military arena. Military intelligence has an offensive side, usually called intelligence, and a defensive side, usually called counterintelligence.

Before defining *intelligence*, it is necessary to define two other terms as they will be used in this article: *war* and *military forces*. *War* is used here as Clausewitz defined it, "the mere continuation of policy [politics] by other means" (Clausewitz 1968, p. 119). It is important to recognize that what is being considered is armed conflict that supports political objectives. *Military forces* are defined as unique organizations created and maintained by a society for the primary purpose of threatening to apply, or actually applying, destructive force against other societies (normally concentrated on their military forces) in support of political objectives.

Intelligence is both a product and a process and, at a minimum, is the systematic, planned, and objective-oriented (i.e., nonrandom) collection, analysis, and dissemination of processed information from either open or denied sources, performed either openly or clandestinely. The very fact that such activity is being undertaken may be either openly acknowledged or kept secret. In all cases the key difference between intelligence and information is the processing involved. Data that has not been processed—that is, has not been analyzed in some fashion—is information, not intelligence. Intelligence is a continual process of mutually supporting, repeating cycles (collection-analysis-reporting) that must be directed, as production is not automatic. The denial of select information to unauthorized persons is both security and counterintelligence. The clandestine collection of information is a proper intelligence function, but covert operations and deception are not intelligence functions. Covert operations are a specific type of operation where a primary objective is to keep the action organization's identity secret, either by maintaining absolute secrecy or by deliberately identifying another organization as the actor. Although covert operations are often associated with the intelligence community, it is important to remember that clandestine collection and covert operations are completely different; only the former is a proper intelligence function.

Deception is a second major military function that is often erroneously considered an intelligence function. Deception is a function planned and executed by the operations officer. In essence, deception is the plan (i.e., the deliberate intent) to take advantage of an enemy's specifically induced or influenced behavior. The inducement of desired behavior requires using intelligence, but intelligence is a supporting element, not the executing element.

General Definition

The term *intelligence* has been misunderstood and misused throughout history because people often define intelligence based more on their own background than on a real understanding of the whole.

OFTEN DEFINED BY PROCESS

Of the three intelligence processes—collection, analysis, and dissemination—the most common definition is based on the most widely discussed: collection. Collection is not intelligence, any more than the other two taken separately are intelligence. Collection itself is only the beginning, although it is the most intriguing and in the popular media the most often portrayed. Spying and espionage are thus erroneously popularized as intelligence. In fact, spying is a form of collection that involves the use of humans as the collecting mechanism; espionage is the act of spying when obtaining denied (i.e., protected) information. The obtaining of this information is also called clandestine collection.

Spies come in two general categories, covert and overt. The covert spy is the type most often thought of in conversation and portrayed in movies and novels. This type of spy is most successful when never detected, and when retired in total anonymity.

While espionage was for centuries the primary intelligence collection method, a new set of systems has clearly become dominant. Much of the imagery, including new and private corporation multispectral capabilities, is based on satellites. While they are expensive to build and operate, satellites have such advantages in terms of access that significant amounts of money and numbers of personnel are now devoted to their development and operation, and the exploitation of their products. Because of this cost and the extraordinarily important access of satellite collectors, intelligence is a significant contributor to space's role as a new theater of military operations.

The second intelligence process, analysis, is rarely used to define intelligence, especially by the public media, because analysts do not make operational decisions, either political or military. The operational decision and the movement of political or military forces are the dramatic efforts. The analyst's contribution is not dramatic enough to capture the public's imagination nor is it the result. Only rarely are intelligence analysts and analyses a major element in the popular media. Two notable film exceptions are the movies *Midway* and *Sink the Bismarck*. In *Midway*, the supremely important work of naval cryptanalysts and order of battle analysts is shown clearly. In *Sink the Bismarck*, the key roles of intelligence analysis and dissemination are shown quite well.

The most common portrayal of analysis is in the detective genre. The detective slowly pieces together the puzzle, but he also has two key advantages over the intelligence analyst. First, the detective performs all three parts of the intelligence process—collection, analysis, and dissemination. Second, the detective is able to take direct action on his intelligence, the chase and arrest, as he is also an operational decision maker.

This third element, dissemination, is the generally forgotten element. This is a major problem, as no matter how thoroughly information is collected and analyzed into intelligence, if the results cannot be transmitted to the proper recipient in a timely fashion, the entire process has been a waste. Dissemination is largely a communications problem, and today this means electronic communications using satellites. In this manner, intelligence has again greatly contributed to space's role as a new theater of military operations.

OFTEN DEFINED BY SOURCE

Intelligence is often defined in terms of source: for example, human intelligence (HUMINT), signal intelligence (SIGINT), and imagery intelligence (IMINT). This definition by source is as misleading and incorrect as defining by process. The three "INTs" are not in themselves intelligence; they are primary sources of information, and as such are primary areas of collection activity. Until analysis is performed, all three provide only potentials that may become intelligence. That is to say, the information contained in an image or in an intercepted signal, or gathered by a human collector, is not yet "intelligence."

OFTEN DEFINED BY CONSUMER ECHELON

Intelligence is often defined in terms of the level of the primary consumer: strategic, operational, or tactical. Strategic refers to that which can in itself dominate a country's overall military situation, cause a major shift in military conditions, and win or lose a war. Operational is the next level down and refers to what is commonly called the theater in geographic terms, and the campaign in operational terms. Tactical is the lowest element and includes everything below the operational level. If one were to look for a historical example, the American Civil War provides all three. The Federal Anaconda Plan, with the objective of physically isolating the Confederate States by blockading the entire Confederate coastline and controlling the Mississippi River, was strategic in nature. Within this, Federal operations under Gen. Ulysses S. Grant to capture Vicksburg were operational in nature, as at the time he commanded only one Federal army and was operating in one theater in order to fulfill the strategic requirement of controlling the Mississippi River. And further, within this campaign, the initial Federal assault on the city's main defenses was a tactical move involving select forces in an attempt to quickly gain the immediate objective, the capture of Vicksburg—which in turn accomplished the operational objective by eliminating the last Confederate stronghold on the river, thus contributing to the strategic objective of physically isolating the Confederacy.

OFTEN DEFINED BY SUBJECT AREA ORIENTATION

Intelligence is often defined in terms of subject area, such as political, economic, diplomatic, or military. At best, subject area orientation is a useful modifier but is in no way a definition of intelligence. Intelligence is intelligence regardless of subject area.

CORRECT GENERAL DEFINITION OF INTELLIGENCE

Having eliminated the common, but incorrect, ways intelligence is defined, the following specific definition is provided. Intelligence is both a product and a process and, at the minimum, is the systematic, planned, and objective-oriented (i.e., nonrandom) collection, analysis, and dissemination of information based on open or denied sources.

The Intelligence Process

The intelligence process, collection-analysis-dissemination, is also often called the intelligence cycle. This cycle is repetitive, planned, objective-oriented, and interdependent.

The cycle is repetitive because the need for intelligence exists before war, during war, and again before the next war. Since there is neither a guarantee that another war will not break out, nor knowledge of when and where it will occur, the cycle must continue. The cycle is also repetitive in that deterrence requires continuous intelligence activity by all interested parties.

The cycle is objective-oriented; that is, intelligence production should never be random or self-generating. Intelligence should always be produced to support decision making. This requirement includes the need to maintain databases, which are the foundation of analysis. Inherent in remaining objective-oriented is planning, but planning itself is not a separate intelligence function.

The processes of the cycle (collection-analysis-dissemination) are interdependent; each element is dependent on the other two for meaning. Collection is required to provide adequate raw material for analysis, analysis is required to transform raw material into an intelligence product, and dissemination is required to give the product utility. All three are also interdependent in that each informs the other as to its own current needs and accomplishments. Thus collection can be continued or redirected, new analysis needs can be determined, and all can know what has been disseminated.

COLLECTION

Collection is the deliberate, planned acquisition of information. The oldest form of collection is HUMINT; spying and reconnaissance are some of the oldest recorded human activities. The newest form of collection is satellite-based imagery and signal collection. This activity has excited a great deal of interest, especially as it relates to nuclear arms control verification and issues. Debate over the ability to monitor nuclear arms control agreements from space has raged for several years, and will likely continue, with much of the debate fueled by media claims on both sides.

In general, three things should be kept in mind when participating in or observing a debate involving intelligence. First, collection alone is not intelligence: analysis must be performed successfully. Second, even when adequate, accurate, and timely intelligence is produced and disseminated, a decision maker must still make a proper decision (intelligence by itself accomplishes nothing). Third, collection is best when it is multisource; some combination of open and denied sources from more than one collection element provides the best opportunity to support superior analysis and detect deception.

ANALYSIS

Analysis, often overlooked, is the central intelligence function, transforming collected information into intelligence. It is the most difficult, and therefore the most challenging, part of the cycle. The difficulty is caused by the combination of two major factors. One is a lack of information—sufficient information rarely exists in a single place; even when it does, there is usually insufficient time to recognize it and incorporate it into truly complete intelligence. A second factor is the difficulty in recognizing and separating useful information from background "noise." The successful Japanese surprise attack on Pearl Harbor in 1941 is an outstanding example of what can happen when these difficulties are not overcome; the American intelligence effort prior to the Battle of Midway is an exception that illustrates what can happen when these factors are overcome. However, intelligence alone never wins; it is the correct use of intelligence by the commander that supports winning.

One of the main reasons that analysis is often not recognized as the central intelligence function is that it is almost impossible to quantify. This limits the ability to measure analysis, and it tends to be considered a zero-sum situation (i.e., the analysis is judged either correct or erroneous). This is not a useful way to look at analysis, which is always based on incomplete or dated information. Sometimes important information is not received in time to be used. Thus, a logical and rational analysis can still draw the wrong conclusions, one of the main reasons that intelligence is often so poorly regarded.

A related reason that analysis is so often not recognized as the central intelligence function is the fact that, historically, senior decision makers have tended to believe themselves to be superior intelligence analysts. This may have worked when one was analyzing manageable volumes of information from only a few sources and the commander could see the entire battlefield. Today, however, collection systems are more complex, sources are manifold, the volume of information is almost unmanageable, and the battlefield has expanded greatly. This combination makes it impossible for senior commanders to function successfully as their own best analysts.

DISSEMINATION

Dissemination is generally the forgotten element of the intelligence process because it is primarily a communications problem. Dissemination of intelligence must compete with all other activities for communications support. The problem worsens as both operations and intelligence continue to develop more data to transmit. For example, intelligence has developed new collection capabilities, such as digital imagery, that require more capable communications systems.

Military Intelligence

Military intelligence is all the above applied to foreign countries, foreign military organizations, and the geography of possible military operations areas.

INTELLIGENCE GIVES WAR ITS RATIONAL FRAMEWORK

The fundamental reason intelligence is so important and central a capability, and therefore a primary function in war, is that intelligence is what makes going to and conducting war a rational act. There are ranges of intelligence forms and qualities; the more sophisticated the military, the more important the role of discrete, formal intelligence analysis may be in war-making decisions. It is also possible for arrogance, anger, and ideology to become overriding factors, but in the overwhelming majority of cases the final engagement decision, whether offensive or defensive, is made with some expectation of victory, and an intelligence assessment is the foundation of that calculation. This victory expectation is based primarily on some success probability calculation, with this calculation's minimum requirements being self-knowledge, knowledge of which way third parties may or may not participate, and knowledge of the enemy's military capability and probable objectives. The last two points require intelligence obtained through collection and analysis.

It should be noted that the victory expectation, while an essential element, is not normally sufficient in itself to decide to go to war. At a minimum, societies must also make some type of cost-benefit ratio calculation that accounts for such things as the war's potential internal political and social consequences. What may make perfect sense in an international context could be too expensive internally. This calculation is necessary to avoid political Pyrrhic victories.

INTELLIGENCE AS A PRIMARY COMMAND FUNCTION

Because intelligence is a primary command function central to the rational conduct of war, it has been the last primary function spun off to professional specialists. Until World War I, Western military forces did not have separate full-time professional intelligence organizations. The U.S. Army did not have a General Staff Military Intelligence Section until 1918 or an all-source integrated intelligence officers branch until the 1960s. Intelligence is designated a command function in the U.S. Army's *Field Manual 100-5: Operations* (U.S. Dept. of Army 1986), which states that "obtaining useful intelligence prior to the initiation of operations is a vital task." Even today the United States Joint Staff does not have a statutory and formally appointed full-time Director J-2 (Intelligence). The director of the Defense Intelligence Agency (DIA) is designated to serve as the J-2, and the DIA, through its directorate, which is assigned current intelligence and Joint Staff support responsibility, provides direct support to the Joint Chiefs of Staff. Additionally, the director of the DIA coordinates and functionally manages the defense intelligence budget through his position as the director of the General Defense Intelligence Budget Staff. Each of the services and all the unified and specified commands do have a senior intelligence staff officer designated.

THE INFORMATION WAR

Since purposeful war, as opposed to anarchy or criminal terrorism, implies objective-oriented, nonrandom, applied violence, such basic decisions as where, when, and how many personnel to move, when to engage and when to disengage, are dominated by intelligence. Otherwise, these basic decisions would be made in near total ignorance during war in general, and combat in particular, and would become random events unconnected to any objective. Such purposeful conduct of war has also always included the desire of commanders to control their own forces, neutralize the enemy's forces, and determine the war's outcome. Successfully accomplishing these three basics is dependent on several things, but knowledge of the enemy ranks high.

Clearly, the contribution of intelligence to campaigns has varied. For instance, Russian intelligence operations in East Prussia in 1914 were clearly inadequate, in stark contrast to the vitally important British integration of ULTRA and the Double Cross System during World War II. Additionally, the intelligence system may be either well organized and formal, or, as in George Washington's time, somewhat rudimentary. Intelligence systems often have been faulty and have produced poor-quality results. But intelligence remains the foundation of a rational conduct of war because intelligence is the source of knowledge about the enemy and the terrain, two of the three primary elements that make up the commander's situational awareness. *Terrain* in this sense refers to more than just the ground over and on which units will operate; rather, it refers to the totality of the operating environment, what the Germans call the *Umwelt*, or the surrounding world situation.

In this context there are really two wars occurring simultaneously, a physical war and an information war. The information war is continuous; it generally intensifies before the physical war, continues during the physical war, and then continues during the peace until the next outbreak of physical war. The information war is both offensive and defensive. The offensive side is intelligence (the gaining of knowledge and understanding) and deception (the planned exploitation of self-generated information deliberately provided to the enemy). The defensive side is counterintelligence and security, which deny true knowledge and understanding.

Military Intelligence's Historically Pervasive Nature

Military intelligence is one of the oldest of political/military functions. The presence and nature of all military forces throughout history have been directly influenced by military intelligence. This influence begins with the most basic decision of all, whether or not to have a military force.

THE NEED FOR AND NATURE OF MILITARY FORCES

An intelligence-based, threat-assessment-driven, self-defense requirement is one of the two fundamental reasons for having armed forces. If there is no threat, no significant military force is required. Any society satisfied with its territory and economy, and not threatened externally, historically either has no military force or has one of no real consequence. This condition is extremely rare, however, as almost all peoples or states either have had external enemies or have been dissatisfied with their condition. A few exceptions are Eskimos, Laplanders, some periods in Pharaonic Egypt, and Japan during parts of the Togukawa Shogunate. Contemporary Iceland and Luxembourg are unusual in that they have virtually no indigenous military. While they have no significant military forces themselves, they are members of a major alliance where an attack on one is considered an attack on all.

The foundation of the first reason for having military forces, a threat perception, is an intelligence-based assessment of potential enemies. This assessment, along with one's own culture, one's geopolitical situation, and the geography of potential operating areas, determines the nature of one's military force.

HISTORICAL EXAMPLES

Since organized war has been written about, intelligence has been considered a key function. And while accurate and timely intelligence has not, in itself, guaranteed victory, its absence has almost always contributed significantly to defeat. This contribution generally has been in the form of an enemy's successful surprise, and the desire to gain surprise and at the same time avoid being surprised has been a primary command objective since the beginning of organized war.

While integrated multisource intelligence operations have appeared irregularly in history (e.g., Walsingham's in Elizabethan England), the widespread development of permanent, professional, all-source military intelligence organizations is relatively new. This is rather puzzling, given that intelligence in political and military events is recorded in some of the earliest documents of civilization. Cryptology, the science of encoding and decoding information, for instance, dates back at least 3,000 years, with China the only high civilization of antiquity not known to have developed any real cryptography. The Old Testament contains numerous stories of Hebrew spying activities in Palestine, and the *Arthasastra* (321–300 B.C.), an Indian classical work on statecraft, describes widespread espionage services in India.

PROFESSIONAL RECOGNITION OF THE IMPORTANCE OF INTELLIGENCE

From the earliest times, the Great Captains, other successful practitioners of the military art, and respected military theorists have recognized intelligence as a key function of war. Some examples will illustrate this point.

Sun Tzu. Sun Tzu's first chapter in *The Art of War* is entitled "Estimates," clearly indicating the fundamental nature of intelligence in war. His second statement in this chapter is, "Therefore, appraise it [war] in terms of the five fundamental factors and make comparisons of the seven elements later named. So you may assess its essentials" (Sun Tzu 1963, p. 63). He then lists these factors as moral influence, weather, terrain, command, and doctrine. Only after this assessment is made is the state committed to war and a general appointed. This assessment is what the Soviets would call a correlation of forces and means calculation.

Sun Tzu's next major point is often quoted, "all warfare is based on deception" (Ibid., p. 66). Deception is based on intelligence and requires fooling the enemy commander, generally by also fooling his intelligence organization, in order to succeed. This applies even if the commander is his own intelligence organization. Sun Tzu further states, ". . .what is supreme in war is to attack the enemy's strategy" (Ibid., p. 77). Thus, while one may be able physically to blunder into an opponent, it is far less likely that one will accidentally attack an enemy's strategy. Attacking an enemy's strategy requires an understanding of one's opponent, and such an understanding is derived from intelligence. He goes on to make the following fundamental points:

> ... know the enemy and know yourself; in a hundred battles you will never be in peril.... When you are ignorant of the enemy but know yourself, your chances of winning or losing are equal.... If ignorant both of your enemy and of yourself, you are certain in every battle to be in peril. (*Ibid.*, p. 84)

Sun Tzu concludes his book with a chapter on the employment of secret agents. He makes the main point very clear when he writes,

> ... the reason the enlightened prince and the wise general conquer the enemy whenever they move and their achievements surpass those of ordinary men is foreknowledge [and] what is called "foreknowledge" cannot be elicited from spirits, nor from gods, nor by analogy with past events, nor from calculations. It must be obtained from men who know the enemy situation. (*Ibid.*, pp. 145–46)

The Strategikon. In the great Byzantine book on military strategy and tactics, *The Strategikon* (attributed to the Emperor Maurice (r. 582–602), intelligence is mentioned prominently several times. In Book 7 it is written, "That general is wise who before entering into war carefully studies the enemy, [and] can guard against his strong points and take advantage of his weaknesses." It is also mentioned that the general is enjoined to ensure that every continuous effort be made "to obtain information about the enemy's movements, their strength and organization, and thus be in a position to prevent being surprised by them."

Machiavelli. Machiavelli, in *The Prince*, makes the point that some capability as a military intelligence analyst is an essential part of being a commander: "... with a knowledge of [the] geography of one particular province one can easily acquire knowledge of the geography of others. The prince who lacks this knowledge also lacks the first qualification of a good commander" (Machiavelli 1981, p. 89). In *The Art of War*, he discusses the importance of avoiding surprise through proper intelligence operations, and states, "... such knowledge (of proper intelligence) is absolutely necessary to anyone anxious to be perfectly instructed in the art of war" (Machiavelli 1965, p. 142).

Marshal de Saxe. Marshal de Saxe, in *My Reveries on the Art of War*, states: "Too much attention cannot be paid to spies and guides. Montecuculli says they are like eyes and are equally necessary to a general. He is right. Too much money cannot be spent to get good ones" (Saxe 1985, p. 291).

Frederick the Great. Frederick the Great, in both *On the Art of War* and *Instruction ... for His Generals*, emphasized intelligence. In *On the Art of War*, he states in the chapter on strategy that,

> above all, you must pay attention to the nature of the country where you wage war ... it is therefore necessary, first of all, that those wishing to formulate a campaign plan have an accurate knowledge of the enemy forces and of the assistance he can expect from his allies. He must compare the enemy forces with his own and with those his friends can furnish him, in order to judge what kind of war he will want to undertake. (Frederick the Great 1966, p. 307)

In his *Instruction ... for His Generals*, he wrote:

> Knowledge of the country is to a general what a rifle is to an infantryman and what the rules of arithmetic are to a geometrician. If he does not know the country he will do nothing but make gross mistakes. Without this knowledge his projects, be they otherwise admirable, become ridiculous and often impracticable. Therefore study the country where you are going to act. (Frederick the Great, 1985, pp. 338–39)

Napoleon. Napoleon in his *Maxims* states in No. 2, "A plan of campaign should anticipate everything which the enemy can do, and contain within itself the means of thwarting him" (Bonaparte 1985, p. 407). No. 79 states, "The first principle of a general-in-chief is to calculate what he must do, to see if he has all the means to surmount the obstacles which the enemy can oppose him and, when he has made his decision, to do everything to overcome him" (*Ibid.*, p. 433).

Jomini. In Article XLII, "Reconnaissance," in his great book, *The Art of War*, Jomini writes,

> One of the surest ways of forming good combinations in war should be to order movements only after obtaining perfect information of the enemy's proceedings. In fact, how can any man say what he should do himself, if he is ignorant of what his adversary is about? As it is unquestionably of the highest importance to gain this information, so it is a thing of the utmost difficulty, not to say impossibility: and this is one of the chief causes of the great difference between the theory and the practice of war. (Jomini 1968, p. 245)

He is stating here that the acquisition of adequate intelligence and the exercise of command using that intelligence are the meaningful differences between actually conducting war and just theorizing about war. He concludes the article with a timeless four-point summary on intelligence:

1. A general should neglect no means of gaining information of the enemy's movements, and, for this purpose, should make use of reconnaissance, spies, bodies of light troops commanded by capable officers, signals, and questioning deserters and prisoners. [Be vigorous in using all possible information sources.]

2. By multiplying the means of obtaining information: for no matter how imperfect and contradictory they may be, the truth may be sifted from them. [Multiple sources are essential.]

3. Perfect reliance should be placed on none of these means. [This should always be borne in mind, with the gross overreliance on ULTRA in Europe that aided the Germans in the 1944 Ardennes surprise being an instructive example.]

4. As it is impossible to obtain exact information by the methods mentioned, a general should never move without arranging several courses of action for himself, based upon probable hypotheses that the relative situation of the armies enables him to make, and never losing sight of the principles of the art. (*Ibid.*, p. 250)

The key points here are that, while intelligence is a key factor, it will never be sufficient in itself, and a commander must maintain operational flexibility and never paralyze himself by continually waiting for just a bit more intelligence before moving.

Clausewitz. In *On War* (Book 1, Chapter 6, "Information in War"), Clausewitz opens the chapter by defining information as ". . . all the knowledge which we have of the enemy and his country: Therefore, in fact, the foundation of all our ideas and actions" (Clausewitz 1968, p. 162).

Triandafilov. The highly respected Russian military thinker, Triandafilov, in his 1929 book, *Nature of the Operations of Modern Armies*, in the section on command and control problems, states:

> . . . it would be erroneous to look upon operational art as some sort of bookkeeping effort, it would be incorrect to convert operational decisions into simple arithmetic multiplication. The material required for every specific case depends not only on the properties of the weapons and arithmetic figures characterizing the length of the front, on the operational and tactical density of the enemy front, how well his positions are fortified, the quality of both the resources and the enemy troops and commanders. These data change too much. The art of the leader is to calculate the operational significance of these changing situational elements correctly and to determine the correct material and personnel resources required to accomplish a given specific mission. (Triandafilov 1929, p. 205)

Clearly inherent and assumed present in this process is adequate intelligence.

Tukhachevsky. Marshal Tukhachevsky, in the 1936 *Red Army Regulations*, begins his command control this way: "The essence of command and control lies in thorough reconnaissance . . ." (Tukhachevsky 1985, p. 167).

Mao Tse-tung. Mao Tse-tung, in *On Guerrilla Warfare*, states that "intelligence is the decisive factor in planning guerrilla operations" (Mao Tse-tung 1961, p. 22). This point should always be kept in mind given that some today consider this to once again be an age of "limited wars."

Changing Nature of Military Intelligence

The nature of intelligence is changing to meet new, more time-sensitive requirements that are being driven by the rapidly increasing lethality of weapons, the increasing operational tempo, and the enormous spatial expansion of the battlefield. Emerging "smart" and "brilliant" weapons, increasing battlefield application of artificial intelligence, improving deception capabilities, and most important of all, increasing battlefield firepower and mobility are all combining to shorten radically decision times at all command levels. This includes air combat and war at sea. As command decision time is continually reduced, the nature of the intelligence requirement is changing; it is now more demanding in terms of time and accuracy. There is less time available to recover from a bad decision, and less ability to absorb rising damage levels.

As intelligence increases in importance it also becomes more and more sophisticated, raising its complexity and cost.

Intelligence Is Often Misunderstood, Misused, and Distrusted

The public misunderstanding and mistrust of intelligence is a peculiarly Western condition, with the American public being one of the most mistrustful. There are several reasons for this extreme position relative to other Western societies; history is perhaps the most important.

U.S. Experience in the Intelligence Field

The American political system is largely derived from the British, and to a lesser extent, liberal French political philosophy and experience. English experience with absolute monarchy, civil war, and the subsequent rise of Parliament clearly established the legislature's ascendancy. However, there is one major difference between the European systems and the U.S. system as far as intelligence is concerned. In Europe, royal intelligence systems were generally continued under democratic control. In the United States, no system existed until the Civil War required extensive military intelligence and counterintelligence, and both capabilities were virtually eliminated after the war. There was an almost total gap in

Intelligence, Military

American military intelligence capability, until World War I required a major domestic counterintelligence effort, largely aimed at preventing sabotage, that was mounted under Justice Department and War Department control. At the same time both the army and the navy had to develop major military intelligence capabilities almost from scratch. Again, the military intelligence systems were largely dismantled after the war. The only notable foreign intelligence effort between 1919 and 1941 was the SIGINT support of the Washington Naval Conference, 1921–22.

The Second World War again required the establishment of major military intelligence and counterintelligence capabilities. While sabotage in the United States was not a major German or Japanese effort, espionage was. The postwar environment, popularly known as the cold war, was the first time in American history that a major postwar intelligence and counterintelligence effort was maintained. Despite the dissolution of the Warsaw Pact and the Soviet empire, this intelligence effort remains massive and has become a major budget expense. The sheer size, complexity, and cost of this effort brings intelligence and counterintelligence to the public's attention. Another major reason for the American public's mistrust of intelligence is that successful intelligence operations require secrecy and security, both of which clash with the American concept of an open society.

The Need for Secrecy and Security

All intelligence operations depend on secrecy and security for success. Therefore, intelligence inherently contains a great irony based on the fact that the more important the operation, the more important are secrecy and security. The irony is that the more successful an intelligence operation, the less is known about that operation; only intelligence failures become widely known. Therefore, the public, whose support is essential, never learns of successes, but only of failures and abuses. This creates a one-sided situation in which, unlike the police or the Federal Bureau of Investigation, the intelligence community cannot present its counterbalancing successes in order to assist informed debate and defend itself. In effect, the Western intelligence communities, and especially the American one, must struggle with a public that has at best a media-generated caricature of intelligence.

Directed Misuse and Abuse of Intelligence Capabilities

Americans have experienced sufficient cause for wariness about how the intelligence community is sometimes used. This wariness is largely based on administration-directed misuse and abuse of intelligence capabilities, the vast majority of which occurred during the turbulent 1960s and early 1970s and were directly related to popular opposition to the Vietnam War. The illegal collection and use of communications intercepts, the filming of demonstrations, and the creation of counterintelligence databases on Americans have been well documented. These activities have had a major impact on the public, and they must always be kept in mind when trying to understand military intelligence and its proper role in American society. Clearly, in peacetime, military intelligence organizations have two proper domestic roles. One is the conduct of personal background investigations on personnel being considered for access to classified information. The second is assisting in the security of classified information and facilities. In wartime there is a proper and larger role for military counterintelligence in defending against espionage and sabotage. The exact boundaries of this role must be determined according to the situation.

Covert Operations and Intelligence

The formal, informal, and improper mixing of covert operations and police functions within intelligence organizations and operations has contributed greatly to the current situation of widespread public distrust of intelligence organizations. As previously stated, covert operations are not an intelligence function. In America, this situation really exists only within the Central Intelligence Agency (CIA), and this combination is by statute. In real terms, however, this mixing of two major and separate functions creates an adverse public environment for the intelligence side of the CIA. While there is significant public understanding of the need for foreign intelligence, there is significantly less support for covert operations. This means that covert operation controversies reflect adversely on intelligence as a whole, thereby significantly and adversely affecting public opinion. In an open society this argues powerfully for a separation of intelligence and covert operations.

The Power of Intelligence

Finally, intelligence is knowledge and knowledge is power, and the misuse and abuse of power are as old as mankind itself. Nothing inherent in intelligence is going to ease this, and it is therefore incumbent on all concerned to keep this in mind and guard against misuse and abuse of intelligence and counterintelligence.

LLOYD HOFFMAN

SEE ALSO: Command, Control, Communications, and Intelligence; Counterintelligence; Covert Action; Deception; Intelligence Analysis, Military; Intelligence Collection; Intelligence Cycle; Intelligence Establishment; Intelligence, General; Intelligence, Human; Intelligence, Imagery; Intelligence, Signal; Operational Art; Satellite Technology Applications; Strategy; Tactics; War.

Bibliography

Barron, J. 1984. *KGB*. New York: Reader's Digest Press.
Bonaparte, N. 1985. Military maxims. In *Roots of strategy*, ed. T. R. Phillips. Harrisburg, Pa.: Stackpole.
Brissaud, A. 1974. *The Nazi secret service*. New York: W. W. Norton.

Clausewitz, C. von. 1968. *On war*, ed. Anatol Rapoport. Baltimore, Md.: Penguin Books.
Cline, R. S. 1981. *The CIA under Reagan, Bush, and Casey*. Washington, D.C.: Acropolis.
Constantinides, G. C. 1983. *Intelligence and espionage: An analytical bibliography*. Boulder, Colo.: Westview.
Corson, W. R. 1977. *The armies of ignorance*. New York: Dial Press.
Dziak, J. J. 1988. *Chekisty*. Lexington, Mass.: Lexington Books.
Eisenberg, D., U. Dan, and E. Landau. 1978. *The Mossad*. New York: Paddington.
Finnegan, J. P. 1984. *Military intelligence: A picture history*. Arlington, Va.: History Office, U.S. Army Intelligence and Security Command.
Frederick the Great. 1966. *The art of war*. Glencoe, Ill.: Free Press.
———. 1985. Instruction of Frederick the Great for his generals. In *Roots of Strategy*, ed. T. R. Phillips. Harrisburg, Pa.: Stackpole.
Garlinski, J. 1979. *The Enigma war*. New York: Scribner's.
Godson, R., ed. 1990. *Intelligence requirements for the 1990's*. Lexington, Mass.: Lexington Books.
Holmes, W. J. 1979. *Double-edged secrets*. Annapolis, Md.: U.S. Naval Institute Press.
Innes, B. 1966. *The book of spies*. London: Bancroft.
Jomini, A. 1968. *The art of war*. Westport, Conn.: Greenwood Press.
Kahn, D. 1967. *The code breakers*. New York: Macmillan.
Kennedy, W. V., ed. 1983. *Intelligence warfare*. New York: Crescent Books.
Laqueur, W. 1985. *A world of secrets*. New York: Basic Books.
Levite, A. 1987. *Intelligence and strategic surprises*. New York: Columbia University.
Machiavelli, N. 1965. *The art of war*. Indianapolis: Bobbs-Merrill.
———. 1981. *The prince*. New York: Penguin Books.
Mao Tse-tung. 1961. *On guerrilla warfare*, ed. S. B. Griffith. New York: Praeger.
Platt, W. 1957. *Strategic intelligence production*. New York: Praeger.
Read, A., and D. Fisher. 1985. *Colonel Z*. New York: Viking.
Roosevelt, K. 1976. *War report of the OSS, Volumes I and II*. New York: Walker.
Rowan, R. W. 1937. *The story of secret service*. New York: Doubleday.
Saxe, M. de. 1985. My reveries on the art of war. In *Roots of strategy*, ed. T. R. Phillips. Harrisburg, Pa.: Stackpole.
Scholars' Guide to Intelligence Literature: Bibliography of the Russell J. Bowan Collection in the Joseph Mark Lauinger Memorial Library. Washington, D.C.: Georgetown University.
Schwien, E. E. 1936. Combat intelligence. *The Infantry Journal*, Washington, D.C.
Sheldon. R. M. 1990. Tinker, tailor, Caesar, spy. Unpublished manuscript.
Stanley, R. M. 1981. *World War II photo intelligence*. New York: Scribner's.
Sun Tzu. 1963. *The art of war*. Ed. and trans. S. B. Griffith. Oxford: Oxford Univ. Press.
Triandafilov, V. K. 1929. *Nature of the operations of modern armies*. Moscow-Leningrad: State Publishing House.
Troy, T. F. 1981. *Donovan and the CIA*. East Haven, Conn.: Aletheia Press (distr. InBook).
Tukhachevsky, M. 1985. Quoted by R. Simpkin in *Deep battle*. Washington, D.C.: Pergamon-Brassey's.
U.S. Department of the Army. 1986. *Field manual 100-5: Operations*. Washington, D.C.: Government Printing Office.
Van Der Rhoer, E. 1978. *Deadly magic*. New York: Scribner's.
Winterbotham, F. W. 1974. *The Ultra secret*. New York: Harper and Row.
Wright, P. 1987. *Spycatcher*. New York: Viking.

INTELLIGENCE, NAVAL

Naval intelligence derives its nature from the special conditions of the world's seas and from the ships, weapons, and other technologies that have evolved to use and protect the sea. It is a demanding environment, covering some 70 percent of the earth and subject to frequently hostile extremes of wind, weather, and visibility. It includes vast, uninhabited surface expanses, considerable depths, air space above, and much adjacent land. Information is required about all this as well as intelligence about any adversary upon the sea or its littorals, where power projection in the national interest is appropriate.

In the past, U.S. naval intelligence has been responsible for protecting the security of the naval establishment against foreign intelligence activity, sabotage, and accidental security breach. This is no longer the case, but in any event, it should be pointed out that security enforcement is, essentially, a police operation. Like any operation, it requires intelligence support, but functionally it is not really an *intelligence* activity, except by fiat.

Intelligence in the History of Seapower

Intelligence has had a role in naval warfare since before the Battle of Salamis in 480 B.C. Themistocles, the Athenian admiral, used reconnaissance information to execute tactics that lured the larger Persian fleet into the narrow Salamis strait, where its numerical advantage was limited against the smaller but more heavily armed Athenian force.

In the sixteenth century, Lord Walsingham, Queen Elizabeth's Secretary and "chief intelligencer," purchased information on the development of the Spanish Armada and other Spanish activity deemed threatening to the British crown.

During the Napoleonic Wars, Adm. Horatio Nelson used patrolling frigates and British consular reports to stay abreast of French fleet sorties and movements.

Naval intelligence came of age in the activity of the British Admiralty in World War I. The considerable success of cryptanalysis efforts of famous "Room 40" was increasingly useful. Only poor intelligence coordination, perhaps, imposed by the Fleet Operations Division, prevented a second Trafalgar at the Battle of Jutland. But lessons were learned, and in World War II, the more desperate battle against German U-boats was finally won—in no small measure through the coordination of the Admiralty Operational Intelligence Centre.

Naval Intelligence in World War II

The wide deployment of shore- and ship-based reconnaissance aircraft, and developments in communication intelligence, had great impact on naval warfare in World War II. The June 1942 Battle of Midway highlighted the crucial importance of intelligence in enabling the U.S. Navy to achieve decisive surprise, and victory, over a much larger Japanese force. U.S. commanders correlated intelligence gleaned from the decrypt of Japanese cyphers with their knowledge of Japanese fleet operations and the results of carrier-launched reconnaissance. American intelligence yielded accurate knowledge of the Japanese objective—Midway. The Japanese, however, with a faulty picture of American carrier deployment in the Pacific, split their invasion armada to support the attack and landing on Midway, thus reducing the defensive support for their carriers and making them more vulnerable to American attack. The resulting loss of four major Japanese carriers to U.S. dive-bombers was the turning point in seapower in the Pacific war.

Intelligence also played a vital role in defeating the German U-boat campaign, in supporting the amphibious landings in both Europe and the Pacific, and in providing a variety of intelligence needed for U.S. carrier task force strikes in the Pacific.

Communications Intelligence

Within the last 100 years, technology has transformed the operational and combat capabilities of navies and has been paralleled by comparable development of naval intelligence capabilities. The advent of wireless radio initiated a revolution in the command and control of navies. It also exposed an Achilles' heel of unwary commanders. Their ships could be located by triangulation and message traffic analysis. More important, crucial secrets could be vulnerable to message intercept and cryptanalysis. Being caught unaware at sea posed particular danger with the development of surprise attack by submarines or aircraft. Operational use of sonar and radar have evolved to reduce such attacks by enabling intelligence to compile acoustic and electronic signature data.

Naval Intelligence Sources

Today, satellite and high-altitude air reconnaissance (including photographic, infrared, and electronic signal collection) provides a large input to naval intelligence. Nevertheless, as in World War II, much "open" information is available from unclassified sources—media, library, academic, technical, and commercial. Library of Congress resources, for example, were heavily exploited in World War II in the production of amphibious landing intelligence and bombing target folders. Sophisticated systems for intelligence have kept apace with other naval technology. Nevertheless, the intelligence process must still transfer a vast amount of information and data into useful intelligence. Increasingly, this process is being facilitated by computer support with information storage and retrieval, cryptanalysis, spreadsheet collation, quantitative and graphic analysis, and employment of "expert systems" of so-called artificial intelligence algorithms.

Fundamental Intelligence Process

What is known, classically, as "the intelligence process" is used to transform "intelligence information" (raw information and data), collected from all relevant sources, into the finished intelligence needed by decision makers: commanders and planners. (This cycle of tasks is, essentially, the same for all intelligence, naval or otherwise.) Once collected, the raw material must then be analytically collated and selected to determine its validity, accuracy, and significance.

The final step in the process, and fully as crucial as any of the others, is that of timely dissemination of intelligence to the users who "need to know." The real intelligence failure of Pearl Harbor, for example, occurred in the Navy Department in Washington, D.C. Excessive, inappropriate security measures for the protection of the easily compromised decrypt of Japanese cypher traffic were imposed by communication officers. Lacking intelligence training and experience, they were unable to match the precious but fragmentary secrets of communications intelligence with the less sensitive into a more complete "all-source" mosaic, and did not permit the analysts in the Office of Naval Intelligence to do so. Worse yet, a variety of crucial evidence of pending attack, unappreciated in Washington, was withheld from Pacific fleet staff in Pearl Harbor, who were sensitive to their vulnerability—a failure, clearly, in dissemination.

Taint of Intelligence by Users

It is important to recognize that the analytical process has been dangerously susceptible to an error-prone influence of intelligence *users*. Given the opportunity, human frailty leads commanders, decision makers, and planners to choose intelligence that supports their decisions or planning. Furthermore, they become increasingly less responsive to subsequent intelligence that casts doubt on their judgment. The surprise achieved at Pearl Harbor, Hitler's surprise offensive in the Ardennes, the failure of Field Marshal Bernard Montgomery's Operation Market Garden in Holland, and the failure of the Bay of Pigs landing in Cuba all reflect the failure of biased intelligence. The first three instances were cases of dominant staff "groupthink" choosing to ignore inconvenient intelligence. The Cuban case is ironic, a CIA venture in operational planning that fell into the same, historic psychological trap. The weakness is human, not functional.

Future

Although the future shows promise of having resolved the threat of communism, a still-troubled world poses a continued need for U.S. force projection. World War II demonstrated the need both for seapower to bring carrier strike and amphibious strength to bear ashore and for supporting intelligence. It seems likely that the need for carrier strikes, naval missiles, and amphibious potential, as demonstrated in Operation Desert Storm, will continue into the foreseeable future. Accordingly, naval intelligence in support of these missions will be needed.

PRESCOTT PALMER

SEE ALSO: Intelligence Analysis, Military; Intelligence, General; Intelligence, Signal; Surveillance, Ocean.

Bibliography

Beesly, P. 1978. *Very special intelligence: The story of the Admiralty's operational intelligence center 1939–45*. Garden City, N.Y.: Doubleday.
———. 1982. *Room 40: British naval intelligence 1914–1918*. London: Hamilton Hamish.
Hinsley, F. H., et al. 1979–88. *British intelligence in the Second World War*, vols. 1, 2, 3. London: Her Majesty's Stationery Office.
Holmes, W. J. 1979. *Double-edged secrets: U.S. naval intelligence in the Pacific during World War II*. Annapolis: U.S. Naval Institute Press.
James, W. 1956. *The eyes of the navy: A biographical study of Admiral Sir Reginald Hall*. New York: St. Martin's Press.
Layton, E. T. 1985. *And I was there: Pearl Harbor and Midway—breaking the secrets*. New York: Morrow.
McLachlan, D. 1968. *Room 39: A study in naval intelligence*. New York: Atheneum.
Pemsel, H. 1977. *Atlas of naval warfare: An atlas and chronology of conflict at sea from earliest times to the present day*. London: Arms and Armour Press.

INTELLIGENCE, SIGNAL (SIGINT)

Signal intelligence (SIGINT) is the result of the collection, evaluation, analysis, integration, and interpretation of information derived from intercepting electromagnetic emissions. It is one of four basic categories of intelligence. The other three are human intelligence (HUMINT), imagery intelligence (IMINT), and scientific and technical intelligence (S&T intelligence).

SIGINT is used to satisfy strategic, operational, and tactical requirements in order to plan, target and deceive the enemy, exercise command/control/communications countermeasures (C^3CM) and electronic warfare (EW), conduct special operations, and successfully employ the full spectrum of weapons throughout the duration of any conflict (see Fig. 1).

Strategic intelligence is that required by national decision makers to formulate national foreign and defense policy in support of the national security strategy. For a major world power these intelligence needs are global and cover all elements of national power. Strategic intelligence needs (as well as operational and tactical needs) vary constantly because of the dynamics of national and international activities. Operational intelligence is used in planning and conducting campaigns within a theater of

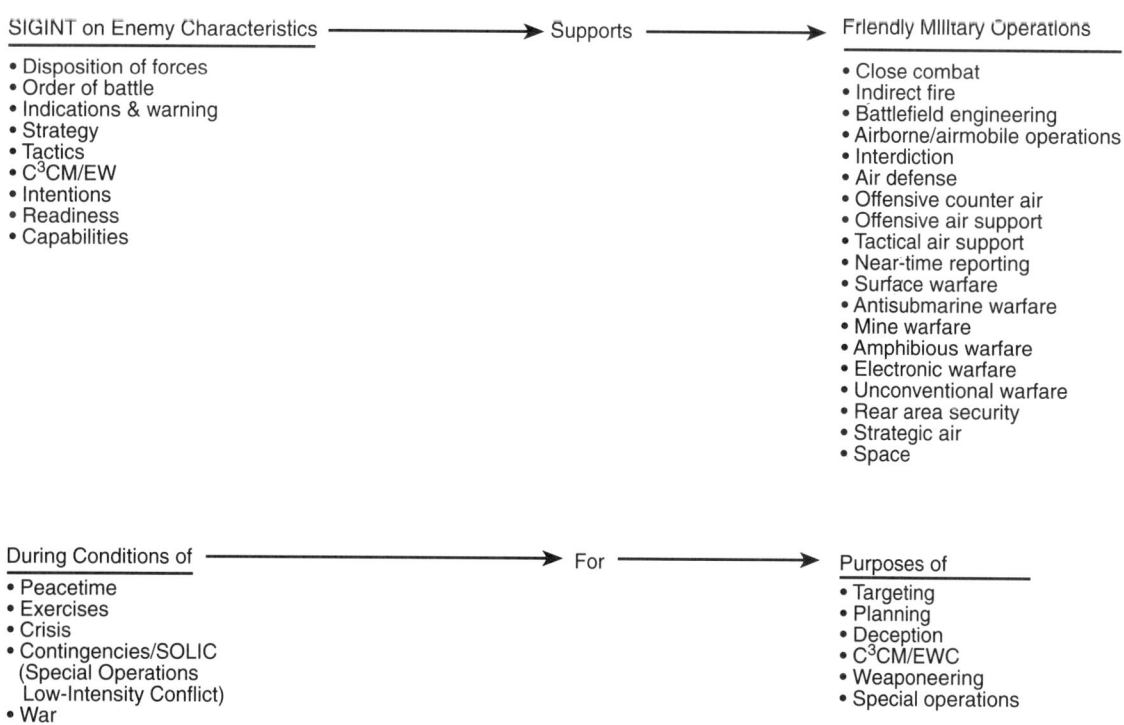

Figure 1. SIGINT support to military operations. (SOURCE: Horgan 1991)

war. SIGINT provides some insight into a target nation's political intentions, economic plans and status, and diplomatic maneuvers. It can also provide indicators and warning (I&W) of forthcoming events or military actions, insight into the opposing commander's intent, and details of his battle plans and troop deployment. Tactical intelligence (sometimes called combat intelligence) is more immediate. It must focus on the needs of the commander in planning and conducting combat.

SIGINT dates almost from the invention of electric telegraphy in the 1830s. During the American Civil War, each side tapped into its opponent's telegraph wires to listen to transmissions and gain information. Between World War I and World War II, the United States broke both Japanese and German codes and the Poles penetrated the German ENIGMA encryption device. The British later built on the Polish success, producing intelligence code-named ULTRA. Intelligence produced by the United States from breaking the Japanese codes was named MAGIC. ULTRA and MAGIC gave the Allies both operational and strategic advantages throughout World War II. Because of Japan's and Germany's inordinate SIGINT efforts and the success of ULTRA and MAGIC, World War II is sometimes known as the SIGINT War.

No one category of intelligence can be depended upon to the exclusion of the others. For example, SIGINT can reveal neither the locations of mined areas, hidden bunkers, gun emplacements, and pillboxes, nor the state of training and morale of enemy troops. Therefore, collecting and processing information into intelligence should be an integrated procedure embodying all the intelligence disciplines. Only an integrated procedure can ensure that all possible avenues to obtaining information are exploited and that the final intelligence product is the best that can be produced.

SIGINT is controlled at the national level. In the United States, it is one of the primary missions of the National Security Agency (NSA). In the former Soviet Union, SIGINT was the responsibility of the Sixteenth Directorate of the KGB and the Sixth Directorate of the Glavnoye Razvedyvatelnoye Upravleniye (GRU or Chief Intelligence Directorate of the Soviet General Staff). The Sixth Directorate was responsible for radio and radio-technical intelligence. Other countries have similar national-level structures to handle SIGINT.

The national SIGINT effort is directed primarily to the strategic level. Subordinate elements of a government and its armed forces support three levels of SIGINT: strategic, operational, and tactical. These three levels of intelligence are always interrelated. Data obtained at the tactical level, when collated with other data, may lead to intelligence of value at the operational or even strategic levels. Similarly, operational or strategic intelligence may be useful at the tactical level. In short, good intelligence requires inputs from multiple sources.

Aspects and Applications of SIGINT

The components of SIGINT are communications intelligence (COMINT), electronics intelligence (ELINT), and foreign instrumentation signals intelligence (FISINT). COMINT is derived from intercepting, monitoring, and collecting information from the enemy's communications systems. ELINT is produced by intercepting and studying the enemy's noncommunications signals, such as radar, and locating the emitters. FISINT is obtained by intercepting foreign electromagnetic emissions, such as telemetry, used with the testing and operational deployment of aerospace surface and subsurface instrumentation. Both COMINT and ELINT are employed at all levels of warfare—strategic, operational, and tactical. FISINT has primarily a strategic application; however, information that is of primarily strategic value may also contain intelligence of tactical or operational value. Monitoring enemy communications systems can produce COMINT of value to a single submarine, aircraft, or army unit, or to national strategy as a whole, depending upon the level at which the surveillance is aimed. ELINT, by identifying new or different enemy fire direction or target acquisition radars, may indicate changes in deployments of field or air defense artillery. The strategic, operational, or tactical application of this intelligence would depend upon the number, types, and locations of these radars. Finally, FISINT may provide early warning of a potential enemy's weapons testing or an increase in aerospace activity.

Search, interception, and direction finding are all used to collect COMINT data. Processing includes transmitter or operator identification, estimation of the range to transmitter/operator, analysis of the signal, traffic analysis, cryptanalysis, and decryption. Cryptanalysis is the art of solving or breaking a code or cipher. Cryptology is the combination of making and breaking codes (cryptography and cryptanalysis, respectively).

Information "external" to the actual message can also produce COMINT. External data include the manner of transmission, identity of the sender, and identity of the recipient. Examples of such data are call signs, frequencies, operator procedure, "conversations" among operators, indicators of where the message is being sent, message precedence, and other procedures. The study of these "externals" is called traffic analysis.

ELINT can be subdivided further into operational ELINT (OPELINT) and Technical ELINT (TECHELINT). OPELINT, as the name implies, deals with current intelligence requirements. It is timely, high-priority intelligence designed to satisfy I&W and current ELINT requirements. OPELINT determines the location and readiness of emitters. It also is used to keep major intelligence databases current, such as the Electronic Order of Battle (EOB) and the ELINT Parameters List. OPELINT reports meet user requirements for parametric data in a machine-readable format that can be used in EW, automatic processing, or displays of certain signals of interest. These reports also include narrative tactical reports.

TECHELINT determines the capabilities and limitations of target foreign electronic emitters and provides the information for a computer database. It collects information on signals characteristics, functions, associations, performance, and technology. TECHELINT is used to estimate the primary functions, capabilities, modes of operation (and malfunctions), and specific roles of noncommunications emitters within a defense network or a complex weapons system.

FISINT includes intelligence information derived from foreign civilian or military electromagnetic emissions. The most significant subcategory of FISINT is telemetry intelligence (TELINT). TELINT is the technical and intelligence information obtained from intercepting, processing, and analyzing foreign telemetry. Access to foreign telemetry is vital to national strategy. For example, telemetry is a vital part of a missile development program. As the missile is being tested, data about its performance are transmitted from the missile to a ground control facility. Telemetry data allow the scientist to assess the missile's operational performance, evaluate its performance capability in comparison to design specifications, and obtain insights into the probable cause(s) for failures.

The SIGINT Cycle

It should be noted that SIGINT is a passive collection discipline: nothing can be intercepted from inactive radars or transmitters. Furthermore, some high-interest communications may be missed because an interception device has not yet been developed. On occasion, the target may send false information, knowing his signals are being intercepted. The SIGINT operative must be aware of these shortcomings and take them into account when collecting, processing, and evaluating signal information.

SIGINT REQUIREMENTS

All intelligence is derived through the same cycle. The cycle starts with a statement of need. This can originate at the national level, from a commander at any echelon, or from an analyst requiring more information to satisfy another request. Often, the volume of requests will exceed the resources available to obtain the desired information. Because some information is of more value than others, analysts review current intelligence collection efforts to determine the new request's priority and whether it requires the development of a collection requirement as well. The remainder of the intelligence cycle is collecting, processing, evaluating, and disseminating the information. The dissemination phase brings the cycle back to the start—the requester.

COLLECTION

To collect SIGINT, one must have a receiver capable of operating on the frequency of the opponent's emitter and of receiving the correct type of signal. Receiving systems are deployed to capture SIGINT at all three levels—strategic, operational, and tactical.

Strategic collection includes both overt and covert efforts. Commercial aircraft may carry covert SIGINT interception devices. Agents disguised as diplomats, governmental officials, or business people may employ covert devices to intercept communications. Fishing or other commercial vessels can also be used to conduct covert SIGINT. Access to the commercial communications of national or international business establishments could supply valuable strategic information.

Overt national-level collection includes specially equipped long-range aircraft and satellites and long-range radio intercept stations, within the nation, its possessions, and its naval units.

Operational intelligence focuses on the theater, army group, field army, or corps. For a fleet, operational intelligence would focus on a large area of the sea. Intelligence originating at the division or lower echelons is considered to have been obtained from the tactical level.

A large array of detection devices mounted on the ground, on naval vessels, and in aircraft is employed at different echelons (e.g., army or corps). This equipment detects and reports specific signaling devices, such as radio transmissions, and radars. In addition, operational intelligence can be supplied from national-level service.

In U.S. doctrine, the area of operations for organizations above the corps level (two to five corps) is 150 kilometers (93 mi.) beyond the forward line of troops (FLOT) and its area of interest extends 1,000 kilometers (620 mi.) beyond the FLOT. (Within the conventional battlefield environment, commanders think in terms of the time and space required to defeat or bypass the enemy force before it can be reinforced. Intelligence requirements stem from areas of influence and areas of interest. The *area of influence* is that part of the battlefield where the commander must be able to acquire targets and bring fire to bear against enemy forces with the weapons under his direction. The *area of interest* is that part of the battlefield that extends beyond the area of influence, in depth and width, to include areas where enemy forces capable of affecting a commander's future operations are located. Both areas are important because they normally contain the targets of intelligence collection efforts. During combat operations, these areas fluctuate, based upon such variables as mission, enemy, terrain, and troops available [see Figure 1 in the article on Tactical Intelligence].)

Because operational planning must occur 72 or more hours before the event, the commander should have information on what the enemy can do to influence the battle at least 96 hours before the event. U.S. commands above the corps level are supported by a military intelligence brigade or similar organization that includes SIGINT collection, processing, and evaluation elements. (This organization compares to the Russian Radio Reconnaissance Regiment [COMINT], Radiotechnical Reconnaissance Regiment [ELINT, COMINT, and direction finding], and a Radiotechnical Brigade supporting a typical front [U.S. theater or army group].)

The corps level usually has an area of operations of up to 150 kilometers (93 mi.) and an area of influence of up to 300 kilometers (186 mi.). The commander at this level needs 72 hours planning. A U.S. corps employs weapons and electronic countermeasures, information from subordinate divisions, intelligence resources from higher echelons, and tactical aerial reconnaissance. Tailored SIGINT service is provided to the corps commander by nonorganic SIGINT producers (not belonging to the corps they are supporting) in response to his requirements. The Russian combined arms army (equivalent to the U.S. corps) has an organic Radio Reconnaissance Battalion (regularly assigned or belonging to the combined arms army) for COMINT and a Radiotechnical Reconnaissance Battalion for COMINT and ELINT. The tank army usually has one Radio Reconnaissance Battalion. In addition to these organic formations, the army commander's SIGINT efforts are supported by subordinate commands, aircraft, and higher echelons in a manner similar to the U.S. corps commander.

The division level has an area of operations that extends 70 kilometers (43 mi.) and an area of interest that extends 150 kilometers. The speed of operations at this level requires that information be processed rapidly to assess enemy dispositions and capabilities. This information includes the location, strengths, and direction of movement of regimental and division command posts and artillery, rocket, air defense, radio electronic combat, and service support units located in or moving to the division's area of influence. The division has organic and subordinate unit collection services, including a military intelligence battalion that has a technical control and analysis element (TCAE) made up of SIGINT and EW assets and companies for communications and jamming, intelligence and surveillance, long-range surveillance, and electronic warfare. The division is aided by corps, army, and air force assets and intelligence. The Russian division has a Reconnaissance Battalion, which includes a Radio/Radar Reconnaissance Company. This company is capable of radio, radar, and air-to-air and air-to-ground interception and direction finding.

Brigade and lower echelons must rely on organic units, ground surveillance radars, plus MI support for intelligence and electronic warfare (IEW).

U.S. and Russian SIGINT organizations and operations are cited as examples. Other nations have similar organizations and procedures for producing and disseminating SIGINT.

Navies, in addition to employing centrally commanded units, provide fleets with their own intelligence centers. Aircraft carriers have a small intelligence unit. Other vessels, down to cruisers, have naval intelligence officers assigned. In smaller vessels, intelligence functions are an additional duty for one of the ship's junior officers. All naval vessels, regardless of which country they represent, possess an array of SIGINT equipment to detect radio, radar, range and direction finding, and other signals. These detection devices are vital to both the defensive and offensive capabilities of any naval vessel.

Likewise, combat aircraft possess an array of sensors and direction-finding and other devices necessary to their ability to fly, fight, and defend themselves. The data these devices obtain must be instantly processed into usable intelligence, otherwise the aircraft could be destroyed by hostile action.

PROCESSING

Processing is the conversion (and reduction) of large amounts of information into a form usable by an analyst. SIGINT often must be processed, evaluated, and disseminated very quickly.

Whether developed on a ship, an aircraft, or a ground system, SIGINT must be processed either by system analysis devices or by humans. The analyst (device or human) integrates, evaluates, and studies information that is often fragmentary and contradictory. Human analysts include experts in a particular field, language translators, cryptanalysts, and traffic analysts.

EVALUATION

To be of value, raw information must be processed, evaluated, and disseminated in a timely fashion. In analyzing material, its reliability, validity, and relevance must be weighed, while the information is placed in its proper context. From all this, the analyst attempts to arrive at a valid conclusion.

DISSEMINATION

SIGINT is usually disseminated electronically to users, but may be published in hard copy and dispatched to them. SIGINT should answer the requirement that generated the search, cause new requirements to be made, or result in another standing intelligence requirement.

The U.S. Joint Chiefs of Staff in JCS Publication 2-0 provide a good summary of intelligence product standards. Intelligence must:

- Be available and accessible in time for effective use.
- Be objective, unbiased, and free from political influence and constraint.
- Be suitable for application.
- Satisfy the needs of commands, staffs, and forces, enabling them to accomplish their missions.
- Be factually correct and convey the true situation.
- Contribute to the understanding of the situation and to the planning, conduct, and evaluation of operations.

Finally, intelligence systems must respond to both existing and contingent operational intelligence requirements (see Fig. 2).

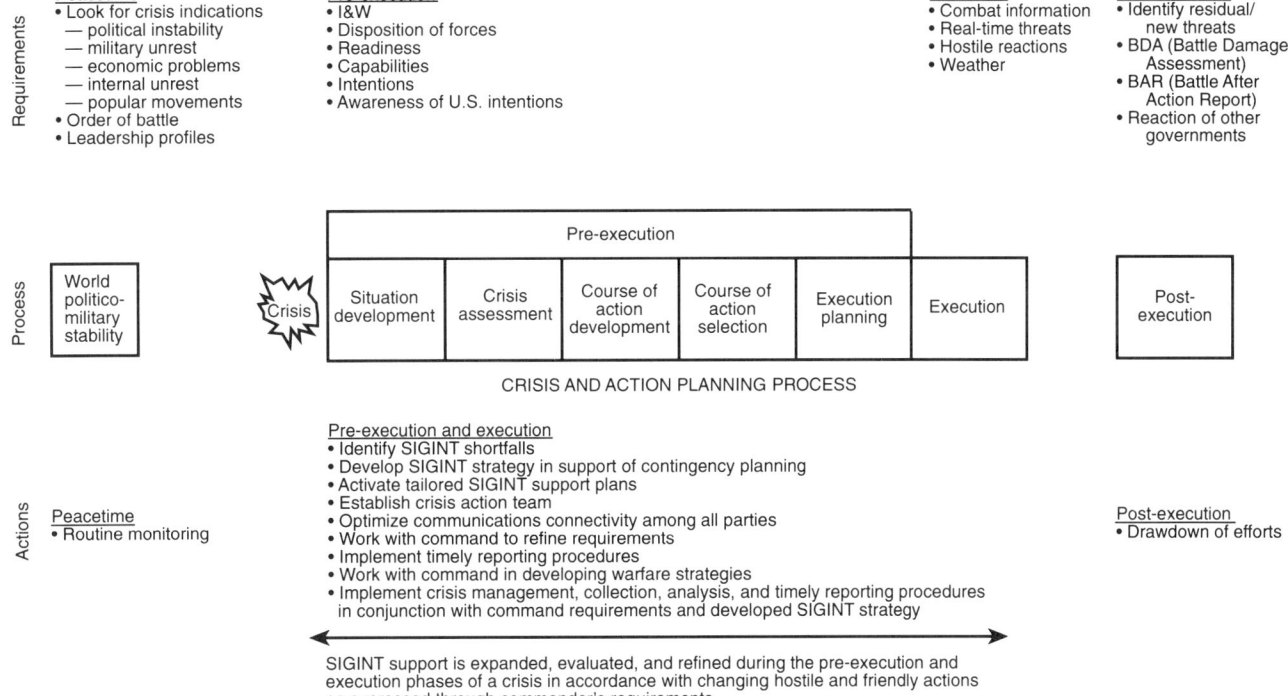

Figure 2. SIGINT response to a crisis. NOTE: This figure depicts the U.S. crisis response procedure, which is fairly typical of procedures used by other nations. (SOURCE: Horgan 1991)

Summary and Conclusion

SIGINT is a dynamic discipline, but its sources are ephemeral. The revelation of a source, or the enemy's capability to detect interception, will result in the loss of that source.

The basic precepts of SIGINT may not change in the future, but the main thrust must change to meet a world order that is rapidly changing. Accordingly, SIGINT efforts will remain at high levels, so that each nation may protect its world-wide interests.

UZAL W. ENT

SEE ALSO: Communications, Signal; Electronic Warfare, Air; Electronic Warfare Technology Applications; Intelligence Analysis, Military; Intelligence Collection; Intelligence Cycle; Intelligence, Strategic; Intelligence, Tactical; Intelligence: Indications and Warning; Radar Technology Applications; Sensor Technology; Strategy, National Security; Surveillance and Target Acquisition Equipment.

Bibliography

Ball, D., and Windrem, R. 1989. Soviet signals intelligence (SIGINT): Organization and management. *Intelligence and National Security* 4(4):621–59.
Carver, G. A., Jr. 1990. Intelligence in the age of glasnost. *Foreign Affairs* 69 (Summer) 147–66.
Christianson, D. 1986. Signals intelligence. In *The military intelligence community*, ed. G. W. Hopple and B. W. Watson. Boulder, Colo.: Westview Press.
Godson, R., ed. 1989. *Intelligence requirements for the 1990s.* Lexington, Mass., and Toronto: Lexington Books.
Horgan, P. 1991. Signals intelligence support to U.S. commanders: Past and present. 1991. USAWC Military Studies Program Paper. Carlisle Barracks, Pa.: U.S. Army War College.
Laqueur, W. 1985. *A world of secrets.* New York: Basic Books.
Strong, J. T. 1986. The defense intelligence community. In *The military intelligence community*, eds. G. W. Hopple and B. W. Watson. Boulder, Colo.: Westview Press.
U.S. Department of the Army. 1987. *Field manual 34-1: Intelligence and electronic warfare operations.* Washington, D.C.: Government Printing Office.
———. 1986. *Field manual 34-10: Division intelligence and electronic warfare operations.* Washington, D.C.: Government Printing Office.

INTELLIGENCE, STRATEGIC

Strategic intelligence is intelligence designed for use in developing strategic-level policies, programs, capabilities, and decisions. This is a simple definition, but to fully understand it requires an understanding of each of its three major parts.

Intelligence is usually defined as either a "process" or a "product." As a process, intelligence generally is described as the targeting, collecting, analyzing, and disseminating of information on foreign entities such as nations, ultra-national organizations, and international outlaws. As a product, intelligence is generally described as "processed information" on these same foreign entities—that is, information that has been carefully evaluated by professional intelligence analysts. This section refers to intelligence as a product: processed information.

The importance of the phrase *designed for use in developing* in the introductory definition cannot be overemphasized. The purpose of intelligence is to help decision makers make decisions. If it is not used for that purpose, then the intelligence and the effort expended in its development are wasted.

One way to categorize intelligence is according to the level of responsibility of the decision makers for whom the intelligence is developed. In military and defense activities, these levels of responsibility are generally described as tactical, operational, and strategic. Hence, *strategic* intelligence assists decision makers at the strategic level. Decisions at the strategic level usually involve the establishment of national policies, the management of national programs to achieve those policies, the creation and deployment of capabilities to carry out those policies, and the issuing of orders in support of those policies.

Strategic intelligence has two main purposes: the first has a mid- to long-term nature, while the second is more immediate. From a mid- to long-term perspective, strategic intelligence assists national security decision makers to properly prepare over the long term to defeat real and potential enemies—those who establish capabilities to interfere with or resist one's own national goals and objectives. The emphasis here is on long-range forecasts. For example, a strategic intelligence report could indicate that an enemy will probably begin a program to double its submarine fleet over the next decade. Such intelligence might cause a defense minister to seek increased funding to improve his nation's antisubmarine capabilities.

From a more immediate perspective, strategic intelligence assists decision makers involved in crises or situations requiring immediate or near-term action to quickly defeat real or potential enemies. In this instance, the intelligence emphasis is on the enemy's present capabilities and intentions. For example, a strategic intelligence report may indicate that foreign naval forces are about to blockade an internationally important waterway. Such intelligence might cause a nation's leader to take immediate unilateral measures or seek near-term international cooperation to prevent such a hostile act.

Strategic intelligence is not always reactive to foreign initiatives. It may be developed at the request, or at the anticipation of a request or need, of national policymakers. For example, a nation's leader who wants to negotiate a trade agreement with a foreign power might request both his commerce and defense ministries to provide assessments of expected responses from the foreign power's civil and military leaders. Both these assessments, one from a civilian organization and one from a defense organization, would qualify as examples of strategic intelligence.

Strategic intelligence is usually "all-source" intelligence: the analysts producing the intelligence evaluate data from many different sources, such as imagery satellites, spies, newspapers, and embassy cables. These data may come from national-level collection assets such as satellites, or from tactical-level assets such as frontline infantry patrols. Strategic intelligence should not be equated or confused with information that comes only from strategic-level collection assets, such as pictures from satellites. A satellite picture, for example, could just as well be part of a tactical intelligence assessment. Strategic intelligence results from the analysis of many different types of data obtained by many different kinds of collection assets located at various levels of intelligence operations. The "all-source" aspect is an important characteristic of strategic intelligence; it is not essential, but it is highly desirable.

Levels of Strategy

Strategic intelligence can be divided into subcategories based on the various levels and categories of strategy: theater military strategy, global military strategy, national defense strategy, national security strategy, and national strategy. Nations may use different names or terms to describe these subcategories, but all major nations engage in most of these strategic activities. The one activity that may not be characteristic of all major nations is global military strategy. This activity is usually only characteristic of those nations that have global military interests or responsibilities and the military capabilities to pursue them. These levels of responsibility are not completely separate and distinct; they often overlap. Their goals are almost always complementary, derivative, and interactive. As a result, strategic intelligence products developed to support one strategy subcategory often can assist decision makers operating at other levels of responsibilities. However, the most effective strategic intelligence products are usually those developed to address specific requirements at a particular level of strategic activity.

The various levels of strategic intelligence and the levels of strategy that they support are described here, beginning at the highest level and following the logic flow to the lower levels. In reality, however, strategic decision making is not an entirely top-down process. Rather, it is a very interactive process in which each level affects the activities of all other levels.

NATIONAL STRATEGY

The development of an overall national strategy is the highest level of strategic activity. At the same time, it is the most unfocused and requires the least intelligence support. The national strategy consists of a general understanding of the nation's most basic social, economic, political, and military goals, an appreciation of resources, and a general will to use these resources to achieve these goals. A foreign threat, if one exists, is generally sufficiently sensed by the populace and the leadership. Strategic intelligence at this level usually tends to describe that threat in the most general terms.

National Security Strategy

Strategic intelligence takes on a more definitive role in the next level of strategy development, that is, the national security strategy. A country's national security strategy is usually developed by national security staff and advisers under the guidance of the nation's leader, often in concert with national security specialists supporting the nation's highest legislative bodies. The national security strategy outlines both the domestic and foreign threats to the achievement of the national strategy. However, unlike the national strategy, the national security strategy begins to formally develop the social, political, economic, diplomatic, and defense strategies necessary to counter those threats.

Strategic intelligence plays an important role in both describing those threats and providing the rationale for the balance of subordinate social, political, economic, diplomatic, and defense strategies being developed to counter those threats. Strategic intelligence products at this level must help the decision makers prioritize the threats, allocate resources, and provide goals and objectives for subordinate strategy developers.

National Defense Strategy

One of the subordinate strategies is the national defense strategy, which is usually developed by a nation's defense minister and civilian and military staffs. It presents the major aspects of the nation's war doctrines and determines how the resources allocated to the defense department will be used to develop the capabilities necessary to carry out these doctrines. Within this level of strategic activity are a myriad of policies and programs such as research and development, the military-industrial base, mobilization, equipment buys and mixes, active and reserve force structures, and foreign deployments/stationings.

Strategic intelligence at this level attempts to comprehensively depict foreign military threats. This intelligence usually focuses on an enemy's existing and future military forces and their capabilities. Extensive study is also done on the support systems for those forces—equipment production capabilities, mobilization capabilities, command and control, training, research and development, and so forth. This intelligence must be detailed, but it must also be strategically significant. For example, it probably is not strategically significant to know that an enemy tank division has 320 tanks instead of 290. It *is* strategically significant, however, to know that a foreign power has 60 tank divisions rather than 40, or that its tank inventory is approximately 50,000 tanks with two-thirds of those having the latest in target-acquisition technology. Such intelligence would help determine the size of one's own tank forces and one's tank production requirement, and give direction to one's own research and development effort.

Global Military Strategy

The next level of strategy that strategic intelligence supports is global military strategy. This is normally developed by the country's highest military staff, usually called a joint or general staff. These global military strategists take into account the men, equipment, organizations, and doctrine provided by the defense strategists and apply them to those areas of the world where the nation has national interests and objectives. Strategic intelligence used in developing a global military strategy focuses on foreign regions critical to a nation's security and provides detailed assessments of how those regions would either hinder or assist the accomplishment of national goals and objectives. Strategic intelligence at this level must help the strategist make decisions in two major areas: first, the effective allocation of combat power between strategic forces and theater forces, and second, the effective allocation of combat power among various theaters.

At this global level, strategic intelligence begins to have a significant operational aspect. Not only does it help the decision maker determine what should be done, but it also contributes to decisions on how military forces will be used to do it. For example, in addition to assessing what type, how many, and where enemy forces might be located, strategic intelligence could provide global strategists with assessments of how these foreign forces might react to the movement of hostile or even nonhostile forces into their area. This level of strategic intelligence should tell the global strategist if a foreign power would attempt to interdict intertheater supply lines and how that interdiction would most likely occur. The focus of this operational aspect of strategic intelligence is not on enemy operational capabilities within a particular theater. Rather, it is on an enemy's capability to prevent the global strategist from projecting the necessary power to the particular theater or between two theaters. Intelligence assessments of a foreign power's operational capabilities within a single theater are the substance of theater military strategy.

Before moving to a discussion of theater intelligence support, the unique case of the globe as a single theater should be addressed. This situation exists as a result of the strategic nuclear strike forces of the United States and the former Soviet Union, which were a significant part of each nation's global strategy. In the context of a full strategic exchange, the globe truly becomes a single theater. Strategists become, in part, tacticians concerned with the mechanics of target acquisition, target-tracking/homing, and target destruction.

Theater Military Strategy

The final area of discussion is strategic intelligence in support of theater military strategy. Theater military strategy exists in the area of overlap between the strategic and operational levels of war. As a result, both strategic and operational intelligence are needed by the commander in wartime decision making. Strategic intelligence helps the commander determine theater military goals and objectives, how many forces are needed in theater, what types of forces, and possibly how and when these forces will be deployed. Operational intelligence in turn will help the commander develop campaign plans to defeat the enemy on the battlefield.

For example, strategic intelligence could inform a theater commander that the enemy needs a 30-division force to successfully attack, but currently only has twenty. Mobilization of ten divisions will take two weeks. As a result of this strategic intelligence assessment, the theater commander may decide to accept a certain degree of risk that the enemy will not attack; therefore, the theater commander might not keep his total force forward deployed or on a high state of alert.

Operational intelligence, on the other hand, would come into play after the enemy had fully mobilized and was about to attack. For example, operational intelligence might assess that the enemy would move along three axes of attack, the main thrust along the center axis, with the 95th tank division in the lead. Such intelligence would help the theater commander develop the battlefield in such a way that he would determine when, where, and under what conditions he would engage the enemy.

Conclusion

In conclusion, strategic intelligence is intelligence designed for use in developing strategic-level policies, programs, capabilities, and decisions. In formulating strategic intelligence, the needs of the specific level of strategic decision maker must be kept in mind. The most effective strategic intelligence is usually all-source intelligence developed by strategic intelligence analysts working in direct support of a specific strategic decision maker.

DENNIS J. QUINN

SEE ALSO: Intelligence, General; Intelligence, Military; Intelligence, Tactical; Intelligence: Indications and Warning; Strategy, National Security.

Bibliography

Berkowitz, B. D., and A. E. Goodman. 1989. *Strategic intelligence for American national security.* Princeton, N.J.: Princeton Univ. Press.
Godson, R., ed. 1990. *Intelligence requirements for the 1990's.* Lexington, Mass.: Lexington Books.
Knorr, K., and P. Morgan, eds. 1983. *Strategic military surprise: Incentives and opportunities.* New Brunswick, N.J.: Transaction Books.
Levite, A. 1987. *Intelligence and strategic surprises.* New York: Columbia Univ. Press.

INTELLIGENCE, TACTICAL

Military intelligence may be categorized, in narrowing order of scope, as strategic, operational, or tactical. Strategic intelligence serves planning at the broadest, usually international, levels; operational intelligence serves planning within a theater of war or on a front as a whole; and tactical intelligence, sometimes known as *combat intelligence,* focuses on the planning and conduct of battle.

In combat situations, modern field commanders must have a clear understanding of both the enemy and the environment. It is the task of tactical intelligence to collect such crucial information through techniques that include ground and aerial surveillance; spying (or human intelligence, HUMINT) and reconnaissance; technical intelligence of various types (e.g., signals intelligence, or SIGINT; imagery intelligence, or IMINT); counterintelligence (CI); interrogation; and sensory data obtained from target-acquisition and night-observation devices. Tactical intelligence also comprises the analysis and effective dissemination of information and decisions based on that information throughout the command structure.

The distinctive purviews of strategic, operational, and tactical intelligence do not prevent them from overlapping. For example, tools that are used primarily in strategic and operational contexts are often employed selectively in tactical situations: maps and charts of terrain, climate, or infrastructure; data on capabilities of enemy systems; political and cultural information; and so on. This is particularly so in low-intensity conflicts (LICs), because such conflicts are frequently motivated and constrained by complex political and cultural considerations. Although tactical intelligence is the beneficiary of substantial amounts of strategic and operational intelligence (i.e., it receives data from other levels), it also provides information used to make strategic and operational estimates through on-the-ground data gathering such as interrogations, descriptions of new enemy equipment or techniques, and so on.

Tactical Intelligence in Western Militaries

Before delving into the division of command responsibilities and the specific development of tactical intelligence in the combat context, it may be useful to see how tactical intelligence conceptualizes the information it must gather, the targets it may confront, and the conditions under which it must operate. Note that the terms used here are those of the United States, although they are generally analogous to those used by most Western military forces (Soviet and Soviet-inspired tactical intelligence is discussed at the end of this article). All military forces use specially trained tactical intelligence and counterintelligence personnel and equipment.

CONCEPTS OF TACTICAL INTELLIGENCE

Modern battle tends to be multidimensional; hence, commanders often must consider its ground, air, and sea aspects concurrently. For example, at a minimum, naval commanders must consider both the sea and air aspects of battle, ground commanders the ground and air aspects.

As one way of managing the multidimensionality of battle, tactical intelligence divides the combat area into three distinctive zones. The first is the *area of operations* (AO), which is the portion of the battle area necessary for the actual military operations of the command involved. The second is the *area of influence*, which is the geographic area in which the commander directly influences operations by maneuver or fire support. Finally, the *area of interest* comprises the two other areas as well as adjacent territory that extends the objectives of current or planned operations into enemy territory. Each of the areas is viewed in terms of width, depth, airspace, and time. The areas may vary in size depending mainly on the size of the command and whether the force is in the attack or defense mode (see Figure 1).

Commanders further classify tactical intelligence information in the following subcategories:

- Order of battle (OB) intelligence—Identification, strength, command structure, disposition of personnel, units, and equipment of a military force
- Technical intelligence—Identification, including the description, capabilities, and limitations of enemy materiel
- Target intelligence—Detection, identification, and location of a target with sufficient accuracy and detail to permit the effective employment of weapons.
- Terrain/weather intelligence—Data and analysis of terrain conditions in light of current and projected weather conditions and the consequences for friendly and enemy movement, communication, and combat

The targets of intelligence fall into four general subcategories: *movers*, or moving elements; (2) *emitters*, such as communications systems; (3) *shooters*, or weapons systems; and (4) *sitters*, or stationary targets, such as command posts or logistical installations.

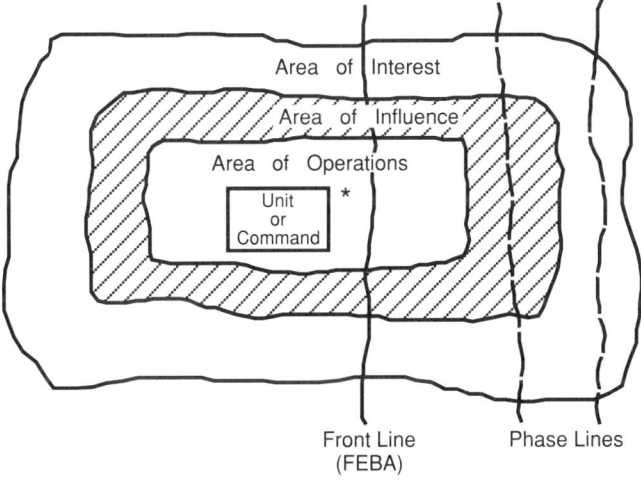

* Note: This applies to an army or air force unit/command or to a naval vessel or task force.

Figure 1. *Battlefield areas.* (SOURCE: Adapted from *U.S. Army Field Manual* 34-1, illus. p. 1–2; and *U.S. Army Field Manual* 34-3, illus. p. 1–3)

COMMAND ORGANIZATION OF TACTICAL INTELLIGENCE

The basic echelons of command are the company (or unit), the battalion, the brigade (or regiment), and the division. Each echelon has a structure of assigned, attached, or supporting intelligence/electronic warfare (IEW) personnel.

Company/unit level. At the company level, tactical intelligence devices from a variety of sources: from *organic* (i.e., officially assigned) on-the-ground resources (troops and patrols) and from technical and supporting elements such as ground surveillance radars (GSRs), fire-support teams, or remotely employed sensors (REMS). For aircraft, the basic tools of collection are the observations of flight personnel and radar or other targeting information. Naval units use sight, radar, sonar, and fire-control equipment. The time-critical nature of combat intelligence at this level means that company or unit commanders serve as the directors, coordinators, producers, and executors of intelligence operations within their units, directing placement of sensors, establishing outposts, and using patrol and troop reports.

Battalion level. At the battalion level, the IEW coordinators are the battalion intelligence and operations officers and their staffs. (In the U.S. Army these officers are known as the S2 and S3, respectively, at the battalion, regimental, or brigade level; as G2 and G3 at the division or higher echelons; and as J2 and J3 on joint staffs.) The S2, the intelligence officer, coordinates combat information and reconnaissance and surveillance operations. The S3, the operations officer, is responsible for electronic warfare (EW) and operations security (OPSEC).

Each frontline battalion in the U.S. Army has a Battalion Information Coordination Center (BICC) working for the S2. The BICC prepares reconnaissance and surveillance plans, information intelligence collection plans, and other plans as required. It employs maneuver companies, scout units, fire-support teams, GSR, and other available units to collect information. It forwards requests to higher or adjacent echelons for information and support that a battalion cannot provide.

The commanders of organic units or of units supporting the battalion are the executors of the battalion. Reconnaissance patrols, the organic battalion scout unit, GSR, REMS, and fire-support teams are employed by the executors to gather information for the S2, S3, and BICC.

Brigade/regiment level. At the brigade or regimental level, the S2 and S3 are again the principal coordinators of IEW operations. The S2 at this level also may have limited counterintelligence duties.

Producers include the BICC, the IEW support element, and a collection and jamming element. BICC responsibilities at this level include:

1. Developing and coordinating the collection plan;
2. Preparing and transmitting tasking messages and reports for information;

3. Developing data for the brigade/regimental S2's intelligence estimate;
4. Developing and maintaining the intelligence database;
5. Processing intelligence;
6. Disseminating combat information and intelligence; and
7. Providing intelligence support to EW and OPSEC.

IEW and collection and jamming support are provided by special military intelligence formations from higher echelons.

These producers combine to provide coordinated IEW support responsive to the commander's needs. Analysis at this level, to include Intelligence Preparation of the Battlefield (IPB—a U.S. Army term), is limited.

Executors include subordinate battalions, direct support field and air defense artillery, engineer elements, and attached or direct support intelligence elements, which include interrogation teams.

The battalions, supporting artillery, and engineer units all perform reconnaissance, surveillance, and target acquisition operations. Artillery radar locates enemy direct fire systems, forward observers collect information, air defense units observe enemy aircraft, and all report information on routes, tactics, types, and numbers. Engineers provide information on terrain, obstacles, and enemy activities. OPSEC is monitored and evaluated by teams or individuals especially trained in OPSEC and CI.

DIVISION

Division coordinators are the divisional intelligence officer (G2) and operations officer (G3), supported by their respective staffs. The G2 may have the added responsibility of formulating division document and personnel security policy. The G3 may formulate similar EW and OPSEC policy and may manage EW and OPSEC operations.

The division G2 section closely coordinates and directs intelligence, CI, special security, weather reporting, and eingeer reporting operations. The G2 operations element coordinates with the G3 operations element and fire-support element to ensure that intelligence and CI operations are integrated to support the commander's scheme of maneuver and fire-support targeting. The G2 also directs IPB and target value analysis efforts within the command.

Division-level producers include intelligence collection and dissemination personnel, information processing personnel (who perform divisional IPB, as well as OPSEC, EW, and weather and terrain analysis), and field artillery and air defense artillery personnel.

The principal executor at division level is an attached military intelligence battalion (in the U.S. Army) or a similar command or elements in other armies (see Figs. 2 and 3).

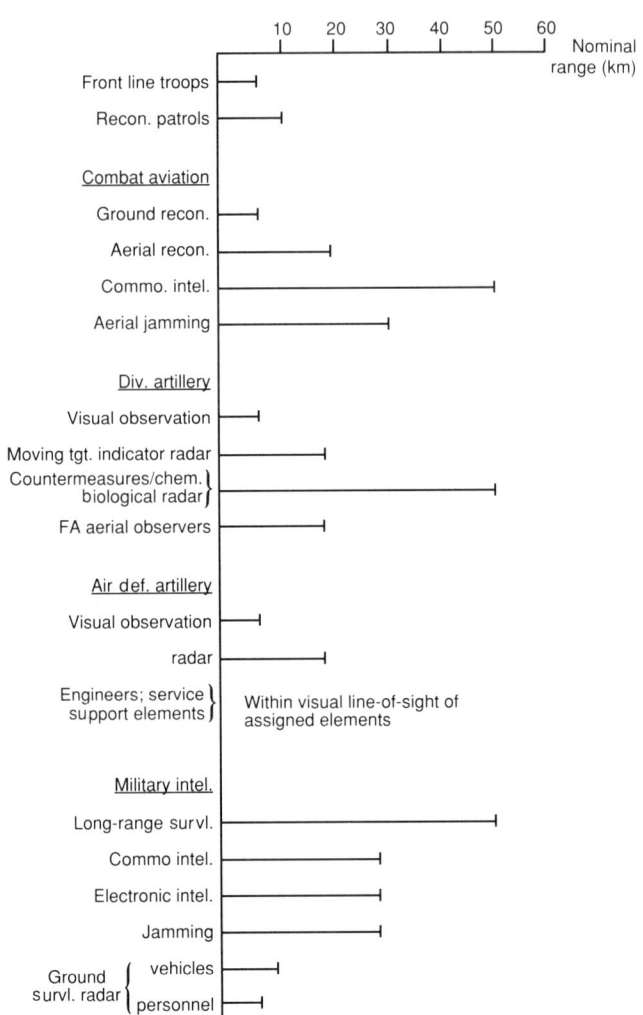

Figure 2. Division information resources. (SOURCE: Adapted from "Intelligence and Electronic Warfare Operations," *U.S. Army Field Manual* 34-1, July 1987, illus. pp. 2-36, 2-37)

DEVELOPMENT OF TACTICAL INTELLIGENCE ON THE BATTLEFIELD

Intelligence preparation of the battlefield is a systematic approach used by the United States to develop combat intelligence. Different national military forces use logical and systematic processes of their own to develop combat intelligence.

IPB is a continual, systematic process of analyzing the enemy, weather, and terrain of a specific geographic area. IPB analysis is based on graphics such as annotated military maps, multilayered overlays, gridded photomaps, microfilm, and large-scale map substitutes, all capable of display on a computer terminal. Battlefield area analysis and intelligence estimates are not replaced by graphics but are converted to them whenever possible.

IPB, according to U.S. Army doctrine, entails five functions: (1) battlefield area evaluation, (2) terrain analysis, (3) weather analysis, (4) threat evaluation, and (5) integration of enemy doctrine with weather and terrain data.

The data developed by the first four of these functions

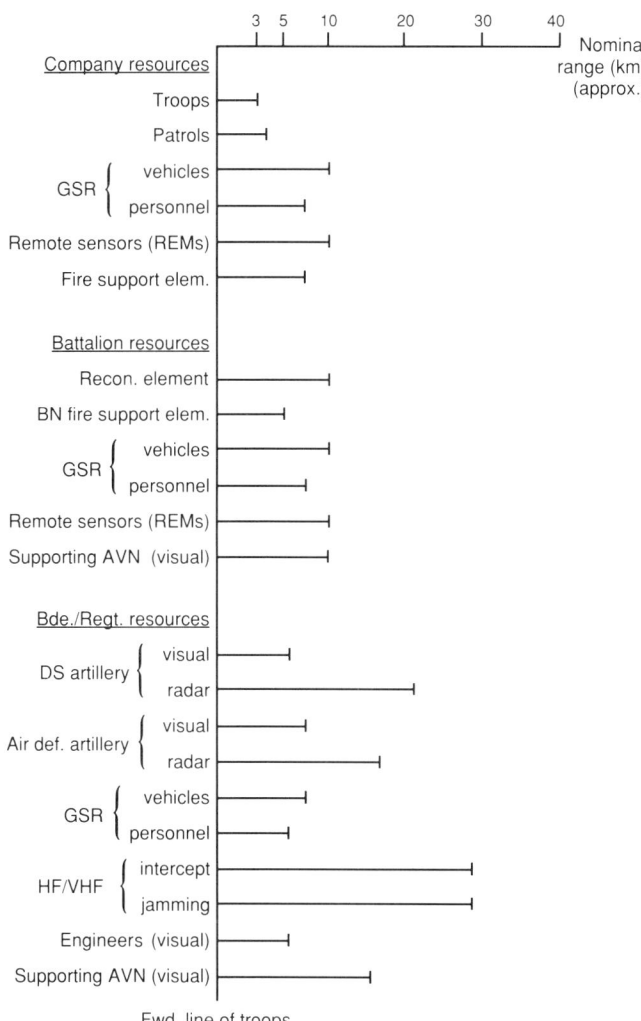

Figure 3. Company-brigade/regiment information resources. (SOURCE: Adapted from "Intelligence and Electronic Warfare Operations," *U.S. Army Field Manual* 34-1, illus. pp. 2-23, 2-25, 2-29)

are reduced to templates, one for each of the functions. The process orients on the AO and area of interest of the command concerned.

IPB Function 1—Battlefield area evaluation. This evaluation process focuses on information about the enemy, terrain, and weather in a specific geographic area (the command's AO and area of interest). The dimensions of these areas are given in terms of width, depth, airspace, and time. The size of the areas is based on an evaluation of the command's mission, the enemy, terrain, troops available, and time. A lower command may recommend the sizes of its areas of operation and interest, but the actual size will be designated by higher headquarters (division, corps, etc.).

IPB Function 2—Terrain analysis. This function addresses the military aspects of the terrain and their effects on both friendly and enemy capabilities to shoot, move, and communicate. Terrain analysis concentrates on several key features.

First, *observation and fields of fire* is primarily visual and electronic line-of-sight analysis. *Concealment* is protection from enemy observation. *Cover* is protection from the effects of enemy fire. *Obstacles* are natural and artificial terrain features that impede, channel, stop, or divert military movement. Since they directly influence mobility, they are the most important considerations in terrain analysis.

The second is *key terrain*—any feature or area that, if seized or controlled, will give a marked advantage in the conduct of operations to either command. The determination of key terrain is dependent on the echelon of command, the mission, the enemy, and the situation. For example, a relatively small hill that dominates an area of operation of a platoon or company is key to those echelons, but may not be significant to a brigade or division.

Third are *avenues of approach*, which are air, ground, or sea routes by which a force can reach an objective or key terrain (or area). Historic examples of sea avenues of approach are Flying Fish Channel, the approach to Inchon Korea (Korean War); and the Straits of Tsushima (Russo-Japanese War, 1904–1905). Avenues of approach are evaluated in terms of their: (1) potential to support maneuver, (2) degree of channelization, (3) obstacles, (4) concealment and cover, (5) observation and fields of fire, and (6) access to the terrain and adjacent avenues.

To make effective use of terrain data, intelligence personnel prepare a terrain factor matrix and a series of overlays to develop a graphic data base of terrain. This matrix facilitates threat integration and guides the selection of terrain and weather factor overlays needed to analyze the terrain. Terrain overlays portray the military aspects of terrain in the AO.

The final step of terrain analysis is to select the avenue of approach that supports friendly and enemy capabilities to move, shoot, and communicate.

The terrain factor matrix matches:

Functions of:	*Against such factors as:*
Observation and fields of fire	Surface configuration (slope) and materials (soils)
Concealment and cover	
Assembly areas	Vegetation
Key terrain	Obstacles (linear)
Ground and air avenues of approach	Weather effects on terrain Transportation
Weapons sites	Built-up areas
DZ (drop zone) and LZ (loading zone)	Surface drainage (hydrology)
Maneuver	
Barriers and fortifications	
Lines of communications/ main supply routes	
Lines of sight	
Communication and EW sites	

IPB Function 3—Weather analysis. Since weather has a significant impact on both friendly and enemy capabilities, weather analysis is critical. In addition, because the weather has a tremendous effect on the terrain, terrain and weather analysis are inseparable. A weather matrix may compare:

Intelligence uses or applications of:	*To factors of:*
Observation and fields of fire	Temperature
Artillery emplacements	Humidity
Concealment	Intervisibility
Camouflage	Surface winds
Ground/air avenues of approach	Precipitation
Cross-country movement	Snow/ice cover
Fording sites	Winds aloft
Air DZ	Cloud data
Helicopter/STOL/VTOL LZ/PZ	Light data
Lines of communication/main supply routes	Severe weather Fog
Nuclear/biological/chemical operations	
Lines of sight (radio/radar)	
Remote sensor emplacement	
Infiltration routes	

IPB Function 4—Threat evaluation. This is a detailed study of enemy forces, their composition and organization, tactical doctrine, weapons and equipment, and supporting battlefield functional systems. The objective of this evaluation is to determine enemy capabilities and how they operate as set out by their doctrine and training. This evaluation is accomplished by using a doctrinal template matrix and doctrinal templates. It also includes an evaluation of high-value targets and troop movement rates, judged within the constraints of the terrain and weather.

The matrix examines types of enemy battlefield weapons, communications, radio electronic combat, reconnaissance, rear services, command and control of division and regiment-level formations, and engineers. These battle systems are applied against types of enemy combat actions (march, meeting engagement, river crossing, attack against a defending enemy, pursuit, hasty defense, and prepared defense).

IPB Function 5—Integration of enemy doctrine. Threat integration is accomplished by the sequential development of situation, event, and decision templates. Since a situation template depicts enemy positions for just one instant in time, a series of templates may be needed to depict how the enemy changes his position during the conduct of an operation. The event template is created after the analyst has decided on the probable enemy course of action. The event template provides the information needed to project what events are likely to occur relative to enemy courses of action. As an enemy moves along an avenue or corridor he will be required by the terrain, weather, and tactics to do certain things at certain times and places. Understanding this, the analyst can select individual areas of interest where he expects certain activities or events of tactical significance to occur. Activity, or lack of it, confirms or denies the enemy course of action. The event analysis matrixing in the U.S. Army is normally performed at division or higher level.

Knowing in advance what the enemy can do and comparing it with what he is doing, the analyst has a basis for predicting what the enemy next intends to do. This information provides the basis for cuing intelligence collection and constructing decision support templates.

Ultimately, the objective of threat integration is to provide options for the commander to defeat the enemy. The Decision Support Template (DST) is developed specifically to aid the commander in decision making. The DST does not dictate decisions to the commander, but it does indicate points where tactical decisions are required. It relates events, activities, and targets of the event template to the commander's decision requirements. The DST is, basically, a combination graphic intelligence estimate and operations plan.

The DST is developed by overlaying the event template, war-gaming enemy courses of action, and placing decision points on the template at points where the commander must decide which course of action to pursue, either to affect the enemy course of action or to change a friendly course of action (such as attack by fire, maneuver, or employment of electronic warfare). Target areas are also selected. There are areas or points where interdiction of enemy forces by maneuver, fires, or jamming will eliminate or reduce a particular enemy capability, cause the enemy to abandon a particular course of action, or require the use of unusual support to continue operations. Some decision points and target areas may be independent, whereas others are associated with one another.

Time phasing is included in this template to illustrate potential enemy movement at his doctrinal rates, as modified by terrain and weather. Potential maneuver corridors are also depicted in the graphics. Other information which may be shown includes key terrain, combat force ratios, or a depiction of how the enemy may have to deploy within each avenue.

In sum, IPB provides one logical and systematic approach to developing military intelligence.

Soviet Intelligence Doctrine

The intelligence doctrine of the former Soviet Union is followed by a number of nations and has affected U.S. and other NATO intelligence and counterintelligence efforts. This doctrine embodies the same three general types of intelligence as U.S. doctrine—strategic, operational, and tactical.

Ground commanders acquire information about opposing forces by means of intelligence collection and target acquisition. The methods they employ to obtain information are basically the same as those employed by other nations. A basic tenet of this doctrine is rapid success. This requires commanders to have timely information on weather, terrain, and the enemy.

The Soviets had an excellent intelligence collection, analysis, and dissemination capability. This capability was organic to all echelons from regiment to front. The results of reconnaissance below the regiment level were passed up to that level, or above, for evaluation.

The most important element of combat support under this doctrine is reconnaissance, which is defined as all measures taken to collect information on:

- Nuclear weapons and other means of mass destruction,
- Formations,
- Organization for combat,
- Intentions, and
- Weather and terrain of the specific area of future operations.

This doctrine emphasizes maximum use of firepower and mobility, which means that target acquisition must be accurate and rapid. Tactical reconnaissance is carried out to varying depths by specialized units as well as by regular troop units.

The various collection methods often overlap and are redundant. Reconnaissance is considered effective only if it is conducted actively, aggressively, and continuously. Timeliness, reliability, and accuracy of information are stressed.

Aviation units organic to Soviet fronts performed air reconnaissance employing visual, photo, infrared, radar, and signal intelligence capabilities. These aircraft were armed and could attack any ground target. High-priority targets could often be engaged within two hours of detection.

The Soviets had extensive radio and radar intercept capabilities. Intercept units were deployed just behind leading maneuver regiments. Their target acquisition radar could intercept electronic emissions approximately 25 kilometers (15.5 mi.) forward of their troops. Very high frequency (VHF) could reach about 40 kilometers, high frequency (HF) groundwave about 80 kilometers, and HF skywave was unlimited. Airborne interception greatly extended these ranges. "Clear" traffic was evaluated and exploited immediately, but decryption was normally very slow.

The Soviet direction-finding capability was equal to that for intercept. Targets within artillery range would be attacked within minutes of location. Front, army, and division artillery units had organic target acquisition units.

Soviet divisions had organic reconnaissance battalions that often operated a day's march ahead of the main body. Regiments on the march used battalion-sized advance guards. The advance guard battalion sent a company forward and that company usually deployed a reinforced platoon forward as a reconnaissance patrol. Engineer, artillery, and chemical reconnaissance elements were cross-attached to leading reconnaissance units in the advance.

A hallmark of the Soviet tactical intelligence effort was aggressive patroling and the employment of outposts, raids, and ambushes (see Fig. 4; compare Fig. 2 and 3).

UZAL W. ENT

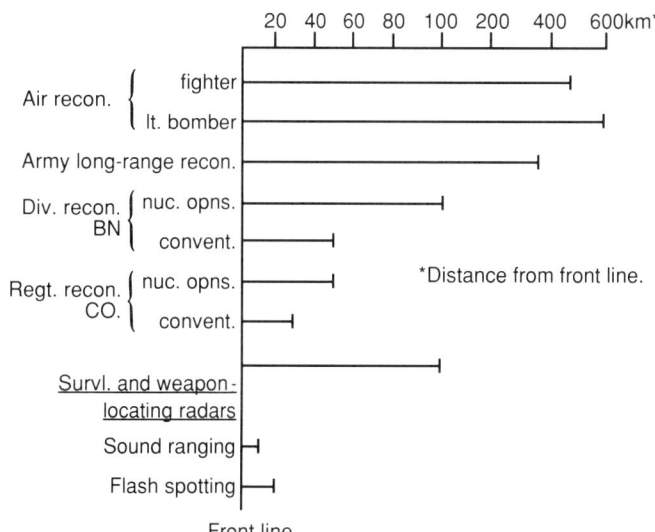

Figure 4. Range of Soviet reconnaissance means. (SOURCE: Adapted from "Intelligence and Electronic Warfare Operations," *U.S. Army Field Manual* 34-1, illus. pp. 4-4)

SEE ALSO: Counterintelligence; Intelligence Analysis, Military; Intelligence Collection; Intelligence Cycle; Intelligence, Human; Intelligence, Imagery; Intelligence, Strategic; Intelligence, Signal; Security Classification; Surveillance and Target Acquisition Equipment.

Bibliography

Campbell, S. 1987. Celtic cross IV. *Military Intelligence*, March, pp. 21, 48. (U.S. Army Intelligence Center and School, Ft. Huachuca, Ariz.)
Eldridge, J. L. C. 1987. Exploiting Soviet vulnerabilities by the brigade and task force. *Military Intelligence*, June, pp. 42–44, 47. (U.S. Army Intelligence Center and School, Ft. Huachuca, Ariz.)
———. 1988. OPFOR in the defense. *Military Intelligence*, January, pp. 38–39. (U.S. Army Intelligence Center and School, Ft. Huachuca, Ariz.)
Ivanov, D. A., V. P. Seval'yev, and P. V. Shemansky. 1969. *Fundamentals of tactical command and control: A Soviet view.* Trans. U.S. Air Force. Soviet Military Thought Series, no. 18. Washington, D.C.: U.S. Air Force.
Levesque, R. W. 1987. LIC doctrine and intelligence. *Military Intelligence*, October, pp. 32–34. (U.S. Army Intelligence Center and School, Ft. Huachuca, Ariz.)
Rundle, S. L. 1987. Battalion task force counter-reconnaissance. *Military Intelligence*, March, pp. 38–40. (U.S. Army Intelligence Center and School, Ft. Huachuca, Ariz.)
Saint, C. E. 1987. Intelligence rquirements at the operational level of war. *Military Intelligence*, March, pp. 6, 48. (U.S.

INTERIOR AND EXTERIOR LINES OF OPERATION

The twin concepts of interior lines of operations and exterior lines of operations (or operations on convergent lines and operations on divergent lines) are simple and straightforward. A military force is operating on interior lines when it is centrally located between two hostile forces that are unable to support each other directly because they are too far apart, because there are intervening terrain features, or because the commanders are simply not skilled enough to coordinate activity against the central force (Fig. 1). A force is operating on exterior lines when two separated forces seek combat by coordination of their activities against a centrally located enemy (Fig. 2). The two situations have much in common, except that when a centrally located force takes the initiative, it is said to be "operating on interior lines," and when two (or more) externally located forces cooperate, or coordinate their activities, against the central force, they are "operating on exterior lines." Note that in this essay the word *operations* is used in two senses: generally it is a term referring to military combat activities at any or all levels of aggregation, such as "operations in mountains," "amphibious operations," or "operations in the face of the enemy." It is in this general sense that the word is used in the term *interior lines of operations*. In relatively modern times, however, the term *operations* is often used to refer to a level of military activity, in contradistinction to the lower level of tactics and the higher level of strategy. In this second sense, the word *operations* is synonymous with the term *operational art*.

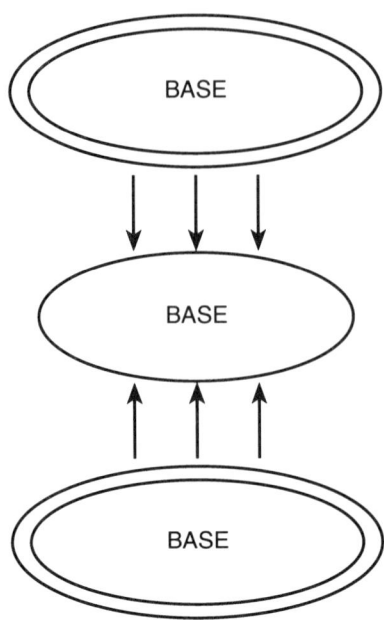

Figure 2. Operations on exterior lines.

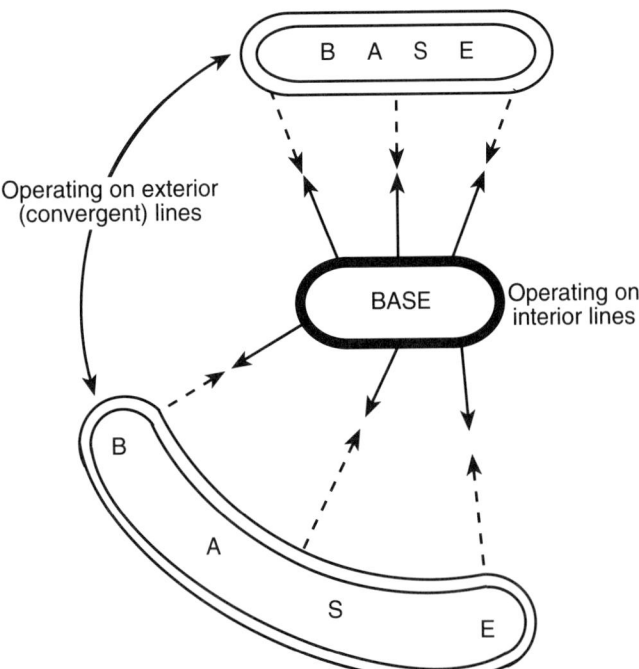

Figure 1. Operations on interior and convergent lines.
NOTE: *In this figure the term "exterior" can be substituted for "convergent."*

Operations on Interior Lines

Operations on interior lines occur when the central force seizes the initiative; this activity can be tactical or operational (or strategic) in nature. It was a favorite maneuver of Napoleon, both strategically and tactically. His first major tactical use of the concept was at the Battles of Montenotte and Dego in 1796. During World War I the German armies also operated strategically on interior lines.

The essence of operations on interior lines is shown in Figure 3. The commander of the central force, threatened by the convergence of two hostile forces (X and Y), usually with a total strength exceeding his own, sends a small fraction of his command (A) to stop or slow down X, one of the converging enemy forces, while massing the bulk of his strength (B) against the other converging force, Y (Fig. 3a). If contingent B defeats, or even merely stops force Y, the central commander will leave a small segment (B') of his contingent B to pursue or hold force Y in check, while turning with most of contingent B to join contingent A in a massive attack against force X. In this way a weaker force can obtain relative local superiority against the overall numerical superiority of the opponent's separate armies, defeating them in detail, or in turn. This was the essence

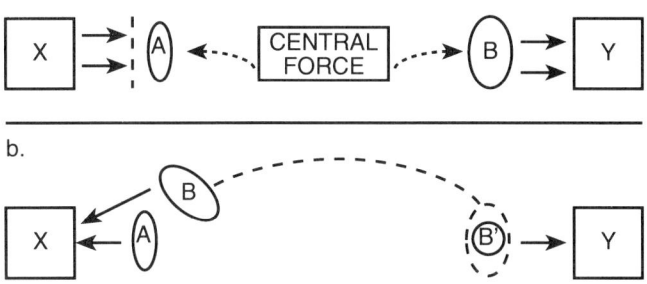

Figure 3. Operations on interior lines.

of the tactics used in 1796 by Napoleon at Montenotte and Dego, and in a number of other battles, in most of which he deliberately inserted his forces between two stronger enemies for the purpose of taking advantage of interior lines.

Napoleon frequently used this concept strategically. He came close to success in using it in each of his last three campaigns: 1813, 1814, and 1815. He used the concept against the armies of Wellington and Blücher at the beginning of the Waterloo Campaign. Eventually he failed in all three of these campaigns because of a combination of superior hostile strength and blunders on the part of some of his subordinates.

Operations on Exterior Lines

Operations on exterior lines occur when converging forces are able to seize the initiative, and coordinate their convergence in such a manner as to prevent, or frustrate, any attempts by the central force to take advantage of its interior lines. Referring to Figure 3a, operations on exterior lines are successful if Force X defeats and drives back the blocking contingent, A, of the central commander, and/or if contingent B is repulsed by Force Y, which then continues its advance toward a junction with Force X.

One of the most successful tactical examples of operations on exterior lines was the victory of the Allies over Napoleon in the Battle of Leipzig in 1813. A classic example of strategic operations on exterior lines was the invasion of France in 1814 by the Allies, eventually overwhelming Napoleon's desperate efforts to use his interior lines. Equally classic was the strategic cooperation of the Western Allies and Soviet forces in the invasion of Germany in 1945.

Complications and Elaborations

Some modern military theorists have related other forms of maneuver with the concept of operations on interior and exterior lines—sometimes successfully, sometimes not. For instance, the concept marks the fundamental start point of what Simpkin (1985, p. 21) quite reasonably refers to as the *maneuver theory*, and which Luttwak (1987, p. 92) less persuasively refers to as the *relational maneuver* approach, which he contrasts with what he calls *attrition-minded* thought.

If the term *line of operations* was inherited from the eighteenth-century "geometry-of-war" theorists such as von Bülow, Lloyd, and Tempelhoff, the distinction between maneuver on exterior lines and maneuver on interior lines was Jomini's legacy.

Although Jomini exerted considerable influence on subsequent military writers, his terminology was not universally accepted. Many writers and staff regulations have other terminology for these two basic forms of maneuver, with stress placed on some intrinsic detailed characteristic of each type.

Other names for the "exterior line maneuver" are "wing or flank maneuver," "maneuver on the enemy's rear" (*maneuver sur les derrières*), or "converging attack." Sometimes, quite incorrectly, "envelopment" or "turning movement" are considered examples of operations on exterior lines. The term *operation on interior lines* is also often associated with other specific forms, such as "dislocation," "penetration," and "breakthrough" in the more general concept of "central maneuver," which is nevertheless different from the attrition or frontal attack (Willoughby 1939, p. 43, 74). Again, it is obvious that these maneuvers are rarely the same as the true operation on interior lines.

The above concepts were born in the framework of land warfare, but they can also be applied in some measure to naval tactics. At least in the pre-airpower and pre-missile era, one of the main goals was the famous "crossing (or capping) the 'T'" maneuver. (This entailed taking the enemy under enfilade fire while avoiding the full effect of the opponent's firepower. It is a naval tactic that has some analogy with the converging attack of land warfare.)

In a very general sense, all classification of maneuver could be applied to higher levels of strategy and grand strategy. Indeed, sea and air lines can be considered as strategic exterior lines on which one can combine seapower, airpower, rapid intervention, and power projection capabilities with air-land operations in a wider grand strategy.

Liddell Hart's concept of the "indirect approach," as unoriginal and obscure as it may be (Aron 1976, vol. 1, p. 210; Bond 1977, pp. 54 ff.), has sometimes been linked with the concept of maneuver on exterior lines. It seems, however, to be a different approach to, rather than a mere system of, the maneuver theory (Simpkin 1985, p. 133). According to this new theoretical approach, the traditional classification of the basic types of maneuver should be replaced by the "strategic and operational deep penetration" concept, considered as the most characteristic, if not the unique, form of maneuver in modern times. In fact, deep penetration may be intended to perform both of the typical goals of the exterior and interior lines maneuvers (i.e., respectively the encirclement and the dislocation of the enemy's force).

Maneuver on Interior Lines and Dislocation

Certain preconditions must exist before an operation on interior lines can be employed. These include the opponent's deployment and operational plan, physical and geopolitical characteristics of the ground, and operational capabilities of one's own forces. An overextension of the opponent's front line and a marching plan based on separate converging, or even parallel, columns over compartmented and hostile terrain, may result in some gaps in the enemy's lines. These gaps can be exploited if one's own army can occupy them faster than the opponent's expectation. Training, speed, and proximity of the army are the fundamental prerequisites for exploiting these gaps. The risk is the possibility that the opponent will succeed in concentrating his separate bodies in a pincer-like maneuver. If that occurs, the strategic interior line maneuver results in a tactical encirclement for the force using interior lines, as happened to the French army at Waterloo.

Such a risk is much smaller in the case of stabilized and continuous fronts, in which the central maneuver can be realized only at the tactical level, even if it may have operational or strategic effects, provided that the eventual success is fully exploited. French military thought has given a special name, "action of dislocation," to the procedure against a part of the enemy forces isolated in the course of the battle, a procedure that may be defined as "exploiting a gap in the line" (Willoughby 1939, pp. 76 f.) in a different way than penetration.

Interior Lines and Breakthrough Operations

At first glance, breakthrough and penetration operations could be regarded as special cases of operations on interior lines. In fact both breakthrough and penetration operations are different from the frontal attacks extended along the entire front line, even if the bloody and ineffective offensives of the First World War were often regarded by the postwar military reformers as if they were nothing but frontal attacks.

But breakthrough and penetration are conceptually different from an interior line operation. The latter is a general maneuver form for all three levels of the art of warfare, while the former are mere tactical procedures, which in principle may be useful not only to a maneuver on interior lines, but also to an operation on exterior lines or a deep penetration maneuver.

Interior Line Advantage

The interior line move almost always is the only resort of the defender who is aiming to prevent, or to stop early, the attacker's converging columns. But deep penetration also may allow the attacker the advantages of interior lines "once the attacker starts to develop a bulge beyond operational depth" (Simpkin 1985, pp. 97 f.).

What military literature calls "the advantage of interior lines" consists of the facilitation of a force's movements and supply, which are in principle faster and safer than those of the enemy. However it is necessary that the distance between the two contingents of the force operating on interior lines be sufficiently greater than the cumulative range of the attacker's fire and that airpower can be neutralized in some way; otherwise the advantage of interior lines would be counterbalanced by the disadvantage of being between two fires of the two enemy forces within supporting distance of each other. As Clausewitz pointed out, the advantages of interior lines grow with their scale, the minimum being at the tactical level and the maximum at the strategic level. Examples of the latter include the cases of Germany in both world wars and Israel in the Arab-Israeli Wars.

In the case of siege war, as well as in the case of encirclement or blockade, the defender enjoys interior line advantages, but lacks both initiative and a line of communications, and thus cannot be said to be operating on interior lines.

Line or Flank Maneuver: Converging Attack and Envelopment

Operations on exterior lines differ from a simple envelopment because the two or more elements of an army maneuvering on exterior lines are widely separated without any reciprocal direct cooperation. Furthermore, concentration of forces is realized in a more sophisticated way (i.e., by coordination of separate bodies and their converging directions, instead of by the physical reunion of the entire army into one body or direct cooperation between the covering and main forces, as in envelopment).

In exterior line operations the physical and geopolitical nature of the terrain, the opponent's deployment and intentions, and the mobility of one's own army are decisive. If the opponent cannot utilize the advantage of interior lines, nor extend his front with reserves, the only possible parry against a converging maneuver is a partial or complete evacuation of the battlefield so that the enemy columns converge on empty space—a "blow in the air" (Willoughby 1939, p. 101).

The converging attack of several columns marching on separate routes concentrates the force on a single point of gravitation, while forcing the opponent to divide his forces in diverging directions. This is fully true at the tactical level, where the concentration of forces nearly coincides with that of fire. The effects of converging firepower upon each side of the enveloped force is theoretically compounded; by contrast, the diverging firepower of the enveloped force has only direct effects. This becomes less and less true as the scale of operations increases, because at the higher levels the converging attack results in an ever-weaker concentration of forces, lacking the cumulative effects of converging firepower.

Flank Maneuver and Deep Penetration

A modern form of flank maneuver is deep penetration, the only possible maneuver on a continuous front. Different from the traditional converging attack, the deep penetration may not only have the goal of enveloping and encircling the opponent's forces, but also the primary goal of dislocating and disrupting it. In the last case, envelopment and encirclement may have a complementary or preliminary role to allow ever-deeper penetration.

Different from the traditional flank maneuver, in deep penetration the main force is not necessarily the biggest, but always the fastest and best of the army. It is a mobile force with logistical autonomy, formed by mechanized, rotary-wing, or airborne units, or by a combination of these. The same strategic and operational characteristics may be recognized in the rapid intervention forces based on power-projection capabilities acting on sea and air lines. The mobile force of the attacker is a true "lever arm," as Simpkin says (1985, pp. 93 ff.). Once hostilities have broken out, an attritional action is first necessary to clear the way for insertion of the mobile force. The latter has to gain local freedom of action by tactical combat combined with semiclandestine and clandestine operations. Then mobile forces can penetrate beyond the opponent's tactical deployment, spreading out in separate columns directed at objectives beyond his defended zone.

The mobile force "turning movement" (what Americans call "enveloping") will become an "envelopment" (in the European sense of the term), perhaps ending in complete "encirclement" of the enemy. To envelop a withdrawing force, the attacker must bar the retirement route with a mobile advanced detachment (as Grant did at Appomattox) or with a heliborne force.

Origin of the Maneuver Theory

Since the late eighteenth century military writers have noted the similarity between Frederick the Great's famous "oblique order" (that facilitated flanking movements) and the tactics used by the Greek general Epaminondas at Leuctra, who refused one flank with a "holding" force to fix the enemy, and a main force for envelopment. Even earlier the ancient Chinese strategist Sun Tzu recommended dividing an army into two forces, the "ordinary" force (cheng) to deceive the enemy and the "extraordinary" force (ch'i) to defeat it (Simpkin 1985, p. 37).

However, the military historian Delbrück (1900–1908, vol. 1, p. 156; vol. 4, p. 318) made a sound and clever distinction between what he called the "wing battle" of ancient military history, which Epaminondas improved, and the "oblique order" of more modern times, where the disrupting wing also has enveloping tasks. According to Delbrück (1900–1908, vol. 4, p. 135) and Pieri (1975, pp. 91 f.) the "wing battle" was theorized first in modern times in 1653, by the Italian general Montecuccoli.

Modern tactics has inherited from ancient military history many other examples of the wing battle. Among these the most famous was undoubtedly Hannibal's double flank maneuver at Cannae in 216 B.C., which inspired the Schlieffen plan against France.

Generally, military writers have followed Jomini's appraisal and attributed to Napoleon a preference for central rather than flank maneuver. Also, it is possible to argue this point on the basis of historical examples (namely, the Napoleonic campaigns of 1796, 1805, 1809, 1813, 1814, and 1815) and by reference to Napoleon's *Maxims of War* (published in 1831), which contained criticism of the effectiveness of the Frederician "oblique order." Napoleon suggested that this was a myth spread by foreign observers who had seen the Prussian army exercises at Potsdam, and also spread deliberately by Frederick the Great to deceive his potential future enemies.

Liddell Hart's well-known prejudice against what he called the "Ghost of Napoleon," whom he made responsible for the bloody frontal attacks of the First World War, may have contributed to the common idea of Napoleon's preference for central maneuver. However, as Willoughby (1939, p. 137) points out, "Napoleon's campaigns contain at least thirty characteristic maneuvers against the hostile rear, but he attempted many others and planned still more."

Clausewitzian Approach to the Theory of Exterior/Interior Lines

Liddell Hart criticized Clausewitz for depreciating the "geometry of war," as well as its chief derivative, the "maneuver theory" (Aron 1976, Bk. 1, pp. 130 f., 210 f., 265, 268). Clausewitz's *On War* (1984, Bk. 7) has been interpreted as suggesting that maneuver is not only different from, but even in some way opposite to, decisive attack, as they are "opening gambits in a game of chess." This is believed to be a misreading of Clausewitz, who emphasized that the purpose of maneuver was to bring about favorable conditions to gain an advantage without which victory might otherwise be difficult or impossible.

With regard to flank maneuvers, Clausewitz (1984, Bk. 5; Bk. 6) wrote: they "are among those prize exhibits of the theorists that are seldom found in actual war." According to Clausewitz, there are two kinds of flank maneuver, one with the entire army united marching on an enemy's rear, the other where an army is divided into a "holding force" and an "enveloping wing." These two kinds of flank maneuvers have different specific risks. The former exposes one's own rear to the enemy, while the latter reduces strength before the battle.

One can find the best Clausewitzian criticism of Jominian maneuver theory in *On War* (Bk. 6), on "convergence of attack" and "divergence of defense." Clausewitz considered maneuver theory self-deluding in imagining that the defender would remain unreactive and motionless, thus granting initiative and freedom of enveloping action only to the attacker. At the strategic level enveloping attack is much less easy than at the tactical level because the locations of the wings are almost always much stronger, as when the defensive front is extended from a sea or neutral country to another. In such a case the attacker cannot operate on converging lines, and his freedom of action is limited. But, Clausewitz said, it is even more limited when the attack has to be converging, as it would be if France and Russia were to attack Germany.

The advantage of attacking the enemy flank or rear decreases as the size of enemy combat units involved becomes greater because an army corps can fight on many sides more easily than can a battalion. Certainly, a converging attack grants strong tactical advantages: improving the effectiveness of the attacker by converging firepower and weakening the defender whose firepower is diverging. These advantages are counterbalanced, however, by the advantages of interior lines paradoxically granted to the defender by the converging attack itself. At the strategic level, Clausewitz points out, the only true superiority granted by converging attack (i.e., that in firepower) becomes uninfluential because "crossfire is not relevant to strategy."

Crisis in Exterior/Interior Lines Operations in World War I

Unlike Liddell Hart's retrospective appraisal, the Clausewitzian criticism of maneuver theory and flank attack was quite uninfluential on military thought, particularly on the Jominian-minded German general staff. The latter put into practice classical flank maneuvers in the 1866 and 1870 campaigns, and planned a similar maneuver on a very large scale against France as early as 1892. General von Schlieffen said that frontal attack is indecisive because it merely drives the enemy to a new position, thus a new operation must be started. The system of converging attack was considered as ideal by the prewar German regulations. Only the 1913 French Regulations considered frontal attacks and tactical breakthroughs as a means of decisive success (Willoughby 1939, pp. 81, 100).

Several military factors changed during the nineteenth century, however, the interaction of which strongly reduced the technical feasibility of flank maneuver. At the strategic level, the advent of massive national armies numbering more than a million men turned natural borders into continuous front lines, as Clausewitz had foreseen. The development of railroads increased the importance of interior lines by increasing the speed and capacity of logistical systems. At the tactical level field fortifications, machine guns, and artillery improved defensive capabilities; and heavy artillery made it possible to deliver converging firepower without converging movements.

At the outbreak of the First World War each opponent tried to envelop the other to avoid expensive frontal attacks. But, both German and French converging attacks and enveloping maneuvers collapsed on the western front in August and September 1914, leading to a rapid extension of the front to its natural extremities. This was followed by stabilization of the front line and stalemate. On the eastern front, a converging attack by the Russians was stopped by the famous central maneuver of Tannenberg, combining a strong defense by the German covering force with a converging attack by the German main force.

The only alternative to trench warfare and bloody attrition warfare was a successful penetration attack followed by close-in central maneuver for the sake of broadening the breakthrough and separating enemy forces. The junction between two enemy armies was the preferred point of penetration. This was especially true if the armies were from different countries because of the wider political implications of a successful offensive. However, these junctions generally were placed in easily defended locations where attacks were difficult to mount. In 1922 Delbrück criticized the geographical location of the 1918 German offensive because General Ludendorff, following the tactical line of least resistance, failed to achieve the political and strategic goal of the campaign—the separation of the English army from the French and the consequent rolling up of the former.

Theory of Exterior/Interior Lines and the Indirect Approach

Even if successful, a breakthrough was rarely followed by appreciable penetration and never by a true central maneuver. The main reasons for this were technical difficulties in advancing artillery to keep up with the infantry and the failure to free troops from other sectors to exploit success. Moreover, as Luttwak (1987, p. 119) wrote,

> The defensive counterconcentration proceeded faster than the advance of foot soldiers against artillery, barbed wire, and machine guns. Indeed, the intellectuals who dominated the postwar French general staff could prove mathematically the counterconcentration superiority of defense over offense, inexorably derived from the *speed* advantage of sideways railway and truck movements along the front over the rate of foot-infantry advance against fire.

Liddell Hart's theory of the "indirect approach" may be considered an interesting attempt to transfer the exterior line concepts from the tactical and operational scale of maneuver theory to the wide scale of grand strategy. Liddell Hart, like other postwar military reformers, believed that traditional infantry armies should be confined to secondary tasks as holding or covering forces, while a decision would be achieved through seapower (according to the traditional British approach to warfare), strategic use of chemical weapons and airpower, power projection capabilities (similar to the failed 1915 Gallipoli amphibious operation), and mobile-mechanized land forces.

In the specific field of land warfare, Liddell Hart stated that his concept of "deep strategic penetration" was influenced by both the Mongol campaigns of the thirteenth century and the operations of William T. Sherman and Nathan B. Forrest in the American Civil War. From Sherman, Liddell Hart derived his concepts of "alternative objectives" (putting the opponent on the "horns of a dilemma," as Sherman said) and "baited gambit" ("to trap the opponent, by combined offensive strategy with defensive tactics") (Bond 1977, pp. 48 f.).

The concepts of "alternative objectives" and "baited gambit" indicate the different goals that are implicit respectively in central (or interior line) and flank (or exterior line) maneuvers. By contrast, the "strategic deep penetration" can hardly be connected with interior lines under the general concept of central maneuver, as Willoughby (1939, pp. 79 ff.) did with regard to penetrations in the First World War.

VIRGILIO ILARI

SEE ALSO: Blitzkrieg; Clausewitz, Karl von; Envelopment; Frederick the Great; Hannibal Barca; Initiative in War; Jomini, Antoine Henri; Lines of Communication; Maneuver; Moltke the Elder; Offense; Operational Art; Sun Tzu; Tactics.

Bibliography

Aron, R. 1976. *Penser la guerre. Clausewitz.* Paris: Gallimard.
Bond, B. 1977. *Liddell Hart. A study of his military thought.* London: Cassell.
Clausewitz, C. von. 1984. *On war.* Ed. and trans. M. Howard and P. Paret. Princeton, N.J.: Princeton Univ. Press.
Delbrück, H. 1900–08. *Die Geschichte der Kriegskunst im Rahmen der politischen Geschichte.* 4 vols. Berlin: G. Stilke.
Luttwak, E. N. 1987. *Strategy. The logic of war and peace.* Cambridge, Mass., and London, U.K.: Belknap Press of Harvard Univ. Press.
Paret, P., ed. 1986. *Makers of modern strategy.* Princeton, N.J.: Princeton Univ. Press.
Pieri, P. 1975. *Guerra e politica negli scrittori italiani.* Milano: Mondadori.
Simpkin, R. 1985. *Race to the swift: Thoughts on twenty-first century warfare.* London: Brassey's.
Willoughby, C. A. 1939. *Maneuver in war.* Harrisburg, Pa.: Military Service.

INTERNAL SECURITY AND ARMED FORCES

The primary purpose of a state's armed forces is to provide for and enhance the state's security. Therefore, they serve as the principal coercive apparatus of the state and have distinctive external and internal functions. First, the armed forces protect the state from external aggressors and promote its interests through the coercion of other states. Second, they protect the state against internal threats. This latter function may be broadly defined as *internal security*, and it consists of those measures taken to free and protect a society from lawlessness, subversion, and insurgency. Such measures are not limited to the sovereign territory of a state, but may also include territories controlled by the state. Nor are these measures the duties only of the state's regular land, sea, and air forces; they may be performed by paramilitary, auxiliary, and militia forces.

Counterinsurgency

Various terms have been used to describe internal threats to a state: guerrilla warfare, internal war, revolutionary war, subversion, and terrorism. These terms have no generally agreed-upon definitions, and they are often used imprecisely, indiscriminately, and sometimes interchangeably. These varying forms of internal political violence may be subsumed under the title of *insurgency*, which is defined as an organized movement aimed at the overthrow of a constituted government through the use of subversion and armed conflict.

Insurgencies have both political and military components. The political wing of an insurgency attempts to mobilize the masses, organize an infrastructure, and create a shadow government. The insurgent military forces seek to isolate the civilian population from incumbent government control, destroy government forces and institutions, and demonstrate the inability of the regime to protect the population and control the situation.

An insurgency develops in several phases. These phases are not unilinear, and they have no set sequence or duration. A particular insurgency may not experience every phase and may, in fact, repeat some phases. The first phase, *latent and incipient insurgency*, consists primarily of subversive activity that may range from the potential threat of subversion to subversion occurring frequently and in an organized pattern. This phase does not involve a major outbreak of violence or uncontrolled insurgent activity. The second phase of insurgency, *guerrilla warfare*, involves the initiation of organized guerrilla warfare or related forms of violence against established authority. During this phase, the insurgent seeks to consolidate political control and to create an infrastructure within liberated areas. The third phase, *war of movement*, is characterized by conventional conflict between organized forces of the insurgents and government forces. The goals of this last phase are the defeat of the government forces and consolidation of power within the political system.

Counterinsurgency consists of military, paramilitary, political, economic, psychological, and civil actions taken to defeat an insurgency. In formulating a viable counterinsurgency strategy, a state must use not only the armed forces, but all the resources at its command, to address the problems of both internal defense and internal development.

An internal defense and development strategy has four

components: balanced development, security, neutralization, and mobilization. The objective of balanced development is to correct those conditions that make a society vulnerable to an insurgency: that is, to balance political, social, and economic development. To do this, the state must identify the root causes of the insurgency and initiate preventive measures. Security includes all those activities designed to protect the population from the insurgent forces and to provide a secure environment for national development. Security operations protect the population and government resources from the insurgents by isolating the insurgent forces and denying them a base of popular support. These operations attempt to create conditions in which the local population can provide for its own security. Neutralization consists of all lawful activities to disrupt, disorganize, and defeat an insurgent organization. The primary targets of neutralization operations are the leadership and control elements of an insurgency. Such operations may range from public disclosure and discrediting of insurgent leaders to full-scale combat operations. Finally, mobilization encompasses those activities intended to motivate and organize popular support for the government. These activities attempt to reinforce existing institutions and create new ones that will respond to the needs of the populace. Mobilization operations are designed to enhance the legitimacy and credibility of the government.

The armed forces perform a number of functions in a counterinsurgency: intelligence, psychological operations, civil affairs, population and resource control, and counterguerrilla operations. Intelligence operations support a counterinsurgency by providing both general and specific knowledge of the area of operations and the insurgent forces. Effective intelligence is, therefore, key to a successful counterinsurgency and must integrate information from all sources. Included under the intelligence function are counterintelligence efforts to protect government information, personnel, and facilities against insurgent intelligence operations. Psychological operations are planned psychological activities directed at enemy, friendly, and neutral audiences to influence attitudes and behavior that affect the achievement of political and military objectives. They serve to mobilize popular support for the government and to undermine the legitimacy and morale of the insurgents.

Civil affairs operations are concerned with the relationship between the military forces and the civil authorities within an area of operations, and may range from military civic action to the exercise of local governmental authority. These operations frequently involve engineering, logistical, and medical projects and may also include the exercise of police and administrative functions. Population and resource control operations are essentially police-type operations. They maintain law and order and ensure public safety. A key objective is to sever the supporting relationship between the population and the insurgent forces.

Counterguerrilla operations are tactical combat operations designed to reduce the guerrilla threat in an area and to provide a secure environment. They include attempts to destroy or capture guerrilla forces, to reduce guerrilla capability for offensive action, and to deny guerrilla access and entry into an area. Counterguerrilla operations incorporate a variety of tactical operations, including strike campaigns, raids, patrols, and ambushes.

All these differing operations have some common features: independent, small-unit operations; centralized control and coordination; flexibility; mobility; patience; and economy of force. When combating an insurgency, the armed forces cannot operate in a vacuum, but must be fully integrated within the broader national strategy.

Modern counterinsurgency theory is drawn from several sources. Among the earliest theorists were the French colonial soldiers Joseph Gallieni and Louis Hubert Lyautey. As a result of experiences in North Africa, Indochina, and Madagascar in the late nineteenth century, they formulated the concept of *tache d'huile* ("oil slick"): a form of progressive pacification that combined the slow methodical expansion of colonial administration with a military presence. France used the *tache d'huile* strategy in Indochina from 1946 to 1954. A product of the French failure in Indochina and the resulting introspection was the concept of *La Guerre Révolutionnaire*. This doctrine saw the world in an integrated political and military context. It also borrowed some Marxist concepts, on the premise that, if the war could be understood from the perspective of Communist doctrine, the enemy's strategy could be frustrated.

The British approach to counterinsurgency was significantly shaped by the Malayan Emergency of 1948–60. Central to their philosophy was the notion that the nature of the insurgency must be understood before it could be eradicated. Sir Robert Thompson's seminal work, *Defeating Communist Insurgency* (1966), was a product of this experience. More recent experiences in Oman and Northern Ireland have further shaped British doctrine.

U.S. counterinsurgency doctrine was strongly influenced by the experience gained advising the Filipino army during the Huk Rebellion of 1946–54. The Philippine and Malayan experiences provided the basis for the counterinsurgency doctrine adopted by the Kennedy administration in 1961. The Cuban Revolution and the threat of other Latin American revolutions gave further impetus to the formulation of American counterinsurgency doctrine. During the 1960s, this doctrine was applied in a variety of situations, primarily in Southeast Asia and Latin America.

An example of the application of U.S. counterinsurgency doctrine was the Strategic Hamlet Program in the Republic of Vietnam during 1962–63, which was essentially a population and resource control measure. Strategic hamlets were areas secured by government forces and the focus of government development and civic action projects. The rural peasantry was relocated to these hamlets to isolate them from the insurgent forces and establish government-controlled enclaves. The program failed for a number of reasons, including rapid overextension, failure to provide adequate security, and the people's resentment at being removed from ancestral lands.

Following the failures in Southeast Asia and the collapse of the Alliance for Progress in Latin America, U.S. interest in counterinsurgency waned. However, the recent emphasis on low-intensity conflict in U.S. military doctrine has revived interest in counterinsurgency and foreign internal defense. While the approaches discussed vary in detail, they all derive from general principles applicable to any counterinsurgency.

Counterterrorism

Terrorism is the unlawful threat or use of force against individuals or property to coerce or intimidate governments or societies to achieve political, religious, or ideological objectives. It has a long and established history and may properly be included within the concept of insurgency as discussed above. However, terrorism has commanded such attention in recent times that it deserves a special comment. The use of terror by separatist and insurgent organizations remains essentially an internal security problem, but the rise of international and transnational terrorism blurs the distinction between internal and external security threats.

As a security function, counterterrorism has a twofold mission: to prevent and deter terrorism and to respond to terrorist acts as they occur. These missions require very specialized skills, methods, and operational techniques. Conventional military units are ill-suited in terms of organization, training, and equipment for counterterrorist operations. Highly trained counterterrorist units that emphasize intelligence, police-type work, and special operations are necessary to defeat a terrorist threat. The prevention and deterrence of terrorism relies on intelligence operations designed to locate the terrorists, penetrate their infrastructure, and detect events before they occur. This requires the integrated use of intelligence from all sources and an effective system of information exchange. Preventing terrorism also requires special operations units to apprehend or pre-empt terrorists before they initiate an operation. In responding to terrorist acts once they occur, intelligence remains an essential element. In these situations there is also a need for operational units trained to neutralize terrorists, conduct assaults, and rescue hostages.

Counterterrorist units are by nature elite special operations units. The very existence of such units is often classified and their organization, equipment, and tactics almost invariably secret. A state may create dedicated counterterrorist units such as the German Border Group 9 (*Grenzschutzgruppe 9*) or the American SFOD-DELTA (DELTA). As an alternative approach, existing special operations units may be tailored to incorporate counterterrorist missions, as the British have done with the Special Air Service. In either case, counterterrorist units are distinct from the conventional military establishment.

Military Occupation

Internal security is not limited to the sovereign territory of a state but may be a function of military occupation. The seizure or acquisition of control is the determining factor in establishing occupation. It may occur in either war or peacetime; while the bases of occupation may differ, the results will be the same. Occupation does not constitute conquest or the transfer of territorial sovereignty as long as the opposing state struggles to regain that territory.

Military occupation frequently occurs in conjunction with an ongoing conventional conflict. The actual occupation begins with the armed forces' first contact with civilians and continues through the establishment of a military government and the consolidation of authority. The occupying power has both significant authority and responsibility. It may supervise or even administer the civil government through existing officials, those it designates, or its own forces. To ensure the proper functioning of government, it may exercise executive and judicial authority, levy and collect taxes, and enact legislation. The occupying power also has the responsibility to maintain order and safety, providing for the protection of the civilian population. If at all possible, it must do so through already existing laws.

In the course of military occupation, the armed forces perform a number of functions similar to those performed in a counterinsurgency: counterguerrilla operations, intelligence, civil affairs, psychological operations, and population and resource control. In a military occupation, however, the nature of the threat is different, resulting in different priorities and objectives for the armed forces. The major security threats during a military occupation are conventional forces that have been cut off or have assumed a stay-behind mission, unconventional warfare units, partisans, and resistance fighters. The objectives of these forces are to disrupt and harass the occupying forces and to support conventional military operations directed against them. Counterguerrilla operations designed to locate, fix, and destroy these forces are a priority security mission. While the scope and scale of these counterguerrilla operations may differ from those directed against an insurgent force, they are essentially the same as counterinsurgency operations.

Civil affairs operations are concerned with establishing a functioning civil government under the supervision or control of military authorities. Given the disruption and dislocation that attends a military occupation, this may be more a process of institution building. Population and resource control operations are police operations designed to maintain order and enforce laws. Both civil affairs and population and resource control operations may be performed by military units directly or through indigenous personnel. The objective of occupying powers is to distance the military from these functions, eventually replacing regular military units with local police and civil authorities.

World War II provides several examples of the armed forces' internal security function during a military occupation. The German occupation of Europe provides a case study of counterguerrilla and security operations, population and resource control, and the exercise of military government. This occupation was ended by military defeat and subsequent occupation by the Allied Powers. During the American occupation of Germany and Japan, major emphasis was placed on establishing a functional civil government and building democratic institutions. Local officials administered the government where possible, thereby facilitating the termination of the occupation. Counterguerrilla operations were not a major concern. Consequently, American occupation forces were reorganized from combat regiments into constabulary force units with a primary mission of population and resource control.

The Soviet occupation of Eastern Europe provides a contrasting case. Following the liberation of this region from German occupation, the Soviet Union installed governments dominated by Communist parties. This resulted in the creation of Communist states supportive of and subservient to the Soviet Union. The formation of the Warsaw Treaty Organization allowed for the continued stationing of Soviet troops in Eastern Europe and, in effect, perpetuated a form of Soviet occupation. This provided a means for suppressing dissent and opposition to Soviet policy, as in the case of Hungary in 1956 and Czechoslovakia in 1968.

A more recent example of military occupation is the Israeli occupation of the West Bank, the Golan Heights, and Gaza. The Israeli military has continued to occupy these territories since the 1967 Arab-Israeli War, performing primarily a population and resource control mission.

Paramilitary Forces

Internal security missions are not restricted to the regular military forces of a state, but may be performed by paramilitary forces. These are forces or groups distinct from the regular armed forces, but which resemble them in organization, equipment, training, or mission. They include national police forces, local defense units, workers' militias, intelligence services, and political or "security" police.

National police are full-time militarized police units under the control of the central government. They are usually quartered in barracks and equipped with light military weapons and military vehicles. Within the Soviet Union, prior to the reorganization that resulted from the August 1991 coup attempt, both the Committee of State Security (KGB) and the Ministry of Internal Affairs (MVD) had paramilitary forces organized along these lines. The border guards of the KGB were organized in regular combat formations and equipped with tanks, armored personnel carriers, aircraft, and ships. The Internal Troops of the MVD were similarly organized and equipped. Both forces were used to secure internal installations and deal with domestic disturbances. Another example of this form of paramilitary organization is the Saudi Arabian National Guard (SANG). Recruited from select Bedouin tribes loyal to the royal family, the SANG is supplied with modern military equipment, including armored vehicles, and is organized for an internal security mission.

Local defense units are rural paramilitary organizations and are frequently referred to as militias. They consist of armed local personnel who reside in their own homes rather than in military barracks. Militia units are typically lightly armed and are often not even uniformed. They are usually deployed in their own area as part-time military defense units in connection with an insurgency. The Vietnam War provides an example of the extensive use of local defense units. South Vietnamese rural paramilitary forces were organized into two tiers: Regional Forces were company-size units that operated at the province level, while Popular Forces were platoon-size units based in hamlets. The Civilian Irregular Defense Group (CIDG) was another distinct rural militia under the direction and control of U.S. Army Special Forces. Begun in 1962 as a village self-defense program, the CIDG was incorporated into the South Vietnamese paramilitary forces upon the departure of U.S. Special Forces from South Vietnam in 1971.

The U.S. Army National Guard (USANG) is a unique institution that has evolved from what were once local defense units. It is an outgrowth of the deeply rooted Anglo-American militia tradition of armed citizens formed into self-defense units under local control. A dual military structure was incorporated into the U.S. Constitution, with provisions for both a national army and state militias under the control of their respective governors. The Dick Act of 1903 brought the state militias under greater federal control and supervision as the National Guard. By 1920 the National Guard, when in federal service, constituted part of the Army of the United States. The USANG is organized and equipped along the same lines as the regular army and is trained primarily for conventional combat. However, these units remain a state force that can be mobilized in the case of civil disturbance or natural disaster. As federal control of the USANG has increased, some states (such as North Carolina) have revived the state militia. In keeping with tradition, these militias are composed of unpaid volunteers who are mobilized when the USANG is called into federal service, and are primarily concerned with emergency preparedness.

Workers' militias are most often found in revolutionary

socialist states. They comprise part-time personnel who are recruited on the basis of political reliability. These units are designed chiefly for political mobilization and domestic security. Members of workers' militias live at home and are partially uniformed and armed. These units are typically mobilized for ceremonial or political occasions, but they can have a military reserve function. Cuban Committees for the Defense of the Revolution and Nicaraguan Committees for the Defense of Sandinism are examples of workers' militias. They coordinate political indoctrination, facilitate mass mobilization, and inform on suspected counterrevolutionaries.

National intelligence organizations are an integral part of a state's internal security system. These intelligence agencies may have independent field operating staffs, special operation units, and other paramilitary units that give them the capability to conduct independent internal security operations. They may also include security police within their organization, as discussed below. Instead of a single intelligence service, most states have a system of intelligence organizations with overlapping functions and responsibilities. Intelligence agencies conduct counterintelligence and counterespionage operations directed at foreign intelligence agents. They also operate against terrorists, insurgents, subversives, and others who are perceived as security threats. Their activities may also include the surveillance of dissidents and the monitoring of contacts between foreigners and their own citizens.

Both liberal democratic states and authoritarian regimes perform these intelligence activities. Germany has several services concerned with internal security: the Military Screen Service (*Militarischer Abschirmdienst*) is responsible for preventing espionage, subversion, and sabotage directed against the armed forces, and the Federal Office for the Protection of the Constitution (*Bundesamt fur Verfassungsschutz*) collects and evaluates intelligence on efforts to disrupt the constitutional order and security of the state. In France, the Directorate for Surveillance of the Territory (*Direction de la Surveillance du Territoire*) is subordinate to the Ministry of the Interior and is primarily charged with counterespionage and counterterrorism. The *Shin Beth*, or SHABAK, is the Israeli counterespionage and internal security service. Its responsibilities include investigation of all forms of subversion, sabotage, and terrorism directed at Israel by internal and external enemies.

The last category of paramilitary forces to be addressed are the political or security police, also referred to as the secret police. These forces differ from other types of police forces in that their principal function is internal security. Political police may not be in uniform, and they make extensive use of part-time personnel. A well-developed network of agents and informants is available to monitor the population. The political police may be run by the interior ministry, the defense ministry, or the state intelligence or security service, or it may be an independent agency. Military intelligence or police units also may operate as security police. Most nations have more than one agency functioning as political police, often in competition with one another.

Political police engage in surveillance, intimidation and coercion, physical control, and detention. They may operate special prisons, labor camps, or detention centers separate from the normal criminal justice system. Political police are an element of authoritarian or totalitarian states and are seldom found in liberal societies. Within Nazi Germany, the Gestapo (*Geheime Staats Polizei*, or Secret State Police) and the SD-Inland (*Sicherheitsdienst-Inland*, the internal party security service) were the principal security police, facilitating internal control and coercion throughout the Reich and occupied Europe. In the case of the Soviet KGB, the security police were an element of the state intelligence agency. Ongoing reform and restructuring will greatly reduce or eliminate these functions. Most of the security services of the former Warsaw Treaty Organization countries were patterned after and had ties with the KGB. While these states have retained some form of security service, they have been placed under greater government control, and ties with the KGB have been severed.

National Security Ideology

National security ideology is a unique set of ideas and principles about national security that has evolved in Latin America. It is derived in part from European geopolitics and French and American writing on counterinsurgency. It was developed in the context of the bipolar conflict between the United States and the Soviet Union. The enemy is perceived to be international communism, but the threat is from internal subversion, not Soviet invasion. This ideology supports a biological conception of the state as an organism with a life and consciousness of its own. The objective of national security is to ensure the well-being and growth of the state; the safety and health of the whole body politic is more important than that of any individual part.

National security ideology assumes a total war in which there are no clear boundaries between war and peace, between military and civilian functions. As in Clausewitz's axiom, war is the continuation of politics by other means. The duty of a soldier is both the external and internal defense of the state. This form of warfare involves all the resources of the state. State security depends upon an integrated system of political, economic, psychological, and military actions, all under the aegis of the military. Anything that disrupts the equilibrium of this organic state, or that fosters political or social decay, is harmful and a threat to security. Underdevelopment is considered to be the root cause of social decay and subversion. Total war requires combating both the agents of subversion (by military action) and the conditions that threaten the social system (by development). Thus, the armed forces no longer have a specialized military mission; instead, they must be both active agents of development and guarantors of security. Counterinsurgency and internal development are necessary elements of internal security. National security mandates that the military control the internal functioning of the state, resulting in a bureaucratic-authoritarian state dominated by the armed forces.

National security ideology is not a monolithic belief system. As a result of differing intellectual traditions, variations of this ideology exist within Latin America. Different interpretations place varying emphasis on military action and development. The Argentine armed forces, for example, were greatly influenced by the French counterinsurgency theorists of the 1950s and 1960s. Writing about their experiences in Indochina and Algeria, the French emphasized military action over political and economic development. Much of Brazil's National Security Ideology was formulated at the *Escola Superior de Guerra* (Superior War College) under the guidance of General Golbery do Couto e Silva. The Brazilian doctrine had a greater balance between national security and national development. In contrast, Peruvian ideology placed a very strong emphasis on development, advocating policies of land reform, nationalization and regulation of multinational corporations, and expansion of the public sector. Developed at the *Centro de Altos Estudios Militares del Peru* (Center for Higher Military Studies), this doctrine was primarily the product of experience gained fighting a rural insurgency. All the variations of national security ideology believe civilian governments are incapable of protecting the security of the state. This leads to greater direct military intervention in political affairs, which results in authoritarian regimes.

Security Assistance

Security assistance consists of those programs designed to help a state establish and maintain an adequate defense posture. For example, the professionalization and modernization of the armed forces improve a state's ability to resist external aggression as well as to provide for internal security. Security assistance programs provide equipment, spare parts, supporting material, and military training and services that enable the armed forces to achieve modernization. While security assistance programs may be directed toward improving the armed forces' capacity for external defense, they may also enhance internal security capabilities. Broadly defined, security assistance programs encompass the regular military forces of a state, and may also include paramilitary, intelligence, and police forces. In addition, they may include development assistance designed to foster a secure and stable environment.

In its current form, security assistance stems from the professionalization of European armies. As these armies modernized, they became models upon which other armies were patterned. In the nineteenth and early twentieth centuries, military missions, primarily French and German, trained the armies of developing states, particularly in Latin America. Such efforts at modernization were designed to improve these armies' external security functions. At the same time, these endeavors improved their capacity for internal security. In the post–World War II era, a number of former colonial powers, notably France and the United Kingdom, provided security assistance to their former dependencies. However, the primary suppliers of security assistance were the United States and the Soviet Union.

Since World War II, U.S. security assistance has been directed at containing communism, promoting stability, and combating internal subversion. In the immediate postwar years, the primary focus was collective security and defense against external aggression. However, the Greek Civil War and the Huk Rebellion demonstrated that internal subversion was still a concern. In the 1960s, emphasis shifted from external security to counterinsurgency and internal development. Security assistance programs were intended to create professional armed forces; the specialization and expertise these forces required, it was believed, left no time for politics. Professional military roles and political roles were theoretically incompatible, making intervention unlikely and the armed forces apolitical. In fact, the professional skills necessary to combat insurgency and promote development foster interdependence of military and political spheres. The net effect is expansion of the political role of the armed forces and greater intervention in politics.

Following decolonization after World War II, a number of newly independent states turned to the Soviet Union and the Eastern bloc. Soviet leaders of the time viewed security assistance as a tool with which to expand their influence and counter the West, and they used it to help revolutionary governments consolidate power (e.g., Angola, Ethiopia, Mozambique, and Somalia). This assistance was used to train elite special units, intelligence services, and security police. A number of Eastern bloc states acted as agents for the Soviet Union, the most notable being Bulgaria, Czechoslovakia, and the German Democratic Republic. A triangular relationship existed in Eastern bloc security assistance: the Soviet Union provided arms, transportation facilities, and strategic influence; the German Democratic Republic provided highly sophisticated technical and administrative expertise; and Cuba provided the necessary manpower. The principal objective of this assistance was to improve the client states' internal security capabilities. With the 1989 unification of Germany, the liberalization of Eastern Europe, and the restructuring of the Soviet Union, security assistance of this nature has become a thing of the past.

Future Developments

The liberalization of Eastern Europe, the changes in the Soviet Union, and the decline in East-West tensions will affect the role of the armed forces in internal security. The Warsaw Treaty Organization has ceased to exist, and Soviet forces are being withdrawn from Eastern Europe and the Baltic States. While it is unlikely that the East European states will totally dismantle their intelligence and internal security services, these organizations will be reduced in scope and less involved in internal repression. They will continue to be concerned with terrorism, ethnic violence, and espionage. It is also likely that these states will reduce security assistance to developing countries. Some states, particularly China and Cuba, are likely to resist liberalization and will not relax their internal security measures or diminish their security assistance.

As a result of the Conventional Forces in Europe (CFE) Treaty, the United States and its NATO allies are less concerned with a conventional conflict in Europe. This should result in a greater focus on low-intensity conflict. Security assistance to those countries threatened by insurgency will continue to be a priority. The link between counterinsurgency and counternarcotics operations is likely to become stronger. There is also the potential for cooperation between the United States and the former Soviet states on such issues as terrorism and regional conflicts.

Ethnic unrest, national allegiances, political discontent, social inequities, and religious strife will continue to pose threats to stability and security. They will manifest themselves in the form of political violence, terrorism, and insurgency and will not be limited to the developing world. Counterterrorism and counterinsurgency will remain a major concern for all states, and internal security will continue to be a principal function of the armed forces.

HENRY AXEL KRIGSMAN, JR.

SEE ALSO: Armed Forces and Society; Civil-Military Relations; Counterintelligence; Coup d'etat; National Guards: International Concepts; Paramilitary Forces; Political Control of Armed Forces; Subversion; Terrorism; Terrorism, Legal Aspects of.

Bibliography

Beckett, I., and J. Pimlott. 1985. *Armed forces and modern counter-insurgency.* New York: St. Martin's Press.
Graves, A., and S. Hildreth. 1985. *U.S. security assistance.* Lexington, Mass.: Lexington Books.
Finer, S. 1988. *The man on horseback: The role of the military in politics.* Boulder, Colo.: Westview Press.
Janowitz, M. 1977. *Military institutions and coercion in the developing nations.* Chicago: Univ. of Chicago Press.
Richelson, J. 1988. *Foreign intelligence organizations.* Cambridge: Ballinger.
Stepan, A. 1973. The new professionalism of internal warfare and military role expansion. In *Authoritarian Brazil*, ed. A. Stepan. New Haven, Conn.: Yale University Press.
Thompson, R. 1966. *Defeating communist insurgency.* London: Chatto and Windus.
Turner, S. 1991. *Terrorism and democracy.* Boston: Houghton Mifflin.

INTERNAL SECURITY FORCES

The need for internal security forces is as old as organized society itself. From the moment societies enacted rules of conduct, some sort of police force became necessary. The Egyptian pharoahs had police, as did the city-states of ancient Greece, both to enforce laws governing the citizenry and to defend the political structure. The Roman Empire developed an urban police force that survives today in Rome as the Vigili Urbani. Later, police forces in Europe were quasi-military in character, capable of defending the cities and their business enterprises against foreign or domestic enemies.

Napoleon gave France the prototype of its present-day paramilitary force, the Gendarmerie, and modernized its existing national police. This system survives not only in France but in many countries that are former French colonies as well as some that were not.

The Allies' victory in World War II brought a new complication to internal security. The victory of democratic principles (except in the Soviet Union and other Communist countries) was widely interpreted to mean that states could not use force against their own citizens even in defense of basic institutions. Nor could the state employ intelligence means within its own borders, so the argument went, because the state exists only through the will of the people. The fallacy of this argument was widely recognized in all states, as a small minority of the citizenry could disrupt society and prevent the exercise of the rights of the majority, obviously an intolerable situation. For this and other reasons, internal security forces were given names, frequently incorporating the word *people's* to identify them with the rights and aspirations of the people at large.

Modern police forces are now armed with more sophisticated weaponry, use modern technology including computerized data banks, and practice those forms of surveillance necessary to identify groups or individuals who threaten internal security. All such police activity is strictly regulated and monitored in most free countries of the world to prevent the abuse of the rights of citizens. Nevertheless, controversy flares from time to time as a result of overaggressive police actions or overly protective reactions of civil libertarians.

In many countries, the rise of dissident groups dedicated to a single political issue, sometimes quasi-religious in character, has led to deliberate defiance of the law. So strong are the beliefs of these groups that they consider their position to be above the law of the state. In a society that encourages freedom of expression, universal agreement on political issues is not expected. Hence, such deliberate defiance can only interfere with the rights of the majority and lead ultimately to a breakdown of law and order. These developments have made it necessary for internal security organizations to concentrate their efforts on intelligence gathering and counterintelligence within the nation to prevent dangerous confrontations.

The forces dedicated to the protection of internal security have been compelled to increase the level of sophistication of their operations while at the same time adopting equipment and firearms comparable to those of the armed forces, making them appear to be paramilitary organizations. In some countries, the armed forces have been given the internal security mission. In a few countries, the armed forces also serve as the national police.

International Police Organizations

The internal security problems of a nation may be rooted within the borders of another nation. Terrorism may be plotted thousands of kilometers away from the target. Because of this, the need for collaboration among the police forces of all nations was recognized early in the twentieth century. The idea of a central information bank was discussed at the first International Congress of Criminal Police in 1914, but it was not implemented due to the outbreak of World War I.

The organization now widely known as Interpol grew out of a second international meeting in 1923 and was reorganized with new governing statutes in 1946 when its headquarters was moved from Vienna to Paris and the name Interpol was adopted as its telegraphic address. Its formal name—International Criminal Police Organization—was adopted in 1956; its headquarters was moved in 1966 to 26 rue Armengaud, 92211 Saint-Cloud, France.

Interpol has no powers of arrest, but its interchange of information is vital to the internal security forces of its 139 member nations. Interpol even collaborates with countries that are not members.

Types of Internal Security Forces

Forces used to protect internal security can be generally classified into three types, although there are variations within each type. The organizations in a few countries do not fit into any of these types.

THE FRENCH MODEL

The first general type of internal security force is the French model, which traces its origins back to Napoleon's Gendarmerie, a highly professional paramilitary police force organized along military lines with its personnel quartered in barracks. It is under military command but collaborates with civil authority throughout the country in defense of national security. Two French agencies with internal security responsibilities are the General Directorate of State Security, which reports on subversive activities, and the External Documentation and Counterespionage Service, which deals with spy infiltration and treasonable activities.

Candidates for enlistment in the French Gendarmerie must meet high standards and must have completed their term of military service, thus providing a military capability in the Gendarmerie.

Many countries that have adopted the French model also maintain a separate national police for major urban areas and a gendarmerie to concentrate on rural areas. Other countries that have adopted the French model for internal security include:

Algeria—a modern French-trained gendarmerie paramilitary force of about 24,000 with a brigade in each of 31 departments of the country. As in France, it is under military command.

Benin—a French-type paramilitary gendarmerie of about 2,000.

Burundi—a gendarmerie of about 1,500 men.

Cameroon—Gendarmerie Nationale du Cameroun is an internal security force of about 4,000.

Congo—People's National Gendarmerie is a French-type paramilitary force of about 1,900.

Guinea—three paramilitary forces including the French-type gendarmerie of about 1,000, a people's militia of about 7,000, and a republican guard of about 1,600.

Madagascar—a national security police force for its provincial capitals and a gendarmerie for rural areas.

Mauritania—a national police for urban areas and a gendarmerie for rural areas.

Morocco—a national security police force and the Royal Moroccan Gendarmerie, a total of about 35,000 paramilitary.

Niger—a national security police force organized in brigades for urban areas and a national gendarmerie, and a paramilitary force for rural areas for a total of about 2,500 paramilitary.

Togo—a paramilitary strength of about 1,500 in a national police force and the Togo National Gendarmerie.

Tunisia—a paramilitary strength of about 5,000 in its national police force and national gendarmerie.

THE ITALIAN MODEL

The second general type is the Italian model with its highly acclaimed and disciplined carabinieri, an arm of the Italian armed forces with the combined duties of military police and internal security force. The carabinieri are organized in a military structure with three divisions controlling nine brigades, which in turn control a total of 24 legions. In addition, there are two separate brigades, one for training and one a mechanized unit. All units are quartered in military barracks. Each provincial capital has a unit to deal with civil disturbances.

For other police duties, Italy also maintains a state police force for highway, frontier, and other duties and a municipal police force (Vigili Urbani) for traffic control and local law enforcement in cities and large towns. The state police represents Interpol in Italy.

Other countries with this type of organizational structure are:

Brazil—both civilian and military police organizations in each state of the republic. Its military police forces are part of the armed forces and are organized in platoons, companies, and battalions.

Chile—Carabinieros de Chile.

Colombia—about 37,500 paramilitary in its carabinieros responsible generally for rural areas and controlled by the Ministry of National Defense.

The Philippines—maintains public order partly through the Philippine Constabulary, which is part of the armed forces with a strength of about 27,000, and partly through the Metropolitan Police Command in Manila, which is dedicated to internal security.

THE SEPARATE ORGANIZATION

The third general type is a model in which the organization is separate from the nation's armed forces, reporting to the Ministry of the Interior or other national agency or, in some cases, to the head of government. Many nations have adopted this model, including the United States and Great Britain, both of which are treated separately below.

In some of these countries, the organization is designed to preserve the status of the armed forces as the defender of the nation against external foes and not for use against its own people (forbidden by law in the United States). Other nations with this organization include the Soviet Union, also treated separately below.

In the Soviet Union and some other Communist states, the organizational structure is designed to protect and defend the national ideology against all opposition, domestic as well as foreign.

UNITED STATES

With 50 states, each having a high degree of autonomy, law enforcement in the United States is of necessity a complex operation. Internal security is a federal responsibility but agencies of the states play important supporting roles. State, county, and municipal police agencies share information and collaborate with federal agencies. The national guard, discussed below, also has a role as a state military force in peacetime to restore order and preserve vital services in any emergency situation, including riots and public disorders as well as natural disasters.

At the federal level, the Internal Security Section is one of the seven main elements of the Criminal Division of the Department of Justice. The investigative arm of the Department of Justice is the Federal Bureau of Investigation (FBI), which has jurisdiction in all domestic security matters including counterespionage and prevention of sabotage. The FBI maintains an extensive data bank and shares its information with other law enforcement agencies throughout the country. The FBI has 59 field divisions with offices throughout the country to conduct its investigative work. Any of the 17,500 police forces at the state or local level can be called upon by the FBI for support and for making arrests. The U.S. armed forces are not authorized to make arrests and, except for the U.S. Coast Guard, are not used as law enforcement agencies. (The Coast Guard is part of the Department of Transportation and has a law enforcement role in drug interdiction and boating safety.)

The Central Intelligence Agency (CIA), established in 1947, has no law enforcement powers. Its director is responsible to the National Security Council, which is headed by the president of the United States. The CIA is responsible for the collection and evaluation of foreign intelligence on all matters of national security. On counterintelligence matters, it must collaborate with the FBI, which has primary domestic responsibility.

The U.S. National Guard has a unique dual status as a state military force under the command of a state's governor and as a major element of the nation's armed forces reserve. In its state status, the national guard can be mobilized by the governor to preserve order and protect property. In the event of war and a general call-up of national guard units into federal service, internal security within states would be provided by state defense forces, now organized and manned by volunteer personnel not eligible for military service by reason of age or disability.

GREAT BRITAIN

After its early history of constables providing police protection in its towns, Great Britain in 1829 established its Metropolitan Police, which is now a sophisticated force utilizing the most modern advances in forensic science and operational techniques. Its area of jurisdiction includes Greater London (except the City, which has its own police) and areas in Essex, Hertfordshire, Kent, and Surrey. It has a strength of about 27,000.

The now widely known Special Branch of the Metropolitan Police has jurisdiction in matters deemed to be dangerous to the security of the state. Also within the Metropolitan Police is the Special Patrol Group, which assists in the suppression of civil disturbances.

The importance of intelligence was recognized in Great Britain more than a century ago. Three main intelligence organizations are now maintained: the Security Intelligence Service, which reports to the Home Secretary and is in charge of internal counterespionage and antisubversive activity; the Secret Intelligence Service, which reports to the Foreign Secretary and is responsible for external intelligence activity; and the Government Communications Headquarters, which is the main intelligence collection agency in charge of signals intelligence.

Within the British armed forces are the regular reserves of about 150,000 and the territorial army of about 75,000, both of which can be called up by a Queen's Order in an emergency.

Soviet Union

After the Russian Revolution of 1917, the Communist government established a strong central police with wide jurisdiction. The organization that evolved was the All-Russian Extraordinary Commission for the Repression of Counterrevolution, Speculation, Sabotage, and Misconduct in Office, best known as the Vecheka, a pronunciation of its acronym VChK, or simply as Cheka. The Cheka consolidated Communist authority through widespread use of its full powers and its dedication to Communist ideology. The Cheka subsequently became the State Political Administration (GPU), the People's Commissariat of Internal Affairs (NKVD), the Ministry of Internal Affairs (NVD), and now the Committee for State Security (KGB).

In the Soviet Union, as in all ideology-based countries, a strong internal security organization was necessary to its stability. The KGB fulfilled this mission in the Soviet Union. Organizationally, the KGB consisted of four chief directorates, four other directorates, and six independent departments, all of which had some role to play in preserving internal security.

The First Chief Directorate was responsible for foreign operations through ten regional departments that oversaw KGB activity throughout the world. The Second Chief Directorate was directly responsible for internal security through twelve departments. The Fifth Chief Directorate maintained surveillance of dissident activity in collaboration with the Second Chief Directorate. The Seventh Directorate was charged with surveillance of foreigners within the Soviet Union and of Soviet citizens who contacted them. All other directorates were responsible for phases of operations of this large government agency.

The KGB also maintained operational control over the military intelligence organ, the Chief Intelligence Directorate (GRU), although the GRU reported to the Armed Forces General Staff.

A separate Soviet paramilitary organization was the Voluntary Society for the Support of the Army, Aviation, and Fleet (DOSAAF). It was a part-time force with an estimated strength of 80 million.

Eastern Europe

Most nations of Eastern Europe, formerly linked closely with the Soviet Union through the Warsaw Pact, traditionally maintained somewhat similar organizations to preserve security of the institutions in their ideology-based states. With breakup of the Warsaw Pact and widespread movements in Eastern Europe toward multiparty political systems, these internal security forces are rapidly evolving into less repressive operations more compatible with those of the Western world. Some of these countries are:

Albania—although not a part of the Warsaw Pact, Albania's new constitution with an elected president, who is not allowed to hold office in a political party, is bringing about gradual reforms in its paramilitary internal security force of 5,000 and in its armed forces and its People's Gendarmerie.

Bulgaria—the end of a tightly controlled communist society greatly reduces the need for 172,500 personnel in its paramilitary force, which includes security police, border guards, and people's territorial militia.

Czechoslovakia—although the communist dictatorship has ended, unrest among ethnic groups demanding autonomy presents a need for internal security forces, traditionally maintained in two paramilitary forces, a people's militia of 120,000 and about 11,000 border troops.

Hungary—rapid movement toward a more open society reduced the need for its state security police numbering about 30,000, half frontier guards and half internal security troops, and its workers' militia, a paramilitary force of about 60,000.

Poland—with a popularly elected president and legislative body and greater tolerance for dissent, it no longer needs a strength of 97,500 in its citizens militia motorized paramilitary force for internal security duties.

Romania—collapse of the long-standing hard-line government and a move toward a more open society reduces the need for the 35,000 men in its paramilitary internal security forces and border guards.

The Army as the Police

In a few countries, the army serves as the national police whether or not there is a separate police force for other law enforcement duties.

In China, for example, although there is a People's Armed Police of 1.83 million, the People's Liberation Army is called upon to enforce national policy against dissent. This was evident in 1989 in events surrounding mass demonstrations in Beijing and elsewhere.

In Israel, the army has been called upon regularly to deal with dissenting Palestinians and occasionally against Israelis demonstrating against the government. However, Israel also has a State Security Department (SHABAK) that reports to the prime minister and is responsible for all internal security duties, although arrests are normally made by members of the police internal security division (LATAM).

In Panama, the 11,000-man national guard had both military and police responsibility until the Noriega government was expelled. The elected government disbanded the national guard and created a new police force.

Other Internal Security Forces

Other countries maintain internal security forces that are generally independent of the armed forces and usually report to the Ministry of the Interior or a similar high-level government department, sometimes with dual responsibility to the Ministry of Defense. Some of those countries are:

Argentina—about 16,000 men in its paramilitary Argentine National Gendarmerie in charge of internal security and border guards.

Austria—a federal gendarmerie, a paramilitary organization that polices rural areas and mans frontier posts; also a mobile militia of eight motorized infantry brigades.

Bahamas—a Security Intelligence Branch in charge of internal security duties within the Royal Bahamas Police Force.

Bangladesh—an armed police reserve of about 36,000 primarily for riot control but also available to support the army if needed. The army strength is about 90,000 plus about 55,000 paramilitary volunteers and the Bangladesh Rifles of about 30,000 border guards.

Belgium—a gendarmerie of 15,900 commanded by an army general. It reports to the Ministry of Defense on military matters, to the Ministry of the Interior on general police work, and to the Ministry of Justice on judicial police work and internal security.

Bolivia—a paramilitary national police of 15,000 and a narcotics police of 6,000.

Burma—two paramilitary units, a people's police force of 38,000 and a people's militia of 35,000. Its army strength is 170,000.

Canada—the famous Royal Canadian Mounted Police, a civil force maintained by the federal government but providing police services under contract with all provinces except Ontario and Quebec. It also provides police services to 187 municipalities in those eight provinces. The RCMP has a strength of about 20,000.

Cape Verde—a paramilitary people's militia for internal security duties with a strength of about 650.

Chad—a paramilitary national military police and a territorial military police with a total strength of about 5,700.

Costa Rica—no armed forces. Its internal security force is a Directorate General of Surveillance and Territorial Security.

Cuba—paramilitary forces of about 15,000 supplemented by about 1.2 million in a territorial militia, all armed.

Dominican Republic—two paramilitary/riot control units, a cavalry squadron, and a machine-gun detachment.

Ecuador—a national police of about 35,000 that can be transferred to the army in an emergency.

Egypt—preserves its security through two internal security organs, the General Intelligence (GI) for data collection and surveillance and the Directorate General of State Security Investigations to investigate political offenses uncovered by GI. Central Security Forces number about 300,000. The armed forces have their own security service.

El Salvador—a 4,000-man paramilitary national guard linked to its armed forces and involved in counterinsurgency operations.

Germany, Federal Republic of—vests its internal security in the Federal Office for Protection of the Constitution, responsible for counterintelligence and surveillance. Federal border protection is performed by a paramilitary police force. Unification of the German Democratic Republic with the Federal Republic of Germany eliminated the need for the separate large internal security forces of East Germany, which included a security police force of 25,000 and 46,500 border troops and an additional 450,000 militiamen organized into combat units.

Greece—a well-organized military reserve system including a national guard of 110,000 for rear-area security and also a paramilitary gendarmerie of 25,000 men.

Haiti—the paramilitary National Security Volunteers, an internal security organ of about 15,000 men, replaced its feared Tonton Macoute.

Honduras—a paramilitary Public Security Force of 5,000 men separate from its armed forces.

India—many paramilitary organizations, including 3,000 men in the National Security Guards, 90,000 in the Central Reserve Police Force (specifically for internal security duties), 90,000 in Border Security Force, 55,000 in Central Industrial Security Force, 70,000 in Railway Protection Forces, and 250,000 in the Provincial Armed Constabulary.

Indonesia—places internal security responsibility on the Command for Restoration of Security and Order plus several paramilitary organs including a militia in which 300,000 men get three weeks' training each year.

Iraq—paramilitary forces in its People's Army, a militia organization of at least 100,000, and security troops of about 6,000 strength.

Jordan—4,000 men in its paramilitary Public Security Force and 15,000 in its People's Army, a militia organization.

Kenya—the police department includes a special unit for government security duties.

Korea, Democratic People's Republic (North)—a 38,000-man paramilitary force in its Ministry of Public Security and a paramilitary militia of about 3 million men.

Korea, Republic of (South)—a paramilitary civilian defense corps of 3.5 million. Its internal security organ is the Agency for National Security Planning.

Malaysia—a contingent of the Royal Malaysia Police Department, an evolution of its paramilitary police force, in each state and territory.

Mexico—its Federal District (Mexico City) police organized along military lines with limited paramilitary roles. The Federal Judicial Police has criminal investigative responsibility along the lines of the U.S. FBI. The Rural Guards is a gendarmerie for rural area police.

Mozambique—National People's Security Service is a paramilitary force responsible for internal security.

New Zealand—a national police under the Ministry of Internal Affairs that cooperates closely with the armed forces when required.

Nigeria—maintains a paramilitary Security and Civil Defense Corps under its Ministry of Internal Affairs as well

Internal Security Forces

as a paramilitary port security police of about 12,000.

Pakistan—internal security is a responsibility of the Security Branch of the Pakistan Police Service. It also maintains about 164,000 men in several types of paramilitary units.

Peru—about 36,000 paramilitary personnel in a section of the national police force known as the Guardia Civil. A separate Republican Guard is a paramilitary force guarding borders and public buildings.

Portugal—security forces are known as the National Republican Guard and are similar to both the French gendarmerie and the Italian carabinieri.

Saudi Arabia—a fully organized national guard with a primary mission of protecting the royal family and vital points in the kingdom. It is under the Ministry of the Interior rather than the Ministry of Defense.

South Africa—maintains a National Intelligence Service, in charge of all internal security matters. It also has extensive paramilitary forces in its police organizations that include a security branch.

Spain—also has a national guard, but the responsibility for maintaining public order and security rests with the national police.

Sweden—a security department within the Swedish state police.

Switzerland—military service is universal and compulsory; state security comes under the federal police, which is supported by canton and municipal police.

Syria—coordinates its internal security through its National Security Council. Duties are performed by the General Intelligence Directorate. The paramilitary Republican Guard is an internal security force for presidential and VIP protection.

Taiwan—a large army with reserves of 1.5 million. The Taiwan Bureau of Investigation is the government internal security police.

Tanzania—defends against insurgency in its rural areas by a Volunteer Defense Corps of about 33,000 with some paramilitary capability.

Venezuela—a well-equipped 20,000-man national guard responsible for internal security.

The Future

The worldwide spread of terrorism and the abhorrence with which it is viewed even by ardent civil libertarians indicate a growing recognition of the need for strong internal security forces with authority to act in time to prevent disasters if possible and to apprehend criminal offenders. At the same time, the public, at least in all free societies, continues to insist that the rights of citizens not be abridged in the absence of reason to suspect threats to the public order. With public support and the great improvement in forensic technology, nations are likely to be better able to preserve their internal security.

STANFORD SMITH

SEE ALSO: Counterintelligence; Gendarmerie; Hostage; Intelligence Collection; Internal Security and Armed Forces; National Guard of the United States; National Guards: International Concepts; Paramilitary Forces; Reserve Components; Terrorism; Terrorism, Legal Aspects of.

Bibliography

Andrade, J. 1985. *World police and paramilitary forces.* New York: Stockton Press.

Association of the U.S. Army. 1988. *Change and challenge—The search for peace in 1988.* Arlington, Va.: Global Assessment AUSA.

International Institute for Strategic Studies. 1988. *The military balance 1988–1989.* London: Brassey's.

Paxton, J., ed. 1989. *Statesman's yearbook 1988–1989.* New York: St. Martin's Press.

Wright, R. K. 1986. *A brief history of the militia and the national guard.* Washington, D.C.: National Guard Bureau.

IRAN, ISLAMIC REPUBLIC OF

Iran, once known as Persia, is one of the oldest nations of the Middle East. Its location between the Caspian Sea and the Persian Gulf has given Iran a rich ethnic and cultural heritage, and its strategic location and valuable oil reserves ensure it an important role in world affairs.

Power Potential Statistics

Area: 1,648,000 square kilometers (636,294 sq. mi.)
Population: 53,766,400
Total Active Armed Forces: 528,000 (0.982% of pop.)
Gross National Product: US$80.0 billion (1990 est.)
Annual Defense Expenditure: US$13 billion (13.3% of GNP; 1991 est.)
Steel and Iron Production:
    Crude steel: 250,000 metric tons (1986)
    Pig iron: 900,000 metric tons (1986)
Fuel Production:
    Coal: 1.262 million metric tons (1986)
    Crude oil: 145 million metric tons (1989)
    Natural gas: 315 billion cubic meters (1986)
Electrical Power Output: 40,000 million kwh (1989)
Merchant Marine: 133 vessels; 4,634,204 gross registered tons
Civil Air Fleet: 42 major transport aircraft; 186 usable airfields (80 with permanent-surface runways); 17 with runways over 3,659 meters (12,000 ft.); 16 with runways 2,440–3,659 meters (8,000–12,000 ft.); 70 with runways 1,220–2,440 meters (4,000–8,000 ft.).

For the most recent yearly statistics, the reader may refer to the following annual publications:

The Military Balance. International Institute for Strategic Studies. London: Brassey's (UK).
The Statesman's Year-Book. New York: St. Martin's Press.
The World Factbook. Central Intelligence Agency. Washington, D.C.: Brassey's (US).

History

Iran, or Persia as it was called before 1920, is an ancient nation; the first Persian state was the empire founded by Cyrus the Great in the mid-sixth century B.C. The most important event in premodern Iranian history, though, is undoubtedly the country's conquest by Muslim Arabs between A.D. 634 and 651. The Arab conquest not only made Iran a Muslim nation, but also tied it into the great Arab civilization of the eighth through eleventh centuries A.D. Moreover, Iran also became one of the great refuges of the Shia Islam, which (unlike the majority Sunni Muslims) has strong messianic and apocalyptic traits, combined with great zeal and a strong tradition of religious martyrdom and sacrifice.

In the ensuing centuries, Iran was ruled by a series of dynasties, some foreign and some domestic, including the Mongol Il-khans (1260–1355), Tamerlane (1387–1405), and the Safavids (1501–1736). During the nineteenth century, rising British and Russian power in India and Central Asia brought Iran under pressure from both nations, but independence was maintained in part because neither power could bear to see the other ascendant. Iran slowly moved into the modern world and gained a measure of modern government with the promulgation of a constitution by Shah Muzaffar ud-Din on his deathbed on 30 December 1907. This event provoked a lengthy period of political instability that ended with the installation of Reza Khan as shah in 1925.

Reza Shah Pahlavi made concerted efforts to modernize the country and expand its economy. Although he enjoyed some success, much of this was achieved with German assistance, and the outbreak of World War II placed the shah in an uncomfortable position. British and Soviet unhappiness with his policies brought about a combined invasion of Iran in August 1941, and Reza Shah was deposed in favor of his son, Mohammed Reza.

The new shah sided with the Western Allies, but otherwise continued his father's policies of modernization. By the mid-1970s, Iran had become a major regional power whose oil exports had financed a powerful military. The Shiite clerics and their allies in the rural population, however, were displeased at increased Westernization, and many Iranians were angry at the shah's lack of toleration for political dissent. Belated moves to liberalize Iran's political life in 1978 came too late, especially after the shah fell ill and left the country in January 1979 for medical treatment. In his absence, the moderate government he had installed fell before a wave of popular Shiite enthusiasm, and the Ayatollah Ruhollah Khomeini, returned from exile, proclaimed an Islamic Republic on 12 February 1979.

The new regime's hostility toward the West, demonstrated by the prolonged American hostage situation (3 November 1979–20 January 1981), as well as its suspicion of the Soviets and its alarming Islamic revolutionary cant, ensured Iranian isolation. This was particularly important after war with Iraq broke out in mid-September 1980.

In the first two years of the Iran-Iraq war, the Iraqis generally maintained the initiative on the ground, but thereafter, superior Iranian manpower, coupled with the rising cost (in casualties) of offensives, led the Iraqis to adopt a defensive and static strategy, letting the Iranians come to them. This the Iranians did, and on several occasions, like the spring 1982 offensives and the Al Dawa offensive against the Faw peninsula in February 1986, scored impressive gains. Other operations, mounted by poorly trained and ill-equipped troops, suffered heavy losses for minimal gains. The continued indecisive nature of the war prompted one commentator to note that the war was "fought with the weapons of World War III, the tactics of World War I, and the mentality of the Middle Ages." Generally, superior Iraqi weapons (and the occasional use of chemical warfare) matched Iranian superiority in manpower.

Despite their lack of success on the ground, Iraq gained air supericority early in the war and used it to some effect, especially in the "tanker war" of attacks on foreign oil tankers in the Persian Gulf. The human casualties and material losses suffered by both sides, coupled with the war's great economic cost, led to near exhaustion by early 1988. The presence of U.S., British, and French naval forces in the Persian Gulf, to escort neutral oil tankers to and from their loading terminals, may have contributed to Iran's eventual acquiescence to a truce. Although this truce, concluded on 20 August 1988, was likened by the Ayatollah to "drinking poison," the Iranians have generally stuck by their parts of the agreement, and now seem determined to reconstruct their war-battered nation.

The death of the Ayatollah Khomeini in the first week of June 1989 marked the end of an era in revolutionary Iran. After a confused power struggle, speaker of the *majlis* (parliament) Ali Akbar Hashemi Rafsanjani emerged as the leading political figure in the republic.

Politico-Military Background and Policy

Under the shah, Iran was a major U.S. ally in the Persian Gulf area, both a check to potential Soviet aggression and a sort of regional policeman. The shah ensured the security of his regime by making loyalty to himself and his family the primary criterion for promotion within the armed forces. The rise to power of the Ayatollah Khomeini in 1979–80 left Iran with two unhappy legacies: an army whose loyalty to the new regime was weak at best, and a large quantity of sophisticated weapons and no adequate source of spare parts or supplies to maintain them.

The Ayatollah's theocratic republic has also reversed Iran's stabilizing role in the region, and in the early 1980s advocated Islamic revolution among its neighbors. The end of the Iran-Iraq war in autumn 1988 has given more moderate elements in Iran greater influence, and Iranian enthusiasm for exporting its revolution seems on the wane. The loss of Western sources of military supplies, and Iranian distrust of the Soviet Union, has led them to other sources of material, including China, Israel, and Brazil, as well as such European nations as France and Sweden.

The loyalty of the armed forces to the shah led the Islamic Republican government to create, during the war with Iraq, the Pasdaran as an alternative military force to the politically unreliable regular forces. The Pasdaran, or Revolutionary Guard, brought under the same command authority as the rest of the armed forces in July 1988, contains land, air, and naval elements and is equipped with material purchased or captured during the war. Even with the end of the war, the Pasdaran will remain as a political counterweight to the regular forces.

Most of the regular forces, some 250,000 out of 355,000, are conscripts. Men are liable for conscription once they reach the age of 21 and serve for 24 to 30 months. Most air force and navy enlisted personnel are volunteers; nearly all conscripts are in the army. The Pasdaran is a volunteer organization, numbering about 250,000 men, and many of its members are below the age of conscription.

Strategic Problems

Iran has three main strategic problems: first, the war with Iraq and the associated hostility toward Iran among the other Gulf states; second, the problem of ethnic and political unrest; and third, the fact that the Gulf War has left Iran with only one real friend, Syria, and no useful military allies.

None of these difficulties is insurmountable, and Iran's position is strengthened by the unlikelihood that any of its neighbors will attack. A moderation of revolutionary cant and the demonstration of sincere compromise would go a long way toward easing tensions in the Gulf, and could perhaps wean supporters away from the Iraqi cause. Similarly, the end of the war with Iraq means that Iran could begin cultivating meaningful relationships with other nations, especially if the government abandons the ultimately futile policy of harboring terrorists. While a reconciliation with the United States is unlikely, some rapprochement with European states is certainly possible.

The ethnic problem is more complex and consequently more difficult to resolve. The Kurds are the largest, most vocal, and potentially most dangerous group, but the current Iraqi policy of using poison gas on their fellows in Iraq strengthens the Kurdish-Iranian alliance. However, the violence in neighboring Soviet Azerbaijan, begun in January 1990, and the continuing civil war in Afghanistan could create problems, as there are Azerbaijanis in Iran, as well as members of several Afghan ethnic groups.

Military Assistance

Iran currently receives no military assistance, although any nation that allows arms sales to Iran is in a sense providing assistance of a sort. The principal Iranian supplier is the People's Republic of China, which has supplied tanks, artillery, aircraft, naval vessels, and missiles. Some equipment and a good deal of spare parts and ammunition have entered Iran through Israel, which also served as a conduit for some of the arms provided by the United States, in the so-called Iran-Contra arms deal.

Defense Industry

Iran produces no major weapons systems of its own and must therefore import nearly all of its heavy equipment, including large-caliber ammunition. Small arms ammunition and assorted "nonlethal" equipment is produced in-country, and there is a limited domestic naval shipyard capacity, reportedly working on one or more small coastal submarines and several small patrol craft. Iranian factories are also capable of undertaking extensive modifications on purchased equipment.

Alliances

Iran is a member of the United Nations but is no longer a member of any other collective security organization. It has no formal alliances, although there are close political ties with Syria, a traditional rival of Iran's enemy, Iraq.

Defense Structure

Supreme authority in the Islamic Republic of Iran rests with the *wali fagih*, or guide, who provides a moral-political conscience for government and nation and is at present the former president Ali Khamenei. The government is headed by a popularly elected president who in turn appoints the prime minister and cabinet, subject to approval by the *majlis*. Command of the regular armed forces and the Pasdaran lay with the Ayatollah until June 1988, when speaker of the *majlis* (and current president) Ali Akbar Hashemi Rafsanjani became supreme commander of the armed forces. This move was a reaction both to Khomeini's age and failing strength and to Iranian military defeats in the spring. To further complicate the issue, the government announced that the Pasdaran would be brought under army command in July 1988, although the truce with Iraq concluded the following month has made the change less important.

Total Armed Forces

Active: 528,000 *Terms of service*: 24 months.
Reserves: Army: 350,000, ex-service volunteers.
ARMY: 305,000 (perhaps 250,000 conscripts)
  est. 3 army HQ.
  4 armd div (each 3 bde).
  7 inf div.
  1 AB bde.
  1 SF div (4 bde).
  Some indep armd, inf bde (incl 'coastal force').
  5 arty gps.
Reserves: 'Qods' bn (ex-service).
Equipment:†
MBT: perhaps 700 incl: T-54/-55, Ch T-59, T-62, some T-72, Chieftain Mk 3/5, M-47/-48, M-60A1.
Light tanks: 40 Scorpion.
Recce: 130 EE-9 Cascavel.
AIFV: P 150+ BMP-1.
APC: perhaps 600: BTR-50/-60, M-113.

Arty: perhaps 1,000.
  Towed: 105mm: M-56, M-101A1; 122mm: D-30, Ch Type-54/Type-60; 130mm: 200 M-46/Type-59; 155mm: 50 M-71, 18 FH-77B, est. 130 GHN-45, est. 50 G-5; 203mm: some 30 M-115.
  SP: 155mm: some 95 M-109A1; 175mm: 30 M-107; 203mm: 10 M-110.
  MRL: 107mm: 40 Ch Type-63; 122mm: 65 BM-21, Ch Type-81, BM-11, RM-70; 230mm: Oghab; 333mm: Shahin 2; 355mm: Nazeat.
Mortars: 81mm; 107mm: M-30 4.2-in.; 120mm: 3,000.
SSM: Scud; local manufacture msl reported under development Iran-130.
ATGW: ENTAC, SS-11/-12, Dragon, TOW, AT-3 Sagger (some SP).
RCL: 57mm: M-18.
AD guns: 1,500: 23mm: ZU-23 towed, ZSU-23-4 SP; 35mm: 92; 37mm; 57mm: ZSU-57-2 SP.
SAM: SA-7, some HN-5.
Aircraft: incl 40+ Cessna (185, 310, O-2A), 2 F-27, 2 Falcon 20, 15 PC-6, 5 Shrike Commander.
Helicopters: 100 AH-1J (attack); 31 CH-47C (hy tpt); 100 Bell 214A; 20 AB-205A; 50 AB-206.

REVOLUTIONARY GUARD CORPS (*Pasdaran Inqilab*): Some 170,000.
Ground Forces: some 150,000; 11 Regional Commands: loosely org in bn of no fixed size, grouped into perhaps 24 inf, 4 armd div and many indep bde, incl inf, armd, para, SF, arty (incl SSM), engr, AD and border defence units, serve indep or with Army; small arms, spt weapons from Army; controls *Basij* (see Paramilitary) when mob.
Naval Forces: some 20,000; five island bases (Al Farsiyah, Halul (oil platform), Sirri, Abu Musa, Larak); some 40 Swedish Boghammar Marin boats armed with ATGW, RCL, machine guns. Italian SSM reported. Controls coast defence elm incl arty and CSS-N-2 (HY-2) Silkworm SSM in at least 3 sites, each 3–6 msl. Now under joint command with Navy.
Marines: 3 bde reported.

NAVY: 18,000 incl naval air and marines†. Bases: Bandar Abbas (HQ), Bushehr, Kharg, Bandar-e-Anzelli, Bandar-e-Khomeini, Chah Bahar.
Principal Surface Combatants: 8:
Destroyers: 3: 1 Damavand (UK Battle) with 4×2 SM-1 (boxed) SSM, 2×2 114mm guns; plus 1×3 AS mor: 2 Babr (US Sumner) with 4×2 SM-1 SSM (boxed), 2×2 127mm guns; plus 2×3 ASTT.
Frigates: 5: 3 Alvand (UK Vosper Mk 5) with 1×5 Sea Killer SSM, 1×3 AS mor, 1×114mm gun; 2 Bayandor (US PF-103).
Patrol and Coastal Combatants: 29:
Missile Craft: 10 Kaman (Fr Combattante II) PFM some fitted for Harpoon SSM.
Patrol, Inshore: 19: 3 Kaivan, 3 Parvin PCI, 3 N. Korean Chaho PFI, plus some 10 hovercraft ((about half serviceable).
Mine Warfare: 3: 2 Shahrokh MSC (1 in Caspian Sea); 1 Harischi MSI; 2 Iran Ajr LST used for minelaying.
Amphibious: 10:
  4 Hengam LST, capacity 9 tk, 225 tps, 1 hel.
  3 Iran Hormuz 24 (S. Korean) LST, capacity 8 tk, 140 tps.
  2 Iran Ajr LST.
  1 Foque LHA, capacity 120t.
  Plus craft: 3 LCT.
Support and Miscellaneous: 13: 1 Kharg AOR, 2 Bandar Abbas AOR, 1 repair, 2 water tankers, 7 Delva spt vessels.

Marines: 3 bn.

Naval Air: 9 armed hel.
ASW: 1 hel sqn with est. 3 SH-3D, 6 AB-212 ASW.
MCM: 1 hel sqn with 2 RH-53D.
Transport: 1 sqn with 4 Commander, 4 F-27, 1 Falcon 20 ac; AB-205, AB-206 hel.

AIR FORCE: 35,000. Some 213 cbt ac (less than 50% of U.S. ac types serviceable); no armed hel.†

FGA: 8 sqn: 4 with some 60 F-4D/E; 4 with some 60 F-5E/F.
Fighter: 5 sqn: 4 with 60 F-14; 1 with 20 MiG-29.
MR: 5 P-3F.
Recce: 1 sqn (det) with some 5 RF-5, 3 RF-4E.
Tanker/Transport: 1 sqn with 4 Boeing 707.
Transport: 5 sqn: 9 Boeing 747F, 11 Boeing 707, 1 Boeing 727, 20 C-130E/H, 3 Commander 690, 9 F-27, 2 Falcon 20.
Helicopters: 2 AB-206A, 39 Bell 214C, 5 CH-47.
Training: incl 26 Beech F-33A/C, 40 EMB-312, 45 PC-7, 7 T-33.
Missiles:
  ASM: AGM-65A Maverick.
  AAM: AIM-7 Sparrow, AIM-9 Sidewinder, AIM-54 Phoenix.
  SAM: 12 bn with 30 Improved HAWK, 5 sqn with 30 Rapier, 25 Tigercat, 35 HQ-2J (Ch version of SA-2). Probable SA-5.

FORCES ABROAD

Lebanon: Revolutionary Guard 1,500.

PARAMILITARY

BASIJ 'Popular Mobilization Army' volunteers, mostly youths: strength has been as high as 1 million during periods of offensive operations. Org in up to 500 300–350-man 'bn' of 3 coy, each 4 pl and spt; small arms only. Not currently embodied for mil ops.
Gendarmerie: (45,000 incl border guard elm); Cessna 185/310 lt ac, AB-205/-206 hel, patrol boats, 96 coastal, 40 harbour craft.†
Kurds: Kurdish Democratic Party armed wing Pesh Merga, est. 12,000

OPPOSITION

Kurdish Communist Party of Iran (KOMALA): strength unknown.
Democratic Party of Iranian Kurdistan (DPIK): perhaps 10,500
National Liberation Army (NLA): est. 4,500. Org in bde, armed with captured eqpt. Iraq-based.

Future

The war with Iraq left the Iranian economy in a shambles, and this has been compounded by the flight overseas of technically skilled but Western-oriented Iranians. The end of the war will allow the government an opportunity to bring both internal peace and prosperity to the nation, but the pressures of revolutionary Islamic populism and the residue of eight years of violent struggle may make that extremely difficult if not impossible. Iran is still deep in the throes of its revolution, and even the death of the Ayatollah Khomeini has brought only a minor reduction of revolutionary rhetoric and behavior.

Iran watched with satisfaction as the U.S.-led coalition shattered Iraq's offensive military capability and with even greater satisfaction the realization of Ayatollah Ruhollah Khomeini's prophecy about the end of communism. Furthermore, Iran has not lost sight of its primary cause: the leadership of the Islamic world. This pre-eminence, however, will not come easily. That battle will be fought between fundamentalist and secular Islamic forces.

DAVID L. BONGARD
TREVOR N. DUPUY

SEE ALSO: Alexander the Great; Graeco-Persian Wars; Gulf War, 1991; Iraq; Middle East; Persian Empire.

Bibliography

Bakhash, S. 1984. *The reign of the ayatollahs*. New York: Basic Books.
Cordesman, A. H. 1987. *The Iran-Iraq war*. London: Jane's.
International Institute for Strategic Studies. 1989. *The military balance, 1989–1990*. London: Brassey's.
Mottahadeh, R. 1985. *The mantle of the prophet*. New York: Simon and Schuster.
Olmstead, A. T. 1960. *History of the Persian empire*. Chicago: Chicago Univ. Press.
Ramazani, R. K. 1986. *Revolutionary Iran: Challenge and response in the Middle East*. Baltimore, Md.: Johns Hopkins Univ.
Sick, G. 1985. *All fall down*. New York: Random House.

IRAQ, REPUBLIC OF

Iraq, almost geographically identical to ancient Mesopotamia, is the cradle of the first known literate human civilization. In or adjacent to the valleys of the Tigris and Euphrates rivers, humans probably first became farmers and herders, evolving from hunters-gatherers, about 8,000 B.C. The first cities were also probably formed in this region at a later date. Iraq has had a troubled history from ancient times to the present, the most recent problems being the Iran-Iraq war in the 1980s and the 1991 war against the UN coalition forces.

Power Potential Statistics

Area: 434,920 square kilometers (167,923 sq. mi.)
Population: 19,854,600 (Population est. is tentative due to unknown casualities in Iran-Iraq and 1991 Gulf wars.)
Total Active Armed Forces: 382,500 (1.927% of pop.)
Gross National Product: US$35 billion (1989 est.)
Annual Defense Expenditure: not available
Iron and Steel Production: not available
Fuel production:
  Crude oil: 138 million metric tons (1989)
  Natural gas: 49.83 million cubic meters (1980)
Electrical Power Output: 20,000 million kwh (1989)
Merchant Marine: 43 vessels: 944,253 gross registered tons
Civil Air Fleet: 64 major transport aircraft; 102 usable airfields (73 with permanent-surface runways); 9 with runways over 3,659 meters (12,000 ft.); 52 with runways 2,440–3,659 meters (8,000–12,000 ft.); 15 with runways 1,220–2,440 meters (4,000–8,000 ft.).

For the most recent information, the reader may refer to the following annual publications:
The Military Balance. International Institute for Strategic Studies. London: Brassey's (UK).
The Statesman's Year-Book. New York: St. Martin's Press.
The World Factbook. Central Intelligence Agency. Washington, D.C.: Brassey's (US).

History

Iraq was the legendary site of the Biblical Garden of Eden, and the Sumerian, Babylonian, Assyrian, Chaldean, and Parthian civilizations flourished there before the time of Christ. Alexander the Great died in Babylon (about 80km [50 mi.] south of Baghdad) in 323 B.C.

The Arabs conquered Iraq in the seventh century A.D. and in the following century the Abassid caliphate established its capital at Baghdad, a center of learning and the arts for several centuries. But, in 1526, the Ottoman Turks conquered Baghdad and made it and Iraq part of their empire.

During World War I the British took Iraq from the Turks. The country was assigned to the British as a mandated territory by the League of Nations in 1920, and was given its independence in 1932 as a constitutional monarchy under the Hashemite family, a branch of which also ruled Jordan.

In 1958, a violent revolution, headed by Gen. Abdul Karim Qasim, seized power in Iraq. The King, Faisel II, and Prime Minister Nuri as-Said, were killed. The Iraqi revolution, following Nasser's overthrow of King Farouk in Egypt, was a significant event that helped the Pan-Arab socialist movement. Qasim, who brutally purged the country of officers and civilians with ties to the old Royalist regime, was himself deposed and killed in early 1963 by the Ba'ath (Arab renaissance) party. This was followed by several coups, with attendant purges, until Saddam Hussein came to power as Chairman of the Revolutionary Command Council (RCC) and President in 1979. Saddam Hussein has continued to hold both positions, as absolute ruler of Iraq, until the present (1990). He promoted grandiose development schemes for Iraq and led it into war with Iran in 1980, a conflict that continued for eight years with terrible losses in personnel and national wealth. The Iraqis were initially successful, but the Iranians gradually pushed them back, almost capturing Basra. The Iraqis prevented this because they were well dug in and they had superior firepower. Late in the war the Iraqis pushed the Iranians, whose capabilities and morale had deteriorated, out of Iraq. There was also a "tanker war" by air attacks in the Persian Gulf, first by Iraq, and then by both nations. Iraqi aircraft hit the USS *Stark* by error in May 1987, killing 37 U.S. sailors. The war ended in an armistice in August 1988.

Politico-Military Background and Policy

Iraq's government is secular and socialist, and as such it is opposed to the Islamic fundamentalist movement in Iran and elsewhere in the Islamic world. Although it has harbored and aided extremist movements in the past, it has become more moderate in recent years (the United States removed Iraq from its list of terrorist nations early in 1982).

Strategic Problems

The principal problems of Iraq, although it has a well-trained and equipped military, are its small population (about 17.5 million) and its limited coastline of about 58 kilometers (36 mi.) Additionally, Iraq has a significant Kurdish minority in the north, which has often been in revolt against the central government. There is also a religious split (50 to 60% of Iraq's Muslims are Shi'a), although there appeared to be no significant religious problem when the Iraqis were fighting Shi'a Iran. But it is a demographic fact that the Iraqi Sunnis, who lead the country, cannot ignore.

Military Assistance

Iraq was a major arms customer of Western Europe, particularly France, before the war. It also received assistance from the Soviet Union. France helped Iraq build a nuclear reactor at Osirah, near Baghdad, which was destroyed in an Israeli airstrike in June 1981.

Early in 1990, press reports stated that Iraq (along with Afghanistan) was receiving armaments from Eastern Europe following the Soviet drawdown there.

Defense Industry

Iraq imports almost all of its weapons and military supplies, although it has a modest capability for refitting. Its chemical weapons, largely mustard gas, were produced in Iraq from Eastern Europe supplies and used during the war against the Iranians and Kurdish rebels in the north.

Alliances

Iraq is a member of the Arab League, OPEC, and OAPEC. In February 1989, Egypt, Jordan, North Yemen, and Iraq formed the Arab Cooperation Council, an Arab common market.

Defense Structure

The President, Saddam Hussein, is commander in chief and holds the rank of field marshal, although he never served in the armed forces. The RCC holds executive authority, with many of its members having military backgrounds. There is a Minister of Defense, with little authority, and command responsibility is vested in the general staff. Saddam is in direct charge of the Baghdad garrison and the Republican Guard.

Total Armed Forces

Active: perhaps 382,500. Terms of service: 21–24 months.
Reserves: 650,000.

ARMY: est. 350,000 (including perhaps est. 100,000 recalled reserves).
7 corps HQ.
24 armd/mech/inf div.
4 Rebublican Guard Force div (2 armd/mech, 2 inf div.).
20+ SF/cdo bde (for COIN ops).
Equipment: It is not possible to give numbers of individual weapons types. Weapons destroyed or captured during the 1991 Gulf War incl: 3,000 tanks, 1,860 armored vehicles and 2,140 artillery pieces.
MBT: perhaps 2,300 incl T-54/-55/M-77, Ch T-59/-69, T-62, T-72, Chieftain Mk 3/5, M-60, M-47.
Recce: perhaps 1,500 incl BRDM-2, AML-60/-90, EE-9 Cascavel, EE-3 Jararaca.
AIFV: perhaps 900 BMP-1/-2.
APC: perhaps 2,000 incl BTR-50/-60/-152, OT-62/-64, MTLB, YW-531, M-113A1/A2, Panhard M-3, EE-11 Urutu.
Towed arty: perhaps 1,000 incl 105mm: incl M-56 pack; 122mm: D-74, D-30, M-1938; 130mm: incl M-46, Type 59-1; 155mm: some G-5, GHN-45, M-114.
SP arty: incl 122mm: 2S1; 152mm: 2S3; 155mm: M-109A1/A2, AUF-1 (GCT).
MRL: perhaps 250 incl 107mm; 122mm: BM-21; 127mm: ASTROS II; 132mm: BM-13/-16.
Mortars: 81mm; 120mm; 160mm: M-1943; 240mm.
SSM: 6 mobile launchers and erectors, 32 launch platforms; 1 Scud, 51 Al-Hussein msl. Totals as declared, pursuant to UN Resolution 687. All are in process of destruction. 20 FROG.
ATGW: AT-3 Sagger (incl BRDM-2), AT-4 Spigot reported, SS-11, Milan, HOT (incl 100 VC-TH).
RCL: 73mm: SPG-9; 82mm: B-10; 107mm.
ATK guns: 85mm; 100mm towed.
Helicopters: est. 350 (120 armed) incl.
  Attack: est. 120 Bo-105 with AS-11/HOT, Mi-24, SA-316 with AS-12, SA-321 (some with Exocet), SA-342.
  Transport: est. 330: hy: Mi-6; med: AS-61, Bell 214 ST, Mi-4, Mi-8/-17, SA-330; lt: AB-212, BK-117 (SAR), Hughes 300C, -500D, -530F.
AD guns: est. 3,200: 23mm: ZSU-23-4 SP; 37mm: M-1939 and twin; 57mm: incl ZSU-57-2 SP; 85mm; 100mm; 130mm.
SAM: SA-6/-7/-8/-9/-14, Roland.

NAVY: est. 2,500. Bases: Basra (currently closed), Umm Qasr.
Frigates: 5: 4 Hittin (It Lupo; new ships retained in Italy since completion) with 1 AB-212 hel (ASW), 2×3 ASTT; plus 8×Otomat SSM, 1×127mm gun; 1 Khaldoum (trg)† with 2×ASTT.
Patrol and Coastal Combatants: Corvettes: 6: 2 Mussa ben Nussair (It Assad, hel version; new ships retained in Italy since completion) 1×AB-212 hel, 2×Otomat SSM; 4 Abdulla ibn abi Sarh (It Assad; new ships retained in Italy since completion) with 6×Otomat, 2×3 ASTT]
Support and Miscellaneous: 1 Agnadeen (It Stromboli) AOR (New vessel laid-up in Alexandria before delivery to Iraq.)

AIR FORCE: 30,000 incl 15,000 AD personnel; total Iraqi air losses cannot be est. 35 ac were lost in air-to-air cbt, over 100 destroyed on ground, 115 flown to Iran; these ac are not included in the totals.
Bombers: est. 6 incl: H-6D, Tu-16, Tu-22.

FGA: est. 130 incl J-6, MiG-23BN, MiG-27, Mirage F1EQ5/-200, Su-7, Su-20, Su-25.
Fighter: est. 125 incl: Ch J-7, MiG-21, MiG-25, Mirage F-1EQ, MiG-29.
Recce: incl: MiG-21, MiG-25.
AEW: incl: Il-76 Adnan.
TKR: incl: 2 Il-76.
Transport: incl: An-2, An-12, An-24, An-26, Il-76.
Training: incl: AS-202, EMB-312, some 20 L-29, some 30 L-39, MB-233, Mirage F-1BQ, PC-7, 30 PC-9.
Missiles:
   ASM: AM-39, AS-4, AS-5, AS-11, AS-12, AS-30L, C-601.
   AAM: AA-2/-6/-7/-8, R-530, R-550.

PARAMILITARY
Frontier Guards.
Security Troops: 4,800.

OPPOSITION
Kurdish Democratic Party (KDP): 15,000 (30,000 more in militia); small arms, some Iranian lt arty, MRL, mor, Sam-7.
Kurdish Workers' Party: strength unknown; breakaway from KDP, anti-Iran, Syria-based.
Patriotic Union of Kurdistan (PUK): 4,000 cbt (plus 6,000 spt). 11 T-54/-55 MBT; 450 mor (60mm, 82mm, 120mm); 106mm RCL; some 200 12.5mm AA guns; SA-7 SAM.
Socialist Party of Kurdistan: est. 1,500.
Supreme Assembly of the Islamic Revolution (SAIRI): claims 2 bde; Iran-based; Iraqi dissidents, ex-prisoners of war.

FOREIGN FORCES
UN (UNIKOM): some 400 incl 300 observers from 33 countries.

Future

Prior to the 1991 Gulf War, Iraq possessed the fourth largest army in the world and was a formidable regional military power. Although the U.S.-led coalition force shattered Iraq's offensive military capability, it did not destroy Saddam Hussein's control over Iraq.

It will be years before the full extent of the Gulf War's political, economic, and environmental impact is known. But Iraq has been removed from the Middle East balance of power for at least a decade; the influence of Syria, Iran, and Turkey will increase; and the influence of Islamic fundamentalism through the region has been enhanced.

WALTER P. WHITE

SEE ALSO: Arab Conquests; Arab League; Gulf War, 1991; Iran; Middle East; Ottoman Empire; Persian Empire.

Bibliography

Chubin, S., and C. Tripp. 1988. *Iran and Iraq at war.* Boulder, Colo.: Westview Press.
Hunter, B., ed. 1991. *The statesman's year-book 1991-92.* New York: St. Martin's Press.
International Institute for Strategic Studies. 1991. *The military balance, 1991-1992.* London: Brassey's
Marr, P. 1985. *The modern history of Iraq.* Boulder, Colo: Westview Press.
Wright, R. 1989. *In the name of God: The Khomeini decade.* New York: Simon and Schuster.

IRELAND (Irish Republic)

Occupying the southern four-fifths of Ireland, the Irish Republic is the westernmost of Europe's nations, and is the only European country whose population is less now (1990) than it was in the early nineteenth century due to emigration and the Potato Famine of 1847–49.

Power Potential Statistics

Area: 70,280 square kilometers (27,135 sq. mi.)
Population: 3,737,000
Total Active Armed Forces: 12,900 (0.345% of pop.)
Gross Domestic Product: US$33.9 billion (1990 est.)
Annual Defense Expenditure: US$458 million (1.6% of GDP, 1990 est.)
Iron and Steel Production: figures unavailable
Fuel Production:
   Coal: 0.054 million metric tons (1986)
   Natural gas: 1,591 million cubic meters (1986)
Electrical Power Output: 14,480 million kwh (1989)
Merchant Marine: 53 vessels; 138,967 gross registered tons
Civil Air Fleet: 23 major transport aircraft; 37 usable airfields (18 with permanent-surface runways); none with runways over 3,659 meters (12,000 ft.); 2 with runways 2,440–3,659 meters (8,000–12,000 ft.); 6 with runways 1,220–2,440 meters (4,000–8,000 ft.).

For the most recent information, the reader may refer to the following annual publications:
The Military Balance. International Institute for Strategic Studies. London: Brassey's (UK).
The Statesman's Year-Book. New York: St. Martin's Press.
The World Factbook. Central Intelligence Agency. Washington, D.C.: Brassey's (US).

History

Britain's conquest of Ireland began in earnest during the Tudor period, culminating in the plantation of the northeast of the country with large numbers of Scottish and English settlers. In 1801 Ireland became an integral part of the United Kingdom. Irish nationalists, however, continued to strive for independence, staging a series of rebellions in the nineteenth and early twentieth centuries. The last of these, the "Easter Rising" of 1916, sparked the War of Independence, or Anglo-Irish War, which began in 1919. Unable to quash the insurrection militarily, and faced with growing domestic and international condemnation for its handling of the conflict, in December 1921 the British government offered Ireland the status of Dominion within the Commonwealth. The six counties of northeastern Ireland, with their Unionist majority, opted to stay in the United Kingdom. In the remaining 26 counties a civil war raged briefly between the "Free Staters," who reluctantly accepted Britain's offer, and the "intransigent Republicans" of the Irish Republican Army (IRA), who wanted to continue fighting for a fully independent Irish Republic. The Free Staters won.

Having effectively left the Commonwealth more than ten years before by gradually eroding the terms of the

1921 Anglo-Irish Treaty, in 1949 the government of the Irish Free State declared the country a republic. To this day, the heirs of the losing side in the civil war refuse to recognize the Irish Republic because it does not have jurisdiction over the six northeastern counties. Some of these people—the new IRA—are now waging war in Northern Ireland, a war that began with the collapse of the Catholic community's civil rights movement in the late 1960s.

Politico-Military Policy and Strategic Problems

The Northern Ireland conflict—known colloquially as "the Troubles"—is the main military security concern of the Irish government. Although the principal purpose of the Permanent Defense Force (PDF) is to defend the state against external aggression, in practice the Defense Forces's main role is to aid the civil power in its struggle against the IRA and related terrorists. This involves the PDF on a daily basis in escorting prisoners, cash, and explosives, manning roadblocks and checkpoints, protecting state installations, and patrolling the border with Northern Ireland.

Defense Industry

Ireland has a very limited domestic defense industry, producing the Timoney wheeled armored personnel carrier.

Alliances

Ireland is not a member of any military alliances. The Irish government unsuccessfully attempted to link Irish unification with a United States offer to join the North Atlantic Treaty Organization (NATO) in 1949. Thus Ireland's military neutrality during the Second World War has been perpetuated into peacetime. Ireland is a member of the EC and the United Nations, and has over 100 personnel assigned to UN missions in Namibia, the Iran-Iraq and Indo-Pakistani borders, Cyprus, the Golan Heights, and Afghanistan. A reinforced battalion with 4 armored cars, 13 Timoney APCs, and 4 120mm mortars, totaling 748 officers and men, is in Lebanon with UNFIL.

Defense Structure

The President of Ireland (the chief of state) is the supreme commander of the PDF, but the Minister of Defense, who is answerable to the Prime Minister, exercises military command. The Defense Minister may consult the Defense Council, which comprises the Minister of State for the Department of Defense, the Secretary of State for Defense, the Chief of Staff, the Adjutant-General, and the Quartermaster-General. The Chief of Staff holds the rank of lieutenant general. The Adjutant-General and the Quartermaster-General each hold the rank of major general.

(For an explanation of the abbreviations and symbols used in the following section of military statistics, see the list of Abbreviations and Acronyms in each volume.)

Total Armed Forces

Active: 12,900 incl 100 women. Terms of service: voluntary, 3-year terms to age 60, officers 56–65.
Reserves: 16,100 (obligation to age 60, officers 57–65). Army: first-line 1,000, second-line 14,800. Navy 300.

ARMY: 11,200.
4 Territorial Commands.
1 inf force (2 inf bn).
4 inf bde:
  2 with 2 inf bn, 1 with 3, all with 1 fd arty regt, 1 cav recce sqn, 1 engr coy:
  1 with 2 inf bn, 1 armd recce sqn, 1 fd arty bty.
Army tps: 1 lt tk sqn, 1 AD regt, 1 Ranger coy.
(Total units: 11 inf bn; 1 UNIFIL bn ad hoc with elm from other bn, 1 tk sqn, 4 recce sqn (1 armd), 3 fd arty regt (each of 2 bty); 1 indep bty, 1 AD regt (1 regular, 3 reserve bty), 3 fd engr coy, 1 Ranger coy).
Reserves: 4 Army Gp (garrisons), 18 inf bn, 6 fd arty regt, 3 cav sqn, 3 engr sqn, 3 AA bty.
Equipment:
Light tanks: 14 Scorpion.
Recce: 19 AML-90, 32 AML-60.
APC: 60 Panhard VTT/M3, 10 Timoney.
Towed arty: 88mm: 48 25-pdr; 105mm; 12 lt.
Mortars: 81mm: 400; 120mm: 72.
ATGW: 21 Milan.
RCL: 84mm: 444 Carl Gustav; 90mm: 96 PV-1110.
AD guns: 40mm: 24 L/60, 2L/70.
SAM: 7 RBS-70.

NAVY: 900. Base: Cork.
Patrol and Coastal Combatants: 7 PCO: 1 Eithne with 1 Dauphin hel; 3 Emer, 1 Deirdre; 2 Orla (UK Peacock).

AIR FORCE: (800): 13 cbt ac, 8 armed hel ; 3 wings (1 trg).
COIN: 1 sqn with 6 CM-170-2 Super Magister.
COIN/Training: 1 sqn with 7 SF-260WE, 1 SF-260 MC ac, 2 SA-342L trg hel.
MR: 2 Super King Air 200, 1 CASA 235.
Transport: 1 HS-125, 1 Super King Air 200, 1 Gulfstream III.
Liaison: 1 sqn with 6 Reims Cessna F-172H, 1 F-172K.
Helicopters: 3 sqn: 1 Army spt with 8 SA-316B; 1 Navy spt with 2 SA-365; 1 SAR with 3 SA-365.

FORCES ABROAD
UN and Peacekeeping:
Afghanistan/Pakistan (OSGAP): 1.
Angola (UNAVEM II): observers.
Central America (ONUCA): 19.
Cyprus (UNFICYP): 8.
Iraq/Kuwait (UNIKOM): 8.
Lebanon (UNIFIL): 1bn + (750); 4 AML-90 armd cars, 10 Sisu APC, 4 120mm mor.
Middle East (UNTSO): 21.

Future

The continuing unrest and conflict in Northern Ireland will remain Ireland's main security concern for some time, barring a sudden settlement that effectively neutralizes the Provisional Wing of the IRA (the Provos). Recent agreements for Irish–United Kingdom cooperation to reduce their mutual security concerns show some promise, but progress is likely to be slow. Ireland also faces sluggish economic growth and continued emigration by young people to Britain and the United States. Still, there are no significant external threats, and the other problems are manageable.

DESMOND DINAN

SEE ALSO: United Kingdom; Western Europe.

Bibliography

Beckett, J. F. 1981. *The making of modern Ireland, 1603–1923.* London: Faber and Faber.
Hunter, B., ed. 1991. *The statesman's year-book, 1991–92.* New York: St. Martin's Press.
International Institute for Strategic Studies. 1991. *The military balance, 1991–1992.* London: Brassey's.
Johnson, P. 1982. *Ireland: A concise history from the twelfth century to the present day.* Chicago, Ill.: Academy Press.
Keatinge, P. 1984. *A singular stance, Irish neutrality in the 1980s.* Dublin: Institute for Public Administration.

ISRAEL, STATE OF

A small nation in the Middle East, Israel was founded in 1948 when the British gave up their League of Nations mandate in Palestine. Israel has fought five major wars since independence, and its armed forces must be accounted the most experienced in the world.

Power Potential Statistics

Area: 20,770 square kilometers (8,019 sq. mi.)
Population: 4,822,000
Total Active Armed Forces: 141,000 (2.924% of pop.)
Gross National Product: US$46.5 billion (1990 est).
Annual Defense Expenditure: US$5.3 billion (13.9% of GNP, 1991 est.)
Iron and Steel Production: 130,000 metric tons (1986)
Fuel Production:
    Crude oil: 0.056 million metric tons (1986)
    Natural gas: 68 million cubic meters (1986)
Electrical Power Output: 17,500 million kwh (1989)
Merchant Marine: 30 vessels; 516,714 gross registered tons
Civil Air Fleet: 27 major transport aircraft; 44 usable airfields (26 with permanent-surface runways); none with runways over 3,659 meters (12,000 ft.); 6 with runways 2,440–3,659 meters (8,000–12,000 ft.); 12 with runways 1,220–2,440 meters (4,000–8,000 ft.).

For the most recent information, the reader may refer to the following annual publications:
The Military Balance. International Institute for Strategic Studies. London: Brassey's (UK).
The Statesman's Year-Book. New York: St. Martin's Press.
The World Factbook. Central Intelligence Agency. Washington, D.C.: Brassey's (US).

History

Palestine has been the homeland of the Jewish people since they first settled there in the second millennia B.C. They resettled the area when they returned from exile in Babylonia in the fifth century B.C., but after the rebellions of 67–72 A.D. and 131–133, the Romans, as a matter of policy, drove the Jews from Israel, spreading them across the Mediterranean basin. A small Jewish population always remained in Palestine, but most Jews lived abroad. They suffered considerable hardship at the hands of Christian Western Europe, but ironically enjoyed good treatment from most Muslim rulers.

Zionism, the modern Jewish nationalist movement, was founded by Theodor Herzl in the 1800s, in large part as a reaction to modern anti-Semitism. Early Jewish immigrants to Palestine faced great hardships, but with perseverance and help from abroad, they established stable communities. Increased Jewish immigration, especially after the British Balfour declaration of 1917 promised a national home for the Jews in Palestine, provoked violent reaction from Palestinian Arabs, who foresaw the alarming prospect of becoming a minority in their homeland. There were particularly severe outbreaks in 1919 and again from 1936 to 1939. The situation became critical after 1945, when survivors of Nazi persecution clamored for admittance to Palestine, and Great Britain, pressured by Jewish guerrillas inside and vocal critics outside Palestine, abandoned their mandate, and left the situation to the United Nations (UN).

By a narrow margin, the UN voted to partition Palestine into Jewish and Arab areas in November 1947, a move opposed by most Arabs. When British forces officially withdrew, the Jews set up the new state of Israel in their area (14 May 1948), declared in a public speech by David Ben-Gurion, a prominent Zionist who became prime minister. Invasion by the armies of Egypt, Syria, Iraq, Lebanon, and Jordan followed almost immediately, in an effort to destroy the new state and rescue Palestine's Arabs. Although poorly equipped, the Israeli army enjoyed a superiority in both numbers and morale, and by the time hostilities ended in January 1949 (without a peace treaty), Israel's survival was assured.

Israel was often the victim of terrorist attacks and cross-border raids from its Arab neighbors. In 1956, Egyptian efforts to close the Suez Canal led to war, and Israeli troops captured the Sinai peninsula in a "lightning" operation from 30 October to 7 November. This coincided with the Franco-British capture of Port Said at the northern end of the Canal, and the resulting international furor not only brought down the British government but compelled Israel to evacuate the Sinai. Further Egyptian provocations, abetted by Jordan and Syria, led to another war from 5 to 10 June 1967, which left Israel in control of the Sinai, the West Bank, and the Golan Heights. These areas, with Arab populations, have not been legally incorporated into Israel, and have proved a focal point for opposition to Israel.

The criticism which Israel netted for its pre-emptive strikes in both these wars impelled the Israelis not to strike first when war threatened again in 1973. Although

Jordan remained neutral, Israel gained the upper hand in the October War (6–23 October) only after heavy fighting and heavy losses on both the Sinai and Golan fronts. Almost five years later, Egypt and Israel signed the Camp David peace treaty (March 1979) sponsored by the United States, but this has so far been the only such peace achieved.

Israeli difficulties with Arab terrorist and guerrilla groups based in Lebanon (which had not had a functional central government since 1975–76) led to the Israeli invasion of Lebanon in June 1982. This was an effort to drive the Palestinian Liberation Organization (PLO), a group led by Yasser Arafat, from its Lebanese bases. Although this was achieved, and the PLO left in August, it made the situation in Lebanon worse, and the PLO returned soon after the Israelis left (May 1983). Israel's continuing security problems in Lebanon have been made more complex with the rise of the *intifada* (popular insurrection) among Palestinians in the West Bank and Gaza in late 1987. Israeli security forces have so far only been able to control the unrest, not end it.

Politico-Military Background and Policy

Surrounded as it is by hostile Arab states which refuse to acknowledge its official existence, Israel's military forces have traditionally played a major role in the formulation of national policy. These problems are aggravated by Israel's small size and lack of strategic depth.

Israel's armed forces are patterned on the Swiss-Swedish model, based on a core of 31,000 long-term professional officers and NCOs. Military service is universal and compulsory for all Jewish and Druze citizens, for periods of 48 months (officers), 36 months (men), and 24 months (women). Obligation begins at age 18 for both men and women; men must serve their three years by age 29, and women their two years by age 26. Men have reserve obligations until age 54, women until age 38 or marriage, whichever happens first. At any one time, there are 110,000 conscripts in service. Christians, Muslims, and Circassians are not subject to conscription, but they may volunteer. Most reservists (there are 504,000, officially) serve four weeks per year, with two weeks' training and two weeks' active duty.

Despite its militia character the Israeli Defense Forces (IDF) has maintained an extremely high level of skill and morale. One of the finest armed forces in the world, the IDF's doctrines have been repeatedly tested and refined in combat. IDF officers, drawn from the ranks, are both highly motivated and highly skilled; it is Israeli doctrine to lead from the front, and this results in the men having great trust in their officers, and in high officer casualty rates in combat. This cost is deemed acceptable because of the undeniable superiority it gives the IDF.

In its early years, Israel was compelled to rely on foreign sources for most heavy military equipment (tanks, artillery, aircraft, naval vessels). The political unreliability of such sources (including the United States) has driven Israel to develop a major domestic arms industry. Israel has also developed its own force of intermediate-range ballistic missiles, to provide a deterrent against long-range chemical or nuclear bombardment of Israeli cities. Although not officially acknowledged, Israel also possesses a nuclear stockpile, variously estimated at between twenty and 100 warheads, for potential use with their missiles. Israeli fears of Arab missile capabilities, especially those developed during the Gulf War (1980–88), has brought about cooperation with the United States in development of an antitactical ballistic missile (ATBM).

Strategic Problems

Israel's strategic problems are twofold: its small size and the presence of hostile states on its frontiers. Israel has attempted to compensate for these weaknesses by maintaining a large, well-equipped, and highly motivated armed force, and has won all of its major wars with its Arab neighbors (1948–49, 1956, 1967, 1973, 1982). Israel has not, however, been able to achieve a lasting peace with any of the neighboring states except Egypt; to that extent, Israeli policy has failed, since the conditions that existed in 1949 have not fundamentally changed.

Israel also faces a long-term threat from the PLO and other Arab guerrilla and terrorist organizations. PLO contact with the United States in late 1988 left Israel isolated in its official refusal to deal with the PLO, and numbers of Israelis have, in violation of the 1986 law prohibiting such contact, met and talked with PLO representatives abroad. The Israeli government's adamant condemnation of the PLO as an outlaw terrorist organization is particularly ironic in light of the activities of some Israeli statesmen (such as former Prime Minister Menachem Begin and current Prime Minister Yitzhak Shamir) with Jewish terrorist organizations operating against the British Mandate during the 1940s. While modern politics have seen numerous transformations from guerrilla leader or terrorist to responsible statesman, a majority of Israelis refuse to consider the possibility that this could happen with Yasser Arafat of the PLO.

Military Assistance

In the late 1950s and 1960s, Israel maintained a close military alliance with France, but this was terminated during the Six-Day War in June 1967. Since that time, the United States has become Israel's chief foreign ally, especially after the October War of 1973. Israel is, in fact, the largest single recipient of U.S. military aid in the world, receiving over one-third of total U.S. funds so expended. While the figures for 1979 reached US$4.0 billion (US$2.7 billion in loans, US$1.3 billion in grants), the figures for 1987 and 1988 included US$1.8 billion in grants each year, with no loans. This aid represents a significant portion of

Israel's defense expenditure, and provides some relief from the economic burden of maintaining the IDF.

Defense Industry

The Israeli defense industry predates the establishment of the state of Israel, when the underground factories of the *Haganah* (the Jewish community's main guerrilla army) manufactured small arms, ammunition, and explosives during the last years of the British Mandate. Although growth was comparatively slow during the 1950s and 1960s, the Israelis developed significant domestic arms production capacity during those years, although it was limited to manufacture of lighter weapons and modifications (often extensive) of materiel purchased abroad. Since the October 1973 war, Israel has undertaken the development and manufacture of its own tank, the Merkava (Chariot; reportedly an excellent system), as well as its own artillery, mortars, armored cars, missiles (including the Gabriel naval SSM and the Shafrir AAM), remotely piloted vehicles (RPVs), and much of the work on the U.S.-engined Kfir jet fighter. A second fighter, the Lavi (Lion) was cancelled in spring 1988 because of excessive cost.

Israel gains considerable foreign exchange from arms sales to developing countries, especially in Africa and Latin America. Soltam howitzers and mortars, and the RBY family of armored vehicles have been particularly popular, and the excellent Israeli Uzi 9mm submachine-gun, famed for its sturdiness and reliability, is in use in dozens of nations, including the United States.

Alliances

Israel is a member of the United Nations, but is often isolated there because of its policies concerning the Palestinians in the occupied territories, and its close friendship with the United States. The United States is Israel's principal foreign ally.

Defense Structure

Control of the IDF is vested in the prime minister and the cabinet. Direct responsibility for administering the armed forces is in the hands of the minister of defense. The Chief of the General Staff is also chief of staff of the army and presides over the triservice unified Defense Forces. The General Staff directs activities and operations of the combat commands (navy, air force, three territorial commands, paratroop command, and armored command), training command, and NAHAL (*Noar Halutzi Lohen* or Pioneer Fighting Youth, a paramilitary organization).

(For an explanation of the abbreviations and symbols used in the following section of military statistics, see the list of Abbreviations and Acronyms in Volume I.)

Total Armed Forces

Active: 141,000 (110,000 conscripts). Terms of service: officers 48 months, men 36 months, women 24 months (Jews and Druze only; Christians, Circassians and Muslims may volunteer). Annual trg as reservists thereafter to age 54 for men, 24 (or marriage) for women.

Reserves: 504,000; Army 494,000; Navy 1,000; Air Force 9,000. Most serve at least one month a year, ideally 2 weeks trg and 2 weeks op duty (border def, security at military installations or administrative duties). Male commitment until 54 in reserve op units may be followed by voluntary service in the Civil Guard or Civil Defence.

STRATEGIC

It is widely believed that Israel has a nuclear capability with up to 100 warheads. Delivery means could include ac, Jericho 1 SSM (range up to 500km), Jericho 2 (tested 1987–9, range est. 1,500km) and Lance.

ARMY: 104,000 (88,000 conscripts, male and female); some 598,000 on mob.

2 corps HQ

3 armd div (2 armd, 1 arty bde, plus 1 armd, 1 mech inf bde on mob).

5 mech inf bde (incl 1 para trained, 1 based on NCO school, 1 on NAHAL (Noar Halutzi Lohem—'Pioneer Fighting Youth', combines military duty with establishing agricultural settlements).

3 regional inf div HQ (border def).

1 Lance SSM bn.

3 arty bn with 203mm M-110 SP.

Reserves:

9 armd div (2 or 3 armd, 1 mech inf, 1 arty bde).

1 airmobile/mech inf div (3 bde manned by para trained reservists).

10 regional inf bde (each with own border sector).

4 arty bde.

Equipment:

MBT: 4,488 incl 1,080 Centurion, 550 M-48A5, 1,000 M-60A1, 600 M-60A3, 488 T-54/-55 (350 mod), 110 T-62, 660 Merkava I/II/III.

Recce: about 400 incl Ramta RBY, M-2/-3, BRDM-2.

APC: 5,900 M-113A1/A2, est. 80 Nagmashot, BTR-50P, 4,400 M2/-3 half track.

Towed arty: 579: 105mm: 70 M-101; 122mm: 100 D-30; 130mm: 109 M-46; 155mm: 300 Soltam M-68/-71, M-839P/-845P.

SP arty: 841: 105mm: 35 M-7; 155mm: L-33, 100 M-50, 530 M-109A1/A2; 175mm: 140 M-107; 203mm: 36 M-110.

MRL: 122mm: BM-21; 160mm: LAR-160; 240mm: BM-24; 290mm: MAR-290.

Mortars: 81mm; 120mm: est. 250; 160mm (some SP).

SSM: 12 Lance, some Jericho 1/2.

ATGW: TOW (incl Ramta (M-113) SP), Dragon, AT-3 Sagger, Mapats.

RL: 82mm: B-300.

RCL: 84mm: Carl Gustav; 106mm: 250 M-40A1.

AD guns: 20mm: 850: incl TCM-20, M-167 Vulcan, 30 M-163 Vulcan/M-48 Chaparral gun/msl systems; 23mm: ZU-23 and 60 ZSU-23-4 SP; 37mm: M-39; 40mm: L-70.

SAM: Redeye

NAVY: 9,000 (3,000 conscripts), 10,000 on mob. Bases: Haifa, Ashdod, Eilat.

Submarines: 3 Gal (UK Vickers) SSC with Mk 37 HWT, Harpoon USGW.

Patrol and Coastal Combatants: 65:

Missile Craft: 22 PFM:

2 Aliya with 4 Harpoon, 4 Gabriel SSM, 1 SA-366G Dauphin hel (OTHT).

3 Romat with 8 Harpoon, 8 Gabriel.

3 Reshef with 2–4 Harpoon, 4–6 Gabriel.
6 Mivtach/Sa'ar with 2–4 Harpoon, 3–5 Gabriel.
2 Shimrit (US Flagstaff 2) PHM with 4 Harpoon, 2 Gabriel.
1 Dvora (with 2 Gabriel.
Patrol, Inshore: 43: 12 Super Dvora PFI (; Dabur PFI.
Amphibious: Craft only; 6 LCT, 1 LCVP.
Support and Miscellaneous: 2: 1 patrol craft depot ship, 1 tpt.

Marines: Naval cdo: 300.

AIR FORCE: 28,000 (19,000 conscripts, mainly in AD), 37,000 on mob; 591 cbt ac (plus perhaps 102 stored), 94 armed hel.
FGA/Fighter: 16 sqn:
4 with 112 F-4E (plus 13 in store); (converting 50 to Phantom 200, some 24 now converted).
2 with 47 F-15 (20 -A, 2 -B, 18 -C, 7 -D);
6 with 149 F-16 (57 -A, 7 -B, 59 -C, 26 -D).
4 with 95 Kfir C2/C7 (plus 75 in store);
FGA: 4 sqn with 121 A-4H/N, plus 14 in store.
Recce: 14 RF-4E.
AEW: 4 E-2C.
EW: 6 Boeing 707 (ELINT/ECM), 1 C-130H, 2 EV-1E (ECM), 4 IAI-201 (ELINT), 4 RC-12D, 6 RC-21D (ELINT), 3 RU-21A.
MR: 5 IAI-1124 Seascan.
Tanker: 5 Boeing-707, 2 KC-130H.
Transport: 1 wing: incl 3 Boeing 707, 19 C-47, 24 C-130H, 10 IAI-201, 3 IAI-1124.
Liaison: 4 Islander, 41 Cessna U-206, 2 -172, 2 -180, 6 Do-27, 9 -28D, 12 Queen Air 80.
Training: 6 Cessna 152, 80 CM-170 Tzukit, 16* F-4E, 5* Kfir TC 2/7, 35 Super Cub, 20* TA-4H, 7* TA-4J.
Helicopters:
Attack: 40 AH-1G/S, 35 Hughes 500MD, 19 AH-64A.
SAR: 2 HH-65A.
Transport: hy: 32 CH-53 (2 -A, 30 -D); med: 12 UH-1D; lt: 55 Bell 212, 40 Bell 206A.
Missiles:
ASM: AGM-45 Shrike, AGM-62A Walleye, AGM-65 Maverick, AGM-78D Standard, Luz, Gabriel III (mod).
AAM: AIM-7 Sparrow, AIM-9 Sidewinder, R-530, Shafrir, Python III.
SAM: 17 bty with MIM-23 HAWK/Improved HAWK, 2 bty Patriot.
Beginning in the late 1970s, the Israelis developed remotely piloted vehicles (RPVs) as important reconnaissance and EW assets. These were proved in combat during the invasion of Lebanon (6–10 June 1982) in action against Syria's SAM-based air defenses in the Bekaa Valley. As a result of this effort, the Israeli air force employs several kinds of RPVs, including the Mastiff 3, Teledyne Ryan 124R, Chukar II, Delilah, and Pioneer, and has shared much of this technology with the United States (and so, indirectly, with NATO).

FORCES ABROAD
Ethiopia: 125 advisers and technicians (probably withdrawn).

PARAMILITARY
NAHAL (Pioneer Fighting Youth): staffed by teenagers and young adults, some of whom have completed military service (NAHAL members have the same military obligations as other Israelis); constructs and operates agricultural settlements in border areas, performing the twofold task of expanding the agricultural economy and providing security in frontier areas. In order to perform their duties, all NAHAL members receive military training.
Border Police: 6,000; 600 Walid 1, equipped with captured BTR-152 APCs.
Coast Guard: 1 U.S. PBR, 3 other patrol craft.

Future

Israel's future, like that of many similar nations, will be strongly affected by the actions of its neighbors. Israel's economy has stagnated in recent years because of high inflation rates (16–20% in 1987–88), and until those rates can be reduced, real prosperity will elude Israel. The most serious problem, however, is the same as it has been for decades: how to reconcile Israel's right to exist as a state with the right of the Palestinian Arabs to autonomy and self-determination. Until Israel reaches some permanent accommodation with both the PLO and its own Palestinian subjects in the occupied territories, real peace in the Middle East will not be achieved.

DAVID L. BONGARD
TREVOR N. DUPUY

SEE ALSO: Arab-Israeli Wars; Egypt; Jordan; Lebanon; Middle East; Syria.

Bibliography

Dupuy, T. N. 1980. *Elusive victory: The Arab-Israeli wars, 1947–1974*. New York: Harper and Row.
Herzog, C. 1982. *The Arab-Israeli wars: War and peace in the Middle East*. New York: Random House.
Horowitz, D., and M. Lissak. 1978. *Origins of the Israeli polity*. Chicago: Univ. of Chicago Press.
Hunter, B., ed. 1991. *The statesman's year-book, 1991–92*. New York: St. Martin's Press.
International Institute for Strategic Studies. 1991. *The military balance, 1991–1992*. London: Brassey's.
O'Brien, C. C. 1986. *The siege: The saga of Israel and Zionism*. New York: Simon and Schuster.
Oz, A. 1982. *In the land of Israel*. London: Fontana Paperbacks.
Sykes, C. 1973. *Crossroads to Israel: 1917–1948*. Bloomington, Ind.: Midlands.

ITALIAN UNIFICATION WARS [1848–66]

During the Italian *Risorgimento*, or rebirth, which occurred between the years 1859 and 1861, Italy evolved from a fragmented and diverse group of kingdoms, duchies, and small states dominated by Austria, into a unified and respected member of the European community. Although complete unification did not occur until after World War I, the forging of a single Italian nation took place within this three-year period. Additions and annexations following this period are simply epilogues.

The freedoms granted by the Napoleonic Code, introduced by French occupiers in the early nineteenth century, awakened Italians to the possibilities of an enlightened government responsive to the basic needs of the people. Further, improvements in finance and communications implemented during the Napoleonic era by French-dominated governments had greatly benefited the Italians.

After the final defeat of Napoleon I in 1815, and the collapse of the Napoleonic organizational structure in the Italian states, there was a regression to fragmentation and autocracy. As provided by the Congress of Vienna of 1815, Austria maintained control over Venetia and Lombardy in northern Italy to prevent French influence from growing too strong there. The repressive Austrian presence in northern Italy, however, gave rise to a strong popular desire for independence and constitutional government.

Anti-Austrian sentiment grew beneath the surface for many years until it broke out in open revolt in Milan on 18 March 1848. The resulting "Five Day Revolt" (March 18–22) ended when Austrian Marshal Josef Radetzky withdrew his forces from the city. The rest of Lombardy and Venetia joined in the revolt and a coalition of Italian forces gathered in northern Italy to confront the Austrians. King Charles Albert of Piedmont (the Kingdom of Sardinia), the most competent and liberal leader in Italy, declared war on Austria on 22 March. The Italian coalition under Charles Albert, despite its numerical superiority, campaigned ineffectively against the Austrians, and suffered a severe defeat at the hands of Radetzky at the Battle of Custozza (24–25 July 1848).

Nevertheless, the Italian independence movement continued to grow. Shortly after the Milan Revolt and the Piedmontese declaration of war, patriots under Daniele Manin declared an independent republic in Venice. In February the following year, Giuseppe Mazzini, with the support of soldier of fortune Giuseppe Garibaldi, declared the formation of a Roman Republic in the Papal States. Despite these events, Radetzky again decisively defeated Charles Albert at the Battle of Novara on 23 March 1849, and Charles Albert abdicated in favor of his son, Victor Emmanuel II.

A French expeditionary force landed at Civitavecchia on 24 April 1849 and moved on Rome, and after an initial repulse, finally forced Garibaldi and his "Legion" to surrender on 29 June. Following this setback, Garibaldi fled to America.

On 9 August Sardinia made peace with Austria, effectively ending the revolution. The Austrians regained firm control in northern Italy, and enacted harsh retaliatory measures, further antagonizing the Italians.

Despite the Austrian success, strong feelings for independence and unification continued to grow among the population. Piedmont, under the leadership of Count Camillo Benso di Cavour, again became the nucleus for the unification movement and for the removal of Austrian influence in Italy.

Although Cavour was trying to build a unified Italian state, he was concerned primarily with expanding the power and prestige of Piedmont. To accomplish this he entered the Crimean War on the side of Britain and France (1853–56), and concluded a secret alliance with Emperor Napoleon III of France at Plombières in 1858. The French promised aid to help drive the Austrians out of northern Italy, but only if this could be accomplished without France being charged with aggression. Following the expulsion of Austria from Italian territory, there would then emerge a federation of four states under the pope's presidency. For her part, France would receive Savoy and Nice from Piedmont.

After the assurance of French aid, Cavour provoked conflict with Austria by stirring up revolts in Lombardy and Venetia. An Austrian ultimatum to disarm was rejected by Cavour, and a brief but bloody war ensued, pitting Piedmont and France against Austria. The decisive Austrian defeat at Solferino (24 June 1859) by the combined French and Piedmontese forces effectively ended major fighting. The Austrian setback at Solferino was a signal for the small north Italian states of Tuscany, Parma, and Modena to overthrow their pro-Austrian leaders.

Napoleon III was alarmed by these revolutions in northern Italy. Hoping to prevent Prussia from entering the war on the side of Austria, he was anxious to end the war quickly and arranged a separate truce with Franz Josef of Austria. On 11 July 1859, at the peace conference held at Villafranca, it was agreed that most of Lombardy (except for the fortress cities of Mantua and Peschiera) would go to Piedmont. Venetia remained under Austrian control, the deposed leaders of Tuscany, Parma, and Modena were reinstated, and the pope was declared president of an Italian confederation.

Piedmont did not participate in the Villafranca meeting, but King Victor Emmanuel reluctantly accepted the terms because he was unwilling to face Austria alone. Cavour then resigned as prime minister. The Treaty of Zürich, between Piedmont and Austria, ratified on 10 November 1859, essentially confirmed the Villafranca agreements.

The idea of an Italian Confederation under the pope was so distasteful to the northern Italian duchies of Parma, Modena, Tuscany, and Romagna that they agreed to be annexed by Piedmont. Napoleon III objected to this, but Cavour reemerged with a compromise solution. He suggested that the previously agreed cession of Nice and Savoy to France be carried out at once in return for Napoleon III's blessing. To this Napoleon agreed and he and Victor Emmanuel II signed the Treaty of Turin formalizing French annexation of Nice and Savoy.

In April 1860, King Francis II of the Kingdom of the Two Sicilies brutally suppressed revolts in Sicily and Naples, arousing indignation throughout Europe. With the covert support of Cavour and King Victor Emmanuel II, Garibaldi and his "Thousand Redshirts" sailed from Genoa and landed at Marsala, in Sicily, on 11 May. He quickly gained broad public support, and incited a revolt against

Francis. Garibaldi defeated forces of the Kingdom of the Two Sicilies at Calatafimi on 15 May and took Palermo on 27 May. He again defeated royal forces at Milazzo on 20 July, and then, with British assistance, Garibaldi crossed the Straits of Messina to the mainland on 22 August. He marched northward, was greeted enthusiastically by the people, and occupied Naples on 7 September after meeting only light resistance.

While Garibaldi was engaged in the Kingdom of the Two Sicilies, unrest in the Papal States provided Cavour with the excuse to send in troops to restore order. Piedmontese forces crossed the border on 10 September and defeated Papal troops at Castelfidardo on 18 September. The Piedmontese then marched southward to link up with Garibaldi. This alarmed Napoleon III because of his desire to maintain the independence of Rome under the pope. French troops were landed and occupied Rome, while naval forces patrolled offshore.

Garibaldi, meanwhile, was able to achieve a victory over the Neapolitan forces at Volturno on 26 October, and then invested the stronghold of Gaeta in early November. The withdrawal of the French fleet in January allowed the Piedmontese navy to arrive and bombard the fortress, forcing its surrender on 13 February 1861.

An all-Italian parliament (with the exception of the papal territories around Rome) was convened on 17 March, and a united Kingdom of Italy was proclaimed with Victor Emmanuel as the first constitutional monarch. Continued French occupation of Rome prevented the inclusion of the Papal States in the new kingdom of Italy. Just three months after seeing his dream come to fruition, Cavour died, exhausted from years of carrying the burden of Italian independence and unification.

Garibaldi and his small army, again with the covert support of Victor Emmanuel, began planning to drive the French from Rome and unite all of Italy. In August 1862 Garibaldi marched northward through Naples toward Rome; however, Victor Emmanuel was not prepared to provoke a conflict with France and sent troops south to stop Garibaldi. The two forces met at Aspromonte on 29 August, where Garibaldi's army was defeated. Garibaldi was wounded and captured, but later released.

Italy now began to plan to recover Venetia from Austria. A treaty was concluded with Prussia in April 1866, and the outbreak of the Austro-Prussian War in June 1866 was followed by Italy's declaration of war against Austria on 20 June. Four days later an Italian army under Victor Emmanuel was defeated by a much smaller Austrian army under Archduke Albert at the Second Battle of Custozza.

Although the Austrians were decisively defeated by the Prussians at Königgrätz, or Sadowa (3 July 1866), and a treaty was concluded at Prague on 23 August, hostilities continued in Italy. Garibaldi won some minor battles in the Alps during July, but withdrew into Italy because Prussia would not agree to the Italian occupation of Trentine Tyrol.

On 20 July the Austrians defeated an Italian naval force at Lissa, in the Adriatic Sea, near Split. A treaty was concluded in Vienna on 12 October between Austria and Italy. Although the Austrians had generally been successful, they accepted the mediation of Napoleon III and ceded Venetia to Italy.

In December 1866 the French occupation force withdrew from Rome. This prompted Garibaldi to try again to force the Papal States into the Italian union. With covert Italian support, he led an invasion in January 1867, while the Italian government attempted to overthrow the papal government by covert and overt means. In October 1867, less than a year after they had been withdrawn, Napoleon again dispatched troops to occupy Rome.

Garibaldi met a French-Papal force at Mentana on 3 November, and was defeated with heavy losses, due primarily to superior French weapons. Garibaldi and his men fled to the Italian border and were arrested by Italian authorities.

The French remained in Rome until the outbreak of the Franco-Prussian War (1870), when they were withdrawn for the defense of France. An Italian army of 60,000 then marched across the border and besieged Rome. Following a brief bombardment, the city was assaulted and taken on 20 September after a short battle. Pope Pius IX, hoping to prevent further bloodshed, ordered his forces to lay down their arms. Rome was then annexed and, following a formal plebiscite, was declared the Italian capital.

ARNOLD C. DUPUY

SEE ALSO: Crimean War; Franco-Prussian War; History, Modern Military; Italy.

Bibliography

Delzell, C. F., ed. 1965. *The unification of Italy, 1859–1861: Cavour, Mazzini, or Garibaldi?* New York: Holt, Rinehart and Winston.
Dupuy, R. E., and T. N. Dupuy. 1986. *The encyclopedia of military history*. 2d rev. ed. New York: Harper and Row.
Leeds, A.C. 1974. *The unification of Italy*. New York: Putnam's.
Mack Smith, D. 1954. *Cavour and Garibaldi*. Cambridge, England: Cambridge Univ. Press.
———. 1956. *Garibaldi: A great life in brief*. New York: Knopf.
Trevelyan, G. M. 1911. *Garibaldi and the making of Italy*. New York: Longmans, Green.

ITALIAN WARS [1494–1559]

Italian wars is the collective name given to a series of wars between France and Spain that were fought primarily in Italy. These wars were characterized in part by the large-scale employment of mercenary soldiers by both sides, a growing use of artillery and other firearms, and the effects of the bastioned trace in fortifications. The end of the wars, brought about by the Treaty of Cateau-Cambrésis, created large numbers of unemployed soldiers in France, and so contributed to the unrest leading to the Huguenot Wars (1560–98).

Background

The collapse of the Hohenstaufen imperial regime in Italy in the early thirteenth century left Italy divided into numerous municipal republics, duchies, and counties that constantly quarreled and fought with each other. In such a chaotic political situation, foreign involvement was inevitable, and French, German, and Aragonese intervention played a significant role in Italian affairs.

France held the County of Asti from 1392, and briefly governed Genoa from 1458 to 1461. Further, the Kingdom of Naples was ruled by the House of Anjou from 1282 to 1442. Sicily, on the other hand, had fallen to Aragón in the aftermath of the Sicilian Vespers revolt against the Angevins in 1282. When King René of Naples died without heirs in 1442, he bequeathed his kingdom to the French throne. The French, deeply involved in the final stages of the Hundred Years' War, were unable to exploit their opportunity, and Aragón instead seized the Kingdom of Naples.

The Early Stages, 1494–1515

By the late fifteenth century, the major powers in Italy were Venice, Milan, Florence, the Papacy, and the Kingdom of Naples, and France, Aragón, and Austria all exerted considerable influence. King Charles VIII of France, who was only 24, determined to invade Italy, both to reassert French sovereignty in Naples and to secure a useful base for launching a crusade against the Turks (as he somewhat disingenuously asserted). He led his army across the Alps in September 1494, and this foreign invasion struck Italy like a thunderbolt. Not only was the ruthlessness and efficiency of French men-at-arms and Swiss pikemen an unwelcome surprise, but the relatively more mobile French artillery was also more effective than its Italian counterparts. The effect of these guns on fortifications was especially shocking, as the Florentine historian Francesco Guicciardini wrote: "So violent was their battering that they could accomplish in a few hours what had previously required many days."

Charles's sudden success caused Ferdinand of Aragón to send a small army under Gonzalo de Córdoba to Italy in spring 1495, in part to succor Ferdinand's cousin King Ferrante of Naples, who had fled to Sicily as the French approached. In addition, Venice, Austria, Spain, Milan, and the Papacy joined in the League of Venice against France. Gonzalo's small army was brushed aside by a French army at Seminara in Calabria (28 June). In the meantime, with enemies gathering to his rear, Charles had realized his dangerous position and, leaving a large garrison in Naples, marched north with most of his army, some 4,100 horse and 7,500 foot. A league army of some 20,000 Italian mercenaries barred his way over the Apennines at Fornovo di Taro (southwest of Parma), but in a furious battle there on 6 July 1495, the French brushed them aside and moved calmly northward toward home.

The French withdrawal was only a pause in the conflict. War continued sporadically into 1496 and resumed in 1499 when France again invaded Naples. Despite some initial success, the French were unable to defeat Gonzalo, who destroyed one French army at Cerignola (21 April 1503) and another in a surprise winter attack at the Garigliano River south of Gaeta (30–31 December). These defeats ended French pretensions in southern Italy.

The League of Cambrai and Milan, 1508–1515

The French, turning their territorial ambitions toward Milan, returned in 1508 as part of the League of Cambrai against Venice. A French army smashed the Venetians at Agnadello (14 May 1509), but the league, led by Pope Julius II, turned on France the following year. A French army under the youthful and energetic Gaston de Foix won a bloody victory at Ravenna (11 April 1512), but Foix's death there spoiled the French success. The following year, a Swiss army routed a French army under Marshal Louis de la Tremoille at Novara (6 June 1513), but a Swiss invasion of France was averted when they accepted a French indemnity. Meanwhile, a Spanish army defeated a Venetian force (Venice had become France's ally) at La Motta (October), and French military prospects were bleak.

Francis I and Milan, 1515–25

Following the Swiss defection, France's other adversaries made peace. After King Francis I ascended the throne, he determined to invade Italy and seize Milan. In a bloody two-day battle at Marignano, Francis drove a Swiss army from the field (13–14 September 1515). This success gave him Milan, recognized by the Treaty of Noyon in 1516. Francis's fear and suspicion of Charles V, whose realm surrounded France on three sides, led to renewed war in 1521. An imperial army under the wily *condottieri* Prospero de Colonna captured Milan. Colonna's skillful use of field entrenchments, coupled with the difficulties French commander Odet de Foix, Marshal de Lautrec, had with his Swiss contingents (unpaid, they threatened to leave unless he launched an assault), gave Colonna a victory at the Battle of Bicocca (27 April 1522). Following the defeat, the battered remnants of Lautrec's Swiss troops left anyway, and Lautrec withdrew to France.

Francis sent a new army to Italy in 1523 under his favorite, Admiral Guillaume Bonnivet. Bonnivet, a soldier of small talent, was surprised and his army routed at La Sesia by Charles de Lannoy, the imperial viceroy of Naples (30 April 1524). Francis frustrated an imperial invasion of the French Riviera by crossing the Alps to the north and descending on Milan in October of that year.

Hastening back to protect its Italian base, Lannoy's army waited while the French army besieging Pavia was weakened by disease and detachments.

Lannoy, joined by Charles, the renegade constable of Bourbon, attacked Francis's army on 24 February 1525, catching the besiegers by surprise. The French army reacted in a haphazard and uncoordinated fashion, and Francis led a locally successful cavalry charge into the imperial center. He realized his predicament only when he was surrounded by infantrymen screaming for his blood, and he was rescued only by the timely arrival of Lannoy himself. A captive of the Spanish, Francis was compelled to sign the ruinous Treaty of Madrid to secure his freedom in February 1526. As soon as he entered France, Francis repudiated the treaty and war resumed (May 1526).

The Hapsburg Valois Struggle, 1526–59

During indecisive campaigning in Italy, Lannoy sent the duke of Bourbon against the pope, who had become a French ally. Bourbon advanced on Rome, but after his death in the first assault his *landsknechts* (German mercenary infantry) went out of control and sacked the city (6 May 1527). The following year, the Marquis de St. Pol's French army was defeated at the Battle (or Rout) of the Landriano (19 June), and Genoa was lost to a revolt led by Andrea Doria. These French reverses led to yet another peace in 1529, but war returned from 1536 to 1538, and broke out again in 1542.

The French sent an army into northern Italy under the command of Francis of Bourbon, prince of Enghien. Enghien brought Marquis del Vasto's imperial army to battle at Ceresole, near Turin, on 14 April 1544. Through rare coordination between infantry and cavalry, the French drove the larger enemy army from the field, although an English-imperial invasion of northern France eclipsed Enghien's success. Peace returned in autumn 1544, but war was renewed in 1547.

In 1552, the French sent yet another army to Italy; this one under the Gascon general Blaise du Monluc. Operating in Tuscany in support of allies there, Monluc was defeated at Marciano by the marquis of Marignano (2 August 1553). Withdrawing into Siena, Monluc led an epic defense of that city against a siege until he was forced to surrender in 1554. Again, the focus of war shifted to northern France, where the last campaigns of the war were waged. French armies were defeated at St. Quentin (10 August 1557) and Gravelines (13 July 1558) before the Treaty of Cateau-Cambrésis ended the conflict in April 1559.

Military Changes During the Wars

Armies had grown larger during the Italian wars, total strength rising from perhaps 12,000 for the French at Fornovo, and much less at Cerignola and the Garigliano, to twice that at Pavia. Alongside the increase in size, the proportion of cavalry had fallen from one-third at Fornovo to about one-eighth at Pavia and one-tenth at Ceresole; moreover, the ratio of heavy to light cavalry had dropped from at least 3:1 at Fornovo to about 1:2 at the later battles. French armies usually contained a larger proportion of cavalry—and a larger fraction of French cavalry was heavy—than did their opponents' armies, in part because the French recognized that one of their strengths was the superiority of their heavy horse, or *gendarmerie*.

Firearms became more common, and the use of artillery became more widespread. While handheld firearms were of minimal import at Fornovo, entrenched Spanish harquebusiers killed the duke of Nemours and broke his attack at Cerignola. A similar combination of entrenchments and firepower at Bicocca wrought fearful execution among the Swiss pike columns, and a Swiss attack at Pavia was handily repulsed by imperial harquebusiers in a wooded thicket on the French left.

Fortresses with bastioned walls proliferated. The ease with which Charles VIII's artillery had battered Italian citadels into submission accelerated the development of low-walled, bastioned fortresses. As sieges took longer (and were frequently no more than blockades, so strong were the new fortresses), the pace of warfare slowed. There were eight major battles in Italy during eleven years of war between 1494 and 1515, five in twelve years from 1515 to 1530, and only two in eighteen years from 1530 to 1559. Fortresses provided armies with safe havens in time of defeat, and as battles decided less and less, they declined in frequency. Warfare became a matter of sieges and "actions," the incessant skirmishes and ambuscades so familiar in the Dutch wars of the late sixteenth century.

MAX GEORGE KELLNER

SEE ALSO: Gonzalo de Córdoba; History, Early Modern Military; History, Medieval Military; Italy; Spanish Empire.

Bibliography

Benedetti, A. 1967. *History of the Caroline wars*. Ed. and trans. D. M. Schullian. New York: Frederick Ungar.

Commines, P. de. 1973. *Mémoires*. 2 vols. Ed. S. Kinser, Trans. I. Cazeaux. Columbia: Univ. of South Carolina Press.

Guicciardini, F. 1969. *The history of Italy*. Ed. and trans. S. Alexander. New York: Macmillan.

Machiavelli, N. 1965. *Chief works and others*. Trans. A. Gilbert. Durham, N.C.: Duke Univ. Press.

Mallet, M., and J. R. Hale. 1984. *The military organization of a renaissance state: Venice, 1400–1619*. Cambridge: Cambridge Univ. Press.

Monluc, B. de. 1964. *Commentaires*. Ed. P. Courteault. Paris: A. and J. Picard.

Oman, Sir C. W. C. 1937. *History of the art of war in the sixteenth century*. London: Methuen.

Taylor, F. L. 1921. *The art of war in Italy, 1494–1529*. Cambridge: Cambridge Univ. Press.

ITALY (Italian Republic)

The Republic of Italy is a strong supporter of NATO and has a key location for southern defense and control of the lines of communications across the Mediterranean to Africa and the Middle East. Italy's defense industry as well as its military forces are key to continued national strength.

Power Potential Statistics

Area: 301,230 square kilometers (116,305 sq. mi.)
Population: 57,322,000
Total Active Armed Forces: 361,400 (0.631% of pop.)
Gross Domestic Product: US$844.7 billion (1990 est.)
Annual Defense Expenditure: US$19.2 billion (2.2% of GDP, 1990 est.)
Iron and Steel Production:
    Crude steel: 25.181 million metric tons (1989)
    Pig iron: 11.9 million metric tons (1986)
Fuel Production:
    Coal: 1.573 million metric tons (1986)
    Coke: 7.0 million metric tons (1986)
    Crude oil: 4.563 million metric tons (1989)
    Natural gas: 16,893 million cubic meters (1989)
Electrical Power Output: 225,000 million kwh (1990)
Merchant Marine: 575 vessels; 7,462,744 gross registered tons
Civil Air Fleet: 125 major transport aircraft; 135 usable airfields (90 with permanent-surface runways); 2 with runways over 3,659 meters (12,000 ft.); 36 with runways 2,440–3,659 meters (8,000–12,000 ft.); 38 with runways 1,220–2,440 meters (4,000–8,000 ft.).

For the most recent information, the reader may refer to the following annual publications:
The Military Balance. International Institute for Strategic Studies. London: Brassey's (UK).
The Statesman's Year-Book. New York: St. Martin's Press.
The World Factbook. Central Intelligence Agency. Washington, D.C.: Brassey's (US).

History

Few regions have had as significant an influence on history and civilization as the Italian Peninsula. The emergence of Rome as a military power in 500 B.C. and its continuance until the fifth century A.D. provided a pattern for future military development. Rome established two major concepts that were used by others: (1) discipline of army units fighting as groups instead of as individual warriors, (2) military engineering works to provide lines of communication for control of vast areas of empire, and to establish permanency.

Following the fall of Rome, factional fighting among city-states and regions continued into the early nineteenth century. This condition facilitated intervention by France, Austria, Germany, and Spain. Yet by the close of the nineteenth century, the efforts of Giuseppe Garibaldi, the dukes of Savoy, and others finally reestablished a trend toward unification. Modern Italian history begins in 1870 with the unification of the entire peninsula under King Victor Emmanuel II of Savoy. This established a constitutional monarchy that lasted until 1922. In 1822 Italy joined the Triple Alliance with Germany and Austria-Hungary, in opposition to the Triple Entente of France, Russia, and England, but in 1915 renounced the alliance in order to join the Allies in World War I.

In 1922 Benito Mussolini established a Fascist dictatorship that eliminated old political parties and many personal liberties, but retained the king as a figurehead. Mussolini effectively used military force to support his policies as required, particularly to expand Italian influence and colonial possessions in the Mediterranean and Africa. Italy joined Germany in declaring war on Britain and France in 1940, and expanded her overseas holdings in Africa and in various Mediterranean islands. Following the Allied invasion of Sicily in 1943, Italy overthrew Mussolini and declared war on Germany, which continued until the Allied victory in 1945. The monarchy was abolished in a 1946 plebiscite when the current Italian Republic was accepted. Since January 1948 a bicameral parliamentary government with different coalitions of several political parties has led Italy politically.

Throughout this period, the state has owned many large enterprises including electricity, telephone and telecommunications, railroads, and airlines. These years have also seen an increase in ties with Western nations, despite the growing popularity of the Italian Communist Party. The lack of natural resources forced Italy to build trade relationships and to create processing and manufacturing industries, which are largely family-owned firms. Principal contributions are precision machinery, motor vehicles, chemicals, electric goods, clothing, military hardware, and oil engineering equipment and services. Italy will continue to process imported materials for value added and thereby seek a balance of trade.

Politico-Military Background and Policy

Despite political instability, the armed forces are traditionally subject to the control of the civilian government of the Republic of Italy. The governments have recognized, since World War II, that stability of the economy and of their continued power is closely related to the economic and military stability of all of Europe. Consequently, loyal support of NATO and of the United Nations has placed Italy in a position of being a major contributing member to world affairs. Throughout this post-war period, countering terrorism of the Red Brigade and of other terrorist organizations has been largely a police matter, not a military activity.

Strategic Problems

Strategic problems for Italy are primarily related to access to raw materials, petroleum, and food. Agreements with neighboring Yugoslavia over the area near Trieste, and the relinquishment of colonial territories after World War II have eliminated most causes for external controversy. The trade relationships with developing nations that need machinery and services provide the means for sustaining the economy. Internal dissatisfaction with potential economic problems if foreign trade is disrupted, and possible terrorist activity in response to supporting international law, appear to be the primary problems facing the republic.

Military Assistance

Italy has received extensive military assistance from the United States since 1950. Post–World War II economic recovery and recent agreements for coproduction of U.S.-designed equipment have reduced the necessity for large amounts of U.S. military assistance. Italy provides 44 bases for American forces and recently agreed to accept a wing of F-16s that had been forced to leave Spain. U.S. foreign military sales credits to Italy in 1988 were US$26.7 million and were projected to be US$75 million in 1989. Much of this increase will concentrate on procuring Patriot air defense systems and upgrading F104G all-weather interceptors.

Defense Industry

The defense industry in Italy covers four major categories of products: ground armaments, aircraft, ship building, and electronics. Markets have expanded beyond developing countries to providing defense products to Western nations as well. Italy is a member of several international consortia (European, South American, and U.S.) for joint development of arms as well as having licenses to produce items designed in other countries. Three major ground combat firms, three aircraft manufacturing consortia, ten major ship-building concerns, and eleven major electronics manufacturers produce equipment for domestic and foreign use.

Alliances

In addition to the United Nations and NATO, Italy is a member of the Organization for Economic Cooperation and Development, the Western European Union, and the Council of Europe, and supports the INTELSAT communications satellite. Additionally, there are several bilateral agreements between the United States and Italy. Italy provides military personnel for two multilateral peacekeeping/observation missions.

Defense Structure

The president is nominally the commander in chief of the armed forces. Civilian control, however, is exercised by the prime minister's cabinet through the minister of defense. The minister of defense receives advice from the secretary of state for defense and the chiefs of staff of the three armed services.

(For an explanation of the abbreviations and symbols used in the following section of military statistics, see the list of Abbreviations and Acronyms in each volume.)

Total Armed Forces

Active: 361,400 (216,000 conscripts). Terms of service: All services 12 months.
Reserves: 584,000. Army 520,000 (obligation to age 45), immediate mob 240,000.
  Navy 36,000 (to age 39 for men, variable for officers to 73).
  Air 28,000 (to age 25 or 45 [specialists]).

ARMY: 234,200 (170,000 conscripts).
Field Army:
3 Corps HQ (1 mtn):
  1 with 1 mech, 1 armd bde, 1 armd cav bn, 1 arty regt.
  1 with 3 mech, 1 armd bde, 1 arty regt.
  1 with 4 mtn bde, 1 armd cav bn, 3 hy arty bn.
1 AD gp: 5 AD arty (1 trg), 4 HAWK SAM bn.
1 hy arty regt.
Avn: 4 wings org in sqn and flt (flt usually has 4 ac or 6 hel):
  7 lt ac flt with SM-1019/Cessna O-1E.
  12 hel sqn, 14 flt with AB-206.
  Multi-role: 12 hel sqn: 9 with AB-205; 2 with AB-212B, 1 with AB-412.
  Med tpt: 4 hel sqn with CH-47.
Territorial Defense:
7 Military Regions.
7 indep mech bde.
Rapid Intervention Force (FIR): 1 AB bde (incl 1 SF bn); 1 mech bde, 1 Marine bn (see Navy), 1 hel unit (Army), 1 air tpt unit (Air Force).
1 amph regt (2 Lagunari bn).
4 armd bn.
1 armd recce bn.
2 inf bn.
1 arty bn.
4 engr bn.
Reserves: On mob: 1 armd, 1 mech, 1 mtn bde.
Equipment:
MBT: 1,220 (25 in store): 300 M-60A1, 920 Leopard (140 in store).
Recce: 6 Centauro B-1 (trials) (CFE HACV).
APC: 3,879 (CFE: 3,339): 2,183 M-113, 1,667 VCC1/-2, 14 Fiat 6614, 15 LVTP-7.
Total arty: 1,952 (101 in store).
  Towed arty: 967: 105mm: 357 Model 56 pack; 155mm: 164 FH-70, 423 M-114; 203mm: 23 M-115.
  SP arty: 283: 155mm: 260 M-109G/-L; 203mm: 23 M-110A2.
  MRL: 227mm: 2 MLRS.
  Mortars: 120mm: 700. Plus 81mm: 1,205.
SSM: 6 Lance launchers.
ATGW: 432 TOW (incl 270 SP), 1,000 Milan.
RL: 1,000 APILAS.
RCL: 80mm: 720 Folgore.
AD guns: 25mm: 50 SIDAM SP; 40mm: 252.
SAM: 126 HAWK, 145 Stinger.
Aircraft: 60: 48 SM-1019, 12 O-1E (target acquisition/utility).
Helicopters: 28 A-109, 6 A-129, 92 AB-205A, 136 AB-206 (observation), 14 AB-212, 17 AB-412, 30 CH-47C.

NAVY: 49,000 incl 1,500 air arm, 600 special forces and 800 marines; (20,000 conscripts). 5 Main Commands: Fleet (Commander also COMEDCENT); Upper Tyrrhenian; Adriatic; Lower Tyrrhenian; Ionian and Strait of Otranto.
Bases: La Spezia (HQ), Taranto (HQ), Ancona (HQ), Brin-

Italy

disi, Augusta, Messina, La Maddalena, Cagliari, Naples (HQ), Venice.
Submarines: 9: 2 Pelosi (imp Sauro) with Type 184 HWT; 4 Sauro with Type 184 HWT; 3 Toti SSC with Type 184 HWT.
Principal Surface Combatants: 32:
Carrier: 1 G. Garibaldi CVV with 16 SH-3 Sea King hel, 4 Teseo SSM, 2 × 3 ASTT (has capability to operate V/STOL ac acquisition in progress).
Cruisers: 2:
  1 Vittorio Veneto CGH with 1 × 2 SM-1 MR SAM, 6 AB-212 ASW hel (Mk 46 LWT); plus 4 Teseo SSM, 2 × 3 ASTT.
  1 Andrea Doria CGH, with 1 × 2 SM-1 MR SAM, 3 AB-212 hel; plus 2 × 3 ASTT.
Destroyers: 3:
  2 Audace DDGH, with 1 × SM-1 MR SAM, 4 Teseo SSM, plus 2 × AB-212 hel, 1 × 127mm gun, 2 × 3 ASTT.
  1 Impavido DDG with 1 SM-1 MR SAM; plus 1 × 2 127mm guns, 2×3 ASTT.
Frigates: 26:
  8 Maestrale FFH with 2 AB-212 hel, 2×533mm DP TT; plus 4 Teseo SSM, 1 × 127mm gun.
  4 Lupo FF with 1 AB-212 hel, 2 × 3 ASTT; plus 8 Teseo SSM, 1 × 127mm gun.
  2 Alpino with 1 AB-212 hel, 2 × 3 ASTT, 1 × ASW mor.
  8 Minerva with 2 × 3 ASTT.
  4 De Cristofaro with 2 × 3 ASTT, 1 ASW mor.
Additional in Store: 1 CGH, 1 DDG, 1 SSC.
Patrol and Coastal Combatants: 18:
Corvettes: 3 Albatros with 2 × 3 ASTT.
Missile Craft: 7 Sparviero PHM with 2 Teseo SSM.
Patrol, Offshore: 4 Cassiopea with 1 AB-212 hel.
Coastal: 4 Bambu (ex-MSC) PCC assigned MFO.
Mine Warfare: 15:
MCMV: 15: 2 Storione (US Aggressive) MSO; 4 Lerici MHC; 7 Castagno (US Adjutant) MHC; 2 Agave MSC.
Amphibious: 2 San Giorgio LPD: capacity 350 tps, 30 trucks, 2 SH-3D or CH-47 hel, 7 craft; plus some 40 craft: about 4 LCU, 22 LCM and 14 LCVP.
Support and Miscellaneous: 34: 2 Stromboli AOR, 2 tugs, 11 coastal tugs, 4 water tankers, 3 trials, 2 trg, 3 AGOR, 6 tpt, 1 salvage.
Special Forces (600) (Comando Subacquei Incursori—COMSUBIN): 6 gp; 2 assigned aslt swimmer craft; 2 raiding ops; 1 underwater ops; 1 SF; 1 school; 1 research.

Marines (San Marco gp) (800): 1 bn gp; 1 trg gp; 1 log gp.
Equipment: 30 VCC-1, 10 LVTP-7 APC, 16 81mm mor, 8 106mm RCL, 6 Milan ATGW.

Naval Air Arm (1,500): 2 cbt ac, 36 armed hel.
FGA: 2* TAV-8B.
ASW: 5 hel sqn with 36 SH-3D, 60 AB-212.
ASM: Marte Mk 2.

AIR FORCE: 78,200 (26,000 conscripts).
FGA: 6 FGA/recce sqn: 3 with Tornado; 1 with F-104S (being modernized); 2 with G-91Y.
CAS: 4 sqn: 2 with AMX; 1 lt attack with MB-339; 1 lt attack/recce with G-91R/R1/R1A (being replaced).
Fighter: 7 sqn with F-104S.
Recce: 1 sqn with F/RF-104G.
MR: 2 sqn with Atlantic (Navy-assigned; to be modernized).
EW: 1 ECM/recce sqn with G-222VS, PD-808.
Calibration: 1 navigation-aid calibration sqn with G-222RM, PD-808, MB-339.
Transport: 3 sqn: 2 with G-222; 1 with C-130H.
Communications: 1 sqn with Gulfstream III, Falcon 50, P-166M, SIAI-208M, PD-808, MB-326, DC-9 ac; SH-3D hel.
Training: 1 OCU with TF-104G; 1 det (Cottesmore, UK) with Tornado; 6 sqn with G-91T, MB-326, MB-339A, SF-260M ac; AB-47G, NH-500 hel.
SAR: 1 sqn and 3 det with HH-3F; 6 det with AB-212.
AD: 8 SAM gp with Nike Hercules; 12 SAM bty with Spada.
Equipment: 449 cbt ac (plus 88 store), no armed hel.
Aircraft:
  Tornado: 81 (71 FGA, 10* in tri-national trg sqn); plus 15 in store.
  F-104: 164; -S: 126 (18 FGA, 84 ftr, 24* trg) plus 20 in store; RF-104G: 18 (recce) plus 16 in store; TF104G: 20 (OCU).
  AMX: 52. 49 (FGA); -T: 3* (trg).
  G-91: 99. -Y: 40 (FGA) plus 15 in store; -T: 59* (trg); plus 15 in store.
  MB-339: 85 (15 tac, 65 (incl 50*) trg, 5 calibration) plus 7 in store.
  Atlantic: 18 (MR).
  MB-326: 56 (liaison).
  Boeing-707: 1 (tkr/tpt). C-130H: 12 (tpt). G-222: 42 (38 tpt, 4 calibration), plus -GE: 1 (ECM).
  DC9-32: 2 (VIP). Gulfstream III: 2 (VIP).
  Falcon 50: 4 (VIP). P-166: 17: (-M: 11, DL3: 6 liaison and trg). PD-808: 18 (ECM, calibration, VIP tpt); SF-260M: 39 (trg). SIAI-208: 36 (liaison).
Helicopters:
  HH-3F: 19 (SAR).
  SH-3D: 2 (liaison).
  AB-212: 35 (SAR).
  AB-412: 4.
  AB-47G: 19 (trg).
  NH-500D: 50 (trg).
Missiles:
  ASM: AS-20, Kormoran, AGM-65 Maverick.
  AAM: AIM-7E Sparrow, AIM-9B/L Sidewinder, Aspide.
  SAM: 96 Nike Hercules, 7 bty Spada.

FORCES ABROAD
UN and Peacekeeping:
Egypt (MFO): 90; 4 PCC.
India/Pakistan (UNMOGIP): 7 observers.
Iraq/Kuwait (UNIKOM): 6 observers.
Lebanon (UNIFIL): 52 hel unit.
Middle East (UNTSO): 9 observers.

PARAMILITARY
Carabinieri (Ministry of Defense) 111,400: Territorial: 9 bde, 24 regt, 101 gp. Trg:1 bde. Mobile def: 2 bde, 1 cav regt, 13 mobile bn, 1 AB bn, avn and naval units.
  Equipment: 48 Fiat 6616 armd cars; 92 VCC2, 119 M-113, 24 M-106 APC; 22 A-109, 4 AB-205, 40 AB-206, 9 AB-412 hel.
Public Security Guard (Ministry of Interior): 80,400: 11 mobile units; 40 Fiat 6614 APC, 3 P-64B, 5 P-68 ac; 12 A-109, 20 AB-206, 9 AB-212 hel.
Finance Guards (Treasury Department): 53,000; 13 Zones, 20 Legions, 128 Gps; 13 A-109, 68 Nardi-Hughes (40 NH-500C, 16 -D, 12 -M) hel; 3 PCI, 65 ⟨, plus 309 boats.
Harbour Control (Capitanerie di Porto) (Subordinated to Navy in emergencies): Some 25 PCI ⟨, 100+ boats.

FOREIGN FORCES
NATO: HQ Allied Forces Southern Europe (AFSOUTH); HQ 5 Allied Tactical Air Force (5 ATAF).

U.S.: 14,100. Army (3,900); 1 AB bn gp; Navy (6,000); Marines (300); Air (3,900); 1 tac, 1 air base gp.

Future

The Republic of Italy may suffer some economic instability with inflation beyond its current 5 percent rate, but efforts to balance the government budget and keep the economy solid will continue, probably successfully. Ties to NATO and the European Community and growing defense foreign sales will be strong contributors to Italy's near future.

JAMES D. BOGGS

SEE ALSO: Atlantic-Mediterranean Area; European Communities; Italian Unification Wars; Western Europe; World War I; World War II.

Bibliography

International Institute for Strategic Studies. 1991. *The Military Balance, 1991–1992.* London: Brassey's.

Hunter, B., ed. 1991. *The statesman's year-book, 1991–92.* New York: St. Martin's Press.

U.S. Central Intelligence Agency. 1988. *World fact book.* Washington, D.C.: Government Printing Office.

U.S. Department of State. 1987. *Background notes—Italy.* Washington, D.C.: Government Printing Office.

J

JACKSON, THOMAS JONATHAN ("Stonewall") [1824–63]

Thomas Jonathan ("Stonewall") Jackson (Fig. 1) was born in Clarksburg, Virginia (now West Virginia), on 21 January 1824, the third of four children and second son of Jonathan Jackson, a lawyer, and Julia Beckwith (Neale) Jackson. His parents died during his early childhood and he was raised by his uncle, Cummins E. Jackson. Thomas added the name Jonathan when almost grown.

Figure 1. Thomas "Stonewall" Jackson. (SOURCE: U.S. Library of Congress)

Early Career

Jackson entered West Point in 1842 handicapped by inadequate prior education. However, he developed a demanding study regimen that enabled him to graduate seventeenth in a class of 59 when he graduated in 1846.

Upon graduation, Jackson was commissioned in the artillery. In the Mexican War he distinguished himself at Vera Cruz, Cerro Gordo, and Chapultepec, winning a succession of brevets to the rank of major. Afterward, he served at Fort Columbus, New York, in 1848–49; then at Fort Hamilton, New York, 1849–51; and finally in Florida. Jackson resigned from the army on 29 February 1852 to accept the professorship of artillery tactics and natural philosophy at the Virginia Military Institute (VMI), Lexington, Virginia.

Middle Years

Jackson was a poor teacher, and his stiff, formal manner and strict adherence to rules and regulations made him the butt of many cadet jokes. At VMI, he continued his strict personal study regimen and developed a strong Presbyterian religious conviction. His devoutness grew over the years.

On 4 August 1853, he married Elinor Junkin, the daughter of a Presbyterian minister, but she died suddenly on 22 October 1854, taking with her the child they were expecting. In 1856, Jackson spent five months in Europe. On 16 July 1857, he married his second wife, Mary Anna Morrison, also the daughter of a Presbyterian minister. This union produced two daughters, one of whom died in infancy.

Civil War Years

Jackson was in charge of the artillery detachment of VMI cadets at the hanging of John Brown at Charles Town, Virginia, on 2 December 1859. Although he did not support secession, Jackson believed that the South should fight for states' rights. When Virginia seceded from the Union in April 1861, he cast his lot with his state. He commanded the VMI corps of cadets, which left for war service on 21 April.

Appointed a colonel in the Virginia forces on 27 April, he was sent to command Harpers Ferry but was soon superseded by Brig. Gen. Albert Sidney Johnston. Jackson was promoted to brigadier general effective 17 June and placed in command of a newly raised brigade.

Jackson earned the sobriquet "Stonewall" during the First Battle of Manassas (First Bull Run), 21 July 1861. Brig. Gen. Barnard E. Bee, rallying his men, shouted: "There is Jackson standing like a stone wall. Let us determine to die here and we will conquer. Follow me." Bee was killed that day and Jackson was wounded in the middle finger of his left hand as he led his brigade of Virginia troops, famous that day forward as the Stonewall Brigade.

On 7 October 1861, Stonewall was promoted to major general of the Provisional Army of the Confederate States. On 4 November, he was given command of the Shenandoah Valley District. Between December 1861 and June 1862, he waged his famous Valley Campaign. His troops (rarely more than 15,000) were equally successful in maneuver and in battle. They tied down some 60,000 Federal troops, which were sorely needed for the campaign against Richmond, the capital of the Confederacy.

Recalled, with his command, to the main Army of Northern Virginia, then operating against the Army of the Potomac just east of Richmond, Jackson was slowed by his own physical exhaustion from the previous campaign and his unfamiliarity with the area. As a result, he failed to attack Federal positions at White Oak Swamp as expected on 30 June. This failure contributed to the disruption of Gen. Robert E. Lee's plan to envelop the Federal army's position. Nonetheless, in the ensuing Seven Days' Battle, the Union army was driven into its fortified base at Harrison's Landing, and Lee turned north to deal with Union general John Pope's Army of Virginia.

Jackson was again given a detached command. After defeating Pope's advanced corps (under Gen. Nathaniel P. Banks) at Cedar Mountain (9 August), he placed his command to the rear of Pope's army and initiated the Second Battle of Manassas, 29–30 August 1862. The arrival of Gen. James Longstreet's corps on Pope's left flank resulted in a crushing defeat of the Union army.

When Lee's army marched north into Maryland in September, Lee entrusted Jackson with five divisions to capture the Federal arsenal at Harpers Ferry. This he did on 15 September, then joined Lee at Sharpsburg. In the ensuing battle of Antietam, Stonewall steadfastly held the greatly outnumbered left wing of the Confederate army against almost overwhelming Union pressure (17 September).

Jackson was promoted to lieutenant general on 11 October 1862 and was given command of the Army of Northern Virginia's Second Corps. He distinguished himself at Fredericksburg on 13 December.

In late April 1863, the Union and Confederate armies were in Virginia's Wilderness region, south of the Rapidan River and west of Fredericksburg. On 1 May, Federal troops had taken up strong positions near Chancellorsville. On 2 May, Jackson led a great turning movement against the Union right flank and rear, wrecking one Union corps and driving the right wing into a constricted perimeter. Returning from reconnaissance forward of his main lines early that evening, General Jackson was mortally wounded when his own men fired into his party, believing them to be Federal troops. His left arm was amputated later that night and a musket ball was removed from his right hand. Moved to Guiney's Station, he died there, on 10 May 1863, of pneumonia. He was buried in Lexington, Virginia.

Assessment

In early life, Jackson suffered from a stomach ailment, causing him to adopt strict dietary and exercise habits the rest of his life. Jackson was five foot ten, slender and wiry, with large hands and feet, a high forehead, brown hair, blue eyes, and a Roman nose. Generally low-voiced and modest as a commander, and particularly in battle, Stonewall was aggressive, brave, determined, disciplined, and energetic. His command was dubbed Jackson's Foot Cavalry for its ability to march long distances in a short period of time. Yet while he asked much of his men, he also paid great attention to their needs and was popular with them. Lee could never replace this remarkable man, and the efficiency of the Army of Northern Virginia was irreparably damaged by his loss.

UZAL W. ENT

SEE ALSO: Civil War, American; Lee, Robert Edward.

Bibliography

Davis, B. 1954. *They called him Stonewall: A life of Lt. General T. J. Jackson, C.S.A.* New York: Rinehart.
Douglas, H. K. 1899. *I rode with Stonewall.* Chapel Hill, N.C.: Univ. of North Carolina Press (Douglas completed the manuscript in 1899, but it was not published until 1940).
Freeman, D. S. 1950. *Lee's lieutenants: A study in command,* vols. 1 and 2. New York: Scribner's.
Henderson, G. F. R. 1898. *Stonewall Jackson and the American Civil War.* 2 vols. London: Longmans, Green.
Vandiver, F. E. 1957. *Mighty Stonewall.* New York: McGraw-Hill.

JAPAN

The dramatic expansion of Japan's international economic activity in the 1980s, which involved Japanese financial institutions and capital flow, as well as trade, is propelling Asia's most dynamic society to examine what role it will play in the international arena during the post–cold war era. While there are considerable uncertainty and debate in Japan about what specific directions and policies the country should adopt, there is growing consensus that this perceived expanded role must be addressed seriously. Compounding the debate is the worsening of relations with the United States regarding trade, technology transfer, and defense burden sharing—issues that present major challenges for the Japanese government.

This article is a brief survey of Japan's national security policy and defense capabilities. The survey first provides a brief overview of the nation's post–World War II history, political-military background, and political-military policy. It then discusses the nation's strategic problems, small but increasingly sophisticated defense industry, alliance relations with the United States, defense structure, and command and control. The survey concludes with a summary description of the major characteristics of the personnel, force structure, and equipment of the ground, maritime, and air self-defense forces.

Power Potential Statistics

Area: 377,835 square kilometers (145,882 sq. mi.)
Population: 124,096,000
Total Active Armed Forces: 246,400 (0.200% of pop.)
Gross National Product: US$2,115.2 billion (1990 est.)
Annual Defense Expenditure: US$20 billion est. (1.0% of GNP, 1990 est.)
Iron and Steel Production:
    Crude steel: 105,681 million metric tons (1988)
    Pig iron: 79.295 million metric tons (1988)
Fuel Production:
    Coal: 11.223 million metric tons (1988)
    Crude oil: 0.689 million metric tons (1988)
    Natural gas: 2,097 million cubic meters (1988)
Electrical Power Output: 790,000 million kwh (1989)
Merchant Marine: 1,019 vessels; 22,396,958 gross registered tons
Civil Air Fleet: 360 major transport aircraft; 157 usable airfields (129 with permanent-surface runways); 2 with runways over 3,659 meters (12,000 ft.); 29 with runways 2,440–3,659 meters (8,000–12,000 ft.); 56 with runways 1,220–2,440 meters (4,000–8,000 ft.).

For the most recent information, the reader may refer to the following annual publications:
The Military Balance. International Institute for Strategic Studies. London: Brassey's (UK).
The Statesman's Year-Book. New York: St. Martin's Press.
The World Factbook. Central Intelligence Agency. Washington, D.C.: Brassey's (US).

History

Japan's postwar national security policy and defense capabilities are based on the legacy of its defeat in the Second World War and the reforms it underwent under the control of the occupation forces between 1945 and 1952.

After losing 3 million lives and suffering atomic bombings on Hiroshima and Nagasaki on 6 and 9 August 1945, the Japanese surrendered on 2 September 1945. The immediate goal of the occupation authorities, led by Supreme Commander for the Allied Powers (SCAP) Gen. Douglas A. MacArthur, was the construction of a demilitarized society, a democratic political system, and a decentralized market economy.

Demilitarization of postwar Japan included demobilization of the imperial army, navy, and air forces, repatriation of Japanese armed forces personnel, dismantling of the war industry, and establishment of a civilian government. A new constitution was drafted by MacArthur and accepted by the Japanese government headed by Shigeru Yoshida. Article 9 of the 1947 Constitution renounced "war as a sovereign right of the nation and the threat or use of force as means of settling international disputes." The same article declared Japan would never maintain "land, sea, and air forces, as well as other war potential."

Democratization of the country was facilitated by a SCAP directive on civil liberties which limited the power of the Ministry of Home Affairs and the centralized police system, released political prisoners, and removed restrictions on fundamental rights. Political parties were also revived. The 1947 Constitution established popular sovereignty and defined the emperor, who had renounced his divinity in January 1946, as the symbol of the state and of national unity. It set up the popularly elected bicameral National Diet as the highest organ of state power, entrusted executive power to a cabinet formed by the parliamentary majority party and responsible to the Diet, and created courts as an independent branch of government, with the power of judicial review entrusted to the Supreme Court. The Constitution also strengthened local self-government.

The intensification of the cold war halted the policy of complete demilitarization of Japan. Two weeks after the outbreak of the Korean War in June 1950, a 75,000-man strong National Police Reserve (NPR) was created for internal security purposes. The paramilitary NPR was reorganized and expanded into a 110,000-man strong National Safety Force (NSF) in October 1952. In July 1954, the NSF was restructured into separate land, sea, and air forces and placed under the authority of the newly created Defense Agency within the Office of the Prime Minister. The three branches were redesignated the Ground Self-Defense Force (GSDF), the Maritime Self-Defense Force (MSDF), and the Air Self-Defense Force (ASDF).

The cold war also affected Japan's re-entry into the international political arena. In September 1951, the San Francisco Peace Treaty was signed by Japan, the United States, and 47 other nations (the Soviet Union refused to sign), and went into effect in April 1952. The treaty restored Japanese sovereignty over the four main islands of Hokkaido, Honshu, Shikoku, and Kyushu and smaller adjacent islands, and forced Japan to surrender all former overseas possessions including the Kuril Islands and southern Sakhalin (the Bonin Islands and the remainder of the Ryukyu Islands, including Okinawa, were administered by the United States from 1945 until they were returned to Japan in June 1968 and May 1972, respectively). The treaty left the status of the Soviet-occupied Kuril Islands unclear because of the Soviet refusal to sign the pact. In October 1956, Tokyo and Moscow signed a joint declaration on the restoration of diplomatic relations and agreed to negotiate a peace treaty, but the dispute over the Habomai, Shikotan, Kunashiri, and Etorofu islands has prevented the two countries from concluding a peace treaty and terminating the state of war.

In a separate treaty signed in September 1951, Tokyo and Washington entered into a mutual security arrangement. The treaty, which went into effect in April 1952, provided for the U.S. bases and facilities and the stationing of American forces in Japan. The pact obligated the United States to defend Japan against aggression but did not impose a reciprocal commitment on Japan. A revised Mutual Treaty of Cooperation and Security was signed in January 1960 and went into effect in June. The treaty became subject to abrogation by either party on one year's notice in June 1970.

Politico-Military Background

Throughout the 1950s Japan undertook to settle its war accounts with other Asian countries by signing a series of separate peace treaties and reparation agreements with them. The normalization of diplomatic relations with Korea took much longer, until 1965. The establishment of diplomatic relations with the People's Republic of China (PRC) was also delayed until August 1978, when Tokyo and Beijing signed a treaty of peace and friendship.

In the 1950s and 1960s, Japan improved its international stature by joining international organizations, including the International Monetary Fund (IMF), the World Bank, the Economic Commission for Asia and the Far East (ECAFE), the General Agreement on Tariffs and Trade (GATT), the United Nations, the International Atomic Energy Agency (IAEA), the Organization for Economic Cooperation and Development (OECD), and the Asian Development Bank (ADB). Japan also hosted the 1964 Summer Olympics.

Domestic political stability was facilitated by the virtually unbroken succession of conservative, pro-U.S. governments in Tokyo, from the first Yoshida government of May 1946–May 1947, through the unification of conservative forces under the banner of the Liberal Democratic Party (LDP) in November 1955, to the Yasuhiro Nakasone government of November 1982–November 1987, and subsequent LDP governments. The LDP favors close and friendly relations with the United States, embraces the Mutual Security Treaty, and supports gradual buildup of the Self-Defense Forces within the framework of Article 9 of the Constitution.

The Socialist Democratic Party of Japan (SDPJ, formerly the Japan Socialist Party (JSP), united in 1955, is the largest opposition party. It opposes constitutional revision, advocates abolition of the SDF, and supports an unarmed, neutral Japan. The SDPJ has moderated its opposition to the Self-Defense Forces and now states that the SDF is "unconstitutional but legal." The Japan Communist Party (JCP), first formed in 1922, is now the third largest opposition party in Japan. It broke with Moscow and Beijing in the mid-1960s and has moved away from Marxist-Leninist dogma since the late 1960s. The JCP is opposed to the SDF as long as Japan–U.S. alliance continues but would favor armed forces in an independent Japan.

The Democratic Socialist Party (DSP), formed in 1959, originally advocated democratic socialism and its triumph over both capitalism and communism. Since the 1970s, however, the party's views have become quite similar to those of the LDP. Its initial support of minimum self-defense measures and its ultimate goal of complete disarmament have been replaced by firmer support of the SDF and the U.S.-Japan Security Treaty. Komeito, or "Clean Government Party," is the second largest opposition party. When it was formed in 1964, it advocated pacifism and opposed Japanese rearmament and the U.S.-Japan Security Treaty. By 1980, however, Komeito had come to accept the bilateral security treaty and begun showing greater support for the current defense buildup.

Politico-Military Policy

Japan's current foreign policy is based on two fundamental premises: the nation is a member of the Western alliance, and it is an Asian-Pacific country. According to the first premise, Tokyo defines its relations with Washington as the cornerstone of its diplomacy and sees those relations as essential, not only to its own peace and security, but to the broader regional and global peace and security. Japan is also deepening its economic and political ties with Western Europe.

Japan supports reduction of East-West tensions and arms control-disarmament efforts in the strategic, tactical, and conventional fields, and welcomed the 1988 INF agreement between the United States and the Soviet Union. In the UN and other multilateral disarmament forums, Japan advocates cessation of all nuclear testing, strengthening of the nonproliferation regime, and early realization of a chemical weapons ban.

A continuing problem between Tokyo and Moscow is the dispute over the so-called "Northern Territories" occupied by the Soviets since the end of the war. Moscow in the 1950s showed a willingness to return two of the four disputed islands (i.e., Habomai and Shikotan islands) in return for conclusion of a peace treaty, but Japan insisted on the reversion of all disputed territories. Moscow subsequently denied even the existence of a territorial dispute with Tokyo, and Japan's position did not change. Beginning in about 1987, against the backdrop of the major diplomatic initiatives of Mikhail Gorbachev globally and in the Asia-Pacific region, some Soviets intimated a softened stance on the issue, but Tokyo remained skeptical. The Gorbachev-Kaifu summit in Tokyo in April 1991 produced Moscow's acknowledgment of the territorial dispute over the Habomai group (Shikotan, Kunashiri, and Entorofu islands), and the Russian government and

Miyazawa government are now negotiating a peace treaty. The outcome of the talks remains uncertain, however, as Moscow has not accepted Tokyo's demand for Russian recognition of Japanese sovereignty over the Northern Territories.

As an Asian-Pacific nation, Japan views the region's peace and stability as critically important to its own security and vice versa. It is committed to the reduction of tensions through diplomacy and sees stable economic development in the region as a means of achieving regional peace and stability.

Japan's expanding economic power enabled the country to become the world's largest provider of official development assistance (ODA) in 1989. Although Japan surrendered the number-one position to the United States in 1990, it is certain that Tokyo will continue to expand its economic assistance to developing countries. In April 1991, Tokyo adopted new ODA guidelines that will gradually increase the government's effort to link its disbursement of aid money to its effort to curb military spending, control nuclear weapons proliferation, restrict arms trade, promote democratization, enhance human rights, and encourage market-oriented economic reforms in recipients of Japanese ODA. In this connection, Japan has in recent years substantially expanded its economic aid to developing countries, including "strategic aid" to Pakistan, Thailand, the Philippines, and Turkey. Upon the completion of its current five-year (1988–93) program of aid expansion, Japan is expected to become the world's largest donor.

Diplomatic and economic approaches to peace and security notwithstanding, Japan recognizes that the maintenance of peace in the present international society requires a balance of power and deterrence. From this perspective, Tokyo is committed to the maintenance of security arrangements with the United States and to the development of effective means of self-defense within appropriate limits.

According to the government's interpretation of Article 9 of the Constitution, Japan is allowed to possess a "minimum self-defense capability"; The use of the SDF for self-defense purposes is constitutional, but Japan cannot dispatch armed forces to foreign territorial land, sea, and airspace for the purpose of using force, and is prohibited from exercising the right of collective self-defense. The "Basic Policy for National Defense" of May 1957 established four principles to achieve the objective of Japan's national defense: to support the United Nations and promote international cooperation; to promote the public welfare and enhance the people's love for the country; to develop progressively the effective defense capabilities necessary for self-defense; and to deal with external aggression on the basis of the Japan-U.S. security arrangements.

During the 1991 Gulf War, the Toshiki Kaifu government failed to win parliamentary approval for a plan to send lightly armed SDF personnel to the Gulf to provide rear support for the multinational coalition forces led by the United States. Tokyo managed to contribute US$11 billion to the coalition forces and US$2 billion in economic and humanitarian aid to the frontline states of Egypt, Jordan, and Turkey, and, following the termination of the war, sent a flotilla of mine sweepers to the Gulf. However, Japan came under international criticism for what some critics called a "checkbook diplomacy" in Tokyo.

Japan has maintained a self-imposed policy of "Three Nonnuclear Principles" since 1967, not to possess, not to produce, and not to permit the introduction of nuclear weapons into Japan. The nation's Atomic Energy Law prohibits it from manufacturing or possessing nuclear weapons. Furthermore, Japan ratified the Nuclear Non-Proliferation Treaty in June 1976 and placed itself under an obligation as a nonnuclear weapons state not to produce or acquire nuclear arms.

In October 1976, the government adopted a "National Defense Program Outline," calling for the "most effective operation" of an "adequate defense capability of its own" and "the smooth functioning" of credible Japan-U.S. security relations. According to the Outline, Japan will "rely on the nuclear deterrent capability of the United States" against nuclear threat. The Outline envisages a 180,000-man SDF to achieve its self-defense goals. It anticipates a GSDF that includes: twelve divisions and two composite brigades deployed regionally in peacetime; mobile operation units composed of one mobile armored division, one artillery brigade, one airborne brigade, one training brigade, and one helicopter brigade; and eight antiaircraft artillery groups. The MSDF is expected to command four escort flotillas designed for mobile operations; ten antisubmarine surface ship divisions; six submarine divisions; two minesweeping flotillas; and sixteen land-based antisubmarine squadrons. The MSDF will be equipped with approximately 60 antisubmarine surface ships, sixteen submarines, and about 220 combat aircraft. The Outline anticipates an ASDF with 28 aircraft control and warning units, ten interceptor squadrons, three support fighter squadrons, one air reconnaissance squadron, three air transport squadrons, one early warning squadron, and six high-altitude surface-to-air missile groups. There will be about 430 combat aircraft. Defense buildup programs, such as the "1984 Mid-Term Estimate" (for FY 1986–90), have since been designed to attain the force levels envisaged by the Outline.

In addition, largely because of U.S. pressure to increase Japan's defense burden but also to meet the Outline's goals, the Nakasone government decided in 1984 to abolish the 1976 policy to limit its defense expenditures to less than 1 percent of the nation's Gross National Product (GNP). As a result, Japan's defense budget slightly exceeded 1 percent of its GNP in FY 1988.

The disintegration of the Soviet Union beginning in 1989 and its dissolution in January 1992 have forced Japanese foreign and security policy makers to review their policies. In 1989, Japan's defense white paper dropped its earlier reference to the "latent threat" from the Soviet Union; in December 1991, the Miyazawa government approved a mere 3.8 percent increase in defense spending, the smallest increase in 32 years; and in the spring of 1992, the government began a major review of the 1976 National Defense Program Outline in view of the end of the cold war and the virtual atainment of the Outline's equipment targets.

Strategic Problems

There are two fundamental facts that define Japan's strategic position: its geographical location and its heavy dependence on foreign sources of raw materials and overseas markets for its exports.

First, Japan is situated in a region (northeast Asia) where the interests of all global powers (i.e., the United States, the former Soviet Union, and the People's Republic of China) meet and often conflict with each other. As well, one of the most volatile conflicts in the world exists in this region, namely the conflict between North and South Korea. Tokyo sees close and friendly relations with the United States as essential not only to Japan's national security but to the peace and stability of the region. Following the Moscow-Seoul rapprochement in 1989, the security situation in northeast Asia improved substantially. However, Tokyo, as others, remains quite concerned over the suspected nuclear weapons development program in North Korea and Pyongyang's apparent attempt to delay acceptance of full IAEA safeguards inspection of nuclear facilities in the country.

The Japanese Defense Agency (JDA) viewed the nuclear and conventional arms buildup in the former Soviet Union as a "growing potential threat" to Japan's national security. The Soviets maintained one-quarter to one-third of their military power in their Far Eastern region. They had deployed in this region one-quarter to one-third of their ICBMs and SLBMs, including SS-18s, SS-N-18s, and TU-95 Bear bombers capable of carrying air-launched nuclear cruise missiles. The Soviet Union also maintained intermediate nuclear forces (INF) in the region, including about 162 SS-20s and about 85 TU-22M Backfire bombers, although land-based INF missiles such as SS-20s, SS-12s, and SS-23s are to be scrapped under the 1988 INF Treaty between Washington and Moscow.

The Soviet ground forces in the Far East included 43 divisions (approximately 390,000 troops), supported by T-72 tanks, armored infantry combat vehicles, and so forth. The Soviet Pacific Fleet, the largest unit of the Soviet navy, comprised about 845 ships (1.9 million tons). Finally, the Soviet air forces in the region comprised about 2,430 combat aircraft, including TU-22M Backfire bombers, MiG-23/27 Floggers, SU-24 Fencers, MiG-31 Foxhounds, and SU-27 Flankers. The Soviet Union had also deployed since 1978 ground troops (the equivalent of a division) on Kunashiri, Etorofu, and Shikotan islands. About 40 Mig-23 Flogger fighters are deployed at Tennei airfield on Etorofu Island.

In response to the Soviet force buildup in the region, Japan stepped up its efforts to improve its antisubmarine, antisurface, and antiair warfare capabilities. A controversial issue in this connection has been whether and under what conditions Japan should or can close its strategic straits to block Soviet naval forces seeking access to the Pacific.

The potentially explosive conflict between North and South Korea is a continuing concern for defense planners in Tokyo. They see the presence of U.S. forces in the South and Washington's firm commitment to the defense of that country as an important deterrent against the outbreak of major armed conflicts on the peninsula. On the diplomatic front, Tokyo has been supportive of a dialogue between North and South Korea for accommodation and peaceful coexistence. It supports international efforts to build a peaceful environment to reduce the tensions on the peninsula. Tokyo actively undertook measures to assist Seoul in hosting a successful International Olympics in 1988.

Tokyo has never considered the People's Republic of China as a security threat. On the contrary, since the early 1970s Japan has taken diplomatic and economic measures to ensure peaceful and friendly relations between the two countries. Japan welcomes the modernization processes taking place in China. Defense planners in Tokyo note Beijing's commitment to military modernization but also recognize that China's top priority remains economic modernization.

The second fundamental fact about Japan's strategic situation is that its extensive, global economic activities both contribute to its status as a global economic power and render it vulnerable to major disruptions in the world economic system. The 1973–74 oil crisis renewed the Japanese sense of vulnerability and made them search for "economic security." In the early 1980s, Japan began discussing the concept of "comprehensive security" that incorporates diplomatic, political, economic, and security efforts at national, regional, and global levels to reduce and, if possible, eliminate multiple sources of threat to the nation's expanding, largely economic, interests.

The Zenko Suzuki government (July 1980–November 1982) pledged that Japan would improve its maritime defense capabilities to defend its sea lines of communication (SLOCs) to a distance of 1,000 nautical miles from its coasts. The public has not fully endorsed the idea. Moreover, countries neighboring Japan remain apprehensive about Japan's military role beyond its immediate territories.

Defense Industry

The defense industry in Japan is small but is improving in both quality and sophistication. Defense production represents only 0.5 percent of Japan's total industrial output.

Defense sales for the top defense contractors constitute an insignificant share of their total sales. In certain defense sectors, however, defense contracts are important. In the aerospace industry, for example, 83.8 percent of the value of Japanese aircraft production in 1985 went to the SDF. The top ten defense contractors account for about 80 percent of the total SDF outlays. Procurement of advanced electronics for defense intelligence and surveillance is becoming an attractive outlet for private companies with potential in this rapidly advancing field.

JDA has emphasized domestic procurement, spending more than 80 percent of its funds domestically each year. Profit and market motivations favor domestic production, as does the private sector's desire to diversify high technology and other defense-related products. Important in this connection is the October 1987 U.S.-Japan agreement concerning a US$8 billion project, dubbed "F-SX," to produce a new generation of fighter support aircraft to replace outdated F-1s and F-4s. After long and difficult negotiations over the selection of the model aircraft, Tokyo decided in October 1987 to opt for coproduction; that is, to use the F-16 design (developed by General Dynamics) as the basis for the new plane's design, but to include a miniaturized active phased-array radar (designed by Mitsubishi Electric) and to build the wings from a single piece of composite materials (developed in Japan) that are radar-absorbent to give the plane "stealth" characteristics. Six prototypes of the plane are scheduled to be built by 1993.

Mitsubishi Heavy Industries (MHI), Kawasaki Heavy Industries (KHI), and Fuji Heavy Industries (FHI) dominate military aircraft production. MHI, Ishikawajima-Harima Heavy Industries (IHI), Mitsui Shipbuilding and Engineering, Hitachi Zosen Corp., and Sumitomo Heavy Industries are the major shipbuilders for the MSDF. MSDF ship orders are limited, but the recent U.S.-Japanese agreement to allow licensed production of the U.S. Aegis escort ship promises expanded production. Similar prospects exist for the production of the U.S. Patriot surface-to-air missile system.

Alliances

Japan defines the U.S.-Japan security system as the foundation of its national security policy and undertakes to maintain smooth and effective operation of the bilateral arrangement. Washington and Tokyo have established four forums for consultation on security matters: the Security Consultative Committee, the Security Subcommittee, the Security Consultative Group, and the Japan-U.S. Joint Committee.

In November 1978, the Consultative Committee acknowledged the "Guidelines for Japan-U.S. Defense Cooperation," calling on the two governments to conduct joint studies to ensure effective attainment of the objectives of the Security Treaty and its related arrangements.

Topics of completed or continuing joint studies include joint defense planning, sea-lane defense, the Japan-U.S. defense coordination center, the exchange of intelligence, common operational preparations, interoperability, and Japanese facilities assistance to U.S. forces in the case of a situation in the Far East outside Japan which will have an important influence on the security of Japan.

Joint military training has also been conducted between Japanese and U.S. forces. The GSDF began combined communication training and command post exercises with U.S. forces in October 1981. Since then, the GSDF has conducted command post exercises and field training exercises every year. The MSDF has participated in joint training centered on antisubmarine and minesweeping training since 1955. The MSDF has also participated in the Rim of the Pacific Exercise (RIMPAC) since 1980. Combined ASDF-U.S. Air Force training began with fighter combat training in November–December 1978. Combined command post exercises started in December 1983 and now occur annually. Finally, all three Japanese defense forces and the U.S. forces conducted their first combined joint field training exercise in October 1986, following the combined joint command post exercise in February the same year.

Cooperation in the field of military equipment and technology has also expanded over the years. The 1954 Mutual Defense Assistance (MDA) Agreement has provided the basis for U.S. grant aid (terminated in 1964), Foreign Military Sales (FMS), and coproduction (since the mid-1950s) of U.S.-developed weapon systems under licensing arrangements. In recent years Japan has produced the P-3C maritime patrol aircraft, the F-15 fighter aircraft, and the Patriot surface-to-air guided missile.

Until recently weapons and weapons technology transfer had been only from the United States to Japan. This was in line with Tokyo's 1967 policy prohibiting arms export. In response to U.S. prodding, Tokyo (in January 1983) exempted the transfer of dual-purpose technologies to the United States from the 1967 policy. Accordingly, in November 1983, an Exchange of Notes was concluded and in December 1985 Detailed Arrangements concerning the transfer of such technologies were concluded. By 1988 Tokyo had approved the transfer to the United States of SAM-related technology, technology related to the construction of naval vessels, and technology related to the modification of U.S. naval vessels.

In March 1985 Washington asked Tokyo to participate in the Strategic Defense Initiative (SDI). Tokyo announced in September 1986 it would cooperate with Washington. Finally, in July 1987, Tokyo and Washington concluded an "Agreement on Japan's Participation in the SDI Research," establishing a framework for future cooperation.

Japan also provides support for the U.S. forces in Japan. Japan is obligated to furnish facilities and areas without any expenses to the U.S. forces, Japan (USFJ) in accordance with the agreements between the Japanese and U.S. governments under the terms of the Status of Forces Agreement. Since FY 1979, Tokyo has expanded its support for the USFJ. In FY 1988, Tokyo budgeted about 79.2 billion yen for the improvement of USFJ personnel's living quarters and other facilities and encumbered another 63.2 billion for subsequent years. Tokyo also spent about 41.1 billion yen to cover labor-related costs for the 21,700 Japanese employees of the USFJ in FY 1988. In addition, Tokyo has shared the rent of USFJ facilities and areas and other related costs, expenses for maintaining the environment of their surrounding areas, and costs related to the retirement of Japanese employees at USFJ facilities. These were budgeted at about 235.8 billion yen in FY 1988.

Defense Structure

Under the Japanese Constitution, Japan's defense forces are placed under strict civilian control. The SDF is under the control of the Diet. The civilian Cabinet makes decisions on legislative and budget bills to be introduced to the Diet for approval, institutes government ordinances, and determines key policies and plans concerning national defense.

In July 1986, a Cabinet-level Security Council was established to replace the 20-year-old National Defense Council and to improve the government's ability to respond quickly to emergencies. Earlier, in December 1980, the government had established a Council of Ministers Concerned with Comprehensive Security within the Cabinet. The Council discusses "economic, diplomatic, and other measures requiring coordination among the adminstrative bodies concerned with ensuring comprehensiveness and coherence from the standpoint of national security."

A civilian minister of state at the Cabinet level is appointed by the Prime Minister as Director General of the JDA and controls SDF affairs. JDA is one of eight external organs of the Prime Minister's Office. The Director General supervises the SDF, and is assisted by Parliamentary and Administrative Vice-Ministers in governing and operating the SDF. Civilian counselors advise the Director General for formulating basic policies relating to the SDF.

Command and Control

The Prime Minister represents the Cabinet and serves as the supreme commander of the SDF. He exercises command and control over the Director General of the Defense Agency, who in turn administers the SDF. Below the civilian command structure is the uniformed SDF command, headed by the Joint Staff Council which includes the Chief of Staff of the GSDF, MSDF, and ASDF. The Joint Staff Council, chaired by the most senior uniformed officer, advises the Director General of the Defense Agency and plans and executes joint exercises. The three branches maintain staff offices that manage operations in their respective branches. While rank differentiations establish echelons of command within the SDF, all three branches are immediately responsible to the Director General and are coequal bodies with the Joint Staff Council and the three staff offices.

The command and control system of the SDF is aided by the central command and communications system, the tactical command and communications system, and other systems. The central command and communications system, built at the Central Command Post of the Defense Agency, transmits commanding orders and various information by connecting the SDF's principal headquarters to the commanding nucleus headed by the Director General. The ongoing computerization of the command and communications system, introduction of an integrated defense digital network (IDDN), and introduction of satellite-aided communications systems will enable the GSDF, MSDF, and ASDF to harmonize their operations, especially in emergencies.

(For an explanation of the abbreviations and symbols used in the following section of military statistics, see the list of Abbreviations and Acronyms in each volume.)

Total Armed Forces

Active: 246,400.
Reserves: Army 46,000; Navy 1,300; Air 1,100.

ARMY: (Ground Self-Defense Force): 156,100.
5 Army HQ (Regional Commands).
1 armd div.
12 inf div (5 at 7,000, 7 at 9,000 men each).
2 composite bde.
1 AB bde.
1 arty bde; 2 arty gp.
2 AD bde; 3 AD gp.
3 trg bde; 2 trg regt.
1 hel bde: 24 hel sqn.
2 ATK hel pl, 1 more forming.
Equipment:
MBT: 1,200: some 347 Type-61 (retiring), some 873 Type-74, some Type-90.
AIFV: some 8 Type-88.
Recce: some 146 Type-82, 51 Type-87.
APC: 300 Type-60, 274 Type-73.
Towed arty: 553: 105mm: 290 M-101; 155mm: some 40 (incl M-1, M-2), 183 FH-70; 203mm: 40 M-115.
SP arty: 296: 105mm: 20 Type-74; 155mm: 200 Type-75; 203mm: 76 M-110A2.
MRL: 130mm: some 120 Type-75 SP.
Mortars: 81mm: 820 (some SP); 107mm: 560 (some SP).
SSM: 50 Type-30, some Type-88.
ATGW: 220 Type-64, some 100 Type-79, some 44 Type-87.
RL: 89mm: 3.5-in M-20.
RCL: 2,900: 75mm; 84mm: Carl Gustav; 106mm (incl Type 60 SP).
AD guns: 120: 35mm: 70 twin; 37mm SP; 40mm SP.
SAM: 180 Stinger, some 48 Type 81 Tan, 200 Improved HAWK.
Aircraft: 19: 17 LR-1, 2 TL-1 (trg).
Helicopters:
  Attack: 57 AH-1S.

Transport: 3 AS-332L (VIP), 20 CH-47J, 55 KV-107, 186 OH-6D/J, 148 UH-1B/H; 33 TH-55 (trg).

NAVY: (Maritime Self-Defense Force): 44,000 (including 12,000 MSDF air). Bases: Yokosuka, Kure, Sasebo, Maizuru, Ominato.
Fleet: surface units org into 4 escort flotillas, of 6–8 DD/FF each; based at Yokosuka (2), Sasebo and Maizuru. Submarines org into 2 flotillas based at Kure and Yokosuka. Remainder assigned to 10 regional/district units.
Submarines: 17:
Tactical submarines: 15:
  2 Harushio with 533mm TT (Jap Type-89 HWT) with Harpoon USGW.
  10 Yuushio with 533 TT (US Mk 37, GRX-2 HWT), 7 with Harpoon USGW.
  3 Uzushio with 533mm TT (Mk 37 HWT).
Other Roles: 2 Uzushio (trg).
Principal Surface Combatants: 66:
Destroyers: 6 DDG:
  2 Hatakaze with 1×SM1-MR Standard SAM: plus 2×4 Harpoon SSM, 1×8 ASROC SUGW (Mk 46 LWT) 2×3 ASTT, 2×127mm guns.
  3 Tachikaze with 1×SM1-MR; plus 1×8 ASROC, 2×3 ASTT, 8×Harpoon, 2×127mm guns.
  1 Amatzukaze with 1×SM1-MR; plus 1×8 ASROC, 2×3 ASTT.
Frigates: 60 (incl 5 training):
FFH: 24:
  2 Shirane with 3×HSS-2B Sea King ASW hel, 1×8 ASROC, 2×3 ASTT; plus 2×127mm guns.
  2 Haruna with 3×Sea King hel, 1×8 ASROC, 2×3 ASTT; plus 2×127mm guns.
  8 Asagiri with 1 Sea King hel, 1×8 ASROC, 2×3 ASTT; plus 2×4 Harpoon SSM.
  12 Hatsuyuki with 1 Sea King, 1×8 ASROC, 2×3 ASTT; plus 2×4 Harpoon SSM.
FF: 36:
  4 Abukuma with 1×8 ASROC, 2×3 ASTT; plus 2×4 Harpoon SSM.
  4 Takatsuki with 1×8 ASROC, 2×3 ASTT; 1×4 ASW RL; plus 2 with 2×4 Harpoon SSM, 1×127mm gun; 2 with 2×127mm guns.
  6 Yamagumo with 1×8 ASROC, 2×3 ASTT, 1×4 ASW RL.
  3 Minegumo with 1×8 ASROC, 2×3 ASTT, 1×4 ASW RL.
  2 Yubari with 2×3 ASTT, 1×4 ASW RL; plus 2×4 Harpoon SSM.
  1 Ishikari with 2×3 ASTT, 1×4 ASW RL; plus 2×4 Harpoon SSM.
  11 Chikugo with 1×8 ASROC, 2×3 ASTT.
  4 Isuzu (trg) with 1×4 ASW RL; plus 4×2 76mm gun.
  1 Katori (trg) with 2×3 ASTT, 1×ASW RL.
Patrol and Coastal Combatants: 13:
Torpedo Craft: 4 Juichi-go PFT with 4×533mm TT.
Patrol: 9 Jukyu-go PCI ⟨.
Mine Warfare: 47:
Minelayers: 1 Souya (460 mines) plus hel deck, 2×3 ASTT, also MCM spt/comd.
Mine countermeasures: 46:
  1 Hayase MCM cmd with hel deck, 2×3 ASTT, plus mine-laying capacity (116 mines).
  25 Hatsushima, 7 Takami MCC.
  6 Nana-go MSI ⟨.
  6 Coastal diver spt ships (ex MSC).
  1 Utone coastal MCM spt.

Amphibious: 6: 3 Miura LST, capacity 200 tps, 10 tk.; 3 Atsumi LST, capacity 130 tps, 5 tk.; Plus craft: 3 LCT, 21 LCM, 22 LCVP.
Support and Miscellaneous: 17: 3 Towada, 1 Sagami AOE, 2 AS, 2 Akizuki trg, 2 trg spt, 6 survey/experimental, 1 icebreaker.

MSDF Air Arm: (12,000); 99 cbt ac (plus 15 in store), 72 armed hel. 7 Air Groups.
MR: 10 sqn: 7 (1 trg) with 59 P-3C (plus 15 in store); 3 with 30 P-2J.
ASW: 6 hel sqn (1 trg) with 72 HSS-2A/B, plus 33 in store.
MCM: 3 hel sqn with 5 KV-107A, 12 S-80.
EW: 1 sqn with 2 EP-2J, 2 EP-3C.
Transport: 1 sqn with 4 YS-11M.
Test: 1 sqn with 3 P-3C, 3 UP-2J, 4 U-36A ac; 2 HSS-2B, 2 SH-60J hel.
SAR: 1 sqn (7 flt) with 7 US-1/1A; 3 rescue sqn with 10 S-61 hel.
Training: 9 sqn with 30 KM-2, 10* P-3C, 22 Queen Air 65, 8 T-5, 23 TC-90/UC-90, 10 YS-11T ac; 10 HSS-2A/B, 12 OH-6D/J hel.

AIR FORCE: (Air Self-Defense Force): 46,300; 422 cbt ac (plus 63 in store), no armed hel.
7 cbt air wings; 1 cbt air unit; 1 recce gp; 1 AEW unit.
FGA: 3 sqn with 70 F-1; 1 with 22 F-4EJ (anti-ship).
Fighter: 10 sqn: 7 with 138 F-15J/DJ (9 more in store); 3 with 72 F-4EJ (being upgraded; 50 more in store).
Recce: 1 sqn with 10 RF-4EJ, 4 more in store.
AEW: 1 sqn with 12 E-2C.
EW: 1 flt with 1 C-1, 4 YS-11.
Aggressor training: 1 sqn with 20 T-2, 2 T-33.
Transport: 5 sqn: 3 with 30 C-1, 15 C-130H, 10 YS-11; 2 heavy-lift hel sqn with 6 CH-47J.
SAR: 1 wing (10 det) with 30 MU-2 ac; 22 KV-107, 6 CH-47J hel. (UH-60J hel being delivered.)
Calibration: 1 wing with 2 MU-2J, 1 YS-11.
Training: 5 wings: 10 sqn: 40* T-1A/B, 50* T-2, 40 T-3, 50 T-4, 10 T-33A (to be replaced by T-4).
Liaison: 11 Queen Air 65, 126 T-33.
Test: 1 wing with C-1, F-4EJ, F-15J.
Missiles:
  ASM: ASM-1.
  AAM: AAM-1, AIM-7 Sparrow, AIM-9 Sidewinder.
Air defense: ac control and warning: 4 wings; 30 radar sites.
  SAM: 6 AD msl gp (18 sqn) with 180 Nike-J (Patriot replacing).
  Air Base Defense Gp with 20mm Vulcan AA guns, Type 81 Tan, Stinger SAM.

PARAMILITARY
Maritime Safety Agency: (Coast Guard) 12,000:
Patrol vessels: Some 335:
  Offshore: 81, incl 2 Mizuho with 2 Bell 212, 8 Soya with 1 Bell 212 hel.
  Coastal: 11.
  Inshore: 243: 3 PFI, 15 PCI, some 225 ⟨.
Miscellaneous: 90 service, 81 tender/trg vessels.
Aircraft: 5 NAMC YS-11Aa, 2 Short Skyvan, 16 King Air, 1 Cessna U-206G.
Helicopters: 32 Bell 212, 4 Bell 206, 2 Hughes 369.

FOREIGN FORCES: 45,100
U.S.: Army (1,800): 1 Corps HQ; Navy (6,300) bases at Yokosuka (HQ 7th Fleet) and Sasebo; Marines (22,000): 1 MEF in Okinawa; Air (15,000): 1 Air HQ, 1 air div some 120 cbt ac.

Future

Japan's current military capacity is very limited in comparison with its economic potential and political influence. This is unlikely to change by a significant degree in the 1990s, for several reasons. Two are of particular importance. First, the Japanese themselves have no real wish to assume the defense burden of the United States in the Far East. Such a course would be costly and might interfere with Japan's vigorous economy. Second, none of Japan's neighbors would welcome a remilitarized Japan, in view of their experiences in the 1930–45 period. China and Korea, particularly, would view a massive rearmament effort in Japan with alarm. These two reservations aside, Japan's modest program of upgrading and re-equipping its armed forces will inevitably continue, as witnessed by two initiatives: (1) the major reorganization of the JGSDF to thirteen divisions to compensate for manpower shortages, and (2) new equipment purchases, including aircraft and new warships, for the JMSDF which is rapidly becoming a blue-water navy with the capability to project naval power far into the Pacific and Indian oceans. For the foreseeable future, Japan will employ its economic power to increase its regional and worldwide political influence.

TSUNEO AKAHA

SEE ALSO: Central and East Asia; Indo-Pacific Area; Japan, Modernization and Expansion of; Russo-Japanese War; Togo, Heihachiro; World War II; Yamamoto, Isoroku; Yamashita, Tomoyuki.

Bibliography

Akaha, T. 1984. Japan's nonnuclear policy. *Asian Survey*. 24:863–67.
Asian Security, 1987. Tokyo: Research Institute for Peace and Security.
Barnett, R. W., 1984. *Beyond war: Japan's concept of comprehensive national security*. Washington, D.C.: Pergamon-Brassey's.
Bunge, F., ed. 1983. *Japan: A country study*. Washington, D.C.: Government Printing Office.
Defense Agency of Japan. 1988. *Defense of Japan, 1988*. Tokyo: Japan Times.
Ellison, H. J., ed. 1987 *Japan and the Pacific quadrille: The major powers in East Asia*. Boulder, Colo.: Westview Press.
Falkenheim, P. L. 1987. *Japan and arms control: Tokyo's response to SDI and INF. Aurora Papers 6*. Ottawa: Canadian Center for Arms Control and Disarmament.
Frost, E. L. 1987. *For richer, for poorer: The new U.S.-Japan relationship*. New York: Council on Foreign Relations.
Galenson, W., and D. W. Galenson. 1986. Japan and South Korea. In *Constraints on strategy: The economics of Western security*, ed. D. B. H. Denoon, pp. 152–94. Washington, D.C.: Pergamon-Brassey's.
Hunter, B. ed 1991. *The statesman's year-book, 1991–92*. New York: St. Martin's Press.
International Institute for Strategic Studies. 1991. *The military balance, 1991–1992*. London: Brassey's.
Institute of Administrative Management. 1986. *Organization of the government of Japan, 1986*. Tokyo: Institute of Administrate Management.
Kimura, H. 1986. The Soviet military buildup: Its impact on Japan and its aims. In *The Soviet Far East military buildup: Nuclear dilemmas and Asian security*, eds. R. Solomon and M. Kosaka, pp. 106–22. Dover: Auburn.
Lawless, B. J., and R. M. Deming. 1987. Japanese defense policy: Needs and realities. In *Essays on Strategy IV*. Washington, D.C.: National Defense University.
McIntosh, M. 1986. *Japan rearmed*. New York: St. Martin's.
Ministry of Foreign Affairs. 1988. *Diplomatic bluebook, 1988 Edition*. Tokyo: Foreign Press Center.
Nish, I. 1986. The United States in East Asia: Japan's perspective. In *The Balance of Power in East Asia*, M. Leifer, ed. pp. 30–43. New York: St. Martin's.
Nishihara, M. 1986. Japan: Regional stability. In *Security interdependence in the Asia Pacific Region*, J. Morley, ed. pp. 65–91. New York: Columbia Univ.
Olson, E. A. 1985. *U.S.–Japan strategic reciprocity: A new international view*. Stanford: Hoover Institution.
Political handbook of the world, 1988. Binghamton, N.Y.: State Univ. of New York.
Reed, R. F. 1983. *The U.S.-Japan alliance: Sharing the burden of defense*. Washington, D.C.: National Defense Univ.
Reischauer, E. 1988. *The Japanese today: Change and continuity*. Cambridge: Harvard Univ. Press.

JAPAN, MODERNIZATION AND EXPANSION OF

The 77 years between the Meiji Restoration and the end of World War II chronicle the emergence of Japan from self-imposed isolation, its transformation into an aggressive, militaristic state, and its near-catastrophic defeat. The evolution of Japan from a preindustrial, agrarian nation into an industrialized, Westernized power in such a short time still elicits admiration, and its aggressive expansionism still evokes bitter memories among the peoples of East and Southeast Asia.

The Meiji Restoration

On 2 November 1868, 265 years of military rule under the Tokugawa Shogunate came to an end. The Shogun Yoshinobu, faced with the continuing civil war launched by pro-imperial *daimyo* (high-ranking feudal lords)—Choshu, Satsuma, Tosa, and Hizen—formally resigned his post and surrendered all executive, legislative, and judicial power to the 14-year-old Meiji emperor (Mutsuhito).

The restoration of imperial power and the far-reaching political, economic, and social changes associated with the reign of the Meiji Emperor were the work of a small oligarchy drawn primarily from the proimperial feudal aristocracy. Spurred on by fears that Japan might suffer the fate of China at the hands of the Western powers, the emperor and his advisers decided to adopt certain elements of Western civilization. The primary objective of the Meiji Era statesmen was succinctly stated in the slogan "*fukoku kyohei*—a rich country, a strong army."

POLITICAL REFORMS

The imperial government's first political reform was the abolition of feudalism. In March 1869, the *daimyo* of Choshu, Satsuma, Tosa, and Hizen voluntarily returned their fiefs to the emperor. In return, the *daimyo* received the governorships of their former domains. Soon thereafter, the remaining *daimyo*, fearing possible government retaliation, also surrendered their fiefs to the emperor.

The emperor, on the advice of Ito Hirombi, decided to follow the Prussian model in the restructuring of the imperial government. According to the Meiji Constitution of 1889, the emperor held supreme power. Although a legislature, the Diet, was created, cabinet members were responsible only to the emperor. The ministers of the army and the navy also were responsible only to the emperor, and candidates for these posts could be nominated only by their respective service high commands.

ECONOMIC REFORMS

Japan's rapid industrialization during the Meiji Era was the result of centralized planning, financing, and control. The Japanese government obtained both foreign assistance and loans. Foreign loans were secured against imperial land revenues and land taxes. Between 1870 and 1900, the Japanese government built and operated steel mills, coal mines, railroads, and textile factories. Eventually the government sold these assets to private corporations.

Although Japan was able to go from a preindustrial to an industrialized society in a short time, industrial development was restricted to heavy industry (iron and steel), coal mining, and textiles. Little attention was paid to the development of light industry. The agricultural sector remained outside the scope of the Meiji reforms.

MILITARY MODERNIZATION

Because a primary reason for Japanese modernization was the threat of Western military power, the imperial government placed military modernization high on its priority list. Initially, the Japanese sought the assistance of the French in the creation of a modern army. In 1875, however, Yamagata Aritomo, the army minister, adopted the German general staff system rather than the French. Eventually the Japanese adopted the German system of military schools, including curricula, as well as the German divisional structure, with organic artillery support. Yamagata also reorganized the Army Ministry along German lines.

The Japanese sought the help of the dominant naval power of the day, Great Britain, in organizing and training the navy. In addition, British shipyards built the first Japanese warships.

Japanese Expansionism During the Meiji Era, 1868–1912

Although military modernization was initially a self-defense measure, the use of military force quickly became a tool of Japanese foreign policy. During the Meiji Era, the Imperial Japanese Army and Navy fought in two foreign wars: The First Sino-Japanese War and the Russo-Japanese War. The cause of both wars was the same: the status of Korea.

THE FIRST SINO-JAPANESE WAR, 1894–95

Disagreements between China and Japan over the status of Korea came to a head when Korean nationalists rebelled against the Korean government. Chinese troops, supporting the Korean government, clashed with Japanese troops. In the ensuing conflict, the numerically superior Chinese forces were soundly beaten by the modern Japanese army. At sea, the Japanese navy sank or captured elements of the Chinese North Seas Fleet, the most modern of China's naval forces.

The First Sino-Japanese War ended on 17 April 1895 with the signing of the Treaty of Shimonoseki. Both China and Japan recognized the independence of Korea. China agreed to pay Japan an indemnity of 200 million taels of silver, approximately US$66 million. Japan also received Taiwan (Formosa), the Pescadore Islands, and a lease to the Liaotung Peninsula. The Chinese also agreed to open the cities of Chungking, Soochow, and Hangchow to Japanese commercial enterprises.

THE RUSSO-JAPANESE WAR, 1904–1905

Having gained the Liaotung Peninsula, the Japanese were forced to return it to China due to pressure from France, Germany, and Russia. This Triple Intervention was shortly followed by German and Russian demands for economic enclaves—concessions—in China. Furthermore, Russia forced China to grant it a lease to the Liaotung Peninsula. The Japanese were outraged. Russia had replaced China as the main threat to Japanese interests. With its control of Liaotung and its penetration of Manchuria, Russia was in a position to dominate Korea and, therefore, threaten Japan.

Imperial Russia had many advantages: its size, population, rich natural resources, and a massive military establishment. Nevertheless, Japan decided to go to war. In order to offset the Russian advantages, the Japanese High Command ordered Adm. Togo Heihachiro to attack the Russian Asiatic Fleet at Port Arthur. The attack came without warning on 9 February 1904. The Russian fleet was bottled up in the harbor, and many ships were sunk or damaged at their moorings. The Japanese army landed and besieged Port Arthur, which finally surrendered on 1 January 1905. The Russian government, in an attempt to strike back, sent the Baltic Fleet halfway around the world. At the Battle of Tsushima Straits (27–28 May 1905), the outnumbered but more modern Japanese fleet outsailed, outmaneuvered, and outfought the Russians and inflicted on them one of the most decisive naval defeats in the history of sea warfare.

Meanwhile, the Japanese also had the best of the long,

drawn-out fighting in Manchuria. The struggle took its toll of both governments. In Russia, the outcry against the war threatened revolution. In Japan, the war threatened bankruptcy. Therefore, when President Theodore Roosevelt offered to mediate, both governments accepted.

According to the terms of the Treaty of Portsmouth, signed 5 September 1905, Russia recognized Japan's paramount interests in Korea. Russia agreed to transfer to Japan, with Chinese approval, the lease to the Liaotung Peninsula, the southern section of the Chinese Eastern Railroad (in Manchuria), and the Russian-operated coal mines. Russia also ceded to Japan the southern half of Sakhalin Island. Both governments agreed to withdraw their forces from Manchuria.

THE ANNEXATION OF KOREA

On 22 August 1910, Japan solved the Korean problem by annexing the kingdom of Korea and placing it under the administration of the Imperial Japanese Army.

Japanese Expansion During the Taisho Era, 1912–26

The Taisho Era can be divided into two periods: the first ten years, 1912–22, during which Japanese expansionism followed the Meiji pattern; and the last four years, 1922–26, during which Japanese policy followed a less aggressive, more cooperative and conciliatory pattern.

THE FIRST WORLD WAR

World War I provided Japan with an opportunity to expand its economic foothold in China. Having declared war on Germany on 23 August 1914, the Japanese attacked German forces in Shantung, China. After securing Tsingtao on 7 November 1914, the Japanese occupied German possessions in the Pacific, the Marshall, Mariana, and Caroline islands. Japan's gains in China and the Pacific were formally recognized by the Treaty of Versailles.

THE LATE TAISHO ERA, 1922–26

The Washington Naval Conference of 1921–22 marked a change in the nature of Japanese foreign policy. In exchange for recognition as one of the five great naval powers and virtual naval control of the western Pacific, Japan agreed to return its concessions in Shantung to Chinese jurisdiction and pledged itself to defend Chinese sovereignty.

For the remainder of the Taisho Era and the first years of the Showa Era, Japan's policy toward China was one of conciliation and emphasis on the benefits of Sino-Japanese cooperation.

Japanese Expansion During the Showa Era, 1926–45

Japan's abrupt return to an aggressive policy of expansionism during the Showa Era can be traced to the economic distress caused by the Great Depression and the rise of militarism. The two were interrelated. The effects of the depression—rising food prices, falling silk prices, rapidly increasing rents, and the worsening condition of the peasants—convinced many members of the military high command that the party politicians who had succeeded the Meiji Era statesmen had sold out to the interests of big business and foreigners. All military factions—Black Dragon, Imperial Way, Control, Cherry Blossom, and the others—agreed that the politicians had abandoned the traditional values of the nation and betrayed the emperor. By 1929, the military had begun to formulate and follow its own foreign and domestic policies, policies often at odds with those of the government. By the late 1930s, the government had become, in effect, a tool of the military.

THE MANCHURIAN INCIDENT, 1931–32

The Kwantung Army—originally stationed in Manchuria to protect Japanese-owned railroads—took control of Manchuria without government approval. By 5 February 1932, the Kwantung Army had occupied all of Manchuria. Manchuria was detached from China and formed into the new state of Manchukuo with Henry Pu-yi, the last Manchu emperor of China, as emperor. Because the Japanese controlled all government functions, Manchukuo was, in reality, a Japanese puppet state.

JAPANESE AGGRESSION IN CHINA PROPER

The successful coup in Manchuria and the inability of the Japanese government to call the army to heel emboldened the militarists. From the creation of Manchukuo on 18 February 1932 to the Marco Polo Bridge Incident of 7 July 1937, the Japanese followed a policy of overt territorial expansionism. They hoped to detach the five northern provinces from China and form them into another puppet state. The Chinese Nationalist government, then embroiled in its anti-Communist extermination campaigns, tried negotiation and conciliation.

In December 1936, as a result of negotiations between Chou En-lai and Chiang Kai-shek, the Communist and Nationalist parties agreed to join forces in a united front against Japanese aggression. Faced with the possibility of a united China, the Japanese army fabricated the Marco Polo Bridge Incident. The Japanese expected the Chinese to request negotiations. Instead, the Chinese chose to fight.

THE SECOND SINO-JAPANESE WAR, 1937–45

The fighting in North China spread rapidly. The Japanese occupied Peking on 28 July and Tientsin the following day. In an attempt to outflank the Chinese defenses, 10,000 Japanese marines landed at Shanghai. Instead of a quick victory, Shanghai developed into a three-month-long battle that allowed the Chinese government to fortify Nanking and make preparations for a withdrawal into the interior. Shanghai fell on 8 November, Nanking on 13

December. With the capture of Nanking, the Chinese capital, the Japanese hoped to end the China Incident. However, Chiang Kai-shek moved his capital to Chungking, and the Chinese fought on.

As the war dragged on, the Japanese government sought ways to end it. Both Britain and the United States condemned Japan's aggression and continued to supply China with war materiel through Burma.

THE GREATER EAST ASIA WAR (WORLD WAR II), 1941–45

American condemnation of Japanese policy in China took concrete form when, in 1940, the United States imposed a series of restrictions on the sale of oil, scrap iron, and other strategic materials to Japan. Faced with the specter of economic strangulation and continued Anglo-American support for China, the Japanese government, after long deliberation, decided to go to war. The plan was to neutralize the threat of American power by destroying the U.S. Pacific Fleet at Pearl Harbor and occupying the Philippines. The conquest of British-controlled Malaya and Burma would not only sever China's supply line but also secure for Japan the rubber and oil reserves of Southeast Asia.

What had begun as an attempt to gain economic control of the resources and markets of China and, therefore, economic security had led Japan into all-out war with China, Great Britain, and the United States. Although initially successful, Japan's limited industrial base and dependence on foreign raw materials severely hampered the war effort. By mid-1943, Japanese forces were in retreat in the Pacific and stalemated in Burma, while in China the fighting continued to drain the strength of the army. By the end of the war, Japan's military and civilian leaders, who had enthusiastically embraced the policy of expansionism, had, in effect, destroyed much of what their predecessors had achieved during the Meiji Era.

LAWRENCE D. HIGGINS

SEE ALSO: Chu-Teh; Giap, Vo Nguyen; Mao Tse-tung; China, People's Republic of; Togo, Heihachiro; Wei Lihuang; Yamamoto, Isoroku; Yamashita, Tomoyuki.

Bibliography

Barnhart, M. A. 1941. *Japan prepares for total war: A search for economic security, 1919–1941*. Ithaca, N.Y.: Cornell Univ. Press.
Beasley, W. G. 1972. *The Meiji restoration*. Stanford, Calif.: Stanford Univ. Press.
Crowley, J. B. 1966. *Japan's quest for autonomy: National security and foreign policy*. Princeton, N.J.: Princeton Univ. Press.
Dower, J. W. 1988. *Empire and aftermath: Yoshida Shigeru and the Japanese experience, 1878–1954*. Cambridge: Harvard Univ. Press.
Francks, P. 1984. *Technology and agricultural development in pre-war Japan*. New Haven: Yale Univ. Press.
Hane, M. 1982. *Peasants, rebels, and outcasts: The underside of modern Japan*. New York: Pantheon Books.
Hattori, T. 1955. *Daitoa Senso Zenshi*. Tokyo: Hara Shobo.
Nobutaka, I. 1967. *Japan's decision for war: Records of the 1941 policy conferences*. Stanford, Calif.: Stanford Univ. Press.
Presseisen, E. L. 1965. *Before aggression: Europeans prepare the Japanese army*. Tucson: Univ. of Arizona Press.
Smith, T. C. 1988. *Native sources of Japanese industrialization, 1750–1920*. Berkeley: Univ. of California Press.
Totman, C. 1980. *The collapse of the Tokugawa Bakufu, 1862–1868*. Honolulu: Univ. of Hawaii Press.

JELLICOE, JOHN RUSHWORTH (1st Earl) [1859–1935]

Admiral of the Fleet Earl Jellicoe had a distinguished career in the Royal Navy from 1872 until 1924. A recognized expert at naval gunnery, Jellicoe was instrumental in modernizing the Royal Navy and superintending the Dreadnought construction program. His name has become enmeshed in the controversy surrounding the Battle of Jutland (31 May–1 June 1916) where Jellicoe commanded the Grand Fleet against the German High Seas Fleet. At this crucial engagement, which involved the two greatest fleets of dreadnoughts ever assembled, Jellicoe failed to defeat the Germans decisively. Afterwards his critics damned his caution as lacking the Nelson touch; his defenders echoed Churchill's comment that "Jellicoe was the only man on either side who could lose the war in an afternoon."

Early Life and Career

John Rushworth Jellicoe was born at Southampton, England, on 5 December 1859 to a family with a long tradition of maritime service. Jellicoe joined the training ship *Britannia* as a naval cadet in 1872, graduated at the top of his term in 1874, and received immediate promotion to midshipman.

Jellicoe's naval career began with service on the frigate *Newcastle* (1874–77) and the battleship *Agincourt* (1877–78) and then attendance at naval schools at Greenwich and Portsmouth. Subsequently Jellicoe served on the *Alexandra* in the Mediterranean and then returned to the *Agincourt* in time to combat the Arabi Pasha rebellion in Egypt in 1882.

Jellicoe next qualified as a gunnery officer, and in 1884 joined the staff of the *Excellent* gunnery school, commanded by Capt. (later Admiral of the Fleet Lord) J. A. Fisher. The following year Fisher became chief of staff of the *Minotaur*, and selected Jellicoe as his staff officer. Jellicoe's proficiency and abilities earned him a deserved place in the "Fishpond," a group of Fisher's proven protégés.

Service as gunnery officer on the *Monarch*, *Colossus*, and *Excellent* followed. In 1889 Fisher requested Jellicoe's transfer to the Admiralty, where he worked for two years before service on the *Sans Pareil*. Jellicoe then served as commander of the Mediterranean Fleet flagship *Victoria*. While Jellicoe was suffering from fever, Adm. Sir George Tryon, Fleet commander, made a tragic blunder in navigation that resulted in the *Victoria* ramming the *Camperdown*. The *Victoria* sank with the loss of nearly 400 men, including Tryon, but Jellicoe was able to swim free from the wreckage.

After recuperating, Jellicoe commanded the new flagship, *Ramillies*, from 1893 to 1896, at which time he returned to England to serve on the Ordnance Committee. He was promoted to captain in January 1897, and the next year became flag captain and chief of staff to Adm. Frederick Seymour in the Far East. Two years later, during the Boxer Rebellion, while Seymour commanded the international naval brigade, Jellicoe was shot in the left lung.

At the Admiralty

Jellicoe, after convalescing, served at the Admiralty where he was responsible for inspecting all ships under construction. A year later he took command of the armored cruiser *Drake*. This tour was interrupted by recall to the Admiralty, where he assisted in the development of the "all-big-gun" battleship *Dreadnought*, which made all earlier battleships obsolete.

In 1907 Jellicoe became second in command of the Atlantic Fleet and the next year returned to the Admiralty as controller and third sea lord. Three years later, Vice Admiral Jellicoe became commander of the Atlantic Fleet, after having supervised an increase in the Royal Navy of some 90 ships, including twelve battleships.

The Grand Fleet

Upon the outbreak of war in 1914, Jellicoe reluctantly assumed command of what was renamed the Grand Fleet. Most authorities and the public expected a major sea battle with the German High Seas Fleet soon after war broke out, but that opportunity never materialized. Jellicoe, however, needed to maintain his fleet in the highest state of readiness in the event the German fleet ventured into the North Sea.

The months of incessant training, drilling, and gunnery practice began to take their toll on Jellicoe's health, and he was increasingly pressured to take the offensive. The German fleet made a number of cautious sorties in an attempt to trap and destroy part of the Grand Fleet. On 30 May 1916 the Admiralty intercepted and decoded messages that warned that the German fleet would probably put to sea the following morning. Jellicoe led his fleet from the harbor that night.

At Jutland, the Grand Fleet's performance was marred from the beginning, primarily by signal errors. When Jellicoe sighted the German fleet, he instantly decided to deploy on his port column and subjected the enemy to withering fire as he crossed the German T on two occasions. As the Germans maneuvered away, Jellicoe, concerned about submarines and approaching nightfall, hesitated to follow. He positioned himself between the enemy and their ports and hoped to renew the action in the morning. The German fleet, however, escaped.

The Battle of Jutland is generally considered to have been inconclusive. The British fleet lost fourteen ships and 6,000 men, compared with the Germans' eleven ships and 2,500 men. Jellicoe's deployment is generally conceded to have been judicious and well executed, but his cautious decision to disengage—as if he had prized the Grand Fleet too highly to subject it to serious risk to win a decisive naval battle—prevented a sound British victory. The results of this battle, however, were due more to the rigidity of the tradition-bound Royal Navy, manifested in a centralization of authority and resultant lack of initiative shown by subordinates. Other factors included mediocre gunnery and defective projectiles. The onus of responsibility fell upon Jellicoe, although the controversy remains to this day.

Aftermath

In November 1916 Jellicoe became first sea lord, and his greatest concern became the German submarine menace and establishment of the convoy system. In the face of political pressure, he became increasingly fatigued and pessimistic. Due primarily to his reluctance to implement the convoy system, Jellicoe was relieved of his last wartime appointment on Christmas Eve 1917, after which he received a viscountcy.

Jellicoe served a successful tour as governor-general of New Zealand from 1920 to 1924 and then retired from the Royal Navy. In 1925 he received an earldom and, upon the death of Lord Douglas Haig in 1928, received the honor of being unanimously elected to fill the presidency of the British Legion. He held office until 1932, and died on 20 November 1935 from a cold caught while attending an Armistice Day ceremony.

Jutland was the apex of Jellicoe's career; the rest a "long, slow anti-climax" (Patterson 1969, p. 258). Jellicoe's renown will be based upon his command of the Grand Fleet against the German High Seas Fleet at Jutland, an action that marked the end of an epoch in naval warfare.

HAROLD E. RAUGH, JR.

SEE ALSO: Battleship; Fisher, John Arbuthnot, 1st Baron of Kilverstone; Fleet; Scheer, Reinhard; Schlieffen, Alfred, Count von; World War I.

Bibliography

Bennett, G. 1964. *The battle of Jutland*. London: Batsford.
Frost, H. H. 1936. *The battle of Jutland*. London: Stevens and Brown.
Gibson, L., and J. E. T. Harper. 1934. *The riddle of Jutland*. New York: Coward-McCann.
Jellicoe, J. 1920. *The crisis of the naval war*. London: Cassell.
MacIntyre, D. 1958. *Jutland*. New York: Norton.

Marder, A. J. 1961–70. *From the Dreadnought to Scapa Flow*. 5 vols. London: Oxford Univ. Press.

Patterson, A. T. 1969. *Jellicoe*. New York: Macmillan.

JOINT OPERATION

A joint operation employs elements from two or more military services of the same nation to accomplish a designated task. Until recent emphasis required a distinction, the term *joint* was often used interchangeably to connote combined (allied) operations, and the term *combined* was sometimes used by the British to designate the multiservice (joint) aspect of amphibious operations. The term *unified* is frequently used in the same context as *joint* in current literature. While joint operations are not new, the conditions of modern military operations compel greater analysis and formal preparation for such undertakings.

The concept of joint operations addresses the need for unity of direction or control over military forces that are distinguished by the specialization of their particular arms (weapons systems). Unity of direction is important where specialized separation of arms exists within branches, or similar subdivisions, of armed services of nations with sophisticated defense establishments. For example, armies have combat components of artillery, infantry, armor, and so forth; navies have surface, air, and submarine forces; and air forces possess bomber, fighter, and transport forces. Intraservice and interservice unity of direction are addressed differently due to the differences in institutional authority of military services and of joint, or unified, national defense establishments. This article concerns interservice operations.

Most modern nations have three main armed services: army, navy, and air force. In addition, several nations have one or two other services, marines and coast guard, which are so integrated into the naval structure, particularly in wartime, that their operational employment with naval forces is not technically considered joint.

Procedures for establishing unity of direction over national armed services differ among nations. The political makeup of governments provides different national authority to implement uniform policy over their respective armed services. Institutional interests and professional perceptions play significant roles in unifying the organization and procedures of traditional armed services.

Historical Precedent

Earliest warfare exhibited specialization in combat forces. Spearmen, archers, cavalry, and numerous other weapons-specialists made up the land armies of ancient forces. Coordinated employment of these mixed military forces often marked the more successful armies.

An additional challenge to achieving unity of effort in military operations was introduced when land forces required support from naval contingents. Early naval support was confined to transport of and resupply of ground forces (or interdiction of the enemy's supply). Naval power was seldom perceived separately from its role in support of the projection of land power. As ships deployed farther from coastal waters, however, functional differences became more pronounced between naval and land army skills. Eventually, institutional responsibilities for the administration of national armies and navies became distinctly separate, as did the perspectives of the respective military commanders.

WARFARE UNTIL WORLD WAR I

Early military operations were not directed by standing unified establishments. Yet numerous land and sea expeditions possessed unity of purpose in the planning and direction of operations. Cooperation was usually achieved through a war council of admirals and generals to devise coordinated (joint) operational plans. The apparent simplicity in directing these joint operations can be attributed to two factors: (1) the forces were small enough that personal leadership and team spirit could prevail, and (2) conflicting institutional interests between the naval and land forces had not yet emerged as dominant issues. If cooperation were lacking, it had to do more with the personalities of the commanders than with any rivalry between the army or navy as institutions.

Joint land-naval operations gradually expanded from the late eighteenth to the early twentieth century. Most amphibious operations were simply "administrative offloadings" and did not require immediate assault against fortified positions. The improvement of naval gunnery, however, allowed the navy to provide artillery fire support to forces within range in coastal areas. Naval bombardment in support of land operations was conducted at the Battle of the Dunes (1658) and in the Crimean War, the American Civil War, and the Dardanelles (Gallipoli) Campaign in World War I, to cite but a few examples.

Revolutionary advancements in naval capabilities—iron construction, steam-powered engines, improved range and accuracy of guns—supported by the advocacy of seapower as a national strategy further separated doctrinal concepts held by land and naval leaders. A new seapower concept was articulated in Alfred Thayer Mahan's theories, which proposed that many national interests could be achieved by the sea arm alone. The seapower theory was used often to support an alternative to armies and, thereby, justify an enlargement of naval strength at the expense of allocation of resources to land forces.

POST–WORLD WAR I

In the United States, following World War I there was a great deal of rivalry between the services and among the specialized arms and branches within the services themselves. The war had been marked by considerable unity of effort in making the air arm, then part of the land forces, responsive to the ground forces. As components of the armies, the new air arm sought more recognition as an independent service, and soon succeeded in doing so: at the end of the war in England, and shortly thereafter in Italy and Germany. The war set in motion other schisms in the military. Armor advocates in the ground forces may not have sought independence, but they certainly proposed dramatic restructuring of combat formations. In navies, naval aviation pushed for dramatic new concepts in the conduct of surface naval combat. The U.S. Marine Corps pursued an identity independent from both the navy and the army.

The introduction of air forces significantly disturbed interservice and intraservice unity. The degree of functional specialization called for in growing army and naval aviation establishments intensified the industrial-technological changes already taking place in armed forces. Claims were made for airpower that were analogous to those made for seapower. Prophets of airpower, such as Douhet (Italy), Trenchard (England), and Mitchell (United States) argued for the distinctive role of air operations in war and for independent air forces. Two major industrial nations established separate air arms prior to 1940: England, the Royal Air Force (RAF); and Germany, the Luftwaffe. In the United States, the Army Air Forces expanded in organizational status and received considerable backing in civilian, industrial, and political circles. The Japanese, U.S., and British navies developed distinct seaborne aviation components.

Traditional army and naval institutions found themselves challenged. Military establishments, particularly in the Western democratic nations, were compelled to debate and to justify their requests for resources. These debates focused ultimately on national budget decisions. Military services sought to support their positions in doctrine—authoritative statements of fundamental principles, which guide military actions. There were at least as many doctrines as there were advocates of a special military arm or service.

Competition for a share of the military budget became a competition of doctrines. This meant that services' roles and missions were in contention, whether they were singular, collateral, or supporting. The services' doctrinal struggles affected cooperative performance, in that precedence of any single joint venture could become an issue in later military doctrinal debate.

Joint relationships between military services were also influenced by management changes in governments, which had been emerging since the middle of the nineteenth century. In earlier times, individuals had influenced the preparation and the direction of military forces, but the growth of national armies and the industrialization of weapons' manufacture introduced a complexity into the supervision of military establishments that was far beyond the capability of single individuals or small committees. The true impact of this change was not fully perceived until the world wars of the twentieth century. Armies and navies devised extensive staff systems to manage the increasing logistical demands of maintaining and directing forces. On the eve of World War II, the Germans held a special advantage. Building upon the heritage of the proven Prussian General Staff, Germany had an integrated, senior military staff (although largely army-dominated) for unified planning and direction of the nation's armed forces.

Although the German High Command was not immune to the parochialism and bureaucratic resistance to change found in most military institutions, it did have an institutional system for structured analysis of military requirements and, thereby, a potential to grasp the changes in technology and to plan and direct integrated employment of specialized arms. Like all national military institutions, the German Command made some initial mistakes that had to be corrected as the war developed. Nevertheless, in the sphere of land operations, the effectiveness of the German High Command was dramatically impressive at the outset of the Second World War.

Joint Operations in World War II

Technological advances in weapons resulted in combat capabilities that propelled joint operations as a paramount military concern in World War II. It became evident that there was some truth in most of the separate service doctrines, but the most meaningful doctrines were those that harmonized the varied services' capabilities with a theater commander's operational objectives.

German Blitzkrieg and Direction of Joint Operations

One of the most striking, modern manifestations of joint operations—an effective blend of land and air forces—was the German blitzkrieg. The concept called for a swift-moving, armor-led thrust supported by tightly coordinated tactical air bombardment, and often integrated with airborne drops. The concept included an early counterair effort and subsequent direct support of the ground offensive by air forces.

German amphibious landings in Norway and the early airborne operations showed a sound concept for the mutual support of the joint arms. The early, spectacular successes probably encouraged later attempts to execute such operations with inadequate supporting resources—such as the airborne operation against Crete. Airborne operations by both sides were probably the most innovative and demanding joint undertakings in the war.

The German High Command did not show as extensive an understanding of strategic air warfare as it did of tactical air. This shortcoming can be partly explained by the German leadership's early perception that the war would be short, and failure to anticipate the development of counterair and offensive air capabilities by the Allies in response to their early reverses. The German armed forces rapidly adjusted their strategic air offensive and air defenses to counter the Allies' efforts. In both cases, German military exertions were compromised by erratic political dictates and were not solely failures of direction by the joint military staff.

Development of Amphibious Operations

Amphibious operations became one of the most distinctive joint operational accomplishments of the war. This complex operation was crucial to the Allies' victories in both the European and Pacific theaters of war. The size and intensity of these operations were without precedent.

Japan, Great Britain, and the United States had conducted amphibious exercises and developed necessary landing craft prior to the war. The U.S. Marine Corps had prepared the basic doctrine upon which the U.S. amphibious operations were based throughout the war. The fundamental tenet of the doctrine addressed procedures for coordinated, supporting firepower and the command and control arrangements while establishing lodgment ashore.

Early British amphibious operations were limited to commando raids due to lack of resources, but they were assimilated easily into the Anglo-American operations in Africa and the Mediterranean. With considerably more resources, the Americans conducted a Pacific theater strategy based primarily on amphibious operations with army and marine corps forces. The Allies were well prepared for the Normandy landing of 1944, the ultimate amphibious invasion and demonstration of joint (and combined) operations up to that time. Although technically an amphibious operation, the Normandy invasion had a distinct supporting air operations component. Local air superiority, deep air interdiction, and airborne delivery of some army units were essential to its success.

One of the serious controversies among the armed services in amphibious operations was the priority of effort given to naval bombardment and direct air support from either naval forces or the U.S. Army Air Forces (USAAF). On more than one occasion in the Pacific area, U.S. naval units withdrew from the provision of direct offshore support in order to engage enemy naval forces. The navy's priority was first to ensure control of the sea. In a similar vein, USAAF and RAF leaders were reluctant to divert their strategic counterair bombing campaign in order to provide concentrated, direct air support to the Normandy landings.

World War II proved how dependent amphibious operations were upon control of the local airspace. This was especially so in noting the reverses in Japanese amphibious offensive and defensive operations that corresponded to their losses in naval aircraft carriers and island airbases.

On a global basis, the Second World War was an air-land-sea war, a fact quickly recognized by the Allied strategic military planners.

Direct Air Support of Ground Operations

Allied air forces began World War II with many of their air leaders committed to strategic bombing and with little experience in providing closely coordinated, air-to-ground firepower support to the armies. Allied unpreparedness for air-ground operations proved costly in North Africa. Allied joint air-ground operations improved throughout the war. The USAAF were greatly assisted by the independent RAF, which provided leadership in the early European tactical air operations. Based on considerable prewar doctrine development, the U.S. Naval and Marine Corps aviation was prepared to give close support to amphibious forces and performed effectively in the Pacific theaters.

The major concepts for direct air support to army forces emerged from the Allies' operations in which highly fluid combat situations limited opportunities for preplanned air strikes. Aided by radio communications, joint fire control and coordination posts were established and air liaison officers were deployed with many ground units. The air liaison officers assisted ground commanders in planning air support requirements, coordinated requests for air support with army artillery, and provided battlefield direction of tactical aircraft tasked to provide direct fire support. A range of tactical air missions evolved as part of most army theater operations. Air missions were categorized as: close air support, interdiction, sector air defense, reconnaissance, and tactical airlift.

World War II produced a wide range of air-land operations for which commanders and staffs were not fully prepared. The principle of central control of theater air assets generally prevailed even though it was challenged by several ground commanders of subtheater-level units. Air space control over combat zones and identification procedures between land and air forces remained serious problems, which led to many tragic errors of friendly forces shooting and bombing one another. Concepts for airborne operations had not been developed to the extent that they had been for amphibious undertakings. Most obvious was recognition that effective interservice operations depended on forceful direction by senior officers with joint strategic vision. The fast pace of modern warfare reduced reliance on battlefield team spirit to make interservice coordination imperative.

Organization for Planning and Directing Joint Operations

Devising a structure for strategic direction of worldwide, joint operations was a challenge to the democratic governments of Great Britain and the United States—nations which carried the major responsibility for the Allied global efforts. Both nations had a tradition of being suspicious of concentrated military power and fostered strongly independent armed services. The British, however, had had a Chiefs of Staff (COS) committee, consisting of the senior officers of each of the three services, for joint planning since 1924. The combined Anglo-American alliance necessitated balanced senior national military structures. Therefore, the United States created, by executive order, the Joint Chiefs of Staff. This joint body served as the senior advisers to the nation's commander in chief and as U.S. members of the Allied Combined Chiefs of Staff.

Concurrently, for the first time, the United States organized military field operating forces on an interservice (joint) basis. This structure was most evident in the United States–dominated Pacific theater of operations. In the European theater, the U.S. joint command was incorporated into the Allied military combined (and inherently joint) command structure.

The Joint Chiefs of Staff and joint theater command structures did not eliminate interservice rivalry (or as some wish to describe it—creative rivalry). Although interservice differences continued during the war years, the effects were abated by the more unifying factor of a national threat and by the abundant U.S. national defense budget. Many of the tensions between services concerned fear of setting precedents that could affect declared service doctrine. Issues developed over marines commanding army forces and vice versa, naval firepower support of the army, the role of USAAF in strategic bombardment versus ground support, and so forth. A significant number of interservice problems that arose, however, were due to the personalities of key personnel, honest misunderstandings, and simple mistakes caused by the confusion of war.

Near the end of the war, it was well accepted, even by the early advocates of airpower, that modern warfare was predominantly executed through joint campaigns and operations. The atomic bombs delivered on Japan by strategic air forces introduced a new era of debate concerning roles and missions of military services.

Post–World War II

During the Cold War era following the Second World War a high level of military activity continued for most of the major nations of the world. Preparedness for nuclear war eclipsed balanced national defense planning, and awarded high priority to those military services and systems that could deliver such weapons. Airpower, at least for a while, held a favored position in defense budgets, but eventually the other services acquired nuclear weapons and gained a viable part of the nuclear mission. Despite the resulting nuclear standoff between the superpowers, military forces continued to be employed in regional, conventional wars and in special operations involving the joint employment of armed forces.

Two major factors influenced joint operations following World War II: the nature of the different governmental systems of the two superpowers, the Soviet Union and the United States; and the increasing complexity of military operations due to rapidly changing developments in weapons technology.

A description of the Soviet military system is covered elsewhere in this encyclopedia. For the purposes of this article, it is only necessary to recognize that the basic concept of central authority inherent in the Soviet political system contributed to unity of direction over their forces. Unity imposed in this manner, however, does not ensure a balanced perspective over the various options permitted by the separate arms. In spite of the very large and modern independent navy and air force, the Soviet unified command's fundamental perception of warfare appeared heavily land oriented. Soviet navy and air forces were not perceived to be as influential in formulating independent doctrine as are their counterparts in the West.

The Soviet doctrine for dedicated aviation support to ground forces was similar to that of the German World War II Blitzkrieg. The Soviets also dedicated land-based air in support of their naval forces. Separate strategic rocket and air forces existed as well as multipurpose air forces for theater (front) army commanders. Soviet doctrine, which emphasized integrated planning and direction of air and ground forces within defined theaters of military operations, appeared to counterbalance the parceling-out of dedicated forces.

The diversity in Western, democratic nations makes it impossible to describe how each addresses joint operations. This article will emphasize developments in the United States since its military posture is the broadest in scope in terms of types of weapons, forces, and geographical commitments. American military organizational and procedural developments for joint military activity are indicative of those adopted by most of the industrialized Western nations. It must be noted that many other Western nations have led in improving aspects of joint operations. This has been particularly the case for nations such as Great Britain and Israel, which have engaged in frequent, limited armed conflicts.

The American Experience

World War II left a legacy of joint armed services cooperation and military organization. Legislative actions quickly recognized the Joint Chiefs of Staff, and created major geographical and unified theater commands with the 1947 National Security Act. The act also created the U.S. Air Force (USAF) as an independent armed service, a civilian secretary to administer a national defense establishment, a National Security Council, and a Central Intelligence Agency. The main thrust of the 1947 Act was to create the structure for unity of direction over the nation's defense establishment. The act also established a framework, if not a mandate, for interagency coordination among civilian and military elements of the executive branch. The 1947 National Security Act, and a series of subsequent amendments, significantly influenced the policy and organization for the nation's military joint actions.

In spite of the unification of defense organization, the American political and military leaderships were reluctant to accept or to promote joint activities below the most senior levels of defense. American supremacy in nuclear weapons encouraged for a period the belief that World War II experience would be irrelevant in the future. The air force and navy clashed over their roles in the strategic delivery of nuclear weapons. Both had the means for strategic deployment (using aircraft carriers or overseas air bases) and the issue was never really resolved even with the introduction of intercontinental bombers and missiles. It remains a shared navy and air force operation linked by a system for joint target planning.

America's loss of the nuclear monopoly and the occurrence of the Korean War (1950–55) were reminders that conventional war was still possible. American forces in Korea were led by military leaders and many unit commanders with World War II experience. This permitted many procedures for joint operations to be taken up from where they were in 1945 and to be developed further. Interservice operational techniques—particularly air-to-ground offensive support—were considerably refined. A structure for Tactical Air Control Systems (TACS) was formed, which remains the basis for most theater air-land operations.

As old problems were solved, new issues arose. Korea saw the intensified struggle between the doctrines of the USAF and Naval/Marine Aviation as to command and control of theater air assets. This remained an unresolved issue throughout America's next conventional war in Vietnam (1960–73).

Vietnam witnessed a high degree of successful joint operations in the theater. Unfortunately, bureaucratic budget and doctrine arguments in the Pentagon often overshadowed the cooperation on the battlefields in southeast Asia. Highly coordinated, supporting air firepower (strategic and tactical) and airlift (fixed-wing and helicopter) became integrated parts of most major in-theater operations.

Besides the major wars of Korea and Vietnam, the United States employed forces in several joint contingency operations. The general analysis of most of these engagements has been critical of the readiness of the army, air, and naval forces in conducting joint operations. The criticism is not directed at the commanders and the forces in the field but, rather, at programming, planning, and preparation of the forces. These observations, along with congressional frustration with interservice confrontations reflected in military advice from the senior programming and budgeting headquarters, have led to military reorganization initiatives by the U.S. Congress.

ORGANIZATION FOR NATIONAL DEFENSE

An important backdrop to U.S. joint operations is the organization and structure for the overall direction of national defense. The U.S. national defense establishment is the product of several amendments to the 1947 National Security Act.

A 1949 Defense Reorganization Act strengthened the authority of the Secretary of Defense and created a Chairman of the Joint Chiefs of Staff (JCS). The Reorganization Act of 1958 created two separate chains of command: an operational chain from the Secretary of Defense, through the JCS, to the combatant commands; and an administrative support chain from the Secretary of Defense to the military departments. Recently, the Reorganization Act of 1986 designated the Chairman of the JCS as the principal military adviser (instead of the JCS corporately). The act provided for more direct involvement of the Chairman, Joint Staff, and Unified/Specified Commanders in the budget process. The act also called for the specific preparation of officers to perform joint duties.

JOINT MILITARY ORGANIZATION AND ARRANGEMENTS

Since World War II, Americans have debated organizing the military for effective unified multiservice planning and operations. There was never a strong suggestion to go as far as Canada did in 1964, in a complete unification of all services (which, some evidence suggests, did not remove fundamental interservice struggles). Most Western democratic nations follow the British and U.S. pattern of a single civilian-headed defense department, or ministry, over the separate army, navy, and air forces.

In the United States, the services exist under the Defense Department along with separate joint military establishments. The joint military establishments have two main parts. One is a senior military advisory and strategic planning establishment in the tradition of the World War II Joint Chiefs of Staff. The other military part is the fighting structure—combatant (for the most part joint) commands.

The Joint Chiefs of Staff (JCS) is made up of the chiefs of the armed services and a chairman (Fig. 1). They are served by a joint staff of officers from all services. They are not commanders, but chiefs, and advisers to the civilian authorities.

The Unified and Specified Command System is designed to ensure unity of effort over the planning and direction of military operations. The system separates responsibility for operations from that for logistical and administrative support, both of which are responsibilities of the separate armed services. Responsibility for the conduct of operations rests with commanders of unified and specified commands, who are designated "Commanders in Chief" (CINCs) and are responsible directly to the Secretary of Defense for mission accomplishment. In turn, the secretary is responsible to the President of the United

Joint Operation

Figure 1. The Joint Chiefs of Staff on 6 August 1991. Pictured from left to right are General Carl E. Mundy, USMC; General Gordon R. Sullivan, USA; General Merrill A. McPeak, USAF; Admiral Frank B. Kelso, USN; General Colin L. Powell, USA, Chairman; and Admiral David E. Jeremiah, USN, Vice Chairman. (SOURCE: U.S. Department of Defense)

States. CINCs normally exercise operational command over their tactical forces through their subordinate service component commanders.

Unified Commands are composed of component forces of two or more services to perform broad, continuing missions. The headquarters of unified commands are made up of integrated joint staffs. Unified commands are generally responsible for a specific geographical area.

Specified Commands are composed of forces primarily from a single service to perform a broad, continuing mission. Their relationship to the Secretary of Defense and the JCS is the same as for unified commands.

A Joint Task Force (JTF) is a force composed of elements from one or more of the services to perform a specific mission or to function as part of the broader mission of a unified command. Some unified commands have standing JTFs. As a rule, a JTF is formed to accomplish a mission of limited duration. They are usually used to conduct special operations in contingencies.

Many other joint structures exist as activities reporting to the Department of Defense or as ad hoc arrangements between services. Various activities coordinate logistical support and doctrinal development, which still remain the responsibilities of the separate services. Some examples are: army–air force agreements focused around tactical air support and the development of rotary-wing systems by the army, and USAF Tactical Air Command and U.S. Army Doctrine Development Command (TAC-TRADOC) working groups, and Joint Assessment and Initiatives Office (1983–86).

TACTICAL PROCEDURES FOR JOINT COMBAT OPERATIONS

As mentioned earlier, World War II forced the development of procedures for joint combat such as amphibious, air-to-ground, and air defense operations. Initially, the responsibility for development and improvement of joint doctrine was with the services, working in coordination to derive mutually agreed positions. The progress made by interservice initiatives in joint doctrine was limited, however, and as a result there has been increased involvement of the Joint Staff and influence from the unified commanders in such matters.

There are a few broad concepts that have been proven in several confrontations and remain basic themes in planning and conducting joint operations:

1. *Coordination and control of offensive firepower* in direct support of ground operations is difficult enough when the army is using its organic artillery, but particularly challenging when using air-delivered support. A primary concern in control of close air support of ground forces is to ensure that enemy troops, not friendly troops, are attacked, or that supporting aircraft are not mistakenly fired upon by friendly air defenses. Planning for such strikes is usually based upon fire support coordination procedures established by the ground commander. Various means employed to guide air strikes are use of ground signal panels and smoke, designation of fire-free zones, and employment of ground and airborne forward air controllers to provide radioed directions to the striking aircraft.

2. *Airspace control* over the combat zone is vested in a single authority designated by the joint force commander. The heavy traffic of air vehicles over the modern battlefield has necessitated management of airspace sectors, much as if it were a major civilian air terminal. The purpose of this control is not only to separate friendly air- and ground-launched missiles, but also to facilitate air defense.

3. *Headquarters' liaison officers* have long been recognized as essential for joint planning when the air, sea, and ground command centers are separated. Many war plans are basically prepared in peacetime and coordinated among separated service component headquarters. Effective execution of these plans depends upon sound understanding of the separate service perspectives during the initial planning. The importance of liaison officers at separate component service combat control centers is also essential. Co-location of service component theater, and subtheater, wartime headquarters (joint command and control centers) is another option being explored in various theater areas.

4. *Amphibious operations* have two distinct command phases. The naval amphibious force commander will retain command of all operations until a lodgment ashore is fully established. At that time, the ground force com-

mander assumes command of forces ashore. With the increased operational ranges of land-based air, most amphibious operations are now triservice ventures.

5. *Special theater assets are centrally managed* and not committed piecemeal. This most frequently applies to air support assets, but the concept extends to other high-value/limited-quantity systems on the modern battlefield. The principle ensures their maximum impact in accomplishing missions critical to the theater as a whole.

6. *Appreciation of the integrated aspects of a theater-level campaign,* as a whole, is represented in many recent concepts of operational art and air-land battle. Such concepts are drawn from some long-held military theories, but are necessarily being recast in terms of modern weapons and service roles.

General Observations

The current environment for addressing joint operations is unsettled. Obviously, expanding the authority of unified defense establishments is accomplished at the expense of the separate services. There are questions as to the need for more unified organization merely to obtain an effective unified approach to operations. Will emphasis on unity constrain the important ability of having a constant review of alternative options in conducting combat operations? Will unified direction suppress initiatives of the separate armed services to fully exploit the potential of their particular combat specialties? In addressing such questions on joint military operations some significant observations are worth noting.

CRITICISMS AND PROBLEMS

Admittedly, even historically successful joint operations have had their share of problems arising from interservice rivalries: jealousy in command arrangements, wasteful duplication of effort, complexity in plans so that every service has a part, and arrogance in one service underestimating the capabilities of another. Each military specialty tends to cultivate fervent attitudes supporting its particular functional area. Such attitudes foster a two-sided dilemma. On the positive side, there is an aggressive effort to exploit the specialty to its fullest potential. The negative side is that it encourages dangerous pursuit of preferred courses of action by a commander in ignorance of other services' perceptions and procedures.

REDUNDANCY: A MILITARY HEDGE

In reality, political strategy undergoes frequent and sudden changes (as a result of such events as turnover of government leaders, international crisis, etc.) and as a result military planners must seek to develop broad capabilities to respond to a variety of contingencies.

In peacetime, efficiency experts decry (but military history frequently vindicates) duplicate efforts. Alternative capabilities often are the "secret weapon" permitting a commander to do what the enemy did not anticipate. The United States has benefited from having available capabilities for one service based upon independent initiatives pursued by another. Examples would be marine corps development of amphibious concepts used by the army in World War II, and tactical air support aircraft (F-4 and A-7) developed by the navy and used by the USAF in Vietnam. Nevertheless, given the value of multiple developments, it is hard to justify lack of interoperable tactical communications equipment, as was reported in the 1983 Grenada operation.

Whatever its value, multiservice approaches need to be guided by a single focus. To some degree, command organization has structured some unity in formal operational planning and direction of joint combat forces. As called for in the recent U.S. Defense Act, however, there remains the challenge to prepare officers with a joint perspective, despite years of specialized experience in their respective services.

ALBERT D. MCJOYNT

SEE ALSO: Amphibious Warfare; Blitzkrieg; Branches, Military; Close Air Support; Combined Arms; Combined Operations; Command, Control, Communications, and Intelligence; Coordination; Doctrine; Identification Friend or Foe; Task Force.

Bibliography

Creswell, J. 1976. *Generals and admirals: The story of amphibious command.* Westport, Conn.: Greenwood Press.
Davis, R. G. 1987. *The 31 initiatives.* Washington, D.C.: Office of Air Force History.
Dupuy, T. N. 1977. *A genius for war, the German army and general staff, 1807–1945.* London: MacDonald and Jane's.
Fergusson, B. 1961. *The watery maze, the story of combined operations.* New York: Holt, Rinehart and Winston.
Fuller, J. F. C. 1970. *The decisive battles of the western world.* 2 vols. London: Paladin.
Galland, A. 1954. *The first and the last, the rise and fall of the German fighter forces, 1938–1945.* Trans. M. Savill. New York: Ballantine Books.
Hansell, H. S., Jr. 1980. *The air plan that defeated Hitler.* New York: Arno Press.
Hittle, J. D. 1949. *The military staff, its development and history.* Harrisburg, Pa.: Military Service.
Millet, A. R. et al. 1986. *The reorganization of the joint chiefs of staff: A critical analysis.* Washington, D.C.: Pergamon-Brassey's.
Momyer, W. W. 1978. *Air power in three wars (WWII, Korea, Vietnam).* Washington, D.C.: U.S. Air Force.
Pogue, F. C. 1973. *Organizer of victory, 1943–1945.* Vol. 3 of *George C. Marshall.* 3 vols. New York: Viking.
Richmond, H. 1941. *Amphibious warfare in British history.* Exeter: A. Wheaton.
Toth, J. E. 1986. *Higher direction of military action.* Washington, D.C.: National Defense Univ. Press.
Turner, G. B., ed. 1953. *A history of military affairs in western society since the eighteenth century.* New York: Harcourt, Brace.

U.S. Department of the Army. 1984. *Field manual 101–5: Staff organization and operations.* Washington, D.C.: Government Printing Office.
———. 1986. *Field manual 100–5: Operations.* Washington, D.C.: Government Printing Office.
U.S. Department of Defense, Joint Chiefs of Staff. 1986. *Unified action armed forces (UNAAF).* JCS Pub. 2. Washington, D.C.: Government Printing Office.

JOMINI, ANTOINE HENRI [1779–1869]

Antoine Henri Jomini, a Swiss of French extraction, was a military historian and theorist who sought to define the principles of warfare systematically. Influenced by the Enlightenment and his observations of the Napoleonic wars, he endeavored to reduce battles, campaigns, and wars to theoretical systems that should govern their conduct. In the process he earned a place in the front rank of military theorists and coined words and concepts still in use.

Early Life and Military Career

Jomini was born into a bourgeois family at Payerne, Switzerland, where his father was syndic (municipal magistrate). Although he was expected to have a business career, he preferred military life. When the Swiss Revolution began, the 19-year-old Jomini left his post in a Paris bank to become first a member of the Swiss General Staff and then, at age 21, commander of a brigade. Subsequently he returned to Paris where he devoted himself chiefly to writing military history.

His first major work, *Traité des grandes opérations militaires*, was published in four volumes (1804–1805). Marshal Ney was so impressed by it that he had Jomini join his staff as volunteer aide-de-camp for the Austerlitz campaign. Napoleon appointed him a colonel in the French service in December 1805, and he became Ney's principal aide. Appreciative of Jomini's grasp of military strategy, Napoleon attached him to his general staff in 1806 for the impending campaign against Prussia. He was present at the battles of Jena and Eylau. For his service at Eylau, he was awarded the Legion of Honor. After Tilsit, he became Ney's chief of staff and was created a baron. Jomini gave Ney valuable advice in the Spanish campaign but vowed to leave French service following a quarrel with the marshal.

Napoleon refused to accept his resignation and compelled him to remain in the French army as a general of brigade, while allowing him to accept an appointment as general in the service of Alexander I of Russia. For a while Jomini held the rank of general in both the French and Russian armies, with the consent of both sovereigns. Because of this conflict of interest he was spared participation in Napoleon's Russian campaign. Jomini did take a command on the line of communication and was engaged there when the retreat from Moscow occurred.

War with Prussia ensued, and Jomini rejoined Ney for the battles of Lützen and Bautzen, where he served as chief of staff of Ney's group of corps. For Jomini's distinguished service at Bautzen, Ney recommended his promotion to general of division. This promotion was blocked by an old enemy, Alexandre Berthier, Napoleon's chief of staff, who erased Jomini's name from the promotion list and had him arrested and censured for failing to file a fortnightly report of strength on time. The charges were trivial and baseless; but the angry, frustrated Jomini joined the allied forces coalescing against France.

In August 1813 the czar appointed him a lieutenant general in the Russian army and a personal aide-de-camp. Jomini's old French comrades considered him treasonous, although in retrospect Napoleon forgave him because he was Swiss, not French. He withdrew from the Allied Army in 1814 when it violated Swiss neutrality by making a passage of the Rhine at Basel. Although Jomini took part in the Congress of Vienna, his influence had been undermined by his futile attempts after Waterloo to have Ney's life spared.

In 1823, after Jomini had spent several years in retirement, Alexander I appointed him full general in the Russian army and called him to Moscow. For the remainder of the decade, he spent his time principally in educating the czarevitch Nicholas and in organizing the Russian staff college, which opened in 1832. His last active service came in 1828 at the siege of Varna, where he was given the cordon of the Alexander order. After retiring from active service in 1829, Jomini spent the next 30 years in scholarly pursuits, chiefly in Brussels. He was called to St. Petersburg to serve as a military adviser to the czar during the Crimean War. Afterwards he settled at Passy near Paris where, until his death in 1869, he wrote treatises, pamphlets, and open letters on military subjects.

Literary Legacy

Jomini's reputation as a military commentator is based on a substantial literary legacy. Consisting of more than 30 volumes, his writings are both historical and theoretical in nature. The histories, issued originally in 27 volumes, discuss the wars of Frederick the Great and the wars of the French Revolution and Napoleon. The Seven Years' War and the revolutionary wars of the 1790s are treated in detail. Four additional volumes published in 1827 cover the political and military careers of Napoleon Bonaparte. But Jomini is best known for a single work entitled *Précis de l'art de la guerre*, published in Paris in 1838. Therein he developed his theories of war and contributed to the vocabulary of military science.

Military Thought

The *Précis* brought to fruition ideas Jomini had originally proposed three decades earlier in his study of the Seven Years' War (*Traité des grande opérations militaires*). In chapters 7 and 14 he speculated on the importance of lines of operation; in chapter 35 he advanced his belief that war was governed by immutable principles. Herein lies the corpus of Jomini's theory. Convinced that war was a rational activity, Jomini wrote the *Précis* in the belief that the central problem in military science was to comprehend and apply the natural laws of strategy. In the *Précis* he gave considerable attention to the definition of strategy, tactics, and logistics and explained their relationship. Noting that battles and campaigns must be meticulously planned, Jomini believed that victory would result from application of these eternal principles of strategy:

1. Massing the forces of an army successively on the decisive points of a theater of war, if possible upon the enemy's communications without endangering one's own;
2. Engaging this mass against fractions of the enemy's army by maneuver;
3. Directing by tactical maneuver one's massed forces to the decisive point of the battlefield, or to that portion of the enemy line which must be taken;
4. Contriving that these masses be brought into action with energy so as to produce a simultaneous effort.

To explain how these principles were to be applied, Jomini devised twelve model battle plans based on geographic formations. In each there was a theater of operations, a zone of operations, and a line of operations. The successful commander should choose a line of operations that would result in his domination of three sides of a rectangular zone of operations. The enemy would be forced either to surrender or to retreat. Because of his reliance on these geometric model battle plans, Jomini sometimes seems to be aloof from the reality of battle.

In summary, Jominian strategy consisted primarily of offensive action by massed forces that pursued vigorously after breaking the enemy's line.

Jomini's Influence

For two generations after Jomini published the *Précis*, it was considered the finest book on war. The *Précis* was studied wherever military leaders were trained or war plans developed. European staff officers held exercises based upon Jominian models and principles. Works by Dennis Hart Mahan, long an instructor at West Point, and Henry W. Halleck, his student, were derived principally from the *Précis*; thus a generation of professional officers entered the American Civil War schooled in Jominian strategy. Jomini's emphasis on the offensive may have affected such battles as Malvern Hill, Fredericksburg, Gettysburg, and Cold Harbor, where attacking armies were shattered by well-placed defenders armed with the rifle musket, a weapon unknown to Jomini. Alfred Thayer Mahan, son of Dennis Hart Mahan, acknowledged Jomini as the single most important influence in his thinking about naval strategy.

There were inherent weaknesses in Jomini's strategic thought (although some authorities have suggested that most of the criticism is debatable). Principal among these criticisms are the following:

- Jomini claimed war to be controlled violence fought according to unchanging principles; neither the validity nor the timelessness of these principles has been proven;
- He failed to understand the importance of strategic mobility, of the element of surprise, or of the role of chance in warfare;
- He urged concentration on one decisive point, with no important alternative objectives;
- He assumed that the enemy would act predictably;
- He tended to undervalue moral and psychological factors in war; and
- He had an inadequate appreciation of how science and technology change warfare.

It is possible, however, to refute each of these criticisms of Jomini by using selections from his own writings. He was neither as philosophical nor as profound as Clausewitz. Nevertheless, his conclusions and theories had much in common with those of Clausewitz.

MAX RAY WILLIAMS

SEE ALSO: Civil War, American; Clausewitz, Karl von; French Revolutionary–Napoleonic Wars; Napoleon I; Principles of War; Science of War; Seven Years' War; Strategy.

Bibliography

Alger, J. I. 1975. *Antoine-Henri Jomini: A bibliographical survey.* West Point, N.Y.: U.S. Military Academy.
Brinton, C., G. A. Craig, and F. Gilbert. 1943. Jomini. In *Makers of modern strategy*, ed. E. M. Earle, pp. 77–92. Princeton, N.J.: Princeton Univ. Press.
Connelly, T. L., and A. Jones. 1973. *The politics of command: Factions and ideas in Confederate strategy.* Baton Rouge, La.: Louisiana State Univ. Press.
Hittle, J. D. 1975. *The military staff: Its history and development.* Westport, Conn.: Greenwood Press.
Jomini, Baron de. 1862. *The art of war.* Trans. G. H. Mendell and W. P. Craighill. Philadelphia, Pa.: Lippincott.
Shy, J. 1986. Jomini. In *Makers of modern strategy*, ed. P. Paret, pp. 143–85. Princeton, N.J.: Princeton Univ. Press.
Weigley, R. F. 1973. *The American way of war, A history of United States military strategy and policy.* New York: Macmillan.
Williams, T. H. 1981. *The history of American wars, from 1745 to 1918.* New York: Knopf.

JORDAN, HASHEMITE KINGDOM OF

The Hashemite Kingdom of Jordan is traditionally one of the most moderate nations in the Middle East. This is due in large part to the political acumen of its monarchs, King Abdullah and his grandson Hussein (reigning since 1953).

Jordan, Hashemite Kingdom of

Power Potential Statistics

Area: 91,880 square kilometers (35,475 sq. mi.)
Population: 4,275,200
Total Active Armed Forces: 101,300 (2.370% of pop.)
Gross National Product: US$4.6 billion (1990 est.)
Annual Defense Expenditure: US$377 million (12.4% of GNP, 1990 est.)
Iron and Steel Production:
  Pig iron: 0.165 million metric tons (1986)
Fuel Production: Processed 2.511 million metric tons of petroleum products (1986)
Electrical Power Output: 3,500 million kwh (1989)
Merchant Marine: 2 vessels; 22,870 gross registered tons
Civil Air Fleet: 19 major transport aircraft; 16 usable airfields (14 with permanent-surface runways); 1 with runways over 3,659 meters (12,000 ft.); 13 with runways 2,440–3,659 meters (8,000–12,000 ft.); none with runways 1,220–2,440 meters (4,000–8,000 ft.).

For the most recent information, the reader may refer to the following annual publications:
The Military Balance. International Institute for Strategic Studies. London: Brassey's (UK).
The Statesman's Year-Book. New York: St. Martin's Press.
The World Factbook. Central Intelligence Agency. Washington, D.C.: Brassey's (US).

History

Before World War I, the histories of Jordan and Israel (or Palestine, as it was known before 1947) were closely intertwined. The area lying east of the Jordan River, some distance from the Mediterranean Sea, has always maintained a more Arab character than the coastal regions, but events that affected one area always affected the other, too.

After Jordan was liberated from Turkish rule during 1917 and 1918 through the efforts of rebel Arab tribesmen, some of them led by T. E. Lawrence ("Lawrence of Arabia"), Jordan became part of the British-administered mandate of Palestine. Jordan (or more properly in those days, Transjordan) was governed separately, since its territory was not part of the Balfour Declaration's Palestinian National Home for the Jews. In 1921, Abdullah ibn Husain, the younger brother of Faisal (who had led the Arab revolt during World War I), became emir of Transjordan, his rule assisted by British advisers. Over the next two decades, Transjordan slowly became independent, also gaining a British-officered army, the Arab Legion.

During World War II, the Arab Legion took part in operations in Iraq and Syria in the spring and summer of 1941, but it was otherwise engaged in its normal peacetime duties of maintaining internal order. In May 1946, the Emirate of Transjordan became the Hashemite Kingdom of Jordan through a treaty with Great Britain. A year later, Jordan's Arab Legion played a major role in the Israeli War of Independence (May 1948–January 1949), preventing Israeli capture of the Arab parts of Jerusalem and holding on to the area later known as the West Bank.

After the war, Jordan assumed government of the West Bank and thereby gained a large population of Palestinian refugees.

King Abdullah was assassinated in Amman on 20 July 1951 and was at first succeeded by his son Talal. Talal, however, suffered from intermittent incapacitating mental illness, and so he in turn abdicated in favor of his son Hussein. Hussein was crowned on 2 May 1953, at the age of 18, and has ruled ever since. Jordan managed to avoid direct involvement in the Arab-Israeli War of 1956, but in the preceding crisis Hussein relieved Gen. Sir John Bagot Glubb (Glubb Pasha) of command of the Arab Legion, a post Glubb had held since 1936, and Hussein's pro-Egypt stance cost him British aid and support.

Jordan did not stay out of the Third Arab-Israeli War (the Six-Day War) of 5–10 June 1967. Outnumbered and in a poor strategic position, the Jordanian forces defending the West Bank were largely destroyed and their remnants driven across the River Jordan in a lightning campaign between 5 and 7 June. This was a great blow to Jordan, not only in terms of materiel but in terms of area and population lost. Despite the Israeli occupation of the West Bank, Jordan continued to pay the salaries of some 6,000 teachers and civil servants there.

The PLO presence in Jordan, and PLO displeasure with King Hussein's relatively moderate policies, led to increasing tension in early autumn 1970. There were PLO clashes with the Jordanian army in February and more serious fighting in early June. In September, King Hussein, fearing for his throne, ordered his army to attack the Syrian-backed PLO, leading to nine days of open warfare (17–26 September). Despite Syrian intervention, the Jordanians were victorious, and the PLO guerrillas accepted Jordan's conditions for remaining in the country. Despite the resulting poor relations with Syria, Jordan sided with the Arabs during the fourth, or October, Arab-Israeli War in 1973. Hussein sent two armored brigades to fight in the Golan under Syrian command and mobilized most of his army along the border to draw off Israeli strength.

Since that war, King Hussein has maintained Jordan as a moderate Arab state. He sent aid to Iraq during the Gulf War of 1980–88, which led to a crisis with Syria in 1980. Hussein surprised the world with his bombshell announcement in September 1988 that he would no longer act on behalf of the Arab residents of the West Bank. Coming as this pronouncement did in the middle of the *Intifada*, the Palestinian revolt against Israeli occupation, it placed greater responsibility on the PLO and the Palestinians themselves, yet it also undercut the public policy of Israel's Labor Party. Aware that peace stood a better chance with Labor in control rather than their more conservative Likud rivals, Hussein took the unusual step in the last week of October (just before the Israeli elections on 2 November) of announcing that he felt Labor was more amenable to a peaceful solution to the Palestinian problem. By acts such as these, Hussein reinforces his well-deserved reputation as a wily political survivor.

Politico-Military Background

Traditionally, the army, either as the Arab Legion or the Jordan Arab Army, has been one of the pillars of the monarchy. Most recruits were formerly drawn from the Bedouin of southern Jordan, among the most loyal to the monarchy. Since the late 1970s, greater numbers of northern Bedouin and some Palestinians (still less than 40 percent, in contrast to their 60 percent of the population) entered the armed forces, but most of the combat troops are still southern Bedouin. Partly because of this close relationship between army and monarchy, but also because of the volatile and often violent politics of the Middle East, Jordan spends a great deal on defense. Since 1973, the defense budget has consumed between 26 percent and 40 percent of all government spending, and between 12 and 16 percent of GNP.

Originally, the armed forces were staffed entirely by volunteers, a system that, coupled with careful reviews of applicants, ensured a loyal army. In 1976, however, the government introduced two-year compulsory military service for all males reaching age 18, although such service could be postponed for educational reasons for up to ten years. Women are not conscripted, although some serve as volunteers, usually in medical and support services.

Strategic Problems

Jordan's strategic problems are related to its relatively small population. People of Palestinian descent now make up nearly three-fifths of the population, and while many of them are well-integrated and productive members of Jordanian society, others owe their first allegiance to Palestinian nationalism and not to Jordan.

Jordan does not have a large population, even accounting for the Palestinians, and its armed forces are, consequently, rather small in comparison to those of Syria, Israel, or Egypt. Jordan is also a small country, most of it desert; it cannot trade space for time in the event of invasion, especially from Israel.

Military Assistance

Up until 1956, Great Britain provided most of Jordan's military aid, including training and equipment. The British influence was particularly strong as long as British officers commanded the Arab Legion. Since that time, British influence has gradually declined, replaced by U.S. sources of funding and materiel; the United States supplied Jordan with $11.75 million worth of military assistance in 1988. Jordan also receives some financial support from Saudi Arabia and the Gulf States; in return, Jordan supplies those countries with military advisers. Many members of the officer corps receive training in the United States or Britain.

Defense Industry

Jordan produces no weapons systems or major military equipment of its own. However, some of the items purchased abroad are modified in-country for domestic use, and items like uniforms, tools, and so forth are produced domestically.

Alliances

Jordan is a member of the United Nations and of the Arab League. It also has close ties with Saudi Arabia and the other Gulf States and has grown close to Iraq since the mid-1980s. Jordan was one of the few countries to support Iraq during the 1991 Gulf War that followed Iraq's invasion of Kuwait in 1990.

Defense Structure

King Hussein is the commander in chief of the Jordanian armed forces. He exercises administrative control through a minister of defense (who would be responsible to parliament if one were in session), but maintains a direct command relationship with both the Jordan Arab Army and the Royal Jordanian Air Force.

(For an explanation of the abbreviations and symbols used in the following section of military statistics, see the list of Abbreviations and Acronyms in each volume.)

Total Armed Forces

Active: 101,300. Terms of service: conscription, 2 years authorized.
Reserves: 35,000 (all services): Army 30,000 (obligation to age 40).

ARMY: 90,000.
2 armd div (each 2 tk, 1 mech inf, 1 arty, 1 AD bde).
2 mech inf div (each 2 mech inf, 1 tk, 1 arty, 1 AD bde).
1 indep Royal Guards bde.
1 SF bde (3 AB bn).
1 fd arty bde (4 bn).
Equipment:
MBT: some 1,131: 260 M-47/-48A5 (in store), 218 M-60A1/A3, 360 Khalid/Chieftain, 293 Tariq (Centurion).
Light tanks: 19 Scorpion.
Recce: 130 Ferret.
AIFV: some 25 BMP-2.
APC: 1,160 M-113, 34 Saracen, some 50 EE-11 Urutu.
Towed arty: 89: 105mm: 36 M-101A1; 155mm: 38 M-114, 11 M-59; 203mm: 4 M-115 (in store).
SP arty: 105mm: 30 M-52; 155mm: 23 M-44, 220 M-109A1/A2; 203mm: 100 M-110.
Mortars: 81mm, 107mm, 120mm: 600.
ATGW: 330 TOW (incl 50 SP), 310 Dragon.
RL: 94mm: LAW-80; 112mm: APILAS.
RCL: 106mm: 330 M-40A1.
AD guns: 408: 20mm: 100 M-163 Vulcan SP; 23mm: 44 ZSU-23-4 SP; 40mm: 264 M-42 SP.

SAM: SA-7B2, 23 SA-8, 20 SA-13, SA-14, Redeye.

NAVY (Coast Guard): 300. Base: Aqaba.
Patrol: 1 Al Hussein (Vosper 30-m) PFI, plus boats.

AIR FORCE: 11,000; 113 cbt ac, 24 armed hel.
FGA: 4 sqn with 62 F-5 (55 -E, 7 -F).
Fighter: 2 sqn with 32 Mirage F-1 (14 -CJ, 16 -EJ, 2 -BJ).
Transport: 1 sqn with 6 C-130 (2 -B, 4 -H), 3 C-212A.
VIP: 1 sqn with 2 Boeing 727, 2 Gulfstream III (VIP) ac; 4 S-76 hel.
Helicopters: 5 sqn:
  Attack: 2 sqn with 24 AH-1S (with TOW ASM).
  Transport: 1 sqn with 10 S-76, 3 S-70; 1 sqn with 12 AS-332M; 1 sqn with 8 Hughes 500D.
Training: 16 Bulldog, 15 C-101, 12 PA-28-161, 6 PA-34-200, *19 F-5 (15 -A, 4 -B).
AD: 2 bde: 14 bty with 126 Improved HAWK.
Missiles:
  ASM: AGM-65 Maverick, TOW.
  AAM: AIM-9 Sidewinder, R-530, R-550 Magic.

FORCES ABROAD
UN and Peacekeeping:
Angola (UNAVEM II): observers.

PARAMILITARY
Public Security Force: The principal police force of Jordan, numbering about 6,000 men. Its function is mainly civil in character, but it has traditionally been headed by an army major general.
People's Army (Civil Defense Force: Founded in the mid-1970s, it is open to males age 16 to 65 and females age 16 to 45 and currently numbers some 225,000 personnel. In the event of war, the People's Army would function as a national guard, serving to maintain internal order and assisting victims of enemy attack.
Palestine Liberation Army (PLA): 1,200; 1 bde (kept under strict control by Jordanian Army).

Future

For over 40 years, Jordan has enjoyed stability and security, due in no small measure to the resourcefulness and courage of its monarchs and the effectiveness of its armed forces. Jordanian security is closely tied to the fate of the Palestinians: If there is a permanent peace settlement between the Arabs and Israel, Jordan will be much more secure; if, on the other hand, Israel or the Palestinians refuse to compromise, Hussein and his eventual successor will be unable to give their nation the real peace they and its citizens desire.

DAVID L. BONGARD
TREVOR N. DUPUY

SEE ALSO: Arab-Israeli Wars; Arab League; Israel; Middle East; Syria.

Bibliography

American University. 1980. *Jordan: A country study*. Washington, D.C.: Government Printing Office.
Glubb, J. B. 1967. *A soldier with the Arabs*. New York: Harper.
Hunter, B., ed. 1991. *The statesman's year-book, 1991–92*. New York: St. Martin's Press.
International Institute for Strategic Studies. 1991. *The military balance, 1991–1992*. London: Brassey's.
Jureidini, P. A., and R. D. McLaurin. 1983. *Jordan: The impact of social change on the role of the tribes*. Boulder, Colo.: Westview Press.
Miller, A. D. 1986. *The Arab states and the Palestinian question: Between ideology and self-interest*. New York: Praeger.

JUNGLE WARFARE

More than three-quarters of the earth's land surface that lies between the latitudes of the Tropic of Cancer (lat. 22°30′N) and the Tropic of Capricorn (lat. 22°30′S) is jungle. These areas receive an average of more than 180 centimeters (70 in.) of rainfall annually. They include the vast basins of the world's great tropical rivers—the Amazon in South America, the Congo in Africa, and the Chindwin-Irrawaddy, Salween, and Mekong rivers in Southeast Asia. Many military operations of the past 50 years have taken place in the Southeast Asian jungle as well as in the Micronesian and Melanesian Island areas of the southwest Pacific. These include operations by the Japanese and by the Allies in World War II and, more recently, United States and allied operations in Vietnam and Cambodia. Jungles are generally important only because their location can provide or prevent access to regions or resources of strategic importance. In addition, however, political circumstances may necessitate military operations in jungle areas (e.g., Vietnam).

The first Japanese operations of World War II were carried out primarily to drive the Americans, British, Australians, Dutch, and French out of the western and southwestern Pacific, and to secure control of the oil resources of Indonesia, which were critical to Japan. Subsequent Allied operations were conducted to regain a strategic presence in the area, deny Japan access to the oil, and stop Japanese expansionism. Operations carried out in places such as Tarawa, Saipan, Buna and others either reduced Japanese control over the area, bypassed and made noneffective a Japanese force, or provided forward basing—especially airfields—for further pursuit of the strategy to regain control and contain the Japanese. Operationally and tactically, this required battles and campaigns in the jungles of the region.

Jungle Environments

TERRAIN

There are two types of jungles: tropical (equatorial) and subtropical. The former generally receive 150 to 350 centimeters (60–140 in.) of annual rainfall and have no marked seasonal variations in temperature, which averages 20°C–30°C (68°F–86°F) day to night year-round. The Amazon and Congo basins, Sumatra, and some South Pacific islands are characteristic.

Subtropical jungles are generally found along the windward coasts of land areas not more than 10° north or south of the equator. There is usually a monsoon climate (i.e., a seasonal change in the prevailing wind direction that brings alternately wet and dry seasons) and vegetation patterns that are different from those in tropical forests. Typical subtropical jungles are found in Vietnam, Burma, the Philippines, Central America, and the Caribbean islands.

Jungle terrain is dominated by the presence of large amounts of water, which in turn generates vegetation. Jungles, tropical and subtropical, include dense forests, grasslands, cultivated areas, and swamps. In primary tropical forests, large evergreen trees form interlocking canopies from a few meters to nearly 100 meters in height. In some jungles, it is possible to operate helicopters under and between these layers. Because tree canopies shut out sunlight, the forest floor is less heavily overgrown than might be expected. However, ground observation is generally limited to no more than 50 meters (55 yd.).

In semitropical jungles a mixture of evergreen and deciduous trees allows more sunlight to reach the forest floor. In turn, this permits development of denser undergrowth than in many tropical forests.

Common to most jungle areas, but especially to subtropical jungles, are swamps (mangrove and palm), savanna grasslands, fields of bamboo 20 meters (65 ft.) high, and rice paddies.

Many Pacific islands are merely the above-surface part of undersea mountains, and thus rapid increases in elevation over short distances are not uncommon. Elevations in island areas vary from sea level atolls like Kwajalein to several-thousand-meter mountains in the Solomon Islands and New Guinea. Roads and tracks, areas flat enough for airfields, and areas to accommodate forward supply bases are generally found along the coasts. Occasionally, for operational or tactical reasons, it may be necessary to occupy the hills in order to control the important lowland areas.

In Southeast Asia, the subtropical areas of Vietnam, Thailand, Burma, and Malaysia feature large alluvial plains (the deltas of the large river systems) and intensive rice culture along the coasts and in the river deltas. Soil conditions vary from delta silt to volcanic mountain structures making movement difficult, especially for vehicles.

CLIMATE AND WEATHER

A benign temperature range and frequent rainfall are the primary characteristics of jungle areas. Southeast Asia and some adjacent island areas have a monsoon climate—a seasonal change of the prevailing wind direction. In the summer, onshore winds pick up moisture from the sea, which is the source of the heavy monsoon rainfalls. Reversing direction in winter, the prevailing winds bring drier air from the interior of the large land masses and, although temperatures remain high—except in mountainous regions—there is some relief from the constant rainfall.

HABITATION

Jungles are heavily populated with many species and varieties of birds, mammals, and reptiles. In larger land areas—Southeast Asia in particular—there are elephants, tigers, large and small simians, lizards, and many kinds of snakes. Human habitation is generally found along the coastlines and waterways, although there is a fairly large population of nomadic tribesmen, the Montagnards, in the highlands of the Indochina Peninsula. Deep jungles have become attractive hidden refuges for dissident revolutionary groups attempting to overthrow established authorities. The Hukbalahaps in the Philippines, Malaysian insurgents in Malaysia, the Vietminh (and later the Vietcong) in Vietnam, and insurgent groups in Borneo have all provided the established governments with a reason to take political-military action against them.

Fighting in Jungles

After the defeat of the Japanese Combined Fleet at Midway in early June 1942, it was believed that the Japanese could only continue offensive action in one area—the southwest Pacific. It therefore became imperative for the Allies to seize the initiative in that area, primarily to protect the lines of communication to Australia, and to stop the spread of Japanese control over oil-rich Indonesia. Given those considerations, it would ultimately be necessary to neutralize the Japanese air and naval base at Rabaul. Thus, within 2 months of the battle at Midway, Allied and Japanese forces were fighting in the mountains and jungles of New Guinea, and not long after, in the rain forests of Guadalcanal. Jungle fighting taxed the resources of all combatants—their physical stamina, courage, and patience. Like the desert, the jungle is neutral—it treats friend and foe alike. Its dense vegetation provides cover and concealment for both sides, its insects and reptiles bother each side equally, and its debilitating climate saps the strength and resolve of all combatants. It is a place, like the desert, where excellent physical conditioning, high-quality small-unit leadership, and tough, demanding, relevant small-unit training are critical—even for survival, let alone for success.

In the first six months of the jungle war, the Allies took the initiative. However, after Guadalcanal and Papua, it took another fourteen months before Rabaul could be isolated by the seizure of the Admiralty Islands. During those months some of the toughest, most demanding tactical battles of the war were fought. It was exclusively an infantry war, foot soldiers on the ground hacking and struggling through jungle and over mountains. The few tanks were only for support of infantry.

Similar combat occurred in Southeast Asia at the start of World War II when the Japanese infantry swept across the Indochina Peninsula, through Malaya, Thailand, and much of Burma, after capturing Singapore from the rear. To regain this ground, and to support China over the famous Burma Road, the Allies found it necessary to re-enter the jungles with infantry and win back meter by meter what had been lost. Battle was difficult in the rain forests, and this left a legacy in the minds of its participants. U.S. perceptions of the World War II Pacific jungle war were at the root of many of the causes for a less than satisfactory outcome for U.S. forces at the tactical and operational levels in Vietnam. In World War II, no armored divisions had tracked across Pacific islands or up forest trails toward Japan, and no leaders of armored units achieved professional fame in jungle battles. Therefore, institutional wisdom saw the jungle as a place where tanks and mechanized units simply could not go. Thus, when deploying forces to Vietnam, after replacing the French in 1954, the United States relied heavily on infantry divisions, initially leaving behind the armored cavalry and tank units that had become organic to infantry divisions following World War II. Provided instead was a new capability in force employment—the helicopter. After experiments with heliborne infantry, many in the U.S. Army believed that airmobile infantry was the answer to the need for mobility so dramatically felt in World War II jungle operations.

To some extent that was true. However, it was quickly apparent that moving troops by helicopter had some inherent disadvantages. Primary among these was a lack of sure knowledge of the terrain and the enemy by the troops and leaders who had been moved quickly by helicopter to respond to a situation. In addition, air mobility of foot infantry left the infantry no more tactical mobility than before, once on the ground. In fact, the enemy often proved more tactically mobile than the heliborne infantry because the enemy lived in the swamps and jungle and learned to use them to his advantage.

The Indochina Peninsula, Vietnam in particular, is not a land totally unsuited to mobile mechanized warfare. More than 45 percent of the terrain can be traveled by mechanized units year-round. Mechanized units—Vietnamese, U.S., and allied—operated successfully in every geographic region. The most severe limitations were encountered in the Mekong Delta to the south, and in the mountain areas of the Central Highlands.

Most of the Mekong Delta is at or below sea level, and rarely more than 4 meters (13 ft.) above. It is wet, extensively cultivated, and crisscrossed with kilometers of water courses—rivers, streams, and canals. Vehicular and foot traffic is restricted to the tops of dikes, dams, and a few raised roads.

In contrast, the jungle-covered Central Highlands to the north (part of the Annamite chain) have peaks as high as 2,600 meters (8,500 ft.) and are heavily forested with tropical evergreen and giant bamboo. Vehicles can travel in most areas after the vegetation is cleared, but movement on a one- or two-vehicle front is vulnerable to an ambush.

The other regions of Vietnam—coastal plain, piedmont, and plateau—include rice paddies, rubber plantations, cultivated fields, and jungle. These areas can be negotiated by mechanized forces about 80 percent of the time. The summer monsoon winds of Indochina blow out of the Indian Ocean June through September, rise against the western slopes of the central mountains, and cause a wet season in the Mekong Delta, the piedmont, and most of the Western Highlands and plateau. From November to February, on-shore northeast winds bring wet season conditions to the northern third of the country.

In 1967, a U.S. Army study team evaluated mechanized and armor operations and opportunities in Vietnam, and concluded that tanks could move in about 60 percent of the country in the dry season, and about 46 percent of the country in the wet season. Lighter armored vehicles (recce vehicles and personnel carriers) could move in about 65 percent of the country year-round.

In addition, it was determined that the U.S. Army armored cavalry squadron (Fig. 1)—a unique unit including organic recce vehicles and self-propelled howitzers and mortars—was the most versatile unit in the inventory. It proved far more cost-effective in terms of casualties and operating expenses than the airmobile infantry with its requirements for extensive helicopter support.

Mobility—strategic, operational, and tactical—is the key to seizing the initiative and so to success in jungle warfare. Modern technology is providing the equipment—helicopters, large transport aircraft, and armored vehicles that outperform their World War II predecessors by orders of magnitude, and more flexible

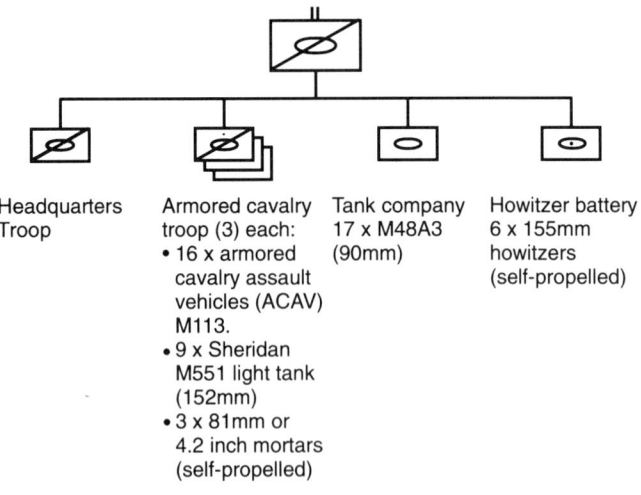

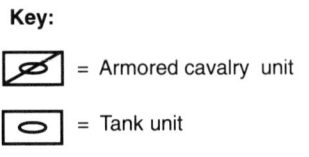

Figure 1. *Armored cavalry squadron (ca. 1969–70).* (SOURCE: U.S. Army circa 1970)

reconnaissance, surveillance, and target acquisition systems—to make mobility at all levels of jungle warfare a reality. Thus, jungle warfare is likely to become more complex than just combat among foot soldiers.

DONN A. STARRY

SEE ALSO: Geography, Military; Vietnam and Indochina Wars; World War II.

Bibliography

Hough, F. O., and J. A. Crown. 1952. *The campaign on New Britain*. Washington, D.C.: Government Printing Office.
Perrett, B. 1989. *Desert warfare: From its Roman origins to the Gulf conflict*. London: Patrick Stevens.
Slim, W. 1956, *Defeat into victory*. London: Cassell.
Starry, D. A. 1977. *Mounted combat in Vietnam*. Washington, D.C.: U.S. Department of the Army.
U.S. Department of the Army. 1982. *Field manual 90–5: Jungle operations*. Washington, D.C.: U.S. Department of the Army.
U.S. War Department, Historical Division. 1945. *The Admiralties: Operations of the 1st Cavalry Division, 29 February–18 May 1944*. Washington, D.C.: Government Printing Office.

K

KAZAKHSTAN

Kazakhstan, one of the fifteen former republics in the Union of Soviet Socialist Republics (USSR), joined ten other Soviet republics in dissolving the Union on 21 December 1991 when they agreed to form the new Commonwealth of Independent States. This dramatic end to the Soviet and Communist state followed several years of dynamic and unprecedented change. For several more years, relations between members of the new Commonwealth and with the rest of the world are likely to continue to change. Over time, new structures and patterns will emerge in economics, trade and commerce, politics and government, finance, manufacturing, religion, and virtually all aspects of human life. New arrangements must be devised for dealing separately as sovereign states and as a Commonwealth with the world outside the boundaries of the former Soviet state. If the history of the Soviet Union since 1985 is any guide, we can expect dramatic surprises and dynamic change.

An important question for the world is how the new states and the Commonwealth will organize and provide for their security. The Soviet Union's armed forces, formerly the largest in the world, are likely to be withdrawn from foreign territory, reduced in size, and divided up between the former republics. Also of great concern is the disposition of the largest arsenal of nuclear weapons, the security of these weapons, the command and control of their potential use, and compliance with arms control agreements entered into by the former Soviet government. The world can only hope these issues are settled amicably.

It will be years before all of these issues are resolved for Kazakhstan and some time before events settle down into more routine and measurable patterns. No accurate description of this new country's policies, defense structure, and military forces was available to be included in this encyclopedia. Only time will reveal the future of Kazakhstan as a separate sovereign state. The reader is thus referred to the historic information contained in the article "Soviet Union," and to the latest annual editions of the *Military Balance*, published by Brassey's (UK) for the International Institute of Strategic Studies; the *Statesman's Year-Book*, published by the Macmillan Press Ltd and St. Martin's Press; and the *World Factbook*, developed by the U.S. Central Intelligence Agency and published commercially by Brassey's (US).

F. D. Margiotta
Executive Editor

KEMAL, MUSTAFA PASHA (Atatürk) [1881–1938]

At the conclusion of World War I, several empires—Russian, German, Austro-Hungarian, and Ottoman—collapsed. During the death throes of the latter, both a new leader (Mustafa Kemal), and a new nation (the Republic of Turkey) emerged in the place of the sultan and his empire. In recognition of Kemal's role in the evolution of the nation, the national assembly gave him the honorific Atatürk—Father of the Turks.

Mustafa Kemal lived through the breakdown of the once powerful Ottoman Empire—an empire by then derisively known as the "Sick Man of Europe." Born in Macedonia, the son of a customs officer, Kemal's home was in the port city of Salonika. Ottoman control over Macedonia faded rapidly as he became a young man. Through those years the "Eastern Question" kept the Ottoman Empire on the brink of crisis, yet for young Kemal good fortune radiated his way. He was selected to attend one of the few institutions—military school—that allowed a young man of humble origins to advance in the closed society of the Ottomans. Kemal graduated with honors as a second lieutenant. His education had been broad and considerably influenced by Western thought.

Reaction and Revolution

As a soldier Kemal quickly advanced in the ranks of the army. The sultan's corruption and autocracy led Kemal to question the status quo. Kemal, nurtured by his reading of Rousseau, Voltaire, Tolstoy, and others, soon advocated reform for the empire. He also resented foreign privilege that he had observed in Constantinople, especially the "Capitulations," which gave foreigners extraterritorial privileges that placed them above Turkish law. His newfound political awareness troubled the sultan's agents, but the Young Turk revolution of 1908 came from the army; Kemal was not in a position to play a key role.

War and Nationalism

The Italian attack on Tripoli in 1911 and the Balkan Wars of 1912 and 1913 foreshadowed the Great War of 1914. For Kemal it was a time of personal loss. As a soldier he experienced defeat in Tripoli, and as a Turk he lost his home to Greece, although he took part in the recapture of Adrianople from Bulgarian troops in July 1913. Events moved into World War I following the assassination of Austrian Archduke Franz Ferdinand during the summer of 1914.

Kemal commanded the 19th Division at Gallipoli and so played a major role in halting the initial British attack there in April 1915. During the ensuing eight months of bitter fighting, Kemal was distinguished as the principal subordinate of German general Liman von Sanders. His performance in these battles won him renown and promotion, although he quarreled with War Minister Enver Pasha over growing German influence in the empire. He commanded the XVI Corps in eastern Anatolia from March to September 1916. Kemal's policy dispute with Enver Pasha continued, resulting in Kemal being placed on indefinite sick leave in late 1917. The deteriorating situation in Palestine led to his recall to command the Seventh Army there in summer 1918. Despite Kemal's best efforts, his meager forces were badly mauled by Allenby's offensive at Megiddo (19–22 September), but he was able to extricate much of his force. Although he re-established a hasty line just north of Aleppo in late October, he was forced to withdraw as British troops approached; when the armistice was concluded in November, his force was in southeastern Anatolia.

The war had convinced Kemal that the future of his people rested with Turkish nationalism. He had witnessed before and during World War I the tide of nationalism within the empire. Kemal recognized that the multinational, multireligious empire led by the Ottomans would fail in the non-Turkish areas. He also concluded that a state based on a single, dominant faith (Islam) would also collapse for the same reason. Kemal had already observed the failure of the faithful to respond to the holy war (*jihad*) that the Ottomans had called during World War I. Last, he believed that a state that linked all Turks together would fail because Europeans dominated the region. Kemal realized that the Anatolian peninsula, peopled by Turks with a common heritage, represented the best opportunity of the Turks to survive a collapse of empire.

War of Independence

On 30 October 1918, the Turks had signed an armistice with the Allies. At the conclusion of the armistice, the Turks believed that an understanding had been reached with the Allies. Greece would not occupy Ismir (Smyrna), and the Allies would not occupy Constantinople. The war had been lost, but it seemed the territorial integrity of the Turkish heartland had been secured. The understanding was only an illusion.

In January 1919, the Allies assembled in Paris to determine the fate of Germany and its allies. The Paris Peace Conference focused on Europe and the competitive claims that the victorious Europeans had in their secret accords. Among them were plans for the dismemberment of the Ottoman Empire and cession of its territory to the prewar allies of Russia, Britain, and France. The Russian revolutions of 1917 caused Russia's claims to be discarded, but Britain and France had not been swayed by Woodrow Wilson's idealistic formula for peace. For the Ottomans, it meant the understandings of the previous October were to be betrayed. For Kemal, however, the betrayal provided an opportunity to ignite nationalism.

In February 1919, Allied troops entered Constantinople, and later that spring, Greece was granted the right to occupy Ismir. As a result, discontent and disorder emerged throughout Anatolia. Allied authorities and their puppet-sultan reacted by sending Kemal to quell the disturbances. The sultan named him inspector general with full power to combine civil and military authority to end the uprisings. Kemal's new position gave him what he had never possessed—legitimate authority. He then used this position to launch the struggle.

On 19 May 1919, Kemal ordered both military and civilian authorities to fight for Turkish dignity and honor. The sultan attempted to recall him, but it was too late. By September, Kemal had held several conferences that affirmed his aim of self-determination for the Turks. Early in 1920, the Allies moved against Constantinople and attempted to establish Armenia as an independent state. The Europeans played into Kemal's hands. Concurrently, the sultan assisted the nationalists when he signed the Treaty of Sèvres on 20 August 1920. In that treaty the sultan's government recognized the end of the Ottoman Empire, the return of the Capitulations, the acceptance of Allied economic controls, the demilitarization of the straits, and Allied intervention. The treaty gave new impetus to Kemal's movement.

Kemal, elected both president of the new national assembly in Ankara and head of the government in April, was faced with a desperate military situation. First, he seized the initiative and launched an offensive against the Armenians, subduing them in a brief campaign and concluding a peace at Alexandropol on 2 December. Meanwhile Kemal had appointed Ismet Pasha (later Ismet İnönü) commander in western Anatolia facing the Greeks. Ismet's success at First İnönü in January 1921 and Second İnönü in late March enabled Kemal's counteroffensive at Sakkaria (24 August–16 September) to be a resounding success.

Kemal spent some ten months reorganizing his forces and consolidating the new Turkish state. He launched his final drive against the Greeks in July 1922, capturing Afyon at the end of August and finally taking Smyrna in a bitter battle from 9 to 13 September. He then began an advance on Istanbul and negotiated the Convention of Mundania (3–11 October), by the terms of which the Allies agreed to withdraw their garrison from Istanbul. The last sultan departed Constantinople on 17 November 1922. The nationalists had won their War of Independence.

Kemal moved quickly to consolidate his triumph. By July 1923, he received international confirmation of his victory in the Treaty of Lausanne. In effect the Treaty of Sèvres was abrogated by the same powers that had imposed it on Turkey in 1920. The odious conditions of Sèvres were removed. On 29 October 1923, the Turkish national assembly created the Republic of Turkey, with Kemal as its president.

Kemalist Programs

The War of Independence had created a new nation. The victory had removed the chains of the Ottoman past and the shackles of imperialism, but Kemal recognized that more was needed to ensure his victory. Kemal's goal was to emulate success, and that meant westernization of the nation.

From the beginning of the republic to his death, the nation experienced radical changes as Kemalist reforms were enacted. Reforms included the abolition of the sultanate and caliphate, the introduction of the secular state, the westernization of the judicial system, the establishment of women's rights (including voting), the replacement of the Arabic alphabet with a Latin version, the acceptance of Western social and cultural lifestyles, and the adoption of Western economic and business standards.

To appreciate the enormity and rapidity of the Kemalist reforms, one not only would have to understand the empire's history, but also have some knowledge of Islam. In Islam there is no separation between secular and religious activities. Kemal's reforms challenged a way of life that had governed Turks for centuries, but he persevered and created a new way of life.

Kemal's program involved six guideposts: nationalism, republicanism, secularization, populism, statism, and reformism. The first three elements of his program had been seen in the war for independence (i.e., the subsequent establishment of the Republic and its secular reforms), but the last three require a brief explanation. Populism in its simplest form meant the abolition of privilege. Every Turk, no matter his station in life, was now equal before the law. Statism involved Kemal's effort to build the economic base of Turkey. In this program the state became a major player in economic development. The last element of his program was reformism. There he sought to continue the evolution of reform in the future. Kemal knew revolutions and their cycles. He warned that, historically, today's reformers become tomorrow's defenders of the status quo. Mustafa Kemal Atatürk tenaciously advocated reform until his death in 1938. His legacy is the modern Republic of Turkey.

J. E. WADE

SEE ALSO: Allenby, Edmund Henry Hynman; Balkan Wars; Ottoman Empire; Turkey; Turkic Empire; World War I.

Bibliography

Berkes, N. 1964. *The development of secularism in Turkey.* Montreal: McGill Univ. Press.
Bodurgil, A. 1974. *Atatürk and Turkey: A bibliography, 1919–1938.* Washington, D.C.: Library of Congress.
Kinross, P. B. 1965. *Atatürk: A bibliography of Mustafa Kemal, father of modern Turkey.* New York: Morrow.
Lewis, B. 1961. *The emergence of modern Turkey.* London: Oxford Univ. Press.
Yalman, A. E. 1956. *Turkey in my time.* Norman, Okla.: Univ. of Oklahoma Press.

KENYA, REPUBLIC OF

A former British colony, Kenya is renowned for the beauty and variety of its animal life, as well as for its agricultural riches. Its rapid population growth rate (nearly 4.0% annually) is one of the highest in the world and a matter of concern to the government.

Power Potential Statistics

Area: 582,650 square kilometers (224,961 sq. mi.)
Population: 26,024,200
Total Active Armed Forces: 23,600 (0.091% of pop.)
Gross Domestic Product: US$8.5 billion (1990 est.)
Annual Defense Expenditure: US$100 million (1.0% of GDP, 1989 est.)
Iron and Steel Production:
  Crude steel: 0.010 million metric tons (1986)
Fuel Production: processed 2.031 million metric tons of petroleum products in 1986
Electrical Power Output: 2,700 million kwh (1990)
Merchant Marine: none
Civil Air Fleet: 14 major transport aircraft; 213 usable airfields (22 with permanent-surface runways); 2 with runways over 3,659 meters (12,000 ft.); 2 with runways 2,440–3,659 meters (8,000–12,000 ft.); 47 with runways 1,220–2,440 meters (4,000–8,000 ft.).

For the most recent information, the reader may refer to the following annual publications:
The Military Balance. International Institute for Strategic Studies. London: Brassey's (UK).
The Statesman's Year-Book. New York: St. Martin's Press.
The World Factbook. Central Intelligence Agency. Washington, D.C.: Brassey's (US).

History

Paleontologists working in Kenya have discovered fossils near Lake Turkana showing that members of genus *Homo*

Kenya, Republic of

lived in the area 2.6 million years ago, thus making Kenya one of the longest-inhabited countries in the world. Kenya's written history begins much later, when Arab traders moving south along the coast encountered Cushitic natives about A.D. 50. Arabs and Persians settled along the coast by the eighth century, and soon after Bantu and Nilotic immigrants also moved into Kenya and the surrounding area. This ethnic mix led to the invention of Swahili, a Bantu-Arabic mixture which served as a *lingua franca* for Arabs and Africans in eastern Africa for centuries.

Arab influences were in part replaced by those of the Portuguese, who arrived in 1498. There are remains of a Portuguese castle at Mombasa, but the Portuguese were displaced by the Omanis in the 1600s, who were in turn displaced by the British in the mid- to late nineteenth century. The Berlin Conference of 1885 gave Kenya to the British, who soon after opened the fertile highland region around Nairobi to European settlement. Kenya served as a major base for operations against German East Africa (now Tanzania) in World War I, and became a colony in 1920. Typically, the British allowed white settlers considerable self-rule, but the natives were barred from politics until 1944.

Between October 1952 and January 1960, Kenya was under a state of emergency because of activities by the Mau-Mau. This group, which drew support from the Kikuyu, conducted a bloody campaign for independence in the early 1950s, but active British countermeasures had largely suppressed the Mau-Mau by mid-1956. British authorities released Mau-Mau leader Jomo Kenyatta from prison in August 1961, and Kenyatta regained his prior role as leader of the independence movement and became head of KANU (Kenyan African National Union), the main political party. As head of KANU he became prime minister on 1 June 1963, and Kenya became an independent state on 12 December. Kenya faced problems from Somali separatist insurgents in the northeast, and in January 1964 weathered a period of unrest with the help of British troops.

Kenyatta's internal policy revolved around the "Kenyanization" of the civil service, the military, and the retail economy; that is, replacing Europeans and Asians with Africans as much as possible. This was tied to his slogan "Harambee," a Swahili word used on work gangs, meaning "pull together." By the time of Kenyatta's death (22 August 1978), these policies had largely achieved their goals. Kenyatta's successor, former vice president Daniel T. arap Moi, continued these policies. Kenya holds regular elections, but since the leftist Kenyan People's Union (KPU) was banned in 1969, KANU has been the only legal party, and the constitution was amended in September 1983 to make Kenya an official one-party state. Elections were last held in March 1988, and are not due again until 1993. Kenya has not been involved in serious armed conflict since independence, although it has expanded its military and acquired additional heavy weaponry and advanced equipment since 1970.

Politico-Military Background and Policy

While Kenya maintains good relations with Western nations, it is also firmly nonaligned. Kenya also enjoys good relations with Ethiopia and Sudan, and since Yoweri Museveni's National Resistance Movement took power there in January 1986, has also had good relations with Uganda. Relations with Tanzania are cool, in part because of the breakup of the East African Community. Relations with Somalia improved after the early 1980s, but remained strained due to Somali interest in the ethnic Somalis living in northeastern Kenya.

The Kenyan army is based on the British King's African Rifles (KAR), but since the mutiny of the KAR 11th battalion (January 1964), the Kenyan government has worked hard to create professional, nonpoliticized armed forces. The Kenyan armed forces are manned by volunteers, who usually sign on for a period of nine years. With a large pool of available manpower, the armed forces and the national police can both be quite selective, and the caliber of personnel in the services is correspondingly high.

Strategic Problems

Kenya's greatest strategic problems, like those of most of its neighbors, are related to economic development and demographic growth. Although an expanding population is often considered a necessary ingredient to a healthy economy, Kenya's rapid population growth rate (the population will double every eighteen years) places a severe strain on public services and on the economy's capacity to absorb new workers. Even with a stringent population control effort, it would take at least a generation for the demographic results to appear.

There are several trouble spots on Kenya's borders. Uganda has undergone severe political upheaval since 1979, and what little stability exists in that country will require great care to take hold. The overthrow of Siad Barre in Somalia in January 1991, leaving political chaos and civil strife, is another nearby trouble spot. The end of the long-running civil war in Ethiopia has eased some regional tensions, but the potential for trouble there or in Somalia is considerable. Last, the continued disaffection southern Sudanese have with their northern-dominated government also affects Kenya; at the very least, refugees will flee to Kenya to escape the fighting.

Military Assistance

Kenya receives military assistance from several countries, including the United States and Great Britain. U.S. aid in 1988 totaled US$6.2 million, a notable reduction from the 1984–86 average of US$21.9 million. Kenya also received some US$776 million in U.S. economic aid through 1986. Britain, as the former colonial power, provided considerable training and logistical support, also selling Kenya most of its heavy weapons and naval vessels, and building the Mombasa naval base. In the years just after independence, many officers were seconded British personnel, and until 1969 the Kenyan army's commander was a British major general. Kenyan personnel have received training in the United States, Britain, Canada, Bulgaria, Ethiopia, and Israel.

Defense Industry

Kenya produces no significant arms of its own, and relies on imports from the United States, Britain, West Germany, Israel, and France.

Alliances

Kenya's long-term security arrangements with the United States and Britain are not formalized by a treaty. Kenya is a member of the United Nations (UN), the Organization of African Unity (OAU), and the Commonwealth.

Defense Structure

The president is, according to the constitution, commander in chief of Kenya's armed forces, and the president is also prohibited from holding any military rank or post. The president exercises his control through a minister of defense. The armed forces are integrated under the commander of the army, who is also commander of the Kenya Military Forces, and who reports directly to the defense minister. Since a failed coup attempt mounted by dissident air force elements (1 August 1982), the air force has been subordinate to the army.

(For an explanation of the abbreviations and symbols used in the following section of military statistics, see the list of Abbreviations and Acronyms in each volume.)

Total Armed Forces

Active: 23,600

ARMY: 19,000
1 armd bde (2 armd bn).
2 inf bde (1 with 2, 1 with 3 inf bn; 1 armed recce, 2 arty bn).
1 engr bde: (2 engr bn).
1 indep air cav bn.
5 inf bn (cadre).
1 AB bn.
1 AA bn.
Equipment:
MBT: 76 Vickers Mk3
Recce: 40 AML-60/-90, 8 Shorland.
APC: 30 UR-416, 10 Panhard M-3.
Towed arty: 105mm: 40 lt, 16 pack.
Mortars: 81mm: 20 L16; 120mm: 10 Brandt.
ATGW: Milan, 8 Swingfire.
RCL: 84mm: 80 Carl Gustav; 120mm: Wombat.
AD guns: 20mm: 50 TCM-20.

NAVY: 1,100. Base: Mombasa.
Patrol and Coastal Combatants: 7:
Missile Craft 6: 2 Nyayo (UK Vosper 56-m) PFM, with 4 Otomat II SSM; 1 Mamba, 3 Madaraka (UK Brooke Marine 37-m/32-m) PFM with 4 x Gabriel II SSM.
Patrol, Inshore: 1 (.

AIR FORCE: 3,500: 28 cbt ac, 38 armed hel.
FGA: 11 F-5 (9 -E, 2 -F).
COIN: 5 Strikemaster Mk 87, 12 Hawk Mk-52.
Transport: 8 DHC-5D, 7 Do-28D, 1 PA-32.
Training: 8 Bulldog 103/127.
Helicopters:
　Attack: 15 Hughes 500MD (with TOW), 8 500ME, 15 500M.
　Transport: 9 IAR-330, 3 SA-330, 1 SA-342.
　Training: 2 Hughes 500D.
Missiles:
　ASM: AGM-65 Maverick, TOW.
　AAM: AIM-9 Sidewinder.

FORCES ABROAD
UN and Peacekeeping:
Iraq/Kuwait (UNIKOM): 8 observers.

PARAMILITARY
Kenya maintains a large and effective national police force. Known as the Kenya Police, this force of 14,000 dates to the early 1900s, and includes the paramilitary General Service Unit (GSU). The GSU was created during the Mau-Mau uprising of the 1950s, and is deployed in company-sized units across the country. Numbering 4,000 officers and men, its purpose is to act as riot police and stop violent demonstrations; its equipment includes vehicles, automatic weapons, mortars, and communications gear. There is also a Police Air Wing, which operates 7 Cessna light aircraft and 3 Bell helicopters, and a Police Naval Squadron, which operates some 12 patrol boats.

Future

Under the present situation in East Africa Kenya is unlikely to face any significant external security threat, but a change of regime or the outbreak of civil war in any of several neighboring countries could lead to Kenyan involvement. Kenyan policies are unlikely to change in the near future, as they have the tacit support of most of the population, but the internal socio-economic problems mentioned above need to be dealt with, and the solutions may not be popular.

DAVID L. BONGARD
TREVOR N. DUPUY

SEE ALSO: Ethiopia; Somalia; Sub-Saharan Africa; Sudan.

Bibliography

American University. 1983. *Kenya: A country study.* Washington, D.C.: Government Printing Office.
Hunter, B., ed. 1991. *The statesman's year-book, 1991–92.* New York: St. Martin's Press.

International Institute for Strategic Studies. 1991. *The military balance, 1991–1992*. London: Brassey's.
Miller, N. 1984. *Kenya: The quest for prosperity*. Boulder, Colo.: Westview Press.
Ominde, S. H. 1985. *Population and development in Kenya*. Portsmouth, N.H.: Heinemann Educational Books.

KESSELRING, ALBERT [1885–1960]

Field Marshal Albert Kesselring was a prominent German field commander of World War II. Although he was actually a Luftwaffe (air force) officer, Kesselring is noted for his superlative performance and conduct of the essentially ground operations of the Italian campaigns of 1943–45. As a result, Kesselring has been lauded as one of the most formidable technicians of war known to the twentieth century.

Early Life and Career

Albert Kesselring was born on 30 November 1885 in Bayreuth, Bavaria, where his father served on the town education committee. He studied at the local Classical Grammar School in 1904, but early in his youth decided on a career as a professional soldier. Kesselring later observed, "I wanted to be a soldier, indeed I was set on it, and looking back I can say that I was always a soldier heart and soul."

In 1904, Kesselring enlisted in the 2d Bavarian Foot Artillery as an officer candidate. Except for attendance at the Military Academy (1905–1906) and the Artillery School (1909–10), he served with his regiment in Munich until the outbreak of World War I.

Kesselring's service during the Great War was primarily as a staff officer. He served successively as adjutant in two Bavarian artillery units until the end of 1917, when he became a general staff officer and served on the eastern front. There, Kesselring conducted local armistice negotiations with a Russian counterpart. At the end of the war, Kesselring was serving on the Sixth Army staff at Lille, and in the immediate postwar years he served in the Bavarian army.

Kesselring was fortunate in being selected to serve in the 100,000-man Reichswehr (limited in size by the Treaty of Versailles), and after service as an artillery battery commander, returned to the staff. He was instrumental in designing the Luftwaffe and the panzer (armored) forces. In 1929, Kesselring was assigned to the Army Personnel Office in Berlin, and the following year he received two promotions, first to lieutenant colonel and then to colonel. In 1931, he returned to the troops as commander of the 4th Artillery Regiment in Dresden.

The advent of Adolf Hitler as chancellor of Germany in January 1933 provided the impetus for the clandestine establishment of the Luftwaffe. Kesselring, one of the first army officers to be transferred to the fledgling German air force, became an air commodore and learned to fly at age 48. The existence of the Luftwaffe was officially announced in March 1935, with Kesselring receiving promotion to major general in 1936 while continuing to urge the inauguration of a general staff for the air arm.

The Luftwaffe general staff was instituted on 1 August 1936, and Kesselring became chief of staff on 15 August 1936. Kesselring became embroiled in controversy over the issue of allocation of resources to strategic or tactical air forces and also over a power struggle with Gen. Erhard Milch, state secretary for aviation. Amid a Luftwaffe structural reorganization, Kesselring resigned as chief of staff only to receive an important command. Kesselring was only biding his time for more favorable political circumstances.

Kesselring's *Luftkreis* III was converted into 1st Group Command on 1 April 1939, which a year later became the First Air Fleet. His command then comprised a majority of the German air forces.

World War II

Kesselring energetically led his forces in support of Army Group North in the invasion of Poland in September 1939. He was largely responsible for eliminating resistance in Warsaw and received the Knight's Cross of the Iron Cross for his achievements during the first month of the war.

The relief of the commander of the Second Air Fleet early in 1940 was fortuitous for Kesselring, who succeeded to this command in the west. In the invasion of Holland in May 1940, Kesselring's command not only supported Army Group B in its blitzkrieg attack, but also conducted revolutionary airborne and glider-borne assaults. After the subjugation of France, the Battle of Britain began, and Kesselring's forces were initially so successful that, in recognition thereof, on 19 July 1940 he was promoted to field marshal.

Kesselring was informed in late 1940 of the plan to invade the Soviet Union, and he commanded the air fleet in support of Army Group Center's main attack when Operation Barbarossa commenced on 22 June 1941. Three months later it was determined that a stronger German presence was needed in the Mediterranean region, and Kesselring was reassigned to Italy in November 1941. He became commander in chief south, making him the superior of Gen. Erwin Rommel, commander of the German Afrika Korps.

With little guidance, Kesselring attempted to develop a strategy for his area of operations, although at this time he considered himself and his command as little more than a conduit for supplies for Rommel. This situation changed dramatically after the Allied invasion of North Africa on 8 November 1942. Stubborn delaying actions were fought in North Africa until the German evacuation on 12 May 1943.

The inevitable Allied invasion of Sicily (which occurred

on 10 July 1943) was complicated for the Germans by the overthrow of Italian dictator Benito Mussolini two weeks later. On 1 September 1943 the Italians concluded an armistice with the Allies.

The Allies invaded the Italian mainland on 9 September 1943. On 21 November 1943, Kesselring was selected, over his now-rival Rommel, to be commander in chief of Army Group C and commander in chief southwest in Italy. By mid-January 1944, he had 21 divisions under his command. He tenaciously and skillfully defended the Italian peninsula against the Allied armies commanded by, first, Gen. (later Field Marshal Earl) Harold Alexander (until 12 December 1944), then by Lt. Gen. (later Gen.) Mark W. Clark.

On 8 March 1945, Kesselring succeeded Field Marshal Gerd von Rundstedt as commander in chief west. At this late stage the German surrender was only a matter of time.

Aftermath

Kesselring was imprisoned by the Allies after the German surrender. He was brought to trial in February 1947 for the murder of 355 civilians in the Ardeatine Caves in March 1944. Found guilty, Kesselring was condemned to death by shooting, but the sentence was later commuted to life imprisonment. He wrote his *Memoirs* (first published in 1953), a self-serving apologia, while in prison. Kesselring was released from captivity in July 1952 when it was discovered he had throat cancer. He died on 20 July 1960.

Field Marshal Albert Kesselring, nicknamed "Smiling Albert" because of his omnipresent grin, had a rare blend of character and competence that permitted him to be the consummate staff officer as well as a dynamic, imaginative, and charismatic commander. At his funeral, Kesselring was fittingly eulogized by his former chief of staff as "a man of admirable strength of character whose care was for the soldiers of all ranks."

HAROLD E. RAUGH, JR.

SEE ALSO: Alexander, Harold Rupert Leofric George; Blitzkrieg; Rommel, Erwin; World War II.

Bibliography

Blaxland, G. 1979. *Alexander's generals.* London: Kimber.
Clark, M. 1950. *Calculated risk.* New York: Harper and Row.
Irving, D. 1973. *The life of Erhard Milch.* London: Weidenfeld & Nicolson.
Jackson, W. G. F. 1967. *The battle for Italy.* London: Batsford.
———. 1972. *Alexander of Tunis as military commander.* New York: Dodd, Mead.
Kesselring, A. 1953. *Memoirs.* London: Kimber.
Linklater, E. 1977. *The campaign in Italy.* London: Her Majesty's Stationery Office.
Macksey, K. 1978. *Kesselring.* New York: David McKay.
Mason, H. M. 1973. *The rise and fall of the Luftwaffe.* London: Cassell.
Shepperd, G. A. 1968. *The Italian campaign, 1943–45.* New York: Praeger.

KHALID IBN AL-WALID
[d. A.D. 642]

Known as "Chaledos" in the Byzantine chronicles, Khalid was an outstanding general, widely regarded as the preeminent war leader of the early Muslim conquests. His prowess in battle earned him the nickname *Sayf Allah* ("Sword of Allah"). He played a central role in the Muslim subjugation of Mesopotamia and Syria in the seventh century.

Background and Early Career

Of noble birth, Khalid was of the *banu-Quraysh*, the Prophet Muhammad's tribe, and a member of one of Mecca's wealthiest and most influential clans. His father, however, was sharply opposed to the nascent Islamic movement. In A.D. 625 the followers of the Prophet in Medina gained a victory over the nonbelieving Meccans at Badr. Seeking revenge the Meccans, including Khalid, who served as a subcommander, confronted the Muslims at Medina and taunted them by pasturing their horses and camels in the Medinans' field of unharvested grain near the hill of Uhud. When the Muslims attacked to drive the interlopers away from their crops, young Khalid led a successful local counterattack. Despite this success the Meccans lost the battle, but the Uhud fight established Khalid's reputation among his Muslim enemies as a tough, natural military genius.

About the time of the Muslim success at the Battle of the Trench (627), or shortly afterward, Khalid embraced Islam and joined the Muslim forces that conquered Mecca in 629. Although his conversion appears to have been opportunistic, Khalid soon demonstrated that he was an efficient, ruthless cavalry commander.

In the "War of Apostasy," or *riddah*, that followed Muhammad's death in 632, the Prophet's successor, Caliph Abu Bakr, dispatched eleven armies throughout Arabia to reestablish Muslim hegemony among tribes that were in revolt, led by "false prophets." In the military operations against the rebels (632–633) the most prominent commander was Khalid, leading the reconquest in the north and east. His ruthlessness shocked the caliph and Omar, who later succeeded Abu Bakr as caliph, but Khalid's success was such that he received only mild rebukes. His victory at Akraba (633) ended the Arab civil war and set the stage for the first expansion of Islam.

Operations in Syria and Palestine

When the traditional Arab border raids along the Arabian-Sassanid Persian (Iraqi) frontier became fused with the emerging Islamic *jihad*, Khalid led a raiding force into Iraq, conquering al-Hira (633). Tribute from al-Hira, along with that imposed on neighboring Sassanid districts, helped offset the loss of trade caused by the *riddah*. At the same time Khalid raided Iraq, another Arab army under Amr ibn al-As invaded Syria and Palestine, the eastern provinces of the Byzantine Empire.

In 634 several more Arab raiding columns entered Syria to reap supposedly easy pickings. The Arabs thrust into southern Palestine in search of booty, avoiding the still-formidable Byzantine *turmae* posted there. (A *turma* was the Byzantine equivalent of the old Roman legion.) However, Byzantine intelligence of the Arabs' operations was excellent, and the Byzantine *turmae* prepared to engage the interlopers. Discovering that the Syrian forces of Byzantine emperor Heraclius were seeking battle, the expeditionary leaders requested reinforcements from Abu Bakr. Having foresworn the military services of apostates, the caliph could find no forces at home to dispatch to Palestine and so sent for Khalid, still raiding in Iraq.

The urgency of Abu Bakr's message caused Khalid with 500 picked men to make a legendary forced march across 500 kilometers (300 mi.) of Syrian desert, reaching the Byzantine eastern frontier in the vicinity of Palmyra in eighteen days. Khalid's first military encounter in Syria was with a strong force of Byzantine auxiliaries, whom he defeated at their camp in Marj Rahit on the plain of Damascus (spring 634). He then joined forces with Amr ibn al-As, and the combined armies inflicted a crushing defeat on a Byzantine *thema* (two or three *turmae*) in July 634 at Ajnadayn on the Gaza-Jerusalem road. Following this victory, Khalid occupied Busra in the Hawran, south of Damascus. Another successful engagement at Fihl (January 635) gave the Arabs control of one of the most strategically important crossings of the Jordan River.

Khalid next moved his army toward Damascus, taking Marj al-Suffar en route. The six-month Muslim siege of Damascus that followed seems to have been brought to resolution by Khalid's conspiring with a disaffected monk within the city walls. Damascus surrendered to Khalid on 4 September 635. Following this, Khalid apparently led the columns that reduced Baalbek, Homs, and Hamah in rapid succession.

Alarmed, Heraclius ordered his general Theodore to marshal a major army—estimated at about 50,000—to deal a decisive blow to Khalid's force. Khalid fell back on a defensive position at the junction of the Yarmuk River within a defile known now as the Deraa Gap, near the southeastern shore of the Sea of Galilee. He positioned his cavalry with their backs to the desert.

When Theodore's Byzantine army attacked (20 August 636), a sandstorm blew up from the desert directly into the eyes of the oncoming Byzantine mass. Khalid's 25,000 men had the wind at their backs. Khalid also was aided by dissension between the Byzantine army's East Roman core and its disaffected Armenian and Arab auxiliaries. Khalid was nominally second in command at this battle—his impetuosity had always irked the new caliph, Omar, who appointed a more manageable commander in chief—but it is believed that Khalid actually directed the maneuvers that resulted in the destruction of the Byzantine army.

After Yarmuk

Khalid's great victory prepared the way for the expulsion of the East Romans from Syria and the consolidation of Arab power in the former Byzantine province (637–645). Khalid was rewarded with the governorship of a part of the new Arab Syrian province, holding this office until his death at Homs, where his tomb can still be viewed.

JAMES J. BLOOM

SEE ALSO: Arab Conquests; Byzantine Empire.

Bibliography

Akram, A. I. 1970. *The sword of Allah: Khalid ibn Al Waleed, his life and campaigns.* Karachi: National Publishing House.
Shaban, M. A. 1971. *Islamic history, A new interpretation* A.D. 600–750. Vol. 1. London: Cambridge Univ. Press.

KING, ERNEST J. [1878–1956]

Ernest Joseph King, Fleet Admiral, U.S. Navy (Fig. 1), was supreme commander of all U.S. naval forces in World War II and principal naval adviser to President Franklin Roosevelt. As naval member of the Joint Chiefs of Staff and the Combined Chiefs of Staff, he participated in all interservice and international planning for the defeat of Germany and Japan.

Early Life and Career

King was born in Lorain, Ohio, on 23 November 1878. In August 1897 he entered the U.S. Naval Academy in Annapolis, Maryland. The following summer he served on the cruiser USS *San Francisco* and came under fire off Havana.

Graduating in 1901 as a passed midshipman, King attended the Naval Torpedo School in Newport, Rhode Island, served as navigator on the geodetic survey ship *Eagle*, was aide to the admiral on the new battleship USS *Illinois*, and was in charge of a division of 40 enlisted men on the cruiser USS *Cincinnati* in 1903 when he was promoted to ensign. King went to the Far East on the *Cincinnati* and was there during the Russo-Japanese War. In October 1905 he married Martha Rankin Egerton. They had six daughters and one son.

While serving on the battleship USS *Alabama* in 1906, King was promoted to lieutenant. That fall he returned to the U.S. Naval Academy as instructor of ordnance, gunnery, and seamanship. In 1909 he went to sea again as flag secretary to Rear Admiral Osterhaus on the USS *Minnesota*. This duty was interrupted by a brief tour as engineer officer on the battleship USS *New Hampshire*. In 1912 King went back to Annapolis as executive officer of the

Figure 1. Adm. Ernest Joseph King. (SOURCE: U. S. Library of Congress)

Naval Engineering Equipment Station. In 1913 he was promoted to lieutenant commander.

World War I

King's first command, in April 1914, was the USS *Terry*, a reserve destroyer. He transferred to a new destroyer, the USS *Cassin*, in July. For a few months he was also aide to Capt. William S. Sims, commander of the Atlantic Fleet Destroyer Flotilla, and then commanded a destroyer division of four ships.

In December 1915 King reported as staff engineer to Vice Adm. Henry T. Mayo, Commander Battleship Force, Atlantic Fleet, and later Commander in Chief, Atlantic Fleet. Promoted to commander in July 1917, King twice traveled to Europe with Mayo during World War I to confer with British officials and inspect U.S. naval units. He received the Navy Cross and the temporary rank of captain.

King reopened the Naval Postgraduate School at Annapolis as commandant in the spring of 1919. His next command was a refrigerator supply ship, the USS *Bridge*. In 1922, with the permanent rank of captain, he took a four-month course at the submarine school in New London, Connecticut, and then was given command of a four-submarine division. He was commander of the submarine base in New London when he came to public attention as commander of the unprecedented operation that, under very difficult conditions, raised the submarine S-51, which had been rammed by a steamer and sunk off Block Island on 24 September 1925.

From submarines King went to naval aviation, as commander of the seaplane tender USS *Wright* from July 1926 to August 1928. He earned his wings at Pensacola, Florida, on 26 May 1927.

When the submarine S-4 was rammed and sunk off Provincetown, Massachusetts, in December 1927, King was abruptly summoned to command the effort to rescue the crew. The attempt had to be abandoned when a winter gale struck, and the rescue became another difficult salvage operation.

After serving briefly as temporary Commander, Aircraft Squadrons, Atlantic, King became assistant to the Chief of the Bureau of Aeronautics, Rear Adm. William A. Moffett. In less than a year he became commander of the Norfolk Air Base in Virginia. In 1930 came a command he had eagerly sought, the aircraft carrier USS *Lexington*. From there he went to the senior course at the Naval War College in Newport. When Admiral Moffett died in the crash of the dirigible *Akron* in April 1933, King, now rear admiral, was made Chief of the Bureau of Aeronautics. Three years later he became Commander, Aircraft, Base Force, with responsibility for land- and sea-based patrol planes. In January 1938 he became Commander, Aircraft, Battle Force, with three aircraft carriers including the *Lexington*, all in the Pacific.

King's career-long hopes of the navy's top post, Chief of Naval Operations (CNO), were dashed in 1939, when Adm. Harold Stark was selected. King went to the Navy's General Board, one of ten admirals at the end of their careers who devoted full time to studying naval problems and advising the secretary of the navy, but his stay was brief.

World War II

In 1940 King accompanied outgoing Secretary Charles Edison and new Secretary Frank Knox on inspection trips. This led to a brief appointment as Commander, Patrol Force, U.S. Fleet. In January 1941, with the rank of vice admiral and soon admiral, King became commander of the new Atlantic Fleet, increasingly involved in assisting Great Britain with convoying and antisubmarine patrols. In August 1941 President Roosevelt traveled on King's flagship, the heavy cruiser USS *Augusta*, to Argentia, Newfoundland, and held meetings with British Prime Minister Winston Churchill.

After the Japanese attack on Pearl Harbor on 7 December 1941, King was called to Washington, D.C. On 30 December Roosevelt appointed him Commander in

Chief, U.S. Fleet (CominCh), in supreme command of all naval operating forces. The division of tasks and responsibilities between CominCh and the CNO soon proved so uncertain that conflicts arose. In March 1942 Roosevelt sent Stark to England as Commander in Chief, U.S. Naval Forces, Europe, and made King both CominCh and CNO. As CominCh King was the principal naval adviser to the president on the conduct of the war and commander of all operating forces. As CNO he was responsible for the preparation, readiness, and logistic support of those forces. No U.S. naval officer had ever been given so much authority and responsibility. Few if any could have exercised it so well.

In 1943 King added another responsibility. To centralize all aspects of the antisubmarine war against German U-boats, he created the Tenth Fleet. Because no one else seemed to be available, he retained command of it himself.

Throughout his career King had been preparing for this wartime job, not only by personal experience in many areas but also by making a serious study of both sea and land warfare, in theory and in practice. As CominCh-CNO and as naval member of the U.S. Joint Chiefs of Staff and the U.S.-British Combined Chiefs of Staff, King contributed with professional expertise to strategic planning for the defeat of both Germany and Japan. He agreed with the decision that Germany must be defeated first, but he insisted that pressure on Japan must be maintained and increased; he saw to it that as much materiel and men as possible went to the Pacific forces and that they advanced steadily north and west. Thus, when Germany surrendered, the defeat of Japan was only a few months behind.

On 13 December 1944, King was given the five-star rank of Fleet Admiral. Shortly after Japan's surrender in September 1945, on King's initiative, the post of Commander in Chief, U.S. Fleet, was abolished. King remained CNO until 15 December, when Fleet Adm. Chester W. Nimitz relieved him. King died in Portsmouth, New Hampshire, on 15 June 1956.

GRACE P. HAYES

SEE ALSO: Nimitz, Chester William; World War II.

Bibliography

Buell, T. B. 1980. *Master of sea power*. Boston: Little, Brown.
King, E. J., and W. M. Whitehill. 1976. *Fleet admiral King: A naval record*. New York: Norton.

KIRGIZSTAN (formerly Kirghizia)

Kirgizstan, one of the fifteen former republics in the Union of Soviet Socialist Republics (USSR), joined ten other Soviet republics in dissolving the Union on 21 December 1991 when they agreed to form the new Commonwealth of Independent States. This dramatic end to the Soviet and Communist state followed several years of dynamic and unprecedented change. For several more years, relations between the members of the new Commonwealth and with the rest of the world are likely to continue to change. Over time, new structures and patterns will emerge in economics, trade and commerce, politics and government, finance, manufacturing, religion, and virtually all aspects of human life. New arrangements must be devised for dealing separately as sovereign states and as a Commonwealth with the world outside the boundaries of the former Soviet state. If the history of the Soviet Union since 1985 is any guide, we can expect dramatic surprises and dynamic change.

An important question for the world is how the new states and the Commonwealth will organize and provide for their security. The Soviet Union's armed forces, formerly the largest in the world, are likely to be withdrawn from foreign territory, reduced in size, and divided up among the former republics. Also of great concern is the disposition of the largest arsenal of nuclear weapons, the security of these weapons, the command and control of their potential use, and compliance with arms control agreements entered into by the former Soviet government. The world can only hope these issues are settled amicably.

It will be years before all of these issues are resolved for Kirgizstan and some time before events settle down into more routine and measurable patterns. No accurate description of this new country's policies, defense structure, and military forces was available to be included in this encyclopedia. Only time will reveal the future of Kirgizstan as a separate sovereign state. The reader is thus referred to the historic information contained in the article "Soviet Union," and to the latest annual editions of the *Military Balance*, published by Brassey's (UK) for the International Institute of Strategic Studies; the *Statesman's Year-Book*, published by the Macmillan Press Ltd and St. Martin's Press; and the *World Factbook*, developed by the U.S. Central Intelligence Agency and published commercially by Brassey's (US).

F. D. MARGIOTTA
Executive Editor

KONEV, IVAN STEPANOVICH [1897–1973]

Marshal of the Soviet Union Konev—front commander in World War II, postwar deputy minister of defense, and commander of groups of Soviet forces in Eastern Europe—was one of the premier Soviet field commanders. Konev also served as a candidate member of the Central Committee (CC) of the Communist party of the Soviet Union (CPSU) from 1929 to 1952 and as a full member of the CC

CPSU from 1952 to 1973. Konev was a delegate to the first through eighth sessions of the Supreme Soviet of the USSR.

Early Life

Konev was born on 16 (O.S.; 28 N.S.) December 1897, in the village of Lodeino (currently in the Podosinov raion of Kirov oblast) to Russian peasants. In 1916 he joined the Russian army and served as a junior NCO in an artillery battalion on the southwestern front. After demobilization in 1918, he joined the Communist party and served first as a member of the *Nikol'skii uezd* executive (*Volgodskii gubernia*), then as its military commissar following his induction into the Red Army in 1918. During the civil war Konev served as the military commissar of an armored train, infantry brigade, and infantry division, and then as a staff officer of the People's Revolutionary Army of the Far East Republic in Siberia opposite Kolchak, Semenov, and the Japanese intervention forces. While a delegate to the Tenth Congress of the Peasant and Workers Party (Bolshevik), Konev participated in crushing the Kronstadt revolt (1920). After the civil war Konev was political commissar of the XVII Maritime Rifle Corps. He graduated from a command training course (1926), then commanded at the regimental (1926–30) and divisional (1930–32) levels. Konev graduated from the Frunze (command and staff) Academy in 1934, then commanded an infantry division (1934–37) and an infantry corps (1937–38). From 1938 to 1940 Konev commanded the Second Detached Red Banner Far East Army; in 1940 Lieutenant General Konev commanded first the Transbaikal, then the North Caucasus, Military Districts.

World War II

Konev's wartime exploits were marked by a bitter rivalry with Marshal Zhukov, which continued after the war. At the outbreak of World War II Konev was commanding the Nineteenth Army, which unsuccessfully defended Smolensk against the German invaders. In September 1941, Colonel General Konev assumed command of the Western Army Group (AG), which suffered serious reverses in the Battle for Moscow. He was relieved of command (10 October 1941) and replaced by Zhukov. A week later the Twenty-ninth, Thirtieth, and Twenty-second armies were combined to reconstitute the Kalinin AG (KAG), and Konev took command. The KAG launched the Soviet counteroffensive at Moscow on 5 December. From August 1942 to February 1943, Konev once again commanded the Western AG. During his command of the Steppe AG (July–October 1943) Konev's troops fought in the pivotal Battle of Kursk (21 July–23 August 1943) and the counteroffensive (Operation Rumiantsev) that liberated Belgorod and Kharkov (3–23 August). Konev commanded the 2d Ukrainian Army Group (2d UAG) (October 1943–May 1944) during the Korsun-Shchevchenkovskii operation (24 January–17 February 1944), which destroyed a large German force formed from Manstein's "South" group of armies and liberated the right bank of the Dnepr River. As commander of the 1st UAG (May 1944–May 1945), Konev was involved in the L'vov-Sandomir offensive (July 1944), the Vistula-Oder operation (12 January–7 February 1945), the Battle for Berlin (16 April–2 May 1945), and the liberation of Prague (one of his subsidiary tank forces occupied the city on 9 May). The L'vov-Sandomir offensive pitted Konev's 1st UAG against the German Northern UAG. Konev's force liberated the western Ukraine and southern Poland, captured L'vov, and established a beachhead across the Vistula River at Sandomir. The Vistula-Oder operation was the largest single Soviet operation in World War II. Fought by Zhukov's 1st Belorussian Army Group (BAG) and Konev's 2d UAG, the Soviets broke out of the Sandomir beachhead to destroy German forces between the Vistula and Oder rivers and created a similar beachhead across the Oder-Neisse River. The Berlin operation, which added a third force (Rokossovskii's 2d BAG to pin down German forces in the north on Zhukov's flank) performed a massive encirclement that surrounded, isolated, and destroyed the defending German Vistula AG and captured Berlin on 2 May. The Allies accepted a German unconditional surrender six days later. Stalin did not assign either commander the responsibility for actually capturing the city; instead he allowed the fierce rivalry between Konev and Zhukov to spur the attack along. Konev was the first to enter Berlin. He was promoted to marshal of the Soviet Union on 20 February 1944 (a rank he held until his death), was twice awarded Hero of the Soviet Union (29 July 1944 and 1 June 1945), and was one of only eleven commanders awarded the diamond-encrusted Order of Victory.

Postwar Career

Konev held a series of senior command positions in the Soviet Union and groups of Soviet forces in Eastern Europe following the war. From 1945–46 Konev was commander in chief (CINC) of occupation forces in Hungary and Austria; as such, he served as CINC of the Central Group of Forces (Hungary) and as supreme commissioner of Austria. In 1946 Konev became CINC, Ground Forces, Soviet Army and deputy minister of the armed forces (MAF). Four years later he was appointed chief inspector of the Soviet Army and deputy minister of defense (MAF was renamed Ministry of Defense in 1950). In 1951 Konev took command of the Carpathian Military District. Under Khrushchev he returned to Moscow in 1955 to become once again CINC, Ground Forces and deputy minister of defense (DMOD). The following year Konev was elevated to first DMOD and CINC, Warsaw Pact. Konev supported Khrushchev's ouster and exile of his rival, Marshal Zhukov, in 1957. In 1960 Konev became chief inspector of the General Inspectorate of the Ministry of Defense

(GIMD), a semiretired position. During the Berlin crisis (1961), however, Konev was brought in to command the Group of Soviet Forces, Germany. He returned to the GIMD in April of 1962, where he served until his death on 21 May 1973; he was buried in the walls of the Kremlin. Konev was named Hero of the Czechoslovakian SSR (1970), the Ukrainian SSR (1970), and the Mongolian People's Republic (1971).

He was the author of *Sorok Piatyi* ['45] (published in 1970) and *Zapiski komanduiushchego frontom, 1943–44* [*Notes of a Senior Commander at the Front 1943–44*], (published in 1972). In an article entitled *Nachalo Moskovskoi bitvy* [*Beginning of the Battle of Moscow*] (*Voenno-istoricheskii zhurnal*, no. 10, 1966) Konev finally rebutted Zhukov's contention that he had been relieved of his command of the WAG in October 1941 when Zhukov replaced him.

DIANNE L. SMITH

SEE ALSO: Civil War, Russian; Manstein, Erich von; World War II; Zhukov, Georgi Konstantinovich.

Bibliography

Bialer, S., ed. 1969. *Stalin and his generals.* New York: Pegasus.
Erickson, J. 1975. *The road to Stalingrad.* London: George Weidenfeld and Nicolson.
———. 1985. *The road to Berlin.* London: George Weidenfeld and Nicolson.
Ziemke, E. 1968. *Stalingrad to Berlin: The German defeat in the East.* Washington, D.C.: U.S. Army Center of Military History.

KOREA, NORTH (People's Democratic Republic of Korea)

The extreme lack of reliable information about North Korea makes any assessment of the situation there problematic. It is fair to begin by recognizing that North Korea is perhaps the most successful modern totalitarian state: It has recovered from the devastation of the Korean War. Its economy, in terms of GNP, has grown at a modest average rate of 3 percent per year. Recently, economic growth has slowed and declined. Signs that North Korea's communist experiment may be flagging include government defaults on foreign loans in the 1970s; a realization that centralized control may not be the best approach to managing growth in increasingly technology-oriented markets; and the difficulty of maintaining a perpetual state of emergency relative to South Korea and the United States.

North Korea, for its area and population, is the most heavily armed country in the world. With a population of 22 to 23 million and a gross domestic product of over US$30 billion, North Korea maintains an armed force of more than 1,000,000 supported by a military budget in excess of US$4 billion (1989). These forces are deployed in about 50 ground combat divisions (or equivalents), 23 major combat ships, and over 800 combat aircraft and helicopters. Nearly all of this force is deployed against South Korea.

Culturally and ethnically homogeneous, North Korea is geographically diverse. It is part of a peninsular extension of northeastern Asia. Its interior is a broad continental base; it is geologically similar to adjacent areas in Manchuria. A major mountain chain extends southward from the northeast, with spurs branching to the west and south. North Korea's west coast on the Yellow Sea features estuaries and tidal flats with tides as high as 6.1 to 9.1 meters (20–30 ft.). On the Sea of Japan, the rugged east coast has tides of only 0.6 to 0.9 meters (2–3 ft.). Two rivers, the Yalu and the Tumen, form the northern border with Manchuria and Russia.

North Korea and its peninsular neighbor South Korea are of great strategic significance in East Asia. Practically separating the island nation of Japan from northern continental China, Korea has been the subject of neighboring interest and control throughout much of its history.

Power Potential Statistics

Area: 120,540 square kilometers (46,541 sq. mi.)
Population: 23,275,600
Total Active Armed Forces: 1,111,000 (4.773% of pop.)
Gross National Product: US$29.7 billion (1990 est.)
Annual Defense Expenditure: US$6–7 billion (20–25% of GNP, 1991 est.)
Iron and Steel Production:
  Crude steel: 4 million metric tons (1982)
  Pig iron: 4 million metric tons (1982)
Fuel Production:
  Coal: 37.5 million metric tons (1986)
Electrical Power Output: 33,000 million kwh (1990)
Merchant Marine: 68 vessels; 465,801 gross registered tons
Civil Air Fleet: major transport aircraft not available; 55 est. usable airfields (about 30 with permanent-surface runways; fewer than 5 with runways over 3,659 meters (12,000 ft.); 20 with runways 2,440–3,659 meters (8,000–12,000 ft.); 30 with runways 1,220–2,440 meters (4,000–8,000 ft.).

For the most recent information, the reader may refer to the following annual publications:
The Military Balance. International Institute for Strategic Studies. London: Brassey's (UK).
The Statesman's Year-Book. New York: St. Martin's Press.
The World Factbook. Central Intelligence Agency. Washington, D.C.: Brassey's (US.)

History

Most Koreans date their national origin to 2333 B.C. when the legendary King Tangun, of "divine origin," ruled over the land. Recorded Korean history, however, dates to the first century B.C. when Chinese settlements dominated northern Korea and influenced the three Han tribes, composed of many clans, in the south.

Historically, Korean culture has been heavily influenced by China. Ethnically, Koreans are a mixture of Han Chinese, Manchurians, and sundry ancient central-Asian groups. They have continually been at the mercy of much stronger neighbors and external events in general. Over the centuries the Koreans have managed to maintain some degree of independence by being exceptionally rebellious against invaders, while playing one strong neighbor off against another. In 1883 Korea signed a treaty of friendship with the United States. This was done with help from China, which was looking for ways to protect its Korean dependency from the growing encroachments of Japan.

During the last century, Japan has been the cause of most of Korea's problems. The Sino-Japanese War of 1894 was fought entirely in Korea. China lost and Korea was forced to accept Japanese influence in its internal affairs. In 1904 Japan went to war with Russia over who would control Korea and Manchuria. Russia lost and in 1910 Korea was annexed by Japan. The harsh Japanese occupation from 1910 to 1945 followed by the devastation of the Korean War (1950–53) dramatically influenced the cultural, political, and economic nature of Korea.

Korea was divided into two occupation zones after World War II, and separate governments were established in 1948. Although North Korea adopted a communist form of government based on the Soviet (Stalin) model, the ensuing years of Kim Il Sung's rule have seen a somewhat greater identification with Chinese communism. North Korea has taken a centrist diplomatic position vis-à-vis China and the former Soviet Union; it frequently played one against the other.

Employing Korea's abundant natural resources and hydroelectric power, heavy industry was developed at the expense of light industry. North Korea has failed to build a competitive industrial economy. At the same time, the collectivization of farming on a village basis was seen by the farmers as preferable to the previous landlord system, and has led to efficient and successful agricultural production. (For a more detailed treatment of Korean history, see "South Korea.")

Politico-Military Background and Policy

North Korea is run as a typical communist police state. After the Korean War, Kim Il Sung instituted the "Juche" doctrine, a combination of nationalism and xenophobia characteristic of Korean isolationism in the centuries before the Japanese occupation in 1910. Kim has also set himself up as the "father of his country" and plans to pass power to his son, thus perpetuating Korean dynastic ideals in a socialist governmental structure. Political and social control is strict, with security and military forces given favored treatment to ensure their loyalty and effectiveness. The disciplined population responds well to nationalistic appeals and to steady, if small, improvements in living conditions. The military is well cared for with most power delegated to trusted military officials, but it is carefully watched by the security services. Some coup attempts have been reported, but none apparently came close to success.

The North Korean armed forces were destroyed during the Korean War, and the country was occupied by Chinese forces for several years thereafter. Conscription was introduced in 1956; the harsh conditions of service provide little free time, home leave, or pay, but abundant food, physical training, drills, and political indoctrination. Most military equipment is of Soviet origin from the 1950s and 1960s.

North Korea has almost always followed an aggressive policy towards its southern neighbor. Infiltration by commandos into South Korea reached a peak in the late 1960s, when more than 500 entries a year were recorded. North Korea began digging tunnels under the demilitarized zone in the early 1970s; eighteen such tunnels are thought to have been built. Other aggressive acts have included assassination attempts against South Korean officials wherever they might be found.

In the mid-1980s a thawing in the relations between North and South Korea was evidenced, and subsequent exchanges have occurred on the subject of reunification, but it has been by no means a smooth period. The current government policy is the so-called "negotiate for reunification"; vague on details, it has nonetheless attracted support of many southerners, but is likely to feature both military confrontation and peaceful negotiation.

Strategic Problems

The difficulties facing North Korea are complex. They relate to many circumstances involving the dissolution of the Soviet Union, China's selection of approaches to becoming a regional power, Sino-Russian rapprochement, North Korea's own internal economic problems, and rapprochement with South Korea. A perceived regional power vacuum from the draw-down of U.S. and Soviet forces in the region and a Japanese commercial and diplomatic tilt toward South Korea are also factors for consideration.

Perhaps the most significant strategic problem facing North Korea is how to achieve reunification with South Korea. Stumbling blocks of its own creation impede progress in this direction, not the least of which is North Korea's strident militarism, recently exacerbated by development of a nuclear weapons production program. A related second-tier issue is the recently flagging economy.

North Korea's strategic significance on the international scene is lessened by South Korea's economic strength, geographic position, and diplomatic ties. Still, North Korea will play a key regional role in the future prospects of China, Japan, Russia, the United States, and the Republic of Korea.

Military Assistance

Military technology, equipment, weapons, and even doctrine and organization have come almost exclusively from

Korea, North

other communist countries, primarily the former Soviet Union and China. From 1970 to 1988, North Korea received US$1.3 billion in aid from communist states. North Korea was a major provider of military aid to Iran in the late 1980s, and sent military advisers to African nations and to Sri Lanka.

Defense Industry

North Korea possesses the heavy industrial base necessary to support manufacture of defense equipment, but it generally lacks the technology to produce sophisticated components and major end-products. Military staples and sundry equipment items are produced for overseas export and internal use. Major weapon systems and mobility platforms have been obtained primarily from the former Soviet Union and China. However, North Korea produces at least two major end-items: an armored personnel carrier, Type M-1973; and a self-propelled antiaircraft weapon system of unknown type.

Alliances

North Korea and South Korea were admitted to the United Nations in 1991. The diplomatic congeniality of its ties with the former Soviet Union and China varies with circumstance.

Defense Structure

North Korea's relatively large and powerful military force comprises army, navy, air force, and paramilitary units. The army is the largest force with approximately 1,000,000 members; the navy has about 41,000 and the air force 70,000. Paramilitary forces include security troops of 200,000, including border guards. There is a worker/peasant Red Guard of about 3.8 million. Most, but not all, equipment is becoming obsolete; some newer items of 1970s and 1980s vintage, in the form of artillery, light tanks, and combat aircraft, are in service. Control is provided through the military command structure, closely controlled by the central government, and monitored by security organizations responsible to the state.

(For an explanation of the abbreviations and symbols used in the following section of military statistics, see the list of Abbreviations and Acronyms in Volume I.)

Total Armed Forces

Active: 1,111,000. Terms of service: Army 5–8 years; Navy 5–10 years; Air Force 3–4 years, followed by compulsory part-time service in the Pacification Corps to age 40. Thereafter service in the Worker/Peasant Red Guard to age 60.

Reserves: Army 500,000, Navy 40,000. Mob claimed in 12 hours; up to 5,000,000 have some Reserve/Militia commitment. See Paramilitary.

ARMY: 1,000,000.
17 Corps (1 armd, 5 mech, 1 inf, 8 all-arms, 2 arty)
25 inf/mot inf div.
15 armd bde.
30 mot inf bde.
3 indep inf bde.

1 Special Purpose Corps: 60,000: 22 bde incl 3 cdo, 4 recce, 1 river crossing regt, 3 amph, 3 AB bn, 22 lt inf bn. 'Bureau of Reconnaissance SF' (8 bn).
Arty Corps:
  Army tps: 8 hy arty (incl MRL).
  Corps tps: 4 bde incl 122mm, 152mm SP, MRL.
Reserve: Pacification Corps: some 1.2m. 26 inf div.
Equipment:
MBT: some 3,500: 200 T-34, 1,600 T-54/-55, 1,500 T-62, 175 Type-59.
Light tanks: 600 Type-63, 50 Type-62, M-1985.
Recce: 140 BA-64.
AIFV: 200 BMP-1/BMP-2.
APC: 4,000 BTR-40/-50/-60/-152, Ch Type-531, N. Korean Type M-1973.
Towed arty: 2,500: 76mm: M-1942; 85mm: D-44/SD-44; 100mm: M-1944; 122mm: M-1931/-37, D-74, Type-54, Type-60, D-30; 130mm: M-46, Ch Type-59; 152mm: M-1937 M-1938, M-1943, Ch Type-66.
SP arty: Some 3,300: 122mm: M-1981, M-1985; 130mm: M-1975; 152mm: M-1974, M-1977; 180mm: M-1978.
MRL: 2,300: 107mm: Type-63; 122mm: BM-21, BM-11 (30 tubes); 130mm: Ch Type-63; 200mm: BMD-20; 240mm: BM-24.
Mortars: 11,000: 82mm: M-37; 120mm: M-43.
SSM: 54 FROG-3/-5/-7; some 15 Scud B.
ATGW: AT-1 Snapper, AT-3 Sagger.
RCL: 107mm: 1,000 B-11.
ATK guns: 37mm: M-1939; 57mm: M-1943; 85mm: D-48 towed; 800 SU-76 and SU-100 SP.
AD guns: 8,000: 14.5mm: ZPU-2/-4 SP; 23mm: ZU-23, ZSU-23-4 SP; 37mm: CH Type-55, M-1939; 57mm: ZSU-57-2 SP, S-60, Ch Type-59; 85mm: KS-12; 100mm: KS-19. N. Korean SP AA, type unknown.
SAM: HN-5A (SA-7 type).

NAVY: 41,000. Bases: East coast: Wonsan (HQ), Chi-aho, Songjin Toejo. West coast: Nampo (HQ), Haeju, Pipaqo. Sagwon-ri. 2 Fleet HQ.
Submarines: 22: 18 Ch Type-031/Sov Romeo with 533mm TT 4 Sov Whiskey with 533mm and 406mm TT; (plus about 40 midget submarines mainly used for SF ops, but some with 2 × TT).
Frigates: 3: 1 Soho with 4 × ASW RL, plus 4 × SS-N-2 Styx SSM, 1 × 100mm gun; 2 Najin with 2 × 5 ASW RL, 1 with 3 × 533mm TT; plus 2 × 100mm guns. 2 SS-N-2 Styx SSM.
Patrol and Coastal Combatants: 366:
Corvettes: 3 Sariwon with 1 × 100mm gun.
Missile Craft: 36: 10 Soju, 6 Sov Osa, 4 Ch Huangfeng PFM with 4 × SS-N-2 Styx; 6 Sohung, 10 Sov Komar PFM (with 2 × SS-N-2.
Torpedo Craft: 173: 3 Sov Shershen with 4 × 533mm TT; some 170 (with 2 × 533mm TT.
Patrol, Coastal: 6 Hainan PFC with 4 × ASW RL.
Patrol, Inshore: some 148: 13 SO-1, 10 Taechong, 12 Shanghai II, 3 Chodo, some 110 (.
Mine Warfare: About 20 MSI (.
Amphibious: craft only; 24 LCM, 7 LCU, about 100 LCI (.
Support and Miscellaneous: 7: 2 ocean tugs, 1 AS, 4 survey.
Coast Defense: SSM: 2 regt: Samlet in 6 sites; Guns: 122mm: M-1931/-37; 130mm: SM-4-1; 152mm: M-1937.

AIR FORCE: 70,000; 732 cbt ac, 50 armed hel.

Bombers: 3 lt regt with 80 H-5.
FGA: 10 regt: 5 with 150 J-5; 3 with 100 J-6; 1 with 40 Q-5; 1 with 20 Su-7, 36 Su-25.
Fighter: 12 regt: 2 with 80 J-5; 2 with 60 J-6; 1 with 40 J-7; 4 with 120 MiG-21; 2 with 46 MiG-23; 1 with 30 MiG-29.
Attack helicopters: 50 Mi-24.
Transport:
  Aircraft: 10 An-24, 5 Il-14, 5 Il-18, 4 Il-62M, 2 Tu-134, 4 Tu-154, 250 Y-5.
  Helicopters: 1 Hughes 300C, 80 -500D, 6 -500E, 100 Mi-2, 60 Mi-8/-17, 30 Z-5.
Training: incl 120 CJ-5, 30 CJ-6, *H-5, 50 MiG-15UTI, *MiG-19U, 10* MiG-21U, Yak-11.
AAM: AA-2 Atoll, AA-7 Apex.
SAM: 4 bde (12 bn, 40 bty) with 72 SA-2 in 45 sites, 2 regt with est. 32 SA-3, 2 regt with est. 72 SA-5.

FORCES ABROAD: Advisers in some 12 African countries

PARAMILITARY
Security Troops (Ministry of Public Security): 200,000 incl border guards.
Worker/Peasant Red Guard: some 3.8m. Org on a provincial/town/village basis. Comd structure is bde–bn–coy–pl. Small arms with some mor and AD guns (but many units unarmed).

Future

The dynamics of change in North Korea are complex; they relate closely to Chinese, Japanese, South Korean, Russian, and U.S. interests. China and the United States are in accord that war between the two Koreas is not in the best interests of the region; South Korea and Japan support this as in their own best interests. The Soviet presence in East Asia had increased in recent years; it is unclear what the future interests of the successor states of the former Soviet Union will be. North Korea would prefer that Japan become a significant trading partner; this seems to be the best political approach to acquiring desperately needed technology. North Korea is not unaware of increasing ties between the United States and China, and of the modernization approach China has taken.

Internally, North Korea is approaching a change in leadership as Kim Il Sung ages. It is not clear what the dynamics of internal forces will be when his son, Kim Chong Il, the designated successor, assumes power. Times of internal political power transfers in authoritarian states can be problematic and unpredictable in their effects.

It is clear that North Korea is nearing a point where some action is necessary to promote future growth and maintain stability. In this regard a reunified Korea under a communist government is most desirable to North Korea. Thus, a combination of actions seem likely: North Korea will expand contact with South Korea; at the same time, relations with China will be strengthened; and a policy of continued confrontation by force will be pursued. These are not inconsistent approaches for Kim Il Sung's government. Should increased internal unrest characterize South Korea, North Korea may stand to gain by exploiting it, either through use of force or politically. North Korea also realizes that any window of opportunity for aggression against South Korea may be disappearing. A pessimistic view allows the possibility of North Korean-instigated fighting for reunification on its own terms.

JAMES FRANCIS DUNNIGAN

SEE ALSO: Central and East Asia; China, People's Republic of; Japan; Japan, Modernization and Expansion of; Korean War; Korea, South; Russo-Japanese War.

Bibliography

Hunter, B., ed. 1991. *The statesman's year-book, 1991–92*. New York: St. Martin's Press.
Scalapino, R. A., and M. Kosaka, eds. 1988. *Peace, politics and economics in Asia: The challenge to cooperate*. New York: Pergamon-Brassey's.
U.S. Central Intelligence Agency. 1992. *The world factbook, 1991–92*. Washington, D.C.: Brassey's.

KOREA, SOUTH (Republic of Korea)

South Korea, with its capital of Seoul, is one of the leading new industrialized countries (NIC) in the world, and occupies an important strategic location on the southern half of the Korean peninsula. Bounded by the Yellow Sea to the west, the Sea of Japan on the east, and separated from Japan by the Tsushima Strait, this country of energetic citizens has accomplished a remarkable recovery since the devastation of the Korean War (1950–53).

With a population of over 44 million and a GNP of US$238 billion, South Korea maintains active armed forces of about 750,000 members and a military budget of over US$10 billion (1989). These forces are deployed in three armies, 27 major combat ships, and over 800 combat aircraft. In fewer than 40 years, South Korea has gone from being a fragment of a Japanese colony to one of the top five military and economic powers in Asia. The bulk of this military power is arrayed defensively against its antagonistic neighbor, North Korea.

Power Potential Statistics

Area: 98,480 square kilometers (38,023 sq. mi.)
Population: 44,338,200
Total Active Armed Forces: 750,000 (1.692% of pop.)
Gross National Product: US$238 billion (1990 est.)
Annual Defense Expenditure: US$10.4 billion (4.5% of GNP, 1991 est.)
Iron and Steel Production: not available
Fuel Production: Coal: 22.7 million metric tons (1988)
Electrical Power Output: 85,000 million kwh (1990)
Merchant Marine: 439 vessels; 7,182,519 gross registered tons
Civil Air Fleet: 93 major transport aircraft; 102 usable airfields (60 with permanent-surface runways); none with runways over 3,659 meters (12,000 ft.); 21 with runways 2,440–3,659 meters (8,000–12,000 ft.); 17 with runways 1,220–2,440 meters (4,000–8,000 ft.).

For the most recent information, the reader may refer to the following annual publications:

The Military Balance. International Institute for Strategic Studies. London: Brassey's (UK).
The Statesman's Year-Book. New York: St. Martin's Press.
The World Factbook. Central Intelligence Agency. Washington, D.C.: Brassey's (US).

History

The long history of Korea springs from a legendary past which holds that a founder of "divine origin"—Tangun, the son of a young maiden who had been transformed from a bear, and of Hwanung, son of the Creator, who breathed on her—became the first king in 2333 B.C. Another legend tells of Chinese influence introduced to Korea in 1122 B.C. by the sage Kija and 5,000 of his followers. The Kija legend rings true, although at a later time; four Chinese colonies were established in northern Korea in 108 B.C.

Three kingdoms dominated much of Korea from the first century B.C. until A.D. 660–668, when the kingdom of Silla (southeastern Korea), with Chinese support, absorbed the kingdoms of Paekche (southwestern Korea), and Koguryo (northern Korea). Silla had unified Korea under one dynasty which continued until a revolt by General Wang Kon forced the last Silla king to abdicate in 935, and Wang Kon established the kingdom of Koryo.

The kingdom of Koryo (from which Westerners derived the modern name Korea) lasted until 1392 and was dominated by Buddhism, although Confucian ideas for government administration continued to have a strong influence. While Koryo generally had good relations with China during this period, invasions by neighboring Chitan (in Manchuria) in 1011 and by the Mongols in 1231 were early indications of the strategic importance of the Korean peninsula. Chitan eventually crumbled, but subsequent Mongolian domination imposed Mongol institutions. In 1274 and again in 1279, Kublai Khan ordered invasions of Japan launched and supported from Korea.

Mongol forces, weakened by warfare with China proper, were finally defeated by a Korean general, Yi Taejo, in 1364. By 1392, General Yi overthrew the last ruler of the decadent Wang dynasty and established his own dynasty which endured until the Japanese annexation of Korea in 1910. The Yi dynasty, recognized by the Ming dynasty in China, adopted an old name for Korea: Choson.

Korean society flourished during the Yi dynasty. Confucian ethics strongly influenced the mores and practices of government officials; there was widespread study of medicine, astronomy, geology, history, and agriculture. In 1403, about 50 years before Johann Gutenberg, Koreans developed movable-type printing presses using Chinese characters.

The latter period of the Yi dynasty was unfortunately characterized by invasions—by the Japanese in 1592, and by the Manchus in 1627—and decaying governmental and political institutions. Increasing corruption and petty differences between ruling factions distanced the central government from the people; weak and venal, Korea became susceptible to the colonial interests of its neighbors, Japan and China, and of Western powers.

Korea continued its policy of isolation until the late nineteenth century. In 1876, Japan forced the Koreans to agree to diplomatic relations; six years later the United States became the first Western power to conclude a treaty with Korea. Similar treaties with other Western countries followed, and by 1886 diplomatic relations existed with every nation with an interest in East Asia.

Japanese influence predominated, and divisiveness between Korean factions led to revolts aided by Japan. The Korean revolts worsened and led to armed conflict between China and Japan in 1884, and again in 1894. Under increasing Japanese pressure, the Yi dynasty sought help from Russia in 1896. This was to no avail; Japan was too strong, and at the end of the Russo-Japanese War (1904–1905) Japan was firmly in control of Korea—formal annexation was proclaimed in 1910.

The period of Japan's annexation of Korea lasted until 1945. At first, Koreans resisted Japanese control by force of arms and, later, by passive resistance. Japan was brutal in its repression of Korean nationalism, even peaceful demonstrations for independence led to violent Japanese response. By the start of World War II, many Korean nationalists had been forced to flee, mainly to the United States, China, Manchuria, and Siberia.

World War II's end also ended Japan's domination of Korea. The Soviet Union declared war on Japan on 8 August 1945, too late to be of much assistance in the fighting, but in time to participate in the surrender of Japanese forces in Korea. U.S. and Soviet occupation forces divided Korea into occupation zones south and north of the 38th parallel.

Although the Allies had agreed prior to the war's end to establish and recognize an independent and united Korea, that agreement was never successfully implemented. A Communist-influenced Korean People's Republic was proclaimed on 6 September 1945 and was opposed by the exiled Korean provisional government (returning later that year) headed by Syngman Rhee and Kim Koo. The U.S. and Soviet militaries remained neutral, maintaining military governance until 1948. In that year efforts by the United Nations to hold national elections and form a government for a united Korea were rebuffed by Soviet-influenced North Korea, and the Republic of Korea (South Korea) was inaugurated on 15 August with Syngman Rhee as its president; on 9 September the Democratic People's Republic of Korea (North Korea) was proclaimed with Kim Il Sung as its premier. In December, the United Nations recognized the Republic of Korea as the only lawful government of Korea (neither country, however, gained UN membership until 1991, when both were admitted).

North Korean raids across the 38th parallel during 1949 were a prelude to the Korean War which began on 25

June 1950 when North Korean forces invaded South Korea. After three years of bitter combat, the Korean War ended inconclusively with an armistice signed on 27 July 1953. The battle lines at the time, somewhat north of the 38th parallel, demarked the continued separation of the two Koreas.

South Korea emerged from the war with its economy in shambles. GNP stood at only US$2.3 billion, and per capita income was only US$87. Despite Western aid and American military presence in South Korea, the decade following the Korean War was marked by internal political turmoil. Although Syngman Rhee was elected to a fourth term in 1960, divisiveness between political factions, the National Assembly, and Syngman, combined with charges of corruption and election rigging and student-led rioting, forced Syngman into exile and the establishment of a caretaker government.

The second Korean republic was formed on 15 June 1960, but succumbed to a military junta on 16 May the following year; martial law was imposed until terminated in December 1962. General Pak emerged as president and eventually instituted political reforms which resulted in national elections in the fall of 1963, with Pak winning a narrow margin of victory.

Also in 1962, the Korean government initiated the first of its Five-Year Economic Development Plans, which continue to be a feature today. These plans provided the framework for South Korea's economic recovery in the 1960s, and rapid growth and industrialization through the 1970s. By the mid 1980s South Korea's economy ranked third in the Pacific Basin behind Japan and Australia, and at an average growth rate of 10 percent was the fastest growing.

Politico-Military Background and Policy

Up until the 1910 Japanese occupation, Korea was a monarchy. Administration was in the Chinese Mandarin style, with a military and civil service chosen by examination. The civil service exam was the more rigorous, leading the civil administrators to look down upon the military. The Mandarins (Yangban) were prone to periods of corruption, which sometimes led to military insurrections. The Japanese occupation eliminated the Korean military, and forced on civil administrators the difficult choice of unemployment or collaboration. When U.S. forces ejected the Japanese in 1945, the new government was formed around a few surviving resistance fighters and many Korean administrators who were tainted by working for the Japanese. This government survived the Korean War, but by the early 1960s was forcibly replaced by the first generation of American-trained military officers. These officers had a tendency toward corruption as practiced by previous, civilian rulers. Despite an anticorruption movement, starting in the late 1970s, popular discontent brought about the first truly democratic government in the late 1980s. The government is based on the U.S. model, but in practice follows the traditional Korean model of strict hierarchy and rigorous respect for authority. In a repetition of an ancient Korean tradition, the civil administrators will be forced to run an efficient and noncorrupt government or risk the return of military control.

South Korea's major ally and trading partner is the United States. Because of the terms of the cease-fire that ended the Korean War fighting in 1953, the war is technically not over and all forces south of the DMZ (Demilitarized Zone) are under the command of the United Nations. As a practical matter, the United States dominates. If North Korea were to invade again, the United States would intervene on South Korea's behalf. The likelihood of other UN members doing so is less certain; this would depend on the circumstances.

South Korea's major diplomatic objective is the reunification of the country. The primary military objective is to prevent North Korea from succeeding at reunification by force. Diplomatic efforts have been frustrated by North Korean insistence on reunification as a communist state. Diplomacy is made more difficult by North Korean persistence in using terrorism to destabilize South Korea.

The focus of South Korean military strategy is the 250-kilometer (155-mi.) DMZ that separates the two Koreas roughly along the 38th parallel. The DMZ is 4 kilometers (2.5 mi.) wide and heavily fortified on both sides. Most of its length runs through mountainous terrain. One sector, 40 kilometers (24.8 mi.) north of Seoul, is most crucial to South Korea's security. This is the relatively flat, narrow corridor leading to Seoul that allows mechanized forces some room to maneuver. Since Seoul is the largest urban area in South Korea, and the nation's capital, nearly half of South Korea's military forces are directed toward the defense of this sector. Thus any future war would likely see this sector as a major battleground. North Korean advances across any other part of the DMZ would be slowed by the terrain, giving the defenders time to bring up reserves.

It is probable that neither of the Koreas is strong enough to invade the other successfully, and both are restrained by their major power mentors.

Strategic Problems

The major strategic issue facing South Korea is reunifying with North Korea, and accomplishing this without war or accepting North Korean terms of a monolithic communist-style central government. A variety of complex and related matters illustrate the context of this strategic matter: South Korea must continue its economic progress, ensure a stable government and help quell potential unrest; the likelihood of a reduction in U.S. forces based in South Korea must be considered; some accommodation with North Korea, and perhaps with China, must be foreseen and its effects determined; the future role of Russia in the region will figure in strategic calculations; and the role of

Japan, the dominant economic power in East Asia, will have an increasing impact on South Korean thought and actions.

Military Assistance

The United States has been the predominant supplier of military equipment to South Korea, although in recent years South Korea has increasingly been able to meet and exceed its own requirements. United States military personnel aid in South Korea's defenses; about 40,000 U.S. service members are stationed in the country. In terms of monetary assistance, the United States has supplied US$3.9 billion, including Export-Import Bank commitments, over the period 1970–85.

Defense Industry

Modern defense-related industrial production became a feature in Korea during the latter part of Japanese occupation when, in the late 1930s, Japan stepped up military activity against China and moved substantial war materials production facilities to the mainland. Although the military-industrial infrastructure survived World War II largely intact, it did not survive the ravages of the Korean War.

The rebuilding of South Korea's economy has been closely linked to its defense industry. Since the early 1970s, South Korea has produced an increasing portion of its own weapons, including artillery, armored personnel carriers, heavy tanks (based on the U.S. M-1), and warships. The major effects of South Korea's deliberately developed defense industry have been an established, diversified, and balanced manufacturing sector capable of sustaining defense requirements. With significant advances into high technology, the defense industry is helping to improve the structural development of other heavy commercial industries.

Alliances and Defense Structure

Korea developed a strong military tradition out of necessity. Surrounded by stronger nations, given to violent internal disputes, and quick to rebel at any real or imagined oppression, there was always work available for organized troops. A professional officer corps was traditionally part of the government, thus providing competent combat leadership. Koreans have always been brave, resolute, and resourceful soldiers. Only when the officer corps became corrupt, as it did from time to time, did the armed forces become ineffective. Korean armed forces were dissolved during the period of Japanese annexation, and Koreans were taken into Japan's military structure as noncombat troops. In 1942, Japan introduced conscription and some Koreans were allowed to become officers. In fact, many of the "Japanese" soldiers who surrendered to U.S. troops in 1945 were actually Koreans.

During the first year of the Korean War, the South Korean Army had mixed success which mirrored the spotty quality of its officers. Although U.S. advisers had set up an officer training school shortly after World War II, only a few thousand junior officers had been trained by 1950. The field grade and senior officers were nearly all former Japanese Army officers or hastily commissioned civilians. The three years of fighting was the crucible that produced a new generation of competent officers and noncommissioned officers (NCOs). These troops confirmed their capabilities when two South Korean divisions served in Vietnam in the 1960s. Over 300,000 troops rotated through these divisions and achieved recognition for their courage and aggressiveness.

More recently, a new generation of officers and troops are entering service who have no knowledge of the Korean and Vietnam wars. The biggest variable is whether or not the South Korean armed forces can maintain a high level of officer competence. Getting the military out of the government may be the key factor in prompting the military to concentrate on maintaining the quality of its leadership.

South Korea's armed forces comprise army, navy, and air force units, with the bulk of the 750,000-member military in the army. The army is organized into three army headquarters (HQs) with a total of eight corps that control two mechanized infantry divisions, nineteen infantry divisions, various independent brigades and battalions, and substantial combat support and combat service support units. In addition there are about 4.5 million reservists under an army HQ with 23 infantry divisions.

South Korea's primary military alliance is with the United States, which has been the case since the Korean War. Without weakening, the alliance has shown signs of evolution; the unified command structure for defense is scheduled to pass for the first time from U.S. commanders to South Korean generals in 1992.

(For an explanation of the abbreviations and symbols used in the following section of military statistics, see the list of Abbreviations and Acronyms in each volume.)

Total Armed Forces

Active: 750,000. Terms of service: conscription, all services, 30–36 months, then First Combat Forces (Mobilization Reserve Forces) or Regional Combat Forces (Homeland Defense Forces) to age 35.
Reserves: 4,500,000; being reorganized.

ARMY: 650,000.
HQ: 3 Army, 8 Corps.
  2 mech inf div (each 3 bde: 3 mech inf, 3 tk, 1 recce bn; 1 fd arty bde).
  19 inf div (each 3 inf regt, 1 recce, 1 tk, 1 engr bn; 1 arty regt (4 bn)).
  2 indep inf bde.
  7 SF bde.
  2 SSM bn with Honest John.
  2 AD arty bde.

2 SAM bde; 3 HAWK bn (24 sites), 2 Nike Hercules bn (10 sites).
1 avn. bde.
Reserves: 1 Army HQ, 23 inf div.
Equipment:
MBT: 1,550: 250 Type 88, 350 M-47, 950 M-48A5.
AIFV: some 530 (KIFV).
APC: some 1,550 incl 450 M-113, 400 Fiat 6614/KM-900/-901.
Towed arty: some 4,000: 105mm; M-101, KH-178; 155mm: M-53, M-114, KH-179; 203mm: M-115.
SP arty: 155mm: 100 M-109A2; 175mm: M-107; 203mm: M-110
MRL: 130mm: 140 Kooryong (36-tube).
Mortars: 5,300: 81mm: KM-29; 107mm: M-30.
SSM: 12 Honest John.
ATGW: TOW.
RCL: 57mm, 75mm, 90mm: M67; 106mm: M40A2.
ATK guns: 76mm: 8 M-18; 90mm: 50 M-36 SP.
AD guns: 600: 20mm: incl 60 M-167 Vulcan; 35mm: 20 GDF-003; 40mm: 80 L60/70, M-1.
SAM: 100 Javelin, some Redeye, 130 Stinger, 110 HAWK, 200 Nike Hercules.
Aircraft: 10 O-1A.
Helicopters:
Attack: 48 AH-1F/-J, 50 Hughes 500 MD.
Transport: 14 CH-47D.
Utility: 144 Hughes 500, 3 KH-4, 30 UH-1B, 70 UH-1H.

NAVY: 600,000 (19,000 conscripts) incl 25,000 marines.
Bases: Chinhae (HQ), Cheju, Inchon, Mokpo, Mukho, Pukpyong, Pohang, Pusan. 3 Fleet Commands.
Submarines: 4: 3KSS-1 Tolgorae SSI (175t) with 2 × 406mm TT; 1 Ge T-209/1400 with 8 × 533 TT.
Principal Surface Combatants: 35:
Destroyers: 9:
7 Chung Buk (US Gearing) with 2 or 3 × 2 127mm guns; plus 2 × 3 ASTT; 5 with 2 × 4 Harpoon SSM, 1 Alouette III hel (OTHT), 2 with 1 × 8 ASROC.
2 Dae Gu (US Sumner) with 3 × 2 127mm guns; plus 2 × 3 ASTT, 1 Alouette III hel
Frigates: 26:
7 Ulsan with 2 × 3 ASTT (Mk 46 LWT); plus 2 × 4 Harpoon SSM.
19 Po Hang with 2 × 3 ASTT; some with 2 × 1 MM-38 Exocet.
Patrol and Coastal Combatants: 83:
Corvettes: 4 Dong Hae (ASW) with 2 × 3 ASTT.
Missile Craft: 11:
8 Pae Ku-52, 3 with 4 Standard (boxed) SSM, 5 with 2 × 2 Harpoon SSM.
1 Pae Ku-51 (US Asheville), with 2 × Standard SSM.
2 Kilurki-71 (Wildcat) with 2 × MM-38 Exocet SSM.
Patrol, Inshore: 68: 32 Kilurki-11 ('Sea Dolphin') 33-m PFI; 36 Chebi-51 ('Sea Hawk') 26-m PFI ((some with 2 × MM-38 Exocet SSM).
Mine Warfare: 9: 1 'Swallow' (mod It Lerici) MHC; 8 Kun San (US MSC-268/289) MSC.
Amphibious: 14: 7 Un Bong (US LST-511) LST, capacity 200 tps, 16 tk; 7 Ko Mun (US LSM-1) LSM, capacity 50 tps, 4 tk; plus about 36 craft; 6 LCT, 10 LCM, about 20 LCVP.
Support and Miscellaneous: 11: 3 spt tankers, 2 ocean tugs, 2 salv/div spt, about 4 survey (civil manned, Ministry of Transport funded).

Naval Air: 24 cbt ac; 35 armed hel.
ASW: 2 sqn:
1 ac with 24 S-2E;
1 hel with 25 Hughes 500MD (ASW);
10 flt with 10 SA-316 hel (ASW), 2 Bell 206 (liaison). Lynx being delivered; to replace SA-316.

Marines: (25,000): 2 div, 1 bde; spt units.
Equipment:
MBT: 40 M-47
APC: 60 LVTP-7.
Towed arty: 105mm, 155mm.
SSM: Harpoon (truck-mounted).

AIR FORCE: 40,000; 405 cbt ac plus 52 in store, no armed hel. 7 cbt, 2 tpt wings.
FGA: 8 sqn: 2 with 48 F-16 (36 -C, 12 -D); 6 with 144 F-5 (44-A, 100 -E) plus 16 in store.
Fighter: 4 sqn with 96 F-4D/E. Plus 36 in store.
COIN: 1 sqn with 23 A-37B, 6 T-28D.
FAC: 20 O-1, 10 O-2A, 25 OA-37B.
Recce: 1 sqn with 18 RF-4C, 10 RF-5A.
SAR: 1 hel sqn with 15 Bell UH-1B, 2 UH-1N.
Transport: 2 wings, 5 sqn:
Aircraft: 2 BAe 748 (VIP), 1 Boeing 737 (VIP), 9 C-54, 1 C-118, 10 C-123J/K, 10 C-130H, 3 Commander.
Helicopters: 7 Bell 212, 3 -412, 5 UH-1D, 5 -H.
Training: 25* F-5B, 35* -F, 25 T-33A, 14 T-37, 20 T-41D.
Missiles:
ASM: AGM-65A Maverick.
AAM: AIM-7 Sparrow, AIM-9 Sidewinder.

PARAMILITARY
Civilian Defense Corps (to age 50): 3,500,000.
Coast Guard: (est. 3,500)
Patrol Craft, Offshore: 13: 1 Mazinger (HDP-1000) (CG flagship); 6 Han Kang (HDC 1150); 6 Sea Dragon/Whale (HDP-600)
Patrol Craft, Coastal: 24: 22 Sea Wolf/Shark; 2 Bukhansan.
Patrol Craft, Inshore: 38: 18 Seagull; 20 (, plus numerous boats.
Helicopters: 9 Hughes 500 D.

FOREIGN FORCES
U.S.: 41,800. Army (31,500): 1 army HQ, 1 inf div, 1 SSM bty with Lance. Air Force (10,300): 1 div: 2 wings: 90 cbt ac.

Future

Continued economic growth for South Korea is likely, but the economy and defense industry are at a new stage of development. Growth rates declined from the early 1970s to the mid-1980s when GNP then rose above 10 percent until slowing to 6.5 percent in 1989. Growth will probably continue above 5 percent in the future, but noninflationary trends are subject to be reversed by continuing worker demands for wage hikes. It is also apparent that the defense industry, which experienced 18 percent growth during 1972–76, and 13 percent during 1977–82, has slowed substantially because of production facilities that currently meet defense requirements while operating at something less than 50 percent capacity. Because of the integrated nature of civil and defense industries, shifts to different products may be achieved, but increases in foreign trade, owing to a rise in regional protectionism, may be problematic.

Diplomatic relations between South Korea and the United States and with Japan came under some strain in 1990 because of disagreements over trade barriers, and such issues will remain a matter in the future. At the same time, President Roh Tae Woo has pursued new international relationships with notable success; diplomatic relations now exist with Czechoslovakia, Bulgaria, Romania, and Mongolia. South Korea also pursued new trade and diplomatic interests with the Soviet Union, with the result of diplomatic recognition and an influx of aid, trade, and investment from South Korea of US$1 billion in 1990–1991, a relationship that will continue with the new Commonwealth of Independent States. Similar efforts with China have succeeded; despite Chinese austerity measures, trade with China surpassed US$3 billion for the second successive year in 1990. Diplomatic recognition by China is less sure; however, it remains a goal of President Roh.

South Korea's leaders have always displayed greater pragmatism than ideology, and, regarding the important issue of reunification, this feature may facilitate compromises with North Korea. A Korean commonwealth is a possible future, with arrangements that would facilitate commerce between north and south. In fact, barter between the two Koreas began in the late 1980s and continues as an unofficial aspect of the relationship between the two nations.

Officially, however, many difficulties remain that separate North and South Korea. The issue of nuclear weapons development in the north is a key aspect of regional security and stability, and has prompted the United States to unilaterally withdraw its nuclear weapons from South Korea, thus removing the excuse North Korea has used to defend its emerging nuclear production capabilities. Moreover, the recent diplomatic talks between both countries has resulted in agreement in principle on some strategic matters. A renunciation of fundamental ideological differences, however, seems unlikely.

Perhaps equally unlikely is that of overt military action between North and South Korea. South Korea has greatly expanded its international relations—North Korea has not. Economic pressures will predominate in future courses of action, and in this regard South Korea has the greater leverage and will maintain this position of strength in negotiations with the north.

JAMES F. DUNNIGAN

SEE ALSO: China; Japan; Japan, Modernization and Expansion of; Korea, North; Korean War; Russo-Japanese War.

Bibliography

Hyde, G. D. 1988. *South Korea: Education, culture and economy*. New York: St. Martin's Press.
Hunter, B., ed. 1991. *The statesman's year-book, 1991–1992*. New York: St. Martin's Press.
Kihl, Y. H. 1984. *Politics and policies in divided Korea*. Boulder, Colo.: Westview Press.
Kim, W. K., and W. J. Rho. 1982. *Korean public democracy*. Seoul: Kyobo.
Koh, B. C. 1984. *The foreign policy systems of North and South Korea*. Berkeley, Calif.: Univ. of California Press.
Kuznets, P. W. 1977. *Economic growth and structure in the Republic of Korea*. Cambridge, Mass.: Yale Univ. Press.
U.S. Central Intelligence Agency. 1991. *The world factbook, 1991–92*. Washington, D.C.: Brassey's.

KOREAN WAR (1950–53)

When Japan surrendered in World War II, a hurried Allied agreement (15 August 1945) established the 38th degree of latitude as an arbitrary dividing line, north of which the USSR would accept surrender of Japanese forces in Korea; Japanese south of the line would surrender to U.S. troops. Following the surrender, the USSR held the 38th parallel to be a political boundary, and dropped the Iron Curtain along it.

After two years of unsuccessful attempts to reach agreement on the unification of Korea, the United States referred the problem to the United Nations (UN). The UN proposed to establish an independent Korean government following free nationwide elections, but the USSR refused to cooperate. Consequently, the Republic of Korea was established in the southern zone (15 August 1947), with Seoul its capital. Declaring the action illegal, the USSR set up the Democratic People's Republic of Korea and organized a North Korean Army (NKA). Soviet troops allegedly evacuated the north in December 1948. U.S. troops completed evacuation of the south in June 1949; a small American military advisory group remained to reorganize a Republic of Korea (ROK) Army. In the following year, a barrage of Communist propaganda, raids, sabotage, terrorism, and guerrilla action was directed against the south, without breaking down the ROK government.

The Opposing Forces

Communist North Korea had a well-trained, Soviet-equipped army with a core composed of 25,000 veterans of the Chinese Communist campaign in Manchuria. The air force consisted of some 180 Soviet planes of World War II type. There were more than 100,000 trained reserves.

The ROK Army—little more than a national police force—consisted of about 100,000 men in eight divisions with little supporting artillery. It lacked medium and heavy artillery, tanks, combat aircraft, and reserves. Naval strength on both sides was negligible.

Operations

1950

On 25 June, North Korean forces crossed the border, driving on Seoul. Surprise was complete; the invaders broke through scattered resistance by elements of the

ROK Army. Their objective was to seize the capital and occupy South Korea, thus presenting the free world with a fait accompli.

In reaction, the UN Security Council met in an emergency session (the USSR boycotted the Council) and called for an immediate end to hostilities and withdrawal of the NKA, asking member nations to assist. Meanwhile, President Truman (27 June) ordered Gen. Douglas MacArthur, commanding U.S. forces in the Far East, to support the ROK defense with air and sea forces. MacArthur effected a naval blockade of the North Korean coast and furnished air support. Reconnoitering the front in person (28 June), as Seoul fell, he reported the ROK Army to be incapable of stopping the invasion even with U.S. air support. Truman authorized the use of U.S. ground troops on 30 June.

Aside from the vessels of the U.S. Seventh Fleet and the Far East Air Force (8½ combat groups), U.S. ground forces in Japan consisted of four understrength divisions organized in two skeleton army corps. Infantry and artillery units were at two-thirds strength in personnel and cannon and short of antitank weapons. Corps troops, such as medium tanks, artillery, and other supporting arms, did not exist. On 30 June, the 24th Division (under Maj. Gen. William F. Dean) began movement piecemeal by sea and air into Korea.

One understrength battalion (two infantry companies) with one battery of artillery joined the ROK Army near Osan on 4 July. The next morning an NKA division with 30 tanks attacked. The ROK troops fled, but the battalion, completely surrounded, held out for seven hours. Then, ammunition exhausted, the survivors cut their way out, abandoning all materiel.

Between 6 and 21 July, the 24th Division partly snubbed the NKA advance down the peninsula, trading terrain for time while the 1st Cavalry and 25th Infantry divisions were being rushed from Japan. A five-day action at Taejon (16–20 July) ended when the NKA assaulted the 24th Division from three directions. Dean, personally commanding his rear guard while the remainder of his division withdrew, was captured. On 22 July his battered troops were relieved by the 1st Cavalry Division, while the 25th Division, on its right, together with reorganized ROK divisions slowed the NKA advance.

In response to a Security Council request to establish a unified command under a U.S. officer, President Truman named MacArthur Commander in Chief, UN Command, on 7 July. Lt. Gen. Walton H. Walker, commanding the U.S. Eighth Army, stabilized his defense on a thinly held line that ran north for 145 kilometers (90 mi.) along the Naktong River from Tsushima Strait to Taegu, and then east for 97 kilometers (60 mi.) to the Sea of Japan. The area embraced the southeast corner of the Korean peninsula, including Pusan, the one available port. In the north, five ROK divisions attempted to contain the invaders. The western flank, where the weight of the incessant NKA attacks fell, was held by U.S. troops, now including two additional infantry regiments and a Marine brigade. The Seventh Fleet protected both sea flanks and harassed NKA movements along the coast, while the Far East Air Force (augmented by an Australian group), together with carrier-based naval air, hammered at NKA lines of communication and furnished much-needed close support. Walker was able to shift a mobile reserve from point to point within the perimeter as the NKA attacks nibbled at his front. Several penetrations of the Naktong River line and a 132-kilometer (20-mi.) NKA advance in the north (August 26) were checked. NKA forces estimated at fourteen infantry divisions supported by several tank regiments, continued a series of uncoordinated assaults all around the perimeter. A three-division attack in the north on 3 September necessitated committing the entire UN reserve (the 24th Infantry Division) north of Kyongju. Arrival of the U.K. 27th Infantry Brigade (14 September) compensated for the withdrawal of the Marine brigade, which was joining the 1st Marine Division for the Inchon landing.

MacArthur's plan for regaining the initiative in Korea and driving Communist forces from the south hinged on an amphibious landing at the port of Inchon, near Seoul. A successful assault there would cut Communist supply lines and compromise their entire military position in South Korea, at the same time avoiding a slow and costly overland advance from the Pusan perimeter.

At dawn on 15 September, the U.S. X Corps (Maj. Gen. Edward M. Almond commanding) began landing over the difficult and treacherous beaches at Inchon. The 1st Marine Division swept through slight opposition, securing Kimpo airport on 17 September. The 7th Infantry Division, following the Marines ashore, turned south, cutting the railroad and highway supplying the NKA in the south, and Seoul was surrounded.

Simultaneously, the Eighth Army broke out of the Pusan Perimeter, the 1st Cavalry Division leading. The NKA, its supplies cut off, and menaced from front and rear, disintegrated. The 1st Cavalry and 7th Infantry divisions met as Seoul was liberated on 26 September.

On 1 October, as directed by the United Nations and President Truman, MacArthur ordered ROK troops north across the 38th parallel. The Eighth Army followed on 9 October, leaving two divisions in the southern area to secure communication lines to Pusan and mop up remnants of the NKA. Pyongyang, the North Korean capital, was overrun (20 October) by a combined airborne landing and overland advance. But a serious military handicap was the injunction prohibiting UN aircraft from flying north of the Yalu River, in order not to provoke the Communist Chinese. The ROK 6th Division reached the Yalu River at Chosan and other ROK units fanned out behind it. By this time, other UN token forces had joined the Eighth Army and were integrated in existing U.S. divisional elements: a

Turkish brigade, and Canadian, Australian, Philippine, Dutch, and Thai battalions.

Reembarked at Inchon on 16 October, the X corps was moved around to Wonsan on the east coast, which had already been captured by the ROK I Corps. A seven-day delay in landing was necessary to sweep the harbor clear of Soviet mines, which had been sown under the direction of Soviet experts with the NKA.

Peking had threatened intervention should UN troops cross the 38th parallel, and heavy concentrations of Chinese Communist troops were reported just north of the Yalu. MacArthur knew neither the full strength nor the dispositions of these troops, nor was he aware that Chinese in considerable numbers had already infiltrated south of the Yalu. On 1 November, the forward ROK divisions were ambushed by these Chinese divisions and a U.S. regiment at Unsan was severely mauled. Walker recalled his leading Eighth Army units and consolidated temporarily along the Chongchon River.

MacArthur's intention was still to advance up the entire front of the peninsula, the X Corps on the east coast, the Eighth Army on the west. The X Corps would turn west at the Yalu, making a sweeping envelopment, and drive all enemy forces south of the border against the Eighth Army. Since the rugged, desolate central massif precluded mutual support, the UN formations acted independently, their control and coordination directed by MacArthur in Tokyo. Almond's X Corps now consisted of the U.S. 1st Marine and 3d and 7th Army infantry divisions and the ROK I Corps (3d and Capital Divisions). The Eighth Army, nine divisions strong, was grouped in three army corps: from left to right the U.S. I and IX and the ROK II. The total combat strength of the entire command was about 200,000 men with perhaps 150,000 more in support functions in the rear. In addition to the ROK corps, some 21,000 more Korean troops were attached to or integrated in U.S. units.

MacArthur had believed that Communist China would not enter the conflict unless Manchuria itself were invaded. Truman, who had met with MacArthur at Wake Island on 15 October, knew this, and the U.S. Central Intelligence Agency held the same opinion. Yet now Chinese troops were in Korea. Since reconnaissance north of the Yalu was prohibited, MacArthur decided the only remaining course was to clarify the situation by a bold advance. Meanwhile, the X Corps had thrust north, widely distributed over an immense front. The ROK Capital Division had reached Chongjin on the coast and the U.S. 7th Division was on the Yalu at Hyesanjin.

MacArthur's "reconnaissance in force" with the Eighth Army began on 24 November. After advancing for 24 hours against practically no opposition, the Eighth Army was suddenly struck a massive blow on 25–26 November, the main effort being directed against its right flank. Some 180,000 Chinese troops, in 18 divisions, shattered and ripped through the ROK II Corps, hit the U.S. 2d Division on the right flank of the IX Corps, and threatened envelopment of the entire army. The 2d Division, attempting to reconstitute its right flank, fell into an ambush at Kunu-ri as the Chinese envelopment trapped its columns while passing through a defile in march order. Some 4,000 men and most of the divisional artillery were lost while trying to fight their way out. Walker threw in his reserves, the U.S. 1st Cavalry Division and the Turkish and 27th Commonwealth brigades. They staved off the envelopment, the Turks in particular taking heavy losses, and the Eighth Army managed to disengage. By 5 December, the Eighth Army, its right flank restored, had completely extricated itself, and the Chinese Communist drive was beginning to lose momentum (Fig. 1). The central and east-coast areas were wide open and a stronger defensive position was essential. Walker accordingly withdrew to the general line of the 38th parallel, slightly north of Seoul and some 130 miles below the 24 November situation. There, as the year ended, the Eighth Army awaited a new Communist offensive.

In the eastern zone, an additional 120,000 Chinese troops, advancing on both sides of the Chosin reservoir, isolated the 1st Marine Division and drove in elements of the 3d and 7th divisions, starting on 27 November. The ROK troops on the coastal flank were hurriedly withdrawn without much molestation. MacArthur then ordered evacuation of the entire force, since the Communist drive, directed on the ports of Hungnam and Wonsan, threatened its piecemeal destruction. The 1st Marine Division, under Maj. Gen. Oliver Smith, consolidated south of the Chosin reservoir. Surrounded by eight Chinese divisions, General Smith, announcing to his division that they were not retreating, but "attacking in another direction," moved southeast on Hungnam, supplied by the Far East Air Force. When the Communists destroyed the one bridge across a gorge otherwise impassable for the division's trucks and tanks, bridging material was flown in by air and the Marine southward "advance" continued. Thirteen days of running fight ended on 9 December when a relief column of 3d Division troops met the Marine vanguard outside the Hungnam perimeter.

Figure 1. Troops of the 24th Infantry Regiment leaving the Kunu-ri area in December 1950 for the Chongchon River bridge to confront the Chinese. (SOURCE: R. Dudley/77th ECC)

Despite continued Communist attacks on both ports, evacuation by air and by sea went smoothly from 5 to 15 December. U.S. Air Force and Navy carrier-plane support, together with naval gunfire, facilitated the final embarkation.

Walker, killed in a jeep accident on 23 December, was replaced by Lt. Gen. Matthew B. Ridgway, who arrived on 26 December. MacArthur gave Ridgway command of all ground operations in Korea, retaining overall ground, air, and sea command.

EARLY 1951

The long-expected Communist assault crossed the 38th parallel at daybreak on 1 January 1951, its main effort in the western zone. Some 400,000 Chinese troops, with an additional 100,000 of the reconstituted NKA, pushed the 200,000-man Eighth Army back almost to Seoul. Then, on 3 January, a heavy penetration in the Chungpyong reservoir area, overran the ROK divisions on both flanks of the U.S. 2d Division, which extricated itself only after serious fighting and the commitment of Ridgway's reserve—the 3d and 7th divisions.

Seoul was evacuated on 4 January, the third time the capital had changed hands. Stubborn resistance of ground troops, plus the Far East Air Force's close support and interdiction of the now-exposed Communist lines of communications, slowly checked the momentum of the drive. The UN position had stabilized by 15 January, some 80 kilometers (50 mi.) south of the 38th parallel, from Pyongtaek on the west coast to Samchok on the east.

Ridgway launched a series of limited-objective attacks, slowly driving north from 25 January to 10 February. A Communist counterattack near Chipyong and Wonju (11–18 February) checked the advance in the center, but on the west UN troops reached the outskirts of Seoul.

Operation Ripper, designed primarily to inflict casualties on the enemy, and secondarily to relieve Seoul and eliminate a large Communist supply base built up at Chunchon, opened on 7 March. The main effort, in the center, forced the Communists back. The Han River was crossed east of Seoul. Patrols of the I Corps found Seoul abandoned on 14 March. Chinese resistance then stiffened, but an airborne drop by a reinforced regiment at Munsan on 23 March forced a general Communist retirement. By 31 March, the Eighth Army was back roughly along the old 38th parallel front. MacArthur and Ridgway decided on further advance, toward the "Iron Triangle"—Chorwan-Kumhwa-Pyonggang, the major assembly and supply area and the communications center for the Chinese.

President Truman relieved General MacArthur of his dual command of UN and U.S. forces in the Far East on 11 April 1951. The president was exercising his prerogative as Commander in Chief. MacArthur was not in sympathy with the policy of limiting the war to the Korean peninsula and had not attempted to conceal his dissatisfaction with the restrictions placed on his operations. Ridgway was appointed in his place and Lt. Gen. James A. Van Fleet was hurried from the United States to command the Eighth Army.

APRIL–DECEMBER 1951

Aware of Communist preparations for a counteroffensive to blunt the threat to the Iron Triangle, General Van Fleet continued forward movement from 12 to 21 April, prepared to fall back, if necessary, to previously prepared defensive positions. There he planned to contain the enemy by heavy firepower and then counterattack.

The first phase of the expected Communist spring offensive began on 22 April; the first assault broke through the ROK 6th Division west of the Chungpyong reservoir. The U.S. 24th Division on the left and the 1st Marine Division on the right promptly restored their respective flanks, but the penetration compromised Van Fleet's general position and he began withdrawal of his left—the I and IX Corps. The Chinese main effort developed against the I Corps, north of Seoul. Hasty withdrawal by the ROK 1st Division exposed the flank of the U.K. 27th Brigade on its right, and a battalion of the Gloucestershire Regiment was cut off. Some 40 men escaped; the remainder were killed or captured. The Communist assault finally lost momentum and came to a pause by 30 April. The Chinese broke contact, retiring beyond UN artillery range. Communist losses in this phase were at least 70,000 men, while Eighth Army casualties were about 7,000.

The second phase of the Communist counteroffensive shifted the weight of their attack to the east. More than 20 Chinese divisions, with NKA divisions on their right and left, struck the right elements of X Corps—the ROK 5th and 7th Divisions. The U.S. 2d Division, next on the left, stood firm, but the ROK III Corps, farther east, went to pieces under heavy assault. The ROK I Corps, on the extreme right, restored its flank against the Communist surge through this wide corridor. The U.S. 2d Division (with French and Dutch battalions attached) and the 1st Marine Division, on the west side, promptly counterattacked. Van Fleet had expected the blow in this area and had already shifted his reserves. Their efforts snubbed the Communist offensive on 20 May. Attacks on the west flank, north of Seoul, and in the center, down the Pukhon River, had been repulsed.

The UN counteroffensive began on 22 May. It was preceded by limited attacks on the far left, anchoring the UN position on the Imjim River north of Munsan. The entire UN front moved north. The ROK Capital and 2d Divisions, on the extreme right, advanced up the east coast with little opposition, reaching Kansong. Advance was slower in the center, but was accelerating by month's end. But the U.S. government ordered Van Fleet to halt. Despite his plea for approval of "hot pursuit" against an enemy on the verge of collapse, the Joint Chiefs of Staff (JCS) refused either increased means or permission for another drive northward. The U.S. government, concerned by Soviet threats, had decided not to risk World War III.

With the advent of the rainy season in early June, Van Fleet decided to establish a defensive belt across Korea, from which he could keep the enemy off balance by a succession of JCS-approved limited-objective moves. Some gains were made at the base of the Iron Triangle and on the southern rim of the "Punchbowl," a fortified circle of hills northwest of Sohwa. Meanwhile, the Communists were organizing in depth to the north.

Soviet ambassador Malik made a cease-fire proposal in the UN on 23 June. It confirmed that the Chinese had been badly hurt in the previous six months' fighting. Estimated enemy losses totaled 200,000 men together with much materiel. UN air attacks also had foiled every attempt to install Communist air bases south of the Yalu. Delegations from both sides met at Kaesong. Negotiations opened in July, and the Communists took advantage of the location of Kaesong (just inside their lines), seizing every opportunity to delay progress wile playing for time to recuperate from their mauling. Meanwhile, clashes between patrols and outposts continued all along the firing line as both sides sought to improve their positions. Negotiations broke down completely in late August.

UN forces resumed limited attacks. Van Fleet's troops cleared the Iron Triangle and the Punchbowl, driving the Chinese back from the Hwachon reservoir and the Chorwan-Seoul railway line. These successes brought prompt Communist requests for resumption of armistice discussions.

Discussions resumed on 12 November at Panmunjon, a village between the lines. General Ridgway ordered offensive operations stopped as talks began.

1952–53

While negotiations dragged out interminably at Panmunjon, minor actions flared continually all along the front. General Ridgway, ordered to replace General Eisenhower in command of NATO forces in Europe, was replaced by Gen. Mark W. Clark in May. The Communists continued building up their strength. By year's end an estimated 800,000 Communist ground troops, three-fourths of them Chinese, were in Korea, while heavy shipments of Soviet artillery were brought in.

Communist wranglings at the Panmunjon conferences centered on the disposition of prisoners of war. About 92,000 UN troops had fallen into Communist hands: 80,000 Koreans, 10,000 Americans, and 2,500 from other UN forces. Consensus of reports by returned POWs indicate that about two-thirds of U.S. prisoners died or were killed in the prison camps. No neutral or Red Cross inspections were ever allowed by the Communists. Some 171,000 Communist prisoners had fallen into UN hands—more than 20,000 of them Chinese. About 80,000 were assembled on the island of Koje, just off Pusan.

Meanwhile, along the front, a succession of attacks and counterattacks cost both sides great losses. In October, the negotiations at Panmunjon again broke off, while the war became a political football in the U.S. presidential election. The American people, tired of the struggle, elected Dwight D. Eisenhower, who had promised to bring about an honorable conclusion to the war.

Unexpectedly, but apparently in tune with internal unrest in the Communist world following Stalin's death (5 March), Premier Kim Il Sung of North Korea and Chinese general P'eng Teh-huai agreed to General Clark's previously ignored proposal for mutual exchange of sick and wounded POWs. They also urged resumption of the Panmunjon conferences. Extensive U.S. plans to renew offensive operations and possibly to extend the war were unquestionably major factors in the Communist change of heart. Operation Little Switch, an exchange of 5,800 Communists for 471 ROKs, 149 Americans, and 64 other UN personnel, took place in April.

In May South Korea's president Syngman Rhee flatly refused to become a party to any agreement that left Korea divided. Massive Communist attacks, mostly against ROK troops, began on 10 June to bring U.S. pressure on Syngman Rhee. On 18 June, Rhee demanded that the UN resume the military offensive. At the same time, he released from his own prison camps 27,000 North Korean POWs unwilling to be repatriated. When Rhee released the prisoners, the Communists, accusing the UN of bad faith, again broke off negotiations and on 25 June launched still another offensive against the ROK sector. Some slight gains were made, but quick shifts of U.S. reinforcements and the inability of the Chinese to exploit their penetrations brought the attack to a halt, with loss of some 70,000 Chinese troops.

Negotiations were resumed on 10 July, and a final armistice was hammered out on 27 July 1953.

The Costs

The Korean War cost the UN 118,515 men killed and 264,591 wounded; 92,987 were captured. (A great many of these died of mistreatment or starvation.) The Communist armies suffered at least 1.6 million battle casualties, 60 percent of them Chinese. An additional estimated 400,000 Communists were nonbattle casualties. U.S. casualties in addition to prisoners were 33,629 killed and 103,284 wounded. Of 10,218 Americans who fell into Communist hands, only 3,746 returned; the remainder (except 21 men who refused repatriation) either were murdered or died. In all, 357 UN soldiers (mainly Korean) refused repatriation. South Korea's toll—which can only be estimated—came to 70,000 killed, 150,000 wounded, and 80,000 captured. Approximately 3 million South Korean civilians died from causes directly attributable to the war.

Fourteen UN member nations besides the United

States took part in this, the first war in which the UN had engaged. Britain and Turkey each contributed a brigade (each equivalent to half of a division). In addition, the United Kingdom furnished one aircraft carrier, two cruisers, and eight destroyers, with Royal Marine and supporting units. Canada sent one brigade of infantry, one artillery group, and one armored battalion; Australia, two infantry battalions, one air fighter squadron and one transport squadron, one aircraft carrier, two destroyers, and one frigate; Thailand, one regimental combat team; France, one infantry battalion and one gunboat; Greece, one infantry battalion and one air transport squadron; New Zealand, one artillery group and two frigates; Netherlands, one infantry battalion and one destroyer; Columbia, one infantry battalion and one frigate; Belgium and Ethiopia, one infantry battalion each; Luxembourg, one infantry company; and the Union of South Africa, one fighter squadron. All these elements bore themselves well and sustained substantial casualties. In addition, from Denmark, India, Italy, Norway, and Sweden came hospital or field-ambulance noncombat units.

The conflict reaffirmed the critical importance of airpower as an essential ingredient of successful combat; it was also a reminder that airpower alone can neither ensure adequate ground reconnaissance nor bring about the final decision in land warfare. The immediate superiority achieved by the UN in the air necessitated bringing in Soviet MiG-15s—then the latest USSR jet fighters, quite superior to America's F-84 and surpassing in some respects the F-86. MiGs, first seen in Korea in late 1950, and presumably all with Russian pilots, increased in number during 1951, but the training and competence of UN pilots—mostly American—compensated for any inferiority in materiel. While UN pilots were never permitted to pursue the MiGs across the Yalu, they were able to neutralize all Communist efforts to establish airbases south of the river. In air-to-air combat, 1,108 Communist planes were destroyed, including 838 MiG 15s; another 177 were probably destroyed, and an additional 1,027 were severely damaged, against a total UN loss of 114 aircraft. As the war drew to a close, U.S. F-86 jets were downing MiG-15s at the rate of 13 confirmed Communist losses to each F-86 shot down. UN plane losses to Communist antiaircraft fire, while giving magnificent close support to ground troops, were 1,213.

The potential of the helicopter as a new means of mobile transportation was clearly demonstrated. It was excellent for reconnaissance, evacuation of wounded, and rescue work.

American command of the sea was one of the principal handicaps to Communist success. The U.S. Seventh Fleet gave valuable gunfire support along the coast and carried out amphibious operations, while U.S. Navy and Marine air units participated in air force interdiction and close support to ground units. The navy's blockade of the peninsula prevented supply of Communist forces by water. Had it been possible to interdict ground-supply channels from Manchuria and Siberia in similar fashion, the war would have been over in short order.

This war was significant on several counts. It was the first major struggle of the nuclear age. While no nuclear weapons were employed, the threat of the atom bomb hung heavy over all concerned and throttled exploitation of success. It was a war between two differing ideologies, a war of stratagem and deceit in which roadbound superior firepower was canceled out by lighter-armed fluidity over desolate, trackless wastes.

TREVOR N. DUPUY

SEE ALSO: China, People's Republic of; Cold War; History, Modern Military; Korea, North; Korea, South; MacArthur, Douglas A.; United Nations.

Bibliography

Appleman, R. 1961. *South to the Naktong, north to the Yalu.* Washington, D.C.: Government Printing Office.
Blair, C. 1987. *The forgotten war: America in Korea, 1950–1953.* New York: Times Books/Random House.
Cagle, M. W., and F. A. Manson. 1957. *The sea war in Korea.* Annapolis, Md.: U.S. Naval Institute Press.
Cumings, B. 1981. *The origins of the Korean War.* Princeton, N.J.: Princeton Univ. Press.
Futrell, J. A. 1961. *The United States Air Force in Korea 1950–1953.* New York: Duell, Sloan and Pearce.
Gugeler, R. A. 1954. *Combat actions in Korea.* Washington, D.C.: Combat Forces Press.
Hermes, W. G. 1966. *Truce tent and fighting front.* Washington, D.C.: Government Printing Office.
James, D. C. 1985. *The years of MacArthur.* Vol. 3 of *Triumph and disaster, 1945–1964.* Boston: Houghton-Mifflin.
Marshall, S. L. A. 1953. *The river and the gauntlet.* New York: Morrow.
Ridgway, M. B. 1967. *The Korean war.* New York: Doubleday.

L

LABOR, MILITARY

A separate labor service or equivalent organization may provide military, civilian, or mixed resources on a single or joint armed service basis to meet peacetime and wartime labor requirements. In war, prisoners of war may also be used for appropriate tasks, provided enough guards are available.

In a theater of operations, very large quantities of equipment and supplies—some fragile, others hazardous—will have to be effectively handled at points of entry, in transit areas, and throughout the lines of communication leading to combat areas. Even if optimum use is made of the mechanical handling equipment positioned in every appropriate location, a substantial labor force will still be required.

If security, local resources, and other circumstances permit, a considerable number of civilians will be recruited, especially for work in rear and base areas; they may require special screening as well as military direction and supervision. If local civilians are available, then combatant military labor will be concentrated for employment in forward areas; organized and allocated as sections, platoons, or companies.

Centralized pooling, control, and administration of labor will be normally exercised at all levels to ensure maximum availability, flexibility, utilization, and economy of resources, and to maintain proper reserves. Demand for labor will be submitted to headquarters, where the staff will allot resources by priorities. Every effort will be made, however, to keep the same labor force on a task that needs continuity of skill and experience.

When a military labor force is maintained in peacetime, apart from its normal tasks and training, its troops may be used for guard duties.

J. H. Skinner

See Also: Lines of Communication; Logistics: A General Survey; Prisoner of War.

Bibliography

Bykofsky, J., and H. Larson. 1957. The transportation corps: Operations overseas. In *U.S. Army in World War II*. Washington, D.C.: Office of the Chief of Military History, Department of the Army.

LANCHESTER EQUATIONS

The Lanchester equations are a mathematical model of combat, developed by F. W. Lanchester in 1914 to show the relationship between the strength of opposing forces and the results of combat. Lanchester equations constitute a system of two differential equations that describe the numerical changes in two opposed forces during a single battle as a function of time.

The results obtained from the solution of these equations relate to fundamental matters; namely the outcome of the combat (winner and loser), the losses sustained by winner and by loser, and the duration of the combat.

The Lanchester equations deal with uninterrupted engagements and are therefore more suitable for the description of a battle than for the description of a war.

Lanchester studied both linear and quadratic forms of his equations, from which he derived a result that he considered very important; namely, justification of the value ascribed to the concentration of forces in modern combat (mass effect).

Lanchester was keenly aware of the extremely simplified nature of his equations compared with the actual conditions of the development of combat. He believed, nevertheless, that such models (which are analogous to the models used in macroscopic physics) could furnish accurate indications of the appropriate structure of forces and the relations between the capacity of forces and the winning or losing of the battle. His major objectives, therefore, related to systems of forces and the conditions of the engagement of forces, rather than to the creation of an exact model of the process of mutual attrition.

These models remained practically unused until World War II, during which they were used by a few operational researchers, such as B. O. Koopman in the United States (1943) and P. M. S. Blackett in England.

These equations, which estimate the losses of opposing forces, are comparable to corresponding equations for chemical reactions, the growth or decrease of competing biological species, and have been applied to social or political phenomena such as studies of the armament race. Thus, analogous models had been the object of numerous chemical, biological, sociological, and mathematical studies during the period between the world wars.

Since World War II the Lanchester equations have been used, often in a form like the original model, as a base for calculating losses (casualties, attrition) in many models of combat, as if their validity in almost all situations, as representations of dynamic process in the real world, had been proven.

Applying those equations to reliable historical numerical data has shown what modifications in them are necessary to make their solutions consistent with the results obtained in the laboratory of history, while at the same time preserving a simple mathematical formula. Some models, which result from this procedure and from judiciously combining a great many fundamental parameters of combat actions, can be described as phenomenological; they can be compared with the models used in other sciences, for example, in macroeconomics.

Another avenue of research has explicitly introduced a number of elements that are not simply a function of the number of men involved and of weapons efficiency—the only essential factors in the original models of attrition. This second line of research has led to the elaboration of Lanchester-type models that are mathematically more complicated, but potentially far more accurate representations of the actual nature of combat, than the original Lanchester equations. Today, the bibliography of derivative studies is enormous, even if only unclassified publications are included.

The Russian school of Operational Research uses the Lanchester systems extensively, using both ordinary differential equations and partial differential equations. This interest in portraying combat by means of differential systems is not surprising considering the vitality of the Russian school of mathematics in these fields.

Especially in the West, this vast output of effort sometimes concentrates on aspects that are not necessarily the most important for understanding the phenomena of combat. It may ignore or downplay factors of the greatest importance—such as the effects of suppression by fire, of shock in the assault or counterattack, of psychological trauma (particularly at the beginning of a conflict), and of tactical maneuver—even when the combat consists of elementary actions in limited space and time. An attempt to establish priorities and identify the most urgent and fundamental topics for research would be needed to improve methods of prediction for future weapon systems and the most effective force structures.

Lanchester

Frederick William Lanchester (1868–1946) did pioneer work on aerodynamics, aeronautics, and automobiles. As his curiosity was universal, we are in his debt for works on economics, management, and fiscal policy as well. He was awarded numerous honors and titles in his lifetime, but his full merit was recognized much later.

Lanchester's article entitled *The Principle of Concentration* was published in the British periodical *Engineering* (2 October 1914). It was the fifth in a series entitled *Aircraft in Warfare: The Dawn of the Fourth Arm*. In 1916 his analysis was resumed and expanded in chapters 5, 6, and 8 of his book entitled *Aircraft in Warfare* (Lanchester 1916). The 1914 article was reprinted in 1956 in *The World of Mathematics* (pp. 2138–57), a publication that recounts the complete development of the N-square laws.

Lanchester Equations and Laws

The equations and laws are well known. They are given below to demonstrate the restrictive hypotheses they postulate and thus to clarify the significance of their results and the limits of their applications.

EQUATION 1—QUADRATIC (N-SQUARE) LAW

Assumptions:

1. Two forces attack each other. Every element (troops and weapons) of each side is within effective range of all the elements of the opposing side.
2. All the elements of the two sides are the same (homogeneous forces), but the performances of the elements may differ from one side to the other; for example, the effectiveness (casualty rate) of the respective weapons systems may differ.
3. Every surviving element has, at all times, information sufficient to enable it to shift its fire immediately from an enemy element that is knocked out to a surviving enemy element.
4. Fire is distributed evenly at all of the surviving enemy elements.

In contemporary notation (different from Lanchester's, which used r for red and b for blue), opposing sides are indicated respectively by the numerals 1 and 2.

t = time ($t \geq 0$)
x_1, x_2 = number of survivors in 1 and 2 at $t > 0$
x_{10}, x_{20} = number of initial units at $t = 0$
k_{12} = average rate at which a unit in 1 destroys a unit in 2
k_{21} = average rate at which a unit in 2 destroys a unit in 1
P_{12} = probability that a projectile fired by a unit in 1 will destroy a unit in 2
r_{12} = average rapidity of fire of a unit of 1 as it fires on the units of 2

P_{21} and r_{21} are symmetrical definitions.

Consequently:

$$K_{12} = r_{12} \times P_{12} \text{ and } K_{21} = r_{21} \times P_{21}$$

This is the Lanchester system:

$$\begin{cases} \dot{x}_1 = dx_1/dt = -K_{21}x_2 \\ \dot{x}_2 = dx_2/dt = -K_{12}x_1 \end{cases}$$

The solution of this system, after eliminating time, is:

$$K_{21}x_2^2 - K_{12}x_1^2 = K_{21}x_{20}^2 - K_{12}x_{12}^2$$

which shows that the difference between the number of survivors on each side conforms to an algebraic function of the second degree proportional to weapons performance symbolized by K_{12} and K_{21} and also proportional to the square of the number of troops initially engaged, hence the term, *square law*.

EQUATION 2—LINEAR LAW

Assumptions:
1. Same as for quadratic.
2. Same as for quadratic.
3. Every unit while firing knows the general area in which the enemy units are deployed, *but does not know the location of individual enemy units; it fires on the whole area of enemy deployment without knowing the result of its fire.*
4. The fire of surviving units is distributed uniformly over the area in which the enemy is deployed.

The symbols for the respective areas are:
A_1 and A_2 = area of engagement for each side
A_{e1} and A_{e2} = area of average effectiveness of each round fired
r_{12} and r_{21} = rapidity of fire by units

Their results:

$$K_{12} = r_{12} \frac{A_{e1}}{A_2} \text{ and } K_{21} = r_{21} \frac{A_{e2}}{A_1}$$

The Lanchester system is, therefore:

$$\begin{cases} \dot{x}_1 = dx_1/dt = -K_{21}x_2x_1 \\ \dot{x}_2 = dx_2/dt = -K_{12}x_2x_1 \end{cases}$$

By eliminating time, the solution of this system is:

$$K_{21}x_2 - K_{12}x_1 = K_{21}x_{20} - K_{12}x_{10}$$

The function expressing the relation between casualties and survivors is linear and is determined by the starting conditions in which the two sides were not aware of the effectiveness of their fire and so were not able to concentrate fire continually on survivors only, hence the term *linear law*.

The quadratic law describes a situation in which each side does have information sufficient to enable it to concentrate fire continually on survivors only. This difference in degree of effectiveness leads to very different conclusions on the value of large units and on the relative casualty rates of the opposing sides.

Historical Background

FISKE'S MODEL

Several authors described in quantitative terms the advantage of concentration of forces before Lanchester. Three of them, J. V. Chase, A. Baudry, and Bradley A. Fiske, are theorists of naval tactics. The best-known is Commander (later Vice Admiral) Fiske, who in 1905 won the Naval Institute prize for his essay "American Naval Policy." In his analysis, he described the cumulative effects of a superior concentration of fire, noting the duration of the intervals between salvos. Fiske concluded that the rate of reduction of the offensive power of the weaker fleet is geometric and not arithmetic, and that the difference between the powers of the opposing fleets increases steadily throughout the combat in favor of the stronger. The logical development of Fiske's procedure contains the elements necessary for deriving the square law.

To Lanchester, however, belongs the credit of giving to a similar intellectual procedure the elegant and easily manipulated form of a simple system of differential equations. It is possible to show clearly how one can advance from Fiske's equations of finite differences to Lanchester's differential system (Engel 1954). It is not known to what extent, if any, Lanchester knew the work of his predecessors, that of Fiske in particular. In general, the great merit of Fiske consists in his having perceived the power of mathematical procedures to contribute to the solution of problems of military operations far in advance of his time and 36 years before the existence of the term *operations research*.

OSIPOV

M. Osipov performed remarkable work almost at the same time as Lanchester. This work remained completely ignored until quite recently by the Western world; in Russian works on operational research, it has merely been cited as historical background.

M. Osipov published, in the Russian journal *Voenniy Sbornik* (Military Collection) between June 1915 and October 1915, a series of five articles with the general title *The Influence of the Numerical Strength of Engaged Sides on Their Losses*. According to the English translator, Robert L. Helmbold (1985), Osipov's work showed deep historical and mathematical thought, attained apparently without knowledge of the work of his Western predecessors and contemporaries such as Fiske and Lanchester. Among the original features of Osipov's work are: the use of exponential and hyperbolic functions to express the reduction of opposing forces as a function of time; the introduction of heterogeneous forces in each side; the intervention of the value of breaking off and abandoning combat well before the values of annihilation are reached; and the application of the most advanced statistical methods of his time for quantifying the historical data. Unfortunately, almost nothing more is known in the West about other Osipov studies (if any), or about the work of Osipov's successors on the theory of combat subsequent to 1915.

L. F. RICHARDSON

Until the renewal of interest in Lanchester's equations about the middle of World War II, there were no important military applications of those equations. However, many works in social science were performed using analogous equations.

Between 1919 and 1953 (the date of his death) Lewis F. Richardson, distinguished English mathematical physicist, did important work on the application of mathematics to international relations, with emphasis on the causes of wars and ways of avoiding them. Among the books on these subjects that Richardson published, two continue to be authoritative: *Arms and Insecurity*, 1947 (2d ed., 1949) and *Statistics of Deadly Quarrels*, 1950.

Richardson's mathematical models (especially those dealing with the armament race) are systems of ordinary differential equations based on a small number of hypotheses entirely analogous to the systems of Volterra for the rivalry between species, and to the Lanchesterian systems for armed competition during combat.

Richardson's simplest system for the armament race is based on only three hypotheses, which are here stated to illustrate the intellectual procedure followed:
1. Each nation tries to increase its armament proportionally to the size of the armament of the other;
2. The national economy is a constraint that tends to limit the rate of armament by an amount proportional to the size of existing forces.
3. A nation will manufacture arms as a result of ambition, resentment, and feelings of hostility, even if the other nation does not directly threaten it.

The three hypotheses lead to a system of two equations, as follows:

$$\begin{cases} dN_1/dt = kN_2 - aN_1 + g \\ dN_2/dt = lN_1 - bN_2 + h \end{cases}$$

in which:
$N_1(t)$ and $N_2(t)$ are the levels of armament of the two nations
$k, a, g, l, b,$ and h = constants
k and l = coefficients of defense or positive reaction (positive constants)
a and b = coefficients of resistance to cost or fatigue (positive constants)
g and h = coefficients of complaints when they are positive and of good will when they are negative

Using the same set of concepts, other authors have followed the work of Lewis Richardson, always from the viewpoint of preparing for war and general strategy. Among these others are W. R. Caspary, Intriligator, T. L. Saaty, and N. Rashevsky.

STUDY OF LANCHESTERIAN SYSTEMS, 1943–50 (WESTERN WORLD)

The rise of interest in Lanchesterian systems took place during World War II. Among the major works are those of the mathematician B. O. Koopman done in the Operations Evaluation Group (OEG) of the U.S. Navy in 1943 and P. M. Morse and G. E. Kimball (1951). Koopman expanded Lanchester's equations by introducing two new factors: (1) chance in combat (random factors); and (2) the rate of industrial production in war.

Introduction of (2) results in the following equations:

$$\begin{cases} dN_1/dt = P - aN_2 - cN_1 \\ dN_2/dt = Q - bN_1 - dN_2 \end{cases}$$

$N_1(t), N_2(t)$ = the respective combatants' forces at the instant t
$P(t), Q(t)$ = the rates of production
a and b = the rates of attrition caused by combat
c and d = the rates of internal wear and tear (losses from accidents, breakdown, wearing out, etc.)

Koopman noted that this system is analogous not only to the systems used in biology to represent the rivalry between species, but also to the systems of chemical kinetics. Among the notable works produced immediately after World War II are those of P. M. S. Blackett in England (1948) and in America those of Helmer, P. M. Morse and G. E. Kimball (1951), T. Oberbeck, N. Rashevsky; Snow, C. Tompkins, and J. H. Engel (1954).

LANCHESTER-TYPE MODELS IN THE SOVIET ARMY

At least since the 1960s, Soviet forces made extensive use of models based on Lanchester-type systems of equations for the conception of systems of forces, for their operational use, and for predicting the results of operations. The landmarks of this growing understanding are found in the work of Chuyev and Mikhaylov entitled *Foresight in Military Affairs*, (1975, English translation 1980).

The Soviet authors cite the historical case of Lanchester's quadratic equation, while noting that Osipov had independently made this discovery. They also give, albeit briefly, other examples that are especially interesting for their involvement with probability theory, and for the study of deployment in space and time using systems of equations containing partial derivatives.

The above-cited work has a bibliography of 75 titles in Russian, including English works translated into Russian. Three of these titles deserve especial attention: P. N. Tkachenko and others, *Mathematical Models of Combat Operations*, Moscow 1969; Y. V. Chuyev, *Operational Research in Military Affairs*, Moscow 1970; and Y. S. Ventsel, *Operational Research*, Moscow 1972.

Indeed, all the cited examples of Lanchester's models refer to those three works, besides noting that Chuyev is co-author of the work on predicting and of an earlier book on cruise missiles.

Verification of the Laws Deduced from Lanchester's Equations

The best way to test the validity of a model is to see if its results seem reasonable and consistent with experience. In many cases the model is adjusted by changing parameters and relationships until it represents the outcome of a particular engagement "reasonably" well.

Such a model can be used, however, only in the set of circumstances for which it has been adjusted. It is, therefore, a trustworthy tool for this limited purpose, but not for extrapolating results and predicting events not yet experienced.

Obviously, it is impossible in peacetime to reproduce all aspects of combat in realistic situations. One of the most fruitful procedures for validating methodology is the comparison with reality that is provided by military history: its interactions during combat; the reactions of human participants; and the effects of coordinated action or the opposite effects of disorganization that can occur within systems of forces. The real objective, however, is something quite different, namely, to obtain an honest adjustment that is robust enough to describe the results of a great many battles. Only consistency of results in a wide variety of situations makes trustworthy predictions about future battles possible.

Many efforts have been made to correlate Lanchester's laws with historical data. Although these efforts sometimes yield valuable lessons about "local," or particular, conditions, they have generally proved to be misleading if regarded as generally, or "universally," valid. Lanchester himself defined the limits of his model, which initially dealt with situations in which contesting forces are of equal quality, affected equally by the conditions of battle, and everything not specified is equal, including morale, tactical skill, and so forth. Clearly such a highly hypothetical situation cannot exist in real life. On the other hand, the hope of reconciling historical facts and theoretical results becomes plausible if the condition of opposing forces is normalized and if the differences in such things as arms systems, environment, and operational circumstances are taken into account; that is, if a system of all-inclusive reference is defined. This has been attempted by Col. T. N. Dupuy after more than fifteen years of original historical analysis. The fruit of this work, the Quantified Judgment Model (QJM), a modified version of Lanchester's equations, results in solutions that give far better representation of quantified historical data and theoretical values. These equations, in their author's notation, are:

$$\begin{cases} dD/dt = - C'(CEV_a)^2 S_a \\ dA/dt = - C''(CEV_a)^2 S_d \end{cases}$$

S_a and S_d = the strength of the respective camps, including factors not only for number, but also for quality and position

CEV_a = combat effectiveness value (the "sum" of human factors, duly quantified and weighted in relative value)

C' and C'' = constants for environment and operations (quantified for every particular battle)

A and D = attacker's and defender's forces

Lanchester's equations do not result in average rates of attrition consistent with historical sources unless the factors incorporated in the model have been adjusted with many factors that affect the outcome of battles. These adjustments are impossible to make for a heterogeneous sample of numerous battles by utilizing only "raw" and simplistic data—as would be done, for example, by considering only the number of survivors that should result from the nominal effectiveness of a given weapons system. This practice, used to evaluate or predict the results of combat with new weapons systems, might be extremely dangerous, giving results that could depart from reality by being many times too large or too small.

Lanchester's model with "normalized" equations is a highly aggregated model, comparable with the models of macroeconomics, whose value depends on the validity of parameters judiciously selected or skillfully adjusted to a large sample of quantified observations, reconstituted for missing data.

Expansion of Lanchester's Original System

Lanchester's original system of two differential equations has proved to be inadequate as an investigative device for solving all problems of military operational research because a great many significant operational factors were omitted in a remarkable effort of abstraction and simplification.

There are at least three ways to give Lanchester's system its maximum practical effectiveness. The first was described in the preceding section on verification. It consists of the elaboration of a modified Lanchesterian system, revealing the general applicability of a highly aggregated system whose normalized parameters are calculated by the QJM using quantified historical data. A second way starts with Lanchester's system. It then progressively adds to that system factors that affect the process of attrition, as well as other fundamental factors that are not determined solely by losses suffered or inflicted, such as rates of advance or retreat, positions gained or lost, and the results of engagements (victory, defeat, draw).

Today the tendency is to create increasingly complex models, either to represent combat of small units with precision, taking into account environment, men's conduct, or elementary maneuvers; or to represent battles of large units, taking into account intelligence, maneuver, and logistics.

In the general effort to understand better the phenomenon of combat, mathematical systems have been developed that approximate ever more closely the complexity of reality.

EXTENSIONS IN LANCHESTER-TYPE SYSTEMS

Heterogeneous forces. These forces vary greatly in quality and belong in sundry categories, especially because of the diversity of weapons systems.

It is necessary to take into account heterogeneous forces and their evolution versus time in order to understand and to measure the values of different components and the possibilities of exchanges and synergies between those components. For example, it is easy to enumerate a dozen or more "protagonists" on each side, even on a comparatively elementary level (e.g., like that of a battalion or regiment on attack or on defense) by counting vehicles of aerial combat (helicopters and ground-support airplanes). It is important to take such factors into account, particularly because the inter-arms factor is more and more obvious even at the smallest level of tactical units.

In general with respect to combat vehicles, mobility—of the vehicles themselves and of units—must be taken into account; fundamental parameters (such as probability of detection, classification, identification, and destruction) depend on speed, sometimes on acceleration, and almost always on the distance between the protagonists.

Combat can end simply because of the exhaustion of the munitions of one side. Limitations due to logistics must be introduced explicitly.

At a fairly high level of forces (in land warfare) it is valuable to introduce the distribution of forces in space and their movements, varying simultaneously with densities and rates of exchanges. This procedure leads to partial differential equations instead of ordinary differential equations (density is a function of space and time).

The mutual support of the various weapons systems on the same side also leads to interactions and adaptations.

These examples show the closely knit nature ("multi-loops") of the interactions between both "protagonists" on the same side and those on opposing sides. They show, too, the nonlinear nature of the equations that describe the unfolding of battle.

Introduction of additional factors. For the most part, classic Lanchesterian systems regard only the effects of complete destruction or removal from combat. Obviously it is reasonable to introduce the other factors that help to determine the outcome of battles, which are equally, if not more, important: the effects of neutralizing (suppression fire), of shock, of capture, and of disorganization, even of the reuse of captured weapons against the original owners (especially in subversive warfare).

Wear and fatigue. Factors of wear of material and fatigue of personnel should be introduced if the combat goes on long enough. The fatigue factor normally works slowly, but can be accelerated suddenly by sundry causes, including delayed reaction to accumulated stress (see above), shock, or psychological trauma.

Reinforcements and relieving forces. Reinforcement is the natural antidote to wear and fatigue. The effect of reinforcement is expressed mathematically by positive time functions in the second member of differential equations. Thus these systems give results very different from the results given by systems with only the decreasing functions, which represent nothing but losses.

Command, control, communications and intelligence. Lanchester-type analytical models can only analyze average trends. The models do not allow for innovative command and control or superior intelligence of enemy intentions, and generally assume a perfect communication system. This causes deterministic models to provide average outcomes that do not necessarily agree with the results of a particular historical battle.

The battle's end. Battles do not usually end with the complete annihilation of the losing side. History shows that this almost never happens. One side gives up before it is completely destroyed, either for material reasons (especially lack of munitions or fuel), because of disorganization, or for psychological reasons. The idea of giving up the fight may occur because of the loss of life, horror at destruction, or the feeling of the uselessness of further fighting.

Consequently, it is necessary to introduce systems factors that represent the breakpoints: the points at which fighting ends; or the thresholds of the end of combat, which generally are not the same as the points where complete destruction takes place. The conditions of victory and of repulse cannot be described solely in terms of purely material factors such as numbers of troops and efficiency of weapons.

Analyses show that the outcome of battle is highly dependent upon the choice of breakpoints for the respective combatants. Definition of rules for the end of combat in complex models is difficult. It is, however, an essential part of the set of rules governing decision making during the successive phases of a battle.

The stochastic aspect (introduction of random factors and processes). Lanchesterian systems base calculations on functions and parameters that represent the average values of stochastic processes during their development and change versus time. For the true representation of an engagement, the distribution of values of each random variable must be examined.

Stochastic models provide more comprehensive information than the corresponding deterministic models, not only about the average value of the factors studied but also about the dispersion of results or the successive states of the system studied.

For every deterministic Lanchesterian model, a corresponding stochastic model can be developed using the same fundamental hypotheses about the mechanisms that govern the evolution of combat. Unfortunately, the average values of the solutions of the stochastic model do not necessarily coincide with the solutions of the corresponding differential system. Taylor has shown, however, that in important cases the difference can be negligible, provided that the number of units on each side is large enough (more than twenty, for example), that the forces were of comparable size at the outset, and that intelligence is not a decisive factor.

In particular, the solutions of the stochastic system differ more and more obviously from those of determinist Lanchesterian systems as the numbers of units decrease, which occurs when the values of annihilation are approached (Koopman 1943).

The richer and more satisfying stochastic systems are closer than the Lanchesterian to the stochastic nature of combat.

When a large number of units is involved (say more than several dozen), however, the use of a stochastic model may exact a heavy price in volume and time of calculations and in experimental values of parameters. It may be impractical even with today's best computers. Then it is necessary to revert to Lanchesterian or other more direct models for approximations.

There is also a hybrid method that introduces random distributions into the parameters of the differential system, transforming it into a stochastic Lanchesterian system that gives average values of the dynamic process and their statistical dispersion. This produces fuller information on the development and outcome of combat, information whose cost in analytical complexity, although greater than that of deterministic Lanchesterian systems, is still much less than that of a completely stochastic model, with values for all the probabilities of transition during changes of the system of opposite forces.

The best choice of model type (deterministic, stochastic, or hybrid) depends on the number of units, the degree of precision sought, and the capacity for calculations. In many practical cases the deterministic model yields valuable results, at least for first approximations.

Finally, no matter how brightly elaborated they may be, neither Lanchesterian, nor stochastic, nor hybrid models can fully account for tactical decisions made in the course of combat. For example, decision to break off a well-prepared attack because of losses sustained may not at first seem an indication of wise and skillful command. Such a decision, however, may have resulted from complicated and little-known considerations. It is therefore possible to question the utility of introducing such decision making into Lanchesterian systems.

In war games the principal factor of decision making is introduced, although kept out of analytical equations. Still, nonquantifiable factors cannot be weighted in the same way as more easily quantified factors. So the investigation is reduced to examining small units as if they were automatons governed by artificial information and by such material factors as are used by Lanchesterian systems.

Models Available in the Early 1990s

Since the original development early in the twentieth century, there has been a dramatic increase in study and work on the Lanchester model. In 1943 the work of B. O. Koopman in the United States and of P. M. S. Blackett in England formed a basis. By 1951 a dozen papers and books on the application of Lanchester equations systems had been written. By 1964 such works numbered 74, and by 1983 there were hundreds of titles (at least 250 of basic importance).

The growth of this literature shows a constantly increasing interest in the Lanchesterian way of constructing models of combat. This is due to two factors: (1) the need for better understanding and better prediction of the unfolding of conflicts on the strategic as well as on the tactical level; and (2) the great number and variety of complicated phenomena with which the Lanchesterian systems can deal.

Use and Limitations of Lanchesterian Systems

A Lanchesterian model is a mathematical approximation of a combat situation. When the situation changes significantly in any respect (such as troops, materials, operational information, commanders' decisions, or morale), the description (the model) must change accordingly. A Lanchesterian model, even an elaborate one, is not self-correcting. A single, static model is incapable of describing from beginning to end the unfolding of sequential combat situations. This fact is especially important in the case of prolonged and complicated conflicts. Accordingly, a battle must be divided into chronological segments, for each of which a revised Lanchesterian model must be devised. This procedure (which may be called "heuristic") is guided by operational experience, historical analysis, logical deduction, and all elements of military knowledge.

Lanchester models are powerful tools for thought and research in the study of particular aspects of combat, and for calculating, when they are inserted into a coherent chain of models many of which are not Lanchesterian models.

MAURICE BRESSON

SEE ALSO: Arms Control and Disarmament: Mathematical Models; Attrition: Personnel Casualties; Models, Simulations, and War Games; Operations Research, Military; Osipov Equations.

Bibliography

Blackett, P. M. S. 1948. Operational research. *Quarterly Journal of the British Association for the Advancement of Science* 5 (April):26–38.

Chuyev, Y. V., and Y. B. Mikhaylov. 1975. *Forecasting in military affairs: A Soviet view* (in Russian). English translation published under the auspices of the United States Air Force, 1980. Washington, D.C.: Government Printing Office.

Dupuy, T. N. 1979. *Numbers, predictions and war: Using history to evaluate combat factors and predict the outcome of battles.* Indianapolis and New York: Bobbs-Merrill.

Engel, J. H. 1954. A verification of Lanchester's law. *Operations Research* 2:163–71.

Helmbold, R. L. 1985. The influence of the numerical strength of engaged sides on their losses. Published in 1915 by M. Osipov (in Russian). Trans. R. L. Helmbold. Unpublished document.

Huber, R. K., L. F. Jones, and E. Reine, eds. 1975. *Military strategy and tactics. Computer modeling of land war problems*. New York and London: Plenum Press.

Koopman, B. O. 1943. *Quantitative aspect of combat*. Office of Scientific Research and Development, Applied Mathematical Panel, Note 6, AMG—Columbia University.

Lanchester, F. W. 1916. *Aircraft in warfare: The dawn of the fourth arm*. London: Constable. Excerpted in vol. 4 of *The world of mathematics*, ed. F. R. Newman, pp. 2138–57. New York: Simon and Schuster.

Morse, P. M., and G. E. Kimball. 1951. *Methods of operations research*. Cambridge: Massachusetts Institute of Technology Press.

Richardson, L. F. 1947. *Arms and insecurity*. Pittsburgh: Boxwood.

———. 1950. *Statistics of deadly quarrels*. Chicago: Quadrangle Books.

Taylor, J. G. 1980. *Force-on-force attrition modelling*. Alexandria, Va.: Military Applications Section of ORSA.

———. 1983. *Lanchester models of warfare*. Vols. 1 and 2. Alexandria, Va.: Military Applications Section of ORSA.

Volterra, V. 1931. *Lecons sur la théorie mathématique de la lutte pour la vie*. Paris: Gauthier-Villars. New printing (1990), Paris: Jacques Gabay.

Weiss, H. K. 1957. Lanchester-type models of warfare. *Proceedings of First International Conference on Operational Research*, Oxford, September, pp. 82–98. Baltimore, Md.: Operations Research Society of America.

LAND FORCES, EFFECTIVENESS AND EFFICIENCY OF

Factors influencing the effectiveness and efficiency of land forces (and criteria for their measurement) are of lasting, vital importance with regard to many problems in defense, such as

- selection and operation of weapon systems,
- allocation of resources in the design and operation of military units within a given or planned force structure,
- planning of force structure,
- assessments of adversarial force structures to derive correct and appropriate evaluations of threats and the means to cope with them in war,
- design and assessment of arms control proposals, and
- the operational planning for and control of military forces in peace and war.

For each of these problem areas, it is sometimes assumed that decision makers hold in their heads an implicit model of how the numerous, interdependent, and stochastic factors influence effectiveness and efficiency of land forces.

In general, one can say:

- technical progress contributes to a dramatic increase in the number of relevant influencing factors;
- selection of effectiveness criteria that satisfy numerous objectives is a difficult problem that includes the risk of making serious mistakes, in particular when considering the problem in too narrow a context;
- great military leaders distinguish themselves by being able to select and promote, from a disturbing variety, essential factors influencing effectiveness;
- computer-based models are being employed apace for purposes of analysis, planning, assessment, and training. A new generation of hardware systems in conjunction with new programming methods and software tools, in particular the employment of expert system technology, promises to contribute to an ever-increasing utilization of model-based analysis of military effectiveness.

Definitions and General Observations

Effectiveness is the degree to which given objectives are satisfied or missions accomplished as production or output. Efficiency is the ratio of that degree of accomplishment to the several possible costs in terms of inputs and throughputs.

A problem of practical significance for military force effectiveness is that objectives (or missions) usually represent only partial objectives or tasks within a complex, multidimensional, hierarchical system of sometimes poorly structured and frequently competing relationships. In addition, services rendered by military forces cannot be assessed by means of free-market mechanisms. Unless one is satisfied with assessments of simple input parameters, such as numbers of available tanks, soldiers, tooth-to-tail ratios, mobilization rate, and so on, one needs to have more or less complex dynamic process models to assess approximately, and off-line, real performance. Assessment criteria applied in a particular situation depend significantly on control level and situations. Also, objectives and missions may differ considerably between peacetime and wartime employments.

Figure 1 is a schematic of the assessment process at

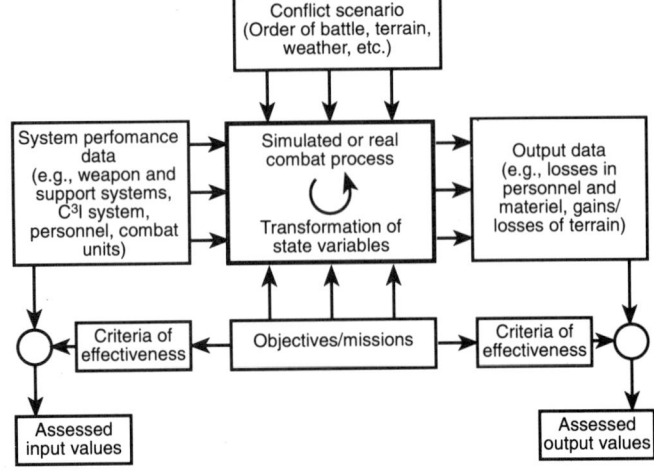

Figure 1. Schematic representation of the assessment process.

various hierarchical levels. Input data are related to the conflict scenario and to the adversary's military forces such as individual weapon and support systems, command, control, communications and intelligence (C^3I) systems, personnel, combat units, and so on. Depending on objectives and missions, real or simulated combat interactions transform state variables of combat systems and provide state changes (e.g., losses in personnel and materiel or the gains/losses of terrain) as output.

From objectives/missions criteria may be derived both input as well as output values for assessing the extent to which objectives are accomplished. Input and output values at two or more hierarchical levels may be aggregated to serve as input values for assessments on a higher hierarchical level.

While input-oriented assessments apply static computational procedures and comparisons (frequently using linear-additive assessment functions) or use merely implicit or fuzzy models, output-oriented assessments are based more often on complex dynamic models of combat by means of which the dynamic interactions in battle are simulated. For analysis, planning, and assessment of problems in defense, such models are highly significant, because in peacetime they represent practically the only way to obtain reproducible information on the effectiveness and efficiency of systems, structures, and concepts. This is because field trials and maneuvers usually do not present very credible models of combat, since safety requirements demand rather artificial circumstances and restrictions.

In what follows, important criteria for the assessment of effectiveness and efficiency of land forces is described. In doing so, distinctions are made between various hierarchical command and control levels on the one hand and input- and output-oriented values on the other.

Table 1 provides examples of important effectiveness criteria at various command and control levels that will be discussed in more detail below.

Effectiveness Criteria on the Weapon System Level

As important characteristics and indicators of performance, input-oriented effectiveness criteria of a typical army weapon system such as a main battle tank or a combat vehicle serve the assessment of such characteristics as firepower, mobility, and protection. Additional factors are related to reliability and availability as well as to certain restrictions (e.g., dimension and weight thresholds for rail and air transport). The degree to which these criteria are met by the weapon system are frequently compared to its cost, which in addition to research and development costs include primarily the procurement and operation costs.

Criteria such as firepower, mobility, and protection cannot be measured as such. Rather, they are the manifestation of some kind of an aggregation process at the basis of which performance parameters are assessed. With regard to mobility, such performance parameters may be a land vehicle's maximum velocity in various types of terrain; its acceleration time from a stationary position to maximum cross-country speed; its range, ground pressure, and the underwater mobility and river crossing capability; admissible gradient angles in terrain, and others. Similarly, the firepower of a combat vehicle is characterized by its type of main and secondary weapons (caliber, initial velocity,

TABLE 1. *Examples of Effectiveness Criteria at Various Command and Control Levels*

LEVEL	INPUT-ORIENTED CRITERIA	OUTPUT-ORIENTED CRITERIA
Weapon system	Firepower Mobility Protection Reliability Availability Cost	Absolute or relative losses Losses over time Force ratios Loss ratios Consumption rates
Tactical (KP, BN, BG)	No. and type and organization of combat units No. and type of direct and indirect-fire weapons Personnel: degree of presence, training status Combat support equipment	Tactical firing rates, kill probabilities, killer-victim scoreboards Relative losses of personnel and materiel Gains or losses in terrain Loss-exchange ratios Cost-effectiveness
Operational (division, corps, army group, theater)	No. and type and organization of combat, support, and combat service support Unit fill and mobilization rates Training status of reserves Data referring to logistics and C^3I systems	Capability to perform given operational missions Survivability and flexibility of combat, combat support, logistics, and C^3I systems Degree to which operational control can be maintained
Strategic/political	Global measures of defense economics Comparative enumerations of individual weapon systems Weapon system categories or unit types Weighted effectiveness indices (e.g., FPS, WEV, ADE)	Local, regional, or global force ratios over time Sustainability Degree of crisis, arms race, conventional and/or deterrence stability

firing rate, range, etc.), type of ammunition and on-board supplies, the fire control system, and the available sensors. Protection afforded by a combat vehicle is determined by its silhouette, means of concealment (such as smoke dispensers), weapon-firing signatures, shape, armor protection as well as means to reduce or control damage (e.g., fire extinguishers, redundant components, active armor).

Input-oriented assessment on the weapon system level carries with it the risk that individual technical characteristics of high performance such as maximum cross-country speed, caliber of the main weapon, thickness of front armor, and availability of additional armor will be considered indicative of a high combat effectiveness that may not materialize under battle conditions. For this reason, an output-oriented assessment based on computer models of battle dynamics is usually preferred. At the weapon system level, these comprise mostly one-sided models; that is, they consider static target arrays that must be neutralized within a given period. To a limited degree, two-sided models, in the form of stochastic duels between two or more opponents, are also being applied to generate output parameters such as mutual draw-down curves for individual weapon systems, the development of force ratios, or loss over time.

It is frequently argued that these simple one- and two-sided models at the weapon system level usually disregard tactical factors (such as movement behavior), the impact of combat support systems (combined arms combat), enemy behavior, terrain effects, and other issues. Furthermore, there is the risk of suboptimization in the sense of taking the support and command and control prerequisites for their successful employment for granted rather than explicitly considering the trade-offs between the respective weapon system under study and its peripheral systems. Thus, it follows that the effectiveness of a weapon system should be assessed in a larger tactical, operational, and finally in a strategic context. This is because individual weapon systems must be assessed in terms of their contribution to the accomplishment of national and alliance security objectives. This ideal complicates the assessment problem. Solutions satisfactory in every respect have yet to be found.

Effectiveness Criteria on the Tactical Level

In addition to assessment of weapon systems in the context of a tactical unit, tactical level evaluations consider the organization of personnel and materiel within combat units with regard to their capability to perform given missions in battle.

Input-oriented parameters for the characterization of tactical level combat units are, among others, unit types, peace- and wartime organization, number and type of direct- and indirect-fire weapons, personnel, training status, availability, communications, and engineering equipment.

Implicit conclusions are drawn from such data with regard to the combat power or capability of a land warfare unit for performing certain tactical missions. These are then contrasted with such input data as the required investment cost, the operating cost, and others.

The application of output-oriented effectiveness criteria on the tactical level requires combat simulation models that imitate the battle of the respective units (including combat support systems) in certain standard scenarios—that is, scenarios defined by combat modes, terrain types, weather conditions, and presumed types of opposing units.

In contrast to the pure duel situations on the weapon system level, these simulations emphasize the weapon system mix (i.e., the combined arms combat in relevant tactical situations and terrain types). Thus, the respective combat simulation models must be able to distinguish between mounted and dismounted combat, account for synergisms of direct- and indirect-fire weapon systems, as well as take into account combat engineers, antiaircraft, communication, and other combat support systems.

Typical output-oriented effectiveness criteria for tactical analysis are:

- killer-victim scoreboards, duel frequencies, tactical firing rates, and kill probabilities versus certain enemy weapon systems;
- relative losses of personnel and materiel;
- relative gains or losses of terrain;
- relative change in force ratios;
- relative loss-exchange ratios;
- required operational depth for the attrition of a given number of attacking enemy units;
- relative cost-effectiveness;
- variance of the above parameters for purposes of risk assessment; and
- sensitivity or robustness of a tactical unit or mix of tactical units, vis-à-vis changes in scenario parameters such as enemy type, terrain type, and weather conditions.

As an example of an output-oriented tactical-level assessment, the reader is referred to Hofmann, Huber, and Steiger (1985), which discusses simulation experiments designed to test cost-effectiveness of battalion-size "reactive" defense modules in comparison with the traditional "active" modules prevalent in general purpose forces. The simulation experiments included four variants of active and more than ten differently designed reactive modules equipped with presently fielded weapons in a variety of circumstances by means of the tactical (battalion/regiment) level model BASIS. The latter is a stochastic, Monte Carlo battle simulation model that permits closed simulations of battalion-size ground forces defending against a sequence of regimental-size attack forces and accounts for organic as well as higher-echelon fire support. It explicitly models each combat vehicle and dismounted infantry down to antitank teams. The effects of each shot are simulated, including the associated visibility degradation. The experiments take place in digital models of several pieces of real estate in Europe's Central Region. The terrain models use a grid size of 50 by 50 meters (55 by 55 yds.) and 10 centimeters (4 in.) altitude resolution and accounts for natural and artificial obstacles and vegetation.

As an example, Figure 2 shows selected results derived from several hundred combat simulation experiments in terms of the relative cost-effectiveness over the relative operational depth required for the attrition of three consecutively attacking Soviet motor rifle regiments in a specific terrain in Bavaria and good visibility conditions. Cost-effectiveness is defined as the ratio of effectiveness to investment cost, with effectiveness being measured in terms of the relative loss-exchange ratio of attacker versus defender. For both cost-effectiveness and cost, the values are multiples of the values that resulted for the active module that turned out to be the most cost-effective of the four active variants tested in the respective scenario.

The reactive modules are battalion-size forces specifically designed for defensive operations in the terrain prevalent within their respective area of operation and incapable of incursions into the opponent's territory, while the active modules are general purpose battalions designed for employment in all combat modes including offense.

The trend line established by the modules E, G, I, and L reflects primarily differences in type and density of antiarmor weapons deployed within the area to be defended.

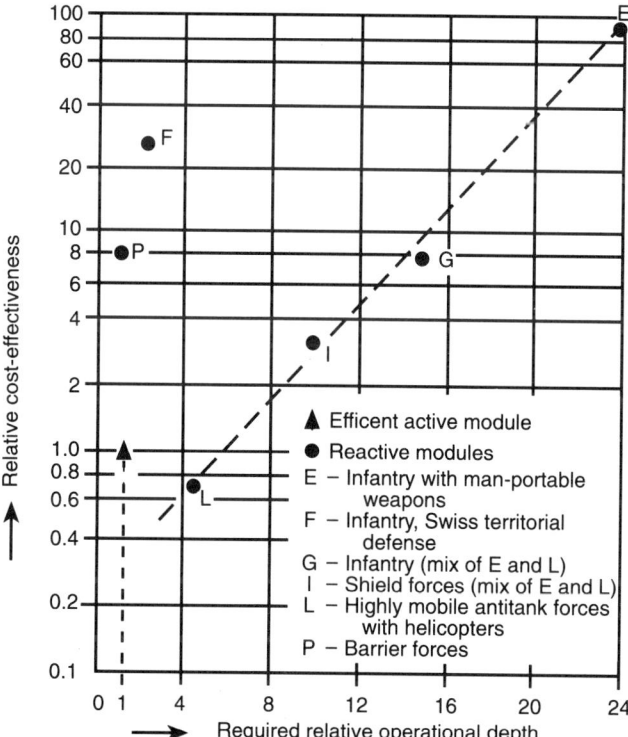

Figure 2. *Relative cost-effectiveness of battalion-size defense modules over the relative operational depth required for the attrition of three consecutively attacking motor rifle regiments under standard scenario conditions.*

At the upper end, E represents infantry teams with limited mobility equipped with man-portable weapons. At the lower end, L represents highly mobile antitank teams equipped with helicopters. G and I have a mix of both. The direct-fire weapon density differs by a factor of more than four between E and L.

In addition to a weapon density 2.5–3 times that of module L, modules F and P are distinguished by a high degree of passive protection afforded through strong points (F) and field fortifications (P) assembled prior to combat from commercially available prefabricated structures stored in the vicinity of their usage. Thus, these modules turned out to be fairly robust against preparatory artillery fire, effects of which are not included in Figure 2.

From the above, it may be seen that the number of factors influencing the tactical level is considerably higher than on the weapon system level. This is due not only to the increased number of sometimes very different weapon systems, but also to the introduction of additional tactical variables such as battle plans and to the impact of combat support elements.

Typical results of combat simulation experiments on this level show that:

- tactics frequently dominate the impact of technical parameters of individual weapon systems,
- variance caused by a change in tactical variables usually exceeds the variance caused by other (technical) variables,
- identical weapon systems exhibit divergent combat-effectiveness parameters depending on the type of tactical unit within which they are employed, and
- effectiveness of a unit is not necessarily increased by introducing a few new high-quality weapons.

In addition to the parameters explicitly incorporated in the model, a comprehensive evaluation often requires the consideration of additional factors not easily modeled or measured. Such factors relate to dirty battlefield effects (impact of dust, dirt, noise, etc.); impact of fatigue; morale; leadership; and so on.

Effectiveness Criteria on the Operational Level

At the operational level—that is, division up to theater—the principal task of a military commander is the spatial and timely deployment of forces in such a manner that interim politico-military objectives are satisfied by maximizing mission accomplishment. In peacetime, the operational-level problem implies an allocation of resources provided to the military so that the prerequisites are created for an optimal employment of forces in wartime. In addition, other objectives usually have to be considered—such as those that may result from the political environment (e.g., alliance policy), stability requirements, the desire for arms control, and others. Thus, in addition to the input- and output-oriented criteria discussed above, a series of further criteria comes into play that result in a more comprehensive assessment appropriate to the operational milieu.

In addition to evaluation of combat and combat support units in the main combat postures, operational-level assessment addresses primarily issues related to mobilization and deployment planning, C^3I, combat service support (including medical services), sustainability, and flexibility.

Accordingly, input-oriented effectiveness criteria reflect equipment quality, unit fill rates, mobilization rates and training status of reserves and combat service support units, the mobilization system (distances between peacetime and wartime deployment positions, street and train capacity), the logistics system (logistic range, supply loads, transportation, maintenance, and repair capability), the C^3I system and its components of intelligence (width and depth of reconnaissance areas of the respective units, number and type of sensors, detection probabilities versus various target types, interference resistance, reliability, etc.), communication (type of system, average message transmission times, redundancies, etc.), and the very command and control elements that are usually described in terms of technical factors (such as vulnerability, reliability, turn around times, etc.).

With respect to output-oriented effectiveness criteria, the complexity of modeling is confusing at this level. Based on the evaluation of expert opinion on "What comprise the significant factors of land combat?," Marshall (1986) has listed more than 100 significant factors ranging from accuracy, advance rate factors, aims and intentions, air superiority, and artillery, among others, to terrain, time, training, visibility, vulnerability, weapon system variables, weapon system interactions, and weather. Marshall concludes that:

> There is little consensus among individual practitioners as to what comprises the principal factors of military combat; indeed, there are significant issues involved in determining just what is a "factor" of combat and what are other components and descriptions in this area.

Many seemingly different terms for "significant/major factors" of combat are similar enough to be considered as one. Their frequency of nomination leads to the following order of priority: firepower, C^3 (including leadership and initiative), environment, human factors, strength (force ratio), intelligence, logistics, mobility, tactics (deployment), miscellaneous, chance, surprise, posture, attacker/defender, and political involvement.

In a tactical and operational context, the authors of *Land Force Tactical Doctrine*—ATP35(A) (NATO 1984, Chapt. 1) distinguish sixteen fundamentals to which armies have adhered in order to achieve their objectives: "Rapidly changing technology and capabilities have altered the emphasis and application of these but the fundamentals themselves remain constant." They are: human factors (leadership, morale, initiative, flexibility, endurance), selection and maintenance of the aim, freedom of action, aggressive action, concentration of effort, economy of force, mobility, maneuver, surprise, intelligence, simplicity, maintenance of forces, flexibility, cooperation, administration, security, and protection.

For main combat modes, additional factors of particular importance are: *Defensive Operations* (intelligence, use of terrain, depth, mutual support, concentration of combat power, maneuver, firepower, electronic warfare, collusion, offensive action, and reserves); *Delaying Operations* (intelligence, maneuver, terrain, time, space, and aggressive actions); and *Offensive Operations* (intelligence, audacity, surprise, concentration, speed, control, depth, security, maneuver). Similar factors are described by Clausewitz in *On War* (1984).

Complexities in modeling are increased by various technical, tactical, operational, and other factors. Variables are interdependent, and their relative importance varies with the scenarios under study. Furthermore, the dynamical battle simulation models on this level need to consider, in addition to the combat processes, the areas of mobilization, combat support, logistics, C^3I, and others. Their processing times must permit simulating a large number of options within a short period of time.

Modeling has progressed considerably in the past few years. Deficiencies remain, however, because military systems analysis emphasizes studies on single weapons systems and tactical levels, which simplifies the modeling. On the operational level, the most common dynamic analysis tool has been the computer-based war game, a technique requiring considerable personnel resources. Therefore, a new generation of fast, closed (without a man-in-the-loop) combat simulation models is being developed. These new models will be implemented on powerful computing systems and will take advantage of modern expert system technologies.

The need for such models is increasingly recognized as many historical examples show that tactical effectiveness (defined as the sum of all capabilities that contribute to the combat unit's lethality and survivability) will not compensate for poor operational effectiveness, which depends on both—that is, tactical effectiveness and commanders' concepts of operations (Hosmer 1988).

With regard to output-oriented effectiveness criteria, the operational level distinguishes the following:

- capability of a given force structure to neutralize forward echelons of an attacking force;
- capability to regain lost territory;
- maximum penetration depth of attacking enemy units prior to restoration of territorial status quo ante;
- fraction of enemy weapon systems neutralized prior to their arrival on the battlefield and the associated time

delays (e.g., for the assessment of follow-on-forces-attack [FOFA] concepts);
- degree of survivability and flexibility of combat, combat support, logistics, and C³I systems or of particular combat functions such as antitank and air defense, movement impairment (barriers);
- degree to which operational control can be maintained in rear combat zones;
- sustainability of various force structures;
- number and robustness of operational decision alternatives for various force structures vis-à-vis given operational options of the enemy; and
- values of variance for the above criteria for the assessment of risk associated with the various decision alternatives.

Effectiveness Criteria on the Strategic Level

In addition to the contribution of the various military services, analyses at the strategic level must consider additional aspects and variables such as civilian defense of a country, available population, industrial potential, alliance structure, the depth of the area of operations, and the length and exposure of lines of communication. Similar to the other levels of analysis, strategic objectives represent guidelines for the selection of effectiveness and efficiency criteria. These objectives are related to:

- improvement of regional or local force ratios;
- maintenance or improvement of crisis, arms race, conventional and/or deterrence stability;
- lifting the nuclear threshold through an improvement of conventional defense capabilities by means of technical, tactical, operational, and structural means within given defense budgets;
- providing for a balanced force structure capable of attaining these strategic objectives; and
- ensuring personnel and materiel preparedness.

Generally accepted dynamic conflict models do not exist at this level, so input-oriented effectiveness criteria dominate assessments. It is the domain of the various static force balance assessments that is frequently discussed in the open literature. These comparisons evaluate force balances in terms of: (1) global measures of defense economics (such as the fraction of gross national product spent on defense, men under arms in time of peace and war compared with the total population of a state, degree of protection provided to the general population and others); (2) comparative enumerations of individual weapon systems (e.g., tanks, armored combat vehicles, antitank weapons), weapon system categories (e.g., armored combat vehicles, air defense systems, artillery systems), or unit types (e.g., tank division, armored infantry division); or (3) weighted effectiveness indices such as firepower scores (FPS), weapons effectiveness values (WEV), weighted unit values (WUV), or armored division equivalents (ADE).

From the foregoing, (2) and (3) may also be applied to operational-level analyses. Attempts are being made to incorporate, among others, command and control and logistics factors into the assessment.

Weighting factors are usually based on judgment and intuitions conditioned by experience. There are, however, serious shortcomings, such as the assumption of utility independence of the weighted parameters, which is obvious from frequent usage of linear and additive utility functions (Hofmann and Huber 1983, p. 153).

Output-oriented effectiveness comparisons on the military strategic level remain based largely on highly aggregated and fairly simple analytical conflict models. For as long as the aggregated input data and implicit assumptions of such aggregated models cannot be qualified by means of a hierarchy of conflict simulation models, results are of limited value. Simple models that are not sufficiently based on empirical evidence incur the risk that the complexity of the underlying problems will be overlooked, leading to faulty interpretations and to pretending an accuracy not found in reality.

Demonstration of variances and risks is considered an essential contribution of military systems analysis on the strategic level. Uncertainty in assessment of force potentials represents not only a source of mistakes (such as inappropriate allocations of considerable resources) but also, depending on the risk attitude of opponents, a probability of crises erupting into military conflict.

Problems in Effectiveness Criteria

The selection of appropriate objectives and effectiveness criteria represents a difficult problem requiring diligence and experience in military systems analysis.

This is especially true due to the lack of sufficiently detailed theater-level models on the one hand and the described disadvantages of aggregated models on the other. Therefore, the present approach is generally one of piecemeal analysis, which involves a risk of suboptimization.

In this context, mistakes frequently made in the criteria selection shall be discussed as per McKean (1970).

IGNORING ABSOLUTE SCALE OF OBJECTIVE OR COST

A common efficiency criterion is the ratio of achievement-of-objective to cost. For example, a problem may concern the selection of an antitank weapon. In a battle on battalion or brigade level, one antitank system may destroy, on the average, 5 tanks at a cost of US$1 million (i.e., an achievement-to-cost ratio of 5 : 1). In the same situation, another weapon system is assumed to neutralize 50 tanks at a cost of US$20 million (i.e., a ratio of 2.5 : 1).

Based on the ratio criterion, the first weapon system would be superior. Nevertheless, this choice might be suboptimal because the capability to neutralize 50 tanks in a certain situation—that is, the capability to thwart the attack of a regiment—can justify the additional cost of US$19 million considering associated tactical and operational effects.

SETTING THE WRONG OBJECTIVE OR SCALE OF OBJECTIVE

A common criterion in most effectiveness problems is to maximize achievement-of-objective for a given cost or conversely to minimize the cost of achieving a specified objective. Choices between these two criteria depend on whether it is the cost or the objective that can be fixed with the greater degree of correctness.

Figure 3 shows the assumed functional relationship between the selected effectiveness criterion (e.g., number of tanks neutralized per unit time) and the required cost for two alternative systems. If we assume the cost restriction to be k_1, system 1 is preferable; for the cost restriction k_2, it is system 2. And vice versa: for a required effectiveness, Z_1, system 1 is preferable, while the effectiveness Z_2 is satisfied by system 2 only. This serves to show how sensitive is the choice of a required degree of effectiveness or how fine is the definition of a cost threshold with regard to choice alternatives. Furthermore, use of fixed-cost or effectiveness thresholds may result in inefficient expenditures or inefficient savings.

This problem is illustrated in Figure 3. If the cost limit k_1 is increased by a relatively small margin Δk, one would select not only a different system, but also obtain a significant increase ΔZ in effectiveness. Thus, cost savings Δk could be considered inefficient. In addition, the convex-concave shape of utility functions typical for most military systems leads to another effect: compensation of cost overruns by means of numerical reductions that result in cost savings Δk that are frequently associated with a significant loss of overall effectiveness.

IGNORING UNCERTAINTY OR RISK

Uncertainty is an essential element in war. Its sources are manifold—for example, occurrence of chance events, lack of situation information, weather, deceptions and dissembly, composition of enemy forces, possible enemy countermeasures, one's own and/or the enemy's objectives, operational employment, and so on.

This means that results based on expected values (e.g., in the form of the mean number of neutralized targets) represent average results, at best, associated with a fairly low probability of occurrence in some cases. It is important that robustness of solutions vis-à-vis results that differ from average results is considered—that is, risks associated with certain decisions. This, in turn, means that assessment models must be capable of generating distributions on the essential assessment parameters rather than merely on expected values. For this reason, modern architectures of battle simulation models attempt to provide deterministic expected values as well as stochastic values. Because of favorable processing times, expected value permits simulating a large number of parametric evaluations, while the more time-consuming stochastic (Monte Carlo) version is applied to determine outcome distributions on interesting parameter constellations.

An example of the assessment of results based on distributional functions in the context of a risk analysis is illustrated in the article entitled Arms Control and Disarmament: Mathematical Models.

IGNORING EFFECTS ON OTHER WEAPON SYSTEMS OR OPERATIONS

A frequent mistake is using criteria that do not measure the effects on other systems or operations. For example, a combat unit that satisfies the requirements for a given defensive mission and also offers the potential for being employed in flank attacks (so that the enemy must provide for additional means in order to protect against the threat to its flanks) has a higher combat potential than a unit that is restricted to a defensive mission only. The same holds true for multirole weapon systems versus single-role systems.

INSUFFICIENT OR WRONG COST DATA

With regard to costs, there are several sources of error, such as the neglect of entire cost categories (e.g., personnel and maintenance costs); systematic underestimation of procurement cost (especially by the producers of systems); considering sunk costs instead of incremental or additional costs (i.e., additional net resource drain that would be incurred by each alternative); neglect of discounting effects (i.e., the fact that expenditures are done at different points in time); as well as disregard for additional resource expenditures that are not easily expressed in monetary terms. For example, assessment of expected personnel losses and the consequences is a largely unsolved problem.

IGNORING THE TIME DIMENSION

The impact of time is usually *the* decisive criterion at all levels of analysis. For example, on the tactical level, the requirement to neutralize a given number of attacking

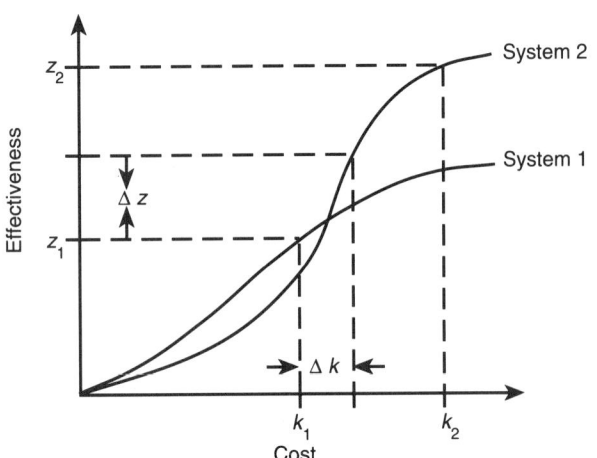

Figure 3. Functional relationship between effectiveness and cost for two alternative systems.

combat vehicles at minimal cost may be quite irrelevant. Capability to do this within a time period of, for example, ten to fifteen minutes might be the decisive criterion for denying the success of an attack.

In summary, the definition of relevant and practical effectiveness and efficiency criteria within the context of a given problem statement requires great diligence and experience in military systems analysis. Frequently, it is only the result of a long analysis process. "The point is that the only way to choose criteria is to undertake analysis, the pitfall is to believe the contrary" (Quade 1970, p. 304).

Future Developments

Selection, procurement, and employment of weapons systems as well as the transformation of technical innovations into appropriate tactical and operational concepts have traditionally presented a great challenge to military leadership. This is because, in peacetime, conservative attitudes usually dominate land forces to the extent that armies are always inclined to prepare for past wars. Thus, tactical requirements and specifications for new weapon and support systems tend to be based on scenarios and employment doctrines that have little relevance. This attitude is of grave consequence at a time when technical progress is accelerating and the potential of new technologies increasing considerably.

For this reason, the tools of analysis must be improved through the development of a new generation of closed simulation models on the levels of operational control (up to theater) that are sufficiently detailed and empirically well founded. Modern concepts of systems analysis and computer science are about to provide the prerequisites for such an improvement. A new generation of extremely capable computing systems that provide multiprocess and parallel processing capability as well as good graphic interfaces, new programming methods and languages, improved software tools, and the employment of expert system technologies comprise the new dynamic assessment environment.

Such models will be employed not just for more adequate assessment of new technologies, new tactical and operational concepts, arms control proposals, and others; they may also be integrated into computer-based operational C^3I systems to be used as command and staff simulators for training commanders and staffs.

If it is correct that tactical and operational variables significantly determine the outcome of battles, as do weapons systems and well-trained troops, then the training of commanders and their staffs by means of simulations attains more significance the fewer opportunities there are to acquire real experience.

HANS W. HOFMANN

SEE ALSO: Arms Control and Disarmament; Combined Arms; Cost Analysis; Defense Decision Making: Analysis and Models; Force Multiplier; Land Warfare; Mathematical Modeling and Forecasting; Measures of Effectiveness; Models, Simulations, and War Games; Span of Control: Military Organizations and Operations; Systems Analysis; Tooth-to-Tail: Land Forces.

Bibliography

Clausewitz, C. von. 1984. *On war*. Ed. and trans. M. Howard and P. Paret. Princeton, N.J.: Princeton Univ. Press.
Dunnigan, J. F. 1983. *How to make war*. New York: Quill.
Helmbold, R. L., and A. A. Khan. 1986. Combat history analysis study effort (CHASE): Progress report for the period August 1984–June 1985. Bethesda, Md.: U.S. Army Concepts Analysis Agency.
Hofmann, H. W., and R. K. Huber. 1983. Systemanalytischer Ansatz zur Zielplanung am Beispiel der Verteidigungsplanung. In *Wirtschaftliche Landesvorsorge im Rahmen der Sicherheitspolitik*, ed. P. Stähly, pp. 149–62. Stuttgart: Haupt Bern.
———. 1988. *On the role of new technologies for conventional stability in Europe*. Report S-8802. Munich: Federal Armed Forces Univ.
Hofmann, H. W., R. K. Huber, and K. Steiger. 1985. On reactive defense options—A comparative systems analysis of alternatives for the initial defense against the first strategic echelon of the Warsaw Pact in Central Europe. In *Modeling and analysis of conventional defense in Europe: Assessment of improvement options*, ed. R. K. Huber, pp. 97–139. London: Plenum.
Hosmer, B. C. 1988. Operational art: The importance of the operational level of war. *Phalanx* 23 (no. 3).
Hughes, W. P., ed. 1984. *Military modeling*. Washington, D.C.: Military Operations Research Society.
Marshall, D. 1986. The factors and structure of combat—A revisit. Paper presented at the 3d International Symposium on Military Operations Research, Shrivenham, Swindon, U.K.
McKean, R. N. 1970. Criteria. In *Analysis for military decisions—The Rand lectures on systems analysis*, ed. E. Quade, pp. 81–91. London: North-Holland Amsterdam.
NATO. 1984. *Land force tactical doctrine—ATP-35(A)*. Military Agency for Standardization. Brussels: NATO.
Quade, E. S. 1970. Pitfalls in systems analysis. In *Analysis for military decisions—The Rand lectures on systems analysis*, ed. E. S. Quade, pp. 300–16. London: North-Holland Amsterdam.

LAND WARFARE

Land warfare is the organized use of military force by a nation-state to achieve political, economic or social objectives. Military force is used directly to defeat the military forces of an adversary or indirectly—by threat—to cause an adversary to modify its political, economic, social, or military behavior. Trained and disciplined individuals are organized into military units that employ a variety of weapons in a coordinated fashion to subdue an enemy military force.

Land warfare is multidimensional. It involves the integration and maneuver of military formations (personnel, vehicles, weaponry, logistics, and communications) and the application of firepower (direct and indirect, air and naval) in a coordinated fashion to exploit an adversary's weaknesses and avoid his strengths, so that assigned objectives are accomplished with minimum expenditure of resources. Battles are orchestrated to accomplish campaign objectives that are aimed at winning wars to accomplish strategic objectives. To achieve those strategic objectives, landpower, seapower, and airpower must be fully interdependent.

Changes in technology have played a central role (and will continue to do so) in developing the doctrine and tactics of land warfare. The development of the bow allowed combatants to distance themselves from each other for the first time. That distance did not change again until the invention of gunpowder. The ability to fire a projectile further separated combatants, and that distance has continued to grow over the last 1,500 years. The 1991 war in the Persian Gulf served notice to land armies worldwide that targets to be engaged will be at even greater distances, and that the human face of the enemy increasingly will be replaced by a blip on an electronic device. Factors that are influencing the conduct of modern land warfare include the electronic battlefield; the increased use and coordination of joint and combined operations; the improved lethality and accuracy of weapons; an array of strategic, operational, and tactical sensors and detectors; the increased maneuverability of forces; the technological sophistication of lesser military powers; and growing requirements for rapid reaction capabilities at the strategic, operational, and tactical levels.

Historical Evolution

The tactics of land warfare emerged from the conflicts of the earliest civilizations. Opponents probably first engaged in personal, brutal, hand-to-hand combat using their bare hands. The first rudimentary handheld weapons were stones, clubs, and stone-head axes. Instruments that thrust objects at adversaries from afar—slings and bows, javelins and spears—further improved lethality while giving some protection by placing distance between combatants. Thus began the cycle of using advancing technology to improve the instruments of war. Improved weaponry meant a gain in advantage over an enemy force, which could lead to victory on the battlefield or, in modern parlance, deter the adversary from entering the fray.

The history and evolution of land warfare revolves around a number of factors: the tactical genius of a few great military leaders; the specific technological breakthroughs with potential military application; the overall modernization and industrialization of societies; and the developing art and science of land warfare as it has grown through analysis of battles, strategic principles, and proven or evolutionary doctrine.

ANCIENT ROOTS OF LAND WARFARE

Although many early societies engaged in land warfare, the Assyrians left one of the earliest records of their weaponry, tactics, and battlefield engagements. As early as 1000 B.C., the Assyrians had made extensive use of military formations of soldiers armed with bows and arrows, spears and slings, on foot, on horseback, and in horse-drawn chariots. Formations of soldiers on foot and horseback would simultaneously launch a mass of projectiles—arrows, spears, and stones—against the enemy and then maneuver in a prescribed formation and direction to complete the task of subduing or destroying the remaining enemy forces.

Two elements of the Assyrians' military prowess continue to be fundamental to an effective military force today. The first factor is the combined use of massive firepower and maneuver of forces to overwhelm and demoralize the enemy. This psychological dimension of warfare at the level of the individual soldier's willingness to fight underlies such battlefield tactics as the employment of the sniper, shock action, blitzkreig, carpet-bombing, and artillery raids. The second factor is related to the Assyrians' use of more readily available iron in lieu of bronze on the tips of arrows and spears and on body armor to improve personal protection from enemy projectiles and swords. The timely recognition and application of technology to land warfare can have a deterrent psychological effect on the enemy, as well as affecting the outcome of the battle.

While the Assyrians and Persians perfected the utility of massed firepower and maneuver, the Greeks countered with extensive use of large, advancing infantry formations of well-disciplined soldiers who were each protected by a large shield and armed with a particularly long, iron-pointed spear. They used this formation to overcome the shock effects of Persian massed fires, similar to those used by the Assyrians, at Marathon in 490 B.C. "With the spears of the first five or six rows of men protruding beyond the front rank to create a deadly hedge of iron points, the phalanx advanced to battle at a steady pace, protected from missiles by heavy shields, and ground [its] way through the enemy formation like a chopping machine" (Kendall 1957).

The Romans improved on the phalanx by making it a more flexible legion of smaller, spaced units and using spears that could be thrown by a soldier bearing a larger shield for greater protection. The legion was modified later to adapt to the need for better leadership and a more professional army in the far-reaching Roman empire. This professional army was recruited from the entire population, organized into 100-man cohorts, led by a professional soldier or centurion, and paid and equipped by the state. This last innovation would continue to come and go throughout the ages.

The ancient civilizations of the Mediterranean region

and of China also contributed great strategists who have had lasting effects on the art and science of warfare. Their contributions to strategy, operational art, and tactics remain relevant to contemporary land warfare doctrine.

THE ANCIENT STRATEGISTS AND GREAT CAPTAINS

As ancient tribal behaviors coalesced into more complex societies, the emerging city-states were charged with securing the interests of their citizens. Individuals served in their respective armies through either conscription or civil obligation. To survive, these ancient societies had to conquer or be conquered. Consequently, they relied heavily on the skills of their military leaders, many of whom today remain central to our understanding of warfare, and land warfare in particular.

Among these ancient generals were Sun Tzu (ca. 500 B.C.), who enunciated well-defined fundamentals of war that underlie modern principles; Alexander the Great (ca. 300 B.C.), who conquered the civilized world from Persia to Egypt to India, adapting firepower, maneuver, and organization to surprise his enemies; Hannibal (ca. 200 B.C.), the Carthaginian, who made the strategic maneuver of crossing the Alps into Italy, thus avoiding the Romans' major force, and subsequently moving his force through foreboding terrain to surprise and defeat a force of Roman legions; and Julius Caesar (ca. 50 B.C.), who demonstrated a genius for adapting the tactics of the legion to the terrain and the enemy's formations and deployments, and engendering a discipline that ensured the commitment of forces to battle at the decisive time and place.

At the close of the last millennium B.C., the ancient great captains had secured their places of prominence in the study of military strategy, art, and tactics that underlie land warfare. The next great captains would not emerge until late into the second millennium A.D.: Gustavus Adolphus of Sweden (17th century); Marlborough of England (18th century); the Prussian, Frederick the Great (18th century); and Napoleon (19th century).

TRANSITION TO MODERN LAND WARFARE

In the first millennium A.D., emerging tactics included the use of horsemen (cavalry), light infantry, and heavy infantry, to be employed in various combinations and weightings. The extent of body armor covering head, chest, arms, and legs varied, as did the size and weight of shields and weaponry (javelin, spear, axe, dagger, sword, bow).

How armies fought also depended on the weaponry and tactics of adversaries. In the western Roman Empire at the start of the millennium, the response to continual barbarian raids on outposts was static defense; in the eastern Byzantine portion of the empire, cavalry prevailed by A.D. 500. The Byzantines codified their doctrine of warfare, focusing on defense in depth and conservation of resources (especially trained and expensive cavalry).

England and the western European tribes and enclaves of civilization were subject to Viking raids. To survive, they organized into protective fiefdoms and made extensive use of fortifications—at first, large earth and timber enclosures and later, walled castle keeps and cities. Siege warfare was common, and the techniques and equipment used to penetrate walls and engage distant targets became increasingly sophisticated.

Feudal armies were usually temporary, brought together for a particular purpose and then disbanded. They varied in size, on the average 5,000 to 10,000 men. Local vassals were eventually replaced by paid soldiers (mercenaries); as the strength of armies became more predictable, longer campaigns could be conducted. Through invasions and counterinvasions among France, England, and later Spain, and as a result of the Crusades and the Hundred Years' War, individuals built power bases and monarchies emerged. At first, the weapons of war and tactics of ancient periods were reinvented. However, with the advent of gunpowder, the relative lethality of infantry and cavalry changed forever; musket replaced bow, and cannon replaced catapult.

The Impact of Technology on Land Warfare

Before the invention of gunpowder, land warfare was fought in a destructive, personal manner, and at closer quarters. As crude weapons were gradually replaced by more effective devices, the nature of land warfare changed. Early man made only relatively simple advances in weaponry over several thousand years. But as man's ingenuity advanced through the first millennium A.D., significant advances in weaponry occurred more frequently.

The short bow, sling, and javelin brought about significant changes in how land wars were fought. Chariots gave archers and javelin throwers a mobile and stable platform from which to launch their projectiles. The use of helmet, breast plate, shin guards, and shield in combination with a spear, axe, or sword, and of lightly equipped soldiers on horseback (cavalry) continued well into the second millennium A.D. The effectiveness of this early weaponry was dependent on the mobility, maneuverability, and discipline of the military formation. Fortified defensive works of earth and stone were essential to the survival of entire city-states and date from earliest periods. These fortresses were besieged by armies using fighting towers, battering rams, catapults, and flame-tipped arrows and other flame devices.

During the first millennium A.D., the role of the soldier on horseback was enhanced by the use of saddle and stirrup, which provided a stable platform and leverage to use weapons while mounted. The cavalryman replaced the two-horsed chariot, which was an economical move in terms of the number of fighters per horse and the forage required for horses. Both the long bow (which required a well-trained archer to launch up to six arrows per minute to a range of 200 yd.) and the cross-bow (which required less training and discipline, had a rate of fire of one to two arrows per minute to a range of 130 yd.) assumed prominent positions for organized use on the battlefield soon after the start of the second millennium A.D. (Macksey 1973). The use of gunpowder emerged soon thereafter.

Gunpowder and the Industrial Revolution

Over the centuries, weapons development focused on improving three elements: range, accuracy, and rate of fire. With the advent of gunpowder-propelled projectiles on the battlefield in the 1400s, science became a tool of the military in the search for ways to inflict more significant casualties on opponents. For the next 400 years, the sophistication and diversification of cannon, artillery, musket, pistol, and rifle revolutionized the conduct of land warfare. With each new invention—the cylindroconoidal bullet, improved explosives, smokeless powders, the fuse, shrapnel, rifling (to increase range and accuracy), breech-loading artillery, the repeating rifle, and the machine gun—rates of fire and the vulnerability of soldiers to long-range fire increased dramatically. The industrial revolution meant these weapons could be manufactured in mass quantities. The railroad, telegraph, and internal combustion engine brought military units a degree of mobility and responsiveness unimagined a century earlier (Garden 1989).

The leap in technology and invention in the nineteenth century continued the trend toward large land armies and greater casualties, as evident in the American Civil War and the Russo-Japanese War. World War I brought the artillery barrage, the machine gun, infantry attacks on entrenched enemy forces, and limited use of chemical weapons. World War I also saw the introduction of the airplane and the first armored vehicle (the tank) to be used in modern warfare to overcome the limitations of trench warfare (Macksey 1973). These two weapon systems have influenced land warfare in an unprecedented fashion. The pace and lethality of military operations was greatly accelerated by these weapon systems and by the application of electronics, including radar, high-speed communications, and encryption, in World War II.

The enhanced ability of opposing military forces to maneuver, employ firepower, and protect their resources brought about new operational methods involving the close coordination of airpower (strategic bombing and close air support) and landpower (rapid-moving armor units and artillery preparations) to engage an opposing force at its most vulnerable location and avoid its strengths. This form of warfare was used with great success by the Germans in World War II, to a lesser extent by the Allies.

Modern Technology

The weapon systems of land warfare that emerged from World War II set the framework for weapons to follow. In spite of its power, the atomic bomb has proven unusable in a ground war. The potential for escalation, side effects of radiation, and political intervention have negated a direct role for nuclear weapons in land warfare. Instead, technology has focused on improvements to traditional applications of energy in weaponry and the vehicles that propel them.

The microminiaturization of equipment components, the microprocessor and computer, the development of sensors and detectors that respond to wide ranges of the electromagnetic spectrum (infrared and laser), electronic countermeasures, near-real-time dissemination of information, ground positioning systems, satellites, lasers, increased lethality of and accuracy of warheads, improved armor and munitions using depleted uranium, and countless related technological adaptations are having a profound effect on warfare. Targets can be detected, tracked, and engaged at ranges that far exceed those of World War II.

The high levels of reliability, protection, and lethality that have been attained can be illustrated by the modern battle tank. The performance of the U.S. Army M1A1 main battle tank in the 1991 war with Iraq reflects technological improvements applied to an established weapon system. After 100 hours of offensive operations, the tank's operational readiness exceeded 90 percent. In a night movement of 300 tanks across open desert, all of the tanks arrived at the destination. Several M1A1 tanks received direct hits from antitank rounds and sustained no damage, attesting to the effectiveness of special armor. The tank's thermal night sight allowed crews to see enemy tanks through smoke from oil well fires, use the laser range finder, maintain gun stabilization on the move, and destroy targets at ranges that exceeded 3,000 meters (9,900 ft.). The state-of-the-art antitank round fired by the 120mm gun of the M1A1 was able to fire through berms protecting enemy tanks and still destroy Soviet-origin T-72 tanks.

Modern land warfare uses technologically advanced systems at the level of the individual soldier to enhance fire support weapons. The individual soldier can be armed with night-vision equipment to allow him to "see daylight." He is also equipped with laser devices to designate targets for engagement by artillery and armed helicopters, and he is armed with individual antitank weapons that can launch smart rounds that stay on target until impact. He can engage enemy helicopters and aircraft using a shoulder-fired, heat-seeking antiaircraft missile and can locate his position within a few feet by using a hand-held global plotting device that receives satellite information.

Artillery weapons fire "smart rounds" that can seek out and destroy tanks from overhead and scatter antipersonnel and antitank mines. Radar systems can "backtrack" the path of enemy projectiles to the artillery location and automatically provide coordinates for counterartillery fires. Aircraft are used in traditional logistical and close air support roles. However, the armed helicopter is able to maneuver and engage enemy armor day or night using smart antitank missiles that respond to laser designation and infrared emissions.

Technology will continue to improve the overall capabilities of traditional land weapon systems. However, the limitations of technology and the man-machine interface will require continual attention. Success in warfare depends heavily on the ability of individuals to use weapons and information systems under battlefield conditions while maintaining the flexibility to adapt and take advantage of changing conditions. The advantages and economies of automation and robotics will have to be balanced with the human ability to process selective information to make good decisions—by a tank gunner's choice of target or a corps commander's timing of an attack.

Conceptual Foundation of Modern Land Warfare

The conduct of land warfare rests on three fundamental piers of analysis: strategy, operational art, and tactics. Successful military strategy achieves national and alliance political aims at the lowest possible cost in lives and national resources. Operational art translates those aims into effective military operations and campaigns. Sound tactics win battles and engagements that produce successful campaigns and operations (Department of the Army 1986).

In addition to these three components, modern land warfare doctrine reflects the precepts of ancient and modern theorists and strategists as reflected in principles of war, and combat power (the ability to fight a war).

The classical principles of war were best articulated by British Major General J. F. C. Fuller during World War I. They remain valid today and are summarized below:

- Direct every military operation toward a clearly defined, decisive, and attainable objective.
- Seize, retain, and exploit the initiative.
- Concentrate combat power at the decisive place and time.
- Allocate minimum essential combat power to secondary efforts.
- Place the enemy in a position of disadvantage through the flexible application of combat power.
- For every objective, ensure unity of effort under one responsible commander.
- Never permit the enemy to acquire an unexpected advantage.
- Strike the enemy at a time or place, or in a manner, for which he is unprepared.
- Prepare clear, uncomplicated plans and clear, concise orders to ensure thorough understanding.

Combat power measures the effect of maneuver, firepower, protection, and leadership, which are defined as follows:

- *Maneuver* is the movement of forces in relation to the enemy to secure or retain positional advantage. It can involve concentrating forces at the critical point to achieve surprise and dominance over enemy forces; it also can be achieved by allowing the enemy to move into a disadvantageous position.
- Maneuver is linked to *firepower* to defeat the enemy's ability and will to fight; however, firepower can also be used independently of maneuver to destroy, delay, or disrupt uncommitted enemy forces.
- *Protection* involves the retention of fighting capabilities so they can be applied at the decisive time and place; this involves actions (such as camouflage, deception, dispersal, and air defense) to counter the enemy's ability to locate friendly forces and use firepower and maneuver against them.
- Competent and confident *leadership* provides purpose, direction, and motivation in land warfare.

Battlefield success is measured by the extent to which it accomplishes the operational goals of a campaign. In turn, the campaign is not successful unless the national security strategic objectives are met. Thus, clear statements of strategic purpose are essential. The effective application of combat power on the battlefield is defined in terms of doctrine.

Modern Battlefield Doctrine

Battlefield doctrine addresses how to use combat power at the operational and tactical levels of warfare. Modern battlefield doctrine prescribes that friendly forces should gain and retain the initiative, act faster than the enemy, synchronize battlefield activities to produce maximum combat power, and operate in depth of space, time, and resources to win the battle.

The linear battlefield dominated the major land wars of the twentieth century. Trench warfare and the use of massive artillery fires in World War I resulted in mutual attrition and high casualties. World War II provided greater opportunities for coordination of mobile forces and firepower to gain the advantage over opposing forces; however, the linearity of the battlefield and heavy casualties still dominated, and rapid victory was elusive. However, the battlefield of the late twentieth century, as demonstrated in 1991 in the ejection of Iraqi forces from Kuwait (Operation Desert Storm), involves fast-paced, fluid operations. Depending on the assigned mission, units conduct close, deep, and rear operations.

Close operations involve units that are committed to battle, including support such as artillery, air defense, reconnaissance, and logistical units. At the operational level, close operations involve the activities of corps comprising divisions. At the tactical level, the subordinate units of the division, the brigades and battalions, fight battles (which can involve deep and rear operations as well). Close operations include maneuver, close combat (including close air support), indirect fire support (artillery or naval gunfire), combat support and combat service support of fighting units, and the necessary command and control (leadership and coordination) to ensure victory.

Modern battlefield doctrine incorporates deep operations to shape the operational situation in which later close operations will occur. Deep operations, which are not new to modern land warfare doctrine, include interdicting enemy supplies, reserves, and communications to minimize their impact on a current or future battle. With the increasing mobility and firepower available to modern tactical-level units, the ability of deep operations to influence the outcomes of battles and ensure early victory continues to grow. Deep operations are undertaken only against those enemy capabilities that can directly affect the conduct of friendly operations. They include deception operations, deep surveillance and target acquisition, interdiction by firepower and maneuver forces, and electronic warfare to disrupt enemy command, control, and communications.

Close and deep operations are closely coordinated with rear operations, which comprise activities to ensure freedom of maneuver and continuity of planned operations or opportunities to exploit success. Friendly reserve forces and fire support are positioned to successfully move and engage enemy formations while remaining secure from enemy observation or attack. Sufficient logistical support and services are conserved, without decreasing the support to currently engaged units, to ensure the sustainment of the reserves if committed.

These categories of operations reflect the fluid and nonlinear nature of the modern battlefield and provide a framework for the development of battle plans. Battle success depends ultimately on the ingenuity of the commanders, the readiness and training of individual soldiers and units, and their ability and flexibility to adapt to rapidly changing situations.

ORGANIZATION FOR LAND WARFARE

The doctrine of the modern battlefield is practiced by organizations within a designated operational area. Military operations are carried out in accordance with the strategic objectives established by national or international command authorities. A definite set of boundaries is established, rules of engagement are defined, and specific objectives are established. Land warfare operations are conducted in a theater of operations in coordination with air and sea forces (joint operations), to include the armed forces of allied countries (combined operations). A joint commander is designated who is responsible for coordinating all activities in the conduct of the campaign, to include all air, sea, and land forces operations. In a joint or combined command, a commander is designated for each of the air, sea, and land components. In a combined operation with forces from other countries, the joint commander may also be the combined commander, or a separate commander can be designated.

In the U.S. military, land forces at the operational level, where objectives are related to the accomplishment of specific strategic objectives, are organized into corps. One or more corps can operate under a theater army commander, who is responsible to the overall joint/combined commander. At the corps level, the operational objectives of the campaign are accomplished through the operations of subordinate tactical divisions: brigades and subordinate battalions and companies fight the battles and engagements.

The types of units assigned to each of these levels of organization—theater, corps, division, brigade—will depend on the overall purpose of the operation, the geographical location, the terrain and weather to be encountered, the sophistication of infrastructure in the theater of operation (i.e., ports, airfields, roadways), constraints of resources and time and, most important, the threat of enemy forces and their capabilities. Each level of organization is task organized to ensure the capabilities and functions necessary to achieve the objectives of the operation. Tactical-level combat units rarely operate as pure organizations; rather, they are augmented (task organized) with other types of units into a task force to take advantage of the combined effects of their capabilities.

Divisions are major ground combat units that have the command and control capabilities (commanders, staff, communications) to effectively plan for and implement operational plans at the tactical level. At this level, tactical missions are assigned, operational orders are developed and implemented, and the battle is fought by assigned units. Operational orders include clear statements of missions to be accomplished, a scheme of maneuver, a visualization of the operation, controls on the movement and firepower of units, assignment of different types of fire support (i.e., close air support, artillery support, naval gunfire support, air defense), priorities regarding transportation, allocation of fuel and ammunition, intelligence information, missions and actions of adjacent friendly units, and time lines for the start of the operation.

The number of personnel and units in a division will depend on the assigned mission, the operational environment, and the capabilities needed to defeat enemy forces. The number of personnel can range from 10,000 to 30,000. The units assigned to the division, like those assigned at the higher corps level and at subordinate brigade, battalion, and company levels, are categorized by their major functions: combat, combat support, and combat service support. Combat units include armor, mechanized, light infantry, air assault, and airborne. Each of these units has unique capabilities that makes it appropriate singularly or in combination with other types of units to conduct a ground operation. Combat support units include artillery, aviation, engineer, and air defense. Combat service support units include medical, transportation, supply, and construction.

Logistical Support

The combat service support units are often neglected in the study of land warfare. Their role is to support the deployment, movement, and sustainment of combat organizations. Functions involved include construction and transport; fuel, water, food, and ammunition resupply; and provision of spare parts and maintenance of equipment and weapon systems. The logistical situation and the support units involved will weigh heavily in decisions regarding the deployment of military forces.

Figure 1 portrays the Kuwait theater of operations that existed at the start of the Operation Desert Storm ground war in February 1991. To emphasize the important role of combat service support and logistical support, only the major combat service support units are displayed. Each of the divisions (displayed to the far right) has its own internal support command to meet short-term supply, maintenance, and medical needs. The corps-level logistical support structure includes a corps support command (COSCOM) for each of the corps (XVIII and VII Corps), which provides a corps support group to support each division; in addition, a corps support group supports other units in the corps area. Each corps support group has a maintenance battalion, a supply and service battalion, and a transportation battalion. To the left are theater-level support units, including ammunition, fuel, spare parts, and transportation.

Land Warfare in the 1990s: Operation Desert Storm

States resort to warfare for many reasons: to promote and spread a political or religious ideology; to gain territory for resources and space for a growing population; to protect national interests in another region; and to conquer. In modern parlance, the reasons for going to war are usually expressed in terms of securing national, regional, or other international interests.

A case study can lend understanding to modern land warfare, and particularly to its complexity in light of the spread of technology and the availability of sophisticated weapons to adversaries. It can also provide some vision of how wars will be fought in the future. The 1990–91 war in the Persian Gulf, which involved Iraq, the United States, and a coalition of other countries, involved a definable threat to the national interests of several countries; international political, economic, and military cooperation and United Nations legitimacy; the use of strategic assets; definable operational campaigns; and the use of advanced weaponry and systems that had not been proven in combat.

The Strategic Setting

Severely strapped for international exchange following an eight-year war with Iran, Iraq pressured Kuwait and other OPEC nations to raise oil prices and to reduce production. It accused the Kuwaitis of digging oil wells on Iraqi territory and extracting more than US$2 billion worth of oil, for which Iraq demanded compensation. Iraq also sought Kuwaiti oil fields near the Iraqi border and demanded a lease of Kuwait islands to gain a seaport on the Persian Gulf.

By late July 1990, Iraq had positioned a large force of troops along the Kuwaiti border. On 2 August, the Iraqi army invaded Kuwait and gained full control of the emirate within one day. That move was promptly condemned by the U.N. Security Council, which demanded the immediate withdrawal of Iraq's forces from Kuwait—a measure that Iraq chose to defy.

By 6 August, with Iraqi forces disposed along the Kuwaiti–Saudi Arabian border and postured for a possible attack on Saudi Arabia, the United Nations authorized worldwide economic sanctions against Iraq. Saudi Arabia, fearing imminent attack, requested assistance in defense of Saudi territory. Thereafter, support from the United States and many other countries moved at a rapid pace. The United States deployed land, air, and sea forces to the Persian Gulf region to deter or defend against an Iraqi invasion of Saudi Arabia. On 7 August, Operation Desert Shield officially began. A 2,300-man contingent of the U.S. Army's 82d Airborne Division immediately deployed by air. On 8 August, Iraq publicly annexed Kuwait and declared it a province of Iraq. The U.N. Security Council took positive and immediate action by approving a resolution that demanded unconditional and immediate withdrawal of Iraq's army.

Operation Desert Shield

The national security objectives of Operation Desert Shield as outlined by the American president were:

- To protect the lives of American citizens
- To deter and, if necessary, repel further Iraqi aggression
- To effect the immediate, complete, and unconditional withdrawal of all Iraqi forces from Kuwait
- To restore the legitimate government of Kuwait

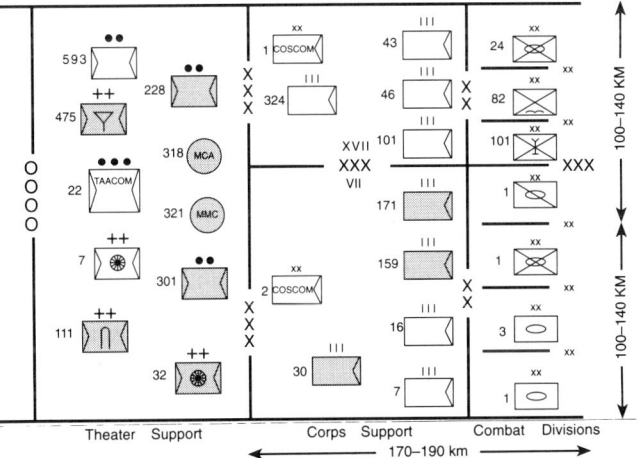

Figure 1. Operation Desert Storm: Kuwait theater of operations, 20 February 1991. (SOURCE: Association of the U.S. Army, 1991b.)

Rapid deployment of forces from all the services proceeded. The first requirement was to deter any further encroachment by Iraqi forces. U.S. naval forces in the region were reinforced and tactical air forces were moved to the theater of operations. Additional mobile light ground forces, including Marine elements and the balance of the 82d Airborne Division, were also moved to the region. Troops of air assault and heavier armored and mechanized divisions, as well as air defense and corps support units, were airlifted to the Gulf, while combat equipment followed in the largest sealift of combat forces since World War II. To provide required combat and combat service support, units from the reserve components of all services were called to active duty. Also, equipment on pre-positioned ships deployed to the region.

Thirty-seven nations sent military forces or medical teams to the region. Ten nations pledged more than US$50 billion to defray the costs of the operation.

All U.S. forces deployed in Desert Shield came under the command of the commander in chief of U.S. Central Command (USCENTCOM). USCENTCOM operated with the Saudi Joint Forces Commander through a Coordination and Communications Center. Initially, British forces were under U.S. operational control and French forces under Saudi control.

Arab and coalition forces were initially positioned behind the Saudi task forces arrayed along the Saudi–Kuwaiti border. The forces that were deployed by air were in place in the first weeks of August; the others arrived in increments through October. The major pacing factor for subsequent movements was the availability of airlift and sealift, with the equipment moving by sea and the troops flying to Saudi Arabia in time to marry up with their equipment. By early November, there was sufficient combat capability to provide an effective defense of Saudi Arabia.

Iraq continued to build its forces in the Kuwaiti theater to more than 400,000 troops, to include mass construction of hardened bunkers, tank traps, mine fields, and miles of earthen walls to reinforce positions along the frontier of Saudi Arabia. On 8 November, it was decided to develop an offensive capability with sufficient combat power to force the Iraqis out of Kuwait. At that point, the U.S. VII Corps from Europe and armored, mechanized, and support units from the United States were ordered to deploy, and the call-up of additional reserve units of all services was started. Toward the end of November, these units began to move, reaching full combat readiness in Saudi Arabia by early February.

In the first 80 days, more than 170,000 people and more than 160,000 tons of cargo were moved to Saudi Arabia by air from the United States. Over 7,500,000 square feet of cargo and equipment were moved by sea. By the time the coalition forces began the offensive on 17 January 1991, the United States had shipped some 460,000 tons of ammunition, 300,000 desert camouflage uniforms, 200,000 tires, and 150 million military meals to sustain the 540,000 soldiers, sailors, airmen, and marines deployed.

OPERATION DESERT STORM

The United Nations established a deadline of 15 January for Iraq to withdraw its forces from Kuwait, and authorized member nations to employ all necessary means to evict them if they did not withdraw. The Iraqis did not withdraw by the deadline. On 17 January 1991, Operation Desert Shield became Operation Desert Storm when the coalition initiated combat operations.

The initial phase, the air campaign, was intensive. The coalition forces employed its air resources—including armed helicopters, cruise missiles, and at least eighteen types of land- and sea-based aircraft—to maximum advantage. Electronic jammers, sophisticated sensors, night vision devices, and precision bombing technology destroyed Iraq's strategic capability.

Coalition air superiority was achieved early in the operation. The campaign was directed against Iraqi ground forces facing coalition units across the Kuwaiti–Saudi Arabian border. Thousands of sorties were flown each day attacking targets of military importance such as missile sites; command and control centers; telecommunications facilities; power generating plants; airfields and runways; aircraft storage shelters; bridges; Iraqi troop positions; and chemical, biological, and nuclear weapon development and production facilities. Air sorties against Iraqi military targets were conducted by U.S., Saudi, Kuwaiti, British, French, Canadian, Bahraini, Qatari, and Italian forces. The air and sea offensive continued for 38 days with a constant, around-the-clock bombardment that brought the war to Iraq.

Despite the high priority given to locating and destroying Iraqi SCUD ballistic missile launchers, missile attacks continued throughout the period, although in decreasing numbers. The SCUDs were intended as terror weapons against civilian targets and were never a serious military threat. U.S. Army Patriot missiles were used for the first time to defeat other missiles in a combat situation.

THE GROUND WAR

Two corps of more than 200,000 troops and thousands of tons of equipment started moving to the western part of Saudi Arabia on 17 January under the cover of air, sea and artillery bombardments. Repositioning for a ground attack into Iraq and Kuwait was underway.

Sufficient fuel, ammunition, spare parts, water, and food were moved as much as 300 miles to establish a 60-day supply in preparation for the coming ground offensive. Special forces teams were inserted deep into Iraq to perform strategic reconnaissance and to report on troop movements. By 16 February, American and coalition forces were in positions spanning a distance of over 300 miles along the Saudi border.

Throughout this phase of the war, numerous feints, probes, and mock attacks were conducted by various elements of the coalition forces. On several occasions, the Navy and Marines rehearsed invasions from the sea and throughout maintained a large presence in Gulf waters off the shores of Kuwait. The American and coalition land forces executed reconnaissance missions all along the fortified borders of Kuwait and Iraq. By concentrating their forces along the southern Kuwaiti border and by fortifying the beaches east of Kuwait City, the Iraqis made it clear that they expected a headlong attack into their most heavily fortified areas.

By mid-February, the emphasis of the air campaign was clearly shifting to inflict maximum damage on Iraqi troop formations and defensive positions, softening them for the pending ground attack. By now, the U.S. Army had over 250,000 troops in the Persian Gulf area. Its combat elements were poised for the attack.

At 4:00 A.M. (Gulf time), 24 February 1991, the coalition forces launched the largest successful ground campaign since World War II. Along a 300-mile front, they rolled into Kuwait and Iraq to engage the world's fourth largest army. One hundred hours later, on 28 February, the coalition declared a temporary cease-fire.

Having maneuvered over 300 miles westward to reposition ground forces composed of two corps for the attack into Iraq and Kuwait, U.S. and coalition forces were positioned as portrayed in Figure 2. XVIII Corps included 82d Airborne, 101st Airborne Division, 3d Armored Cavalry Regiment, the 24th Infantry Division (Mechanized), and the French 6th Light Armored Division. Further east, VII Corps included the 1st and 3d Armored Divisions, the 2d Armored Cavalry Regiment, and the British 1st Armored Division. Near the confluence of the Saudi-Iraqi-Kuwaiti borders, were the 1st Infantry Division (Mechanized) and the 1st Cavalry Division. To their right was a pan-Arab force consisting of Saudi, Kuwaiti, Egyptian, and Syrian units at the western edge of the Saudi-Kuwaiti border. The 2d Marine Division and a brigade of the 2d Armored Division were positioned to the east of the pan-Arab force; the 1st Marine Division anchored the right flank. Two additional Saudi task forces were prepared to advance up the Persian Gulf coast, while Marines of the 4th and 5th Expeditionary Brigades were poised for amphibious operations off the Kuwaiti coast. Earlier, a number of Special Forces teams had been inserted deep in Iraq to track enemy movements and especially to locate SCUD missile sites.

The ground war started with two simultaneous attacks, one in the east, where pan-Arab forces and U.S.

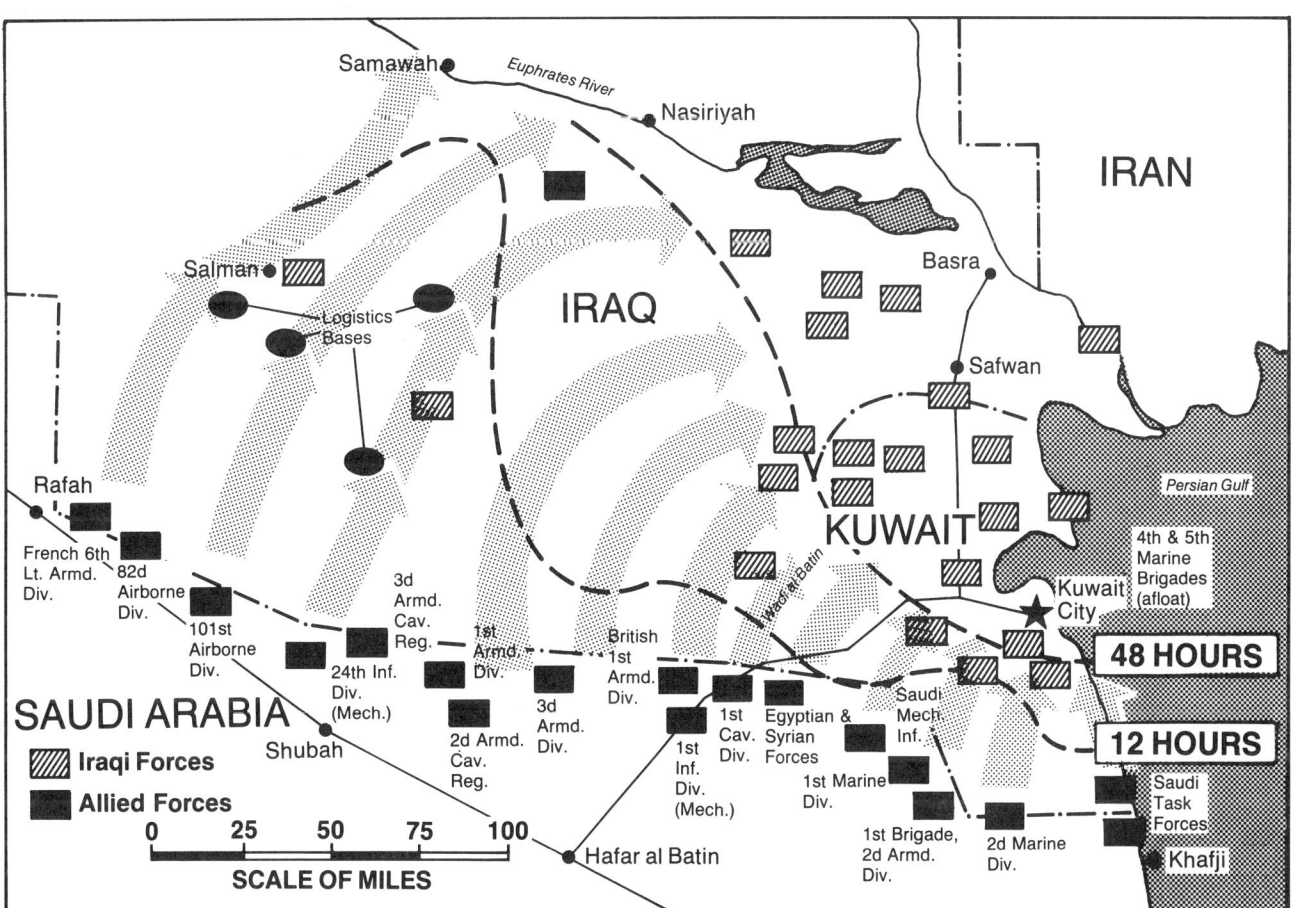

Figure 2. Operation Desert Storm: U.S. and coalition ground forces positioned for attack. (SOURCE: Association of the U.S. Army, 1991b.)

Marines breached the first line of Iraqi defenses and drove up the coast toward Kuwait City, and the other 300 miles west, consisting of the French 6th Light Armored Division and one brigade of the 82d Airborne Division attacking 90 miles into Iraq to seize the airfield at Salman and establish a security screen for the western flank. At the same time, the Marines in the Gulf, aided by intense naval gunfire, feinted an assault against Iraqi forces dug in along Kuwait's coast. Similarly, two brigades of the 1st Cavalry Division attacked about a dozen miles up the Wadi al Batin against sporadic resistance.

At 8:00 A.M., 23 February, the 101st Airborne Division launched the largest air assault operation in military history more than 70 miles into Iraq and then continued the attack to the Euphrates River. That afternoon, the 3d Armored Cavalry Regiment and the 24th Infantry Division (Mechanized) attacked north into Iraq.

In VII Corps' sector, the 1st Infantry Division breached Iraqi defensive positions and attacked north, followed by the British 1st Armored Division. The 2d Armored Cavalry Regiment and the 1st and 3d Armored Divisions attacked rapidly, bypassing hundreds of enemy positions. Along the coast, Saudi-led coalition forces breached defensive barriers and joined the Marines in the attack on Kuwait City. By the end of 24 February, all major coalition forces were engaged.

On 26 February, elements of the XVIII Airborne Corps and VII Corps maneuvered to the east to trap and destroy what was left of the Iraqi forces. Having driven almost 100 miles into Iraq, the 24th Infantry Division and the 3d Armored Cavalry Regiment turned toward Basra to cut off retreating Iraqi forces. VII Corps units turned east to attack the Iraqi reserve. During the night, VII Corps, which now included the 1st Cavalry Division, conducted a coordinated attack and destroyed two Iraqi divisions. The 1st and 2d Marine Divisions also reached the outskirts of Kuwait City and fought for control of the international airport.

On 27 February, Kuwait was liberated. By the time a suspension of offensive combat operations was declared (at 8:00 A.M., 28 February 1991, Gulf time), U.S. and coalition forces had destroyed or rendered ineffective 43 Iraqi divisions, captured more than 80,000 prisoners, and destroyed or damaged 4,000 tanks, 2,100 artillery pieces, 1,800 armored personnel carriers, seven helicopters, and 103 Iraqi aircraft.

The United Nations worked out the details of the formal cease-fire agreement, which was accepted by Iraq on 6 April and proclaimed on 10 April. Included in the U.N. action was the authorization of a 1,440-member observer team to oversee a newly created demilitarized zone (DMZ) between Kuwait and Iraq. Operation Desert Storm, preceded by Operation Desert Shield, became Operation Provide Comfort as some 13,000 coalition military personnel—including about 9,000 U.S. troops—turned their attention to giving food, shelter and medical care to the refugees.

As U.S. and coalition forces redeployed, their troops in the DMZ were relieved by U.N. forces; the refugee support effort in the south was accepted by the Saudis. In the north, Operation Provide Comfort—support for the Kurdish refugees, much of which was rendered through Turkey—became the responsibility of the U.S. European Command.

Desert Shield and Desert Storm were the largest operational tests of modern military forces and doctrine since World War II.

The Future of Land Warfare

Land warfare, as the Desert Shield and Desert Storm operations indicate, involves the coordinated use of all military resources at a nation's disposal to achieve specific national objectives. The strategic, operational, and tactical settings of land warfare are shaped by airpower and seapower. Once strategically deployed and operationally positioned, the landpower component is charged with carrying the fight to the enemy land forces and dislodging them from the area of contention. Throughout, airpower and seapower continue to play a major role in maintaining supremacy in the skies and on the seas.

Ancient and medieval history and the technological impact of the industrial, atomic, and electronic ages have forged today's military doctrine. Operation Desert Storm reconfirmed many of the principles of war and provided some vision of land wars yet to be fought. Some of the parameters of future land warfare might be as follows:

- Early deployment of military forces, particularly land forces, to the area of contention is the clearest signal of national resolve in a crisis. An accompanying clear statement of the purpose and objectives of the military operation provides a framework for strategic and operational planning and judicious use of national resources.
- Developing nations have access to sophisticated, high-tech weaponry and, especially in the face of overwhelming opposing forces, are prepared to employ them to achieve national goals and resolve disputes with neighbors.
- Land forces are the major military resource of most countries. Therefore, nations will continue to use land warfare as the predominant instrument of military force, particularly at lower levels of conflict.
- A vast array of sensors is now available to detect, locate, and engage targets. Future land warfare may see most enemy capabilities engaged and rendered ineffective long before they can be maneuvered and employed.
- Operational doctrine will be broadened to more thoroughly encompass activities short of war, such as peacekeeping, refugee support and security, and environmental disaster relief.
- The interdependence of nations' interests and the num-

ber of nations with sophisticated armed forces point toward greater use of limited, multinational coalitions in future land wars. Temporary, opposing regional coalitions and more permanent international forces will probably be necessary to pool limited and complementary resources to mount a military operation.

The accuracy and lethality of weapons available to opposing military forces will continue to improve. Successful and economical employment of these expensive and sophisticated weapons will depend on the skill of the individuals and crews that employ them. While modern technology will provide "smart weapons," simplicity of use will drive their success in battle engagements.

JAMES D. BLUNDELL

SEE ALSO: Airborne Land Forces; Alexander the Great; Armor; Army Aviation; Artillery; Art of War; Attack; Caesar, Julius; Clausewitz, Karl von; Close Air Support; Coalition Warfare; Combat Power and Potential; Combined Arms; Combined Operations; Conventional War; Envelopment; Fire and Movement; Firepower; Force Multiplier; Fuller, J. F. C.; Infantry; Limited War; Low-intensity Conflict; Maneuver; Mechanized Warfare; Napoleon I; Operational Art; Principles of War; Tactics; Tank; Trench Warfare

Bibliography

Association of the U.S. Army. 1991a. *The U.S. Army in Operation Desert Storm: An overview.* Arlington, Va.: AUSA Institute of Land Warfare.
———. 1991b. *Operations Desert Shield and Desert Storm: The logistics perspective.* Arlington, Va.: AUSA Institute of Land Warfare.
Bellamy, C. 1987. *The future of land warfare.* New York: St. Martins Press.
Gabriel, R. A. 1990. *The culture of war: Invention and early development.* Westport, Conn.: Greenwood Press.
Garden, T. 1989. *The technology trap: Science and the military.* London: Brassey's.
Jones, A. 1987. *The art of war in the western world.* Oxford: Oxford Univ. Press.
Kendall, P. 1957. *The story of land warfare.* Westport, Conn.: Greenwood Press.
Liddell Hart, B. H. 1991. *Strategy.* New York: Penguin Books.
Macksey, K. 1973. *The Guinness history of land warfare.* Enfield, England: Guinness Superlatives.
U.S. Department of the Army. 1986. *Field manual 100-5: Operations.* Washington, D.C.: Government Printing Office.

LAOS (Lao People's Democratic Republic)

The Lao People's Democratic Republic, a communist state with its capital at Vientiane, is a landlocked country bordered on the east by Vietnam, the south by Cambodia, the west by Thailand and Burma, and by China on the north. Created in its present form by external powers, notably France and Japan, the current government was formed on 2 December 1975. Present-day Laos is strategically significant from a geographical perspective: as a part of Indochina it is nominally a close ally of that subregion's leader, Vietnam; the Mekong River, which forms much of the border with Thailand (a U.S. ally), flows into northern Laos from China, and exits from its southern border into Vietnam and the Mekong Delta.

Laos is one of the world's poorest nations, with a 1990 per capita income of US$150, high unemployment and inflation, and an economy that is based largely on subsistence agriculture. Most of the country is ruggedly mountainous with some plateaus; agriculture is limited to about 4 percent arable land and 3 percent pastures. Unlike other Southeast Asian states, Laos has few towns. The largest is Vientiane; other important towns are Luang Prabang, the southern Mekong River town of Pakse, and Savannakhet. The country suffers from occasional floods, deforestation, soil erosion, and periodic droughts.

Power Potential Statistics

Area: 236,800 square kilometers (91,429 sq. mi.)
Population: 4,183,400
Total Active Armed Forces: 52,600 (1.257% of pop.)
Gross Domestic Product: US$600 million (1990 est.)
Annual Defense Expenditure: not available (3.8% of GDP, 1987 est.)
Iron and Steel Production: none
Fuel Production: none
Electrical Power Output: 1,100 million kwh (1990)
Merchant Marine: none
Civil Air Fleet: major transport aircraft: not available; 51 usable airfields (9 with permanent-surface runways) none with runways over 3,659 meters (12,000 ft.); 2 with runways 2,440–3,659 meters (8,000–12,000 ft.); 13 with runways 1,220–2,440 meters (4,000–8,000 ft.).

For the most recent information, the reader may refer to the following annual publications:
The Military Balance. International Institute for Strategic Studies. London: Brassey's (UK).
The Statesman's Year-Book. New York: St. Martin's Press.
The World Factbook. Central Intelligence Agency. Washington, D.C.: Brassey's (US).

History

Historically, external political and cultural influences in Laos, as in other Southeast Asian countries, have come from India. In the mid- to late-thirteenth century, however, Mongol invasions of southern China and Southeast Asia influenced the region, and forced the southern migration of the Lao-Lum and Lao-Tai (Lao branches of the Thai) peoples.

Laos originated in 1353 when Prince Fa Nqhua broke from the Khmer Empire and formed the Kingdom of Lan Xang ("land of a million elephants"). Lan Xang's power peaked during the reign of King Suligna Vong Sa (1637–99); thereafter it declined until 1707, when Laos became divided into the three competing states of Vientienne, Luang Praban, and Champa Sak.

In the nineteenth century, Laos avoided becoming a part of Thailand by asking for French assistance; it was made a French Protectorate in 1893, acquiring its present frontiers in 1907. When the Japanese invaded in 1941, French authority was suppressed, and at the end of World War II an independence movement known as Lao Issara (Free Laos) set up a government under Prince Phetsarath. When the French returned in 1946, this government collapsed and the leaders of the movement fled to Thailand. Under French control, a constitutional monarchy was established in 1947, and in 1949, Laos became an independent sovereign state in the French Union.

Support for rebellion and independence remained strong, however, and led to a condition of civil war that lasted twenty years. In 1953 a complex conflict ensued involving three parties: a pro-Western organization under General Phoumi Nosavan; a neutralist group led by Prince Souvana Phouma; and the Pathet Lao forces of Souvana Phouma's half-brother, Souphanouvong, which were aligned with the Viet Minh and occupied much of northern Laos.

In 1961, after North Vietnam commenced its incursion of the South via the Ho Chi Minh Trail through Laos, a fourteen-power conference in Geneva forced the establishment of a neutralist coalition under Souvana Phouma. This had little effect. Fighting resumed between the Pathet Lao (supported by North Vietnam) and the Royal Lao Government (supported by U.S. forces and Thai mercenaries), continuing until 1973 when an agreement and a protocol were signed. This resulted in another coalition government led by the two half-brothers Souvana Phouma and Souphanouvong. With the Vietnamese Communist victory in Vietnam in 1975, however, the Pathet Lao took full control in Laos, and by December of that year Souvana Phouma had to relinquish all power to Souphanouvong. The Lao king and queen subsequently died in "reeducation" labor camps; loyalist resistance has continued to the present.

Politico-Military Background and Policy

Laos's relevant politico-military past begins in 1975. The governmental structure is somewhat less monolithic than other communist states, and is based on a president who is head of state, with a Supreme People's Assembly that is appointed by a National Council of Ministers. The cabinet is totally communist, but the council includes several neutralists and non-communists.

The Laotian military of less than 60,000 members is in no position to pursue aggressive policies outside its borders. Neighboring Vietnam, a co-communist state, has about one million in active service; Thailand, with its U.S. connections, maintains a military force approaching 300,000. Thus, Laotian military policy has focused on maintaining internal security, and in this regard has been generally effective.

Largely dependent on foreign aid for survival, Laos and its four million inhabitants are intent on pursuing a course that will secure their status among other nations in the region. Its governmental policies are directed toward improving economic conditions and gradually extending its contacts abroad. Diplomatically, Laos frequently follows Vietnam's lead, but there are substantial cultural and ethnic differences between the two; increasingly, Laos has followed its own direction in foreign affairs.

In the mid-1980s the United States and Laos moved quietly to establish diplomatic relations. In September 1985, Vice Foreign Minister Soulivong Prasithdeth became the first Communist official from Indochina to visit with U.S. diplomats in Washington, D.C., since 1975. Later, diplomatic representation between these two countries began at the chargé d'affaire level, which resulted in Laos being taken off the so-called enemy list by the United States, an essential step toward creating a trading environment.

More recently, the Lao leadership has moved toward a free-market economy, has provided greater autonomy for some state-run enterprises, and has made limited moves to increase private sector activities. Since 1988, foreign companies have been permitted to participate in Laotian joint ventures and business activities; few, however, have done so. The exception to this is Thailand. The Thais, sometimes bending their accords with ASEAN partners, have increased their commerce and trade with Laos. Laos, sensitive to Vietnam's position, has nonetheless responded in kind; there is some trade also with Malaysia.

A significant aspect of Laos's economic and military policy is one that affects Indochina in general—the loss of the Soviet Union as a major provider of aid, and as a trading partner. In this regard the gradual lessening of central control and expansion of foreign contacts is understandable, and Laos may be setting an example for its communist neighbors.

Strategic Problems

In 1960, U.S. President Eisenhower called Laos "the key to Southeast Asia," and "the crucial domino," in reference to the so-called domino theory of Communist expansion in the region. The opposite condition also pertains: Southeast Asia is the key to Laos. Laos is affected by the problems and policies of China to the north, Vietnam to the east, Thailand and Burma on the west, and—in particular—by Cambodia in the south. Laos may never have started a war, but it has been fought over by Thai and Vietnamese peoples, the French and Japanese, and assaulted by the United States, Thailand, and South Vietnam during the Vietnam War.

In addition to its strategic geographic position, Laos faces its most urgent strategic problem at home: improving the economy to relieve its heavy dependence on foreign aid. Efforts to do so have involved modifying central government policies, and extending its contacts with non-communist nations abroad. This in turn has created strategic political and diplomatic situations that have required deft management and negotiation.

A problem of international concern involves illicit drugs: Laos was once considered to supply 90 percent of the world's opium. It is now third, behind Burma and Afghanistan.

A number of U.S. servicemen, mostly pilots, became missing in action (MIA) in Laos during the Vietnam War. The problem of their identification and recovery of their remains (or the determination if there are, or were, any who remained as prisoners of war) is a problem that is being addressed with the United States and with Vietnam's approval.

Of interest is that China, no particular friend of Vietnam, represents a potentially strategic threat on Laos's northern border. In mitigation, Laos has to some extent closer ethnic ties to the southern Chinese. In addition, any relationship with the United States will take into account Chinese policy.

Perhaps the greatest external threat has to do with Cambodia and the nature of the solution to that country's troubles with the Khmer Rouge and Vietnam. ASEAN efforts to negotiate a peaceful settlement and the creation of a consensual government have had some diplomatic successes within the region, but the Cambodian people are less sanguine about any future involving the Khmer Rouge. Instability, or open hostilities, along Laos's southern border and the vital Mekong River is an undesirable feature.

Military Assistance and Alliances

Until 1990, Laos received a substantial amount of military aid and economic assistance from the USSR and Eastern European communist countries (US$995 million from 1970 to 1989). Since then most military assistance has been provided by Vietnam, largely in the form of about 5,000 economic construction troops stationed in the country.

Laos is more or less allied with Vietnam as a part of Indochina and neighboring communist states. Laos is also a member of the United Nations. Future alliances, with as much emphasis on economic cooperation as on military defense, are probable.

Some degree of foreign economic assistance has also come from other sources: Mongolia, which built a provincial hospital for the Lao; Australia helped construct an important bridge over the Mekong; and Sweden has provided aid.

Defense Industry

Laos has no major or minor defense industrial capability. In addition, the country's industrial infrastructure is poor: there are no railways; the road network is sparse and mostly unpaved; and the Mekong River is the major transportation route.

(For an explanation of the abbreviations and symbols used in the following section of military statistics, see the list of Abbreviations and Acronyms in each volume.)

Total Armed Forces

Active: 52,600. Terms of service: conscription, 18 months minimum.

ARMY: 50,000. 4 Military Regions.
 5 inf div.
 7 indep inf regt.
 3 engr (2 construction) regt.
 5 arty, 9 AD arty bn.
 65 indep inf coy.
 1 lt ac liaison flt.
Equipment:
MBT: 30 T-54/-55, T-34/85.
Light tanks: 25 PT-76.
APC: 70 BTR-40/-60/-152.
Towed arty: 75mm: M-116 pack; 105mm: 25 M-101; 122mm: 40 M-1938 and D-30; 130mm: 10 M-46; 155mm: M-114.
Mortars: 81mm; 82mm; 107mm: M-2A1, M-1938; 120mm: M-43.
RCL: 57mm: M-18/A1; 75mm: M-20; 106mm: M-40; 107mm: B-11.
AD guns: 14.5mm: ZPU-1/-4; 23mm: ZU-23, ZSU-23-4 SP; 37mm: M-1939; 57mm: S-60.
SAM: SA-3, SA-7.

NAVY: est. 600.
Patrol Craft, River: some 40 ⟨.

AIR FORCE: 2,000; 34† cbt ac; no armed hel.
FGA: 1 regt with some 30 MiG-21.
Transport: 1 sqn with 5 An-24, 2 An-26, 2 Yak-40.
Helicopters: 1 sqn with 2 Mi-6, 10 Mi-8.
Training. †4 MiG-21U.
AAM: AA-2 Atoll.

PARAMILITARY
Militia Self-Defense Forces: village 'homeguard' org for local defense.

OPPOSITION: Numerous factions/groups. Total armed strength est. 2,000.
Largest group United Lao National Liberation Front (ULNLF).

FOREIGN FORCES
Vietnam: 5,000: mostly economic construction tps.

Future

The re-education labor camps are now closed, and some measure of reconciliation is taking place; some Laotian exiles have returned home from the United States and from Thailand. These returnees, mostly intellectuals and skilled workers, are viewed as welcome and important to developing the economy. There is a significant rapprochement with Thailand, to the benefit of both countries. Low-key diplomatic relations with the United States and trade delegations to France, Germany, and other non-communist nations, combined with a lessening of central governmental control of the economy, signal a willingness to expand commercial and political horizons, and become less dependent on foreign aid.

Laos has substantial undeveloped resources, including a high potential (some already developed) for hydroelectric power. Development and economic improvement will come through cooperative arrangements with Western, non-communist states, and Thailand will play a key role. This, so long as a peaceful solution to the Cambodian problem can be assured.

CHARLES F. HAWKINS
DONALD S. MARSHALL

SEE ALSO: Burma; Cambodia; China; South and Southeast Asia; Thailand; Vietnam.

Bibliography

Adams, N. S., and A. Adams, eds. 1970. *Laos: War and revolution.* New York: Harper and Row.
Brow, M., and J. J. Zasloff. 1986. *Apprentice revolutionaries: The communist movement in Laos, 1930–1985.* Stanford, Calif.: Hoover Institute Press.
International Institute for Strategic Studies. 1991. *The military balance 1991–1992.* London: Brassey's.
Natkiel, R., and J. Pimlott, cartographer and ed. 1988. *Atlas of warfare.* New York: W. H. Smith.
Scalapino, R. A., and M. Kosaka, eds. 1988. *Peace, politics and economics in Asia: The challenge to cooperate.* New York: Pergamon-Brassey's.
Stuart-Fox, M. 1986. *Laos: Politics, economics and society.* London: Frances Pinter.
U.S. Central Intelligence Agency. 1991. *The world factbook, 1991–1992.* Washington, D.C.: Brassey's.
U.S. Department of State. 1986. *Laos background notes.* Washington, D.C.: Government Printing Office.

LATIN AMERICAN WARS OF INDEPENDENCE

The Latin American wars of independence were a series of early nineteenth-century conflicts in the areas of the Western Hemisphere governed by France, Spain, and Portugal. Although there were some linkages among them, it is helpful to analyze separately the independence wars of Haiti, of Mexico and Central America, of Spanish South America, and of Brazil. At the conclusion of these wars, only a few French Caribbean islands, Spanish Cuba, and Puerto Rico remained colonies of the Latin nations of Europe. From 1861 to 1865, Santo Domingo, for a time under Haitian rule and for a time independent, came briefly under Spanish rule again before peacefully regaining independence.

The military aspects of the wars can be described generally as struggles for independence fought between masses of Indian and mestizo (mixed Spanish-Indian) troops led on the one hand by small numbers of Spanish military professionals, and on the other by *criollos*, first- or second-generation Spaniards born in the colonies. The most noteworthy campaigns were those carried out by the Venezuelan Simón Bolívar and the Argentine José de San Martín, both of whom made epic crossings of the Andes Mountains to take the Spaniards by surprise.

The overall pattern of the Spanish American wars of independence was an initial period of political confusion stemming from Napoleon's occupation of the Iberian Peninsula in 1808, followed by declarations of independence first from Napoleon's puppet government in Madrid, and then from Spain as the mother country. These declarations were buttressed by a string of early patriot victories. But, after Spain threw off the Napoleonic yoke, she reconquered many of her American possessions, and this led to an extended period of conflict between Spaniards and rebels until the rebels gained final victory and independence in the early and mid-1820s (Fig. 1).

Figure 1. The Western Hemisphere at the end of the wars of independence (1826). (SOURCE: Jack Child, 1988)

Background of the Latin American Wars of Independence

GENERAL CAUSES

The general causes of the wars of independence lie in French, Spanish, and Portuguese neglect of their American colonies, their economic exploitation of them, and their feeling that the *criollos* in the colonies were politically, economically, and culturally inferior to their cousins on the Iberian Peninsula. Serious rebellions had been put down brutally in the late eighteenth century, and memories of these outbreaks were still fresh when winds of change began blowing in the Americas, fueled by the American and French revolutions, British-Spanish rivalry, and the ideas of the Enlightenment brought in by Freemasons and liberals.

THE INDEPENDENCE OF THE ISLAND OF HISPANIOLA

The first American colony to achieve independence after the United States was the black republic of Haiti on the island of Hispaniola, although the special circumstances of that event held little precedent for the other colonies in the hemisphere. The French Revolution had abolished slavery, and in Haiti, in the 1790s, this led to a civil war involving black former slaves (led by François Toussaint L'Ouverture), mulattoes, and French settlers. Napoleon's expeditionary force under General Leclerc managed to capture Toussaint, but it was eventually exhausted by disease and the firm resistance of the former slaves. It was only after a decade of bloodshed that Haiti's independence was achieved in 1804 on the western half of the island of Hispaniola. Santo Domingo, on the eastern half of the island, was to be under the intermittent control of Haiti, France, and Spain for another 40 years before achieving independence as the Dominican Republic in 1865.

THE IMPACT OF NAPOLEON'S INVASION OF IBERIA

The spark that set off the process of independence for the rest of Latin America was Napoleon's invasion of Spain and Portugal in 1808. In Portugal, the ruling Braganzas were able to escape the French armies and make their way to their colony of Brazil under the protection of British warships. From Brazil, they continued their rule and eventually returned to Portugal, thus avoiding the sharp splits that characterized the Spanish-American independence process. In Madrid, King Charles IV was forced to abdicate. His son Ferdinand was imprisoned by Napoleon, who placed his brother Joseph Bonaparte on the Spanish throne (1808–13). Spanish resistance to Napoleon was strong in the south of the peninsula, where a junta ruled in the name of Ferdinand. This political division provided Spain's American colonies with the opportunity to form their own local juntas and declare themselves independent from Napoleonic Spain. Many of the backers of this process expected that this would lead to full independence, which it eventually did after fifteen years of bitter struggle.

Mexico and Central America

The Viceroyalty of New Spain included what is today Mexico, Central America, and large portions of the southwestern United States. In 1808, the viceroy declared himself independent of Napoleon and established a government in the name of Ferdinand VII. This, however, was not good enough for the *criollos*, who had been conspiring for greater autonomy. On 16 September 1810, the priest Miguel Hidalgo launched the Mexican war of independence by ringing the bells of his parish church and uttering the "Cry of Dolores" against the Spaniards in the name of the Mexican Virgin of Guadalupe. Father Hidalgo organized a government in Guadalajara and led masses of Indians in battles against the Spanish authorities in Mexico City. Unfortunately, he proved to be a more effective spiritual and political leader than a military campaigner, and he was defeated by the viceroy's forces and captured early in 1811. He was condemned and excommunicated by the Inquisition. Shortly afterward he was tried and shot by the civil government. A fellow rebel priest, José Morelos, took up Hidalgo's banner and convened a congress that adopted a constitution and declared independence (November 1813). But Morelos, too, was defeated in battle and then executed by the Spaniards. It was left to the former Spanish professional soldier Augustín de Iturbide, who joined the patriot cause in 1820, to defeat the Spaniards, which he did in 1821, and consolidate the independence of a sovereign Mexican nation.

In Central America, there were several local independence movements in the period from 1811 to 1814, but these were crushed by the Spaniards. Central America finally achieved independence from Spain as a result of the victories of Iturbide, who invited the peoples of the old Capitancy-General of Guatemala (today's Central America) to join his Mexican Empire. But when Iturbide was overthrown in 1823, the Central Americans seceded from Mexico and established a federation of the "United Provinces of Central America," which lasted for a decade before splitting up into today's five independent Central American nations.

Spanish South America

The military struggle for independence by Spanish South America was essentially a strategic pincer movement focusing ultimately on the Spanish stronghold of Peru and led from the north by the Venezuelan Simón Bolívar and from the River Plate area in the south by the Argentine José de San Martín. Both campaigns came after the 1810 declarations of rebellion against Napoleonic Spain and were characterized by initial victories that were followed by defeats as Spain temporarily regained her colonies, and then final independence in the mid-1820s.

Southern South America

The *criollos* of Buenos Aires had considerable confidence in their ability to make good their independence because they had defeated British invasions in 1806 and 1807, when the Spanish authorities abandoned the *criollos* to their fate. As a result, when news of Napoleon's invasion of Spain and of the incarceration of Ferdinand VII reached Buenos Aires, there was a firm determination to use the events in Madrid as the basis first to declare a break from Napoleonic Spain (25 May 1810) and then to effect a complete break from the mother country (9 July 1816). Gen. Manuel Belgrano was a key figure in this early period, but he suffered military reverses after Ferdinand VII was restored to the throne. This was also the period when Paraguay split off in its own independence movement, when the Uruguayan general José Gervasio Artigas was struggling against encroachments from both Argentina and Brazil, and when the Chilean patriot Gen. Bernardo O'Higgins was decisively defeated by the Spanish and had to retreat across the Andes Mountains into Argentina.

After 1812, military leadership of the southern South American independence movement passed into the able hands of José de San Martín. His strategic concept was to establish himself in western Argentina, then cross the Andes and strike the Spanish forces in Chile by surprise. From there he would head north to the center of Spanish authority in Peru, while Bolívar similarly headed south from his base in Venezuela. In early 1817, San Martín's forces, working with the Chilean patriots under General O'Higgins, crossed the Andes in an unprecedented feat and caught the Spaniards off guard, defeating them at Chacabuco (12 February 1817). A year later, at the Battle of Maipú (5 April 1818), he and O'Higgins consolidated the independence of Chile and began the campaign against the Spaniards in Peru. A fleet was organized with the help of Englishman Lord Thomas Cochrane. In July 1821, after victories on land and sea, San Martín entered Lima and proclaimed its independence. Important Spanish garrisons, however, remained in control of large sections of the region now encompassing Peru, Ecuador, and Bolivia. In July 1822, San Martín and Bolívar met privately for two days at Guayaquil, Ecuador, to plan the final campaign against the Spaniards. There is much mystery surrounding these meetings between the two liberators, but when the conversations ended San Martín withdrew from the scene and went into European exile, leaving to Bolívar the direction of the final stages of the wars of independence.

Northern South America

The *criollos* of Caracas, like those of Buenos Aires, had declared their independence of Napoleonic Spain in 1810, and Simón Bolívar, the son of wealthy landowners, had led Venezuelan forces in a series of early victories against the Spanish. His War to the Death against the Spanish general José Boves was particularly bloody, and when the resurgent Spanish captured Caracas in 1814, Bolívar was forced to flee first to Bogotá (Colombia) and later to the island of Jamaica. Left behind to fight in the plains of the Orinoco Basin was Gen. José Antonio Páez, who led his irregular cavalry forces in a series of running battles against the Spanish.

Bolívar returned to Venezuela early in 1817 and began the second phase of the struggle against the Spanish. By late 1818, his forces, allied with those of Páez, were in control of most of the Orinoco Valley. This permitted the patriots to declare independence formally at Angostura (20 November 1818) and to begin the campaign against the Spanish forces still holding Caracas. The final defeat of the Spanish in Venezuela was assisted by substantial numbers of foreign volunteers, mainly British. With Venezuela in his hands, Bolívar, like San Martín, planned a hazardous crossing of the Andes and a surprise attack on the Spanish garrison in what is today Colombia. He achieved this at the critical Battle of Boyacá (7 August 1819), finally expelling the last Spanish forces from northern South America at the Battle of Carabobo, on 25 June 1821. The following year, one of Bolívar's lieutenants, Gen. Antonio José de Sucre, obtained the independence of Ecuador at the Battle of Pichincha, 22 May 1822, fought on the slopes of a volcano just outside the capital city of Quito.

After the fateful meeting between San Martín and Bolívar in Guayaquil, Bolívar was left as supreme military commander of patriot armies in South America. Scattered Spanish forces held on, however, until decisively defeated by Bolívar and his generals first at the Battle of Junín (6 August 1824) and then at Ayacucho (9 December 1824).

Brazil

There were no wars of independence in Brazil, which separated from Portugal gradually and with little violence. The Portuguese Braganza dynasty had established itself in Brazil after Napoleon invaded the peninsula, and when Portuguese King John VI returned to Portugal, his son Pedro stayed behind. When Brazil obtained its independence from Portugal on 7 September 1822, it was as an empire, with the Braganzan Pedro I as its first sovereign.

Jack Child

See Also: Bolívar, Simón; Central America and the Caribbean; San Martín, José de; South America; Spanish Empire.

Bibliography

Anna, T. E. 1983. *Spain and the loss of America*. Lincoln: Univ. of Nebraska Press.
Bethell, L., ed. 1984. *The Cambridge history of Latin America*. Vol. 3: *From independence to c1870*. Cambridge: Cambridge Univ. Press.
Domínguez, J. I. 1980. *Insurrection or loyalty: The breakdown of the Spanish-American empire*. Cambridge: Harvard Univ. Press.

Herring, H. 1968. *A history of Latin America.* 3d ed. New York: Knopf.
Humphreys, R. A., and J. Lynch, eds. 1966. *The origins of the Latin American revolutions.* New York: Knopf.
Keen, B. 1986. *Latin American civilization: History and society, 1492 to the present.* 4th ed. Boulder, Colo.: Westview Press.
Lynch, J. 1973. *The Spanish American revolutions, 1808–1826.* New York: Norton.
Madariaga, S. de. 1947. *The fall of the Spanish American empire.* New York: Macmillan.
Prago, A. 1970. *The revolutions in Spanish America.* New York: Macmillan.

LATVIA

Latvia, one of the fifteen former republics in the Union of Soviet Socialist Republics (USSR), declared its independence from the Union on 4 May 1990. On 6 September 1991, the Soviet government formally recognized the independence of Latvia and the two other Baltic republics, which the USSR annexed in 1940. This dramatic establishment of Latvian sovereignty followed several years of dynamic and unprecedented change in the Soviet Union. For the next several years, relations between Latvia and the rest of the world are likely to continue to change. Over time, new structures and patterns will emerge in economics, trade and commerce, politics and government, finance, manufacturing, religion, and virtually all aspects of human life. New arrangements must be devised for dealing as a sovereign state with the new Commonwealth of Independent States (which is itself undergoing continual change) and with the world outside the boundaries of the former Soviet state. If the history of the Soviet Union since 1985 is any guide, we can expect dramatic surprises and dynamic change.

An important question for the world is how Latvia and the Commonwealth will organize and provide for security. Most of the Soviet Union's armed forces stationed in Latvia are likely to be withdrawn. Also of great concern is the disposition of nuclear weapons, the security of these weapons, the command and control of their potential use, and compliance with arms control agreements entered into by the former Soviet government. The world can only hope these issues are settled amicably.

It will be years before all of these issues are resolved for Latvia and some time before events settle down into more routine and measurable patterns. No accurate description of this new country's policies, defense structure, and military forces was available to be included in this encyclopedia. Only time will reveal the future of Latvia as a separate sovereign state. The reader is thus referred to the historic information contained in the article "Soviet Union," and to the latest annual editions of the *Military Balance*, published by Brassey's (UK) for the International Institute for Strategic Studies; the *Statesman's Year-Book*, published by the Macmillan Press Ltd and St. Martin's Press; and the *World Factbook*, developed by the U.S. Central Intelligence Agency, and published commercially by Brassey's (US).

F. D. MARGIOTTA
Executive Editor

LAW, MILITARY

Because the armed forces of all states possess lethal weapons, it is essential that discipline be maintained and enforced against individual servicemen by military courts applying military law. Thus, the British *Manual of Military Law* states that the object of military law is "to provide for the maintenance of good order and discipline among members of the army." To do this, military law creates purely military offenses, such as absence without leave or insubordination to a superior officer, but it may also enforce the ordinary criminal law through military courts when the offenders (and usually, the victims) are servicemen. The reason often given for such a wide jurisdiction is that the system of military law established for the armed forces is designed to operate in peace as well as in war. The armed forces of some states, such as those of the United States and of the United Kingdom, are frequently stationed outside their own territories, and therefore outside the jurisdiction of their own domestic courts. In these circumstances, the criminal law of the state to which the armed forces belong can be enforced against servicemen only by the military courts of that state. In contrast, the domestic law of some states (e.g., Sweden) permits criminal offenses committed by servicemen to be tried only by its ordinary criminal courts and not by its military courts.

The military law of any state will be structured around the legal system and the traditions of that state. It is likely also to reflect the nature of military service in that country. If the state depends on an all-volunteer force, its military law is likely to equate more fully with the rights and obligations of civilians, whereas compulsory military service may lead to a harsher regime for individual servicemen. In the United Kingdom, the move from compulsory military service to an all-volunteer force has led to a rethinking of the structure of military law.

Military law may appear to be a legal system within a legal system. It is important, therefore, that the civilian courts exercise some control over military courts. This they may do through the appeal system (whereby a serviceman convicted before a military court is given the right of appeal to a civilian court), through the review of procedures adopted by military courts, or through the right of an individual serviceman to petition a supranational body (as under the European Convention on Human Rights). All three options operate within the military legal system of the United Kingdom.

Although it is within the sole competence of a state to determine its own military law, the international law of war (also known as the law of armed conflict or humanitarian law) applies to all states bound by the treaties from which it derives its force. These range from the Hague Conventions of 1907 through the Geneva Conventions of 1949 and the 1977 Additional Protocols. By 1 May 1991, 164 states were party to the Geneva Conventions of 1949, while 102 states were party to the First Additional Protocol 1977 and 92 states were party to the Second Additional Protocol of 1977. In addition, a number of rules of customary international law bind all states. These are very basic rules, such as the principle that combatants must distinguish between military objectives and civilian persons and objects and attack only the former.

Military Law: The Military Legal System

A court-martial has jurisdiction only over those subject to military law: all members of the regular forces, reservists while on duty, and civilians who are within the limits of an officer commanding a military unit and who come within a list set out in the Fifth Schedule to the Army Act 1955. This last category comprises the families of servicemen living with them and those, such as teachers and administrators, who accompany the armed forces abroad. Subjecting civilians to military law enables them to benefit from the NATO Status of Forces Agreement of 1951. Under the law of the United States, by way of contrast, civilians are not subjected to military law and can be tried only by the courts of the United States or, should they be accompanying the U.S. forces abroad, the courts of the state in which the crime was committed.

The British Army, Royal Air Force, and Royal Navy have separate legal systems; the marked differences among them reflect the historical growth of each service. (In other countries, a uniform system of military justice is common.) In the British army, a soldier may be tried by a general or a district court-martial for an offense within the Army Act 1955, depending upon the seriousness of the offense. These range from misconduct in action, insubordination, and mutiny, to criminal offenses under Section 70 of the Act. A general court-martial comprises five officers and a district court-martial, three. In wartime, a field general court-martial may be established with a minimum of only two officers. In some armed forces, noncommissioned officers may serve, but it is a virtually universal rule that all or most of the members of the court must be officers. The procedure is similar to a civilian court trying a criminal offense, and the rules of evidence applying in that court will also apply in a court-martial, but there are a number of important differences. In a court-martial trial, there is no jury; under British military law, the verdict is reached by a majority of the members. A British Army or RAF court-martial verdict and sentence (but not that of a Royal Navy court-martial) are subject to confirmation by the military officer who convened the court. The officer may take legal advice and a convicted soldier is entitled to petition him not to confirm the verdict or sentence. Errors that may have occurred during the court-martial may be corrected at this stage instead of by the Court-Martial Appeal Court, a purely civilian court. (A number of countries possess a court of appeals from courts-martial trials composed of either military officers appointed directly to it, or a combination of civilian judges and military officers.) A serviceman may appeal the verdict to the Court-Martial Appeal Court, but the sentence imposed may not be changed if the verdict is subsequently confirmed. The military authorities, acting through court-martial trial and through the confirmation procedure, are deemed the best arbiters of the army's need to enforce discipline.

Civilians subject to military law when the British armed forces are serving abroad may be tried by a Standing Civilian Court when the offense charged is not a serious one. For serious offenses, they may be tried by court-martial on the same basis as servicemen and if convicted may appeal both sentence and verdict to the Court-Martial Appeal Court.

In a number of states, a commanding officer may impose military punishment for a purely military offense. This is a faster and simpler procedure than convening a court-martial. Article 15 of the Uniform Code of Military Justice in the United States permits a commanding officer to impose nonjudicial punishments; Article 16 deals with the types of courts-martial, including a summary court-martial consisting of one commissioned officer. In the British Army, a soldier may have a charge of theft or a purely military offense dealt with by his commanding officer; these procedures are intended to be speedy and limited. Once the charge is established, the soldier may be ordered to pay a fine (based on units of a day's pay) or he may be placed in detention for up to 28 days. The Royal Navy has more extensive powers of summary disposal due to the difficulty of convening a court-martial at sea.

A committee set up by the British government reported in 1946 that "In the matter of legal safeguards, citizens should be no worse off when they are in the Forces than in civil life unless considerations of discipline or other circumstances make such a disadvantage inevitable." The rights of a serviceman are not identical with those of a civilian, but it is commonly argued that any differences are essential to maintain discipline in the armed forces. Four such differences are the continuing punishment of homosexual acts between consenting adult servicemen, the imposition of the death penalty for certain military offenses, the ban on trade union activity, and the limited rights of employment in the armed forces. By way of example, the position in the British armed forces will be considered.

Although it is no longer an offense for consenting adults to commit homosexual acts in private, it continues to be an offense for a serviceman; in 1990, 74 were so convicted by court-martial or were discharged administratively. The Ministry of Defence reasons that "the nature of service operations means that absolute trust and confidence within and between all ranks is essential and any activities, such as homosexual practices, which might disrupt such trust and confidence must be avoided . . . [homosexuality] makes people liable to blackmail and hence may present a security risk."

The death penalty has been abolished in most European countries, but all countries retain this penalty for certain serious military offenses committed during wartime. In 1991, the British House of Commons attempted to abolish the death penalty altogether in the armed forces, but the attempt to do so was defeated. The reason commonly given for its retention is the need to apply the ultimate deterrent to servicemen who might consider committing such offenses.

The ban on trade union activity is also widespread among the armed forces of the world. Some countries, such as the Netherlands and Sweden, permit servicemen's organizations to exist, but these organizations are largely the product of conscript armies. All ban the taking of industrial action by members of the armed forces. Such action would normally be considered as mutiny, a serious military offense. Allied with this is the restriction of employment rights of servicemen compared with the civilian. In the armed forces of some states (e.g., Great Britain) a serviceman who is dismissed is unable to bring an action before the courts for wrongful dismissal or for redundancy. However, if no remedy were available to servicemen who believed that they had been wrongly treated, discipline would inevitably suffer. It is therefore common to provide some form of complaints machinery, in the form of either a serviceman's ombudsman (as in Australia) or a right to make a complaint to a commanding officer (as in the British armed forces). In the latter case, the courts may exercise jurisdiction should the complaints procedure be carried out in an improper manner (R. v. Army Board of the Defence Council, ex parte Anderson, 1991).

The Laws of War Since World War II

The distinction between military law and the international law of war has been discussed above. It may be difficult to decide whether an armed conflict is "international" and therefore whether the international laws of war apply to it. The Vietnam War is a good example: the United States took the view that it was an international armed conflict between two separate states, North and South Vietnam; the North Vietnamese considered it a civil war and refused to concede that the laws of war applied. In the Falklands/Malvinas conflict in 1982 and the Gulf war in 1990–91, the international laws of war clearly applied. However, individual servicemen remained subject to their own military law and could have been tried by court-martial for any offenses committed against their fellow servicemen or (subject to any status-of-forces agreement) against civilians. In addition, a serviceman may be liable, under either his own military law or international law, for committing any war crimes (defined as breaches of the laws and customs of the international laws of war).

Conclusion

Any system of military law must strike a balance between the need to enforce discipline in the armed forces and the rights of servicemen. Different armed forces will take differing stances on this balancing act, depending on their own legal tradition and on whether recruits are drawn from volunteers or from conscripts. Should the armed forces of a state become involved in an international armed conflict, the laws of war will be imposed in addition to the obligations already placed on the serviceman by his military law.

PETER ROWE

SEE ALSO: Command; Discipline; Geneva Conventions; Hague Conventions; Legal Assistance; NATO Policy and Strategy; Rules of Engagement: Legal Aspects; War Crimes.

Bibliography

McCoubrey, H. 1990. *International humanitarian law.* Aldershot: Dartmouth.
Rowe, P. 1987. *Defence: The legal implications.* London: Brassey's.
Ministry of Defence. 1972. *Manual of military law.* 3 vols. London: Her Majesty's Stationery Office.

LAW OF THE SEA AND PIRACY

The law of the sea is one of the most dynamic arenas of international law. The very nature of the oceans as a medium for international activity signifies their importance. Nations must share in their use. The oceans play a vital role in defining the nature of international relations. Commercial-economic, geopolitical, and military relations are all affected by our use of the world's oceans. As a result, the law of the sea is central to the economic prosperity and security of the Western world.

This article covers the historical development and sources of the law of the sea, its basic doctrinal content, and a few specific areas of the law influenced by recent attempts to reach universal agreement on their content. These areas include: the territorial sea, archipelagic states, international straits, and exclusive economic zones. This is followed by consideration of several areas where more progress and international cooperation are needed before a satisfactory consensus can be attained. Included here are: restrictions on innocent passage in territorial seas, delineating straight baselines used for measuring territorial seas, and historical water claims.

Historical Development

As nation-states emerged (during the 16th and 17th centuries), customary practice became the dominant source of international law. Traditional or customary practices evolved into international law by states engaging in, and consenting to be bound by, such practices. Today, the sources of international law include customary practice, international conventions, judicial decisions, legal writings, and what is referred to as general principles of international law. Although the law of the sea has historically drawn on all these, it still depends primarily on customary law and its reflection in international convention.

Customary Law and Early Attempts at Codification

The customary law that has developed over the past four centuries has attempted to regulate the delicate balance between coastal state controls and maritime interests. Customary international law has defined the essential principles that comprise the law of the sea, including high seas freedoms, rights in territorial seas, flag state jurisdiction over ships, and jurisdiction over marine resources.

The other principal source of the law of the sea is conventional law, reflected in written multilateral agreements among participating states. There have been several modern attempts to codify the law of the sea in such agreements.

The first United Nations Conference on the Law of the Sea (UNCLOS I) convened in Geneva in 1958. UNCLOS I sponsored four conventions: the High Seas; the Continental Shelf; the Territorial Sea and Contiguous Zone; and the Living Resources of the High Seas conventions. These conventions failed to resolve a number of important, but contentious, issues, such as the maximum breadth of the territorial sea, the breadth of the continental shelf, coastal state rights in environmental protection, and conservation of living resources beyond the territorial sea.

The second United Nations Conference on the Law of the Sea (UNCLOS II), held in 1960, failed by one vote to achieve agreement on the maximum width of the territorial sea.

Third United Nations Conference on the Law of the Sea (UNCLOS III)

UNCLOS III, convened in 1973, was the most ambitious effort to codify the law of the sea. It addressed the entire spectrum of maritime issues and produced a comprehensive code (the 1982 LOS Convention) that is widely accepted (although not yet formally in force) as an authoritative statement governing the traditional uses of the seas. Except for provisions dealing with deep seabed mining, the 1982 LOS Convention largely reflects customary or emerging customary law.

Doctrinal Aspects of the Law of the Sea

The law of the sea comprises a set of rules that regulate access to and use of the world's oceans. The following selection of doctrinal rules illustrates how the law of the sea strikes a delicate balance between coastal state jurisdictional claims and maritime interest in free use and access to the seas.

Internal Waters

Internal waters refers to waters landward of the baseline (normally the low-water mark along the coast) from which the territorial sea is measured. The coastal nation exercises complete sovereignty over these waters, and other nations have no right to navigation or overflight without consent of the coastal state.

Territorial Sea

The territorial sea is the belt of ocean area seaward of the baseline and extending seaward a maximum of 12 nautical miles (NM). The coastal nation exercises sovereignty over such waters, including the seabed, the subsoil, and the airspace above. Other nations have the right of innocent passage for surfaced ships and transit passage for all ships and aircraft through territorial seas in international straits, in accordance with specific rules prescribed in the law of the sea.

Contiguous Zone

The contiguous zone is a belt of ocean area adjacent to the territorial sea and extending up to 24 NM from the baseline, in which the coastal nation may exercise control necessary to prevent or punish infringement of its customs, fiscal, immigration, or sanitary laws and regulations within its territory or territorial sea.

Exclusive Economic Zone (EEZ)

The exclusive economic zone is adjacent to the territorial sea, extending up to 200 NM from the baseline. In this zone the coastal state has sovereign rights for the purposes of exploring, exploiting, conserving, and managing the natural resources of the waters, seabed, and subsoil; other nations may undertake resource-related activities only with the consent of the coastal state. The coastal state may exercise jurisdiction with regard to establishment of artificial islands, marine scientific research, and protection of the environment.

High Seas

The High Seas are all parts of the sea that are not included in the exclusive economic zone, in the territorial sea, or in the internal waters of a coastal nation.

Continental Shelf

The continental shelf comprises the seabed and subsoil of the submarine area extending beyond the territorial sea, throughout the natural prolongation of the land territory, to the outer edge of the continental margin (or generally to a distance of 200 NM from the baseline where the continental margin falls short of 200 NM, or according to a complex formula in certain situations). The coastal state exercises sovereign rights for purposes of exploring and exploiting natural resources on the continental shelf; other nations may undertake such activities only with the consent of the coastal state.

Archipelagic States

Archipelagic states are nations made up of one or more archipelagoes and other islands—but no continental land mass. Archipelagoes are groups of islands and their interconnecting waters and natural features that together form an intrinsic geographical, economic, and political entity. An archipelagic state may delineate a system of straight baselines enclosing the outermost islands; waters within such lines are called archipelagic waters. Normal routes through and over such waters are termed archipelagic sea-lanes, which are subject to continuous and expeditious navigation (surface and submerged) by all ships, and overflight by aircraft.

High Seas Freedoms

Outside the territorial seas and exclusive economic zones lie high seas areas where all nations enjoy free access to and use of the oceans and resources. High seas freedoms include, but are not limited to: freedom of navigation and overflight; fishing; laying of submarine cables and pipelines; construction of artificial structures, islands, or installations; and marine scientific research. Such freedoms must be exercised with due regard for the interests of other nations to enjoy the same freedoms. The high seas freedoms that relate to navigation and overflight, and the laying and maintenance of submarine cables and pipelines may also be exercised in the EEZ.

Innocent Passage

Innocent passage applies only to navigation through the territorial sea. All ships have the right to transit such seas on the surface so long as such transit is continuous and expeditious and not prejudicial to the peace, good order, or security of the coastal state. There is no right of overflight by aircraft over the territorial sea. Coastal states may regulate only certain aspects of innocent passage, such as prescribing traffic schemes for safety, and conservation of resources.

Transit Passage

Transit passage refers to navigation (surface and submerged) and overflight through international straits, that is, straits that are at some point completely lapped by territorial seas and that are susceptible to international navigation between one part of the high seas or EEZ and another part of the high seas or EEZ. All ships and aircraft have the right of transit passage so long as the transit is continuous and expeditious and not prejudicial to the peace, good order, or security of the coastal state. Strait states may enact laws and regulations, relating to transit passage, in the following areas: safety of navigation and regulation of maritime traffic; prevention, reduction, and control of pollution; fishing; and customs (fiscal, immigration, and sanitation). Such laws and regulations may not discriminate among foreign ships or in their application have the practical effect of denying, hampering, or impairing the right of transit passage as set forth in Part III of the 1982 LOS Convention.

Impact of the 1982 LOS Convention

The following sections describe the impact of the LOS Convention on four aspects of the law of the sea.

Territorial Sea

The concept of the territorial sea arose with the development of nation-states in the sixteenth century. However, questions regarding the nature of a state's jurisdiction in its territorial sea, as well as the breadth of such seas, have been subject to dispute.

The nature of jurisdiction in the territorial sea concerns the extent of coastal state control. By the 1930s the notion of complete sovereignty, subject to certain transit rights (e.g., innocent passage), was generally accepted.

The issue of the territorial sea's breadth has been more contentious. Early rules focused on criteria such as the limits of visibility, fixed distances, or the range of coastal defense armaments (i.e., the cannon shot or 3-NM rule). These rules eventually resulted in the 3-NM limit which was widely accepted by the late nineteenth and early twentieth centuries as the customary maximum breadth.

Subsequently, and especially since World War II, however, states have asserted broader territorial sea claims, generally of 12 NM. One of the principal objectives of UNCLOS III was agreement on the maximum breadth of territorial seas. By the mid-1970s, the 12-NM limit was emerging as the customary rule of international law. The 1982 Convention helped solidify the process by recognizing a maximum breadth of 12 NM.

The 1982 Convention also solidified the allocation of rights and responsibilities in the territorial sea. It is more specific about the respective rights and duties of coastal and noncoastal states than the 1958 Convention on the Territorial Sea and Contiguous Zone. For example, the 1982 Convention enumerates the kinds of activities of a ship that are inconsistent with innocent passage.

Archipelagic States

States composed of far-flung island groups, such as Indonesia and the Philippines, have long advocated the need to recognize their special archipelagic status. They have argued that the political integrity and security of such island territories require a different treatment for the waters between their islands than for the waters off the coasts of continental states. Prior to UNCLOS III, however, the principle of the archipelagic state was not widely accepted.

Maritime nations traditionally rejected the concept because it threatened their freedom of navigation and overflight through and over important shipping routes, which were regarded as high seas.

By the time the 1982 Convention was completed, archipelagic states and maritime nations had accepted an allocation of their respective rights and responsibilities. Maritime nations have accepted the proposition that archipelagic states exercise sovereignty over the waters enclosed by their archipelagic straight baselines and over the belt of territorial sea adjacent to those baselines. By the same token, the archipelagic states recognized that those waters are subject to navigational rights and freedoms: archipelagic sea-lanes passage for all ships and aircraft on, over, and under routes through archipelagic waters normally used for international navigation, as well as innocent passage for vessels on the surface in all other archipelagic waters.

The 1982 LOS Convention endorsed this practice by recognizing the archipelagic status of states meeting certain geographic criteria, that is, a baseline limitation of 100 NM and requirements that the states comprise islands only, and that their water-to-land ratios are between 1 : 1 and 9 : 1. With these criteria, and with agreement on navigational rights through such states, there has been a high degree of compliance with the archipelagic provisions in the convention by nations claiming archipelagic status.

Transit Passage Through International Straits

An international strait is commonly described as a natural passageway of water connecting one area of the high seas (or EEZ) with another area of high seas (or EEZ) that can be used for international navigation. In many instances, these passageways form vital links between strategically important ocean areas. Transit passage, defined above, applies to international straits that may be at least partly overlapped by the territorial seas of the bordering coastal state or states. The trend toward 12-NM territorial seas has increased the importance of this right by increasing the number of straits overlapped by territorial seas.

The right of transit passage is part of customary international law and is reflected in the 1982 LOS Convention. Arguments have been made that this right was a creation of UNCLOS III, but state practice predating the conference indicates otherwise. For example, free navigation and overflight by military aircraft through and over the straits of Bab-el-Mandeb and Hormuz, without coastal state permission, have taken place for over 25 years, notwithstanding the overlapping territorial sea claims of the littoral states.

The 1982 LOS Convention did, nevertheless, have a positive impact with respect to transit passage. It not only recognized the right, but gave it added support and definition. The convention sets out the specific rights and duties of coastal nations and maritime nations. For example, coastal states may prescribe sea-lanes for surface ships where necessary for navigational safety; ships and aircraft, on the other hand, must proceed without delay through straits and may not engage in, or threaten, the use of force against the bordering states.

Exclusive Economic Zones

The relatively new concept of EEZ emerged as an outgrowth of the post–World War II trend toward assertions of resource-related jurisdiction seaward of the territorial sea. The rationale for EEZs is the perceived need by coastal nations for greater control over fish and other resources in coastal regions that are susceptible to exploitation by other nations. The EEZ provides a satisfactory means by which resource-related jurisdiction can be exercised without extending the territorial sea, thereby preserving other high seas freedoms outside 12 NM.

This trend in customary practice, which coincided with the UNCLOS III process, is likewise reflected in the 1982 LOS Convention. States generally have endeavored to conform their claims—both in terms of geographic limits and jurisdictional control—to the language and requirements of the convention.

Subject Areas Less Directly Influenced by the Convention

The preceding sections focused on four areas in which the 1982 Convention helped solidify state practice. There remain, however, other substantive areas concerning traditional uses of the ocean where the convention's impact has been less effective. What follows is a short discussion of two such areas: restriction on innocent passage in territorial seas, and claims of excessive straight baselines used to measure territorial seas.

Restrictions on Innocent Passage

An important area in which the 1982 LOS Convention has not been as successful as desired is the eradication of unlawful restraints on innocent passage. As of April 1991, there are approximately 38 nations with some requirement for prior notice of, or prior permission for, warships conducting innocent passage in their territorial seas. These requirements are unfounded in customary international law, and their legitimacy is not reflected in the 1982 LOS Convention.

The 1982 Convention amplifies the definition of innocent passage with a list of specific activities which, if undertaken by any vessel, would be inconsistent with innocent passage. This list also contains activities uniquely associated with warships—reflecting the fact that they are entitled to conduct innocent passage under the same circumstances and conditions as other vessels. Neither customary international law nor the 1982 Convention advances the notion of prior permission or notice as legitimate conditions of coastal state control over innocent passage.

STRAIGHT BASELINE AND HISTORICAL WATER CLAIMS

In recent years there has been a proliferation of straight baseline and historical water claims. These claims, when excessive, result in unlawful restrictions on navigation and overflight.

Normally, the territorial sea is measured from the low-water mark along the coastline. But when the coastline is deeply indented and cut into, or where there is a fringe of islands along the coast in its immediate vicinity, international law permits the use of straight baselines to join appropriate points along the coast. From these lines the breadth of the territorial sea may then be measured.

Historical waters—often claimed in the geographic context of bays—are waters over which the coastal state asserts that it has historical title. This occurs when the coastal state exercises sovereignty over the waters for a considerable amount of time, and where other states acquiesce in such exercise of authority.

Waters landward of straight baselines, and waters comprising historical waters, are largely treated as internal waters and thereby subject to the complete sovereignty of the coastal state.

Approximately 60 nations have claimed straight baselines, and an estimated 30 historical bays and other historical water areas have been claimed. Unfortunately, the 1982 LOS Convention provides little or no specific criteria with which to judge the validity of these claims. As a result, there continues to be a need for the international community to formulate specific criteria for assessing the legitimacy of such claims.

Conclusion

The law of the sea is an integral part of the overall fabric of international relations. It is a dynamic body of law that has followed the general evolution of international law and, in some respects, provided a model for its development.

There have been numerous attempts to codify this body of law, most recently in UNCLOS III. Many of the long-recognized tenets of customary law were adopted by the UNCLOS III process and are reflected in the 1982 LOS Convention. This document provides a single comprehensive source of authority for essentially all law of the sea matters that concern the traditional uses of the oceans.

It has been less influential, however, in stemming excessive coastal state claims, such as restricting innocent passage and asserting rights to historical waters. Nevertheless, the 1982 LOS Convention represents the greatest milestone yet in the development and safeguarding of the law of the sea. However, as with all law, its development is a day-to-day process. The reality is that all nations must remain ever vigilant in diplomacy and practice in ensuring that the law remains responsive to, and supportive of, the needs of the world community as a whole.

Piracy

The term *piracy* conjures up visions of pirates of old sailing over the horizon with the skull and crossbones. But even today, pirates threaten commerce on the seas. They are active in such places as the Gulf of Thailand, the Straits of Malacca, and off the western coast of Africa.

Piracy today has also been confused with other types of threats on the oceans. For example, it is sometimes used to describe armed conflict between states or state-sponsored terrorism at sea, activities which are more properly classified under different regimes of international law.

To help alleviate such confusion, the following paragraphs describe the background and technical meaning of the term *piracy* in international law.

Piracy is an international crime, one of those few acts that is considered to violate the laws of nations (others include genocide, slavery, war crimes, and torture). Pirates are international criminals subject to the universal jurisdiction of the world's nation-states.

Nations of the world have often used the term *piracy* and *pirate* in their domestic legislation to define and proscribe a wide range of heinous activity (usually occurring at sea) that goes beyond that included under international law as part of piracy. Under customary international law, the definition of *piracy* is robbery or forcible depredation on the high seas. A more recent definition, derived from the 1958 Convention on the High Seas, is contained in Article 101 of the 1982 LOS Convention, and states:

> Piracy consists of any of the following acts:
> (a) any illegal acts of violence or detention, or any act of depredation, committed for private ends by the crew or the passengers of a private ship or a private aircraft, and directed:
>   (i) on the high seas, against another ship or aircraft, or against persons or property on board such ship or aircraft;
>   (ii) against a ship, aircraft, persons or property in a place outside the jurisdiction of any State;
> (b) any act of voluntary participation in the operation of a ship or of an aircraft with knowledge of facts making it a pirate ship or aircraft;
> (c) any act of inciting or of intentionally facilitating an act described in sub-paragraph (a) or (b).

This definition contains two essential limitations: the piratical acts must involve two vessels or aircraft, and the piratical acts contemplate private objectives or gains as opposed to governmental objectives or gains. There is some authority, and considerable juridical sentiment, for a crime of piracy to exist where only one vessel is involved, but the generally accepted definition does not extend that far. A more substantial problem is over the issue whether the "private ends" requirement excludes acts, otherwise piratical, committed for public purposes by, for example, a politically motivated group.

While precedent exists for classifying acts undertaken for political gain as piracy under domestic criminal law, it is generally accepted under international law that piracy does not include criminal acts undertaken for purely political motives.

The "private ends" requirement of the 1958 High Seas Convention was not changed in the 1982 LOS Convention. Thus, situations that are perpetrated for political purposes, like the 1961 forcible takeover of the Portuguese cruise ship *Santa Maria* in the Caribbean by the insurgent Captain Galvao, and the 1985 terrorist hijacking of the Italian liner *Achille Lauro* are not international piracy. They involved no second ship and were not for "private ends." Nor is belligerent action undertaken by states in armed conflict properly termed piracy.

Traditional piracy continues today and remains a violation of the law of nations. Its definition, however, is limited and does not include many modern-day forms of violence at sea. Terrorism and armed conflict must be addressed under other international legal regimes. Today, the international focus is not on expanding the definition of piracy, but rather on dealing with these other types of activity.

WILLIAM L. SCHACHTE, JR.

SEE ALSO: Terrorism, Legal Aspects of; United Nations.

Bibliography

Alexander, L. M. 1986. *Navigational restrictions within the new LOS context: Geographical implications for the United States.* Peace Dale, R.I.: Offshore Consultants.

Brittin, B. H. 1986. *International law for seagoing officers.* 5th ed. Annapolis, Md.: U.S. Naval Institute Press.

Churchill, R. R., and A. V. Lowe. 1983. *The law of the sea.* Manchester, U.K.: Manchester Univ. Press.

Crockett, C. H. 1976. Toward a revision of the international law of piracy. *De Paul Law Review* 26:78–99.

Ellen, E. F., ed. 1986. *Violence at sea.* Paris: ICC Publishing.

Harlow, B. A. 1983. Freedom of navigation in a post–UNCLOS III environment. *Law and Contemporary Problems* 46(2):125–35. Durham, N.C.: Duke Univ. School of Law.

McCullough, L. A. 1986. International and domestic criminal law issues in the Achille Lauro incident: A functional analysis. *Naval Law Review* 36:53–108. Newport, R.I.: Naval Justice School.

Negroponte, J. D. 1986. Who will protect the freedom of the seas? In *The law of the sea—What lies ahead?* Proceedings of the 20th annual conference of the Law of the Sea Institute, ed. T. A. Clingan, pp. 126–30. Honolulu: Univ. of Hawaii.

O'Connell, D. P. 1975. *The influence of law on sea power.* Annapolis, Md.: U.S. Naval Institute Press.

———. 1982. *The international law of the sea.* Oxford: Clarendon Press.

Oxman, B. H. 1983. The new law of the sea. *American Bar Association Journal* 69:156–62.

Reagan, R. W. 1983. Exclusive economic zone of the United States of America. Proclamation No. 5030. *Federal Register* 48:10605. Washington, D.C.: Government Printing Office.

Schachte, W. L. 1986. Violence at sea. In *The law of the sea— What lies ahead?* Proceedings of the 20th annual conference of the Law of the Sea Institute, ed. T. A. Clingan, pp. 432–44. Honolulu: Univ. of Hawaii.

Smith, R. W. 1986. *Exclusive economic zone claims: An analysis and primary documents.* Dordrecht, The Netherlands: Martinus Nijhoff.

United Nations. 1983. *The law of the sea: United Nations convention on the law of the sea.* New York: United Nations.

U.S. Department of Defense. 1987. *The commander's handbook on the law of naval operations (NWP-9).* Philadelphia, Pa.: Naval Publications and Forms Center.

LEADERSHIP

Leadership may be defined simply as the process of influencing others. Given the indispensable role of leadership in all cultures throughout history, people have long sought to understand what makes some leaders more effective than others. Nevertheless, the "secret" of good leadership continues to elude explanation.

Part of the difficulty lies in the variety of influence processes that may be called leadership. Valuable insights about political leaders might have no relevance to a gang leader organizing fellow prisoners; a leadership technique that works for the minister of a church might fail for an environmental activist. Yet, all these people are leaders.

This article deals specifically with organizational leadership, since a focused discussion may permit more precise conclusions about the essential nature of leadership.

Organizational leadership is the process of influencing human behavior to accomplish the organization's goals. This definition can apply to any organization including the military. An organizational leader is appointed by the organization, and is not necessarily the same person who would emerge from the group as the group's leader, or who might be elected by the subordinates.

Leadership and Management

The difference between leadership and management has been discussed at great length. The continuing debate is fueled in part by cultural biases about the meaning of these words that favor "leaders" over "managers." The word *leadership* conjures mental images of a charismatic and visionary war hero, while the word *management* implies a desk-ridden planner, and we constantly hear pleas for more leaders and fewer managers.

Yet leadership appears to be a subset of management. Leadership concerns interpersonal influence processes; management refers to the coordination and application of all resources, including people. A successful military leader must practice both management and leadership. A useful discussion will examine the difference between the skills of leadership and management, which are not inherent in character or personality, but instead can be learned. Soviet Marshal Georgi Zhukov noted that commanding troops "embraces a wide range of military-political, moral, material, and psychological factors," and that study of all of these factors is critical to success.

Approaches to Leadership

The notion that inherent character attributes enhance leadership potential led British Field Marshal Sir Archibald P. Wavell to conclude that "no amount of learning will make a man a leader unless he has the natural qualities of one." These traits have been the subject of considerable study and speculation. The most often cited characteristics include physical and moral courage, integrity, loyalty, dependability, ambition, unselfishness, enthusiasm, intelligence, endurance, initiative, self-discipline, tact, and physical attractiveness. Yet there are effective leaders who lack many of these traits, and ineffective leaders who have these traits in abundance. There are many glaring exceptions to anyone's list of necessary leader characteristics: Napoleon was short, American Civil War General James Longstreet graduated near the bottom of his West Point class, Alexander the Great killed his best friend in a fit of drunken rage, and Admiral Ernest J. King had an extremely abrasive personality. Yet all these men were extraordinarily effective military leaders.

Noting this incongruity, social scientists have criticized the historically based studies of leadership, arguing that a complete picture cannot be gained merely by cataloging the characteristics of the leader. Applying the scientific method of inquiry, they have attempted to capture leadership in the form of testable, supportable principles. Greatly influenced by behavioral psychology, they have sought to understand not what an effective leader *is*, but what an effective leader *does*.

Patterns of leader behaviors frequently are referred to as leadership "styles." The most-often discussed leadership styles are "task-oriented" and "people-oriented." A leader using a task-oriented style tends to structure his and the subordinates' roles toward accomplishment of the mission, often making decisions for the group. A leader using a people-oriented style is friendly and supportive, listens to subordinates, and may also seek their input. These behaviors are not mutually exclusive opposites; in fact, leaders frequently have a capacity for displaying varying degrees of both styles simultaneously.

Some leadership theorists have argued that the "one best style" of leadership is that in which the leader demonstrates high degrees of both styles simultaneously (Yukl 1989). Unfortunately, evidence to support this claim has been scarce. While it has been generally demonstrated that people-oriented behaviors consistently improve subordinates' satisfaction, neither style, nor any combination of the two, can generate group productivity or effectiveness.

Still, even a thorough understanding of what the leader is (through characteristics or traits) and what the leader does (through observable behaviors) does not give a complete picture of leadership. Recent leadership theories have included "situational" variables such as the task, the subordinates, and the effect of the external environment.

These situational theories and models of leadership suggest that effective leader behavior depends on such factors as how difficult or routine the task is, how motivated the subordinates are, and how well prepared they are to accomplish the task. For instance, a highly effective drill instructor might fail miserably if he uses the same repertoire of behaviors to organize a battalion picnic that proved so successful in pushing soldiers through the obstacle course. For an effective leader, then, the question is not which style or behaviors to use, but under what circumstances to employ certain leader behaviors.

Subordinate Reactions to Leadership

Leadership is a process of interpersonal interaction; therefore, it is imperative to understand how the subordinates perceive themselves, their task, and their leader. Changing a soldier's behavior is relatively easy; military leaders are generally imbued with enough power and authority to get subordinates to comply with the leader's orders to avoid punishment or to obtain status, promotion, recognition, and so forth. But is compliance enough? Xenophon noted 2,300 years ago that "willing obedience always beats forced obedience." Obviously, if the subordinates are committed to the leader or the mission, the likelihood of success increases. Therefore, an effective leader wants to do more than change behaviors; he wants to change attitudes. As General Eisenhower put it, "I would rather try to persuade a man to go along, because once I have persuaded him, he will stick. If I scare him, he will stay just as long as he is scared, and then he is gone."

Recent theories of "transformational" leadership attempt to understand how leaders induce followers to move beyond mere compliance. These theories suggest that the best leaders are able to persuade their followers to transcend basic selfish needs and emotions, often by articulating a vision and setting an example that appeals to higher values such as liberty, justice, fraternity, patriotism, or equality. Subordinates' reactions to a leader's attempt to influence them depend on whether they perceive the leader to be relying on position power or personal power.

Position power is a function of the leader's ability to issue rewards and punishments due to the leader's role-vested authority in the organizational hierarchy; more simply, it is the expectation that the leader will be obeyed because he, by virtue of his rank, can demand obedience. Such power is not limited to a military commander's authority to convene a court-martial or award a medal. For instance, Alexander the Great graciously rewarded his victorious soldiers with extended leaves and a general cancellation of debts. The Vietcong lacked the resources of either Alexander or their American foe. However, relying upon the importance to fellow soldiers of recognition and esteem, they frequently punished itinerants through *kiem thao*, public group criticism. During the 1990–1991 occupation of Kuwait, Iraqi leaders became increasingly reliant on the threat of summary executions for any soldiers who attempted to desert their positions.

Unlike position power, *personal power* is not a derivative of the leader's rank or position. Personal power is based on the leader's expertise or skill, as well as the leader's ability to cultivate the trust, respect, and admiration of the followers. The word *charismatic* is frequently applied to a leader who is able to influence followers with some set of personal characteristics or behaviors, although there is still much debate about exactly what charisma is and whether leaders can consciously develop it.

Marshal Aleksandr Suvorov, the founder of the modern Russian Army, displayed tremendous personal power. To the consternation of his fellow officers, Suvorov talked, ate, and slept with his soldiers. In the eighteenth-century Russian Army, few officers were willing to talk to the common soldier, even if they could (since most officers had been raised to speak French rather than Russian). As a result, Suvorov quickly endeared himself to the Russian soldiers; his lifetime of brilliant campaign successes was ample testimony to his soldiers' dedication.

The Israeli Defense Forces (IDF) have institutionalized a unique model for ensuring that leaders have personal power. Unlike most of the world's military organizations, the IDF promotes leaders only "from the ranks." Hence, the leader at any level—squad, platoon, regiment—is the best soldier in that unit. Furthermore, the basic tenet of leadership for an Israeli leader is to lead from the front, which may help account for the enormous respect that the Israeli soldiers almost always have for their leaders (as well as the extremely high casualty rates among officers and NCOs).

Position power and personal power elicit different responses from subordinates. Leaders may depend on position power to get public compliance to their influence. At least while they are supervised, subordinates will behave so as to avoid punishment or gain reward. This type of influence is good only as long as the leadership can supervise and sanction, for the soldiers' hearts will not be in the mission. Many of the Iraqi soldiers who carried out their duties during the 1991 Gulf War only under the threat of execution surrendered after offering only brief token resistance to the multinational force confronting them.

If the leader has personal power, followers are more likely to trust and accept that leader's influence, either because the leader knows best or because the subordinates identify with the leader. In such cases, the subordinates will surpass mere compliance—they will be committed to achieving the organization's goal. Consider the esteem that the Confederate soldiers of the American Civil War had for their commander Gen. Robert E. Lee. The Confederate soldiers' attachment to their commander in chief was nowhere more evident than at the battle of Spotsylvania Courthouse, where a Union attack threatened to overwhelm the Confederate position. General Lee, sensing the precarious situation, rode forward to personally lead a desperate counterattack. The distraught Confederate soldiers, fearing for their commander's safety, shouted, "General Lee to the rear!" Lee relented, and the counterattack went forward without him.

One method of gaining commitment is ensuring the followers understand and believe in what they are doing. Gen. George Washington considered it essential that leaders "impress upon the mind of every man, from the first to the lowest, the importance of the cause and what it is they are contending for." Mao Tse-tung also considered the commitment of his followers essential. When he found that his soldiers' lack of education was hindering his ability to convey his vision of a communist China, Mao had women wear placards of Chinese characters on their backs. While his Army marched, many soldiers in the column would eventually memorize enough to be able to read and write. The former Soviet army's reliance on the *zampolit* (political officer) is, at least in part, a recognition of how imperative it is that soldiers understand *why* they serve, fight, and die.

Cultural Differences in Leadership

Societal or cultural background frequently dictates the nature of the relationship between leader and follower. The relative importance of status between leaders and followers may vary from culture to culture. Many countries have an entrenched class system that closely parallels the social status of different ranks. (For example, in certain traditional European models, officership is the prerogative of the privileged class.) In countries such as North Korea, a harsh totalitarian political system socializes people to willingly submit to the authority of the leader. In the armies of these societies, subordinates are often less likely to respond effectively to a democratic or participative leader style.

In a culture with a high level of collectivism, the leader is more capable of transcending subordinate self-interest for the good of the organization. The fearsome Mangoday of Genghis Khan's army turned the tide of pitched battles with their shocking suicidal tactics. More recently, Japanese kamikaze pilots in World War II willingly went to their certain deaths to fulfill the ultimate warrior ethic.

Leaders must understand and accept cultural differences when working with a different culture's military organization. For instance, to the credit of Baron von Steuben, this professional Prussian officer did not attempt to shape the young Continental American army to mimic the European conscript armies of the time. Frederick the

Great had a highly proficient system of discipline based on the notion that the troops needed to fear their leaders more than they feared the enemy. Von Steuben, however, had to adjust his training after noting that American conscripts expected to be treated with much more dignity than their European counterparts, had a zeal for individualism, and wanted to know what they were fighting for.

Leaders in Action

It is dangerous to draw conclusions about effective leadership solely on anecdotes and historical snapshots of successful leaders. The most obvious limitation is that examples can be found to support any and every belief about what makes an effective leader, so such examples can prove nothing. Also, historical examples are generally available only for the leadership of successful high-level wartime combat commanders; the leadership stories of lower-ranking leaders are less frequently available. There is also less information available about successful leadership back from the action of the front line. Little is known about the many excellent NCOs and officers who died doing their job, never having the chance to become famous and merit biographical recognition. Nevertheless, these examples can illustrate and help give context to what might otherwise be a rather theoretical and unexciting discussion.

It is virtually a prerequisite for any discussion of successful combat leadership to document heroic leadership in battle, but setting the example through selfless sacrifice can also be demonstrated away from the battlefield. While commander of a tank brigade in World War I, Col. George S. Patton, Jr., was observing 37mm gun practice. A round exploded in one gun's muzzle, wounding some soldiers. The next round fired from this gun exploded in the breech, killing the gunner. Fearing that the troops would lose confidence in their equipment, Colonel Patton went to the gun and personally fired the next three rounds without incident.

Effective leaders also must possess and demonstrate tactical and technical expertise. Followers are much more likely to respond favorably to a leader who will win battles and keep soldiers alive. Field Marshal Erwin Rommel distinguished himself as one of Germany's finest small-unit leaders on the French, Italian, and Romanian fronts in World War I. When he captured his lessons in the 1937 book *Infantrie Greift an* (Infantry Attacks), his demonstrated tactical genius caught the eye of the German leadership, to include Hitler himself. Rommel's expertise was genuine; in combat from 1940 until his death in 1944, he built a singular reputation that distinguishes him as one of the great military leaders of all time.

Good leaders also establish challenging goals and standards and build effective teams. One of Adm. Ernest King's first leadership challenges as a new ensign was to take over a division of 40 sailors on board the USS *Cincinnati*. Morale was quite low, and the desertion rate was extremely high. King took advantage of an impending gunnery competition to challenge his sailors. The sailors, responding with vigor to King's enthusiasm and self-confidence, won the competition. Morale, cohesion, and pride in King's division soared, and young Ensign King's reputation as a winner was launched.

Effective leaders are in touch with their subordinates. Although the claim is almost certainly exaggerated, Julius Caesar reportedly could recognize every man in his legions by name. During inspections and speeches, Napoleon also showed the ability to single out campaign veterans by name, asking about their families and other personal details (although an adjutant's advance briefing aided his recognition). Good leaders also have sought to understand the concerns of their soldiers. According to Shakespeare, an incognito King Henry V made rounds of the English campfires the night before the battle of Agincourt, seeking to learn the soldiers' thoughts and fears.

Beyond understanding the troops, the effective leader must provide for their well-being. Although this is often interpreted as providing the troops with material comforts, Rommel pointed out that "the best form of 'welfare' for the troops is first-class training, for this saves unnecessary casualties." Taking care of soldiers also means ensuring that they have the support they need to do their job. A young Col. George C. Marshall, as a division operations officer in World War I, discovered that his division's system for publishing orders to lower commands was so slow that the battalions, companies, and platoons of the division seldom had time to conduct a reconnaissance prior to attack time. Marshall directed his staff to send orders over the phone, instead of by messenger. He weighed the increased security risk against the advantage an extra two hours would give subordinate commands in preparing for the next day's operation. Marshall's action demonstrated not only a concern for subordinates, but a willingness to take initiative.

There are many examples of leaders giving inspirational speeches—*feldherrnrede*—before battle, or at critical moments; Napoleon, always sensitive to the human element of command, noted that "a man does not have himself killed for a few half-pence a day or for a petty distinction. You must speak to the soul in order to electrify the man." Although assembling large bodies of soldiers before battle may now be tactically impractical, modern commanders can take advantage of other means to communicate their confidence and high expectations in soldiers. Gen. H. Norman Schwarzkopf, commander of Allied efforts during the Persian Gulf War, had an inspirational note delivered to each American soldier at the onset of the ground offensive. In this message, Schwarzkopf incited his soldiers to be "the thunder and lightning of Desert Storm."

Leadership Challenges in Future Warfare

Leaders in future wars will be confronted with a proliferation of weapons of mass destruction (nuclear, chemical, and biological); advances in the acquisition capability, lethality, range, and accuracy of weapons systems; sophisticated electronic warfare and counterwarfare measures; increased mobility of weapon systems; and an enhanced capacity to fight at night and in any weather. The increased lethality of warfare and the reduction in military forces due to both economic and political initiatives will disperse military forces on the battlefield more than ever before. Furthermore, as technology has made the "deep" battle a reality, it can no longer be assumed that combat is restricted to a relatively shallow battle zone at the forward edge of the battle area.

It is important for leaders to understand that social dynamics play a critical role: soldiers in imminent danger seek the companionship of fellow soldiers. Such support will be more difficult to ensure on the dispersed battlefield of the future. This isolation will be heightened when soldiers must fight at night or wear nuclear, biological, or chemical protective gear.

Both isolation and uncertainty may be amplified by the loss of communications with other units, a real possibility as battlefield distances increase and electronic warfare is employed. Uncertainty also will be fed by the increased mobility of troop and weapons carriers, which will increase the likelihood of surprise and impel more abrupt shifts on the already fluid battlefield.

The anxiety produced by increased isolation, lack of mental or physical down-time, heightened uncertainty, and reduced reaction time will place a tremendous strain on the individual soldier. The future battlefield leader will have to be able to shield his soldiers from these numbing effects by building a high degree of unit cohesion, which can be strengthened by increasing the soldiers' interaction and communication, establishing unique norms and symbols, creating a sense of unit identity, providing tough challenges that require teamwork to succeed, and rewarding unit success while punishing unit failure.

The future leader also must understand how to manage stress. Redundancy of functions (to include leadership) should be worked into unit training to ensure that every soldier receives adequate rest periods. Soldiers must have the opportunity to interact, to talk about their fears with one another. The unit must be well trained, even overtrained, in peacetime, so that proficient performance becomes a reflex reaction. Furthermore, the more similar the peacetime training conditions are to combat, the greater the likelihood that soldiers will not be overwhelmed by the reality of combat.

There is a relationship between unit cohesion and the ability to withstand stress. During the 1973 October War, Israel reacted to the Arab surprise attack by sending reserve tank crews to the front lines before the units had a chance to organize into their normal combat teams. These tank crews were in combat before they had even met each other. When later compared with organic tank crews, these rapidly assembled units had a profoundly higher rate of stress-related psychiatric casualties (Gal 1986).

Because of the reduced reaction time necessary from more-dispersed forces and the less-reliable communications between them, junior leaders (company-grade officers and NCOs) must be developed and prepared to make independent decisions. On the fluid modern battlefield, it is impossible for a commander to issue specific orders for every possible contingency; subordinates must have the willingness and ability to take action without requesting guidance or permission from headquarters. This philosophy of *auftragstaktik*, as it was called by the Prussian Army of a century ago, encourages junior leaders to seize the initiative in accordance with their commander's intent. It is unlikely that a leader in combat will suddenly be rendered able to implement *auftragstaktik*, build a cohesive team, and manage stress. These capabilities are best developed before the battle. As a Chinese proverb suggests, "the more you sweat in peace, the less you bleed in war."

Future Trends in Leadership

Leadership theorists have begun to recognize that senior- and junior-level leaders may require different skills. Senior-level leaders must become adept at managing increasingly complex organizational systems. While face-to-face, direct leadership remains important, senior leaders must work through levels of subordinate leaders to influence subordinates indirectly. Thus, necessary leadership skills will not be the same in all levels of an organization. Organizations will have to adapt by ensuring that effective junior leaders receive the additional developmental experiences and training needed to become effective senior leaders.

As technology has flourished, the military increasingly needs better-educated soldiers who can effectively operate complex systems. Furthermore, the current size reductions in many armed forces will make recruitment more selective. As a result, incoming members of the armed forces probably will be better educated than in the past. The soldiers of the future may have different capabilities, needs, and aspirations and will require well-educated and technically competent leadership. Leaders themselves will have to balance competing demands to assume the specialized management functions of a technical leader along with the more traditional roles of troop leadership.

With smaller armed forces and a move toward all-volunteer staffing, the balance between those recruits who view the military as an occupation and those who view it as a profession becomes a key issue for leaders. Those who view the military as a profession would value intangible rewards such as duty, country, sacrifice, or patriotism; those who view it as an occupation would expect financial compensation and a quality of life comparable to that provided by a civilian occupation. Leaders should note that soldiers who subscribe to one model will be motivated by different rewards than those who take the opposite view.

Finally, the moral-ethical dimension of leadership has received increasing attention. Leaders influence their command's ethical climate by the way they administer rewards and punishments, establish goals and competitions, issue guidelines, and manage stress (U.S. Military Academy 1988). When subordinates act unethically, leaders can reasonably be held responsible for the ethical climate which they established or tolerated.

A great deal remains to be learned and understood about the complex phenomenon of military leadership. The significance of good leadership at all levels of a military organization remains clear, however, and leaders can benefit from reflecting on both what is known and what is yet to be known about effective leadership.

KEVIN S. DONOHUE
LEONARD WONG
STEVEN M. JONES

SEE ALSO: Command; Generalship; Management.

Bibliography

Bass, B. M. 1990. *Bass and Stogdill's leadership handbook.* New York: Free Press.
Gal, R. 1986. Unit morale: From a theoretical puzzle to an empirical illustration: An Israeli example. *Journal of Applied Social Psychology* 16:549–64.
Fitton, R. A., ed. 1990. *Leadership: Quotations from the military tradition.* Boulder, Colo.: Westview Press.
Hunt, J. G., and J. D. Blair, eds. 1985. *Leadership on the future battlefield.* McLean, Va.: Pergamon-Brassey's.
Matthews, L. J., and D. E. Brown, eds. 1989. *The challenge of military leadership.* McLean, Va.: Pergamon-Brassey's.
Taylor, R. W., and W. E. Rosenbach, eds. 1984. *Military leadership: In pursuit of excellence.* Boulder, Colo.: Westview Press.
U.S. Department of the Army, 1990. *Field manual 22-100: Military leadership.* Washington, D.C.: Government Printing Office.
U.S. Military Academy, Department of Behavioral Sciences and Leadership. 1988. *Leadership in organizations.* Garden City Park, N.Y.: Avery Press.
Yukl, G. 1989. *Leadership in organizations.* Englewood Cliffs, N.J.: Prentice Hall.

LEAHY, WILLIAM DANIEL
[1875–1959]

Adm. William D. Leahy, U.S. Navy, served as a personal adviser to President Franklin D. Roosevelt during World War II (Fig. 1). In that capacity he had tremendous influence over America's wartime policies. He and Gen. George Marshall were so successful in the performance of

Figure 1. William Daniel Leahy. (SOURCE: U.S. Library of Congress)

their duties that many subsequent presidents came to rely on the U.S. military as a source of advisers.

Early Life and Career

Leahy was born on 6 May 1875 in Hampton, Iowa, and was graduated from the U.S. Naval Academy in Annapolis, Maryland, in 1897. As a naval cadet during the Spanish-American War he was stationed aboard the battleship *Oregon* (BB-3), where he fought at the Battle of Santiago Bay in July 1898. He was then transferred to the second-class battleship *Texas* in October 1898. In 1899 he was sent to the Far East and saw action in the Philippine insurrection and the Boxer Rebellion, serving aboard the gunboat *Castine* and on the supply ship *Glacier* (AF-4), and as the commanding officer of the gunboat *Mariveles* in 1902. Leahy's duty assignments in the following years included: the new cruiser *Tacoma* (C-18); the protected cruiser *Boston* while it patrolled off the Panama Canal when that waterway was being built and later while *Boston* provided assistance to victims of the San Francisco earthquake of 1906; science instructor at the Naval Academy; navigator on the armored cruiser *California* (ACR-6) from 1909 to 1911; staff officer to fleet commander Adm.

Chauncey Thomas on the same ship from 1911 to 1912; and chief of staff to Adm. W. H. H. Southerland on the flagship *West Virginia* (ACR-5) during the Nicaraguan intervention of 1912. He then served in various shore duty billets.

Prior to U.S. entry into World War I, Leahy at first continued his shore duty and then became commander of the dispatch gunboat *Dolphin* (PG-24) in the West Indies in 1915. While aboard, he participated in the occupation of Santo Domingo in 1916, in demonstrations off Mexico, and, following the U.S. declaration of war against Germany (6 April), in searches for German supply craft during the spring of 1917. He then became executive officer of the *Nevada* (BB-36), and served while it conducted wartime patrols out of Norfolk, Virginia. In April 1918, Leahy took command of the *Princess Matoika*, a troop transport that ferried elements of the American Expeditionary Force to Europe. It was during his command of this ship that he became friends with Franklin D. Roosevelt, who was then assistant secretary of the navy.

Senior Assignments in the U.S. Navy

Following the war, Leahy rose rapidly in the navy. He commanded the *St. Louis* (CA-18) while it was the flagship of U.S. naval forces off Turkey when that nation was torn with strife in 1921. He served as commander of the Atlantic Fleet's Mine Squadron One, and in 1922 as commander of the Control Force of submarines and smaller craft. From 1923 to 1926, he directed the officer personnel division of the Bureau of Navigation and then commanded the *New Mexico* (BB-40) before his promotion to rear admiral in 1927. While Leahy would have additional sea assignments, his administrative skills were his greatest strength, and it was in shore assignments that he had his greatest effect. From 1927 to 1931, he served as chief of the Bureau of Ordnance. Following a tour as chief of the Bureau of Navigation from 1933 to 1935, he was commander of Battleships Battle Force as a vice admiral, and commander of Battleships Force in the *California* (BB-44) as a rear admiral from 1936 to 1937. He served as chief of naval operations from 1937 to 1939 and virtually ran the navy in the absence of the secretary of the navy, Claude A. Swanson, who was ill during this period. Leahy was 64 (mandatory retirement age) when he retired as an admiral in July 1939.

Leahy as Statesman

Although Leahy never again served in the navy's chain of command, he served in several positions that had great effect on both naval and national policy. President Roosevelt made him governor of Puerto Rico in September 1939 and U.S. ambassador to the Vichy government of unoccupied France in November 1940—an assignment that required the greatest diplomatic skills, given German control over French foreign policy.

After America entered World War II, Leahy was recalled to active duty in July 1942 and was appointed to a newly created position, chief of staff to the president. He also served as chairman of the Joint Chiefs of Staff. He was promoted to fleet admiral in December 1944 and accompanied Roosevelt to all the strategic conferences of the Allies, including the Yalta conference in 1945.

After Roosevelt's death in April 1945, Truman retained Leahy as his chief of staff and he continued in his advisory role during the early years of the cold war. He retired at 73 in 1949 and wrote his memoirs, *I Was There: The Personal Story of the Chief of Staff to Presidents Roosevelt and Truman*, in 1950. He died on 20 July 1959.

Leahy's Significance

Admiral Leahy's historic significance lies in his diplomatic and administrative skills. During World War II he served as the senior military member of the small group of politicians and military men that dictated wartime military policy. His personal relationship with President Roosevelt gave him great influence in wartime decisionmaking. Additionally, because Leahy was considered so successful as a presidential adviser the practice of using high-ranking military officers in high-level government positions was continued. Such officers, regarded as competent and apolitical and offering no political threat, have served in many sensitive positions such as directors of the Central Intelligence Agency and the National Security Agency, and as national security advisers to presidents.

BRUCE W. WATSON

SEE ALSO: World War I; World War II.

Bibliography

Adams, H. H. 1985. *Witness to power: The life of fleet admiral William D. Leahy*. Annapolis, Md.: U.S. Naval Institute Press.
Leahy, W. D. 1977. *Diaries, 1939–1949*. Washington, D.C.: Library of Congress Photoduplication Service.
———. 1950. *I was there: The personal story of the chief of staff to Presidents Roosevelt and Truman*. New York: Whittlesey House.
Spiller, R. J. 1989. *American military leaders*. New York: Praeger.
Thomas, G. E. 1973. *William D. Leahy and America's imperial years, 1893–1917*. New Haven, Conn.: Gerald E. Thomas.

LEAVE, MILITARY

Traditionally, military personnel are considered available for duty at all hours seven days a week except when granted official leave. Such a leave of absence is generally regarded not as an entitlement but as a privilege to be granted or denied by commanders, depending on the needs of the service.

The word *leave*, then, can be defined as "permission to

be absent from duty," and is thus a generic term. What civilians call a *vacation*, enlisted military personnel customarily call *furlough*. Navy people use the word *liberty* when giving a sailor permission to leave his ship while in port, and a soldier may be given a *pass* to be absent from his post or duty station for a short period, usually less than 72 hours.

Leave policies have evolved over the years just as military forces themselves have evolved. At one point in time, no leave was ever granted for any reason. Today, leave may be granted professional military personnel under policies similar to civilian vacation plans.

Rationale for Leave

Governments and military forces have adopted laws and regulations granting leave to provide personnel with a specified number of days for rest and relaxation away from the arduous duties of military service. These policies also provide for the accrual of leave days, since personnel may not be able to take leave because of military requirements that include peacetime deployments, wartime combat, and overseas postings.

The rationale behind such policies is that they enhance military effectiveness. The United States Department of Defense Directive on this subject states: "Experience has shown that vacations and short periods of rest from duty, as well as authorized absences to attend to emergency situations, provide benefits to morale and motivation that are essential to maintaining maximum effectiveness. Therefore, an aggressive leave program is an essential military requirement."

Historical Background

Historians have noted that military institutions reflect the nature of the society of their time and place. Engels wrote that nothing was more dependent on economic conditions than an army or navy.

Alfred Vagts wrote that "each stage of social progress or regress has produced military institutions in conformity with its needs and ideas, its culture as well as its economics"(Vagts 1937).

Max Jahns, another historian, wrote that forms of war corresponded to those of the political economy. The medieval economy, based on land ownership, produced armies of barons and knights. Under the feudal system, if a war was to be waged, the knight's followers simply mustered at their master's summons. When the campaign was over, the members of the feudal army returned to their homes. Under this system, there was no requirement to grant leave, and there was no professional military until about the eleventh century.

In the Middle Ages, citizens of towns and city-states fought battles to develop and protect trade. Soon the burghers found idle men willing to do the fighting for a price. This was a manifestation of early capitalism. The senior officers of these armies still were members of the nobility, even though most of the troops were mercenaries.

The development of new weapons brought the need for more troops, but when the fighting ended, the mercenaries were paid off and sent home. As with the earlier followers of the knights and barons, military service was continuous, without any leave expected or granted.

In *A History of Militarism* (1937), Vagts noted that the nobility were not fully reconciled to the new weapons, but they nevertheless remained entrenched in the officer positions. Thus, the quality of the armies deteriorated. The eventual solution to this state of affairs was to require military education for officers. And over a long span of time, this produced a professional officer corps. With the military education requirement, the need for leave to attend the required schooling became a necessity, as did, eventually, a general leave from duty at periodic intervals.

Finally, the militia idea developed, under which citizens served when needed, either by conscription or as volunteers. Generally, the longer the committed term of service, the greater the likelihood of some system to grant leave.

Before World War I, most European armies were royal forces loyal to a monarch. The armies of other countries evolved from armed revolution. In these and most developing countries, the "freedom fighters" mixed with the military professional, becoming the professional military forces that earned and generally received the privilege of occasional leave from duty.

The British Navy as a Case History

The evolution of the British navy from pre-Tudor times (800–1485) to the present provides a case history of the development of the modern professional configuration. When needed, the old navy was constituted by the monarch by drafting existing ships of any useful kind with their crews and adding a leader and his followers (lords and the men from their manors). When the need was over, the fighters went home. Under this system, leave was impractical and not expected.

By the seventeenth century, England saw the need for a permanent navy, and some steps were taken to make life acceptable for ordinary seamen. But after peace came in 1674, England could not afford to maintain a large naval force. Out of this situation, England developed a professional navy officer system with the commissioned officer appointed by the Crown but only as an officer of a particular ship. The officer drew full pay when posted to a ship. Otherwise, he was at home on half-pay while awaiting a posting to another ship. This was retainer pay or a form of leave.

By 1713, the Royal Navy was well established as a full-fledged, honorable profession for officers, like the army, church, and law.

At the end of the eighteenth century, England still had no full-time enlisted seamen. Instead, it was relying on a pool of merchant sailors and fishermen. Living conditions for seamen during these times were so poor that men would not serve willingly. This led to impressment, with press gangs rounding up men for service. In *The History of the British Navy* (1989), Michael Lewis noted that this system kept these seamen "underpaid, overcrowded, harshly disciplined and practically prisoners" for indefinite periods. Under these conditions, leave could not be granted for fear of desertion. Even shore leave in distant places was risky.

In 1853 the English government inaugurated long-term service in the navy. The Bluejacket came into existence as a volunteer and was the nucleus for wartime expansion of the navy through its newly established Fleet Reserve. For officers, modern equipment and weapons required specialization, and creation of the "retired list" brought the officer corps into its niche in the modern professional navy. As in other modern professional military forces, periodic leave for rest and relaxation was expected and granted.

Types of Leave

Leaves of absence take many forms, depending upon such factors as the military status of the individual (career professional, reservist or conscript, and length of service), specific purpose of the leave, and the period of the proposed absence.

ORDINARY LEAVE

Sometimes called annual leave, ordinary leave fulfills the basic purpose of leave policies. It is based on a system that allows accrual on a specified basis; for example, 2½ days per month for American military personnel on extended active duty. Commanders are directed to establish annual leave programs that provide opportunities for all eligible personnel to use their leave on a regular basis. Under the U.S. system, members accrue 30 days of leave entitlement per year and are encouraged to take at least one extended two-week leave.

Commanders can exercise wide leeway in encouraging use of ordinary leave during traditional national holidays, after long periods of arduous duty or protracted deployment from home station or port, or when there is evidence of family problems for a service member.

Ordinary leave, therefore, is the fundamental basis of any national leave policy. As a practical matter, it may be limited to the standing active duty force and may not apply to conscripts or reservists on short active duty tours. The latter groups may accrue leave under a national policy that limits its use to the end of the specified period of active duty.

EDUCATIONAL LEAVE

With the increased complexity of warfare, education requirements increased, particularly for officers, and these requirements often can best be met through attendance at civilian schools. Such attendance necessitates the granting of leave for the period required to attend such schools. Educational leave is granted for military personnel to pursue a program of education compatible with the individual's military career. Normally, educational leave cannot exceed two years. To be eligible, individuals must meet service requirements such as length of service and commitment for future service. In the United States, this commitment for future service is two months for each month of educational leave, and the member is entitled to basic pay but not allowances while on this leave. In case of national emergency, educational leave can be terminated immediately.

CONVALESCENT LEAVE

A service member who is hospitalized for a prolonged period may be granted convalescent leave for up to 30 days while awaiting final disposition of his case—that is, return to duty, limited service, or discharge. Convalescent leave is always limited to the minimum deemed essential in relation to the diagnosis and medical prognosis.

REST AND RECUPERATION (R&R) LEAVE

Rest and recuperation leave is a term that applies only to leave granted to personnel serving in areas of imminent danger where dependents are not allowed and unofficial travel is controlled. Under these circumstances, military personnel may be granted R&R to specified areas with military transportation provided. American military personnel serving in Vietnam were granted one R&R per year. Travel time was not charged against ordinary leave accrual.

ENVIRONMENTAL AND MORALE LEAVE (EML)

Sometimes, military personnel must serve in areas of such adverse environmental conditions that the full benefits of the ordinary (annual) leave program are offset. Under these conditions, service members may be granted environmental and morale leave (EML).

Eligibility is based on such conditions as extraordinarily difficult living conditions (geographic isolation, poor housing, lack of cultural or recreational facilities), notably unhealthful conditions (high incidence of disease and epidemics or lack of public sanitation), and conditions that cause excessive physical hardships (climate, altitude, or dangerous conditions affecting physical or mental health).

Under U.S. Department of Defense regulations, service members and their dependents may use military transportation to the designated EML areas, including the continental United States.

Reenlistment Leave

Because enlisted personnel are committed to serve for only a specified number of years, continued service depends upon reenlistment at the end of each committed term of service. At such times, personnel often desire a break to visit family and friends. Reenlistment leave may then be granted for up to 90 days, depending on the amount of leave time already accrued.

Reenlistment leave is charged against leave accrued during the enlistment being completed and against leave that may accrue during the next enlistment, the latter not to exceed 30 days.

Leave in Conjunction with Change of Station

On the occasion of a permanent change of station (PCS), military personnel are normally accorded a leave. It is a convenient time both for the service member to visit family and friends and for the losing and gaining military units to minimize disruption of unit activity. This type of leave is chargeable to the member's leave account.

This leave is particularly valuable for members who have just completed basic training and are moving to their first duty station, for navy personnel moving from one ship to another, for members moving to or from tours of duty without family member accompaniment, and for enlisted members who have just completed officer candidate training and are moving to their first assignment as an officer.

Liberty (Pass)

In peacetime, military personnel are normally free from the end of working hours each day until the beginning of working hours the next day and on weekends. Unless operational conditions dictate otherwise, the individual's military identification card is sufficient documentation for absence from the duty station. If security requirements or operational conditions dictate otherwise, a commander may require additional documentation in the form of a pass to control such authorized absences.

In the U.S. armed forces, liberty passes may not exceed 72 hours and may not be used for consecutive 72-hour periods. If extended beyond 72 hours, the additional time is charged against the member's accrued leave.

Special Liberty

A form of leave known as special liberty may be granted for a period of three or four days after significant service of unusually long working hours, arduous duty away from home station, or duty in an isolated location. It may also be used as a special recognition for exceptional performance. Special liberty may not be combined with other forms of leave so as to result in an extended period of absence that is otherwise charged as ordinary leave.

Administrative Absence

Absences that contribute to the military mission may be granted to service members as administrative absence and are not chargeable against accrued leave. Administrative absences are granted for attendance at professional meetings that have a direct relationship to the service member's military duties, for participation in sanctioned competitive sports events, for attendance in response to a subpoena or summons as a witness in a legal proceeding, or for participation in other official or semiofficial programs that enhance the member's value to his military service.

Required Leave

If a service member is convicted by a court-martial and sentenced to be dismissed or to receive a punitive discharge, he may be required to take leave without pay pending the outcome of any appeal of the conviction. This avoids returning a convicted individual to his or her unit. Under these circumstances, the convicted member may use accrued leave with pay for the number of days accrued, but if the appellate review takes longer, the member must then take leave without pay until the review is completed.

Terminal Leave

When an individual approaches the date of retirement, separation, or release from active duty, terminal leave may be granted without requiring the individual to return to the duty station on the final day of active service. Documentation must be completed before the terminal leave begins.

Under U.S. law, military personnel discharged, retired, or released from active duty under honorable conditions (or their survivors in the case of deceased members) are entitled to reimbursement in a lump-sum cash payment for not more than 60 days of unused accrued leave.

Trends Affecting Leave Policies

A worldwide survey of military forces reveals an increasing number of countries relying on either an all-volunteer force or a mix of volunteers and conscripts with emphasis on the professional career force. Approximately 50 countries have all-volunteer active duty forces, among them are the United States, the United Kingdom, Canada, Australia, Pakistan, Japan, Burma, and the Philippines.

This emphasis on the professional force increases the importance of such enlightened personnel policies as leave that contributes to morale and esprit de corps and also leave that increases the professional competence of career personnel.

Another factor affecting leave policies is the increased reliance, in many countries, on reserve personnel. As the number of reserve forces increases, the size of the standing professional force decreases. It is, therefore, necessary

Leave, Military

to have policies that will encourage individuals to make a career in active duty.

Future

As in the past, military forces reflect the social and economic society they represent. Throughout the world, governments today recognize the individual rights of their citizens and consider no one group more deserving of fair and equitable treatment than those who have sworn their loyalty to their country and serve in its military forces. Because of this, governments recognize the importance of a realistic leave policy and the need for clearly stated laws and regulations to provide for most, if not all, of the types of leave discussed above.

W. STANFORD SMITH

SEE ALSO: Compensation, Military; Conditions of Service; Morale.

Bibliography

International Institute for Strategic Studies. 1988. *The military balance 1988–89*. London: International Institute for Strategic Studies.
Keegan, J. 1983. *World armies*. 2nd ed. Detroit, Mich.: Gale Research.
Lewis, M. 1989. *The history of the British navy*. Fair Lawn, N.J.: Essential Books.
U.S. Department of Defense. 1985. Directive 1327.5, *Leave and liberty*. Washington, D.C.
Vagts, A. 1937. *A history of militarism*. New York: Norton.

LEBANON, REPUBLIC OF

Ravaged by civil war from 1975 to 1990, Lebanon, with its capital of Beirut, has suffered damage to its economic infrastructure and its position as a Middle East entrepôt and banking hub. Lacking any real government during this time, power lay with the religiously and ethnically based political parties and their "militias," which were in fact full-time armies. The situation was further complicated by the stationing of Syrian troops and an extensive Palestine Liberation Organization (PLO) presence.

Arab League intervention and cooperation in negotiations by rival parties established a legitimate government in November 1989. The subsequent defeat and ouster of a last resistance effort in October 1990 has enabled the Lebanese government to reunite its capital city and begin a phased plan to disarm the militias and re-establish authority throughout Lebanon.

Power Potential Statistics

Area: 10,400 square kilometers (4,015 sq. mi.)
Population: 2,674,400
Total Active Armed Forces: not available
Gross Domestic Product: US $3.3 billion (1990 est.)
Annual Defense Expenditure: US$168 million (7.3% of GDP, 1991 est.)
Iron and Steel Production:
  Crude steel: less than 80,000 metric tons (average 1976–86)
Fuel Production: none
Electrical Power Output: 3,870 million kwh (1989)
Merchant Marine: 60 vessels; 257,220 gross registered tons
Civil Air Fleet: 15 major transport aircraft, 8 usable airports (6 with permanent-surface runways); none with runways over 3,659 meters (12,000 ft.); 3 with runways 2,440–3,659 meters (8,000–12,000 ft.); 2 with runways 1,220–2,440 meters (4,000–8,000 ft.).

For the most recent information, the reader may refer to the following annual publications:
The Military Balance. International Institute for Strategic Studies. London: Brassey's (UK).
The Statesman's Year-Book. New Yokr: St. Martin's Press.
The World Factbook. Central Intelligence Agency. Washington, D.C.: Brassey's (US).

History

Lebanon has a long recorded history, dating back to the second millennia B.C. The cities of Sidon and Tyre were famous in the ancient world as bases of Phoenician merchants and as manufacturing centers for the purple dye that made those cities wealthy. Lebanon was also part of the domains of the European Crusaders in the twelfth and thirteenth centuries. Lebanon became independent from the French-administered League of Nations mandate on 22 November 1943. Lebanese troops fought against the Israelis in the first Arab-Israeli War (May 1948–January 1949), but thereafter remained neutral.

Lebanon's major problem since independence has been the presence of Palestinian refugees and the guerrillas based in the refugee camps. Efforts to control their activities failed, and the guerrillas were often able to undermine civil authority. In March 1975, scattered fighting broke out, mostly in the southern and central part of the country, and expanded until the Syrian-led Arab Deterrent Force restored some degree of order in November 1976.

The ensuing precarious peace was disrupted by an Israeli invasion in retribution for PLO guerrilla attacks on Israel (14–21 March 1978), but the Israelis only advanced as far as the Litani River. A larger scale operation was mounted four years later, prompted by the unchecked growth of a PLO state-within-a-state ("*Fatah*land") in southern Lebanon. The Israelis crossed the border on 6 June 1982, and within a week had swept aside the Syrian army, defeated the PLO's armed forces, and nearly reached Beirut. Further fighting brought the Israelis up to Beirut and drove the PLO from southern Lebanon. From 26 June to 3 September the Israelis besieged West (Muslim) Beirut, ending their operations only when the PLO evacuated by sea after an extensive U.S. diplomatic effort, and a U.S.-French-Italian peacekeeping force was installed to maintain order.

Civil war, which had never really died away in the

preceding eight years, broke out afresh when Christian Phalangist militiamen and Lebanese army forces attempted to replace the Israeli forces that had withdrawn from the Chouf Mountains south of the Awali River in September 1983. The resulting crisis led to suicide bombing attacks by pro-Iranian terrorists on the U.S. Marine barracks and the French headquarters on 23 October; partly in retaliation, the United States conducted naval and aerial bombardments of Syrian and Shiite Amal militia positions in the Chouf Mountains (November–December). That stage of the Lebanese crisis really ended only when the multinational peacekeeping force was withdrawn in February–March 1984.

Since late 1982, the PLO has re-established itself in Lebanon, but it lacks its former degree of autonomous power. The central government, generally supported by the Syrians, controls mere portions of the countryside, and most of the country is in the hands of the various militias. Pres. Amin Gemayel took office following his brother Bachir's assassination on 14 September 1982, after less than a month in office. An attempt to hold national elections (the first since 1972) produced a severe governmental crisis in May–June 1988 and resulted in the creation of rival Christian and Muslim governments. In November 1988, the Muslim government appointed a new commander for the army, thereby ensuring that the army lost most of its remaining national identity. Gen. Michel Aoun, Christian prime minister and commander of Christian army elements, declared a "war of national liberation" against occupying Syrian forces in April 1989, and severe fighting ensued in Beirut for over a month. Renewed efforts to arrange a national settlement led Saudi Arabia to host a meeting of 63 of the surviving members of the Lebanese parliament at Taif (early October), which resulted in a compromise agreement between Christian and Muslim political factions. The parliament elected René Moawad president on 5 November, but he was assassinated on 17 November. The parliament elected Elias Hrawi as Moawad's successor, but General Aoun refused to recognize Hrawi. Aoun's renegade elements forced a stalemate in Beirut, and the continued fighting throughout most of 1990 damaged what remained of the industrial and commercial infrastructure. In September 1990 constitutional modifications allowed the creation of the Second Republic; this enabled the implementation of the political reform provisions of the Taif accords. In October 1990 the Lebanese army, with assistance from Syrian forces, succeeded in defeating General Aoun's militia. Aoun took refuge in the French embassy. In November and December the militias withdrew from Beirut, and a new Government of National Reconciliation was announced on 24 December. Syria and Lebanon signed a treaty of friendship and cooperation in May 1991, and Syrian troops continue in partial occupation of Beirut and Lebanon.

Politico-Military Background and Policy

In theory, the armed forces exist to protect the government as a whole. The breakdown of the old political consensus in early 1975 also split the army (most of which sided with the Christians), and the bulk of air force and navy personnel and assets likewise sided with the Christians. Neither Hrawi's government nor Aoun's regime could have ruled effectively; a settlement had to be reached if peace and "normalcy" were ever to return to Lebanon. The armed forces are manned by volunteers.

Strategic Problems

The ever-present potential for national disunity is a problem that overshadows all others. However, with unity restored, Lebanon could be in a potentially awkward position between Syria, which regards Lebanon as a natural addition to its territory, and Israel, which devoutly wishes Lebanon to remain neutral and uncommitted.

Alliances and Military Assistance

Before 1975, Lebanon bought most of its military equipment from the United States and France, but it had no formal defense alliances although it was a member of the Arab League and the United Nations. Since the civil war, the armed forces have received very little new equipment, and much of what was on hand has been siphoned off by the militias. Many of the militias have purchased arms abroad, and some have been supplied by Israel or Syria.

National Armed Forces

ARMY: some 17,500. 11 bde (−) incl Presidential Guard, SF.
 Equipment:
MBT: some 175 M-48A1/A5, 70 T-54/-55.
Light tanks: 32 AMX-13 (with 75mm or 105mm guns).
Recce: 50 Saladin and 10 Ferret.
APC: 250 M-113, Saracen, 40 VAB-VTT, AMX-VCI.
Towed Arty: 105mm: 15 M-101A1; 122mm: 18 M-102, M-1938/D-30; 130mm: 15 M-46; 155mm: 10 Model 50, 18 M-114A1, 36 M-198.
Mortars: 81mm: 150; 120mm: 30.
ATGW: ENTAC, Milan, 20 BGM-71A TOW.
RL: 85mm: RPG-7; 89mm: M-65.
RCL: 106mm: M-40A1.
AD Guns: 20mm, 23mm: ZU-23; 30mm: towed; 40mm: 15 M-42 SP.

NAVY: Some 500†. Base: Juniye.
Patrol Craft: Inshore: 1 Byblos ⟨. Plus about a further 14 PCI ⟨ of doubtful operational status.
Amphibious: Craft only; 2 Sour (Fr Edic) LCT†.

AIR FORCE: Some 800.
Equipment: (Numerous ac destroyed 1989–90; operational status of remainder is doubtful):
Fighters: 3 Hunter (2 F-70, 1 T-66).
Helicopters:
 Attack: 2 SA-342 with AS-11/-12 ASM;
 Transport: (med): 4 AB-212, 6 SA-330; (lt): 2 SA-313, 3 SA-319.

Training: 3 Bulldog, 3 CM-170.
Transport: 1 Dove, 1 Turbo-Commander 690 A.

PARAMILITARY

Internal Security Force: (Ministry of Interior): 9,000 (being reorganized); 30 Chaimite APC.
Customs: 2 Tracker inshore patrol craft ⟨.

Militias

Most militias are being disbanded and heavy weapons handed over to the National Army. Reports suggest that large quantities have been placed in hidden stockpiles.

CHRISTIAN:
Lebanese Forces Militia: est. 18,500 (incl The Phalange). 35,000 all told.
Equipment:
MBT: some 125 T-55, M-48.
Light tanks: some 5 AMX-13.
APC: M-113.
Arty: some 100: 105mm; 122mm: 60; 130mm; 155mm: 10.
Mortars: 60mm, 81mm, 120mm.
RL: RPG-7
AD: 12.7mm, 14.5mm, 23mm guns.
Patrol craft: 1 Tracker, 2 Dvora PFI ⟨.

MUSLIM
Amal (Shi'a, pro-Syria): 10,200 active; some 15,000 all told. Most eqpt has been handed over to the National
Hezbollah ('The Party of God'; Shi'a, fundamentalist, pro-Iranian): est. 3,500 active; 15,000 all told.
Equipment incl: APC, arty, RL, RCL, ATGW, (AT-3 Sagger) AA guns.

DRUZE
Progressive Socialist Party (PSP): 8,500 active; perhaps 15,000 all told. Eqpt being handed over to the Syrian Army.
Equipment:
MBT: 70 T-34, T-54/-55
APC: BTR-60/-152

SOUTH LEBANESE ARMY (SLA): est. 2,500 active, (mainly Christian, some Shi'a and Druze, trained, equipped and supported by Israel, occupies the 'Security Zone' between Israeli border and area controlled by UNIFIL)
Equipment:
MBT: 40 M-4, 30 T-54/-55.
APC: M-113.
Towed arty: 122mm: M-1938; 130mm: M-46; 155mm: M-198.

Foreign Forces

UNITED NATIONS (UNIFIL): some 5,900; Contingents from Fiji, Finland, France, Ghana, Ireland, Italy, Nepal, Norway and Sweden.

Syria: 30,000.
  Beirut: elm 1 armd bde, elm 7 SF regt.
  Metn: elm 1 mech bde.
  Bekaa: corps HQ, 1 inf, possibly elm 1 mech inf bde.
  Tripoli: 2 SF regt, elm PLA.

Iran: Revolutionary Guards: some 1,500 including locally recruited Shi'ia Lebanese; may be withdrawn.

Palestine Liberation Organization (PLO): All significant factions of the PLO and other Palestinian military groups are listed here irrespective of the country in which they are based. The faction leader is given after the full title. Strengths are estimates of the number of active 'fighters', these could be trebled perhaps to give an all-told figure. PLO in Lebanon have agreed to hand over their heavy and medium weapons.
Fatah: 4,500.
PLF (Palestine Liberation Front, Al-Abas): est. 300.
Fatah (dissidents, Abu Musa): 1,000.
PFLP (Popular Front for Liberation of Palestine, Habash): 900.
PFLP (GC) (Popular Front for Liberation of Palestine, (General Command), Jibril): 500.
Saiqa (al-Khadi): 600.
PSF (Popular Struggle Front, Ghisha): est. 500.
DFLP (Democratic Front for Liberation of Palestine, Hawatmah): est. 1,000.
FRC (Fatah Revolutionary Council, Abu Nidal): est. 500.

Future

A consequence of the fighting in Beirut during much of 1990 was that Lebanon's growth rate dipped to an estimated −15 percent. However, Lebanon's future may be brighter than at any time since 1975. If the Syrian-backed Lebanese government can maintain order, the economic and industrial infrastructure of Lebanon can begin to recover. Rebuilding Beirut can provide a major stimulus to the economy in the years ahead.

Since the ouster of Aoun, a return to full-scale war seems unlikely. However, small-scale clashes between militias have been a feature, and successful disarmament of these forces is necessary to achieve and maintain order. There is also the feature of PLO presence to consider, and the fact that a real peace settlement between the PLO and Israel, while being fitfully negotiated, has yet to occur.

One problem of long standing for Lebanon seems to have been resolved; another of longer standing may be resolved given time, patience, and understanding. In the meantime, Lebanon will gradually recover economic and internal stability if it can maintain a careful political balance.

DAVID L. BONGARD
TREVOR N. DUPUY

SEE ALSO: Arab-Israeli Wars; Crusades; Israel; Middle East; Syria.

Bibliography

Faris, H. A. 1982. *Beyond the Lebanese civil war: Historical issues and the challenges of reconstruction.* Washington, D.C.: Georgetown Univ. Center for Contemporary Arab Studies.

International Institute for Strategic Studies. 1989. *The military balance 1989–1990.* London: Brassey's.

Khalidi, W. 1979. *Conflict and violence in Lebanon: Confrontation in the Middle East.* Cambridge: Harvard Univ. Center for International Affairs.

Rabanovich, I. 1984. *The war for Lebanon, 1970–1983.* Ithaca, N.Y.: Cornell Univ. Press.

LEE, ROBERT EDWARD [1807–1870]

Born at Stratford, Westmoreland County, Virginia, on 19 January 1807, Robert E. Lee was the fifth child and third son of Henry ("Light Horse Harry") Lee, a distinguished cavalry officer of the American Revolutionary War, and his wife, Lucy Grymes Lee.

When his father died in 1818, Robert assumed the responsibility of caring for his partially disabled mother and an ailing sister. Scholastically, he excelled in mathematics.

Early Military Career

Lee graduated number two in the 1829 class of the United States Military Academy at West Point, New York, without a single demerit. He was breveted second lieutenant of engineers.

He subsequently served at Fort Pulaski, Georgia; as assistant engineer at Fort Monroe, Virginia (May 1831–November 1834); as assistant in the chief engineer's office in Washington, D.C. (November 1834–July 1837; helped survey the Ohio-Michigan border in 1835); as superintending engineer for St. Louis harbor and the Upper Mississippi and Missouri rivers (July 1837–October 1841); and at Fort Hamilton, New York (October 1841–August 1846).

On 30 June 1831 he married Mary Ann Randolph Custis, daughter of George Washington Parke Custis, grandson of Martha Washington. They were destined to have seven children.

Mexican War service provided Lee with lessons in the value of audacity and of a trained staff; of the relationship between careful reconnaissance and sound strategy; of the strategic possibilities of flanking movements; of the relationship of communications to strategy; and of the value of fortification. All of these lessons—acquired under the tutelage of Gen. Winfield Scott—proved of inestimable value to him in the forthcoming Civil War. He was cited by his superiors for bold reconnoitering, effectively emplacing artillery under heavy enemy fire, accurate reports, and well-thought-out recommendations. Slightly wounded at the battle of Chapultepec on 13 September 1847, he was sent home.

Lee had been promoted to first lieutenant in 1836 and captain in 1838. Brevet promotions during the Mexican War advanced him to colonel in 1848. From November 1848 to August 1852 he supervised construction at Fort Carroll, Baltimore, Maryland. Against his wishes, he was appointed superintendent of West Point. Lee did not feel experienced enough for this singular honor.

He was a diligent and efficient superintendent. He tightened academy discipline and academic standards, but also devoted personal attention to individual cadets.

In March 1855 Lee happily accepted appointment as lieutenant colonel of the 2d Cavalry, but was actually assigned to court martial duty most of the next two years.

When his father-in-law died in October 1857, Lee took leave from the army. Responsibilities as estate executor and caring for an ailing wife kept him from duty until 1859. That year he was sent to Harpers Ferry, Virginia (today, West Virginia), to quell an insurrection led by the abolitionist John Brown.

Civil War

Lee did not favor secession of the southern states from the Union, but he believed his principal allegiance was to his native state of Virginia.

Between February 1860 and February 1861, he commanded the Department of Texas. At the outbreak of civil war he was recalled to Washington and on 18 April was offered field command of the U.S. Army by General-in-Chief Scott. However, Virginia seceded, and Lee resigned his army commission.

Lee was then 54 years old. He stood 5 feet 11 inches tall and weighed just under 170 pounds. His physique was sound, his eyesight unimpaired, and he possessed great endurance (Fig. 1).

He was a disciplined professional soldier with good strategic sense, a master of reconnaissance, an excellent topographer, and had a superior knowledge of fortification. He had never, however, experienced defensive warfare, he had difficulty delegating work to others, and he lacked a good understanding of logistics.

Lee was nominated to command all Virginia troops on 23 April. In August he was appointed general in the Confederate States Army, to rank from 14 June, and appointed military adviser to President Jefferson Davis.

Between November 1861 and March 1862 he organized South Atlantic seaboard defenses. On 31 May 1862 he was appointed commander of the Confederate Army in Virginia, which he named "The Army of Northern Virginia."

In June, employing interior lines and brilliant maneuvering, Lee won a series of battles near Richmond, known as the Seven Days. In August he won the Second Battle of Manassas (Second Bull Run). However, when he moved north in September, he was checked by the Federals under Gen. George B. McClellan at Sharpsburg (Antietam) in Maryland. In December, Lee's well-positioned troops repulsed repeated Union assaults at Fredericksburg.

At Chancellorsville in May 1863, Lee defeated the Union army by dividing his army and enveloping the

Figure 1. Robert E. Lee. (SOURCE: U.S. Library of Congress)

Union right wing. However, the price was high: Lt. Gen. Thomas Jonathan ("Stonewall") Jackson was lost, mortally wounded when shot by his own men. Lee moved north again in June but was decisively defeated at Gettysburg, Pennsylvania (1–3 July).

In 1864, Lee's army won tactical victories, but the persistent brilliance of Union general U. S. Grant, as well as the numerical, logistical, and transportation superiorities of the Union doomed Lee and the Confederacy to defeat. The end for the Army of Northern Virginia came on 9 April 1865 at Appomattox Court House, Virginia.

After the war, Lee became president of Washington College, Lexington, Virginia (today Washington and Lee University). He died 12 October 1870 and was buried in Lexington.

Conclusion

Lee successfully took chances with his army many times during the Civil War. A number of battles were won by the timely arrival of reinforcements at the critical point on the battlefield. However, he could never censure subordinates whose caution, independence, or disagreement with his ideas altered, or ruined, planned operations. He inspired confidence and was just and kind to all. The respect and admiration Robert E. Lee earned from friend and foe alike has grown with the passage of time.

UZAL W. ENT

SEE ALSO: Civil War, American; Grant, Ulysses Simpson; Jackson, Thomas Jonathan ("Stonewall"); Scott, Winfield.

Bibliography

Connelly, T. L. 1977. *The marble man: Robert E. Lee and his image in American society.* New York: Knopf.
Davis, B. 1956. *Gray fox: Robert E. Lee and the Civil War.* New York: Rinehart.
Dowdey, C. 1965. *Lee.* Boston and Toronto: Little, Brown.
Freeman, D. S. 1934–35. *R. E. Lee: A biography.* 4 vols. New York and London: Scribner's.
———. [1942] 1950. *Lee's lieutenants: A study in command.* 3 vols. New York: Scribner's.
Taylor, W. H. [1877] 1962. *Four years with General Lee.* Reprint. New York: Bonanza Books. Bloomington, Ind.: Indiana Univ. Press.

LEGAL ASSISTANCE

The capacity of the military to perform, in large part, is based upon the effectiveness and morale of its military members. Indeed, low morale and inefficiency may result from personal legal problems affecting soldiers and sailors. To avoid these consequences, effective and efficient legal assistance—by way of both preventive law programs and prompt advice or other legal action—is provided in one form or another by the various military services to its military personnel. The term used to describe this benefit in the United States military is *legal assistance*.

Some countries have limited legal assistance for military personnel, while others provide no such service. A global survey of legal assistance is not feasible because many services are provided *ad hoc*, and often are not reported.

The United States, more than any other country, has institutionalized legal assistance as part of military life. This is primarily because the United States military has one of the largest standing military forces, with more troops stationed in more diverse areas of the world than any other nation. In addition, the societal emphasis of dispute resolution through the legal system appears to be much greater in the United States than in other countries. A historical and functional analysis of the U.S. military's legal assistance programs discloses substantially all of what other countries, in the aggregate, may have adopted.

Early History

In 1941, the American Bar Association (A.B.A.) created the Special Committee on War Work to address the legal problems of military members, veterans, and their families. By 1943, the army and the navy had established legal assistance offices in all army posts and all naval districts. An integral part and a requirement of the executive mandate establishing those offices was liaison with the A.B.A. and other civilian bar organizations. The A.B.A. Committee on War Work promoted working relationships between state bar associations and the newly established legal assistance offices.

The legal assistance function of military and naval judge advocates gradually became fully institutionalized. As the complexities of the law have made deeper and deeper inroads into the day-to-day lives of the members of the military community, the need for greater and more competent legal assistance has increased. The A.B.A. recognized an increased responsibility on it to establish an ongoing liaison with the military legal assistance function. To that end, after World War II ended and the immediate mission of the A.B.A. Committee on War Work passed, the A.B.A. Standing Committee on Legal Assistance for Military Personnel (the "LAMP" Committee) was established. That committee's primary mission is to function as a link between the civilian bar and the military legal assistance activities so as to aid in providing the best quality legal assistance to military and naval personnel.

Statutory Authority

Although legal assistance became an integral part of the mission of the judge advocate general in each military department, it did not receive the mantle of law until 1984, when the U.S. Department of Defense Authorization Act of 1985, Public Law 98-525, included the following provision (10 United States Code 1043):

Legal Assistance
(a) Subject to the availability of legal staff resources, the Secretary concerned may provide legal assistance in connection with their personal civil legal affairs to:
    (1) members of the armed forces under his jurisdiction who are on active duty;
    (2) members and former members under his jurisdiction entitled to retired or retainer pay or equivalent pay; and
    (3) dependents of members and former members described in clauses (1) and (2).
(b) Under such regulations as may be prescribed by the Secretary concerned, the Judge Advocate General (as defined in section 801(1) of this title) under the jurisdiction of the Secretary is responsible for the establishment and supervision of legal assistance programs under this section.
(c) This section does not authorize legal counsel to be provided to represent a member or former member of the armed forces, or the dependent of a member or former member, in a legal proceeding if the member or former member can afford legal fees for such representation without undue hardship.
(d) The Secretary concerned shall define "dependent" for the purposes of this section.

Congress was careful not to create a mandate that legal assistance be provided; but rather, it is provided "as available." The consequence is that Congress does not separately fund for legal assistance. Providing legal assistance is not supported by budget requirements and legal assistance is wholly dependent on resource availability in excess of budgeted requirements for military and naval legal offices. Despite the demonstrable need for legal assistance, fiscal constraints suggest that future legal assistance services will remain without separate funding. As a consequence of ever-shrinking military budgets, this may result in fewer and fewer active duty resources.

Legal Assistance Service Providers

The requirements for being authorized to provide legal assistance vary from service to service. The minimal common denominator is that the person must be a lawyer who is a member of the bar of a federal court or of the highest court of a state or territory of the United States.

The U.S. Navy limits legal assistance attorneys to members of the Navy Judge Advocate General's (JAG) Corps who have been designated as such by the judge advocate general or his designee. Policy guidance for U.S. Marine Corps legal assistance comes from the navy judge advocate general. The regulations regarding legal assistance officers in the U.S. Marine Corps somewhat mirror the U.S. Navy requirements in that legal assistance officers must be designated by the commandant of the Marine Corps, through its Judge Advocate Division.

The U.S. Air Force considers all of its judge advocates as legal assistance officers without further designation and regardless of assignment. Each such officer is deemed capable of providing legal assistance as an ancillary duty to his or her other assigned responsibilities.

The U.S. Army is in the midst of a transition from a regulation similar to the navy regulation, which designated members of the JAG Corps who were authorized to provide legal assistance, to a new regulation similar to the air force regulation, which designates all judge advocates, regardless of assignment and regardless of whether they are active duty or reserve, as authorized to provide legal assistance. In addition, the army general counsel has authority to provide legal assistance through civilian attorneys of the Department of the Army to the extent it is consistent with other duties or responsibilities. In foreign countries the army authorizes legal assistance by civilian employees of the United States who are licensed or otherwise professionally qualified under local law.

Persons Eligible for Legal Assistance

The scope of eligibility in part is defined in the federal statute, 10 U.S.C. §1034, *id*. Congress contemplated that legal assistance would be provided to active duty personnel, their dependents, and retired members and their dependents. Regulations from the different services have expanded and particularized the scope of eligible persons. These include reservists on active duty for more than 29 days and their dependents; civilian personnel on orders for duty outside the United States, and their dependents who accompany them; discharged persons confined in military confinement facilities; reservists incident to pre-mobilization requirements, and since Desert Storm, reservists and their dependents for a limited period of post-mobilization counseling; members of allied forces and their dependents in the United States when serving with the armed forces of the United States; dependents of active duty personnel who died while on active duty; and such other persons authorized by the judge advocate general of each service or other designated authority.

Since legal assistance is subject to the availability of resources, at a minimum, the services attempt to provide legal assistance to all active duty personnel (approximately 1.6 million). Beyond that, certain services have prioritized availability based on the category of the client. Other services leave the matter to the discretion of the judge advocate general of the particular service.

Scope of Services

Domestic relations. Domestic relations is one of the most complicated areas in which military legal assistance operates. The nature of the circumstances that arise in this field creates the most severe morale problems. The multi-jurisdictional and international aspects to this field often require expertise in the local law of more than one state and/or country. As well, domestic relations law is one of the most rapidly changing fields of law in the United States, and internationally. Further, in most settings both the service member and his dependent(s) need assistance, and often look to the same facilities to fulfill that need. Finally, many of these cases require court action with parties who cannot afford individual representation.

The primary role of the legal assistance officer is office advice on the implications of divorce, legal separation, annulment, custody, paternity, domestic violence actions, and spousal and child support obligations. This *may* include negotiating and preparing property settlement agreements, court documents to be submitted by the client *pro se*, and other professional functions short of court appearances.

Wills and estates. Other than complex estate planning, legal advice and wills are routinely provided.

Adoption and name changes. Advice and preparation of pleadings are usually provided.

Nonsupport and indebtedness. Advice and negotiations with third parties are provided.

Taxes. Other than advice on business or investment matters, tax advice will be provided. Tax returns will not be prepared or signed.

Landlord-tenant. Legal advice and assistance, and with some limitation, lease negotiation and drafting may be provided.

Civil suits. Office advice and negotiations with third parties may be provided. Appearances in court are not permitted except through the Expanded Legal Assistance Programs of the various services, which will be discussed hereinafter.

Soldiers' and Sailors' Civil Relief Act. This legislation provides a wide range of protection to military members, both as to debts, other obligations, and the prosecution of claims in court. Counseling, preparation of documents, and correspondence are usually provided.

Criminal matters. Limited general advice on minor offenses in civilian courts may be provided. Serious criminal matters in civilian courts are not within the scope of legal assistance.

Military administrative and personnel matters. Matters arising from disciplinary or judicial proceedings under the Uniform Code of Military Justice or administrative discharges are not normally a legal assistance function; services in this area are usually provided by military defense counsel. Other matters not directly adverse to the government, however, may come within the legal assistance function. Examples are disability discharges, actions to correct military records, and other appeals to the secretary of the military department not arising out of the Uniform Code of Military Justice, security clearances, letters of reprimand, bars to re-enlistments, and line-of-duty determinations. When and if these matters go beyond administrative matters and reach the arena of litigation, however, the legal assistance function ceases.

Other services. Advice and assistance may be provided as to powers of attorney, real estate, bankruptcy, contracts, consumer affairs, insurance, immigration, naturalization, citizenship, torts, and other areas not inconsistent with service regulations or adverse to the interests of the U.S. Government.

In-court services. Although in-court services to legal assistance clients, as a general rule, are not permitted, the U.S. Army, Navy, and Marine Corps have established an

Expanded Legal Assistance Program (ELAP). Subject to criteria established by agreements with or orders or statutes of local or state bars, and further subject to availability of resources, legal assistance officers, whether or not members of the local state bar, may represent active duty personnel and their dependents in court.

Generally, the criteria established for in-court representation is that the service member or dependent is unable to afford to pay fees to private counsel. Typically, this will apply to the lowest enlisted grades. Others showing hardship will also be permitted ELAP services.

Unlike the other services, the army has a reserve arm to its ELAP program. To the extent approved by the local or state bar and the army judge advocate general, reservists are authorized to appear in local courts on behalf of qualifying service members and dependents. Their active duty counterparts serve as associate counsel to reservists who are not members of the local bar. Of course, if the active duty legal assistance officer is a member of the local bar, he may appear without the need for local reserve counsel.

Preventive and Premobilization Law Programs

The preventive law programs of the different services are designed to diminish legal problems confronting military personnel and their dependents, and thereby enhance readiness. They operate through preventive law services such as unit education, training, legal assistance checkups, and briefings. In addition, there are education programs to ensure awareness of counseling services, the import of legal advice before acting in important matters, knowledge of one's rights, privileges and responsibilities, and the import of predeployment preparation as to legal matters.

Premobilization counseling of reserves is assumed to be part of their reserve training. During the mobilizations of Desert Shield and Desert Storm, however, it became clear that units that were geographically remote from installations with legal assistance officers had to rely on the civilian legal community and reserve judge advocates in their local area to meet day-to-day legal needs related to mobilization. Indeed, even when activated, members of reserve units tend to be assigned to the same geographic area, and upon an incident such as a SCUD missile attack, the casualties tend to come from the same locale in the United States. This creates a geographic concentration of survivors and dependents who have specialized local law needs. Because of the remoteness of dependents to active duty legal services, civilian and reserve lawyers become critically important.

Although regulations relating to premobilization legal assistance to reservists and to reservists as casualties while on active duty are limited, the role of the A.B.A. through its LAMP Committee, and equivalent activities in state and local bars, reflects the essential link of the legal assistance function in the military to the civilian bar and reserve judge advocates. The legal assistance role was engendered through that link in 1943, and remains viable because of it.

Referrals, Fees, Conflicts, and Malpractice

Legal assistance attorneys regularly refer cases to other government and civilian attorneys based upon expertise, work load, potential conflict situations, convenience to the client, or when the services sought are prohibited by regulation. When referring work to civilian counsel, the military legal assistance officer must be sure to avoid the appearance of showing favoritism to one or a few selected attorneys, while at the same time attempting to provide competent civilian counsel. Regulations bar any economic relationship between a referring attorney and civilian counsel relative to such referrals. Likewise, legal assistance counsel cannot later represent a member or dependent on a matter in which he or she provided military legal assistance.

In the event a legal assistance officer becomes subject to a malpractice action, pursuant to Public Law 100-448, § 15(a)(2), 10 United States Code §1054, the United States will defend and pay any award resulting from such action. Indeed, such action is deemed to be a tort action against the United States and is treated as such.

Statistical View

Between active duty, reserve, and retired personnel and dependents entitled to legal assistance, the total community consists of approximately four million people. During 1990, the services provided assistance to almost two million clients, many of whom required assistance as a result of Desert Shield and Desert Storm. The U.S. Army saw an increase of 48 percent in legal assistance between 1989 and 1990.

There are approximately three hundred full time legal assistance officers in the U.S. military. A statistical count of part-time legal assistance officers is not possible because the legal assistance function is an ancillary duty for all judge advocates in a number of services.

U.S. Prospects

Except for the U.S. Coast Guard, where legal assistance is being revitalized after a number of years of having been all but eliminated, it is anticipated that legal assistance will be scaled down due to budgetary constraints. However, the trend appears to be that all judge advocates and law specialists will become legal assistance officers who, in field conditions, armed with laptop computers and comprehensive legal assistance computer programs, will be able to deliver top quality service at all times and places.

Other Countries

The country that is most closely modeled after the U.S. legal assistance program (although on a much smaller scale) is Canada. With 70 lawyers and 40 reservists, the legal assistance function is primarily in the areas of powers of attorney, wills, and notarial services, but these services are somewhat expanded in the five Canadian legal offices in Germany. The other British Commonwealth nations provide less legal assistance, and in England, where there is no meaningful legal assistance, substantially all matters are referred to civilian counsel.

France, Germany, and Israel provide no legal assistance; Korea provides military legal assistance on a limited basis, and has been known to provide it to civilian populations in remote agricultural areas. The Soviet military has military lawyers, but the legal assistance concept is very limited. Botswana implemented a program in 1988 modeled after U.S. Army regulations. Nigeria has a limited program that will be expanding in scope, but it does not include dependent coverage.

Conclusion

The role of the military lawyer in the personal problems of military members that are not service connected is extremely narrow, if it exists at all, in most nations of the world. To the extent that a given country seeks to embrace or expand the legal assistance concept within its own force, the experience of the United States will be the guidepost as to scope, eligibility and dedication of resources. This is so because of the long history and relative import the United States places on the need for quality legal assistance so as to enhance combat readiness and morale.

STEPHEN C. GLASSMAN

SEE ALSO: Civil-Military Relations; Discipline; Family Dependent Programs; Law, Military; Morale.

Bibliography

U.S. Congress. House. Conference Committee. 1985. *Department of Defense Authorization Act*. 98th Cong., 2d sess. 10297–98. H.R. Rep. 98-1080.

U.S. Congress. Senate. Committee on Armed Services. 1985. *Omnibus Defense Authorization Act*. 98th Cong., 2d sess. 217, 307 (1984). S. Rep. 500 (Sec. 157, Authorization for Legal Assistance Services, Legislative Provisions Adopted, part F - Miscellaneous).

U.S. Dept. Army. 1989. *Army Regulation. 27-3: Legal Assistance*. Chapters 1–4. Washington, D.C.: Government Printing Office.

U.S. Dept. Navy. Manual of the Judge Advocate General of the United States Navy. Chapter 7: *Legal Assistance*. Washington, D.C.: Government Printing Office.

———. 1943. Navy Department Bulletin. *Legal Assistance for Naval Personnel* 3 (1): R-1164.

———. 1943. Navy Department Bulletin. *Office of the Judge Advocate General: Administration of legal assistance offices* 3(6): R-1397.

LIBYA (Socialist People's Libyan Arab *Jamahiriya*)

A thinly populated country in North Africa, Libya is often in the news because of the policies of its leader, former Col. Muammar Qadhafi, which have led to clashes with the United States, most recently in January 1989.

Power Potential Statistics

Area: 1,759,540 square kilometers (679,358 sq. mi.)
Population: 4,765,600
Total Active Armed Forces: 85,000 (1.784% of pop.)
Gross National Product: US$24 billion (1989 est.)
Annual Defense Expenditure: US$2 billion est. (11.1% of GNP, 1987 est.)
Iron and Steel Production: 10,000 tons ferrous metals in 1986
Fuel Production:
  Crude oil: 53 million metric tons (1989)
  Natural gas: 29,000 million cubic meters (1982)
Electrical Power Output: 13,600 million kwh (1990)
Merchant Marine: 30 vessels; 807,539 gross registered tons
Civil Air Fleet: 59 major transport aircraft; 123 usable airfields (53 with permanent-surface runways); 7 with runways over 3,659 meters (12,000 ft.); 31 with runways 2,440–3,659 meters (8,000–12,000 ft.); 44 with runways 1,220–2,440 meters (4,000–8,000 ft.).

For the most recent information, the reader may refer to the following annual publications:
The Military Balance. International Institute for Strategic Studies. London: Brassey's (UK).
The Statesman's Year-Book. New York: St. Martin's Press.
The World Factbook. Central Intelligence Agency. Washington, D.C.: Brassey's (US.)

History

During the ancient era, Libya was under the domination of one or more foreign powers, including the Phoenicians, Carthaginians, Greeks, Romans, Vandals, and Byzantines. One of the most significant events in Libyan history was the invasion and conquest of the country by Muslim Arabs in A.D. 642–43. Over the centuries, the natives adopted Arab language and culture, and retained these even when the country fell under the control of the Ottoman Empire in the sixteenth century. Ottoman control soon waned, and Libya was to a great extent autonomous after 1600; the corsairs of Tripoli were, together with the pirates from Tunis and Algiers, the scourge of the western Mediterranean. By the late nineteenth century, Libya had become a backwater, and the Turkish garrisons along the coast held little sway over inland nomads.

Italy invaded Libya during its war with Turkey (September 1911–October 1912), and gained Libya as a colony after defeating Turkey. It took the Italians nearly twenty years and several major campaigns to subdue the interior. In World War II, Libya was conquered by the British in late 1942, and remained under British administration until Libya became an independent nation on 24 December 1951. The independence process was completed through the offices of the United Nations (UN), and Libya was made a kingdom under the rule of Idris I, former Emir of Cyrenaica (the area around Benghazi and Tobruk). The discovery of oil in 1959 improved the country's fortunes, but King Idris was deposed in a military coup on 1 September 1969.

The new government was led by then-Col. Muammar Qadhafi, chief of the Revolutionary Command Council (RCC). The new government made public its steadfast support of liberation movements in the developing nations, and moved to close U.S. and British military bases in Libya (March–July 1970). Qadhafi convened a General People's Congress on 3 March 1977 to proclaim the Socialist People's Libyan Arab *Jamahiriya* (Republic), a new state based on the tenets of Qadhafi's political philosophy, described in his "Green Book of Revolution." In theory, power in modern Libya rests in the hands of the Popular Congress, but in practice much authority rests with civil and military hierarchies, and the revolutionary committees.

Libya's (and Qadhafi's) advocacy of "national liberation" movements has involved Libya in terrorist activities. U.S. anger over Libyan involvement in a bombing incident which killed U.S. servicemen led to a reprisal air raid on Tripoli (12 April 1986). Libya also gave considerable aid (including troops) to rebel forces in Chad (1981–86 and 1991). This reached a peak between August 1983 and November 1984. Renewed Libyan intervention in 1985–86 resulted in an embarrassing defeat for Qadhafi when local Chadian forces with U.S. and French equipment routed Libyan forces in a series of running battles (August–September 1987). Libya has also been involved in minor clashes with U.S. forces in the Mediterranean (the first in August 1981), one of which (24 March 1986) resulted in the loss of 1 corvette sunk and a second badly damaged.

Politico-Military Background and Policy

Broadly, Libyan foreign policy has two goals: Arab unity and the elimination of Israel (and Western influence) from the region. Libya shelters, trains, and equips guerrillas and terrorists from many nations in order to further these goals. Qadhafi has been wary of committing Libyan forces, with the exception of Chad, to support foreign policy goals. Libya has also engineered unsuccessful coup attempts against neighboring governments, including Morocco, Egypt, Tunisia, and Sudan (where the effort may have succeeded).

The armed forces are manned through selective conscription of physically fit males over age 17. Conscripts serve for a term of two to four years, and an estimated 40 percent of each age-class are called up for service. After conscript service, men have a reserve liability to age 49.

Strategic Problems

Libya's greatest strategic weakness is its relatively small population and its concentration along the coast and in the few other areas of arable land. Libya's policies have earned it few friends, although as an Arab state it gains some benefit from Arab solidarity, and it has had close, if unofficial, security ties to the former Soviet Union and other East European countries.

Military Assistance

Libya received considerable military assistance from the Soviet Union and its allies, although Libya purchased most of the materiel from Soviet sources. The Libyans have purchased US$15 billion in Soviet arms and $10 billion in Eastern European weapons since the early 1970s. It has also had some East bloc advisers and training personnel.

Defense Industry

Libya depends on imports for nearly all of its heavy equipment, including tanks, aircraft, and naval vessels. There have been rumors of a Libyan nuclear weapons project, but Libya at this point still has no nuclear reactor, and no country seems eager to sell it such equipment. In late 1988, the United States made public its evidence that Libya had constructed a chemical weapons plant some 80 kilometers (50 mi.) south of Tripoli, assisted by several West German companies. The Libyans deny this, claiming the plant makes fertilizers, but have not yet explained why there are SAMs and antiaircraft gun emplacements around the plant, which is also protected by guards and barbed wire.

Alliances

Libya has entered into several unions with other Arab nations, most recently with Morocco (1984). Although often touted as major steps toward Arab unity, these alliances lack real substance and are little more than symbolic. Libya negotiated a peace and friendship accord with the USSR in 1983, but this was not signed and did not become an official treaty. Libya is a member of the United Nations (UN), OAPEC, the Arab League, OPEC, and the OAU.

Defense Structure

Officially, the secretariat of the General People's Congress serves as a cabinet. This secretariat is headed by the secretary general, an office held by Colonel Qadhafi until his "retirement" in 1980, and currently held by Omar al-Muntasir. Precise responsibility and chain of command is not clear, since Qadhafi retains a great deal of power despite his lack of official title and position.

Total Armed Forces

Active: 85,000. Terms of service: selective conscription, 2–4 years.
Reserves: People's Militia, some 40,000.

ARMY: 55,000. 4 military districts.
28 bde (11 armd, 11 mech, 5 inf, 1 National Guard).
42 tk bn.
48 mech inf bn.
19 para/cdo bn.
53 arty, 14 AD arty bn.
7 SSM bde.
Equipment:
MBT: 2,150 (incl 1,200 in store): 1,500 T-54/-55, 350 T-62, 300 T-72.
Recce: 270 BRDM-2, 380 EE-9 Cascavel.
AIFV: 1,000 BMP-1.
APC: 850 BTR-50/-60, 90 OT-62/-64, 40 M-113, 100 EE-11 Urutu.
Towed arty: some 720: 105mm: some 60 M-101; 122mm: 270 D-30, 60 D-74; 130mm: 330 M-46.
SP arty: some 370: 122mm: 130 2S1; 152mm: 60 2S3, DANA; 155mm: 160 Palmaria, 20 M-109.
MRL: some 650: 107mm: Type 63; 122mm: BM-21/RM-70, BM-11.
Mortars: 82mm; 120mm; 160mm: M-43; 240mm.
SSM launchers: 40 FROG-7, 80 Scud B.
ATGW: 3,000: Vigilant, Milan, AT-3 Sagger (incl BRDM SP), AT-4 Spigot.
RCL: 106mm: 220 M-40A1.
AD guns: 600: 23mm: ZU-23, ZSU-23-4 SP; 30mm: M-53/59 SP; 40mm: L/70; 57mm: 92 S-60.
SAM: SA-7/-9/-13, 24 quad Crotale.
Helicopters:
  Transport: 18 CH-47.
  Liaison: 5 AB-206, 11 SA-316.
Deployment:
Aouzou Strip: est. 2,000; 2 mech bn, 2 tk bn.

NAVY: 8,000 incl Coast Guard. Bases: Tarabulus, Benghazi, Darnah, Tubruq, Sidi Bilal, Al Khums.
Submarines: 6 Al Badr (Sov Foxtrot) with 533mm and 406mm TT.
Frigates: 3: 1 Dat Assawari (UK Vosper Mk 7) with 2×3 ASTT; plus 4 Otomat SSM, 1×114mm gun; 2 Al Hani (Sov Koni) with 4×ASTT, 2×ASW RL; plus 4 SS-N-2C SSM.
Patrol and Coastal Combatants: 45:
Corvettes: 7: 4 Assad al Bihar (It Assad) with 4 Otomat SSM; plus 2×3 ASTT (A244S LWT). 3 Ean al Gazala (Sov Nanuchka) with 2×2 SS-N-2C Styx SSM.
Missile Craft: 24: 9 Sharara (Fr Combattante II) with 4 Otomat SSM; 12 Al Katum (Sov Osa-II) with 4 SS-N-2C SSM; 3 Susa ⟨ with 8SS-12M SSM.
Patrol Craft, Inshore: 14: 4 Garian, 3 Benina, 7 ⟨.
Mine Warfare: 8 Ras al Gelais (Sov Natya MSO).
Amphibious: 5: 2 Ibn Ouf LST, capacity 240 tps, 6 tk, 1 hel; 3 Sov Polnocny LSM, capacity 180 tps, 6 tk; plus 2 LCT.
Support and Miscellaneous: 4: 1 log spt/dock, 1 trg, 1 salvage, 1 diving spt.

Naval Aviation: 31 armed hel.
Helicopters: 2 sqn: 1 with 25 Mi-14 (ASW); 1 with 12 SA-321 (6 ASW, 6 SAR).

AIR FORCE: 22,000 (incl Air Defence); (some Syrian pilots, Soviet, N. Korean and Pakistani instructors); 409 cbt ac, 45 armed hel (many ac in store, number nk)
Bombers: 1 sqn with 5 Tu-22.
FGA: 7 sqn:
  20 MiG-23BN, 8 MiG-23U, 22 Mirage 5D/DE, 10 Mirage 5DD, 8 Mirage F-1AD, 10 Su-24, 45 Su-20/22.
Fighter: 9 sqn: 50 MiG-21, 100 MiG-23, 12 -23U, 55 MiG-25, 3 MiG-25U, 12 Mirage F-1ED, 6 -BD.
COIN: 1 sqn with 30 J-1 Jastreb.
Recce: 1 sqn with 6 Mirage 5DR, 7 MiG-25R.
Transport: 2 sqn: 11 An-26, 12 Lockheed (7 C-130H, 2 L-100-20, 3 L-100-30), 20 G-222, 16 Il-76, 15 L-410.
Helicopters:
  Attack: 35 Mi-24, 10 Mi-35.
  Transport: hy: 18 CH-47C; med: 7 Mi-8, 50 Mi-4; lt: 10 Mi-2, 4 SA-316.
Training: 89 Galeb G-2 ac; 20 Mi-2 hel; other ac incl 2 Tu-22, 70 L-39ZO, 77 SF-260WL.
Missiles:
  ASM: AT-2 Swatter ATGW (hel-borne).
  AAM: AA-2 Atoll, AA-6 Acrid, AA-7 Apex, AA-8 Aphid, R-530, R-550 Magic.

Air Defense Command:
'Senezh' AD comd and control system.
3 bde with SA-5A: each 2 bn of 6 launchers, some 4 AD arty gun bn; radar coy.
3 Regions: 2 bde each 18 SA-2; 2–3 bde each 12 twin SA-3; est. 3 bde each 20–24 SA-6/-8.
Some 2,000 Soviet personnel reportedly man the SA-5 complexes. Expatriates form a large proportion of the technical support staff.

PARAMILITARY
There are 4 paramilitary groups in Libya, although one is largely ceremonial (People's Cavalry Force), and another is the Customs/Coast Guard, which is under navy control and operates 14 fast inshore patrol craft and 7 small coastal patrol boats drawn from navy inventories.
Liwa Haris al Jamahiriya (Revolution Guard Corps): a popular militia force of some 3,000; equipped with materiel drawn from army inventories, including T-54/-55/-62/-72 tanks, armored cars, APCs, multiple rocket launchers, ZSU-23-4 self-propelled 23mm antiaircraft guns, and SA-8 SAMs.
Islamic Pan-African Legion: some 2,500 men, many are political exiles from other Saharan and North African countries; organized into 1 "brigade" each of armored troops, infantry, and parachute/commando forces. Some 700–1,000 legion personnel may be in the Sudan, but reports are unreliable. Legion equipment, also drawn from army inventories, includes many T-54/-55 tanks, some EE-9 Cascavel armored cars, and BTR-50/-60 APCs.

Future

Under its present government, which shows no signs of losing control of the country, Libya is unlikely to change its policies. The Qadhafi regime will almost certainly continue to provide refuge and support to terrorists. Libya will also continue to interfere in the internal affairs of its neighbors, although Qadhafi's willingness to antagonize nearby countries seems to have declined since the early 1980s. It can only be hoped that Qadhafi may become more inclined to negotiation and compromise, and less willing to face the dangers of confrontation.

DAVID L. BONGARD
TREVOR N. DUPUY

SEE ALSO: Arab Conquests; Arab League; Colonial Empires, European; North Africa; Organization of African Unity.

Bibliography

American University. 1979. *Libya: A country study*. Washington, D.C.: Government Printing Office.

Cooley, J. K. 1983. *Libyan sandstorm: The complete account of Qaddafi's revolution*. New York: Holt, Rinehart, and Winston.

Deeb, M., and M. J. Deeb. 1982. *Libya since the revolution: Aspects of social and political development*. New York: Praeger.

Hunter, B., ed. 1991. *The statesman's year-book, 1991–92*. New York: St. Martin's Press.

International Institute for Strategic Studies. 1991. *The military balance, 1991–1992*. London: Brassey's.

Paxton, J., ed. 1988. *The statesman's year-book, 1988–1989*. New York: St. Martin's Press.

LIMITED WAR

Although the literature on limited war was extensive in the 1950s and 1960s, this does not mean that: (1) no limited wars had been fought before that; (2) theories on limited war were developed in those two decades only; (3) it has been possible to define unequivocally the strategic value of limited war theories, or even that (4) a clear and universally accepted definition of limited war has been found. Usually, however, *limited war* is defined as a war in which neither side seeks the total annihilation of the opponent. Since 1945 this has meant war short of nuclear war.

It is significant that modern limited war theory was developed mainly by British and American scholars. And it is equally significant that each limited war fought by U.S. forces has heavily influenced the evolution of limited war theory. This was true for the Korean War as well as for the Vietnam War. In both cases the United States went to war with a war theory not suitable for the kind of warfare it was going to wage. Hence, that theory had to be developed in the aftermath of those painful experiences. Neither the Korean War nor the Vietnam War were the first or last limited wars fought in the twentieth century. But they were the first wars in the nuclear age fought by a superpower without using nuclear weapons. It has yet to be seen if the Soviet Union will develop a limited war theory of its own after its experience in Afghanistan.

While limited war theory has been greatly elaborated upon during the 1950s, 1960s, and 1970s, ideas about limited war have been included in thoughts about the nature of war throughout recorded history. Fortunately, most wars have been limited in one way or another. The total destruction of Carthage by the Romans is a unique event in history.

Limited wars must not end with the annihilation of the loser; the victor must settle for more limited objectives. As outlined by all theoreticians of limited war, this limitation must be deliberate, and consequently must lead to the limitation of means, not only in favor of the loser, but also to achieve an optimum peace at the least possible cost. Limitation in warfare can therefore be viewed as the proof of wisdom in statesmanship and strategic planning. As the Chinese military thinker Sun Tzu said about 500 B.C.: "Generally, in war the best policy is to take a state intact, to ruin it is inferior to this. To capture the enemy's army is better than to destroy it. . . . To subdue the enemy without fighting is the acme of skill." Or, as Sir Basil H. Liddell Hart wrote in 1925 (*The Future of War*): "The aim of a nation in war is, therefore, to subdue the enemy's will to resist, with the least possible human and economic loss to oneself." Warfare during the period between these two historical statements was limited in most cases, but not always deliberately so. Limitation in aims, and more often in means, occurred merely because of a lack of power or technical abilities to achieve the total destruction of the enemy. The crusades and the Thirty Years' War serve as examples.

Because limitation was not deliberate, and was not chosen as a result of political decision making, there was little development of limited war concepts during those centuries. Only in the eighteenth century were concepts of limited war developed—motivated primarily by the fear of ruining the costly and valuable professional armies of the time.

At the beginning of the nineteenth century, the development of mass armies during the French Revolutionary—Napoleonic wars led to a decline in thinking about limited war. Clausewitz's statement that "war is an act of violence and there is no limitation in using it," was meant to emphasize the necessity for political control and congruence of means and ends in the reality of the political-military process of a war. He showed, in fact, that there is a wide range of limitations on the theoretically unlimited violence of war. This period marks the beginning of the age of total war, leading to the American Civil War and to World Wars I and II. Only a few military thinkers, for example, the Austrian Archduke Charles in the nineteenth century and Sir Basil Liddell Hart in the period between the two world wars, tried to argue against *guerre à l'outrance*. However, to many war leaders of the nineteenth and twentieth centuries, the prospect of total victory seemed to be attractive enough to risk total defeat by committing all the resources of the nation in war. Finally, at the end of World War II, the nuclear bomb (the "ultimate weapon," the "weapon to end all wars"), seemed to be the solution. The risk of total annihilation should deter all wars and bring peace forever.

Reality destroyed this overly optimistic view within a few years after World War II. Civil and revolutionary wars in China, Indonesia, Greece, the Philippines, Indochina, Nigeria, Malaya; the Kashmir Dispute; the first Arab-Israeli War (1947–49); and eventually the Korean War clearly showed the impossibility of deterring all wars by threatening use of the nuclear bomb. It even proved impossible when a nation in possession of the nuclear bomb was directly involved. This fact, together with the loss of a nuclear monopoly by the United States upon the introduction of nuclear weapons into the Soviet inventory, implying that nuclear deterrence might lead to mutual annihilation, prepared the way for modern limited war theory. A number of military thinkers before the Korean War had foreseen the implicit risks of relying on nuclear weapons to deter or to end a war. Among these was Sir Basil Liddell Hart, who as early as March 1946 expressed his hope that "international agreement will recognize the great destructive power of new weapons and lay down rules for limiting their use." Similar thoughts about the limitation of warfare were aired by a number of strategists such as Brodie, Kissinger, Osgood, and Halperin. Most of them had two main concerns: how to avoid the escalation of a local conventional war into an all-out nuclear war between the two superpowers, and how to contain the expansion of communism. Different ideas on how to solve these two interrelated problems were the basis for the theory of limited war, which is more a set of ideas than a single formulated theory.

Meaning of Limited War

The basis for the modern limited war theory is the maxim expressed by Clausewitz that war is not an end in itself, but an instrument of policy. Since a policy leading to mutual destruction is inconceivable, wars in the nuclear age must be limited, at least when nuclear powers are involved or might become involved. In general, there are three fields of limitation: limitation of objectives, limitation of means, and limitation of scope.

LIMITATION OF OBJECTIVES

Limitation of objectives is the first and most important feature of limited war. Objectives form the basis for national policy and must be supported by the national will and the resources of the nation. Only if the objectives are limited can the war itself be kept limited. The limitation of objectives must be deliberate; it must guide, as well as be guided by, policy, and not by the mere inability to seek unlimited objectives.

The Second World War was an unlimited war because it was fought for an unlimited objective (i.e., unconditional surrender of the Third Reich and Japan). Vietnam was a limited war because the United States had limited objectives: to repulse the North Vietnamese aggression, and to foster the survival of South Vietnam. In practice, limiting objectives means a willingness to settle for a negotiated peace and to accept a compromise, an outcome only partly satisfactory. Clausewitz's classic definition that the object of war is to impose one's will upon the enemy must be constrained. The object of a limited war must be a better peace (Liddell Hart 1967).

LIMITATION OF MEANS

Directly connected with the limitation of objectives, and generally the most obvious limitation, is the limitation of means. While limitation of objectives might not be apparent, or might not be believable to the adversary, restraints in using force are easy to recognize and to prove. Again, restraints must be deliberate and not motivated only by the lack of means. They must also be massive. A small margin of limitation is not enough to make a war limited. World War II serves as an example. Despite the fact that chemical weapons, while available to both sides, were not used, one can hardly call this war limited. Since all the resources and all the forces of the nations involved were brought to bear, the nonuse of one single kind of weaponry did not make any difference.

Limitation of means has two aspects: one political-economic, the other military-technological. If a nation at war is dedicating all political and economic efforts to the single purpose of fighting (and winning) the war, one can hardly speak of a limitation of means. Limitation of means in this respect only occurs if the state is able to maintain its normal political and economic activities—at least to some extent. The Falklands/Malvinas campaign is a case in point.

The military-technological aspect of the limitation of means is more difficult to define. The question whether the term *limited war* can be used for a war with nuclear weapons has been discussed at great length. In his first book on the subject, *Nuclear Weapons and Foreign Policy* (1957), Kissinger presented the case for a limited nuclear war strategy. Others have rejected this idea. Nonetheless, there is as yet no consensus as to the extent (weapons yields, or selection of targets) to which the use of nuclear weapons can be "allowed" within a limited war. Even a concept for a limited strategic war, with the possible use of strategic nuclear weapons against military and selected civil targets was conceived. The only limitation suggested in this case was to avoid massive strategic bombing of cities with nuclear weapons. This is the basis for the only definition of limited war on which all theorists are agreed: Limited war is a war short of strategic war with bombing of cities. The majority of scholars, however, have also agreed that a limited war must not include any strategic nuclear exchanges between the United States and the Soviet Union. This provision is one of the rationales for the strategy of flexible response adopted by the North Atlantic Treaty Organization (NATO). Below this threshold there is a second distinct borderline, perhaps the most significant distinction between limited and total war: the use or nonuse of any nuclear weapons. On the one hand, it has been widely argued that this might be the most important constraint in keeping a war limited. On the other hand, this position has been challenged with the argument that the effects of properly targeted small-yield nuclear weapons do not exceed the effects of conventional weapons; therefore their use would not make any difference. The difference, according to this argument, lies in the targeting policy. The basic criterion would be to not target civilian assets or targets on the home territory of a nuclear power.

LIMITATION OF SCOPE

Limitation of the geographic scope has always occurred in warfare and will probably occur in the future. While both world wars were fought worldwide, they were not waged everywhere. If the term *geographically limited war* is to make sense, the war must be restricted to only a part of the territories of the belligerents. Again, this limitation should be deliberate and massive, as it was in the Falklands/Malvinas War.

The second possibility for a limitation of scope is the number of states involved in a war. Thus, the only really unlimited war would be a real world war, but neither of the world wars fought so far, nor any conceivable in the future, has included all nations.

This consideration has led to the suggestion that the term *local war* should be used for wars in a limited area or with a limited number of belligerents, not including one of the superpowers. This might make sense from the point of view of a superpower, but it seems that the rules for, and the problems of, limited warfare also apply to most local wars.

There are no generally accepted delineations among the kinds of wars discussed above. Most likely there are no clear-cut boundaries either, and thus the nature of these wars might change during the course of a war. Further clarification of this issue could be one of the future tasks of limited war theory.

Problems of Limited Warfare

The basic and most important problem of limited warfare is keeping the war within the limits envisaged. This normally cannot be done by unilateral declaration or behavior. Furthermore, sovereign states sometimes find it difficult to settle for limited objectives in a war; to fight with limited means; and to adhere to these limitations subject to the pressures of public opinion, unexpected enemy moves, and an unfavorable course of the war. The momentum of the war itself might endanger the self-imposed bounds.

Obviously, there is an absolute requirement for communication or agreement between the adversaries, a clear and understandable explanation of purposes and objectives to the public, and tight political control of the military side of the war.

Communication between the belligerents can be open or tacit. Its purpose is—explicitly or implicitly—to come to an agreement on limits and rules of conflict. As Thomas Schelling (1963) has suggested, a kind of "bargaining" might take place. One side may declare limits or indicate by its behavior that it would not go beyond certain limits (e.g., by not using all means at hand or by leaving some sanctuaries to the forces of the adversary). The enemy, when accepting these limits, may reciprocate by open agreement or abstentions on his part. In case of disagreement, the enemy may answer by letting the war become more violent and less restrained. This kind of mutual, deliberate escalation can be regarded as "tacit bargaining" and may go on until a level of warfare has been reached that both sides regard as an equilibrium between the "costs" and the possible benefits of the war.

The problem is that both sides must understand the messages of their adversaries and must have the desire to observe the "rules." Otherwise, escalation might easily get out of control and lead to disaster. As Schelling (1963) pointed out, competitive escalation in a war has some similarities with the "chicken" game played by teenagers in the United States. In this game two cars are driven toward each other at top speed, both astride the centerline of the road. The first driver to pull over "chickens out" and loses the game. Obviously in this game, as well as in an escalation process, success is with the player willing to run higher risks. Dangerous and apparently insane as this behavior is, Schelling has correctly stressed its essential rationality, since one's objectives might be achieved in this way.

Successful escalation (i.e., an escalation that leads to an accepted level of conflict short of an intolerable level of war) is only conceivable if the objectives are not intolerable either. This again demonstrates the close interrelation between objectives and means in limited war.

Limiting the means between adversaries of comparable strength is only possible if limited objectives are pursued. Goals short of victory are not easily explained to the public. They are a kind of negative goal (e.g., not to lose a country to communism, or to defend freedom). Hence these objectives are often put into question. Doubts may arise as to whether or not the objectives are worth fighting a war for, or allocating the resources and means necessary for that kind of war. In a limited war, both political and public interest are not necessarily focused on the war at all times. Therefore, ongoing international relations as well as domestic politics have considerable bearing on its conduct. As happened in the Vietnam War, this influence might lead to the complete abandonment of the initial objectives of the war.

Yet the decision-making process during the war may be swayed in the other direction, to abandon limited objectives and to increase the violence and the pace of the war. As history suggests, any society has a certain level of tolerance it can accept in a war. Beyond that level, a society is tempted to risk everything in order to gain total victory or to avoid total defeat. This threshold is different for each society and for each war. It is influenced by ideology and religion, by the internal structure of the society, and by the actual course of the war.

The notion that it might be possible to gain much more with just a little more effort may cause the incentive to override any restrictions. This may also be caused by the opposite perception. A belligerent who believes that he has been driven back against a wall might try to cope with his problem by escalating the war into an all-out war. Hence, both approaches need to be avoided. Neither total victory nor total defeat should be perceived as possible by any of the opponents. This requires continual monitoring and reassessment of the status of the war, and a continual effort to keep the delicate balance of justifiable limited objectives and means, both in the eyes of one's own public and in the perception of the enemy.

The same kind of monitoring and reassessment is necessary to prevent the war from gaining too much momentum. As Clemenceau said: "War is too serious to leave to the generals." An example commonly cited is the Korean War and the argument between President Truman and General MacArthur ("There is no substitute for victory.") over the future course of the war, which eventually led to MacArthur's dismissal. But it is not always the generals who press for unlimited objectives. Halperin (1963) cited a Republican minority report to the U.S. Congress issued after the MacArthur hearings: "Our policy must be to win. Our strategy must be devised to bring about decisive victory."

Another problem in limited warfare is caused by third-party involvement. While it might be easy to limit objectives, means, or scope as long as the war is only between two opponents, things become more complex with each additional adversary that gets involved in the war. This not only automatically enlarges the scope of the war, it could heavily influence efforts to keep objectives and means limited. Third-party involvement may be directly military, or it could be in the form of political, economic, or logistic support. In any case, the objectives of the third party must be taken into account and will influence the process of decision making. The more important the involvement of the third power, the more influence it will have on the real decisions (its influence may become greater than that of the original belligerents), thus possibly leading to a slow erosion and finally elimination of limitation in the war. Coalition warfare, difficult enough in unlimited wars, demands the utmost of statesmanship in limited wars.

Third-party involvement may work in the other direction as well. As McClintock (1967) observed, "Another characteristic of twentieth-century limited war is the fact that international organization has played an important role, either in settling these armed conflicts or as serving as a useful palliative and sally port of belligerents in need of saving face." More than that, third parties (international organizations or states) may, through deeds and through avoidance, provide for the limitation of the war. Not delivering weapons is one way to moderate a war. Another is to mediate between the two adversaries. As O'Brien (1981) described for the 1973 October War, it can even be in the interest of both superpowers to limit a conflict, and therefore to work together in the process of moderation and limitation.

Criticism of Limited War

As with any theory in the realm of social science, limited war theory has attracted a flood of criticism that questions the usefulness of the theory as a whole and brings to light perceived flaws in the theory.

The Soviet Union always considered *local wars*—its term for wars in which it was not directly involved—as a kind of imperialistic aggression. It even regarded (as G. D. Arbatov put it in 1974) "the idea to introduce and accept rules, courses of action and artificial limitations as illusionary and impracticable." Especially considering a war between NATO and the Warsaw Pact, Soviet military theory for some time flatly denied the feasibility of a limitation of war. According to Arbatov: "It is hardly [sic] to conceive, that a nuclear war, once unleashed, can be kept within certain bounds and it would not escalate into an all-out war." From this point of view, war between the two blocs, between socialism and capitalism, must lead to the final victory of the first. It is interesting to observe that, by 1989, the originally formulated position that such a war is unavoidable due to the class nature of the conflict between the two systems was no longer stressed by Soviet writers. Instead, the newly expressed political position was that war was no longer a means of politics. This political position has not yet been transferred to the technical side of military doctrine. Yet, Gorbachev's idea of "reasonable sufficiency" of military forces, sufficient for the defense and incapable of launching decisive offensives, may form the future nucleus for a new doctrine of limited war for the new states that once formed the Soviet Union.

The second major criticism of limited war theory as a whole has come from Western analysts. They have claimed that a limited war strategy might undermine the strategy of deterrence pursued by the two superpowers. They suggested that only the prospects of global nuclear destruction prevented the outbreak of a war between the two blocs for at least four decades. If, based on a limited war theory, a war could be considered as not inevitably leading to a nuclear war, the reluctance to wage war might decrease and thus weaken the strategy of deterrence. Under a limited war theory, this argument continues, war could also again be seen as a viable means of politics. It could be regarded as a usable technique for achieving political objectives, to be used whenever deemed appropriate and controllable in a way so as to preclude escalation to total war. Therefore, the probability of wars would increase.

It was Robert McNamara who, in the 1960s, while defending his strategy of flexible response, rejected these arguments by emphasizing that "one cannot fashion a credible deterrent out of an incredible action." He argued that a deterrent threat must be a rational instrument of policy, implying that the same must be true for the war theory forming the fundament of the deterrent threat.

Otherwise, deterrence would not be credible. Except in very ominous circumstances, massive nuclear retaliation would be an incredible response and hence a poor deterrent; whereas limited war as a credible response would be much more of a deterrent in most of the conflict scenarios conceivable. Therefore, a limited war strategy would complement the strategy of nuclear deterrent and enhance overall deterrence.

Since McNamara's tenure as U.S. Secretary of Defense (1961–68), flexible response has been the cornerstone of United States and NATO strategy. But still the relation between nuclear deterrence and the capability to fight and control a limited war leaves more than enough space for strategic thinkers and policymakers. Views as expressed in the report of the Commission On Integrated Long-Term Strategy, *Discriminate Deterrence*, may mark a new beginning of limited war thinking, helping to implement a strategy for a future multipolar world. In addition to that, the process of arms control, in the nuclear as well as in the conventional realm, may cause further adjustments of limited war theory.

The third fundamental argument against limited war theory is that it was only developed to implement U.S. policy and that it is not applicable to any conflict scenario between other states, with the possible exception of war scenarios between nations possessing nuclear weapons and nonnuclear countries. This argument is difficult to reject, because limited war theory was evolved by American (plus a few British and French) strategists. The use of limited military means to achieve limited goals by the United Kingdom in the Falklands/Malvinas War and by France in the Chad War seems to prove this point.

On the other hand, the Gulf War between Iraq and Iran could be viewed as proof of the usefulness of limited war theory for wars between less powerful states. It might be protested that a war with millions of dead is not a "limited" one. But the war was started for limited objectives; it was limited in scope, and there were some restraints in the use of force despite the fact that chemical weapons were used by one side and some bombing of civilian targets ("war of the cities") occurred. The increase and decrease in the level of violence, which could be observed several times during the war, could be seen as deliberate escalation and deescalation in order to settle, through tacit bargaining, for acceptable conditions to terminate the war. Of course, it could be argued that some of the limits, especially in the level of violence, were not deliberate, but were merely dictated by the availability of military means. As of early 1991, the available information is inadequate to prove this point, but it is evident that the opponents seem to have achieved a peace settlement without having reached a single objective of the war, and without the total defeat of either of them—a typical feature of a limited war.

The group of arguments against more practical aspects of the limited war theory includes the question on how to control a limited war. In the 1960s, it was argued that the whole structure of reasoning implied a level of rationality on the part of decision makers that was quite unrealistic. Because statesmen were liable to human weakness too, it was highly dangerous to assume that they could conduct a war as rationally and coolly as they could play a game of chess.

It is true that decision making is not always a matter of complete, strictly target-oriented rationality. As Graham T. Allison outlined in his book *Essence of Decision* on the Cuban Missile Crisis, decision making resembles a bargaining process heavily influenced by governmental politics and the output of bureaucratic organizational actions. But even when accepting this, no critic of political decision making would deny the usefulness or even the necessity of a rational framework—like the limited war theory—that influences the decision makers or other "players" in the process. Furthermore, historical examples from Korea to the Falklands/Malvinas provide evidence of the practicality of controlling the level of violence in a war and keeping it limited—presupposing a political decision to do so.

Some of the arguments against the merits of limited war theory are connected with the problem of limitation of means. One may argue that it depends on the point of view as to whether the means of warfare used are considered limited or not. In the Vietnam War, the limited use of force by the United States might have been considered unlimited by the North Vietnamese, who, from their side, used all military means at their disposal, and did not in any way restrict the use of force, which was still, by the yardstick of a superpower, a limitation of means.

This brings about two more points of criticism: the problem of asymmetry of means (or objectives) of the two adversaries, and the case of using limited means to attain unlimited objectives.

Again, the Vietnam War serves as an example. Both sides pursued different kinds of goals and used means that were considered limited by one side only. On the one hand, the U.S./South Vietnam side aimed at regaining the *status quo ante* and at repelling the North Vietnamese attack—a clearly limited objective. On the other hand, North Vietnam longed for the reunification of Vietnam under its terms, meaning the extinction of South Vietnam as a sovereign state—a total objective.

Asymmetry of means in a limited war is especially crucial in a war between states of different power. While the more powerful side could consider the war limited because it is not forced to use all forces at its disposal, the other, less powerful side, may be pressed to use virtually all of its military inventory and national resources. The limited war of one side can be a total war for the other side. One side may fight for limited objectives, such as limited territorial gains, whereas the other side may be forced to fight for sheer survival.

The last argument to be discussed is whether or not a

war with limitations that result only from the nonavailability of means and from geographical restrictions can be called limited, and whether or not the limited war theory can be applied. In other words, is a war limited when the limitations of means and scope are not deliberate and the objectives are unlimited, at least by implication?

The first Arab-Israeli War (1948–49) was limited in all aspects except for the objectives; but the limitations did not occur as a matter of policy, they were dictated by circumstances. The objectives were rather unlimited, with both sides wrestling for total supremacy in Palestine. The means were limited because no more were at hand, at least on the side of the Israelis, who had mobilized their complete society and economy. One other typical feature of a limited war did not occur—the war was not terminated by settling for a compromise between the opponents; rather, the war ended under pressure by third parties.

Obviously, the problem in this case is again connected with the point of view. Seen from "outside," especially from the standpoint of a great power, a war might seem limited. Seen from "inside," the same war might seem unlimited, using up all available means and pushing the nation involved to the brink of extinction.

Conclusion

In modern war theory, limited war is seen as a political process conducted with military means. The aim of this process is not to win at any cost but rather not to lose, and to fight in such a way that the enemy will not escalate the level of violence too far and will settle for a compromise peace.

Limited war, as opposed to total war, must be conducted by an adherence to self-imposed, deliberate, and massive limitations, especially of objectives and means. As of today, no one had found a singular and simple answer to delineate the borderline between limited and unlimited means. Because of the problems involved in a nuclear escalation process, most scholars agree that a limited war must not be fought with nuclear weapons.

Nevertheless, limited wars can take on different forms that cover a broad spectrum of conflicts with or without direct or indirect involvement of a superpower. At one extreme of this spectrum are unlimited or total wars. On the other end can be found a congeries of conflicts—the realm of subconventional war and low-intensity conflicts.

The limited war theory has its bearing on strategy, especially the strategy of nuclear deterrence. It has been formulated to enhance deterrence, and it remains to be seen how the theory of limited war will influence the future interrelated processes of arms control and strategy formulation, and how it will, in turn, be influenced by those.

The fact that most wars in human history have been limited, be it because of the prevailing circumstances or of the wisdom of statesmen and generals, has secured the survival of mankind so far. In the nuclear age, limitation of wars is more necessary than ever. If only because limiting war is the only way to avoid a nuclear Armageddon should war become unavoidable, limited war strategies must be devised and adapted to the political development of the world. Limited war is much worse than peace but much better than nuclear annihilation.

HEINZ KOZAK

SEE ALSO: Arms Control and Disarmament; Cold War; Conventional War; Deterrence; Demonstration; General War; Gunboat Diplomacy; Hague Conventions; Nuclear Employment Policy and Planning; Nuclear Theory and Policy; Spectrum of Conflict; Strategy, National Security; Total War; Unconventional War; War.

Bibliography

Allison, G. T. 1971. *Essence of decision: Explaining the Cuban Missile Crisis*. Glenview, Ill.: Scott, Foresman.
Brodie, B. 1959. *Strategy in the missile age*. Princeton, N.J.: Princeton Univ. Press.
Deitchman, S. J. 1966. *Limited war and American defense policy*. Cambridge, Mass.: Massachusetts Institute of Technology Press.
Halperin, M. E. 1963. *Limited war in the nuclear age*. New York: John Wiley and Sons.
Howard, M., ed. 1979. *Restraints on war, studies in the limitation of armed conflict*. Oxford: Oxford Univ. Press.
Kahn, H. 1965. *On escalation*. London: Pall Mall Press.
Kissinger, H. A. 1957. *Nuclear weapons and foreign policy*. New York: Harper.
Knorr, K., and T. Read., eds. 1962. *Limited strategic war*. New York: Praeger.
Liddell Hart, B. H. 1967. *Strategy*. 2d ed. New York: Signet.
McClintock, R. 1967. *The meaning of limited war*. Boston: Houghton Mifflin.
O'Brien, W. V. 1981. *The conduct of just and limited war*. New York: Praeger.
Osgood, R. E. 1979. *Limited war revisited*. Boulder, Colo.: Westview Press.
Schelling, T. C. 1963. *The strategy of conflict*. Oxford: Oxford Univ. Press.

LINES OF COMMUNICATION

Lines of communication are the routes that connect an operating military force with the bases that support it. Along these paths supplies and reinforcing military forces move. The efficient organization of lines of communication is a basic requirement of military planning. A nation may lose a war even though it has good lines of communications; but it cannot win a war without them.

Actual lines of communication are shaped by a number of factors and do not fit a standard pattern. The lines are composed of various types of support bases that are connected by land, sea, and air transportation depending upon geography and military requirements.

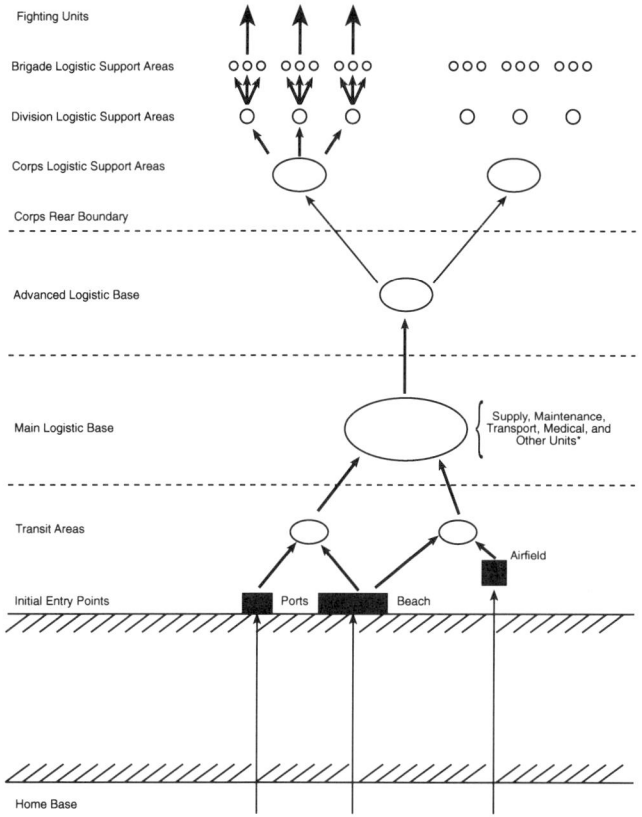

Figure 1. Lines of communication in an overseas theater.

Initial Deployment

The movement of a force from its home base to a theater of operations may be hazardous, for a part or all of the trip. Vessels may need a protective escort of surface and air forces, and air transport may need fighter protection against enemy attack. If the theater is sufficiently far away, the deploying force may have to replenish its fuel and other supplies at some intermediate supporting facility.

The force will enter the campaign area through initial entry points that could be some combination of ports, airfields, beaches (in the case of an amphibious operation), or drop zones (for an airborne operation). A temporary base would form around each entry point, and the tactical situation may require these bases to be made secure quickly so that combat operations can be supported while additional forces and supplies arrive. Force calculations must make allowances for additional troops to guard all logistics installations against attack and sabotage. Indeed, protection of bases is an operational requirement at every stage of a campaign.

An initial lodgment of forces that is accomplished using air transportation will usually be supported by follow-up echelons arriving by sea with the heavier equipment and bulk stocks of materiel that cannot be transported by air. Because of its limited lifting capacity, airlift can seldom be used to maintain a large force over a protracted period, and a seaport or harbor would have to be secured and brought into use as soon as possible. If a port or harbor is not available, then specialized vessels and amphibious vehicles would have to be used to move men and materiel ashore. Also, artificial harbors or jetties may have to be constructed to enable normal shipping to discharge cargoes. At the very least, protected anchorages are required, and their use may impose time and operating constraints.

Unless a force of reliable, skilled workers is available locally, the incoming force will have to provide its own labor; a protecting force will also be required if the landing zone is not secure from enemy action. Specialized logistic units, supplemented by military and possibly civilian labor, will undertake such tasks as freight identification, handling, and storage; operation of cranes and mechanical handling equipment; control of harbor activities; and engineering and construction work.

Supply Depots in the Early Phase of Operations

In the initial stages of an operation, stores and equipment may have to be distributed directly to combat units from a port, beach, or airfield. This must be regarded as a temporary expedient, however, for entry points usually become congested and are, therefore, ideal targets for air strikes. Because of this congestion, they almost always operate on the verge of chaos. As quickly as possible, supply depots (dumps) should be established forward of the initial entry points to clear congestion and ease control of logistic activities.

Once established, these dumps can be used to unscramble the masses of materiel that pour into a lodgment, for no matter how well planned ship and aircraft loads may be, unloaded bulk freight is invariably unsuited for immediate delivery to users. Within the forward depots, supplies may be sorted and prepared for transporting to receiving units.

Main Logistic Base

As soon as the tactical situation permits, at least one main logistic base should be established and its facilities gradually built up. Reinforcement troops and substantial logistic stocks will be held here, dispersed and yet grouped together for protection and control.

When fully developed, a base supporting a large force would consist of a number of facilities, including a full range of supply depots that hold and dispense things like vehicles, munitions, food, and fuel. Also included would be maintenance workshops, transportation units, hospitals, and special service units that take care of such things as mail delivery and payment of personnel. Finally, a headquarters command would be required to administer the activities of the base.

Advanced Logistic Bases

As the combat forces move forward and the in-theater lines of communication lengthen, it becomes necessary to establish advanced logistic bases. At first, these depots are relatively small and may hold only a limited range and amount of materiel. As a force continues its forward movement, however, the advanced bases become larger and one or more of these may eventually reach the size of a main logistic base, with materiel being moved forward from the initial entry points, through supply depots near the entry points, to these large forward bases.

As the lines of communication advance still farther toward the area where units are engaged in combat, a succession of logistic support bases is established. Stocks in these forward bases are usually unloaded from transportation vehicles and stored on the ground; but if the base is very near the fighting, stocks may be kept loaded on vehicles or even supplied by air.

Planning and Organizing

Careful consideration of logistical support must be included in the planning for each of the major stages of an operation: force commitment, deployment, combat, and support. Planners should bear in mind that at each stage, the provision and movement of men and materiel require a reliable communications network without which the lines of communication cannot operate effectively.

As with all operational plans, allowance has to be made for the unexpected and unforeseen, for a sudden break in a crucial line of communications could be disastrous. Flexible contingency options have to be considered and incorporated into the operations plan. If necessary, logisticians can improvise to overcome unforeseen difficulties, but adequate planning will anticipate shortages and schedule slippages and provide for a margin of safety to compensate for possible problems.

The initial logistic planning should involve a number of considerations. These include such things as the careful integration of tactical and logistic support plans; the order of priority for developing combat and logistic support units; the distance to the overseas theater of operations, routes to it, and the speed of commitment; the availability and accessibility of sealift and airlift resources; and the climatic and terrain conditions in the theater. Another consideration is the availability of resources and facilities in the area of the campaign. Included here would be the transportation infrastructure, food supplies, and sources of electricity and water. Also important is the attitude of the local people. If they are hostile, there will be a threat of sabotage and guerrilla actions against which logistic facilities must be defended.

Conclusion

Although logistics is the backbone of military operations, students of war rarely give it the attention it deserves. Without proper planning and the establishment of adequate lines of communication, a military force will surely fail disastrously in virtually any operation it undertakes. As Karl von Clausewitz put it: "There are countless campaigns in which nothing happened, which missed their goal and squandered their resources to no purpose, and the excuse for it all was difficulties of supply."

N. T. P. Murphy

SEE ALSO: Interior and Exterior Lines of Operation; Logistics: A General Survey; Logistics, NATO; Logistics, Naval; Strategy; Theater of War.

Bibliography

Clausewitz, C. von. 1976. *On war*. Ed. and trans. M. Howard and P. Paret. Princeton, N.J.: Princeton Univ. Press.
Creveld, M. van. 1977. *Supplying war: Logistics from Wallenstein to Patton*. Cambridge, U.K.: Cambridge Univ. Press.
Eccles, H. E. 1959. *Logistics in the national defense*. Westport, Conn.: Greenwood Press.
Jomini, H. 1971. *The art of war*. Westport, Conn.: Greenwood Press.

LIN-PIAO [1907–1971]

This article provides a brief glimpse into the life of Lin-Piao, a Chinese general and strategist, who played a major role in the military and political life of modern-day China.

Early Years: 1907–1934

Lin-Piao (Lin Biao) was born on 5 December 1907 to the family of a dye-house owner in Huanggang County, Hubei Province. He joined the Chinese Socialist Youth League in 1923, and in 1925, inspired by the 30 May Anti-Imperialist Movement, he threw himself into the student movement. He initiated the formation of the Gongjin Reading Society, and organized students to read progressive books and periodicals. He was elected by the Students' Union of Hubei Province as a delegate to the Congress of the Nationwide Students' Unions held in Shanghai. In winter of the same year (1925), the local Party organization recommended him to the Fourth Class of the Huangpu (Whampoa) Military Academy in Guangzhou (Canton). He subsequently joined the Chinese Communist Party (CCP) there. After his graduation in July 1926, he was assigned to Ye Ting's Independent Regiment, which belonged to the Fourth Army of the National Revolutionary Army, and took part in the Northern Expedition as a platoon leader. He participated in the Nanchang Uprising in August 1927 and after its failure followed Chu-Teh and Chen Yi to the borders of Fujian, Guangdong, and Jiangxi and was promoted to company commander. In January 1928, he took part in the South Hunan Uprising and then went to the Jinggang Mountains with the army. Serving in the Fourth Army of the Chinese Workers' and Peasants' Red Army, he became commander of the 1st Battalion of the 28th Regiment and later became the commander of the regiment itself. He took part in the struggle against the Kuomintang's (KMT) "encirclement and suppression" of the Red Army in the Jinggang Mountains Revolutionary Base. In the spring of 1929, he went to South Jiangxi and West Fujian with the main force of the Fourth Army as the commander of the First Column. He was promoted to commander of the Fourth Army the following year, and later to commander of the First Army Corps. He led his troops in the campaigns of Changsha, Ganzhou, Zhangzhou, Naxiong Shuikou, and Yihuang of Le'an. He also took part in the defense against the KMT's five "encirclement and suppression" campaigns in the Central Revolutionary Base Areas (1930–34).

Military Exploits: 1934–49

Lin-Piao was subsequently elected a member of the First and Second Executive Committees of the provisional government of the Chinese Soviet Republic and member of the Central Revolutionary Military Committee. In October 1934, he led his troops in the retreat from the Central Revolutionary Base and joined in the Long March. En route he took part in the Xiangjiang Campaign and fought in the battles of Tucheng, in the forced crossing of the Dadu River, and in the taking of the Luding Bridge. In September 1935, he became the assistant commandant of the Shaanxi-Gansu Detachment, made up of the First and Second Red Armies, and commander of the First Column. When the Shaanxi-Gansu Detachment was renamed the First Front Army on arriving at the Shaanxi-Gansu Revolutionary Base, he was its commander in chief and then led his army in the Zhiluozhen and East Expedition campaigns. In 1936 he was made the president and political commissar of the Red Army Academy of Resistance Against Japan and, later, the president of the Military and Political Academy.

After the outbreak of the War of Resistance Against Japan (World War II) in July 1937, he became the commander of the 115th Division of the Eighth Route Army, and together with Nie Rongzhen, he led the troops on the North China front. In the first battle fought at Pingxing Pass, they badly mauled the Itagaki Division of the Japanese army and won the first important Chinese victory in the Anti-Japanese War. When marching past Xixian County, Shanxi Province, on 2 March 1938, he was accidentally wounded by KMT sentries of Yan Xishan's troops and returned to Yan'an for medical treatment. That same year he went to the Soviet Union for treatment and recuperation and acted as the representative of the Chinese Communist Party Central Committee's Delegation in Moscow. Returning to Yan'an in January 1942, he became the vice-president of the Central Party School of the CCP.

After the end of the victorious Anti-Japanese War in 1945, he was sent to North China and became commander in chief of the Northeastern People's Self-Governing Army, commander in chief and political commissar of the Northeastern Democratic United Army, commander and political commissar of the Northeast Military Area Command and Northeastern Field Army, and secretary of the Party's Northeast Bureau. He participated in planning and commanding the Linjiang Campaign; the summer, fall, and winter offensives in the Northeast in 1947; and the Liaoning-Shenyang Campaign, thereby playing a major role in the liberation of the northeast. He also led the reconstruction of the Liberated Areas in the northeast. In 1948 he led his troops down into the Shanhaiguan area and was appointed secretary of the Front Committee during the Beijing-Tianjin Campaign and, together with Luo Ronghuan and Nie Rongzhen, exercised united command of the main forces of the Northeast Field Army and the troops of the North China Military Area Command. In April 1949 he led his troops to the mid-south of China in a strategic pursuit. He was made successively the commander of the Fourth Field Army (formerly the Northeast Field Army), commander of the Mid-China Military Area Command (later changed to Mid-South Military Area Command), and the first secretary of the Party's Mid-China Bureau (later changed to the Mid-South Bureau). He participated in commanding the campaigns of Hengbao, Guangdong, and Guangxi.

Political Advancement and the Cultural Revolution: 1949–71

With the founding of the People's Republic of China on 21 September 1949, Lin-Piao was appointed chairman of the Mid-South China Army and Government Committee and the Mid-South China Administrative Committee. Later he filled other positions such as vice-chairman of the People's Revolutionary Military Committee of the Central People's Government, vice-premier, and vice-chairman of the National Defense Council. In September 1955 he received the rank of field marshal of the People's Republic of China. In September 1959 he became the defense minister and then vice-chairman of the Central Military Committee, taking charge of its day-to-day work. From June 1945 on, he was a member of the Party Central Committee; from April 1955 on, a member of the Politburo of the CCP Central Committee; and from May 1958 on, a member and vice-chairman of the Standing Committee of the CCP Central Committee.

During the Cultural Revolution, Lin-Piao formed a counterrevolutionary clique with Chen Boda, Huang Yongsheng, Wu Faxian, Ye Qun, Li Zuopeng, Qiu Huizuo, and others. Hand in glove with Jiang Qing's counterrevolutionary clique, he schemed against and persecuted other party and state leaders, incited a movement for "overthrowing everybody and unleashing a comprehensive civil war," and plotted to usurp the supreme powers of the party and the state. On 8 September 1971, Lin-Piao issued an order for a coup, planning to assassinate Mao Tse-tung and set up another central authority. On 13 September 1971, when the conspiracy was uncovered, he tried to flee abroad, taking with him his wife, Ye Qun; his son, Lin

Liguo; and other members of his clique. The plane carrying Lin-Piao's party crashed at Undurkhan in the People's Republic of Mongolia, killing all on board.

Final Comment

On 20 August 1973, the CCP Central Committee decided to expel Lin-Piao's name from the party. On 25 January 1981, he was condemned as the prime culprit of the counterrevolutionary clique by a Special Court of the Supreme People's Court of the People's Republic of China.

JIANG FENG-BO

SEE ALSO: China, People's Republic of; Chu-Teh; Mao Tse-tung; Taiwan; Unconventional War; War: The Soviet View; World War II.

Bibliography

Institute of Military History of the Military Academy of the People's Liberation Army. 1987. *Military history of the Chinese People's Liberation Army.* Vols. 1–3. Beijing: Military Science Press.

Research Office of the Party's History of the Party Central Committee. 1987. *Chronicles of the Chinese Communist Party.* Beijing: People's Publishing House.

LITHUANIA

Lithuania, one of the fifteen former republics in the Union of Soviet Socialist Republics (USSR), declared its independence from the Union on 11 March 1990. On 6 September 1991, the Soviet government formally recognized the independence of Lithuania and the two other Baltic republics, which the USSR annexed in 1940. This dramatic establishment of Lithuanian sovereignty followed several years of dynamic and unprecedented change in the Soviet Union. For the next several years, relations between Lithuania and the rest of the world are likely to continue to change. Over time, new structures and patterns will emerge in economics, trade and commerce, politics and government, finance, manufacturing, religion, and virtually all aspects of human life. New arrangements must be devised for dealing as a sovereign state with the new Commonwealth of Independent States (which is itself undergoing continual change) and with the world outside the boundaries of the former Soviet state. If the history of the Soviet Union since 1985 is any guide, we can expect dramatic surprises and dynamic change.

An important question for the world is how Lithuania and the Commonwealth will organize and provide for security. Most of the Soviet Union's armed forces stationed in Lithuania are likely to be withdrawn. Also of great concern is the disposition of nuclear weapons, the security of these weapons, the command and control of their potential use, the compliance with arms control agreements entered into by the former Soviet government. The world can only hope these issues are settled amicably.

It will be years before all of these issues are resolved for Lithuania and some time before events settle down into more routine and measurable patterns. No accurate description of this new country's policies, defense structure, and military forces was available to be included in this encyclopedia. Only time will reveal the future of Lithuania as a separate sovereign state. The reader is thus referred to the historic information contained in the article "Soviet Union," and to the latest annual editions of the *Military Balance,* published by Brassey's (UK) for the International Institute for Strategic Studies; the *Statesman's Year-Book,* published by the Macmillan Press Ltd and St. Martin's Press; and the *World Factbook,* developed by the U.S. Central Intelligence Agency, and published commercially by Brassey's (US).

F. D. MARGIOTTA,
Executive Editor

LOGISTIC MOVEMENT

In the military context, *movement* is a term describing the act of moving personnel and materiel—passengers and freight—in peacetime, during transition to war, and in war. Tactical movement is confined to the immediate battle area. To the rear, strategic and in-theater movement is one of the principal logistic functions of armed forces whether it is undertaken by sea, air, rail, road, inland waterway, or by a combination of these means.

In peacetime, units on rotation or change of situation and stocks of materiel are moved between stations at home and overseas. Routine turnover of individuals or groups of individuals (including conscripts in some armies) is an extensive and frequent task in which air trooping has largely replaced sea trooping. Unit and individual movements may include families as well as troops in some forces and in some circumstances.

During transition to war, reinforcement formations, units, and individuals, as well as materiel, have to be moved speedily to their operational areas. On arrival, formations and units may have to collect their heavier equipment from pre-positioned stockpiles well in the rear of their combat areas or battle positions. At the same time noncombatants and dependents return home from overseas stations in areas that are potential combat zones.

With the onset of active hostilities the scale of movement activities intensifies still further, now under the threat of enemy action. Larger quantities of a wide range of military equipment and commodities, many and varied personnel and equipment replacements for losses incurred in battle, and the remaining reinforcements are moved to the theater or area of operations. By that time the casualty and equipment evacuation systems may have begun to function in the opposite direction.

This article concentrates primarily on the logistic movement of land forces and supporting elements; naval and air forces are dealt with elsewhere. Even in this restricted context, however, it must be acknowledged that individual armies organize and execute logistic movement in different ways, using their own techniques and technology. Therefore this is a generalized description of the function, unrelated to any specific national or allied system or nomenclature, or to any particular operation of war.

The second main term is *movement system*. Figure 1 illustrates a typical theater of operations where long lines of communication link the home base with the combat area.

Personnel and materiel are conveyed from mounting areas, ports, and strategic airfields in the home base to sea and air entry points in the theater of operations. Once unloaded from ships and aircraft they are normally held temporarily in marshaling or staging areas near ports, beaches, and airfields before being moved forward to force and/or corps support areas. During this process certain formations and units collect their stockpiled equipment from rear supply depots. They then move onward to their operational areas by appropriate means. The diagram shows, symbolically, only one railway line and two main roads when in reality many more, as well as inland waterways, might be used.

At each stage, and at every major nodal point and formation support area, along the lines of communication there are movement control units or detachments to regulate a system that is complex, often congested, and sometimes vulnerable. More problems may well arise if allied, multinational forces use common or overlapping lines of communication; then movement planning, coordination, and control requirements will increase.

Movement control is the third principal term. It describes the planning, routing, scheduling, coordination, and control of logistic movement. A possible deployment pattern of movement control units and detachments is outlined in Figure 1. Wherever they are deployed—in sea, air, or inland surface movements—these elements ensure that the system works effectively. Moreover, movement control officers are responsible for the executive control of personnel and materiel—passengers and freight—passing through the system. Again, there is no standard organization: some armies have a separate movement service such as the Russian Commandant's Service; some link the functions of movements and transport within a single logistic service; in others paramilitary or military police undertake movement control.

Movement staffs in headquarters also play a key role and are normally employed at all levels as far forward as an army corps HQ; in NATO they form part of the G4 (logistics) staff.

These staffs plan logistic movement in peacetime, in peace for war, during emergencies, and in war. Movement priorities are determined by the commander's policy and requirements. Movement staffs arrange the transportation, reception, and staging of personnel and materiel according to his priorities. They allocate routes along the lines of communication and produce movement instructions, schedules, and tables. They regulate and monitor the execution of plans and instructions, and rapidly adjust them if breaks in the movement system or other problems occur. They also have a vital national coordinating role and also a multinational role when allied forces are involved.

Movement staffs and movement control units combine their efforts to ensure, through centralized planning and effective control, that the system works flexibly, is well regulated—it maintains an even flow of personnel and materiel according to set priorities—and uses all transportation means efficiently and economically. They must have excellent communications and liaison: with each other; with commanders and their staffs; with all transportation agencies; with ports, airfields, rail terminals, road traffic control organizations (military and civilian police); and with personnel-holding and stockholding units in mounting areas and in formation support areas. The way the system operates is largely dependent on the *movement plan*.

The movement plan is agreed upon at the highest level and at each echelon in the military structure. It incorporates the movement requirements of forces, formations, and other tactical commands. When agreement has been reached, the logistic commander will publish it as an ad-

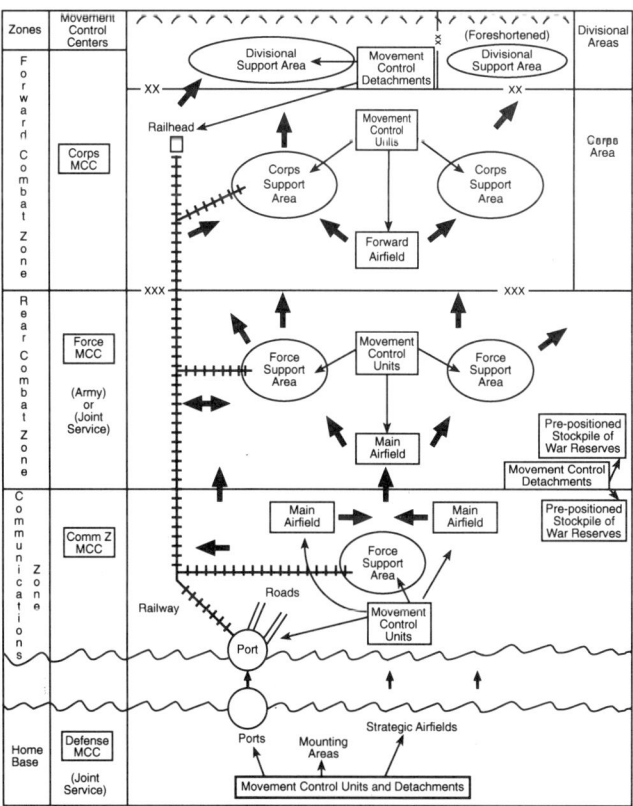

Figure 1. A typical theater movement system.

dendum to his logistic plan for peace, emergencies, or war.

A network of movement control centers (MCC)—shown on the left-hand side of Figure 1—implement, review, and revise the movement plan as it affects their particular part of the operation. In the illustrative scenario used here, the Defense MCC located in the home base is a joint service organization, although single services may well have their own principal movement agencies integrated or linked directly with the main defense center. The DMCC is normally part of the national ministry of defense. It is likely to be manned by movement staffs and representatives from other government departments and civilian bodies, particularly those that provide additional shipping, aircraft, and other transportation to supplement military resources in peacetime, emergencies, or war. The Force MCC is the main center within the theater of operations. It is either organized and operated on a joint service basis or, in the scenario under discussion here, run by the army with representatives from the other armed services, the host nation, other local bodies, and their movement agencies.

Another main term is *movement resources*. Many factors influence the availability, readiness, and selection of these resources including: the movement requirement; the distance and nature of the link between the home base and theater of operations; the type and environment of operations; the facilities provided by the host nation or by other local agencies in a developed or undeveloped country; and, not least, the capacity of transportation means. All movement resources will be matched against the stated movement requirement; in this way the movement plan is initially made and subsequently updated. From this plan sea, air, rail, and road movement tables will be produced that classify personnel (passengers) and materiel (cargo or freight) for movement purposes. They also specify the type or mode of conveyance, the composition of the personnel draft or materiel load, and detailed instructions and data concerning a particular movement task.

The composition and configuration of materiel loads are important considerations. They may be organized on a tactical or combat, commodity or composite basis; special regulations govern the stowage and handling of dangerous or hazardous items. Some military loads will consist of single major pieces of equipment (vehicles, weapon systems, radars, cranes), others of multipackaged commodities that may be palletized or placed in large containers. The availability and utility of materials-handling equipment are also important considerations at every stage in the movement system.

Mounting areas in the home base will ideally be sited near ports and strategic airfields. Maritime movement resources require good ports to embark passengers and cargo. They may need shore facilities in passage, and may possibly transship some passengers and cargo to improve tactical readiness en route: this form of transshipment was usefully carried out off Ascension Island during the voyage of 12,800 kilometers (8,000 mi.) from the United Kingdom to the Falkland Islands/Islas Malvinas in 1982. Maritime movement resources require protected ports, jetties, anchorages, or beaches at which to off-load in the theater of operations. Military ships and craft may be designed to discharge at sea, at improved shore facilities, or to land on beaches, but additional or requisitioned merchant shipping is not likely to be as flexible. A hostile operational situation may also require alternative sea entry points should the main ones close or be denied.

Airlift resources are classified in two ways: *strategic*—between the home base and a distant theater of operations; and *tactical*, or intra-theater (not to be confused with *tactical* in the army sense, which applies to combat resources alone). It may be practicable to fly passengers and cargo by strategic transport aircraft directly to the forward airfield (located in the corps area in Fig. 1). It may be feasible to land only on the main airfields in the rear combat zone and communications zone; tactical transport aircraft will then fly personnel and materiel to airfields and landing strips in the forward combat zone, with helicopters providing priority, additional and flexible airlift resources mainly to operational areas.

Most passengers and cargo will be unloaded after landing, but some troops and materiel may be air-dropped by parachute into battle positions, particularly in isolated areas. Strategic transport aircraft may need maintenance, tactical reloading, and refueling stops en route to the theater of operations. In-flight refueling substantially enhances long-haul airlift, as in the Falklands/Malvinas campaign.

Rail movement resources will depend largely on the availability, utility, and capacity of host nation or other local facilities, not least on the ever-vital condition and positioning of rolling stock. Given appropriate rail communications and an amenable operational situation, through-running of trains from entry points to forward railheads in corps and divisional areas, or to force and corps support areas, will be an invaluable, high-tonnage movement resource. Some armies retain a railway operating capability likely to constitute a localized supplement to existing civilian rail facilities.

Road movement resources feature prominently for land forces. In this illustrative scenario, it is assumed that the developed country has a good network of routes with many roads and bridges of suitable strength to sustain heavy-density and heavy-weight traffic in both directions, thus permitting optimum tonnages to be moved. It is also assumed that a large proportion of the heavier, logistic road movement resources will be provided by the host nation as either indigenous military or civilian transport; if this is possible, it reduces appreciably the numbers of trucks and drivers that need to be moved to the theater of operations from outside. A less well-developed country may have a more basic route network with roads and bridges of lower weight-bearing classification, with restricted access to double-flow traffic, and only reasonable access to single-flow usage. When local road movement resources are scarce, this limitation will proportionately increase the amount of transport imported by the force.

In either case—developed or undeveloped local facilities—a system of main supply routes will be established. Along these most military road traffic will travel, normally under movement authorization and control, and usually supervised by traffic control elements. These may comprise traffic regulating headquarters, traffic control posts, and mobile patrols provided by the military police, possibly reinforced by local civilian traffic police.

A regulated convoy of vehicles will move from an assembly or staging area through a movement control checkpoint where loads and drivers will be checked and documented. It will then proceed to a designated start point, and before reaching the release point at the other end of its journey the convoy may pass through a series of such checkpoints, as well as through traffic control posts previously mentioned. Well-regulated road movement will preplan, publish, and monitor the pass time of the convoy, the running time per vehicle, and the total time the convoy should and does take to complete the move at a specified speed, density, and rate of flow. Time is allowed for halts.

Inland waterways—rivers, lakes, and canals—also may provide another useful movement resource. Indeed, in undeveloped countries, river movement may be a vital, routine means of transportation. Although relatively slow, it can represent a valuable capacity supplement, especially for large tonnages of bulky cargo.

Movement resources of all types make a significant contribution to the successful buildup of the military force in a near or distant theater or area of operations, in a developed or undeveloped country. The movement plan, and the collective part played by movement staffs and movement control units in implementing the plan, greatly influences the rate at which prescribed force levels are built up. It helps first to introduce a military presence, gain a tactical foothold, or deploy for war, and then establish an effective, operational organization. The movement system must also work in the reverse direction, principally by evacuating personnel and equipment casualties by road, rail, air, or sea.

J. H. SKINNER

SEE ALSO: Casualties: Evacuation and Treatment; Lines of Communication; Logistics: A General Survey; Logistics, NATO; Logistics, Soviet and Warsaw Pact; Reinforcements; Replacements: Personnel and Materiel; Stockpiles; Supply; Theater of War; Transportation.

Bibliography

Brown, K. N. 1987. *Strategics, the logistics-strategy link.* Washington, D.C.: National Defense Univ. Press.
Paulson, R. M., and T. T. Tierney. 1971. *Logistics and technology: Some thoughts about future military implications.* Santa Monica, Calif.: Rand Corp.
Van Creveld, M. 1977. *Supplying war: Logistics from Wallenstein to Patton.* New York: Cambridge Univ. Press.

LOGISTICS: A GENERAL SURVEY

Armed forces around the world use a number of definitions of the term *military logistics* which differ in presentation of content as well as mode of expression. For instance, the separate article on logistics in NATO contains a detailed version. As an introduction to this general survey of logistics, an abridged description is herewith offered: Military logistics is the business—and big business at that—of planning, preparing, and providing materiel support for forces, thus enabling them to live and move, train in peacetime, mobilize and deploy in an emergency, and to fight in war or keep the peace.

The military term *materiel* is a key word in this and other definitions. It is used to describe the vast range and quantities of equipments and commodities in the inventories of armed forces—inventories that have increased in length and complexity commensurate with all the impressive advances in military technology of recent years. They include items of materiel ranging from capital equipments such as ships, aircraft, tanks, and all types of vehicles and weapon systems, through replacement assemblies and spare parts, to missiles and munitions, explosives and fuel, food and water, constructional stores, medical supplies, clothing and personnel equipment, maps, and other necessaries. To aid identification and management, military materiel is normally grouped into categories or classes of supply.

Logistics has two primary tasks, and both emphasize the central theme of materiel support. The first involves the development and production, the procurement and storage, the distribution, and eventually perhaps the disposal of the equipments and commodities used by modern military forces. The second task concerns the provision of an extensive variety of logistics services, listed in Figure 1. Each service has operational and nonoperational functions which may differ in situations of war or peace. For example: the supply function may vary from replenishing the ammunition stocks of fighting troops engaged in battle, to replacing furniture in barracks; constructional engineering tasks may vary from building a fuel tank-farm in a field location just behind the fighting troops and then connecting it to a fuel pipeline, to maintaining accommodation in peacetime military garrisons. There are separate articles on all the logistic services listed in Figure 1.

One of the principal military functions included in Figure 1 is *support*. Although logistics is a main support function, as are military administration and the provision of communications and information, there is a problem of military terminology. Specifically, armed forces tend to

Logistics: A General Survey

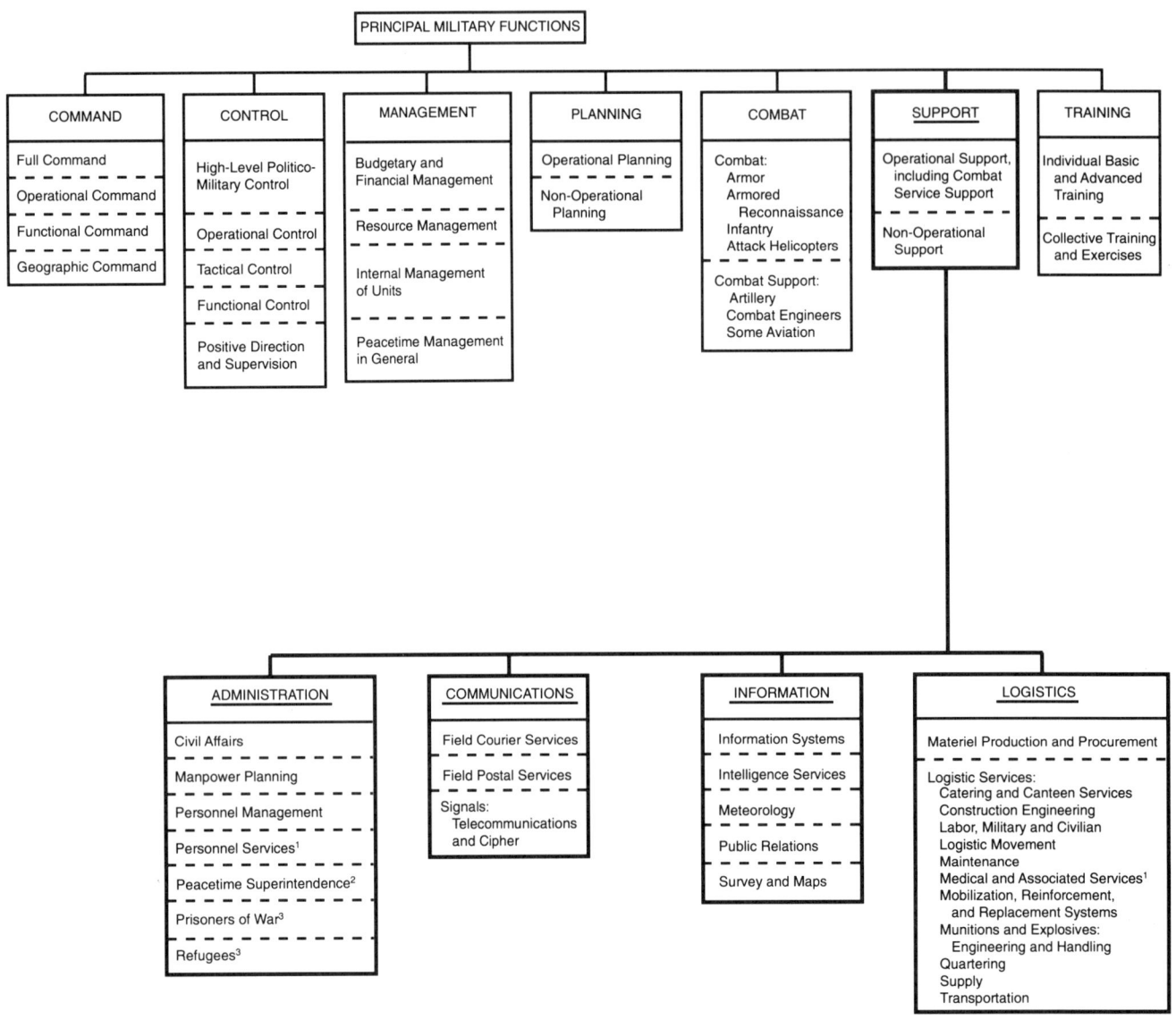

Figure 1. Principal military functions.

classify their functions differently, and some regard logistics as an integral part of military administration. This article, in common with most of today's practice, treats military logistics and military administration separately, thus recognizing the continuous and sustained growth in popularity of the term *logistics* and its eminent importance as a main support function. This treatment in no way depreciates the value of military administration, which relates principally to people, while logistics mainly concerns materiel.

Operational logistics sustains forces engaged in combat, counterinsurgency, internal security, peacekeeping and disaster-relief operations, or training exercises. That part of operational logistics that directly supports and is in close contact with the fighting is often described as combat service support. Nonoperational logistics, on the other hand, covers a multiplicity of support tasks that are all part of the domestic routine of forces living and working in peacetime garrisons and barracks, at home or overseas.

Another way of describing operational logistics is to differentiate between levels or lines of support as outlined in Table 1.

By linking these levels of support by lines of communication, long distances can be spanned from home bases to troops deployed in distant theaters or areas of operations. The practical framework suggested in Table 1 can be used flexibly to provide the most suitable and most effective logistic structure to fit the circumstances. For

TABLE 1. *Levels or lines of support*

LEVELS	LINES	EXPLANATION	ALTERNATIVE DESCRIPTION
Immediate logistic support	First line	Support integral or organic to user/consumer units	Unit support
Primary logistic support	Second line	Direct support given to user/consumer units by primary logistic support units that are normally mobile and in field locations	Direct or field support
Secondary logistic support	Third line	Support given to primary by secondary logistic support units that are sometimes mobile and usually in field locations	General support
Tertiary logistic support	Fourth line	Support given to secondary by tertiary logistic support units and installations from invariably static, permanent locations	Base support

example, fighting units have organic supply, maintenance, and medical support elements and, in certain circumstances, may receive support direct from home or overseas bases without it being necessary to introduce the intermediate levels. There are other permutations.

The terms *field* and *base logistics* are also quite commonly used by the military. Field logistics encompasses the first three levels listed in Table 1, whereby materiel and services are provided in the field: from temporarily acquired or erected infrastructure, from hastily built or converted shelter, or, literally, from fields or forests. In an operational or combat environment, camouflage and concealment, local defense, and security can be as important as the technical provision of logistic support. Base logistics, as the term suggests, is provided from home bases, from main bases in overseas theaters, or from advanced bases in overseas areas of operations. It is usually, but not always, provided from permanent accommodation in a secure environment. Field logistics involves mobile, as well as static, stocks and service support, whereas base logistics invariably features large, static installations. These and other terms are illustratively used in some hypothetical operational scenarios presented later in this article.

Another way of describing logistics is to differentiate between the producer, or wholesaler, and the user or consumer, both linked by the supplier or retailer. To elaborate: to the defense producer, the military supplier is the intermediary who calculates materiel requirements, who procures and holds stocks in base depots or field storage, and who distributes the user's needs either automatically or on demand. To the military user, the supplier and the producer together satisfy his legitimate needs at the right time, in the right place, and in the proper quality and correct quantity.

Historical Development

The English word *logistics* has been in American military use for well over 100 years, and has been gradually adopted by most other English-speaking defense communities and armed forces. More recently, the term has also become popular in civilian commerce and industry. It derives from the Greek adjective *logistikos* meaning skilled in calculating. (Throughout history, the tasks of equipping, supplying, and moving armies have involved arithmetical calculation, some of considerable complexity. Napoleon is reputed to have described these calculating skills as "not unworthy of a Newton or a Leibniz.") In Roman and Byzantine times, the *logista* was an official who employed this skill as part of his vocation. Moving forward through history to Louis XIV's time, the French Army introduced the appointment of *maréchal-général des logis*, who was responsible for quartering troops. (The French verb *loger* means to billet or lodge, and the adjective *logistique*, when used in the military context, originally described the quartering service.)

In its broader and more modern sense, the term *logistics* first gained prominence in Jomini's theory of war. This was published in 1838, after the author had seen service with Napoleon. In expounding the trinity or interdependence of strategy, tactics, and logistics, Jomini described logistics as the means and the arrangements that enable strategy and tactics to be put into effect. His contemporary, Clausewitz, stressed that a wise commander, even if the resources of the province are quite sufficient, will not neglect to form magazines behind him as a provision against unforeseen events, so as to be able more readily to concentrate his strength at certain points. Both statements remain true today, although the scope and magnitude of modern military logistics have greatly expanded.

Unlike Jomini, who considered logistics to be of equivalent importance to strategy and tactics, Clausewitz regarded logistics as having a subservient role to the other two. This difference of emphasis is interesting because it still prevails in certain cases. In some armed forces, there is a definite acceptance by commanders, staffs, and units of the interdependence and interactivity of strategy, tactics, and logistics, to the point that all three have for long been equivalently represented on the general staff—the operational planning staff present at all levels of headquarters in military forces. C. Barnett describes the British machinations of the late nineteenth century: whether or not to adopt and employ the "continental" general staff structure and system that has been, or shortly would be, introduced in the main European armies and the U.S. Army at that time (*Britain and Her Army: 1509–1907* [London: Allen Lane, 1970], pp. 310, 336). In the end, the "continental" version was rejected; the British Army developed its own staff structure and system, but recently has partially introduced a general staff along the lines of the "evolved continental" model. B. von Schellendorf outlines the "continental" structure employed at that time (*The Duties of the General Staff* [London: His Majesty's Stationery Office, 1905]; see also Dupuy 1977).

In other armed forces, logistics has been separated with no similar effective integration of logistics within the general staff at every level of the force structure. To illustrate this difference, the Prussian and, later, the German Army incorporated the *quartiermeister-general* and his supply, transportation, and quartering staffs as part and branch of the *grosgeneralstab*, from the elder von Moltke's time onward, whereas the British Army's quartermaster-general is not part of the general staff—he reports directly and independently to a government minister—and there is no single chief of staff at each organizational level who directs and coordinates all aspects of operational planning on behalf of his commander.

Attitudes, like those of Jomini and Clausewitz, are hard to change. Tradition, rather than logical practicality, can govern military organization and method, for good and bad reasons, and disparities of attitude and approach toward logistics are reflected in other ways. For instance, authoritative published works on military logistics remain few and far between. Eccles (1959, 1965); Huston (1966); Thorpe (1986), including Falk's introduction; and van Creveld (1977) provide excellent commentaries on logistics, although some are now a little dated. In particular, Falk gives a brief, handy, abridged summary of the historical evolution of logistics.

The deficiency of publications moved Fuller, before the Second World War, to observe: "Surely one of the strangest things in military history is the complete silence about the problems of supplies [logistics]. In ten thousand books written on war not one is to be found on the subject, yet it forms the basis on which rests the whole structure of war: it is the very foundation of tactics and strategy" (G. C. Shaw, *Supply in Modern War* [London: Faber and Faber, 1938], p. 9; Preface by J. F. C. Fuller). Jomini would certainly have agreed wholeheartedly with Fuller's concluding statement. They and others appreciated and recognized that the practical art of moving armies and keeping them supplied comprises "90 percent of the business of war" (A. C. P. Wavell, *Speaking Generally* [London, 1946], pp. 78–79). In another instance, Wavell's experiences during World War II led him to conclude:

> The more I have seen of war, the more I realise how it all depends on administration and transportation (what our American allies call logistics). It takes little skill or imagination to see where you would like your army and when; it takes much knowledge and hard work to know where you can place your forces and whether you can maintain them there. A real knowledge of supply and movement factors must be the basis of every leader's plan: only then can he know how and when to take risks with these factors; and battles and wars are won by taking risks. (*Ibid.*)

Since World War II, several notable works have emphasized the importance of logistics. One of them, by van Creveld (1977, p. 231), contains the assertion that, because there is a continuing dearth of published, authoritative logistical material, most available military history tends to place the cart before the horse. When reviewing van Creveld's work, Michael Howard illuminates deficiencies in a penetrating style all of his own:

> The traditional and disastrous distinction between "teeth" and "tail" arms, between executive and engineering branches, between flying and technical officers which has done so much to destroy the efficiency of our [British] armed forces in the twentieth century extends to writings about war as well. Military historians devote as little attention as they decently can to logistical matters before going on to describe the splendours and miseries of battle and generalship that make up the core of the drama. (*Sunday Times*, London, 19 Feb. 1978)

Historical commentary on strategy, tactics, and command may well be devalued to a considerable extent because of a lack of essential logistical dimensions. Yet, we must be cautious of publications that deal with logistics in an isolated manner. Military history should accurately and objectively record that the conduct of wars, campaigns, and battles is a *combined* effort. Strategists and tacticians, commanders and logisticians, soldiers and civilians, all share the trials and tribulations when working together to achieve success in war. Just as all military forces and units—whatever their function—cooperate and collaborate to accomplish their missions and tasks to achieve victory. If operations are viable and feasible, because their logistic support is ready and sufficient, then less needs to be written of logistics than if operational risks—premeditated risks—have to be taken because of logistical constraints or inadequacies. When logistics has a profound influence on military policies, plans, and priorities, or when logistics moderates or restricts operations, then these facts and their associated circumstances are notable. Perhaps these notable facts and circumstances are too infrequently recorded in adequate detail. If this is so, then the cart may well have been placed before the horse, as van Creveld (1977) claims.

When reviewing the past and anticipating the future, it is worth remembering that logistics is a principal means of creating and sustaining combat power. It can be a key determinant of operational success or failure as is suggested here:

> Well planned and well executed logistics will provide the key to victory in battle, and plays a key part in combat readiness in war. The obverse is true. Neglected logistics can quickly reduce what may appear to be a potent military force to a rag-bag of sick, dispirited men, trained as fighters, but unable to

move, shoot or communicate. It is a certain prescription for defeat, disastrous morale, and lack of confidence in their commanders, supporting staffs, and services, and, ultimately and maybe very quickly, in their weapons, equipment and ability to fight. (Alexander 1986 p. 43)

For centuries in the past, fighting men foraged and pillaged for food; they requisitioned quarters and commandeered the means of movement. Logistics, or its equivalent then, was considered to be a way of waging punitive war, as well as a means of supporting war. Lands were laid waste, bare of resources and their people starved. Some armed forces still employ these methods very much as part of their operational doctrine today.

In early times, individuals or small bands of combatants seldom strayed far from sources of potable water, daily provisions, or other essentials. Early warriors were admirably self-reliant, self-contained, and self-sufficient, much more than most soldiers are today. However, as these small bands grew larger in size, when they proliferated in number, and when in ancient and medieval times they began to take the shape of modern fighting units and modern fighting formations, their operational aspirations became more ambitious and their combat capabilities more powerful. Thus, their need for sustained and responsive logistics also grew. This newfound capacity to rove and fight further afield produced the requirement for specialized logistics, dedicated to the support of combat.

When conditions allowed, military forces continued to live off the land, often at the local population's and the enemy's expense. But it was the gradual, progressive introduction of an increasing variety of more sophisticated weaponry, transport, and equipment that demanded a proportional and parallel increase in the quantity and the quality of logistic support. During the last century and a half, the advent first of railways, steamships, telegraphs, and telephones, then wheeled and tracked motor vehicles, a miscellany of maritime and aerial craft, and a diversity of powerful and lethal forms of generating energy, has given nations and their armed forces the capacity to wage war much more vigorously and destructively. With the dramatic development of the technology of warfare, so logistics has become recognized as a key military function and a key organizational component of modern armed forces.

Timely, effective logistic support has always depended upon the ability of commanders and their staffs, their units, and their logisticians to anticipate requirements. The techniques of forecasting and calculating have been more successfully applied when logisticians have been able to work in close contact with those they support, and when they have been able to keep themselves briefed on the probable course of combat, abreast of operations, and party to changes of plan and priorities. Early logisticians were usually attached to the military tactical headquarters on the field of battle, or were grouped together as a supporting echelon as close as practicable to the fighting. In either case, they were firmly under the commander's direction and in easy communication with his subordinates. Logisticians also had to cooperate closely with the authorities of the state wherein they were operating or through which they were passing, from whom they acquired provisions, fodder, and other necessities. Sometimes their procurement methods were brutal and arbitrary; often payment or compensation was by no means automatic.

Whatever the composition of the military force, and whatever its operational doctrine and methods, logisticians and logistic support echelons had the weighty, continuous, unavoidable responsibility of anticipating and providing the everyday needs of those whom it was their duty to sustain. Whether troops were under siege, whether they were conducting siege operations or were on the march, whether they were attacking or defending, advancing or retreating, their requirements had to be met. Any failure to do so had obvious repercussions. Force requirements were prestocked in field dumps along the route, held in static or rolling magazines (van Creveld describes the rise of the magazine system [1977, pp. 17–26]), or carried in wagon columns behind the fighting troops. Thus, the ancient, professional business of military logistics evolved and developed into the modern complex of supporting services, facilities, and systems that forces have come to know, expect, and appreciate. In principle and in concept, logistics has changed little. Yet, in keeping with the growth of modern military forces and the complexity of modern technology, the function has grown enormously in scope and magnitude.

The introduction of standing armies and, in some cases, their stationing abroad meant that logistics had to be a substantial, active function in peacetime, as well as in wartime. The domestic needs of garrisoned troops had to be met, units had to be kept equipped for their war roles, and formations had to remain operationally ready. Furthermore, logistic plans had to be made and implemented for the swift, smooth mobilization of reserves to reinforce standing armies. Stockpiles of materiel had to be quickly activated, and national economies and industries placed on a war footing. Moreover, logistic planning and the rapid buildup of logistic support played an increasingly crucial part in the concentration and deployment of the total force. Once an emergency was declared or operations of war started, then logistical reinforcement and replenishment, replacement and refurbishment, came into their own. Whenever and wherever troops were engaged in armed conflict, it was necessary to sustain them at a specified level of combat intensity for a prescribed duration. Also, allowances had to be made for changes of plan and other unforeseen eventualities. Again, in principle and in concept, little has changed. Flexible response has for long been a familiar logistical characteristic.

Logistics has therefore become the challenging military vocation it is today: a professional occupation demanding special training, qualification, and expertise. Logisticians of all ranks require a high degree of conscientiousness, dedication, diligence, and skill. It is possibly their consistent, persistent attention to detail that creates a type of backstage, engine-room mentality, and, because of this, conveys a somewhat unglamorous image to other soldiers. Yet, it is their duty to be ever-present, to advise commanders, staffs, and units when operations can be supported or when logistic shortages require an adjustment to plans or priorities. It is also their duty to advise when operations are not sustainable or are not viable for logistical reasons; this is not to be misinterpreted as some gloomy preoccupation with obstructing operational intentions, for logisticians must be essentially "can-do" advisers. Logisticians may see themselves principally as resourceful perfectionists, but they also have to be practical realists. They must be prepared to accept that sometimes taking risks with logistics is prudent, as well as necessary. It is easier for them to accept this when they know their advice is invited and welcomed, evaluated, and heeded. Close, confident working relationships with their commanders, general staff planners, and the units dependent on their support encourage logisticians to give of their best.

Their best is undoubtedly needed, because advances in military technology, especially weapon systems, have created operational capabilities of voracious logistical appetite. This has paradoxical implications. As combat forces have become increasingly dependent upon a continuous connection with their logistic support, so their operational mobility can be inherently degraded, in spite of the constant efforts of operational commanders and planners to find ways round this restricting trend. As logistics now contributes ever more significantly to the development of operational strategies, doctrines, and concepts, and to the conduct of modern warfare, so the financial costs of providing sufficient logistics unerringly escalate. Moreover, the operational costs of *not* providing enough escalate as well.

The potentially huge logistic bill that nations and alliances may have had to face may well have played a part in deterring general total nuclear war, and also encouraged force reductions in central Europe. Military powers will have to continue to invest heavily in logistics in order to retain an operationally viable and credible limited-localized war capability: a combat capability that effectively deters or defeats the maverick opportunist or adventurer who threatens peace, democracy, and foreign territorial integrity; a logistic capability that effectively creates a force and sustains operations in difficult physical conditions and, for some, far from the home base, as in the recent Gulf War.

Therefore, military logistics remains vital, and Rommel explained why, in the context of his own desert campaign:

> The first essential condition for an army to be able to stand the strain of battle is an adequate stock of weapons, petrol and ammunition. In fact, the battle is fought and decided by the quartermasters before the shooting begins. The bravest men can do nothing without guns, the guns nothing without ammunition; and neither guns nor ammunition are of much use in mobile warfare unless there are vehicles with sufficient petrol to haul them around. Maintenance must also approximate quantity and quality to that of the enemy. (B. H. Liddell Hart, ed., *The Rommel Papers* [New York, 1953], p. 328)

It is with the knowledge that logistics will continue to have a crucial role in future military operations that the discussion now turns to logistic principles—principles that can usefully assist with planning and implementing future support.

Principles

There is no set of commonly accepted global logistic principles. They differ, just as the principles of war differ, from one nation's armed forces to another's. Variations stem from doctrines, procedures, and, of course, from terminology and language. In any case, long-established principles should be periodically reviewed to ensure their currency, and for the reasons Huston stated:

> No one aspect of the Army's logistical experience can be singled out as most valuable in providing guidelines for the future, for the future is, as always, uncertain. One thing can be forecast with assurance—the continuation of change. But it may also be assumed that, however far-reaching the changes, there must always be links with the past. Any general conclusion drawn from history as a whole must include the principle of change and the principle of continuity. (Huston 1966, pp. 689–90)

All history—not only military history—entails the study of past events, their relationship with the present, and their relevance as lessons for the future. Principles are constructed in similar fashion and serve the same purpose. Those concerned with continuity and change, for instance, apply equally to all military functions, not just to logistics. While there is no room in any respect for inertia, there is, as in all enterprises, a need for stability mixed with dynamism. With this in mind, a review of the literature suggests some threads of commonality among established logistic principles, as well as some requirements for updating. Six principles are proposed and each will be examined in turn:

1. Unity of purpose
2. Preparedness
3. Viability
4. Economy
5. Responsiveness
6. Resourcefulness

Every logistic principle must support and correlate with the principles of war adopted by armed forces, taking those followed by the American and British as a baseline. Also, in selecting logistic principles, the interdependence and interactivity of logistics with strategy, tactics, and operational command must be taken into account. Logistic principles are, after all, but guidelines to help plan and provide combat power at critical times and places, in forms most conducive to operational success. They are neither theoretical abstractions nor rules for rigid compliance. They should help to teach and test, check and improve, monitor and evaluate concepts, such as military logistics. Not only are principles strongly governed by circumstances, they can also be contradictory. For example, the suggested logistic principle of viability presupposes sufficiency, yet the need for economy, implicit in another principle, indicates that sufficiency may not always be attainable. The American slogan "first with the most" may therefore have to be amended to "first with the best available."

There also may be some overlap, duplication, and repetition among a set of six principles, such as those suggested here. In addition, certain characteristics or *constants* would seem to apply to all or most of the six proposed principles. These constants are summarized in Table 2.

Having discussed the usefulness of principles and having established eight constants that commonly apply, we can now examine each of the six principles proposed.

UNITY OF PURPOSE

The main logistic services are listed in Figure 1. Although each is a separate service, together they form a cohesive part of the corporate logistic organization and effort. All should be integrated—that is, combined and coordinated—under appropriate command or control, at every level within a nation's military force structure, both in peace and war. Together, these services deliver to military forces the means of livelihood and functional capacity, and, when necessary, the materiel and services to support combat operations.

At the higher levels of military organization—in ministries or departments of defense and in superior military headquarters—the corporate integration of logistic services and the corporate delivery of logistic support require unity of politico-military control, unity of civil-military management, and unity of military direction. The goal is corporately agreed-upon policies, plans, and priorities, with logistic and strategic objectives firmly set and pursued, regularly monitored, and objectively evaluated.

At lower levels—in force, formation, and tactical headquarters—a similar degree of integration of operational effort and activities is necessary: unity of command of a particular region or sector of operations, unity of operational planning, and unity of assigned missions and tasks. All enable strategic and tactical objectives to be achieved with the logistic resources available.

The tendency, at any organizational level, to give a commander or a chief of staff some lesser responsibility for the logistic support of forces—lesser, for instance, than he has for strategic and tactical matters—should be avoided. Full powers of command and comprehensive operational planning contribute valuably to the quest for unity of purpose within forces and formations and facilitate the guidance and conduct of logistic support.

Any tendency logisticians may have—when undertaking support tasks—to promote separate, vested interests in preference to common, corporate interests must be resisted and discouraged. Through firm leadership, good training, and varied military employment, logisticians should find it easier to view the broad perspective of military requirements, rather than restrict themselves solely to logistical routine and detail of equal importance. Logisticians must not hide, or be hidden, in watertight compartments. If they are, then unity of purpose may well be jeopardized; whether they work in a single service, joint service, or multinational environment; whether they are civilian producers or military providers; whether the situation is operational or nonoperational; whether it is peace or wartime. Overcompartmentalization and internal rivalry are best eschewed.

Logisticians should maintain profitable producer-user links, acting as the military intermediary between those who design and develop, produce, or procure materiel, and those who use the provided equipment and commodities. The close working relationships of logisticians with commanders, staffs, dependent units, and producers of materiel should extend to all areas of their work. Mutual understanding and confidence, fruitful cooperation, and enterprising teamwork engender and strengthen unity of purpose in general, not just in logistical affairs.

PREPAREDNESS

If corporate operational planning involves logistics, what is planning? In this context, it may be described as a

TABLE 2. *Eight constants*

Information	This concerns the prompt, accurate provision of logistic intelligence drawn from the wealth of raw data that modern information systems generate, particularly computer systems.
Timeliness	This is key to any activity and to the application of all six logistic principles suggested.
Continuity Momentum Endurance Resilience	Cumulatively, these concern the *staying power* necessary in logistic support, as well as in most other military activities and actions.
Flexibility	This implies the capacity readily to accept and cope with change, without sacrificing stability.
Simplicity	This reflects in the ability to be clear, concise, and uncomplicated without impairing comprehension, efficiency, or attention to essential detail.

continuous process of matching military aspirations with intentions, commitments with capabilities, requirements with resources, availabilities with constraints. Plans seek to optimize economic, industrial, and logistic capacity, actual operational readiness, and potential combat effectiveness. However, as planning is principally a prediction, and since neither foresight nor military judgment is infallible, so logistic plans should include a safety margin for unforeseen and unforeseeable eventualities. They are bound to arise both in peacetime emergencies or in war.

By predicting likely contingencies and by anticipating the expected course of operations, general staff planners and their logistic advisers together endeavor to forecast force requirements. They may devise an operational structure of in-being, in-place forces and first add regular (permanent) and reserve (mobilizable) reinforcements, then battle replacements. Equipped, trained reinforcements may then be needed to complete formation orders of battle or unit establishments. Replacements of similar readiness may also be required to make up for equipment losses due to breakdown and, once war starts, due to battle damage. Through foresight, military judgment, and persevering, painstaking staff work, through reliable estimates of casualty and consumption rates, so the logistic plan begins to materialize. Like all plans, it requires constant updating.

Operational plans are best made, reviewed, and revised under the direction of one chief of staff at each structural level within a military force. He tasks and coordinates general staff and logistic planners, and acts on behalf of the government minister or military commander to whom he reports. He ensures that all considerations and requirements—logistics included—are competently, cohesively, and completely assessed. Logistic preparedness is a product of this universal, continuous planning process.

Assured preparedness greatly eases the necessary expansion of forces and logistic infrastructure if mobilization is necessary. There should be no fundamental alteration of logistic support organizations and systems during transition to war. A well-planned, well-regulated, smooth transition significantly assists forces to convert from their peacetime operational readiness status to a combat effectiveness status. In the event that some logistical adjustments are necessary, they are best undertaken swiftly and decisively. Adjustments caused by lack of foresight, lack of planning, or lack of preparedness are avoidable and could delay or dislocate operations or the support of operations. *Semper paratus* prevails.

VIABILITY

Before a force is committed to combat, that force must itself be viable and its operations feasible and sustainable. To aid viability (Russian: *zhivuchest*), force structures contain an appropriate mix and balance of constituent components and an appropriate ratio of combat to logistic elements. The mix, balance, and ratio may vary by type of operations, physical environment, and other considerations. For instance, at the end of World War II, the ratio of combat troops to logistic troops stood at 1:5 in certain American formations. It is difficult, however, to be precise about these parameters and measurements—is a forward repair team a combat or a support element? Nevertheless, precision is necessary as a contribution toward producing credible logistics.

When armed forces lack logistic credibility, both support and confidence will suffer. Given the absence of this credibility, adversaries may well be encouraged to embark upon some military opportunism or adventurism that they might not otherwise have contemplated. When armed forces are deprived of feasible, credible, logistic sufficiency, they do not have the resources to succeed. In short, operationally viable forces require sustainable logistic support.

Logistic sustainability does not alone guarantee operational viability. Other factors include: sound strategies and tactics; effective planning and force structure; good equipment; experienced command and inspiring leadership; proper training and readiness; a powerful combat capability with adequate reinforcement and replacement capacity; high morale and belief in a cause. However, logistic sustainability is a vital ingredient of successful military operations. It is achieved by combining resource sufficiency and forward impetus so that continuous, mainly rear-to-front support constantly—and, whenever possible, automatically—satisfies operational requirements, without operational commanders, formations, and units having to look frequently over their shoulders. Viable, sustainable logistic support enables commanders and their forces to achieve strategic and tactical objectives and to accomplish their operational missions and tasks, and gives them the freedom of action to wrest the initiative, exploit success, and avert disaster.

Logistically viable forces are likely to be balanced, collective groupings of combined-arms-services teams. These teams are capable of conducting and supporting combat at a specific level of intensity for a prescribed period of operations—and for a part of the time without resupply, because viability assumes a degree of self-sufficiency. Logistic sustainability follows from this. It is measured and provided using three quotients: *days* of combat; *rates* of battle attrition and casualties, expenditure, and consumption; and *estimates* of replacement equipments and replenishment commodities. Logistic sufficiency is, therefore, the matching of estimated or determined requirements with the resources available. Resources include those provided from the rear by logistic organizations and systems, and those held by the fighting formations and units themselves. Self-sufficiency ensures operational flexibility and ensures against breaks in, or disconnections with, rear support.

Logistic viability is more effectively ensured under operational conditions when support facilities are adequately dispersed and protected, whether the facilities are production plants or base installations, field support units or stockpiles of operational reserves. These potential targets, like lines of communication, are vulnerable to enemy attack. They are also liable to congestion and disruption. Dispersion and protection prolong their viability as logistic assets without running counter to certain principles of war, namely concentration and economy of effort. As in all things, moderation should prevail. Modern communication and information systems now considerably assist the control, management, and deployment of dispersed resources in time and space. Security—another principle of war—has to be optimized in operational areas that increasingly tend to have no clearly delineated, convenient front lines. Therefore, neither security nor logistic viability should be unacceptably reduced by creating a multiplicity of logistic bases, support areas, and lines of communication. Too much dispersion may degrade the logistic sufficiency and sustainability of a force at a given time and in a given area. Too little dispersion and too little protection may lose that force valuable, irreplaceable assets, making it logistically and operationally unviable.

Economy

The well-established principle of logistic economy is open to misunderstanding and hints at rationing. To enforce economies does not imply a deliberate reduction of resources to such an extent that logistic support is no longer viable, sufficient, or sustainable. Just as effectiveness is a product of efficiency and economy working in tandem, so the constant drive for logistic effectiveness involves correction of inefficiencies and elimination of any unnecessary overlap or duplication of responsibilities and waste of precious resources. Logistic economy therefore means using resources—services, facilities, and stocks—most productively and sparingly. Ensuring that resources are effectively used by military forces may ameliorate the adverse effects of materiel and service constraints—often externally and suddenly imposed on forces. Logistic economy also means striking a judicious balance between over- and underinsurance: by providing not too little, not too much, but just enough resources. It additionally means striking a judicious balance between role-effectiveness and cost-effectiveness by economizing without unduly restricting the performance of a combat or support task.

Designing and producing military equipment of optimum reliability, maintainability, and endurance can reduce equipment downtime, which is the period that equipment is inoperable and therefore not available for use, and can also improve user confidence in equipment performance. Regular inspection and preventive maintenance also enhance equipment serviceability. Should equipment become defective or battle damaged, a forward repair capability optimizes battleworthiness and serviceability. Should equipments have to be evacuated to supporting rear maintenance facilities, skilled attention can reduce battle casualty replacement requirements. All these measures help to promote role-effectiveness and cost-effectiveness.

Logistic economy is also gained by long production runs of military equipment to reduce unit costs or by using suitable civilian models to save expensive procurement of special-to-military versions, when the latter are not really essential. In addition, cosmetic modifications, made during production or during the in-service life of equipment, are not always cost-effective, whereas functional modifications are role-effective. Furthermore, prolonging the in-service life of equipment may postpone its expensive replacement. On the other hand, delaying replacement may inflate future maintenance costs. The timely introduction of a new equipment, complete with an initial scaling of assemblies and spare parts, may well be more role-effective and more cost-effective.

Logistic economy is further gained by selecting the right equipments for development and purchase, by specifying the correct degree of sophistication and ruggedness, and by choosing the best manufacturer. Informed decisions have to be made between: multirole and single-purpose models; proven design and the latest technology; a well-established, highly specialized, comparatively narrow production base and a more speculative, more competitive, broader base; long lead times and off-the-shelf, supply-in-time procurement; and between repairable and discardable items. Moreover, equipment standardization and production specialization are likely to be logistically, as well as operationally, beneficial to both national and multinational forces.

Logistic economy is additionally gained by rationalizing and improving in-service support. Large logistic bases and installations can absorb extravagant resources purely for their own upkeep, when cheaper alternatives will suffice. Greater use of civilians and contractors in peacetime may well free military assets for other purposes, provided operational readiness is not penalized. Transport aircraft can be a quicker, more flexible means of logistic movement. However, if these aircraft consume too large a proportion of their fuel cargo in flight, a slower, less flexible, more economical form of transportation may be preferred. Operational stocks and war reserves held for a protracted period within the supply system may in practice inaccurately reflect updated, estimated, or actual requirements. The peacetime scale of equipment engineering and maintenance often exceeds the needs of war. More interoperability, mutual support, and cross-servicing may be economically achievable within military alliances.

Appropriate joint service and international integration is logistically economical and efficient and may also be operationally effective. Centrally agreed-upon equipment policies and programs, centrally directed and coordinated producer logistics, and centrally established resource management can benefit nations and military alliances.

Single-service major-user management of specific ranges of materiel items in common inventories and the rationalization of base logistic support can reduce overheads. Controlled competition between defense manufacturers for declining orders, and between armed forces for constrained resources, can also be advantageous. The escalating logistic bill compels the military to consider every effective measure to integrate logistics as a means of achieving prudent, practical economies.

RESPONSIVENESS

In an effectively responsive supply system, logistic support installations and echelons *push* predictable requirements forward in prescribed daily quantities, mostly automatically; and dependent units *pull* their unpredictable needs from the rear by demanding specified items. In an effectively responsive maintenance system, mobile specialist teams perform forward repair, if practicable. When tasks are beyond their capability or of lesser priority, repairable equipments are backloaded to workshops where defects or damage are rectified and the equipments returned to units. In an effectively responsive personnel casualty evacuation and treatment system, skilled first aid and medical attention are administered to the wounded as soon as practicable. If forward treatment is not appropriate or available, casualties are evacuated by waiting ambulance transport to field and general hospitals, and, when fully recovered, personnel normally return to their units. Other logistic support systems are similarly responsive to operational requirements and equally sensitive to combat priorities.

Responsive logistic services and facilities anticipate changes—they do not only react when changes have occurred. By increasing support in one sector and reducing support in another, adjustments are flexibly made to meet changing operational and tactical requirements. Thus, adjustments respond to a commander's developing plans and priorities. If it is necessary to introduce centralized control of logistic support—in a particular sector, at a specific level within a force, or for a stated period of time—this is effected in the interests of improving flexible response and not to undermine or fetter an operational commander's overall authority and responsibility.

Experience shows that operational plans made in peace need to be constantly reviewed, and those made in war require to be continuously revised. Indeed, peacetime plans seldom fit the real circumstances or meet the actual requirements of war. Responsive logisticians may anticipate and adjust, but they must also allow for unforeseen events and requirements. For logisticians to be responsive is commendable; for them to be able to maintain responsiveness over the period of a war, campaign, or battle is excellent. Both objectives call for unrelenting dedication to duty and attention to detail. The challenges and lack of rest logisticians and their units experience during peacetime exercises are but a glimpse at their task in war.

RESOURCEFULNESS

If the suggested principles of viability and economy appear to be somewhat contradictory, those of responsiveness and resourcefulness are surely complementary.

Logisticians, as well as all soldiers, may not respond effectively to the dangers and difficulties of war unless they are resourceful. The most versatile draw retrospectively on past experiences to deal as competently as they can with unexpected challenges. They may have to refurbish embattled formations or units in the best way practicable, enabling these forces to resume operations. They may have to cannibalize a battle-damaged tank, enabling another to be repaired to battleworthiness standards, or modify equipment so that it can, perhaps at short notice, adequately perform an operational role different from that for which it was designed. Forces may have to live off the land, use local resources, or depend on other unconventional forms of supply.

Resourcefulness derives from practical aptitude and demands a positive attitude of mind. Imagination and ingenuity help the resourceful innovator to keep the initiative in times of crisis. However, developing the right aptitudes and attitudes in peacetime is a problem; therefore, resourcefulness in war often has to be instantaneously intuitive.

Some argue that, because advanced technology breeds specialists, soldiers who are not specialists are no longer as self-reliant and self-contained as they once were. Armies that traditionally emphasized *equipping the man* are now preoccupied with *manning equipment*. When the equipment breaks down, the unresourceful nonspecialist may blame technology and seek a replacement. An example often quoted concerns the division of responsibility and labor between tank crews and maintenance specialists. In peacetime, crew training may be limited to regular servicing and minor adjustments with specialists attending to repairs. In war, when specialists may not be available to help, it may be a matter of life and death for crews to make-do-and-mend and keep their tanks in action. Practical preparation of nonspecialists is thus likely to improve their resourcefulness in a crisis, just as first aid training has saved the lives of countless wounded soldiers in battle.

Resourcefulness is not only an individual gift; resourceful logistics can be collectively acquired. For instance, logistically self-sufficient units and formations, combined-arms-services teams, possess integral self-reliance and self-containment. Potentially, they are more resourceful and less dependent upon specialized rear-area support. Working together, commanders, the general staff, and logisticians have ample scope and an undeniable need to plan and encourage resourceful support of military operations.

Summary of Principles

Table 3 summarizes the six suggested principles of logistics. The key words might usefully be read in conjunction with the constants, listed in Table 2. These principles relate to separate articles dealing with logistics in limited war, logistics in NATO, and logistics in Soviet and Warsaw Pact forces. They also relate to the four hypothetical, illustrative scenarios that follow.

Scenario A

The logistic support of general total nuclear war is explained in fuller detail in separate articles; therefore, only some key considerations are outlined in this general survey of logistics.

Scale and Scope

This first scenario depicts very large multinational, conventional forces backed by nuclear, chemical, and biological weapons. It envisages operations on a major scale, conducted by combined forces in a number of theaters and areas of operations. It envisages central, allied operational command of balanced, collective forces: a mix of national contingents using common tactical and logistical doctrines, concepts, and operating procedures. It also envisages integrated logistics of full scope and magnitude to support the forces deployed and the operations to be undertaken.

TABLE 3. *Summary of Suggested Logistic Principles*

Unity of Purpose	Corporate effort
	Functional interdependence and integration
	Mutual understanding and confidence
	Cooperation and teamwork
Preparedness	Foresight and military judgment
	Determination of requirements
	Coordination of planning and plans
	Operational readiness
Viability	Feasibility and credibility
	Sufficiency
	Sustainability
	Dispersion and protection
Economy	Elimination of inefficiency and waste
	Role-effectiveness versus cost-effectiveness
	Rationalization, standardization, and specialization
	Integrated support
Responsiveness	Forward impetus and momentum
	Local, centralized control of resources
	Proactive, as well as reactive, support
	Dedication to duty
Resourcefulness	Versatility
	Improvisation and innovation
	Self-reliance, self-containment, and self-sufficiency
	Development of aptitudes and attitudes

As far as land forces and joint land/air operations are concerned, allied commanders at corps level and above and national commanders at divisional level and below have full authority over assigned logistics, and are fully responsible for the effective logistic support of their force, formation, or unit. Logistic integration and the reasonable degree of equipment standardization already achieved have enhanced the prospects of interoperability, mutual support, and cross-servicing among the national contingents that make up the combined force. Single channels of command and comprehensive operational planning, local logistic control, and compatible management information systems have also enhanced capabilities at all structural levels to determine operational requirements, to take quick account of changed operational priorities, and to monitor and regulate day-to-day, hour-by-hour logistic support.

The plan provides for operational stocks of materiel in-place in theaters and area of operations. They are sufficient to sustain intense, conventional combat for a prescribed period of time. The plan also provides for stockpiles of war reserves either in-place or available in logistic bases. It additionally provides for logistic resupply from home or other bases to replenish materiel usage, to replace materiel wastage, and to prolong conventional or nuclear operations. The planned system is outlined in Figure 2.

Synchronization and Coordination

Intelligence indicators, alert measures, and the synchronized mobilization of multinational reserve forces all signal the start of transition to war. Existing in-place forces move to their operational sectors or battle positions, and they are joined by out-of-area, regular reinforcements. Concurrently, logistic support is activated and the initial outloading of stocks begins. These and the deployment of mobilized reserve elements add to the complexity and vitality of a massive movement operation, the plan for which has been agreed upon and rehearsed in peacetime. The movement plan incorporates all sea, air, and overland dispatch to, from, or within a theater of operations such as the one hypothetically illustrated in Figure 2. It also incorporates allied and host-nation transportation and movement control facilities. The implementation of the plan is synchronized and coordinated by allied commanders.

Replenishment and Resupply

Initially, operations depend on in-theater or in-area stocks which are outloaded from depots and deployed in close support of forces. War reserves of materiel are designed to cover requirements for additional tasks and abnormally high expenditure, consumption, or casualty rates after war begins. In addition, strategic stocks of raw materials and components held in peacetime enable the defense industries of allies taking part in the operations to step up production so that resupply can begin as soon as practicable.

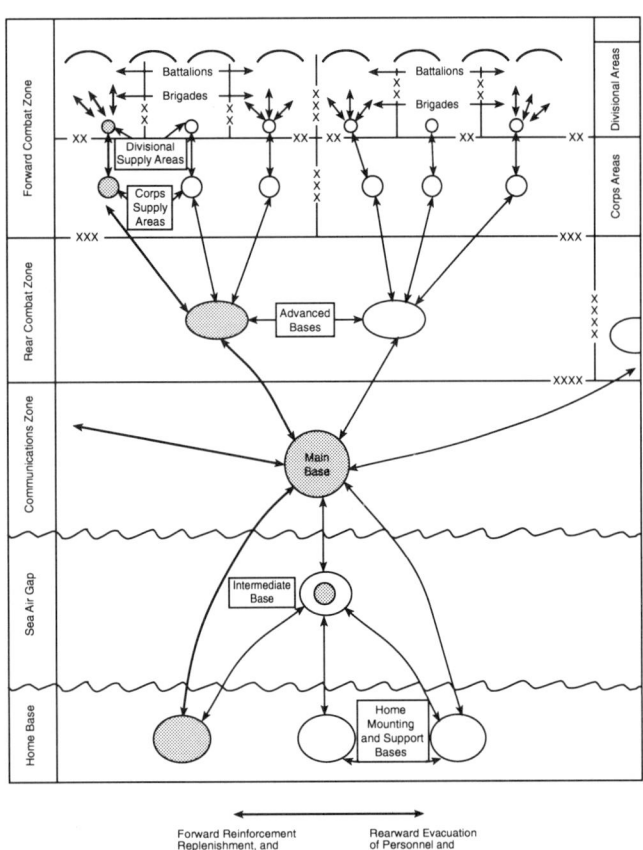

Figure 2. Scenario A: A developed theater of operations.

The replenishment and resupply systems are then complete, and continue to operate for as long as possible.

HOST NATION SUPPORT

The state or states forming the theaters and areas of operations have agreed to provide specific host nation support to allies contributing forces. This support is incorporated within the logistic buildup and deployment planning, and, following implementation of the plans, continues for as long as possible.

FULL LOGISTIC SYSTEM

Reference back to Figure 1, Table 2, and Figure 2, respectively, will recall the range of logistic functions and services available, the levels or lines of support, and the framework for an illustrative theater of operations. The full logistic system spans the facilities and distances from forward troops to home production plants, for as long as the lines of communication remain open. If the system fails at any point, friendly forces have to make do with the logistic resources they already possess, conserve them, and protect them from enemy attack.

SELF-SUFFICIENCY

Fighting formations and units are reasonably self-sufficient. They can operate, if the need arises, for several days without supply replenishment. They have facilities for repairing equipment and treating personnel casualties. Equipment and personnel decontamination and specialized medical support may be required should nuclear, chemical, or biological weapons be used. They have limited resources for dealing with massed casualties and have neither suitable nor sufficient road transportation to travel far for resupply.

Scenario B

The second hypothetical scenario illustrates a very much smaller scale of military deployment to assist the defense of a state that is under the threat of invasion by a hostile, well-armed neighbor. It outlines the operational setting, again concentrating on logistics. The commentary is written from the perspective of the supporting ally who has strong forces and a developed operational capability.

A small, oil-rich state suddenly faces a threat of armed aggression from the north. It is an arid, desert, and quite primitive country with good roads, reasonable tracks, a developed port complex, and a prestigious international airport with facilities for strategic transport aircraft. There is another airfield near the principal oil field town in the center of the state. This airfield is capable of receiving and servicing medium-range tactical transport aircraft, helicopters, and fighters. A major surface oil pipeline connects the installations around the town with the port at the capital on the coast, where there is also a large refinery. The pipeline conveys crude oil to the port and delivers refined fuel to storage facilities at the town's airfield and the capital's international airport. The capital has modern infrastructure, including a modern, underutilized general hospital, and excellent port facilities run by the oil company (see Fig. 3).

The state is within strategic transport aircraft range of the supporting ally's home bases and a week's fast sailing from its home ports. A contingency plan for reinforcement and defense exists, and, after a rapid review and some revision of that plan, both allies agree to the following main tasks:

1. Indigenous land forces are to deploy as shown in Figure 3. The separate royal guard is to continue to protect vulnerable points, including the capital, port, and international airport, as well as key dignitaries.
2. Indigenous naval and air forces, without reinforcements, are, respectively, to continue to protect coastal waters and the harbor from their base in the port, and to provide tactical air reconnaissance, offensive air support, medium-range transport aircraft, and logistic helicopters from their bases at the international airport and the inland airfield.
3. The supporting ally's reinforcement land forces are to deploy near to the northern frontier and, initially, to conduct aggressive patrolling.
4. The supporting ally is to supplement and strengthen the state's operational command and planning capabilities.

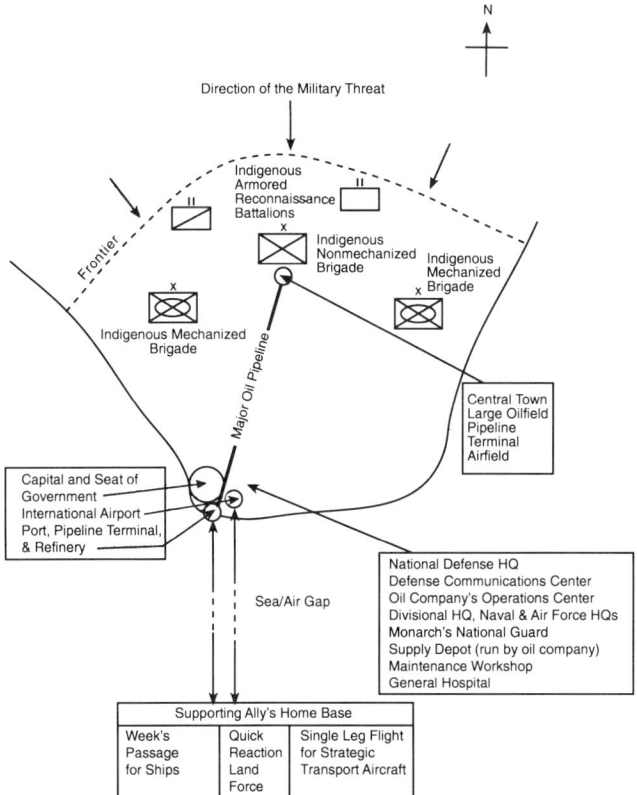

Figure 3. Scenario B: Opening situation.

5. The supporting ally is to provide logistic support for all indigenous and reinforcement land forces, less the self-contained royal guard. Some support is available from existing facilities and from local resources. The remainder is to be transported to the state by sea and air.
6. The supporting ally is to mount strategic air reconnaissance and early warning sorties and build up intelligence activities in the area.

The indigenous and reinforcement land forces are similarly organized. They use common main battle equipments, operating procedures, and terminology. The supporting ally's language is spoken extensively by the educated military and local population. The major oil company is jointly owned by the small state and the supporting ally, and the majority of its senior management and technical staff comprises seconded nationals or expatriates from the supporting ally. In addition, a number of the supporting ally's officers and technicians fill key appointments in the indigenous naval and air forces. The state's navy and air force are small but powerful. They are well trained and equipped, with sufficient logistic support. Further military details follow.

The national defense headquarters is located with other government departments in the capital. The commander in chief, a member of the royal family, has a small, efficient joint staff and a large, inexperienced civilian complement to assist him. He commands all forces, including the royal guard. Indigenous land forces, as well as the royal guard, are trained only for internal security duties.

The national defense headquarters has a good communications center, network, and facilities. The oil company runs a small operations center which regulates its own large transport fleet, the pipeline, the refinery, and port facilities. It is located at the company's head office adjacent to national defense headquarters, and has room for expansion.

A small divisional headquarters commands the indigenous land forces, but not the royal guard. It is currently sited with the national defense headquarters. The divisional commander is experienced and competent, but his general staff is inadequately trained and he has no logistic advisers. His division is undergoing intensive training at present and consists of:

- Two mechanized infantry brigades comprising one tank and two mechanized infantry battalions, one artillery battery, and signal and logistic support companies
- One nonmechanized infantry brigade of three rifle battalions, one artillery battery, signal and logistic support companies, and a troop-carrying transport battalion
- Two armored reconnaissance battalions equipped with tracked vehicles and helicopters, and some organic support.

The logistic support organic to these formations and units needs no reinforcement, only training and specialist advice and assistance. To date, they have received support direct from static installations in the port area because no dedicated divisional level logistic support exists. The supply depot provides a composite materiel service and is run by the oil company on an agency basis. The well-equipped maintenance workshop nearby is run by military expatriates. The general hospital admits military and civilian patients. The oil company can provide a substantial number of tanker trucks and other transport vehicles. Fuel, food, potable water, and labor are adequately available in the capital and central oilfield town. The logistic assessment indicates the following main indigenous deficiencies: inadequate logistic command and planning capabilities, organization, systems, and training; no divisional logistic support; an inadequate military supply depot and no local resources team; no movement planning and control facilities.

The indigenous navy and air force can handle the reception and turnaround of military shipping and strategic transport aircraft.

Following the supporting ally's assessment of all operational and support requirements, it undertakes to provide:
1. A deputy commander in chief to be based at the national defense headquarters with the following main responsibilities: to act as chief of staff; to direct and coordinate the joint staff on behalf of the commander in chief; to task the defense communications center, network, and facilities; to establish a defense movement control center within the oil company's existing oper-

ations center and to be responsible for its tasking and performance; to reinforce the joint logistic planning staff; and to set up and chair a logistic management committee with the oil company's managing director and key military officers and civilian officials as members
2. An increment for divisional headquarters comprising: a deputy commander; a chief of staff; several general staff officers, including a complete G4 branch to undertake operational logistic planning; and certain specialist logistic advisers and their staffs
3. Two reinforcement armored brigades with organic armored reconnaissance, combat, combat support, and combat service support units
4. A reinforcement logistic support force consisting of a headquarters, three battalions, a local resources team, and staff for the new defense movement control center
5. Logistic support for all land forces, other than resources available locally
6. An operational reserve in the supporting ally's home base, should it be needed

Four additional operational planning arrangements are among those agreed upon for immediate implementation. They are:
1. Reinforcements for the joint defense headquarters, the divisional headquarters, the local resources team, the staff for the defense movement control center, and key advanced parties are to arrive as soon as possible to start planning.
2. The national defense headquarters is to undertake in-area joint logistic planning of all support provided by an advanced base to be established around the capital, the port, and the international airport. The J4 branch is to be responsible for joint logistic planning, including joint movement planning.
3. Divisional headquarters is to undertake in-area land force and land/air operational planning, with its G4 branch dealing with logistic support and logistic training within the complete formation.
4. The supporting ally is to undertake all planning required to move reinforcements to the state and to sustain in-area operations by air and sea.

Once the initial reinforcements have arrived and planning has started, several additional decisions are made. Divisional headquarters is to move immediately to the central oilfield town. It will locate at the airfield and establish a forward base there. It will extend its command to include the two reinforcement armored brigades, divisional troops, and the forward base. Air force headquarters is to provide a detachment of air staff for incorporation within divisional headquarters to assist with the planning and control of land/air operations and air logistic support. Air force headquarters is also to provide aero-medical casualty evacuation teams positioned in the forward and advanced bases.

The commander of the reinforcement logistic support force is to set up his headquarters in the advanced base, which he will command. He will deploy his force as follows:
1. A multifunctional logistic battalion is to be located in the *forward* base and will consist of
   - A maintenance company with a field workshop and supply platoon
   - A medical support company with a field hospital, ambulance platoon, and supply section
   - Several constructional engineering teams with requisite plant and equipment
   - An administrative company with labor and local defense elements
2. A supply battalion is to be located in the forward base and will be composed of
   - An equipment and vehicle company with integral stock management, stock maintenance, handling, and transportation facilities
   - Two commodity companies, each with the facilities as above
   - An ordnance engineering platoon
   - Catering and canteen services
   - A map supply section
   - An administrative company as in paragraph 1.
3. A supply battalion is to be located in the *advanced* base and will comprise elements similar to those listed in paragraph 2, plus a local resources supply platoon and a reserve of road transport to supplement the oil company's vehicle fleet. In addition, individual logistic reinforcements are provided to supplement the existing maintenance workshop and existing general hospital staffs.
4. Reinforcements are to be provided to run the newly established defense movement control center, including detachments to be positioned in the forward base, at the advance base headquarters, in the port, and at the international airport.

Having outlined the initial plan for in-area operations and support we should now return to the supporting ally's home base where the main reinforcement process and the logistic buildup is about to begin.

PHASE 1—FINAL PREPARATIONS

The supporting ally's quick reaction force is ready for deployment at short notice. It consists of two component parts: air-portable elements and heavier elements that have to be transported by sea. The air-portable elements already have light scales of equipment and their immediate logistic support stocks. Additional items appropriate to desert operations are issued, such as special camouflage equipment and paint.

When the composition of the reinforcements is decided, the movement plan is made. The formations and units of the quick reaction force involved in the operation are returned from leave and placed on movement alert. The movement plan is implemented and the order given by the commander of the quick reaction force (deputy commander in chief designate) to move from peacetime locations to staging areas at home air bases and sea ports. Final checks and adjustments are made to the movement plan and detailed instructions issued. The airlift and sealift of reinforcements commence.

PHASE 2—AIRLIFT

Air-portable elements are flown from one or more home air bases to the international airport of the state being reinforced. The movement plan allocates elements to sorties and provides loading instructions and manifests.

The strategic transport airlift requires over-flying rights and air traffic control clearance, with emergency provision for landing and refueling en route. The aircraft return to home bases for additional passengers or cargo.

PHASE 3—SEALIFT

Troops, weapons, vehicles, equipment, and bulk stocks are embarked at one or more home ports. Sealift begins simultaneously with the start of the airlift. Transportation of these heavier elements is by naval logistic ships, auxiliary vessels, and civilian freighters. They disembark at the port of the state being reinforced.

Sailing time is one week for the fastest ships. The movement plan allows for a staggered arrival which is convenient for the receiving port.

PHASE 4—BUILDUP

The national defense headquarters's in-area movement plan provides reception arrangements for reinforcements arriving by air and sea. Some are immediately flown or moved by road to the forward base, their operational deployment areas, or other employment locations. Others staying in the advanced base are moved by road to a transit area and then to their employment locations.

Urgent logistic requirements arrive by air, but the main buildup of operational stocks and war reserves follows by sea. In-area deployment to stockholding depots, units, or echelons is mainly undertaken by road. The logistic annexes of the divisional and joint operational plans respectively prescribe stocks to be held at all levels within the force.

Figure 4 illustrates the situation in the area of operations when all four reinforcement phases are complete. The following details of logistic support explain the services and systems provided:

1. In-area supply is normally by road from the supply depot in the advance base, via the forward base and brigade supply areas, and then to combat units. Air supply is used for urgent, air-portable requirements or when road transportation is impracticable. As well as the multipurpose helicopter, air supply involves medium-range transport aircraft operating into and out

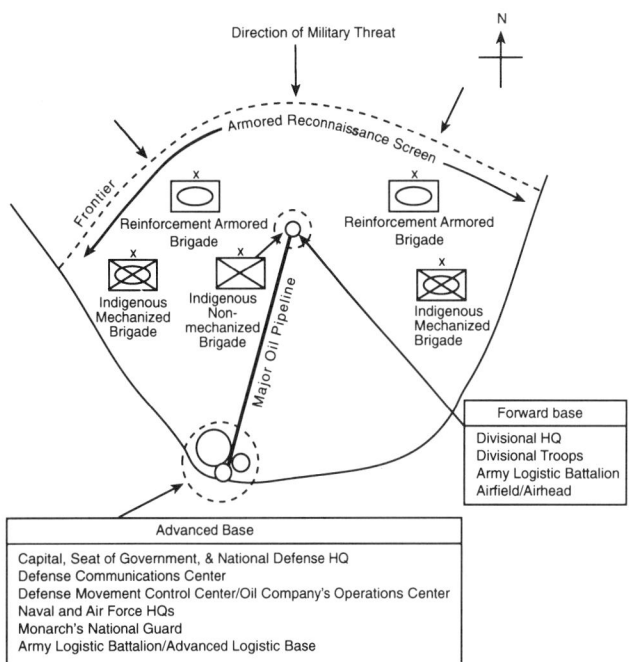

Figure 4. Scenario B: Reinforced situation.

of the forward base. For example, at the airhead there, cargoes are loaded onto aircraft for air-dropping by parachute or free-fall delivery to units.
2. Forward repair is supplemented by the small field workshop in the forward base and the larger maintenance workshop in the advanced base.
3. Sick and wounded are treated forward and, if necessary, evacuated by road or air ambulance to the small field hospital in the forward base and the larger general hospital in the advanced base.
4. Constructional engineering tasks are undertaken by military teams in forward areas and range from building temporary accommodation to digging wells. Tasks in the forward and advanced bases are carried out by civilian contractors. Adequate materials are available locally.
5. The oil company provides fuel from its pipeline and refinery. Some stocks of petrol, diesel, and lubricating oils are held in cans, but the main means of supplying fuel to brigades and units is by tanker truck. Food and certain other commodities are also available locally. Forward troops receive fresh and refrigerated rations regularly. Maps, mail, newspapers, and canteen supplies reach them through the normal supply system. Civilian labor is plentiful.
6. An additional task is to train indigenous logistic support personnel and units.

The reinforcement operations, the strong defensive posture, the active and aggressive patrolling along the frontier areas, and the logistic buildup all serve to deter an attack. After six months, the reinforcement forces return to their home bases, leaving behind a logistic training team and certain logistic stocks. The supporting ally updates the contingency plan based on the experience gained.

Scenario C

Whereas the previous scenario describes a military emergency when reinforcement forces are invited to intervene and where local resources are plentiful, the next presents a different situation. Foreign insurgents have infiltrated a state and have forced the legal government to flee to a friendly neighboring country. The intruders have now set up a puppet administration in the capital and are conducting a campaign to gain the support of the local population by brutal coercion. In this scenario, the supporting ally receives the request to intervene, from a government in exile, in a country that has precious little local resources to offer a military force, and where certain elements of the population are hostile or unsupportive to military intervention. However, the request has been made, not only to intervene, but also to defeat the insurgents, drive them out of the state, restore the rightful government, and re-establish law and order. The supporting ally agrees to assist and to mount an operation as soon as practicable.

The country is jungle covered (see Fig. 5). It has only primitive communications between an unwelcoming hinterland and the small coastal capital. There are jungle tracks linking riverside villages with hunting, farming, and logging areas, but, until the insurgents arrived, the two main rivers were the main arteries for trade and contact. Now riverboat trade has ceased. The local paramilitary police force is loyal to the legal government and is doing its best to remain in close touch with the hinterland tribespeople and the capital's business community. It is also in contact with exiled government ministers and officials, providing all the intelligence it can on the local situation. The police force has managed to hide the riverboat company's craft in creeks where the two main rivers join at the capital and port. It appears that most of the inland villagers have either fled to the jungle or are subjugated. The insurgents have set up several operating bases in the jungle and regularly visit villages to acquire food, medical supplies, and information. The puppet administration is evidently endeavoring to impose a 24-hour curfew in the capital, without much success. The military intervention is clearly also a *hearts and minds* operation.

The capital's port area has a single jetty with facilities to unload one reasonably sized freighter, and a deep-water harbor where ships may anchor while their cargoes are unloaded by lighters onto the quayside. There is one airfield on the outskirts of the capital, with an all-weather strip and facilities suitable for medium-range tactical transport aircraft. The paramilitary police force had some helicopters until the president, ministers, senior officials, and their families fled in them. Several of the main villages on both rivers have helicopter landing sites, with several other landing sites within jungle areas.

The supporting ally has a large overseas base in the region, within medium tactical transport aircraft range of the capital's airfield and two days' sailing time for ships to reach its port. The operation is planned and mounted along lines similar to those described in Scenario B. The following commentary outlines the main events, again concentrating on logistics.

Phase 1—Intervention

Two parachute battalions drop on the airfield, secure it and establish an operating base. They quickly make contact with the local paramilitary police force; secure the port, harbor, and riverboats; then secure vital points in the capital and capture the insurgent leaders. The battalions have sufficient logistic support to last 72 hours; however, their operation has been so successful that it enables the transport aircraft to fly in without delay.

The remainder of the air-portable elements of a parachute brigade are airlanded. They establish a forward base around the airfield, with brigade headquarters, the third parachute battalion, combat support and combat service support units, and initial logistic support stocks located there.

Air-portable elements of a nonmechanized infantry brigade are flown in and relieve the parachute brigade in the capital and forward base. The parachute brigade regroups as a mobile, operational reserve. The main airlifted logistic buildup begins.

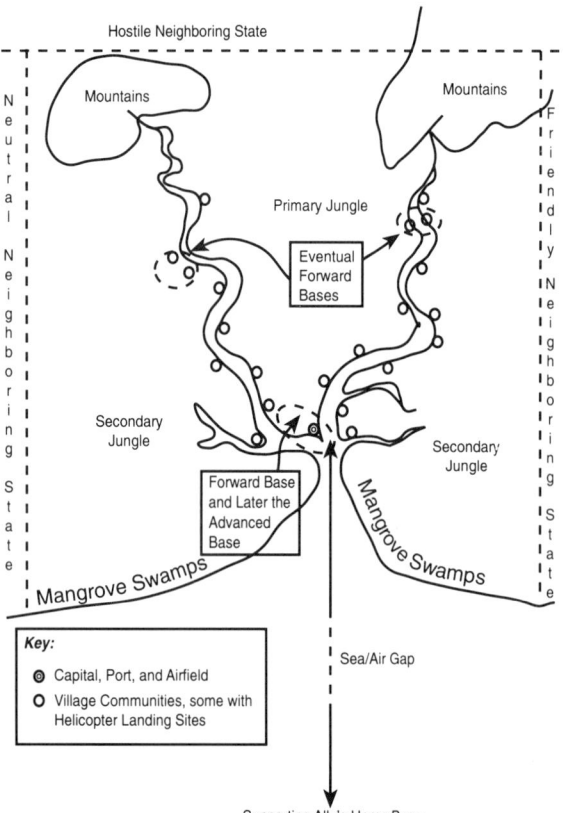

Figure 5. Scenario C.

Logistic helicopters, the advanced parties of divisional headquarters, and key support units arrive in the forward base.

The local paramilitary police force re-establishes its authority in the capital and imposes strict curfews there, in the port and harbor areas, and in the lower reaches of the two main rivers.

The legal government returns and the president appoints the supporting ally's divisional commander as director of operations. The divisional commander establishes an operations and security committee with military, police, and government representation which also coordinates logistics.

PHASE 2—EXPLOITATION

The parachute brigade captures and secures the main villages upriver and opens several helicopter landing sites. It leaves one parachute battalion in the forward base as a mobile, operational reserve. It also opens and stocks some battalion bases in the best village locations.

The local police and tribespeople help to build defensible perimeters around the main villages and helicopter landing sites.

The infantry brigade, transported by rivercraft and helicopter, systematically and progressively clears both main rivers, and links up with the parachuted forces. It leaves one infantry battalion in the capital to support the police. Logistic stocks are then ferried forward by river as well as by air.

The seatail arrives in the harbor and port. Divisional headquarters, divisional troops, and the two brigades are now complete with their heavier elements. The logistic buildup is accelerated.

Combat and constructional engineers improve all the main facilities in the area of the capital, build harbor and river defenses, and improve existing and develop new upcountry bases.

A second nonmechanized infantry brigade arrives by air and sea. The force is complete and the logistic buildup continues.

PHASE 3—CONSOLIDATION

Having regrouped and reorganized his force, the divisional commander/director of operations allots each infantry brigade a main river sector to clear and secure, with one infantry battalion remaining in the capital. He orders the parachute brigade to mount deep penetration patrols into the jungle and mountainous areas along the border with the state that sent in the insurgents. One parachute battalion remains as the mobile force reserve.

A brigade base is established in both river sectors, and an all-weather airstrip is built in each. These bases now become forward bases with enough logistic support and stocks for the two infantry brigades to be assured of reasonable self-sufficiency.

The original forward base around the capital now becomes the advanced base for the operations, and divisional headquarters, divisional troops, and the mobile force reserve are also sited there.

The local police and local guides provide invaluable assistance, tactically and logistically, to the conduct of operations in the hinterland. Gradually, well-controlled and -coordinated insurgent-free areas are being extended outward from firm, secure operational and logistic bases. These bases are established at patrol/platoon, company, battalion, and brigade levels.

Logistic replenishment of forward bases continues, the in-area buildup of stocks is now complete, and resupply from the supporting ally's overseas base maintains in-area operational stocks and war reserves.

Business is again starting to flourish in the capital. River trading resumes, the military force provides some logistic assistance for the hinterland tribespeople, the paramilitary police force is expanded and trained, the operations and security committee is working well, and life for a growing portion of the local population is returning to normal.

Without repeating relevant details for Scenario B, this commentary is concluded with some further logistical elaboration. Jungle operations present some severe challenges. Combat and combat support units operate mainly on foot, carrying light equipment and their immediate logistic stocks. Their heavier equipment and larger support stocks have to be transported over the jungle by air and moved into jungle clearings whenever practicable by helicopters. If no air transport is available, loads have to be carried by river craft or by troops, porters, or pack animals along jungle tracks, often over difficult terrain and in bad weather. The constantly hot, humid climate can also cause health and hygiene problems. The sick invariably outnumber the wounded in this type of operation. Acclimatization and many other preventive health and hygiene measures are vital, as in most extreme climates. Skilled first aid is also vital, as in all combat operations. Casualty evacuation can be difficult and hazardous—by air, by river, or on foot—in the quest to move the seriously ill and wounded to the nearest doctor or, if necessary, further back to a field or general hospital.

The all-pervading dampness encountered in jungle conditions can result in inoperable equipment. The relatively small number of maintenance specialists, combined with movement problems, may well inhibit forward repair and equipment evacuation. As operations of this type critically depend on each item of equipment, wherever it is deployed, logistic support must effectively and by priority overcome these problems. Likewise, there are rarely enough sappers for the wide variety of constructional engineering tasks they are expected to complete in an area of operations such as the one depicted here. These tasks are, of course, additional to normal combat engineer missions. Improved shelter, tracks, bridges over rivers, and water supplies can lift morale considerably. Mail, news, and canteen supplies are also substantial comforts, especially for troops operating in small groups in an alien jungle environment, and items such as these are likely to be moved forward as part of the normal supply system—however tenuous the system.

Daily replenishment is the aim and is achieved whenever and wherever practicable. If bad weather, the unavailability of helicopters, or the inaccessibility of troop locations makes air supply impracticable, then jungle bases hold a small reserve of stocks. Otherwise, a simple operating procedure enables each patrol or unit to obtain daily replenishment according to a standard, agreed-upon list of commodities. Each commodity on the standard list has a simple code, used when it is necessary to add or subtract items, which is quickly transmitted by radio. This system of management by exception is efficient and economical and minimizes radio traffic. Daily supply packs are made up to standard or variable requirements in rear bases and airdropped, airlanded, or moved by surface means. Daily supply is the system predominantly in use in this scenario. Air support is, however, the key to its full effectiveness and success. Without transport aircraft and helicopters, the delivery of daily supply packs can cause severe logistic problems. Air logistic support is a joint land/air operation that is best planned, coordinated, and controlled by the divisional commander in this scenario, with authority for local tasking of certain logistic helicopters also vested in brigade headquarters.

Scenario D

The three preceding scenarios have dealt with logistics in general war, a threatened limited-localized war of much lesser scale, and counterinsurgency operations. The last scenario illustrates support of an internal security situation when military forces come to the aid of a civil power in a remote island state. The island is heavily populated and largely undeveloped, with mineral resources of critical strategic and commercial importance to the supporting ally. The latter has a vital telecommunications, radio intercept, and satellite tracking installation on the island's single, high mountain not far from the capital. In addition, the supporting ally has enjoyed sole mining and export rights of the mineral resources, and a considerable number of its nationals manage the mines and mining services on the island. Apart from these nationals and a privileged minority of the local population, most islanders are uneducated, have a low quality of life, lack basic amenities, and are overtly discontented (see Fig. 6)

In an effort to increase its income and improve the lot of the ordinary people, the island's government negotiates and signs a deal that will enable other countries to dump large quantities of highly toxic waste in unused mine workings. This signals widespread dissidence, sabotage, and terrorism. The internal security situation is fast deteriorating and there are bomb fatalities occurring in commer-

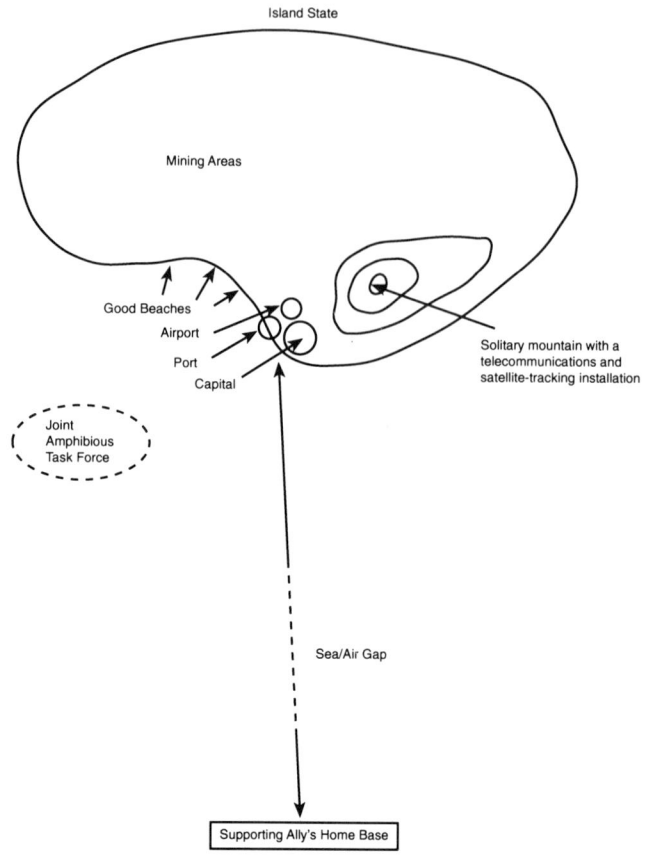

Figure 6. Scenario D.

cial, retail, and higher-class residential areas. Strikes disrupt the mining industry and all public services, and the one port and the one airport are closed. The island's government revokes the toxic waste dumping agreement, imposes curfews, and threatens to declare martial law. The local police force is stretched to the breaking point.

The supporting ally anticipated trouble, and has recently reviewed and revised its contingency plan for protecting its nationals, both resident and working on the island; protecting its mining interests and other vital installations; and evacuating its nationals should the need arise. It has no overseas base within operating distance of the island state, so has positioned a joint amphibious task force nearby, exercising and monitoring events closely. The supporting ally agrees to mount an internal security operation from the joint amphibious task force in order to help restore law, order, and normality on the island.

The task force consists of an aircraft carrier, maritime protection vessels, logistic ships, and a hospital ship. It carries afloat marines and infantry trained and equipped for riot control and other internal security duties, some special forces and intelligence elements, several explosive ordnance disposal (bomb disposal) teams, as well as appropriate logistic support for a force deployed on the island. The following description concentrates on the main military actions and particularly focuses on logistics.

The rear admiral commanding the joint task force has

direct working links with the island state's government and with his home base. He has responsibility for planning and conducting the internal security operations on the island and their support from his task force. He completes his exercise and moves his force nearer to the island.

Naval helicopters transport marines, infantry, and their immediate logistic support to an operating and logistic base established at the airport. The land-based element quickly builds up and deploys to selected vital or vulnerable points, commanded by a brigadier general who sets up his headquarters at the airport base. He remains responsible for land-based operations and support to the admiral and has priority call on sea-based reinforcements, logistic support, and helicopters. The land-based commander also provides direct liaison with the island's police force and public authorities. He gradually widens the scope and capacity of his internal security operations, builds up his land-based logistics, and progressively establishes a network of smaller operating and logistic bases.

Meanwhile, island nationals of the supporting ally, together with threatened local civilians, are moved into refuge areas for protection. Some are moved into transit areas to await transport to the airport or to the naval task force. The land-based military forces help the police and the managers of public services, communications, and utilities to get essential facilities working again as quickly as possible.

The port, however, remains closed and is the center of militant political and trade union activity. It is badly damaged, and its sea and land access is obstructed. The joint amphibious task force has logistic ships capable of beaching and discharging their roll-on/roll-off cargoes over the good beaches indicated in Figure 6. From early in the operations, logistic ships have been unloading reinforcements, the heavier equipments, and bulk logistic stocks to strengthen the land-based forces. This means of entry substantially supplements the air bridge from task force to island and allows helicopters to be used for tactical purposes and for transporting urgent logistic requirements from ship to shore or on over-island sorties. Helicopters returning to their land or afloat bases carry personnel casualties for treatment on the hospital ship or vital air-portable equipments for task force repair. Non–air-portable repairable equipment is evacuated from the island across the beaches and onto logistic ships.

The strike is broken at the airport and the facilities there start receiving strategic transport aircraft flown from the supporting ally's home base, staging and refueling in-flight en route. High-priority air-portable reinforcements and logistic requirements are thereby transported to the island. Families are evacuated on returning flights.

When the land-based force is complete, the land force commander becomes directly responsible to the state's government for internal security operations. The admiral and the residue of his joint amphibious task force remain in island waters as a reserve, with a large proportion of the naval helicopters disembarked and land-based under the operational control of the land-force commander.

Eventually, the port is repaired and opened, the sea bridge links the island with the supporting ally's home base, and the logistic buildup is completed.

The amphibious task force, less its land-based elements, then withdraws. When the situation on the island returns to normal, the land force thins out and a smaller internal security garrison of principally infantry and support elements remains for a few months. The garrison retains sufficient specialized logistic personnel and facilities to meet its operational requirements; it obtains, by then, most of its domestic needs from island sources. Apart from a contingency reserve of military materiel kept on the island for use by any reinforcements—should they be requested and airlifted there in the future—all other surplus logistic resources are returned to the supporting ally's home base by sea.

In this scenario as in others, the helicopter proves its worth tactically and logistically. In its logistical role, the helicopter carries internal or external cargo. Its crew is adept at reaching isolated locations to land or winch down loads. Ground- or ship-based supporting teams are specially trained to prepare loads and to task logistic helicopters. Weather, limited flying hours, serviceability, and availability all govern helicopter operations, but, with sound servicing and handling, the helicopter introduces invaluable flexibility and speed of reaction to operational environments.

There have been other significant improvements in the field of logistic movement. Bulk stocks of commodities are placed on pallets, in special containers, or on a combination of pallets loaded into containers for transportation by sea, air, rail, road, or inland waterway. For example, a single pallet may contain a number of complete artillery rounds—shell, propellant charge, fuze, and primer; pallets and containers may carry multi-item or single-item stocks for a particular purpose, mission, or user; the loading, unloading, or transloading of pallets or containers is considerably assisted by roller-conveyors mounted on the floor of the mode of conveyance—hold of a ship, floor of an aircraft, railway wagon, or road vehicle. Pallets and containers require specialized handling equipment and skilled operators. Provided these are available, substantial savings in time and manual effort are made by utilizing these and similar logistic movement means most effectively.

The computer, telecommunications, and such specific aids as telefacsimile have revolutionized the methods of monitoring, regulating, and controlling logistics. Management information systems of impressive and ever-increasing usefulness help commanders, planners, logisticians, and users to make the best operational use of available logistic resources. Rapidly advancing technology is having as much impact on logistic doctrine, concepts, procedures, and techniques as any other facet of military activity. This is so, whether we are examining the logistic support of global total nuclear war (Scenario A) or some form of armed conflict of lesser magnitude and intensity (such as Scenarios B–D); whether we are examining the logistic support of land or air forces, maritime logistics, or a combination of operational environments.

Conclusion

The challenges of *pure* logistics (forecasting and determining requirements) and *applied* logistics (planning, preparing, and providing support) remain as difficult today as in the past. Tailoring support to sustain a small, quick-reaction force is just as challenging and vital in the context of a particular operation as the accumulation of huge stockpiles of resources during a protracted period of cold war. It is possible that the days of vast, permanent, expensive, base logistic installations are numbered as greater dependence is judiciously placed upon the swifter means of logistic movement, "just-in-time" supply practices, and the management information systems now available. The days of "shoestring" logistics featuring rationed resources and financial stringencies are never likely to be numbered. The resources devoted to military support are always likely to be less than most conscientious logisticians relish.

Sound military judgment will continue to be crucially applied when deciding the *extent* of viable, yet affordable, logistic support needed to sustain military operations or foreseen contingencies. Sound logistic experience and expertise will continue to be crucially applied when deciding the *organization* and *methods* required to achieve effective operational readiness and effective combat service support. The principle of continuity provides a stable platform from which to provide operational logistics, while the principle of change underlines the constructive part that imaginative, innovative, new ideas should play in determining the most effective form of logistic support for the future.

Before the Second World War, when experimentation and preparedness were lacking, the military inclination to remain the same, despite the pressing need for change, drew this pithy comment from Liddell Hart: "There's only one thing more difficult than getting a new idea into a military mind, and that's getting an old one out" (J. C. T. Downey, *Management in the Armed Forces: An Anatomy of the Military Profession* [London: McGraw-Hill, 1977], p. 19). Now that the change from globalized cold war places renewed emphasis on the potential continuity of localized hot wars, now that the quest to reduce the overbearing logistic bill for standing forces stresses the desirability of seeking new support concepts without jeopardizing the essential need for resourceful logistics responsive to all military calls, so there should be an inquisitive demand for fresh ideas and some experimentation. Military commitments and requirements are compelling certain nations to review their defense policies, priorities, programs, and capabilities, given declining military budgets.

In conclusion, this general survey of logistics offers some reminders of the past that would seem relevant to the future. They may help in the review and revision of military support functions so that logistics remains appropriately geared to possible future developments.

SOME CONCLUDING REMINDERS

Huston (1966) makes this claim in his masterful work on logistics, *The Sinews of War:* "Logic would suggest—and military planners would prefer to believe—that logistic plans stem from strategic plans" (p. 424). He then proceeds to remind us of the succession of instances during World War II when high-level strategic decisions were based on logistical limitations, more than on any other consideration—hence the need to assess all factors and blend all considerations when making operational plans and command decisions, and when making all preparations and meeting all requirements for support.

Van Creveld (1977) reminds us in his more recent, major work, *Supplying War:* "The aim of military organization is not to make do with the smallest number of supporting troops, but to produce the greatest possible fighting power" (p. 225). Hence the need to replace a fanciful preoccupation with "teeth-to-tail" ratios—of minimal descriptive and functional value—with force structures that create and maintain a mix, balance, and ratio of combat and support elements in meaningful, effective harmony.

Logistical lessons from past campaigns should be objectively recorded and descriptively presented; and, if they remain relevant to the future, they should be carefully taught and assiduously learned. Only by satisfying all these requirements will we be sure that evaluated lessons from the past are employed appropriately in the future, and are not neglected or overlooked.

We are reminded to avoid the mistake of automatically attributing the logistic performance of combat and support forces during peacetime training to the way they may acquit themselves in a real war. Only if major peacetime exercises realistically test their logistic performance may we draw valid conclusions about operational readiness and combat effectiveness.

Military commanders and senior operational planners—who may lack all-around operational experience, through no fault of their own—need to receive sound logistical advice. Only by obtaining the best advice are they able to make informed military judgments and effective decisions. Advice on logistics also has to be relayed to, and understood by, involved politicians in many situations of military emergency or war.

We must be wary of lax, complacent, and unimaginative logisticians who may well offer "broad-brush" generalities as solutions or sanction waste and other inefficiencies in order to keep an uncontentious, low profile during their sometimes undistinguished and long professional careers. Only if logisticians are prepared to be demonstrably single-minded and to "rock the boat" when conditions warrant, will deficiencies be corrected and improvements made.

Arbitrary cuts in peacetime may well expose yawning gaps in logistic support in an emergency or war. It will invariably be too late and too difficult to rectify deficiencies when the chips are down.

Finally, we are reminded that the continuity of change is a constant phenomenon of military logistics, as of any worthy enterprise, and this serves to emphasize the importance of planning and managing change now and in the future.

J. H. SKINNER

SEE ALSO: Administration, Military; Assistance, Military; Assistance, Mutual; Budget and Finance, Military; Canteen and Shopping Services; Casualties: Evacuation and Treatment; Combat Service Support; Consumption Rates, Battlefield; Cross-servicing; Engineering, Constructional; Food and Catering Services, Military; Host Nation Support; Infrastructure, Military; Labor, Military; Lines of Communication; Logistics, Limited War; Logistics, NATO; Logistics, Soviet and Warsaw Pact; Maintenance; Maritime Logistics; Mobilization; Munitions and Explosives: Engineering and Handling; Procurement, Military; Quartering, Military; Reinforcements; Replacements: Personnel and Materiel; Resource Management; Stockpiles; Supply; Sustainability and Viability; Transportation.

Bibliography

Alexander, G. M. 1986. Military logistics. *Journal of Defense and Diplomacy* 4(6).
Beaumont, R. A. 1985. Beyond teeth and tail: The need for new logistical analogies. *Military Review* 1985 (March).
Brown, K. N. 1987. *Strategics: The logistics-strategy link.* Washington, D.C.: National Defense Univ. Press.
Dupuy, T. N. 1977. *A genius for war: The German army and the general staff: 1807–1945.* Englewood Cliffs, N.J.: Prentice Hall.
Eccles, W. E. 1959. *Logistics in the national defense.* Harrisburg, Pa.: Stackpole.
———. 1965. *Military concepts and philosophy.* New Brunswick, N. J.: Rutgers Univ. Press.
Freedman, L. 1986. Logistics and mobility in modern warfare. *Armed Forces* 1986 (February).
Huston, J. A. 1966. *The sinews of war: Army logistics 1775–1953.* Army Historical Series. Washington D.C.: Office of the Chief of Military History, United States Army.
International Institute for Strategic Studies (IISS). 1969. *Military logistic systems in NATO: The goal of integration—Part 1: Economic aspects.* London: IISS.
———. 1970. *Military logistic systems in NATO: The goal of integration—Part 2: Military aspects.* London: IISS.
Kelley, R. C. 1977. Applying logistic principles. *Military Review* 1977 (September).
Kennon, J. E. C. 1983. Logistics and the Royal Navy. *Journal of the Australian Naval Institute* 1983 (February).
Meixner, D. 1895. *Historischer Rückblick auf die Verpflegung der Armeen in Felde.* 2 volumes. Vienna.

North Atlantic Treaty Organization. 1989. *NATO logistic handbook.* Brussels: NATO.
Peilow, B. F. 1987. Should logistics replace administration as a principle of war? *Naval Review* 1987 (July).
Reed, J. 1988. A look at current and future logistic systems. *Armada International* 1985 (July).
Thorpe, G. C. 1986. *Pure logistics: The science of war preparation.* Washington, D.C.: National Defense Univ. Press.
van Creveld, M. 1977. *Supplying war: Logistics from Wallenstein to Patton.* Cambridge: Cambridge Univ. Press.

LOGISTICS, AIR FORCE

Logistics is a twentieth-century word that is often carelessly defined. The precise definition, in a military context, includes all aspects of combat from acquisition to personnel, including hospital and cemetery services. To the air forces of the world, logistics has been particularly troubling. Air forces use thousands of gallons of fuel; large, very heavy engines, which last only a few thousand hours; and literally hundreds of support personnel along the various flight paths. In fact, during the past ten years the Falklands/Malvinas War, the Afghan Civil War, and the 1991 Gulf War have confirmed old logistics problems as well as opened new ones. References in this article must, by necessity, use the United States Air Force as providing the most instructive air logistics model. Not only is it the world's largest active air force, its worldwide routes and support systems confirm it as a good choice.

In 1985, lacking a viable and current statement of logistics doctrine, the U.S. Air Force established a committee, which two years later produced the first statement of logistics doctrine in twenty years. To accommodate the engineering support community and the war fighting emphasis of the time, the document was called *Air Force Manual (AFM) 1-10, Combat Support Doctrine.* It provides a combat support perspective that states, "Combat support exists to meet combat operational needs" and that it is "the art and science of creating and sustaining combat capability." The second chapter deals with the process of the support life cycle: resources are added to a mix of combat needs, to result in combat capability. The life cycle consists of several stages—definition, acquisition, maturation, distribution, integration, preservation, restoration, and disposal. The concluding chapter describes the eight principles of combat support, which are "objective, leadership, effectiveness, trauma/friction, balance, control, flexibility and synchronization." These principles provide direction, explain influences, and focus on perfecting combat support.

In the summer of 1991, the U.S. Air Force began the consolidation of two commands—logistics and acquisition—into a new materiel command. In the third quarter of 1992, the process was to be completed and the Air Force Materiel Command established.

Several terms must be identified and defined at this

juncture; the first two are *wholesale* and *retail*. Base-level work is usually termed retail, while all else is almost always wholesale. Wholesale supply and maintenance occur at the depot or contractor level or at some place beyond the base operating level. The Defense Logistics Agency now stocks millions of air force consumable items. The General Service Agency is another important wholesale agency for everything from furniture to rental cars. For instance, an F-16 that needs wholesale maintenance would be shipped or flown to Hill Air Force Base (AFB), Utah—the maintenance center that has responsibilities for (among other things) the F-16. Retail logistics covers everything consumed at the lowest unit level or location.

Two other terms used are *push* and *pull* logistics. In World War II, the United States pushed supplies and maintenance into the theater without a demand summary. In the Korean War, many items were pulled into the theater as a result of a generated demand, even though the United States was still basically pushing. Both systems have their advantages and disadvantages. In recent times they have been joined by a new term, *pre-positioned logistics*, which means that items (spares like tires, fuel tanks, and batteries) are stored directly in the theater close to the equipment that the items will support. New computer systems and a developing science of requirements usage makes it possible to predict how much of a resource will be used and what parts will break with what frequency. In the past, units have stored what is easy to store—the classic case being that of the British, who collected 10,000 fuel tanks in Bahrain in 1960. However, once the demand channel opened, it never closed, and today Bahrainis still use fuel tanks as watering troughs and window boxes.

Air Force logistics may be divided into five functions: maintenance, supply, transportation, contracting, and planning. A sixth division, acquisition, is part of combat support under the new Air Force Materiel Command, but is not necessary to the traditional logistics definition as practiced by the U.S. Air Force. A more detailed discussion of these functions follows.

Maintenance

As one of logistics' most important functions, maintenance is also key to any successful combat operation. There can be as many levels of maintenance as the circumstances require. The first level is what practitioners call "on-equipment," and is performed on the aircraft or platform while it is in operation. The second level is, simply enough, "off-equipment"; this is a more interesting proposition. In the last 40 years, off-equipment maintenance has been studied and restudied. Questions asked include: Should this maintenance be centralized or decentralized? Should it be primarily efficient or generally effective?

Decentralized maintenance was promoted by General Creech, who instituted it first at Tactical Air Command (TAC). It was an important step forward in military effectiveness. Work would be performed at the lowest levels, and the process would be personalized—from the man with the wrench to the crew chief to the pilot, work would be done with pride. But it emerged that the cost of testing equipment, specialized tools, and the like was very high. The efficient solution may have been to use an intermediate TAC facility (at Langley AFB, for instance), and to rely for backup upon a depot that is primary for the aircraft being serviced. With three levels of off-equipment maintenance, both efficiency and effectiveness are served.

However, efficiency proponents argue that only two tiers are necessary—base level and depot. They argue that this approach concentrates expertise, and the depot should become extraordinarily efficient. At the same time, an effort to concentrate world-class troubleshooters at base level would prove militarily effective. At this writing, the Strategic Air Command (SAC), which in 1992 was to become part of the Air Combat Command, has tested a hybrid regional system that would balance effectiveness and efficiency. A SAC tanker, for instance, could be fixed at its home base, at a regional maintenance center, or at the assigned depot.

Regardless of the course of action taken in 1992, maintenance costs are a huge part of the readiness budget. As a result, those who acquire weapon systems for the service have been urged by the users to design more reliability and sustainability into the systems. Design costs are high and at some point may exceed lifetime maintenance costs. The military acquirer must balance the design/sustainability equation with the requirements of the operator. The future should prove challenging to military maintenance personnel.

Supply

Another definition of *logistics* often used by airmen is "the creation and sustained support of weapons and forces...." This definition emphasizes the need for rational and timely supply systems.

The wholesale supply system encompasses the former Air Force Logistics Command (now Air Force Materiel Command) at Dayton, Ohio, the Defense Logistics Agency at Cameron Station, Virginia, and the General Services Agency in the District of Columbia. For purposes of this article, the air force supply system will get all our attention. The command is headquartered at Dayton in a complex employing more than 25,000 military and civilian personnel. The complex is further divided into three areas; one that is not logistical deals primarily with bomber and tanker operations, another is dedicated to the acquisition functions, and the third centers on logistics functions.

The logistics function also commands a series of five wholesale logistics centers operating across the nation. These centers are responsible for purchase, storage, warehousing, distribution, technical assistance and maintenance/repair/overhaul of systems. At least 75,000 people are employed at these centers. Each center has specialized weapons systems assignments. For example, the F-15 is at Warner-Robbins, Georgia, and the F-16 is at Hill AFB, Utah. Each center maintains an adequate but not excessive inventory of every part that may be required. If Eaker AFB needs an F-16 avionics box, which they do not stock, they would request it from the Ogden Air Logistics Center at Hill AFB. Repair performed at Hill could take at least ten times longer than home base repair. Thus, most logisticians prefer the replacement mode, rather than remove and repair.

Most real supply is retail and most serving personnel are very familiar with their base supply organization. Base supply is chartered to provide supplies, equipment, and petroleum products to the base. They are "pull" oriented. Users requisition supplies from them, and they in turn requisition supplies from the wholesale suppliers. They must maintain intricate software systems to account for the myriad of items listed under the base supply account. Because of the need for both accountability and service, the standard base supply organization has a chief of supply and five branch chiefs responsible to him. The first branch chief is in charge of management and systems, which includes procedures, funds, documents, training, and inventory systems. The next chief controls operations support, which encompasses mission support, demand processing and repair cycle support. The chief of the third section, materiel management, oversees stock control, equipment management, retail sales, mobility, and munitions management. The fourth chief manages storage and distribution, which includes receiving, pickup, and delivery, as well as storage and issue and bench stock support. The last chief commands fuels, which includes quality control and inspection, fuel operations, accounting and administering of fuels, mobility, training for fuels management, and a cold storage section that deals with cryogenics problems.

During the 1991 Gulf War, attention was focused on the operations support branch. When a unit is assigned to a war zone or crisis, this branch manages the war readiness spares kit (WRSK) and the base-level spares support (BLSS), which are essential to providing the unit the supplies expected to be used in a wartime deployment. In addition, the unit receives constant software upgrades through the management and systems branch, which improves the ability of supply personnel to forecast requirements. The thoroughly modern supply officer is usually armed with a bar code reader, a laptop computer, and a host of software programs.

Transportation

Both wholesale and retail transportation systems exist, but in the restructured air force, wholesale transportation is in another command: the Air Mobility Command. Traditional wholesale transportation consists of airlift and sealift.

Retail transportation, however, is very much in the logistics or combat support domain. The base-level chief of transportation, often a civilian, has three sections: traffic management, vehicle operations, and vehicle maintenance. All together these sections coordinate movements of people, property, and freight throughout their areas of responsibility.

The traffic management office (TMO) handles a host of arrangements. They provide moving vans to move personnel from one base to another, they write authorizations for commercial airline, railroad, or bus tickets to move personnel from temporary or permanent duties elsewhere, and they work shipments of military cargo throughout the system.

Vehicle operations or vehicle management is the primary organization providing four-wheel support of the base. From cars to trucks and to school buses, this section operates and schedules vehicle support, evaluates drivers, and trains replacements.

The goal of vehicle maintenance is a base filled with safe and serviceable vehicles to answer the military mission. This section must ensure adequate inspection of vehicles, workable schedules of service and maintenance, and an adequate supply of parts for broken vehicles. Because bases have fork lifts and fire engines as well as a host of more orthodox vehicles, separate work centers may be set up for the various pieces of equipment. In successful vehicle maintenance, much coordination is required among the work centers and base-wide interdependence is usually the standard. Finally, the planning office within transportation is usually in the chief of transportation's purview. Arrangements are made in that office to meet the mobility needs of a military deployment.

Contracting

Perhaps the most difficult logistics function for personnel to understand is that of contracting. Again, there are wholesale and retail realms in contracting. Wholesale contracting was managed by Air Force Systems Command, which has become one of two subsections of Air Force Materiel Command. Retail contracting is the main focal point in this article.

This function contracts through civilian channels to provide goods and services not available through normal channels. The chief of contracting or the base contracts chief monitors the contracting cycle. The cycle originates in a demand, progresses to a purchase order, results in (usually) a fixed-price contract award, and ends up in a delivery. The user comes in with a solid description of the item needed, and the contracting office must determine the character of this need. Is it one-time only? Have these items been bought before? Are there many sources? Are there affirmative action or minority contracts involved?

Within contracting are usually four branches. The first, and usually the smallest, systems management, governs computer activity and routine purchases of prepriced and standardized items. The supply section, or supplies contracting section, contracts for all "nonissue"-type supplies. They will usually state a need, ask for open bids, and accept an offer. That offer, without negotiation, is the basis for a firm fixed-price contract. Negotiation comes into play when there are only a few bidders, a difficulty with schedules, a local shortage, a strike or work stoppage, or confusion over the supplies required. In such a case, a cost reimbursement or cost-plus-profit contract may be awarded.

The services section, or services contracting branch, buys electricity, construction, special services, or other utilities. Buyers set up these contracts on an annual or semiannual basis, often basing them on letter contracts followed up by a definitive arrangement.

Finally, the contract administration branch maintains a file on all awarded contracts. Often they must follow up on late delivery claims, terminations, and other actions that are necessary to keep the base functioning in a timely fashion. Every base organization will interact with the base contracting function.

Planning

Because coordination of logistics areas is so essential to successful military operations, a base-level logistics planning office is necessary. Planners must coordinate emergency actions and ensure that they can and will take place. Logistics planners work continuously to remain an integral part of campaign tactics and war strategies. They consider units deployed, support required, mobility, and logistics flexibility. Evidence of their work is depicted in the logistic annex to the war plans.

As with the other logistics functions, the planning operation functions at both wholesale and retail levels. At the wholesale level, the director of Logistics Plans and Programs on the Air Staff inputs logistics requirements directly into the Joint Operations Planning Systems (JOPS). The basis of that input largely rests with the Air Force War and Mobilization Plan (WMP). Annexes to these plans include unit type codes (UTC). The UTC tracks the passage of units to the retail or base-level part of logistics planning. The UTC also appears in the Time Phased Force Deployment Data (TPFDD). The TPFDD comprises a listing of units (TPFDL) and a transportation requirement (TPTRL). These are the most essential documents to the logistics planner, who now must calculate all deployment and employment logistics support requirements. At the retail level, the logistics planner works through AF Forms 2511 and 2512, and must ensure that planning from the wholesale down through the base or retail level is adequately coordinated, so that the mission can be accomplished. Every logistics function is related to the actions of the logistics planner.

Acquisition

Logistics and acquisition are coordinated and related functions under the Air Force Materiel Command. Until recently, these were two separate commands, but they have now been rejoined as they were in the early postwar period. As previously mentioned, combat support doctrine describes acquisition as the first step in the process of combat support.

Basically, acquisition requests resources and converts them into war fighting systems. The process is detailed and precise, with more checkpoints than any other system in the government. At the wholesale level, the national command authority determines a useful strategy for the nation. The secretary of defense guides the services through a defense planning guide which provides them information upon which to build a program objective memorandum (POM). These POMs are negotiated into program decision memoranda (PDM) by the services and in turn are set down as budget estimate submissions (BES). After more negotiations, a budget is submitted and eventually approved by Congress. The acquirers direct this budget toward programs in a host of different stages in the acquisition process—from B-2s in the production stage to the stealth transporter in the concept exploration stage.

The stages of the acquisition process are designated by milestones. They are as follows:

Mission need/statement	Milestone 0
Concept exploration	Milestone 1
Demonstration/validation	Milestone 2
Full-scale development	Milestone 3
Production/deployment	Milestone 4
Operational readiness and support	Milestone 5
System operational effectiveness review	

For a new system to pass from Milestone 0 to Milestone 5 is a process of eight to twenty years. There are numerous checks and balances throughout with most interest groups represented.

There is now an undersecretary in the Defense Department responsible for weapons system acquisition: the USDA. There is, in each service, a service acquisition executive (SAE). A program executive officer (PEO) is responsible to the SAE; all program managers (PM) are responsible to the PEO, who is usually a two-star general. The authority for all acquisition programs lies in this chain of command.

Members of this chain meet regularly as programs pass from milestone to milestone. A Joint Requirement Oversight Council (JROC) meets before Milestone 0. The Defense Acquisition Board (DAB) approves a program through Milestone 0, and after another service report called the Integrated Program Summary, pushes the program to Milestone 1. At this point, the DAB has produced a documented report called the Acquisition Decision Memorandum (ADM). The process gets repeated through the various milestones. The system is layered, it is bureaucratic, and it is redundant, but because weapons systems acquisition must be driven by strategic plans based on combat capability, the acquisition process may be the most important function within the combat support community.

Under the reorganization scheme of 1992, cradle-to-grave combat support will be under one overall command: Air Force Materiel Command (AFMC). Centralization beyond what is organizationally necessary, however, will be avoided because the total quality management (TQM) concept has made deep inroads into the combat support community, which is opposed to further centralization. Air force leadership has announced strong support for TQM.

New Directions in Logistics Management

LOGISTICS MANAGEMENT CENTER

Primarily to avoid overreliance on outside consultants and others to solve major logistics problems, the Air Staff almost twenty years ago established the Logistics Management Center (LMC) at Gunter AFB, Alabama. Workers were tasked at the time to help coordinate wholesale/retail interactions and solve base-level logistics problems.

This small organization is made up of approximately 100 mostly junior analysts, operations researchers, and pure (as opposed to applied) researchers. Their success stories are legion and have received universal application. They led the air force into bar coding, computer modeling of spare parts usage, and laptop load planning of deploying aircraft.

The LMC reports directly to the deputy chief of staff for logistics on the Air Staff; previously, it operated under Air Force Logistics Command.

ENGINEERING

For the past ten years, the deputy chief of staff for logistics and engineering was also responsible for engineering and services. He was assigned a directorate whose job was to monitor base-level civil engineering. Under the 1992 air force reorganization, civil engineering and services reverts to independent status.

AIR FORCE JOURNAL OF LOGISTICS

In considering the strategy-tactics-logistics model, the air force finally recognized that logistics is a legitimate subset of the military profession. Logistics had its own body of knowledge, training programs, and certification procedures; therefore, it should have its own professional journal, and in 1979 the *Air Force Journal of Logistics* was unveiled. The *Journal* reinforces the professional community of logistics. It appears quarterly through the Government Printing Office as AFRP 400-1.

Conclusion

Logistics to the modern air force is a most expensive proposition. Because it consumes literally billions of dollars it is the most regulated and watched subset of the military profession. A change from the current model in the United States is unlikely, however, because the United States must plan for a host of wartime and peacetime logistics contingencies. From a typhoon in Bangladesh to an attack on Baghdad, the logistics system must respond in a flexible and timely manner to a host of national demands.

TED KLUZ

SEE ALSO: Airpower; Combat Service Support; Combat Support; Logistics: A General Survey; Logistics, NATO; Logistics, Soviet and Warsaw Pact; Organization, Air Force; Technology Acquisition and Development; Technology and the Military.

Bibliography

Brodie, B., and F. M. Brodie. 1973. *From crossbow to H-bomb.* Bloomington: Indiana Univ. Press.
Eccles, H. E. 1965. *Military concepts and philosophy.* New Brunswick, N.J.: Rutgers Univ. Press.
Holley, I. B., Jr. 1983. *Ideas and weapons.* Hamden, Conn.: Archon Books.
Thorpe, G. C. 1986. *Pure logistics.* Washington, D.C.: National Defense Univ. Press.
Van Creveld, M. 1979. *Supplying war.* London: Cambridge Univ. Press.

LOGISTICS, LIMITED WAR

Limited war is armed conflict short of general war, exclusive of incidents, involving two or more nations. General war is armed conflict between major powers, in which the total resources of the belligerents are employed and the survival of at least one belligerent as an independent nation is in jeopardy.

Limited wars include guerrilla war—military and paramilitary operations conducted in enemy-held or hostile territory by irregular, predominantly indigenous forces—and unconventional war—a broad spectrum of military and paramilitary operations conducted in enemy-held, enemy-controlled, or politically sensitive territory.

Guerrilla war and unconventional war are often considered synonymous, but, to the professional, unconventional war connotes a less well-defined enterprise. It includes evasion, escape, subversion, sabotage, and other operations of a low-visibility, covert, or clandestine nature. These interrelated aspects of unconventional warfare may be prosecuted singly or collectively by predominantly indigenous personnel, usually supported and directed in varying degrees by external sources.

Limited wars are now precisely classified by professionals in "the spectrum of conflict" (see Fig. 1) as "low inten-

Logistics, Limited War

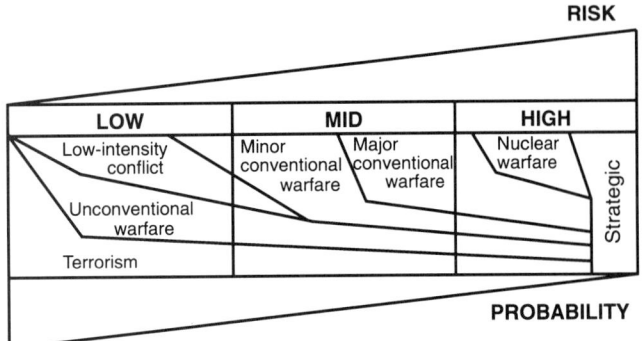

Figure 1. The spectrum of conflict.

sity," which includes unconventional war and terrorism; and "mid-intensity," which is subdivided into minor and major conventional war.

Each of the wars in the gradations of the spectrum of conflict has its own type of supply line, and whether that war thrives or not depends on what supplies are provided. The supply line is vital to military operations and is that part of the military called logistics—the science of planning and carrying out the movement, supply, and maintenance of military forces. In a low-intensity conflict, logistics may provide terrorists with no more than a hand grenade and a plane ticket. On the other hand, as an example of a major conventional conflict, America provided 300,000 individual line items of supply in Vietnam, from Bailey bridges (interchangeable latticed-steel elements rapidly pinned together for crossing rivers) to bridgework that dentists use to repair and reconstitute a damaged jaw. Logistics is a "womb-to-tomb" endeavor, but also includes a recycling effort. Modern battlefields are often policed and cleaned up. For example, during four years of the Vietnam War, the U.S. Army cleaned up almost 848,000 tons of scrap and extracted almost US$400 millions' worth of reusable property.

The reader should understand that there have been innumerable limited wars throughout history, and that the logistical needs and requirements of these wars have been diverse. The aims or purposes of the war, the geography and climate, the technology of the adversaries, all combined to make each situation unique. Therefore, a useful way to understand the logistics of limited war is to examine some classic limited wars.

The American Revolution (1775–81)

George Washington, unlike many revolutionary heroes, was not a guerrilla. He was British-trained, and, as a general, he was a strict adherent to the British way of war.

The American Revolution started for logistic reasons. General Gage, in command of British forces, directed his troops to destroy colonial militia stores at Concord, Massachusetts. As they did so, rebel riflemen confronted them at Concord Bridge and war began.

As a forerunner of a logistics system for the War of Independence, colonial militiamen supplied themselves with the basics: horses, guns, and gunpowder. Soldiers who could not afford their own supplies received weapons from their hometowns with the understanding that they would pay for them later. It was not unusual for the hometown to be a logistic base. As Ralph Waldo Emerson noted, Concord, a town of 1,300 people, "furnished 67 men, paying them, itself . . . with every levy to the end of the war. For these men, it was continually providing shoes, stockings, shorts, coats, blankets, and beef" (R. W. Emerson, *The Complete Writings of Ralph Waldo Emerson* [New York: William H. Wise, 1921]).

At the start of the war, each of the thirteen colonies had its own logistic system. The thirteen systems slowly and painfully evolved into one, which sought unified standards to become more efficient than the individual colonial variations.

Overseas supplies came via ship from France, but the British fleet captured two out of every three ships. On the other hand, American privateers, much to the consternation of the Royal Navy, regarded British merchantmen as wartime inventory.

Fortunately for the rebels, gunpowder and cannon had been produced in the Colonies for well over a century before the start of the American Revolution, and small-arms factories were abundant. Washington's vital shortages included gunpowder, lead, and cartridge paper. Also in short supply were salt, as a food preservative, and shoes.

The standard weekly issue for every hundred men was three pounds of candles, more than twice that weight of soap, and six times the weight of candles in salt. More often than not, the shortage experienced by the first American army was the one that has persisted and is endemic in all armies: not supplies, but transportation. The refrain about the war being lost for want of a horse would most aptly apply to the pack horse.

With the British in control of the seas and the navigable rivers of America, Washington was compelled to rely for lines of communication on waterways too shallow for the British warships. Logistics in the colonies was hampered for both sides by the lack of roads (there were mostly trails). Thus, although the colonists relied on shallow rivers for lines of communications, they had to risk storing some of the American army's supplies in stockades located along the banks of the deep, navigable rivers (Hudson, Delaware, Susquehanna). The combined effect of those American-built forts, designed to hamper British logistic support of its inland forces, and underwater rails designed to rip the hulls of British ships, further disrupted Crown strategy. Lesser rivers—the Catawba, the Dan, and the Pee Dee—proved, with their shallowness, rocks, rapids, and falls, to be obstacles to the British ships, but not to colonial boats. These rivers and colonial amphibious engineers were a main logistic strength of Washington's forces. Although Washington's Christmas adventure across the Delaware to capture surprised Hessians at Trenton has become legendary, more significant to his mission was the earlier rescue of his army by his amphibious engineers, who transported 9,500 of his troops across New York's East River in a safe retreat from Long Island.

The days-of-supply criterion of the U.S. Army today was the basic logistic factor for Washington. At York, Pennsylvania, west of the Susquehanna River, Washington ordered four months' supplies for 10,000 troops; at Lancaster, two months'. Dispersed sites had fewer days-of-supply for fewer men, which accommodated the flexible mobility that fitted Washington's maneuvers. Supply discipline in America's first war appears to have been far stricter than in some of its later ones, thanks to Baron Friedrich Wilhelm von Steuben, Washington's Prussian-born inspector general. Each soldier had an account book, and failure to account for supplies resulted in a deduction from his pay.

Other logistic features, characteristic of all of America's wars, began with Washington. These include incentive payment to contractors and interest-free loans on capital investments; price controls; and exemption from service for logistic work that supports the field army.

Unlike the U.S. Army today, which feeds its troops in combat with a steady flow of rations, Washington's soldiers were on four-day ration cycles, which were suspended during battle. As a result, if rations started on the day a battle began, as they did at the Battle of Bemis Height, the quick pursuit of a defeated enemy, in that case General Burgoyne, was not possible because the troops were tired and hungry. After replenishment, however, the Americans were able to overtake Burgoyne and force his surrender at Saratoga.

The British had supply problems, too. The surrender of Gen. Charles Cornwallis at Yorktown was due essentially to the presence of the French navy, which neutralized the ability of the British, logistically and tactically, to support its forces. The well-timed allied sea-land cooperation in the days before telegraph and radio is a credit to the skill of the French-American alliance in synchronizing a master stroke.

Thus, in sum, the American army's basic logistic factors and terminology began with General Washington. Days-of-supply, ton-mile costs of transport, and pounds-of-supply became a lasting part of America's logistic primer. Depot sites were based at transportation nodes. For Washington, these nodes were located at waterway junctions, usually in places too shallow for British ships to navigate.

The logistics of the American Revolution were based on water. The British controlled the sea; time and space utility lay with the British fleet. While it took a month for a courier to carry a message from Washington's Northern army to his Southern army, it took but a week with a favorable wind for the British fleet to deliver troops and supplies in quantity over the same distance. It was only when the French navy offset the British advantage at sea that the war went Washington's way. Thus did the waging of strategic warfare at sea against sea lines of communication (SLOCs) alter the outcome of a limited war on land. The same strategy is applicable to general war and is a feature of quarantine as a war preventive. Pres. John F. Kennedy dramatically and successfully invoked such a measure during the Cuban Missile Crisis with the Soviet Union (1962).

U.S.-Mexican War (1846–48)

The United States' limited war with Mexico was won as much by logistic acumen and the technology that flowed from it as by its application. Generals Zachary Taylor and Winfield Scott maneuvered brilliantly, avoiding battle and thus avoiding attrition of their army in a foreign country, and fighting skillfully with confidence.

With army-navy cooperation, the Americans stifled Mexican logistics by closing Mexican ports on the Pacific coast and in the Gulf of Mexico, and by driving overland from Vera Cruz into the heart and capital of Mexico.

The political results of this limited war had a great impact. It meant the accession of territory in Texas, brought on the cession of New Mexico and California, led to the Gadsden Purchase, and achieved the Atlantic-to-Pacific Manifest Destiny of the United States.

Beginning of Modern Logistics: Crimean War (1853–56)

Friedrich Engels, after observing the Russian logistic influences that prevailed in the Crimean War, said that the Crimean War was a hopeless struggle between a nation with primitive techniques of production and nations that were up to date. He also understood the need for the allies to end the war on the peninsula, because the sheer size of Russia would make a further invasion, as both Napoleon I and Hitler discovered, a giant logistic trap.

Britain had logistic troubles from the start—a lack of troop ships, cavalry horses, and ships to carry horses. Setting up an attenuated 4,850-kilometer (3,000-mi.) logistic system for the guns and ammunition provided for embarrassment when they arrived at the objective area and there were not enough pack animals to carry them. The logistic chagrin of this limited war for the British, French, Turks, and Sardinians was matched in one respect a century later by that of American, Australian, Korean, Thai, and Philippine allies in their limited war in Vietnam: excessive delay in the unloading of ships.

The private, charitably funded medical support and hospitalization of British soldiers was a disgrace for which the

Turkish town of Scutari became a synonym. Florence Nightingale became a symbol of the healing arts, bringing hygiene, the Victorian privacy of amputation screens, and all that human kindness could offer to the Tommies, then known as "the wretches."

The French had far superior logistics. Efficient transport corps and a medical service staffed by nuns were exemplary. The British, however, also had unacclaimed feminine support; about six wives per company were permitted with the troops, and they provided support in the form of laundry, cooking, and practical nursing.

The managerial support for logistics in London was inept. For example, the British Commissariat fed troops abroad; Parliament provided the Master of Ordnance, who fed troops at home. The Transport Corps and the Staff Corps, which had performed admirably during the Duke of Wellington's campaigns in the early 1800s, had been eliminated by the British government, which was wary of a military strong enough to attempt a coup. The duke's transport organization ("wagon train") had served in both directions: troops and materiel had been brought to the battle, dead and wounded had been brought out, withdrawal had been achieved, and inventories had been balanced and shifted. The Staff Corps, which would have planned and supervised logistic mobility, was nonexistent for the Crimean War.

With the first use of a railroad in war, the traction motor was applied on rails to roll the big siege guns, which devastated Sevastopol, into place. Steam-propelled trains presaged the Austro-Prussian and Franco-Prussian limited wars and three great general wars: America's Civil War and World Wars I and II. It was the dominating logistic influence of railroads on both strategy and tactics that vastly extended the reach of Pres. Abraham Lincoln's army in the American Civil War (1861–65), and that united Germany, allowing it to increase its military strength.

The Crimean War, politically designed to prevent Russia, the largest country in the world, from gaining territory, was notable for the logistic reforms on both sides. England and Russia, particularly, modernized their mediocre military ability to produce, deliver, and sustain in distant places food and clothing supplies, medical treatment, and hospitalization. Weaponry, too, was reevaluated and updated.

For the British, the Crimean War meant army reformation. The War Office was established, supply and commissariat authority was excised from the Treasury, a central clothing factory (instead of the colonels in command) was assigned responsibility for purchasing uniforms, and Enfield became the unified producer of small arms. In addition, because of the presence of Florence Nightingale, as well as artists, photographers, and reporters with access to the telegraph, Scutari was exposed as a slaughterhouse, giving rise to the Medical Service Corps and making the world aware of the need for medical service reform.

Although the Crimean War was a limited war, its effects were significant. As a result of the Crimean War, Russia was kept out of the Middle East, the Black Sea was declared neutral, the Danube was recognized as an international waterway, and Serbia became independent.

Modern war and modern logistics began due to the public exposure created by the advent of the war correspondent and the use of the telegraph in the Crimean War. Public scrutiny and outrage forced the military to improve its organization and performance, especially its logistics. As much as the military today criticizes freedom of the press on the battlefield in the name of security and surprise, it must be remembered that it was the media that brought about a wide array of necessary reforms.

Austria-Prussian War of 1864 and the Franco-Prussian War of 1870–71

Germany's superior logistic infrastructure of interior lines was established by Field Marshal Helmuth von Moltke, Chief of the Prussian and, later, the German, General Staff, who declared that he wanted "no more fortresses built, only railroads." These railroad lines, later augmented by the autobahn, led to the German army's prodigious ability to concentrate and shift its striking power. In 1866, Prussia's short Seven Weeks' (limited) War with Austria presaged a cataclysmic future.

Moltke showed his and the German General Staff's logistic talent in the Austrian war. He used five railroads to mobilize and concentrate his troops; Austria was confined to one. Germany thus formed its strategy for the next war with France, and later the Schlieffen Plan, which Moltke's nephew "Moltke the Younger" purportedly bungled.

Following the success of the Seven Weeks' War against Austria in 1866 with equally brilliant use of railroads against the French (who outnumbered the Prussians in 1870), Moltke the Elder took advantage of his intelligence sources about French railroads. While the German railroads provided for swift Prussian troop concentration, the French railroad system forced Napoleon III's forces to be split, and slowed French mobilization efforts. After another seven weeks of ground war and his march into Paris, Moltke took twenty more weeks to overcome French control of the sea, which had kept the French logistic door open.

Although Moltke left tactics to his generals, he did participate in the planning of strategy for the Prussian army. He also retained control of logistics, even when it led to quarrels with the Iron Chancellor, Prince Otto von Bismarck. Moltke argued that the railroads should be used to carry supplies to the front, but Bismarck decided instead to use them to move siege trains, with their big guns, into a position to shell Paris.

Indian Wars and the Custer Massacre (1876–77)

The "logistics of nature" on the Great Plains of the American West contributed to the American Indians' most renowned battlefield success, the massacre of Custer's forces at the Little Big Horn.

Compelled by their horses' need for grass and water and their own need for nutriment from the buffalo, the Plains Indians were forced to disperse and were less vulnerable to the concentrated firepower and conventional Civil War attack pattern of the American army.

The buffalo was a logistic bonanza; the herds were a vast mobile logistic base with ready resources. Buffalo meat provided food, the skin and hide provided tepees, clothes, moccasins, swaddling dress, and winding sheets. Buffalo hair provided rope, bridles, and fly swatters, and the stomach was used as a water bladder. Bones were used to make needles, and buffalo chips as fuel for cooking and heating.

Nature had provided the Plains Indian with a logistic advantage. The ample supply of buffalo, elk, deer, berries, and fish enabled the Indians to travel light, the ultimate guerrilla of very limited war.

Features of war that trouble commanders in the field change little. The cost-benefit equation and the prospect of an empty gun was present in the Plains Indian Wars as it was a century later in Vietnam and Afghanistan. A cavalry captain, campaigning with Gen. George Crook before Crook defeated Geronimo and the Apaches, described the concern for cost and ammunition wastage that prevailed in Custer's day as it does today. As the South Vietnamese and the Nicaraguan Contras became painfully aware, limited wars are limited by their cost and their budget, and ammunition is invariably the limiting factor.

During the Sioux-Cheyenne War against the U.S. Army, General Custer's Seventh Cavalry had a pack train in its rear echelon. With protective troops, it required the attention of 20 percent of Custer's command (130 soldiers), a factor that contributed to Custer's disastrously split force being destroyed by the tribes.

The Indians also had the advantage in versatility in weaponry. Custer's men had 100 rounds per rifle and 24 rounds per revolver, but no alternative weapons. There is evidence that Custer's troops and the Indians under Crazy Horse ran out of ammunition at the same time. The Indians, however, then used their alternative weapons—bows and arrows, knives, and tomahawks—to massacre Custer's troops.

Boer War (1899–1902)

Two limited wars at the turn of the last century, one at the southern tip of Africa (Boer War), the other at the eastern edge of Asia (Russo-Japanese War), presaged the general war called World War I, or the Great War.

The Boer War brought to the fore the fatal effect of the optional air burst, point detonation of artillery on masses of men by men who pulled lanyards and were miles away from the carnage of falling shells. With smokeless powder and the accurate 500-yard range of rifles, men advancing together could not see the hidden operator of one of the deadliest hand-held rifles invented: the Krag Jorgensen, progenitor of the Garand, the modern Russian AK-47, and the American M-16.

Wars often have similarities, even when a half-century apart. General the Earl H. H. Kitchener, British commander in the Boer War, laid miles of barbed wire with hundreds of watchtowers, resembling Robert McNamara's electronic fence in America's Vietnam War. Kitchener's 80 columns, who marched back and forth across the vast country "scouring and sweeping," were replicated by Gen. William Westmoreland's "search and destroy" missions, and the "body count" was used by both as a measure of success.

The bloody, not-so-limited, second *fin de siècle* Russo-Japanese War (see next section) dramatically exhibited the machine gun as an effective producer of mass death. The logistic systems in these limited wars had to supply huge amounts of ammunition and nations quickly developed mass production lines for explosives. The use of the machine gun for defensive purposes, however, was not recognized during this war.

Limited wars are tactical and logistical laboratories for general wars. The Boer War brought on British hegemony and the Union of South Africa. The Russo-Japanese War provided Japan with territory on the mainland of Asia, curtailed Russian expansion in the Far East, and triggered the end of czardom and its replacement by Communism.

Technology, Tactics, and Logistics

RUSSO-JAPANESE WARS OF 1904–1905 AND 1935

Russian logistics in the Russo-Japanese War was entirely dependent on the Trans-Siberian Railroad, a line that necessitated crossing Lake Baikal, the deepest freshwater body in the world. Trains were ferried across the lake in the short Siberian summer, and tracks were laid on the ice in the winter. Despite trains that broke through the ice, the railroad upheld Russian logistics. The czar's forces extended a railway branch to each of the three armies and averaged eight trains a day along each branch.

The causes of the Russo-Japanese War were Russia's eastward expansion, which aimed at absorbing Korea, and the rise of Japanese imperialism. The Japanese realized that they would be no match for the Russians once the gap in the Trans-Siberian Railway at Lake Baikal had been bridged, so they decided to strike while geostrategy favored them. As long as the railroad gap existed, reinforcement of the Russian troops in Manchuria would be slow. The Japanese, racing against the Russian effort to close the Lake Baikal logistic gap, paid a high price. The Japanese took much higher casualties—in some battles, a 6:1 disadvantage. Thus, again, geostrategic logistical lines of communication (LOCs) played a key role in a limited war.

For 40 years after the Treaty of Portsmouth (15 September 1905), Russia was required to grant Japan's "paramount interest" in Korea, to leave Manchuria to Chinese (and later Japanese) hegemony, and to cede railroads and the southern half of Sakhalin Island. Their long-sought warm water port, Port Arthur, lost in battle, was lost again in Portsmouth. Of the two Latin axioms of international law, *status quo ante bellum* (the way it was before the war) is seldom applied to the loser at a peace table, while *uti possidetis* (belligerent parties keep what they acquire) holds firm.

In 1935, Russia and Japan were again at war but by this time, the Trans-Siberian railroad bypassed Lake Baikal to the south, thus removing the lake as an obstacle, and Communist Russia reversed the czarist loss. Moreover, Soviet Russia's cause was in the hands of Georgii Konstantinovich Zhukov. While in 1904 the Japanese had brought the machine gun to the modern battlefield, in 1935 Zhukov gave notice to the world of a new combination of war machines—he combined armor with airpower. The old triumvirate of infantry, cavalry, and artillery was transformed into a quartet of power: infantry and artillery with tactical air and armor on treads.

As a result of their decisive defeat in 1935, the famed Japanese Kwangtung Army in Manchuria was reluctant to attack the Red Army throughout World War II. This hesitancy enabled the Trans-Siberian railroad to serve as a logistic lifeline to Western Russia from America.

History lists the Murmansk ocean run and the Persian Gulf rail corridor as the American supply lines to Russia. What is largely ignored is that the Japanese let American ships, usually carrying aircraft and aircraft parts, enter Vladivostok during the war. This was a combined ocean-rail third logistic source for Russia.

Sino-Japanese War of 1937

A worldwide depression was the background for the Sino-Japanese War of 1937. Deprived of trade with the United States by American isolationists, the Japanese sought to offset their deprivation by military improvisation—the annexation of Manchuria—thus becoming an island parasite of militarily languid China. The unintended national logistic consequences of this limited war resulted in a general war in East Asia and the Pacific Ocean.

This limited war started favorably for the Japanese in the north and along the coast, but within one year it stalled. China's internal strife was temporarily halted as Chiang Kai-shek, Mao Tse-tung, and the warlords united against the common enemy in a limited war of stubborn and steady attrition. Instead of becoming a source of supplies for the Japanese invaders, China forced Japan to use its limited island resources.

After being forced to use much of their own sparse natural resources as a result of the Chinese resistance, the Japanese, for logistic reasons, were forced southward to Indochina and appropriation of the Dutch East Indies, to sign a neutrality pact with the Soviets, and, after the fall of France, to join the Axis powers in 1941.

Thus the limited war with China, which drained Japanese resources, led step by step to the faulty Japanese concept of a one-strike victory at Pearl Harbor. The Japanese were not aware that the Americans could be just as resilient as the Chinese, and that the Americans possessed the resources and inventiveness necessary to recover from Pearl Harbor and move to the offensive.

Three Indochina Wars (1950–75)

Revolution of 1950–54

The first Indochina War (1950–54) was a revolutionary war. What Ho Chi Minh viewed as bringing forth a new country was viewed as something else by Dwight D. Eisenhower, who backed the French and partition.

Ho's general, Vo Nguyen Giap, commander in chief of the People's Army, was undersupplied by the Chinese, and was forced to loot the French to make up for logistic shortages. After successful guerrilla warfare early in the war, Giap ventured into conventional warfare, and subsequently suffered severe losses from General Jean de Lattre de Tassigny in the Red River delta. After these losses, Giap shifted to logistic warfare, cutting the French lines of communication. The French, with superior firepower and mobile teams, could move about at will, but they could not stay where they went because Giap's guerrillas closed the supply lines.

The decisive French defeat at Dien Bien Phu was a strangulation. Attempted aerial delivery, as the perimeter of the fortress shrank, resulted in more supplies being dropped to Giap's forces than to the besieged French. Toward the last day, a trio of twin-tailed "flying boxcars," flying against skillful antiaircraft fire, accidentally dropped almost 20 tons of artillery and mortar ammunition, which was intended for and sorely needed by the French, to the Viet Minh.

The monsoon season, antiaircraft weaponry, and tunneling as an assault technique contributed mightily, along with Giap's wiliness, to the humiliating French defeat.

Second Indochina War (1961–72)

The second Indochina War, unlike the first, was both a revolutionary war (in that it was fought between countrymen) and a conventional, international, large-scale war (in that it was fought between adversaries with major-power allies).

When the second Indochina War started in late 1961, there were 900 U.S. military personnel present. That number peaked at 542,000 at the end of 1969, then declined to zero when the Military Assistance Command,

Vietnam (MACV) was inactivated in 1973. Whether viewed as a success or failure, it is certain that this war provided the United States with an innovative logistic laboratory that pioneered improvements.

During the Vietnam War, there was a dramatic increase in the speed of delivery of supplies, equipment, and troops. It featured the first use of container ships with their own cranes, which could off-load 10,000 tons and, in 24 hours, back-load retrograde. Break-bulk vessels, on the other hand, required five or six days for such an operation, not including the waiting lines to get pier space at Cape St. Jacques and along the Saigon River. U.S. Army logisticians had portable deLong piers (the pivot of the container ships) installed at the ports of Cam Ranh Bay and Da Nang.

The U.S. Air Force invested in the largest freight and troop-carrying aircraft in the world, the C-5, and in the smaller C-141. What would otherwise have taken weeks to cross the Pacific by sea—the combat delivery of crucial, spare parts and, if need be, an M-48 tank—could be delivered in hours.

The delivery system that provided unexcelled logistic and troop mobility within the theater of operations was based on the helicopter, on the portable Bailey bridge, and on computer management, which cross-leveled supplies.

Considering the usual havoc of war and human failings, the United States developed a delivery system that had the growth and force of an avalanche and the grace of beneficence. It included hot meals and ice cream supplied to squads in the field. Even more important for the infantry—ARVN (Army of the Republic of Vietnam), U.S., ROK (Republic of Korea), Australian, and Thai forces—was the medical evacuation of the wounded by helicopter (MEDEVAC). These aerial ambulances, with daring pilots, crews, and medical corpsmen, often flew into a hostile landing zone (LZ) to pick up the wounded and, bypassing the triage tent, delivered them to a hospital for surgery in a matter of minutes, well within the "six golden hours" from wound to surgery that military surgeons say gives the best assurance of survival.

The ARVN, having fought with the French, knew the vital statistic that put the odds in their favor when fighting as allies with Americans. The casualty ratio of killed in action to wounded in action (KIA:WIA) in Indochina for the French Union Forces was 75:65; that is, for every thirteen casualties, seven died and six survived. For the U.S. Army, the KIA:WIA ratio was about 1:4; that is, of five casualties, there was one death and four wounded survivors. In large measure, this was due to superb military medical care and a MEDEVAC system that did not discriminate among the casualties.

Communications also took a leap forward for the U.S. Army in Vietnam. It was reported that improvements in communications paralleled or even exceeded improvements in mobility. Brigade commanders were able to see and talk to all their platoon leaders. Division commanders had secure voice radios and instantaneous communications up and down the chain of command to every tactical unit and to other services or allies.

The logistic system of the U.S. Army, however, did have deficiencies. Supply, maintenance, storage, and warehousing skills were inadequate in the theater of operations, and, in the continental United States, the supply of qualified military personnel was not sufficient.

Refrigeration for dry battery storage, armored vehicles for truck convoy security, dropsides for ease of loading and unloading of trucks with forklifts, computerized quick-reaction inventory control units, the use of commercial off-the-shelf supplies and equipment (particularly for engineer construction), the sanitizing and entomological demands for retrograde equipment—all were subjects for correction within the U.S. Army's logistic laboratory, which professionals considered a serendipitous positive outcome of the Vietnam War.

Lt. Gen. Joseph M. Heiser, the U.S. Army's logistical expert, asked tough and thoughtful questions:

> Why multi-fuel engines (with their confounded maintenance troubles)? . . . Did we need communications retransmission capability in tanks (more maintenance headaches)? . . . Did we train sufficient marksmen to justify precision rifles? Would not lightweight, cheaply constructed, mass fire weapons serve better than expensive, maintenance-significant items? . . . Must vehicles be waterproof for their occasional submerging? Would lightweight, plastic flotation bags serve better? (J. Heiser, *Logistics Support* [Washington D.C.: U.S. Dept. of Army, 1972])

The North Vietnamese Army (NVA) and Viet Cong (VC) logistic system was dependent on the sanctuary countries (usually observed as such by the allies) of Laos, Cambodia, and, except for selected aerial targets, North Vietnam itself. It was also dependent on the ability to excavate and tunnel and to live in underground compounds that included warehouses, canteens, hospitals, and barracks; on salvage and scavenger organization; and on innovation, which adapted battle-damaged or abandoned American equipment as precious raw materiel (e.g., confiscated artillery shells were routinely converted into mines or used in mortars).

The North Vietnamese gradually shifted from a logistic system dependent upon thousands of peasants and their bicycles to a large and sophisticated logistic system dependent on cargo trucks. The system had large truck parks, fueling bases for trucks along the routes, and major truck maintenance and repair facilities.

As war weariness, cost (US$24 billion a year by 1968), public pressure, and politics began to have an effect, the United States began to withdraw its forces (more than half a million men) and to "Vietnamize" the war. As U.S.

troops withdrew from ground combat, North Vietnamese forces under General Giap confidently struck with a three-pronged conventional war offensive across the demilitarized zone (DMZ), into the central highlands, and across the Cambodian border, aiming at Saigon.

Fortified by U.S. naval gunfire and airpower, and logistically supported by the United States, the South Vietnamese fought aggressively against the North. They were particularly professional in their resilient showing against what became known as Giap's 1972 Easter offensive. The ARVN showed that it could perform well without U.S. ground troops' support.

Massive Soviet and Chinese logistic support, which included fuel pipelines extending the length of the Ho Chi Minh trail, enabled Soviet-made tanks to invade the South. Although the ARVN, aided by U.S. B-52s and naval gunfire on the China Sea flank and to the rear of the NVA as Giap's troops broke through the DMZ, lost an entire division, they did not lose their composure. They held stubbornly, and pushed the enemy back to the DMZ. Resilient in the highlands, refusing to retreat from An Loc, the gateway to Saigon, the South Vietnamese compelled General Giap to withdraw.

THIRD INDOCHINA WAR (1973–75)

The third Indochina War ended with the demise of the Republic of Vietnam, the result of logistic anemia. As in all war, the supply of materiel is critical. The flow of supplies from the Soviet Union and China to North Vietnam increased to support Giap's next campaign. The flow of supplies to South Vietnam, however, which started as the "Peace with Honor" American promise to provide one-for-one replacement of combat consumption or losses of materiel, was a promise never kept.

Nor was the promise kept of U.S. B-52 bomber and naval gunfire support in the event of enemy peace violations. Enemy violations were committed blatantly and confidently, with the expectation of no hostile U.S. reaction. The North Vietnamese correctly interpreted the meaning of Watergate and the political paralysis of the American presidency. Congressional mandates, known as the War Powers Act, declared that the president, without Congressional declaration of war or specific authorization, could not introduce U.S. armed forces into hostilities or into situations of imminent hostilities. The limited war in Southeast Asia lost its logistic support when it lost the public funds to back it.

Limited War: Britain's Battles

Since World War II, Great Britain has experienced five limited wars:
- The Protestant-Catholic conflict in Northern Ireland has gone on longer than the Thirty Years' War, and is destined to continue.
- The United Nations Korean War (1950–53) was the most intense (actually mid-intensity) limited war and negotiated peace that Britain has engaged in since World War II.
- Malaysia was a ten-year intervention in a civil war (1950–60).
- The 1956 Suez fiasco, in company with the French, was short.
- The fight (1963–66) by British regulars, Gurkhas, Australians, and New Zealanders to prevent Indonesia from annexing Brunei, Sabah, and Sarawak in Borneo was a skillful triumph.
- The 1982 improvised and successful invasion of the Falklands/Malvinas was brief.

NORTHERN IRELAND

The retail logistics of the outlawed and sporadically supported Irish Republican Army (IRA) permits a terrorism with a lower annual murder count than that of the city of Detroit in the United States. With the mass media as a gauge, however, the British conflict in Ireland is far less acceptable. British logistics is standard, ample, and of a quality to be expected in a small close-to-home operation without unusual demands. The IRA logistics provides just enough to exasperate the conventional British military and to excite the press, and to challenge the craft of the London-Dublin-Belfast politicians.

SUEZ

Logistic limitations in the 1956 Suez venture—the lack of troop airlift, amphibious landing craft, ocean and air terminal facilities; the lack of a likely staging area for the relatively limited military task of taking over the Suez Canal; and the adjunct task of taking over the government from Gamal Nasser—all contributed to the British-French defeat. Hostility from U.S. president Eisenhower also played an important role.

BORNEO

The Borneo operation against Indonesia (1963–66), in contrast, was a logistically illuminating, if patience-testing, success. The use of fixed- and rotary-wing aircraft to support bases from which British troops fanned out on their jungle and mountain patrols proved the key to this small, tropical conflict with large political results. Radio-detection devices that located and targeted the Indonesia enemy were a decisive advantage. The foiling of Indonesian annexation and the ousting of President Sukarno were the results, as was the acceptance of Malaysia as a sovereign state.

FALKLANDS/MALVINAS WAR

Like the Indonesian conflict, the Falklands/Malvinas War (1982) was on the edge of the old British protectorate, but otherwise, these conflicts were very different. The Indonesian war occurred in the tropics, involved infantry squads finding and fixing the enemy in jungle camouflage and heat, and came to a slow conclusion. The Falklands/

Malvinas War occurred in a cold climate; involved ground, air, and naval forces with sophisticated weapons in Antarctic waters; and ended in weeks.

The Falkland Islands, also known as the Islas Malvinas, have a land area about half that of Massachusetts parceled among 200 islands (two large ones); they were, and remain, a British Crown colony in the South Atlantic. Logistics played a significant role in this war; only a slight change in the British or Argentine logistics could have resulted in the Argentines retaining control of the islands.

Strategic logistics favored the Argentines. Britain's line of communications from the Falklands was 13,000 kilometers (8,000 mi.) with a seaport and airport pivot at Ascension Island, about 4,800 kilometers (3,000 mi.) from the objective area. The Argentine line of communications was only 400 kilometers (250 mi.) long.

Each side in the Falklands/Malvinas conflict disastrously misread the other. Britain thought Argentina would not invade. After Argentina invaded, Argentina thought Britain would not assault. Both countries were stunned by the adversary. That the Argentine junta would seek the traditional way of armed aggression to divert the populace from the unrest of unemployment, inflation, and draconian injustice was to be expected. That Britain would tolerate for a moment a military affront from a second-rate power was a fantasy.

Politics immediately interfered with British logistics: compelled to sail hurriedly on waves of emotion, the Royal Marines were unready for an across-the-beach assault. Fortunately, Ascension Island provided a respite halfway between Africa and South America where the fleet, its auxiliaries, and the ground forces were able to put their logistics in good order.

Logistic support from the U.S. Air Force on Ascension Island was pivotal. Moreover, U.S. Secretary of State Alexander Haig made sure that sophisticated intelligence apparatus and tanker aircraft were at the service of the NATO ally.

The Argentines, thwarted by the Royal Navy's exclusionary zone, set up an aerial logistic pipeline. This failed, however, when British airpower bombed the airfields, making them unusable.

Another critical logistic problem for the Argentine forces was the inability of Argentine land-based aircraft to refuel in the air and thus the inability to reach the British fleet with bombs or rockets. Even when they did reach the fleet and successfully dropped their bombs on British ships, the bombs frequently failed to explode. After having suffered a missile attack, the fleet was careful to remain beyond the range of the Argentine air force.

Whenever the Royal Navy was forced to come within range of Argentine airpower, it took heavy losses, including the loss of a ship carrying all of Britain's heavy-lift helicopters. As a result, the troops had to march in frigid wind to a landing zone where the choppers would have carried them.

The British logistic system was perilously marginal. The logistic shortcomings of Her Majesty's forces might have resulted in a different outcome if the Argentine ground forces, faced with a pending and embarrassing lack of ammunition, had not surrendered early.

Despite the British logistic shortcomings, the Falklands/Malvinas, as a logistic testing ground, accented the British application of state-of-the-art technology. It displayed a logistic system that developed and made capital investment in the right hardware, stocked it, maintained it, and had it in place at the crucial moment. Foresight gained the victory.

The old Argentine heavy cruiser *Belgrano* was sunk by a British nuclear attack submarine. Night vision devices on British weaponry zeroed in on Argentine infantry, which was then compelled to move and fire in the dark. On the other hand, the Exocet missiles on Argentine aircraft endangered British ships and entirely depleted the logistic mobility and morale value of the heavy-lift helicopters

Inventive and innovative logistics led to the transformation of British bombers into aerial tankers. Helicopters were unhindered by the roadless country and partially compensated for the more numerous Argentine troops, who had to rely on movement by land.

The British also had the vertical/short takeoff and landing (VSTOL) tactical aircraft on site and the agility to parachute onto and take the strategic airfield at Goose Green. All of these factors, plus the flexibility of amphibious assault craft, the gallantry of troops, and the quality of the combat leadership, figured in the quick achievement of this limited war's political aim.

UN Logistics: The Korean War (1950–53)

The Korean conflict was the first United Nations (UN) war, the first war in which jet fighters clashed, and, from the first to last, a war of logistics. The official American army historian called it "the first American war where materiel mobilization actually received the first emphasis."

In 1950 a U.S. State Department disclosure that Korea was not within the American sphere of vital interest had seemed to North Korea an invitation to invade without reprisal from South Korea's (Republic of Korea, ROK) ostensible protector. On 25 June 1950, the North Korean People's Army, the Inmun Gun, comprising seven infantry divisions and an armored brigade with Soviet T-34 tanks, backed by 200 Russian YAK fighters, burst across the 38th parallel to attack the South Korean army of eight divisions. The ROK army did not have what was thought a conventional necessity of the time, a blitzkrieg package; that is, it had no tanks and no combat aircraft. It took three days for the ROK capital to fall and five days for the

United States to commit its army. The U.S. forces consisted of an understrength 24th Division of poorly prepared occupation troops from Japan, augmented by troops from the guardhouses.

After the United States announced intervention, the aim of the Inmun Gun was to crush the ROK forces and push the remnant 24th Division into the sea as quickly as possible—before the U.S. 8,000-kilometer (5,000-mi.)-long logistic pipeline could start up, and before the eighteen other UN members who had promised troops could fulfill those promises.

Fortunately for the UN forces, excess World War II stocks (which Japan had prepared for the invasion that never happened) were available in the islands. Its stockpiling was inspired by hereditary fear: before World War II, Japan's foreign policy view was that "Korea was a dagger aimed at the heart of Japan." By 1950, Japan itself, with remarkable resilience, was producing up to a half a billion tons of steel annually, exceeding the steel production of any other nation in the Far East; it was also producing locomotives and trucks. As a logistic fortress, Japan had the experience of a defeated nation that understood expedience.

According to Gen. Douglas MacArthur, the race between North Korean advances and American build-up was touch-and-go after twelve weeks. The whole complexion of the war changed almost overnight, however, when MacArthur, with the ROK and U.S. Army forces pinned down in the Pusan perimeter, made a surprise amphibious assault at Inchon, west of Seoul.

The move on Inchon proved to be a stroke of military genius. The U.S. Eighth Army in Korea broke out of the Pusan perimeter, and crushed the Inmun Gun. In little over a month, MacArthur's U.S. forces had taken Pyongyang, the North Korean capital, and were speeding toward the Yalu River. Then Mao Tse-tung's general, Chou En-lai, crossed the Yalu with 30 divisions (about 300,000 Communist Chinese) and pierced the center of the UN line.

Confusion reigned among the UN forces in their retreat. Vast quantities of equipment were abandoned. This created a problem for stateside logisticians—how to replenish the forces rapidly? Certain kinds of consumables (such as rations and fuel) are automatically forwarded on the basis of the number of personnel to feed and the number of engines. For the most part, however, needs are "requisitioned." For the Far East, most items moved by sea, and it took on the average 120 days from the day of requisition to the day of receipt in the Far East depots.

To reduce the time, Pentagon logisticians ignored procedure and requisition. They consulted unit tables of organization, which listed all the basic supplies, tools, and weaponry, and ordered those lists for priority delivery to all combat units in retreat. Thus, the U.S. units were refitted in time to brace and hold, and then retake the original position of the ROK forces, fulfilling successfully the limited aim of the limited war.

There were also successes in logistics that presaged the future. American statesmen and war planners did not, as they had in the past, mobilize troops before their weapons and supplies were ready; troops were synchronized with means, weapons, and supplies. Inter-theater airlift, which would prove vital in the 1972 Arab-Israeli war and in the 1972 Easter offensive of the North Vietnamese Army, came into its own in the Korean War. When the war started, the U.S. Air Force lifted about 70 tons a month to Japan. In three months the airlift was 100 tons a day. Of course, with the cost of airlift at US$4,000 a ton versus US$38 a ton for sealift, only preciously needed cargo was airlifted. The high price fades in importance when the combat troops need whole blood, bazookas and rockets, airdrop equipment, ammunition, and recoilless rifles.

Coalition UN warfare created novel ration requirements for American logisticians, who were the major providers. Religion, taste buds, and habit led to a menu in combat that delighted some and left others unsatisfied. Pork was taboo for Muslims, beef for Hindus. Indians, Thais, and Filipinos wanted rice and stinging spices; the Dutch favored milk and cheese. Ethiopians wanted their food saturated with hot sauce.

Overkill and Logistic Folly: Grenada (1983)

The American invasion of Grenada on 25 October 1983 is a good example of bad logistics in a limited war. The strategy was sensible: to uphold the Monroe Doctrine and contain Communist inroads in the Caribbean and the Americas. The tactics, however, were poor.

Some concession to the fog of war is tolerable, but the events in Grenada were inexcusable. The four military services of the United States, in an excessive investment of forces, were used for what should have been a marine corps or ranger operation.

The invasion was launched in response to a call from a British diplomat for help, with reinforced requests from Caribbean island nations, to stabilize Grenada, eliminate its leftist leanings, and ensure the safety of American citizens.

Preinvasion intelligence was abominable. The Marines invaded without suitable maps; they used a tourist brochure and a map of the island not updated since 1895, until captured maps were available.

Ship-to-shore coordination of naval gunfire, so well executed and devastating in Vietnam, was inept. Army radios could not communicate with navy radios. A credit card phone call to the army's base in North Carolina placed a fire support request to be transmitted to the fleet.

No drinking water was available, nor was more than a day's food ration. There was an absence of transport and a shortage of ammunition for the army. This scenario—the result of incredible logistic shortsightedness—was traceable to the military stress on secrecy at the highest echelons and to the short time between decision and invasion.

It is reported that the J-4, the Chief of Logistics of the Joint Chiefs of Staff, was inexplicably excluded from knowledge, planning, and direction of the operation.

In some instances, good sense triumphed over foolish orders. When the navy directed that army helicopters that landed on the deck of a navy ship with wounded could not be refueled (because there was no interservice fund for fuel repayment agreement), the ship's captain ignored the order.

Despite the logistic travesty of this three-day limited war, the political objective was achieved. The medical students were all rescued unharmed, a large cache of ammunition and weapons was captured, 200 or more Cubans were deported, and the jumbo jet airfield under construction (suspiciously large for tourist traffic) was reduced to a cost-benefit size more compatible with trade.

The logistic flaws in Operation Urgent Fury were mainly attributable to poor intelligence and an obsession with operational security (OPSEC). The rigid insistence on such secrecy ensured surprise of the enemy, but it also surprised the experts in the invading forces, whose business it was to provide the soldier with the necessities for battle.

War in Afghanistan (1979–89)

At the start of the Soviet invasion of Afghanistan, Soviet MiGs, helicopters, tanks, and a reinforced conventional army supported by local government troops swept through the country capturing strategic points and key cities. The opposing warrior tribesmen fled to the mountains or migrated, with 2 million countrymen, to sanctuary in Pakistan.

Gradual improvements in the quality and range of logistic support from the United States and China enabled the tribesmen to administer such shock to the Soviet army that, after nine years of limited war, the USSR withdrew.

Old British Enfield rifles were replaced with Chinese-made replicas of the Soviet AK-47, and Japanese and American trucks and jeeps appeared. U.S. rocket-propelled grenades (RPGs), Chinese recoilless rifles, multiple rocket launchers, ZU-011 antiaircraft weaponry, and U.S. shoulder-launched Stinger surface-to-air missiles (SAMs) were supplied to the mujaheddin.

The primitive Afghan tribesmen proved adept fighters when equipped with modern weapons. The "nice-to-have" supplies of a first-rate power were unnecessary. A tribesman's ration was bread, rice, and onions, with occasional meat. He wore a wool hat instead of a helmet, sandals instead of boots. He slept on the ground in the open air, and the supplies of ammunition were kept in dry caves.

While walkie-talkies and inflatable rafts for crossing rivers were some of the earliest modern supplies to reach the Afghan resistance fighters, and the sophisticated heat-seeking 70-pound shoulder-held Stinger missiles were credited with turning the war around in favor of the resistance, American mules also played an important role. About 700 mules were flown from Texas in Flying Tiger Boeing 747 jets, each jet capable of carrying 114 mules. The mules proved invaluable, carrying supplies and ammunition through the Khyber Pass to the Hindu Kush. The Afghan rebels could have used more mules; in one offensive against Asadābād Heights, rebel troops could not continue their heavy bombardment for lack of mules to transport enough shells and rockets from Pakistan.

The export of American mules to war is an old tradition. During the Boer War, Kitchener was supplied with thousands of American mules. Mules were used in America's war with Mexico and in the Spanish-American War. In World War I, they pulled supplies along narrow-gauge railroads to the front, and were so effective that German soldiers were ordered to shoot mules before men. In World War II, 350,000 mules were used to carry 300-pound packs; and mules were transported to Burma in gliders.

The wide-ranging logistic support (from SAMs to mules) by the United States (US$4.8 billion in weaponry and supplies) and China during the nine-year war enabled the Afghan resistance to make life so difficult and uncomfortable for the occupying Soviets that they withdrew. Although the limited war in Afghanistan was long and costly (in lives) to the mujaheddin, the results pleased the mujaheddin, the Americans, and the Chinese.

Rapid Expenditure of Supplies

Israel has participated in five near-high-intensity limited wars from 1948 to 1982. The 1973 war, in which Jordan, the Palestinian Arabs, Syria, and Egypt tried to squeeze Israel into extinction, holds lessons from a logistic point of view about superpower interplay and the consumption of materiel by both sides.

The ability of the United States and the Soviet Union to supply and air-deliver replacements for critical losses (e.g., tanks) to the adversaries enabled Israel, with its superior leadership, to gain victory, and Egypt and Syria to minimize their embarrassment in defeat.

The expansion of the second Pakistan-India war (1965) from Kashmir into the Punjab, with eight divisions on each side, was also influenced by countries not directly engaged. The war ceased when the United Kingdom and the United States stopped the flow of supplies to the adversaries.

In both the previous examples, the suppliers saw rapid destruction of billions of dollars of inventory in a matter of weeks, for which the client states could not pay.

Future of Logistics in Limited War

A limited war differs from a general war in that it is limited; it is an attempt to avoid the exhaustion of a general war.

From the limited wars discussed here, it is easy to project some facets of limited war into the future. The major military powers may continue to introduce troops and techniques into conflicts in developing nations. One can also make certain statements about past and future limited wars:

- Seapower as a main or winning factor is evident from the American Revolutionary War and the U.S.-Mexican War.
- The impact of the modern media and increasing public influence on limiting war may encourage negotiation and treaty by the withdrawal of logistic support, as happened in the last Indochina War.
- The advantage of interior lines and modern transport, which, when combined, allow for rapid concentration of forces, is apparent from Moltke's wars.
- The logistic advantage of warriors in certain primitive environments (mostly mountainous and jungle) is evident from the American Indian and other guerrilla wars (early Indochina, Malaysian, Afghanistan).
- Limited wars, to the professional, are places to test and develop tactics, technology, logistics, and leadership, as shown in the three Indochina wars and in Operation Desert Shield.
- The fog of war will continue to be a factor in war, as shown by the Grenada conflict.
- Logistic support of opposite sides by other nations will continue (e.g., the war in Afghanistan). In such wars, the US$3.5 billion-a-year business of selling arms will also continue, mainly as a competition among the United States, the republics of the former USSR, France, Italy, Britain, Sweden, and China.

It is also safe to project that advances in technology will affect all of these features:

The 34-pound shoulder-fired Stinger missile and a few hundred mules played key parts in the Soviet army withdrawal from Afghanistan. The Stinger, with a range of 10 kilometers (6.2 mi.), is also credited with turning the tide of battle in the favor of anti-Marxist guerrillas in Angola and with encouraging Cuban troop withdrawal.

The U.S. AWACS (airborne warning and control system) and U.S. satellite intelligence provided to the Iraqis on Iranian targets gave a decided advantage to the Iraqis.

Tear gas and rubber bullets are deterrents that may be used as benign weaponry in certain low-intensity trouble areas.

Weaponry designed to spot and destroy anything equipped with radar is being developed. Electromagnetic rail guns with superhard bullets able to penetrate the toughest metal are being developed for use against tanks.

Aseptic packaging that prevents spoilage without refrigeration or chemicals is available. It sustains the purity of rations and thus the morale of the troops.

Vulnerability looms for logistic systems that are almost totally reliant on computers. Computer crime is growing, and there are worrisome military implications. Potential enemies are raiding military computer systems in ways that disrupt or paralyze altogether logistic computer systems. Computer saboteurs can run magnets across disks, erase data, or inject a latent "virus" with software control that, in war or threat of war, can erupt and cripple or virtually destroy the computer-dependent logistic system. The potential for mischief is infinite. Planes, trucks, ships, and trains can be stopped, or computer-directed to crash, by traffic-controlling computer devices. In the U.S. defense inventories, requisitions enter computers by federal stock numbers that are easily changed by guerrilla programming. A requisition for an electric detonator can be easily changed to one for an electric light bulb.

Biological diseases (viruses) and nerve agents may be part of the weaponry in future general wars, in wars ranging from low to high intensity. While the evidence of "yellow rain" in Laos is questioned by scientists, the U.S. investment in research and development during the Reagan administration for protection against chemical and biological warfare increased by more than five times since its start in 1980, from US$63 million to US$334 million. Genetic engineering is also in this program. While such an investment is relatively small for the Pentagon, the growth indicates rising concern and fear about its application. Despite an international ban on nerve gases since 1925, and a virtual moratorium since World War I, they are likely to be used in the future, as they were in the Iran-Iraq war. Mustard gas and hydrogen cyanide will probably account for death and lingering affliction in future limited wars, with the logistic problems of inventory control, handling, and distribution of hazardous materials. White phosphorus, or "Willie Peter," used for smoke screens and (along with napalm) to burn flesh, will continue to be used extensively, as it was in the Vietnam War, and will be widely available in national inventories of chemical weapons.

Because of scientists' ingenuity at finding new ways for humans to exterminate one another, war will not merely be an exchange of explosives. Computers that can perform a billion computations per second could be the basis for the neutralization of intercontinental ballistic missiles. Genetic engineering and herbicides give rise to the idea of a spray that can paralyze an army or a populace. Spraying machines in war, as in Vietnam and in Malaysia, may have only seen their infancy. The battlefield of the future may be a toxic wasteland degrading a soldier's immune system (and that of his family), and promising an early death. Herbicides used by the British in Malaya and the Americans in Vietnam were of general tactical utility in both attack and defense. As the vegetation died, fields of fire were opened.

In future limited war, one can also expect improved electronic countermeasures (ECM), which jam enemy radar and fool it with stealth technology. Night vision devices will be commonplace. Laser sights will ensure that if the rifle trigger is squeezed when the laser spot is on the target, the hit is guaranteed.

Fear, augmented by long-range and medium-range ballistic missiles with nuclear warheads, makes limited war the only sensible survival choice for mankind. Notwithstanding mankind's survival standoff, ballistic missiles, particularly tactical ones, will proliferate and will be applied to limited war. Sixteen countries are known to have, or are obtaining and developing, ballistic missiles. China is selling them to Saudi Arabia and Syria. India is producing its own, while Pakistan is on the verge of producing a ballistic missile with a range of 950 kilometers (600 mi.). The United States is funding half the cost of the Arrow, an antitactical missile produced in Israel.

The unlikelihood of nuclear war waged by the major powers does not eliminate the likelihood of the use of nuclear weapons. The portable nuclear bomb, the SADM (Special Atomic Demolition Munition), also called a "backpack nuke"— a little less than a yard square, with a warhead weighing 56 pounds—is already in the U.S. inventory and, probably, in other forces. Miniaturization is predictably in progress. The one-kiloton or smaller nuclear explosive may be used in limited war, to make a big crater where there once was an airfield or ball-bearing factory, to change the course of a river, or to close the entrance to a harbor with rubble.

JOHN E. MURRAY

SEE ALSO: Afghanistan, Soviet Invasion of; Grenada, U.S. Intervention in; Logistics: A General Survey; Special Operations Forces; Unconventional War.

Bibliography

Blair, C. 1987. *The forgotten war: America in Korea.* New York: Times Books.
Braestrup, P. 1984. *Vietnam as history.* Washington, D.C.: Univ. Press of America.
Clausewitz, K. von. 1943. *On war.* New York: Random House.
Gorce, P. M. de la. 1963. *The French army: A military history.* New York: George Braziller.
Hanks, R. J. 1980. The Boer war. *Strategic Review,* spring.
Hastings, M., and S. Jenkins. 1983. *The battle of the Falklands.* London: W. W. Norton.
Jomini, H. de. 1962. *The art of war.* Westport, Conn.: Greenwood Press.
Osanka, F. 1962. *Modern guerrilla warfare.* New York: Free Press of Glencoe.
Paret, P., ed. 1986. *Makers of modern strategy: From Machiavelli to the nuclear age.* Princeton, N.J.: Princeton Univ. Press.

LOGISTICS, NATO

During the 40 years of the North Atlantic Treaty Organization's (NATO's) existence as a defensive alliance, its membership has grown from twelve to sixteen free and sovereign states—nations with different political, economic, social, and military philosophies and organizations. These nations have agreed on the nature of the military threat they face in Europe and bordering seas and the strategy they will use to meet that threat, and have chosen the allied commanders who will exercise authority over the operational forces assigned to NATO should war occur. Despite their agreement on these important principles, the member nations have always maintained that the logistic support of NATO's forces remains "a national responsibility." In this respect, NATO differs substantially from the Warsaw Pact.

The declared doctrine of national responsibility for logistics within NATO, in war as well as in peace, has led to severe misgivings among some of NATO's supporters. They feel that this doctrine affects adversely NATO's readiness and capability to conduct and sustain military operations on the scale envisaged. Many military planners now consider that the levels of effective logistic support needed for NATO to provide a viable defense can only be met by collective and integrated effort by member nations. They further suggest that the weaknesses in NATO's logistic preparation will deteriorate further as nations develop different perceptions of the threat they face.

The preparation of plans for national forces to fight side by side in battle is comparatively straightforward. It is far more difficult to persuade independent nations to agree to a common standard of logistic support by which those national forces will be sustained. The question is not simply one of providing substantial resources. This is only part of the problem; the major difficulty lies in the planning and procedures necessary to support the complex semi-integrated organizations that make up NATO's military structure in wartime. The current procedures are often practiced on exercises, but this does not prove their efficiency in wartime.

Definition of Logistics in NATO

Because of the need to provide a basis of logistics that is common to all its member nations and that is inclusive of all their logistic philosophies, NATO has defined *logistics* broadly as:

> the science of planning and carrying out the movement and maintenance of forces. In its most comprehensive sense, those aspects of military operations which deal with:
> a. design and development, acquisition, storage, movement, distribution, maintenance, evacuation and disposition of materiel;

b. movement, evacuation and hospitalisation of personnel;

c. acquisition or construction, maintenance, operation and disposition of facilities;

d. acquisition and furnishing of services.

The textbook definition from the NATO *Handbook of Logistics* is comprehensive. It is drawn so widely that NATO has had to use three terms to delineate the separate functions involved; *production logistics, consumer logistics,* and *civil emergency planning.* Each of these is dealt with by a different part of the NATO organization and is described below.

PRODUCTION LOGISTICS

Within NATO, *production logistics* covers research, design, development, and manufacture of materiel, and the acceptance of materiel into service. It therefore includes the functions of standardization and interoperability, contracting, quality assurance, procurement of spare parts, reliability and defect analysis, safety standards for equipment, specification and production processes, trials and testing, codification, materiel documentation, configuration control, and modifications. The lead authorities for these responsibilities at NATO headquarters in Brussels are the Defence Support Division of the International (Civil) Staff (IS) and the Armaments and Standardisation Division of the International Military Staff (IMS).

CONSUMER LOGISTICS

Consumer logistics within NATO covers the reception of materiel by the armed forces; its storage, transportation, and maintenance (which includes serviceability and repair); its operation in combat; and its eventual disposal. This category comprises the functions of stock control, provision or construction of facilities (except those involved with production logistics), movement control, reliability and defect reporting, safety standards for storage, and transportation, handling, and related training. The lead authorities for consumer logistics at NATO headquarters are the Logistics Directorate of the IS and the Logistics and Resources Division of the IMS.

CIVIL EMERGENCY PLANNING

Civil emergency planning is not strictly part of NATO's military logistics, but, since so much of the alliance's logistic planning depends upon it, this function must be included in any discussion on NATO logistics. It includes all defense-oriented responsibilities and activities of national government departments (excluding foreign affairs and domestic military matters). More specifically, it covers the expanded responsibilities and activities required in crisis and war for the continuity of the machinery of government; the mobilization of national resources, including energy, manpower, transport systems, production, food and agriculture, raw materials, and telecommunications; and civil defense measures. The lead authority at NATO headquarters is the Civil Emergency Planning Directorate of the IS.

COORDINATION AND CONCURRENT ACTION

These three areas of logistics constantly overlap; very few aspects of logistic planning do not involve at least two of them. Thus, a logistic proposal may be suggested by a NATO nation or commander; once it has been agreed upon and accepted in the appropriate international committee, its implementation must be coordinated through all the staff branches concerned—operational, administrative, military, civil, international, or national. Effective implementation of such agreements depends not only on good coordination, but also on concurrent action by all the military and civil staffs concerned. This process is often complicated and lengthy.

Principles of NATO Logistics

Each of the member nations of NATO has evolved its own logistic principles, organizations, and practices over the years, as a result of its own history and military experience. These national logistic practices vary widely and reflect the economic strength of individual nations as well as their perceptions of the type of war they expect to fight and the threat they plan to meet.

NATO's logistic principles were formulated to provide a set of standards for logistics as they apply in the particular case of NATO. They reflect the special factors that have to be accounted for in international military cooperation. They do not replace the normal principles of logistics; they are the additional rules of logistic planning that are necessary in an alliance.

It is agreed that economy of logistic force is the basic logistic principle. Cooperation and collaboration are required for more efficient and economical use of logistic resources. Logistic interdependence requires "guaranteed satisfaction" of other national force logistic requirements equivalent to one's own. Provision of logistic resources to meet NATO operational plans is a national responsibility, while determination of logistic requirements and multinational logistics are NATO responsibilities. Logistic practices must be the same in peace as in war.

Moreover, materiel and services should be standardized, and interoperability must compensate for any lack of standardization. Logistic information will be fully and reliably exchanged within NATO, and constant satisfactory logistic readiness must be maintained. Logistic plans must be based on updated combat operational plans and immediately convertible to combat logistic requirements at the moment of enemy attack or threat. Mobility and dispersion must replace voluminous static storage of combat supplies and equipment. NATO logistic facilities must be configured for passive defense in peacetime to ensure survivability in war. Duplication of common logistic functions must be minimized within the Alliance through such measures as specialization and single management. Finally, logistic procedures should be standardized and harmonized to provide flexibility among nations in logistic support of NATO forces.

Logistic Responsibilities within NATO

The precise division of responsibilities for logistic planning in NATO is often clouded by the very large number of countries concerned. One of NATO's greatest difficulties is that the size and complexity of the organization, as well as its international structure, serves to obscure and weaken the chain of responsibility. NATO has no written constitution, and clear accountability is sometimes difficult to identify and achieve.

The basic problem is to reconcile the overall needs of the alliance with the concept of the sovereignty of independent nations. A NATO instruction can easily be regarded as interference in the management of forces of individual nations. The collective will exists, but the means to put it into practice are sometimes lacking. The relationship between NATO's internationally agreed-upon decisions and the military forces of the member nations who implement them might be likened to membership in a private club. Member nations join of their own accord and remain free to leave whenever they wish—even though they realize that their membership contributes to the common good. While they remain members, they obey the rules of the club; in a sense, they surrender a degree of national freedom of action, since an element of control over their national forces is transferred to NATO, especially when those forces are placed under NATO's operational command.

The general principles of logistic responsibility were agreed upon as long ago as 1952 by the highest NATO authority, the North Atlantic Council. A regrettable fact is that the passage of time has weakened the resolve of some nations to adhere to the spirit of that agreement: "The responsibility for logistic support to national component forces will, in general, remain with the responsible authorities of the nations concerned. The responsibility for coordination will, however, rest with the Supreme Commander and his major subordinate commanders at the appropriate levels."

Logistic Coordination

Coordination is often used to describe the sometimes limited authority exercised by NATO commanders. Because of language differences within the Alliance, some nations considered that the word meant giving firm orders or directions; others felt it meant only the passing of information. It was therefore necessary to agree upon a definition of coordination; thus, within NATO, the term is held to mean

> the authority granted to a commander or individual assigned authority for coordinating specific functions or activities involving forces of two or more countries, of two or more services, or of two or more forces of the same service. He has the authority to require consultation between agencies involved or their representatives, but does not have the power to compel agreement. In case of disagreement between the agencies involved, he should attempt to obtain essential agreement by discussion. In the event that he is unable to obtain essential agreement, he shall refer the matter to the appointing authority.

It has been argued that this is too weak an interpretation of the term, and that such a meaning achieves little. Nations should have sufficient commitment to NATO to provide allied commanders with more authority. However, the definition epitomizes both the way an alliance of independent nations works and the inherent weakness, in military terms, that stems from lack of unified control. The current understanding of where logistic responsibility lies in NATO is discussed below.

National Logistic Responsibilities

Nations are responsible for planning and providing the logistic support of their forces to meet allied requirements, furnishing allied commanders with sufficient information on their national logistic plans and resources, and providing the maximum logistic assistance to the forces of other nations. They also are responsible for making logistic support arrangements that are in line with NATO plans and policy.

The aim is straightforward. However, the armed forces of the NATO nations are structured in dissimilar ways, and nations adopt different attitudes toward NATO proposals. These differences are not confined to military affairs. Some nations have defense commitments outside the NATO area, and their forces are not all necessarily earmarked for use within NATO in wartime. This affects the emphasis of their military strategy and planning. Other nations have political or constitutional limitations placed upon their commitment to NATO in peacetime, which, in turn, could weaken their contribution to the alliance in war.

NATO Headquarters' Logistic Responsibilities

Much of the impetus for logistic planning in NATO comes from the military side as a response to changes in strategic policy, tactical doctrine, or logistic concepts. However, on basic issues of finance, international agreement, and multinational cooperation on such matters as weapon production and fuel supplies in war, much of the logistic work at NATO headquarters is carried out by the civilian IS.

The second factor that governs this aspect of logistic planning is the NATO system of consultation. Although every member nation has representatives at NATO headquarters at Brussels, they are unable to decide or agree on every proposal put to them. Working papers and draft proposals are therefore disseminated from Brussels to national capitals. The IS and IMS at NATO headquarters share in the initial processing and preparation of these papers. The IS civil officials are NATO employees and follow a policy line independent of national views. The IMS comprises military officers seconded to NATO; while employed in that capacity, they also adopt an independent viewpoint.

IS Logistic Responsibilities

The primary role of the IS is to provide a secretariat for NATO committees, but, in practice, its duties go far beyond this. Proposals for logistic policies and plans can originate in individual nations, national military headquarters, allied commanders, or their staffs. They all have to be examined, prepared, processed, and circulated for discussion.

The IS handles these tasks, but it also has a more positive role: to evaluate national positions and the logistic readiness of national forces and to submit recommendations arising from its own corporate knowledge and experience of NATO. The IS should therefore be considered as an important source of information and planning for those topics within its area of responsibility.

Of the five divisions of the NATO IS, three have particular responsibilities for logistics. The Defence Planning and Policy Division is responsible for preparing "force goals" (these are planning objectives for nations) and developing and preparing "ministerial guidance" (this is the current NATO planning policy agreed upon by nations at ministerial level). The Defence Support Division is responsible for coordinating production logistics—harmonizing concepts and requirements for future air, land, and maritime requirements; harmonizing procurement and replacement plans to facilitate greater standardization and commonality; encouraging coordination of national research, development and production programs; coordinating national research, development, production, and logistic capabilities, to use resources more efficiently; organizing systematic exchanges of information to facilitate international cooperation in the sphere of defense equipment and its support; and developing procedures to systemize and rationalize armaments planning. Finally, the Division of Infrastructure, Logistics and Civil Emergency Planning, as its name implies, is concerned with consumer logistics and civil emergency planning.

IMS Logistic Responsibilities

The IMS, which is manned by military personnel, is responsible to the Military Committee. Apart from this, its organization is basically similar to the IS. Three divisions of the IMS have particular responsibilities for logistics. The Plans and Policy Division assesses the proposals put forward by NATO commanders, evaluates national responses to the NATO defense planning questionnaire, and reviews the status of nations' logistic readiness. The Armaments and Standardisation Division deals with armament planning activities and has the important task of reconciling and coordinating plans for bringing new equipment into service in line with alliance needs. This division represents the interests of the Military Committee at production logistics meetings, and it is also the channel through which the major NATO commanders submit proposals for new weapons and equipment. If a proposal is approved by the Military Committee, this division is responsible for processing it to the IS for action by the production logistics branches. The Logistics and Resources Division has primary responsibility for dealing with consumer logistics on behalf of the Military Committee. It coordinates, manages, and advises on such issues as war reserve levels, medical planning, and host nation support. It services the senior NATO logisticians' conference and represents the Military Committee on the senior civil emergency planning committee.

Major NATO Commander Logistic Responsibilities

Three senior allied officers are designated as major NATO commanders: the Supreme Allied Commander Europe (SACEUR), the Supreme Allied Commander Atlantic (SACLANT), and the Commander-in-Chief Channel (CINCHAN). Each has wide logistic responsibilities but comparatively little logistic authority. This again emphasizes one of NATO's fundamental deficiencies: that even the most senior allied operational commanders have insufficient direct authority over the logistic support of the forces that come under their operational command in crisis or war. In effect, their logistic authority is limited to: definition of force proposals, publication of stockpile planning guidance, specification of projects forming part of the commonly funded infrastructure program, and prioritization of national supplies, such as ammunition and fuel.

The major NATO commanders specify the levels of logistic support they wish to see in national forces. Although nations may agree on these levels in principle, their implementation depends entirely on the nations' willingness to comply with the NATO commanders' wishes. This is a constant problem in NATO's planning to resist attack.

Nations unable or unwilling to meet the required level of logistic support often defend their action by restating the principle that logistics is a national responsibility. They interpret this to mean that national logistics is not the direct concern of a NATO commander. But every commander must ensure that his forces have adequate logistic support to guarantee their operational effectiveness.

Logistics is not and cannot be the sole responsibility of nations. It is the task of nations to provide logistic support for their forces in accordance with agreed-on NATO plans. Until this concept is accepted and implemented, NATO's logistic status will remain suspect.

SUMMARY OF LOGISTIC RESPONSIBILITIES IN NATO

Logistic policies and requirements in NATO originate from NATO commanders, staffs at NATO headquarters, and national staffs. If they are agreed on by nations, they should then be implemented by nations, under NATO coordination. Many organizations, departments, divisions, and headquarters are involved, but none works in isolation.

Logistic information is gathered, planning is undertaken, and consultation is conducted both up and down as well as across the chains of national and NATO command. Thus, the commander of a national army in Europe will submit a logistic proposal directly to his superior allied commander, copying it to his own national ministry of defense. That ministry of defense will consult with other nations or the civil IS at NATO headquarters; at the same time, SACEUR will discuss the logistics proposal with the Military Committee, which will have been briefed by the IMS. Parallel and concurrent action such as this saves time during the consultation process and speeds up international cooperation.

Logistic Committees in NATO

There are many committees in NATO, and they have various titles: conferences, coordination boards, panels, research groups, or working groups. There is no special significance in these titles, and all the committees have the same basic structure: (1) a chairman, normally a senior official of the NATO headquarters concerned, (2) members representing the nations concerned; (3) representatives from the immediate superior and subordinate NATO headquarters; (4) representatives from other parts of the NATO organization whose opinion or advice may be required; and (5) a secretary from either the NATO Secretariat or the same branch/division as the chairman.

Although different committees may address the same basic topic, the degree of detail discussed will vary with the level of the committee. For example, a senior committee at NATO headquarters concerns itself with national stocks of ammunition, while a committee at a subordinate level examines ammunition holdings in field formations. Both levels provide specialist advice to their sponsoring authority and make recommendations for implementing plans and policies agreed on by them.

Most logistic decisions in NATO are made through this committee structure, whether the committee is concerned with production logistics, consumer logistics, or civil emergency planning.

Production Logistics

The logistic ideal in NATO would be for nations to have one standard rifle, one standard frigate, and one standard combat aircraft. However, in an alliance of sixteen sovereign nations, political as well as economic factors make this impracticable.

In the early days of NATO, there was considerable equipment commonality among allied forces, since so many of them used American equipment surplus from World War II. However, as European nations recovered economically and restarted their own armament industries, they tended to produce their own versions of weapons and equipment, with their own specialized spare parts and logistic systems to support them.

Although general agreement was reached within NATO on strategy and battle tactics (mainly through the efforts of a series of competent Allied commanders), nations developed their operational formations around the weapons they possessed. Since they tended to place more reliance on weapons that met their own particular needs, their view of the correct combination of tanks, infantry, artillery, and other units reflected national perceptions, rather than those of the NATO commander under whom they would fight in war.

NATO has accepted the economic reality of diversification of production among nations. The current emphasis on production logistics within NATO is therefore on partnership agreements and coordination, rather than on dependence on a single source of production of equipment or weaponry. The aim is to satisfy and reconcile both the collective needs of the alliance and the economic needs of NATO's member nations. Progress in this area has been slow, but some positive results have been achieved in two areas: agreements on production, and agreements on standardization.

AGREEMENTS ON PRODUCTION

The reconciliation of national and alliance aims is met by encouraging nations to manufacture identical equipment in more than one country; purchase equipment produced by another nation; produce one part of "a family of weapons" (for example, one nation undertakes production of a short-range version of a weapon while others produce medium- and long-range versions); and set up a joint international production agency for equipment.

AGREEMENTS ON STANDARDIZATION

Because of the complexity of modern weaponry and equipment, standardization of materiel among the sixteen alliance nations seems an impossible objective. NATO therefore encourages agreements among nations to ensure that national equipment is compatible with that of other nations; wherever possible, equipment of one nation's forces is interoperable with that of others; and equipment components and spares are interchangeable. The list is not complete, but it illustrates the variety of forms that collaboration can take.

PRODUCTION LOGISTICS COMMITTEES

The senior committee that deals with production logistics is the Conference of National Armament Directors (CNAD), which reports directly to the North Atlantic

Council. The CNAD is responsible for promoting more effective use of available resources through cooperative equipment programs and increased standardization and interoperability of weapon systems. It covers all research, development, and production of equipment for NATO forces.

The CNAD meets in full international session twice a year and holds routine meetings with representatives from the national delegations to NATO headquarters whenever required. While the meetings are used to make formal decisions, information flows constantly between the IS and nations. The CNAD provides a forum for the exchange of information on operational concepts, national equipment programs, and appropriate technical and logistic matters where cooperation can benefit nations. It also promotes discussion on longer-term research activities to provide guidance on future military needs through advances in technology or new scientific discoveries.

Because of the wide area covered by production logistics, there are many working groups and subcommittees subordinate to the CNAD. These can be categorized under two heads: committees dealing with equipment projects, and those dealing with topics common to all aspects of production logistics.

The main committees dealing with equipment projects are: the Naval Armaments Group, the Army Armaments Group, the Air Force Armaments Group, the Defense Research Group, and the Tri-Service Group on Communications and Electronic Equipment.

The committees undertaking activities of general concern to all the groups in the CNAD are the Group of National Directors on Codification (of equipment); the Group of National Directors on Quality Assurance; the Group of Experts on Safety Aspects of Transportation; the NATO Group on Acquisition Practices; the Group on Rationalisation and Design Principles, Test and Safety Criteria for Explosives Materials and Explosive Stores; and the Group on Materiel Standardisation. Their titles give some indication of the wide variety of work concerned with production logistics within NATO.

NATO Projects and NATO Production and Logistics Organizations

There are three common terms for the types of collaborative projects in NATO. A *NATO project* is any form of NATO armament production cooperation in which two or more member nations participate and participants commit themselves to report progress annually to the CNAD until the equipment is produced or the project is otherwise terminated. It must provide for the admission of other NATO countries, subject to original participants accepting reasonable and equitable conditions.

A *NATO project steering committee* comprises national representatives established by intergovernmental agreement between two or more member nations to coordinate, execute, and supervise an equipment procurement program that has qualified as a NATO project. Examples are the NATO Sea Sparrow, involving Belgium, Denmark, Germany, Italy, Norway, and the United States, and the NATO Sea Gnat, with Denmark, the United Kingdom, and the United States participating.

NATO production and logistic organizations (NPLOs) are special development and production management organizations set up as required. Examples are the NATO Hawk Production and Logistic Organization (NHPLO) and the NATO Multi-Role Combat Aircraft Development and Management Organization (NAMMO).

Phased Armaments Programming System

The Armaments Planning Review conducted periodically within the alliance seeks to identify opportunities for armaments cooperation. Based on national information and the advice of NATO military authorities, suitable cooperative development candidates are then presented to nations.

In 1977, the Phased Armaments Programming System (PAPS) was introduced to provide a systematic and flexible framework for production cooperation. Recognizing the sovereignty of member nations in making decisions on equipment, PAPS also utilizes the modus operandi and structure of NATO to fulfill the most pressing military equipment requirements, adapt to political, economic, and technical realities, and establish and maintain broad cooperation throughout the life cycle of the equipment. The concept of PAPS is straightforward and follows the principle that there are a finite and fairly consistent number of points (milestones) in the life of a weapon system where the nature of the program changes. These are:

1. Evaluation of mission need
2. Prefeasibility studies
3. Feasibility studies
4. Project definition
5. Design and development
6. Production
7. In-service acceptance

At each of these points, decisions must be made regarding alternative courses of action—in particular, on whether nations will continue to participate through the next phases.

During the early stages of a typical weapons system the detail and form of the information exchanged between nations, as well as the nature and timing of these exchanges can be crucial to achieving the aims of the system. Although the options diminish progressively even during these phases, it is important to exploit whatever potential exists to involve the maximum number of interested nations.

The main advantage of PAPS is flexibility. Each project can be conducted according to its merits and needs. The key elements of PAPS are that nations discuss their requirements early and that they agree on a common specific requirement as the basis for each stage.

Consumer Logistics

Consumer logistics, which has been already defined in this article, encompasses the in-service life of all materiel, whether it is complete equipment, a component part, or an item of supply. It covers receipt of the item into service and its storage, transport, use, maintenance, modification, and eventual consumption or disposal out of service.

Senior NATO Logistician's Conference

The Senior NATO Logistician's Conference (SNLC) deals with consumer logistics at the high level and comprises civilian, military, international (NATO), and national representatives. It is the only committee or body that reports to both the North Atlantic Council and the Military Committee. It meets in joint military/civil session with two cochairmen: the Assistant Secretary General for Infrastructure, Logistics, and Civil Emergency Planning and the Deputy Chairman of the Military Committee.

Some indication of the range of subjects covered by the SNLC can be seen in its terms of reference, which include

1. To review, assess, and evaluate, in conjunction with other NATO authorities and bodies as appropriate, the overall consumer logistics organizations, plans, procedures, and capabilities with a view to enhancing the performance, efficiency, sustainability, and combat-effectiveness of alliance forces
2. To provide the focal point and forum for the consideration of consumer logistics matters
3. To address consumer logistics problems of the alliance, particularly those that have resisted solution elsewhere, as well as those that are not within the purview of other NATO bodies
4. To provide consumer logistics advice and assistance to the North Atlantic Council and the Military Committee in developing the policy needed to resolve consumer logistic matters and problems, reporting at the same time any areas of concern and forseeable consequences thereof
5. To recommend procedures to the North Atlantic Council and Military Committee for implementing consumer logistic policy
6. To direct the work of any subordinate bodies that may be created in specific consumer logistics fields of endeavor

The committee examines a wide scope of consumer logistics that may affect NATO's overall military efficiency. It has the authority to investigate any aspect of consumer logistics not covered elsewhere, and it has an element of executive authority. The terms of reference reflect the realization that many of NATO's logistic problems spread across the normal functional areas of its organization and need coordination across the entire military/civil field.

Subordinate Consumer Logistic Committees

Because the main emphasis of consumer logistics in NATO is on military effectiveness—practical problems on and behind the battlefield—several subordinate committees or bodies dealing with consumer logistics function in the NATO military headquarters and report to the Military Committee. Two are of particular importance.

The Allied Command Europe Logistic Coordination Centre (LCC), located at SACEUR's headquarters, is the senior logistic forum in Allied Command Europe (ACE). It provides a permanent link for consultation and coordination in crisis or war on logistic requirements between the Allied nations and SACEUR. Its task can include any logistic problem arising in NATO Headquarters, at Brussels, in the Supreme Headquarters Allied Powers Europe (SHAPE), or anywhere in ACE that affects SACEUR's assigned or earmarked forces. The LCC takes action on reports of national surpluses and deficiencies that the national ministries of defense note and process requirements for emergency logistic assistance between nations using agreed-on procedures.

The second subordinate committee is the SACLANT/CINCHAN Logistic Coordination Board (LCB). It was established to provide a forum for liaison among NATO commanders and nations on maritime logistic matters. Its responsibilities include publishing, coordinating, and reviewing SACLANT's Logistic Planning Guidance; reviewing and studying maritime logistic problems and making subsequent recommendations of action to the SNLC; and developing and promoting maritime interoperability, standardization, and cross-servicing. It also reviews levels of logistic coordination, reallocation, and planning within the areas for which SACLANT and CINCHAN are responsible; reviews maritime logistics and medical standardization agreements (STANAGS) with follow-up recommendations for change to the NATO Military Agency for Standardization (MAS); and develops and updates standard logistic criteria and medical plans to conform to SACLANT/CINCHAN plans.

Because of the mixture of national forces to be found at every NATO level of command, multinational logistic conferences take place down to army corps level. They work on such topics as cross-servicing and mutual assistance among national contingents; detailed movement plans; and location of supply, ammunition, and fuel stocks.

Civil Emergency Planning

The aim of national defense plans in the civil sector is to enable nations to proceed as smoothly as possible from a peacetime to a wartime footing, and then to continue to function effectively. It is self-evident that nations' plans for resource management in war provide a stronger economic base if they are coordinated and directed toward a common goal. This is the philosophy of civil emergency planning within NATO.

It is sometimes said within NATO and among member nations that civil emergency planning is a separate issue from logistics. There are various reasons for this. The main one is that the purpose of civil emergency planning is to look after the civil population of Europe, not to maintain its armed forces. Another reason sometimes given is that civil emergency planning is so wide a subject that it would be impossible to include it with logistics in any manageable organizational structure. However, it is important enough to be included in any review of NATO logistics.

The NATO concept of flexible response to attack rests as much on civil preparedness as it does on military readiness. The civil resources of NATO nations have to be mobilized rapidly to sustain the civil population as well as to provide support for the rapid mobilization, deployment, reinforcement, and resupply of allied military forces in Europe. The ships, aircraft, and road transport controlled by NATO's armed forces provide only a fraction of the transport capacity needed in war. The majority comes from civil sources, as will the fuel, food, and industrial support required. This fact cannot be too strongly emphasized.

SENIOR CIVIL EMERGENCY PLANNING COMMITTEE

The Civil Emergency Planning Director at NATO Headquarters provides staff support for the Senior Civil Emergency Planning Committee (SCEPC). SCEPC meets monthly in a permanent session, with representatives drawn from the national delegations to NATO; it meets in full plenary sessions twice a year, with representatives attending from national capitals.

The civil emergency planning staffs perform tasks based on the Ministerial Guidance issued every four years. From this, work programs are produced that identify tasks, priorities, and targets in specific areas. At similar intervals, nations are asked to respond to the Civil Emergency Planning Questionnaire on governmental readiness, energy, food and agriculture, civil air transport, industry, ocean shipping, inland surface transport, civil defense, civil communications, and manpower. A full report on the state of civil preparedness, with an assessment by the NATO military authorities, is submitted to the North Atlantic Council every four years.

PLANNING BOARDS

SCEPC coordinates and guides the activities of eight specialist planning boards and committees: the Planning Board for Ocean Shipping (PBOS), the Planning Board for European Inland Surface Transportation (PBEIST), the Civil Aviation Planning Committee (CAPC), the Food and Agriculture Planning Committee (FAPC), the Industrial Planning Committee (IPC), the Petroleum Planning Committee (PPC), the Civil Defense Committee (CDC), and the Civil Communications Planning Committee (CCPC).

The work of these NATO planning boards and committees on civil preparedness is conducted in parallel with the planning branches of national ministries and the NATO staffs. Wartime requirements are assessed, shortfalls are identified, and remedial action is planned. Military demands on civil resources are identified by joint civil/military meetings held in national capitals and coordinated where appropriate by the planning boards in conjunction with the NATO military staff. The flow of information among national civil ministries is constant, especially among government transport agencies, who already cooperate on such matters as trains that cross national borders, port facilities, and airport usage.

NATO CIVIL WARTIME AGENCIES

NATO Civil Wartime Agencies (NCWAs) are set up in periods of tension or emergencies. They deal with specific problems as they arise, and continue the consultative and coordination roles performed by the peacetime planning boards.

The NCWAs have the same basic structure: a directing body in which all participating nations are represented and an international staff of experts (air traffic controllers, regional controllers on railways, etc.). They are not official government committees sitting in isolation; the members are specialists in their field and include representatives from the NATO military authorities to ensure that they are quickly apprised of military problems and needs. They play a vital role in NATO logistics.

The powers and functions of the NCWAs vary. Like NATO itself, they are not supranational, but operate on the basis of international cooperation. When, as in the case of the Defence Shipping Authority (DSA), assets are made available to them, they can be said to have executive authority. In other areas, such as the NATO Refugee Agency (NRA), their task remains one of liaison, consultation, and coordination. They work with national organizations including such bodies as national oil boards and national shipping authorities.

The Defence Shipping Authority is a wartime agency that manages the NATO pool of ocean-going shipping. It allocates vessels to member nations with the aim of meeting their civil and military requirements. The DSA has two branches, in Europe and America, and consists of the Defence Shipping Council, the Committee of National Representatives, a Mediterranean Shipping Group, Route Management Organizations located for the convenient management of their respective liner routes, and an international staff headed by a managing Director reporting directly to the Defence Shipping Council.

The Agency for the Co-ordination of Inland Surface Transport in Central Europe (ACTICE) coordinates the assignment, relocation, improvement, maintenance, and repair of all inland surface transport in Central Europe. It works through five specialized commissions: the Road Transport Commission, the Inland Waterways Commission, the Railroad Transport Commission, the Ports and Beaches Commission, and the POL Transport Commission.

The Agency for the Co-ordination of Transport in the Mediterranean (ACTIMED) was formed to deal with the particular geographic conditions of the Mediterranean area and sea transport among the nations bordering it.

The NATO Civil Aviation Agency (NCAA) is responsible for mutual assistance, exchange of information, and the coordination of civil aviation efforts. It consists of the NATO Civil Aviation Board (NCAB), a body of national representatives empowered to take decisions on behalf of their governments on policy issues, and the Bureau for Co-ordination of Civil Aviation (BOCCA), which coordinates the use of civil aircraft made available by member nations and the arrangements for civil aviation support of military reinforcement operations.

The Central Supply Agency (CSA) coordinates the plans of member nations to ensure continued availability and equitable distribution of essential supplies (food, agriculture, and industry).

The NATO Wartime Oil Organization (NWOO) ensures the continued availability and equitable distribution, for both military and civil purposes, of essential supplies of oil and petroleum products in support of the alliance. It has two branches in crisis or war, one in America colocated with CSA and DSA branches, and the other in Europe located with the European DSA branch. It has a NATO Oil Executive Board at each location and operational staff.

Finally, the NATO Refugee Agency (NRA), located in Europe, coordinates actions of member nations on refugee problems of international significance.

It took some time for the importance of civil emergency planning to be recognized, since the multiplicity of factors involved in planning the military logistics of an alliance of sixteen nations tended to obscure the significance of civil involvement. Nevertheless, an appreciation of the importance of civil emergency planning is vital to an understanding of NATO logistics. It forms the basic support structure without which NATO cannot operate effectively in war.

Areas of Coordinated Logistic Planning

The previous sections have concentrated on the organization and structure of logistic planning in NATO. Those that follow describe some of the problems NATO logisticians face and the methods by which they would operate in war, and illustrate the importance of cooperation and joint planning between national and international civil and military organizations.

HARMONIZATION OF THE COMMUNICATIONS ZONE

The NATO definition of the *communications zone* is the rear part of the theater of operations (behind but contiguous to the combat zone) which contains the Lines of Communications (LOC), establishments for supply and evacuation, and other agencies required for the immediate support and maintenance of the field forces.

General. Every logistician knows the importance and the difficulties of planning a communications zone for a single national force. When, as in the case of NATO, some sixteen nations are involved, the problems become even more complex. In Central Europe, for example, one country acts as host nation to the armed forces of six reinforcing nations.

The result is that, in NATO, each communications zone will have in front of it combat forces from different nations organized in different ways, with different weapons and equipment, and with differing priorities of support. The communications zones themselves will contain support units from various countries, each with different resupply and maintenance systems, and dependent on a variety of national and international civil agencies. The coordination and cohesive organization of these functions and facilities have been given the term "harmonization of the communications zone."

Background. In the first two decades of NATO, when the "trip-wire" philosophy was followed, the problems of the malpositioning of NATO's forces and their supplies was secondary to the nuclear response. When it became clear that NATO would have to expand its conventional capabilities to fight a war based on "flexible response," the problems assumed a new importance. The numbers of men and tonnages of supplies involved meant that far greater importance had to be given to the communications zone, and that NATO commanders had to coordinate the logistic support plans of nations far more closely than they had done previously. This coordination by NATO commanders may take the form of setting priorities, if these are in doubt; assessing national plans; and, if appropriate, consulting with host and reinforcing nations during their bilateral negotiations. This last factor is still a cause of controversy in NATO, because it seems to infringe upon the sovereignty of nations. However, it was exactly this sort of coordination by NATO commanders that was agreed upon by nations in the North Atlantic Council resolution of 1952.

Responsibilities. In 1978, a NATO study (the Long Term Defence Programme) listed 56 recommendations for the harmonization of the communications zone. This is some measure of the magnitude of the problem. However, it is possible to summarize the recommendations and, in so doing, to describe the elements of this harmonization process. Nations plan the movement of their combat forces to arrive at the right place and at the right time; the movement and correct location of stocks and facilities; the organization, movement, and functioning of their logistic support forces; and the provision of logistic support to allies and/or the receipt of support from them. In these respects, NATO commanders have the authority to: advise on priorities, when these are in doubt; assess national plans and bilateral agreements between nations; and request alterations, amendments, or additions to the plans where this is necessary to accord with NATO policies.

Both these groups of functions and powers depend upon a third element: a clear understanding and agreement of the relationship in peace between NATO commanders and the national civil and military authorities responsible for providing the forces allotted to NATO in war. Only by the close coordination of these functions is NATO able to create a robust logistic harmonization zone in its different regions.

A great deal of work has been done in this complex area. For example, logistic cooperation has now reached the stage where it is commonplace for British ammunition to travel by Belgian railway trucks through two different countries before being unloaded by German labor. This type of logistic activity can be planned. What is not yet properly appreciated is the full extent of logistic cooperation that will be necessary at lower levels. Despite much hard work by NATO staffs, national forces are still reluctant to have their logistic assets transferred to forces of another nation. Such questions as joint use of petrol depots or field hospitals must be agreed on at the brigade and divisional level. Practical problems of this type do not lend themselves to planning in many cases, but they demonstrate how the efficiency of the communications zone depends on cooperation and coordination at every level, not just in international committees.

REINFORCEMENT PLANNING

The theory of reinforcement planning—of moving national forces and equipment to where they will be needed in war—is easy to understand. Practiced in an alliance like NATO, the matter is more complex, and includes:

- The establishment of operational lines of communication to sustain military forces, while providing for the joint use of facilities required to sustain the civil population during crisis and war.
- The integration of multinational civil and military requirements.
- The coordination of peacetime planning and the control of available multinational civil and military movement resources in wartime.

In turn, essential functions of the lines of communication (LOC) include reception and movement planning; engineering support; security and defense; movement and coordination; medical support and hospitalization; coordination of changes in plans; identification of losses of critical resources; re-evaluation of priorities; monitoring of the progress of reinforcing formations and critical materiel; and ensuring coordinated reaction to the demands of the civil and military authorities. The effectiveness of reinforcement planning depends on a network of organizations: forces using the lines of communication, host nations, NATO military authorities, NATO Civil Wartime Agencies, and the civil and transport resources of allied nations.

As one example of reinforcement planning, the U.S. Air Force is committed to reinforcing Europe with 1,200 to 1,500 aircraft in a matter of days. While the transit of the aircraft and crews is a task for the military movement staffs, their operational effectiveness depends on the provision of bases and facilities from which they can operate, as well as on sufficient logistic resources to enable them to do so. However, there are insufficient resources to move the logistic backup of men and materiel for such a large force within the same timeframe. This problem has been met within NATO by the greatest single asset in NATO logistic planning: host nation support.

HOST NATION SUPPORT

Host nation support (HNS) is the civil and military assistance rendered in peace and war by a host nation to allied forces and NATO organizations located in or in transit through the host nation's territory. In the case of the U.S. reinforcement aircraft, reception facilities have been made available at 70 allied air bases. Some 85 storage locations have been provided in twelve host nations for pre-positioned fuels, ammunition, and spare parts to enable the aircraft to operate until normal resupply comes into operation.

The concept of HNS, and NATO's dependence on it, emphasizes the error of regarding logistics as being solely a national responsibility—that is, the sole responsibility of a nation providing combat forces. The distances involved, the short time-scale and the large number of combat troops needed to produce an effective alliance defense simply do not allow simultaneous movement of the necessary logistic support. By pre-positioning equipment and materiel, many of the reinforcement problems of NATO can be solved. HNS is the mechanism by which it is achieved.

The general guidelines for planning HNS can be summarized as follows: The first stage consists of the initial agreement or Memorandum of Understanding between the nations concerned; accepting the need for a bilateral agreement and authorizing further work. At the second stage, Technical Arrangements are made—more detailed negotiations and subsequent agreements between staffs of the ministries of defense and other government departments of the two nations, covering quantities and types of support to be provided, and taking into account legal and financial arrangements. The third stage, Joint Lines of Communication/HNS planning, involves detailed consultation between the national and NATO staffs to achieve an agreed-upon joint plan, including specific quantities, timings, and alternatives, and the preparation of detailed implementation plans.

Providing support for a reinforcing nation can reduce the combat potential of the host nation itself. Furthermore, allowance must be made for the needs of the host nation's civilian population in war. NATO has found that the correct balance depends on complex and delicate negotiation, but its successful conclusion adds immeasurably to the effectiveness of alliance logistics.

MOVEMENT AND TRANSPORTATION

The need to coordinate NATO movement planning stems from three factors: The first of these is the change from NATO's philosophy of "trip-wire" nuclear response to that of "flexible response." The latter, based on conventional combat until the nuclear threshold is reached, entails the movement of far greater numbers of reinforcements than had been planned previously. Second, in peacetime only a small part of NATO's forces are located in the area they would defend in war. In a crisis, the great majority of NATO assigned and earmarked forces have to move to deployment areas within short time limits. Finally, in the initial stages of reinforcement, transport facilities are unlikely to be adequate to meet all the civil and military needs of the alliance simultaneously.

Movement requirements. In 1977, NATO Ministers directed that reinforcements should reach potential areas of conflict either before aggression takes place, or, if warning time is very short, in time to influence the initial course of hostilities. As a result, the three major NATO commanders conducted a study into the problems of reinforcing Europe and agreed on the concept known as the Rapid Reinforcement Plan. This aimed at easing the trans-Atlantic movement problems by pre-positioning equipment in Europe and by providing services from host nations to reduce the number of logistic personnel accompanying the incoming forces.

Even with this assistance, the commitment is still a heavy one. Initial movement involves approximately 1 million metric tons of stores from the United States, and 130,000 men and 100,000 metric tons of stores from the United Kingdom. Follow-up movement involves approximately 500 shiploads per month to support the armed forces, and 85 million metric tons per month for the economic survival of Europe.

Although the trans-Atlantic commitment is the largest single task, it is only part of the alliance movement plan. The armed forces of European nations travel shorter distances to their deployment areas, but this is offset by problems of terrain, availability of transport, movement control in densely populated areas, and the provision of transport facilities for other incoming alliance formations.

Movement planning in NATO requires agreement on sound procedures, planning, and the coordination of all available transport resources. Movement planning to achieve these objectives involves not only national planning staffs, but will require international coordination by NATO commanders and NATO Civil Emergency Planning Boards/Committees.

Movement planning organizations. Each nation has movement and transport staffs responsible for the coordinated (civil/military) planning of the movement requirements of their forces, as well as for such other NATO forces for which they act as host nation.

At SHAPE, the movement staff in the Logistic Division deals with policy and guidance for movement planning and ACE-wide coordination, in conjunction with NATO subordinate commanders and the NATO Civil Emergency Planning agencies. They work closely with SACLANT and CINCHAN on the priority of movement of forces and with the NCWAs on the allocation of resources.

The three NCWAs with specific responsibilities for movement are the Defence Shipping Authority, the NATO Civil Aviation Agency, and the Agency for the Co-ordination of Inland Surface Transport in Europe. Their functions are described in the following paragraphs.

1. Movement by Sea. In an emergency or war, 95 percent of the materiel and supplies for reinforcing nations (Canada, Portugal, United Kingdom, and United States) will move by sea. NATO governments have committed ships to a NATO pool from which allocation will be made to meet alliance needs. This pool of ships will be managed by the Defence Shipping Authority, which is staffed by shipping experts from NATO nations. Nations are responsible for the operation of shipping allotted to them.

2. Movement by Air. Although in principle nations should meet their own airlift requirements, arrangements have been made to make the fullest use of alliance resources. Aircraft available for use are declared to NATO commanders on a planned or opportunity basis. There is a set procedure for making such allocations and, while the provision of military aircraft is dealt with by the staff at NATO military headquarters, requests and subsequent allocation of civil aircraft are dealt with by the NCAA (NATO Civil Aviation Agency).

3. Inland Surface Movement. In wartime, NATO's inland surface transport resources (railways, road transport, and canals) remain under national control for all practical purposes. The emphasis is therefore not on allocation of resources to nations, but on the coordination of the resources they already hold. In peacetime, this is carried out by the Planning Board for European Inland Surface Transport (PBEIST), which has separate subcommittees for Northern, Central, and Southern Europe.

Road Transport is planned and operated by nations, using standardized NATO convoy rules. Where necessary, it will be governed by the regulations of the nations through which it passes. If road transport assistance is needed, requests are passed to the host nation, based on bilateral agreements between the nations concerned or using standard NATO request procedures.

Rail transport is already closely coordinated among the European railway organizations, and through-running of trains is normal. In crisis or war, changes will be made to priorities rather than to operating methods. The railway authorities work closely with national planning staffs and, through them, with the NATO military authorities. PBEIST has a subcommittee on rail transport which national representatives attend. The Central Region nations have arranged to coordinate the use of freight wagons in wartime by the creation of a Central European Wagon Pool.

Inland waterways and port operations are dealt with by a special PBEIST subcommittee dealing with ports, beaches, and inland waterways. The Rapid Reinforcement Plan has increased the importance of efficient use of ports, and NATO has initiated studies to match civil and military requirements to the ports and port facilities available.

SUPPLY AND MAINTENANCE

The movement of a reinforcement brigade from the United States to its combat location in West Germany involves several national and international agencies and illustrates clearly the extent of cooperation and joint planning within the NATO alliance. In the areas of supply and maintenance, however, different factors apply.

The distinction between the two fields of activity is a subtle one. Movement planning in NATO depends on a variety of movement agencies using their normal procedures and organization to cooperate on joint action to meet an emergency. The basis of supply and maintenance planning in NATO is to alter or adjust national logistic practices and methods in peacetime, so that no changes will be necessary in war. This entails laborious, lengthy consultation and negotiation to achieve agreed-on standard supply and maintenance procedures. It attracts little publicity, but it is a vital factor in the logistic strength of the alliance.

Rationalization procedures. The overall aim is to improve NATO's collective logistic effort by rationalization. This term is used in NATO for

> any action that increases the effectiveness of defence resources committed to the Alliance. Rationalisation includes consolidation and reassignment of national priorities to higher Alliance needs, standardisation, specialisation, mutual support or greater co-operation. Rationalisation applies to both weapons/materiel resources and organisational and procedural matters.

While most of these terms are generally understood, the complexity of modern weapon systems has made it necessary for NATO to define standardization. It is now taken to mean the process of developing concepts, doctrines, procedures, and designs to achieve the most effective levels of compatibility, interoperability, interchangeability, and commonality in the fields of operations, logistics, and materiel.

While this definition may appear at first superfluous, it has proved necessary in an organization that covers the wide field of NATO logistics. The lengthy process of negotiation and consultation within the alliance on maintenance procedures or weapon and equipment research makes it vital to distinguish between spare parts that are identical or compatible or interchangeable. Depending on their purpose, research, money, and time can be saved if the lesser level of need can be accepted.

The negotiations on such detailed issues can be tedious and time-consuming, but they are the building blocks of NATO logistics. Slow, steady work in these areas draws the logistic procedures and practices of the alliance armed forces ever closer together.

By agreeing on standard methods of identifying equipment and terminology, and by reaching accord on standard ammunition and procedures for requesting logistic assistance, NATO has started to create a common logistic structure across the armed forces of its member nations.

Military Agency for Standardization. The bulk of standardization work in NATO is dealt with by the Military Agency for Standardisation (MAS) and its Service Boards, or in CNAD and its subcommittees. In general terms, MAS deals with standardization of doctrine, procedures, and equipment in current use. It is user-oriented and concentrates on aspects of direct concern to NATO's armed forces. CNAD and its subcommittees look principally at future weapons and equipment and are oriented to research, development, and production. MAS is located at NATO Headquarters and works under the authority of the Military Committee with the aim of rationalizing tactical concepts, procedures, materiel, training, and terminology among member nations. It maintains close liaison with technical groups working on equipment standardization and operates through a system of working parties. These comprise national representatives who meet to formulate, discuss, and negotiate a specific topic for standardization. When agreement is reached, it is registered by MAS and published as a STANAG (Standardization Agreement) or as an AP (Allied Publication). There are now some hundreds of these publications covering every aspect of military administration, from an agreed-upon list of drugs used in military hospitals to a common procedure for refueling ships at sea.

Classes of supply. Over the years, nations have developed their own systems for classifying military materiel. This led to obvious difficulties in logistic cooperation and an agreed-upon simple method of grouping was clearly necessary. The need was particularly important for passing logistic information in the field to an allied commander. If logistic resources could not be readily identified and grouped together, it would be difficult for a formation commander to ascertain what his logistic situation was. Because of this, a simple five-class system has been adopted, commonly known as the "NATO Five Classes of Supply":

Class I: Items that are consumed by personnel or animals at an approximately uniform rate, irrespective of local changes in combat or terrain conditions (e.g., food and forage).

Class II: Supplies for which allowances are established by

tables of organization and equipment (e.g., clothing, weapons, tools, spare parts, vehicles, etc.).

Class III: Fuel and lubricants for all purposes, except for operating aircraft or for use in weapons such as flame throwers (e.g., gasoline, fuel oil, greases, coal, etc.); for air forces, Class IIIA comprises aviation fuel and lubricants.

Class IV: Supplies for which initial issue allowances are not prescribed by approved issue tables (e.g., fortification and construction materials, and additional quantities of Class II items such as vehicles).

Class V: Ammunition, explosives, and chemical agents of all types.

Codification. The NATO Codification System is an agreed-upon standard procedure for identifying, classifying, and stock-numbering supply items for NATO nations. It is designed to achieve maximum effectiveness in logistic support and to facilitate the data management of materiel.

The system has been agreed upon by all members of the alliance for identifying equipment and supplies, especially common equipment for NATO projects and equipment used by two or more countries and procured from another. The system, based on the U.S. Federal Catalogue system, is governed by the NATO Group of National Directors on Codification of Equipment, one of the cadre groups of the CNAD. The basis of the system is that each "Item of Supply" has a single item name, a single uniform classification, a single uniform identification, and a single uniform NATO stock number. In practice, the user normally works from two main elements: the item name and the NATO stock number. Approved item names are used whenever possible and are listed in national handbooks, which give item names with definitions and item name codes. The U.S. Handbook, for example, contains 28,000 approved item names as well as 11,000 colloquial names. A NATO stock number consists of thirteen digits and is based on the uniqueness of an item of supply. An example of a NATO stock number (NSN) is 1005-13-123-4567, where

- 1005 is the NATO Supply Classification Code
- 13 is the code for the National Codification Bureau
- 123-4567 is the nonsignificant number issued by the National Codification Bureau

Quality assurance. As soon as NATO nations started to implement collaborative equipment programs in which the manufacture of components was undertaken by industries of different countries, it became clear that uniform standards of quality control were required. The task of drawing up these procedures was given to the Group of National Directors of Quality Assurance, another of the CNAD cadre groups. It has established basic doctrine and policy in a series of NATO Allied Quality Assurance Publications (AQAPs). These are not intended to supersede national policy, but to provide a uniform framework within which NATO nations and agencies can place contracts for materiel with the confidence that supplier nations will follow adequate quality assurance standards.

In parallel with their work on quality assurance, the committee is also responsible for NATO Requirements for Reliability and Maintainability. In NATO, *reliability* is the ability of an item to perform a required function under stated conditions for a specified period of time. *Maintainability* is the ability of an item, under stated conditions of use, to be retained in or restored to a specified condition when maintenance is performed by personnel having specified skill levels under stated conditions and using prescribed procedures and resources. These standards are published in Allied Reliability and Maintainability Publications (ARMPs).

Storage. In addition to the normal facilities in use by national contingents, NATO has developed its own storage program. The first general category is storage for a NATO project or organization. Under this category, nations may either accept responsibility for storage of NATO equipment or share the costs, dependent on their participation in the project. Such arrangements are normally subject to specific agreement for each project.

The second category of storage consists of the four types of storage facilities funded by the NATO common infrastructure program (which is explained later): (1) FSTS—Forward Storage Sites; (2) POMSS—Pre-positioned Organisational Materiel Storage Sites; (3) TR Sites—Theatre Reserve Sites; and (4) AS—Ammunition Storage Sites. FSTSs provide facilities for the peacetime storage of combat-essential and fast-moving supplies for land forces (including barrier and other materials) in the forward areas of the combat zone to ensure immediate availability of combat support for assigned and earmarked forces. They are not confined to reinforcing forces from outside Europe. They stem from the recognition that the storage depots of nations were poorly located to support conventional operations. A program was therefore agreed upon to deploy materiel depots to more appropriate locations to support the type of conflict now envisaged.

The other three types of storage come under the Reinforcement Support Category: facilities to support alliance forces coming from countries outside the European mainland. The overall aim of these facilities is to ensure the fastest possible combat readiness of these reinforcing forces, by pre-positioning equipment and supplies in the area of planned deployment.

Maintenance. In common with supply, maintenance is a basic logistic task that is the responsibility of nations. In cases of war, maintenance units—like all field support units—within each army corps come under the operational command of their parent formation, part of the NATO integrated operational force structure. In rear of each Army Corps, units remain under national command.

The opinion is sometimes voiced that in a future conflict there will be insufficient time to repair any equipment. This is a possibility, but recent experience has shown that for every four tanks hit in combat, three can be repaired. This proved to be a battle-decisive factor in the Israeli campaign of 1973.

An efficient maintenance organization is an essential component of NATO's defensive structure, and the terms of reference of NATO commanders include the right to full information about maintenance and resupply facilities in their command and the encouragement of nations to make bilateral and multilateral agreements on cross-repair.

Aircraft cross-servicing. Although certain NATO aircraft are built as collaborative projects, many different national models exist, and a great deal of work has been accomplished in standardizing components and maintenance. This is reflected in the large number of STANAGs (over 175) covering NATO aircraft. One major advance is the aircraft cross-servicing system now in operation in NATO. In general terms, the system enables aircraft of one NATO nation to be serviced at the airfields of another. Because of the different technical equipment required (specialized refueling equipment, calibration sets, etc.), the system does not allow all NATO aircraft to be serviced at all NATO airfields. However, it does indicate which airfields can provide facilities for specified aircraft, through regularly updated publications.

Medical Support

Although medical planning is considered in NATO to be part of logistics, there are certain important distinctions. Logistics deals primarily with equipment; medical planning deals primarily with men. Medical units have protection under the Geneva Conventions; logistics units do not. And lack of logistics can prevent a war being won, but lack of medical services can take away the will of men to fight it.

Medical planning in NATO includes cooperation between the civil and military departments of alliance nations and NATO staffs as well as detailed standardization measures. Although specialized medical research and standardization are carried out in many areas of NATO, the main responsibility lies with the four medical working parties in the NATO MAS.

A series of medical STANAGs and APs has been published across a wide range of medical activities. Some of these are published in each of the ten NATO languages and can therefore be used at the field and hospital level.

Executive Organizations and Logistic Agencies

Throughout this review of NATO logistics, two words—*cooperation* and *coordination*—have occurred repeatedly. This is unavoidable, because it is through these means that NATO works. However, certain areas of NATO logistics need more direct control. In these cases, NATO organizations have been established to manage specific projects or functions. These organizations have different tasks and authority, but are all responsible to NATO. Two of particular logistic importance are NAMSO and CEPS.

NATO Maintenance and Supply Organization

The NATO Maintenance and Supply Organization (NAMSO) is the principal logistic organization in NATO. It is structured at two levels: a legislative level comprising a Board of Directors supported by committees, and an executive level consisting of the NATO Maintenance and Supply Agency (NAMSA). The Board, directly responsible to the North Atlantic Council, provides policy guidelines to NAMSA and oversees implementation of its policy.

NAMSA's main task is to assist nations by supplying spare parts and providing maintenance and repair facilities to support various weapon systems. This assistance is available whenever two or more nations have the same system in their national inventories and choose to use NAMSA's facilities to provide support. The objective is to promote materiel readiness, improve logistics efficiency, and effect substantial savings.

NAMSA tailors its support to nations' needs and acts as a central agency to prevent duplication of effort. By adopting competitive bidding procedures and centralizing stocks of spares, it is able to provide an economic repair and supply facility for the equipment it supports. Its main sites are at Cappellen in Luxembourg and at Taranto in Italy, while other sites cater to specialized repairs.

Central Europe Pipeline System

The Central Europe Pipeline System (CEPS) provides the facilities for bulk distribution of fuel to NATO forces in Central Europe. Financed by NATO infrastructure funds, it crosses a number of national borders and has eight national users. It is part of the NATO Pipeline System, which has seven other networks in Turkey, Greece, Italy, Denmark, Norway, Portugal, and the United Kingdom. Like NAMSO, CEPS has its own management structure and is classified as a NATO Production and Logistics Organization (NPLO).

CEPS is one element in the NATO Petroleum, Oil, and Lubricants (POL) organization, which has several committees, reflecting the importance of those commodities to both the civil and military sides of the alliance. These committees look at various aspects of POL, including

- Civil preparedness within NATO to meet oil problems
- Bulk distribution and storage of fuels for military use (i.e., the NATO Pipeline System)
- Air base, naval base, and unit support
- Types of military fuels and their relationship to weapon systems, equipments, and vehicles
- Standardization, interchangeability, and research concerning POL and related products

Infrastructure

As well as the special management organizations created to deal with specific projects, there are many other areas in NATO where a cooperative approach is far more economical than individual national efforts. Where this can be achieved by common funding, the NATO infrastructure system is used. *Infrastructure* in the NATO context means those fixed installations that are necessary for the deployment and operation of NATO assigned forces. Examples are airfields, signals installations, military headquarters, fuel pipelines, radar installations, missile sites, and support facilities for reinforcement forces.

Under the infrastructure system, common financing of projects is agreed upon according to a cross-sharing formula, which is reviewed at intervals. Under this formula, nations make financial contributions to approved construction projects over a fixed number of years. The procedure of agreeing on the scope of projects to be included in the program is lengthy and complex, but the result is that airfields and communications installations can be constructed for NATO use in countries that would otherwise be unable to pay for them.

Future of NATO Logistics

There are three major areas of concern in NATO logistics: the authority of NATO commanders over nations' logistic assets; the practice of cross-servicing and mutual support by field formations; and the future of equipment production in NATO. As long as nations are reluctant to allow NATO commanders to exercise control over their logistic assets, NATO's logistic posture is suspect. Coordination and cooperation are excellent principles for peacetime negotiation, but they cannot replace the authority of a commander in war. The exigencies of combat may overcome nations' reluctance to accept this authority, but the danger is that they may not recognize their mistake until it is too late.

The navies and air forces of NATO nations have practiced mutual support and cross-servicing for many years. However, it is still exceptional for land forces to do so. It seems of little use for NATO committees to spend months agreeing on a procedure whereby armies can use each other's ammunition, if the procedure is never put into practice. Until nations make cross-servicing and mutual logistic support a normal aspect of peacetime military routine, they cannot expect to gain the benefits of it in war.

NATO nations are prepared to fight together, to make financial sacrifices to ensure their joint security, and to agree that their national forces accept orders from a foreign commander in war. They have collaborated in many joint equipment projects. In recent years, they have come to recognize the advantages of the United Kingdom and the United States developing mines for use in the Atlantic while European nations develop weapons for the shallow seas around the English Channel. This is a start. Is there any reason, apart from national pride, why they cannot extend this philosophy? For example, an agreement whereby the United States manufactures all the combat aircraft for NATO, Germany makes all the tanks, the United Kingdom makes all the frigates, and so on, would offer immeasurable advantages. The savings in national defense costs, the advantage of an assured market, the simplicity of spares supply, and the maintenance of equipment make such an idea very attractive. All that is lacking is the political will. There are many improvements still to be made in NATO's logistic philosophy, as well as its practical application.

N. T. P. MURPHY

SEE ALSO: Assistance, Mutual; Civil Emergency Planning; Cross-servicing; Host Nation Support; Lines of Communication; NATO; NATO Policy and Strategy; Standardization and Interoperability.

Bibliography

Gabriel, C. A. 1984. Tactical air reinforcement for Europe. *NATO's Sixteen Nations*, special issue, February.
Lawson, R. L. 1984. Wartime host nation support. *NATO's Sixteen Nations*, February/March.
NATO Information Service. 1984. *Facts and figures: The North Atlantic Treaty Organization.* Brussels: NATO.

LOGISTICS, NAVAL

Naval logistics encompasses the art and science of planning for and providing the goods and services necessary for creating and sustaining naval forces at sea, whether in peace or war. It deals specifically with the resources needed to ensure that ships and their crews are combat ready, and with the processes necessary for identifying, obtaining, and delivering those resources.

At the tactical level, naval logistics deals with the readiness of combat forces to fight. Therefore, the responsibility for overall logistics status lies with command—that is, the commanding officer of a ship or aircraft squadron, or the commander of a naval task force.

At the theater level, naval logistics includes the entire infrastructure of command and control (C^2), the transportation network, stock points, storage, and production that provide operating naval forces the wherewithal for war at locations required to execute a strategic plan.

Maritime logistics are the resources and processes conducted on or adjacent to the oceans that sustain a nation and all of its armed forces in wartime. Maritime logistics encompasses naval logistics plus the movement of commodities across the seas to sustain the fighting forces and populace of a nation at war, including merchant shipping, manpower, port facilities, and organizations such as the U.S. Naval Control of Shipping Office, which organizes and monitors the movement of merchant shipping in wartime, whether in convoy or sailing independently. In addition, logistics over the shore (LOTS) provides for the special characteristics and requirements of support for an amphibious landing. Maritime logistics are not directly addressed herein, but many of its additional resources and functions may be inferred by extrapolation.

Overview

An understanding of naval logistics is best approached at two levels. First, the logistic needs of a ship, its crew, and its weapons will be examined. It is the ship, operating independent of land-based facilities, with its ability to use most of the earth's surface as its highway, that makes navies unique. These capabilities also bring with them special logistical liabilities. Second, and on a grand scale, a navy requires large expenditures in order to provide the resources necessary to keep the ships operational. The decisions about how much to spend in support, and the types of support provided, determine the ultimate ability of a navy to sustain readiness over great distances in the face of an enemy.

Logistics deals with resources and processes. Fuel, food and water, munitions, and supplies are examples of logistics resources, stored aboard ship, consumed in operations, and requiring resupply if the ship is to continue its mission. There are four major processes involved with naval logistics. Three of them—requirements determination, procurement (acquisition), and distribution—deal with the resources listed above. The fourth, maintenance, centers on the ship, aircraft, and their equipment. Of the four, three are similar to processes that take place in land combat. Distribution, however, has many aspects unique to naval logistics.

Logistics as a Function of Command

In 1950, Capt. Henry E. Eccles, U.S. Navy, in his seminal work *Operational Naval Logistics* (1950), said: "The determination of requirements for military operations is a responsibility and prerogative of military command." Eccles drew on his study of U.S. Navy wartime experience to conclude that logistics could not be effective when it was managed under a separate command structure. The readiness of his ship is the captain's responsibility, and the status of his weapons, fuel, combat systems, and personnel will determine both the options available to him and the risks he may decide to take in a combat situation. The number of tactical alternatives available to a commander will be seriously reduced when sufficient assets are not available. Prominent among a commander's many concerns before combat is the need to top off his resources.

At the theater level, the same close involvement of the commander with logistics is necessary in the preparation of strategic plans and their effective execution. Logistics, just as much as the numbers of ships and aircraft in the order of battle, determines the limits in time and distance within which naval operations can be conducted.

First in importance are the weapons systems; having as many missiles, gun shells, and chaff canisters as the ship is capable of holding will allow it to survive, endure, and prevail. Speed and endurance—for maneuvering, for repositioning, for defense—are vital capabilities for a fleet commander to have at his disposal. Consequently, the importance of refueling is never as critical as it is in the period prior to combat.

In a book published in the United States in 1917, Lt. Col. Cyrus Thorpe of the U.S. Marine Corps wrote: "Strategy and Tactics provide the scheme for the conduct of military operations: Logistics provides the means therefor." It is incumbent on the commander to know what means he has available, and to plan strategy and tactics within the capabilities those means provide.

The Logistics Pyramid

As with logistics on land, the ultimate consumer of a resource is at the apex of a very large, complex pyramid of logistic support organizations (Fig. 1). In order for each ship in a fleet to be ready to fight today, it and its support had to have been planned for and designed years earlier, built, and equipped and outfitted with thousands of pieces of hardware, from pumps to radar antennas, that may themselves have undergone parallel design and production processes. The crew also had to have been recruited and trained in the myriad of technical and nontechnical tasks required to keep a modern warship of the "electronic battlefield" ready to sail and fight.

Once joined, the ship and crew require additional time to train for and exercise their missions. Equipment, with all the attendant manuals, parts, tools, and special test equipment, has to be maintained satisfactorily; the design

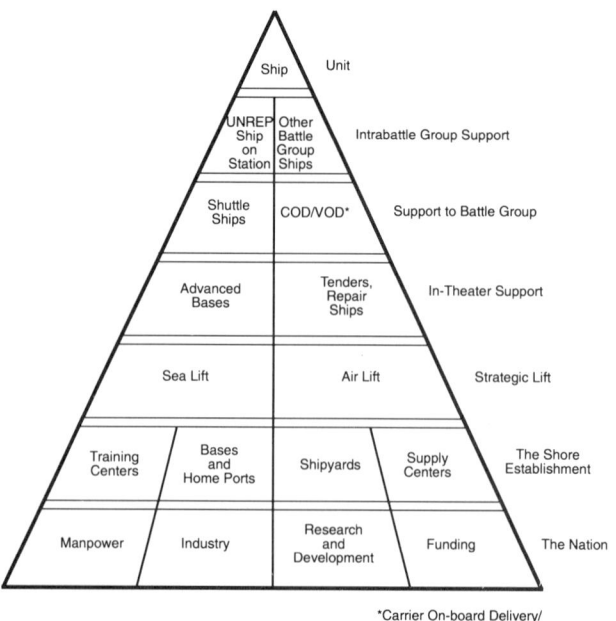

Figure 1. Logistics pyramid.

of the maintenance program, specifications for the replacement parts, and the writing and updating of the manuals keep numerous engineers busy. Fuel, food, supplies, and weapons need to be delivered to the ship on a routine basis, keeping shore-based support activities busy obtaining and storing these resources until needed by the ship. Hull work and major equipment overhaul mandates that shipyards and repair shops be available.

After training, the ship may be required to deploy to another location. For many navies and ships, the locations available are limited by the number of refueling ports accessible to them. Access to overseas bases has long been a key to projecting power at sea; denial of access has effectively reduced the capability of many forces. Those navies that have the capability to replenish at sea become free to project power over a greater surface of ocean, although at the considerable cost of maintaining a fleet of specialized ships that perform the replenishment function.

The pyramid has at its base the nation itself, from which come the economic resources needed to support everything above it. Those resources are substantial, for to deploy a ship requires not only the current expenditures for the resources the ship uses in its deployment but also includes the costs of building and maintaining the ship and recruiting and training the crew, the costs of maintaining the underway replenishment ships and the forward bases to resupply the ship while it is deployed, and the costs of maintaining the homeport facilities the ship will need when the deployment is complete.

Logistics on the Shipboard Level

Each warship needs fuel for propulsion, food and water for the crew, ammunition and missiles for the weapons systems, and a variety of supplies for maintenance and repair, damage control, and day-to-day life at sea. Inherent in the design of the ship is both the rate of consumption and the capacity to store items. A comparison of usage rate and storage capacity indicates how long a resource will last. A ship that uses 3,800 liters (1,000 gal.) of marine fuel per hour at 16 knots, with a usable storage capacity of 456,000 liters (120,000 gal.), has its endurance defined as five days and its maximum range at 16 knots as 1,920 nautical miles; no amount of exhortation or planning will change these. Still, the length of time any specific resource will last is affected by the ship's mission, the difficulty of resupply, and the numerous other weight/space/capability tradeoffs made in ship design.

In the age of sail, the factors limiting a warship's endurance were typically food and water for the crew. When a ship was captured in war, that ship's stores were frequently a high priority. The advent of steam brought about the need to fuel the power plant, first with coal then with liquid fuel. With the notable exception of nuclear-powered vessels, the amount of fuel on board and the fuel consumption rate remain the most critical factors in determining the length of time a ship can sustain itself in peacetime. In combat, the rate of ammunition expenditure may become the important factor, since in modern warfare a vessel can expend an entire magazine of missiles in a matter of minutes.

Logistics for a Navy

The infrastructure required for supporting a navy can be vast, even for a small force. The "shore establishments," as such organizations are frequently called, include bureaus for the design, procurement, maintenance, and support of ships, aircraft, and their many specialized equipments; personnel recruiting and training stations; shipyards, with the drydocks and specialized shops and equipment needed for ship repair; naval bases, with their fueling and loading piers, ammunition loading facilities, specialized maintenance equipment, medical facilities and hospitals, and recreation areas for the crew; and depots for food, repair parts, supplies, and ammunition. There is then an extra layer of resources dedicated to supporting the shore establishment itself, to maintain the shipyards, depots, bases, and other facilities. In fact, the share of a naval budget that must be spent on the shore establishment may be a significant part of a nation's entire naval outlay. Obtaining exact figures is difficult, but one may get an idea by considering that during the period 1971–80, when comparatively little weapons procurement was being done, the U.S. Navy's direct operating and support costs averaged just 50 percent of the total budget.

Logistics Processes

Requirements determination involves the answers to the questions: What is needed? How much? When will it be needed? From where should it be obtained? The answers are integral to the decisions of command. Restricted opportunity to refuel may limit the traveling speed, force a detour to a friendly base, or prevent the use of zigzags and other tactical measures. In wartime, more difficult decisions may be faced, especially since the replenishment rate for missiles is at least an order of magnitude slower than the usage rate.

Acquisition (also known as procurement) is the process of obtaining the necessary resources in a condition usable to the naval force being supported. In peacetime, this usually means purchase of materials and equipment that meet military standards. In combat, it may involve obtaining resources by force as well as by purchase.

Distribution is the movement of the resource from the place it is acquired to the end user, maintaining acceptable condition throughout. Distribution entails packaging, transportation, materials handling, warehousing, and inventory management. Worldwide distribution systems exist, with the important military goal of reducing the time needed for replenishment of the combat unit.

Maintenance is the process of ensuring that a ship and its equipment does not degrade significantly despite its operating environment, or returning the equipment to a normal operating state once it has degraded. At sea, hull and machinery maintenance take on critical importance because of the corrosive nature of the constant salt spray; the chipping, painting, lubrication, and constant testing that is the toil of many young sailors is as vital today as mending the sails was 200 years ago. And electronic maintenance practices are designed to be accomplished despite the constant movement of the ship.

The crew must be "maintained" as well. Providing food and clothing, medical and dental care, recreation or diversion during off-duty hours, and the occasional letter from home are all important in keeping a crew physically and psychologically fit during extended periods at sea.

Naval distribution systems involve the unique capability to replenish a ship while it is at sea. A significant amount of marine engineering has gone into developing the equipment and techniques that enable this last step of the distribution chain—delivery to the customer—to be made. Transfer while ships are traveling in tandem, 36 meters (120 ft.) apart, linked by the high-tension wires that support the weight of the load being transferred, has been developed to a fine art in the U.S. Navy and was a significant factor in the success of the U.S. Pacific Fleet in World War II. The capability to perform underway replenishment under less than ideal conditions is an absolutely vital part of maintaining a true blue-water navy.

Naval Logistics History

Many historians draw attention to the connection between naval logistics and the outcome of a battle or campaign. Naval historians have regarded maritime logistics—the movement of goods and services over the oceans—as even more influential as to policy, strategy, and the outcome of war. Still, naval historians have preferred to center on people, places, tactics, and grand strategies. There are two reasons for this. First, histories deal primarily with what was: the order of battle, the movements of the forces, the decisions made in the heat of the moment that affected the outcome. These are frequently definable from messages, ships' logs, interviews, and the writings of those who took part. The contributions of logistics (and logisticians) are much harder to find, record, and weigh in the balance; they frequently took place weeks, months, sometimes years before the fateful events, in bureaus and shipyards, storehouses and magazines. Indeed, poor logistics decisions may have occasionally prevented any engagement at all, and therefore there is no "history" to report.

A modern example of this is the failure of the USS *Saratoga*'s task group to reinforce Wake Island in December 1942. The group included the oiler USS *Neches*, built in World War I, and capable of just 12½ knots. The destroyers were unable to refuel en route on the course that had been set despite only moderate seas. A change in course to facilitate refueling took them away from Wake, costing 24 hours; the task group arrived too late to inhibit the invasion and was ordered back to Pearl Harbor for fear the Japanese carriers that had attacked might still be in the area. What might have been an American success, had the *Saratoga* hit the Japanese transports in the middle of the landing operations was instead a nonevent. Wake fell to the Japanese. Despite U.S. strategic planning for a "logistics" war, the necessary implementing detail was missing. Underway replenishment, although practiced in fleet exercises, had had no engineering attention brought to bear for the development of rigs that could be used in moderate seas. A suitably fast fleet oiler was a deficiency the importance of which had been undervalued (Miller 1985).

A second reason for their neglect is that logistics systems are evolutionary not revolutionary. As a long-term process, there are thousands of little events that finally add up to some significant difference that evokes attention. The "evolved" logistics system may be reported in place at some crucial time in history, so we know it has evolved, but its path to that point is a mystery. An example, the evolution of which can be traced, is the magnificent World War II shipyard at Pearl Harbor, whose work after the Japanese attack on 7 December 1941 was unceasing and much more than it was ever conceived to be capable of. Of particular note was the battle damage repair of the USS *Yorktown* in late May 1942 after it was damaged in the Battle of the Coral Sea. The carrier was only able to bring its air wing to bear at the crucial Battle of Midway because the shipyard workers made it capable in three days, instead of the three *months* originally thought required. That shipyard was viable when needed because of many prior decisions made to build capabilities in Pearl Harbor, not the least of which was the decision to move the fleet to Hawaii in 1940, compelling a buildup of all of the base's support capabilities. On the other side of the coin, the failure of Admiral Nagumo to reattack on 7 December 1941 and destroy the installation completely was one of the great blunders of the war. Histories have aptly pointed out the vulnerability of fuel tanks and ammunition depots as sustaining elements of the U.S. Pacific Fleet, but few have noted the serendipitous extent of the installations. Those crucial decisions and the evolution of the base are simply taken for granted in the standard histories.

For perspective on the history of naval logistics, therefore, one must resort to a series of vignettes. The snapshots that follow are as space permits so that in flipping from one to another a pattern may be discerned.

DAYS OF FIGHTING SAIL

In the days of fighting sail, ships were limited by the endurance of their crews. The great voyages of the Persians, the Polynesians, the Vikings, the Portuguese, and later the Spanish and English were a testament to the endurance of early sailors and the self-sufficiency of their craft. For warships, ammunition was a limiting factor but rarely were other stores. Historian Duncan Ballantine (1947) wrote:

> The warships of the eighteenth century and earlier, it is true, were more capable of casting off the shackles that bound them to land than are the warships of today. But the qualities of mobility and self-reliance they enjoyed were the result mainly of technical factors in their construction and operation. Only the wind was provided them by the environment of the sea. For the rest they were capable of endurance at sea because they could carry with them a relatively higher ratio of essential subsistence than can the ships of modern times. Provisions could be stocked for a six-month voyage. Fresh water for three months could be taken aboard, and was easy to replenish. Ammunition was more slowly expended. The needs of the ship, in short, were simple; its logistics was therefore of the simple and primary sort—namely the building into the ship itself of a maximum cruising range.

The great sailing vessels of the sixteenth through nineteenth centuries still required some forms of specialized support—material for sails, lines, spars, and caulking (known as naval stores)—many of which were unique to ships and naval warfare. Shipyards to build and repair ships sprang up quickly in the countries that wanted a merchant fleet, a navy, or both. Mariners often deplored how quickly the yards were closed when the crisis passed or the advantage of trade faltered. Specialized shipbuilding crafts grew in England and, from them, in America's New England. France, the Netherlands, and Spain built merchant fleets that spanned the globe. As trade expanded, so did the need to develop ports or bases—places where the ships could go to pick up or deliver cargo but that also had the stores and supplies the vessels needed.

In this context, the voyage of the great French admiral Pierre Andre de Suffren is all the more striking. Suffren managed to operate in the Indian Ocean from 17 December 1781, when he left the French port of Isle of France (Mauritius), until 6 October 1783, when he departed his station near India for the long voyage home after a peace treaty had been signed. During that period, he never entered a friendly naval port. He participated in five significant engagements with the British admiral Sir Edward Hughes, and he retook the Dutch trading port of Trincomalee, in Sri Lanka. He received meager support from scattered transports that managed to get past the English ships interdicting the sea-lanes, but for the most part he improvised. Part of his tale is retold by Alfred Mahan in *The Influence of Sea Power on History, 1660–1783* (1890, p. 451):

> To appreciate thoroughly this contrast between the two admirals, it is necessary also to note how differently they were situated with regard for material for repairs. After the action of the 6th, Hughes found at Madras spars, cordage, stores, provisions, and material. Suffren at Cuddalore [his anchorage] found nothing. To put his squadron in good fighting condition, 19 new topmasts were needed, besides lower masts, yards, rigging, and so on. To take the sea at all, the masts were removed from the frigates and smaller vessels and given to the ships-of-the-line, while English prizes were stripped for the frigates. Ships were sent off to the Straits of Malacca to procure other spars and timber. Houses were torn down on shore to find lumber for repairing the hulls. The difficulties were increased by the character of the anchorage, an open roadstead with frequently heavy sea, and by the near presence of the English fleet; but the work was driven on under the eyes of the commander-in-chief, who . . . inspired the working parties by his constant appearance among them. . . . It was indeed to this activity on the Coromandel coast that the success at Trincomalee was due. The weapons with which Suffren fought are obsolete; but the results wrought by his tenacity and fertility in resources are among the undying lessons in history.

NINETEENTH CENTURY

During the nineteenth century, however, the great endurance of these sailing warships was seen to be less desirable than another military capability: maneuverability in action. Yet the transition was not swift for practical reasons. Although the age of steam arrived at sea with Robert Fulton's *Clermont* in 1807, the early ships had many problems. First was the problem of fuel for their boilers. Wood as a fuel was too bulky, so coal had to be used. Even coal took up a lot of space, and since the coal bunkers needed to be near the boiler rooms, and the boilers and propulsion machinery had to be below the waterline to reduce its vulnerability to damage in combat, gun locations and the entire internal arrangement of the ship had to be changed.

The necessity of having numerous fires onboard presented a great hazard to the ships, which were still made of wood. Coal was dirty as well, and a great deal of extra effort was required to clean a ship whenever coal was loaded. Early sailors, used to the days when their primary function was working the rigging—from which they could see the sea and sky—intensely disliked work on the mechanical propulsion equipment or stoking the boilers in hot, sweaty rooms below the waterline. The boilers required distilled water, with few impurities allowed. The early steam plants had constant leaks, requiring replacement of the water to keep the boilers operating. Since water storage was limited, development of reliable water-distilling equipment became important as well.

These first ships were propelled by paddle wheels installed along the sides of the ship, which was both a defensive and offensive liability. These wheels were extremely vulnerable to attack from the traditional broadside and reduced the number of places where guns could be placed. Strategically, these new ships had a severely reduced action radius, even if they had both sail and steam available (Beach 1986). The U.S. Navy ordered its first steam vessels in 1817 but continued to build sailing ships. Among the first steamships built for the open ocean was the *Mississippi*, which Comdr. Matthew Perry chose as his flagship on his visit to Japan in 1854. Steam propulsion was a technological marvel with which Perry wished to impress the Japanese.

The objectionable paddle wheels were made obsolete by the screw propeller in 1845, when the English propeller-driven brig *Rattler* demonstrated its superiority by towing the paddle-wheel steamer *Alecto* around during a stern-to-stern test of their relative might. From that time on, development of propeller-driven, steam-powered naval vessels was assured, accompanied by great new logistical problems of providing for their fuel and endurance. Imparting a radius of action comparable to that of a sailing fleet would not be achieved for a hundred years, when a combination of underway replenishment techniques and nuclear power restored (at considerable expense) the sailing ship's strategic mobility.

Coal was the fuel of choice in the nineteenth century, and merchant colliers were used to bring coal to ships operating away from their bases. In the U.S. war with Mexico in 1842, and later in the American Civil War, colliers brought coal from New England to the Gulf of Mexico, where ships were forced to come to a stop in a protected harbor to allow the collier to transfer the coal. By the end of the century, the British had coaling and repair stations around the globe, supporting their vast empire. Mahan (1890) wrote passionately about the need for coaling stations to support naval operations. Implicitly, he understood that logistics was at the very heart of strategy, for he wrote strongly on these aspects of strategic issues.

> These remarks [about the necessity of bases], always true, are doubly so now since the introduction of steam. The renewal of coal is a want more frequent, more urgent, more peremptory, than any known to the sailing-ship. It is vain to look for energetic naval operations distant from coaling stations. It is equally vain to acquire distant coaling stations without maintaining a powerful navy; they will but fall into the hands of the enemy. But the vainest of all delusions is the expectation of bringing down an enemy by commerce-destroying alone, with no coaling stations outside the national boundaries. [p. 329]

There were other significant logistic issues that arose during the latter half of the nineteenth century. The machinery of the steam engines had to be kept in working order, and the introduction of metal hulls required different maintenance and repair techniques. Ballantine (1947) comments:

> The significance of logistic factors in the wars of the eighteenth century has perhaps been obscured by the great increase in their importance which followed the introduction of the ironclad, steam-propelled warship in the nineteenth century. For the United States that technical revolution, which got underway in the 1880's, was doubly significant, for it was accompanied by a revolution of strategic doctrine which intensified the problem of naval logistics. The warship of the "New Navy" had lost even the few qualities of self-reliance of its predecessor. And since the United States had then no foreign establishments either colonial or military, its warships were, as Mahan wrote, "like land birds, unable to fly far from their own shores." Without coal they were helpless; and in turn the necessity of providing space for bunkers cut down the space available for stores and ammunition. Wooden hulls could go several years without cleaning; iron hulls fouled rapidly and must be scraped and painted at least once a year. Larger crews increased the requirement for subsistence and reduced available storage space. Water for the boilers was as essential as coal. The greater firepower of modern guns raised the problem of replenishment of ammunition. Spare parts and machinery repairs beyond the power of the ship to perform itself were now necessary.

The great naval buildup at the end of the nineteenth century and the beginning of the twentieth was accompanied by major changes in hull design, naval gunnery, and the first sniff of the electronic revolution as the "wireless" went to sea. Before World War I, however, the problems of coal still had a major impact on the use of fleets worldwide. There had been coal-burners to circumnavigate the globe—the Spanish ironclad steamer *Numancia* did it in 1865—but fleet actions were limited. Comdr. George Dewey's defeat of the Spanish fleet in Manila was preceded by an extended stay in Hong Kong, loading up with coal and provisions, rendezvousing with an ammunition ship that had been hurried across the Pacific, and buying two British colliers to accompany his fleet to the Philippines. Had he not won his battle and seized the Spanish base at Cavite, he would have been swinging at the end of a long rope, for the British closed Hong Kong to him once the war started, and the nearest home base was 11,200 kilometers (7,000 mi.) away. On the other side of the globe, Comdr. Winfield Schley forced the invasion of Guantanamo, Cuba, because he needed a protected harbor in which to coal during his blockade of Santiago.

Early Twentieth Century

American engineer Spencer Miller spent the years 1898 to 1913 developing coaling-at-sea systems for U.S. and foreign navies. Admiral Dewey first advised Miller to develop a coaling-at-sea system in which all the equipment could be carried aboard a battleship. Dewey's concept would allow the battleship flexibility in that it could then coal at sea from any collier, even a captured one. Miller's first prototype was installed in the battleship *Illinois* (BB7). The battleship would tow the collier, with a sea-anchor tensioned highline. Coaling-at-sea speed was 6 to 7 knots and the transfer rate was nearly one 3,000-pound bag per minute. This system was later used by the Russian fleet in its transit from the Baltic to Tsushima (Miller 1985).

The Russian czar Nicholas II unintentionally exposed the vulnerability of the coal-burners when he sent 10 battleships and 30 smaller vessels 28,800 kilometers (18,000 mi.) from the Baltic toward Japan in 1905. Even using Miller's new rig to transfer coal at sea, the lack of ports for coaling and voyage repairs wore on the ships and the crew. They arrived at the Straits of Tsushima in terrible condition, and were promptly defeated by a prepared fleet close to its own bases.

In 1907, President Theodore Roosevelt sent his "Great White Fleet" of twelve battleships and supporting vessels on a cruise around the world. They had with them a supply ship, a refrigerator ship, and a repair ship. The Navy Department had made arrangements for coal to be delivered in many of the ports they visited, but sometimes the arrangements fell through. The British were suspected of wishing the failure of the voyage, and colliers arranged for Auckland, Sydney, Melbourne, and Albany (a coaling station at Australia's southwest corner) failed to fulfill their contracts. Rear Adm. Charles S. Sperry, commanding the fleet, had to negotiate with the Australians for low-grade coal that was not properly suited for naval ship use. In addition, the dependence on foreign sources may have helped produce another problem: in Trinidad and other ports they found sticks of dynamite in the coal.

Even food became a problem before the voyage was over: the fleet's supply ship was called into relief action at the great earthquake in Messina in January 1909, the liner bringing supplies sank en route, and provisions obtained from a French dealer were found to be contaminated. The British garrison at Gibraltar gave them enough bully beef and hardtack for their voyage back to America (Hart 1965). All in all, the voyage showed what could be done by warships in peacetime when friendly ports were available and there was no need to replace ammunition spent in battle or to repair the type of damage only battle brings. But it could not have been done in wartime.

World War I

When World War I started, oil was just starting to be used as fuel. The U.S. Navy had only two battleships using oil exclusively. Merchant shipping adapted to oil-burning engines first, because they allowed more cargo space, greater speed, and quicker turnaround in port. As the facilities and technology advanced, fuel oil for warships was adopted. In 1918, the United States built 811 oil-burning vessels; by 1932, the conversion was complete.

Conversion to oil opened new possibilities for naval use. Oil was cleaner to handle than coal, and the possibility of transferring it from ship to ship by hose was investigated. During the war, the U.S. Navy's second oiler, *Maumee* (AO2), with then Lt. Chester Nimitz aboard as executive and engineering officer, refueled destroyers 34 times while on station for three months south of Greenland. Nimitz developed a way of rigging ships together that allowed them to transfer fuel in daytime at 5 knots. This technique was later practiced in fleet exercises prior to World War II. In 1939, while Nimitz was a rear admiral commanding Task Force Seven, he developed a method for aircraft carrier refueling at sea as well (Miller 1985). At the time, however, the techniques were viewed as emergency procedures and not part of the day-to-day routine.

Another significant logistic development occurred during World War I that planted an important seed for later growth: the use of destroyer tenders. Tenders and supply ships had been auxiliaries to combat fleets of sailing ships for hundreds of years, but the level of expertise required for repair work on the metal hulls, machinery, and gun systems of the modern ships was far more complex and demanding. Prior to the war, most strategists, perhaps influenced by Mahan, felt that developed bases, with their drydocks and extensive machine shops, were necessary to support the fleet. Yet even though repair ships had been developed, they were conceived as only capable of providing voyage repairs and minor maintenance.

Two destroyer tenders, the *Dixie* and *Melville*, stationed at Queenstown, Ireland, from 1917 to 1919, changed the thinking. The level of repairs done was astonishing and led to the idea that afloat tenders and repair ships could have enduring wartime value. After the war, when the U.S. fleet was transferred to the Pacific, the Base Force was established, encompassing most of the service vessels and auxiliaries: the repair ships, oilers, fresh- and frozen-food ships, fleet tugs, and target repair ships. The submarine and destroyer tenders remained separate, however, although Base Force ships supported them. By 1940, the Base Force had increased to 51 vessels, including one floating drydock that could hold a destroyer, 14 oilers, 2 repair ships, and a hospital ship. When the U.S. Pacific Fleet moved its homeport to Hawaii in May 1940, the Base Force moved as well (Carter 1951).

WORLD WAR II

Well before World War II, farsighted Americans had foreseen that the war in the Pacific would be one of logistics. Nimitz and others had developed theories about how to do this while studying at the U.S. Naval War College in the 1920s. While at the college in 1923, Nimitz wrote:

> To bring such a war to a successful conclusion Blue [Note: "Blue" and "Orange" were used to designate the major protagonists in a conflict scenario; Blue was always the United States; Orange was Japan] must either destroy Orange military and naval forces or effect a complete isolation of Orange by cutting all communications with the outside world. It is quite possible that Orange resistance will cease when isolation is complete and before steps to reduce military strength on Orange soil are necessary. In either case the operations imposed on Blue will require the Blue Fleet to advance westward with an enormous train, in order to be prepared to seize and establish bases enroute. . . . The possession by Orange of numerous bases in the western Pacific will give to her fleet a maximum of mobility while the lack of such bases imposes on Blue the necessity of refueling at sea, or of seizing a base from Orange for this purpose, in order to maintain even a limited degree of mobility. [Hattendorf, et al. 1984]

The first eighteen months of the war in the Pacific, however, were characterized by shortages. Despite the planning and strategic concepts developed prior to the war, the actuality was different. Even though American industry started to increase production in 1940, the benefits went primarily to the European theater, not to the Pacific. For the Battle of Midway, Admiral Nimitz had to exhort the Pearl Harbor shipyard to fix the *Yorktown* in three days, as previously mentioned; he also had to scramble for parts to have the aircraft repaired as well as for the ammunition they were to carry. The early war demonstrated that there was a vast difference between peacetime plans and wartime execution.

Three important organizational developments during the war would have far-reaching consequences for naval logistics. The first was the creation of construction forces dedicated specifically to the needs of naval base building. The second was the development of the floating base. The third was the creation of a full service underway replenishing group. Each of these were vital to the strategy of the Pacific campaign.

The Naval Construction Battalions (known as SEABEEs) were established by Congress in January 1942 in recognition of the need for naval base construction in advanced areas where civilian contractors could not—or would not—work. During the war, the SEABEEs swelled to almost 250,000 men and 12,000 officers, and their accomplishments included helping to build more than 400 bases in all theaters of operation, including the naval operating base in Guam, which was at one time the largest in the world; assembling pontoons and causeways for amphibious assaults; handling underwater demolition and cargo offloading; and performing combat engineer functions for the marines. Their ability to quickly build an effective base was legendary, starting to build airfields and other facilities on many islands before the fighting was over. The planning process for base building was raised to a fine art by the end of the war with the inception of the Advance Base Functional Capability (ABFC) concept: a strategic planner could select the functional capabilities a planned base should have from a "menu" of possible ABFCs covering all conceivable base functions. Each ABFC was tied directly to lists of equipment, materials, skills, and shipping requirements, thus facilitating the accomplishment of the desired task.

By late 1943, however, the fleet was advancing faster than the SEABEEs could build their forward support bases. The Base Force, renamed the Service Force in 1942, had grown steadily during the war. An idea first advanced in 1904 by a navy civil engineer finally came to the fore: put the facilities the fleet requires aboard ships and barges and float them to a harbor or atoll that can provide calm anchorage. In October 1943, one of the Service Force's squadrons was sent to Funafuti in the Ellice Islands to furnish logistic support for the fleet; in February 1944, the famous Squadron TEN was formed and sent to Majuro in the Marshalls for the same purpose. These were not just tenders or repair ships; these squadrons became the complete substitute for an established land base, with the single exception that they did not provide long-term ship repairs. By June 1944, Service Squadron TEN had 4 destroyer tenders; 6 repair ships; 3 repair shop barges; 6 drydocks; 13 ammunition barges; 15 storage barges for freight, spare parts, ground tackle, radio, medical, torpedo, marine stores, and so on; 23 oil and gasoline storage barges; 15 old tankers or Liberty ships for storage and local services; 6 large concrete supply barges; 11 water barges; 5 YP cold storage vessels; and 15 tugs, along with assorted smaller and special service vessels (Carter 1951). The squadron provided a massive amount of logistical support to the fleet and had one great advantage over fixed bases: it was mobile. It did not take months to build a base only to become useless when the fighting passed further on. The mobile base moved from Majuro to Eniwetok in June 1944, to Ulithi in October, and finally, in March 1945, to Leyte.

By 1944, U.S. forces were advancing steadily. The base infrastructure so necessary for Japanese control of the Western and South Pacific was falling apart as their lines of communication were interrupted, first by U.S. submarines, then by the presence of the U.S. fleets operating in the areas, supported by the bases built by the SEABEEs and by Squadron TEN, in the central Pacific campaign; and by the presence of the army forces and army air forces in the South Pacific, supported by fleet elements. The logistics war was being won by the United States.

In 1945, the island-hopping campaign reached its climax: Iwo Jima and Okinawa were the last great stepping-stones to the home islands of Japan. They were, however, 3,200 kilometers (2,000 mi.) from the nearest U.S. base. Refueling at sea was now being done regularly, but the massive quantities and varieties of ammunition used in the air war and island bombardment would require ships to leave battle for at least ten days at a time if they were to replenish their ammunition in port. Adm. Raymond Spruance directed his staff to develop a method for transferring ammunition at sea, and it was done within three weeks. The *Shasta* (AE6) issued ammunition at the rate of 16 tons per hour to the *Bennington* (CV20) on 23 February 1945, while moving at a speed of 10–12 knots (Miller 1985).

From then until the end of hostilities in August, the 15 aircraft carriers, 7 battleships, and 100 other combatants of Fast Carrier Task Force 58 (and later Task Force 38) were replenished at sea by Squadron SIX, an underway replenishment group made up of 38 oilers, ammunition, and supply ships. The oilers were in fact makeshift multiproduct issue ships for the destroyers, as they carried ammunition, dry stores, medical stores, replacement personnel, depth charges, aircraft engines, and drop tanks along with their normal load of fuel for the ships, aviation gasoline, and lubricating oil. Ships stayed at sea for as long as three months at a time as the battle was brought to the Japanese homeland. Peak demand for fuel oil was 220,000 barrels per day and required 39 oilers to be in regular shuttle between Ulithi and the fleet. During May and June 1945, the aircraft carriers were able to launch strikes on Japan for two days in a row, after which they would withdraw to a safe distance (out of reach of kamikaze attack) at night. On the third day, they would go through the lines of oilers, ammunition ships, and supply ships and replenish. That night they would return to their operating area and strike again at dawn the next day (Carter 1951).

In his 1948 *History of the United States Navy*, American naval historian Dudley W. Knox summed up the progress made during the Pacific campaign as follows:

> The form into which American sea power evolved in the Pacific was highly novel, immensely powerful, and quite extraordinary. Always sea power has been regarded as including several major components; naval ships, merchant shipping, home and overseas bases, home specialized industries and the like, not to mention essential personnel including Marines . . . [a] primary alteration of form of sea power that developed in the Pacific was the method of logistic support. As Nimitz's theater of control progressed closer and closer to Japan, the necessity for continuous fleet support of forward areas became more and more imperative. The length of the lines of communications to the most advanced permanent base at Pearl Harbor rapidly increased until the fleet could not return there for replenishment without leaving our frontal footholds in danger. The alternative was a system of mobile service squadrons of ships and floating facilities. They moved forward, step by step, to newly conquered anchorages to be used as temporary advanced bases. They served the fleet there or at sea, when and where needed, with the logistic support which is the foundation of naval operations. The methods of refueling and resupply in the open sea amounted to a "secret weapon" of great effectiveness. By such resourceful measures the speed of progress to Japan was enormously increased.

Post–World War II Developments

At the end of the war, the techniques developed were written down, but nothing new was done as the nation stood down from war. Five years later, the lessons learned were dusted off as aircraft carriers took to the waters off Korea. The advent of jet aircraft, which require much more fuel than their predecessors, heightened the awareness that better methods had to be found for transferring fuel and material at sea. The World War II methods were for day-use only and required cooperating seas. Besides, they were slow in that they required each aircraft carrier to make separate approaches to three different replenishment ships. Necessity bringing invention, adjustments were made to improve the rigs, and long-range development of improved methods was started.

Major changes took place after Korea. Spurred on by Chief of Naval Operations Arleigh Burke, improved rigs were developed that not only increased the transfer rate but also allowed transfer at speeds of 20 knots or more, techniques for night replenishment were worked out, and most significantly, a true, highly capable, multiproduct ship was developed. Called the Fast Combat Support Ship (AOE), the new class carries fuel for both ships and aircraft, food, ammunition, and a range of other supplies and is capable of moving with a carrier task group at 26 knots. A single AOE carries enough material to completely support a modern aircraft carrier task group for a week or more; this was first demonstrated with the aircraft carrier groups on Yankee and Dixie stations off Vietnam in the late 1960s and early 1970s (Hooper 1972).

A recent development was the refinement of vertical replenishment (VERTREP): the use of helicopters to deliver ammunition and supplies. First developed in the late 1950s, it is now a standard capability for naval auxiliaries. It allows much greater delivery flexibility and safety. In many places, VERTREP accounts for more than 50 percent of the stores, ammunition, and food transferred between ships.

The Falklands/Malvinas War in 1982 and U.S. operations in the Indian Ocean since 1980 have again pointed out the importance of replenishment at sea. To support the fight in the Falkland Islands/Islas Malvinas British merchant ships were pressed into service and modified, allowing the British fleet to operate 12,800 kilometers (8,000 mi.) from home without a friendly port for support (Villar 1984). And the continued presence of a U.S. carrier battle group in the North Arabian Sea has been maintained by the U.S. Navy auxiliaries, which have the ability to deliver virtually anything warships need while remaining under way.

RECENT DEVELOPMENTS

A logistics history is not complete without mentioning the effects of two revolutions in naval warfare: the use of nuclear power and the electronics revolution.

Nuclear power frees a ship from the bonds that oil dependence brings. Nuclear-powered task forces have circled the globe like no other ship since the sailing vessels of the nineteenth century, but with considerably more striking power. The first instance of nuclear-powered ships in combat occurred on 2 December 1965, as the USS *Enterprise* (CVAN-65) and USS *Bainbridge* (DLGN-25) took positions at Yankee Station off North Vietnam. Nuclear power for propulsion has increased the amount of internal space available for internal stores and allows an aircraft carrier to carry considerably more aviation fuel.

The other revolution is not so kind. Electronics in the form of sensors (radars, sonars, and the like) and automated, sophisticated weapons systems have mushroomed since World War II—as has the cost of keeping them in operational order. They demand a higher level of training in operator and repair personnel; they require many expensive, sometimes sensitive and delicate, repair parts and test equipment; and they need more room—to allow access for repair under way—and for storage of test equipment and increased numbers of repair parts. The electronics extend to naval aircraft as well. The cost of ships and aircraft in the 1980s is dominated by the cost of the electronics and the systems required to maintain them. The importance of keeping them working has increased the importance of a link with the shore establishment. Today's carrier task forces rarely operate for any length of time without setting up an airhead so that vital repair parts can be sent to the fleet via regularly scheduled flights of special cargo aircraft, known as Carrier Onboard Delivery (COD) aircraft. These airhead operations even make it possible for special technicians to travel wherever and whenever they are needed—worldwide.

History suggests two things. First, the new and undesirable tie to the shore brought about by electronics can be overcome if need be, as Suffren overcame his difficulties and stayed on station in the Indian Ocean 200 years ago, and as Spruance stayed on station off Japan 45 years ago. But history also indicates that the solution will await the next war.

Logistics Support for Naval Units

NAVAL SHIP DESIGN

Designing a ship is an extraordinarily complex process of tradeoffs between capabilities and cost. The ideal ship might have a 10,000-nautical-mile range at 35 knots, turn on a button, carry all of the latest sophisticated electronic equipment and weapons systems, be 100 percent reliable in all weather and sea states, and be cheap enough so that a fleet can be built. But the reality is that these characteristics are competitors in the design process, and most of them will be modified by the time the ship is launched. Many of the factors, as they change, will affect the ship's eventual ability to support itself and to be supported at sea.

A number of military characteristics are considered in ship design, some of which have direct logistics impacts. *Speed* is one of the most basic and is normally considered in two ways: as the burst speed available for tactical maneuvers, and as the cruise speed at which the maximum range is achieved. Naval vessels generally have a cruise speed of around 15–17 knots and burst speeds of around 30 knots. In standard hulls, going beyond the cruise speed becomes increasingly inefficient as fuel consumption at twice cruise speed can easily be four times as great. Speed for rapid deployment of forces, then, is an option only as long as sufficient fuel is available along the route. The need for speed in a design may limit the number or type of propulsion plants that may be used; they in turn will help determine the range and endurance of the ship.

Range is the distance a ship can go without refueling. Range at best cruising speed and range at burst speed are both considered in warship design. For any given propulsion plant, range will be determined by the quantity of fuel carried in the ship's tanks. A range characteristic for a fleet escort ship might specify that it should be able to travel 8,000 nautical miles at its cruising speed and then enter port with a 30 percent reserve left in its tanks. If range is vital—because of a proposed primary mission for the ship—the ship design will have to accommodate fuel tanks large enough to provide the desired range. Even without a need for propulsion, however, a ship will use fuel in making electricity to run the equipment necessary for supporting the crew and maintaining its weapons systems.

Next comes *endurance*, which in peacetime is the length of time the ship can remain at sea without replenishment of its consumable stores—food, personal supplies, and the like. Endurance for large naval ships will frequently be on the order of 30–60 days, and will be limited by the ship's storage capacity for food. Food-storage facilities are generally designed into the hull using planning factors, such as ".1 cu. ft. of storage capacity per man-day for dry stores." For a 30-day endurance, a ship with a 250-man crew would then need 21 cubic meters (750 cu. ft.) of dry store space. By the time aisles and other access space and storage aids are added in, a dry storeroom might have to be about 3 meters by 5.4 meters (10 ft. by 18 ft.). Equally important, it should have good access for loading the stores and should be handy to the galley. These factors are typical of those that restrict and constrain the configuration of the internal spaces.

Water is a very special commodity on a ship, and lack of it may severely limit a ship's endurance. Major naval vessels use water of very high purity in their propulsion systems. Water cools selected electronic and other equipment, and fresh water sustains the crew. If the ship's water-processing equipment (evaporators or distilling plants) breaks down, endurance may be limited to a week or less. Water use and production are watched closely at sea, and many sailors have had to live with the hardships of fewer or shorter showers so the propulsion plant would have enough feedwater for its steam cycle.

Combat endurance is a function of the number of engagements in which a ship can participate with a full store of weapons aboard. Combat endurance is a very difficult figure to specify, because quantification of weapons use depends on the scenario an analyst uses. In general terms, however, medium- or high-intensity combat operations will result in weapons being depleted before other consumable stores—and quite possibly before fuel.

Availability is the statistical likelihood that the ship or any system will be in a state of operational readiness, and includes the reliability of the system (mean time between failures—MTBF) and the mean time to repair the system (MTTR) after it fails. Reliability of equipment is a design characteristic for every electronic and mechanical item on board; the need to be reliable despite a salt atmosphere and the constant motion of the ship accounts for a great deal of the high cost of naval hardware over similar equipment in a civilian application.

System reliability is sought through a variety of methods. Redundancy of equipment provides for system reliability but at an increased cost in dollars, space, and weight that the extra equipment needs. Vessels designed for combat may rely on redundancy to ensure that vital ship control or other functions are available even after battle damage is sustained.

MTTR is a more complex design issue. Early in the design process, concepts about the maintenance plan for the ship must be made, for they affect many subsequent decisions. Shall a system by maintainable at sea, or shall it be designed to be repaired only in port? In his 1986 book *Modern Warship Design and Development*, Norman Friedman discusses the basic problems with the maintenance of modern electronics at sea. This maintenance requires that the equipment be accessible while under way, which increases the volume required for each piece of equipment and reduces the weight density of the ship; it requires manning ships with trained technicians, which in turn requires improved habitability to aid retention in peacetime; and it requires stowage for the many maintenance manuals, equipment, and repair parts. The alternatives, however, can be equally costly: accept a higher risk that a system will be out of commission or provide a lot of redundancy. The U.S. Navy has opted for having each warship maintain its own systems at sea to the greatest extent possible; the Soviet navy tends to do its maintenance at shore facilities in port or while tied up to one of the many repair ships they have built to compensate for each warship's lack of organic capability.

The last characteristic with direct logistics impact is *habitability*—the ability of men and women to endure the shipboard environment for extended periods of time without any significant degradation of their ability to do their job. Increased habitability entails larger berthing and messing spaces, improved ventilation and air-conditioning, and reduced noise and vibration. In the last 30 years, habitability standards on U.S. ships have been greatly upgraded: berthing areas are larger; there is effort to minimize stress from heat, cold, and noise; more varied and healthy menus are provided; and recreational facilities have been added. Improved habitability has enabled the U.S. Navy to retain more of its highly trained young technicians and has fostered a more productive work environment at sea. More habitable ships required more space dedicated to personnel; this forced increases in the hull size of U.S. warships, which have in turn caused other design tradeoffs.

All these factors have an intricate impact on combat readiness. Design goals for high speeds and long ranges require large fuel tanks and a larger hull, thereby changing stability and seakeeping characteristics. Arrangement of tanks and machinery affect the arrangement of other spaces and may limit options for making weight/space tradeoffs among payload (the ship's sensor and weapons systems), habitability requirements, storage for food and other supplies, and access space for equipment.

In many navies, warships are designed with the ability to replenish at sea. Capabilities for refueling and transfer of ammunition, food, and other supplies through vertical (helicopter) or alongside replenishment are vital if the navy is to stay at sea for weeks on end. These capabilities must be designed in, because underway replenishment equipment and operations have their own set of requirements. The stations from which replenishment and refueling will be accomplished must be clear of entanglements from weapons or sensor systems; access to and from the stations must be adequate to allow material to be moved away from the stations quickly, or there must be staging areas to prevent bottlenecks; helicopter landing or drop zones must be free of masts and antennas and allow for various angles of clear approach. These capabilities complicate the arrangement of topside equipment, but replenishment at sea has so proven its worth as to be a paramount design consideration.

Logistics in Ship Operations

Military missions may require the ship to travel all over the globe. Monitoring the day-to-day status of fuel, water, and weapon system readiness is a continuing concern of the commanding officer. In peacetime, this may mean visiting a friendly port or asking for a rendezvous with an underway replenishment ship. The underway replenishment operation, despite its frequency, can be hazardous. In some navies, refueling at sea has been done via the astern method, in which the delivering ship trails a fuel hose for the receiving ship to pick up. This is a slow process; the ships are limited to a single hose hookup. The preferred method is alongside refueling, which allows multiple-hose—and multiple-product (ship and aviation fuel)—refueling at the same time. In the case where a ship is replenishing from a multiproduct ship like an AOE, it is possible to receive fuel, stores, food, and ammunition all in the same evolution, using multiple refueling hoses, two connected replenishment dry stores/ammo rigs, and vertical replenishment simultaneously.

Fleet Replenishment at Sea

Modern fleets are organized to fight in task forces, which are composed of one or more capital ships plus escorts (including submarines and aircraft) that perform scouting and screening functions. Sometimes there is also one or more logistic support ships. In the U.S. Navy, the battle group, the basic building block, is formed around one capital ship (aircraft carrier or battleship). One or more battle groups are then combined to accomplish a task. Battle groups provide tactical support to each in other combat environments, of course, as well as mutual logistic support, even without a logistics ship in company. Typical battle group support functions include sharing supplies, repair parts, and maintenance technicians. In addition, capital ships generally have more and larger repair shops onboard, extensive medical and dental facilities, and carry extra fuel for the smaller ships in the formation.

U.S. Navy Logistics Support Ships

Specific logistics ships travel with the battle group if the mission requires it. In the U.S. Navy, a battle group will include either a Fast Combat Stores Ship (AOE) or a combination of an Ammunition ship (AE) and a Replenishment Oiler (AOR). The latter pair are not as capable as an AOE, being limited to about 20 knots. On the other hand, the assignment of two ships instead of one provides compensatory operational and scheduling flexibility in some situations.

The *Sacramento*-class AOE is the finest underway replenishment vessel in history. This ship is 241 meters long (795 ft.), with a full-load displacement of 53,600 tons. It can travel in excess of 26 knots and can carry 177,000 barrels of fuel, 2,150 tons of ammunition, 750 tons of provisions and dry stores, and has 1,275 cubic meters (44,625 cu. ft.) of storage space for refrigerated stores. Manned with approximately 600 officers and crew, it also carries two UH-46 Sea Knight helicopters for vertical replenishment. It can replenish two ships at a time—on each side—and frequently does so with an aircraft carrier to port and a smaller escort to starboard. It has seven cargo transfer stations, three double-hose fuel-issuing stations, and three single-hose delivery stations. It can increase the peacetime endurance of an aircraft carrier battle group, including up to seven escorts, by more than one week. The United States originally built four of this class during the 1960s, and has a similar class, starting with the USS *Supply* (AOE-6), which began construction in 1987.

The U.S. Navy also has seven smaller multiproduct replenishment ships (AORs) in the Wichita class. Shorter and slower than an AOE, these ships have a full displacement load of 38,100 tons. These vessels also provide a substantial underway replenishment capability to deployed U.S. battle groups. In addition, the United States has seven 16,000-ton Fast Combat Stores Ships (AFS) to deliver food, consumable stores, personnel support items, and a wide range of repair parts. Single-product underway replenishment ships include twelve ammunition ships, with five more planned to replace the five oldest ships in the early 1990s. The largest of the ammunition ships are those of the seven-ship Kilauea class, which displace almost 20,000 tons; this class entered service between 1968 and 1972. Five 27,500-ton oilers in the Cimarron class entered service in the 1980s; these are already being lengthened, and after alteration in the early 1990s will displace nearly 38,000 tons. The U.S. Navy also has several other large and capable auxiliaries that support its overseas fleet deployments. Six 20,000-plus-ton destroyer tenders in the Samuel Gompers class, the last four of which entered service in the 1980s, plus four older destroyer tenders, routinely deploy with the Sixth and Seventh fleets on a rotating basis. The United States also has five submarine tenders of similar size in the L. Y. Spear class, and seven older submarine tenders. Two old (1940s) repair ships remain on active service.

In the 1970s, a number of support ships started operations under the Military Sealift Command (MSC), and now fill an important role in fleet operations. These ships furnish fuel and stores whenever needed, and operate functionally like their naval counterparts. These include one ammunition ship of the Kilauea class and three combat stores ships. Manned by a civilian crew, they carry a small naval component. The MSC also has several support ships under its control, including two 1,000-bed hospital ships placed in service in 1987 after being converted from merchant tankers.

Soviet Navy Logistics Support Ships

The navy of the former Soviet Union also developed an underway replenishment force, although it was not considered equal to that of the U.S. Navy. The USSR deployed six ships in the 24,500-ton Boris Chilikin class of replenishment oiler in the 1970s, and the 36,000-ton multiproduct replenishment ship *Berezina* in 1978. The latter is roughly equivalent to a U.S. AOR. Soviet warships refueled underway astern or alongside, depending on the rigs installed on the ship being serviced, and also transferred stores alongside. They accomplished replenishment tied to a tender or support ship in a friendly anchorage more often than their Western counterparts; whether this was to facilitate maintenance or simply to avoid the hazards of underway replenishment is not known.

The Soviet navy had more major auxiliary support ships than any other navy in the world. Among their many ships were 12 submarine tenders in the Ugra and Don classes; the 17,000-ton missile transport *Aleksandr Brykin*, which entered service in 1987, and 17 smaller missile transports; 2 12,000-ton submarine repair ships, which entered service in 1984 and 1985, and 43 smaller repair ships of various classes; 24 oilers; 16 water tankers; 33 naval cargo vessels; 8 provisions ships; 10 ammunition ships; and 2 hospital ships. The large number of repair ships were necessary adjuncts to Soviet fleet movements because of the aforementioned lack of maintenance and repair capability on the warships themselves.

UNDERWAY REPLENISHMENT SHIPS OF OTHER NAVIES

Of other navies of the world, the British Royal Navy has the most capable auxiliary force. Operating as the Royal Fleet Auxiliary (RFA), these ships provide for excellent strategic mobility. The Royal Navy underway replenishment fleet is headed by the two-ship 22,750-ton Fort Grange-class ammunition, food, and spares replenishment ships, and three 36,000-ton Olwen-class replenishment oilers. In addition, Great Britain is currently planning to build six to twelve more underway replenishment ships in the Fort Victoria class, which will be 31,500 tons fully loaded and a fully capable multiproduct ship issuing fuel, stores, and ammunition via connected and vertical replenishment. The first two of these have been ordered. The Royal Fleet Auxiliary is fully capable of helping the Royal Navy maintain a naval presence anywhere in the world.

The French navy has a substantial underway replenishment force, including its four 17,800-ton Durance-class multiproduct ships, with another being built. The Netherlands has two 17,000-ton multiproduct ships, Italy has two smaller unrep ships, and the Federal Republic of Germany has plans to build five 12,000- to 14,000-ton multiproduct ships. The usefulness of multiproduct underway replenishment (unrep) is seen in the building and conversion programs of navies around the world. Canada has three 22,000- to 25,000-ton full-capability multiproduct ships. In South America, Brazil is having a 12,000-ton cargo vessel, the *Almirante Gastao Motta*, converted for multiproduct unrep. Across the Atlantic, the Republic of South Africa has two such ships, the newest being the 12,500-ton *Drakensburg*, which entered service in late 1987. In the Indian Ocean area, both Iran and India have multiproduct unrep ships: the 33,000-plus-ton Iranian *Kharg* was converted from a British Olwen-class oiler, and entered service in 1980. In the Western Pacific, the navy of the People's Republic of China has three Fuqing-class 21,700-ton multiproduct ships that entered service in the 1980s; the Australian navy has its 18,000-ton *Success*, which started service in 1986; and the Japanese currently have two ships and are building two more like the 15,800-ton *Towada* (Couhat, et al. 1988).

UNDERWAY REPLENISHMENT

Underway replenishment takes precise ship handling to maintain the ships at a constant distance apart: coming too close risks a collision and being too far can snap the transfer rigs. Speeds of 12 knots are considered ideal for control alongside. Higher replenishment speeds are now routinely practiced when the formation is on the move, but in seas above state 3, the motion of the sea running between the ships will cause choppy waves and increase the suction tending to pull the ships together. The length of time required to accomplish the desired transfer is dependent on the transfer rates of the rigs and the staging areas and material-handling equipment available on both ships. Because of the demands on shiphandling—and in wartime the added vulnerability to attack—commanding officers of ships desire the evolution to be as quick as possible. An hour alongside is typical, but some transfers can extend much longer, occasionally over twelve hours at a time, straining the shiphandling, engineering, and material-handling personnel to their limits.

Support from Outside the Battle Group

CARRIER ON-BOARD DELIVERY (COD)

Support from outside the battle group can come in a variety of ways. Daily COD flights to a U.S. carrier battle group will be scheduled when there is a suitable airfield from which to stage material. COD flights bring mail, critical repair parts, and passengers to the ships at sea. Material destined for ships other than an aircraft carrier will have their material and passengers delivered onward by helicopter. Repair parts, notably for the critical electronics that suffer failure for which onboard capabilities are not sufficient, can be flown from great distances in a short time, obviating the need for the ship to return to port. The COD requires only a small "beach detachment" of personnel, maybe three to five men at an air terminal, who receive material, sort and assign priority to, and load it on the COD aircraft for the daily flight to the battle group.

SHUTTLE SHIPS

A second means of supporting the battle group is with shuttle ships. Merchant or military oilers can bring fuel to an AOE or AOR in the vicinity of a battle group, allowing them to replenish their supply without leaving the group. AFS ships usually shuttle from port with stores and fresh provisions, and whatever heavy cargo the battle group may require. With such logistic systems, the U.S. Navy routinely keeps carrier battle groups at sea for 90 days at a time with no degradation of operational readiness.

ADVANCED BASES

Advanced bases also support a fighting fleet. Providing ships and aircraft repairs beyond what can be done at sea and occasional replenishment in a less stressful environment, overseas bases are still prized as a necessity for some types of support. These bases may be as simple as a protected anchorage where a tender or repair ship may tie up and provide support to ships operating nearby, to the elaborate shore facilities found at the Subic Bay Naval Base in the Republic of the Philippines, which, as of early 1991, was a primary support site for the U.S. Seventh Fleet. The facilities include an extensive supply and fuel depot, an ammunition magazine, ship and aircraft repair facilities, a hospital, and numerous recreation activities. The base, like others, is serviced by long-haul transportation routes, both air and water, with material from major supply centers in the United States.

HOMEPORTS

Last, and best loved, of all elements in the naval logistics chain is the homeport. Ships at times need very specialized support in their life cycle. Hull work, replacement of propellers and heavy internal machinery, alterations and improvements to weapon and sensor systems, and general maintenance in depth are all usually done during designated periods in homeport, where ship systems can be shut down and taken apart. Homeports usually provide this full range of needs that a ship may require. Extensive repair and drydocking facilities, access to supplies and material of all kinds, and training facilities for the crew, with housing for the families of the crew are all typical homeport functions.

WAYNE P. HUGHES, JR.
MARK L. MITCHELL

SEE ALSO: Mahan, Alfred Thayer; Maritime Logistics; Naval Auxiliary and Support Ships; Navy; Nimitz, Chester William; Seapower; Spruance, Raymond Ames.

Bibliography

Ballantine, D. S. 1947. *U.S. naval logistics in the Second World War.* Princeton, N.J.: Princeton Univ. Press.
Beach, E. L. 1986. *The United States Navy—A 200-year history.* Boston: Houghton Mifflin.
Carter, W. R. 1951. *Beans, bullets, and black oil.* Washington, D.C.: Government Printing Office.
Couhat, J. L., and B. Prezelin, eds. 1988. *Combat fleets of the world 1988–89: Their ships, aircraft, and armament.* Annapolis, Md.: U.S. Naval Institute Press.
Eccles, H. E. 1950. *Operational naval logistics.* Washington, D.C.: Government Printing Office.
———. [1959] 1981. *Logistics in the national defense.* Reprint. Westport, Conn.: Greenwood Press.
Friedman, N. 1987. *Modern Warship design and development.* New York: Mayflower Books.
Hart, R. A. 1965. *The great white fleet.* Boston: Little, Brown.
Hattendorf, J. B., B. M. Simpson III, and J. R. Wadleigh. 1984. *Sailors and scholars: The centennial history of the U.S. Naval War College.* Newport, R.I.: Naval War College Press.
Hooper, E. B. 1972. *Mobility, support, endurance: A story of naval operational logistics in the Viet Nam War, 1965–1968.* Washington, D.C.: Government Printing Office.
Knox, D. W. 1948. *A history of the United States Navy.* Rev. ed. New York: G. P. Putnam's Sons.
Mahan, A. T. 1890. *The influence of sea power on history 1660–1783.* Boston: Little, Brown.
Miller, M. 1985. Standby for Shotline. *USNIP* Vol. 111/4/986, pp. 75–79.
Thorpe, G. C. [1917] 1988. *Pure logistics: The science of war preparation.* Reprint. Washington, D.C.: Government Printing Office.
Uhlig, F., Jr. 1986. *Vietnam—The naval story.* Annapolis, Md.: U.S. Naval Institute Press.
Villar, R. 1984. *Merchant ships at war—The Falklands experience.* Annapolis, Md.: U.S. Naval Institute Press.

LOGISTICS, SOVIET AND WARSAW PACT

[This article was written prior to the 1991 dissolution of the Warsaw Pact and the political events that have since occurred in the Soviet Union. It is included because of the valuable historical insight provided into Soviet logistical doctrine.]

The former Soviets insisted that the maintenance of "viability" of forces in battle depends to a considerable extent on the ability of the "rear" (*tyl*) to furnish the requisite support. An important feature of this is the ability to maintain adequate supplies of ammunition and fuel. The Russian word *tyl* embodies two wide concepts. The first comprises the homeland (*rodina*) with all its industrial production capacity, human and material resources from which the state draws its military strength. The second refers to the entire rear services organization of the armed forces and its various branches, including supply, technical maintenance, medical support, sustained operation of lines or communication (LOC), and so on.

The rear services must be organized, equipped, and trained in a manner that enables them to operate in accordance with the principles of military art. They themselves must be capable of maintaining the rapid tempo of operations, a flexibility of maneuver, effecting an efficient concentration of effort together with preserving their own combat effectiveness. The rear services must, by their actions, enable the combat forces to operate effectively in accordance with these principles. Since the military organization of the other Warsaw Pact countries mirrored that of the Soviet armed forces, reference here is made only to the latter.

Establishing the Principles of the Rear

Modern operational analysis concentrated mostly on historical examples from the third period of the Great Patriotic War (GPW), where the high-speed offensives and widespread employment of mobile groups were considered more relevant to the likely conditions of a future battlefield. The basic factors that were identified as being important lessons for a future war follow.

1. The organization of the rear—the composition, structure, and grouping of its services and systems—was the most important factor contributing to the effectiveness of rear support; it permitted the timely and flexible maneuver of forces.
2. Also vital were the concentration of resources on the main axes and their location as close as possible to the forces being supplied or supported.
3. As the operations for deep penetration developed, major efforts to regroup forces became necessary, putting a great burden on the transport system in the rear and on command and control of rear assets.
4. Rear support of mobile groups came to be extremely important as these formations were used increasingly to split up enemy groupings and achieve deep penetration. Forward location of dumps and stockpiles was essential for successful resupply. Supply by air was employed where possible. Operational control groups were set up to provide greater flexibility of command.
5. As the volume of supplies required increased, the principle of centralizing stock control became correspondingly more important. Sufficient stocks had to be established for the total requirement of troops for the whole operation. Centralized control made for greater flexibility and was most effective as a means of reducing the impact of enemy air interdiction. Full authority of control was vested in the deputy commander of the rear.
6. Centralization of control was required for effective support whereby larger, rearward supply elements must distribute resources to subordinate rear elements or direct to combat forces, with senior commanders taking full responsibility for timely delivery.
7. Railways were the most important means of strategic and operational transport. However, as the war went on, the importance of railway transport in the operational rear was reduced and replaced with vehicle, pipeline, and air transport. At front level, the repair of railways and coping with the different gauges of track required a special study.
8. Prediction of supply requirements became more important as the rate of advance and depth of penetration increased, intensifying the strain on logistic assets and communications.

Some details of characteristics of supply and expenditure during the GPW are given in the following section. As a result of this experience, modern Soviet rear support was based on a set of carefully formulated principles that directed the development of rear services. First, the rear services had to be organized, equipped, and trained in a manner that enabled them to operate in accordance with the principles of the military art. That is to say, they had to be capable themselves of maintaining a rapid tempo of operations and flexibility of maneuver, effecting an efficient concentration of effort, and preserving their own combat effectiveness. They had to operate, in other words, like the rest of the armed forces. Second, the rear services, by their actions, had to enable the combat forces to operate effectively in accordance with these principles. This second principle necessitated the formulation of additional principles specific to the conduct of the rear. Adherence to these was considered essential if the effective rear support of an operation was to be assured. These principles, which laid down broad guidelines for the organization of rear support, were:

1. The organization of the rear must reflect the character of the war and the nature of the fighting.
2. Reserves must be echeloned in depth and deployed before the war starts.
3. The higher command is responsible for supplying subordinate formations, units, and subunits.
4. All available forms of transport must be used.
5. Equipment repair assets must be deployed to those areas with the greatest number of repairable vehicles and/or weapons.
6. The medical services must be deployed as close as possible to the areas with the largest number of casualties.
7. Foraging for local supplies must be undertaken wherever possible.

These principles were basic to the organization of the Soviet logistic system. Adherence to the first principle demanded primarily a logistic system fully capable, in particular, of supporting a high-speed offensive in nuclear or conventional conditions. But the system also had to be or become capable of supplying Soviet troops in other types of war, such as a defensive battle on the Chinese border, the war in Afghanistan, or a longer period of conventional war with members of the North Atlantic Treaty Organization (NATO).

The second principle, requiring the advanced creation and echeloned emplacement of reserves, meant that, in preparation for any high-speed offensive, it had to be expected that the Soviet Army would establish supply bases well forward along all axes of advance to effect the necessary support. This was of particular importance in terms of identifying an imminent Soviet operation and predicting the main axes of troop movement, requiring a large-scale campaign of camouflage, deception, and concealment. The Soviet invasion of Czechoslovakia in 1968 was preceded by a large-scale logistics exercise: "Neuman."

Misunderstanding of the third principle, the responsibility of higher commands for delivery to lower rear elements, caused some misjudgment of Soviet logistic capability in the past. The rear service element of a Soviet division was often criticized as being too small to be able to support the division effectively. In fact, it did not have the purpose of supporting the division's regiments and battalions. The division received support from army or front. The bulk of logistics resources were held at army and front level, and were under the control of those higher formation commanders. This method of organization gave the senior commander a high degree of flexibility in deciding which axes of advance to support and which to abandon, enabling him to quickly concentrate his effort on the chosen axes. This centralized control was always considered by the Soviets as one of the principal strengths of their logistic system.

The fourth principle of the rear, demanding full use of all forms of transport, reflected a Soviet appreciation, based on their experience in the GPW, of the problems of a general shortage of transport and a lack of road space. The Soviet Army was well aware that the operational vulnerability of its long lines of communication and supply from the Soviet heartland to potential combat zones such as Germany, could cause problems particularly in trying to maintain fuel and ammunition supplies to a rapidly advancing army. The experience of expenditure rates of stocks by combatants in the recent Mideast wars reinforced Soviet appreciation of this problem.

The fifth and sixth principles, bringing the repair or treatment facilities to the casualty (human or mechanical) rather than vice versa, must be seen in the context of an offensive on relatively narrow axes; there would be no secure "rear area" and no stable "forward edge of the battle area" but rather "spurs and re-entrants" of friendly and hostile forces. Repair and medical aid would be complicated not only by this factor but also by the nature of the battle, where periods of rapid and possibly isolated advance into the enemy position with few casualties would alternate with periods of great localized destruction due to the use of weapons of mass destruction or a high density of modern improved conventional weapons. There would not, therefore, be an even and steady attrition rate. Nor would it be feasible, due to lack of fuel and the vast distances involved, to drag vehicle or personnel casualties back to a "safe" rear base given the volume and concentration of losses that could be expected. In such circumstances, it would be easier to move highly mobile, if limited, repair and medical facilities to the casualties.

The principles governing the actions of these repair and medical facilities would be to give, as a first priority, immediate attention to those casualties only lightly affected, so that they could return to battle as quickly as possible. Vehicle casualties, incapacitated and requiring major attention, would be left for higher formation workshops. Human casualties would be evacuated to centrally located field hospitals. No attempt would be made in the field to repair damaged vehicles or weapons where the work would require more than a few hours' attention. This Soviet attitude toward repair was often the root of another misconception: that the Soviet Army disposed of divisions, tanks, and men in a "use once and throw away" manner; in other words, casualties were left to die, damaged vehicles abandoned, and divisions ground away to nothing and completely replaced. Although this did happen in the first period of the GPW, when retreating from the Wehrmacht, it was never Soviet policy or doctrine.

The Russians did not consider it either flexible or cost-effective to wear divisions and armies down completely during an offensive. Furthermore, Soviet studies clearly demonstrated that it was essential to reduce the attrition rate during an operation so as to maintain the viability of the formation. Consequently, the main task of the medical and maintenance services was to return to action the maximum number of lightly wounded soldiers and damaged pieces of equipment during the battle to keep up the strength—and therefore the viability—of the operation. German sources refer constantly to the very high recovery rate, during the 1944–45 period, of Soviet tanks and to the ubiquitous presence on the battlefield of armored recovery teams and medical personnel. This does not mean, however, that the Soviets would try to keep all their subunits and formations up to strength by "topping" up with reinforcements. The badly mauled regiment would either be rationalized into a smaller force broken down into companies to merge with other reduced regiments to reinforce them, or be reorganized into a composite formation. To maintain the momentum of the offensive, once a formation suffered a loss of effectiveness through casualties, its leading position in the advance and reorganized formation might become the second echelon or reserve, or it might be switched to a minor axis as the offensive continues.

The final principle of the rear is that of making use of local resources. Foraging for food and fuel was of great importance, particularly if the length of the battle exceeded the expected norm. A Soviet rear service unit in peacetime practiced self-provisioning.

The Mathematics of Sustainability Norms

From 1945 to the late 1980s, Soviet military doctrine sought to reduce the uncertainties of the modern battlefield and to establish standards of activity through comprehensive statistical calculations, thereby ensuring an objective and common approach to the planning of future operations and campaigns. The Soviet concept of sustainability, in common with the whole basis of military art, was founded upon a firm statistical base formed by exhaustive research and mathematical and scientific analysis not only from the GPW, but also from the results of the more recent local wars (Arab-Israeli conflicts, the Falklands/ Malvinas, and Afghanistan), exercises, training, and the introduction of new weapon systems and equipment. From this base standards were produced that had the force and authority of regulations—that is, law. They were known as *normativy* or *normy*; in English, simply "norms."

The *Great Soviet Encyclopedia* describes a norm as "the minimum of something, as established by a rule or plan; for example, a time norm or a sowing norm." The *Soviet Military Encyclopedia* is more detailed. It states that the word *norm* originates from the Latin *normatio*, meaning "to regularize," and for military purposes, norms were subdivided into a number of groups: operational and tactical (spatial and temporal); expenditure; and supply.

Spatial and Temporal Norms

The first type of norms are those scales that characterize the spatial and temporal factors of operations, including tactical tasks of combat forces and the terrain on which they act. Spatial operational-tactical norms take the form of depth-of-battle tasks; dimensions of zones, areas, and sectors of combat operations; areas of grouping of forces; battle order, formations, and groupings along the front and in depth; and scale of redeployments and regrouping.

Temporal norms are concerned with the time taken to complete a task, march, or maneuver. They were worked out taking into account the fighting strengths and capabilities of one's own forces and those of the enemy, battle experience, experience of operations and tactical exercises, degree of preparedness and training of personnel, results of special research, terrain conditions, time of year, and time of day. Norms were calculated with regard to the level of training of personnel for every type of weapon or equipment and their method of use on the battlefield. Norms were recalculated and again finalized according to the degree of a weapon's development or changes in tactics or training. Some examples of operational tactical norms pertaining to the GPW are shown in Tables 1 to 5.

The data in Table 1 show the increase in density of personnel and weapons on the breakthrough sector between the first and third periods of the GPW, which had grown, on average, two to three times in infantry, four to ten times in artillery, and six to ten times in tanks and self-propelled (SP) guns. The higher densities, especially during the breakthrough battle, allowed a decisive superiority over the enemy on the axis of the main thrust. In real terms, this amounted to three to five times more in infantry and some six to eight times more in tanks and artillery, demonstrating the move toward greater "viability."

The scale of offensive operations increased, as shown in Table 2. As front offensives developed in size and scope, these changes were reflected in the dimensions of the rear areas. Table 3 shows these changes. It was on the basis of these changes that future rear support was planned.

Tables 4 and 5 show the structuring of fronts and combined-arms armies (CAA) at the start of the Vistula-Oder operation and the rate of advance of tank armies (TA) in the third period of the GPW.

These tables give a general picture of the operational norms that can be expected to apply in theaters of operations with a depth of front operations of 300 kilometers (186 mi.), a speed of advance between 35 and 50 kilometers (21–31 mi.) per day, and an overall superiority in the breakthrough sector of between 4.2:1 and 6.5:1.

Table 1. *Average Density of Forces and Weapons on Breakthrough Sectors*

Forces/ Weapons	Average density on 1 km of breakthrough sector during the GPW		
	1st period	2d period	3d period
Rifle troops (km per one rifle division)	4–5	2.5–3	1.2–2.5
Guns and mortars	20–80	120–220	200–300
Tanks and SP guns	3–12 (inf. sup. 3–6)	18–40 (inf. sup. 10–20)	70–85 (inf. sup. 12–30)

Table 2. *Scale of Front Offensive Operations*

Item	1st period	2d period	3d period
Sector of advance in km	300–400	75–250	200–250
Depth of operations in km	70–80	100–200	200–300+

Table 3. *Changes in the Depth of the Rear Area During the GPW*

Period during GPW	Depth of operational rear in kilometers			
	Front	Army	Formation*	Overall
Prewar period	Up to 500	75–125	50–75	650–750
During GPW on basis of most important operations:				
Defense	180–250	100–150	30	310–400
Advance				
Prep. of ops	150–250	50–100	15–20	215–370
During ops	200–300	150–200	—	350–500

* Includes regimental rear area (8–12 km).

TABLE 4. *Structuring of Fronts and Combined Arms Armies—Vistula-Oder Operation*

Fronts	1st Ech.	2d Ech.	Mobile Groups	Reserves	Armies	Front Width in km	Breakthrough Sector in km
1st Belorussian	47A	3SA	1GTA	7GCC	47A	14	4
	1A		2GTA		1A	53	—
	61A		2GCC		61A	30	4
	5SA				5SA	12	6
	8GA				8GA	30	7
	69A				69A	54	7
	33A				33A	36	6
Total armies:	7	1	2 TAs, 1 Cav Corps	1 Cav Corps	7		
1st Ukraine	6A	59A	3GTA	7G-MechC	6A	94	—
	3GA	21A	4TA	1GCC	3GA	12	2
	13A				13A	11	11
	52A				52A	10	10
	5GA				5GA	3	13
	60A				60A	110	3
Total armies:	6	2	2	2	6		
Grand total:	13	3	4	3 Corps	13		

TABLE 5. *Rates of Advance of Tank Armies in the Third Period of GPW (from a Selection of Operations)*

Campaign	Army	Depth of Adv. in km	Max. Rate km per 24 hours
Lvov-Sandomir	1GTA	400	60
	3GTA	300	60
	4TA	350	55
Yassi-Kishinev	6TA	300	60
Vistula-Oder	1GTA	610	75
	2GTA	705	90
	3GTA	480	50
	4TA	400	60
East Prussia	5GTA	250	50
Berlin	1GTA	110	20
	2GTA	130	25
	3GTA	130	50
	4GTA	170	50

NORMS OF EXPENDITURE OF MATERIEL RESOURCES

The second group of norms relevant to the military sphere are *normy raskhoda*: norms of expenditure of materiel resources. These are concerned with the accounting of supplies in units of mass or volume or as individual items in their expenditure by servicemen, weapon systems, subunits, units, formations, and armies. Again, norms were laid down by the Soviet Ministry of Defense on the basis of researched and calculated data. For instance, the basic norm of consumption of fuel (diesel, gasoline, oil, and lubricants) was laid down in liters or kilograms for each vehicle, usually for 100 kilometers (62 mi.) of movement or for one hour of operation or static running. When special conditions (difficult terrain, bad weather) prevailed, a supplement was added. Norms of expenditure were laid down for ammunition in *boyevoy komplekt* (BK, or units of fire), fuel in *zapravkiy* (refills), and rockets in batches. Norms of expenditure for each nature of ammunition and fuel for battle were calculated beforehand on the basis of the envisioned action or operation. For example, norms of fuel consumption for tanks in an offensive battle were calculated according to the planned depth of the operation, taking into consideration terrain, weather, and coefficients of maneuverability. As a rule, norms of expenditure also took into account the availability of materiel resources.

NORMS OF SUPPLY

The third type of norms are *normy snabzheniya*: norms of supply. These are the amount of materiel resources laid down for supply to servicemen, subunits, units, or formations and designated for use in a specific period of time. Included in this category are spare parts, types of instruments, materiel stores, ammunition, petroleum/oil/lubricants (POL), and rations. They are closely linked to norms of expenditure.

STATISTICS RELATING TO NORMS OF EXPENDITURE AND SUPPLY DURING THE GPW

Table 6 shows the scale of material supplies in front-line units and those held in the operational rear according to prewar opinions.

Tables 7 and 8 cover the provision of materiel resources in the Voronezh front for the Belgorod-Kharkov operation in July/August 1943, including ammunition, fuel, rations, and forage available.

Tables 9 and 10 indicate material resources for the commencement of the Vistula-Oder operation in February 1945 and the Berlin operation in April 1945.

TABLE 6. *Scale of Materiel Supplies in Front-Line Units and in the Operational Rear According to Prewar Opinions*

Types of supplies and their echeloning	Materiel		
	Ammo (BK)	Fuel (fills)	Rations (24 hr.)
Mobile (i.e., in front-line units)			
In rifle (cavalry) divisions	1.5	3.0	5
In tank (motorized) divisions	2.0	2.5	5
Previous balance (expendable)			
In army depots/dumps			
In the advance	0.75–1.5	Up to 2.0	2–5
In the defense	0.75–1.0	Up to 1.0	3–4
On front depots	8–10	Up to 10	Up to 30

TABLE 7. *Rear of the Voronezh Front in the Belgorod-Khar'kov Operation—Provision of Material Resources July/August 1943 (Ammunition Resupply as of 3 August 1943 [BK])*

Type of ammo	Army							
	1TA	5GTA	5GA	6GA	27A	38A	40A	47A
Mortar bombs, all calibers	2.55	2.85	1.3	1.15	1.6	1.95	1.8	2.03
45mm	4.0	3.5	2.1	2.2	1.9	3.1	2.0	2.0
76mm	2.3	3.1	2.4	2.5	1.4	2.3	2.75	1.6
122mm	1.5	3.2	1.3	1.0	0.5	1.1	2.1	0.5
152mm	–	2.7	–	2.2	–	–	4.4	–

TABLE 8. *Voronezh Front Rations and Forage Available as of 3 August 1943 for the Belgorod-Khar'kov Operation, Measured in "Day's Stocks"*

	Availability			
Commodity	In front-line units	In front depots	Total in front	In transit from depots and central bases
Bread/grain	10.3	0.7	11.0	5.0
Maize	7.0	0.6	7.6	6.0
Meat	16.4	1.9	18.3	7.6
Oil/fats	13.3	2.3	15.6	13.0
Sugar	21.9	1.1	23.0	12.0
Tobacco	20.5	1.5	22.0	11.0
Forage	24.3	7.7	32.0	3.0

TABLE 9. *Amassing of Materiel at Start of Vistula-Oder Operation*

	Categories of materiel resources							
	Ammo BKs		Fuel fills			Rations		
Fronts	Inf. wpns.	Arty.	Avn.	Gasoline	Diesel	Grain	Groats	Fat	Sugar
1 Belorussian	1.5/2.5	3.1/9.8	14.1	4.3	3.4	140	65	33	66
1 Ukraine	1.5/2	3.5/4	6.4	5.1	4.6	21.7	20	28.8	36

TABLE 10. *Stockpiled Supplies at the Start of the Berlin Operation in April 1945*

	Artillery ammunition BKs					Fuel fills		
Fronts	76–100 mm	122 mm	152–203 mm	AA arty.	Mors.	Avn.	Gasoline	Diesel
2 Belorussian	1.8	1.25	2.2	2.8	1.2	6.8	3.1	5.2
1 Belorussian	2.9	3.0	3.0	3.0	4.25	8.7	5.8	5.0
1 Ukraine	1.85	2.7	2.95	2.75	2.0	6.5	4.7	5.0
Average	2.18	2.32	2.7	2.85	2.48	7.3	4.53	5.06

Tables 11 and 12 provide some details of tank army logistics during the GPW. The faster the advance of forces, the lower the ammunition expenditure and fuel consumption. Table 13 illustrates the part played by captured fuel stocks.

TABLE 11. *Ammunition Expenditure in Offensive Operations of Tank Armies*

		Ammo expenditure (BK)					
Operation	Army	Rifle ammo.	MG ammo.	82mm mor.	120mm mor.	76mm arty.	76–85mm tank
Belgorod-Kharkov	1TA	0.35	0.7	1.74	1.93	0.43	1.7
Korsun-Shevchenko	2TA	1.2	1.0	1.4	1.0	1.6	0.8
Lvov-Sandomir	4TA	0.8	1.0	1.9	2.5	1.0	0.9
Lublin-Brest	2TA	0.95	1.1	1.1	1.1	1.1	1.4
Vistula-Oder	1GTA	0.5	0.6	0.6	0.4	0.5	0.7
	2GTA	0.5	0.6	0.2	0.5	0.4	0.65
	3GTA	0.6	0.8	0.6	2.5	1.4	1.5
	4TA	0.54	1.25	0.57	1.11	1.22	0.59
Berlin	1GTA	0.38	0.98	1.26	3.7	1.64	1.9
	2GTA	0.7	0.8	0.8	2.1	1.55	1.9
	3GTA	0.64	0.62	1.3	1.7	0.85	1.06

TABLE 12. *Fuel Consumption by Tank Armies in Offensive Operations*

		Diesel		Gasoline		Aviation fuel KB-70		Selected totals	
Operation	Army	Tons	Fills	Tons	Fills	Tons	Fills	Tons	Fills
Orel	2TA	232	2.9	656	5.0	112	4.3		
	3GTA	1,100	3.7	2,224	9.6	190	6.4		
	4TA	458	2.5	1,568	4.3	113	2.8		
Belgorod-Kharkov	1TA	561	3.6	2,071	9.8	329	6.8	2,961	20.2
Kiev	3GTA	459	2.4	1,014	3.6	101	3.5		
Korsun-Shevchenko	5GTA	500	2.2	850	2.4	39	2.4		
Proskurov-Chernovit	1TA	500	3.3	1,090	5.0	75	3.6		
Lvov-Sandomir	1GTA	1,720	7.5	3,090	11.2	235	5.7	5,045	24.4
	3GTA	1,435	8.6	3,077	10.2	303	7.7		
	4TA	960	7.0	2,467	8.6	301	9.8		
Lublin-Brest	2TA	948	3.5	1,915	5.3	152	4.2	3,015	13.0
Vistula-Oder	1GTA	1,175	3.9	2,535	6.5	382	6.0		
	2GTA	885	3.0	2,182	4.0	218	4.5	3,285	11.5
	3GTA	1,920	6.0	3,519	7.6	392	6.7		
	4TA	1,214	4.7	1,739	6.7	249	4.1		
East Prussian	5GTA	857	3.4	1,951	5.9	209	4.5		
East Pomeranian	1GTA	408	1.7	984	2.6	98	2.0		
	2GTA	470	2.3	1,834	3.6	107	2.5		
Berlin	1GTA	525	2.7	1,192	3.2	148	3.5	1,865	9.4
	2GTA	800	4.1	2,149	4.8	157	4.0	3,106	12.9
	3GTA	654	2.6	1,746	3.9	147	3.5	2,547	10.9
	4GTA	458	2.5	1,568	4.3	113	2.8		

TABLE 13. *Captured Fuel as a Percentage of Fuel Consumed by Tank Armies in Offensive Operations (in metric tons)*

Operation	Army	Overall fuel consumption (metric tons)	Incl. captured fuel	Captured fuel as a % of total consumption
Belgorod-Kharkov	1TA	2,961	962	23.0
Lvov-Sandomir	1TA	5,045	94	1.8
Lublin-Brest	2TA	3,015	921	30.5
Yassi-Kishinev	6TA	1,800	1,288	71.5
Vistula-Oder	2GTA	3,285	822	25.0
Berlin	1GTA	1,865	840	45.4
	2GTA	3,106	568	18.0
	3GTA	2,547	386	15.8

ARTILLERY AMMUNITION: ACCOUNTING AND ALLOCATION

The discipline of a military doctrine and the establishment of fixed norms in order to apply effective fire tended to make the Soviet more thorough and precise in terminology than his Western counterpart. As mentioned above, ammunition was accounted for in BKs and fuel in *zapravkiy*. A BK consisted of a certain number of rounds pertaining to a particular type of weapon: for a 122mm howitzer, it consisted of 80 rounds per barrel. A BK was used as a planning figure for administrative accounting. It had some relevance to the load-carrying capacity of the supply vehicle and average expenditure rates in war. The following extract from the administrative part of a fire plan shows rounds per gun for an artillery battalion.

Preparation for attack	1.25 BK (80 + 20 rounds 122 mm)
Support of attack	0.3 BK (approx. 24 rounds 122 mm)
Accompaniment of attack in depth	0.9 BK (approx. 72 rounds 122mm)
Total	2.45 BK

Likewise, an artillery battalion defensive fire plan accounting showing rounds per gun follows.

Interdiction of advance and deployment	0.4 BK (32 rounds)
Repulse of attack	0.6 BK (48 rounds)
Support of defense in depth	0.6 BK (48 rounds)
Support of counterattack	0.4 BK (32 rounds)
Total	2.0 BK (160 rounds)

The divisional ammunition holdings in terms of BKs are shown in Table 14; Table 15 indicates number of rounds per BK.

TABLE 14. *Breakdown of Divisional Ammunition Holdings of BK*

Ser.	Echelon	82mm	120mm	BM-21	Arty.
1	Co./bty. tpt.	1.25	0.5	0.75	1.0
2	Bn. tpt.	–	0.5	0.375	0.5
3	Regt. tpt.	2.0	2.0	–	3.0
4	Div. tpt.	2.0	2.0	1.0	1.0
Total div. holdings		5.25	5.0	2.125	5.5

TABLE 15. *Artillery and Mortars, BK Weights*

Ser.	Weapon	Rounds per unit of fire	Weight per BK (metric tons)
1	120mm mor.	80	1.8245
2	122mm D-30	80	3.19
3	122mm 2S1	80	3.19
4	152mm D-20	60	4.65
5	152mm 2S3	60	4.65
6	122mm BM-21	160	11.907
7	82mm mor.	120	0.58

Organization of Army Rear Services

LOGISTIC RESPONSIBILITIES

There was no equivalent in the Soviet Army of an ordnance corps or a maintenance corps. Supply and maintenance responsibilities were borne by the service using that equipment. In a unit or formation headquarters, there was a deputy commander for each arm of service, while the rear services, under the deputy commander for rear services, had responsibility for the centralized direction of logistic resources.

The deputy commander for rear services was also chief of the rear area and was responsible for the security and efficient running of the rear area of his unit or formation. This responsibility included local defense and security of lines of communication in the tactical rear, traffic control, and prisoner handling. Details of these responsibilities were:

1. Artillery troops were responsible for the supply of all ammunition, including chemical and nuclear, down to regimental level. Transport allocation was coordinated with the rear services.
2. The commander of chemical troops supervised the supply of chemical munitions but was not responsible for their transportation, which was an artillery matter.
3. Artillery commanders reported ammunition states every twelve hours through artillery channels to the next higher formation. Reports were compiled there and ammunition issues planned depending on the extent of the demanding unit's depletion, the stocks available at the issuing formation, and the supply capability of the next higher formation.

Priority was given to units that would have to exert the greatest effort in the immediate future. Transport was supplied by the higher echelons. Artillery weapons and equipment states were sent every 24 hours.

Tank and armored fighting vehicle (AFV) ammunition demands up to divisional level went to the chief of the rear. Demands were made in multiples of BKs.

Table 16 illustrates in tabulated form the responsibilities of commanders and logistic staffs.

LOGISTIC SUPPORT AVAILABLE TO THE FIGHTING TROOPS

For supply and maintenance requirements, a subunit, unit, or formation had organic subunits or units, plus additional rear service elements that the senior commander had subordinated to the group to enable it to fulfill its mission. (These organic elements are listed below.) However, not only could extra vehicles be allocated, but units could also be stripped of vehicles and equipment if desired. Second- or third-line formations might have incomplete or ad hoc equipment tables, depending on transport availability. As the new Kamaz trucks came into service at divisional and regimental level replacing the Zil (3.2 metric tons [3.5 tons]) and Ural 375 (4.5 metric tons [5 tons]) together with the Kraz 260 at army and front level replacing the Kraz 255B, the lift capacity of unit and formation rear services was considerably enhanced.

Battalion level. A combat arms battalion had a supply platoon of five 5-ton trucks and trailers and three light trucks. In BMP and tank battalions, there were an additional three fuel bowsers. A maintenance section of up to two workshop vehicles and a medical section with a soft-skinned ambulance were also integral to the unit.

Regimental level. A regiment had a rear service battalion consisting of: a headquarters with a control section (Commandants Service); a transport company for fuel ammunition, and cargo with 40 load-carrying vehicles, (4.5 and 7.3 metric tons [5 and 8 tons]) with trailers and 20 fuel bowsers; a mobile maintenance company for the repair of vehicles and weapons with three tank armored recovery vehicles (ARVs), four workshop vehicles, and twelve 20 × 5 or 8-ton trucks; a support platoon, including mobile kitchens, with eight 5-ton trucks; and a medical company with four ambulances, four 5-ton trucks, and three motorized stretcher vehicles. In addition, each subunit in the regiment had its own transport element sufficient to carry personnel, weapons, and immediate supplies.

Divisional level. A division had a support battalion, a maintenance battalion, a medical battalion, and a substantial traffic control element at rear headquarters. In addition, it had a helicopter squadron. Details are:

1. The divisional support battalion had four transport companies—two designated solely for ammunition, one for fuel, and one for spares and general stores. Each ammunition transport company had 60 × 5 or 8-ton trucks and trailers. The fuel company had 80 bowsers and trailers. The ammunition company's trucks, however, could carry packed fuel in cans or pillow tanks if required. The battalion had its own control, maintenance, and repair facilities of some fifteen trucks and ten trailers, two light recovery vehicles, and ten workshop vehicles. It also had smoke-generating vehicles and ambulances. In addition, it deployed up to four field kitchens and a field bakery and was responsible for the armored cars, trucks, and motorcycles of the Commandant's Service (rear area security and movement control).

2. The divisional maintenance battalion deployed as three workshop companies for the repair of AFVs, SP guns, and other artillery equipment. Each company was equipped with up to two cranes, twelve workshop vehicles, six to ten trucks and trailers, reconnaissance vehicles, and a number of heavy lift cargo vehicles. A light equipment repair platoon of six workshop vehicles, the battalion's own support platoon of six ARVs, and artillery tractors with two amphibious load carriers completed its complement. In addition, in the division, each combat support element (engineers, chemical, signals; but not artillery) had its own transport element for immediate supplies.

Army level and above. An army support regiment (brigade) had some 360 heavy (10-ton) trucks and trailers and 240 large fuel bowsers and trailers. Fronts would have an additional three such regiments (brigades); on strategic axes within the theater of operations (TVD), more of these brigades would be in evidence. In the past, almost all such higher logistic transport could be expected to be requisitioned from civilian sources. However, over the last fifteen years, it was Soviet policy to invest in reserves of transport, as it had always amassed reserves of weapons systems.

One of the crucial problems the Soviet Army would have faced in close country was the shortage of real estate in which to deploy. As the rear support requirement increased to meet the norms required for logistic sustainability, the increasing number of logistic vehicles would have lowered a formation's operational viability by presenting a concentration of targets. Hence the developments to increase the lift capacity by phasing out the Zil 131 and Ural 375 trucks. Replacements included the Kamaz 7.3 metric ton (8 ton), while the lift capacity of the Kraz 255 would be upgraded 6.8 to 9 metric tons (7.5 to 10 tons) and the Kraz 260 by 100 percent by using trailers. Still, these load figures were for cross-country usage; road performance would have entailed a greater lift capacity.

A divisional rear support element, while sufficiently

TABLE 16. *Logistical Responsibilities from Front to Battalion Level*

Commodity/ Support service	Responsibility			
	Front	Army Division	Regiment	Battalion
Provision and repair of: artillery weapons (including mortars, rocket launchers, antitank missiles, and AA weapons); small arms; all ammunition; spare parts	Chief of Artillery (assisted by the head of artillery supply)			
Supply and repair of all armored vehicles, including tanks, APCs, and SP artillery	Chief of Armored Troops	Deputy Commander Rear Services	OC Maintenance Company	Battalion Technical Company
Supply and repair of all soft-skinned vehicles	Chief of MT Troops			
Supply and repair of engineer equipment	Chief of Engineer Troops	Commander Engineer Battalion	Commander Engineer Company	
Supply and repair of signals equipment	Chief of Signals Troops	Commander Signals Battalion	Commander Signals Company	
Supply and repair of chemical and NBC defense equipment	Chief of Chemical Troops	Commander Chemical Battalion	Commander Chemical Defense Company	
Fuel and lubricants supply and foraging	Chief of Fuel and Lubricants		Fuel Supply Officer	
Rations and water (Intendance Service)			Food Service Officer	
Catering			Catering officer	
Medical services (incl. medical supplies)			Regimental doctor	

light to enable it to be highly mobile and maneuverable, still had a respectable transportation capacity. The centralization of large transport reserves at army and front level provided the flexibility for a higher command to quickly transfer logistic support from one axis to another in response to the battle situation. Army and front maintained large dumped stocks for the replenishment of second-echelon troops advancing from inside the Soviet Union. Movement of supplies to these dumps was by rail and pipeline. Movement of supplies forward from these dumps to the rear of the fighting troops was the responsibility of army and front transport organizations.

LOGISTIC SUPPLY IN TIME OF WAR

From the above it could be assumed that, as the Soviet army had an abundance of cargo vehicles and large stocks of both fuel and ammunition, it would have little difficulty in supporting a short war. By drawing from state resources in the USSR and Eastern Europe, it would also have been capable of supporting a war for a longer period. From a Soviet point of view, however, the problem was still difficult. It was not the possession of an adequate volume of materiel and supplies that guaranteed success, but getting them to the right place at the right time. Their own experience in the GPW, recent exercises, and a study of the Arab-Israeli Wars imposed upon the Soviets an appreciation of the problems of delivery. Enhanced mechanization increases fuel consumption; consequently, improvement in fuel supply was probably the rear services' greatest single task in the 1980s. The main feature of the Soviet delivery system was that by maintaining large reserves and exercising the "delivery forward" system of responsibility, large variations in logistic requirements could be met.

On the battlefield, the rear was organized into three echelons: tactical or forward troop rear (*voyskovoy tyl*) at division and below; operational rear; and deep (strategic) rear in the TVD. Supplies were moved in bulk, mainly by rail and pipeline, into the army rear, where dumps were established or replenished. In the army rear, the materiel was re-sorted and dispatched to the troop rear by road transport or occasionally by tactical fuel pipeline, or dispersed in dumps tactically for the replenishment of reinforcements (2d echelon troops) before commitment to battle.

In the operational-strategic rear, lines of communication and supply would probably have been guarded by Ministry of the Interior (MVD) troops, KGB border guards, and troops of the Warsaw Pact country through which the lines passed. The front chief of the rear would organize a special control service (*dispetcherskayha sluzhba*) to ensure the coordination of supply; direction would be enforced by a large traffic control network of the Commandant's Service assisted by the Military Vehicles Inspectorate (VAI).

As a general principle, in the forward troop rear, supply was provided two steps at a time down the chain of command. Army vehicles would deliver not only to the divisional rear but also to the regimental rear when the battle permitted. Divisional transport would deliver not only to the regimental rear but also directly to battalions and, in the case of artillery, often directly to gun positions.

RAILWAYS

The Soviet railway system was always the main form of transport for all types of freight and passengers. The lack of adequate roads and the inhospitable terrain and climate coupled with a poor safety record in civilian air travel resulted in the railways remaining the most used means of travel, one of the most efficient systems in the world. Heavy reliance on rail transport would have been a feature of Soviet involvement in a future war.

The movement of troops and military equipment, repair and construction of lines, stations, protection, creation and removal of obstacles, and other railway-related engineering tasks were the responsibility of the Railway Troops, a special subsection of the Transport and Road and Railway Construction Troops. Originally formed in 1851 to protect the Moscow-to-Petersburg line, their role increased in importance. In peacetime, they supplemented the work of the civilian railway staff in all aspects of rail transport. Large numbers of railway troops were employed full time in the construction of the Baykal-Amur mainline (BAM). The movement of troops by rail during the GPW was required not only between fronts and from the rear but also within fronts.

SUPPLIES IN THE FIELD

The location of dumps and stockpiles is dictated by the conditions of battle. Being highly mobile, divisions did not create dumped stockpiles but instead maintained mobile holdings, keeping as much as possible on wheels. At army and front level, replenishment depots were maintained, often near to or at railheads. There was a strategic pipeline system to front level and a front-army pipeline system for POL that also supplied airfields directly.

At divisional level, mobile replenishment stocks were held at convenient road junctions. Fuel could be stored in static rubber reservoirs, although this was more widespread at army level. Stock holdings were under the command of a deputy commander for resupply, subordinate to the chief of the rear of the division.

At regimental level, the same principles applied, but the deputy commander of the rear himself commanded a supply distribution center.

THE FUNCTIONING OF THE REAR IN BATTLE

A Soviet unit of formation headquarters normally had four elements, as listed below.

1. The forward command post comprised the commander and a small staff including a senior deputy, signals officer, artillery liaison officer, and air liaison officer. It was established to bring command close to the focal point of the battle, improving the commander's feel for the battle. In the advance, it would move well forward in the march order; in the attack, it would be with the first echelon.

2. The main command post was under control of either the first deputy commander or the chief of staff in the absence of the commander. This was a larger organization involving all heads of arms coordinated by the chief of staff. The function of the main command post was to wield effective detailed control in accordance with the commander's orders, coordinating the movement of all arms units, their deployment on the battlefield, and the maintenance of their combat capability and supply.

3. The reserve command post usually found adjacent to a subordinate headquarters was where one of the deputy commanders with a small personal staff was located. When this was not formed, a subordinate headquarters was designated as the alternative command headquarters.

4. The rear control point was responsible for controlling all rear services and supply and for the running of the unit or formation rear area. This was usually a mobile headquarters element under the deputy commander for rear services (chief of the rear) and his assistants. On the march and in battle it moved behind the main body at the head of the rear service elements.

Details of location of rear supply and support elements are given in Table 17.

LOGISTIC SUPPORT BY AIR

It was the task of the Military Transport Aviation (VTA) to support ground, sea, and air operations. The Soviets were well aware of the value of logistic support by air. The 8,000 metric tons (8,800 tons) of supplies delivered to six GTA during the Manchurian operation is often quoted as an example.

Aeroflot, the Soviet Civil Airline, was under command of a senior air force general and would come under military control in the event of hostilities. Almost all conscripts rotated to the Group of Soviet Forces Germany (GSFG) were carried every six months by Aeroflot with no disruption to its civil air schedules. Military control in peacetime provided an unparalleled mobilization capacity. Consequently, the Soviet civil air fleet had to be included in calculations of air lift capacity in event of war.

CONTROL OF THE REAR

Control of the rear in field formations was exercised from the *Tylovoy Punkt Upravleniye* (TPU), a mobile rear headquarters commanded by the deputy comander for the rear. Subordinate to him were chiefs of service and an artillery deputy for supply. Staff procedures for the rear support of a regiment or division were as follows:

Normal procedure. The rear commander (deputy commander for the rear) went to the regimental or divisional CP (daily or whenever required), received battle orders from the chief of staff or commander, and returned to the TPU. He then worked out a time and movement scale, informed subunit rear commanders, and issued a warning order. At the orders group with his chiefs of service and deputy artillery chief for supply, he explained the tasks of

TABLE 17. *Location of Rear Supply and Support Elements*

1ST ECHELON SUBUNITS/UNITS	APPROX. DEPTH FROM BATTLE ZONE	
Company	*Offensive*	*Defensive*
First-aid post	–	50–100 m
Ammo. distribution points	–	100–150 m
Rations and water point	–	up to 1 km
Battalion:		
Battalion	1–2 km	1.5–3 km
Ammo.	3 km	2–3 km
Rations and field kitchen	3 km	2–4 km
Technical observation post	1–2 km	2–4 km
Regiment		
Regimental medical post	5–7 km	6–10 km
MT platoon	5–7 km	up to 20 km
POL point	10–15 km	10–20 km
Ammo. distribution point	10–15 km	10–20 km
Rations	10–15 km	10–20 km
Tank and MT repair and evacuation group	up to 8 km	up to 10 km
Damaged vehicle collection pound	5–7 km	6–10 km
2d echelon regiments	16–20 km	all rear service elements
Division		
Div. medical post	10–14 km	up to 20 km
MT repair & depot	10–14 km	up to 20 km
Tank repair & depot	20–40 km	35–60 km
Artillery & small arms repair & depot	20–40 km	35–60 km
MT bn.	20–30 km	20–40 km
Engineer bn.	20–40 km	35–60 km
Ammo. dump	25–30 km	35–50 km
POL dump	25–30 km	35–50 km
Rations dump & field bakery	25–30 km	35–50 km
Baths/laundry/water	25–30 km	40–45 km
2d echelon divisions	40–70 km	

the rear, evaluated the logistic state of the rear, and prepared his proposals for the rear support of the operation. Draft orders for the rear elements were then drawn up. He returned to the regimental or divisional CP and presented his draft proposals for the endorsing signature of the commander. He returned to the TPU and worked out final details of the rear support plan, checking allocation of tasks and organizing control. The detailed plan was confirmed with his chiefs of services, artillery supply officer, and the regimental or divisional chief of staff. A final confirmatory report was made to the regimental or divisional commander.

This had been the normal or standard procedure, but it came to be considered too lengthy. The following procedures were being established for the sake of speed, but they required greater knowledge, preparation, flexibility, and better training of staff officers.

The rapid or parallel procedure. The rear commander, accompanied by his artillery supply officer and chief of fuel supply, attended a briefing of combat unit and subunit commanders. While they were preparing their combat plans, the rear commander, together with his two subordinates, rapidly worked out proposals for rear support with the help of pre-prepared calculation tables and proforma. Using these, the rear commander issued his proposals for the support of the regiment of division at the same time as combat subunit commanders received their orders. When this had been approved by the regimental or divisional commander, the rear commander issued carbon copies to combat troop commanders so that the rear commander's orders on logistic support were included at the same time they issued battle orders to their subunits. Thus, rear support planning was completed at the same time as combat unit operational planning instead of waiting for the completion of the latter in order to commence logistic planning. The rear commander returned to his TPU to brief his remaining subordinates.

Supervision. The Soviets considered conformity with orders to be vital so that the rear commander knew the exact situation should a sudden change of plan be implemented, even if, to obtain such knowledge, it meant halting columns on the march to check their composition. The Commandant's Service and VAI could help with rapid mobile checks, and helicopters at divisional level provided a useful method of supervising column movements and exercising control.

Traffic control. Although it was not strictly a function of the rear alone, it is impossible to discuss rear control without mention of the *Komendantskaya Sluzhba* or Commandant's Service. This organization carried out traffic control and many of the disciplinary measures exercised by military police in Western armies, but without wide investigative and judicial responsibilities. The Commandant's Service provided dispatchers and traffic regulators on all roads in the combat zone and at forward railheads where the duties of railway troops ceased.

Route availability. This was a key element in the viability of rear support itself. Traffic management was a specialization demanding, in Soviet eyes, a great deal of calculation and supervision. Western analysis of this vital subject (which was subjected to considerable study during World War II) concluded that it took over six times as much road capacity to move a formation's vehicles as to move that formation's daily supplies, and over a hundred times as much rail capacity to move the formation by rail as it did to move the formation's supplies by rail.

Soviet doctrine for rear area security. The Soviet concept of "viability of the rear" (*zhivuchest tyla*) included defensive measures against a whole range of threats, such as strikes by conventional and mass destruction (OMP) weapons, all forms of aerial attack, and assaults by deep-penetration ground forces. Defense against all of these threats followed a general pattern, that of tight centralized control of resources with individual units responsible for their own local defense. The Soviet air defense umbrella did not exempt units from the duty of deploying their own defenses. A common element in Soviet concepts of defense against all the above threats was the importance of *maskirovka*: camouflage, concealment, deception, and security.

The high command of forces within the TVD. Established in the 1980s, it would probably have been responsible for overall coordination of theater rear service troops, with a chief of rear service troops located at this level. One of his major preoccupations would be to coordinate the operations of the security forces of the Soviet Union's Warsaw Pact allies. These would include regular troops and worker's militia. It is unlikely, however, that the Soviet Union would have left the defense of its lines of communication entirely in the hands of its allies. MDV and KGB troops could have been deployed outside Soviet borders for greater security. Both organizations had large numbers of trained reserves available.

DELIVERY PROCEDURES

Moving forward behind the combat troops, the rear support elements would off-load supplies, either at a preplanned site, such as a dump, or directly to AFVs or onto a gun position. They would return with casualties and perhaps prisoners of war (POWs). Having delivered these to a central casualty collecting point, field dressing station, or POW control agency, they would return to the divisional rear or regimental dispatch point to reload. In highly fluid operational situations, instead of off-loading at a dump where supplies must be reloaded onto other vehicles, a supply transfer area was allocated where army supply vehicles unloaded their cargo directly onto divisional and regimental transport. During an assault river crossing, it was usual to replenish tanks prior to snorkel preparations.

Special features of ammunition supply. As 90 percent of ammunition was packaged or palleted, ammunition vehicles had to have mechanical handling devices with them, or have them available at the guns, dumps, or supply transfer points. Provision of these specialized equipments was the responsibility of the artillery supply officer. The following were problems associated with ammunition handling.

1. There was a tendency for vehicles to be loaded so hastily during tests and exercises that loads varied from 50 to 90 percent, making it difficult for staff planners to work on an 88 percent capacity load.

2. Delays at ammunition dumps occurred frequently. Once ammunition was grounded, quickly organizing its reissue became difficult, as did observing the need for camouflage and protection.

3. Artillery ammunition requirements (*potrebnost v boyepripasakh*) were calculated on the basis of three types of reserve held by subunits. These were defined as the "running expenditure" (*tekushchiy raskhod*)—that is, rounds kept in/on the gun; the "mobile reserve" (*podvizhnny zapas*) on the subunit transport; and the "supplementary reserve" (*dopolnitelnyy zapas*), normally more than the battery or battalion transport could carry without additional transport. Antiaircraft ammunition was categorized in a different manner.

4. Expenditure rates were highest during the preparation and close support of the attack phases. Expenditure rates decreased as the guns supported the advance into the depth of the enemy position. As a general rule, delivery of ammunition within a regiment was undertaken by divisional or army transport to the regimental artillery dump and moved from there to the gun lines by regimental transport. It could be delivered directly to the gun lines if the tactical situation permitted. On occasion, however, the artillery battalion transport would return to the regimental dump to collect ammunition. Shells were delivered to the gun lines packed in batches and this required unpacking, sorting, degreasing, and stacking in weapon pits. Plans for ammunition delivery started as soon as reconnaissance identified enemy firing positions. Ammunition stockpiled at a firing position may have had to be left there when the guns moved, because unit transport could only carry the mobile reserve. It was advisable, the Soviets stated, to allot more transport resources to artillery units in support of an attack to cover this eventuality. On redeployment, guns were to move with a full mobile reserve of ammunition.

Fuel supply. Soviet forces believed that the demand for fuel in a future war would constitute over 50 percent by weight of all supplies moved during an offensive phase. During the 1941–45 war, the rate of consumption rose rapidly from 5,200 metric tons (5,720 tons) per day for the Orlov-Kursk operation in 1943 to 8,800 metric tons (9,680 tons) for the Berlin operation. This demand continued to increase, and present-day operations by a front would have required 20,000 metric tons (22,000 tons) or more per day. Captured fuel stocks would have made a significant contribution to meeting this requirement. Furthermore, quality and purity of fuel became more important with the greater sophistication of vehicles and aircraft. The mobility required of the fuel supply system was as great as that required for a tank unit. There were several areas in the supply chain where the performance of the fuel supply service was critical:

1. Ensuring the discreet, efficient mass issue of fuel from permanent operational and strategic rear dumps in a very short space of time with little or no warning.

2. Siting and establishing effective field fuel dumps with due regard to camouflage, protection, access, and traffic control together with provision of discreet assembly areas for units to be refueled, both in the rear of advancing units and formations and for the mass refueling of second-echelon formations.

3. Ensuring the rapid and efficient transfer of fuel at formation dumps from bowsers to cans and tanks.

4. Ensuring that units and formations carried sufficient fuel in cans, and logistic support elements provided replenishment facilities in battle, to give maximum flexibility of maneuver.
5. It was the duty of the fuel supply system to train drivers to reduce fuel consumption and wastage, thus reducing demand.
6. Bulk fuel distribution from permanent storage involved rapid off-loading from fixed storage tanks into railway bowsers, vehicle bowsers, and vehicle-carried rubber reservoirs. Such static storage sites had light signals, loudspeakers, or intercoms so that a bowser driver could drive to a pump outlet, state his demand by signal or intercom, fill his vehicle, and drive off with the minimum of preparation and delay.
7. The task of siting field fuel pumps (*polevoy zapravochniy punkt* or PZP) was a complex one. Bowsers and vehicles carrying rubber reservoirs from permanent fuel dumps had to find a secure and convenient location, deploy as many filling points as possible, and organize maximum traffic access. At this level (formation rear or operational rear), the role of the rubber-fabric reservoir increased rapidly. These reservoirs were various sizes: 4,000; 6,000; 25,000; 50,000; 150,000; and 260,000 liters (4,228–274,820 qt.). They could be filled with gasoline, diesel, paraffin, lubricating oil, or water. Their disadvantages were the ease with which they could be punctured, and the fact that they were slower to fill than metal tanks. One URAL 375 and trailer in bowser configuration could carry 9,500 liters (10,042 qt.) of fuel or 20,000 (21,140 qt.) in four rubber reservoirs. A fuel dump could be established within 90 minutes and redeployed in 80 minutes. Formations marching from the rear would not rely on prepositioned supply dumps the whole time but would bring some of their fuel in mobile columns moving ahead of AFVs to refuel them at each tactical halt.
8. Army dumps supplying fuel to divisions faced the same problem with regard to transloading that afflicted ammunition resupply. In the fuel supply system, this was often complicated by the need to load from bowser to can or barrel.
9. In the forward area, refueling was completed in three ways: by AFVs coming to a forward regimental or divisional fuel dump; by fuel being taken directly to AFVs on the combat line; or by a combination of both.
10. To reduce demand, as much extra fuel as possible was carried in spare drums which were jettisoned in action, or in external cans mounted on the outside of vehicles.

Fuel pipelines. For movement of fuel forward to army dumps and airfields, the tactical field pipeline provided a valuable contribution. To examine the tactical fuel pipeline system in detail, it is convenient to divide the subject into five parts: storage means at either end of the pipeline, pumping stations, the pipe itself, fuel, and pipeline-laying troops.

1. At the start of the pipeline at a railhead fuel depot in the strategic rear, fuel would be pumped from permanent storage tanks or collapsible rubber-fabric pillow tanks replenished from railway tankers, or from the railway tankers themselves. At the pipeline terminal, fuel would be pumped into smaller pillow tanks, or into conventional or temporary bowsers.
2. There was a requirement for an initial pumping station plus auxiliary pumping stations at regular intervals to maintain the necessary pressure. These stations were usually mounted on a wheeled chassis and towed behind a vehicle that carried the operating personnel, spare parts, and tools.
3. The PSG-75 mobile fuel pumping station was designed for unloading and transloading fuel from rail tankers into refueling vehicles, tankers, and storage systems, and for pumping fuel within a depot. It was also used as the main pump for the PMTP-100 field pipeline. It was motorized on a VAZ 452 vehicle and consisted of an SFSN-60m centrifugal self-priming pump, an FGN-60 filter, suction and pressure lines with shut-off valves, instrument panels, and control levers. It is thought that the Soviets used thin-walled flanged aluminum piping of 100mm, 150mm, and possibly 200mm bore.
4. The pipeline system could pump all types of fuel, including aviation spirit, gasoline, paraffin, and diesel oil. The Soviets specified two types of diesel oil: D-Z R-20 and U-20. These fuels could be pumped through consecutively without the pipeline being cleared or pumping being halted. Productivity with the lowest viscosity diesel oil was up to 100m^3 per hour; a 150mm-bore pipe would give a rate of flow of approximately 1.5m/sec. (1.6 yd./sec. or 5.7 kph). If necessary the pipeline system could also be used for pumping water.
5. Pipeline-laying troops (*truboprovodniye voyska*) came under command of the Directorate of Rear Services. A pipelaying regiment consisted of three or four battalions of 350 to 400 men. Each battalion, in turn, consisted of three companies and included troops responsible for constructing and guarding the pipeline, with technical personnel to supervise construction and operate the system. A company was responsible for a pumping station plus 12 kilometers (7.2 mi.) of pipeline. A regiment, therefore, could expect to be responsible for 96 to 128 kilometers (60–80 mi.) of pipeline. An army formation could be expected to have one such regiment. Other East European armies, particularly the NVA, also numbered pipeline-laying troops among their forces.

The Quartermaster's Service: Food, Clothing, and Finance

At first sight, the relevance and importance of this service to the viability of a high-speed offensive may not be so obvious despite the importance of its functions to domestic garrison life and administration. Yet the tasks of this service have a role in helping to ensure wartime efficiency.

Peacetime norms established rations at 2 kilograms (4.4 lb.) per day, but wartime stringency might reduce this considerably at times. Food had the lowest supply priority, save perhaps for medical stores, so the Intendance Service had to make provision for foraging in order to ease the strain when the ration issue had been consumed. A five-day ration supply was the normal quoted figure at the start of an operation. For the 1945 Manchurian operation, troops received fourteen-day rations initially and no more was supplied until the successful completion of the operation.

It is here that the peacetime practices would have borne fruit. All garrison units in the USSR (and in other Groups of Forces) were expected to run unit farms to provide a proportion of their rations. In military districts, certain state farms were controlled by the army and their produce supplied direct to the military authorities. Units submitted their food requirement to their military district or group every November. The shortfall between their own production and what was supplied was filled by the Intendance Service by local purchase. Several procurement exercises for formation food supply services were conducted, training the staff in identifying areas with potential food resources. Moreover, all units were established with cooks trained in coping with unprepared foodstuffs—grain and meat "on the hoof"—thus exploiting the advantages of foraging. Despite earlier comments about low priority, the importance of a reliable food supply was not underestimated by the army staff, as its important bearing on combat capability was recognized. Serious efforts also were made to improve the quality of food both in garrison and on exercise.

In the field the army was well equipped for cooking. Equipment included: the PAK-200 field kitchen supplied to tank regiments mounted on a Zil 131; the KC-30D supplied to airborne regiments, which operated from exhaust gases of a GAZ-66D; the KP-200 for use in cold climates mounted on an MTL; and the PKhZ bakery mounted on an articulated truck. This last piece could bake bread on the move.

Accounting and finance in units played an important role in Soviet garrison life. Thrift in all matters, from food to fuel, was a constant demand that impinged on every Soviet soldier. A unit was constrained by an annual financial budget for expenditure on supplies, maintenance, and construction. The task of the unit finance officer, once the annual audit and estimate were completed, required constant monitoring of departments to try and reduce expenditure on shirts, shells, or rubles; preventing waste; praising economizers; and curbing the tendency to overindent. Officers and soldiers who, through negligence, caused or allowed losses to occur, equipment to be damaged, or material to be wasted unnecessarily were fined. Soldiers whose vehicles used more fuel than the norm for the task had to pay for it out of their own pockets at double the market price. Such a tight financial rein was undoubtedly good training for economy in war.

FORAGING

Foraging is the exploitation of resources found in the civilian economy, and "trophies" are equipment and supplies taken from the enemy. However, their place in the Soviet plan merits special mention. Except on special and relatively rare occasions, the use of trophy weapons and ammunition would not figure highly in Soviet plans because it was too difficult a resource to predict. Still, captured fuel, food, transport, accommodation, and water would figure in Soviet planning. Capturing fuel stocks, including civilian assets, might be the only way a forward detachment or elements of an operational maneuver group (OMG) could be replenished. Soviet air force units, deploying forward, were equipped to exploit NATO airfield support systems. Civilian earthmoving and heavy plant equipment would be a useful addition to Soviet engineers. The Soviet Army in 1945 depended on captured fuel stocks for its mobility. The Wehrmacht had a serious fuel shortage in 1945 yet in the Berlin operation almost half of one GTA's fuel was obtained from captured stocks, and in the Yassi-Kishinev operation over 70 percent of six TAs' fuel represented captured stocks.

Technical Support: Maintenance and Repair

The importance attached by the Soviets to the repair and restoration of equipment and personnel casualties and their rapid return to the battlefield was often not fully appreciated in Western military circles. The Soviets maintained this to be a most important factor contributing to the continued viability of formations.

The Soviet principles of operational art and tactics that shaped their ground forces' capacity to fight a high-speed offensive required battalions, regiments, and divisions to be highly mobile, possess great firepower, and use surprise. They could not afford to be burdened with a heavy, organic, rear support, since this would reduce their mobility. Equally, they could not do without this support. The Soviets sought to reconcile these conflicting requirements through weapon designs that stressed reliability, simplicity, and, in the case of vehicles, long range without refueling; echeloning tactics that enabled fatigued units and formations to be relieved and restored to combat effectiveness while maintaining unremitting pressure on the enemy; and organizational measures that provided for only light organic rear support up to divisional level while maintaining substantial rear support at a higher formation level that could be concentrated on major axes.

PRINCIPLES

Certain principles specific to the conduct of the rear in war were applicable for the restoration of viability of mauled units and formations. These were: the need for advanced creation and echeloned positioning of reserves; the responsibility of higher commands for provision of support to lower elements; and the need to move the repair facility forward to the areas of greatest casualties.

The first of these principles demands that formations create in advance and maintain at each level (division, army, and front) specially designated reserves of equipment and vehicles to supplement those that are damaged. The second principle stresses the centralized nature of the restoration work, underlining the fact that units are not solely responsible for their own restoration. The third principle, while reinforcing the second, at the same time emphasizes that repair/treatment facilities will not be distributed evenly over the battlefield but will be concentrated forward specifically to those areas where they can be the most effective. German sources in the last war underestimated the repair facilities available to Soviet forces on the main axes of an offensive because they failed to appreciate this last principle.

As mentioned earlier, the Soviets made a number of studies of their wartime experience and depended on statistical analysis to predict the level of loss rates and maintenance requirements that might apply in future conflicts. While absolute numbers of AFVs in a TA might change, the Soviets considered that percentages, ratios, and trends were likely to remain the same. It was on these figures and the conclusions to be drawn from them that Soviet attention focused. It was the collection, analysis, and use of data that enabled them to plan a repair system which would contribute to the maintenance of a high weapon system availability; and if it had been achieved, the level of availability would have been higher than their Western counterparts'. NATO armies, as a rule, do not collect and analyze these data to the same degree. NATO repair systems are, therefore, reduced to guessing the scales of spare parts and repair facilities to provide for their armed forces. Furthermore, Soviet forces had a higher ratio of "teeth" repair and recovery elements organic to combat and some combat support arms than NATO armed forces, with considerable savings in resources and greater mobility on the battlefield.

The ability to predict servicing and maintenance requirements made it possible, when procuring weapons systems, to include in the budget a fixed sum to cover servicing and maintenance costs for the whole life of the weapon system, with an annual review increment based on a national statistic to take account of the rate of inflation. Such fixed life-cycle costings assisted budgetary calculations and promoted greater stability in weapons procurement processes.

REPAIR AND RESTORATION ON THE BATTLEFIELD

The principles listed above and the procedures that resulted from their application affected the repair and maintenance of all weapons, equipment, and the treatment of human casualties. When looking to the battlefield, it was the experience of tank formations of 1941–45 that provided the best model for operational analysis. The most significant conclusion that Soviet analysts reached is that during an operation, new equipment would in practice be introduced into formations only during long marches, during periods of preparation, or during pauses between phases of an operation. It would be very rare for new equipment to be fed into Soviet formations at any level or echelon during the course of the battle. It was not realistic to rely on major reinforcement of formations during a battle as a means of maintaining their viability. With few exceptions, the only reliable sources of reinforcement with major items of equipment during the operation would be from operational reserves, or from damaged equipment repaired on the battlefield at higher formation level.

The Soviet Army did not start the war in 1941 holding this concept. It was their experience during the large-scale offensives of 1943–45 that convinced them of its validity. In some operations, loss and repair rates reached staggering proportions. During the L'vov-Sandomir operation, the number of tanks and SP assault guns repaired and returned to battle in three GTAs by the end of the operation exceeded the total number of AFVs in three GTAs at its start. During that operation, each tank and SP gun had been knocked out on average two or three times and then had undergone repair to fight again. An analysis of Soviet loss rates during eleven different operations reveals several interesting points. These operations were not selected at random, nor are they representative of the war as a whole. They were selected and assessed by Soviet analysts because of their special relevance to modern battle conditions. Total losses (vehicles not repairable at operational level) of AFVs during a fifteen- to twenty-day operation totaled about 25 percent of the initial number of available AFVs, while overall loss rates (total number of vehicles "knocked out" or rendered unserviceable by mechanical failure) ran at 82 percent of the initial number available. Daily percentage loss rates of AFVs ran at something over 5 percent, of which 1.7 percent were total write-offs, 0.6 percent required heavy repair and therefore could not be repaired at formation level, and 3 percent were returned to battle after light or medium repair at unit or formation level. Loss rates were far from uniform or steady; instead, they fluctuated widely depending on the nature of the army's engagements, the intensity of the action, the density of the enemy's antitank defenses, and the level of Soviet crew training. During the last year of the GPW, of all AFV losses due to shaped-charge projectiles, 30 percent were total write-offs, 15 percent required major (not battlefield) repair, 25 percent required medium repair in army and front workshops, and 30 percent needed light repair. These statistics were held to be relevant to modern conditions when attempts were made to predict the effect of modern light antitank weapons on Soviet AFVs.

The Soviets maintained that the overall figures from 1941–1945 showed that 60 percent of all AFVs joining an army just before an operation and 85–90 percent arriving with fighting forces during an operation came direct from repair workshops. Thus, repair was and would remain the prime way of replacing battlefield losses. In the latter stages of the GPW, where a typical operation involved a high rate of advance, delivery from factory via depot to formation was rare. Most of the new tanks delivered to formations came from operational reserves constituted for that specific purpose.

THE ORGANIZATION OF REPAIR AND MAINTENANCE

A modern Soviet motor rifle or tank battalion had a light maintenance team consisting of up to three soft-skinned workshop vehicles under the command of a technically qualified NCO. In a regiment, the deputy commander for technical services had a repair and maintenance company consisting of an armored car, three ARVs (5 in a tank regiment), and up to twenty soft-skinned workshop and transport vehicles. At divisional level, the repair battalion included a recovery team (six ARVs and an amphibious transport vehicle), a transport and support platoon (10–12 trucks plus trailers, a light equipment repair platoon of 6 workshop vehicles), and three repair and maintenance companies. The latter were specialized for the repair of AFVs, SP guns, and other artillery. They each included ten to fifteen workshop vehicles, vehicle-mounted cranes, and up to twelve transport vehicles.

At higher formation (army and front) levels, repair or maintenance regiments deployed as several general repair battalions (similar to those at divisional level) or as specialized field workshops.

In peacetime, units used their organic heavy equipment, such as their tanks, very rarely. For reasons of economy, in order to keep operational equipment as little used as possible, units often bused their personnel to permanent training regiments where they used equipment dedicated for training purposes. The heavy use that these training vehicles received gave the formation repair teams ample opportunity to practice repair techniques. Within the unit, vehicle and weapons crews were taught to service and maintain the vehicle and undertook basic repairs (such as track-link replacement on a tank or APC). Within garrisons, unit and divisional repair teams would, in peacetime, complete far heavier repair tasks than on the battlefield. Both in garrison and on the battlefield, weapon and vehicle crews assisted the repair teams in their task.

The conscript system limited the training of soldiers. There were relatively few technically qualified regular NCOs. Most of the major maintenance and repair was done under the supervision of, and often personally by, technically proficient junior officers. This was particularly the case where the repair of high-technology equipment was involved.

REPAIR AND MAINTENANCE IN BATTLE

Planning the technical support of an operation began the moment the army commander passed his warning order to his chief technical deputy. The plan would be based on the availability of repair facilities, the most effective means of their employment, the ability to supplement capacity by use of local resources, the ability of unit personnel to effect running repairs, the expected losses and types of mechanical breakdown, and the availability of spares. The levels of recovery and repair are shown in Figure 1.

Inventiveness and resourcefulness were always the attributes of a Soviet repairman. The reduced maintenance requirement that was a concomitant to the simplicity of the Soviet equipment design was complemented by a traditional Soviet ability to improvise, to make do with essentials only, and to cope with adverse weather conditions. The Soviets had three classifications of repair:

Running or light repair. This was a quick repair that could be undertaken at regimental- or divisional-level mobile workshops, or even by the vehicle crew themselves. It involved changing a damaged or worn-out component,

(1) *Company*	(2) *Battalion*	(3) *Battalion Repair*
Technical deputy. Carries immediate spares to assist quick repairs. Refers other repairs to battalion.	Technical deputy at Technical Inspection (observation) Post. Identifies damaged equipment and assesses extent of damage. Directs them to battalion repair and evacuation group.	Attempts only repairs that will return the vehicle or weapons to the battle within 60 minutes. Other damaged equipment will be dragged off the routes and left for the regiment.
(4) *Regimental*	(5) *Damaged Vehicle Collection Point*	
The regiment maintenance and repair company commander in the Technical Inspection (observation) Post. Identifies and assesses damaged equipment and directs his recovery vehicles to assemble the equipment at the DVCP nearby.	The regimental repair company. Will attempt repairs that will enable the equipment to rejoin the regimental order of battle (3 hrs max. repair). The remainder will be left in the DVCP for the attention of the division.	
(6) *Divisional Maintenance Battalion*	(7) *Army and Front Repair Assets*	
Major repair resources are deployed onto the axes with the greatest collection of damaged equipment. Repair is attempted by workshops specializing in artillery, trucks, armored vehicles, and light equipment. Serious repairs (+3 hours) are left for attendance by army.	Often deployed in the divisional rear areas; specialize in repair of equipment or components; can cannibalize weapons for supplies of spares.	

Figure 1. Soviet repair and recovery structure.

making technical adjustments, light welding, or simple mechanical work. Three hours was the maximum time spent on a repair at this level during an operation.

Medium repair. In the last war, this was carried out at corps and army level. In the 1980s, it was done at army and front level. Medium repair involved the replacement of major components or modules on an AFV and servicing. It was the task of the divisional evacuation and repair teams to recover vehicles and bring them to Damaged Vehicle Assembly Points (in GPW designated SPAM, now SPPM). The mobile repair bases moved to the vicinity of these assembly points. The guiding principle of operation was to repair first those vehicles with the least damage. During the GPW, corps and army repair teams performed many "running" repairs on vehicles that battalions, regiments, and divisions were unable to complete due to lack of time. Having carried out all medium repairs possible within the limits of time and the availability of spares, the mobile tank repair base would pass on to the next Damaged Vehicle Assembly Point, leaving behind the more badly damaged vehicles for the army repair workshops.

Heavy or major repair. By the third period of the 1941–45 war, this was no longer attempted at the operational level. Vehicles or weapons needing complete overhaul were recovered to static workshops. Under modern conditions, this would have required a destroyed tank or gun to be recovered to a factory in Eastern Germany or any other Warsaw Pact country for rebuilding, cannibalization, or scrap.

Similar procedures existed for the maintenance and repair of aircraft. Very little work was undertaken at the forward airfields. Aircraft were returned to higher formation repair facilities for anything more than routine servicing or minor repairs.

In the initial stages of operations during the GPW, army repair battalions and evacuation-recovery companies operated forward with divisions and corps, reinforcing their effort and undertaking running or medium repairs. Only during a slow advance did army repair bases undertake major repairs. During a high-speed offensive operation, there were too many running and medium repairs to be done and insufficient time available to make major repairs worthwhile. In many cases, tanks needing major repair were cannibalized for spares to enable more light or medium repairs to be undertaken. The rest were evacuated to front repair workshops but not returned during the operation.

As a general rule, during a high-speed offensive operation, army repair facilities would almost all be deployed at lower formation (divisional) level and would operate to maintain the viability of battalions, regiments, and divisions; this would also be the primary task of their own, organic maintenance subunits. Holding most repair and maintenance facilities under army control created the advantage, in effect, of keeping them in reserve for allocation and deployment to wherever they were most needed on the battlefield.

THE LOCATION OF REPAIR FACILITIES

During the Vistula-Oder operation, Damaged Vehicle Assembly Points (SPAM) were established 40 to 50 kilometers (24–30 mi.) apart, behind the advancing forces. Each point was the base for a mobile repair-recovery group consisting of a tank repair company, a mechanical repair platoon, a recovery-tractor company, and a spares detachment with fuel, communications, and reconnaissance vehicles. Each point was in operation for six to seven days. So rapid was the rate of advance of the main forces that a SPAM established just to the rear of the battle area was 150 kilometers (90 mi.) behind the forward edge of the battle area (FEBA) by the time it came to move forward again. Drawing from the important lessons of the Vistula-Oder operations, the Soviets applied them to the modern-day organization of their support. The first was to have available within formations at all levels a dedicated reserve of AFVs that could be rapidly deployed to restore viability to a damaged unit or lower formation. The second was to ensure the deployment of repair bases well forward, relocating them frequently so that lines of communication did not become too extended.

At front level, the most successful organizational innovation of the last war was the mobile tank component repair workshop (PTARZ). This consisted of several specialist repair workshops mounted in vehicles. Behind the advancing troops of each front, PTARZs would seek out the areas where most damaged and unrepaired vehicles had been collected by SPAMS. PTARZs would then remove and repair major components of the damaged vehicles, such as engines, and either reconstruct entire tanks from cannibalized and repaired components or send forward replacement components and assemblies as spares for use by army repair teams. These specialist repair workshops were so successful that their establishment at front and army level would have been likely in a future war. The high level of technology in all arms of service would require these workshops to be established on a special-to-arm basis dedicated to repairs of tanks, guns, electronics, and engineering equipment.

Medical Support

There is no doubt that the attitude of past and modern-day Soviet military commanders toward casualties was different from that which many Western officers appeared to assume. Western military commanders of the last war were constantly criticized by Soviet historians and strategists for their failure to follow through on military operations for fear of a high casualty rate. The Soviets accepted that casualties in any future war would be high, especially in an offensive campaign involving weapons of mass destruction. Many Western soldiers equated this Soviet willingness to accept casualties with a careless attitude toward human life. Consequently, they made the erroneous assumption that the Soviet Army had no medical support and that casualties were left to die on the battlefield. Certainly medical services, as part of rear services, did not receive the same priority as combat arms when it came to defense expenditure. Nevertheless, the Soviets always recognized that the ability to treat the wounded quickly and effectively increased battle performance by returning them to battle more quickly; it also improved morale and, therefore, fighting capability. As a consequence, a great deal of attention was devoted to medical care of soldiers on and off the battlefield. The study of casualty rates was very important as a means of establishing the viability of a formation in the offensive.

MEDICAL SERVICES

The Soviet Military Medical Service formed a separate corps of the Soviet Army. For command and organization it was subordinate to the Ministry of Defense Directorate of Rear Services. In medical matters it was subordinate to the Ministry of Health of the USSR, being the senior branch of that ministry. The personnel of the Military Medical Service were doctors, medical assistants, medical orderlies, and assistant orderlies. Doctors, always of officer rank, could either be specialists or general practitioners. To join the Military Medical Service, a candidate doctor—one who is not already qualified—had to gain entry by a competitive examination to the Kirov Red Banner Military Medical Academy for a six-year course to qualify as a Military Doctor in a general or specialist category.

An important person in the organization of the Medical Services was the medical assistant (*feldsher*), who was either a long-service NCO (*praporshchik* or ensign) or a junior officer. Highly trained in the minor surgical and medical tasks of a doctor, as well as in dispensing, his task was to relieve the doctors by dealing with minor ailments and the daily administrative routine. Medical orderlies (*santarnyy instruktor*), usually *praporshchiks*, were responsible for the basic training of first aid, field hygiene, administration of medical units and subunits, and for battlefield tasks: lifesaving, first aid, and the organization of primary evacuation. Women were employed as nurses and medical assistants throughout the Military Medical Service. During the GPW, they were frequently frontline doctors. Unskilled or semi-skilled conscripts drafted into the Medical Service as assistant orderlies (*pomoshchiki*) undertook the minor, routine tasks.

The Military Medical Service was responsible in peace and war for all medical and dental care in the Soviet Army. Within this broad framework, their priorities were defined as:
1. Development, organization, and implementation of medical protective measures of troops to warn them against or reduce the effects of NBC (nuclear, biological, chemical) weapons.
2. Organization and provision of medical aid and the treatment of the sick and wounded.
3. Organization and provision of constant medical supervision in military life, working conditions, and training of troops to prevent or provide early warning of illness and inhibit the spread of disease. This involved: responsibility for training and supervising soldiers in personal hygiene and first aid—including medical anti-NBC measures and some supervision of personal NBC decontamination; organization and running of all medical centers, first-aid posts, hospitals; organization of battlefield evacuation and all medical treatment; organization of all medical supplies from source to user; and supervision of water and food supplies, feeding arrangements, and sanitation.

When executing its responsibilities in specialist fields such as supervision of decontamination or education in hygiene, the Medical Service cooperated with other arms (for decontamination, this would involve the chemical defense troops for hygiene education), and with the party political organizations.

BASIC PRINCIPLES OF ORGANIZATION IN THE FIELD

The medical service of a front in war and a military district or group of forces in peacetime was run by a Military Medical Directorate. This body managed the various hospitals in its area for diagnosis, specialist treatment, and transport of the sick or wounded back to the USSR. In addition, a wartime front had an organic medical regiment that was responsible for establishing a field hospital and organizing the casualty evacuation system.

Each army had an organic medical regiment. This regiment was of variable size, depending on the size and the role of the formation. It was expected to establish one or more reserve field hospitals at army level as directed by the front military directorate. The regiment was also responsible for casualty evacuation from army to front.

The basic unit of the Medical Service was a medical battalion; there was one in each division. Most of the battalion's resources were allotted to establishing a field hospital with a capacity of up to 200 patients every 24 hours, providing general nonspecialized surgical operations and up to 60 beds for those casualties either too weak to be moved farther or so slightly wounded as to be sure of quick recovery.

At regimental (brigade) level there was a Regimental Medical Post run by the organic medical platoon. The function of this post was to act as a dressing station, treating light casualties and preparing the others for evacuation to division. It had a capability for minor operations and transfusions. The post was commanded by a senior regimental medical officer, usually a major, and might have two more doctors, together with medical assistants, assistant orderlies, nurses, and attached ambulance drivers.

A battalion had a Battalion Medical Post under the command of a medical assistant, with an orderly, assistant orderlies, and drivers. Each motor rifle company had an orderly and one or more assistants.

In wartime, medical service units and subunits took to the field in the form of medical aid posts or field hospitals. On the march, they were the leading element of the rear services, moving as close as practicable to the fighting units, so as to effect treatment as rapidly as possible. A company medical orderly and assistants moved with the main body of the company either in their own vehicle or in a company APC.

The battalion medical post would establish itself 1.5 to 3 kilometers (.9–1.8 mi.) behind the front line, a regimental medical post about 6 kilometers (3.6 mi.) behind the fighting in an offensive battle, and 9 kilometers (5.4 mi.) in defense, while a divisional field hospital was usually established about 12 kilometers (7.2 mi.) back when in the attack and 20 kilometers (12 mi.) back in defense—in all cases, farther forward than logistic supply depots.

Medical supplies were sent via medical channels. There was no separate supply organization, rear units being responsible for the supply to forward units.

The first principle of medical treatment was the prevention of disease and the maintenance of hygiene among the troops. Company or battalion medical personnel were to be present at morning inspections and were to hold sick parades. They were responsible for checking the siting and construction of latrines, testing the purity of the water supply, and supervising the preparation of food and the storage of rations. After combat they were responsible for organizing the burial of the dead. At all levels, great attention was paid to the implementation of antiepidemic measures.

From experiences in the civil war and the 1941–45 war, the Soviet Military Medical Service developed a system of "treatment and evacuation by stages," which constituted the main characteristic of its battlefield organization. The aim of this system was to provide continual medical treatment at the various stages of medical evacuation, combined with the evacuation of the sick and wounded—as far as diagnosis permitted—to specialized medical facilities, where they could receive more effective medical care. Though this system could function in any type of war, it was in fact tailored to provide appropriate medical support in an offensive operation. To help rationalize their treatment and evacuation system, the Soviets defined five classes of medical assistance: first aid, premedical treatment, preliminary medical treatment, full medical treatment, and specialized medical treatment. First aid was rendered on the spot by the casualty himself, a comrade, or the company medical staff. It was limited to commonsense techniques to prevent further pain or injury to the wounded man. Premedical treatment was given by the company medical staff or at the battalion medical post and was restricted to dressing or redressing wounds, improving splints, checking the application of tourniquets, and administering chemical antidotes. Preliminary medical treatment was not normally given until the patient reached the regimental medical post. It was classified by degree of urgency as: immediate, when the life of the patient was in danger; or delayed, when further treatment could be postponed without undue risk to the patient. Full medical treatment was rendered when the patient reached the divisional medical field hospital or specialized hospitals farther to the rear. The category was once again subdivided with reference to the degree of urgency: immediate, to save the patient's life; urgent, when delay was undesirable; or nonurgent, when delay would not be detrimental to health.

Conclusion

It is possible that some authorities in the past have underestimated the Soviet capacity for sustained combat because of the failure to understand the nature of the Soviet system. There can be no doubt that, with the benefit of a firm statistical base relating to logistical expenditure during the GPW, together with the ability to update from the results of other conflicts and local wars (including the experience from Afghanistan) and the strong bias of weapon and equipment design toward reliability and maintainability, sustainability and viability of the Soviet ground forces would have continued to be enhanced. The degree of viability of a Soviet offensive in Central Europe, however, would have depended on how well the Warsaw Pact generated a limited number of forces for the Soviet strategic operation, conducted defensive operations where appropriate, maintained rear area security, and provided supplies. Good performance in all these respects would have been required not only for a short, sharp operation but also under the less favorable conditions of a long, drawn-out conflict. During a more protracted operation, the strains of "enforced coalition" might have surfaced with serious and debilitating effects, particularly to the security of the logistic arteries flowing from the *tyl* of the Soviet Union.

The above describes the Soviet logistics system as it was through 1988; very significant logistical changes were in the offing when the Soviet Union collapsed in 1991. The half-completed withdrawal of Soviet forces from Hungary, Czechoslovakia, and Poland (and subsequently from Germany); the adoption of a new Soviet defensive military doctrine and posture (with a counteroffensive capability retained); and the overall reduction of Soviet military forces and arms would have markedly changed logistical requirements and planning, including those for force generation, mobility, and mobilization. The factors of speed, volume, and priority of resupply were different for USSR-interior defensive operations than for foreign, exterior offensive operations. Standardizations and new logistical norms for mobility inside the USSR would have become more important than in the past, lines of communication ipso facto shorter and presumably more defensible. The likely transformation of the Soviet armed forces from a large, conscription-based force to a smaller, professionalized force with territorial components would have changed many of the logistical parameters of the past. Civil war, maintaining public order, and reacting to increasing internal chaos and disorder would have placed added demands on already perturbed and constrained supplies of consumables for military operations inside the USSR.

C. W. BLANDY

SEE ALSO: Attrition: Personnel Casualties; Consumption Rates, Battlefield; Infrastructure, Military; Lines of Communication; Logistics: A General Survey; Maintenance; Logistic Movement; Organization, Army; Replacements: Personnel and Materiel; Resource Management; Soviet Military Doctrine; Supply; Sustainability and Viability; Warsaw Pact; Warsaw Pact Policy and Strategy.

Bibliography

Great Soviet Encyclopedia. 1973. Vol. 30. New York: Macmillan.
Donnelly, C. N., et al. 1986. The sustainability of Soviet ground forces, Soviet Studies Research Centre (SSRC), Royal Military Academy, Sandhurst, Research Paper.
Malygin, N. A. *Sovershenstvovaniye Operativnovo Tyla.* Moscow: Voyenizdat.
Peredel'skiy, G. E. 1984. *Artilleriskikh Divizion v Boyu.* Moscow: Voyenizdat.
Radziyevskiy, A. I. 1979. *Tankovyy Udar.* Moscow: Voyenizdat.
Soviet Military Encyclopedia. 1983. Vol. 5. Moscow: Voyenizdat.
Voyenno-Istoriocheskiy Zhurnal (VIZH). 1965, January and April; 1976, April (Moscow).
Wardak, G. D., comp. 1989. *The Voroshilov lectures: Materials from the Soviet General Staff Academy.* Vol. I: *Issues of Soviet military strategy.* Washington, D.C., National Defense Univ. Press.

LOW-INTENSITY CONFLICT: THE MILITARY DIMENSION

The term *low-intensity operations* was introduced in the 1960s, and from it, the term *low-intensity conflict* (LIC) emerged in the 1980s. It refers to a situation between peace (when no arms are used) and the full-scale use of weapons in conventional war.

During such a situation there is extensive diplomatic, economic, and psychological activity, but its primary feature is the covert or overt use of military force short of war. The military may have noncombat missions, such as training assistance, advisory assistance, or peacekeeping, but the missions may also be combat oriented, such as in counterinsurgencies, reprisals, or limited contingencies that call for so-called "special operations."

Because low-intensity conflict is not a uniform phenomenon but rather a series of dissimilar incidents, the part that military power can or should play in connection with those incidents cannot be precisely defined. As the forces that created a low-intensity conflict situation become more violent and confrontational, the opportunities for military involvement become clearer. These may appear in categories that include: insurgency, counterinsurgency, peacekeeping, and contingencies calling for "special operations."

Low-intensity conflict can involve activities ranging from diplomatic, economic, and psychological through subversion, insurgency, and guerrilla warfare. Clandestine and covert actions to seize or to release hostages may take place, and acts of terrorism—used as psychological weapons—may be carried out. Special operations forces may be used in any or all of these roles and in such capacities can contribute either to the success of insurgents or to that of the forces mobilized to fight insurgency.

Insurgency

Resistance to the governing structure of a country becomes most critical when dissidence and subversion produce active insurgency. While still technically within a low-intensity conflict context, insurgent movements may spawn guerrilla forces that add a substantial military dimension to the struggle. If guerrilla formations develop to proportions that permit them to engage conventional military units on equal terms, the insurgency may move from the low-intensity conflict arena to that of conventional warfare. Guerrilla warfare is a product of those forces that have combined to oppose the targeted government on every level at which it is seen to be vulnerable: psychological, political, economic, or military.

Both the political and military leaders of a nation threatened by insurgency must analyze and understand its root causes. If the targeted government becomes aware of and is sensitive to the rumblings of discontent among its population prior to open antigovernment violence, that government may reduce the threat through political, social, and economic reforms. Without a dynamic information campaign to convince the public that the contemplated reforms are real and imminent, however, the effect of even extensive ultimate reforms upon the incipient insurgency may be negligible.

Although a government may genuinely seek answers to problems known to spawn dissidence, subversion, and insurgency, corrective measures that are theoretically possible may not be feasible. Deep-seated racial, ethnic, religious, and social differences, which have resulted in conflict and turmoil for centuries, often defy correction even if patience, logic, and law are invoked hand-in-hand over protracted periods. Crushing economic woes of a one- or two-commodity country can sometimes be alleviated only by major changes in the foreign policies and trading patterns of several other countries. Those countries may be unwilling to modify their national lifestyles in order to relieve the internal stresses of the afflicted country and thereby to save it from internal upheaval.

Histories of major insurgent movements indicate that

dissatisfaction with entrenched regimes is frequently so deep and fundamental that no concessions short of abdication will satisfy the insurgent leadership. If hostility toward the government, as expressed by dissident subversive propaganda, is shared by any considerable element of the population, efforts to suppress the insurgency are not likely to be successful.

Social and political systems that permit free dissent provide a much less favorable climate for subversion and insurgency than those in which voiced opposition is likely to be punished. On the other hand, some totalitarian systems have developed the mechanics of repression into a fine art. These may prove adequate for indefinitely holding in check even widespread potential for insurgency. Such repressive systems are often brittle, however, and subject to catastrophic disintegration. As was demonstrated in Eastern Europe in 1989, the majority who have been held vassal to the few may decide to endure the risk, pain, and trauma of breaking their chains.

Counterinsurgency

Counterinsurgency is a term applied to political, economic, social, military, and paramilitary measures that indigenous governments and associates use to forestall or defeat revolutionary war. It applies also to similar measures that occupying powers use to prevent or defeat resistance movements. The role of military power in defeating an insurgency may range from substantial to supporting or even to peripheral. Counterinsurgency measures call for the use of almost every resource available to the government that is the target of the insurgency, including:
1. intelligence and counterintelligence,
2. psychological operations,
3. police,
4. military forces, and
5. diplomacy.

Subversion

Successful counterinsurgency planning requires in-depth information about the subversive infrastructure. This infrastructure provides the insurgency with inspiration, intelligence, recruits, training, planning, and logistics. Neutralization of the subversive infrastructure by the counterintelligence systems of the government will normally receive a high priority. The cellular structure of a well-implanted resistance movement, however, makes it difficult to neutralize; vigorous government efforts may net minor and local successes without seriously damaging the overall underground movement.

A desperate government may try to inhibit the growth of subversion with repressive measures against the general population; the aim being to make cooperation with the resistance movement more hazardous. Such measures have achieved mixed results historically, and generally create more dissidence than they suppress.

Psychological Programs

In free and open societies, psychological operations against domestic audiences range from difficult to impossible. In societies where free press and freedom of speech are looked upon as part of the fabric of the nation, government attempts to develop information programs, either for or against movements with strong political overtones, are not feasible.

Police-Military Cooperation

A well-trained, disciplined police force can be the first line of defense against civil disorder. If it works meticulously within the framework of the law and demonstrates attitudes that are viewed as fair and impartial, it can sometimes prevent escalation of minor incidents into major confrontations. Good police intelligence systems normally acquire intimate knowledge of the personalities, rumors, and activities in crowded urban ghettos, where living conditions may be one of the factors that gives rise to and nurtures subversion. If an insurgency grows to proportions that are beyond the capabilities of the police, military assistance may be called upon. A smooth transition from police authority to that of a militia or the regular armed forces can be facilitated by provisions of law, enacted and tested prior to their invocation in an actual emergency. The lack of such a smooth transition procedure in China in 1989 set the stage for the Tienanmen Square incident in Beijing.

Conventional military forces are not normally well suited to engage in counterinsurgency operations without substantial modifications. Changes in orientation, training, personnel selection procedures, organization, and equipment are dictated by the requirements of missions against an enemy whose tactics are unorthodox and whose targets frequently are less purely military than they are psychological and political. Coordination of military counterinsurgency activities with those of police and other nonmilitary governmental agencies requires mutually accepted and practiced operating procedures that may vary considerably from standard military models.

Military Counterguerrilla and Antiguerrilla Operations

Destruction of guerrilla movements through the complex formulas applicable to counterinsurgency can differ substantially from antiguerrilla tactics used by military forces in the field under wartime conditions. In the first instance, to supplement and complement tactical operations against guerrillas, popular support for the guerrilla movement must be eroded through carefully planned and executed civic actions, propaganda campaigns, and governmental reforms. In the latter case, troop commanders may hold the civilian population responsible for guerrilla attacks and take drastic measures in reprisal. The long-term and lasting benefits that stem from viewing guerrilla movements as part of a political and social phenomenon that cannot be addressed solely through the use of military strength may be considered of secondary importance by a troop commander who sees his primary mission as one best accomplished through fire and movement.

Tactical operations by conventional military forces against guerrillas call for:
1. mobility comparable to or greater than that of the guerrillas,
2. small-unit leadership of exceptionally high quality,
3. intelligence not always available through purely military collection systems, and the
4. ability to live in the field with minimal resupply for protracted periods.

Military Civic Action

Urgently needed civic assistance programs using conventional military forces, skills, equipment, and supplies tend to project the image of a benevolent government and to challenge subversive propaganda. But extensive use of military personnel for activities that normally lie within the civilian sector can also serve to create friction with labor unions and elsewhere within the economy. This in turn can give certain elements within the indigenous society perceived excuses for antigovernment attitudes. Military forces in civic action roles are most effective when their presence, guidance, and material support serve to inspire the local populace to address and solve their own problems. In matters of counterinsurgency, the actions and attitudes of the police and the military are generally seen by the indigenous population as reflecting those of the government whose orders are being implemented. Thus, in confronting an actual or incipient insurgency, the message sent to the people through the security forces is vital.

Population and Resource Control

In addition to their intelligence and counterintelligence capabilities, the police and military establishments are major instruments through which population and resource control may be exercised. Inhibition of freedom to travel—through such means as requirements for special passes, movable roadblocks, check points, and curfews—assists security forces in making the environment more inhospitable for insurgents. Rationing of items needed to carry on normal living and business activity may aid in cutting the flow of public assistance to underground organizations and force them to surface in search of supplies. Both population and resource control can become counterproductive from a psychological point of view if they are viewed as illegal or arbitrary, or if they are designed so that they harass the innocent while seeking the guilty. If martial law is put into effect as a counterinsurgency measure, its impact on the general public can be lightened by repeated government explanations of the reasons why certain civil liberties are being denied temporarily and promises that they will be restored as soon as national security is no longer in jeopardy. The success of population and resource control as a counterinsurgency measure depends on the selection and training of the security forces and their sensitivity in dealing with people under conditions of stress.

Insurgency and Counterinsurgency—Role of Foreign Assistance

Insurgency movements, both in the underground stage and later as active, organized guerrilla forces, may seek and receive assistance from foreign sources. In such cases the external counterinsurgency strategy of the targeted government will include not only physical interdiction measures but also vigorous diplomatic initiatives. The latter may urge censure and economic penalties against the foreign supporters of the indigenous insurgency while soliciting international approval, funds, military equipment, and supplies for its own counterinsurgency campaigns.

The impact of foreign assistance upon an indigenous counterinsurgency effort is a function of several important factors. Among these is the extent to which foreign aid and advice are the products of in-depth understanding of the history, political aims, anthropology, and mores of the recipient country.

It is unlikely that the strategy, tactics, organizational concepts, and equipment reflected in the skills and philosophical orientations of the foreign military forces will be appropriate for the highly specialized counterinsurgency needs of the recipient country. Motivation for foreign sources to provide assistance for governments confronted with incipient or actual insurgency generally falls within categories that can be described as strategic, economic, traditional, or idealistic.

The internal stability of a potential client nation may attract a patron nation's aid to ensure the patron's own security. Continued access to world markets and to strategic raw materials important both to national defense and to economic vitality may even suggest counterinsurgency assistance to governments whose ruling philosophies are only marginally compatible with those of the aid donor. Politically active ethnic groups within the societies of potential sources of foreign counterinsurgency assistance can either facilitate or negate the flow of aid to governments that have requested it. Idealistic or unrealistic parameters limiting the amount or the manner in which counterinsurgency assistance is used may stem from the foreign providers' attempts to project their own political and social systems as models for the government under attack.

In rare cases, the strategic aims and political objectives

of the indigenous government under insurgent pressure coincide with those of the foreign sources of assistance. Most often, however, foreign counterinsurgency "advisers" are accepted primarily as conduits through which supplies, equipment, and funds may be acquired. Pressures from foreign aid donors aimed at influencing the political end products of indigenous counterinsurgency efforts succeed or fail in direct proportion to the sophistication and dexterity of the diplomatic efforts through which they are applied.

A domestic insurgency represents a condition that only the indigenous people themselves can solve satisfactorily. Outside help may be solicited and may be instrumental in enabling the indigenous government to remain in power; but the conditions placed upon provision of assistance from foreign sources may serve to produce a hybrid product unsuited to either the political or the social fabric of the recipient country. If foreign assistance to a beleaguered indigenous government is the decisive factor in defeating an insurgency, it may demonstrate that the base for the insurgency among the people was neither wide nor deep.

Historical Examples of Counterinsurgency

Examination of the counterinsurgency successes achieved by the British in Malaya (1948–57); the government of the Philippines under President Magsaysay (1946–54); and in Greece (1945–49) provides source material from which a number of common characteristics can be extracted. The British success in Malaya was chiefly due to easy ethnic identification of the insurgents (Chinese) and well-executed programs that isolated the guerrillas from the indigenous population. Magsaysay's victory was characterized by brilliant civic action and propaganda programs. Generally the success of the counterinsurgents in Greece has been attributed to the fact that the guerrillas' access to external support was effectively severed.

On the other hand, the failures of counterinsurgency measures in Cuba (1953–59); in South Vietnam (1959–65); and in Portuguese Africa and Zimbabwe (Rhodesia) during the same periods illustrate some characteristics of unsuccessful counterinsurgencies. The inability of the Batista regime to deal with Fidel Castro's insurgent movement had much to do with the blatant corruption of the government and lack of inspirational leaders working against the guerrillas. South Vietnamese forces failed to defeat the Vietcong because of political turmoil, superb communist guerrilla organization, and mediocre government leadership. Both the Portuguese government in Angola and Mozambique, as well as the government in Rhodesia, lacked popular support and were therefore doomed.

Military Aspects of LIC

The role military forces play in low-intensity conflict situations may be quite minor. Most likely, more insurgencies have been quelled by rapid, effective political action than by military repression. On the other hand, external military assistance to an insurgent movement may be limited and, in some cases, counterproductive if that aid is seen as intrusive by the indigenous population. Intelligent, responsive, political measures, however, coupled with rapid, pervasive military operations may combine to defeat the best of insurgent movements. Although subversion and psychological programs are not necessarily military in nature, they may have military aspects. Military civic action and population and resource control efforts are likely to involve military forces and can be critical to the success of a counterinsurgency campaign.

The ongoing insurgent and counterinsurgent efforts in Afghanistan, Ethiopia, El Salvador, Cambodia, and Angola readily illustrate the persistence of LIC and the roles of military forces in these conflicts.

WILLIAM P. YARBOROUGH

SEE ALSO: Assistance, Military; Covert Action; Hostage; Military Aid to the Civil Power; Paramilitary Forces; Psychological Warfare; Special Operations Forces; Spectrum of Conflict; Subversion; Unconventional War.

Bibliography

Alexander, Y., ed. 1979. *Terrorism: An international journal* 3 (1 and 2). New York: Crane Russak.
Bacevich, A. J., J. D. Hallums, R. H. White, and T. F. Young. 1988. *American military policy in small wars: The case of El Salvador*. Institute for Foreign Policy Analysis. Cambridge, Mass., and Washington, D.C.: Pergamon-Brassey's.
Collins, J. M. 1987. *Green Berets, Seals, and Spetsnaz*. McLean, Va.: Pergamon-Brassey's.
Kitson, F. 1971. *Low intensity operations*. Harrisburg, Pa.: Stackpole Books.
Levytsky, B. 1972. *The uses of terror*. New York: Coward, McCann and Geoghegan.
Molnar, A. R., et al. 1965. *Human factors considerations of undergrounds in insurgencies*. Washington, D.C.: American Univ.
Navarre, H. 1956. *Agonie de L'Indochiné*. Paris: Librairie Plon.
Sully, F. 1968. *The age of the guerrilla*. New York: Parents Magazine Press.

LOW-INTENSITY CONFLICT: THEORY AND CONCEPT

Low-intensity conflict (LIC) is a term that covers a broad area of military and nonmilitary operations below the level of conventional combat between regular military forces. The U.S. Joint Chiefs of Staff define low-intensity conflict as "a limited political-military struggle to achieve political, military, social, economic, and psychological objectives. It is often protracted and ranges from diplomatic, economic, and psychological pressure through terrorism to insurgent war. Low-intensity conflict is generally characterized by constraints on the geographic area, weaponry, tactics, and level of violence" (U.S. Department of Defense 1985).

Within the conflict spectrum, low-intensity conflict is distinguished from high-intensity conflict and mid-intensity conflict. As the definitions of high- and mid-intensity conflicts are confined to war between regular armed forces, low-intensity conflict corresponds to the threat or use of force below the level of regular conventional or nuclear conflict. This does not, however, exclude possible escalation of a low-intensity situation into higher stages of the conflict spectrum.

Concept of LIC

In general terms, low-intensity conflict covers both nonmilitary (diplomatic, economic, and/or psychological) and military operations in the gray area between peace and open, conventional warfare. It applies to two distinct sets of phenomena. One encompasses the various forms of insurgency and counterinsurgency, guerrilla warfare, and terrorism. The other set of phenomena comprises several forms of military pressure—gunboat diplomacy (show of force) and clandestine and overt use of force below the level of warfare. In this context LIC would include insurgencies and counterinsurgencies and actions for their support such as the Afghan resistance, interventions by Belgian forces in the Congo crisis, Cuban and South African forces in Angola, French forces in Chad, U.S. forces in Lebanon, the Turkish intervention in Cyprus, some aspects of the Vietnam conflict, the deployment of Western naval forces to the Persian Gulf in 1986, and peacekeeping operations.

It should be noted that the term *low intensity* was applied by Kitson (1971) not in the context of "conflict," but under the heading of low-intensity operations. He used *low intensity* to define certain types of military operations characterized by a low level of force (counterinsurgencies and peacekeeping operations).

Sarkesian (1988) proposed re-introducing the term *unconventional conflict* to mean conflict ranging between noncombat employment of military power on the lower side and conventional war. *Unconventional war* would include *special operations* on the one side, and *low-intensity conflict*, reduced to revolution/counterrevolution, on the other. This approach closely corresponds to the term *sublimited unconventional war* as used in the 1960s, clearly separating low-intensity conflict from (conventional) *limited war* as the next rung of the conflict spectrum.

Low intensity would then characterize certain operations within the framework of unconventional conflicts as well as other unilateral military operations, which threaten or use force below the level of conventional warfare, and also operations that take place outside armed conflicts, for example, military assistance not related to an actual conflict, peacekeeping operations, or "armed neutrality" operations.

The Phenomenon of LIC

Low-intensity conflict is multidimensional and occurs on the tactical/operational, strategic/political, and sociopolitical levels.

Tactical/Operational Level. On the tactical/operational level, low-intensity operations differ in several aspects from normal, or conventional, military operations. In insurgency/counterinsurgency campaigns, they are mostly identical with guerrilla/counterguerrilla warfare. Guerrilla warfare is a typical indirect approach employing irregular or unconventional use of force in a widespread, sporadic, low-intensity fashion rather than in massive concentration. It does not envisage territorial gains but rather is intended to drive off an adversary (Liddell Hart 1974)

Modern terrorism resembles guerrilla warfare in many tactical specifics, and in the final objectives, but differs in crucial elements. Guerrilla warfare at its core remains a military strategy, inherently driving to escalate from unconventional to conventional warfare. Terrorism, on the other hand, aims at weakening the opponent psychologically. Terrorism is based on the belief that demonstrating the vulnerability of a society will lead to the collapse of the political order. Thus, terrorism does not attack military objectives but aims for nonmilitary, civilian targets.

Countering both guerrilla warfare and terrorism requires low-intensity operations, emphasizing tactical dispersion in a "fine but closely woven net over the widest possible area" (Liddell Hart 1974), as well as intelligence and close civil-military relations (Kitson 1971). Counterinsurgencies often approach police-type operations more than traditional military operations. In the same vein, the "paracriminal" character of terrorism by itself calls for police-type counteroperations.

The same is true for strike operations, evacuations, and similar actions, intended either to punish, to end a breach of law, or to restore law and order. It is also true for peacekeeping operations, which police a certain area (e.g., against the breach of an armistice agreement), as well as armed neutrality operations, which police an area against the spillover of actual conflict. As these operations do not envisage genuine combat actions and the full employment of armed force, they are also properly characterized as low-intensity operations. From a tactical perspective, low-intensity operations are sometimes characterized by their police-type elements.

Strategic/Political Level. On the strategic/political level, the emergence of modern low-intensity forms of operations, as well as strategies of unconventional conflict, may be traced to two distinct factors. On the one hand, insurgency and guerrilla warfare increasingly have become the "weapon of the weak" (This has been true throughout history). Terrorism has been employed frequently where even guerrilla warfare has failed, for example, in the Middle East after unsuccessful Palestinian guerrilla campaigns between 1967 and 1969; and in Europe, where domestic terrorists tried unsuccessfully to mobilize the masses for insurgencies before turning to terrorize them.

On the other hand, unconventional warfare and low-intensity operations have been related to the global rivalry of great powers because of the nuclear stalemate, which burdens even conventional warfare with uncalculable risks. Liddell Hart (1974) predicted that guerrilla war may be increasingly developed as a form of aggression suited to exploit the nuclear stalemate, leading to a new type of *camouflaged war*, which has since been characterized as *covert warfare;* that is, low-intensity operations of an offensive type. This, in turn, has promoted the search by the global powers for military answers leading to concepts of military assistance or of assistance in internal defense operations, and similar types of low-intensity operations.

In a similar way, global powers see the requirement for unilateral military operations below the level of conventional war, including reactive or even preemptive strikes or similar operations of power projection for promoting strategic/political objectives, for preventing adverse developments, or for punitive actions.

Sociopolitical Level. On the sociopolitical level, low-intensity conflict is primarily characterized by insurgency and counterinsurgency. There, the underlying conflict takes place primarily within a society. At its core, this type of LIC is either revolutionary or counterrevolutionary—the promotion or prevention of change in the social or political system being its main objective. It aims at "milieu goals" in contrast to "possession goals" in traditional warfare, that is, the changing of the sociopolitical framework rather than the gaining or losing of territory. Sarkesian (1988) has pointed out that in low-intensity conflicts the center of gravity of the conflict (in Clausewitzian terms) is rarely in the armed forces of the antagonists, but in the political-social milieu of the indigenous system.

Historical Roots

In the modern concept of low-intensity conflict there are two distinct historical roots. One is the root of insurgency, and the other is the root of gunboat diplomacy.

INSURGENCY AND COUNTERINSURGENCY

The concept of insurgency has two roots: (1) the phenomenon of uprisings against an occupying power (nationally motivated uprising), and (2) the phenomenon of revolutionary civil war (socially motivated uprising). The nationally motivated uprising can be traced at least back to the time of the Roman Empire; the best-known example is the Jewish uprising in Palestine, A.D. 66–74. Another nationally motivated uprising was the Spanish "guerrilla" or "small war" against Napoleon. Nationally motivated uprisings continue in this century. Socially motivated uprisings can be traced from the plebian uprisings in ancient Rome through the peasants' uprisings in Medieval Europe (which in Germany escalated to conventional warfare until they were defeated in 1525), and on to the major revolutions of this century.

Earlier Phenomena. Early examples in more recent times include the colonial uprisings and "small wars," which most of the major powers experienced during the nineteenth century; for example, the Sepoy mutiny in British India in 1857–58. The United States experienced the "Indian wars" during its expansion westward, and imperial Russia had similar experiences in Siberia and other Asian parts of the Russian Empire. Later, the phenomenon shifted toward support for resistance movements. The British were instrumental in the Arab revolt against Turkish domination in World War I. In World War II, it became part of Churchill's war strategy against Germany to organize resistance movements in Europe (Liddell Hart 1974; Foot 1980). Armed insurgency by the Yugoslav partisans proved very successful, growing from unconventional to conventional warfare by the end of the war. Soviet partisans in the occupied territories operated in coordination with the regular forces, as did Mao Tse-tung's guerrillas in the Far East.

Ideology and the "Struggle for Liberation." The resistance movements of World War II merged the social and national roots of uprising. Marxism had given an ideological momentum to social uprising, which proved successful in Russia in 1917. Marxist-Leninist doctrine stressed the revolutionary role of the urban proletariat, but Mao Tse-tung in 1927 launched a rural guerrilla campaign that contradicted most major Marxist tenets. Mao then merged the social revolutionary motive with a nationalistic motive in the struggle against the Japanese occupation, which after World War II left him with the political and military structures required to overcome the noncommunist Chinese government as well.

Mao's successful guerrilla campaigns were the model for numerous subsequent guerrilla campaigns. In most of these, the nationalistic motive of anti-colonialism blended with either orthodox Marxist or at least some "socialist" ideology. This was true of the Vietminh insurgency in Southeast Asia, of the Algerian insurgency, and of other anti-colonial insurgencies in Africa and Asia during the 1950s and early 1960s. Even Castro's domestic Cuban revolution contained some anti-colonial elements stemming from the previous domination of Cuba by the United States.

Conversely, counterinsurgency became important to the Western military forces. Counterinsurgency campaigns were waged by the British in Malaya, and by the French in Algeria. The Vietnam conflict led to U.S. participation in counterinsurgency and retained these elements, although it escalated into conventional war with the increasing involvement of North Vietnamese forces. At that time, the concept of counterinsurgency gained a prominent place in Western military theory and practice.

Then, in the late 1960s, insurgency turned away from rural areas to the cities. Revolutionaries in Latin America promoted the idea of the "urban guerrilla." These ideas, soon copied in Europe, particularly in Germany, Italy, and France, and to a certain extent even in the United States, led to the emergence of domestic terrorism. This method was adopted by other groups with mixed social and nationalist motivations, including parts of the Palestinian resistance movement, the Irish Republican Army (IRA) and its different factions, and the Basque ETA. In turn, antiterror operations and counterterrorism gained some prominence in the military training and doctrine of the regular military forces of nations under terrorist attack.

Gunboat Diplomacy

The second major root of low-intensity operations is gunboat diplomacy, or "show of force." A unilateral threat or use of force was frequently employed by European powers and the United States to protect their citizens and promote their political objectives against less powerful states in colonial and semicolonial status during the eighteenth, nineteenth, and twentieth centuries.

Legal Aspects of LIC

Unconventional warfare is a deviation not only from conventional warfare with regard to its strategic, operational, and tactical concepts, but also from the legal conventions regulating war. Two problems exist. One is the problem of *just war* (the legitimization of the use of force). The other problem is the legal regulations for warfare, concerning mainly participation in war (the problem of the lawful combatant) and the conduct of war (the problem of means and methods of warfare).

Legal Development

In the period of the classical "law of war," neither colonial wars nor internal insurgencies were considered wars in the legal sense. The threat or use of force as exercised in gunboat diplomacy conformed with the inherent right of states to declare war (*ius ad bellum*) and thus to use military force on a lower level. This did not change with the creation of the League of Nations or with the Kellogg-Briand Pact of 1928, which was aimed at outlawing war. The traditional laws of land warfare, as codified in the 1907 Hague convention, provided no regulation of and thus no protection for those engaging in insurgencies or armed resistance. Combatant status was accorded only to regular forces or to militia and volunteer forces under certain criteria.

After World War II, there were two major developments regarding war. One was the Charter of the United Nations, which prohibited any threat or use of force, except for the purpose of lawful self-defense and as sanctioned by the UN Security Council. The other development concerned the changing nature of warfare and its legal regulation. The Geneva Conventions of 12 August 1949 included members of organized resistance movements in the category of lawful combatants. Two Additional Protocols to the Conventions were adopted in 1977. The First Additional Protocol is devoted to international armed conflicts, including all cases of anticolonial insurgencies. Other cases, however, are regarded as noninternational armed conflicts and covered by the Second Additional Protocol.

These protocols are characterized by two principal features. First, they avoid any reference to the legal status of war; they refer rather to armed conflict. Second, they emphasize elements typical of insurgencies; for example, distinguishing combatants from noncombatants by their actions rather than by formal status, regulating the conduct of armed forces of any kind instead of insisting that only states are entitled to organize armed forces, and explicitly prohibiting measures terrorizing the civilian population.

Few Western powers have ratified the protocols. Their main content, however, would be applicable even among nonsignatories of the protocols, as they serve to clarify existing understandings rather than create new law. The protocols cover most unconventional warfare or low-intensity operations because they are applicable to any kind of use of military force without reference to a legal status of "war." On the other hand, terrorists could not invoke the protocols because they clearly treat terrorism as a breach of the law.

Legal Problems

There remain several problems with the UN Charter's prohibition of the use of force in some low-intensity operations. As these operations cross the threshold of the threat or use of force, they could be legitimized only inasmuch as they constitute acts of justified self-defense against an armed attack. On the other hand, the United Nations' 1974 definition of aggression extends not only to conventional military operations but also to the sending of armed bands, groups, insurgents, or mercenaries by a state or in its name, which employ armed force. The legal framework for military actions countering irregular actions remains blurred, both with regard to its justification as acts of self-defense against an act of armed aggression and with regard to the legal status of those participating in strike operations as long as they do not constitute an "armed conflict" in the meaning of the Additional Protocols, or as long as the state for which they operate has not ratified the protocols.

Future Prospects for LIC

Low-intensity conflict, in its various forms, is likely to grow in importance. Although the threat or use of armed force has been limited legally to defensive purposes, it can nevertheless occur as a breach of law in an offensive fashion. As long as the international order cannot provide centralized law enforcement agencies or other effective means to prevent or counter armed conflict, states will not abandon the option to take the law into their own hands and resort to unilateral or combined actions, including the threat and use of armed force. The better deterrence works on both the nuclear and the conventional levels, the more strongly will states be inclined to use surrogate, unconventional strategies in the threat or use of force. This may be true for either promoting or containing insurgencies for both "milieu goals" and political/ideological/economic influence, for unilateral actions of power projection, intervention, covert and overt operations not involving the employment of regular forces, and the use of regular forces for "surgical strikes" and similar operations.

The number of states has increased considerably, and their capabilities for promoting and supporting insurgencies, and for power projection, have grown as well. Both components of LIC are likely to be employed by a growing, rather than a declining, number of states. The emerging pattern of growing differentiation into regional and subregional power struggles inside as well as outside the framework of the East-West conflict may enhance the proliferation of low-intensity conflicts. Thus, LICs are likely to become more frequent in interstate power relations.

The growing number of actors other than states on the international and intranational level indicates continued growth in low-intensity conflict. These new actors include multinational companies, nationalist or similar movements, and terrorist or transnational criminal organizations, such as drug syndicates. These actors usually cannot resort to regular military forces and strategies but have to rely on irregular forces. By necessity they will have to use unconventional strategies and low-intensity operations. The reemergence of religion, ideology, nationalism, separatism, and even tribalism as factors in domestic as well as international politics—especially on the regional level—may further contribute to these trends. An increase in LICs of this type may provoke growing capabilities, as well as actual employment, of interventionist strategies by regional, as well as global, powers that are adversely affected by them. That may be true despite improved relations between the global powers and their willingness, in principle, to disengage from regional conflict locking them against each other, for such trends cannot induce them to ignore regional developments entirely.

Finally, low-intensity operations not involving active participation in actual conflicts also appear to be on the rise. This is true for peacekeeping operations precisely because of the global powers' disengagement from some regional conflicts, employing the United Nations and its peacekeeping forces as a stabilizing element as well as a face-saving device. It is also true for operations outside the framework of properly established peacekeeping, where assets and interests have to be protected against the spillover from actual conflicts without becoming involved in them; in other words, the rediscovery of armed neutrality. The more likely actual conflicts on the regional level become, and the less inclined global powers are to let themselves be dragged into actual participation while maintaining their presence in the region, the more this specific form of low-intensity operation will be employed in the future.

HEINZ VETSCHERA

SEE ALSO: Limited War; Low-intensity Conflict: The Military Dimension; Special Operations; Unconventional War.

Bibliography

Clauseqitz, C. von. 1976. *On war*. Ed. and trans. M. Howard and P. Paret. Princeton, N.J.: Princeton Univ. Press.
Filiberti, E. F. 1988. Defining the spectrum of conflict. *Military Review*, April, pp. 34–43.
Foot, M. R. D. 1978. *Resistance*. London: Granada/Paladin.
Liddell Hart, B. H. 1974. *Strategy*. Chapter 23, Guerrilla War. New York: Signet.
Kitson, F. 1971. *Low-intensity operations: Subversion, insurgency, peace-keeping*. London: Faber and Faber.
Mao Tse-tung. 1967. *On protracted war*. Peking: Foreign Languages Press.
Motley, J. B. 1985. A perspective on low intensity conflict. *Military Review* 65(1):2–11.
National Defense University. 1986. *Proceedings of the low-intensity warfare conference*. Washington, D.C.: Fort McNair.
Sarkesian, S. 1986. *The new battlefield: The United States and unconventional conflicts*. Westport, Conn.: Greenwood Press.
———. 1988. The myth of U.S. capability in unconventional conflicts. *Military Review*, September, pp. 3–17.
U.S. Department of Defense. 1985. Memorandum SM-793-85 of the Joint Chiefs of Staff.

LUDENDORFF, ERICH [1865–1937]

Erich Ludendorff is one of the most tragic figures in military history. A soldier of great intellect and thorough competence, Ludendorff became the dominant personality in the German war effort in World War I. Besides determining national policy, Ludendorff also supervised and implemented changes in tactical doctrine that brought the German Army to a level of tactical excellence that has seldom, if ever, been equaled. But, as one biographer has aptly stated, Ludendorff was a "tormented warrior," a victim of his own terrible intensity. Without others around him to help steady him, his behavior often exhibited emotional instability, and his judgment would become distorted.

Career Prior to World War I

Erich Friedrich Wilhelm Ludendorff was born near Posen (Poznan), Prussia on 9 April 1865. His family was middle-class landowners; unlike many of his Prussian military contemporaries, he was not an aristocrat. Tragedy touched him early in his life when his father became bankrupt.

Ludendorff became a cadet at the age of 12 and quickly displayed a great capacity for hard work, a characteristic that he retained all his life. Commissioned in the infantry, he was selected to attend the Kriegsakademie, the school of the German General Staff, in 1893.

From 1904 to 1913 he served with the Great General Staff in Berlin. Ludendorff became deeply involved with the issue of army manpower, and he pressed vigorously for increases. So doggedly did he pursue the issue that in 1913 he was posted to a regimental command out of Berlin, probably to get him away from the capital and away from sensitive policy issues. During his posting in Berlin, Ludendorff also met Margarethe Pernet, a divorced housewife with four children, who, drawn to this intense army officer, married him. Vivacious and beautiful, Margarethe helped Ludendorff display a warmer, more human side. He was devoted to her children, who warmly returned his affection.

World War I

With the advent of the greatest war in Europe since Napoleon, Ludendorff was a staff officer with the Second Army as it swept into Belgium, part of the great turning movement of the so-called Schlieffen Plan. Taking over a brigade whose commander had been killed, Ludendorff boldly directed the unit to take the critical Belgian fortress of Liège. With a combination of bluff and cool bravery, he persuaded the confused Belgian defenders to surrender the citadel. For this action, in which he demonstrated both great courage and great tactical skill (it was also his first experience under fire), he was awarded the nation's highest decoration.

Events on the eastern front now unexpectedly demanded Ludendorff's attention. As the bulk of German forces swept through Belgium in the attempt to defeat France quickly, the German Eighth Army in the east was to hold off the Russians until victorious forces from France could later be diverted to that theater. However, the commander of the Eighth Army began to panic as two Russian field armies massed on the borders of East Prussia. The German high command decided to appoint a new commander and a new chief of staff to that field army. General Paul von Hindenburg was called out of retirement to be the commander, and Ludendorff was named the chief of staff. The two men, who had never previously met, greeted each other at a railway station in Germany, and headed east on a special train. Thus began one of the most successful partnerships in military history. Hindenburg's calm, steady, and humane demeanor was the perfect complement to Ludendorff's brilliant but high-strung mind.

The Hindenburg-Ludendorff combination enjoyed immediate success. By concentrating first (near a town named Tannenberg) against one Russian field army, decisively defeating it, and then hitting the other army, the German Eighth Army drove the Russian forces out of East Prussia and inflicted severe casualties on them. By the end of 1914, as the war settled down to a long struggle, the Hindenburg-Ludendorff team (or H-L team, as Winston Churchill would call it) was placed in command of all German forces in the east. For the next two years these German forces hammered the Russians, driving them back deeper into Russia. In August 1916, frustration with German operations in the west caused the Kaiser to select the Hindenburg-Ludendorff team to direct the entire German war effort, with Hindenburg named as the chief of the general staff and Ludendorff as the first quartermaster general, a name he chose himself, not wishing to be a "deputy" or "second."

The Tactical Revolution

As Ludendorff now moved west to concentrate most of his efforts on the most crucial front, the western front, he must have been tempted to advocate "eastern" solutions. To his great credit, he set out to learn about the peculiar conditions of the western front, refraining from arrogantly imposing his own solutions from the experience in the east.

In adjusting to the tactical conditions of the western front and in directing changes in German tactics from 1916 to 1918, Ludendorff was at his most brilliant. By soliciting opinions and observations from the fighting units, Ludendorff's small operations staff developed new tactics; disseminated them to the units; refined them based on feedback; and then trained, organized, and equipped the army to apply them. Ludendorff was not the inventor; more important, he understood that tactical creativity existed *not* in a remote headquarters, but in the fighting units, and he led the corporate effort to *discover* solutions to vexing tactical problems.

In 1917 the German forces in the west were on the defensive. Applying a new tactical concept, of the elastic defense-in-depth, the Germans absorbed the tremendous allied offensives and responded with mauling counterattacks. Many German observers remarked that Germany would not have survived the battles of 1917 had not Ludendorff managed the tactical changes so successfully.

For offensive tactics (which the Germans applied in 1918), the Germans developed the concept of the sudden attack in depth, using carefully planned concentrations of artillery to disrupt the enemy, followed closely by infiltrating infantry units probing deeply, bypassing when necessary, constantly keeping the adversary off balance. Under the direction of Ludendorff, the German principles of offense and defense recognized that tactics had become decentralized in execution, that the small unit, even down to the squad, was a key element. The uniformly equipped rifleman was no longer the basic element of the tactical unit; soldiers with specialized weapons now had to be organized in the proper mix at the small-unit level. Special assault troops, called stormtroopers, were trained. The tactics also stressed the importance of combined arms, particularly the combination of infantry and artillery.

Strategic Errors

While Ludendorff was truly in his element directing the tactical innovation of the German Army, his influence in directing the entire German war effort led him into areas well beyond his expertise. The military talent that Germany possessed in abundance was not matched by political talent. Into this vacuum the dominant personality of Ludendorff penetrated, but the role exceeded even his capacity. The Hindenburg-Ludendorff team, dominated by Ludendorff, directed such diverse endeavors as the war economy and U-boat policy. In the latter case, their policy of unrestricted submarine warfare brought the United States into the war. When an opportunity for a negotiated peace emerged in 1917, Ludendorff ended it by making unrealistic territorial demands.

In order to destroy the Western powers before the full weight of the United States would tip the scales, Ludendorff decided to make one last offensive in the west in the spring of 1918. The great offensives that ensued were classic examples of German organizational ability. The new offensive tactics proved quite effective, as the Germans acquired more territory than had any army in the west since 1914. But the precarious position of Germany, with dwindling military resources and a suffering population at home, demanded a flawless strategic direction. This Ludendorff did not provide. He did not concentrate the German effort, nor did he reinforce success. Beginning in June 1918, the opposing allies, initially thrown into confusion, now rallied with counteroffensives of their own. Soon Ludendorff, completely exhausted, told Hindenburg that Germany should seek an armistice. Ludendorff found himself absorbing most of the blame for Germany's failure and resigned in October 1918. He eventually went to Sweden to write his memoirs, but before he left Germany he uttered the famous line that Germany had not been defeated, but had been "stabbed in the back" (referring to naval mutiny and political, revolutionary unrest that led a new German government to ask for an armistice).

Later Life

The strain upon Ludendorff during the war had been immense. In essence, he had directed the entire nation in 1917 and 1918; he had also suffered personal loss, for two of his stepsons had been killed in the war.

Returning to Germany in 1919, Ludendorff became involved in politics. (Fig. 1 shows Ludendorff in 1921.) He seized upon extreme ideas of German ethnic "purity" and became an advocate of the newly formed National Socialist German Workers Party. In 1923 Ludendorff partici-

Figure 1. Gen. Erich Ludendorff is surrounded by high officers of the German Army at the funeral of ex-King Ludwig and the former queen of Bavaria, 1921. (SOURCE: U.S. Library of Congress)

pated in the abortive Beer Hall Putsch with Adolf Hitler, was tried for treason, and was acquitted. Continuing in politics, he was elected to the Reichstag (parliament), but was disappointed by a dismal showing when he ran for president in 1925.

In his last years, Ludendorff retreated into his own world of disappointment and hatred. He divorced Margarethe in 1926 and married Dr. Mathilde von Kemnitz, whose bizarre ideas now appealed to him. He and his new wife indulged in pagan beliefs and racial hatred. Ludendorff became increasingly hostile to Hitler, who ironically continued to try to win the old general's approval. Estranged from his old commander, Hindenburg, Ludendorff nonetheless wrote to him when Hindenburg, as president of Germany, made Hitler Chancellor. Ludendorff—with surprising prescience—warned that Hitler was a demagogue who would bring ruin to Germany. Just before his death in 1937, Ludendorff wrote *Total War*, in which he inverted Clausewitz's dictum subordinating war to politics, arguing that a military dictator should run the nation in war. (He appeared oblivious to the fact that his own experience disproved this thesis.)

Assessment

Ludendorff was an extremely able soldier, but he required the calming presence of others who were less intense. In politics, he was not in his element. While one can sympathize with his assuming great responsibility during the war because of an absence of competent political direction, his strange conduct after the war revealed a character that had lost its balance. In tactics, however, he was one of the most brilliant figures in military history. His conceptual tactical accomplishments in 1917 and 1918 have few parallels. Through his efforts, the theoretical foundations were laid for twentieth-century tactics. In the interwar years, his successors on the German General Staff added the benefits of mobility (mechanization and airpower) and radio communication to his tactical innovations of 1917 and 1918. The practitioners of blitzkrieg were the heirs of Ludendorff.

TIMOTHY T. LUPFER

SEE ALSO: Hindenburg, Paul von; Prussia and Germany, Rise of.

Bibliography

Barnett, C. 1963. *The swordbearers*. New York: Morrow.
Craig, G. A. 1964. *The politics of the Prussian army*. London and New York: Oxford Univ. Press.
Dupuy, T. N. 1970. *The military lives of Hindenburg and Ludendorff*. New York: Franklin Watts.
Goerlitz, W. 1953. *The German general staff*. New York: Praeger.
Goodspeed, D. 1966. *Ludendorff, genius of World War I*. Boston: Houghton Mifflin.
Ludendorff, E. 1934. *The general staff and its problems*. 2 vols. Trans. F. A. Holt. New York: Dutton.
———. 1919. *Meine Kriegserrinnerungen 1914–1918*. Berlin: Mittler.
Lupfer, T. T. 1981. *The dynamics of doctrine: The changes in German tactical doctrine during the First World War*. Fort Leavenworth, Kans.: Combat Studies Institute.
Middlebrook, M. 1978. *The Kaiser's battle*. London: Penguin.
Parkinson, R. 1978. *Tormented warrior*. New York: Stein & Day.
Tschuppik, K. 1932. *Ludendorff: The tragedy of a military mind*. Trans. W. H. Johnston. New York: Houghton Mifflin.
Wynne, G. [1940] 1976. *If Germany attacks*. Reprint. Westport, Conn.: Greenwood Press.